Birkhäuser

Cornerstones

Series Editor
Steven G. Krantz, *Washington University, St. Louis, MO, USA*

More information about this series at http://www.springer.com/series/7161

Robert B. Burckel

Classical Analysis in the Complex Plane

Birkhäuser

Robert B. Burckel
Department of Mathematics
Kansas State University
Manhattan, KS, USA

ISSN 2197-182X ISSN 2197-1838 (electronic)
Cornerstones
ISBN 978-1-0716-1963-6 ISBN 978-1-0716-1965-0 (eBook)
https://doi.org/10.1007/978-1-0716-1965-0

Mathematics Subject Classification (2020): 30-01, 30-03, 01A05

This book is published under the imprint Birkhäuser, www.birkhauser-science.com by the registered company Springer Science+Business Media, LLC
The registered company address is: 1 New York Plaza, New York, NY 10004, U.S.A.

CONTENTS

FOREWORD

Complex analysis is one of the *grandes dames* of modern mathematics. Beginning with the fifteenth-century efforts of Scipione del Ferro to solve cubic equations and proceeding to important developments by Bombelli and Cardano in the sixteenth century, Descartes in the seventeenth century, de Moivre and Euler in the eighteenth century, Cauchy and Riemann and Weierstrass in the nineteenth century, and Ahlfors and Nevanlinna and a host of others in the twentieth century, the subject continues to evolve and grow. It is safe to say that complex variables is one of the most active areas of mathematical research today.

In fact, the corona problem and harmonic measure are cutting-edge areas of today's mathematical research. Value-distribution theory, conformal mapping, and the Cauchy transform maintain considerable interest. Complex dynamical systems is one of the hottest fields around, and some of the leaders of world mathematics (including two Fields Medalists) ply their trade in that venue. There are few 500-year-old disciplines that command such attention.

And the connections of complex variables with physical applications have certainly contributed to the robustness of the field. Engineering, fluid mechanics, mathematical physics, and many other fields use complex variables routinely. Even though complex numbers were known early on as "imaginary" just because people could not decide how to justify the existence of a square root of -1, they are now a very real part of scientific life.

There have been—especially in the twentieth century—a good many texts and monographs on function theory. The paradigmatic book was the one first published in 1953 by Lars Ahlfors. It can be argued that virtually every other modern text is derivative from that one, although each has its own particular emphases and point of view. Some texts are for the undergraduate audience (mostly engineering students), and have a very applied bent. Others are for beginning graduate students, and tend to be rather austere and pure. More advanced monographs will explore particular byways in the field, such as potential theory, or subharmonic functions, or the corona problem or conformal mapping.

The book by Robert Burckel stands alone in the vast literature of complex variables. It presents the subject from an authoritative historical perspective, and describes the ideas at a level of detail and insight that is rarely encountered in the mathematical literature—especially the textbook literature. Just as an example, it is a rare book indeed that says very much about

Painlevé's theorem that a conformal mapping of a smooth domain extends smoothly to the boundary. Burckel not only discusses the result, but gives detailed bibliographic references. This is just about the only text that treats the Aumann-Carathéodory theorem. Burckel gives an extremely unusual—but quite enlightening and gratifying—proof of the Riemann mapping theorem. He has a detailed discussion of the theorems of Julia, Carathéodory, and Wolff. He provides a nice treatment of value-distribution questions.

This book is simultaneously a monograph/reference and a textbook. It has lovely and detailed exercise sets (with loads of hints for the timid), lots of in-text exercises, loads of didactic insight. Every page was lovingly written to convey the author's passion for, and lifelong commitment to his subject.

There are a variety of different ways to teach the subject of complex variables. Among the points of view or wellsprings that one might adopt are:

- The Cauchy theory;
- Conformality;
- The argument principle;
- Partial-differential equations;
- topologies on function spaces.

It can be said that the point of view of Burckel's book is unabashedly traditional function theory. This text follows in the tradition of Cauchy and Riemann (although in a considerably more rigorous vein) in developing key ideas about *holomorphic functions*. And, no matter what your point of view, those really are the grist of the subject. A complex analyst makes his/her living by bending holomorphic functions.

The book is very much written in the Bourbaki style. There are no figures, and few examples. The reader is expected to battle the subject out for himself, but with copious aid from the author. One of the extraordinary features of this book is a Bibliography of over 150 pages. That is longer than many a freestanding book. And this Bibliography is an exemplary resource for anyone doing scholarly work in the subject. There are many references to primary sources in the eighteenth- and nineteenth-century literature, and few omissions. Practically every twentieth-century monograph on complex variables is listed. Where appropriate, Burckel is careful to also provide Math Reviews numbers for bibliographic citations. It would take years to assemble a Bibliography of this sort.

I can honestly say that Robert Burckel's book has profoundly influenced my view of the subject of complex analysis. It has given me a sense of the historical flow of ideas, and has acquainted me with byways and ancillary results that I never would have encountered in the ordinary course of my work. Burckel's sense of history and precision is both moving and inspiring. The care exercised in each of his proofs is a model of clarity in mathematical writing. Certainly Robert Burckel is no friend of handwaving. Any budding mathematician could use this text as a role-model in the stylistic arts of our profession.

The book is long, indeed encyclopaedic. But this is a necessary feature of the kind of treatment that Burckel essays. Anyone teaching a course from this book would cherry-pick topics, and leave certain portions of the material for the students to enjoy on their own. But that is part of the joy of a book of this kind: it presents a rich tapestry of ideas that have developed over half a millennium, and gives the reader a profound sense of how they came about. Anyone in the field should have this book on his/her bookshelf just as a resource and an inspiration. I keep it close to hand and refer to it frequently. It has certainly saved me many a tedious trip to the library.

Some books are long because they attempt to cover a great many topics, or because they attempt to bring the neophyte reader up to speed in some current research area, or because they provide copious (elementary) background material. Burckel's book is none of these. His goal is to cover the most fundamental areas of his subject *in depth*, and with precision and rigor. He certainly succeeds admirably in that quest. This book would provide the beginning graduate student a definitive source for studying for the qualifying exams; it also is a great leaping-off spot for the beginning of a research career.

Robert Burckel's text is not a "first book" in the subject of complex analysis, nor is it a "last book". It is a companion for a long journey, one that you can depend on and take strength from. We can be grateful for a scholarly effort of this kind, and only hope that it serves as a model for future writers.

Steven G. Krantz

PREFACE

This is a text-monograph which derives from a 1979 book (*An Introduction to Classical Complex Analysis*, ISBN 3-7643-0989-X; **MR** 81d:30001) of the author on the same subject. Some portions of the earlier text have been taken over with minimal modification; however, about half of it has been extensively rewritten and significantly expanded. The promised second volume of that book is still on the drawing board! Since it will make frequent references to its predecessor(s), it seemed desirable to harmonize this new work with volume 1 as much as possible. To that end, the structural outline (chapter and section titles) has been retained. Occasionally procrustean tactics were needed to accomplish this, making the geography of the text less transparent; I hope that an unusually robust index will compensate (and even that readers' interest will be piqued by occasional exotic entries such as "Hamlet" or "Puritans").

The title is now more descriptive: "In the plane" signals *one* variable, in $\mathbb{C}$, not on a Riemann surface. However, "analysis" in $\mathbb{C}$ inextricably involves topology, and before serious analysis begins, the book examines some of the deeper topological issues, such as the Invariance of Domain and the Jordan Curve Theorem. As before, "classical" refers to the subject matter and style, although the latter is more baroque than that of the classic master Edmund (*Satz-Beweis*) Landau. Just as Rachmaninoff and Sibelius enriched humanity by extending the luxuriant oasis of 19th-century music well into the 20th, there is an abundance of post-Weierstrassian material that is classical in this extended sense and finds a place here. [One cannot pass directly from the Classical to the Modern era; there must be a Romantic interlude. See HENLE [1996], in whose well thought-out taxonomy the mathematics in this book is rather Romantic than Classical.]

The book is almost self-contained logically, but the role of the instructor in providing heuristics, motivation, overview and a realistic syllabus remains significant. The prerequisites are rather modest; they are detailed in Chapter 0 but consist mainly of an ϵ-δ familiarity with (and attitude towards) analysis, at about the level of RUDIN [1976]. (However, complete mastery of that canonical work is by no means necessary.)

Perhaps the book's most distinctive features are the historical notes (over 70 pages) and bibliography (over 2,200 works). ["A man will turn over half a library to make one book," as SAMUEL JOHNSON noted.] The aim remains to present this impressive body of knowledge as a work of art, and to promote the view of C. G. J. JACOBI: "*...mais un philosophe comme lui* [Fourier] *aurait dû savoir que le but unique de la science, c'est l'honneur de l'esprit humain...*" (see **MR** 99c:01015). But the richness of its techniques and the depth and multiplicity of its results recommend it to the pragmatic reader as well. The expositional challenge is to achieve logical rigor without sacrificing beauty. Two archetypes of this synthesis are classical ballet and the gothic cathedral, and the latter suggests which goal has precedence in [rare] cases of incompatibility.

As the book's size suggests, it is designed more for the explorer who wants to know the whole terrain (and "pause to smell the flowers"), less for the military commander interested only in major roads and production facilities. On one hand, basic results are often reached in several different ways, with full backward and forward references tying them together. And the numerous off-road topics treated avoid overly specialized material. Also contributing significantly to this size is the detail in which proofs are done – as a safety net for the reader who may (occasionally) need it; moreover, outside of the modest prerequisites detailed in chapter 0, no unproven facts are used in the text. On the other hand, sequestered among these details are some subtleties and touches of elegance. I hope these features will allay readers' anxiety over the book's unusual length. All prospective readers should know that many of the *major* themes here are presented with unrivaled concision and clarity in 160 pages by SARASON [2007], a book I recommend as a *vade mecum* for every student. For the reader with more experience, wondering how my text differs from others, the table of contents is supplemented by a few pages of "highlights."

AMS-TEX-ing my manuscript was an onerous undertaking and I am grateful to Pranab Banerjee (for doing chapters 0–III), to Kuppusamy Ravindran (for doing the lion's share, chapters IV–XII), to Joseph Kincaid (for doing the bibliography and a year-long revision of the whole) and to his son Thomas (for the indexing). Significant early revisions, plus a laborious updating of computer code were ably and cheerfully done by Sheree Walsh, amid many other professional and familial obligations. For reading the whole manuscript, spotting several subtle slips and many typos, and for quite a few constructive suggestions I am indebted to Miroslav Engliš and Mark Shaw. During the long gestation of the book the author was privileged to spend three years at the Mathematical Institute of the University of Erlangen-Nürnberg, and he thanks all his friends there, especially Prof. Dr. Heinz Bauer (†), for the hospitality and ideal working conditions they provided. The staff of the interlibrary loan department at Kansas State University's library have my thanks for their valiant efforts over many years to supplement its chronically underfunded and inadequate resources. The good advice and intercession of Steven Krantz and the understanding and support of a succession of Birkhäuser editors, from Ann Kostant to Danielle Walker, Loren-Marie Durr, and Christopher Tominich were essential to the realization of the book. Equally essential was my wife's help with the foreign-language literature.

My final and deepest thanks are for my family, the loving matrix that nurtured whatever I have accomplished; but there, of course, words are inadequate: Irene (†), Arthur (†), Glenna, and Nicholas — there could be no better parents, wife, and brother.

R. B. Burckel
June, 2021

Apologia Pro Libro Suo

I hold every man a debtor to his profession,
from the which as men of course do seek to
receive countenance and profit, so ought they
of duty to endeavor themselves by way of amends
to be a help and ornament thereunto.

Francis Bacon,
(1561 — 1626)

Notre devoir, . . . , n'est-il pas de regrouper,
en vue d'un simple effort de conservation,
les éléments essentiels de notre culture,
afin qu'un jour nos descendants les retrouvent
intacts à l'aube d'une nouvelle Renaissance?

André Weil,
Oeuvres Scientifiques
vol. 1, p. 359

Für uns Mathematiker ist es die Aufgabe,
sie [die Mathematik] wie ein Heiligtum zu hüten,
damit einst *alles* menschliche Wissen überhaupt
der gleichen Präzision und Klarheit teilhaftig wird.

David Hilbert,
Mathematische Annalen
104 (1931), p. 494

HIGHLIGHTS & UNUSUAL FEATURES

Chapter 0

1. The algebraically motivated introduction of $\mathbb{C}$ along with its historical significance in §3 is unique. There too the historical evolution of $\mathbb{R}$ is discussed, but its actual construction is out-sourced. A scrupulous effort is made to never exceed the (modest) prerequisites detailed in this chapter.

Chapter I

1. Existence of roots in $\mathbb{C}$ is secured by a novel, elementary topological argument, without the complex exponential function, 1.26.

2. Lemma 1.27 is trivial but surprisingly useful in the many local-to-global arguments in this subject.

3. Corollary 1.34 is an indispensable tool for rigorous treatment of the Jordan Curve Theorem and multiple connectivity. (It's not in any prior texts, although REMMERT [1991] adopted it after my 1979 book.)

4. 1.47 is an efficient, self-contained (one might even claim "natural") exercise set doing log and exp in $\mathbb{R}$ (later obviated by the complex exponential, but good "Fingerübungen" and consolidation for neophytes of things supposedly "known").

Chapter II

1. Weierstrassian construction of primitives without integrals, 2.12.

2. The contact with real analysis in §4 is more extensive than customary. It includes the weak (i.e., f' hypothesized to be continuous and only rectangular contours) Cauchy Integral Theorem and pre-π Integral Formula, together with two beautiful proofs of the Fundamental Theorem of Algebra (= FTA), "known" but cast in optimal form (e.g., no use made of the complex exponential or cyclometric functions). This is Exercises 2.20 and 2.21.

3. In Exercise 2.22 readers are guided through a very unusual development of the complex logarithm via $\int_1^z \frac{d\xi}{\xi}$, overcoming all topological subtleties, with no resort to power series. (This exercise is acknowledged as optional, since all is later done from scratch via series.)

4. Extensive Notes on weakening the requirements for holomorphy, on the founding Trinity, and on sources for the history of our subject.

Chapter III

1. After the canonical differentiation of power series, a short minimally computational one is given (Remark following 3.6).

2. Quick proof of the rearrangement theorem for power series and its use to prove their differentiability (Exercise 3.12 and the Remark following it.)

3. *Motivation* for the exponential power series (opening paragraphs of §3).

4. The "Remarks on Geometry" following definition 3.16 showing that isogonal (resp., isotropic) real-differentiable functions are holomorphic (or anti-holomorphic) are unusually detailed.

5. The Remarks following 3.22 give an interesting analytic derivation of Archimedes' estimate for π, and 3.28 outlines short, elementary proofs of the irrationality of e and π.

6. Gauss' probability integral and the Fresnel integrals are evaluated by a single elementary real-variable method in Exercise 3.24.

7. The long, multi-part Exercise 3.35 develops Weierstrass' double limit theorem, the Montel-Vitali-Porter compactness theorems, the uniqueness and open-mapping theorems and the FTA by pure power-series methods. (All are later proved anew by Cauchy's methods.)

8. §4 does power sums, Bernoulli numbers and functions, and the (historically) first Fourier series efficiently.

9. §6 begins with a "Hier stehe ich; ich kann nicht anders" recognition of the consequences of the exponential function not being univalent, followed by several different realizations of logarithms in slit planes and a historical note on Henry Briggs' neat pre-calculus computation of logarithms.

10. In Remark 3.43 the "full answer" to when continuous logarithms exist in open or in compact sets is previewed.

11. The chapter closes by defining the Γ-function directly, without developing the theory of infinite products, and then proving a subtle iterated limit for Euler's constant

$$\gamma = \lim_{x \downarrow 1} \sum_{n=1}^{\infty} \left(\frac{1}{n^x} - \frac{1}{x^n} \right),$$

Exercise 3.46(ix). (A recurring theme in the book is the significance of permuting two limit operations in analysis. Cf. **MR** 96e:01013.)

Chapter IV

1. Theorem 4.0 produces continuous logarithms in a very elegant,efficient manner and is used incessantly. (The application of it following Corollary 4.1 to "fair division" is unusual and cute.)

2. The index (winding number) is defined from the start for all continuous curves — by "lifting" to an exponential representation, and contact made with integrals immediately thereafter (Corollary 4.3 ff.).

3. A novel analytic proof of the rather non-trivial fact that the figure-8 curve distinguishes nullhomologous from nullhomotopic is sequenced in Exercise 4.13. Because of the equivalence of these two notions in simply-connected regions, such an example is illuminating. (Proofs in algebraic topology involve covering spaces and are less elementary.)

4. An abstract monodromy theorem to generate continuous logarithms (destined for repeated use) gets a methodologically revealing proof in 4.18.

5. The full story of continuous self-maps of the circle is told in 4.22 and is vital for later developments. Exercise 4.23 contains immediate and elegant consequences. Some of these and numerous others in this chapter are "ars gratia artis", not needed for the development of complex analysis.

6. The keys to (civilized) analytic treatments of the Jordan Curve Theorem and multiple-connectivity are Theorems 4.29 and 4.30. I have arranged it so that only twice in the text, here and in the Runge theory, is "tiling" by squares employed. In both cases the intricate boundary behavior can be handled rigorously yet unoppressively.

7. §5 contains the "optimal" proof of the Jordan Curve Theorem and its relative, the Theorem on Theta Curves (4.47), needed for later extensions of the Riemann map to the boundary. Exercise 4.49 is a nice application to defining the orientation of a homeomorphism in the plane.

8. §6 is an especially luxuriant display of applications, including the FTA (and a not well-known consequence, 4.54), Brouwer's fixed-point theorem (4.58(ii)), an argument principle for (merely) continuous functions (4.59), the open-mapping theorem for univalent continuous functions in $\mathbb{C}$ (4.61), Borsuk's Antipodal Theorem (4.62), Poincaré's Vector-field [or "Hairy Ball"] Theorem (4.63), and Kakutani's Orthogonal Triads Theorem (4.65). The latter two are exceptionally elegant.

9. In §7 and §8 the first half of the giant "Hauptsatz of the Cauchy Theory" (=10.2, the fifteen odd equivalences of simple-connectivity) is proved via the key factorization theorem of §4. Thus Chapter IV reaches (somewhat farther than) all the deeper topological facts about $\mathbb{C}$ needed (but often not acknowledged) for classical analysis, without the reader having had any prior knowledge of algebraic topology. However, the relation to algebraic topology is brought out in §3 of chapter X. (See, especially Remark (v) following 10.30.)

10. The chapter Notes contain a *very* detailed history (and alternative proofs) of the FTA.

Chapter V

1. Corollary 5.2 (to the canonical proof of Cauchy-Goursat) gets the integral theorem almost instantly for convex regions, even on the boundary. This is

very useful in an alternative march to the integral formula, sketched in the Remarks following 5.2 and the Chapter Notes.

2. The (circumferential) mean-value property of holomorphic functions (5.5) is made something of a workhorse in the sequel, and the Remarks following it have considerable interest.

3. The Poisson formula is derived directly from the mean-value property in a very natural (but not so common) way, unusual care being given the necessary change of variables which brings the Poisson kernel into existence.

4. After the solution of the Dirichlet problem in the disk (5.20), seven significant applications to Fourier theory are succinctly presented (no background knowledge presupposed), destined to be used later in various "alternative" proofs.

5. In 5.23(iii) a surprising 1-radius converse of Gauss' mean-value theorem for harmonic functions is presented, together with a striking counterexample (5.24(v)).

6. 5.26(iv) is an application of maximum-modulus to an attractive result of Erdös and Radó about generalized "lemniscates".

7. In analogy with the Cauchy product of series and Leibniz' formula for $(f \cdot g)^{(n)}$, the computation (due to Faà di Bruno) of $(f \circ g)^{(n)}$ is treated succinctly in Exercise 5.31(ii).

8. For those complacent about creating anti-derivatives by integration, Exercise 5.32(i) and (ii) might serve as a vaccination.

9. Two strong (but elementary) forms of Morera are given in 5.39(ii) and (iii) [and another in the discussion preceding 3.17].

10. In the treatment (via several concrete cases) of harmonic majorization (§5) unusual detail is offered in support of the claimed boundary behavior, and the applications (5.50, 5.56, 5.60) are out of reach of many introductory texts.

11. §6 (uniqueness theorems) contains some unusual examples of that property of holomorphic and harmonic functions: 5.64(i), (ii); 5.66(i); 5.67(iii) and (especially) (v).

12. The introduction to 5.73 outlines a very fruitful program to be carried out in Exercise 5.73 which gives a constructive (interpolatory) proof of the uniqueness theorem and can even be exploited to prove the Vitali-Porter convergence theorem (5.74) [which also gets proved by quite different means in 3.35(v) and in 7.5].

13. 5.76 is a strong boundary-uniqueness result of T. Radó.

14. The "local theory" (§7) is pursued in two independent ways: one emphasizing the open-map theorem and factorization (5.83), and one based on Cauchy integral formulae and index ideas (5.88).
The first proof (of several) that f holomorphic and univalent implies f' zero-free and f^{-1} holomorphic (5.78) is noteworthy for being "bare-hands" (e.g., no power series).

15. Useful (later) implicit function and factorization theorems (5.93, 5.96, 5.98) round out this longest (of all 13) chapter.

Chapter VI

1. 6.1–6.5 manage to contain all the basic Schwarz-lemma lore, including the characterization of self-maps of $D(0,1)$ that are isometric in the pseudo-hyperbolic metric. J. Wolff's "Schwarz lemma on the boundary" (= existence of invariant horodisks) destined for use later, gets an elegant, elementary proof in 6.6.

2. Blaschke's uniqueness condition is derived in a novel way from an inequality of Löwner and Radó: 6.9(iii).

3. §2 is devoted to *proper* maps, described by the colorful neologism *compactifugal.* Its striking theorems (in very few basic books) 6.14, 6.15, 6.17 see significant service later, for example, in the complete characterization of proper holomorphic maps between annuli (6.18) and between disks (6.16). An idea in the proof of 6.17 can replace certain complicated analytic continuation arguments in proving the monodromy theorem.

4. The famous circle of ideas of Carathéodory, Julia, Wolff, *et al.* about angular derivatives on the boundary of self-maps of half-planes and disks is optimally laid out in 6.27, 6.28.

5. §5 draws conclusions from the elementary principle of subordination, including strong Hadamard-Borel-Carathéodory type inequalities (6.35), the coefficient bounds $|c_n| \leq |c_1|$ for univalent convex maps of $D(0,1)$ (6.38), and the covering result $g(D(0,1)) \supset D(g(0), 1/2)$ for every convex holomorphic function g in the unit disk $D(0,1)$ with $|g'(0)| \geq 1$.

Chapter VII

1. This chapter immediately addresses the big convergence and compactness theorems, getting both Vitali-Porter and Montel (7.5, 7.6) in an unusual way. In 7.14 several convergence results get a unified proof.

2. Osgood's theorem that mere pointwise convergence almost entails local uniform convergence is proved succinctly in 7.18, and the question of whether inversion is a locally uniformly continuous operation (involved later in constructive proofs of the Riemann mapping theorem) is treated in 7.19.

3. In §2 the convergence theorems are used to prove "differentiability behind the integral sign" under weak hypotheses (7.21), and also Dixon's elegant homology version of Cauchy's integral formula (7.22), leading to its homotopy forms (7.23) [in which the intermediate curves in the homotopy need not be rectifiable].

4. The earlier compactness theorems are used to deduce local uniform boundedness of sequences of univalent functions and prove covering theorems of the Bloch-Landau variety (7.28, 7.29(iii)). Explicit (although not best-possible) constants are obtained, leading to a nice proof of the Little Picard Theorem, without Schottky or a modular function (7.29(v)).

5. In §3 Weierstrass' Theorem (every discrete set is a zero-set) is directly proved for arbitrary regions, without the apparatus of infinite products. (I think the proof is well motivated. Later, in the Remarks following 11.40 of chapter XI, a very elegant but little-known proof of the special case where the region is $\mathbb{C}$, from Weierstrass' *Werke*, is presented.) Then Blaschke products and Frostman's characterization of them (7.39) are done.

6. §4 applies the compactness and convergence theorems to the sequence of iterates of a bounded holomorphic function to obtain well-known (but in few elementary texts), elegant results like 7.43, 7.44, 7.47 and 7.51, which is Carathéodory's "Starrheit" characterization of simple-connectivity in terms of self-maps with a fixed-point.

7. After a diversion with the "tower of powers" of i (7.53), the serious business of completely characterizing the orbit behavior of self-maps of disks and annuli is carried out in §5. See 7.54, 7.55, 7.56 and 7.59 (very elegant covering-space techniques of Maurice Heins), respectively.

Chapter VIII

1. Opens by proving the polynomial and rational approximation theorems of Carl Runge. Two useful forms of each are proved (8.9, 8.11 and 8.10, 8.12, respectively). The proofs are functional analysis inspired, but one still needs a tiling argument to get started, with attendant subtle combinatorics. These are kept antiseptic (proof of 8.1) and a novel proof (8.3) shows that the non-canceling sides form *closed* polygons.

2. Following a 1933 idea of Joseph Walsh, Cauchy's theorem for simply-connected (the definition in this book is: no bounded complementary components in $\mathbb{C}$, i.e., no "holes") regions is derived in one line from Runge (8.13), providing one important link in the grand equivalence (10.2).

3. Two other immediate but significant applications of Runge are to a Rouché theorem in which the boundary is arbitrary (not necessarily a curve) and a decomposition theorem which generalizes the Laurent expansion: 8.18 and

8.24, respectively. Another application (8.25) is to a beautiful summability theorem of E. Borel. [None of this material is in any other basic text.]

4. For many applications it is important to surround a hole K in a region G with a closed curve Γ in G which winds once around each point of K. By invoking Jordan's theorem the drudgery and inelegance usually associated with the construction of such a curve is avoided and it becomes a corollary of the Runge construction (8.21). Alternative versions that can be purchased without the J-curve theorem and are adequate for most applications are pointed out (following 8.21). The first application is another link in the 10.2 chain, relating winding number 0 to connected complement (8.22).

5. In §4 some special kinds of polynomial and rational approximation theorems are proved, for example (8.27), every continuous map f of $\overline{D(0,1)}$ into $D(0,1)$ which is holomorphic in $D(0,1)$, the open unit disk, is an infinite convex sum of finite Blaschke products.

6. Torsten Carleman's beautiful extension of Weierstrass' uniform polynomial approximation theorem to all of $\mathbb{R}$ is proved in 8.32: For every continuous complex-valued function f on $\mathbb{R}$ and continuous positive function ε on $\mathbb{R}$ there is an entire function g such that $|f(x) - g(x)| < \varepsilon(x)$ for all real x. The techniques are elementary but basic and of significant pedagogical value. For example, use is made of W's original exponential approximate identity.

7. Via Carleman's theorem the Dirichlet problem in a half-plane is reduced to the case of a boundary function which vanishes at infinity, and via the Cayley transform this version is reduced to the disk problem solved already in chapter V. *Kurz und bündig*: see 8.33.

8. §6 develops Poisson's formula in the half-plane under various restrictions on the (continuous) boundary function, but using no Riemann-Stieltjes integrals or Helly's (compactness) theorem [pursuant to a promise in the preface pertaining to prerequisites]. The results are 8.34(viii) and 8.38. The Poisson kernel in the disk is *characterized* by its positivity and null boundary limit at all but one point (Remark (ii) following 8.37). The Carathéodory-Julia-Wolff theorem on angular boundary derivatives in the half-plane is deduced *anew* from these integral representations (8.39), as is a beautiful growth theorem for subharmonic functions in a half-plane due to the brothers Nevanlinna (8.41). Finally, the integral representation formulas lead painlessly to the solvability of the D-problem, and even a Poisson formula, in an infinite strip (8.43).

Chapter IX

1. A very detailed presentation of the constructive proof of Riemann's mapping theorem (due to Carathéodory & Koebe) opens this chapter. Concrete estimates for the rate of convergence of the sequences constructed are pro-

vided, and there is no passage to unspecified subsequences. However, much shorter proofs which *do* involve Montel's compactness theorem are outlined in the exercises (9.9).

2. In §4 the homeomorphic extendibility of Riemann's map to the closure of any Jordan region (theorem of Osgood and Carathéodory) is proved. It is a bit arduous but avoids Lebesgue theory and transforms geometric arguments into comparably transparent analytic ones. I believe it maximizes elementariness and rigor, without grievously compromising elegance.

3. Consequences of the Osgood-Carathéodory theorem are drawn; for example, a very general argument principle (9.15 and 9.18), and the fact that a continuous function on the closure of a Jordan region which is holomorphic in the region and one-to-one on its boundary is one-to-one throughout (9.16). And, of course, the Schönflies extension theorem, normally a brutal workout in point-set topology, is now painlessly available (9.19).

4. §6 treats special cases of Carathéodory's work relating geometric convergence of simply-connected regions to the local uniform convergence of their Riemann maps in order to get Walsh's approximation theorem, intermediate to Runge's and Mergelyan's: Every function which is continuous on the closure of a Jordan domain and holomorphic in its interior is approximable by polynomials uniformly over the whole domain (9.26). A very general "Cauchy theorem on the boundary" naturally emerges (9.27), as does Painlevé's removable-arc theorem (9.28).

5. In §7 the equivalence of Laplace's equation (under minimal differentiability hypotheses) with being locally the real parts of holomorphic functions is finally proved. Then Perron's method is used (following an unusual path of O. D. Kellogg's 9.41) to solve the Dirichlet problem in the plane whenever there are no one-point boundary components (9.44). An interesting and elementary recent example of a simply-connected region in the plane for which the D-problem is *not* solvable is analyzed (9.47).

6. Maurice Heins' elegant proof of the equivalence of the existence of harmonic conjugates with simple-connectivity (another link in the 10.2 chain) is proved in §8 (9.50). There too a rigorous derivation of a Poisson-integral solution of the D-problem in a Jordan region is outlined (9.52).

7. In the final section the Schwarz-Christoffel formula for a rectangle is derived. I think all the subtle boundary matters are handled rigorously yet elegantly.

Chapter X

1. Opens with a satellite view of all roads leading into Rome, the theorem which unifies all aspects of Cauchy's theory: 15 equivalences of the central concept of simple-connectivity. Together with 5 others (see 10.3(i)) it is the

most complete such list I know of. The Monodromy Theorem (10.3(iv)) is treated unorthodoxly, and effectively marginalized.

2. The beautiful theorem of E. Heinz (f continuous, holomorphic off its zero-set implies f holomorphic) gets a somewhat novel proof in 10.10. Proper maps (from chapter II, §2) re-appear and in view of the Hauptsatz we are able to prove the striking local-to-global result 10.12 (see also the remark following it).

3. In §2 we once again (cf. 6.18 and 6.24) find all conformal maps between annuli and even locate the conformally invariant radius-ratio in the structure of the *ring* of bounded holomorphic functions (10.15, 10.16). A covering-space argument (*in situ*, no outside topology results called on) is used to give the first of three proofs that every doubly-connected region is conformal to an annulus (10.20). [No textbook I know of proves this — unimpeachably.]

4. What amounts to a homology basis for a finitely-connected region Ω is constructed in 10.26, leading to the theorem on harmonic conjugates for such regions (10.27) and the counting of holes by the codimension of the subspace of derivatives in $H(\Omega)$, as well as that of the subspace $\operatorname{Re} H(\Omega)$ in the real-valued harmonic functions on Ω (10.29), and the rank of the quotient *group* of the zero-free functions in $H(\Omega)$ modulo its subgroup of exponentials in $H(\Omega)$ (10.30). In remarks following 10.30 these spaces are related to the first homology group of Ω. The discussion is *ab ovo*: modest knowledge of vector spaces and group theory is expected but none of algebraic topology.

5. A new proof of another characterization of simple-connectivity (due to Aumann and Carathéodory) is given in §4. It affirms that if the group of conformal automorphisms of a bounded region Ω which fix some point $z_0 \in \Omega$ is infinite, then Ω is conformal to a disk. The proof shows many important techniques in action.

6. Finally, §5 gives two quite different (again, technique-rich, topological/analytic) proofs of the fundamental mapping theorem for doubly-connected regions — one of the book's high-points and very distinctive features, I feel.

Chapter XI

1. This chapter is full of applications, to other areas of analysis. It opens (naturally) with Cauchy's formula in the annulus, realized in ten lines without the usual inelegance of integration over keyhole contours. Then the Laurent decomposition (via a technique used to prove a more general result in 8.24) and its bilateral series representation are derived. The local theory is done rather economically but completely (11.5 - 11.7). The determination of orders and residues in 11.5(iv), (v) seems to often be overlooked, as does the elementary fact 11.8(i).

2. In §2 some elementary facts about rational functions are derived: Their partial-fraction decomposition is proved (11.13) and those which are of modulus 1 or are positive on the unit circle are determined (11.14, 11.15). And the circle-preserving meromorphic functions in $\mathbb{C}$ are shown to be just the Möbius maps (11.20(vii)). Rational functions whose Taylor series at 0 have integer coefficients are quotients of polynomials with integer coefficients (11.23). When a holomorphic function is known to have just one singularity on its circle of convergence and that a pole, how to locate that singularity in terms of the Taylor coefficients is shown in 11.26.

3. The most general form of the residue theorem is proved in 11.29. Among the applications made are the evaluations of the Fresnel and Gauss (probability) integrals, as well as the famous Gauss sums (11.32). Dirichlet's Fourier series treatment of the latter is an easy application of the short Fourier analysis digression in chapter V. See Remark (iv) following 5.20. Similarly a basic theta-function identity is proved both via residues (11.33(vi)) and via Fourier analysis (11.33(vii)); the latter involving as well the famous *Poisson summation formula*, 11.33(vi).

4. The multi-part Exercise 11.34 is a spectacular application of contour integrals and residues to prove a beautiful and very explicit summability result of Ernst Lindelöf's, which expresses f throughout a convex domain in terms of its Taylor series at any point.

5. The multi-part 11.37 is a pretty efficient derivation of the partial-fraction development of cosec^2, thence that of cot, thence Euler's infinite product for the sine. This last is somewhat novel, requiring no preliminary study of convergence conditions for infinite products. The methods are function-theoretic but alternative Fourier derivations are also within reach, and so are presented (11.37(xv)). In 11.38 a special case of the Whittaker-Shannon-Kotelnikov Sampling Theorem is proved, the method being closely related to the foregoing.

6. Mittag-Leffler's "snuggling-up" theorem is proved, in which not only the singular part, but a partial sum of the regular part of a holomorphic function are being specified at each point of a set S with no cluster point in the region Ω (11.41). Using Runge, the proof is rather short. This is a wonderful tool for studying the ideals in the ring $H(\Omega)$: We prove (11.44) that any subset of $H(\Omega)$ which has no common zero contains a finite or countable set $f_1, f_2, \ldots$ such that $\sum f_n g_n = \mathbf{1}$ for some $g_1, g_2, \cdots \in H(\Omega)$. It follows that every finitely-generated and every closed (in the local uniform convergence topology) ideal in $H(\Omega)$ is principal, whence every maximal ideal that is closed is a point evaluation.

7. Meromorphic functions in region Ω are defined locally and the global char-

acterization (imparting the name) that they are the quotient field of the integral domain $H(\Omega)$ is proved in 11.45 (from Weierstrass' theorem on specifying zeros). The appropriate Argument Principle and Rouché Theorem (involving zeros *and* poles) is proved in 11.49.

8. 11.49(vii) and (viii), and 11.50 are interesting results that utilize methods from this chapter.

9. 11.51-11.52 (together with the remarks following it) is a wonderful display of the techniques of this chapter to characterize the class of meromorphic functions in $\mathbb{C}$ which map $\mathbb{R}$ into $\mathbb{R}$ and the upper half-plane into itself: See 11.52(ii).

10. Several different and very instructive proofs of the solvability of the Dirichlet problem in annuli are given in §7 and a Poisson formula for the annulus is developed. The form of harmonic functions in annuli and punctured disks (11.55) and some less common removable singularity theorems for the latter in 11.57, 11.60 close out the chapter.

Chapter XII

1. This chapter opens with Jacobi-Jensen type inequalities (12.4(i)), followed by a long march (12.8-12.14) to *Miranda's theorem*, which furnishes for each non-negative integer n and $0 < r < 1$ a constant $C_n(r)$ bounding on the circle of radius r *every* holomorphic function f in the unit disk for which $f^{(n)} - \mathbf{1}$ and f are zero-free and $|f(0)| \leq 1$. The proofs are clever and computational but quite accessible, important because this is a gentle introduction to Nevanlinna theory.

2. Normality results (12.27) flow at once from Miranda, as does a striking theorem of Pólya & Saxer: f entire and $f \cdot f' \cdot f''$ zero-free is equivalent to f being the exponential of a linear function. All solutions in entire functions of $f^n + g^m = \mathbf{1}$ are found.

3. Schottky's famous theorem (12.15) is a special case of Miranda, and it furnishes an "elementary" proof (meaning one that does not require a modular function) of Picard's Great Theorem, as well as the existence of Julia directions at every essential singularity.

4. Striking normality theorems for *univalent* functions are proved in 12.31, 12.32, and in §5 limit theorems in sectors and strips; 12.33, 12.41, 12.42 are good examples.

5. All the important results from chapter VII on iterates of self-maps of bounded regions are quickly generalized to regions whose complements contain at least two points. Theorem 12.50 is the grand summary. It falls just a little short of Maurice Heins' definitive 1941 result. Following it a beautiful, not so well known "rigidity" theorem of Behnke & Zumbusch (1936) is proved:

Every non-constant holomorphic map of $\mathbb{C} \setminus \{0, 1\}$ into itself is a conformal automorphism.

6. With §7 the chapter and thus the book ends. Due to the great importance of Schottky's theorem, a second proof (due to Ostrowski, who wrote a whole monograph on the subject) is presented here. This proof furnishes *explicit* constants: if f is holomorphic in the unit disk and $(f - \mathbf{1}) \cdot f$ is zero-free, then for each $0 < r < 1$

$$\sup_{|z| \leq r} |f(z)| \leq \exp\left(\frac{169}{1-r} \log\left(\frac{e}{1-r}\right) \max\{1, |f(0)|\}\right)$$

(one of the more dramatic results in the whole book!). An application to a recent deep result of Mañé, Sad and Sullivan (12.57) is the last hurrah.

Chapter 0
Prerequisites and Preliminaries

As indicated in the Preface, only very modest specific knowledge is required of the prospective reader. Those (salient) *facts* that will be used in the text without proof I attempt to state fully below with references (along with a little history and polemics). What is really expected of the reader is a level of mathematical maturity comparable to that fostered by books like STROMBERG [1981], RUDIN [1976] and E. LANDAU [1950], but by no means a mastery of the contents of these outstanding classics. A junior or senior mathematics major at any college or university nowadays should therefore be adequately equipped. The recent user-friendly text ROSS [2013] is an ideal source for review and proofs of the facts simply recorded in this chapter.

§ 1 Set Theory

We'll be using the language and notation of set theory: $\in$, $\forall$, $\exists$, $\cup$, $\cap$, $\subset$, etc. When $P(x)$ is a proposition about x and S is a set, the set of all x in S for which $P(x)$ is true is denoted by $\{x \in S : P(x)\}$. When $P(x)$ is the statement "x does not belong to A", then $\{x \in S : P(x)\}$ is denoted by $S \setminus A$. As usual, the empty set is denoted by $\varnothing$. The function on S which is 1 at points of $A \subset S$ and 0 at points of $S \setminus A$ is called the *indicator function* of A and denoted by 1_A (terminology and notation which unfortunately do not indicate the domain S). As is customary, the function constantly equal to c is also denoted by c, often in bold type if c is 0 or 1. The reader is assumed to have modest acquaintance with naive (i.e., couched in a language and logic which have not been fully formalized), axiomatic set theory. Primarily he should know the bare rudiments of cardinal arithmetic. For five decades the standard testament at this level has been HALMOS [1960]. This pretty little basic book is warmly recommended to the reader, even though in the present work only a small fraction of its contents will be utilized. Consult that work in particular or the handbook of HOWSON [1972] for the standard terminology

R. B. Burckel, *Classical Analysis in the Complex Plane*, Cornerstones,
https://doi.org/10.1007/978-1-0716-1965-0_1

like *map, function, injective, bijective, surjective, one-to-one, onto, countable,* etc.

The mathematical text uses quotation marks sparingly. For example, the universal usage "In what follows, x denotes an element of X" is favored over "In what follows, 'x' denotes an element of X," even though the latter is the grammatically correct form, since it is 'x' that is doing the denoting (of the element x). This is the notorious *use* versus *mention* distinction, which mathematicians typically ignore without dire consequences – although as youngsters most had been puzzled why, if 4/8 is 1/2, the denominator of 4/8 is not the denominator of 1/2. In the same spirit, "the function $f(x)$" is understood as "the function f," and not as the constant function whose value is that of f at input x (cf. MENGER [1960]). This is convenient when it is a matter of a recipe that does not have a function name, e.g., "the function $x^3 + 3x + 1$." Another popular alternative to naming the function is the use of a special arrow: "the function $x \mapsto x^3 + 3x + 1$," which however still has the disadvantage of leaving the variable unquantified. The arrow is read "x maps to $x^3 + 2x + 1$." In this book a special effort is made to see that in all displayed formulas which are to be cited later, the variables *are* quantified: Readers should not have to break their train of thought just to track down for what x's a referenced formula from five pages earlier is valid.

The formulas "$a := b$" and "$b =: a$" mean that a is being (or has been) *defined* by this equality. Occasionally, when quantifiers would disrupt the flow of a sentence, an expression such as "$f(x) \equiv g(x)$" serves instead. The symbol "$\subset$" ; means subset inclusion, but not necessarily proper inclusion. Sometimes the end of a proof by contradiction is signalled by the colliding arrows $\longrightarrow\!\!\!\!\longleftarrow$.

The rubric $f\colon S \mapsto T$ or $S \overset{f}{\mapsto} T$ means that f is a function with *domain* or *initial space* S and *range* T. If $A \subset S$, then $f(A) := \{f(x) : x \in A\}$ is called the *image of A under f* and $f|A$ or $f|_A$ traditionally denotes the *restriction of f to A*. If $B \subset T$ (and not necessarily $B \subset f(S)$), then $f^{-1}(B) := \{x \in S : f(x) \in B\}$. This is called the *inverse image* or *pre-image (=Urbild) of B under f*; it may be void. When B is a singleton, $B = \{t\}$, we write simply $f^{-1}(t)$. The points of $f^{-1}(0)$ are called the *zeros* of f [not to be confounded with *roots* of f, which would be *functions* g satisfying $g^n = f$ for some positive integer n]. When f is one-to-one, the symbol f^{-1} is also used for the function on $f(S)$ inverse to f; so in this case, for each $t \in f(S)$, $f^{-1}(t)$ also means the unique $s \in f^{-1}(\{t\})$. In case of zero-free complex-valued functions, I write $\frac{1}{f}$ or $1/f$ for the function whose value at each x is $1/f(x) = (f(x))^{-1}$. I will never use "f^{-1}" to denote $1/f$. The composition of two functions f and g whose domains and images are properly related is denoted by $f \circ g$. For a positive integer n, the n-fold self-composite $f \circ f \circ \cdots \circ f$ (called the nth *iterate*; of f) is denoted $f^{[n]}$, whereas f^n is the n-fold product $f \cdot f \cdot \cdots \cdot f$.

The cartesian product of the sets A and B is denoted by $A \times B$. The cross is hardly ever used to signify multiplication, which is usually indicated by juxtaposition. Occasionally in long formulas dots are used to indicate multiplication and bold dots are used with the inner product in euclidean spaces. (See below.)

If $\mathcal{S}$ is a family of sets, we write $\bigcup \mathcal{S}$ (or $\bigcap \mathcal{S}$) for the union (or intersection) of all the sets in $\mathcal{S}$. If $\mathcal{S} := \{S_\alpha : \alpha \in I\}$, we write $\bigcup_{\alpha \in I} S_\alpha$ (or $\bigcap_{\alpha \in I} S_\alpha$). When I is the integers or the natural numbers, we write $\bigcup_{n=-\infty}^{\infty} S_n$ or $\bigcup_{n=1}^{\infty} S_n$, etc. Other obvious variations on this notation are also employed. E.g., $\bigcup_{P(\alpha)} S_\alpha$ and $\bigcup\{S_\alpha : P(\alpha)\}$ for the union of all the S_α such that the statement $P(\alpha)$ is true. We follow the convention that the union of a void family of subsets of a set S is $\varnothing$ and the intersection of a void family of subsets of S is all of S.

§ 2 Algebra

I use freely the language of algebra, like *semigroup, group, ring, algebra, field, integral domain* (for which consult any contemporary algebra text or HOWSON [1972]) but almost no facts from algebra. The first isomorphism theorem for (commutative) groups is used implicitly once (Chapter IV), as is the fact that a permutation is a product of disjoint cycles (Chapter VIII). I assume that the reader has a passing acquaintance with vector spaces, their dual spaces and their linear transformations (matrices). Once (Chapter XI) I invoke the row-rank equals column-rank theorem and from time to time a dimension argument intervenes, often implicitly. Quotient constructions in groups and vector spaces occur briefly in Chapters IV and X (where a free abelian group is also utilized). The analog of bases and dimension in vector spaces, that of independent generators and rank in torsion-free abelian groups, also comes up briefly there. A few elementary properties of determinants are used in §2 of Chapter XI. Indeterminates and formal power series do not come up (in earnest), so I write simply "polynomial" for what is more properly called a "polynomial function", that is, a function of the form $p(z) = \sum_{j=0}^{n} c_j z^j$, n a non-negative integer, c_j and z complex numbers (see §3), and $z^0 := 1$.

§ 3 The Battlefield

$\mathbb{N}$ shall denote the natural numbers: $1, 2, 3, \ldots$; $\mathbb{Z}$ (for *Zahl*) the integers: $\mathbb{N} \cup \{0\} \cup -\mathbb{N}$; $\mathbb{Q}$ (for quotient) the *ratio*nal numbers: $\{m/n : m, n \in \mathbb{Z}, n \neq 0\}$. $\mathbb{N}$ is brought into being from (Zermelo-Fraenkel) set theory and a "successor" function, HALMOS [1960], or directly via Peano's axioms. $\mathbb{Z}$ is manufactured from an equivalence relation on $\mathbb{N} \times \mathbb{N}$, and in turn $\mathbb{Q}$ (the quotient field of the integral domain $\mathbb{Z}$) is built from an equivalence relation on

$\mathbb{Z} \times \mathbb{N}$. We use $\mathbb{R}$ to denote the real numbers, which are constructed from $\mathbb{Q}$ via Dedekind cuts (in a mere $4\frac{1}{2}$ pages in RUDIN [1976]). For the (consummate) details of this whole *Aufbau*, starting with Peano's axioms, see E. LANDAU [1951]. An alternative, more algebraic construction of $\mathbb{R}$ from $\mathbb{Q}$ (due to Eduard Heine, Georg Cantor and Charles Méray) via equivalence classes of Cauchy sequences in $\mathbb{Q}$ is presented in detail in COHEN and EHRLICH [1963]. Weierstrass too had a (protracted) go at grounding $\mathbb{R}$, but it was less satisfactory–see **MR** 92e:01048. There is also a natural construction of $\mathbb{R}$ directly from $\mathbb{N}$ via (formal) decimal expansions. All three methods are treated in BURRILL [1967], and F. BACHMANN [1939] surveys various constructions of $\mathbb{R}$. But there are numerous more exotic ones, e.g., **MR** 50#292 (ultrafilters), 52#362 (wreath products), 54#4983 (bilateral sequences), 84j:26002 (continued fractions), 88i:11006,7 (infinite products), 90e:00002 (sequences), 90k:26003 a,b (special series), 95m:12007 (cuts), 2004m:00004 (formal logic), 2006b:20079 (near-endomorphisms of $\mathbb{Z}$), and **Zbl** 276.26002 (Egyptian fractions), 913.12007 (completions). It is the "received view," (*pace* Hermann Weyl, *et al.*), codified in the expression *the real line*, that $\mathbb{R}$ so constructed adequately reifies the linear continuum. In a sense this initially motivated and finally justifies the whole algebraic enterprise — the "arithmetization of analysis."

Since 1960 various enlargements of $\mathbb{R}$ have claimed to represent the continuum more effectively; for example, the non-standard reals, ordered fields $^*\mathbb{R} \supset \mathbb{R}$, in which numbers greater than every $n \in \mathbb{N}$ (and hence positive numbers less than every $1/n$) exist. The $^*\mathbb{R}$ are sometimes grandiloquently described as "infinitesimal-enriched" continua, and more prosaically as "thick" continua. They will play no role in this book, but for a good overview of them see part C of EBBINGHAUS, *et al.* [1990] and **MR** 97a:00004. The inestimable efficiency of $\mathbb{R}$ over $\mathbb{Q}$ when doing analysis is claimed by some for $^*\mathbb{R}$ over $\mathbb{R}$ as well. But (save for a coterie of devotees) analysts have been reluctant to embrace all the formal logic and semantics needed to build $^*\mathbb{R}$. Most readers will have assimilated the dominant view of $\mathbb{R}$ but might be interested in the efforts of dissenters to chronicle the parallel persistence and interpretation of "infinitesimals" since Cauchy, in light of their belief that "history is written [distorted] by the victors." BŁASZCZYK, *et al.* [2013] is one of their recurrent "minority reports".

At the other extreme, some (notably Kronecker) maintained that analysis had no business trying to ground geometry at all; it properly ought to be (and could be!) carried on using *only* rational numbers. A modern constructivist version of analysis, from this point of view, begun by Errett Bishop, has had considerable success; see E. BISHOP [1967] and BISHOP & BRIDGES [1985].

The prospective reader is *not* expected to have made any of the above pilgrimages to $\mathbb{R}$, but merely to know what lies atop the holy mountain: not an Elysian field, but an order-complete one. Prescinding from the existence question (which such a hajj settles), this entity is fully described by 15 properties (axioms): the customary 10 axioms for a field (a commutative group under addition, with a commutative multiplication which distributes through addition and with respect to which the non-zero elements are a group), 4 axioms for a total order compatible with the addition and multiplication (meaning that the sum and product of positive elements is positive), and an axiom guaranteeing that among upper bounds of sets there is always a least one (see §5). This axiom evidently involves second-order logic, that is, quantification over not just the aborning "numbers", but over *sets* of them as well. One effect of it is to disqualify $\mathbb{Q}$ and the constructive reals (as too small) and ${}^*\mathbb{R}$ (as too large). In fact, these axioms are categorical: to within "isomorphism" they have at most one model; and the proof of this is very easy. (An unusual set of 11 axioms for $\mathbb{R}$, which are independent as well as categorical, is offered in OMAN [2009].) Note, however, that when $\mathbb{R}$ arrives this way, fully developed, $\mathbb{N}$ loses its Promethean role and must be *defined*, as the additive semigroup generated by 1, of course. And $\mathbb{Q}$ is then defined as the smallest subfield of $\mathbb{R}$ that contains $\mathbb{N}$; it is a proper subset of $\mathbb{R}$ but is dense with respect to the ordering. Chapter 2 of EBBINGHAUS, *et al.* [1990] provides an unusually readable and succinct account of both the constructive and the axiomatic generation of $\mathbb{R}$, while Moritz Epple's 32-page chapter in JAHNKE, *et al.* [2003] recounts their historical evolution and attendant philosophical issues in more detail. But all these brilliant (not uncontroversial) achievements of the human mind aside, a traditional "operational" knowledge of $\mathbb{R}$ (that needed to understand §4-§7 below) suffices for reading the present book.

Realizing that the negative numbers in $\mathbb{R}$ do not have square roots in $\mathbb{R}$ might lead one to enlarge $\mathbb{R}$ by adjoining *new* numbers that fulfill this role, and one then realizes that as soon as a square root i of -1 has been "created", $\sqrt{-x}i$ gives a square root for every *negative* $x \in \mathbb{R}$. From the modern perspective we further realize that this minimal enlargement of $\mathbb{R}$, consisting of all sums $x + yi$ with $x, y \in \mathbb{R}$, is a vector space over $\mathbb{R}$, in which 1 and i are linearly independent (since $i \notin \mathbb{R} \cdot 1$). Therefore this can all be brought about by envisioning $\mathbb{R}$ as the horizontal axis in the cartesian plane $\mathbb{R} \times \mathbb{R}$ and choosing any vector off that axis to play the role of i, the canonical candidate being $(0, 1)$. Of course, there is an obvious way to multiply numbers of the form $x + yi$, since $i^2 = -1$. We thus arrive at the "complex number system $\mathbb{C}$," *kurz und bündig*, in a form adequate for reading this book. But this "natural" evolution was not the historical one. Our forefathers were not comfortable "creating" new numbers. They felt that "$\sqrt{-1}$" was out there somewhere in the Platonic universe; only its metaphysical status and its algebraic legitimacy had to be clarified. Another issue for them, one that despite Euler's seminal work in 1749 continued unsettled after the grounding of $\mathbb{C}$ by Gauss and others, was exponentiation of complex numbers: What is $(a+bi)^{c+di}$ and can it be expressed as $x+yi$? Gradually this was recognized as an analysis issue, rather than one which can be settled merely by the algebraic structure of $\mathbb{C}$. (Its elucidation, and the prosecution of analysis in $\mathbb{C}$ generally is what this book is about.) The absence of an ordering and a trichotomy law (articulated by Euler) added to the reluctance to accord full numberhood to the complexes.

In precising the preceding it is fruitful to adopt a more structural but more open-ended approach, driven by other properties than square roots for negative numbers, which we want the enlarged number system to have. In the end, these properties each determine one and the same $\mathbb{C}$, testimonial to its naturalness, if not to its Platonic apriority. The formal and official development of $\mathbb{C}$ in the following paragraphs will elaborate this.

The *complex* numbers are a 2-dimensional number system $\mathbb{C}$ arising from the introduction of a multiplication [signalled below by juxtaposition] in the $\mathbb{R}$-vector space $\mathbb{R}^2 := \mathbb{R}\times\mathbb{R}$. A natural requirement on such a multiplication is that it make $\mathbb{R}^2$ an $\mathbb{R}$-algebra having $\varepsilon_1 := (1,0)$ as multiplicative identity. Setting $\varepsilon_2 := (0,1)$, this means that for all $a,b,c,d\in\mathbb{R}$

$$(0)\qquad (a,b)(c,d) = (a\varepsilon_1 + b\varepsilon_2)(c\varepsilon_1 + d\varepsilon_2) = ac\varepsilon_1 + bd\varepsilon_2\varepsilon_2 + (ad+bc)\varepsilon_2.$$

Therefore the multiplication is necessarily commutative and is already completely determined except for specifying what vector (x,y) the product $\varepsilon_2\varepsilon_2$ should be. One further natural requirement, that the length of any product of vectors equal the product of their lengths, determines this. For that requirement entails that $x^2+y^2=1$ and, because $\varepsilon_1+\varepsilon_2$, $\varepsilon_1-\varepsilon_2$ each have length $\sqrt{2}$, the length of $(\varepsilon_1+\varepsilon_2)(\varepsilon_1-\varepsilon_2) = \varepsilon_1 - \varepsilon_2\varepsilon_2 = (1-x)\varepsilon_1 - y\varepsilon_2$ is 2. That is, $2 = \sqrt{(1-x)^2+y^2} = \sqrt{1-2x+x^2+y^2} = \sqrt{2-2x}$, whence $x=-1$, $y=0$, $\varepsilon_2\varepsilon_2 = -\varepsilon_1$. The multiplication in $\mathbb{C}$ must therefore be

$$(1)\qquad (a,b)(c,d) = (ac-bd, ad+bc).$$

Cf. Richmond [1951], Fried and Varga [1956], D. J. Hansen [1964] or p. 66 of Ebbinghaus, *et al.* [1990]. The easily confirmed algebraic identity

$$(2)\qquad (ac-bd)^2 + (ad+bc)^2 = (a^2+b^2)(c^2+d^2)$$

testifies that the product (1) does indeed treat lengths multiplicatively. It is an interesting fact that the multiplicativity of length which (2) encodes underlay one of the early geometric adumbrations of complex numbers, François Viète's calculus of triangles. He "multiplied" the right triangle with legs a,b and hypotenuse c times the one with data α,β,γ by requiring that the resultant right triangle have hypotenuse $c\gamma$. He knew the algebraic identity

$(a^2+b^2)(\alpha^2+\beta^2) = (a\alpha - b\beta)^2 + (a\beta + b\alpha)^2$ for expressing the product of sums of squares as a sum of squares. See §5, chapter 5 of BAŠMAKOVA and SMIRNOVA [2000], and **MR** 57#12047 or **MR** 2003j:01006.

The intervention of the geometric concept of length in this algebraic enterprise can be avoided. We can instead ask (as Hermann Hankel did in 1867) that the multiplication in $\mathbb{R}^2$ make it an $\mathbb{R}$-algebra which is also an *integral domain* (no two non-zero elements have product 0, a condition weaker than multiplicativity of length) and which possesses a multiplicative identity ε_1^*. Extend $\{\varepsilon_1^*\}$ to a basis $\{\varepsilon_0^*, \varepsilon_1^*\}$. Then there exist $c, d \in \mathbb{R}$ such that $\varepsilon_0^*\varepsilon_0^* = c\varepsilon_1^* + 2d\varepsilon_0^*$, that is, $(\varepsilon_0^* - d\varepsilon_1^*)^2 = (c+d^2)\varepsilon_1^*$. If $k := c + d^2$ is non-negative, then neither of $\varepsilon_0^* - d\varepsilon_1^* \pm \sqrt{k}\varepsilon_1^*$ is 0, yet their product is 0. Thus $k < 0$, and then $\varepsilon_2^* := (1/\sqrt{-k})(\varepsilon_0^* - d\varepsilon_1^*)$, which is linearly independent of ε_1^*, since ε_0^* is, satisfies $\varepsilon_2^*\varepsilon_2^* = -\varepsilon_1^*$. (Cf. FUCHS and SZELE [1952] or p. 68 of EBBINGHAUS, *et al.* [1990], with a less transparent version in PASCH [1904].) In the basis $\{\varepsilon_1^*, \varepsilon_2^*\}$ multiplication has the same form as above:

$$(3) \qquad (a\varepsilon_1^* + b\varepsilon_2^*)(c\varepsilon_1^* + d\varepsilon_2^*) = (ac - bd)\varepsilon_1^* + (ad + bc)\varepsilon_2^*.$$

The other geometric feature of multiplication (1) is the *direction* it imparts to the product vector: If $u := (a, b)$ is a vector of unit length and $v := (c, d)$ any vector, a trivial computation reveals that uv is the vector $cu + du^\perp$ whose coordinates c, d relative to the basis $u, u^\perp := (-b, a)$ are the coordinates c, d of v relative to ε_1, ε_2; that is,

(4) $\qquad uv$ is the result of *rotating* v by u.

Conversely, the validity of (4) just for $u = v = \varepsilon_2$ entails $\varepsilon_2\varepsilon_2 = -\varepsilon_1$, and so from (0) it follows that the multiplication must be that in (1). Moreover, every rotation is (1)-multiplication by some unit-length vector u, because generally every self-map of $\mathbb{R}^2$ which fixes $(0, 0)$ and preserves distances between vectors has either the form $(x, y) \mapsto (a, b)(x, y)$ or the form $(x, y) \mapsto (a, b)(x, -y)$ for some unit-length vector (a, b) and the multiplication defined by (1). *Very* elementary proofs of this will be found in GAUGHAN [1962] and on p. 85 of EBBINGHAUS, *et al.* [1990]; cf. the hints to 6.5(iv) below. Since $(x, y) \mapsto (x, -y)$ is a non-identity $\mathbb{R}$-linear transformation which fixes the non-zero vector ε_1, it is *not* a rotation; hence no rotation $(a, b)(x, -y)$ of it is either.

Because property (4) and multiplicativity of length are so useful, their progenitor (1) is universally taken to be the multiplication in $\mathbb{R}^2$ that *defines* $\mathbb{C}$, which thereby becomes not merely an $\mathbb{R}$-algebra, but a *field*, the multiplicative inverse of the non-zero element (a, b) being $((a^2 + b^2)^{-1}a, -(a^2 + b^2)^{-1}b)$.

Multiplicativity of length is a surprisingly strong hypothesis on an algebra multiplication, as the famous Gelfand-Mazur theorem illustrates: any $\mathbb{C}$-algebra equipped with a multiplicative norm necessarily possesses a multiplicative identity and is the set of all $\mathbb{C}$-multiples of it. See DORAN [1980].

When the elements of $\mathbb{R}^2$ are written as column vectors, (1) shows that the $\mathbb{R}$-linear transformation of multiplication by (a, b) is represented by the matrix $\left(\begin{smallmatrix} a & -b \\ b & a \end{smallmatrix}\right)$. Consequently, the set of all such two-by-two real matrices with the usual matrix multiplication is $\mathbb{R}$-algebra isomorphic to $\mathbb{R}^2$ with multiplication (1). $\mathbb{C}$ could therefore have been *defined* to be this subalgebra of matrices and its multiplication would already have been in place!

By the same token, when $\mathbb{R}^2$ is viewed as the set of all *functions* from $\{1, 2\}$ into $\mathbb{R}$, then with the usual operations on functions it becomes a commutative unital $\mathbb{R}$-algebra. But the multiplication does not treat lengths multiplicatively and abounds with zero-divisors. Consequently, $\mathbb{R}^2$ with this "coordinatewise" multiplication has not found great utility.

The mapping $(x, y) \mapsto (x, -y)$ is the only unital $\mathbb{R}$-algebra endomorphism of $\mathbb{C}$ other than the identity mapping—a fact which is trivial to confirm. It is an example of an *involution* in a ring and is called (*complex*) *conjugation*. The image of z is denoted $\bar{z}$ (in some of the literature z^*) and called the (*complex*) *conjugate* of z. The number $(0, 1)$ is called the *imaginary* unit and is universally denoted (following Euler) by i; its real multiples are called *purely imaginary* numbers. [Naturally "i" is henceforth unavailable for any other jobs, in particular, as an index of summation.] The $\mathbb{R}$-vector-space equality $z = (x, y) = x\cdot(1, 0) + y\cdot(0, 1)$ is usually written $(x, y) = x + yi$, that is, $\mathbb{R}$ is identified with its isomorphic copy $\mathbb{R}\cdot(1, 0)$ in $\mathbb{R}^2$. We write $|z|$ for $\sqrt{z\bar{z}}$, the *non-negative* square root of $z\bar{z} = x^2 + y^2$. (See §6 below for existence.) This number, the *norm* or *length* of the vector z, is also called the *modulus* or the *absolute value* of z. x is called the *real part* or *abscissa* of z, y the *imaginary part* (despite its being a real number!) or *ordinate* of z and we write $x =: \operatorname{Re} z$, $y =: \operatorname{Im} z$. Notice that multiplicativity of length and inversion in the field $\mathbb{C}$ are quite transparent in terms of conjugation: $|zw|^2 = zw\overline{zw} = zw\bar{z}\bar{w} = |z|^2|w|^2$, and from $z\bar{z} = |z|^2$ follows $z^{-1} = |z|^{-2}\bar{z}$.

The terms "real" and "imaginary" are relics from the days when these concepts had not yet been formalized, were imperfectly understood, especially their ontological status, and were generally rather mysterious. (Some of the bizarre pre-Gaussian views of complex numbers are sampled on pp. 208-9 of PIEPER [1991].) They have been de-mythologized over the years, so there is now neither any need nor any zeal to eradicate these so characteristically unmathematical terms from the official lexicon. Of course, no scintilla of their common parlance meaning still attaches to them. Gauss himself saw in them a barrier to understanding and (1832) suggested "lateral" for "imaginary," "direct" for "positive real" and "inverse" for "negative real"; but he recognized it was already too late to effect such a change in usage. In subsequent development of mathematics, in which *all* numbers are recognized as products of the imagination (see below), the expression "imaginary number" has become more pleonastic than problematic. It is otherwise with "imaginary *quantity*" (no longer in use), which has a definite oxymoronic character (and whose incoherence is ironically illustrated in the paper **MR** 2007k:01007).

A 21st-century reader may be slightly amused or slightly annoyed at the terminology, and wonder about its genesis. Probably he finds the definition of $\mathbb{C}$ as $\mathbb{R} \times \mathbb{R}$ with the multiplication (1) a mathematically routine matter and is not aware of the long and painful historical birth of $\mathbb{C}$. In fact, this one phenomenon is a singular chapter in the history of mathematics, because it chronicles the steps to the modern abstract, axiomatic view of mathematics. First, "imaginary" numbers were "discovered" in the 16th century as Tartaglia, Cardano and Bombelli attempted to solve cubic polynomial equations. Up to that point algebra had been the science of quantity; numbers "measured" how much or "counted" how many, they had no independent existence. (The procrustean persistence of this point of view retarded the enfranchisement of negative as well as complex numbers.) Yet the operations on numbers could be carried out in a larger context (and some combinations of the resulting "impossible" numbers were real and eventuated in real answers). So maybe algebra should instead be the science of these operations, a foreshadowing of Hilbert's formalism by the 19th-century British school of Robert Woodhouse, George Peacock, Duncan F. Gregory, Augustus De Morgan, George Boole, *et al.* (see **MR** 43#4612, 86d:01015, 98d:01023, 2003k:01012, 2003m:01032, 2007h:01004 and especially chapter VII of Cauchy's revolutionary 1821 "textbook"). Continental efforts went toward finding a model (*avant la lettre*), usually geometric, in which to interpret "imaginary" numbers. The period 1797–1831 saw this completed, independently by several men (Caspar Wessel in Denmark, Jean Robert Argand in France, Gauss in Germany). A synthesis occurred with the work of the Irishman William Rowan Hamilton. He felt that in the "science of symbols" the symbols had to symbolize *something* and the geometric "models" did not fill this ontological need, not to mention that founding algebra on geometry was philosophically unsatisfactory. So he introduced ordered pairs and operations on them. As he also isolated the key rules that algebraic operations should obey (done as well by Martin Ohm in Germany), we see the concept of a field being born. In calling these "numbers," we have the final emancipation of number from measuring and the "real" world, a liberation that took centuries of struggle—although a *logically* satisfactory grounding of $\mathbb{R}$ was still to come. [At this point the reader who hasn't dozed off should be hearing the majestic last movement (*Des Helden Weltflucht und Vollendung*) of Richard Strauss' epic opus 40—or is opus 35 more appropriate?] These ideas did not immediately carry the day, mathematicians and textbook writers continuing for some decades to entertain "unreal" attitudes toward $\mathbb{C}$. For example, Cauchy came on board as late as 1847—actually, in CAUCHY [1847] he constructed his own $\mathbb{C}$-worthy vessel, namely the quotient of the polynomial ring $\mathbb{R}[X]$ modulo the ideal generated by X^2+1. By the late 19th century, however, the pendulum had swung further, and all numbers outside of $\mathbb{N}$ were recognized as "artificial," a point of view memorably articulated by Leopold Kronecker in "Die ganzen Zahlen hat der liebe Gott gemacht, alles andere ist Menschenwerk." But in fact, Georg Cantor, Richard Dedekind, Gottlob Frege, Giuseppe Peano and subsequent generations of mathematicians came to regard even $\mathbb{N}$ as a creation of the human mind (although we encounter the adjective 5 in the world, the noun 5 raises ontological questions). For example, $\mathbb{N}$ may be created from $\varnothing$ via a successor function, suggesting a radical revision of Kronecker that incorporates Plato's idea [in the *Sophist*] of being as power: "God created the empty set and empowered man to succeed." Less reverentially phrased, mathematics is thus *die Lehre von der Leere*. Much has been written on this subject, often under the title "The number concept" (*Der Zahlbegriff*)—e.g., **MR** 39#2574, 47#3134, 83d:00015, 85d:01003, 98f:01030, 99b:01003. The preceding sketch is a crude condensation of NAGEL [1979], to which the philosophically inclined reader is enthusiastically referred (see also §11.4, 11.5 in DUNMORE [1992], J. J. GRAY [1992] and §7.2, 18.1 in FERRARO [2008]); a more poetic account is given by LORIA [1917] and MAZUR [2003]. For the lighter side of things see SLAUGHT [1928], AIGNER [1978], CREMER [1982] and WILDBERGER [1999].

From its earliest birth-pangs $\mathbb{C}$ had an air of the "Philosopher's Stone": It seemed increasingly likely that $\mathbb{C}$ had the potential to "transmute" not just its generating polynomial $X^2+1=(X+i)(X-i)$, but all "base" polynomials with complex coefficients, into "noble" products of linear factors. And successive generations strove to realize this. Since a really unimpeachable proof is hardly possible until $\mathbb{C}$ itself is well grounded, the evolution of such proofs and of $\mathbb{C}$ proceeded in tandem. The profound fact that $\mathbb{C}$ has this property, known as *algebraic closure*, was once (rightly) considered "the fundamental theorem of algebra" (FTA), a distinction it has lost in the vast panorama of modern algebra. Nowadays very elegant yet elementary proofs exist, employing nothing more than ideas discussed in this preliminary chapter. For them (plus history) see 2.21 and the notes and references in chapter IV. An interesting consequence of algebraic closure is the finality it implies for the progression of ever larger number systems: That stops at $\mathbb{C}$, in the sense that $\mathbb{C}$ is the only proper field extension of $\mathbb{R}$ which is finite-dimensional (or even just algebraic) over $\mathbb{R}$. See R. M. YOUNG [1988] for the easy proof of this claim (which goes back to Gauss).

For history of $\mathbb{C}$ see §§1–11 of STUDY and CARTAN [1908], chapter 18 of PIEPER [1991], LÜTZEN [1998], or the monumental FLAMENT [2003] and LÜNEBURG [2008]. But the best account of $\mathbb{R}$ and $\mathbb{C}$ which covers their historical evolution plus technical details is the first three chapters of EBBINGHAUS, *et al.* [1990].

The order relation in $\mathbb{N}$ is transmitted up through $\mathbb{Z}$, $\mathbb{Q}$ and into $\mathbb{R}$ (LANDAU, *loc. cit.*). We write $\mathbb{Z}^+$, $\mathbb{Q}^+$, $\mathbb{R}^+$ for the non-negative elements of $\mathbb{Z}$, $\mathbb{Q}$, $\mathbb{R}$, respectively. For a, $b \in \mathbb{R}$ we write $a \le b$ and $b \ge a$ if $b - a \in \mathbb{R}^+$, $a < b$ and $b > a$ if $b - a \in \mathbb{R}^+ \setminus \{0\}$, i.e., if $b - a$ *is positive.* When $a \le b$ we write

$$\begin{aligned} [a,b] &:= \{x \in \mathbb{R} : a \le x \le b\}, && \text{called a closed interval ,} \\ (a,b) &:= \{x \in \mathbb{R} : a < x < b\}, && \text{called an open interval ,} \\ [a,b) &:= \{x \in \mathbb{R} : a \le x < b\}, && \text{called a half-open (on the right) interval ,} \\ (a,b] &:= \{x \in \mathbb{R} : a < x \le b\}, && \text{called a half-open (on the left) interval.} \end{aligned}$$

In each case a is called the *left endpoint*, b the *right endpoint* of the interval. There is no order in $\mathbb{C}$ but an interval notation is handy there too and will be introduced in Chapter I. We rely on context to resolve the ambiguity in the notation (a, b)—is it an interval or an ordered pair of real numbers? The problem seldom arises because we don't often use ordered pair notation for complex numbers and then not in the presence of open intervals. The author asks readers' (further) indulgence for his inability to break old habits and adopt the superior Bourbaki interval notation:] for (and [for). Another deviation from the Bourbaki canon about which the reader needs to be alerted (although it is widespread) is that the definition of $\mathbb{N}$ adopted above excludes 0.

For any a, $b \in \mathbb{R}$ we also write

$$\begin{aligned} [a,\infty) &:= \{x \in \mathbb{R} : a \le x\} \\ (a,\infty) &:= \{x \in \mathbb{R} : a < x\} \\ (-\infty,b] &:= \{x \in \mathbb{R} : x \le b\} \\ (-\infty,b) &:= \{x \in \mathbb{R} : x < b\} \end{aligned}$$

with the obvious verbalizations. [No ontological significance attaches to the symbols ∞, $-\infty$ here (as yet).]

§ 4 Metric Spaces

Recall that a *metric space* is a (non-empty) set, together with a *distance function* or *metric* $d \colon X \times X \mapsto [0,\infty)$ which satisfies for all x_1, x_2, $x_3 \in X$

(1) $\quad d(x_1,x_2) = d(x_2,x_1)$ (symmetry)

(2) $\quad d(x_1,x_2) = 0 \Leftrightarrow x_1 = x_2$ (reflexivity)

(3) $\quad d(x_1,x_3) \le d(x_1,x_2) + d(x_2,x_3)$ (triangle-inequality).

Formally, a metric space is a pair (X, d) as above but one speaks loosely of "the metric space X". If $x_0 \in X$ and $r \in [0, \infty)$, the set $\{x \in X : d(x, x_0) < r\}$ is called the (open) *ball of radius* r *and center* x_0. It is denoted by $B(x_0, r)$. A set in a metric space is *bounded* if it lies wholly in some ball, *open* if it is a union of open balls and *closed* if its complement is open. The *closure* (or *adherence*) of a set S is the intersection of all the closed supersets. It is evidently closed. We denote it by $\overline{S}$. The union of all the open subsets of S is called the *interior* of S and is denoted $\overset{\circ}{S}$; it is obviously an open set but may well be empty (see the convention at the end of §1). Note that $\overset{\circ}{S} = X \setminus \overline{(X \setminus S)}$. A *neighborhood* of a point $x \in X$ is any set which contains some $B(x, r)$, $r > 0$. Generally N (neighborhood), V (*voisinage* or vicinity) or U (*Umgebung*) are used to designate such sets. Most often they are open but this is not part of the definition here (though it is in some books).

The function $d(z, w) := |z - w|$, z, $w \in \mathbb{C}$ possesses properties (1)–(3) above and is the standard (euclidean) metric in $\mathbb{C}$. Here balls are called *disks* (to be more suggestive) and we use $D(z_0, r)$ instead of $B(z_0, r)$, calling this set the (*open*) *disk of radius* r *and center* z_0. Notice that $\overline{D}(z_0, r) = \{z \in \mathbb{C} : |z - z_0| \leq r\}$, called the (*closed*) *disk of radius* r *and center* z_0. In a general metric space however $\overline{B}(x_0, r)$ lies in, but need not coincide with, $\{x \in X : d(x, x_0) \leq r\}$. (E.g., $X = \mathbb{N}$, r a positive integer.) The set $C(z_0, r) := \{z \in \mathbb{C} : |z - z_0| = r\}$ is called the *circle of center* z_0, *radius* r. For the set $C(0, 1)$ the notation $\mathbb{T}$ (for torus) and for $D(0, 1)$ the notation $\mathbb{D}$ will be used. The set $\{z \in \mathbb{C} : r < |z - c| < R\}$ is called the *annulus of center* c, *inner radius* r *and outer radius* R. It is defined for every $c \in \mathbb{C}$ and $0 \leq r < R \leq \infty$. The notation $A(c; r, R)$ is used for this set, or simply $A(r, R)$ if $c = 0$. Two other sets which come up often enough to deserve special names are the open *upper half-plane* $\{z \in \mathbb{C} : \operatorname{Im} z > 0\}$ and the open *right half-plane* $\{z \in \mathbb{C} : \operatorname{Re} z > 0\}$; we suggestively denote them $\mathbb{U}$ and $\mathbb{H}$, respectively. Sometimes we write $\mathbb{C}_-$ for $\mathbb{C} \setminus (-\infty, 0]$, $\mathbb{C}_+$ for $\mathbb{C} \setminus [0, \infty)$.

For each positive integer n we let $\mathbb{C}^n := \mathbb{C} \times \cdots \times \mathbb{C}$, the cartesian product of n copies of $\mathbb{C}$. Formally this is the set of all functions z from $\{1, \ldots, n\}$ into $\mathbb{C}$, but we don't think in those terms. Rather we write z_j for $z(j)$, $1 \leq j \leq n$, and we display z as the *ordered* n*-tuple* $(z_1, \ldots, z_n)$. This is a vector space over $\mathbb{C}$ (coordinate-wise operations). A conjugate-bilinear form is defined in $\mathbb{C}^n$ by

$$(z_1, \cdots, z_n) \bullet (w_1, \cdots, w_n) := \sum_{j=1}^{n} z_j \overline{w}_j .$$

This is called the *dot*, or *scalar*, or *inner product* in $\mathbb{C}^n$ and together with its obvious linearity properties it satisfies the so-called *Cauchy-Schwarz Inequality*

$$(4) \qquad |(z_1, \cdots, z_n)\bullet(w_1, \cdots, w_n)|^2 \leq \sum_{k=1}^{n} |z_k|^2 \cdot \sum_{j=1}^{n} |w_j|^2,$$

thanks to the directly and easily confirmed *Lagrange identity*:

$$\left|\sum_{k=1}^{n} z_k \overline{w}_k\right|^2 = \sum_{k=1}^{n} |z_k|^2 \cdot \sum_{j=1}^{n} |w_j|^2 - \frac{1}{2}\sum_{k=1}^{n}\sum_{j=1}^{n} |z_k w_j - z_j w_k|^2.$$

A further useful consequence of this identity is that equality holds in (4) and some $w_j \neq 0$ if and only if the number $c := z_j/w_j$ satisfies $z_k = cw_k$ for all k, that is, $(z_1, \ldots, z_n) = c(w_1, \ldots, w_n)$. The inner product defines a *norm* or *length* function $\|\cdot\|\colon \mathbb{C}^n \mapsto [0, \infty)$ by means of

$$\|z\| := \sqrt{z\bullet z}.$$

In this language (4) reads simply

$$(5) \qquad |z\bullet w| \leq \|z\|\|w\| \qquad \forall z,\ w \in \mathbb{C}^n.$$

The usual *euclidean metric* in $\mathbb{C}^n$ is that defined by the above norm:

$$d(z, w) := \|z - w\| \qquad \forall z, w \in \mathbb{C}^n.$$

It is, of course, via (5) that we verify property (3) [(1) and (2) being obviously true for this d].

A *sequence* in a metric space X is (formally) a function x from $\mathbb{N}$ or $\mathbb{Z}^+$ into X, with $x(n)$ usually written as x_n. [Functions from $\mathbb{Z}$ to X are called *bilateral* sequences.] We denote the sequence x variously by $\{x_n\}_{n=1}^{\infty}$, $\{x_n\}_{n=0}^{\infty}$ (as the case may be) or simply $\{x_n\}$. A sequence $\{x_n\}$ is *Cauchy* or *fundamental* if for each $\varepsilon > 0$ there exists an $N(\varepsilon) \in \mathbb{N}$ such that $d(x_n, x_m) < \varepsilon$ whenever n, $m \geq N(\varepsilon)$. The sequence $\{x_n\}$ *converges* to $p \in X$ if for each $\varepsilon > 0$ there exists an $N(\varepsilon)$ such that $d(x_n, p) < \varepsilon$ for all $n \geq N(\varepsilon)$. Obviously such a p is unique; it is called the *limit* of the sequence. We write $\lim_{n\to\infty} x_n = p$ and sometimes $x_n \to p$ as $n \to \infty$, or simply $x_n \to p$. The sequence $\{x_n\}$ is called *convergent* if such a p exists, *divergent* otherwise. (Notice that a convergent sequence is necessarily Cauchy and a Cauchy sequence is necessarily bounded.) A metric space (X, d) is called *complete* if every Cauchy sequence in it is convergent.

If $k\colon \mathbb{N} \mapsto \mathbb{N}$ $\left(\text{or } \mathbb{Z}^+ \mapsto \mathbb{Z}^+\right)$ is any increasing (see §5) function and x is a sequence, then the composite $x \circ k$ is called a *subsequence* of x. With the above conventions we write k_n for $k(n)$ and x_{k_n} for $x(k_n) = x(k(n))$.

The concept of a *series* in a euclidean space $\mathbb{C}^m$ is very simple. But, as with "real" and "imaginary," some ambiguous terminology has become historically entrenched, and all must learn to live with it. A *series* s is a sequence having the form $s_n := \sum_{k=1}^n x_k$ for all n, where $x\colon \mathbb{N} \mapsto \mathbb{C}^m$ is another sequence. x_k is called the kth *term* or (better) *summand*, s_n the nth *partial sum* of the series s. (But of course, x_k is also the kth term of the sequence x!) The symbol $\sum_{k=1}^\infty x_k$, sometimes simply $\sum x_k$, does double duty: It serves as a name for s and, in case $\lim_{n\to\infty} s_n$ exists, it serves as a name for this limit too. The crowning element of confusion: *Every* sequence $y\colon \mathbb{N} \mapsto \mathbb{C}^m$ is a series. One has only to define $x\colon \mathbb{N} \mapsto \mathbb{C}^m$ by $x_1 := y_1$, $x_{k+1} := y_{k+1} - y_k$ for all $k \in \mathbb{N}$ in order to have y be the series $\sum x_k$. (What are the terms of the sequence [series] y?) Similar conventions are common in other contexts. For example, with $x : \mathbb{Z} \mapsto \mathbb{C}^m$ a bilateral sequence, the nth partial sum s_n is $\sum_{k=-n}^n x_k$ and when $\lim_{n\to\infty} s_n$ exists, either $\sum_{\mathbb{Z}} x_k$ or $\sum_{k=-\infty}^\infty x_k$ designates it.

The series $\sum x_k$ is said to be *absolutely* or *normally convergent* if the associated series $\sum \|x_k\|$ converges in $\mathbb{R}$, and due to the least-upper-bound property of $\mathbb{R}$ (see §5) the latter is equivalent to the boundedness of $\{\sum_{k=1}^n \|x_k\| : n \in \mathbb{N}\}$. The completeness of $\mathbb{C}^m$ confirms what this language suggests: absolute convergence implies convergence. For the Cauchyness of the sequence of partial sums of $\sum x_k$ follows from that of $\sum \|x_k\|$ via the triangle-inequality.

An *isolated point* of a set S in a metric space X is a point $x \in X$ such that $B(x, r) \cap S = \{x\}$ for some $r > 0$. A *cluster point* of S is a point $x \in X$ such that every ball centered at x meets S. A non-isolated cluster point is called a *limit point* or an *accumulation point.* (The reader is cautioned that there is no unanimity in the literature concerning the use of these last three terms.) The cluster points of S are just the limits of sequences in S; the set of all of them coincides with $\overline{S}$. The *boundary* or *frontier* of S, denoted ∂S or *bdry*(S), is the set $\overline{S} \setminus \overset{\circ}{S} = \overline{S} \cap \overline{(X \setminus S)}$. It consists of those $x \in X$ every neighborhood of which meets both S and $X \setminus S$.

A metric space X is *compact* if whenever a class $\mathcal{O}$ of open subsets of X satisfies $X \subset \bigcup \mathcal{O}$ (we say $\mathcal{O}$ *covers* X), then $X \subset \bigcup \mathfrak{F}$ for some finite $\mathfrak{F} \subset \mathcal{O}$ (we say X *admits a finite subcover* from $\mathcal{O}$). This is equivalent to the condition that every sequence in X possess a convergent subsequence. (See RUDIN [1976], pp. 36–40.) If we take complements of open sets, the definition of compactness is that the space have the *finite-intersection property*: Whenever $\mathcal{C}$ is a class of closed subsets of X such that $\bigcap \mathfrak{F} \neq \varnothing$ for each finite $\mathfrak{F} \subset \mathcal{C}$, then $\bigcap \mathcal{C} \neq \varnothing$. See PIER [1980] and KALB [1987] for the history of this important concept.

If (X, d) is a metric space and $\varnothing \neq Y \subset X$, then $(Y, d\,|\,Y \times Y)$ is a metric space. Notice that for $y_0 \in Y$ the ball of center y_0 and radius r *in* Y is the intersection with Y of the corresponding ball in X. $S \subset Y$ is called *relatively open* (or *closed*) if it is open (or closed) in the metric space $(Y, d \,|\, Y \times Y)$. This is equivalent to $S = T \cap Y$ where T is open (or closed) in (X, d). Y is called a *subspace* of X. It follows that the metric space $(Y, d \,|\, Y \times Y)$ is compact if and only if every cover of Y by open subsets of X admits a finite subcover and from this it follows that Y must be closed if it is compact. A subset Y of X is called *pre-compact* or *conditionally compact* or *relatively compact* if $\overline{Y}$ is compact. The two most important features of $\mathbb{C}^n$ are that it is a complete metric space and a subset of it is compact if and only if closed and bounded (*Heine-Borel* theorem; see the remark following 1.8). Then the aforementioned sequential characterization of compactness means that every bounded sequence in $\mathbb{C}^n$ contains a convergent subsequence (*Bolzano-Weierstrass* theorem). Thus, non-empty compact subsets of $\mathbb{R}$ contain smallest and largest elements.

Though the terminology is paradoxical, there is nothing logically wrong with a subset Y of a metric space being both closed and open. (*Every* subset of the metric space $\mathbb{Z}$ has this property, for example.) For such a set the neologism *clopen* is employed, though many linguistic purists prefer the less colorful term closed-and-open (or open-and-closed).

A metric space (X, d) is called *connected* if the only clopen subsets of X are $\varnothing$ and X. A *disconnection* of a metric space X is a pair of disjoint sets X_1, X_2 which cover X and are each non-empty and open (equivalently, by looking at complements, each closed). Then, evidently, X is connected if and only if it possesses no disconnection. This concept is explored in some depth in Chapter I and together with compactness is of fundamental importance for all that follows.

We often face the following situation: X is a countable set, Y a metric space and for each $n \in \mathbb{N}$ a map $f_n \colon X \mapsto Y$ is given. And we want a subsequence $\{f_{n_j}\}_{j=1}^{\infty}$ of $\{f_n\}_{n=1}^{\infty}$ such that $\{f_{n_j}(x)\}_{j=1}^{\infty}$ converges in Y for each $x \in X$. A sufficient condition for such a subsequence to exist is the relative compactness in Y of the orbit $\{f_n(x) : n \in \mathbb{N}\}$ of each $x \in X$. The procedure for obtaining a subsequence of the desired kind is called the (*Cantor*) *diagonal process*; for its (elementary) details see RUDIN [1976], Theorem 7.23 or STROMBERG [1981], Theorem (3.144).

§ 5 Limsup and All That

In the Dedekind construction of $\mathbb{R}$ (i.e., that in E. LANDAU [1951]) the order relation is absolutely primal, and one of the first things proved about $\mathbb{R}$ so constructed is its *order completeness*: for every non-empty subset of $\mathbb{R}$ which is *bounded above* (i.e., lies in $(-\infty, b]$ for some $b \in \mathbb{R}$) there is a *least upper bound* (i.e., there is a least such b). When $\mathbb{R}$ is not manufactured but only axiomatically summoned, this property must be postulated: $\mathbb{R}$ is a (the unique) field with an order relation that behaves in the expected way with respect to addition and multiplication and is complete in the above sense (so-called *Least Upper Bound Axiom*). [For some of the surprisingly many logically equivalent ways to express the completeness of $\mathbb{R}$, see **MR** 3365375.] In any case, it is from this fundamental property that the metric completeness of $\mathbb{R}$ (described in §4) and the compactness of closed-and-bounded sets (as well as the connectedness of intervals, to be proved in 1.8) is derived. [The extension of these properties to $\mathbb{C}$, thence to $\mathbb{C}^n$, is routine.] The most immediate consequence of this property of $\mathbb{R}$ is that its order is *Archimedean*, meaning that its subset $\mathbb{N}$ is not bounded. For if $\mathbb{N}$ were bounded, its least upper bound b would satisfy $n+1 \leq b$ for all $n \in \mathbb{N}$, whereupon $b-1$ would be a smaller upper bound. At the other extreme, every positive real number exceeds $1/n$ for some $n \in \mathbb{N}$. Thus, given $a, b \in \mathbb{R}$ with $a < b$, there is $n \in \mathbb{N}$ with $1/n < b - a$. Letting k denote the largest integer such that $k \leq na$, we then have $a < (k+1)/n < b$, showing that $\mathbb{Q}$ is *dense* in $\mathbb{R}$.

The order relation in $\mathbb{R}$ can be rendered unconditionally complete by adjoining a least element $-\infty$ and a greatest element $+\infty$ (often written just ∞). These are two (distinct) points not in the set $\mathbb{R}$. The order relation in $\mathbb{R}$ is extended to $\overline{\mathbb{R}} := \mathbb{R} \cup \{\pm\infty\}$ by the decree

$$-\infty < x < +\infty \qquad \forall x \in \mathbb{R}.$$

The algebraic structure of $\mathbb{R}$ is not extended (or sometimes only partially extended) to $\overline{\mathbb{R}}$. In $\overline{\mathbb{R}}$ the above axiom reads: every subset of $\overline{\mathbb{R}}$ has a least upper bound. The abbreviation l.u.b. S for the least upper bound of a set S is common. The usual convention about vacuous hypotheses leads to l.u.b. $\varnothing = -\infty$. A synonym for least upper bound is *supremum* and the supremum of S is noted $\sup(S)$. *Greatest lower bound*, g.l.b. S, *infimum* and $\inf(S)$ are defined in the obvious way. A few self-explanatory notational variants prove quite convenient. E.g., $\sup_{\alpha\in I}\{x_\alpha\}$ for $\sup\{x_\alpha : \alpha \in I\}$ or $\sup_{n\geq k}\{x_n\}$ for $\sup\{x_n : n \geq k\}$.

For a non-void subset S of a metric space (X, d) the element

$$\sup\{d(x,y) : x, y \in S\}$$

of $[0, +\infty]$ is called the *diameter of* S, sometimes noted $\operatorname{diam}(S)$. For any point $x \in X$ the *distance from* x *to* S is denoted $d(x, S)$ and is defined as

$$d(x,S) := \inf\{d(x,y) : y \in S\}.$$

(See 1.32.)

For a sequence $\{x_n\}_{n=1}^\infty \subset \overline{\mathbb{R}}$ we define $\overline{\lim}_{n\to\infty} x_n = \limsup_{n\to\infty} x_n$ (called the *limit superior* or *upper limit* of the sequence $\{x_n\}$) as $\inf_{n\geq 1}\big[\sup_{k\geq n} x_k\big]$.

Analogously $\underline{\lim}_{n\to\infty} x_n = \liminf_{n\to\infty} x_n$ (called the *limit inferior* or *lower limit*) is defined as $\sup_{n\geq 1}\bigl[\inf_{k\geq n} x_k\bigr]$. With the obvious notion of sequential convergence in $\overline{\mathbb{R}}$ (based on the order topology, rather than the absolute-value metric), we have

$$(*) \qquad \underline{\lim}_{n\to\infty} x_n \leq L \leq \overline{\lim}_{n\to\infty} x_n$$

for every subsequential limit L of the sequence $\{x_n\}$, and the sequence $\{x_n\}$ itself converges to $L \in \overline{\mathbb{R}}$ if and only if equality holds in $(*)$. It follows easily from the definition that $\underline{\lim}_{n\to\infty} x_n$ is the smallest subsequential limit of $\{x_n\}$ in $\overline{\mathbb{R}}$ and $\overline{\lim}_{n\to\infty} x_n$ is the largest.

For a sequence $\{x_n\} \subset \mathbb{C}$ we write $x_n \to \infty$ if the sequence $\{|x_n|\}$ converges to $+\infty$ in $\overline{\mathbb{R}}$. This is sometimes verbalized as "x_n *converges* to ∞" and sometimes as "x_n *diverges* to ∞"!

If $S \subset \overline{\mathbb{R}}$, $f, g\colon S \mapsto \overline{\mathbb{R}}$, then $f > g$ means that $f(x) > g(x)$ for all $x \in S$, and f is *positive*, written $f > 0$, if $f(x) > 0$ for *all* $x \in S$. The function f is called *non-decreasing* or *order-preserving* or *isotonic* [*increasing*] if $f(x) \leq f(y)$ $[f(x) < f(y)]$ whenever x, $y \in S$ and $x < y$. Sometimes *strictly increasing* is used for increasing. *Non-increasing, antitonic, decreasing*, etc., have then the obvious meanings. The term *monotonic* is generic for all these properties. [This is the site of another lost linguistic battle: we ought to be saying "nowhere-decreasing" or "weakly increasing"; after all, a case can be made that the sine function is non-decreasing—it is not a decreasing function! Similarly, although every $x \in \mathbb{C} \setminus \mathbb{R}$ is non-negative, universal usage condones "consider non-negative x" when "consider non-negative real x" is meant.]

§ 6 Continuous Functions

If (X_1, d_1), (X_2, d_2) are metric spaces, $S \subset X_1$, $f\colon S \mapsto X_2$, $b \in X_2$, a a limit point of S, we say that $f(x)$ *converges to* b *as* x *converges to* a if for every $\varepsilon > 0$ there exists a $\delta = \delta(f, a, \varepsilon) > 0$ such that $d_2(f(x), b) < \varepsilon$ whenever $x \in S$ and $0 < d_1(x, a) < \delta$. Because a is a limit point of S, there always are such x and consequently at most one such b. It is called the *limit of* f *at* a and denoted $\lim_{x\to a} f(x)$. Notice that $\lim_{x\to a} f(x) = b$ is equivalent to the implication

$$\{x_n\} \subset S \setminus \{a\} \;\&\; \lim_{n\to\infty} x_n = a \Rightarrow \lim_{n\to\infty} f(x_n) = b.$$

[This equivalence, and that between sequential and covering compactness, is "elementary," but in fully formalized set theory it necessarily involves the Axiom of Choice. See **MR** 33#3909.]

We say that f is *continuous at* $a \in S$ if either a is an isolated point of S or a is a limit point of S and $\lim_{x\to a} f(x) = f(a)$. For a subset T of S, we say f *is continuous on* T if f is continuous at each point of T. This is (generally) different from (stronger than) $f|T$ being a continuous function on its domain. That f be continuous on S is equivalent to the requirement that $f^{-1}(U)$ be a relatively open subset of S whenever U is an open subset of X_2. If f is continuous on S and for each $\varepsilon > 0$ the $\delta(f, a, \varepsilon)$ can be chosen the same (say, $\delta(f, \varepsilon)$) for each $a \in S$, then f is called *uniformly continuous on* S. If f is continuous at each $a \in S$ and S is compact, then f is uniformly continuous on S. Indeed, for each $\varepsilon > 0$, $n \in \mathbb{N}$ the set $C_n := \{(x, y) \in S \times S : d_1(x, y) \leq 1/n$ & $d_2(f(x), f(y)) \geq \varepsilon\}$ is closed in the compact space $S \times S$, and $C_n \downarrow \varnothing$. Hence (by the finite-intersection property of compacta) for some N, $C_N = \varnothing$, that is, $x, y \in S$ & $d_1(x, y) \leq 1/N \Rightarrow d_2(f(x), f(y)) < \varepsilon$. A family $\mathfrak{F}$ of continuous functions on S is called *uniformly equicontinuous on* S if in addition, for each $\varepsilon > 0$ and all $f \in \mathfrak{F}$ a common value $\delta(\varepsilon)$ can be selected for the $\delta(f, \varepsilon)$.

f is called an *open map* if $f(V)$ is relatively open in $f(S)$ for every relatively open subset V of S.

If (X, d) is a metric space, $S \subset X$, $f \colon S \mapsto \overline{\mathbb{R}}$ and a a limit point of S, we define

$$\overline{\lim_{x\to a}}\, f(x) = \limsup_{x\to a} f(x) := \inf_{r>0} \left[\sup f(B(a, r) \cap S)\right].$$

Similarly we define $\underline{\lim}_{x\to a} f(x) = \liminf_{x\to a} f(x)$. It is evident that $\overline{\lim}_{x\to a} f(x)$ is the maximum of the elements $\overline{\lim}_{n\to\infty} f(x_n) \in \overline{\mathbb{R}}$ over all possible sequences $\{x_n\} \subset S$ which converge to a. When X is $\mathbb{R}$, $\overline{\lim}_{x\to\infty}$, $\overline{\lim}_{x\to-\infty}$, etc., have the obvious meanings.

Well-known elementary facts about continuous functions which we shall use: the composite of two continuous functions is continuous. The functions $x \mapsto \|x\|$, $x \mapsto \lambda x$ ($\lambda \in \mathbb{C}$) are continuous from $\mathbb{C}^n$ to $\mathbb{R}$ and $\mathbb{C}^n$, respectively; the functions $(x, y) \mapsto x\bullet y$ and $(x, y) \mapsto x + y$ are continuous from $\mathbb{C}^n \times \mathbb{C}^n$ to $\mathbb{C}$ and $\mathbb{C}^n$, respectively. The function $(x, y) \mapsto xy$ is continuous from $\mathbb{C} \times \mathbb{C}$ to $\mathbb{C}$ and $z \mapsto 1/z = z^{-1}$ is continuous from $\mathbb{C} \setminus \{0\}$ into $\mathbb{C} \setminus \{0\}$. With the composition law the continuity of all the elementary real- and complex-valued functions is thus assured.

The inverse of a continuous, one-to-one function need not be continuous (those functions for which the inverse *is* continuous are said to be *bicontinuous* and are called *homeomorphisms*), but when the domain and range are each open subsets of $\mathbb{R}$ or of $\mathbb{C}$, continuity of the inverse follows: Open subsets of $\mathbb{R}$ are treated in Exercise 1.46 (easy) and open subsets of $\mathbb{C}$ (hard) in Chapter IV. One other important case is also easy: if X is a compact metric space, Y any metric space and $f \colon X \mapsto Y$ is continuous and one-to-one, then $f(X)$ is compact and f^{-1} is continuous on it. Indeed, any closed subset C of X is compact, so $(f^{-1})^{-1}(C) = f(C)$ is compact (see below), hence closed in $f(X)$. [But note that having the compactness hypothesis instead on the image space is inadequate to insure continuity of the inverse; the example $X := \{0\} \cup [1, \infty)$, $Y := [0, 1]$, $f(0) := 0$, $f(x) := 1/x$ for $x \geq 1$, illustrates this.]

The restriction of a continuous function to a subspace is continuous. However, if Y is a subspace of X and f is defined on X, then $f|_Y$ may be continuous at a point y of Y without f being continuous at y. On the other hand, if A and B are two closed subsets of X and $f|A$, $f|B$ are each continuous functions, then $f|A \cup B$ is a continuous function. This elementary fact is used repeatedly in the sequel to build up a continuous function from continuous pieces; for its trivial proof see LIPMAN [1960].

On the line $\mathbb{R}$ there are the notions of *left-* and *right-hand continuity*. We use the notation $x \uparrow a$ or $x \to a^-$ to indicate x approaching a in $(-\infty, a)$, i.e., from the left. Similarly for $x \downarrow a$ or $x \to a^+$. Generally to indicate x approaching a from inside some set S (in any metric space) we write $\lim_{x \to a, x \in S}$.

The continuous image of a compact space is compact, and the continuous image of a connected space is connected. (For a proof of the latter, see 1.4.) When $X = [a, b]$ and $f(X) \subset \mathbb{R}$, this latter result amounts to the *Intermediate-Value Theorem* for continuous functions: if y is a number between $f(a)$ and $f(b)$, then $y = f(x)$ for some x between a and b. This property ensures that every positive real number has a positive n^{th} root, i.e., is the nth power of some positive real number (for each $n \in \mathbb{N}$). Cf. 1.47.

If (X, d) is a metric space, we denote by $C(X)$ the set of all bounded continuous complex-valued functions on X, by $C_{\mathbb{R}}(X)$ the real-valued functions in $C(X)$. $C(X)$ is an algebra over $\mathbb{C}$ and $C_{\mathbb{R}}(X)$ is an algebra over $\mathbb{R}$, both with "point-wise" operations, e.g., fg is the function $(fg)(x) := f(x)g(x) \quad (\forall x \in X)$. The sets $C(X)$, $C_{\mathbb{R}}(X)$ are usually made into metric spaces by introducing the *supremum* or *uniform norm*

$$\|f\|_\infty := \|f\|_{C(X)} := \sup |f(X)|$$

and the accompanying *uniform metric*

$$d_\infty(f, g) := \|f - g\|_\infty.$$

It is a fundamental and elementary fact that $C(X)$ is then complete and $C_{\mathbb{R}}(X)$ is a closed subset (hence also complete) of $C(X)$.

§ 7 Calculus

The elements of differential and integral calculus of real- and complex-valued functions on $\mathbb{R}$ as presented (from first principles with unsurpassed clarity) in chapters 5 and 6 of Ross [2013] will be taken for granted. (Differential forms and integration in multi-dimensional Euclidean space or on manifolds are not needed in this book, and the word "differential" is always an adjective, never a noun.) Only piecewise-continuous functions are ever integrated and neither Lebesgue nor Riemann-Stieltjes integrals are employed. When $f : \mathbb{R} \mapsto \mathbb{C}$ is piecewise continuous and $\lim_{r\to\infty} \int_{-r}^{r} f$ exists (in $\mathbb{C}$), we denote this limit by $\int_{-\infty}^{\infty} f$ or $\int_{\mathbb{R}} f$. An obvious sufficient condition for its existence, and even the existence of the asymmetric $\lim_{r,R\to+\infty} \int_{-r}^{R} f$ is that $M := \sup\{\int_{-r}^{r} |f| : r \in \mathbb{R}\}$ be finite. (For then $\lim_{r\to+\infty} \int_{-r}^{r} |f| = M$ and whenever $0 < r < R < +\infty$, $|\int_{-R}^{R} f - \int_{-r}^{r} f| \le \int_{-R}^{R} |f| - \int_{-r}^{r} |f| \le M - \int_{-r}^{r}$.) Similar meaning attaches to $\int_a^b f$ if f is continuous on (a, b) and $\int_A^B f$ has a limit in $\mathbb{C}$ as $A \downarrow a$ and $B \uparrow b$ (so called *improper integrals*). Mainly we need the two forms of the *Fundamental Theorem of Calculus* (FTC): If g is continuous and F is continuously differentiable on $[a, b]$, then $\int_a^x F' = F(x) - F(a)$ and $G(x) := \int_a^x g$ satisfies $G'(x) = g(x)$ for all $x \in [a, b]$. (The standard proof of the latter is reprised in 2.11.) *Any* function G satisfying $G' = g$ is called a *primitive* (=*Stammfunktion* in German) or an *antiderivative* of g. We often cite the *Mean-Value Theorem* for differentiable *real-valued* functions on a subinterval of $\mathbb{R}$, and the corollary thereof which says that a function with a positive derivative is increasing, as well as the *Chain Rule* for differentiating composites. (See 2.3 for the canonical proof of the latter in one variable and the discussion beginning §4 of chapter II for the two-variable version.) *Integration-by-parts* is used, as are the *Cauchy-Schwarz* inequality for integrals and the most elementary facts about the relation of uniform convergence to integration. Partial differentiation with respect to the kth coordinate variable (be it real or complex) is denoted by D_k. However, if f is a function of one complex variable, and $F(x, y) := f(x + iy)$ the corresponding function of two real variables, we follow custom and write $D_k f(z)$ for $D_k F(\operatorname{Re} z, \operatorname{Im} z), k = 1, 2$. Ordinary differentiation on the line is usually denoted by $'$, rarely is d/dx used. The nth iteration of this differentiation operator on a function f is denoted $f^{(n)}$.

No knowledge of the transcendental functions log, exp, sin, cos is supposed, as these will be developed (in $\mathbb{C}$) *ab ovo*.

§ 8 *De Figuris*

In the preface of his 1755 book *Institutiones calculi differentialis* (= Ser. I, Vol. 10, p. 9 of his *Opera Omnia*) the great Leonhard Euler wrote:

Hic autem omnia ita intra Analyseos purae limites continentur, ut ne ulla quidem figura opus fuerit, ad omnia huius calculi praecepta explicanda.

Similar sentiments were expressed by Joseph-Louis Lagrange in the foreword of his famous 1788 treatise *Mécanique Analytique* ("On ne trouvera point de Figures dans cet Ouvrage...") and there are no diagrams in Cauchy's famous 1821 textbook or in the Bourbaki treatise. ("The Bourbaki were Puritans, and Puritans are strongly opposed to pictorial representations of truths of their faith" is how one of them provocatively put it — **Zbl** 915.01010.) I follow these illustrious predecessors (*sans* ideological baggage) because I feel that the traditional figures, to the extent they are enlightening at all, are best generated by the reader. Significant pictorial insights probably require computer graphics (e.g., **MR** 2007h:30001), although the work of NEEDHAM [1997] is unusually perspicuous and should be consulted. A new development is the use of color in graphing and encoding the behavior of complex-valued functions of a complex variable; see **MR** 3024399.

§ 9 On Problems and Exercises

Most of the "exercises" in this book are really just continuations of the text, with less fulsome proofs (called "hints"). Sometimes calculations are relegated there, as are special results which will be needed as mortar later in the structure; in any case, most are teleological, and there are no *Fingerübungen.* Students will therefore want to supplement their study here with one of the many available problem collections listed in the bibliography, e.g., KNOPP [1948], [1953], VOLKOVYSKII, *et al.* [1965], PÓLYA and SZEGÖ [1972], [1976], KRZYŻ [1971], FEYEL and PRADELLE [1973], EVGRAFOV, *et al.* [1974], TAUVEL [1994], HERZ [1996], DE SOUZA and SILVA [2004], or ALPAY [2011].

Chapter I
Curves, Connectedness and Convexity

We want to develop here as expeditiously as possible some routine facts about the topology of the plane centering on connectedness; these are essential in all that follows. A deeper exploration of the topology of the plane occurs in Chapter IV, after the exponential function has been developed (in Chapter III) and the useful concept of the index of a curve is available.

§ 1 Elementary Results on Connectedness

Throughout this section M will be an arbitrary metric space.

Theorem 1.1 *If $\mathfrak{F}$ is a collection of connected subsets of M, one of which meets every other one, then $\bigcup \mathfrak{F}$ is connected.*

Proof: Suppose $F_0 \in \mathfrak{F}$ meets every set in $\mathfrak{F}$. It suffices to show that the only relatively clopen subset S of $\bigcup \mathfrak{F}$ which meets F_0 is $\bigcup \mathfrak{F}$. Every $F \in \mathfrak{F}$ which meets S must be a subset of S, because meeting $M \setminus S$ would disconnect F. Thus $F_0 \subset S$. But then every $F \in \mathfrak{F}$, since it meets F_0, also satisfies $F \subset S$.

Theorem 1.2 *If S is a connected subset of M, then any set T satisfying $S \subset T \subset \overline{S}$ is also connected.*

Proof: If V is any non-void relatively clopen subset of T, then $V \cap S$ is a relatively clopen subset of S. Moreover $V \cap S \neq \varnothing$, for the closure of S in T is $\overline{S} \cap T = T$; that is, S is dense in T, and V is open in T. It follows from the connectedness of S that $V \cap S = S$. But then the relatively closed subset V of T contains also the closure in T of S, viz., T itself.

Exercise 1.3 *Prove the following useful extension of* 1.1. *If $\mathfrak{F}$ is a collection of connected subsets of M and for every pair of its sets the closure of one meets the other, then $\bigcup \mathfrak{F}$ is connected.*

R. B. Burckel, *Classical Analysis in the Complex Plane*, Cornerstones,
https://doi.org/10.1007/978-1-0716-1965-0_2

Hints: Enlarge each $F \in \mathfrak{F}$ to the set $\widetilde{F}$ of all points which $\overline{F}$ has in common with any of the sets in $\mathfrak{F}$. By 1.2 each $\widetilde{F}$ is connected, and the family $\widetilde{\mathfrak{F}}$ of all of them is eligible for 1.1. Note that $\bigcup \widetilde{\mathfrak{F}} = \bigcup \mathfrak{F}$.

Theorem 1.4 *If X, Y are metric spaces and $f\colon X \mapsto Y$ is continuous, then $f(X)$ is connected if X is.*

Proof: If V is a proper, non-void, relatively clopen subset of $f(X)$, then by continuity of f the set $f^{-1}(V)$ is a proper, non-void, clopen subset of X.

Exercise 1.5 (i) (Painlevé, 1902) *Show that if $X_1 \supset X_2 \supset \cdots$ are closed, connected subsets of M at least one of which is compact, then $X = \bigcap_{n=1}^{\infty} X_n$ is connected.*

Hints: Suppose X is disconnected, i.e., it is covered by two non-empty, disjoint, closed subsets C_1, C_2. By compactness there is a $\delta > 0$ such that $|z_1 - z_2| \geq \delta$ for all $z_j \in C_j$ $(j = 1, 2)$, so the open sets $U_j = \bigcup\{B(z, \delta/2) : z \in C_j\}$ $(j = 1, 2)$ are disjoint. Since they cover X, we have $X_n \subset U_1 \cup U_2$ for some n by the finite-intersection property of compacta. [The sets $X_n := \mathbb{C} \setminus [-n, n]$ illustrate the need for the compactness hypothesis here.]

(ii) *Show that if $X_1 \supset X_2 \supset \cdots$ are closed connected subsets of $\mathbb{C}$ and $X = \bigcap_{n=1}^{\infty} X_n$ is compact and non-void, then some X_n is compact and X is connected.* (Cf. W. M. Whyburn [1935].)

Hints: Take open V with $X \subset V \subset \overline{V}$ compact. Then $V \supset X = X \cap \overline{V} = \bigcap_{n=1}^{\infty}(X_n \cap \overline{V})$. These sets being compact, it follows $X_n \cap \overline{V} \subset V$ for some n. Then $X_n \cap \overline{V}$ is clopen in X_n, hence not a proper subset.

Call a function f on a metric space M *locally constant* if it is constant in a neighborhood of each point of M. Clearly this is equivalent to $f^{-1}(w)$ being an open subset of M for each $w \in f(M)$, leading to

Exercise 1.6 *Show that a metric space is connected if and only if every locally constant function on it is constant.*

Hints: If V_1, V_2 disconnect the metric space M, then the indicator function of V_1 is locally constant but not constant. On the other hand, if $f : M \mapsto \mathbb{C}$ is locally constant, then the sets $f^{-1}(w)$, $w \in f(M)$, are open, disjoint and cover M; so if M is connected, exactly one of them is non-void.

Exercise 1.7 *Prove that if F_1, $F_2\colon M \mapsto \mathbb{C} \setminus \{0\}$ are continuous, M is connected and $F_1^2 = F_2^2$, then either $F_1 = F_2$ or $F_1 = -F_2$. What happens if the F_j are allowed to take the value zero?*

Hints: Note that M is the disjoint union of two closed subsets $(F_1+F_2)^{-1}(0)$ and $(F_1 - F_2)^{-1}(0)$.

§ 2 Connectedness of Intervals, Curves, and Convex Sets

Theorem 1.8 *If a, $b \in \mathbb{R}$ and $a \le b$, then the set $[a, b]$ is connected.*

Proof: We are to show that there is no proper, non-void, relatively clopen subset A of $[a, b]$. With such an A the set $B := [a, b] \setminus A$ is another such and one of A or B contains a. Let, say, $a \in A$. Since A is relatively open in $[a, b]$, there is some $a < x \le b$ such that $[a, x] \subset A$. Let c denote the supremum of such x. Then by definition of c we have $[a, x] \subset A$ for every $x < c$, and so since A is closed, $[a, c] \subset A$. If $c < b$, then since $c \in A$ and A is relatively open in $[a, b]$, there exists $c < d < b$ such that $[c, d] \subset A$; and then $[a, d] = [a, c] \cup [c, d] \subset A$, contrary to the maximality of c. It follows that $c = b$, so $[a, b] = [a, c] \subset A$, contrary to the propriety of A.

Remark: The preceding method also furnishes a simple proof, directly from the least-upper-bound property, that closed, bounded intervals in $\mathbb{R}$ are compact: If open sets cover $[a, b]$, the $x \in [a, b]$ such that $[a, x]$ is covered by finitely many constitute a non-void relatively clopen subset A of the connected set $[a, b]$. For an abstraction and history of this method see HATHAWAY [2011], and CLARK [2019].

Corollary 1.9 *Any (non-void) interval in $\mathbb{R}$ is connected.*

Proof: If I is an interval and $x \in I$ is arbitrary, then

$$I = \bigcup_{\substack{a,b \in I \\ a \le x \le b}} [a, b].$$

An appeal to the last result and 1.1 finishes the proof.

Exercise 1.10 *Establish the converse of the last corollary, namely that every connected subset of $\mathbb{R}$ is an interval.*

Hints: Given connected $C \subset \mathbb{R}$, let $\alpha := \inf C$, $\beta := \sup C$, elements of $[-\infty, \infty]$. Evidently $C \subset [\alpha, \beta] \cap \mathbb{R}$. If $x \in (\alpha, \beta)$, the open sets $(-\infty, x)$ and (x, ∞) both meet C non-voidly (why?), so they cannot cover that connected set. Hence $x \in C$. In summary, $(\alpha, \beta) \subset C$.

Definition 1.11 (i) A *curve* is a continuous function γ from a compact interval $[a, b]$ in $\mathbb{R}$ into $\mathbb{C}$. $\gamma(a)$ is called the *initial* point and $\gamma(b)$ the *terminal* point of γ, and γ is called *closed* if $\gamma(a) = \gamma(b)$. When there is no possibility of confusion, we sometimes permit the linguistic abuse of writing simply γ for the set $\gamma[a, b]$. This set is called the *range* or *image* or *trajectory* (=*Spur* in German) of γ. A closed curve is also called a *loop*. A one-to-one curve is called an *arc*. A loop $\gamma\colon [a,b] \mapsto \mathbb{C}$ is called *simple* if γ is one-to-one on $[a,b)$.

(ii) A curve $\gamma\colon [a,b] \mapsto \mathbb{C}$ is *continuously differentiable* or *smooth* if it is differentiable in (a,b), has left- and right-hand derivatives at b and a, respectively, and the derivative function γ' is continuous on $[a,b]$; for closed curves there is the additional requirement that the one-sided derivatives at a and b coincide; that is, the periodic extension of γ to $\mathbb{R}$ should have continuous derivative everywhere.

(iii) A curve $\gamma\colon [a,b] \mapsto \mathbb{C}$ is *piecewise continuously differentiable* or *piecewise smooth* if there exist $a = t_0 < t_1 < \cdots < t_n = b$ such that γ is continuously differentiable on each $[t_{j-1}, t_j]$, without any equalities being required among the one-sided derivatives at $t_1, \ldots, t_{n-1}$.

(iv) For a closed curve $\gamma\colon [a,b] \mapsto \mathbb{C}$ define $0{*}\gamma\colon [a,b] \mapsto \mathbb{C}$ to be the constant curve $\gamma(a)$, $(-1){*}\gamma\colon [a,b] \mapsto \mathbb{C}$ by $(-1){*}\gamma(t) := \gamma(b+a-t)$, $n{*}\gamma\colon [a,b] \mapsto \mathbb{C}$ for each positive integer n by $n{*}\gamma(t) := \gamma(a+n(t-t_{j-1}))$ for $t \in [t_{j-1}, t_j]$, $t_0 := a$, and $t_j := a + (b-a)j/n$, $j = 1, 2, \ldots, n$; and for every negative integer k define $k{*}\gamma$ to be $(-k){*}(-1){*}\gamma$.

(v) A *Jordan-curve* is a homeomorph in $\mathbb{C}$ of $\mathbb{T}$; the term *topological circle* is also used for such a set.

Remarks: The notation in (iv) will seldom be needed. It is undeniably clumsy, but popular use of $-\gamma$ or γ^{-1} to designate $(-1){*}\gamma$ is worse than clumsy. Definition (v) is traditional (hence inviolable?), but according to it and (i) a Jordan-curve is *not* a curve. The gratuitous hyphen is intended to remind the reader of this. Since $\mathbb{T}$ is the *range* of a curve (cf. 3.16), every Jordan-curve is too. On the other hand, as most readers will know, the whole square $[0,1] \times [0,1]$ is also the range of a curve (G. Peano, 1890). But this interesting fact, however counter-intuitive, in no way jeopardizes the utility of definition (i) for our subject.

Definition 1.12 (i) If $a_0, \ldots, a_n \in \mathbb{C}$, then $[a_0, a_1, \ldots, a_n]$ is used to denote both the piecewise smooth curve $\gamma\colon [0,n] \mapsto \mathbb{C}$ defined by

$$\gamma(t) := a_{k-1} + (t-k+1)(a_k - a_{k-1}), \quad k-1 \le t \le k, \quad 1 \le k \le n$$

and the range of this curve. Any such curve is called a *polygon* , more precisely, an *n-gon* . When $n = 1$ the term *interval* is also used. Furthermore we let $(a_0, \ldots, a_n]$, $[a_0, \ldots, a_n)$ and $(a_0, \ldots, a_n)$ denote the restriction of the above γ to $(0,n]$, $[0,n)$ and $(0,n)$, respectively, as well as the ranges of these restriction functions.

(ii) If $a, b, c, d \in \mathbb{R}$, $a < b$, $c < d$, and $R := (a,b) \times (c,d)$, then

$$\partial R := \partial \overline{R} := [a+ci, b+ci, b+di, a+di, a+ci].$$

Remarks: Note that here the convention in 1.11(i) regarding ranges is consistent with the usual definition of topological boundary. See §4, Chapter 0. As the Remarks following 2.8 suggest, much of what we do with polygons does not depend on their being parameterized in this canonical way, and it is often convenient to replace an n-gon γ by the re-parameterized version $\gamma(nt)$, $0 \leq t \leq 1$. This is what is intended, for example, by expressions such as "an arbitrary polygonal curve $\gamma\colon [0, 1] \mapsto \mathbb{C}$."

Definition 1.13 A non-void subset S of $\mathbb{C}$ is *polygonally connected* if each pair of points of S lies on some polygon wholly contained in S. (Compare 1.21.)

When this happens, a one-to-one polygon can always be selected, a technical fact which will be quite useful in Chapter IV. This follows from

Exercise 1.14 *Let S be a connected set which is a finite union of (closed) intervals. Then any two points of S lie on a polygonal arc in S.*

Hints: Go by induction on the (minimal) number of intervals in S. The result is clearly true if S is comprised of a single interval. If the result is true whenever S is comprised of fewer than n intervals, consider a connected set S' which is the union of n intervals $I_1, \ldots, I_n$. Consider any two points a', $b' \in S'$, say $a' \in I_n$. We want to find a polygonal arc in S' joining a' to b'. There is nothing to prove if $b' \in I_n$. So we suppose $b' \notin I_n$. Let S denote the union of all the connected subsets of $I_1 \cup \cdots \cup I_{n-1}$ which contain b'. By 1.1, S is connected and if I_j meets S for some $j \leq n-1$, then by 1.1 and 1.8, $S \cup I_j$ is connected. It follows that S is a union of some of the intervals $I_1, \ldots, I_{n-1}$ and is disjoint from each of the others. Let S'' denote the union of the latter. If I_n were disjoint from S, then S and $I_n \cup S''$ would constitute a disconnection of S'. We infer that I_n meets S. Letting $I_n := [a_n, b_n]$, it follows that one of $[a', a_n]$ and $[a', b_n]$ meets S. We take a to be the first point of this interval which lies in S. By the induction hypothesis there is a polygonal arc p in S joining b' to a. Then $p \cup [a, a']$ is a polygonal arc in S' joining b' to a'.

Definition 1.15 (i) If $S \subset \mathbb{C}$, an arc with one endpoint on ∂S and all other points in S is called an *end-cut in S*. An arc with both endpoints on ∂S but all other points in S is called a *cross-cut in S*.

(ii) A point $z \in \partial S$ is *accessible from S* if it is an endpoint of an end-cut in S.

Exercise 1.16 *For any open subset U of $\mathbb{C}$ the accessible points are dense in ∂U.*

Hints: Let $z_0 \in \partial U$ and $\varepsilon > 0$. Since z_0 is a boundary point, there exists $w \in U \cap D(z_0, \varepsilon/2)$. There obviously exists (elementary compactness argument) at least one point $z_1 \in \partial U$ which is at a minimum distance δ from w: $|z_1 - w| = \inf\{|z-w| : z \in \partial U\} =: \delta > 0$. Then $|z_1 - w| \le |z_0 - w| < \varepsilon/2$, whence $|z_1 - z_0| \le |z_1 - w| + |w - z_0| < \varepsilon$. By minimality of δ, $[w, z_1)$ does not meet ∂U. Since it is connected and meets U, it must lie wholly in U. Therefore $[w, z_1]$ is an end-cut, and z_1 is an accessible point of ∂U lying in $D(z_0, \varepsilon)$.

Definition 1.17 (i) A subset S of $\mathbb{R}^n$ is *starlike* (*star-shaped=sternförmig*) with respect to the point z, which is called a *star center*, if $[z, w] \subset S$ for every $w \in S$. We say simply that S is *starlike* if it is starlike with respect to some point.

(ii) A subset S of $\mathbb{R}^n$ is *convex* if it is starlike with respect to each of its points, that is, if $[z, w] \subset S$ for every $z, w \in S$.

(iii) If S is a subset of $\mathbb{R}^n$, the *convex hull of* S, noted $\operatorname{co} S$, is the intersection of all convex subsets of $\mathbb{R}^n$ which contain S.

(iv) A real-valued function f defined on a convex subset C of $\mathbb{R}^n$ is called a *convex function* if its epigraph $\{(x, y) : x \in C, y \ge f(x)\}$ is a convex subset of $\mathbb{R}^{n+1}$.

Exercise 1.18 *A* convex combination *of elements of a subset S of $\mathbb{R}^n$ is any finite sum $\sum_{j=1}^N \lambda_j x_j$ where $x_j \in S$, $0 \le \lambda_j \le 1$ and $\sum_{j=1}^N \lambda_j = 1$.*

(i) *Show that S is convex if and only if it contains all convex combinations of its elements.*

Hints: Use induction on N.

(ii) *Show that $\operatorname{co} S$ is convex and consists of all convex combinations of elements of S.*

All that we will need about convex sets is summarized in the following catch-all theorem:

Theorem 1.19 (i) *If S is a bounded subset of $\mathbb{C}$, then $\operatorname{co} S$ is bounded and has the same diameter as S.*

(ii) *If K is a compact subset of $\mathbb{C}$, then $\operatorname{co} K$ is compact.*

(iii) *Let C be a nonvoid, closed convex subset of $\mathbb{C}$. For each $z \in \mathbb{C}$ there is a point in C nearest to z. Moreover, any choice $N(z)$ of such points satisfies*

$$(*) \qquad |N(z) - N(w)| \le |z - w| \qquad \forall z, w \in \mathbb{C},$$

so that $N(z)$ is unique and N is a continuous map of $\mathbb{C}$ onto C which coincides with the identity in C.

(iv) *If C is an open convex subset of $\mathbb{C}$, then for any $z_0 \in C$, $z_1 \in \overline{C}$ the half-open segment $[z_0, z_1)$ lies in C.*

(v) *If C is an open convex proper subset of $\mathbb{C}$ which contains 0 and a is any point in the closed set $\mathbb{C} \setminus C$ which is nearest to 0, then*

$$\operatorname{Re}\big[\bar{a}(z-a)\big] < 0 \qquad \forall z \in C.$$

Proof: (i) Let d be the diameter of S. It evidently suffices to show that $|x-y| \le d$ for $x, y \in \operatorname{co} S$. For each $w \in S$ we have $|w-z| \le d$ for all $z \in S$, that is, S lies in the convex set $\bar{D}(w,d)$. Therefore $\operatorname{co} S \subset \bar{D}(w,d)$, that is, $|w-y| \le d$ for every $y \in \operatorname{co} S$ and every $w \in S$. This says that S lies in the convex set $\bar{D}(y,d)$, whence $\operatorname{co} S \subset \bar{D}(y,d)$, for every $y \in \operatorname{co} S$; that is, $|x-y| \le d$ for every $x \in \operatorname{co} S$.

(ii) Suppose $n > 3$, $z_1, \ldots, z_n$ are n distinct points in K, $t_1, \ldots, t_n$ are positive and $t_1 + \cdots + t_n = 1$. The $n-1$ vectors $z_j - z_n$ $(1 \le j \le n-1)$ in $\mathbb{R}^2$ are linearly dependent over $\mathbb{R}$ (since $n-1 > 2$), so there exist $a_1, \ldots, a_{n-1} \in \mathbb{R}$ not all 0 such that

$$\sum_{j=1}^{n-1} a_j(z_j - z_n) = 0.$$

Set $a_n := -\sum_{j=1}^{n-1} a_j$ and choose k so that $|a_j|/t_j \le |a_k|/t_k$, $1 \le j \le n$.

This inequality shows that $a_k \ne 0$ (since not all a_j are 0). Furthermore, replacing all the a_j by their negatives if necessary, we can suppose that $a_k > 0$. Set then

$$c_j := t_j - t_k a_j/a_k, \quad 1 \le j \le n$$

and note that $c_k = 0$ and

$$c_j \ge 0, \quad \sum_{j=1}^{n} c_j = 1 \quad \text{and} \quad \sum_{j=1}^{n} t_j z_j = \sum_{\substack{j=1\\ j\ne k}}^{n} c_j z_j.$$

An obvious inductive argument then shows that

$$\sum_{j=1}^{n} t_j z_j \in \{ax + by + cz : a, b, c \in [0,1], x, y, z \in K, a+b+c = 1\}.$$

The latter set, call it C, is a continuous image of the compact set $\{(a,b,c,x,y,z) \in [0,1]^3 \times K^3 : a+b+c = 1\}$ and so is compact, while the sum on the left is the generic point of $\operatorname{co} K$ by 1.18(ii). Since evidently $C \subset \operatorname{co} K$, we have finally $\operatorname{co} K = C$.

(iii) Existence is a trivial compactness argument: If we set $d := \inf\{|z-c| : c \in C\}$, then there exist $c_n \in C$ with $|z - c_n| \to d$. In particular, $\{c_n\}$ is a bounded subset of C and so has a cluster point c in C. Evidently $|z-c| = d$. Consider now any z, $w \in \mathbb{C}$ and any choices $N(z)$, $N(w)$ of nearest points to them in C. Let $\delta := \operatorname{Re}\big[(z - N(z))\overline{(N(w) - N(z))}\big]$ and $0 < t < 1$. Then by convexity of C and the characteristic property of $N(z)$

$$\begin{aligned}
|z - N(z)|^2 &\le \big|z - [(1-t)N(z) + tN(w)]\big|^2 \\
&= \big|z - N(z) - t[N(w) - N(z)]\big|^2 \\
&= |z - N(z)|^2 - 2t \operatorname{Re}\big[(z - N(z))\overline{(N(w) - N(z))}\big] \\
&\qquad + t^2|N(w) - N(z)|^2 \\
&= |z - N(z)|^2 - t\big[2\delta - t|N(w) - N(z)|^2\big],
\end{aligned}$$

whence

$$2\delta - t|N(w) - N(z)|^2 \le 0.$$

This holds for all $0 < t < 1$. Let $t \downarrow 0$ to conclude that $2\delta \le 0$. That is,

$$\text{(1)} \qquad \operatorname{Re}\big[(z - N(z))\overline{(N(w) - N(z))}\big] \le 0 \qquad \forall z, w \in \mathbb{C}.$$

Reversing the roles of z and w here gives

$$\begin{aligned}
&\operatorname{Re}\big[(w - N(w))\overline{(N(z) - N(w))}\big] \le 0, \quad \text{that is,} \\
\text{(2)} \qquad &\operatorname{Re}\big[(N(w) - w)\overline{(N(w) - N(z))}\big] \le 0.
\end{aligned}$$

Adding (1) and (2) gives

$$\begin{aligned}
&\operatorname{Re}\big[(z - w + N(w) - N(z))\overline{(N(w) - N(z))}\big] \le 0, \\
&(N(w) - N(z))\overline{(N(w) - N(z))} \le \operatorname{Re}\big[(w - z)\overline{(N(w) - N(z))}\big], \\
&|N(w) - N(z)|^2 \le |w - z||N(w) - N(z)|.
\end{aligned}$$

The desired inequality $(*)$ then follows. If $N^*(z)$ is another candidate for nearest point to z, apply $(*)$ with $w = z$ and $N^*(z)$ in the role of $N(w)$, to see that $N^*(z) = N(z)$.

(iv) There is an $r > 0$ such that $D(z_0, r) \subset C$. Any $z \in (z_0, z_1)$ has the form $z := tz_1 + (1-t)z_0$ for some $0 < t < 1$. Since $z_1 \in \overline{C}$, there exists $z_2 \in C$ with $|z_1 - z_2| < \left(\frac{1-t}{t}\right) r$. Then $z_3 := z_0 + \frac{t}{1-t}(z_1 - z_2) \in D(z_0, r) \subset C$. Hence by convexity the point $tz_2 + (1-t)z_3$ lies in C. But this point is just z.

(v) Note that $a \neq 0$. Fix $z \in C$ and set $\delta := \operatorname{Re}\left[\bar{a}(z-a)\right]$. For any $t > 0$ look at the point $w := a - t(z-a)$. If it were in C, then convexity of C would lead us to the contradiction

$$a = \frac{t}{1+t} z + \frac{1}{1+t} w \in C.$$

Hence $w \in \mathbb{C} \setminus C$ and so, as a is nearest to 0 in $\mathbb{C} \setminus C$,

$$\begin{aligned} |a|^2 \leq |w|^2 &= \left[a - t(z-a)\right]\left[\bar{a} - t(\bar{z} - \bar{a})\right] \\ &= |a|^2 - t\left[2\delta - t|z-a|^2\right]. \end{aligned}$$

Therefore

$$2\delta - t|z-a|^2 \leq 0 \qquad \forall t > 0,$$

which means that $\delta \leq 0$. That is,

$$\operatorname{Re}\left[\bar{a}(z-a)\right] \leq 0 \qquad \forall z \in C.$$

But since C is open, the point $z^* := z + ta$ lies in C if $t > 0$ is sufficiently small. Hence

$$\begin{aligned} &0 \geq \operatorname{Re}\left[\bar{a}(z^* - a)\right] = \operatorname{Re}\left[\bar{a}(z-a)\right] + \operatorname{Re}\left[\bar{a}(ta)\right], \quad \text{so} \\ &\operatorname{Re}\left[\bar{a}(z-a)\right] \leq -t|a|^2 < 0. \end{aligned}$$

Theorem 1.20 *Every curve is connected.*

Proof: 1.4 and 1.8.

Corollary 1.21 *Every polygonally connected set is connected.*

Proof: If S is polygonally connected, fix $x \in S$ and for every $z \in S$ let p_z be a polygon in S which contains x and z. p_z is connected by the preceding theorem and evidently $S = \bigcup \{p_z : z \in S\}$, so by 1.1, S is connected.

Corollary 1.22 *Every starlike set and (hence) every convex set is polygonally connected and (hence) connected.*

Corollary 1.23 *Center-punctured disks, annuli and circles in $\mathbb{C}$ are connected.*

Proof: We treat the case of open disks and annuli centered at the origin. The reader can easily extend this to all the other cases. Specifically,

$$\mathbb{D} \setminus \{0\} = \left(\mathbb{D} \setminus (-1, 0]\right) \cup \left(\mathbb{D} \setminus [0, 1)\right).$$

The two sets on the right-hand side are starlike (with respect to any point in $(0,1)$ or $(-1,0)$, respectively) and so connected. Also they have points in common (for example, $\frac{1}{2}i$). Therefore their union is connected, by 1.1.

Now if $0 \le a < b < \infty$, then $\psi(t) := tb+(1-t)a$ maps the interval $(0,1)$ continuously onto the interval (a,b), so $z \mapsto \psi(|z|)\cdot z/|z|$ maps $\mathbb{D}\setminus\{0\}$ continuously onto the annulus $A(a,b)$. Now cite 1.4.

Take $a = b$ in (an argument similar to) the above to see that the circle $\{z \in \mathbb{C} : |z| = a\}$ is connected.

Theorem 1.24 *If Ω is an open connected subset of $\mathbb{C}$ and A is a subset of Ω which has no limit point in Ω, then $\Omega\setminus A$ is open and connected.*

Proof: Since A has no limit point in Ω, each point of $\Omega\setminus A$ is the center of a disk which does not meet A. Therefore $\Omega\setminus A$ is open. Suppose U_1, U_2 are disjoint, non-void, open subsets of $\Omega\setminus A$ such that $\Omega\setminus A = U_1\cup U_2$. For each $a\in A$ there is an $r_a > 0$ such that $D_a := D(a,r_a)\subset\Omega$ and $D_a\cap A = \{a\}$. Each subset $D_a\setminus\{a\}$ of $\Omega\setminus A$ is connected (by 1.23) and so lies entirely in one of the U_j. Let $A_j := \{a\in A : D_a\setminus\{a\}\subset U_j\}$ $(j=1,2)$. Then $V_j := U_j\cup\bigcup\{D_a : a\in A_j\}$ $(j=1,2)$ are disjoint, non-void, open subsets of Ω, that is, a disconnection of Ω.

Exercise 1.25 (i) *If C is a proper subset of a circle, then $\mathbb{C}\setminus C$ is connected.*

Hints: Let the circle be $\mathbb{T}$ and select $p\in\mathbb{T}\setminus C$. Note that

(1) $$\mathbb{D}\subset\overline{\mathbb{D}}\setminus C\subset\overline{\mathbb{D}},$$

(2) $$\mathbb{C}\setminus\overline{\mathbb{D}}\subset(\mathbb{C}\setminus\overline{\mathbb{D}})\cup\{p\}\subset\mathbb{C}\setminus\mathbb{D} = \overline{\mathbb{C}\setminus\overline{\mathbb{D}}}.$$

Now $\mathbb{D}$ is connected (1.22) and so is $\mathbb{C}\setminus\overline{\mathbb{D}}$, being the image of the connected set $\mathbb{D}\setminus\{0\}$ under the continuous map $z\mapsto 1/z$. Thus (1), (2) and 1.2 yield

(3) $\overline{\mathbb{D}}\setminus C$ and $(\mathbb{C}\setminus\overline{\mathbb{D}})\cup\{p\}$ are each connected.

But then, since p lies in both sets, their union $\mathbb{C}\setminus C$ is connected, by 1.1.

(ii) *If Ω is an open, connected subset of $\mathbb{C}$ and D is an open disk such that $\bar{D}\subset\Omega$, then $\Omega\setminus\bar{D}$ is connected.*

Hints: Let D_1 be an open disk concentric with D such that

$$\bar{D}\subset D_1\subset\Omega.$$

Suppose U and V are a disconnection of $\Omega\setminus\bar{D}$: U and V are non-empty, open subsets of $\Omega\setminus\bar{D}$ and $U=\big(\Omega\setminus\bar{D}\big)\setminus V$. The subset $D_1\setminus\bar{D}$ of $\Omega\setminus\bar{D}$ is connected (1.23) and so lies wholly in one of U or V, say,

$$D_1\setminus\bar{D}\subset U.$$

Then

$$U\cup\bar{D}=U\cup D_1,\quad \text{an open subset of } \Omega.$$

We have

$$V=\big(\Omega\setminus\bar{D}\big)\setminus U=\Omega\setminus\big(U\cup\bar{D}\big),$$

so V and $U\cup\bar{D}$ are disjoint, non-empty, open subsets of Ω which cover Ω, contradiction.

Exercise 1.26 (Existence of roots in $\mathbb{C}$)

(i) *Show that for each $w\in\mathbb{C}_-$ the number*

$$z:=\frac{w+|w|}{|w+|w||}\sqrt{|w|}$$

is the unique element of $\mathbb{H}$ whose square is w. Infer that for every $n\in\{2^k\colon k\in\mathbb{N}\}$ every complex number is the nth power of another. (Cf. hints to 3.44(ii).)

Hints: The motivation for this definition of z is the rotation significance of multiplication and the parallelogram significance of addition in $\mathbb{C}$ (see §3 of chapter 0). Calculation, keeping in mind $|w|^2=w\overline{w}$, confirms that $z^2=w$. Every $w\in\mathbb{C}\setminus\mathbb{C}_-=(-\infty,0]$ is the square of $\sqrt{|w|}i$. The rest is induction.

(ii) *Show that for every $n\in\mathbb{N}$ every complex number is the nth power of another.*

Hints: In view of (i) it suffices to treat odd n. Consider the continuous power function $p(z):=z^n$. If $u\in\mathbb{T}\setminus p(\mathbb{T})$, then also $\bar{u}\in\mathbb{T}\setminus p(\mathbb{T})$ and $u\neq\pm1$. Thus the disjoint open sets $\{z\in\mathbb{C}:\operatorname{Re}z<\operatorname{Re}u\}$, $\{z\in\mathbb{C}:\operatorname{Re}z>\operatorname{Re}u\}$ cover $\mathbb{T}\setminus\{u,\bar{u}\}\supset p(\mathbb{T})$ and meet $p(\mathbb{T})$, in -1 and 1, respectively. They therefore disconnect $p(\mathbb{T})$. Since the latter is connected, thanks to (1.23) and (1.4), we have a contradiction, which shows that no such u exists. So $\mathbb{T}\subset p(\mathbb{T})$, whence $\mathbb{C}\subset p(\mathbb{C})$.

A devious device discovered by Dedekind deserves detailing here, as it supplies a second elementary solution to (ii): Given *odd* $n\in\mathbb{N}$ and $\pm1\neq u\in\mathbb{T}$, use (i) to write $u=w^2$. Then $w\neq\overline{w}$, so the polynomial $p(z):=iw(z+i)^n-i\overline{w}(z-i)^n$ has the (odd) degree n. Moreover, $p(x)=\overline{p(x)}$ for all real x, so p has a real zero y. Consequently, $\left(\frac{y-i}{y+i}\right)^n=w/\overline{w}=w^2=u$.

§ 3 The Basic Connectedness Lemma

Open, connected, non-empty subsets of $\mathbb{C}$ occur so frequently in the sequel that it is convenient to have a generic name for them. The literature is not uniform on this point: the term "domain" is sometimes used for such sets, but more often they are called *regions*, and that term will be adopted here.

Many arguments in function theory involving connectedness of the domain Ω can be viewed as showing that the "good" part of Ω has a certain "uniform openness" property. It is worth codifying this argument once and for all:

Theorem 1.27 (Basic Connectedness Lemma) *Let U be an open subset of $\mathbb{C}$ and $\varnothing \neq A \subset U$ have the property that for some $\theta > 0$*

$$a \in A \quad \textit{and} \quad D(a,r) \subset U \Rightarrow D(a,\theta r) \subset A.$$

Then A is relatively clopen in U. In particular, if in addition U is connected, then $A = U$.

Proof: Evidently A is open in U, since U is open. To see that $U \setminus A$ is also open in U, let $b \in U \setminus A$ and pick $R > 0$ so that $D(b,R) \subset U$. It follows that if $D\big(b, \theta R/(1+\theta)\big)$ meets A, say in a, then we have

$$(*) \qquad |a-b| < \frac{\theta R}{1+\theta}.$$

We have for $r := R - |a-b| > 0$

$$D(a,r) \subset D(b,R) \subset U.$$

But then $D(a,\theta r) \subset A$ and from $(*)$ we get

$$|a-b| < \theta R - \theta|a-b| = \theta r,$$

which says that $b \in D(a,\theta r) \subset A$, contradiction.

Corollary 1.28 *Every region is polygonally connected.*

Proof: Let Ω be a region. It suffices to show that for each fixed $a \in \Omega$ the set A of z in Ω for which there is a polygon in Ω with initial point a and terminal point z is all of Ω. This set is non-void ($a \in A$) and if $z \in A$ and $r > 0$ is such that $D(z,r) \subset \Omega$, then $D(z,r) \subset A$, because if $[a,\ldots,z]$ is a polygon in Ω and $w \in D(z,r)$, then $[a,\ldots,z,w]$ is also a polygon in Ω. Thus A meets the hypothesis of the Basic Connectedness Lemma with $\theta = 1$.

Exercise 1.29 *Any two distinct accessible points on the boundary of a region Ω lie on a cross-cut in Ω.*

Hints: let a, b be distinct accessible boundary points of Ω. Let A, B be endcuts in Ω with a the initial point of A, b the terminal point of B. By ignoring an appropriate initial segment of B, we can suppose A and B disjoint. By 1.28 and 1.14 there is an arc $P\colon [0,1] \mapsto \Omega$ with initial point on A and terminal point on B. Let t_a be the largest $t \in [0,1]$ such that $P(t) \in A$. We have $t_a < 1$ since $A \cap B = \varnothing$ and $P(1) \in B$. Let t_b be the smallest $t \in [t_a, 1]$ such that $P(t) \in B$. Then A, $P(t_a, t_b)$, and B are mutually disjoint and, since each of the functions involved here is one-to-one, this implies that $A \cup P[t_a, t_b] \cup B$ contains an arc which contains a and b. It is evidently a cross-cut from a to b.

§ 4 Components and Compact Exhaustions

Recall that a *component* of a set is a maximal connected subset.

Theorem 1.30 *Let X be a non-empty, proper subset of $\mathbb{C}$. Then*

(i) *Each component of X is relatively closed.*

(ii) *Distinct components are disjoint.*

(iii) *Each connected subset of X lies in a unique component.*

(iv) *X is the (disjoint) union of its components.*

(v) *The boundary of X contains the boundary of each component of X and of each component of $\mathbb{C} \setminus X$.*

If X is open, then:

(i)′ *Each component of X is open and there are at most countably many of them.*

Proof: (i) follows from the fact that a component is a maximal connected set while its relative closure is still connected (by 1.2).

(ii) If C_1 and C_2 are components with a point in common, then $C_1 \cup C_2$ is connected by 1.1 and so by the maximality of C_1 and C_2 we see that $C_1 = C_1 \cup C_2 = C_2$.

(iii) If C is a connected subset of X, then the union of all the connected subsets of X which contain C is connected by 1.1 and evidently is maximal among connected subsets of X, i.e., is a component.

(iv) Since single points are connected sets, (iv) follows from (iii).

(v) Since $\partial(\mathbb{C} \setminus X) = \partial X$, it suffices to treat components C of X. Consider $x \in \partial C$. Then $x \in \overline{C} \subset \overline{X}$, so for every $r > 0$, $D(x,r)$ meets X. If $x \notin$

∂X, then some $D(x,r)$ does not meet $\mathbb{C} \setminus X$, so $C \cup D(x,r) \subset X$. Since $x \in \overline{C}$, $C \cap D(x,r) \neq \varnothing$ and consequently (1.3), $C \cup D(x,r)$ is connected. By maximality of C then, $C \cup D(x,r) = C$, $D(x,r) \subset C$ and $x \notin \partial C$. —×—

(i)′ Suppose X is open and C is a component. Then by (v), $\partial C \subset \partial X \subset \mathbb{C} \setminus X \subset \mathbb{C} \setminus C$. That is, $C \cap \partial C = \varnothing$, making C an open set. Thus C contains $a + bi$ for some rational a, b. As there are only countably many such points and components are disjoint, there can be only countably many C.

Theorem 1.31 *Let U be an open subset of $\mathbb{C}$. Then there exist compact sets K_n such that*

(1) $K_n \subset \overset{\circ}{K}_{n+1}$.

(2) $\bigcup_{n=1}^{\infty} K_n = U$.

(3) *Every bounded component of $\mathbb{C} \setminus K_n$ contains a component of $\mathbb{C} \setminus U$. In particular, if $\mathbb{C} \setminus U$ has no bounded components, then each set $\mathbb{C} \setminus K_n$ is connected.*

Proof: We may assume $\varnothing \neq U \neq \mathbb{C}$. For each positive integer n form

(4) $$C_n := \{z \in \mathbb{C} : d(z, \mathbb{C} \setminus U) \geq 1/n\} \subset U,$$

where $d(z, \mathbb{C} \setminus U) := \inf\{|z - w| : w \in \mathbb{C} \setminus U\}$. Notice that

(5) $$\mathbb{C} \setminus C_n = \{z \in \mathbb{C} : d(z, \mathbb{C} \setminus U) < 1/n\} = \bigcup_{a \in \mathbb{C} \setminus U} D(a, 1/n)$$

is obviously open, so C_n is closed. Form

(6) $$K_n := C_n \cap \overline{D}(0,n).$$

Then K_n is compact and lies in U. Moreover it is obvious that (2) holds: given $z \in U$, pick positive integer $n \geq |z|$ so that $D(z, 1/n) \subset U$. Then $z \in K_n$.

To prove (1), we set $r_n := 1/n - 1/(n+1)$ and show that $D(z, r_n) \subset K_{n+1}$ for every $z \in K_n$. Indeed, given such a z and $w \in D(z, r_n)$, we have

$$\begin{aligned} |w| < |z| + r_n &\leq n + r_n, \quad \text{since } z \in K_n \subset \overline{D}(0,n), \\ &\leq n + 1, \end{aligned}$$

so $w \in \overline{D}(0, n+1)$. But also for any $a \in \mathbb{C} \setminus U$ we have

$$\begin{aligned} |w - a| \geq |z - a| - |w - z| &\geq \frac{1}{n} - |w - z|, \quad \text{since } z \in C_n \\ &> \frac{1}{n} - r_n = \frac{1}{n+1}, \quad \text{since } w \in D(z, r_n). \end{aligned}$$

It follows that $w \in C_{n+1}$ and so finally $w \in \bar{D}(0, n+1) \cap C_{n+1} = K_{n+1}$.

It remains to look after (3). Notice that

$$\begin{aligned} \mathbb{C} \setminus K_n &= \big[\mathbb{C} \setminus \bar{D}(0,n)\big] \bigcup [\mathbb{C} \setminus C_n] \quad \text{by (6)} \\ &= \big[\mathbb{C} \setminus \bar{D}(0,n)\big] \cup \bigcup_{a \in \mathbb{C}\setminus U} D(a, 1/n) \quad \text{by (5).} \end{aligned} \tag{7}$$

The set $\mathbb{C} \setminus \bar{D}(0,n)$ is the image of the center-punctured disk $D(0,1/n) \setminus \{0\}$ under the continuous function $z \mapsto 1/z$ and so is connected [by 1.23 and 1.4]. It therefore lies wholly in a connected component of $\mathbb{C} \setminus K_n$, which is consequently the *unique* unbounded component of $\mathbb{C} \setminus K_n$. Now let C be a bounded component of $\mathbb{C} \setminus K_n$. Since C is the union of all the connected subsets of $\mathbb{C} \setminus K_n$ which meet it, we see from (7) that on the one hand C does not meet $\mathbb{C} \setminus \bar{D}(0,n)$ and on the other hand

$$C = \bigcup_{a \in \mathbb{C}\setminus U} D(a, 1/n) \cap C = \bigcup_{a \in C \cap (\mathbb{C}\setminus U)} D(a, 1/n).$$

In particular, $C \cap (\mathbb{C} \setminus U) \neq \varnothing$. Being non-void, this set meets some component K of $\mathbb{C} \setminus U$. But then K is a connected subset of $\mathbb{C} \setminus U \subset \mathbb{C} \setminus K_n$ (by (2)) which meets C and hence lies in C.

Exercise 1.32 *Let S be a non-empty subset of $\mathbb{R}^n$, $g\colon S \mapsto [0, \infty)$ a bounded, continuous function. For each $x \in \mathbb{R}^n$ define*

$$d(x) := d_g(x) := \inf\{\|x - y\| g(y) : y \in S\},$$

where $\|\cdot\|$ is the euclidean norm in $\mathbb{R}^n$.

(i) *Show that d is a continuous function on $\mathbb{R}^n$.*

Hints: Let M be a bound for g. Then for any x_1, $x_2 \in \mathbb{R}^n$, $y \in S$ we have $d(x_1) \leq \|x_1 - y\| g(y) \leq \big[\|x_1 - x_2\| + \|x_2 - y\|\big] g(y) \leq M\|x_1 - x_2\| + \|x_2 - y\| g(y)$, that is, $d(x_1) - M\|x_1 - x_2\|$ is a lower bound for $\big\{\|x_2 - y\| g(y) : y \in S\big\}$.

Consequently,

$$d(x_1) - M\|x_1 - x_2\| \leq d(x_2).$$

Reversing the roles of x_1 and x_2, we are therefore led to

$$|d(x_1) - d(x_2)| \leq M\|x_1 - x_2\|.$$

(ii) *For the function $g = 1$ we have $d_1^{-1}(\{0\}) = \bar{S}$.*

Theorem 1.33 (Hausdorff [1914], p. 303) *Let X be a compact subset of $\mathbb{C}$ and C a component of X. Then C is the intersection of all the relatively clopen subsets of X which contain it.*

Proof: If $\mathcal{C}$ is the class of all relatively clopen subsets of X which contain C, then $\mathcal{C}$ is not void ($X \in \mathcal{C}$) and the closed set $K := \bigcap \mathcal{C}$ contains C. Because C is a maximal connected set, it suffices to show that K is connected in order to conclude that $K = C$. This we do by *reductio ad absurdum*. If K is not connected, there exist disjoint non-void closed sets K_1, K_2 such that $K = K_1 \cup K_2$. Now there are disjoint open $V_j \supset K_j$. In fact, if $d_j(z)$ denotes the distance from z to the closed set K_j, i.e., $d_j(z) = \inf\{|z - w| : w \in K_j\}$, then by 1.32, d_j is continuous and so the function $f := d_1 - d_2$ is continuous on $\mathbb{C}$. Evidently $K_1 \subset f^{-1}(-\infty, 0)$ and $K_2 \subset f^{-1}(0, \infty)$. Now $\bigcap \mathcal{C}$ is compact and lies in $K = K_1 \cup K_2 \subset V_1 \cup V_2$ and so some finite-intersection of elements of $\mathcal{C}$, i.e., another element of $\mathcal{C}$, lies in $V_1 \cup V_2$. Say, $C_0 \in \mathcal{C}$, $C_0 \subset V_1 \cup V_2$. Then $C_0 \cap V_1 = C_0 \cap (X \setminus V_2)$. Since C_0 and V_1 are relatively open, while C_0 and $X \setminus V_2$ are closed, this shows that the set $C_0 \cap V_1$ is relatively clopen. Similarly for $C_0 \cap V_2$. Now not both these sets, whose union C_0 contains $K \supset C$, can meet C, because C is connected. Let, say, $C \subset C_0 \cap V_1$. Then $C_0 \cap V_1 \in \mathcal{C}$ and so $C_0 \cap V_1 \supset K$ by definition of K. Therefore $V_1 \supset K$, so $V_2 \cap K = \varnothing$, a contradiction, since $V_2 \cap K \supset K_2$.

Corollary 1.34 (Šura-Bura [1941]) *Let X be a closed subset of $\mathbb{C}$ and C a compact component of X. Then C is the intersection of all the compact, relatively open subsets of X which contain it. If X has only finitely many compact components, then each of them is relatively open in X.*

Proof: It suffices to produce for each open subset U of $\mathbb{C}$ with $C \subset U$ a compact relatively open subset of X which contains C and lies in U. Cover C with finitely many open disks whose closures lie in U and let V be their union. Thus $C \subset V \cap X \subset \overline{V}$ compact $\subset U$. If $\mathcal{C}$ is the family of all compact, relatively open subsets of $\overline{V} \cap X$ which contain C, then $C = \bigcap \mathcal{C}$ by the preceding theorem. By the finite-intersection property of compacta, since $\bigcap \mathcal{C} = C \subset V$, some intersection of finitely many members of $\mathcal{C}$, which is again a member K of $\mathcal{C}$, must already fall into the open set V. Thus $C \subset K \subset V \cap X \subset U$. K is compact and K is relatively open in $\overline{V} \cap X$, whence in $V \cap X$, whence in X, since V is open in $\mathbb{C}$.

Being clopen in X, K must wholly contain any component it meets. Hence, being bounded, it is disjoint from every unbounded component. If there are only finitely many other compact components, then for U we may take the complement in $\mathbb{C}$ of their union, and K will be disjoint from them too. In this case then K must coincide with C.

Our next result will play a role in one of the myriad characterizations of simple-connectivity.

Theorem 1.35 *Let Ω be a proper open subset of $\mathbb{C}$, $\phi\colon \mathbb{D} \mapsto \Omega$ a continuous surjection such that $\phi^{-1}(K)$ is compact for each compact $K \subset \Omega$. Then $\partial\Omega$ is connected if Ω is bounded and otherwise $\partial\Omega$ has no bounded component. [Notice that the hypothesis is satisfied by any* homeomorphism *ϕ of $\mathbb{D}$ onto Ω.]*

Proof: Let $D_n := D(0, 1 - 1/n)$, $A_n := \mathbb{D} \setminus \bar{D}_n$ and $X_n := \overline{\phi(A_n)}$. Since ϕ is continuous, it follows from 1.2, 1.4 and 1.23 that each X_n is connected. Set $X := \bigcap_{n=1}^{\infty} X_n$.

Suppose that some X_{n_0} is compact. Then by 1.5(i), $\bigcap_{n=n_0}^{\infty} X_n = X$ is connected. Moreover, $\Omega \subset \phi(\bar{D}_{n_0}) \cup X_{n_0}$, a union of compact sets, shows that Ω is bounded in this case.

Next, suppose that X is non-empty and has a bounded component C. Then by 1.34 C lies in a compact subset K of X which is also relatively open in X, that is, $X \setminus K$ is closed. There exists then an open subset V of $\mathbb{C}$ such that

$$K \subset V \subset \bar{V} \text{ compact } \subset \mathbb{C} \setminus (X \setminus K).$$

We have

$$\bigcap_{n=1}^{\infty} (X_n \cap \bar{V}) = X \cap \bar{V} = K \cap \bar{V} = K \subset V.$$

Therefore, since the sets $X_n \cap \bar{V}$ are each compact, the finite-intersection property of compacta implies that not all of them meet the closed set $\mathbb{C} \setminus V$. So

$$X_{n_0} \cap \bar{V} \subset V \quad \text{for some } n_0.$$

Then the set

$$W := X_{n_0} \cap \bar{V} = X_{n_0} \cap V$$

is relatively clopen in X_{n_0}, contains $X \cap V \supset K \supset C$ and so is non-void. Since X_{n_0} is connected, this clopen subset must exhaust it. But then $X_{n_0} = W$ is compact, and by the preceding paragraph X itself is connected and Ω is bounded.

Finally, we show that X is in fact $\partial\Omega$. We have $\partial\Omega = \bar{\Omega} \setminus \Omega = \overline{\phi(\mathbb{D})} \setminus \phi(\mathbb{D}) \subset \overline{\phi(\mathbb{D} \setminus D(0, 1 - 1/n))} = X_n$ for all n. On the other hand, if $w \in X = \bigcap_{n=1}^{\infty} X_n \subset \bar{\Omega}$, then for each n, $|w - \phi(z_n)| < 1/n$ for some $z_n \in A_n$. Since such z_n are eventually outside every compact subset of $\mathbb{D}$, the hypothesis on ϕ implies that the $\phi(z_n)$ are eventually outside every compact subset of Ω. It follows that $w \notin \Omega$, i.e., $w \in \bar{\Omega} \setminus \Omega$. Thus $X = \partial\Omega$, which is not void since $\Omega \neq \mathbb{C}$. The results of the preceding paragraph therefore complete the proof.

Theorem 1.36 *Let K be a compact, connected subset of $\mathbb{C}$, $x \in K$. Then x is a limit point of each component of $K \setminus \{x\}$.*

Proof: Suppose contrariwise that C is a component of $K \setminus \{x\}$ and $x \notin \overline{C}$. Then C, being closed in $K \setminus \{x\}$, is closed in K, hence compact. Since $x \notin C$, there is then an $r > 0$ sufficiently small that $\overline{D}(x,r) \cap C = \varnothing$. Evidently C is then a component of the compact set $K \setminus D(x,r)$. By 1.33 there is a compact set U which is relatively open in $K \setminus D(x,r)$, which contains C and which lies in the open set $\mathbb{C} \setminus \overline{D}(x,r)$. Then $K \setminus D(x,r) \setminus U$ is relatively closed in $K \setminus D(x,r)$ and so is compact, while

$$K \setminus U = \left[K \cap \overline{D}(x,r)\right] \bigcup \left[K \setminus D(x,r) \setminus U\right].$$

This representation shows that $K \setminus U$ is compact. Therefore U disconnects K.

§ 5 Connectivity of a Set

The theorems of this section are technical results which will be needed later (*inter alia*, in Chapter X). They require only material presented so far and it is a good idea to dispose of them now. This will mean one less distraction when the proofs of these later results, often lengthy enough in their own right, are carried out. The impatient reader may skip this material and return to it later as needed.

Definition 1.37 For a subset A of $\mathbb{C}$ a *hole* in A is any *bounded* component of the complement $\mathbb{C} \setminus A$. A subset A of $\mathbb{C}$ is called *simply-connected* if it has no holes, *doubly-connected* if it has exactly one hole. In general we will say A has *connectivity* $n \in \mathbb{N} \cup \{\infty\}$ if it has exactly $n-1$ holes. This definition makes *no* requirement that A itself be a connected set!

Exercise 1.38 (i) *Show that an open subset of $\mathbb{C}$ is simply-connected if either* (a) *its boundary has no bounded component or* (b) *it is bounded and has a connected boundary. Infer from this and* 1.35 *that any open subset of $\mathbb{C}$ which is homeomorphic to $\mathbb{D}$ is simply-connected.*

Hints: Suppose open $U \subset \mathbb{C}$ satisfies (a) or (b) and that C is a component of $\mathbb{C} \setminus U$. 1.30(v) insures that C meets ∂U in some point z. Let B denote the component of ∂U which contains z. By (1.1) $B \cup C$ is a connected set, and it lies in $\mathbb{C} \setminus U$. Maximality of C then entails $B \cup C = C$. In case (a) this proves C is unbounded. In case (b) it means that $\partial U = B \subset C$, and therefore by disjointness of components, C must be the only component of $\mathbb{C} \setminus U$, and it is unbounded because U is bounded in case (b).

(ii) *Show that $A \subset \mathbb{C}$ is simply-connected if and only if $S \subset A$ for every bounded $S \subset \mathbb{C}$ such that $\partial S \subset A$.*

Hints: ($\Rightarrow$) From $\bar{S} \setminus \mathring{S} = \partial S \subset A$ follows $\bar{S} \cap (\mathbb{C} \setminus A) = \mathring{S} \cap (\mathbb{C} \setminus A)$, so this is a relatively clopen subset of $\mathbb{C} \setminus A$. It therefore contains every component of $\mathbb{C} \setminus A$ which it meets. As $\mathbb{C} \setminus A$ has only unbounded components, this means its bounded subset $\bar{S} \cap (\mathbb{C} \setminus A)$ meets no component, hence is empty.

($\Leftarrow$) According to 1.30(v) any component S of $\mathbb{C} \setminus A$ satisfies $\partial S \subset A$. Hence S cannot be bounded, else the contradiction $S \subset A$ would follow from the present hypothesis. As it is bounded and $\mathbb{C} \setminus A$ has no bounded components, $\bar{S} \cap (\mathbb{C} \setminus A)$ is empty.

(iii) *Show that an open subset U of $\mathbb{C}$ is simply-connected if and only if $\mathbb{C} \setminus U$ contains no non-empty, compact, relatively open subset.*

Hints: ($\Rightarrow$) Such a subset would be a union of components of $\mathbb{C} \setminus U$ (why?).

($\Leftarrow$) If $\mathbb{C} \setminus U$ had a compact component C, 1.34 would supply a set K of the forbidden kind.

Theorem 1.39 ("Filling the Holes") *For each non-empty open $U \subset \mathbb{C}$ the union of U with all its holes is an open, simply-connected set, which is connected if U is.*

Proof: Each component of $\mathbb{C} \setminus U$ is closed (1.30(i)). Denote by $\mathcal{C}$ the set of all the bounded ones, the holes in U. Each $C \in \mathcal{C}$ is compact, so by 1.34 there is a compact, relatively open subset V_C of $\mathbb{C} \setminus U$ which contains C. Now V_C meets each (closed) component of $\mathbb{C} \setminus U$ in a compact, relatively open subset of it, hence would disconnect it if that component were unbounded. That is,

$$(*) \qquad V_C \subset \bigcup \mathcal{C}.$$

But then

$$U^* := U \cup \bigcup \mathcal{C} \subset U \cup \bigcup_{C \in \mathcal{C}} V_C \overset{(*)}{\subset} U \cup \bigcup \mathcal{C} = U^*.$$

Equality thus prevails here, making U^* open, since $V_C = W_C \cap (\mathbb{C} \setminus U)$ with W_C open in $\mathbb{C}$ makes each set $U \cup V_C = U \cup W_C$ open in $\mathbb{C}$. Evidently the components of $\mathbb{C} \setminus U^*$ are just the unbounded components of $\mathbb{C} \setminus U$, which means that U^* is simply-connected. (It is called the *simply-connected hull* of U.) Since $\bar{U} \cap C \supset \partial C$ by 1.30(v), and ∂C is non-void, it follows from 1.3 that if U is connected, so is each set $U \cup C$, and then too the union U^* of all these sets.

Theorem 1.40 *Let Ω be a region, C_1, C_2 two distinct holes in Ω. Then there is an unbounded, connected, closed subset of $\mathbb{C} \setminus C_1$ which contains C_2.*

Proof: C_1, C_2 are disjoint and compact, and each meets $\overline{\Omega}$. Pick any point $c_2 \in C_2 \cap \overline{\Omega}$ and take a convex neighborhood N_2 of c_2 which is disjoint from C_1 and pick $b_2 \in N_2 \cap \Omega$. Consider then two cases:

Case I: Ω is bounded. Then $\mathbb{C} \setminus \Omega$ has (exactly) one unbounded component C_3. Pick $c_3 \in C_3 \cap \overline{\Omega}$ and a convex neighborhood N_3 of c_3 disjoint from C_1. Pick $b_3 \in N_3 \cap \Omega$ and let (1.28) Γ be a curve in Ω from b_2 to b_3. Consider then

$$C_2 \cup [c_2, b_2] \cup \Gamma \cup [b_3, c_3] \cup C_3.$$

This is an unbounded, connected, closed set which contains C_2 and is disjoint from C_1.

Case II: Ω is unbounded. Then use 1.28 to construct a curve Γ in Ω joining b_2 to some $\omega \in \Omega$ with $|\omega| > \sup |C_1|$.

Then

$$C_2 \cup [c_2, b_2] \cup \Gamma \cup [1, \infty)\, \omega$$

is an unbounded, connected, closed set which contains C_2 and is disjoint from C_1.

Theorem 1.41 *Let Ω be a region in $\mathbb{C}, C_1, \ldots, C_n$ distinct holes in Ω. Then there exists an open subset U_1 of Ω in which C_1 is a hole but none of $C_2, \ldots, C_n$ lies in a hole. In particular, if $C_1, \ldots, C_n$ are all the holes in Ω, then C_1 is the unique hole in U_1.*

Proof: The preceding theorem provides unbounded, connected, closed sets K_j with $C_j \subset K_j \subset \mathbb{C} \setminus C_1 \quad (j = 2, \ldots, n)$. Let $U_1 := \Omega \setminus (K_2 \cup \cdots \cup K_n)$. Then

$$\mathbb{C} \setminus U_1 = (\mathbb{C} \setminus \Omega) \cup K_2 \cup \cdots \cup K_n.$$

Denote by $\mathcal{C}$ the set of unbounded components of $\mathbb{C} \setminus \Omega$. Obviously each K_j and each $C \in \mathcal{C}$ lies in an unbounded component of $\mathbb{C} \setminus U_1$. If a bounded component C of $\mathbb{C} \setminus \Omega$ touches any K_j, then it lies in the unbounded component of $\mathbb{C} \setminus U_1$ which contains K_j. Any other bounded component of $\mathbb{C} \setminus \Omega$ is necessarily also a bounded component of $\mathbb{C} \setminus U_1$. In particular, C_1 is a bounded component of $\mathbb{C} \setminus U_1$, that is, a hole in U_1.

Exercise 1.42 *Let U be an open subset of $\mathbb{C}$ and $C_1, \ldots, C_M$ finitely many of the holes in U. Show that there exists compact $K \subset U$ such that $C_1, \ldots, C_M$ lie in distinct holes in K.*

Hints: Choose bounded, open $V_j \supset C_j$ such that $V_1, \ldots, V_M$ are pairwise disjoint. Apply 1.34 to select compact $K_j \subset \mathbb{C} \setminus U$, relatively open in $\mathbb{C} \setminus U$ and satisfying $C_j \subset K_j \subset V_j$. Now $\mathbb{C} \setminus (U \cup K_1 \cup \cdots \cup K_M) = (\mathbb{C} \setminus U) \setminus (K_1 \cup$

$\cdots \cup K_M)$ is relatively closed in $\mathbb{C} \setminus U$ (since the K_j are relatively open there) and so is closed in $\mathbb{C}$. Therefore $U \cup K_1 \cup \cdots \cup K_M$ is open in $\mathbb{C}$. Consequently there exist U_j open in $\mathbb{C}$ such that

(1) $\qquad K_j \subset U_j$ open $\subset \overline{U}_j$ compact $\subset U \cup K_1 \cup \cdots \cup K_M$

and

(2) $\qquad$ the $\overline{U}_j$ are mutually disjoint.

Define

$$K := \partial U_1 \cup \cdots \cup \partial U_M.$$

This set is compact and is disjoint from $K_1 \cup \cdots \cup K_M$ by (1) and (2), but lies in $U \cup K_1 \cup \cdots \cup K_M$ by (1). Therefore $K \subset U$. (1) and (2) also show that $U_j \subset \mathbb{C} \setminus K$. Since $\overline{U}_j \setminus U_j = \partial U_j \subset K$, we have $U_j = U_j \cap (\mathbb{C} \setminus K) = \overline{U}_j \cap (\mathbb{C} \setminus K)$, so U_j is open and relatively closed in $\mathbb{C} \setminus K$. Hence U_j contains any component of $\mathbb{C} \setminus K$ which it meets. As $C_j \subset U_j$ and the U_j are bounded and disjoint, it follows that $C_1, \ldots, C_M$ lie in different bounded components of $\mathbb{C} \setminus K$.

Finally, a very technical result which will only be needed in volume II:

Theorem 1.43 *There is a countable family $\mathcal{K}$ of simply-connected, compact subsets of $\mathbb{C} \setminus \overline{\mathbb{D}}$ with the property that every simply-connected, compact subset of $\mathbb{C} \setminus \overline{\mathbb{D}}$ is a subset of some element of $\mathcal{K}$.*

Proof: In the space of continuous complex-valued functions on $[0, 1]$ which are 1 at 0 there is a countable set $\{f_1, f_2, \ldots\}$ which is uniformly dense. Indeed, by an easy uniform-continuity argument one sees that the piecewise linear functions with rational complex "vertices" are such a class.

For each pair of positive integers n and m define

$$(1) \quad U(n,m) := D\left(0, 1 + \frac{1}{m}\right) \cup \bigcup_{t \in [0,1]} D\left(f_n(t), \frac{1}{m}\right) \cup \left[\mathbb{C} \setminus \bar{D}\left(0, |f_n(1)| - \frac{1}{m}\right)\right].$$

This is an open, connected set. For

$$(2) \qquad f_n[0,1] \cup D(f_n(t), 1/m), \quad t \in [0,1]$$

is connected, being the union of connected sets with a point in common. Similarly, since $f_n[0,1]$ meets both the disk $D(0, 1 + 1/m)$ (in $f_n(0) = 1$) and the set $\mathbb{C} \setminus \bar{D}(0, |f_n(1)| - 1/m)$ (in $f_n(1)$), the set

$$(3) \qquad D(0, 1 + 1/m) \cup f_n[0,1] \bigcup \left[\mathbb{C} \setminus \bar{D}(0, |f_n(1)| - 1/m)\right]$$

is connected. Then the union of the set (3) and all the sets in (2), viz., (1), is connected. Define

$$K(n,m) := \mathbb{C} \setminus U(n,m). \tag{4}$$

Thus $K(n,m)$ is a compact set which is disjoint from $\overline{\mathbb{D}}$ and has connected complement.

Now let K be any simply-connected, compact subset of $\mathbb{C} \setminus \overline{\mathbb{D}}$. We will show that $K \subset K(n,m)$ for some n, m. Choose $r > 1 + \sup|K|$. Then 1 and r are two points in the open connected set $\mathbb{C} \setminus K$. So by 1.28 there is a continuous $\gamma\colon [0,1] \mapsto \mathbb{C} \setminus K$ with $\gamma(0) = 1$, $\gamma(1) = r$. Choose $m \in \mathbb{N}$ so large that

$$D(0, 1 + 1/m) \subset \mathbb{C} \setminus K, \tag{5}$$

and

$$\frac{2}{m} < \operatorname{dist}(\gamma[0,1], K). \tag{6}$$

Then choose n so that

$$\sup_{t\in[0,1]} |f_n(t) - \gamma(t)| < 1/m. \tag{7}$$

It follows from (7) and (6) that

$$\bigcup_{t\in[0,1]} D(f_n(t), 1/m) \subset \mathbb{C} \setminus K \tag{8}$$

and also that

$$\begin{aligned} \sup|K| < r - 1 = \gamma(1) - 1 &\le |f_n(1)| + |\gamma(1) - f_n(1)| - 1 \\ &< |f_n(1)| + 1/m - 1 \\ &\le |f_n(1)| - 1/m \quad (\text{if } m \ge 2), \end{aligned}$$

that is,

$$K \subset D(0, |f_n(1)| - 1/m). \tag{9}$$

It follows from (5), (8) and (9) that $K \subset K(n,m)$.

§ 6 Extension Theorems

The first result of this section will be useful in Chapter IV when we want to produce logarithms. It is classical and indispensable in analysis.

Theorem 1.44 (TIETZE [1914]) *C is a non-empty closed subset of metric space M and $f\colon C \mapsto \mathbb{R}$ is continuous. Then there exists continuous $\phi\colon M \mapsto \mathbb{R}$ such that $\phi = f$ on C and* $\sup|\phi(M)| = \sup|f(C)|$.

Proof: (URYSOHN [1925]) Suppose first that f is bounded and consider any positive real $m \geq \sup|f(C)|$. Let d be a metric for M and for any $A \subset M$ define

$$d(x, A) := \left\{ \begin{matrix} \inf\{d(x,a) : a \in A\} & \text{if } A \neq \varnothing \\ 1 & \text{if } A = \varnothing \end{matrix} \right\}, \quad x \in M.$$

This is a continuous function of x whose zero-set is the closure of A (cf. 1.32). The sets

$$C_- := f^{-1}[-m, -m/3] \quad \text{and} \quad C_+ := f^{-1}[m/3, m]$$

are closed, disjoint subsets of C. Hence for every $x \in M$ at least one of $d(x, C_-)$ and $d(x, C_+)$ is positive. Therefore

$$\psi(x) := \frac{d(x, C_-) - d(x, C_+)}{d(x, C_-) + d(x, C_+)}, \quad x \in M$$

is a well-defined, hence continuous, function and it satisfies

(1) $\psi(M) \subset [-1, 1]$

(2) $\psi(C_-) = \{-1\}$ if $C_- \neq \varnothing$, $\psi(C_+) = \{1\}$ if $C_+ \neq \varnothing$.

Consider $\phi_1 := m\psi/3$. By (2) and the definition of C_-, $|f - \phi_1| \leq 2m/3$ on C_-. Similarly on C_+. Since $|\phi_1| \leq m/3$ (by(1)) and f maps $C \setminus (C_- \cup C_+)$ into $[-m/3, m/3]$ the inequality $|f - \phi_1| \leq 2m/3$ holds on $C \setminus (C_- \cup C_+)$ as well. In summary

$$|f - \phi_1| \leq 2m/3 \text{ on } C \text{ and } |\phi_1| \leq m/3 \text{ on } M.$$

Repeat this construction with $f - \phi_1$ in the role of f and $2m/3$ in the role of m. Inductively, we get continuous $\phi_n\colon M \mapsto \mathbb{R}$ which satisfy

$$\left| f - \sum_{j=1}^{n} \phi_j \right| \leq (2/3)^n m \text{ on } C \text{ and } |\phi_n| \leq (2/3)^{n-1} m/3 \text{ on } M.$$

The series $\sum \phi_n$ therefore converges uniformly on M to a (continuous) function ϕ that satisfies $|\phi| \leq \sum (2/3)^{n-1} m/3 = m$ on M, and

$$|f - \phi| = \lim_{n\to\infty} \left| f - \sum_{j=1}^{n} \phi_j \right| \leq \lim_{n\to\infty} (2/3)^n m = 0 \text{ on } C.$$

In case f is unbounded, apply the preceding to get a continuous extension Φ of the function $F := f/(\mathbf{1} + |f|)$ which satisfies $|\Phi| \le \mathbf{1}$. Replacing Φ with $\Phi_1(x) := \Phi(x)/(1 + d(x, C))$ gives a continuous extension of F which in addition satisfies $|\Phi_1| < \mathbf{1}$. A little algebra shows that the function $\phi_1 := \Phi_1/(\mathbf{1} - |\Phi_1|)$ is a continuous extension of f.

Exercise 1.45 *Prove that* 1.44 *remains valid if* $\mathbb{R}$ *is replaced by* $\mathbb{C}$.

Hints: Get a continuous extension f_1 of ϕ by extending $\operatorname{Re}\phi$ and $\operatorname{Im}\phi$ separately. If $\sup|\phi(M)|$ is finite, we may assume it is 1 and set $f := f_1/\max\{\mathbf{1}, |f_1|\}$. [With finer arguments an extension can be constructed having range in the convex hull of $\phi(M)$.]

Exercise 1.46 *Let a, b, A, $B \in \mathbb{R}$ with $a < b$, $A < B$ and ϕ be a continuous bijection of (a, b) onto (A, B). Show that ϕ is strictly monotonic and extends to a homeomorphism of $[a, b]$ onto $[A, B]$ with endpoints mapping to endpoints.*

Hints: (M. Lloyd) The kernel of the problem is to show that ϕ is monotonic. To this end, look at the convex set $T := \{(x, y) : a < x < y < b\}$ and the continuous real-valued function $\psi(x, y) := \phi(y) - \phi(x)$ on it. Since ϕ is injective, ψ is zero-free. Hence either the connected set $\psi(T)$ lies in $(0, \infty)$, which means that ϕ is increasing, or $\psi(T)$ lies in $(-\infty, 0)$, which means that ϕ is decreasing.

Our closing illustration of extension techniques is one of the standard developments of the exponential and logarithmic functions in $\mathbb{R}$. Although the results are known to most readers and are wholly subsumed under the later construction of these functions in $\mathbb{C}$ (cf. 3.14 and 2.21), an organic and elementary treatment of the theory in $\mathbb{R}$ has pedagogical value and illustrates a characteristic feature of our enterprise, expressed quite generally by T. S. Eliot in the lines

> "We shall not cease from exploration
> And the end of all our exploring
> Will be to arrive where we started
> And know the place for the first time."

Exercise 1.47 [Exponentiation in $\mathbb{R}_+$; cf. remarks following 3.14.]

(i) *Show that for each $n \in \mathbb{Z}\backslash\{0\}$ the function $p_n(x) := x^n$ maps $(0, \infty)$ bijectively and continuously. Using* 1.46 *deduce that p_n^{-1} is continuous.* [After (xv) we will be able to verify an iterative construction of $p_n^{-1}(x)$ involving only integer powers of x. See (xvii).]

(ii) *For $m \in \mathbb{Z}$, $n \in \mathbb{N}$ and positive real x define $p_0 := \mathbf{1}$ and*

$$x^{1/n} := p_n^{-1}(x), \qquad x^{m/n} := p_m(p_n^{-1}(x)) = (x^{1/n})^m.$$

Show that the latter is well defined (that is, depends only on the ratio m/n) *and is a monotonic continuous function of* x.

(iii) *Formulate and substantiate an algebraic theory of rational exponents on positive bases.*

(iv) *Show that for each positive base* $b \neq 1$, $r \mapsto b^r =: E_b(r)$ *is a strictly monotonic function on* $\mathbb{Q}$.

(v) *Show that* $\lim_{s \downarrow x} E_b(s) = \lim_{r \uparrow x} E_b(r)$ *for every positive* $b \neq 1$, $x \in \mathbb{R}$, *and this number is* b^x *if* $x \in \mathbb{Q}$.

Hints: Monotonicity ensures that both limits exist. Since $E_b(s)/E_b(r) = E_b(s-r)$ by (iii), to show the two limits are equal, it suffices to consider $x = 0$ and approach from the right. In turn, by monotonicity, it suffices to show that for $n \in \mathbb{N}$, $\lim_{n\to\infty} b^{1/n} = 1$. If $\varepsilon > 0$ and $n \in \mathbb{N}$ exceeds b/ε, then $(1+\varepsilon)^n = \sum_{j=0}^{n} \binom{n}{j}\varepsilon^j \geq 1 + n\varepsilon > b$, so $b^{1/n} < 1 + \varepsilon$.

(vi) (Laws of real exponents). *For positive* $b \neq 1$, $x \in \mathbb{R}$ *define* b^x *to be the limit in* (v). *Show that* $E_b(x) := b^x$ *defines a continuous isomorphism of* $(\mathbb{R}, +)$ *onto* $((0,\infty), \cdot)$, *called* the exponential function with base b.

Hint: If $x_n \to x$, there exist rational r_n with $|r_n - x_n| < 1/n$ and $|b^{r_n} - b^{x_n}| < 1/n$. Then $r_n \to x$ and by definition $b^x = \lim b^{r_n}$.

(vii) *Infer the further laws of exponents*

$$(ab)^x = a^x b^x, \qquad (b^x)^y = b^{xy} \qquad \forall x, y \in \mathbb{R}, \text{ positive } a, b.$$

Hint: For the second equality, suppose $1 \neq b > 0$; $r \in \mathbb{Q}$ with $r \to x$; $m \in \mathbb{Z}$, $n \in \mathbb{N}$, $s := m/n$. The power function $p_s := p_m \circ p_n^{-1}$ is continuous, so $(b^r)^s = p_s(b^r) \to p_s(b^x) = E_{b^x}(s)$, while $(b^r)^s = E_b(rs)$ [by (iii)] $\to E_b(xs)$. Hence $E_{b^x}(s) = E_b(xs)$. Let $s \to y$.

(viii) *Show that for real* $b > 1$

$$(b^r - 1)r^{-1} < (b^s - 1)s^{-1} \qquad \forall \text{ rational } s > r > 0.$$

Hint: If $r = p/n$, $s = q/n$ ($p, q, n \in \mathbb{N}$), replace b by $b^{1/n}$ and reduce to the case $r = p$, $s = q \in \mathbb{N}$. In that case it suffices to consider $s = r + 1$ and to divide both products by $b - 1$.

(ix) *Prove that the inequality in* (viii) *holds for all real* $s > r > 0$.

(x) *Prove that* $\lim_{r\to 0}(b^r - 1)r^{-1}$ *exists in* $\mathbb{R}$.

Hints: Consider $b > 1$. The limit $r \to 0^+$ comes from (ix). Use this fact, continuity of E_b at 0, and the identity $(b^{-r} - 1)(-r)^{-1} = E_b(-r)(b^r - 1)r^{-1}$ to settle $r \to 0^-$.

(xi) *Prove that E_b is differentiable on $\mathbb{R}$ and that $E_b' = E_b'(0)E_b$. Deduce from non-constancy of E_b that $E_b'(0) \neq 0$.*

(xii) *Using* (xii) *and the chain rule, show that for any non-zero real α, $E'_{b^\alpha}(0) = \alpha E_b'(0)$ and deduce that there is a unique $b_1 > 1$ such that $E'_{b_1}(0) = 1$ and $E'_{b_1} = E_{b_1}$.*

(xiii) (Logarithms). *For positive $b \neq 1$ return to* (vi) *and denote the inverse of E_b, an isomorphism from $\cdot$ to $+$, by L_b. Show that for all $x > 0$*

$$E_b'(0)L_b'(x) = 1/x, \quad \textit{and}$$

$$E_b'(0)L_b(x) = \lim_{\varepsilon\to 0} \frac{x^\varepsilon - 1}{\varepsilon} = \lim_{n\to\infty} n(x^{1/n} - 1).$$

Hints:

$$\frac{L_b(x+\varepsilon) - L_b(x)}{\varepsilon} = \left[\frac{E_b(L_b(x+\varepsilon)) - E_b(L_b(x))}{L_b(x+\varepsilon) - L_b(x)}\right]^{-1}.$$

Since L_b is continuous at x (by 1.46) and E_b is differentiable at $L_b(x)$, it follows that $L_b'(x)$ exists and equals $1/E_b'(L_b(x)) = 1/E_b'(0)E_b(L_b(x))$. Using (vii), if $x \neq 1$

$$\frac{x^\varepsilon - 1}{\varepsilon} = \frac{E_b(\varepsilon L_b(x)) - 1}{\varepsilon L_b(x)} L_b(x) \longrightarrow E_b'(0)L_b(x) \quad \text{as} \quad \varepsilon \to 0.$$

(xiv) *For the b_1 of* (xii) *prove that*

$$E_{b_1}(x) = \lim_{n\to\infty}\left(1 + \frac{x}{n}\right)^n \qquad \forall x \in \mathbb{R}.$$

In particular, b_1 is $\lim_{n\to\infty}(1 + 1/n)^n$. (Cf. 3.15 and 3.21.)

Hint: $L_{b_1}(1 + x/n)^n = x \cdot [L_{b_1}(1 + x/n) - L_{b_1}(1)] \div (x/n) \to x \cdot L'_{b_1}(1) = x$ by part (xiii). Now use the continuity of E_{b_1}.

(xv) *Using* (vii) *show that for all $x, b \in \mathbb{R}$, $b > 0$*

$$E_b(x) = E_{b_1}(xL_{b_1}(b))$$

and infer the differentiability of the generalized power function $b \mapsto b^x$ *on $(0, \infty)$. In fact, using* (xii) *and the Chain Rule, the derivative with respect to b is seen to be xb^{x-1}.*

(xvi) *Given real $a > 0$ and integer $n > 1$, consider the function*

$$F(x) := [(n-1)x + ax^{1-n}]/n, \qquad x \geq a^{1/n}.$$

Show that

$$(*) \qquad a^{1/n} < F(x) < x \qquad \forall x > a^{1/n}.$$

Hints: (Ken Ross) The second inequality is immediate. After writing $b := a^{1/n}$, $t := b/x < 1$, and using the fact that $F(b) = b$, routine algebra confirms the first inequality to be equivalent to each of the following: $b^n\left(\frac{1}{b^{n-1}} - \frac{1}{x^{n-1}}\right) < (n-1)(x-b) \Leftrightarrow b(1-(b/x)^{n-1}) < (n-1)(x-b) \Leftrightarrow t - t^n < (n-1)(1-t) = n(1-t) - 1 + t \Leftrightarrow (1-t)\sum_{j=0}^{n-1} t^j = 1 - t^n < n(1-t)$.

(xvii) *Inequality* $(*)$ *enables iteration of* F. *Show that for every* $x > a^{1/n}$ *the sequence of iterates* $\{F^{[k]}(x)\}$ *decreases to a limit* $L(x) \geq a^{1/n}$, *and in fact, equality holds here.*

Hint: $F^{[k+1]}(x) = F(F^{[k]}(x)) \to F(L(x))$ and $F^{[k+1]}(x) \to L(x)$, making $L(x)$ a fixed-point of F, which $(*)$ asserts is impossible were $L(x) > a^{1/n}$. [It should be noted that this is all genuinely elementary, as the $F^{[k]}(x)$ involve no fractional exponents and are rational functions of a.]

Notes to Chapter I

The results of this chapter belong to the realm of general point-set topology and all of them are valid in the appropriately general topological space (most in any [Hausdorff] topological space; some require local compactness or connectedness). For some of the history of some of the concepts here consult the notes at the end of HAUSDORFF's books [1914], [1957]. On the important concept of connectedness, see especially Freudenthal's notes on pp. 486–488 of BROUWER [1976], and WILDER.

In almost any function theory book (e.g., CONWAY [1978], pp. 8–10; see also the preliminaries to 4.63 below) the reader will find a discussion of the stereographic projection of the plane onto the sphere $\mathbb{C}_\infty := \{(x_1, x_2, x_3) \in \mathbb{R}^3 : x_1^2 + x_2^2 + x_3^2 = 1\}$ minus its north pole $(0,0,1)$. One regards $\mathbb{C}$ as a subset of $\mathbb{C}_\infty$ and does all of function theory on the sphere. This has several advantages (not the least being esthetical), since now the point "at infinity" does not enjoy a privileged role (topologically, at least). The statements and proofs of many results thus acquire greater cohesion and uniformity. This point of view is essential in the theory of Riemann surfaces and differentiable manifolds and in the study of meromorphic functions. A penalty is, however, exacted, which I have elected not to pay; namely, the consideration of cases returns when one engages in algebraic operations. $\mathbb{C}_\infty$ is called the *Riemann sphere*—see §8 of OSGOOD [1901].

Simple-connectivity for subsets S of $\mathbb{C}$ is usually defined as $\mathbb{C}_\infty \setminus S$ being connected. It is obvious that our definition agrees with this one when S is bounded, and via 1.34 it is easy to demonstrate when S is open. (For closed S, see p. 133 of GAIER [1987].) For arbitrary S, however, the two definitions do not agree. To see this, consider a connected subset C of $\mathbb{C}_\infty$ with an "explosion point" x, i.e., a point $x \in C$ such that $C \setminus \{x\}$ is totally disconnected (meaning that each component is a single point). In 1921 B. Knaster and K. Kuratowski proved that such exotica indeed exist; see page 145 of STEEN and SEEBACH [1978]. We may situate C in $\mathbb{C}_\infty$ so that $x = (0,0,1)$ and let $S := \mathbb{C}_\infty \setminus C$, a subset of $\mathbb{C}_\infty \setminus \{(0,0,1)\} = \mathbb{C}$. Then of course this set has a connected complement in $\mathbb{C}_\infty$, but its complement in $\mathbb{C}$ is the totally disconnected set $C \setminus \{(0,0,1)\}$ and so has infinitely many bounded components.

A stronger form of 1.5(ii) occurs in (9.9) of SAKS and ZYGMUND [1971]. A much more general version of 1.14 is true: any two points on any curve lie on an arc which lies on the curve. For a proof of this surprisingly non-trivial fact (due to R. L. MOORE [1916/17]) see TIETZE [1919], MAZURKIEWICZ [1920], KAMIYA [1931] or G. T. WHYBURN [1932] or p. 62 of ČERNY [1992]. Contained in the proof of 1.19(ii) is a famous result of CARATHÉODORY [1911] (p. 200): if $S \subset \mathbb{R}^k$, then every element of $\operatorname{co} S$ is a convex combination of $k+1$ or fewer points of S. The proof in the text is due to STEINITZ [1913] (p. 153).

In spite of 1.28 the converse of 1.21 is false. Any two points in the set $S := (0,1] \times \{0\} \cup \{1/n : n \in \mathbb{N}\} \times [0,1]$ lie on a 3-gon in S, and the point $(0,1)$ is in $\overline{S}$. So the union of that point with S is a connected set S_1, by (1.2). But no curve in S_1 joins $(0,1)$ to the point $(1,0) \in S$. For if $\gamma : [0,1] \mapsto \mathbb{C}$ is such a curve, then $\operatorname{Re}\gamma[0,1] = [0,1]$, so for each $n \in \mathbb{N}$ there is $t_n \in [0,1]$ with $\operatorname{Re}\gamma(t_n) = \frac{1}{2}[\frac{1}{n} + \frac{1}{n+1}]$ and consequently $\operatorname{Im}\gamma(t_n) = 0$. A subsequence of $\{t_n\}$ converges to some t_0. Then $\gamma(t_0) = (0,0)$, a point not in S_1. —×—

1.33 was also proved in MULLIKIN [1922], p. 147. In Hausdorff's version X can be any compact metric space. The proof in the text clearly works in that generality too. This version of the theorem implies 1.34 when applied to the closure of X in $\mathbb{C}_\infty$. Šura-Bura dispensed with the metric hypothesis.

The idea of the proof of 1.35 comes from HEINS [1962b] (p. 150) and that for 1.40 is from SAKS and ZYGMUND [1971] (p. 210), who achieve most of the goals of our 1.33 in a different way with their (9.6). See also DE VITO [1957] for 1.39. Definition 1.37 is due to HAUSDORFF (p. 351 of his book [1914]) and supersedes a less tractable one which Carathéodory had made earlier. For 1.31 see p. 334 of KURATOWSKI [1945].

There is another proof of 1.44 in DIEUDONNÉ [1969] (pp. 89, 90). It is quite similar to one in BORSUK [1937] and to the second of two proofs which HAUSDORFF [1919] offers. Both of Hausdorff's proofs are elegant and interesting; the first involves (theorem of Hans Hahn) inserting a continuous function between an upper semicontinuous function and a lower semicontinuous function. (See also pp. 276–277 and 281–283 of HAUSDORFF [1957].) This first proof of Hausdorff's, together with an unusual one by Harald Bohr, are detailed as multi-part exercises in the classic text STROMBERG [1981]. For other proofs see R. RADO [1931] and VIZZINI [1932]. Theorem 1.44 was proved for $M = \mathbb{R}^n$ by LEBESGUE [1907]. The plane version with C compact can already be found in PHRAGMÉN [1900]. And a version of 1.24 was proved by PHRAGMÉN [1885/86]: if P is a totally disconnected closed subset of $\mathbb{C}$, then $\mathbb{C} \setminus P$ is connected. Our proof of 1.44 is from SAKS and ZYGMUND [1971]. Urysohn actually proved 1.44 in any topological space X in which every pair of disjoint, non-empty closed sets C_1, C_2 lie in disjoint open sets (such an X is called *normal*) by constructing, without the aid of a metric, a continuous map $\psi\colon X \mapsto [-1, 1]$ with $\psi(C_j) = (-1)^j$ $(j = 1, 2)$. Such functions provide the ϕ_n from which ϕ was constructed in the proof of 1.44. The elegant, simple form into which this proof has evolved will be found in MANDELKERN [1993] and OSSA [1998].

Chapter IV will illustrate the importance of extending zero-free functions so as to remain zero-free, which is not always possible if continuity is also wanted. An elegant, elementary treatment of that important theme is MORTINI and RUPP [2012].

Elementary treatments of the real exponential and logarithmic functions, somewhat different from 1.47 but also involving no integration, appear in KEMENY [1957] and M. KNESER [2004]. See also WULF [1897], HURWITZ [1911] and chapter 2 of E. LANDAU [1950]. A maximally elementary but thorough and pedagogically valuable account is PATYI [2015].

Chapter II
(Complex) Derivative and (Curvilinear) Integrals

About half of this short chapter is comprised of very basic and very easy material, likely familiar to most readers in one form or other, but I record it here for later reference and to fix notation and terminology. However, §5 is rather novel and even *cognoscenti* might enjoy looking at it.

§ 1 Holomorphic and Harmonic Functions

Classical complex analysis, the subject of this book, is primarily the study of holomorphic functions:

Definitions 2.1 If f is a complex-valued function defined in a neighborhood of the point z_0, we say f is *(complex-)differentiable* at z_0 if the quotient of complex numbers $(f(z) - f(z_0))/(z - z_0)$ has a limit in $\mathbb{C}$ as z approaches z_0. When it exists, this limit is called the *derivative of* f *at* z_0 and is noted $f'(z_0)$. If f is (complex) differentiable at each point of an open subset U of $\mathbb{C}$, we say f *is holomorphic in* U and by $H(U)$ denote the set of all such functions. This alternative terminology is primarily to distinguish complex-differentiability from real-differentiability (of $\mathbb{R}^2$-valued functions in $\mathbb{R}^2$—see §4). We denote by $H^\infty(U)$ the subset of all functions in $H(U)$ which are bounded on U. A function which is holomorphic in $\mathbb{C}$ is called *entire*. The terms *univalent, injective* and *schlicht* are synonyms for one-to-one. A holomorphic function which is one-to-one is called *conformal*, and a conformal mapping of U onto U is called a *conformal automorphism* of U; the set of these is denoted by $\mathrm{Aut}\,(U)$. [It will be proved in 5.77 and 5.78 that conformal mappings have open image sets and holomorphic inverses. So, from that point on conformal will be seen to be equivalent to *biholomorphic* and $\mathrm{Aut}\,(U)$ to be a group under composition, but these facts are not needed in the meantime.]

Remarks: Often the term conformal is bestowed on holomorphic functions which are only locally univalent, i.e., univalent in some neighborhood of each

R. B. Burckel, *Classical Analysis in the Complex Plane*, Cornerstones,
https://doi.org/10.1007/978-1-0716-1965-0_3

point. We shall see later (5.89) that this is equivalent to the derivative being zero-free. The rationale for this term may be found in the discussion following 3.16. It should also be noted that earlier British usage for entire is *integral.* Sometimes *analytic* is also used for holomorphic, but equally often that term has a more technical use. (See chapter VI of SAKS and ZYGMUND [1971].) The terms *monogenic* and *regular*, now largely archaic, are other synonyms for holomorphic. But by now we are teetering on the edge of Ockham's razor. The reason for the multitude of synonyms is the multitude of disparate properties which are all equivalent (as we shall gradually see) to holomorphy; historically each acquired its name before the grand equivalence was recognized. The most fundamental fact in this subject, which we will uncover in two stages, 3.39(ii) and 5.3, is that:

> *Every holomorphic function is $\mathbb{C}$-differentiable arbitrarily often and is represented in a neighborhood of each point by its Taylor series.*

Definition 2.2 A function f defined in a subset U of $\mathbb{C}$ is called *harmonic in U* if U is a union of open sets in each of which f is the real part of some holomorphic function; that is, for each $a \in U$ there is an $r > 0$ and a function F_a holomorphic in $D(a, r)$ such that $f = \operatorname{Re} F_a$ in $U \cap D(a, r)$.

Remarks: This is not the customary definition. Later the myriad equivalences of this important concept will be established. One of them is the traditional definition of harmonic: f satisfies Laplace's partial-differential equation. (See 9.29, 9.31.) A complex-valued function F is defined to be harmonic if each of $\operatorname{Re} F$ and $\operatorname{Im} F$ is harmonic.

Although one of the simplest of functions, $f(z) := \bar{z}$, is not complex-differentiable at any point z_0 [because the difference quotient is 1 whenever $z - z_0 \in \mathbb{R}$ and is -1 whenever $z - z_0 \in i\mathbb{R}$], the expected elementary results prevail (and thus, e.g., $H(U)$ and $H^\infty(U)$ are algebras). Their simple proofs are rendered completely transparent when we note (following Carathéodory) that differentiability of f at z_0 is obviously equivalent to the demand that there exist a function F defined in a neighborhood of z_0 which is continuous at z_0 and satisfies $f(z) - f(z_0) = (z - z_0)F(z)$ for all z in this neighborhood; $F(z_0)$ is then $f'(z_0)$.

Theorem 2.3 *Let f, g, h be complex-valued functions with f and g defined in a neighborhood of z_0 and differentiable at z_0 and h defined in a neighborhood of $f(z_0)$ and differentiable at $f(z_0)$. Then*

(i) *f is continuous at z_0.*

(ii) *$f + g$ and $f \cdot g$ are each differentiable at z_0 and*

$$(f + g)'(z_0) = f'(z_0) + g'(z_0), \quad (f \cdot g)'(z_0) = f'(z_0)g(z_0) + f(z_0)g'(z_0).$$

(iii) *The composite $h \circ f$ is defined in a neighborhood of z_0, is differentiable at z_0 and $(h \circ f)'(z_0) = h'(f(z_0)) \cdot f'(z_0)$. The latter formula is called the* Chain Rule.

Proof: Let F, G, H defined in neighborhoods U, V, W of z_0, z_0, $f(z_0)$ be the functions associated to f, g, h by the preceding observation. The equation

$$f(z) - f(z_0) = (z - z_0)F(z) \qquad \forall\, z \in U$$

and the continuity of F at z_0 immediately yield (i). From the identity

$$\begin{aligned}(fg)(z) - (fg)(z_0) &= \big[f(z) - f(z_0)\big]g(z) + \big[g(z) - g(z_0)\big]f(z_0)\\ &= (z - z_0)\big[F(z)g(z) + G(z)f(z_0)\big] \qquad \forall\, z \in U \cap V\end{aligned}$$

and (i) the less trivial half of (ii) follows. By (i) a subneighborhood V_0 of z_0 in V is mapped into W by f. Then

$$\begin{aligned}h\big(f(z)\big) - h\big(f(z_0)\big) &= \big[f(z) - f(z_0)\big]H\big(f(z)\big)\\ &= (z - z_0)F(z)H\big(f(z)\big) \qquad \forall\, z \in V_0.\end{aligned}$$

The function $F \cdot H \circ f$, which is continuous at z_0, therefore witnesses the differentiability of $h \circ f$ at z_0 and furnishes the claimed value for its derivative.

Remark: Notice that the proof of (iii) is valid *verbatim* for the chain rule for functions on intervals in $\mathbb{R}$ and for the mixed version in which f is a complex-valued function on a real interval and h is (complex-) differentiable at $f(z_0)$. It is only a question of the ambient space in which U, V, W are supposed to be neighborhoods.

Exercise 2.4 *Let f be holomorphic in an open set U. Define $U^* := \{\bar{z} : z \in U\}$ and f^* on U^* by $f^*(z) := \bar{f}(\bar{z})$. Show that f^* is holomorphic in U^*.*

Hint: For z, $z_0 \in U^*$ with $z \neq z_0$, we have $\bar{z}$, $\bar{z}_0 \in U$ and

$$\frac{f^*(z) - f^*(z_0)}{z - z_0} = \overline{\left[\frac{f(\bar{z}) - f(\bar{z}_0)}{\bar{z} - \bar{z}_0}\right]}.$$

Then f^* is differentiable at z_0 with derivative $\overline{f'(\bar{z}_0)}$.

Exercise 2.5 *Consider the functions*

$$\phi(z) := \frac{z+1}{z-1} = 1 + \frac{2}{z-1}, \quad z \in \mathbb{C}\backslash\{1\}, \quad \psi(z) := \frac{1}{2}\left(z + \frac{1}{z}\right), \quad z \in \mathbb{C}\backslash\{0\}.$$

(i) *Show that ϕ is univalent, its range is $\mathbb{C} \setminus \{1\}$ and $\phi^{-1} = \phi$.*

(ii) *Show that for each $a \geq 0$ the disk $D\left(\frac{a}{a+1}, \frac{1}{a+1}\right) \subset \mathbb{D}$ is mapped by ϕ onto the half-plane $-a - \mathbb{H}$.*

Hint: By "completing the square," show $\operatorname{Re}\phi(z) < -a \Leftrightarrow |z - \frac{a}{a+1}|^2 < \left(\frac{1}{a+1}\right)^2$.

(iii) *Show that ϕ maps $[-1, 1)$ onto $(-\infty, 0]$ and (consequently) maps $\mathbb{C} \setminus [-1, 1]$ onto $\mathbb{C}_- \setminus \{1\}$.*

(iv) *Show that ψ maps $\mathbb{C} \setminus \overline{\mathbb{D}}$ biholomorphically onto $\mathbb{C} \setminus [-1, 1]$.*

Hints: For $z \in \mathbb{H}$ set $s(z) := z^2$ and note that in

$$\mathbb{C} \setminus \overline{\mathbb{D}} \overset{\phi}{\mapsto} \mathbb{H} \setminus \{1\} \overset{s}{\mapsto} \mathbb{C}_- \setminus \{1\} \overset{\phi}{\mapsto} \mathbb{C} \setminus [-1, 1]$$

all maps are one-to-one and onto, surjectivity of s being due to 1.26(i). Then check that $\psi(z) = \phi([\phi(z)]^2)$. In view of this representation of ψ, to check holomorphy of ψ^{-1} it suffices to confirm that of s^{-1}. The latter inverse is given by 1.26(i), which shows it to be continuous. Hence if $w_0 \in \mathbb{C}_-$ and $w \in \mathbb{C}_- \setminus \{w_0\}$ converge to w_0, then

$$\frac{s^{-1}(w) - s^{-1}(w_0)}{w - w_0} = \frac{1}{s^{-1}(w) + s^{-1}(w_0)} \to \frac{1}{2s^{-1}(w_0)}.$$

(v) *Show that ψ maps $\mathbb{U}$ biholomorphically onto $\mathbb{C} \setminus \{x \in \mathbb{R} : |x| \geq 1\}$.*

Hints: Let I denote the identity map in $\mathbb{C}$, $s := I^2$. Check that $\phi \circ \phi^2$ is well defined in $\mathbb{C} \setminus \{0, 1\}$ and that

(1) $$\psi = \phi \circ \phi^2 = \phi \circ s \circ \phi \quad \text{throughout } \mathbb{C} \setminus \{0, 1\}.$$

Confirm that in (2) and (3) the mapping is bijective:

(2) $$\mathbb{R} \setminus \{1\} \overset{\phi}{\longmapsto} \mathbb{R} \setminus \{1\}.$$

(3) $$(-\infty, -1] \cup (1, \infty) \overset{\phi}{\longmapsto} [0, 1) \cup (1, \infty) = [0, \infty) \setminus \{1\}.$$

Check that in (4)–(8) all mappings are holomorphic and bijective:

(4) $$\mathbb{C} \setminus \{x \in \mathbb{R} : |x| \geq 1\} \overset{\phi}{\longmapsto} \mathbb{C}_+.$$

(Use (3) and (i).)

(5) $$-\mathbb{U} \overset{\phi}{\longmapsto} \mathbb{U} \overset{s}{\longmapsto} \mathbb{C}_+.$$

(For the first arrow use (2) and (i).)

(6) $$\mathbb{U} \overset{\phi^2}{\longmapsto} \mathbb{C}_+ \overset{\phi}{\longmapsto} \mathbb{C} \setminus \{x \in \mathbb{R} : |x| \geq 1\}.$$

(Use (5), (4) and the equality $\phi = \phi^{-1}$ from (i).)

(7) $$\mathbb{C} \setminus \{x \in \mathbb{R} : |x| \geq 1\} \overset{-\phi}{\longmapsto} \mathbb{C}_- \overset{s^{-1}}{\longmapsto} \mathbb{H} \overset{-iI}{\longmapsto} -\mathbb{U} \overset{\phi}{\longmapsto} \mathbb{U}.$$

(For s^{-1} recall the hints to (iv).) Summarizing (1), (6) and (7),

(8) $$\mathbb{U} \overset{\psi}{\longmapsto} \mathbb{C} \setminus \{x \in \mathbb{R} : |x| \geq 1\} \xmapsto{\phi\circ(-is^{-1})\circ(-\phi)} \mathbb{U}.$$

Finally, compute the composition of maps in (8): $\psi \circ \phi \circ (-is^{-1}) \circ (-\phi) \overset{(1)}{=} \phi \circ s \circ \phi \circ \phi \circ (-is^{-1}) \circ (-\phi) \overset{(i)}{=} \phi \circ s \circ (-is^{-1}) \circ (-\phi) = \phi \circ (-is^{-1})^2 \circ (-\phi) = \phi \circ (-(s^{-1})^2) \circ (-\phi) = \phi \circ (-I) \circ (-\phi) = \phi \circ \phi = I$.

(vi) *The expression for the inverse of* $\psi|\mathbb{U} =: \Psi$ *found in the* (v) *hints is very cumbersome. Other arguments deliver a better recipe for* Ψ^{-1} *(once its existence has been secured, as in* (v)*, say). Prove that*

(9) $$\Psi^{-1}(w) = w + is^{-1}(1 - w^2) \qquad \forall w \in \mathbb{C} \setminus \{x \in \mathbb{R} : |x| \geq 1\} =: \Omega.$$

Hints: Check that $1 - w^2 \in \mathbb{C}_-$ for all $w \in \Omega$, and so for such w and $z \neq 0$

$$\psi(z) = w = \frac{1}{2}\left(z + \frac{1}{z}\right) \Leftrightarrow (z - w)^2 = w^2 - 1 = (is^{-1}(1 - w^2))^2.$$

Thus we have two continuous functions on Ω

(10) $$f(w) := w + is^{-1}(1 - w^2) \text{ and (by(v)) } \Psi^{-1}(w),$$

which satisfy

$$(f(w) - w)^2 = (is^{-1}(1 - w^2))^2 = (\Psi^{-1}(w) - w)^2 \qquad \forall w \in \Omega,$$

and this expression is never 0 [since $s^{-1}(1-w^2) \in \mathbb{H}$]. Check that at $w = 0 \in \Omega$ both functions in (10) have value i, then cite 1.7 to conclude that they must be identical, confirming (9). [If one could prove directly that f maps Ω onto $\mathbb{U}$, then this argument would settle the existence issue in (v) as well. Is a formula like (9) available for the inverse in (iv)?]

Remark: The conformal map of the open upper half-plane onto the open unit disk given by $z \mapsto \phi(iz) = (z - i)/(z + i)$ is called the *Cayley transform* and the map ψ is often called the *Joukowski (Zhukovskii) transform* or the *Kutta transform*. They are important in applications and we will make good use of them later. §51, vol. I of MARKUSCHEWITSCH [1965] and 10.25(ii) below elaborate a bit further on the mapping properties of ψ. If C is a circle centered in $\mathbb{U}$, passing through -1 and surrounding 1, then $\psi(C)$ is a Jordan-curve J (called a *Joukowski profile*) which resembles the cross-section of an aircraft wing with trailing edge at -1, and ψ maps the outside of C conformally onto the outside of J (cf. 11.50(iii)). This enabled N. E. Joukowski and W. M. Kutta, at the dawn of aeronautical engineering, to reduce the analysis of fluid flow around an airfoil to the easier problem of flow around a cylinder. See §31, vol. II of the cited book by Markuschewitsch.

Exercise 2.6 *For each $a \in \mathbb{D}$ define f_a by*

$$f_a(z) := \frac{z-a}{1-\bar{a}z}.$$

Show that f_a is one-to-one and continuous on $\overline{\mathbb{D}}$, holomorphic in a neighborhood of $\overline{\mathbb{D}}$, that $f_a(\mathbb{D}) = \mathbb{D}$, $f_a\big(\overline{\mathbb{D}}\big) = \overline{\mathbb{D}}$ and that the inverse of f_a on $\overline{\mathbb{D}}$ is f_{-a}.

Hints: Evidently f_a is holomorphic in $\mathbb{C} \setminus \{1/\bar{a}\} \supset \overline{\mathbb{D}}$ in case $a \neq 0$ and entire if $a = 0$. Direct computation leads, for $z \in \mathbb{D}$, to

$$\left|\frac{z-a}{1-\bar{a}z}\right|^2 = 1 - \frac{\big(1-|z|^2\big)\big(1-|a|^2\big)}{|1-\bar{a}z|^2} < 1.$$

It follows that $f_a(\mathbb{D}) \subset \mathbb{D}$. Another simple calculation confirms that for all $z \in \overline{\mathbb{D}}$, $f_{-a}(f_a(z)) = z$. Thus f_a is one-to-one in $\overline{\mathbb{D}}$. Applying all the conclusions so far secured to $-a$ in the role of a, we learn that for all $z \in \mathbb{D}$, $f_{-a}(z) \in \mathbb{D}$ and $f_a(f_{-a}(z)) = z$. Thus $f_a(\mathbb{D}) \supset \mathbb{D}$. We now have $f_a(\mathbb{D}) = \mathbb{D}$. By continuity then $f_a\big(\overline{\mathbb{D}}\big) = \overline{\mathbb{D}}$ and all claims about f_a and f_{-a} are established.

Remark: In 6.2(ii) it is proved that, conversely, every conformal automorphism f of $\mathbb{D}$ has the form $f = uf_a$, for some $a \in \mathbb{D}$ and some $u \in \partial\mathbb{D}$. Such functions f and the ϕ of 2.5 belong to a family called *Möbius maps*.

§ 2 Integrals along Curves

Exercise 2.7 *Suppose $a =: t_0 < t_1 < \cdots < t_n := b$ and $f\colon [a,b]\backslash\{t_0,\ldots,t_n\} \mapsto \mathbb{C}$ is continuous and bounded. Then, however one chooses to define f at the t_j, the extended function (call it f again) is Riemann-integrable and the function $F(t) := \int_a^t f$ is continuous on $[a,b]$.*

Using this elementary fact we can make a very useful

Definition 2.8 Let $\gamma\colon [a,b] \mapsto \mathbb{C}$ be a piecewise smooth curve, $f\colon \gamma[a,b] \mapsto \mathbb{C}$ a continuous function. Then $\int_a^b |\gamma'|$ is called the *length of* γ and is denoted by $l(\gamma)$, while $\int_a^b (f \circ \gamma) \cdot \gamma'$ is called the *integral of* f *along* γ and is denoted by $\int_\gamma f$ or sometimes $\int_\gamma f(\zeta)d\zeta$. A modest extension of this notion will prove very convenient. Given piecewise smooth curves $\gamma_1, \ldots, \gamma_n$ ($n \in \mathbb{N}$), not necessarily distinct, we call the n-tuple $\gamma := (\gamma_1, \ldots, \gamma_n)$ a *chain* and define $\int_\gamma f := \sum_{j=1}^n \int_{\gamma_j} f$ for every function f which is continuous on $\gamma_1 \cup \cdots \cup \gamma_n$. [The suggestive notation $\gamma_1 \oplus \cdots \oplus \gamma_n$ for γ is used in many books.]

Remarks: My attitude towards curves and integrals is frankly utilitarian. There is no need to integrate over curves more general than piecewise smooth ones, so we will not. There is, however, some esthetic loss in *defining* the length of γ to be $\int |\gamma'|$. One ought to call a curve $\gamma\colon [a,b] \mapsto \mathbb{C}$ *rectifiable* if there is a finite M such that the length $\sum_{j=1}^n |\gamma(t_j) - \gamma(t_{j-1})|$ of the polygon $[\gamma(t_0), \ldots, \gamma(t_n)]$ inscribed in γ is not greater than M whatever be the points $a \le t_0 \le t_1 \le \cdots \le t_n \le b$. In such cases call the least upper bound of the lengths of these inscribed polygons the *length* of γ. It is then a theorem (and an easy one, see 2.13(ii) below) that if γ is piecewise smooth, then it is rectifiable and its length in the latter sense equals $\int |\gamma'|$. Only this integral intervenes in our subsequent considerations and I chose for it its traditional name, the length of γ; the theorem just cited is the rationale for this. It should be noted, however, that what is being rectified is the path traced by a point that is moving in time, with position vector $\gamma(t)$; $l(\gamma)$ is the total (linearized) distance such a point travels, and backtracking or multiple traversals of parts of the image increase it. Only if γ is one-to-one does $l(\gamma)$ have any claim to measure the length of the image set $\gamma[a,b]$, and generally to even give meaning to the latter is a non-trivial task. Thus, for example, $\gamma(t) := (2t-1)^2$, $0 \le t \le 1$, is a smooth curve having image $[0,1]$ but length 2.

Our definition of $\int_\gamma f$ appears very natural, *vis-á-vis* the Riemann theory of integration, upon noting (see 2.13(iii)) that the Riemann sums

$$\sum f(\gamma(t_j))\big(\gamma(t_j) - \gamma(t_{j-1})\big)$$

converge to $\int_\gamma f$ whenever the partitions $a =: t_0 < t_1 < \cdots < t_n := b$ satisfy $\max_j(t_j - t_{j-1}) \to 0$. And in view of this it is not surprising that this definition leads to both forms of the Fundamental Theorem of Calculus (2.10(i) and the proof of 2.11). Herein lies its value for the theory of functions. But it also has geometric significance and physical interpretations: The real (imaginary) part of the integral $\int_\gamma f$ is the circulation along (the flux across) the curve γ of the flow field $\bar{f}$ (called the Pólya vector field associated with f). See BRADEN [1987], GLUCHOFF [1991] and NEEDHAM [1997].

Again, esthetic considerations might lead one to want to have $\int_\gamma f$ defined for any curve γ (lying in the domain of the holomorphic function f). After Cauchy's theorem (in a disk) is proved in chapter V, this can easily be done by approximating γ uniformly with piecewise smooth curves (whether γ is rectifiable in the above sense or not). See PERRY and YOUNGS [1947], REDHEFFER [1969] and the Remarks following 7.23. On the other hand, if γ is rectifiable and f is merely continuous on the range of γ, then the Riemann sums above still converge (to the Riemann-Stieltjes integral $\int_a^b (f \circ \gamma)\, d\gamma$) and with a further natural hypothesis on γ both forms of the Fundamental Theorem of Calculus are valid. See GAIER [1998].

The difficult task of defining the integral of f over a curve thought of merely as the *set* $\gamma[a,b]$, or of analyzing the invariance of our definition under different parametrizations of the set $\gamma[a,b]$, is not worth pursuing. Essentially only two forms of the latter show up in the sequel: For a piecewise smooth bijection $\tau\colon [c,d] \mapsto [a,b]$ we have $\int_{\gamma\circ\tau} f = \pm \int_\gamma f$, according as $\tau(c) = a$ or $\tau(c) = b$ (cf. 1.46). If γ is a *closed* curve, the integral is also invariant under change of initial point, that is, $\int_\gamma f = \int_{\gamma_0} f$ if Γ is the $(b-a)$-periodic extension of γ to $\mathbb{R}$ and γ_0 is the restriction of Γ to any closed interval of length $b-a$. The reader can easily confirm these claims; in the sequel they will be routinely used without comment. For example, in case τ and γ are smooth, let F be an antiderivative of $(f\circ\gamma)\gamma'$. Then $F\circ\tau$ is an antiderivative of $(f\circ\gamma\circ\tau)(\gamma\circ\tau)'$, so $\int_{\gamma\circ\tau} f = F(\tau(d)) - F(\tau(c)) = \pm[F(b) - F(a)] = \pm\int_\gamma f$. Similarly, $l(\gamma\circ\tau) = l(\gamma)$. It may however be of interest to note that every piecewise smooth curve $\gamma : [a,b] \mapsto \mathbb{C}$ can be re-parameterized to be smooth, merely by "coming to a stop" at each point where one-sided derivatives disagree. Thus if $a = t_0 < t_1 < \cdots < t_n = b$ and γ is smooth in each $[t_{j-1}, t_j]$, we find a smooth increasing bijection τ_j of $[t_{j-1}, t_j]$ satisfying $\tau_j'(t_{j-1}) = \tau_j'(t_j) = 0$ and redefine γ in each $[t_{j-1}, t_j]$ to be $\gamma\circ\tau_j$. [It's unclear to the author the extent to which this artifice could eliminate merely piecewise smooth curves from the theory of holomorphic functions, or the value of doing so. On the other hand, some texts build into the definition of smooth curve that the derivative be zero-free. The reader needs to beware.]

Exercise 2.9 (i) *If γ is a piecewise smooth curve and f is continuous on the range of γ and M is a bound for $|f|$ there, then*

$$\left|\int_\gamma f\right| \le M \cdot l(\gamma).$$

(ii) *If* $a_0, \ldots, a_n \in \mathbb{C}$ *and* f *is continuous on* $[a_0, \ldots, a_n]$, *then* $l([a_0, \ldots, a_n]) = \sum_{j=1}^{n} |a_j - a_{j-1}|$, $\int_{[a_0,\ldots,a_n]} f = \sum_{j=1}^{n} \int_{[a_{j-1},a_j]} f$, $\int_{[a_0,a_1]} f = -\int_{[a_1,a_0]} f$, *and if* $a_1 \in [a_0, a_2]$, *then* $\int_{[a_0,a_2]} f = \int_{[a_0,a_1]} f + \int_{[a_1,a_2]} f$.

Theorem 2.10 *Let* U *be an open subset of* $\mathbb{C}$, $F \in H(U)$ *and suppose that* F' *is continuous on* U. [This last hypothesis will later be shown to be redundant.] *Then for any piecewise smooth curve* γ *in* U *with initial point* z_0 *and terminal point* z_1

(i) $$\int_\gamma F' = F(z_1) - F(z_0).$$

If in addition γ *is closed, then*

(ii) $$\int_\gamma F' = 0.$$

If U *is connected, there is for each* z_0 *and* z *in* U *a polygon* γ_z *in* U *with initial point* z_0 *and terminal point* z *(by* 1.28*). For any such polygons*

(iii) $$F(z) = F(z_0) + \int_{\gamma_z} F' \qquad \forall z \in U.$$

(When U is starlike with respect to z_0, the interval $[z_0, z]$ is an appropriate γ_z.)

(iv) *If* U *is connected and* $F' = 0$ *throughout* U, *then* F *is constant in* U.

Proof: If the domain of γ is $[a, b]$ and $a =: t_0 < t_1 < \cdots < t_n := b$ are such that γ is continuously differentiable on each $[t_{j-1}, t_j]$, then

$$\begin{aligned}
\int_\gamma F' &= \sum_{j=1}^{n} \int_{t_{j-1}}^{t_j} (F' \circ \gamma)\gamma' \\
&= \sum_{j=1}^{n} \int_{t_{j-1}}^{t_j} (F \circ \gamma)' \qquad \text{(Chain Rule)} \\
&= \sum_{j=1}^{n} [(F \circ \gamma)(t_j) - (F \circ \gamma)(t_{j-1})] \quad \text{(Fundamental Theorem of Calculus)} \\
&= (F \circ \gamma)(t_n) - (F \circ \gamma)(t_0) = F(\gamma(b)) - F(\gamma(a)) \\
&= F(z) - F(z_0) \qquad \text{(by definition of } z \text{ and } z_0\text{)}.
\end{aligned}$$

This establishes (i), and (ii), (iii), and (iv) are trivial corollaries of (i). In the spirit of Weierstrass we can also prove (iv) without appeal to integration theory. Suppose $F' = 0$ throughout the region U and consider $D(z_0, r) \subset U$. For any z in this disk the function $g(t) := F\big(tz + (1-t)z_0\big)$ is differentiable on $[0, 1]$ and its derivative is 0. Apply the Mean-Value Theorem to $\operatorname{Re} g$ and $\operatorname{Im} g$ to infer that g is constant. In particular, $F(z) = g(1) = g(0) = F(z_0)$. This proves that F is locally constant, and since U is connected it follows that F is constant (1.6). An unusual, though still elementary, integration-free proof is offered by DESBROW [1996].

There is a converse to part (ii):

Theorem 2.11 *Let Ω be a region and f a continuous function in Ω with the property that $\int_\gamma f = 0$ for every polygonal loop γ in Ω. Then there is a $\mathbb{C}$-differentiable function F in Ω such that $F' = f$.*

Proof: According to 1.28, Ω is polygonally connected. Fix $z_0 \in \Omega$ and for each $z \in \Omega$ let γ_z be a polygon in Ω with initial point z_0 and terminal point z, and define

$$F(z) := \int_{\gamma_z} f.$$

Since this definition implicitly appeals to the Axiom of Choice, the continuity of F, let alone its differentiability, would seem unlikely. Nevertheless, we can show that at each $z_1 \in \Omega$, $F'(z_1)$ exists and equals $f(z_1)$. Pick $r > 0$ so that $D(z_1, r) \subset \Omega$. If $z \in D(z_1, r) \setminus \{z_1\}$, we may consider the piecewise smooth loop $\gamma\colon [0, n+m+1] \mapsto \Omega$ defined by

$$\gamma(t) := \begin{cases} \gamma_{z_1}(t) & 0 \le t \le n \\ z_1 + (t-n)(z - z_1) & n \le t \le n+1 \\ \gamma_z(n+m+1-t) & n+1 \le t \le n+m+1, \end{cases}$$

where $[0, n]$, $[0, m]$ are the parameter intervals of γ_{z_1} and γ_z, respectively. After some trivial changes of variable we see that

$$\int_\gamma f = \int_{\gamma_{z_1}} f + \int_{[z_1,z]} f - \int_{\gamma_z} f = F(z_1) - F(z) + \int_{[z_1,z]} f.$$

By hypothesis this integral is zero, that is,

$$F(z) - F(z_1) = \int_{[z_1,z]} f = \int_0^1 f(z_1 + t(z - z_1))(z - z_1)dt,$$

whence

$$\left|\frac{F(z) - F(z_1)}{z - z_1} - f(z_1)\right| = \left|\int_0^1 [f(z_1 + t(z - z_1)) - f(z_1)]dt\right|$$
$$\le \sup_{0 \le t \le 1} |f(z_1 + t(z - z_1)) - f(z_1)|.$$

From the continuity of f at z_1 the desired conclusion is now immediate.

The reader who steals a glance at 2.15(iii) will see an example of a holomorphic function f in $\Omega := \mathbb{C} \setminus \{0\}$, namely, $f(z) := 1/z$, such that $\int_\gamma f \neq 0$ for some polygonal loop γ in Ω. According to 2.10(ii), f has no primitive in Ω, a fact which is realized in 2.21(xxii) by other means. This contrasts with the situation in $\mathbb{R}$: *every* continuous function on a connected subset of $\mathbb{R}$ has a primitive (gotten by integration, of course). On the other hand, for certain Ω the conclusion of 2.11, with the same proof, follows from an apparently weaker hypothesis on f, namely

Scholium: *If the open set Ω is starlike with respect to the point z_0, and f is a continuous function on Ω which satisfies*

$$(*) \qquad \int_{[z_0,b,c,z_0]} f = 0 \qquad \textit{whenever } co\{z_0, b, c\} \subset \Omega,$$

then $F(z) := \int_{[z_0,z]} f$ defines a function in $H(\Omega)$ such that $f = F'$.

Proof: In the preceding proof take $\gamma_z := [z_0, z]$, for each $z \in \Omega$. Central to that proof is the equality

$$F(z) - F(z_1) = \int_{[z_1,z]} f \qquad \forall z \in D(z_1, r)$$

whenever $D(z_1, r) \subset \Omega$. This follows from the weaker hypothesis $(*)$ once we show that $\text{co}\{z_0, z_1, z\} \subset \Omega$ for each such z. A typical point $w \in \text{co}\{z_0, z_1, z\}$ has the form $w = \lambda_0 z_0 + \lambda_1 z_1 + \lambda_2 z$ for some $\lambda_j \geq 0$ with sum 1. If $\lambda_1 + \lambda_2 > 0$, the point

$$z' := \frac{\lambda_1}{\lambda_1 + \lambda_2} z_1 + \frac{\lambda_2}{\lambda_1 + \lambda_2} z \in [z_1, z] \subset D(z_1, r) \subset \Omega,$$

and therefore $w = \lambda_0 z_0 + (\lambda_1 + \lambda_2) z' \in [z_0, z'] \subset \Omega$, as wanted.

Primitives can be found by other means than integration. For example, for every polynomial a primitive can immediately be written down (by anti-differentiating). More is true: It can be shown by a very elementary argument employing only the Mean-Value Theorem that every uniform limit of polynomials has a primitive, which is itself a uniform limit of polynomials. This follows from 2.12 by selecting the polynomial anti-derivatives so that all vanish at a common point a in the region.

Theorem 2.12 *Let Ω be a region, $\{F_n\}$ a sequence in $H(\Omega)$ such that $\{F'_n\}$ converges uniformly on each compact subset of Ω to a function f, and for some $a \in \Omega$ the numerical sequence $\{F_n(a)\}$ converges in $\mathbb{C}$. Then $\{F_n\}$ converges uniformly on each compact subset of Ω to a function $F \in H(\Omega)$ such that $F' = f$.*

Proof: (Cf. GODEFROY [1904]). Let C denote the set of $z \in \Omega$ such that the sequence $\{F_n(z)\}$ converges. Since $a \in C$, this set is not empty. Consider any $z \in C$ and any $r > 0$ such that $\bar{D}(z,r) \subset \Omega$. For all w in this disk and all $n, m \in \mathbb{N}$

$$(1) \qquad |F_n(w) - F_m(w) - [F_n(z) - F_m(z)]| \leq 2|w - z| \sup_{\bar{D}(z,r)} |F'_n - F'_m|.$$

This follows from forming the functions $t \mapsto (F_n - F_m)(z + t(w-z))$ on $[0,1]$ and applying the Mean-Value Theorem to their real and imaginary parts. Since the numerical sequence $\{F_n(z)\}$ converges, and $\{F'_n\}$ converges uniformly on the compact set $\bar{D}(z,r)$, (1) shows that the sequence $\{F_n\}$ is uniformly Cauchy in $\bar{D}(z,r)$, hence uniformly convergent there. One consequence of this is that $\bar{D}(z,r) \subset C$. Thus the hypotheses of the Basic Connectedness Lemma 1.27 are met (with $\theta = 1$) and we can conclude that $C = \Omega$. But then the above argument shows that the sequence $\{F_n\}$ converges uniformly in a neighborhood of each point of Ω.

Denote the limit function by F. Given $\varepsilon > 0$, there exists $N = N_\varepsilon \in \mathbb{N}$ such that

$$(2) \qquad \sup_{\bar{D}(z,r)} |F'_n - f| \leq \varepsilon/6 \qquad \forall n \geq N$$

and so

$$(3) \qquad \sup_{\bar{D}(z,r)} |F'_n - F'_m| \leq \varepsilon/3 \qquad \forall n, m \geq N.$$

Letting $m \to \infty$ in (1) and using (3) we get

$$(4) \qquad |F_N(w) - F_N(z) - [F(w) - F(z)]| \leq 2|w - z| \cdot \varepsilon/3 \qquad \forall w \in \bar{D}(z,r).$$

Since F_N is differentiable, there exists a positive $\delta(\varepsilon) < r$ (depending as well on N and z) such that

$$|F_N(w) - F_N(z) - (w - z)F'_N(z)| \leq |w - z|\varepsilon/6 \qquad \forall w \in D(z, \delta(\varepsilon)),$$

which in view of (2) gives

$$(5) \qquad |F_N(w) - F_N(z) - (w-z)f(z)| \le |w-z|\varepsilon/3 \qquad \forall w \in D(z, \delta(\varepsilon)),$$

From (5), (4) and the triangle inequality we learn that

$$|F(w) - F(z) - (w-z)f(z)| \le |w-z|\varepsilon \qquad \forall w \in D(z, \delta(\varepsilon)).$$

As this holds for every $\varepsilon > 0$, it shows that F is $\mathbb{C}$-differentiable at z, with $F'(z) = f(z)$. Here z is any point in $C = \Omega$, so $F \in H(\Omega)$ and $F' = f$.

Remark: In older literature $f \colon \Omega \mapsto \mathbb{C}$ is called *integrable in* Ω if it has a primitive, i.e., if $f = F'$ for some $F \in H(\Omega)$. The pre-Cauchy view of integration as being nothing but anti-differentiation is perhaps responsible for this terminology. On the other hand, even though every continuous function on Ω is Riemann integrable over every piecewise smooth curve in Ω, it is not unreasonable to withhold the designation "integrable" unless the integrals along all paths connecting two points agree, since as 2.10, 2.11 show, that condition is what allows the integrals to constitute a primitive.

For a subset S of $\mathbb{C}$ and a bounded function $f \colon S \mapsto \mathbb{C}$ let

$$\omega_f(\delta) := \sup\{|f(z) - f(w)| : z, w \in S,\ |z-w| \le \delta\}$$

for each $\delta \ge 0$. The function $\omega_f \colon [0,\infty) \mapsto [0,\infty)$ so defined is called the *oscillation of* f, or in case f is continuous on S, the *modulus of (uniform) continuity of* f.

Exercise 2.13 (i) *Let S be a subset of $\mathbb{C}$ and $f \colon S \mapsto \mathbb{C}$ a bounded function. Show that ω_f is continuous at 0 if and only if f is uniformly continuous on S.*

(ii) *Let $\gamma \colon [0,1] \mapsto \mathbb{C}$ be a piecewise smooth curve. Show that*

$$\sum_{j=1}^{n} |\gamma(t_j) - \gamma(t_{j-1})| \le l(\gamma)$$

whenever $0 = t_0 < t_1 < \cdots < t_n = 1$, and that, given $\varepsilon > 0$, there is a partition for which this sum exceeds $l(\gamma) - \varepsilon$.

Hints: By adding more points t_j the left-hand side is not decreased, so we can suppose that all the points of discontinuity of γ' are among the t_j. Then $\gamma(t_j) - \gamma(t_{j-1}) = \int_{t_{j-1}}^{t_j} \gamma'$ and $l(\gamma) = \sum_{j=1}^{n} \int_{t_{j-1}}^{t_j} |\gamma'|$. Since γ' is uniformly continuous on each $[t_{j-1}, t_j]$, by inserting more subdivision points we can suppose that the oscillation of γ' on each $[t_{j-1}, t_j]$ is less than $\varepsilon/2$. Then from two applications of the triangle-inequality

$$\begin{aligned}\int_{t_{j-1}}^{t_j} |\gamma'| &\le \int_{t_{j-1}}^{t_j} |\gamma' - \gamma'(t_j)| + |\gamma'(t_j)|(t_j - t_{j-1}) \\ &= \int_{t_{j-1}}^{t_j} |\gamma' - \gamma'(t_j)| + \left|\int_{t_{j-1}}^{t_j} \gamma'(t_j)\right| \\ &\le \int_{t_{j-1}}^{t_j} |\gamma' - \gamma'(t_j)| + \left|\int_{t_{j-1}}^{t_j} (\gamma'(t_j) - \gamma')\right| + \left|\int_{t_{j-1}}^{t_j} \gamma'\right| \\ &< \varepsilon(t_j - t_{j-1}) + |\gamma(t_j) - \gamma(t_{j-1})|.\end{aligned}$$

(iii) *With γ, $t_0, \ldots, t_n$ as above, set $\delta := \max_{1\le j\le n}(t_j - t_{j-1})$. Show that if f is a continuous function on the range of γ and all the points of discontinuity of γ' are among the t_j, then*

$$\left|\int_\gamma f - \sum_{j=1}^{n} f\circ\gamma(t_j)[\gamma(t_j) - \gamma(t_{j-1})]\right| \le \omega_{f\circ\gamma}(\delta)l(\gamma).$$

Hint: Use the fact that

$$\int_\gamma f = \sum_{j=1}^{n} \int_{t_{j-1}}^{t_j} (f\circ\gamma)\gamma'.$$

§ 3 Differentiating under the Integral Sign

We shall use (and prove), as we progress, several successively stronger theorems on differentiating behind the integral sign. (See, e.g., 7.21.) The easiest and most useful of these is codified in 2.14. When $n = 1$ the integral that occurs there is called the *Cauchy transform* of g and is an example of the *convolution product* $z \mapsto \int g(\xi)k(\xi - z)\,d\xi$ of two integrable functions. The factor $k(\xi - z) := 1/(\xi - z)$ in this role is known as the *Cauchy kernel.* One of the foundational results of our subject affirms (roughly) that a continuous function g is holomorphic if and only if it can be represented locally by such transforms, and indeed by the transform of g itself.

Theorem 2.14 *Let γ be a piecewise smooth curve, g a function continuous on the range of γ. Then the functions*

$$G_n(z) := \int_\gamma \frac{g(\zeta)}{(\zeta - z)^n}\, d\zeta, \qquad n \in \mathbb{N},$$

are (infinitely) differentiable in the complement of the range of γ and satisfy

$$(*) \qquad G_n' = nG_{n+1}.$$

Proof: We base the proof on a simple algebraic identity which we derive first. Consider two unequal, non-zero complex numbers a and b.

$$\begin{aligned}\frac{1}{b^n} - \frac{1}{a^n} &= \frac{1}{b^n a^n}(a^n - b^n) = \frac{a-b}{b^n a^n}\sum_{j=0}^{n-1} b^{n-1-j}a^j \\ &= \frac{a-b}{a^{n+1}}\sum_{j=0}^{n-1}\left(\frac{a}{b}\right)^{j+1} = \frac{a-b}{a^{n+1}}\sum_{j=1}^{n}\left(1+\frac{a-b}{b}\right)^j \\ &= \frac{a-b}{a^{n+1}}\sum_{j=1}^{n}\sum_{k=0}^{j}\binom{j}{k}\left(\frac{a-b}{b}\right)^k.\end{aligned}$$

$$\begin{aligned}(1)\quad \frac{1}{a-b}\left(\frac{1}{b^n} - \frac{1}{a^n}\right) - \frac{n}{a^{n+1}} &= \frac{1}{a^{n+1}}\sum_{j=1}^{n}\left[\sum_{k=0}^{j}\binom{j}{k}\left(\frac{a-b}{b}\right)^k - 1\right] \\ &= \frac{1}{a^{n+1}}\sum_{j=1}^{n}\sum_{k=1}^{j}\binom{j}{k}\left(\frac{a-b}{b}\right)^k \\ &= \frac{1}{a^{n+1}}\sum_{j=1}^{n}\sum_{l=0}^{j-1}\binom{j}{l+1}\left(\frac{a-b}{b}\right)^{l+1} \\ &= (a-b)\cdot\frac{1}{a^{n+1}b}\cdot\sum_{j=1}^{n}\sum_{l=0}^{j-1}\binom{j}{l+1}\left(\frac{a-b}{b}\right)^l.\end{aligned}$$

If $|a-b| \le r$ while $|a| \ge r$ and $|b| \ge r$, then this gives

$$\begin{aligned}\left|\frac{1}{a-b}\left(\frac{1}{b^n} - \frac{1}{a^n}\right) - \frac{n}{a^{n+1}}\right| &\le \frac{|a-b|}{r^{n+2}}\sum_{j=1}^{n}\sum_{l=0}^{j-1}\binom{j}{l+1} \\ (2)\qquad &\le \frac{|a-b|}{r^{n+2}}\sum_{j=1}^{n}(1+1)^j \le \frac{2^{n+1}}{r^{n+2}}|a-b|.\end{aligned}$$

Consider now any z, z_0 not on γ, with $z \neq z_0$. We have

$$(3) \quad \frac{G_n(z) - G_n(z_0)}{z - z_0} - nG_{n+1}(z_0)$$
$$= \int_\gamma g(\xi)\left[\frac{1}{z - z_0}\left(\frac{1}{(\xi - z)^n} - \frac{1}{(\xi - z_0)^n}\right) - \frac{n}{(\xi - z_0)^{n+1}}\right] d\xi.$$

Now choose $r > 0$ so small that $D(z_0, 2r)$ is disjoint from γ and consider only $z \in D(z_0, r)$ with $z \neq z_0$. Then $|\xi - z_0| \geq r$ and $|\xi - z| \geq r$ for any $\xi \in \gamma$ and so for the integrand above, (2) provides the upper bound

$$|g(\xi)||z - z_0| \cdot \left(\frac{2}{r}\right)^{n+2}.$$

From this and (3) follows (recalling 2.9(i))

$$\left|\frac{G_n(z) - G_n(z_0)}{z - z_0} - nG_{n+1}(z_0)\right| \leq M|z - z_0|\left(\frac{2}{r}\right)^{n+2} l(\gamma),$$

where M is the maximum of $|g|$ on γ. Now let z approach z_0 to see that $G_n'(z_0)$ exists and equals $nG_{n+1}(z_0)$.

Exercise 2.15 *Let γ be a piecewise smooth loop, n an integer not equal to* 1.

(i) *Show that*

$$\int_\gamma \frac{d\xi}{(\xi - z)^n} = 0 \qquad \forall z \in \mathbb{C} \setminus \gamma.$$

Hint: The integrand is a derivative, so 2.10(ii) may be cited.

(ii) *Show that the function*

$$z \longmapsto \int_\gamma \frac{d\xi}{\xi - z}$$

is constant in each component of $\mathbb{C}\setminus\gamma$ *and is* 0 *throughout the unbounded component.*

Hint: Use (i), 2.14 and 2.10(iv). Due to 2.9(i), the integral converges to 0 as $|z| \to \infty$.

(iii) *For $r > 0$ let $S_r := (-r, r) \times (-r, r)$. Show that*

$$\int_{\partial S_r} \frac{d\xi}{\xi} = 8i \int_0^1 \frac{du}{u^2 + 1}.$$

Hint: (PRINGSHEIM [1896a] and [1920]) Set $v := -r(1+i)$ and notice that $\partial S_r = [v, iv, -v, -iv, v]$ (see 1.12) and that its component intervals are

$$[v, iv] =: \gamma, \quad [iv, -v] = i\gamma, \quad [-v, -iv] = -\gamma, \quad [-iv, v] = -i\gamma.$$

Since $(i^k\gamma)'/i^k\gamma = \gamma'/\gamma$ ($k \in \mathbb{Z}$), it follows from 2.9(ii) that

$$\int_{\partial S_r} \frac{d\xi}{\xi} = \sum_{k=0}^{3} \int_{i^k\gamma} \frac{d\xi}{\xi} = 4\int_\gamma \frac{d\xi}{\xi} = 4\int_0^1 \frac{(iv - v)}{v + t(iv - v)}\, dt$$
$$= 4\int_{-1}^1 \frac{du}{u - i} = 4\int_{-1}^1 \frac{u + i}{u^2 + 1}\, du = 8i\int_0^1 \frac{du}{u^2 + 1},$$

using the facts that $u/(u^2+1)$ is an odd and $1/(u^2+1)$ is an even function.

§ 4 Elementary Real-variables Methods

Recall the definition of real-differentiability for a function $f = (u, v)\colon U$ open $\subset \mathbb{R}^2 \mapsto \mathbb{R}^2$. Such a function is *real-differentiable* at a point $z_0 := (x_0, y_0) \in U$ if there exists an $\mathbb{R}$-linear map $L\colon \mathbb{R}^2 \mapsto \mathbb{R}^2$ which provides a first-order approximation to f near z_0, in the sense that

$$(*) \qquad \lim_{z \to z_0} \frac{|f(z) - f(z_0) - L(z - z_0)|}{|z - z_0|} = 0.$$

If we first consider $z = (x, y_0)$ and let $x \to x_0$, then consider $z = (x_0, y)$ and let $y \to y_0$, we see that $D_1 f(x_0, y_0), D_2 f(x_0, y_0)$ each exist and that the matrix of L in the basis $\begin{bmatrix} 1 \\ 0 \end{bmatrix}, \begin{bmatrix} 0 \\ 1 \end{bmatrix}$ of $\mathbb{R}^2$, thought of as column vectors, is

$$(**) \qquad \begin{bmatrix} D_1 u(z_0) & D_2 u(z_0) \\ D_1 v(z_0) & D_2 v(z_0) \end{bmatrix},$$

called the *Jacobian matrix* of f at z_0. Thus, if such an L exists, it is unique; it's called the *derivative of f at z_0* and sometimes denoted $Df(z_0)$.

As in §1, $(*)$ is equivalent to the existence of a function $F : U \mapsto \mathbb{R}^2$ which is continuous at z_0 and satisfies $F(z_0) = 0$ and

$$f(z) = f(z_0) + L(z - z_0) + (z - z_0)F(z) \qquad \forall z \in U.$$

And as before, this formulation is particularly conducive to proving the chain rule. If $\gamma : [a, b] \subset \mathbb{R} \mapsto U$ is differentiable at t_0 and $\gamma(t_0) = z_0$, then $\mathbb{R}$-linearity of L gives:

$$\frac{f(\gamma(t)) - f(\gamma(t_0))}{t - t_0} = L\left(\frac{\gamma(t) - \gamma(t_0)}{t - t_0}\right) + \frac{\gamma(t) - \gamma(t_0)}{t - t_0} F(\gamma(t)).$$

Trivially every $\mathbb{R}$-linear transformation on $\mathbb{R}^2$ is continuous, so if $t \to t_0$ we see that

$$(f \circ \gamma)'(t_0) \text{ exists and equals } L(\gamma'(t_0)).$$

This equality can be expressed in complex numbers via $(**)$ as

$$(f \circ \gamma)'(t_0) = D_1 f(\gamma(t_0)) \operatorname{Re} \gamma'(t_0) + D_2 f(\gamma(t_0)) \operatorname{Im} \gamma'(t_0).$$

19th-century mathematicians seemed to know what they wanted the (total) differential of a function of two real variables to accomplish (in particular, to yield this form of the chain rule), but only very late, in an 1893 textbook of Otto Stolz, was the (very natural) condition $(*)$ recognized as just the right hypothesis which would secure the desired properties, that is, which should be made the *definition* of (total) differentiability. See **MR** 58#15979. This definition makes sense in any normed vector space, where it is called *Fréchet differentiability*. Note that when f is real-valued, $p, v \in \mathbb{R}^2$ and $\gamma(t) := p+(t-t_0)v$, the preceding equality is the familiar directional derivative formula

$$D_v f(p) = (D_1 f(p), D_2 f(p)) \bullet v.$$

Actually this *two-dimensional chain rule* is not only a consequence of $\mathbb{R}$-differenti-ability of f, but conversely it implies the latter. Precisely, if there exists an $\mathbb{R}$-linear transformation L such that $f \circ \gamma$ is differentiable at t_0 with $(f \circ \gamma)'(t_0) = L(\gamma'(t_0))$ for *every* γ as described, then $(*)$ holds and f is $\mathbb{R}$-differentiable at z_0. See Exercise 5, p. 157 of DIEUDONNÉ [1969]. However, f may fail to be $\mathbb{R}$-differentiable at z_0 and yet $f \circ \gamma$ be differentiable at t_0 for every differentiable curve γ with $\gamma(t_0) = z_0$ and $\gamma'(t_0) \neq 0$, as the function $f(x, y) := x^3/(x^2 + y^2)$, $f(0, 0) := 0$ illustrates.

The relation of $\mathbb{R}$-differentiability to $\mathbb{C}$-differentiability is pretty obvious. If f is $\mathbb{C}$-differentiable at z_0, then multiplication by the complex number $f'(z_0)$ on $\mathbb{C}$ defines an $\mathbb{R}$-linear mapping L on $\mathbb{R}^2 = \mathbb{C}$ which satisfies $(*)$. Of course, this L is really $\mathbb{C}$-linear. Conversely, if f is $\mathbb{R}$-differentiable at z_0, and the associated $\mathbb{R}$-linear map L is actually $\mathbb{C}$-linear, then f is $\mathbb{C}$-differentiable, with $f'(z_0)$ being the number $L(1, 0) \in \mathbb{R}^2 = \mathbb{C}$. For in this case the mapping L is multiplication on $\mathbb{C} = \mathbb{R}^2$ by the complex number $L(1, 0)$: $L(w) = L(w \cdot 1) = w \cdot L(1)$ for all $w \in \mathbb{C}$.

Any $\mathbb{R}$-linear transformation L is represented by a real matrix $\begin{bmatrix} a & b \\ c & d \end{bmatrix}$ with respect to the natural basis. As just noted, L is $\mathbb{C}$-linear if and only if it is multiplication by the number $L(1, 0)$. The condition on the matrix which expresses this is evidently

$$d = a \quad \text{and} \quad c = -b.$$

Hence f is $\mathbb{C}$-differentiable at z_0 if and only if it is $\mathbb{R}$-differentiable at z_0 and in addition satisfies

$$D_2 v(z_0) = D_1 u(z_0) \quad \text{and} \quad D_1 v(z_0) = -D_2 u(z_0),$$

two real equations which can be combined into the single complex equation

$$\text{(C. R.)} \qquad D_2 f(z_0) = i D_1 f(z_0).$$

Although it originates with d'Alembert (see p. 13 of OSGOOD [1901]), this is universally known as the *Cauchy-Riemann* equation. If f is $\mathbb{R}$-differentiable at z_0 and γ is a smooth curve with $\gamma(t_0) = z_0$ and $\gamma'(t_0) \neq 0$, then the "curvilinear" derivative

$$(***) \quad \lim_{t \to t_0} \frac{f(\gamma(t)) - f(\gamma(t_0))}{\gamma(t) - \gamma(t_0)} \text{ exists and equals } \frac{(f \circ \gamma)'(t_0)}{\gamma'(t_0)} = \frac{L(\gamma'(t_0))}{\gamma'(t_0)},$$

and of course this limit is $f'(\gamma(t_0)) = f'(z_0)$ if f is $\mathbb{C}$-differentiable at z_0. For $\gamma(t) := z_0 + t - t_0$ [resp., $\gamma(t) := z_0 + (t - t_0)i$] this limit is $D_1 f(z_0)$ [resp., $-iD_2 f(z_0)$]. Hence two different curves γ having the same $(***)$ limit represents a generalized (C. R.) equality, and in fact the consequence is the same: If γ_1, γ_2 are smooth curves, $\gamma_1(t_0) = \gamma_2(t_0) = z_0$ and the vectors $\gamma_1'(t_0)$, $\gamma_2'(t_0)$ are $\mathbb{R}$-linearly independent (i.e., neither is 0 and their ratio is not in $\mathbb{R}$), then the equality of the limits in $(***)$ for $\gamma = \gamma_1$ and $\gamma = \gamma_2$ implies (C. R.), and therewith the $\mathbb{C}$-differentiable of f at z_0. To see this, set $v = v_1 + iv_2 := \gamma_1'(t_0)$, $w = w_1 + iw_2 := \gamma_2'(t_0)$, and $\begin{bmatrix} a & b \\ c & d \end{bmatrix} = Df(z_0)$. Then after some trivial algebra the equality of the two limits in $(***)$ becomes

$$(a + ic)\left(\frac{w_1}{w} - \frac{v_1}{v}\right) = (d - ib)\left(\frac{-iw_2}{w} + \frac{iv_2}{v}\right).$$

The two right-hand factors are clearly equal, and to divide them away and get the desired equalities $d = a$, $c = -b$, we only have to show they're non-zero. More routine algebra confirms this:

$$\operatorname{Im}\left[wv\left(\frac{w_1}{w} - \frac{v_1}{v}\right)\right] = \operatorname{Im}[vw_1 - v_1 w] = v_2 w_1 - v_1 w_2 = \operatorname{Im}(v\bar{w}) = |w|^2 \operatorname{Im}\left(\frac{v}{w}\right),$$

which is non-zero since $v/w \notin \mathbb{R}$.

Returning to the component functions $u = \operatorname{Re} f$, $v = \operatorname{Im} f$, (C. R.) can be written in terms of gradients as

$$\nabla v = i \nabla u.$$

Since gradients provide normal vectors to level-sets, this equation affirms that the level-sets $\{z \in U : v(z) = v(z_0)\}$ and $\{z \in U : u(z) = u(z_0)\}$ intersect orthogonally at z_0, provided $\nabla u(z_0) \neq 0$. This orthogonality of level-sets will later be seen to be but a special instance of an angle-preservation property (elucidated after 3.16) which holomorphic functions enjoy.

In summary, (C. R.) is a necessary condition for $\mathbb{C}$-differentiability:

Exercise 2.16 (i) *If f is $\mathbb{C}$-differentiable at z_0, then $D_1 f(z_0)$, $D_2 f(z_0)$ exist and satisfy* (C. R.). *In fact, $D_1 f(z_0) = f'(z_0) = -iD_2 f(z_0)$.*

Hint: The hypothesis is that the quotient

$$\frac{f(z_0 + h) - f(z_0)}{h}$$

has a limit as the complex number h approaches 0. Consider first $h \in \mathbb{R}$, then $h \in i \cdot \mathbb{R}$.

(ii) *If f is holomorphic in a region Ω and* $\operatorname{Re} f$ *is constant, then f is constant.*

Hint: Equating real and imaginary parts in (C. R.) shows that $D_1(\operatorname{Im} f) = -D_2(\operatorname{Re} f)$. Therefore, $f' = D_1 f = D_1(\operatorname{Re} f) - iD_2(\operatorname{Re} f)$. The latter is 0 if $\operatorname{Re} f$ is constant. Cite 2.10(iv).

(iii) *If f is holomorphic in a region Ω and $|f|$ is constant, then f is a constant.*

Hint: If the constant is $c \neq 0$, then $\bar{f} = c^2/f$ is $\mathbb{C}$-differentiable, and so therefore are the functions $f - \bar{f}$ and $(f + \bar{f}) \cdot i$. Each of the latter has real part $\mathbf{0}$ and so is constant by (ii) .

(iv) *Two zero-free holomorphic functions f and g in region Ω have the same direction at every point if and only if each is a positive constant multiple of the other. What can be said under the alternative hypothesis $|f| = |g|$?*

Hint: $f/|f| = g/|g|$ means that f/g is a real-valued holomorphic function. [5.62 and 1.24 combine to eliminate the "zero-free" hypothesis.]

The converse of 2.16(i) is false. For example, $f(z) := \left|\operatorname{Im}(z^2)\right|^{1/2}$ satisfies $D_1 f(0) = D_2 f(0) = 0$, but $\left|f(z)/z\right| = 1$ on the diagonals. However, under additional hypotheses there is a converse. The weakest such hypotheses are discussed in the chapter notes, but the most useful criterion is perhaps the elementary

Theorem 2.17 (Otto Hölder). *Let $z_0 \in U$ open $\subset \mathbb{C}, f\colon U \mapsto \mathbb{C}$ be a function for which $D_2 f(z_0)$ exists, while $D_1 f$ exists throughout U and is continuous at z_0. Then f is $\mathbb{R}$-differentiable at z_0. If, in addition,* (C. R.) *holds, then f is $\mathbb{C}$-differentiable at z_0 and $f'(z_0) = D_1 f(z_0)$.*

Proof: Let L be the $\mathbb{R}$-linear transformation whose matrix with respect to the natural basis in $\mathbb{R}^2$ is $(**)$ above. Thus

$$(0) \qquad L(h+ik) = hD_1 f(z_0) + kD_2 f(z_0) \qquad \forall h,k \in \mathbb{R}.$$

Consider $r > 0$ such that $Q_r := z_0 + [-r,r] \times [-r,r] \subset U$. In what follows h, k are numbers in $[-r,r]$. Generally the mean-value theorem fails for complex-valued functions, so we apply it to $u := \operatorname{Re} f$ and $v := \operatorname{Im} f$ separately, to find real numbers a, b (dependent on z_0, h, k) satisfying $|a|, |b| \le |h|$ and

$$(1) \qquad u(z_0 + h + ik) - u(z_0 + ik) = hD_1 u(z_0 + a + ik)$$

$$(2) \qquad v(z_0 + h + ik) - v(z_0 + ik) = hD_1 v(z_0 + b + ik).$$

From (0), (1) and (2)

$$\begin{aligned} &f(z_0 + h + ik) - f(z_0) - L(h+ik) \\ &\quad = f(z_0 + h + ik) - f(z_0) - hD_1 f(z_0) - kD_2 f(z_0) \\ &\quad = f(z_0 + h + ik) - f(z_0 + ik) - hD_1 f(z_0) \\ &\qquad + f(z_0 + ik) - f(z_0) - kD_2 f(z_0) \\ &\quad = hD_1 u(z_0 + a + ik) - hD_1 u(z_0) + ihD_1 v(z_0 + b + ik) - ihD_1 v(z_0) \\ &\qquad + f(z_0 + ik) - f(z_0) - kD_2 f(z_0). \end{aligned}$$

Consequently,

$$\begin{aligned} (3) \qquad &\big|f(z_0 + h + ik) - f(z_0) - L(h+ik)\big| \\ &\qquad \le |h| \sup_{z,w \in Q_r} \big(\big|D_1 u(z) - D_1 u(z_0)\big| + \big|D_1 v(w) - D_1 v(z_0)\big|\big) \\ &\qquad\quad + \big|f(z_0 + ik) - f(z_0) - kD_2 f(z_0)\big|. \end{aligned}$$

Given $\varepsilon > 0$, there exists $r = r_\varepsilon > 0$ so that

$$(4) \qquad \sup_{z \in Q_r} \big|D_1 u(z) - D_1 u(z_0)\big| < \varepsilon, \qquad \sup_{w \in Q_r} \big|D_1 v(w) - D_1 v(z_0)\big| < \varepsilon$$

and

$$(5) \qquad \big|f(z_0 + ik) - f(z_0) - kD_2 f(z_0)\big| \le \varepsilon |k| \quad \text{whenever } |k| \le r$$

due to the continuity at z_0 of $D_1 f = D_1 u + iD_1 v$, and the existence at z_0 of $D_2 f$, respectively. From (3), (4) and (5) it follows that

$$\big|f(z_0 + h + ik) - f(z_0) - L(h + ik)\big| \le 2\varepsilon|h| + \varepsilon|k| \le 3\varepsilon|h + ik|$$

holds whenever the real numbers h, k satisfy $|h + ik| \le r_\varepsilon$. This proves that f is $\mathbb{R}$-differentiable at z_0. The rest follows from the earlier discussion of (C. R.).

Remarks: The popular example $f(x, y) := xy(x^2 + y^2)^{-1}$ for $(x, y) \neq (0, 0)$, $f(0, 0) := 0$ shows the existence of $D_1 f$, $D_2 f$ throughout U together with the equality $iD_1 f(z_0) = D_2 f(z_0)$ is insufficient to insure even the continuity of f at z_0. On the other hand, continuity of $D_1 f, D_2 f$ is not necessary for $\mathbb{R}$-differentiability of f. Elementary functions can be constructed to illustrate this, but the quickest example anticipates the trigonometric functions: Define $\phi : \mathbb{R} \mapsto \mathbb{R}$ by $\phi(0) := 0$, $\phi(x) := x^2 \sin(1/x)$, $x \neq 0$. It is everywhere differentiable, with $\phi'(0) = 0$. However, ϕ' is not continuous at 0. Therefore the function $f(x, y) := (\phi(x), \phi(y))$ is $\mathbb{R}$-differentiable at $(0, 0)$, the linear map L being $\mathbf{0}$, and $D_1 f, D_2 f$ exist throughout $\mathbb{R}^2$ but neither is continuous at $(0, 0)$.

Exercise 2.18 (D. R. Curtiss [1901/02], Pingya [1998]) *One can modify the mean-value argument in 2.17 to prove the $\mathbb{R}$-differentiability of f at z_0 when the hypothesis is changed to read: All first-order partial derivatives of u and v exist at z_0, and one of $D_1 u, D_2 u$ and one of $D_1 v, D_2 v$ is continuous at z_0.*

Exercise 2.19 *Let a, b, c, $d \in \mathbb{R}$ with $a < b$, $c < d$, $R := (a, b) \times (c, d)$.*

(i) (Leibniz' Rule) *Show that if $\phi : \overline{R} \mapsto \mathbb{C}$ is a continuous function and $D_2\phi$ exists and is continuous on $\overline{R}$, then the function $\Phi(y) := \int_a^b \phi(x, y)\,dx$ is differentiable in $[c, d]$ with $\Phi'(y) = \int_a^b D_2\phi(x, y)\,dx$.* (Cf. 7.21.)

Hint: When ϕ is real-valued, use the mean-value theorem of the differential calculus and the uniform continuity of $D_2\phi$ in $\overline{R}$. Alternatively, note that for distinct $y, y_0 \in [c, d]$

$$\begin{aligned}
&[\Phi(y) - \Phi(y_0)][y - y_0]^{-1} - \int_a^b D_2\phi(x, y_0)\,dx \\
&\qquad = \int_a^b (y - y_0)^{-1} \int_{y_0}^{y} [D_2\phi(x, t) - D_2\phi(x, y_0)]\,dt\,dx.
\end{aligned}$$

(ii) (Fubini) *Show that if ψ is a continuous function on $\overline{R}$, then*

$$\int_a^b \int_c^d \psi(s, t)\,dt\,ds = \int_c^d \int_a^b \psi(s, t)\,ds\,dt.$$

Hints: Show that $\phi(x,y) := \int_c^y \psi(x,t)\,dt$, $(x,y) \in \overline{R}$, is continuous and apply part (i) to get

$$D_y \int_a^b \left[\int_c^y \psi(s,t)\,dt\right] ds = \int_a^b \left[D_y \int_c^y \psi(s,t)\,dt\right] ds = \int_a^b \psi(s,y)\,ds.$$

By an application of the fundamental theorem of calculus to this continuous function of y

$$\int_c^d \left[\int_a^b \psi(s,y)\,ds\right] dy = \int_a^b \left[\int_c^d \psi(s,t)\,dt\right] ds - \int_a^b \left[\int_c^c \psi(s,t)\,dt\right] ds.$$

(iii) (H. A. Schwarz) *Suppose* $\Phi\colon \overline{R} \mapsto \mathbb{C}$, $D_1\Phi$, $D_2\Phi$ *and* $D_2D_1\Phi$ *exist, and the latter is continuous on* $\overline{R}$. *Show that* $D_1D_2\Phi$ *exists and coincides with* $D_2D_1\Phi$ *throughout* $\overline{R}$.

Hints: Define $\psi := D_2D_1\Phi$ and

(1) $$\Psi(x,y) := \int_a^x \int_c^y \psi(s,t)\,dt\,ds, \quad (x,y) \in \overline{R}.$$

The uniform continuity of $D_2D_1\Phi$ on the (compact) set $\overline{R}$ not only justifies invoking the fundamental theorem of calculus to get

$$\int_c^y \psi(s,t)dt = D_1\Phi(s,y) - D_1\Phi(s,c) = \frac{d}{ds}\left[\Phi(s,y) - \Phi(s,c)\right],$$

but it also insures the continuity of this function on $\overline{R}$, thereby justifying a second appeal to the fundamental theorem to get

(2) $$\Psi(x,y) = [\Phi(x,y) - \Phi(x,c)] - [\Phi(a,y) - \Phi(a,c)].$$

Use (ii) and the fundamental theorem of calculus to conclude from (1) that $D_1D_2\Psi = \psi$. Then use (2) to see that $D_2\Phi(x,y) = D_2\Psi(x,y) + D_2\Phi(a,y)$, whence $D_1D_2\Phi$ exists and equals $D_1D_2\Psi$.

(iv) (Green's Theorem) *Suppose* f *is continuous on* $\overline{R}$ *and that* D_1f, D_2f *exist and are continuous on* $\overline{R}$. *Show that*

$$\int_{\partial R} f = \int_a^b \left[\int_c^d (-D_2f(x,y) + iD_1f(x,y))\,dy\right] dx.$$

Hints: Split the sum on the right-hand side, use (ii) on the second summand and then unravel all iterated integrals via the fundamental theorem of calculus into line integrals. [Note that continuity of f is a consequence of the other hypotheses, via a slight strengthening of 2.17.]

(v) *Suppose that f is holomorphic in a neighborhood of $\overline{R}$ and that f' is continuous on $\overline{R}$. Show that $\int_{\partial R} f = 0$.*

Hint: (C. R.) + (iv).

(v)* *The method in* (v) *can also deal with smooth distortions of rectangles: $R := [0,1] \times [0,1] \subset U$ open, and $h : U \mapsto \mathbb{C}$ has continuous partial derivatives of first and second order. Suppose that F is holomorphic in a neighborhood of $h(R)$ and that F' is continuous on $h(R)$. Show that*

$$\int_{h\circ\partial R} F = 0.$$

[In 7.23(iii) these hypotheses are significantly weakened.]

Hints: (VÝBORNÝ [1979].) For notational ease denote the *curve* ∂R by γ. In view of 2.17 the 2-dimensional chain rule (at the beginning of this section) is valid and gives

$$\begin{aligned}\int_{h\circ\partial R} F = \int_{h\circ\gamma} F &= \int_0^4 (F\circ h\circ\gamma)(h\circ\gamma)' \\ &= \int_0^4 (F\circ h\circ\gamma)[(D_1h\circ\gamma)\operatorname{Re}\gamma' + (D_2h\circ\gamma)\operatorname{Im}\gamma'].\end{aligned}$$

In view of definition 1.12, this says that

$$\begin{aligned}\int_{h\circ\partial R} F = &\int_0^1 F(h(x,0))D_1h(x,0)dx - \int_0^1 F(h(x,1))D_1h(x,1)dx \\ &+ \int_0^1 F(h(1,y))D_2h(1,y)dy - \int_0^1 F(h(0,y))D_2h(0,y)dy\end{aligned}$$

Continuity of F' lets the Fundamental Theorem of Calculus transform this to

$$\begin{aligned}\int_{h\circ\partial R} F = &\int_0^1\int_0^1 -D_2(F(h(x,y))D_1h(x,y))dydx \\ &+ \int_0^1\int_0^1 D_1(F(h(x,y))D_2h(x,y))dxdy. \qquad (3)\end{aligned}$$

Since F is holomorphic, the product and chain rules give

$$\begin{aligned}D_1((F\circ h)D_2h) &= (F\circ h)D_1D_2h + D_1(F\circ h)D_2h \\ &= (F\circ h)D_1D_2h + (F'\circ h)D_1hD_2h, \quad \text{and} \qquad (4)\\ D_2((F\circ h)D_1h) &= (F\circ h)D_2D_1h + D_2(F\circ h)D_1h \\ &= (F\circ h)D_2D_1h + (F'\circ h)D_2hD_1h \\ &= D_1((F\circ h)D_2h) \quad \text{by (4) and (iii).} \qquad (5)\end{aligned}$$

Finally, equality (5) and Fubini (ii) make the right-hand side of (3) equal 0.

(vi) *Ω is open and starlike with respect to 0, $f \in H(\Omega)$ and*

$$g(z) := \int_0^1 f(tz)\,dt, \quad z \in \Omega.$$

Assuming that f' is continuous in Ω, show that $g \in H(\Omega)$ and

$$g'(z) = \int_0^1 tf'(tz)\,dt \qquad \forall z \in \Omega.$$

Hints: For each $t \in [0,1]$ the function $f_t(\xi) := f(t\xi)$ is holomorphic in Ω, with $f_t'(\xi) = tf'(t\xi)$. Fix $z \in \Omega$ and consider $r > 0$ such that $\bar{D}(z,r) \subset \Omega$. Then $[z,w] \subset \Omega$ for every $w \in \bar{D}(z,r)$ and by 2.10(i)

$$f(tw) - f(tz) = \int_{[z,w]} f_t' = \int_0^1 tf'\big(t(sw + (1-s)z)\big)(w-z)\,ds,$$

and so

$$\frac{g(w)-g(z)}{w-z} - \int_0^1 tf'(tz)\,dt = \int_0^1 t \int_0^1 \big[f'\big(t(sw+(1-s)z)\big) - f'(tz)\big]\,ds\,dt.$$

The function f' is uniformly continuous on each compact subset of Ω, and the points $t(sw + (1-s)z)$ with $s,t \in [0,1]$ and $w \in \bar{D}(z,r)$ constitute such a set. Hence, given $\varepsilon > 0$, r may be chosen small enough that the bracketed expression will have modulus less than ε for all such s, t, w.

(vii) *With the notation and hypotheses of* (vi), *show that*

$$F(z) := z\int_0^1 f(tz)\,dt, \quad z \in \Omega$$

defines a holomorphic function whose derivative is f. (Cf. 5.2.)

Hint: Note that $zf'(tz) = \frac{d}{dt} f(tz)$ and in the equation

$$zg'(z) = \int_0^1 tzf'(tz)\,dt$$

of (vi) integrate by parts to get $zg'(z) = f(z) - g(z)$.

(viii) *Use* (vii) *and* 2.10(ii) *to generalize* (v) *above: If Ω is a starlike open subset of $\mathbb{C}$, $f \in H(\Omega)$ and f' is continuous in Ω, then $\int_\gamma f = 0$ for every piece-wise smooth loop γ in Ω.*

(ix) (Converse of (v)): *Suppose that $f, D_1 f, D_2 f$ are continuous in the open set $U \subset \mathbb{C} = \mathbb{R}^2$ and that $\int_{\partial R} f = 0$ whenever $\overline{R} \subset U$. Use Green's theorem from* (iv) *to show that f satisfies* (C. R.) *in U.*

Hints: Given $a + ic \in U$, we have $\overline{R} \subset U$ whenever b, d are sufficiently near a, c, respectively. Hence

$$0 = \int_a^b \left[\int_c^d (-D_2 f(x,y) + i D_1 f(x,y)) dy \right] dx.$$

Therefore

$$- D_2 f(a,c) + i D_1 f(a,c) =$$
$$\frac{1}{(b-a)} \frac{1}{(d-c)} \int_a^b \left[\int_c^d (D_2 f(x,y) - D_2 f(a,c) + i D_1 f(a,c) - i D_1 f(x,y)) dy \right] dx.$$

The right-hand side converges to 0 as $b \downarrow a$ and $d \downarrow c$.

(x) *Suppose $D_1 f$ and $D_2 f$ are continuous in the open set $U \subset \mathbb{C} = \mathbb{R}^2$ and that* (C. R.) *holds throughout U. Prove the $\mathbb{C}$-differentiability of f in U by using integration arguments like the preceding.* (Cf. 2.17.)

Hints: After a translation we may suppose that the typical point z_0 of U at which existence of $f'(z_0)$ is to be proved is $z_0 := 0$. Pick $r > 0$ so that $D(0,r) \subset U$. Then for any $z = (x,y) \in D(0,r)$

$$\begin{aligned} f(z) - f(0) - z D_1 f(0) &\\ &= f(z) - f(0) - x D_1 f(0) - iy D_1 f(0) \\ &\overset{\text{(C. R.)}}{=} f(x,y) - f(0,y) - x D_1 f(0) + f(0,y) - f(0) - y D_2 f(0) \\ &= \int_0^x [D_1 f(s,y) - D_1 f(0)] ds + \int_0^y [D_2 f(0,t) - D_2 f(0)] dt. \end{aligned}$$

Consequently,

$$\begin{aligned} |f(z) - f(0) - z D_1 f(0)| &\\ &\leq |x| \sup_{|w|<r} |D_1 f(w) - D_1 f(0)| + |y| \sup_{|t|<r} |D_2 f(0,t) - D_2 f(0)| \\ &\leq |z| \sup_{|w|<r} |D_1 f(w) - D_1 f(0)| + |z| \sup_{|w|<r} |D_2 f(w) - D_2 f(0)|. \end{aligned}$$

[Evidently, as regards $D_2 f$, only the continuity of each map $y \mapsto D_2 f(x,y)$, not that of $D_2 f$, is needed.]

(xi) *f is $\mathbb{C}$-differentiable in $S := [0,1] \times [0,1]$ and ε is positive. By an ε-decomposition of S is meant a decomposition into finitely many closed squares S_j whose sides are parallel to the coordinate axes and whose interiors are disjoint, such that each S_j contains a point z_j satisfying*

$$(6.j) \qquad |f(z) - [f(z_j) + f'(z_j)(z - z_j)]| \leq \varepsilon |z - z_j| \qquad \forall z \in S_j.$$

Prove that S has such a decomposition (GOURSAT [1900]).

Hints: If not, then when S is divided into four congruent subsquares, at least one will have no ε-decomposition. Proceed inductively to get a decreasing sequence of closed squares S_n having side-length 2^{-n} and no ε-decomposition. The finite-intersection property of compacta supplies a point $z_0 \in \cap S_n$. As S_n has no ε-decomposition and $z_0 \in S_n$, there must be $z_n \in S_n \setminus \{z_0\}$ with

$$|f(z_n) - [f(z_0) + f'(z_0)\ (z_n - z_0)]| > \varepsilon |z_n - z_0|.$$

Division by $z_n - z_0$ then contradicts differentiability of f at z_0, since $z_n \to z_0$.

(xii) *With f and S_j, having side-length ℓ_j, as in* (xi), *show that*

$$(7.j) \qquad \left| \int_{\partial S_j} f \right| \leq 4\varepsilon \ell_j \cdot \sup_{z \in S_j} |z - z_j| \leq 8\varepsilon \ell_j^2.$$

Hints: Since the bracketed expression in (6.j) integrates to 0 [2.10(ii)], we have $\int_{\partial S_j} f = \int_{\partial S_j} (f(z) - [f(z_j) + f'(z_j)(z - z_j)] dz$, and because of inequality (6.j) and 2.9(ii), the latter integral is majorized as claimed.

(xiii) *Continuing with* (xii), *show that $\sum_j |\int_{\partial S_j} f| \leq 8\varepsilon$.*

Hint: Since the S_j lie in S and have disjoint interiors, $\sum_j \ell_j^2 = \sum_j \text{area}(S_j) \leq \text{area}(S)$.

(xiv) *For f, S as in* (xi), *does* (xiii) *justify the conclusion $\int_{\partial S} f = 0$?* (Cf. DE LA VALLÉE POUSSIN [1914].)

Hint: Although the details are a combinatorial headache, one can *prove* the (obvious?) equality $\int_{\partial S} f = \sum_j \int_{\partial S_j} f$. *Granted this*, it follows from (xiii) that $|\int_{\partial S} f| \leq 8\varepsilon$, an inequality that holds for every $\varepsilon > 0$.

Remarks: As will be seen, $\int_{\partial R} f = 0$ (gotten as in (v) by Cauchy in 1846) leads to the remarkable fact that f is infinitely often differentiable and even (3.6) of power-series type in R. Cauchy suspected that the continuity hypothesis in (v) was unnecessary, but this was only confirmed in 1900, 43 years after his death. Modulo a (non-)trivial detail, the argument in 2.19(xiv) proves this. A combinatorially simpler proof, also featuring the method of descent, will be given in 5.1 (which the reader could easily turn to right now!). Neither of these proofs involves Green. But one can adapt these arguments to prove Green under the weaker hypothesis that $D_2f - iD_1f$ merely be Riemann double-integrable over R. Evidently (v) without the continuity hypothesis is a corollary of this improved Green's theorem. See BOCHNER [1955b], P. J. COHEN [1959], CRAVEN [1964], GRUNSKY [1966], ACKER [1996] and GREENLEE [2004/05]. See also **MR** 2005e:30069.

The aforementioned power-series representation of a holomorphic function comes from expanding the kernel in a certain integral representation of the function (as in 3.8), which in turn is derived from Cauchy's Integral Theorem 2.19(v) by an ingenious idea that the next exercise sketches, but which gets a fuller treatment in chapter V. See also 3.39 and the remarks following it. It is noteworthy that by itself 2.19(v) is able to provide a brief, elegant proof of the fundamental theorem of algebra, essentially Gauss' third proof freed of reference to the transcendental functions sine and cosine (cf. also the remarks following 5.5).

Exercise 2.20 *Let $a, b, c, d \in \mathbb{R}$ with $a < b, c < d$, $R := (a, b) \times (c, d)$, $\overline{R} \subset U$ open $\subset \mathbb{C}$.*

(i) *Suppose $z_0 \in R$, $F : U \setminus \{z_0\} \mapsto \mathbb{C}$ is holomorphic, is bounded near z_0, and F' is continuous. Show that $\int_{\partial R} F = 0$.*

Hints: For $r > 0$ sufficiently small and $S_r := (-r, r) \times (-r, r)$, we have $z_0 + \overline{S}_r \subset R$. Extend the sides of the square $z_0 + S_r$ and thereby pave R as the union of this square and eight rectangles disjoint from it. To each of the latter 2.19(v) is applicable. When the integrals over the boundaries of these rectangles are written as sums of integrals over intervals and added, significant cancellation occurs, resulting in

$$0 = \int_{\partial R} F - \int_{z_0 + \partial S_r} F.$$

Since F is bounded near z_0 and $l(z_0 + \partial S_r) = 8r$, it follows from 2.9(i) that the second integral converges to 0 with r.

(ii) *Suppose f is holomorphic and f' is continuous in U. Show that*

$$f(z_0) \int_{\partial R} \frac{d\xi}{\xi - z_0} = \int_{\partial R} \frac{f(\xi)}{\xi - z_0} d\xi \qquad \forall z_0 \in R.$$

Hint: For each $z_0 \in R$ part (i) is applicable to the function defined by $F(z_0) := f'(z_0)$, $F(z) := (f(z) - f(z_0))/(z - z_0)$ for $z \in U \setminus \{z_0\}$. An alternative proof method is the Fubini argument employed in 3.39(ii).

(iii) *Using* 2.15(iii) *and the paving technique in the hints above, show that*

$$\int_{\partial R} \frac{d\xi}{\xi - z_0} = 8i \int_0^1 \frac{du}{u^2+1} \qquad \forall z_0 \in R.$$

(iv) *Combine* (ii) *and* (iii) *to prove the following weak version of* Cauchy's Integral Formula: *If f is holomorphic and f' is continuous in a neighborhood of $\overline{R}$, then*

$$f(z) = \frac{1}{2pi} \int_{\partial R} \frac{f(\xi)}{\xi - z} d\xi \qquad \forall z \in R,$$

where $p := 4\int_0^1 \frac{du}{u^2+1}$. *Infer* (2.14) *that f is $\mathbb{C}$-differentiable of all orders!*

(v) *Suppose P is a complex polynomial of degree n with no zeros in $\mathbb{C}$. Use the (obvious) fact that* $\lim_{|z|\to\infty} zP'(z)/P(z) = n$ *to deduce that*

$$\lim_{r\to\infty} \int_{\partial S_r} \frac{P'}{P} = 8n \int_0^1 \frac{1}{\gamma},$$

where $S_r := (-r, r) \times (-r, r)$ *and* $\gamma := [-1 - i, 1 - i]$.

Hints: Set $f := P'/P$. Note that $\gamma' = 2$ and $\operatorname{Im}\gamma = -1$. As in the hint to 2.15(iii), the intervals comprising the polygonal loop ∂S_r are $r\gamma$, $ir\gamma$, $-r\gamma$, $-ir\gamma$, so 2.9(ii) together with the fact $\gamma' = 2$ gives

$$\int_{\partial S_r} f = 2\sum_{k=0}^{3} \int_0^1 \frac{i^k r\gamma f \circ (i^k r\gamma)}{\gamma}.$$

Since $|\gamma| \geq |\operatorname{Im}\gamma| = 1$, we have $|r\gamma| \to \infty$ uniformly on $[0, 1]$ as $r \uparrow \infty$. Consequently, $\lim_{|z|\to\infty} zf(z) = n$ implies that each integrand on the right converges to n/γ, uniformly on $[0, 1]$ as $r \uparrow \infty$.

(vi) (FTA) *From* (v) *and Cauchy's theorem* 2.19(v) *infer that every nonconstant complex polynomial has a zero in* $\mathbb{C}$.

Hints: If P is a complex polynomial of degree n which has no zeros in $\mathbb{C}$, then $f := P'/P$ is holomorphic and f' is continuous in $\mathbb{C}$. Hence by 2.19(v) every integral on the left-hand side of the equality in part (v) above is 0. From that equality and the fact that $\operatorname{Im}\gamma = -1$ therefore follows

$$0 = \operatorname{Im}\left(8n \int_0^1 \frac{1}{\gamma}\right) = 8n \operatorname{Im} \int_0^1 \frac{\bar{\gamma}}{|\gamma|^2} = 8n \int_0^1 \frac{1}{|\gamma|^2},$$

whence $n = 0$.

(vii) *Infer the* FTA *directly from* (iv).

Hints: If P is a zero-free polynomial, then $f := 1/P \in H(\mathbb{C})$ and $f' = -P'/P^2$ is continuous. Hence according to (iv)

$$0 \neq f(0) = \frac{1}{2pi} \int_{\partial S_r} \frac{f(\xi)}{\xi}\, d\xi \qquad \forall r > 0, \tag{1}$$

where S_r and p are as in (v) and (iv). From 2.9 then

$$\begin{aligned} |f(0)| &\leq \frac{1}{2p} l(S_r) \cdot \max_{\xi \in \partial S_r} \left| \frac{f(\xi)}{\xi} \right| \leq \frac{1}{2p} 8r \max_{\xi \in \partial S_r} \frac{|f(\xi)|}{r} \\ &\leq \frac{4}{p} \max_{\xi \in \partial S_r} |f(\xi)| \qquad \forall r > 0. \end{aligned} \tag{2}$$

If P has positive degree, $\lim_{|\xi| \to \infty} |P(\xi)| = \infty$ is clear, so the right-hand side of inequality (2) converges to 0 as $r \to \infty$, contradicting (1).

Remark on the FTA: The principal ingredients in the preceding two proofs are the growth estimate of a polynomial via its leading term, the Cauchy-Riemann equation, and Fubini's theorem about iterated integrals. For esthetic (not to mention pedagogical) reasons one would like a proof of this basic (algebraic!) result that does not involve differentiation and integration, but merely field properties of $\mathbb{C}$. There are several ingenious such minimalist proofs, admittedly inspired by the view from higher up the mountain. Here is one, from VAGGIONE [1997], that uses just one property of $\mathbb{C}$, that closed and bounded sets are compact. In the axiomatic founding of $\mathbb{R}$ this is readily seen to be equivalent to the Least Upper Bound Axiom. So this proof is certainly minimal with respect to prerequisite knowledge and tools employed.

Exercise 2.21 (i) *Show that for each $k \in \mathbb{N}$ there exist $u_r \in \mathbb{T}$ such that $u_r^k = i^r$ for $r = 1, 2, 3$.*

Hints: For appropriate integers $l \geq 0$ and odd $m \geq 1$, $k = m2^l$. Then m is congruent to $r \in \{1, 3\}$ modulo 4. Repeated use of 1.26(i) produces $u \in \mathbb{C}$ such that $u^{2^l} = i$. Then

$$\begin{aligned} u^k &= (u^{2^l})^m = i^m = i^r \\ (u^2)^k &= (u^k)^2 = (i^r)^2 = (-1)^r = -1 = i^2, \quad \text{since } r \text{ is odd,} \\ (u^3)^k &= (u^2)^k u^k = (-1)^r i^r = (-i)^r = i^{-r} = i^{4-r}. \end{aligned}$$

Hence some ordering of u, u^2, u^3 does the job.

(ii) *$a \in \mathbb{C}$, $\varepsilon > 0$ and f maps $D(a, \varepsilon)$ into the left half-plane. Show that if $\lim_{z\to a}(z-a)^{-k}f(z)$ exists in $\mathbb{C}$ for some $k \in \mathbb{N}$, then it equals* 0.

Hints: Call the limit $L \in \mathbb{C}$. Then for $u \in \mathbb{T}$, and non-zero $r \in \mathbb{R}$ we have $L = \lim_{r\to 0}\,(ru)^{-k}\; f(a+ru)$, that is, $u^k L = \lim_{r\to 0} r^{-k} f(a+ru)$. From $\operatorname{Re} f \leq 0$ we thus learn that $\operatorname{Re}(u^k L) \leq 0$, for every $u \in \mathbb{T}$. The validity of this inequality merely for $u = 1$ and the three numbers in (i) already forces $L = 0$.

(iii) (Maximum Modulus Theorem for Rational Functions) *Let p, q be polynomials such that for some $a \in \mathbb{C}$, $\varepsilon > 0$, $0 \notin p(D(a, \varepsilon))$ and the rational function $r := q/p$ satisfies $|r(z)| \leq |r(a)|$ for all $z \in D(a, \varepsilon)$. Show that r is constant.*

Hints: Suppose contrariwise that

(1) $$r \text{ is not constant.}$$

Then the polynomial $Q := p(a)q - q(a)p$ is not identically 0. Since it has a zero at a, there is $k \in \mathbb{N}$ and a polynomial P such that $P(a) \neq 0$ and $Q(z) = (z-a)^k P(z)$. (See the hint to 4.52(i).) Thus

$$(z-a)^{-k}(r(z) - r(a)) = (z-a)^{-k}Q(z)/p(a)p(z) = P(z)/p(a)p(z)$$

and consequently

(2) $$\lim_{z\to a}(z-a)^{-k}(r(z) - r(a)) = P(a)/p^2(a) \in \mathbb{C} \setminus \{0\}.$$

For appropriate $u \in \mathbb{T}$, $|r(a)| = ur(a)$. Therefore by hypothesis, for $z \in D(a, \varepsilon)$, $|ur(z)| = |r(z)| \leq |r(a)|$, whence

$$\operatorname{Re} ur(z) \leq |r(a)| = ur(a) = \operatorname{Re} ur(a).$$

That is, the function $f := ur - ur(a)$ satisfies

$$\operatorname{Re} f(z) \leq 0 \qquad \forall z \in D(a, \varepsilon).$$

This inequality and (2) contradict (ii), which proves (1) untenable.

(iv) (FTA) *Every non-constant polynomial p with complex coefficients has a zero in $\mathbb{C}$.*

Hints: Evidently $|p(z)| \to \infty$ as $|z| \to \infty$, so for some $N \in \mathbb{N}$, $|p(z)| > |p(0)|$ whenever $|z| \geq N$. The disk $\bar{D}(0, N)$, being compact, contains an a such that $|p(w)| \geq |p(a)|$ for all $w \in \bar{D}(0, N)$. Hence $|p(z)| \geq |p(a)|$ for all $z \in \mathbb{C}$. Therefore if p is zero-free, $r := 1/p$ is a non-constant rational function satisfying $|r(z)| \leq |r(a)|$ for all $z \in \mathbb{C}$, which contradicts (iii).

§ 5 Complex Logarithms, *avant la lettre*

This chapter closes with a long multi-part exercise adapted from LEBORGNE [1991]. It illustrates one way to bring the twin functions logarithm and exponential to birth in $\mathbb{C}$ and is optional, since in the next chapter we shall officially pursue a quite different development. This one gives primacy to the logarithm, defining it by an integral that is a natural analog of the integral $\int_1^x dt/t$ which defines the real-valued logarithm on $(0, \infty)$. Whereas in the next chapter we give primacy to the exponential. The less enthusiastic reader may wish to skip this section. On the other hand, since this approach makes the indispensable exponential function available without a theory of series, following it lets the reader who is philosophically so disposed, or simply in a hurry, go right to the Cauchy theory of Chapter V without passing through Chapter III.

Exercise 2.22 *Recall the notation* $\mathbb{C}_- := \mathbb{C} \setminus (-\infty, 0]$ *and define*

(1) $$L(z) := \int_{[1,z]} \frac{1}{\xi}\, d\xi = \int_0^1 \frac{z-1}{1+t(z-1)}\, dt, \qquad z \in \mathbb{C}_-.$$

(i) *Show that* $\mathbb{C}_-$ *is starlike with respect to the point* 1, *making* L *well defined.*

(ii) *Show that* L *is holomorphic in* $\mathbb{C}_-$ *and that*

(2) $$L'(z) = 1/z \qquad \forall z \in \mathbb{C}_-.$$

Hint: This is all a consequence of 2.19(vii), but a direct treatment is easy. For $w, z \in \mathbb{C}_-$ with $w \neq z$, a little computation reveals that

$$\frac{L(w) - L(z)}{w - z} = \int_0^1 \frac{dt}{[1 + t(w-1)][1 + t(z-1)]}.$$

When $w \to z$, this integral converges to $\int_0^1 [1 + t(z-1)]^{-2}\, dt$, which equals 1 if $z = 1$ and otherwise evaluates to

$$-\frac{1}{z-1} \cdot \frac{1}{1 + t(z-1)}\Bigg]_{t=0}^{t=1} = \frac{1}{z}.$$

(iii) *Show that*

$$L(z) = \int_{[1,1+i\,\mathrm{Im}\, z, z]} \frac{1}{\xi}\, d\xi \qquad \forall z \in \mathbb{C}_-$$

and infer that on the line $\operatorname{Im} z = b$, $b \in \mathbb{R}$,

$$L(z) = \int_0^b \frac{y+i}{1+y^2}\,dy + \int_1^{\operatorname{Re} z} \frac{x-ib}{x^2+b^2}\,dx.$$

Hint: Use (ii) above and 2.10(i).

(iv) *Use* (1) *to show that*

$$L(\bar{z}) = \overline{L(z)} \qquad \forall z \in \mathbb{C}_-.$$

(v) *Prove that*

$$L(rz) = L(r) + L(z) \qquad \forall z \in \mathbb{C}_-,\ r > 0,$$

$$L(1/z) = -L(z) \qquad \forall z \in \mathbb{C}_-,$$

$$L(wz) = L(w) + L(z) \qquad \forall\, w, z \in \mathbb{H}.$$

Hints: For fixed positive real r, the function $z \mapsto L(rz)$ is holomorphic in $\mathbb{C}_-$ and has derivative $r \cdot 1/rz = 1/z = L'(z)$. Hence by 2.10(iv) it differs from $L(z)$ by a constant. Similarly, $z \mapsto 1/z$ maps $\mathbb{C}_-$ onto $\mathbb{C}_-$ and $z \mapsto L(z) + L(1/z)$ has derivative 0. Lastly, $z \mapsto wz$ maps $\mathbb{H}$ into $\mathbb{C}_-$ if $w \in \mathbb{H}$ (routine algebra – see the hints to 3.44(iii)), and $z \mapsto L(wz) - L(z)$ has derivative 0 in $\mathbb{H}$.

(vi) *Show that*

$$\operatorname{Re} L(z) = L(|z|) \qquad \forall z \in \mathbb{C}_-.$$

Hints: From (iv) and (v)

$$2 \operatorname{Re} L(z) = L(z) + L(\bar{z}) = L(z) + L(|z|^2/z) = L(z) + L(|z|^2) + L(1/z) = 2L(|z|).$$

(vii) *Show that* L *maps* $(0, +\infty)$ *strictly increasingly onto* $\mathbb{R}$.

Hint: For $r \in (0, +\infty)$ we have from (2)

$$L(r) = \int_1^r \frac{dx}{x},$$

which is a strictly increasing function that satisfies $L(1/r) = -L(r)$, by (v). Moreover, (v) also shows that $L(2^n) = nL(2)$, so $\lim_{n\to\infty} L(2^n) = +\infty$.

(viii) *Set* $p := 4\int_0^1 \frac{dt}{1+t^2}$ *and define* $\Theta := \operatorname{Im} L$. *Prove that* Θ *maps* $\mathbb{R} - i$ *strictly increasingly onto the interval* $(-p, 0)$ *and maps* $\mathbb{R} + i$ *strictly decreasingly onto the interval* $(0, p)$.

Hints: Because of (iv) it suffices to prove the second half of this claim. For $t \in \mathbb{R}$, taking $b = 1$ in (iii) yields

$$\Theta(t+i) = \int_0^1 \frac{1}{1+y^2}\,dy + \int_1^t \frac{-1}{x^2+1}\,dx = \frac{p}{4} - \int_1^t \frac{dx}{x^2+1}.$$

Note that by the change of variable $x = 1/u$

$$\lim_{t\to+\infty} \int_1^t \frac{dx}{x^2+1} = \int_0^1 \frac{du}{1+u^2} = \frac{p}{4},$$

and

$$\lim_{t\to-\infty} - \int_1^t \frac{dx}{x^2+1} = \int_{-1}^1 \frac{dx}{x^2+1} - \lim_{t\to-\infty} \int_{-1}^t \frac{dx}{x^2+1}$$
$$= \int_{-1}^1 \frac{dx}{x^2+1} + \int_{-1}^0 \frac{du}{1+u^2} = \frac{3p}{4}.$$

(ix) *Show that $\Theta(z) = 0$ iff $z \in (0,\infty)$ and that otherwise $\Theta(z)$ has the same sign as $\operatorname{Im} z$. Confirm the formulas*

$$\Theta(z) = \int_0^{\operatorname{Im} z/\operatorname{Re} z} \frac{du}{1+u^2} \qquad \forall z \in \mathbb{H}.$$
$$\Theta(w) = 2\int_0^{\operatorname{Im} w/(|w|+\operatorname{Re} w)} \frac{du}{1+u^2} \qquad \forall w \in \mathbb{C}_-.$$

Hints: $\Theta = 0$ on $(0,\infty)$ by (vii). If $y := \operatorname{Im} z > 0$, then $w := \frac{1}{y} z \in \mathbb{R} + i$, and so by (v) and (viii)

$$\Theta(z) = \Theta(yw) = \Theta(y) + \Theta(w) = 0 + \Theta(w) \in (0,p).$$

Suppose both $\operatorname{Re} z > 0$ and (keeping (iv) in mind) $\operatorname{Im} z > 0$. If we write $t := \operatorname{Re} z/\operatorname{Im} z$, then $w = t + i$ and, as shown in the hints to (viii),

$$\Theta(z) = \Theta(t+i) = \frac{p}{4} - \int_1^t \frac{dx}{x^2+1} = \int_0^1 \frac{du}{1+u^2} + \int_1^{1/t} \frac{du}{1+u^2},$$

by definition of p and the change of variable $x = 1/u$. For the second formula, use 1.26(i) to express w as z^2 for some $z \in \mathbb{H}$, then apply the first formula and part (v).

(x) *Prove that L is one-to-one in $\mathbb{C}_-$ and maps $\mathbb{C}_-$ onto $\mathbb{R} \times (-p,p)$.*

Hints: If $z,\ w \in \mathbb{C}_-$ and $w \neq z$ but $w = rz$ for some $r > 0$, then by (v)

$$L(w) = L(r) + L(z) \neq L(z),$$

since $r \neq 1$ means (by(vii)) that $L(r) \neq 0$. Next consider $z,\ w \in \mathbb{U}$ with $w \neq rz$ for any $r > 0$. Then for the positive numbers $\alpha := \frac{1}{\operatorname{Im} w}$, $\beta := \frac{1}{\operatorname{Im} z}$ we have $\alpha w,\ \beta z \in \mathbb{R} + i$ and these numbers are unequal. Hence by (viii), $\Theta(\alpha w) \neq \Theta(\beta z)$. In view of (v) and the fact that $L(\alpha)$, $L(\beta)$ are real, this yields $\Theta(w) \neq \Theta(z)$. Hence $L(w) \neq L(z)$. The case where $z \neq w$ are both in $\mathbb{C}\backslash\overline{\mathbb{U}}$ reduces to the preceding via (iv), and the cases where they are in different components of $\mathbb{C}\setminus\mathbb{R}$ or just one is in $(0,\infty)$ are settled by (ix). The first three lines of the (ix) hint show that $\Theta(\mathbb{U}) = (0,p)$, and similarly (or from (iv)) $\Theta(-\mathbb{U}) = (-p,0)$. Since $\Theta(0,\infty) = \{0\}$, we thus see that $\Theta(\mathbb{C}_-) = (-p,p)$. But (ix) also shows that $\Theta(rz) = \Theta(z)$, so (vi) and (vii) can be used to prove the surjectivity claim.

(xi) *The map* $E_0\colon \mathbb{R}\times(-p,p) \mapsto \mathbb{C}_-$ *which is inverse to* L *is continuous.*

Hints: Suppose $z_n,\ z \in \mathbb{C}_-$ and $L(z_n) \to L(z)$. By (vi) we have

$$L\big(|z_n|\big) = \operatorname{Re} L(z_n) \to \operatorname{Re} L(z) = L\big(|z|\big).$$

Now from (vii) we infer that $\big[L|_{(0,\infty)}\big]^{-1}$ is continuous (cf. 1.46). Therefore it follows that

$$|z_n| \to |z|. \tag{3}$$

As in the hints to (ix), write $z_n = w_n^2$, $z = w^2$, with $w_n, w \in \mathbb{H}$. From (v) and $L(z_n) \to L(z)$ we infer $L(w_n) \to L(w)$; in particular,

$$\Theta(w_n) \to \Theta(w). \tag{4}$$

Now $x \mapsto \int_0^x (1+u^2)^{-1}du$ is a continuous, strictly increasing function on $\mathbb{R}$, so it has a continuous inverse (cf. 1.46). It follows then from (4) and (ix) that

$$\frac{\operatorname{Im} w_n}{\operatorname{Re} w_n} \to \frac{\operatorname{Im} w}{\operatorname{Re} w}. \tag{5}$$

From (5) and (3)

$$\operatorname{Re} w_n = |w_n|\left[\left(\frac{\operatorname{Im} w_n}{\operatorname{Re} w_n}\right)^2 + 1\right]^{-\frac{1}{2}} \to |w|\left[\left(\frac{\operatorname{Im} w}{\operatorname{Re} w}\right)^2 + 1\right]^{-\frac{1}{2}} = \operatorname{Re} w.$$

In turn, from this and (5) follows

$$\operatorname{Im} w_n \to \operatorname{Im} w.$$

So finally, $w_n \to w$ and $z_n = w_n^2 \to w^2 = z$, which is what needed to be proved.

(xii) *The map* $E_0 \colon \mathbb{R} \times (-p, p) \mapsto \mathbb{C}_-$ *which is inverse to* L *is holomorphic and satisfies* $E_0' = E_0$.

Hints: Let $w_n, w \in \mathbb{R} \times (-p, p)$ with $w_n \to w$ and $w_n \neq w$ for all n. Then $E_0(w_n) \neq E_0(w)$ for all n and by (xi) $E_0(w_n) \to E_0(w)$. Hence

$$\frac{L\big(E_0(w_n)\big) - L\big(E_0(w)\big)}{E_0(w_n) - E_0(w)} \to L'\big(E_0(w)\big) = \frac{1}{E_0(w)}, \quad \text{by (ii).}$$

Taking reciprocals, we get

$$\frac{E_0(w_n) - E_0(w)}{w_n - w} \to E_0(w).$$

(xiii) *Show that*

$$L(z) = L(-z) + ip \qquad \forall z \in \mathbb{U}.$$

Hints: The function $F(z) := L(-z) + ip$ is holomorphic in $\mathbb{U}$ and satisfies $F'(z) = -1/(-z) = L'(z)$. Hence F differs from L by a constant. To find that constant, note that by (iii)

$$L(i) = \int_0^1 \frac{y+i}{1+y^2}\,dy + \int_1^0 \frac{x-i}{x^2+1}\,dx = 2\int_0^1 \frac{i}{1+y^2}\,dy = \frac{pi}{2},$$

and by (iv), $L(-i) = \overline{L(i)} = -pi/2$.

(xiv) *Use* (ix) *and* (x) *to see that* $L(\mathbb{U}) = \mathbb{R} \times (0, p)$ *and then use* (xiii) *to prove that*

$$E_0(w - ip) = -E_0(w) \qquad \textit{whenever } 0 < \operatorname{Im} w < p.$$

(xv) *Use* (xiv) *to show that a well-defined extension of* E_0 *to a* $2ip$*-periodic function* E *on* $\mathbb{C}$ *is gotten by the rule*

$$E(w) := (-1)^k\, E_0(w - ikp) \quad \textit{whenever } w \in \mathbb{R} \times \big((k-1)p, (k+1)p\big),\ k \in \mathbb{Z}.$$

(xvi) *Show that for* $w \in \mathbb{C}$, $E(w) = 1$ *if and only if* $w \in 2ip\mathbb{Z}$.

Hints: Since E_0 maps into $\mathbb{C}_- \subset \mathbb{C} \setminus \{-1\}$, definition (xv) makes it clear that if $E(w) = 1$, then $w \in \mathbb{R} \times \big((k-1)p, (k+1)p\big)$ for some even k, say $k = 2n$. Thus $w_0 := w - 2nip \in \mathbb{R} \times (-p, p)$ and $E_0(w_0) = 1$. Consequently, $w_0 = L\big(E_0(w_0)\big) = L(1) = 0$, whence $w = 2nip$.

(xvii) *Show that E is holomorphic in $\mathbb{C}$, satisfies $E' = E$ and has range equal to $\mathbb{C}_- \cup (-\mathbb{C}_-) = \mathbb{C} \setminus \{0\}$.*

(xviii) *Show that E is a homomorphism of $(\mathbb{C}, +)$ onto $(\mathbb{C} \setminus \{0\}, \cdot)$ and as a consequence of* (xvi) *its kernel is $2ip\mathbb{Z}$.*

Hint: Given $w \in \mathbb{C}$, the function $F(z) := \frac{E(w+z)}{E(z)}$ is entire and satisfies $F' = \mathbf{0}$ by (xvii). Hence $F = F(0) = E(w)$.

(xix) *Derive the following properties of E:*

$$E|_{\mathbb{R}} \text{ is positive and strictly increasing;} E(\mathbb{R}) = (0, +\infty), \tag{6}$$

$$E(\overline{w}) = \overline{E(w)} \qquad \forall w \in \mathbb{C}, \tag{7}$$

$$\big|E(w)\big| = E(\operatorname{Re} w) \qquad \forall w \in \mathbb{C}. \tag{8}$$

Hints: $L|_{(0,+\infty)}$ maps $(0, +\infty)$ increasingly onto $\mathbb{R}$ by (vii), and $E|_{\mathbb{R}}$ is its inverse. If $w \in \mathbb{R} \times (-p, p)$, then $z := E_0(w) \in \mathbb{C}_-$ and $w = L(z)$. Consequently $\overline{w} = \overline{L(z)} = L(\overline{z})$ by (iv) and so $E(\overline{w}) = E\big(L(\overline{z})\big) = \overline{z} = \overline{E(w)}$. Equation (7) follows in general by continuity and $2ip$-periodicity. Finally, note that

$$|E(w)|^2 = E(w)\overline{E(w)} = E(w)E(\overline{w}) = E(w + \overline{w}) = E(2 \operatorname{Re} w) = \big[E(\operatorname{Re} w)\big]^2.$$

(xx) *Show that*

$$E(w) = \sum_{k=0}^{n} \frac{w^k}{k!} + \frac{w^{n+1}}{n!} \int_0^1 (1-t)^n E(tw)\, dt \qquad \forall w \in \mathbb{C},\ n \geq 0 \tag{9}$$

and infer that

$$\text{the series } \sum_{k=0}^{\infty} \frac{w^k}{k!} \text{ converges to } E(w) \qquad \forall w \in \mathbb{C}. \tag{10}$$

Hints: Prove (9) by induction on n, using integration by parts and the fact that $E' = E$. Then by (6) and (8)

$$\left| E(w) - \sum_{k=0}^{n} \frac{w^k}{k!} \right| \leq \frac{|w|^{n+1}}{n!} \int_0^1 (1-t)^n \big|E(tw)\big|\, dt \leq \frac{|w|^{n+1}}{(n+1)!} E\big(|\operatorname{Re} w|\big).$$

Finally, from $n > m \Rightarrow n! = n(n-1)\ldots m(m-1)! > m^{n-m+1}(m-1)!$ infer that $n > m \geq |w| \Rightarrow$

$$\begin{aligned} \frac{|w|^{n+1}}{(n+1)!} &\leq \frac{|w|}{n+1} \frac{|w|^n}{m^{n-m+1}(m-1)!} = \frac{|w|}{n+1} \left(\frac{|w|}{m}\right)^n \frac{m^{m-1}}{(m-1)!} \\ &\leq \frac{|w|}{n+1} \frac{m^{m-1}}{(m-1)!}. \end{aligned}$$

(Taking $m \geq 2|w|$ evidently reduces these majorants by the factor 2^{-n}.)

(xxi) *Use integration by parts to confirm that for all $z \in \mathbb{C}_-$, $n \in \mathbb{N}$*

$$(11)\quad L(z) = -\sum_{k=1}^{n} \frac{(1-z)^k}{k} - (1-z)^{n+1} \int_0^1 \big(1-t(1-z)\big)^{-n-1} (1-t)^n \, dt$$

and infer that

$$(12)\quad \text{the series } -\sum_{k=1}^{\infty} \frac{(1-z)^k}{k} \text{ converges to } L(z) \text{ for each } z \in D(1,1).$$

Hints: For $n = 1$ integration by parts on the right-hand side of (11) produces (1). If $|1 - z| = r < 1$, the integral in (11) is majorized by

$$\int_0^1 (1 - tr)^{-n-1} (1 - t)^n \, dt < \int_0^1 (1 - tr)^{-1} \, dt < (1 - r)^{-1}.$$

(xxii) *Show that the holomorphic function $z \mapsto 1/z$ has no primitive in $\mathbb{C} \setminus \{0\}$ and infer that E has no holomorphic right inverse; that is, no $F \in H(\mathbb{C} \setminus \{0\})$ satisfies $E(F(z)) = z$ for all $z \neq 0$.*

Hints: If such an F exists, $F'(z) = 1/z$ by the chain rule. Then (2) above and 2.11(iv) say that $F - L$ is constant in $\mathbb{C}_-$. The equality in (xiii) is therefore valid with F in the role of L. Let $z \in \mathbb{U}$ converge first to 1, then to -1 in (xiii) and use continuity of F in $\mathbb{C} \setminus \{0\}$ to get first $F(1) = F(-1) + ip$, and then $F(-1) = F(1) + ip \longrightarrow\!\!\!\!\!\!\times\;$.

Exercise 2.23 *Show that on $\mathbb{R}$ the function E in 2.22 coincides with the (exponential) function E_{b_1} in 1.47. That is, $E(x) = E_{b_1}(x) = b_1^x$ for all $x \in \mathbb{R}$.*

Hint: We have $E(0) = E_{b_1}(0) = 1$, and $(E/E_{b_1})' = 0$, thanks to 2.22(xii) and 1.47(xii).

Notes to Chapter II

Sometimes it is convenient to speak of a function $f\colon A \mapsto \mathbb{C}$ being holomorphic in A without A being open. There would be two possible definitions, a local one and a global one: (i) For each $a \in A$ there is an open neighborhood V_a of a and an $f_a \in H(V_a)$ such that f coincides with f_a in $A \cap V_a$. (ii) There is an open neighborhood V of A and $F \in H(V)$ such that f coincides with F in A. We will not use this language, but the two definitions are equivalent, a not entirely trivial fact: see MINDA [1973] (and 5.71 below). For more general results see ROGERS and ZAME [1982].

Though adumbrated by Gauss and Poisson, the idea of the complex contour integral in 2.8 was codified by Cauchy in his famous 1825 memoir. See STÄCKEL [1900], [1901], ETTLINGER [1921–22], YUŠKEVIČ [1947] and chapter 4 of SMITHIES [1997]. Eighteenth-century mathematicians sometimes (but with suspicion) made an "imaginary substitution" in an integral (that is, a $\mathbb{C}$-valued change of variable, 11.33 and the generalized Fresnel integral in Remark (ii) following 11.32 being historically significant examples, and $x := it$ in $\int_{-\infty}^{\infty}(1+x^4)^{-1}dx$ being one reason for suspicion); this can be seen as foreshadowing the contour integral, and it was still the prevalent view in Cauchy's early work. See pp. 10–23 of SMITHIES [1997].

Much work has gone into efforts to weaken the conditions in 2.17 and these conditions can indeed be considerably relaxed. The natural global converse of 2.16(i) is:

> *If f is continuous, $D_1 f$, $D_2 f$ exist and satisfy* (C. R.) *throughout an open set U, then $f \in H(U)$.*

But it is not significantly easier to prove than the following:

Theorem (LOOMAN [1923], RIDDER [1930a], WILKOSZ [1933], MENCHOFF [1936]) *If f is continuous on U, $D_1 f$ and $D_2 f$ each exist at all but countably many points of U and the equation* (C. R.) *holds almost everywhere in U (in the sense of plane Lebesgue measure), then f is holomorphic in U.*

In fact there is nothing exceptional about the horizontal and vertical directional derivatives here: MENCHOFF [1935] showed that if f is continuous on U and through each $z_0 \in U$ there pass two distinct lines $L_1(z_0)$, $L_2(z_0)$ such that

$$\lim_{\substack{z \to z_0 \\ z \in L_1(z_0)}} \frac{f(z) - f(z_0)}{z - z_0} \quad \text{and} \quad \lim_{\substack{z \to z_0 \\ z \in L_2(z_0)}} \frac{f(z) - f(z_0)}{z - z_0}$$

exist and coincide, then f is holomorphic in U. For some remarkable counterexamples when the continuity hypothesis of the theorem is absent, see PELLING [1979].

The requirement of differentiability at every point in the definition 2.1 of holomorphy can also be relaxed. 5.41, 5.42 and 9.28 are examples; here is another:

Theorem (POMPEIU [1905a], LOOMAN [1922/1924], RIDDER [1930a]) *If f is continuous in U, $\overline{\lim}_{h\to 0}\big|(f(z+h) - f(z))/h\big|$ is finite for all but countably* *many z in U and f' exists almost everywhere in U, then f is holomorphic in U.*

For details of these results, which are heavily real-variable in nature, the reader may consult MENCHOFF [1936], pp. 195–201 of SAKS [1937] or the exhaustive (but terse) treatise of TROKHIMCHUK [1964]. For a somewhat more elementary account see pp. 43–50 of NARASIMHAN [1985], MEIER [1951], [1960], and ARSOVE [1955]. For a survey of results and history (together with a very accessible proof of the first theorem above) see DISNEY, GRAY and MORRIS [1975] and GRAY and MORRIS [1978]; in TROKHIMCHUK [1962] the reader will find a longer survey and some proofs. See also TAR [1969], [1971], [1973], [1978], MEIER [1970], PONOMAREV [1970], M. N. BERENŠTEĬN [1971], SINDALOVSKIĬ [1987], the survey DOLZHENKO [1992], and HOWARD [1990], RAO [1990], TELYAKOVSKIĬ [1994] for recent results. For one-to-one functions even weaker conditions suffice to ensure analyticity. See pages 5, 6 of MONTEL [1933b] and the references there, as well as MENCHOFF [1936], [1937] and BRODOVICH [1970], [1974], [1991], [1993], [1997]. Chapter 1 of MARKUSCHEWITSCH [1976] is also devoted to these questions and contains a proof of the first theorem above.

In view of these two theorems it is perhaps surprising that there exists a compact $E \subset \mathbb{C}$ having 2-dimensional Lebesgue measure 0, and a bounded, continuous $f\colon \mathbb{C} \mapsto \mathbb{C}$ which is holomorphic in $\mathbb{C} \setminus E$ but not in $\mathbb{C}$ (nor in any neighborhood of any E-point). Such a function was constructed by URYSOHN [1923]; see also ZALCMAN [1968b].

The uniqueness conclusion 2.10(iv) follows from even (slightly) weaker hypotheses than in the preceding theorem: If f is continuous in U and the limit $\rho(z) := \lim_{h\to 0}\left|[f(z+h)-f(z)]h^{-1}\right|$ exists and is finite for all but countably many $z \in U$ and $\rho(z) = 0$ for a dense set of z, then f is locally constant. See TROKHIMCHUK and SAFONOV [1999].

As noted in the text, (C. R.) is equivalent to $\nabla v = i\nabla u$, from which it follows that $\nabla v \perp \nabla u$. But the orthogonality condition is obviously weaker than the equality. If f and its partial derivatives are continuous, then $\nabla v \perp \nabla u$ together with $|\nabla v| = |\nabla u|$ almost gets us back to (C. R.): B. WALSH [1979] shows that under these conditions either f or $\bar{f}$ is holomorphic. See also R. K. WILLIAMS [1973].

The choice of $\mathbb{C}$-differentiability (2.1) as the *terminus a quo* of the theory to be erected here, though natural, is somewhat ahistorical. That theory did not spring fully-armed from the head of Zeus, but condensed gradually out of the primordial vapors. There are at least three distinct lines of development: the theory of Augustin Cauchy (1789–1857), based on integrals of the type 2.14, has an analytic character; that of Bernhard Riemann (1826–1866), based on the equation (C. R.) and closely connected with potential theory, is geometric (Riemann surfaces); and that of Karl Weierstrass (1815–1897) [and Charles Méray (1835–1911)] based on power series, is purely arithmetic. Each of these has its strengths and weaknesses. For example, the (unadulterated) Weierstrass theory did not deal so well with mapping problems and the beautiful theorem 9.7 eluded it. On the other hand, it was the most logically rigorous and philosophically satisfactory of the three because it started with only the arithmetic of complex numbers (i.e., rational functions) and added just one more ("natural") ingredient—uniform convergence. Moreover, this theory seemed better able to handle functions of several variables. It is not clear where integration (a convergence issue, to be sure) fits into Weierstrass' program philosophically, but he and his more ardent disciples tried to circumvent it, or to substitute for it various (uniform) averaging processes. The issue and its resolution are treated in A. KNESER [1920] and PRINGSHEIM [1920], [1921] (as well as in the latter's book).

As is customary nowadays, the account to be presented here imperiously conscripts the best features from each of these. The strategy consists in exploiting first one (until it encounters a chasm it cannot bridge) and then another of these points of view, and sometimes multiple bridges are constructed. And, as with any anabasis, extensive logistical preparation is necessary. This is the work of the next two chapters.

It is nevertheless of interest to try to formalize the relations between the three points of view, and one way to do so is to prove the equivalence of the following statements about continuous $\mathbb{C}$-valued functions f on open disks U in $\mathbb{C}$:

(0) $D_1 f, D_2 f$ *exist and satisfy* (C. R.) *throughout* U.

(i) f *is represented by a convergent power series in* U.

(ii) f *is* $\mathbb{C}$*-differentiable at each point in* U.

(iii) $\int_\gamma f = 0$ *for every triangle (or every rectangle)* γ *in* U.

(iv) f *is the* $\mathbb{C}$*-derivative of a* $\mathbb{C}$*-differentiable function in* U.

These equivalences *are* proven in due course, but submerged in other details. Proving them together and directly would be an appropriate opening salvo in a more concise exposition than this book aims for. However, readers who are interested in such a coherent picture can have it right now; namely, they can prove (i)⇒(ii)⇒(iii)⇒(iv)⇒(i) by citing, respectively, 3.6; 5.1; the scholium to 2.11; 2.20(iv) (applied to a primitive of f) and 3.8; all of which material is accessible at this point. According to 2.16, (ii)⇒(0), and according to 2.17, (0) *with the added hypothesis that* f *be* $\mathbb{R}$*-differentiable in* U implies (ii). But, as remarked earlier, the full implication (0)⇒(ii) regrettably involves techniques outside the scope of this book.

For short biographies of these architects of the theory, which, besides personal details, survey their scientific contributions, see FREUDENTHAL [1971a,b], K.-R. BIERMANN [1971], [1966], MITTAG-LEFFLER [1923], and BEHNKE [1966]. In FOUËT [1904] the reader will find the three theories compared, with many references to the original works. A deeper comparison of the three schools will be found in the work of NEUENSCHWANDER [1981], and a shorter overview of them in BOTTAZZINI [1994]. HURWITZ [1898] compares the Weierstrass and Riemann approaches, and the master's own view can be gleaned from WEIERSTRASS [1925]. MANNING [1975] gives the historical evolution of Weierstrass' theory and ULLRICH [1989] surveys Weierstrass' famous Berlin lectures on function theory (WEIERSTRASS [1988]), while NEUENSCHWANDER [1996] reconstructs Riemann's important Göttingen lectures. A consistent development of function theory along Weierstrassian lines is given by PRINGSHEIM [1925], [1932] and can be glimpsed in our later multi-part Exercise 3.35. SMITHIES [1997] details the evolution of Cauchy's ideas. His end-of-chapter summaries are particularly valuable for the general reader.

A comprehensive general history of function theory has only recently been written, by BOTTAZZINI & GRAY [2013] and J. J. GRAY [2015], but a short one is MARKUSCHEWITSCH [1955], [1981]; and BOTTAZZINI [1986], [1987], LAUGWITZ [1999], JAHNKE, *et al.* [1999] and ROY [2011] cover function theory as part of a larger program. The 170-page Introduction that Bottazzini composed for the 1992 edition of CAUCHY [1821] is also a useful historical resource and *explication de texte*. The period 1700–1900 is briefly described in VERLEY [1978]. A good historical survey for the general reader is chapter 27 of M. KLINE [1974]. The encyclopedia articles of BIEBERBACH [1921], LICHTENSTEIN [1919], OSGOOD [1901], and PRINGSHEIM and MOLK [1909] are rich sources of historical information, as is the long *DMV* report of BRILL and NOETHER [1892–93], the monograph of TIMTSCHENKO [1892–1899], and GRATTAN-GUINNESS [1970], which also inform on the pre-history of our subject (as does FERRARO [2008]). R. NEVANLINNA [1966], GAIER [1990] and HAYMAN [1994] describe post-Weierstrassian developments. See also the congress lectures of HURWITZ [1898], JULIA [1932], and BOTTAZZINI and GRAY [1996]. The extensive historical notes interlacing the two volumes REMMERT [1991], [1998] make valuable and enjoyable reading.

In FERRAR [1930] is the following elementary extension of 2.12: If $\{F_n^{(N)}\}_{n\in\mathbb{N}}$ converges locally uniformly in Ω and $\{F_n(z)\}_{n\in\mathbb{N}}$ converges in $\mathbb{C}$ for N different points $z \in \Omega$, then $\{F_n\}$ converges locally uniformly in Ω. Theorem 2.12 and its proof remain valid for $\mathbb{R}$-valued functions on open intervals of $\mathbb{R}$.

On the relation between (i), (ii) and (iii) in 2.19 see SEELEY [1961] and FISHER and SHILLETO [1986]. LEWIN [1987] proves (i) and (ii) elementarily but under notably weaker hypotheses. Schwarz' work is on pp. 275–284, vol. II of his *Gesammelte Mathematische Abhandlungen*; there he also offers an alternative proof that does not involve the integral calculus, as well as a counter-example to (iii) when the continuity hypothesis is dropped. (A simpler example than his is $f(0,0) := 0$, $f(x,y) := xy^3/(x^2+y^2)$ for $(x,y) \neq (0,0)$.). Actually, in (iii) Schwarz hypothesized continuity of $D_1\Phi$ and $D_2\Phi$ as well, but the proof does not seem to need this. Exercises 27 and 28, p. 242 of RUDIN [1976] also illustrate the failure of (iii) and (i) when the continuity hypotheses are weakened. For a local version of (iii) see R. NEVANLINNA [1952] or (better) 9.41 of RUDIN [1976]. The interesting history of this important result is set out in KOETSIER [1990].

A version of 2.19(v)* is already found in Cauchy's first memoir (1814); see pp. 28, 29 of SMITHIES [1997] and ETTLINGER [1921–22]. He considered, *inter alia*, $h(x,y) := (x, xy)$, which maps the square R onto a triangle.

The development 2.22 may have its origins in MÉRAY [1890]. There a multiple-valued logarithm is defined throughout $\mathbb{C} \setminus \{0\}$ by integrals, and the exponential function is accessed via a holomorphic inverse-function theorem (cf. 5.93), so the whole apparatus of the Cauchy theory is needed, in contrast to the "bare-hands" treatment here. The second formula in (ix) is from BAKER and SLUIS [1997], while the series in (12) goes back to Isaac Newton, 1664 and Nicolas Mercator, 1668.

As intimated in the text, Definition 2.2 is a much too parochial view of harmonic functions; for one thing, it only covers dimension 2. Laplace's equation (see the remark following 9.29) is meaningful in $\mathbb{R}^n$ for $n > 2$ and its solutions are called harmonic functions. Their theory is as rich and varied as that of holomorphic functions in $\mathbb{C}$, and need not be subordinated to the latter, as will be done in this book. A very readable modern treatment with an enviable light touch is that of AXLER, BOURDON, and RAMEY [2001]. On the other hand, a theory of harmonic functions in $\mathbb{R}^2$ which is more systematic and complete than that in the present book, but which also exploits the close connection with holomorphy, will be found in RANSFORD [1995].

Chapter III

Power Series and the Exponential Function

§ 1 Introduction

As with integrals, power series (though fascinating and from Weierstrass' point of view the *essence* of our subject) are a tool here and are not pursued extensively. For more on series the reader should consult BROMWICH [1926], KNOPP [1951] or DÖRRIE [1951]. To make our definitions and get started only one simple result is needed:

Theorem 3.1 $\lim_{n\to\infty} n^{1/n} = 1$ *and for any* $a > 0$, $\lim_{n\to\infty} a^{1/n} = 1$.

Proof: Set $x_n := n^{1/n} - 1$. Since this is non-negative, for $n \geq 2$

$$n = (1 + x_n)^n = \sum_{j=0}^{n} \binom{n}{j} x_n^j \geq \binom{n}{2} x_n^2 = \frac{n(n-1)}{2} x_n^2.$$

Therefore

$$0 \leq x_n \leq \sqrt{\frac{2}{n-1}}$$

and so $x_n \to 0$. If $a \geq 1$, then $1 \leq a^{1/n} \leq n^{1/n}$ for all $n \geq a$, so $a^{1/n} \to 1$. If $0 < a < 1$, apply this conclusion to $1/a$. (PRACHAR [1975] is an alternative approach.)

Remarks: We are, of course, assuming known the theory of rational roots and powers on $[0, \infty)$. This is an easy consequence of the completeness of $\mathbb{R}$. (See §6 of chapter 0.) Irrational powers on $[0, \infty)$ can be dealt with at the same level of elementariness (see 1.47) but will not be needed. Everything is finally subsumed under Theorem 3.14.

R. B. Burckel, *Classical Analysis in the Complex Plane*, Cornerstones,
https://doi.org/10.1007/978-1-0716-1965-0_4

§ 2 Power Series

Definition 3.2 An infinite series of the form $\sum_{n=0}^{\infty} c_n(z-z_0)^n$, where $c_n, z, z_0 \in \mathbb{C}$, is called a *power series*, more precisely a *power series about* z_0, or a *power series centered* at z_0. (No convergence claim is being made.)

Theorem 3.3 *Let $c_n, z_0 \in \mathbb{C}$ and consider the power series*

$$\sum_{n=0}^{\infty} c_n(z-z_0)^n. \tag{1}$$

Then there exists an $R \in [0, \infty]$ such that:

(i) *The series (1) converges absolutely for all $z \in \mathbb{C}$ with $|z - z_0| < R$ and uniformly in $D(z_0, \rho)$ for any $\rho < R$.*

(ii) *The terms of the series (1) are unbounded (and hence the series diverges) for every $z \in \mathbb{C}$ with $|z - z_0| > R$.*

(iii) *The extended real number R is the reciprocal of the extended real number $\overline{\lim}_{n\to\infty} \sqrt[n]{|c_n|}$ and is called the* radius of convergence *of the series (1).*

(iv) (Ratio Test) *If the c_n are eventually all non-zero, then R is not less than $\underline{\lim}_{n\to\infty} |c_n/c_{n+1}|$ and coincides with it if the limit exists.*

Proof: Define R to be the least upper bound in $[0, \infty]$ of the $r \in [0, \infty)$ for which $\{|c_n|r^n\}$ is a bounded sequence. If $|z - z_0| > R$, then the sequence $\{c_n(z - z_0)^n\}$ is unbounded, while if $\rho < R$ and we pick r satisfying $\rho < r < R$, then by definition of R we shall have that $\{|c_n|r^n\}$ is a bounded sequence, say $|c_n|r^n \le M < \infty$ for all n. Then for all z in $D(z_0, \rho)$

$$|c_n(z-z_0)^n| = |c_n|r^n \cdot \left[\frac{|z-z_0|}{r}\right]^n \le M \cdot \left(\frac{\rho}{r}\right)^n$$

and since $\rho/r < 1$, the assertion (i) follows from convergence of geometric series.

Now if $0 < r < R$, then $\{|c_n|r^n\}$ is bounded, say by $M(r) \in (0, \infty)$:

$$|c_n|r^n \le M(r)$$
$$|c_n|^{1/n} \le \frac{1}{r}[M(r)]^{1/n} \qquad \forall n \in \mathbb{N}.$$

Let $n \to \infty$ and cite 3.1 to conclude that $\overline{\lim}_{n\to\infty} |c_n|^{1/n} \le 1/r$ and since r is arbitrary in $(0, R)$,

$$\overline{\lim_{n\to\infty}} |c_n|^{1/n} \le 1/R. \tag{2}$$

Equality necessarily holds in (2) if $R = \infty$. If $R < \infty$ and $R < r < \infty$, then by definition of R the sequence $\{|c_n| r^n\}$ is not bounded and so in particular infinitely many of its terms exceed 1 and therefore $|c_n|^{1/n} r > 1$ for infinitely many n. It follows that $\overline{\lim}_{n\to\infty} |c_n|^{1/n} \geq 1/r$. Again as $r \in (R, \infty)$ is arbitrary, it follows that

$$\overline{\lim_{n\to\infty}} |c_n|^{1/n} \geq 1/R, \tag{3}$$

which with (2) establishes the claim (iii). A glance ahead to 3.9 confirms (iv).

Example 3.4 The radius of convergence of the power series $\sum_{n=0}^{\infty} (1/n!) z^n$ is infinite. For if $n \geq N+1 \geq 2r$, then $n! = n(n-1)\ldots(N+1)N! \geq (N+1)^{n-N} N!$ and consequently

$$\sum_{n=N+1}^{\infty} \frac{r^n}{n!} \leq \sum_{n=N+1}^{\infty} \frac{r^n}{(N+1)^{n-N} N!}$$
$$= \frac{r^N}{N!} \sum_{k=1}^{\infty} \left(\frac{r}{N+1}\right)^k \leq \frac{r^N}{N!} \sum_{k=1}^{\infty} \left(\frac{1}{2}\right)^k = \frac{r^N}{N!}.$$

Of course, this proves that the original series converges absolutely for every $z \in \mathbb{C}$, but it also provides an upper bound for the remainder:

$$\left| \sum_{n=N+1}^{\infty} \frac{z^n}{n!} \right| \leq \frac{|z|^N}{N!} \qquad \forall z \in \mathbb{C}, N \geq 2|z|.$$

Lemma 3.5 *If $c_n, z_0 \in \mathbb{C}$, then the two series $\sum_{n=0}^{\infty} c_n (z-z_0)^n$ and*

$$\sum_{n=1}^{\infty} n c_n (z-z_0)^{n-1} = \sum_{n=0}^{\infty} (n+1) c_{n+1} (z-z_0)^n$$

have the same radius of convergence.

Proof: Since $\sum_{n=0}^{\infty} n c_n (z-z_0)^n = (z-z_0) \cdot \sum_{n=1}^{\infty} n c_n (z-z_0)^{n-1}$, it suffices to show that the two series $\sum_{n=0}^{\infty} n c_n (z-z_0)^n$ and $\sum_{n=0}^{\infty} c_n (z-z_0)^n$ have the same radius of convergence. And by part (iii) of the preceding theorem, for this it suffices to show that $\overline{\lim}_{n\to\infty} |n c_n|^{1/n} = \overline{\lim}_{n\to\infty} |c_n|^{1/n}$. This in turn follows from 3.1: Since $n^{1/n} \to 1$, the subsequential limits of $\{|n c_n|^{1/n}\}_{n\in\mathbb{N}}$ and those of $\{|c_n|^{1/n}\}_{n\in\mathbb{N}}$ coincide.

Theorem 3.6 *Let c_n, $z_0 \in \mathbb{C}$ and let the radius of convergence R of the series $\sum_{n=0}^{\infty} c_n(z-z_0)^n$ be positive. Then the function f defined in the open disk $D(z_0, R)$ by*

$$(1) \qquad f(z) := \sum_{n=0}^{\infty} c_n(z-z_0)^n$$

is infinitely differentiable and for every positive integer k its kth derivative $f^{(k)}$ satisfies

$$(2.k) \quad f^{(k)}(z) = \sum_{n=k}^{\infty} n(n-1)\cdots(n-k+1)c_n(z-z_0)^{n-k} \qquad \forall z \in D(z_0, R).$$

In particular, $c_k = f^{(k)}(0)/k!$ for every k, so a power series with positive radius of convergence (and therefore, e.g., every polynomial function) determines its coefficients uniquely.

Proof: It suffices to prove that f is differentiable in $D(z_0, R)$ and satisfies equation (2.1). For then we may, via 3.5, induce on k. Also, by considering the function $F(z) := f(z+z_0)$ in $D(0, R)$, we may assume without loss of generality that $z_0 = 0$. Let g denote the function defined (by 3.5) in the disk $D(0, R)$ by the right-hand side of the equation (2.1). Fix $z \in D(0, R)$ and choose $0 < r < R - |z|$. Then for $0 < |w| < r$

$$(3) \qquad \frac{f(z+w)-f(z)}{w} - g(z) = \sum_{n=2}^{\infty} c_n \left[\frac{(z+w)^n - z^n}{w} - nz^{n-1} \right].$$

According to the binomial theorem the square-bracketed term is

$$\frac{1}{w} \sum_{k=1}^{n} \binom{n}{k} w^k z^{n-k} - nz^{n-1} = \sum_{k=2}^{n} \binom{n}{k} w^{k-1} z^{n-k}$$

and therefore it is majorized by

$$\begin{aligned} \sum_{k=2}^{n} \binom{n}{k} |w|^{k-1} |z|^{n-k} &\leq \sum_{k=2}^{n} \binom{n}{k} |w| r^{k-2} |z|^{n-k} = |w| r^{-2} \sum_{k=2}^{n} \binom{n}{k} r^k |z|^{n-k} \\ &\leq |w| r^{-2} \big(|z| + r\big)^n. \end{aligned}$$

Consequently, (3) gives

$$(4) \qquad \left| \frac{f(z+w)-f(z)}{w} - g(z) \right| \leq \frac{|w|}{r^2} \sum_{n=2}^{\infty} |c_n| \big(|z| + r\big)^n \qquad \forall 0 < |w| < r.$$

The sum here is finite, because $|z| + r < R$, by definition of r, and the series converges absolutely in $D(0, R)$. Letting $w \to 0$ in (4) therefore proves that $f'(z)$ exists and equals $g(z)$, as claimed.

Remark: It might be noted that 3.6 is an immediate consequence of 2.12, applied to the polynomials $f_n(z) := \sum_{k=0}^{n} c_k(z - z_0)^k$, $n \in \mathbb{N}$. Because of 3.5 the sequence $\{f_n'\}$ converges locally uniformly in $D(z_0, R)$ to the function $g(z) := \sum_{k=1}^{\infty} kc_k(z-z_0)^{k-1}$. Thanks to 2.12, the limit f to which the sequence $\{f_n\}$ converges is differentiable and $f' = g$. On the other hand, the maneuvers employed in the proof of 3.6 yield a very important rearrangement theorem for power series, of which 3.6 itself is a trivial corollary. See 3.12(iii) and the remark following 3.12(vi).

Exercise 3.7 *If the series* $f(z) := \sum_{n=0}^{\infty} c_n z^n$ *has radius of convergence* $R > 0$ *and* $f'(0) = c_1 \neq 0$, *then for some* $0 < r \leq R$, f *is one-to-one in* $D(0, r)$.

Hints: If $0 < r < R$ and $z, w \in D(0, r)$ then, as in the proof of 3.6,

$$f(z) - f(w) = c_1(z - w) + (z - w) \sum_{n=2}^{\infty} c_n \sum_{k=1}^{n} w^{k-1} z^{n-k},$$

$$|f(z) - f(w)| \geq |c_1||z - w| - |z - w| \sum_{n=2}^{\infty} |c_n| \sum_{k=1}^{n} |w|^{k-1} |z|^{n-k}$$

$$\geq \left[|c_1| - \sum_{n=2}^{\infty} |c_n| \sum_{k=1}^{n} r^{n-1} \right] |z - w|$$

$$= \left[|c_1| - \sum_{n=2}^{\infty} n|c_n| r^{n-1} \right] |z - w|.$$

By 3.5 the series $\sum_{n=1}^{\infty} n|c_n|z^{n-1}$ has radius of convergence R, so for sufficiently small $0 < r < R$, $\sum_{n=2}^{\infty} n|c_n|r^{n-1} < \frac{1}{2}|c_1|$ and it follows that

$$|f(z) - f(w)| \geq \tfrac{1}{2}|c_1||z - w|.$$

[A very short proof using only the existence and continuity of f', but not the power series, appears near the end of the 7.29(i) hints. The full story here (to be made precise and proven later) is that the restriction of f to some neighborhood of 0 has an inverse given by a power series centered at $c_0 = f(0)$.]

Exercise 3.8 *In the notation of* 2.14 *show that each* G_n *is of power series type near each point of* $\mathbb{C} \setminus \gamma$, *i.e., for each* $a \in \mathbb{C} \setminus \gamma$ *there is a power series about* a *which agrees with* G_n *in some neighborhood of* a. *Deduce, along the way, a new proof of* 2.14.

Hints: (after Cauchy) From $1/(1-w) = \sum_{k=0}^{\infty} w^k$ for all $w \in \mathbb{D}$ and 3.6 follows by repeated differentiation that

$$\text{(1)} \qquad \frac{1}{(1-w)^{n+1}} = \sum_{k=n}^{\infty} \binom{k}{n} w^{k-n} \qquad \forall w \in \mathbb{D}.$$

Given $a \in \mathbb{C} \setminus \gamma$, choose $R > 0$ so that $D(a, R) \subset \mathbb{C} \setminus \gamma$. Then for any $0 < r < R$, $z \in D(a, r)$ and $\xi \in \gamma$ we have $|(z-a)/(\xi-a)| < r/R < 1$, and so

$$\begin{aligned}\frac{1}{(\xi - z)^{n+1}} &= \frac{1}{(\xi-a)^{n+1}} \frac{1}{\left(1 - \dfrac{z-a}{\xi-a}\right)^{n+1}} \\ &\overset{(1)}{=} \frac{1}{(\xi-a)^{n+1}} \sum_{k=n}^{\infty} \binom{k}{n} \left(\frac{z-a}{\xi-a}\right)^{k-n},\end{aligned}$$

with convergence uniform in such z and ξ. Consequently, for $z \in D(a, r)$

$$\begin{aligned} nG_{n+1}(z) &= \int_\gamma \frac{ng(\xi)}{(\xi-z)^{n+1}}\, d\xi = \sum_{k=n}^{\infty} \int_\gamma \frac{g(\xi)}{(\xi-a)^{n+1}} n \binom{k}{n} \left(\frac{z-a}{\xi-a}\right)^{k-n} d\xi \\ \text{(2.}n\text{)} \qquad &= \sum_{k=n}^{\infty} \left(\int_\gamma \frac{g(\xi) d\xi}{(\xi-a)^{k+1}} \right) n \binom{k}{n} (z-a)^{k-n} \\ &= \frac{d}{dz} \left[\sum_{k=n-1}^{\infty} \left(\int_\gamma \frac{g(\xi) d\xi}{(\xi-a)^{k+1}} \right) \frac{n}{k-n+1} \binom{k}{n} (z-a)^{k-n+1} \right], \\ &\quad \text{by 3.6,} \\ &= \frac{d}{dz}[G_n(z)], \end{aligned}$$

by equation $(2.n-1)$ and the fact that

$$\frac{n}{k-n+1} \binom{k}{n} = \binom{k}{n-1}.$$

Exercise 3.9 *Show that if a_n is positive for all (sufficiently large) n, then*

$$\varliminf_{n\to\infty} \frac{a_{n+1}}{a_n} \le \varliminf_{n\to\infty} \sqrt[n]{a_n} \le \varlimsup_{n\to\infty} \sqrt[n]{a_n} \le \varlimsup_{n\to\infty} \frac{a_{n+1}}{a_n}.$$

Hints: (Cf. CAUCHY [1821], p. 54.) Call the last limit superior $a \in [0, \infty]$. It suffices to prove the last inequality. For if $a = 0$, this finishes our work, while if $a > 0$, this result can be applied to the sequence $\{1/a_n\}$.

In proving the last inequality we may assume $a < \infty$. For an $\varepsilon > 0$ there is an $N = N(\varepsilon)$ such that

$$\frac{a_{n+1}}{a_n} < a + \varepsilon \qquad \forall n \geq N.$$

Multiply these inequalities for $n = N, \ldots, m-1$ to obtain

$$\frac{a_m}{a_N} < (a+\varepsilon)^{m-N} \qquad \forall m > N,$$

$$\sqrt[m]{a_m} < (a+\varepsilon)\left[\frac{a_N}{(a+\varepsilon)^N}\right]^{1/m} \qquad \forall m > N.$$

It follows from this and 3.1 that $\overline{\lim}_{m\to\infty} \sqrt[m]{a_m} \leq a + \varepsilon$.

Exercise 3.10 *Show that for each positive integer k*

$$\lim_{n\to\infty} \sqrt[n]{\binom{kn}{n}} = \frac{k^k}{(k-1)^{k-1}}.$$

Hint: The corresponding limit for ratios is routine, so use the preceding exercise.

Exercise 3.11 (Euler Summability) *If $a, a_0, a_1, \ldots \in \mathbb{C}$ and $\lim_{n\to\infty} a_n = a$, then*

$$\lim_{n\to\infty} 2^{-n} \sum_{k=0}^{n} \binom{n}{k} a_k = a.$$

Hints: Since

(1) $$\sum_{k=0}^{n} \binom{n}{k} = (1+1)^n = 2^n,$$

we have

(2) $$a - 2^{-n} \sum_{k=0}^{n} \binom{n}{k} a_k = 2^{-n} \sum_{k=0}^{n} \binom{n}{k} (a - a_k).$$

Now divide and conquer: given $\varepsilon > 0$, choose $k_0 = k_0(\varepsilon)$ so that

(3) $$|a - a_k| < \varepsilon/2 \qquad \forall k \geq k_0;$$

then use the fact that $n^{k_0}/2^n = [(n^{1/n})^{k_0}/2]^n \overset{(3.1)}{\longrightarrow} 0$ to pick $n_0 = n_0(\varepsilon) \geq k_0$ so that

$$\frac{n^{k_0}}{2^n} < \frac{\varepsilon}{2Mk_0} \qquad \forall n \geq n_0$$

for $M := 1 + \sup\{|a - a_k| : k = 0, 1, 2, \ldots\}$. Then for $k = 0, 1, \ldots, k_0$

$$2^{-n}\binom{n}{k} = \frac{n(n-1)\cdots(n-k+1)}{2^n k!} \leq \frac{n^k}{2^n} \leq \frac{n^{k_0}}{2^n} < \frac{\varepsilon}{2Mk_0} \qquad \forall n \geq n_0,$$

and so

$$(4) \quad 2^{-n}\sum_{k=0}^{k_0-1}\binom{n}{k}|a - a_k| \leq M\sum_{k=0}^{k_0-1} 2^{-n}\binom{n}{k} < M\sum_{k=0}^{k_0-1}\frac{\varepsilon}{2Mk_0} = \frac{\varepsilon}{2} \qquad \forall n \geq n_0.$$

From (2), (3) and (4)

$$\begin{aligned}\left|a - 2^{-n}\sum_{k=0}^{n}\binom{n}{k}a_k\right| &\leq 2^{-n}\sum_{k=0}^{k_0-1}\binom{n}{k}|a - a_k| + 2^{-n}\sum_{k=k_0}^{n}\binom{n}{k}\frac{\varepsilon}{2} \\ &< \frac{\varepsilon}{2} + 2^{-n}\sum_{k=0}^{n}\binom{n}{k}\frac{\varepsilon}{2} \overset{(1)}{\leq} \frac{\varepsilon}{2} + \frac{\varepsilon}{2} = \varepsilon \qquad \forall n \geq n_0.\end{aligned}$$

Exercise 3.12 (Cauchy [1821], p. 541). *Let $c_{n,m} \in \mathbb{C}$ be such that $\sum_{n=1}^{\infty}[\sum_{m=1}^{\infty}|c_{n,m}|] < \infty$. Show that then*

$$\text{(i)} \quad \sum_{m=1}^{\infty}\left[\sum_{n=1}^{\infty}|c_{n,m}|\right] = \sum_{n=1}^{\infty}\left[\sum_{m=1}^{\infty}|c_{n,m}|\right] = \sup_{M,N}\left\{\sum_{m=1}^{M}\left[\sum_{n=1}^{N}|c_{n,m}|\right]\right\}.$$

Consequently,

(ii) *for each n the series $\sum_{m=1}^{\infty} c_{n,m}$ converges (say to s_n), for each m the series $\sum_{n=1}^{\infty} c_{n,m}$ converges (say to t_m) and the series $\sum_{n=1}^{\infty} s_n$, $\sum_{m=1}^{\infty} t_m$ are (absolutely) convergent and have the same limit.*

Hints: Let L denote the (finite) number in (i), S_n, T_m the (finite) numbers $\sum_m |c_{n,m}|$, $\sum_n |c_{n,m}|$, and set

$$\delta_{N,M} := L - \sum_{n=1}^{N}\left[\sum_{m=1}^{M}|c_{n,m}|\right].$$

Show that $\sum_{n=N+1}^{N'} |S_n| \le \delta_{N,M}$ and $\sum_{m=M+1}^{M'} |T_m| \le \delta_{N,M}$ for all N and M and all $N' > N$, $M' > M$. In the inequality

$$\left|\sum_{n=1}^{N}\sum_{m=1}^{M'} c_{n,m} - \sum_{m=1}^{M}\sum_{n=1}^{N'} c_{n,m}\right| \le \sum_{m>M}^{\infty}\sum_{n=1}^{\infty} |c_{n,m}| + \sum_{n>N}^{\infty}\sum_{m=1}^{\infty} |c_{n,m}|$$
$$= L - \sum_{m=1}^{M}\sum_{n=1}^{\infty} |c_{n,m}| + L - \sum_{n=1}^{N}\sum_{m=1}^{\infty} |c_{n,m}| \le 2\delta_{N,M}$$

valid whenever $N' > N$ and $M' > M$, first let $M' \to \infty$ and then let $N' \to \infty$ to see that $\left|\sum_{n=1}^{N} s_n - \sum_{m=1}^{M} t_m\right| \le 2\delta_{N,M}$.

(iii) *The power series $f(z) := \sum_{n=0}^{\infty} b_n z^n$ has radius of convergence $R > 0$, $a \in D(0,R)$ and $r := R - |a| > 0$. Show that for each integer $m \ge 0$ the series $\sum_{n=m}^{\infty} b_n \binom{n}{m} a^{n-m}$ converges to a complex number a_m and that the series $\sum_{m=0}^{\infty} a_m (z-a)^m$ converges to $f(z)$ for all $z \in D(a,r)$.*

Hints: Fix $w \in D(0,r)$, define

$$c_{n,m} := \begin{cases} b_n \binom{n}{m} a^{n-m} w^m & \text{if } 0 \le m \le n \\ 0 & \text{if } 0 \le n < m \end{cases}$$

and note that, by the binomial expansion

$$\sum_{n=0}^{\infty}\sum_{m=0}^{\infty} |c_{n,m}| = \sum_{n=0}^{\infty}\sum_{m=0}^{n} |b_n| \binom{n}{m} |a|^{n-m} |w|^m = \sum_{n=0}^{\infty} |b_n| \big(|w| + |a|\big)^n$$

which is finite since $|w| + |a| < r + |a| = R$. Now cite (i).

(iv) (Cauchy product) *A_n, $B_n \in \mathbb{C}$, $C_n := \sum_{m=0}^{n} A_m B_{n-m}$ for integers $n \ge 0$. Show that if $\sum |A_n| + \sum |B_n| < \infty$, then $\sum |C_n| < \infty$ and $\left(\sum A_n\right)\left(\sum B_n\right) = \sum C_n$.*

Hint: Apply (ii) with $c_{n,m} := A_m B_{n-m}$ if $n \ge m$, $c_{n,m} := 0$ if $n < m$.

(v) *Show by example that convergence of $\sum C_n$ in* (iv) *can fail if $\sum A_n$, $\sum B_n$ converge but neither absolutely.* (See, however, 3.13(vii).)

Hint: (Cauchy [1821], p. 134). For $A_0 = B_0 := 0$, $A_n = B_n := (-1)^{n-1}/\sqrt{n}$ ($n \in \mathbb{N}$) we have $|C_n| = \sum_{m=1}^{n-1} \frac{1}{\sqrt{m}\sqrt{n-m}} \ge \sum_{m=1}^{n-1} \frac{1}{\sqrt{n-1}\sqrt{n-1}} = 1$ for all $n \ge 2$. Cauchyness, hence convergence of $\sum A_n$ follows from

$$\left|\sum_{n=2M-1}^{2N} A_n\right| = \sum_{k=M}^{N} \left(\frac{1}{\sqrt{2k-1}} - \frac{1}{\sqrt{2k}}\right)$$
$$< \sum_{n=2M}^{2N} \left(\frac{1}{\sqrt{n-1}} - \frac{1}{\sqrt{n}}\right) < \frac{1}{\sqrt{2M-1}} \qquad \forall\, M < N \in \mathbb{N}.$$

(vi) (Newton's square-root power series). *Using* (iv) *and the customary definitions*

$$\binom{x}{0} := 1, \ \binom{x}{n} := \prod_{k=1}^{n} \frac{x-k+1}{k}, \qquad n \in \mathbb{N}, x \in \mathbb{C},$$

show that the series $q(z) := \sum_{n=0}^{\infty} \binom{1/2}{n}(z-1)^n$ *converges absolutely for* $z \in \overline{D}(1,1)$ *to a square-root of* z *which also lies in* $\overline{D}(1,1)$. Cf. 3.35(xiv).

Hints: Evidently

(1) $$\left|\binom{1/2}{n}\right| = (-1)^{n-1}\binom{1/2}{n}$$

and this number does not exceed 1, for every $n \in \mathbb{N}$. So the series for $q(z)$ converges for all $z \in D(1,1)$. Notice that $\binom{2x}{n}$ and $\sum_{k=0}^{n}\binom{x}{k}\binom{x}{n-k}$ are each polynomials in x of degree n, and if x is an integer greater than n, then they are the coefficient of z^n in $(1+z)^{2x}$ and $[(1+z)^x]^2$, respectively. The difference polynomial having thus infinitely many zeros is therefore identically 0 (cf. 4.52(ii)). The equality of these two polynomials at $x = 1/2$ and (iv) imply that

(2) $$q^2(1+z) = \sum_{n=0}^{\infty}\binom{1}{n} z^n = 1+z \qquad \forall z \in \mathbb{D}.$$

For $0 < t < 2$, $q(t)$ is real and $q^2(t) = t$ by (2). Hence by 1.7

(3) either $q(t) = -\sqrt{t} \quad \forall t \in (0,2)$ or $q(t) = \sqrt{t} \quad \forall t \in (0,2)$.

Using (1) we have

(4) $$q(z) = 1 - \sum_{n=1}^{\infty}\left|\binom{1/2}{n}\right|(1-z)^n \qquad \forall z \in D(1,1).$$

When $z = t \in (0,1]$ this is clearly an increasing function, a fact which decides for the second alternative in (3), whereupon (4) yields

$$\sum_{n=1}^{N}\left|\binom{1/2}{n}\right|(1-t)^n \le \sum_{n=1}^{\infty}\left|\binom{1/2}{n}\right|(1-t)^n = 1 - q(t) = 1 - \sqrt{t} < 1$$

for all $N \in \mathbb{N}$, $t \in (0,1]$. Letting first $t \downarrow 0$, then $N \uparrow \infty$ on the left-hand side gives

$$\sum_{n=1}^{\infty}\left|\binom{1/2}{n}\right| \le 1.$$

Then (4) shows that

$$|q(z) - 1| \le \sum_{n=1}^{\infty} \left|\binom{1/2}{n}\right| |1 - z|^n < \sum_{n=1}^{\infty} \left|\binom{1/2}{n}\right| \le 1 \qquad \forall z \in D(1,1)$$

and insures that the series converges absolutely for all $z \in \bar{D}(1,1)$, and the limit lies in $\bar{D}(1,1)$.

Remark: Note that when (iii) is written as

$$[f(z) - f(a)]/[z - a] - a_1 = (z - a) \sum_{m=2}^{\infty} a_m (z - a)^{m-2},$$

it makes clear that $f'(a)$ exists and equals $a_1 := \sum_{n=1}^{\infty} n b_n a^{n-1}$, thereby furnishing a short, new proof of 3.6. The mechanism of proof here makes clear why this result is sometimes called the *rearrangement theorem* (a designation, alas, also used for a famous theorem of Riemann about permuting the terms of a conditionally convergent numerical series).

Exercise 3.13 (i) *Let X be a non-void set, $a_k : X \mapsto \mathbb{C}$, $b_k : X \mapsto [0, \infty)$ with $b_k \downarrow \mathbf{0}$ uniformly. Set $A_n := \sum_{k=0}^{n} a_k$ and suppose that $\{A_n\}$ is uniformly bounded on X. Show that $\sum_{k=0}^{\infty} b_k a_k$ is uniformly convergent on X.*

Hints: Since $a_k = A_k - A_{k-1}$, a little juggling with indices of summation leads to (*Abel's partial summation formula*):

$$\sum_{k=p+1}^{q} b_k a_k = \sum_{k=p+1}^{q} A_k (b_k - b_{k+1}) + A_q b_{q+1} - A_p b_{p+1}, \qquad 0 \le p < q.$$

If M is a bound for $\{A_n\}$ on X, we then get

$$\begin{aligned} \left| \sum_{k=p+1}^{q} b_k a_k \right| &\le M \sum_{k=p+1}^{q} (b_k - b_{k+1}) + M b_{q+1} + M b_{p+1} \\ &= M(b_{p+1} - b_{q+1}) + M b_{q+1} + M b_{p+1} = 2M b_{p+1}, \end{aligned}$$

showing that the sequence $\{\sum_{k=0}^{n} b_k a_k\}$ is uniformly Cauchy on X.

(ii) *If $b_k \in \mathbb{R}$ and $b_k \downarrow 0$, then the series $\sum_{k=0}^{\infty} b_k z^k$ converges throughout $\bar{\mathbb{D}} \setminus \{1\}$ and uniformly on each compact subset.*

Hints: For any $0 < r < 1$ and all $n \ge 0$ we have

$$\left| \sum_{k=0}^{n} z^k \right| = \left| \frac{z^{n+1} - 1}{z - 1} \right| \le \frac{2}{|z - 1|} \le \frac{2}{r} \qquad \forall z \in \bar{\mathbb{D}} \setminus D(1, r);$$

now cite (i).

(iii) *The hints to* (i) *deliver the same conclusion under the alternative hypothesis that* $\{A_n\}$ *converges uniformly on* X*, the non-negative sequence* b_k *is decreasing, and* b_1 *is bounded.*

Hint: By changing a_0, arrange that $A_n \to \mathbf{0}$.

(iv) (Abel's Convergence Theorem) *If* $\sum_{k=0}^{\infty} c_k z^k$ *converges uniformly for* $z \in B \subset \mathbb{T}$*, then it converges uniformly for* $z \in B^* := \{rb : b \in B, 0 \le r \le 1\}$*, the radial extension of* B*.*

Hint: Apply (iii) with $X := [0,1] \times B$, $a_k(r,b) := c_k b^k$, and $b_k(r,b) := r^k$.

(v) (Stolz's Convergence Theorem) *For some* $b \in \mathbb{T}$ *the series* $\sum_{k=0}^{\infty} c_k z^k$ *converges at* $z = b$*. Show that for all* $a, c \in \mathbb{D}$ *the series converges uniformly in* $\operatorname{co}\{a,b,c\}$ *and infer that it therefore defines a function in* $\mathbb{D} \cup \{b\}$ *which is continuous in each such convex set.*

Hints: Note that the series converges throughout $\mathbb{D}$ by 3.3(i), and is continuous in $\mathbb{D}$. Now it suffices to consider $b = 1$ and $c = \bar{a} \in \mathbb{D} \setminus \mathbb{R}$. (Why?) The inequality

$$(0) \qquad (\operatorname{Im} z)(1 - |\operatorname{Re} a|) \le |\operatorname{Im} a|(1 - \operatorname{Re} z)$$

trivially holds for $z = 1$, a and $\bar{a}$. Multiplying these inequalities by $\lambda_j \in [0,1]$ satisfying $\lambda_1 + \lambda_2 + \lambda_3 = 1$, and adding the results, confirms (0) for all $z \in \operatorname{co}\{1, a, \bar{a}\} = \Delta$. Thus, for $T := |\operatorname{Im} a|/(1 - |\operatorname{Re} a|)$ we have

$$|\operatorname{Im} z| \le (1 - \operatorname{Re} z)T \qquad \forall z \in \Delta.$$

Writing $x := \operatorname{Re} z$, we have for $1 \ne z \in \Delta$

$$\frac{1 - |z|}{|1 - z|} \ge \frac{1 - \sqrt{x^2 + (1-x)^2 T^2}}{(1-x)\sqrt{1 + T^2}} =: \frac{1}{\sqrt{1+T^2}} \cdot \frac{\phi(1) - \phi(x)}{1 - x}.$$

Since $\lim_{x \uparrow 1} \frac{\phi(x) - \phi(1)}{x - 1} = \phi'(1) = 1$, this quotient is bounded away from 0 for $1 \ne z \in \Delta$. That is,

$$(1) \qquad \sup_{1 \ne z \in \Delta} \frac{|1 - z|}{1 - |z|} =: M < \infty.$$

For non-negative integers n, k write $R_{n,k} := \sum_{j=n}^{n+k} c_j$. Thus, given $\varepsilon > 0$, there exists $n_\varepsilon \in \mathbb{N}$ such that

$$(2) \qquad |R_{n,k}| < \varepsilon \qquad \forall k \ge 0, n \ge n_\varepsilon.$$

For such an n, manipulating indices (as in the hints to (i)) confirms that

$$\begin{aligned}\sum_{j=0}^{k} c_{n+j}z^j &= \sum_{j=0}^{k-1} R_{n,j}(z^j - z^{j+1}) + R_{n,k}z^k \\ &= (1-z)\left[\sum_{j=0}^{k-1} R_{n,j}z^j + R_{n,k}\frac{z^k}{1-z}\right], k \in \mathbb{N}, 1 \neq z \in \mathbb{C}.\end{aligned}$$

Consequently, by (2) for all $k \in \mathbb{N}$ and $1 \neq z \in \mathbb{C}$

$$\begin{aligned}\left|\sum_{j=n}^{n+k} c_j z^j\right| &= |z^n|\left|\sum_{j=0}^{k} c_{n+j}z^j\right| \le |z|^n|1-z|\left[\sum_{j=0}^{k-1}\varepsilon|z|^j + \frac{\varepsilon|z|^k}{|1-z|}\right] \\ &= |z|^n\varepsilon|1-z|\left[\frac{1-|z|^k}{1-|z|} + \frac{|z|^k}{|1-z|}\right] \\ &\le \varepsilon|1-z|\left[\frac{1-|z|^k}{1-|z|} + \frac{|z|^k}{1-|z|}\right] \qquad \forall k \in \mathbb{N}, z \in \mathbb{D} \\ &= \varepsilon\frac{|1-z|}{1-|z|} \le \varepsilon M \quad \text{by (1)} \qquad \forall k \in \mathbb{N}, 1 \neq z \in \Delta.\end{aligned}$$

This proves that the series $\sum c_j z^j$ is uniformly Cauchy in $\Delta \setminus \{1\}$, hence in Δ.

(vi) *The function f in* (v) *is generally not continuous in $\mathbb{D} \cup \{b\}$.*

Hints: Consider the polynomials $p_n(z) := (z^{3^n} - z^{2\cdot 3^n})/n$, $n \in \mathbb{N}$. The (power) series $f(z) := \sum_{n=1}^{\infty} p_n(z)$ converges for all $z \in \mathbb{D} \cup \{1\}$. Let us show that f is not continuous at 1. Suppose $u_k \in \mathbb{T}$ satisfying $u_k^{3^k} = i$ and $\lim_{k\to\infty} u_k = 1$ are available. Then $0 \le r < 1$ and $n > k$ entail

$$p_n(ru_k) = (r^{3^n} i^{3^{n-k}} - r^{2\cdot 3^n}(-1)^{3^{n-k}})/n.$$

Since $i^{3^{n-k}} = \pm i$ and $(-1)^{3^{n-k}} = -1$, this means

$$\operatorname{Re} p_n(ru_k) = r^{2\cdot 3^n}/n,$$

and consequently, for every $k \in \mathbb{N}$,

$$\operatorname{Re} f(ru_k) = \operatorname{Re}\sum_{n=1}^{k} p_n(ru_k) + \sum_{n=k+1}^{\infty} r^{2\cdot 3^n}/n,$$

which implies that $\lim_{r\uparrow 1} \operatorname{Re} f(ru_k) = \infty$. If $r_k \uparrow 1$ sufficiently rapidly, we then have $z_k := r_k u_k \in \mathbb{D} \to 1$ and $\operatorname{Re} f(z_k) \geq k$. Where do such u_k come from? After studying the complex exponential in 3.14, we would simply *define* $u_k := e^{\pi i/2\cdot 3^k}$. Alternatively, using only ideas developed so far, the following elementary fact can be inductively exploited (beginning with $w := i$) to ensure that such u_k exist (without, however, producing them!):

Each $w \in \mathbb{T}$ has a cube root $u \in \mathbb{T}$ which satisfies $|u-1| \leq \frac{2}{3}|w-1|$.

To see this, let $\varepsilon := (-1+\sqrt{3}i)/2$ and u_1 be any cube root of w, as furnished by 1.26(ii). Then $u_2 := \varepsilon u_1$ and $u_3 := \bar{\varepsilon} u_1$ are its other two cube roots. Setting $\beta := \operatorname{Im} u_1$, $\alpha := \operatorname{Re} u_1$, we have $\operatorname{Re} u_2 = (-\alpha - \sqrt{3}\beta)/2$ and $\operatorname{Re} u_3 := (-\alpha + \sqrt{3}\beta)/2$. Consider $u \in \{u_1, u_2, u_3\}$ and set $x := \operatorname{Re} u$, $y := \operatorname{Im} u$. A brief computation, using $u\bar{u} = 1$ gives

$$(*) \qquad |u^2+u+1| = |u||u+1+\bar{u}| = |2x+1|.$$

If $|\alpha| \leq 1/2$, then $|\beta| = \sqrt{1-\alpha^2} \geq \sqrt{3}/2$, we choose $u := u_2$ or u_3 in order to have

$$2x+1 = -\alpha + \sqrt{3}|\beta| + 1 \geq -1/2 + \sqrt{3}\sqrt{3}/2 + 1 = 2.$$

If $\alpha > 1/2$, we choose $u := u_1$ and have $2x+1 > 2$. If $\alpha < -1/2$, we choose $u := u_2$ or u_3 in order to have

$$2x+1 = -\alpha + \sqrt{3}|\beta| + 1 \geq -\alpha + 1 > 3/2.$$

So in all cases $(*)$ yields $|u^2+u+1| > 3/2$, whence

$$|w-1| = |u^3-1| = |u-1||u^2+u+1| \geq \frac{3}{2}|u-1|.$$

(vii) (Cauchy product again) a_n, $b_n \in \mathbb{C}$, $c_n := \sum_{j=0}^{n} a_j b_{n-j}$ *and all three series* $\sum a_n$, $\sum b_n$, $\sum c_n$ *converge, to* A, B, $C \in \mathbb{C}$, *respectively. Use* (v) *above and* 3.12(iv) *to show* (*as was done by* Abel) *that* $C = AB$.

Hint: $\sum a_n z^n$, $\sum b_n z^n$ converge for all $z \in \mathbb{D}$.

§ 3 The Complex Exponential Function

The most important power series in analysis will now be introduced and studied. Besides the interesting numerical results which we will derive from it in this chapter, it will serve as the basis for a rigorous logical treatment of the idea "the number of times a curve winds around a point". This will be essential to the whole Cauchy theory of holomorphic functions and also will be one of the main tools in the rather complete topological analysis of the plane offered in Chapter IV.

As soon as one seeks solutions of the differential equation $y'(z) = y(z)$ in $\mathbb{C}$ via power series $y(z) = \sum c_n z^n$, the recursion $nc_n = c_{n-1}$ on the coefficients emerges, from which it follows that $c_n = c_0/n!$, and this series really converges and satisfies $y'(z) = y(z)$ for all $z \in \mathbb{C}$. Alternatively, if one seeks a non-zero power series $f(z) = \sum c_n z^n$ which maps $(\mathbb{C}, +)$ homomorphically into $(\mathbb{C}, \cdot)$, then an easy argument using 3.12(iv) (and only the weaker property $f(2z) = f^2(z)$) confirms that $c_n = c_1^n/n!$ for all n. This is the motivation for Theorem 3.14, but there are other approaches. Since the function $y \mapsto e^{iy}$ is, up to a constant, the unique (continuous) homomorphism of the topological group $\mathbb{R}$ onto the topological group $\mathbb{T}$ (see 3.25), it can be introduced "naturally". One has only to show somehow, without the exponential, that these groups are locally isomorphic. Similarly with the real exponential $x \mapsto e^x$. This elegant, but arduous, program is carried out in BOURBAKI [1966], chapter V, §§3 and 4, and chapter VIII, §2. Cf. also DZEWAS [1968], BEHREND [1956] and §5 of our Chapter II.

A more geometric approach consists of measuring arclength on $\mathbb{T} \cap \overline{\mathbb{U}}$ via 2.8 and the parametrization $(t, \sqrt{1-t^2})$, $t \in [-1, 1]$, using it to define the inverse sine and cosine functions (cf. 5.82(v)), using $e^x(\cos y + i \sin y)$ as the definition of the exponential function at $x + iy$, and finally using (C. R.) to confirm that this function is entire and coincides with its complex derivative. Here $x \mapsto e^x$ is the inverse of the function $y \mapsto \int_1^y dt/t$, $y \in (0, \infty)$ or is defined in 1.47. Series can be avoided altogether. See SPIVAK [1967]. Alternatively, one can avoid improprieties at $t = \pm 1$ by using the (rational) parameterization $((1-t^2)/(1+t^2), 2t/(1+t^2)) = (1+it)/(1-it)$ of $\mathbb{T} \cap \overline{\mathbb{H}}$, and the arclength formula will define the inverse tangent. See EBERLEIN [1966].

Theorem 3.14 *The function E defined for all $z \in \mathbb{C}$ (cf. 3.4) by*

$$E(z) := \sum_{n=0}^{\infty} \frac{1}{n!} z^n$$

has the following properties:

(i) $E' = E$.

(ii) *For all $z_1, z_2 \in \mathbb{C}$, $E(z_1 + z_2) = E(z_1)E(z_2)$.*

(iii) *There is a positive number π such that $E\left(\frac{\pi}{2} i\right) = i$ and*

$$\forall z_1, z_2 \in \mathbb{C}, \qquad E(z_1) = E(z_2) \iff \frac{z_1 - z_2}{2\pi i} \text{ is an integer.}$$

(iv) *$E|\mathbb{R}$ is strictly increasing and $E(\mathbb{R}) = (0, \infty)$.*

(v) *The mapping $x \mapsto E(ix)$ maps $\mathbb{R}$ onto the unit circle $\mathbb{T} = \{z \in \mathbb{C}: |z| = 1\}$.*

(vi) *$E(\mathbb{C}) = \mathbb{C} \setminus \{0\}$.*

Note: (ii), (iii) and (vi) say that E is a homomorphism of the additive group $\mathbb{C}$ onto the multiplicative group $\mathbb{C} \setminus \{0\}$ with kernel $2\pi i\mathbb{Z}$.

Proof: That E is differentiable and that

$$E'(z) = \sum_{n=1}^{\infty} n \cdot \frac{1}{n!} z^{n-1} = E(z)$$

for all z follows from 3.6 and 3.4.

Given $c \in \mathbb{C}$, define a function $f: \mathbb{C} \mapsto \mathbb{C}$ by

$$f(z) := E(c - z).$$

Then by (i) (and the Chain Rule), the function f is differentiable throughout $\mathbb{C}$ and satisfies $f' = -f$. Therefore the function $F := f \cdot E$ satisfies $F' = \mathbf{0}$, whence (by 2.10) $F = F(0)$, that is, for all $z \in \mathbb{C}$

$$E(c - z)E(z) = F(z) = F(0) = E(c)E(0) = E(c).$$

Taking $c = z_1 + z_2$ and $z = z_1$, we get (ii). [A more natural and more direct alternative proof follows from 3.12(iv). Even better perhaps is to use (i), 3.12(iii) and 3.6 to write $E(z) = \sum_{m=0}^{\infty}(E(a)/m!)(z - a)^m = E(a)E(z - a)$.]

For positive real x we have from the series definition of E

$$E(x) > 1 + x,$$

so

$$\lim_{x \to \infty} E(x) = \infty; \tag{1}$$

while from (ii) follows that for all $z \in \mathbb{C}$

$$1 = E(0) = E(z)E(-z) \tag{2}$$

and then (1) yields

$$\lim_{t \to -\infty} E(t) = \lim_{-t \to \infty} \frac{1}{E(-t)} = \lim_{x \to \infty} \frac{1}{E(x)} = 0. \tag{3}$$

Since conjugation is continuous and $\mathbb{R}$-linear on $\mathbb{C}$,

$$\overline{E(z)} = E(\bar{z}) \qquad \forall z \in \mathbb{C}. \tag{4}$$

In particular $E(\mathbb{R}) \subset \mathbb{R}$ and in fact, bearing in mind (2) that E is never zero, for any $x \in \mathbb{R}$

$$E(x) = E(x/2 + x/2) = [E(x/2)]^2 > 0.$$

Therefore $E(\mathbb{R})$ is a connected subset of $(0, \infty)$, which by (1) and (3) must be all of $(0, \infty)$. Moreover, $E'(x) = E(x) > 0$ for all $x \in \mathbb{R}$ shows that $E|\mathbb{R}$ is strictly increasing. Hence (iv) is proved.

Define functions $C, S \colon \mathbb{R} \mapsto \mathbb{R}$ by

$$C(x) := \operatorname{Re} E(ix), \qquad S(x) := \operatorname{Im} E(ix).$$

Now the derivative of the function $x \mapsto E(ix) = C(x) + iS(x)$ is, by the (mixed) chain rule, $iE'(ix) = iE(ix) = iC(x) - S(x)$, so

$$\begin{aligned} &C'(x) + iS'(x) = iC(x) - S(x), \\ &-C' = S, \qquad S' = C. \end{aligned} \tag{5}$$

Also from (4) and (2) we have for all $x \in \mathbb{R}$

$$\begin{aligned} &C^2(x) + S^2(x) = |E(ix)|^2 = E(ix)\overline{E(ix)} = E(ix)E(-ix) = 1, \\ &C^2 + S^2 = \mathbf{1}. \end{aligned} \tag{6}$$

The definition of C gives

$$\begin{aligned} C(x) &= \frac{E(ix) + \overline{E(ix)}}{2} = \frac{E(ix) + E(-ix)}{2} \\ &= \sum_{n=0}^{\infty} \left[\frac{1 + (-1)^n}{2}\right] \frac{i^n x^n}{n!} = \sum_{k=0}^{\infty} \frac{i^{2k} x^{2k}}{(2k)!} \\ &= \sum_{k=0}^{\infty} \frac{(-1)^k x^{2k}}{(2k)!} = 1 + \lim_{N\to\infty} \sum_{k=1}^{2N} \frac{(-1)^k x^{2k}}{(2k)!} \\ &= 1 + \lim_{N\to\infty} \sum_{n=1}^{N} \left[\frac{(-1)^{2n-1} x^{4n-2}}{(4n-2)!} + \frac{(-1)^{2n} x^{4n}}{(4n)!}\right] \\ &= 1 + \sum_{n=1}^{\infty} \left[\frac{x^2}{(4n)!} - \frac{1}{(4n-2)!}\right] x^{4n-2}. \end{aligned}$$

Observe that for $0 < x \le 2$ and $n \ge 1$

$$\frac{x^2}{(4n)!} \le \frac{4}{(4n)(4n-1)(4n-2)!} < \frac{1}{(4n-2)!}$$

and so

$$\begin{aligned} C(x) &= 1 + \left[\frac{x^2}{4!} - \frac{1}{2!}\right]x^{4-2} + \sum_{n=2}^{\infty}\left[\frac{x^2}{(4n)!} - \frac{1}{(4n-2)!}\right]x^{4n-2} \\ &< 1 + \left[\frac{x^2}{4!} - \frac{1}{2!}\right]x^{4-2}. \end{aligned}$$

Therefore $C(\sqrt{3}) < -\frac{1}{8}$. Since $C(0) = \operatorname{Re} E(0) = 1$, we have that $C\big[0, \sqrt{3}\,\big]$ is a connected subset of $\mathbb{R}$ which contains positives and negatives and hence zero; that is, the compact set $\big[0, \sqrt{3}\,\big] \cap C^{-1}(0)$ is not void. Let $\pi/2$ denote its least element:

$$\pi/2 := \min\Big(\big[0, \sqrt{3}\,\big] \cap C^{-1}(0)\Big). \tag{7}$$

Thus

$$0 < \pi/2 < \sqrt{3} \quad \text{and} \quad C(\pi/2) = 0. \tag{8}$$

[The notation π, meant to suggest "perimeter" (of a circle having unit diameter), anticipates the remarks following 3.16.] An insertion-of-parentheses argument similar to the above shows that

$$C(x) > 1 - \frac{x^2}{2!} + \frac{x^4}{4!} - \frac{x^6}{6!} =: P(x), \quad x \in \big(0, 3\sqrt{10}\big].$$

This polynomial is decreasing on $\big[0, \sqrt{6}\big]$ (look at its derivative), so

$$C(x) > P(x) \ge P(3/2) > 0 \qquad \forall x \in (0, 3/2]\,.$$

We infer that $\pi/2 > 3/2$, i.e.,

$$\pi > 3. \tag{8$'$}$$

From (8) and (6) we get

$$S(\pi/2) = \pm 1. \tag{9}$$

However by (5) and the definition of π

$$S' = C > 0 \quad \text{in } [0, \pi/2)\,, \tag{10}$$

so that S is strictly increasing in $[0,\pi/2]$. Therefore $S(\pi/2) > S(0) = 0$ and from equation (9) we get

$$S(\pi/2) = 1. \tag{11}$$

Putting (8) and (11) together:

$$E\Big(\frac{\pi}{2}i\Big) = C(\pi/2) + iS(\pi/2) = i, \tag{12}$$

whence

$$\begin{aligned} &E(2\pi i) \overset{\text{(ii)}}{=} \Big[E\Big(\frac{\pi}{2}i\Big)\Big]^4 = 1, \\ &E(z+2\pi i) = E(z)E(2\pi i) = E(z) \qquad \forall z \in \mathbb{C}. \end{aligned} \tag{13}$$

Now suppose $z := x + iy$ $(x, y \in \mathbb{R})$ is such that $E(z) = 1$. Then $1 = |E(z)| = |E(x)E(iy)| = E(x)|E(iy)| = E(x)$, since $E(\mathbb{R}) \subset (0,\infty)$ and $|E(iy)| = 1$ by (6). Since $E(0) = 1$ and $E|\mathbb{R}$ is strictly increasing, it follows that $x = 0$. To prove that $y/2\pi$ is an integer it is enough, in view of (13), to show that $E(it) \neq 1$ if $0 < t < 2\pi$. Given such a t, we have $0 < t/4 < \pi/2$ and, since C and S are both positive in $(0,\pi/2)$ (by (10)), we have

$$C(t/4) > 0, \qquad S(t/4) > 0. \tag{14}$$

Writing $z^4 - 1 = (z+1)(z-1)(z+i)(z-i)$, shows that $\pm 1, \pm i$ are the only fourth roots of 1. Inequalities (14) show that $E(it/4) = C(t/4) + iS(t/4)$ is none of these numbers. Therefore $E(it) = [E(it/4)]^4$ is not 1. This proves (iii).

Now if $u^2 + v^2 = 1$ and $u \geq 0, v \geq 0$, then $C(\pi/2) = 0 \leq u \leq 1 = C(0)$ and so by continuity of C there is a $t \in [0,\pi/2]$ such that $C(t) = u$. But S is non-negative in $[0,\pi/2]$ (by (10)), so (6) gives

$$\begin{aligned} S(t) &= \sqrt{1 - C^2(t)} = \sqrt{1-u^2} = v, \\ E(it) &= u + iv. \end{aligned}$$

If $u^2 + v^2 = 1$ and $u < 0, v \geq 0$, pick t according to the previous recipe so that $E(it) = v - iu$ and then have by (12)

$$E\Big(i\Big(t + \frac{\pi}{2}\Big)\Big) = E(it)E\Big(i\frac{\pi}{2}\Big) = (v - iu)\cdot i = u + iv.$$

If $u^2 + v^2 = 1$ and $v < 0$, pick t according to one of the previous two recipes so that $E(it) = -u - iv$ and then have

$$E(i(t+\pi)) = E(it)E(i\pi) = (-u - iv)\,(-1) = u + iv.$$

The assertion (v) is thus proven. [Note that the argument in 1.26(ii) gives a quick alternative proof of (v).]

If $w \in \mathbb{C}\setminus\{0\}$, then $|w| \in (0,\infty) = E(\mathbb{R})$, so $|w| = E(x)$ for some $x \in \mathbb{R}$; while $|w/|w|| = 1$, so $w/|w| = E(iy)$ for some $y \in \mathbb{R}$ by the result (v). Thus

$$E(x+iy) = E(x)E(iy) = w$$

and we have $\mathbb{C} \setminus \{0\} \subset E(\mathbb{C})$, thus completing the proof of (vi).

Remarks: If n is a positive integer, then $E(1/n)$ is a positive real number whose n^{th} power is $E(1)$, that is, $E(1/n)$ is the unique positive n^{th} root $[E(1)]^{1/n}$ of $E(1)$. If m is any integer, then

$$E(m/n) = [E(1/n)]^m = [E(1)]^{m/n}.$$

Consequently, if we follow the universal custom and let e denote the number $E(1)$, then for every rational r

$$E(r) = e^r.$$

[3.14 obviates (most of) 1.47; but readers familiar with that theory, in which e^r is defined for $r \in \mathbb{R} \setminus \mathbb{Q}$ as $\lim e^{r_n}$, $r_n \in \mathbb{Q}$ convergent to r, will see that $E(r) = e^r$ then holds as well for such r, by continuity of E.]

This equality explains another universal notation, namely e^z, as a name for $E(z)$, in view of which the homomorphic property of E extends a familiar "law of exponents." Thus

Definition 3.15 $e := \sum_{n=0}^{\infty} 1/n!$ and $e^z := \sum_{n=0}^{\infty} z^n/n!$ for every $z \in \mathbb{C}$. This function is called the (*complex*) *exponential* function and sometimes also denoted exp. Its restriction to $\mathbb{R}$ is called the *real* exponential function.

Remark: Only very crude rational approximations to e are ever needed in the sequel. For example, since $n! > 3^{n-1}$ for all $n \geq 5$,

$$2\tfrac{43}{60} < e = 2\tfrac{43}{60} + \sum_{n=6}^{\infty} \frac{1}{n!} < 2\tfrac{43}{60} + \sum_{n=6}^{\infty} \frac{1}{3^{n-1}} = 2\tfrac{1171}{1620}.$$

Definition 3.16 For any $a \in \mathbb{C}$ and $r > 0$ the *circle of center a and radius r* is the simple loop defined by $t \mapsto a + re^{2\pi it}$, $t \in [0,1]$. We denote this curve by $C(a,r)$ or $\partial D(a,r)$; this is consistent with the definition in Chapter 0 and the convention of 1.11(i), thanks to 3.14(v).

Remarks on Geometry. The image of the simple loop $\gamma(t) := e^{it}$, $t \in [0,2\pi]$, is $\mathbb{T}$. What we learned in the course of the proof of 3.14 about the increasing and decreasing characteristics of $\operatorname{Re}\gamma(t) = C(t)$ and $\operatorname{Im}\gamma(t) = S(t)$ and about their signs as t advances through $[0, 2\pi)$ shows that $\gamma(t)$ "traverses $\mathbb{T}$ bijectively counterclockwise." For any fixed $\theta \in [0, 2\pi)$ the length of the arc counterclockwise from 1 to $e^{i\theta}$ is therefore (see Definition 2.8) $\int_0^\theta |\gamma'| = \int_0^\theta |i\gamma| = \int_0^\theta \mathbf{1} = \theta$. This is the customary interpretation of θ as the counterclockwise *angle* from the vector 1 to the vector $e^{i\theta}$, or the *argument* or *amplitude* of the complex number $e^{i\theta}$. [The pure geometric concept of "angle" seems insusceptible to precise definition, so the preceding in effect conflates it with that of "numerical-measure-of-angle", expedient and noncontroversial here, though *perhaps* otherwise in the real world – see DREXLER and KRYSTEK [2016].] Note too that since γ maps every half-open interval of length 2π bijectively on $\mathbb{T}$, $C = \operatorname{Re}\gamma$ is never 1 on the open interval $(0, 2\pi)$; similarly, $S = \operatorname{Im}\gamma$ is never 1 on the open interval $(\pi/2, 2\pi + \pi/2)$. Hence 3.14(iii) can be augmented by the observation that *neither* of the functions C, S has a positive period less than 2π. Cf. 3.30(iii).

According to the remarks on the parameter invariance of arclength which open §2 of Chapter II, the (bijective) rectangular coordinatization of the semicircles $\mathbb{T} \cap \overline{\mathbb{U}}$ and $\mathbb{T} \cap \overline{\mathbb{H}}$ mentioned in the prelude to 3.14 should, like γ, assign arc-length π to these semicircles. Of course, the (differentiable) changes of variable in the arclength integral that explicitly demonstrate this are $t := C(\theta)$, $\theta \in [0, \pi]$ and $t := S(\theta/2)/C(\theta/2)$, $\theta \in [-\pi/2, \pi/2]$, respectively. (Cf. 3.22.)

If $a, b \in \mathbb{C} \setminus \{0\}$, then $(a/|a|)/(b/|b|) = e^{i\theta}$, for a unique $\theta \in [0, 2\pi)$ which we will call the "angle from vector b to vector a." In terms of Definition 3.27 and the inner product $\bullet$ in $\mathbb{R}^2$,

$$\cos\theta := \operatorname{Re}(e^{i\theta}) = \operatorname{Re}\left(\frac{a}{|a|}/\frac{b}{|b|}\right) = \operatorname{Re}\left(\frac{a}{|a|}\frac{\overline{b}}{|b|}\right) = \frac{a}{|a|} \bullet \frac{b}{|b|},$$

which, incidentally, delivers a one-line proof of the *Law of Cosines*: The triangle with sides a, b and $c := a - b$, satisfies $|c|^2 = (a-b) \bullet (a-b) = |a|^2 + |b|^2 - 2|a||b|\cos\theta$.

If $\gamma_1, \gamma_2 : [0, 1] \mapsto U$ open $\subset \mathbb{C}$ are smooth curves such that $\gamma_1(0) = \gamma_2(0) = z_0$ and $\gamma_1'(0) \cdot \gamma_2'(0) \neq 0$, and if f is holomorphic in U with $f'(z_0) \neq 0$, then the image curves $\Gamma_j := f \circ \gamma_j$ satisfy

$$\text{(1)} \qquad \frac{\gamma_1'(0)}{|\gamma_1'(0)|} \bigg/ \frac{\gamma_2'(0)}{|\gamma_2'(0)|} = \frac{\Gamma_1'(0)}{|\Gamma_1'(0)|} \bigg/ \frac{\Gamma_2'(0)}{|\Gamma_2'(0)|},$$

as a routine calculation with the chain rule confirms. This says that the angle between (the tangents to) the curves γ_1, γ_2 at their intersection z_0 equals that between the image curves Γ_1, Γ_2 at their intersection $f(z_0)$. This is the "angle-preservation" feature implicit in the terminology "(locally) conformal mapping."

Conversely, a weak form of this "angle-preservation" property implies holomorphy. Recall the discussion at the very beginning of §4, chapter II and consider a real-differentiable $f\colon U$ open $\subset \mathbb{R}^2 \mapsto \mathbb{C}$. For any smooth curve γ in U, $\Gamma := f\circ\gamma$ is a differentiable curve and the chain rule in $\mathbb{R}^2$ gives

$$\Gamma'(t) = D_1 f(\gamma(t))(\operatorname{Re}\gamma)'(t) + D_2 f(\gamma(t))(\operatorname{Im}\gamma)'(t)$$

or, after a little algebra,

$$(2)\ \ \Gamma'(t) = \frac{1}{2}\big[D_1 f(\gamma(t)) - iD_2 f(\gamma(t))\big]\gamma'(t) + \frac{1}{2}\big[D_1 f(\gamma(t)) + iD_2 f(\gamma(t))\big]\bar{\gamma}'(t).$$

For the special case where $z_0 \in U$, $u \in \mathbb{T}$ and γ is the line segment $t \mapsto z_0 + tu$, equation (2) when $t = 0$ is called the *directional derivative* of f at z_0 in the direction u, often denoted $D_u f(z_0)$. That equation when $t = 0$ now reads

$$D_u f(z_0) = \frac{1}{2}\big[D_1 f(z_0) - iD_2 f(z_0)\big]u + \frac{1}{2}\big[D_1 f(z_0) + iD_2 f(z_0)\big]\bar{u}$$

or, conveniently abbreviating the factors of $u, \bar{u}$ as A, B, respectively,

$$(3.u)\qquad D_u f(z_0) = Au + B\bar{u}.$$

The equality (1) expressing that at z_0 the function f preserves the angle between $u, w \in \mathbb{T}$ reads

$$(1.u.w)\qquad \frac{u}{w} = \frac{D_u f(z_0)}{|D_u f(z_0)|}\bigg/\frac{D_w f(z_0)}{|D_w f(z_0)|}.$$

$$\left\{\begin{array}{l}\textit{Suppose that there are } u, v, w \in \mathbb{T}, \textit{ no two collinear,}\\ \textit{such that } f \textit{ preserves the angle between } u \textit{ and } w, \textit{ and}\\ \textit{the angle between } v \textit{ and } w.\end{array}\right.$$

Then $(1.v.w)$ says

$$\frac{D_v f(z_0)}{v} = \rho\frac{D_w f(z_0)}{w},$$

whence via $(3.v)$, $(3.w)$

$$\frac{Av + B\bar{v}}{v} = \rho\frac{Aw + B\bar{w}}{w},$$

for some positive ρ. If $B \neq 0$, this says that $A/B + \bar{v}^2 = \rho(A/B + \bar{w}^2)$. That is, $\bar{v}^2 \in \mathcal{L} := \{-A/B + t(A/B + \bar{w}^2) : t \in \mathbb{R}\}$. For fixed w this set is either a straight line or a point, so it meets $\mathbb{T}$ at most twice. But our hypothesis about z_0 is that there is another point $u \in \mathbb{T}$, not collinear with v or w, which fulfills the role of v here. Thus $\bar{u}^2$, $\bar{v}^2$, $\bar{w}^2$ are distinct (why?) and lie in $\mathbb{T} \cap \mathcal{L}$, a contradiction which proves $B \neq 0$ untenable. We infer that $B = 0$, which says that (C. R.) holds at z_0. Being $\mathbb{R}$-differentiable at z_0, f is consequently $\mathbb{C}$-differentiable at z_0.

R. K. Williams [1973] considered the alternative hypothesis that f is continuous with continuous first-order partial derivatives in U and that at each $z_0 \in U$ the orthogonal rays $z_0 + \mathbb{R}^+$ and $z_0 + i\mathbb{R}^+$ are mapped orthogonally by f, and likewise the orthogonal rays $z_0 + (i \pm 1)\mathbb{R}^+$, from which he deduced that in each component of U either f or $\bar{f}$ must be holomorphic.

Besides being "winkeltreu" (=*isogonal*) in the preceding sense, holomorphic mappings f manifest another kind of "troth" at points z_0 where $f'(z_0) \neq 0$: They are "streckentreu" in the sense that all infinitesimal vectors at z_0 are elongated by the same positive factor independent of their directions; length distortion at z_0 is *isotropic*. More generally and more precisely

$$\lim_{z \to z_0} \frac{|f(z) - f(z_0)|}{|z - z_0|} \quad \text{exists in } \mathbb{R} \text{ and is non-zero,} \tag{4}$$

this limit being just $|f'(z_0)|$ in case f is holomorphic. Property (4) almost characterizes holomorphy: Consider a real-differentiable $f \colon \Omega$ region $\subset \mathbb{R}^2 \mapsto \mathbb{C}$. For each $z_0 \in \Omega, u \in \mathbb{T}$ we have (3.u),

$$|D_u f(z_0)| = |Au^2 + B|. \tag{5}$$

The set $\mathcal{C} := \{Au^2 + B : u \in \mathbb{T}\}$ is a circle of radius $|A|$ centered at B.

$$\begin{cases} \textit{Suppose } (4) \textit{ holds just for linear approach to } z_0 \textit{ along} \\ \textit{three non-collinear directions } u, v, w \in \mathbb{T} \end{cases}$$

That is, $|D_u f(z_0)| = |D_v f(z_0)| = |D_w f(z_0)| \neq 0$. Call this number r. Then (5) says that $Au^2 + B$, $Av^2 + B$, $Aw^2 + B \in \mathcal{C} \cap r\mathbb{T}$. Since two different circles intersect at most twice, it follows that either these points are not distinct, whence $A = 0$ and $B \neq 0$, or the circles $\mathcal{C}$ and $r\mathbb{T}$ coincide, whence $B = 0$ and $A \neq 0$. We have thus proved, following Menchoff [1936], that

$$D_1 f(z_0) = \pm i D_2 f(z_0) \neq 0. \tag{6}$$

Now suppose that the preceding hypothesis is satisfied at every z_0 in region Ω. Suppose *additionally* that $D_1 f$ and $D_2 f$ are continuous in Ω. Then from (6) holding for all $z_0 \in \Omega$ and 1.7 it follows that

either $\quad D_1 f(z_0) = -i D_2 f(z_0) \qquad \forall z_0 \in \Omega$

or $\quad D_1 f(z_0) = i D_2 f(z_0)$ [equivalently, $D_1 \bar{f}(z_0) = -i D_2 \bar{f}(z_0)$] $\quad \forall z_0 \in \Omega$.

The first alternative implies that f is holomorphic in Ω, while the second implies that $\bar{f}$ is holomorphic in Ω—see 2.17.

In proving that these geometric properties entail holomorphy one would like to dispense with or weaken the *a priori* assumption that f is real-differentiable, inferring this too from the geometric properties. Formulation (4) of Streckentreue suggests such a possibility. But formulation (1) of angle-preservation involves actual derivatives, because implicit in it is the definition of the angle between curves γ_1, γ_2 as the argument of the ratio of their normalized *tangent* vectors. An alternative definition of this angle (not requiring differentiability) is the limiting argument of the ratios of normalized *chords* in the curves:

$$\lim_{t\to 0} \frac{\gamma_1(t)-\gamma_1(0)}{|\gamma_1(t)-\gamma_1(0)|} \Big/ \frac{\gamma_2(t)-\gamma_2(0)}{|\gamma_2(t)-\gamma_2(0)|}.$$

This suggests a definition of Winkeltreue more general than (1), which [like (4)] involves neither $\mathbb{R}$-differentiability nor curves. It is that the *arguments* of the difference quotients of f at z_0 converge; more precisely, that

$$(1)' \qquad \lim_{z\to z_0} \frac{\frac{f(z)-f(z_0)}{z-z_0}}{\left|\frac{f(z)-f(z_0)}{z-z_0}\right|} \quad \text{exists for each } z_0 \in U.$$

These ideas are treated in REMAK [1914], BOHR [1918], RADEMACHER [1919a] and MENCHOFF [1936], [1937]. Bohr hypothesized only that f is continuous and one-to-one in region Ω and was able to prove (by lengthy real-variable arguments) that if f satisfies (4) at every $z_0 \in \Omega$, then either f or $\bar{f}$ is holomorphic in Ω. The example $\Omega := \mathbb{D}$, $f(z) := z$ (resp., $\bar{z}$) if $\operatorname{Im} z \geq 0$ (resp., $\operatorname{Im} z < 0$) shows the need for the global univalence hypothesis. Later MENCHOFF [1931] improved on Bohr by showing that the limits in (4) need only be hypothesized to exist and be equal (possibly 0) along three different directions at each z_0. TELYAKOVSKIĬ [1994] showed that (1)′ with no further hypothesis on f entails holomorphy. A qualitative study of the angle distortion that occurs when f does not satisfy either equality in (6) is undertaken in PULFORD and WALDEN [2003].

Now that the exponential function is available, we can give polar-coordinate and circular formulations of some rectangularly formulated facts from Chapter II.

First:

If g is real-differentiable at the non-zero point $z_0 = \rho_0 e^{i\theta_0}$, then all relevant partial derivatives exist and satisfy

$$(\text{C. R.})^\circ \qquad (iD_1 g - D_2 g)(z_0) = ie^{i\theta_0} D_\rho g(\rho_0 e^{i\theta_0}) - \frac{1}{\rho_0} e^{i\theta_0} D_\theta g(\rho_0 e^{i\theta_0}).$$

To see this, compute D_ρ, D_θ using the curves $\gamma(\rho) := \rho e^{i\theta_0}$, $\gamma(\theta) := \rho_0 e^{i\theta}$ in the chain rule formulated in the discussion with which §4 of Chapter II begins.

From (C. R.)° follows the *polar form* of the Cauchy-Riemann equation

$$\text{(C. R.)}^{\circ\circ} \qquad i\rho_0 D_\rho g(z_0) = D_\theta g(z_0),$$

a *necessary* condition for $\mathbb{C}$-differentiability of g at z_0 (2.16), which is also sufficient under the real-differentiability hypothesis here. Thanks to 2.17 the (stronger) hypothesis that $D_1 g, D_2 g$ exist near and be continuous at z_0 implies (C. R.)°. Symmetrically, such a hypothesis imposed instead on $D_\rho g, D_\theta g$ as functions of (ρ, θ) also implies (C. R.)°, but the proof of this is more subtle, as the chain rule in this case proceeds via a real-differentiable local inverse of the complex exponential function (gotten, for example, explicitly from 3.19 or implicitly from 5.79). The existence of holomorphic inverses for the exponential function will be taken up in §6, but the preceding discussion is one (heavy-handed) way to get global inverses from local ones. For example (see 3.17 and 3.18), define $g : \mathbb{C}_- \mapsto \mathbb{C}$ by

$$g(\rho e^{i\theta}) := \log(\rho) + i\theta \qquad \text{whenever } \rho > 0 \text{ and } -\pi < \theta < \pi.$$

Evidently, $D_\rho g, D_\theta g$ exist, are continuous functions of (ρ, θ), and satisfy (C. R.)°°. Under the above local-holomorphic-inverse hypothesis it follows that g is holomorphic in $\mathbb{C}_-$. Note that $e^{g(z)} = z$ for all $z \in \mathbb{C}_-$. (See 3.43(ii).)

Similarly, the exponential function makes possible a disk version of Green's theorem 2.19(iv), namely:

If f has continuous first-order partial derivatives in a neighborhood of $\overline{D}(a, r)$, then

$$\text{(G)}^\circ \qquad \int_{C(a,r)} f = \int_0^{2\pi} \left[\int_0^r (-D_2 f + iD_1 f)(a + \rho e^{i\theta}) \rho d\rho \right] d\theta.$$

Take $g(z)$ above to be $f(a + z)$. As in 2.19, the proof is a straight-forward application to (C. R.)° of Fubini (2.19(ii)), the fundamental theorem of calculus, and a few integrations by parts. Then the disk analog of 2.19(v) follows from (G)° and 2.16:

If f is holomorphic in a neighborhood of $\overline{D}(a, r)$ and f' is continuous there, then

$$\text{(C)}^\circ \qquad \int_{C(a,r)} f = 0.$$

As in 2.19(ix), this has a converse:

If f has continuous first-order partial derivatives in an open set U and (C)° *holds for each $a \in U$ and each sufficiently small $r > 0$, then f is holomorphic in U.*

This follows from 2.17 once it is confirmed that f satisfies (C. R.) in U. That in turn follows from (C)° and (G)°, together with another elementary fact:

$$\lim_{r \downarrow 0} \frac{1}{\pi r^2} \int_0^{2\pi} \left[\int_0^r h(a + \rho e^{i\theta}) \rho d\rho \right] d\theta = h(a)$$

whenever h is continuous in a neighborhood of a. Of course, in the present context h is the function $iD_1 f - D_2 f$. (Cf. hints to 2.19(ix).) Note that the argument uses only the weaker (than (C)°) hypothesis

(C)°° $$\lim_{r \downarrow 0} r^{-2} \int_{C(a,r)} f = 0 \qquad \forall a \in U.$$

In fact, (C)°° and the mere continuity of f imply its holomorphy: Looman [1922/24], Wolff [1922/24], and p. 196 of Saks [1937]; see also C. Müller [1954] and G. Springer [1957].

According to 3.14 the real exponential function is strictly increasing and has range exactly $(0, \infty)$. Consequently there is a function from $(0, \infty)$ onto $\mathbb{R}$ which is inverse to the real exponential. The notation and terminology are again traditional (cf. 1.47):

Definition 3.17 The (*natural*) *logarithm* function, noted log, is the function with domain $(0, \infty)$ and range $\mathbb{R}$ which is inverse to the real exponential.

Most of the properties of this function are immediate consequences of 3.14. For example,

Scholium 3.18 Log is strictly increasing and, since its range is $\mathbb{R}$, we have

(1) $$\lim_{x \downarrow 0} \log x = -\infty, \quad \lim_{x \uparrow \infty} \log x = \infty.$$

Log is continuous (cf. 1.46) and (therefore) differentiable:

$$\log'(x) = \lim_{y \to x} \frac{\log y - \log x}{y - x} = \lim_{y \to x} \frac{1}{\left(\dfrac{e^{\log y} - e^{\log x}}{\log y - \log x} \right)} = \frac{1}{e^{\log x}}$$

(2) $$= \frac{1}{x},$$

since the exponential function is its own derivative. Since the real exponential is an isomorphism of the additive group $\mathbb{R}$ onto the multiplicative group $(0,\infty)$, its inverse log is an isomorphism of the multiplicative group $(0,\infty)$ onto the additive group $\mathbb{R}$, so

$$\log(xy) = \log x + \log y \qquad \forall x, y \in (0,\infty). \tag{3}$$

Equation (3) essentially *characterizes* the function log; see 3.25(iii). We noted in the proof of 3.14 that for positive x we have $e^x > 1 + x$. Therefore $x > \log(1+x)$ and so

$$\log y < y - 1 \qquad \forall y > 1.$$

Since also $\log 1/x = -\log x$ from (3), it follows that, in spite of (1), we have for any positive real number r

$$|x^r \log x| = \left|\frac{2}{r} x^r \log x^{r/2}\right| = \frac{2}{r} x^r \log \frac{1}{x^{r/2}}$$
$$< \frac{2}{r} x^r \cdot \frac{1}{x^{r/2}} = \frac{2}{r} x^{r/2}, \quad \text{if } x < 1;$$
$$\lim_{x \downarrow 0} x^r \log x = 0. \tag{4}$$

Now the exponential function is not one-to-one in $\mathbb{C}$ but is locally so, and we have the following important theorem on the (local) existence of holomorphic right-inverses:

Theorem 3.19 *If $c \in \mathbb{C} \setminus \{0\}$, then there is a function L $(:= L_c)$ defined in $D(c, |c|)$ by a convergent power series such that $e^{L(z)} = z$ and $L'(z) = 1/z$ for all $z \in D(c, |c|)$. Also, the series for L converges uniformly on each compact subset of $\bar{D}(c, |c|) \setminus \{0\}$ and extends L to a continuous function on $\bar{D}(c, |c|) \setminus \{0\}$ which satisfies $e^{L(z)} = z$ there.*

Proof: The preceding discussion shows that $\log(1-x)$ is an $\mathbb{R}$-differentiable function of $x \in \mathbb{R} \cap \mathbb{D}$ with derivative $-1/(1-x) = -\sum_{k=0}^{\infty} x^k$. This suggests considering the $\mathbb{C}$-valued function $l(z) := \sum_{k=1}^{\infty} z^k/k$, which is holomorphic in $\mathbb{D}$ and has $\mathbb{C}$-derivative

$$l'(z) = \sum_{k=1}^{\infty} z^{k-1} = \frac{1}{1-z} \qquad \forall z \in \mathbb{D}. \tag{1}$$

Hence, for the function

$$F(z) := (1-z)e^{l(z)}, \quad z \in \mathbb{D},$$

the chain rule gives

$$F'(z) = (1-z)l'(z)e^{l(z)} - e^{l(z)} \overset{(1)}{=} 0 \qquad \forall z \in \mathbb{D}.$$

It follows from 2.10(iv) that

$$F(z) = F(0) = e^{l(0)} = e^0 = 1 \qquad \forall z \in \mathbb{D};$$

that is, upon recalling the definition of F,

$$1 - z = \frac{1}{e^{l(z)}} = e^{-l(z)} \qquad \forall z \in \mathbb{D}. \tag{2}$$

By 3.13(ii) the series $\sum_{k=1}^{\infty} z^k/k$ converges uniformly on each compact subset of $\overline{\mathbb{D}} \setminus \{1\}$, so it defines a continuous function, which we will continue to denote by l, on $\overline{\mathbb{D}} \setminus \{1\}$. By continuity (2) holds as well for all $z \in \overline{\mathbb{D}} \setminus \{1\}$.

Now given $c \in \mathbb{C} \setminus \{0\}$, for each $w \in D(c, |c|)$ apply (2) to

$$z := \frac{c-w}{c} \in \frac{-1}{c}[D(c,|c|) - c] = \frac{-1}{c}D(0,|c|) = \mathbb{D}$$

and get

$$\frac{w}{c} = 1 - \frac{c-w}{c} = 1 - z = e^{-l(z)} = e^{-l((c-w)/c)},$$

$$w = ce^{-l((c-w)/c)} \qquad \forall w \in D(c,|c|). \tag{3}$$

Now by 3.14, $c = e^b$ for some $b \in \mathbb{C}$, so if we define L in $D(c,|c|)$ by

$$L(w) := b - l\left(\frac{c-w}{c}\right), \qquad w \in D(c,|c|),$$

then L is a power series centered at c and by (3) satisfies

$$w = e^{L(w)} \qquad \forall w \in D(c,|c|),$$

as desired. The continuous extension of l to $\overline{\mathbb{D}} \setminus \{1\}$ furnishes the desired continuous extension of L to $\overline{D}(c,|c|) \setminus \{0\}$ and completes the proof.

Scholium: $\sum_{k=1}^{\infty}(-1)^{k+1}/k = \log 2 = -\log(1/2) = \sum_{k=1}^{\infty} 1/k2^k$. [Take $b := 0$ above.]

Corollary 3.20 *The function defined by the series* $\sum_{k=1}^{\infty} 2z^{2k-1}/(2k-1)$ *maps the open unit disk* $\mathbb{D}$ *univalently onto the strip* $\mathbb{R} \times (-\pi/2, \pi/2)$. (Cf. 3.45(v).) *Its inverse is the holomorphic mapping* $w \mapsto (e^w - 1)/(e^w + 1)$.

Proof: Recall (proof of 3.19) that

$$l(z) := \sum_{k=1}^{\infty} \frac{1}{k} z^k$$

converges for all $z \in \overline{\mathbb{D}} \setminus \{1\}$ and satisfies

$$e^{-l(z)} = 1 - z. \tag{1}$$

Replacing z with $-z$ and dividing the two versions of (1), gives

$$e^{l(z)-l(-z)} = \frac{1+z}{1-z} =: \psi(z).$$

Therefore

$$L(z) := l(z) - l(-z) = 2\sum_{k=1}^{\infty} \frac{z^{2k-1}}{2k-1}, \quad z \in \overline{\mathbb{D}} \setminus \{\pm 1\}$$

satisfies

$$e^L = \psi. \tag{2}$$

According to 2.5 ψ maps $\mathbb{D}$ univalently onto $\mathbb{H} := (0, \infty) \times \mathbb{R}$. The univalence of L therefore follows from (2). Now the exponential maps $\mathbb{R} \times \{m\pi/2 : m$ is an odd integer$\}$ into $i \cdot \mathbb{R} \subset \mathbb{C} \setminus \mathbb{H}$. Therefore (2) shows that we have

$$L(\mathbb{D}) \subset \bigcup_{n \in \mathbb{Z}} \mathbb{R} \times ((2n-1)\pi/2, (2n+1)\pi/2).$$

As $L(\mathbb{D})$ is connected [1.4 and 1.22], it follows from the last inclusion that $L(\mathbb{D}) \subset \mathbb{R} \times ((2n-1)\pi/2, (2n+1)\pi/2)$ for a single n. Then $L(0) = 0$ shows that $n = 0$. Moreover, if $w \in \mathbb{R} \times (-\pi/2, \pi/2)$, then $e^w \in \mathbb{H} = \psi(\mathbb{D}) = e^{L(\mathbb{D})}$, that is, $e^w = e^{L(z)}$ for some $z \in \mathbb{D}$. By periodicity, $L(z) - w = 2\pi k i$. But $L(z) - w \in \mathbb{R} \times (-\pi, \pi)$. Therefore $k = 0$, $w = L(z)$ and we have shown that $L(\mathbb{D}) \supset \mathbb{R} \times (-\pi/2, \pi/2)$. For w in this strip, $z := L^{-1}(w) \in \mathbb{D}$ satisfies $e^w = e^{L(z)} \overset{(2)}{=} (1+z)/(1-z)$, whence $(e^w - 1)/(e^w + 1) = z = L^{-1}(w)$.

There is an important limit expression for e^z which we will examine now. The whole theory can be derived very elegantly from this limit. Particularly noteworthy is the Mephistophelean use of it to secure the surjectivity result $E(i\mathbb{R}) = \mathbb{T}$. See VAN YZEREN [1970] and M. KNESER [2004].

Theorem 3.21 $\lim_{n\to\infty}(1+z/n)^n$ *exists and equals* e^z *for each* $z \in \mathbb{C}$ *and the convergence is uniform on any compact set. In fact, for all* $z \in \mathbb{C}$ *and all positive integers* m

$$\left|e^z - \left(1+\frac{z}{m}\right)^m\right| \le e^{|z|} - \left(1+\frac{|z|}{m}\right)^m \le \frac{|z|^2 e^{|z|}}{2m}.$$

Proof: (Cf. Biehler [1887] and A. J. Macintyre [1949].) For $m \ge 2$, the binomial expansion gives

$$\left(1+\frac{z}{m}\right)^m = \sum_{k=0}^{m}\binom{m}{k}\left(\frac{z}{m}\right)^k = 1+z+\sum_{k=2}^{m}\left(1-\frac{1}{m}\right)\cdots\left(1-\frac{k-1}{m}\right)\frac{z^k}{k!}$$

and so

$$(1)\quad e^z - \left(1+\frac{z}{m}\right)^m = \sum_{k=2}^{m}\left[1-\left(1-\frac{1}{m}\right)\cdots\left(1-\frac{k-1}{m}\right)\right]\frac{z^k}{k!} + \sum_{k=m+1}^{\infty}\frac{z^k}{k!}.$$

Two applications of (1) show that

$$(2)\quad \left|e^z - \left(1+\frac{z}{m}\right)^m\right| \le \sum_{k=2}^{m}\left[1-\left(1-\frac{1}{m}\right)\cdots\left(1-\frac{k-1}{m}\right)\right]\frac{|z|^k}{k!} + \sum_{k=m+1}^{\infty}\frac{|z|^k}{k!}$$
$$= e^{|z|} - \left(1+\frac{|z|}{m}\right)^m.$$

Notice that for $2 \le k \le m$, by an easy induction argument,

$$\left(1-\frac{1}{m}\right)\cdots\left(1-\frac{k-1}{m}\right) \ge 1-\sum_{j=1}^{k-1}\frac{j}{m} = 1-\frac{k(k-1)}{2m}.$$

Therefore (2) yields

$$(3)\qquad e^{|z|} - \left(1+\frac{|z|}{m}\right)^m \le \sum_{k=2}^{m}\frac{k(k-1)}{2m}\cdot\frac{|z|^k}{k!} + \sum_{k=m+1}^{\infty}\frac{|z|^k}{k!}.$$

Now $m \ge 1$ means that $m+1 \ge 2$, $(m+1)m \ge 2m$. Consequently $k(k-1) \ge 2m$ for $k \ge m+1$ and we can estimate the second sum in (3) thus:

$$e^{|z|} - \left(1+\frac{|z|}{m}\right)^m \le \sum_{k=2}^{m}\frac{k(k-1)}{2m}\cdot\frac{|z|^k}{k!} + \sum_{k=m+1}^{\infty}\frac{|z|^k}{2m(k-2)!}$$
$$= \frac{1}{2m}\sum_{k=2}^{\infty}\frac{|z|^k}{(k-2)!} = \frac{1}{2m}\sum_{n=0}^{\infty}\frac{|z|^{n+2}}{n!} \qquad (\text{set } n = k-2)$$
$$= \frac{|z|^2}{2m}e^{|z|}.$$

Remarks: (i) There is an alternative proof of the limit in 3.21 which is short and transparent but a bit less elementary, and affords no estimate of the speed of convergence. Let L be the function L_1 of 3.19. For integers $n > N > 0$, $z \in \overline{D}(0, N) \setminus \{0\}$ implies $1 + z/n \in D(1, 1)$. Consequently, by 3.19

$$nL(1 + z/n) = z \cdot \frac{L(1 + z/n) - L(1)}{z/n} \to zL'(1) = z \quad \text{as} \quad n \to \infty,$$

uniformly in such z, whence $(1+z/n)^n = [\exp L(1+z/n)]^n = \exp(nL(1+z/n)) \to \exp(z)$. [Similarly, $(1 + z_n/n)^n \to e^z$ whenever $z_n \to z$.]

(ii) According to 3.21 the numbers $a_n := n!\,n^{-n}$ satisfy $a_{n+1}/a_n = (1 + 1/n)^{-n} \to 1/e$, in view of which 3.9 tells us that $a_n^{1/n} = n^{-1}(n!)^{1/n} \to 1/e$. After a little algebra, involving $n^{-1}\sum_{j=1}^{n} j = \frac{1}{2}(n + 1)$, another interesting feature of e comes to light:

$$\frac{1}{2}e = \lim_{n\to\infty} \frac{\text{Arithmetic Mean of } \{1, 2, \ldots, n\}}{\text{Geometric Mean of } \{1, 2, \ldots, n\}}.$$

(A generalization to means of $\{1^s, 2^s, \ldots, n^s\}$ $(s > -1)$ was proved by different [elementary] means by XU [2010].)

Although the constant π is ubiquitous in this subject (see CASTELLANOS [1988]), one seldom needs estimates of it finer than (8) and (8)$'$ in the proof of 3.14. Nevertheless, we will look at a few of the myriad series and integrals by which it can be explicitly calculated.

Theorem 3.22 (Gregory-Leibniz series) $\displaystyle \frac{\pi}{4} = \int_0^1 \frac{dx}{1 + x^2} = \sum_{n=0}^{\infty} \frac{(-1)^n}{2n + 1}.$

Proof: The quadratic polynomial $X^2 - i$ has exactly two zeros in $\mathbb{C}$, $\pm(1 + i)/\sqrt{2}$. Since, in the notation of the proof of 3.14,

$$\left[C\left(\frac{\pi}{4}\right) + iS\left(\frac{\pi}{4}\right)\right]^2 = E\left(i\frac{\pi}{4}\right)^2 = E\left(i\frac{\pi}{2}\right) = i$$

and C, S are non-negative in $\left[0, \frac{\pi}{2}\right]$, it must be that

(1) $$C\left(\frac{\pi}{4}\right) = S\left(\frac{\pi}{4}\right) = \frac{1}{\sqrt{2}}.$$

Since C is an even function and is never zero in $[0, \pi/2)$, by definition of π, we may form the function $T := S/C$ on $(-\pi/2, \pi/2)$ and have from equations (5) of the proof of 3.14 that

(2) $$T' = \frac{S'C - SC'}{C^2} = \frac{C^2 + S^2}{C^2} = 1 + (S/C)^2 = 1 + T^2 > 0.$$

Therefore T is strictly increasing. Since $T(0) = 0$ and $T(t) \to +\infty$ as $t \uparrow \frac{\pi}{2}$, the range of the odd function T is $\mathbb{R}$. Now $T^{-1}\colon \mathbb{R} \mapsto (-\pi/2, \pi/2)$ exists and by (2)

$$(3) \qquad \int_0^x \frac{1}{1+t^2}\,dt = \int_{T^{-1}(0)}^{T^{-1}(x)} \frac{T'}{1+T^2} = \int_{T^{-1}(0)}^{T^{-1}(x)} 1 = T^{-1}(x) \qquad \forall x \in \mathbb{R}$$

and in particular, because of (1),

$$(3)' \qquad \int_0^1 \frac{1}{1+t^2}\,dt = T^{-1}(1) = \frac{\pi}{4}.$$

For every $N \in \mathbb{N}$, $t \in \mathbb{R}$

$$\sum_{n=0}^{N-1} \left(-t^2\right)^n = \frac{1-\left(-t^2\right)^N}{1-\left(-t^2\right)} = \frac{1}{1+t^2} - \frac{\left(-1\right)^N t^{2N}}{1+t^2}$$

and therefore

$$\begin{aligned} \left|\sum_{n=0}^{N-1} \frac{\left(-1\right)^n x^{2n+1}}{2n+1} - \int_0^x \frac{1}{1+t^2}dt\right| &= \left|\int_0^x \left[\sum_{n=0}^{N-1} \left(-t^2\right)^n - \frac{1}{1+t^2}\right] dt\right| \\ (4) \qquad &= \left|\int_0^x \frac{\left(-1\right)^{N-1} t^{2N}}{1+t^2} dt\right| \qquad \forall x \in \mathbb{R}. \end{aligned}$$

The trivial inequalities $\frac{1}{2}t^{2N} \le t^{2N}(1+t^2)^{-1} \le t^{2N}$ for $t \in [-1,1]$ let us estimate the last integral in (4), below and above, yielding for $x \in [-1,1]\,, N \in \mathbb{N}$.

$$(5) \qquad \frac{|x|^{2N+1}}{4N+2} \le \left|\sum_{n=0}^{N-1} \frac{\left(-1\right)^n x^{2n+1}}{2n+1} - \int_0^x \frac{1}{1+t^2}dt\right| \le \frac{|x|^{2N+1}}{2N+1}.$$

In particular, using (3)

$$(6) \qquad T^{-1}(x) = \sum_{n=0}^{\infty} \frac{\left(-1\right)^n x^{2n+1}}{2n+1} \qquad \forall x \in [-1,1]\,.$$

Now equations (3)′ and (6) prove the theorem, but (3)′ and (5) also furnish the estimate

$$(7) \qquad \frac{1}{4N+2} \le \left|\sum_{n=0}^{N-1} \frac{\left(-1\right)^n}{2n+1} - \frac{\pi}{4}\right| \le \frac{1}{2N+1} \qquad \forall N \in \mathbb{N}.$$

Remarks: (i) See 3.37(vii), 3.38(viii) and 11.37(viii) [with $z := 1/2$] for different proofs. This series converges excruciatingly slowly, as (7) shows; for one converging to π with geometric rapidity, see 5.82(ii). After checking (see 3.30(ii) Hints) that $T(\pi/6) = 1/\sqrt{3}$, one can also take $x := 1/\sqrt{3}$ in (6) and get the even more rapidly converging representation

$$\frac{\pi}{6} = \frac{1}{\sqrt{3}} \sum_{n=0}^{\infty} \frac{(-1)^n}{2n+1} \cdot \frac{1}{3^n}. \tag{8}$$

In 3.45(v) we will see that for each $x \in \overline{\mathbb{D}} \setminus \{-i, i\}$ the series (6) converges to a solution w of the equation $\tan w = x$. (See def. 3.27.)

(ii) Besides using the partial sums of the series or Riemann-sum approximations to the integral, there is a third (more efficient?) way to exploit 3.22 for rational approximations to π. Each $n \in \mathbb{N}$ determines a polynomial p_n with integer coefficients such that $x^{4n}(1-x)^{4n} = p_n(x)(1+x^2) + (-4)^n$. (One checks that $p_1(x) := x^6 - 4x^5 + 5x^4 - 4x^2 + 4$ works and the rest is induction.) Therefore $\int_0^1 x^{4n}(1-x)^{4n}(1+x^2)^{-1}dx$ is a rational number minus $(-4)^{n-1}\pi$. This integral is positive and less than $[\max\{x(1-x) : 0 \leq x \leq 1\}]^{4n} = 4^{-4n}$. A rational approximation to π closer than $4^{-4n-(n-1)} = 4^{-5n+1}$ emerges. In case $n = 1$, $\int_0^1 x^4(1-x)^4(1+x^2)^{-1}dx = 22/7 - \pi$. Since

$$\frac{1}{630} = \int_0^1 x^4(1-x)^4 dx > \int_0^1 x^4(1-x)^4(1+x^2)^{-1}dx > \int_0^1 x^4(1-x)^4 2^{-1}dx = \frac{1}{1260},$$

we get $1/630 > 22/7 - \pi > 1/1260$, and so

$$3 + \frac{10}{71} < \frac{22}{7} - \frac{1}{630} < \pi < \frac{22}{7} - \frac{1}{1260} < 3 + \frac{1}{7}.$$

This is from Dalzell [1944], with more in Lucas [2009]. The extreme inequalities are Archimedes' famous estimate; his geometric construction can be found on pp. 31–34 of C. H. Edwards [1979].

(iii) Yet a fourth way to utilize 3.22 is to express $\pi/4 = \arctan(1)$ as a rational linear combination of $\arctan(\varepsilon_j)$ for small ε_j (for which the series is more rapidly convergent). One of these goes back to Euler: $\pi/4 = 4\arctan(1/5) - \arctan(1/70) + \arctan(1/99)$. There are numerous such relations; see pp. 274–275 of Berggren, Borwein and Borwein [2004], pp. 59–65 of Eymard and Lafon [2004] and pp. 90–98 of Castellanos [1988].

(iv) In 1755, Euler found a series which, in contrast to (6), converges to $T^{-1}(x)$ throughout $\mathbb{R}$:

$$T^{-1}(x) = \frac{x}{1+x^2} \sum_{k=0}^{\infty} \frac{2^{2k}(k!)^2}{(2k)!} \frac{1}{2k+1} \left(\frac{x^2}{1+x^2}\right)^k \qquad \forall x \in \mathbb{R}. \tag{9}$$

To confirm it, change the variable of integration in (3) via $t =: xC(\theta)$, $x \neq 0$, $\theta \in [0, \pi/2]$, to get

$$(1+x^2)/(1+t^2) = 1/[1 - \frac{x^2}{1+x^2} S^2(\theta)] = \sum_{k=0}^{\infty} \left(\frac{x^2}{1+x^2}\right)^k S^{2k}(\theta).$$

After commuting the uniformly converging series in $S(\theta)$ with the integration, the kth coefficient is $\int_0^{\pi/2} S^{2k+1}(\theta)d\theta$; that it equals $2^{2k}(k!)^2/(2k+1)!$ is confirmed by induction, reduction of k being accomplished by an integration by parts. [The Yin & Yang relation between (9) and 5.82(iv) is noteworthy.]

(v) Since $(T^{-1})'(t) = 1/(1+t^2)$, we have

$$\frac{1}{2}(T^{-1}(x))^2 = \int_0^x T^{-1}(t)/(1+t^2)\, dt.$$

Use (9) on the integrand, permute integration and summation, and make the change of integration variable $u := t^2/(1+t^2)$. You get a combinatorially remarkable companion to (9):

$$(9)^2 \qquad (T^{-1}(x))^2 = \sum_{k=0}^{\infty} \frac{2^{2k+1}(k!)^2}{(2k+2)!} \left(\frac{x^2}{1+x^2}\right)^{k+1} \qquad \forall x \in \mathbb{R}.$$

Exercise 3.23 *With the convention $0^0 := 1$, the function x^{-x} is continuous on $[0,1]$. (Cf. 3.18(4) and recall the customary definition $x^{-x} := e^{-x \log x}$ for $x > 0$.) Prove the satisfyingly symmetric formula*

$$\int_0^1 x^{-x}\, dx = \sum_{n=1}^{\infty} n^{-n}.$$

Hints: $x^{-x} = \sum_0^\infty (-1)^n (x \log x)^n / n!$. Interchange summation and integration and evaluate $\int_0^1 x^n \log^n(x)\, dx$ by repeated integration-by-parts.

Exercise 3.24 *Consider $f(x) := \int_0^1 (1+t^2)^{-1} e^{-x^2(1+t^2)} dt$ and $g(x) := \int_0^x e^{-t^2}\, dt$, $x \in [0, \infty)$.*

(i) *Show that $0 < f(x) < e^{-x^2} f(0)$ for all $x > 0$.*

(ii) *Use 2.19(i) to show that $f'(x) = -2e^{-x^2} g(x)$ for all $x \geq 0$; that is, $f' = (-g^2)'$.*

(iii) *Prove that $\int_0^\infty e^{-t^2}\, dt = \sqrt{\pi}/2$.*

Hints: Integrate in (ii) from 0 to x, let $x \to \infty$ and use (i); evaluate $f(0)$ via 3.22.

(iv) *Show that* $\int_0^\infty t^k e^{-\pi t^2} \cos(2\pi x t)\, dt$ *and* $\int_0^\infty t^k e^{-\pi t^2} \sin(2\pi x t)\, dt$ *exist for each* $x \in \mathbb{R}$, $k \in \mathbb{N} \cup \{0\}$, *and they define functions* $C_k(x)$, $S_k(x)$ *which satisfy* $C_k' = -2\pi S_{k+1}$. (Cf. 2.19(i) and glance ahead to definition 3.27)

Hints: The power series shows that $e^x \geq x^{k+1}/(k+1)!$ for all $x \geq 0$, $k \in \mathbb{N} \cup \{0\}$. Hence $t^k e^{-\pi t^2} \leq (k+1)!/\pi^{k+1} t^{k+2}$, an inequality whose validity for all $t \geq 1$ clearly insures the existence of the relevant improper integrals. For real $\alpha < \beta$

$$\left| \frac{\cos\beta - \cos\alpha}{\beta - \alpha} + \sin\alpha \right| = \frac{1}{\beta - \alpha} \left| \int_\alpha^\beta (\sin\alpha - \sin t) dt \right| \leq \sup_{t \in [\alpha, \beta]} |\sin\alpha - \sin t|$$

$$= \sup_{t \in [\alpha, \beta]} \left| \int_\alpha^t \cos \right| \leq \beta - \alpha.$$

If we apply this inequality in the relevant integrand, we see that for all real $x \neq x_0$

$$\left| \frac{C_k(x) - C_k(x_0)}{x - x_0} + 2\pi S_{k+1}(x_0) \right| \leq \int_0^\infty 2\pi t^{k+1} e^{-\pi t^2} |2\pi x t - 2\pi x_0 t|\, dt$$
$$= 4\pi^2 C_{k+2}(0) |x - x_0|.$$

(v) *Identify the function* C_0 *of* (iv) *as being* $C_0(x) = \frac{1}{2} e^{-\pi x^2}$ *for all* $x \in \mathbb{R}$.

Hints: Using (iv) and integration by parts deduce that $e^{\pi x^2} C_0(x)$ has derivative **0**. The constant value $C_0(0)$ of the function $e^{\pi x^2} C_0(x)$ is found via (iii).

(vi) *Show that* $\int_{-\infty}^\infty e^{-\pi t^2} e^{-2\pi i x t}\, dt = e^{-\pi x^2}$ *for all* $x \in \mathbb{R}$.

(vii) *In the integral defining* f *replace* $-x^2$ *by* ix^2 *and in the integral defining* g *replace* $-t^2$ *by* it^2. *Call the new functions* f_i, g_i, *respectively, and carry through for them the program in* (ii) *to get*

(1)
$$g_i^2(x) = \frac{\pi i}{4} - i e^{ix^2} \int_0^1 \frac{e^{ix^2 t^2}}{1 + t^2}\, dt \qquad \forall x \in \mathbb{R}.$$

Infer from this and 3.22 that

(2)
$$|g_i^2(x)| \leq \pi/2 \qquad \forall x \in \mathbb{R}.$$

(viii) *Use integration by parts to show that*

$$\int_0^1 \frac{e^{ix^2 t^2}}{1 + t^2}\, dt = \frac{1}{2x} g_i(x) + \frac{1}{x} \int_0^1 \frac{2t}{(1 + t^2)^2} g_i(xt) dt \qquad \forall x > 0.$$

Infer from this and (2) that $\int_0^1 (1+t^2)^{-1} e^{ix^2t^2}\,dt \to 0$ *as* $x \to \infty$*; then from the latter and (1) infer that*

(3) $$\lim_{x\to\infty} g_i^2(x) = \frac{\pi i}{4}.$$

(ix) *From (3) and continuity of* g_i *infer that*

$$\int_0^\infty e^{it^2}\,dt := \lim_{x\to\infty} g_i(x) \text{ exists and equals } \sqrt{\frac{\pi}{2}}\left(\frac{1+i}{2}\right).$$

Hints: From (3) we know that $\operatorname{Im} g_i^2(x) > \pi/8$ for all large x. Consequently $\operatorname{Im} g_i(x)$ is of constant sign for large x. Since $\sin \geq 0$ in $[2k\pi, (2k+1)\pi]$ for all $k \in \mathbb{Z}$ (see definition 3.27 and the proof of 3.14),

$$\begin{aligned}
0 &\leq \sum_{k=0}^{n-1} \int_{2k\pi}^{(2k+1)\pi} \sin t\left(\frac{1}{\sqrt{t}} - \frac{1}{\sqrt{t+\pi}}\right)dt \\
&= \sum_{k=0}^{n-1}\left(\int_{2k\pi}^{(2k+1)\pi} \frac{\sin t}{\sqrt{t}}\,dt + \int_{(2k+1)\pi}^{2(k+1)\pi} \frac{\sin t}{\sqrt{t}}\,dt\right) \\
&= \int_0^{2n\pi} \frac{\sin t}{\sqrt{t}}\,dt = \int_0^{\sqrt{2n\pi}} 2\sin(t^2)\,dt = 2\operatorname{Im} g_i(\sqrt{2n\pi}) \qquad \forall n \in \mathbb{N}.
\end{aligned}$$

Therefore the eventual constant sign of $\operatorname{Im} g_i(x)$ is $+$. Setting $A := \sqrt{\frac{\pi}{2}}(\frac{1+i}{2})$, we have $A^2 = \frac{\pi i}{4}$ and (3) reads

$$\lim_{x\to\infty} [g_i(x) - A][g_i(x) + A] = 0.$$

As $\operatorname{Im}[g_i(x)+A] \geq \operatorname{Im} A = \sqrt{\pi/8}$ for all large x, this product can only converge to 0 if its first factor does.

Remarks: The results (iii) and (ix) will be derived by complex-analysis and Fourier analysis methods in 11.32 and its Remark (iii). The real-variables techniques here are adapted from Anonyme [1889], Bloch [1922], R. M. Young [1991], and especially Výborný [1991], where $\int_0^\infty e^{-ct^2}\,dt$ is treated for arbitrary non-zero $c \in \overline{\mathbb{H}}$, thereby getting (iii) and (ix) in one stroke. For history see footnotes 1433–1436 in Burkhardt [1914].

Exercise 3.25 (i) *Show that if* f *is differentiable in an open, connected subset* Ω *of* $\mathbb{C}$ *or* $\mathbb{R}$ *and* $f' = cf$ *for some constant* c*, then*

$$f(z) = f(a)e^{-ca}e^{cz} \qquad \forall a, z \in \Omega.$$

Hint: Let $g(z) := e^{-cz}$ and note that $(f \cdot g)' = 0$ throughout Ω.

(ii) *Find all continuous homomorphisms f of the additive group $\mathbb{R}$ into the multiplicative groups $\mathbb{T}$, $\mathbb{C} \setminus \{0\}$ and $\mathbb{R} \setminus \{0\}$.* Cf. 4.23(iii).

Hints: Since f is a homomorphism, $f(0) = 1$. By continuity then there exists $a > 0$ such that $\int_0^a f = b \neq 0$. The homomorphic property $f(x+t) = f(x)f(t)$ (all $x, t \in \mathbb{R}$) then implies that

$$bf(x) = f(x)\int_0^a f(t)\,dt = \int_0^a f(x+t)\,dt = \int_x^{x+a} f = \int_0^{x+a} f - \int_0^x f.$$

From this and the fundamental theorem of calculus we learn that the function f is not only continuous but is differentiable, and

$$bf'(x) = f(x+a) - f(x) = f(x)[f(a)-1] \qquad \forall x \in \mathbb{R},$$

so $f' = cf$, with $c := [f(a)-1]/b$. This leads via (i) to $f(x) = f(0)e^{cx} = e^{cx}$. It is thus obvious that f maps into $\mathbb{T}$ if and only if $\operatorname{Re} c = 0$, and maps into $\mathbb{R}$ if and only if c is real.

(iii) *Show that every continuous homomorphism h of the multiplicative group $\mathbb{R}\setminus\{0\}$ into the additive group $\mathbb{R}$ has the form $c\log\circ|\ \ |$ for some $c \in \mathbb{R}$.*

Hints: According to (ii) there is a $c \in \mathbb{R}$ such that $e^{h(e^x)} = e^{cx}$ for all $x \in \mathbb{R}$. For $x = \log y$ this gives $h(y) = cx = c\log y$, for all $y > 0$. But $0 = h(1) = h((-1)(-1)) = 2h(-1)$, so $h(-y) = h(-1) + h(y) = h(y)$ for all $y \in \mathbb{R} \setminus \{0\}$.

(iv) *Show that* (iii) *remains valid if at every occurrence $\mathbb{R}$ is replaced by $\mathbb{C}$.*

Hint: Apply (iii) to $\operatorname{Re} h$ and $\operatorname{Im} h$ restricted to $\mathbb{R} \setminus \{0\}$ to come up with $c \in \mathbb{C}$ such that $h(x) = c\log x$ for all $x > 0$. Since continuous homomorphism h maps $\mathbb{T}$ onto a compact additive subgroup of $\mathbb{C}$, $h(\mathbb{T}) = \{0\}$. Thus $h(z) = h(|z|) + h(z/|z|) = h(|z|)$, for all $z \in \mathbb{C} \setminus \{0\}$.

Exercise 3.26 *Let $\phi\colon [0,1] \mapsto \mathbb{C}$ be continuous and satisfy $\phi(1) = \phi(0)$. Show that there exists a continuous $\Phi\colon \mathbb{T} \mapsto \mathbb{C}$ satisfying*

$$(*) \qquad \Phi\big(e^{2\pi i t}\big) = \phi(t) \qquad \forall t \in [0,1].$$

Hints: By 3.14(v) and the $2\pi i$-periodicity of the exponential function, the map $t \mapsto e^{2\pi i t}$ sends $[0,1)$ one-to-one onto the circle $\mathbb{T}$. Hence we can define Φ on $\mathbb{T}$ by the decree

$$(1) \qquad \Phi\big(e^{2\pi i t}\big) = \phi(t) \qquad \forall t \in [0,1)\,.$$

The condition $\phi(1) = \phi(0)$ ensures that in fact $(*)$ holds. Suppose now that Φ is not continuous at some $z_0 \in \mathbb{T}$. This means that there exists an $\varepsilon > 0$ and a sequence $\{z_n\} \subset \mathbb{T}$ such that $z_n \to z_0$ and

$$(2) \qquad |\Phi(z_n) - \Phi(z_0)| \geq \varepsilon, \quad n \in \mathbb{N}.$$

For each $n \in \mathbb{N}$ there is a unique $t_n \in [0, 1)$ such that $z_n = e^{2\pi i t_n}$ and there is a subsequence of $\{t_n\}$ which converges to some $t_0 \in [0, 1]$. Passing to that subsequence, we can simply suppose that $\{t_n\}$ itself converges to t_0. Then $z_n = e^{2\pi i t_n} \to e^{2\pi i t_0}$, and so $z_0 = e^{2\pi i t_0}$. But then

$$\Phi(z_n) = \phi(t_n) \to \phi(t_0) \overset{(*)}{=} \Phi\big(e^{2\pi i t_0}\big) = \Phi(z_0),$$

in violation of (2).

It is customary to introduce functions cos, sin, tan and cot by

Definition 3.27 The functions cos and sin are defined for all $z \in \mathbb{C}$ by

$$\cos(z) := \frac{e^{iz} + e^{-iz}}{2} = \sum_{n=0}^{\infty} \frac{(-1)^n z^{2n}}{(2n)!}$$

$$\sin(z) := \frac{e^{iz} - e^{-iz}}{2i} = \sum_{n=0}^{\infty} \frac{(-1)^n z^{2n+1}}{(2n+1)!}.$$

The function tan is defined as sin/cos. Since the zeros of cos are evidently the odd integer multiples of $\pi/2$, the domain of tan is $\mathbb{C} \setminus \pi(\mathbb{Z} + 1/2)$. The function cot is defined as cos/sin, with the appropriate domain, viz., $\mathbb{C} \setminus \pi\mathbb{Z}$ (cf. 3.30(iii)). The closely related entire functions $z \mapsto \cos(iz) = \frac{1}{2}(e^z + e^{-z})$ and $z \mapsto -i\sin(iz) = \frac{1}{2}(e^z - e^{-z})$, the even and odd parts of the exponential function, are designated $\cosh(z)$ and $\sinh(z)$ [in French literature, $ch(z)$ and $sh(z)$].

Remarks: The functions C and S in the proof of 3.14 were just cos and sin restricted to $\mathbb{R}$ and are known collectively as the *circular* or *cyclometric* or *trigonometric* functions. The functions cos and sin are infinitely differentiable throughout $\mathbb{C}$ and $\cos' = -\sin$, $\sin' = \cos$. For each $x \in \mathbb{R}\setminus\mathbb{Z}$, $y \in \mathbb{R}\setminus\{0\}$ they map the lines $\frac{1}{2}\pi x + i\mathbb{R}$, $\mathbb{R} + iy$ onto semi-hyperbolas and ellipses with foci ± 1, as the reader can easily confirm. The (only occasionally encountered) name cis for the function $z \mapsto e^{iz}$ is suggested by the identity $e^{iz} = \cos(z) + i\sin(z)$. The functions cos, sin, and tan map $[0, \pi]$, $[-\pi/2, \pi/2]$, and $(-\pi/2, \pi/2)$ bijectively onto $[-1, 1]$, $[-1, 1]$, and $\mathbb{R}$, respectively. The inverses of these restrictions are sometimes denoted by $\cos^{-1}$, $\sin^{-1}$, and $\tan^{-1}$ (a slight abuse of language), but preferably by arccos, arcsin, and arctan. The rationale for the latter notation is the fact (see the Remarks following 3.16) that every $x+iy \in \mathbb{T}_+$ is $\cos\theta + i\sin\theta$ for a unique $\theta \in (0,\pi)$, and this θ is the length of the counterclockwise *arc* of $\mathbb{T}$ with initial point 1 and terminal point $x+iy$; so $\theta = \cos^{-1}(x)$ is an arclength.

Cosh and sinh are known as *hyperbolic trigonometric* functions. Notice that cosh is an even function which effects a strictly increasing mapping of $[0,\infty)$ onto $[1,\infty)$, and sinh is an odd function which effects a strictly increasing mapping of $\mathbb{R}$ onto $\mathbb{R}$. The inverses of these restrictions are denoted arcosh and arsinh, respectively. Just as $t \mapsto (\cos t, \sin t)$, $t \in [-\pi/2, \pi/2]$, parameterizes the right half of the circle $x^2 + y^2 = 1$, the function $t \mapsto (\cosh t, \sinh t)$, $t \in \mathbb{R}$, parameterizes the right half of the "equilateral" hyperbola $x^2 - y^2 = 1$. But the parameter t appears to lose its arclength significance in the latter case. Rather it is now an area: The sector bounded by the circle and its two radii $[0, x \pm iy]$ $(x > 0, y > 0)$ has area $\arccos(x) = \arcsin(y)$, while the analogous region bounded by the hyperbola and its "radii" $[0, x \pm iy]$ has area $\operatorname{arcosh}(x) = \operatorname{arsinh}(y)$, as the reader can routinely compute. These facts elucidate the use of the adjective "hyperbolic" and the fact that the prefix "ar" is intended to suggest "area".

In the spirit of *ars gratia artis* I want to briefly divert from business to offer the reader a glimpse into "the BOOK," Paul Erdös' famous fantasy repository of proofs that the Almighty itself judges to be *ne plus ultra*. The idea of this proof goes back to Charles Hermite, in 1873, but the exposition here follows AIGNER and ZIEGLER [2010]. T. W. JONES [2010], ZHOU and MARKOV [2010] and ZHOU [2011] analyze the mechanism of the π-proof and its evolution. They also show how these ideas can prove the irrationality of $\cos(r)$ and $\tan(r)$ for every non-zero rational r. Other elementary proofs will be found in HANČL [1986], DESBROW[1990] (via complex analysis), LACZKOVICH [1997], and HUYLEBROUCK [2001].

Theorem 3.28 *π^2 is irrational, and e^m is irrational for every $m \in \mathbb{N}$.*

Proof: A central role in both proofs is played by the $2n$-degree polynomial

$$(1) \qquad p(x) := x^n(1-x)^n/n!,$$

$n \in \mathbb{N}$ to be specified later. The properties needed are

$$(2) \qquad 0 < p(x) < 1/n! \qquad \forall x \in (0,1),$$

$$(3) \qquad p^{(k)}(0), p^{(k)}(1) \in \mathbb{Z} \qquad \forall k \in \mathbb{Z}_+.$$

Since $0 < x(1-x) < 1$ for $x \in (0,1)$, (2) is clear. Evidently the polynomial $q(x) := x^n(1-x)^n$ has the form $\sum_{j=n}^{2n} c_j x^j$ with coefficients $c_j \in \mathbb{Z}$ (use, say, induction on n), and so $q^{(k)}(0) = 0$ if $k < n$ or $k > 2n$. If, on the other hand, $n \le k \le 2n$, then $q^{(k)}(x) = \sum_{j=k}^{2n} j(j-1)\dots(j-k+1)c_j x^{j-k}$, so $q^{(k)}(0) = k!c_k$, an integer multiple of $n!c_k$. Hence $p^{(k)}(0) = q^{(k)}(0)/n! \in \mathbb{Z}$. And since $p(x) = p(1-x)$, $p^{(k)}(1) = (-1)^k p^{(k)}(0) \in \mathbb{Z}$.

Now suppose first that for some $m, a, b \in \mathbb{N}$

$$e^m = a/b. \tag{4}$$

Since the power series for e^{m^2} converges, we may pick $n \in \mathbb{N}$ so that

$$(m^2)^n/n! < 1/am. \tag{5}$$

With the p in (1) form

$$P := \sum_{k=0}^{2n} (-1)^k m^{2n-k} p^{(k)}.$$

Since $p^{(2n+1)} = \mathbf{0}$, we see that

$$\begin{aligned} P' = \sum_{k=0}^{2n-1} (-1)^k m^{2n-k} p^{(k+1)} &= \sum_{j=1}^{2n} (-1)^{j-1} m^{2n-(j-1)} p^{(j)} \\ &= -m \sum_{j=1}^{2n} (-1)^j m^{2n-j} p^{(j)} = -m[P - m^{2n}p]. \end{aligned}$$

Hence

$$\frac{d}{dx}[e^{mx}P(x)] = me^{mx}P(x) + e^{mx}P'(x) = m^{2n+1}e^{mx}p(x).$$

Consequently, the mean-value theorem supplies $x \in (0,1)$ such that $e^m P(1) - e^0 P(0) = m^{2n+1}e^{mx}p(x)$. That is, in view of (4),

$$N := aP(1) - bP(0) = bm^{2n+1}e^{mx}p(x), \tag{6}$$

which is an integer, due to (3) and the definition of P. From (2) and (6)

$$0 < N \le bm^{2n+1}e^m/n! \overset{(4)}{=} m^{2n+1}a/n! \overset{(5)}{<} 1,$$

inequalities which contradict the fact $N \in \mathbb{Z}$ and thus prove that (4) is untenable.

Turning to the π^2-claim, its proof being similar, a few computations can be left to the reader. Suppose then that for some $a, b \in \mathbb{N}$

$$\pi^2 = a/b, \tag{7}$$

and select $n \in \mathbb{N}$ such that

$$a^n/n! < 1/\pi. \tag{8}$$

With the p in (1) form

$$P := b^n \sum_{k=0}^{n} (-1)^k \pi^{2(n-k)} p^{(2k)}$$

and calculate that

$$P'' = -\pi^2 P + b^n \pi^{2n+2} p.$$

Use this to further calculate

$$\begin{aligned}\frac{d}{dx}\Big[\pi^{-1}P'(x)\sin\pi x - P(x)\cos\pi x\Big] &= \Big[\pi^{-1}P''(x) + \pi P(x)\Big]\sin\pi x \\ &= b^n\pi^{2n+1}p(x)\sin\pi x\end{aligned}$$

and from this and the mean-value theorem get an $x \in (0,1)$ such that

$$N := P(1) + P(0) = b^n \pi^{2n+1} p(x)\sin\pi x. \tag{9}$$

Because of (7), $b^n\pi^{2(n-k)} \in \mathbb{N}$ for each $k = 0, 1, \ldots, n$ and therefore (3) and the definition of P tell us that $P(0)$, $P(1) \in \mathbb{Z}$. Hence $N \in \mathbb{Z}$. But from (9) and (2)

$$0 < N \le b^n\pi^{2n+1}/n! \overset{(7)}{=} a^n\pi/n! \overset{(8)}{<} 1,$$

inequalities which contradict $N \in \mathbb{Z}$ and thus prove that (7) is untenable.

I think readers will agree: a proof "from the BOOK."

Exercise 3.29 (i) $|\sin x| \le |x|$ *for all real* x.

Hint: For each $x \neq 0$ the Mean-Value Theorem provides an x' such that

$$\left|\frac{\sin x}{x}\right| = \left|\frac{\sin(x) - \sin(0)}{x - 0}\right| = |\sin'(x')| = |\cos(x')| \le 1.$$

(ii) $\cos x \ge 1 - 2x/\pi$ *for all* x *in* $[0, \pi/2]$.

Hints: The function $g(x) := \cos x + 2x/\pi$ satisfies $g(0) = g(\pi/2) = 1$. Also $g' = 2/\pi - \sin$. From equation (10) in the proof of 3.14, sin increases on $[0, \pi/2]$ from $\sin(0) = 0$ to $\sin(\pi/2) = 1 > 2/\pi$. Therefore g' has a unique zero $x_0 \in [0, \pi/2]$, $g' > 0$ on $[0, x_0)$ and $g' < 0$ on $(x_0, \pi/2]$. It follows that g is increasing on $[0, x_0]$, decreasing on $[x_0, \pi/2]$ and so $\min g[0, \pi/2] = g(0) = g(\pi/2) = 1$.

Exercise 3.30 (i) $\sin x \geq 2x/\pi$ *for all* $x \in [0, \pi/2]$.

Hints: The function $h(x) := \sin x - 2x/\pi$ satisfies $h(0) = h(\pi/2) = 0$ and $h' = -2/\pi + \cos$. Now $\cos' = -\sin$ is negative on $(0, \pi/2]$, so h' decreases from $h'(0) = -2/\pi + 1 > 0$ to $h'(\pi/2) = -2/\pi < 0$. As in the hints of 3.29(ii), it follows that $\min h[0, \pi/2] = h(0) = h(\pi/2) = 0$. (For refinements of (i), see **MR** 2006m:26027.)

(ii) $|\sin(\pi z)| > 1/2$ *for all* $z \in \mathbb{C} \setminus \bigcup_{n \in \mathbb{Z}} D(n, 1/4)$.

Hints: Consider the hyperbolic sine function

$$h(t) := \frac{1}{2}\left(e^t - e^{-t}\right) = \sum_{k=0}^{\infty} \frac{t^{2k+1}}{(2k+1)!}, \qquad t \in \mathbb{R}.$$

Evidently $h(|t|) \geq |t|$. By elementary algebra, if $x, y \in \mathbb{R}$ and $z := x + iy$, then $|\sin(\pi z)|^2 = \sin^2(\pi x) + h^2(\pi y)$. Suppose $|\sin(\pi z)|^2 \leq 1/4$. Then

$$\text{(1)} \qquad \sin^2(\pi x) \leq 1/4 \quad \text{and} \quad h^2(\pi y) \leq 1/4.$$

Since $(\sqrt{3} + i)/2$ and $e^{\pi i/6}$ each lie in $\mathbb{U} \cap \mathbb{H}$ and each is a cube root of i, they must be equal. So $\sin(\pi/6) = 1/2$. Since sin is an odd function, strictly increasing on $[-\pi/2, \pi/2]$, it follows that $t \in [-\pi/2, \pi/2]$ & $|\sin t| \leq 1/2 \Rightarrow |t| \leq \pi/6$. For the x in (1) choose $n \in \mathbb{Z}$ such that $|x - n| \leq 1/2$. Then $\pi x - n\pi \in [-\pi/2, \pi/2]$ and $|\sin(\pi x - n\pi)| = |\sin(\pi x)| \leq 1/2$, so $|\pi x - n\pi| \leq \pi/6$. Furthermore, since

$$(\pi y)^2 = |\pi y|^2 \leq h^2(|\pi y|) = h^2(\pi y),$$

inequality (1) yields $(\pi y)^2 \leq 1/4$ and so

$$\text{(2)} \qquad |z - n|^2 = (x - n)^2 + y^2 \leq \left(\frac{1}{6}\right)^2 + \frac{1}{4\pi^2}.$$

But by $(8)'$ in the proof of 3.14, $\pi > 3$, so (2) gives

$$|z - n|^2 < \left(\frac{1}{6}\right)^2 + \left(\frac{1}{6}\right)^2 < \left(\frac{1}{4}\right)^2.$$

(iii) *Show that the only zeros in* $\mathbb{C}$ *of* sin *are those in* $\pi\mathbb{Z}$ *and the only periods in* $\mathbb{C}$ *of* sin *are those in* $2\pi\mathbb{Z}$.

Hints: $\sin z = 0 \Leftrightarrow e^{2iz} - 1 = 0 \Leftrightarrow 2iz \in 2\pi i\mathbb{Z}$. From the (routinely confirmed) identity

$$\sin(z + 2p) - \sin(z) = 2\cos(z + p)\sin(p) \qquad \forall z, p \in \mathbb{C}$$

it follows that $2p$ is a period of sin if and only if $\sin(p) = 0$.

Exercise 3.31 $\int_0^{\pi/4} e^{-R\cos 2x}\,dx \leq \pi/4R$ *for all* $R > 0$.

Hint: By 3.29(ii), $e^{-R\cos 2x} \leq e^{-R(1-4x/\pi)} = e^{-R}e^{(4R/\pi)x}$, so

$$\int_0^{\pi/4} e^{-R\cos 2x}\,dx \leq \left[e^{-R}\frac{\pi}{4R}e^{(4R/\pi)x}\right]_{x=0}^{x=\pi/4} < e^{-R}\cdot\frac{\pi}{4R}\cdot e^{4(R/\pi)(\pi/4)}.$$

Exercise 3.32 *Suppose* $\lim_{N\to\infty}\sum_{n=-N}^{N} c_n z^n$ *exists uniformly on* $C(0,r)$, *for some* $r > 0$. *Denote this limit by* $f(z)$.

(i) *Prove* Parseval's Formula

$$\frac{1}{2\pi}\int_0^{2\pi} \left|f\left(re^{it}\right)\right|^2 dt = \sum_{n=-\infty}^{\infty} |c_n|^2 r^{2n} := \lim_{N\to\infty}\sum_{n=-N}^{N} |c_n|^2 r^{2n}.$$

Hints: The functions $S_N(z) := \sum_{n=-N}^{N} c_n z^n$ satisfy

$$\begin{aligned}\left|S_N\left(re^{it}\right)\right|^2 &= \left[\sum_{n=-N}^{N} c_n r^n e^{int}\right]\left[\sum_{k=-N}^{N} \bar{c}_k r^k e^{-ikt}\right]\\ &= \sum_{n=-N}^{N}\sum_{k=-N}^{N} c_n\bar{c}_k r^{n+k} e^{i(n-k)t},\end{aligned}$$

and so

$$\int_0^{2\pi} \left|S_N\left(re^{it}\right)\right|^2 dt = \sum_{n=-N}^{N}\sum_{k=-N}^{N} c_n\bar{c}_k r^{n+k}\int_0^{2\pi} e^{i(n-k)t}\,dt = \sum_{n=-N}^{N} 2\pi|c_n|^2 r^{2n}.$$

By hypothesis, $S_N\left(re^{it}\right) \to f(re^{it})$ uniformly in $t \in \mathbb{R}$ for $N \to \infty$.

(ii) *Show that if* $c_n = 0$ *for all* $n < 0$, *then*

$$\frac{1}{2\pi}\int_0^{2\pi} \left[\operatorname{Re} f\left(re^{it}\right)\right]^2 dt = (\operatorname{Re} c_0)^2 + \frac{1}{2}\sum_{n=1}^{\infty} |c_n|^2 r^{2n}$$

and infer that if additionally $|\operatorname{Re} f| \le A$, *then*

$$2(\operatorname{Re} c_0)^2 + \sum_{n=1}^{\infty} |c_n|^2 r^{2n} \le 2A^2.$$

Hints: Define

$$c_n^* := \begin{cases} c_n & \text{if } n > 0 \\ c_0 + \bar{c}_0 & \text{if } n = 0 \\ \bar{c}_{-n} & \text{if } n < 0 \end{cases}$$

and note that $c_k^* c_{-k}^* = |c_k^*|^2$ for all k. Hence

$$2 \operatorname{Re} S_N\big(re^{it}\big) = \sum_{n=0}^{N} c_n \big(re^{it}\big)^n + \sum_{n=0}^{N} \bar{c}_n \big(\overline{re^{it}}\big)^n = \sum_{k=-N}^{N} c_k^* r^{|k|} e^{ikt}$$

$$\big[2 \operatorname{Re} S_N\big(re^{it}\big)\big]^2 = \sum_{j=-N}^{N} \sum_{k=-N}^{N} c_k^* c_j^* r^{|k|+|j|} e^{i(k+j)t}, \quad \text{whence}$$

$$\int_0^{2\pi} \big[2 \operatorname{Re} S_N\big(re^{it}\big)\big]^2 dt = \sum_{k=-N}^{N} c_k^* c_{-k}^* r^{2|k|} \cdot 2\pi = 2\pi \sum_{k=-N}^{N} |c_k^*|^2 r^{2|k|}$$

$$= 2\pi \left[(2 \operatorname{Re} c_0)^2 + 2 \sum_{n=1}^{N} |c_n|^2 r^{2n} \right].$$

Now let $N \to \infty$

(iii) *For every* $j \in \mathbb{N}$ *the coefficient* c_j *in* (ii) *can be expressed as an integral of* $\operatorname{Re} f$ *over* $C(0, r)$. *Find explicit formulae.*

Hint: Multiply the above sum for $2 \operatorname{Re} S_N(re^{it})$ by e^{-ijt}, integrate and let $N \to \infty$ to learn that $\pi r^j c_j = \int_0^{2\pi} e^{-ijt} \operatorname{Re} f(re^{it})\, dt$.

(iv) *Consider* $R(z) := \sum_{j=-n}^{n} c_j z^j$ $(n \in \mathbb{N}, c_j \in \mathbb{C})$, $M(r) := \max\{|R(z)| : |z| = r\}$ *and* $m(r) := \min\{|R(z)| : |z| = r\}$ *for* $r > 0$. *The useful inequalities*

$$m^2(r) \le \sum_{j=-n}^{n} |c_j|^2 r^{2j} \le M^2(r) \qquad \forall r > 0$$

are a simple consequence of (i). *But they can be proved without recourse to either integrals or the exponential function, by a clever averaging technique that has broader applicability.*

Hints: (suggested by Oswaldo Rio Branco de Oliveira) Fix $u \in \mathbb{T}$ such that $1 \notin \{u^m : 1 \leq m \leq 2n\}$. Then define $K := \max\{|1-u^m|^{-1} : 1 \leq m \leq 2n\}$ zand note that

$$(*) \qquad \left|\sum_{k=0}^{N-1} u^{km}\right| = \left|\frac{1-u^{Nm}}{1-u^m}\right| \leq 2K \quad \text{for} \quad 1 \leq m \leq 2n, N \in \mathbb{N}.$$

Direct calculation delivers, for $z \neq 0$ and $k \in \mathbb{Z}$

$$|R(u^k z)|^2 = \sum_{-n\leq j\leq n} |c_j|^2 |z|^{2j} + 2 \sum_{-n\leq j<l\leq n} \operatorname{Re}[\bar{c}_j \bar{z}^j c_l z^l u^{k(l-j)}].$$

Summing over k gives

$$\begin{aligned} N\sum_{-n\leq j\leq n} |c_j|^2|z|^{2j} &\leq \sum_{k=0}^{N-1} |R(u^k z)|^2 + 2\sum_{-n\leq j<l\leq n} |c_j c_l||z|^{j+l}\left|\sum_{k=0}^{N-1} u^{k(l-j)}\right| \\ &\overset{(*)}{\leq} \sum_{k=0}^{N-1} |R(u^k z)|^2 + 4K \sum_{-n\leq j<l\leq n} |c_j c_l||z|^{j+l}. \end{aligned}$$

Taking $z \in C(0,r)$, this yields

$$N\sum_{-n\leq j\leq n} |c_j|^2 r^{2j} \leq NM^2(r) + 4K \sum_{-n\leq j<l\leq n} |c_j c_l| r^{j+l} \qquad \forall r > 0, N \in \mathbb{N}.$$

Divide by N and let $N \to \infty$. ("O frabjous day! Callooh! Callay!") Obvious sign changes in this argument lead to the $m^2(r)$ inequality.

(v) *Prove* Gutzmer's Inequalities:

$$m^2(r) \leq \sum_{n=-\infty}^{\infty} |c_n|^2 r^{2n} \leq M^2(r).$$

Hints: This too follows trivially from (i), but in Weierstrass' approach to function theory a proof without integration theory was desirable, and follows routinely from (iv): Given $\varepsilon > 0$, there is an $N(\varepsilon) \in \mathbb{N}$ such that the partial sums $S_N(z) := \sum_{n=-N}^{N} c_n z^n$ satisfy

$$|S_N(z) - f(z)| \leq \varepsilon \qquad \forall z \in C(0,r), N \geq N(\varepsilon).$$

The numbers $M_N := \max\{|S_N(z)| : |z| = r\}$ consequently satisfy

$$M_N \leq \varepsilon + \sup\{|f(z)| : |z| = r\} = \varepsilon + M(r),$$

and (iv) delivers

$$\sum_{n=-N}^{N} |c_j|^2 r^{2n} \le M_N^2 \le (\varepsilon + M(r))^2 \qquad \forall N \ge N(\varepsilon).$$

Let $N \to \infty$ and then let $\varepsilon \downarrow 0$.

Exercise 3.33 *The series* $\sum_{n=0}^{\infty} c_n z^n$ *converges to* $f(z)$ *for all* $z \in D(0,R)$ *and some* $0 < R \le \infty$.

(i) (Liouville) *Deduce from either* 3.32(i) *or* (v) *that if* f *is bounded and* $R = \infty$, *then* f *is constant.*

(ii) *Deduce from either* 3.32(i) *or* (v) *that if* $|f(z)| \le |f(0)|$ *for all* $z \in C(0,r)$ *and some* $0 < r < R$, *then* f *is constant.*

Hint: Parseval gives

$$\sum_{n=0}^{\infty} |c_n|^2 r^{2n} \le \frac{1}{2\pi} \int_0^{2\pi} |f(0)|^2 \, dt = |c_0|^2,$$

whence $\sum_{n=1}^{\infty} |c_n|^2 r^{2n} \le 0$.

(iii) (Maximum-Modulus Principle for functions of power series type). $F : \Omega \mapsto \mathbb{C}$ *is bounded in the region* Ω, *each point of which is the center of a disk in which* F *is represented by a power series. Show that if there is some* $a \in \Omega$ *such that* $|F(a)| = M := \sup |F(\Omega)|$, *then* F *is constant.*

Hints: Rotating F lets us assume $F(a) = M$. Consider the set $S := F^{-1}(M)$. By continuity of F it is relatively closed in Ω. If $c \in S$, we can apply (ii) to infer that $F = M$ throughout some whole disk centered at c. Thus S is open. Finally, $S \ne \varnothing$, since it contains a.

Exercise 3.34 (i) *Deduce from* 3.32(i) *or* (v) *the so-called* Cauchy estimates: *If* $c_{-n} = 0$ *for all* $n \in \mathbb{N}$, *and if we set*

$$M(r) := \sup_{|z|=r} |f(z)|,$$

then for every non-negative integer k

$$\frac{|f^{(k)}(0)|}{k!} = |c_k| \le \frac{M(r)}{r^k},$$

and equality holds for some k *if and only if* f *is a monomial of degree* k.

Hint: Since the left-hand side of 3.32(i) is majorized by $M(r)^2$, an equality $|c_k| = M(r)/r^k$ entails that all other coefficients must be 0.

(ii) *Deduce from* (i) *that if* $F(z) := \sum_{k=0}^{\infty} c_k z^k$ *converges throughout* $\mathbb{C}$ *and for some* $n \in \mathbb{N}$ *and some* $r_j \uparrow \infty$, $|F(z)/z^n| \le M < \infty$ *on* $\cup C(0, r_j)$, *then* F *is a polynomial of degree at most* n.

Hint: We can assume $\left|\sum_{k=0}^{n} c_k z^{k-n}\right| \le |c_n| + 1$ for all $|z| \ge r_1$. Then the power series $f(z) := z^{-n}\left[F(z) - \sum_{k=0}^{n} c_k z^k\right] = \sum_{k=1}^{\infty} c_{k+n} z^k$ converges throughout $\mathbb{C}$ and every $M(r_j)$ in (i) is majorized by $M + |c_n| + 1$, whence every coefficient in f is 0.

(iii) *Formulate and prove, using* 3.32(ii), *the analog of* (i) *for the function* $A(r) := \sup_{|z|=r} |\operatorname{Re} f(z)|$ *and the generalization* (cf. 6.37 and its notes) *of* (ii) *to the case where only* $z^{-n} \operatorname{Re} f(z)$ *is bounded on large circles.*

(iv) *Suppose* $f(z) := \sum_{k=0}^{\infty} c_k z^k$ *converges throughout* $\mathbb{C}$ *and for some* M, $N \in \mathbb{Z}$ *and some finite constant* C *satisfies*

$$(*) \qquad |f(z)| \le \frac{C|z|^M}{|\operatorname{Re} z|^N}$$

for all large $z \in \mathbb{C} \setminus i\mathbb{R}$. *Show that* f *is a polynomial of degree not greater than* $max\{0, M - N\}$. (Cf. 5.26(iii).) *How does the conclusion change if* $(*)$ *is satisfied by* $\operatorname{Re} f(z)$ *instead of* $f(z)$?

Hints: (Stone [1948]) For each $k \in \mathbb{N} \cup \{0\}$ and all large r form

$$I(r, k) := \frac{1}{2\pi i} \int_{C(0,r)} z^{-k-1} \left(1 + \frac{z^2}{|z|^2}\right)^N f(z)\, dz.$$

Since

$$\left|1 + \frac{z^2}{|z|^2}\right| = \left|\frac{z(\bar{z} + z)}{z^2}\right| = \frac{2|\operatorname{Re} z|}{|z|},$$

it follows from $(*)$ that

$$|I(r, k)| \le 2^N C r^{M-N-k} \to 0 \quad \text{as } r \to \infty, \text{ for } k > M - N.$$

On the other hand, we may directly multiply

$$z^{-k-1}\left(1 + r^{-2} z^2\right)^N = \sum_{j=0}^{N} \binom{N}{j} r^{-2j} z^{2j-k-1}$$

by the power series $\sum_{n=0}^{\infty} c_n z^n$ and get for the integrand of $I(r,k)$ a series that converges uniformly in $z \in C(0,r)$ and in which the coefficient of z^{-1} is the number

$$\sum_{j=0}^{\min\{N,k/2\}} \binom{N}{j} r^{-2j} c_{k-2j}.$$

Only this coefficient survives the term-by-term integration, and as $r \to \infty$ it converges to c_k.

(v) *Suppose that $c_0 \in \mathbb{R}$ and that $\sum_{k=0}^{\infty} c_k z^k$ converges for all $z \in \mathbb{C}$. Call the sum $f(z)$ and for each $r \geq 0$ set $M(r) := \sup\{|f(z)| : |z| = r\}$ and $A(r) := \sup\{|\operatorname{Re} f(z)| : |z| = r\}$. Show that*

$$M(r) \leq \sum_{k=0}^{\infty} |c_k| r^k \leq 2\sqrt{2} A(2r) \qquad \forall r \geq 0.$$

Hints: For all $r > 0$ the Cauchy estimates in (iii) give

$$|c_k| \leq \frac{\sqrt{2}A(2r)}{(2r)^k}, \quad k = 0, 1, 2, \ldots$$

whence

$$\sum_{k=0}^{\infty} |c_k| r^k \leq \sum_{k=0}^{\infty} \frac{\sqrt{2}A(2r)}{(2r)^k} r^k = \sqrt{2}A(2r) \sum_{k=0}^{\infty} \frac{1}{2^k}.$$

The following mammoth, multi-part exercise proposes power-series proofs (Weierstrassian) for several important facts that the text will later develop in different ways.

Exercise 3.35 *A sequence of power series $f_n(z) := \sum_{k=0}^{\infty} c_{n,k} z^k$ $(n \in \mathbb{N})$ is given, each convergent in $\mathbb{D}$ and bounded by a common $M \in \mathbb{N}$.*

(i) (Cf. 7.9.) *Suppose that*

$$(*) \qquad \lim_{n\to\infty} c_{n,k} =: c_{0,k} \in \mathbb{C} \text{ exists for each } k.$$

Show that the series $\sum_{k=0}^{\infty} c_{0,k} z^k$ converges in $\mathbb{D}$ and defines a holomorphic function f_0 to which the f_n converge locally uniformly as $n \to \infty$.

Hints: (Divide and conquer.) Letting $r \uparrow 1$ in 3.34(i) shows that $|c_{n,k}| \leq M$ for all $n \in \mathbb{N}$, $k \geq 0$, hence for $n = 0$ too. Therefore the f_0-series converges in $\mathbb{D}$. Given $0 < r < 1$ and $\varepsilon > 0$, choose N so large that $\sum_{k>N} 2Mr^k < \varepsilon/2$; then choose n_ε so large that $\sum_{k=0}^{N} |c_{n,k} - c_{0,k}| < \varepsilon/2$ for all $n \geq n_\varepsilon$. This will ensure that $|f_n(z) - f_0(z)| < \varepsilon$ for all such n and all $z \in \overline{D}(0,r)$.

(ii) *Replace the boundedness hypothesis on the sequence* $\{f_n\}$ *by the hypothesis that it converge locally uniformly to some* f. *Show that* $(*)$ *holds and* $f(z) = \sum_{k=0}^{\infty} c_{0,k} z^k$ *for all* $z \in \mathbb{D}$. *Prove further that* $\{f'_n\}$ *converges locally uniformly to* f'. Cf.5.44.

Hints: Consider $0 < r < 1$. Uniform convergence of $\{f_n\}$ in $\bar{D}(0,r)$ clearly implies the existence of an $M(r) < \infty$ such that $|f_n| \leq M(r)$ in $\bar{D}(0,r)$ for all $n \in \mathbb{N}$. From 3.34(i)

$$|c_{n,k} - c_{m,k}| \leq 2^k \sup\{|f_n(z) - f_m(z)| : |z| = 1/2\} \qquad \forall n, m, k$$

and this implies that for each k the sequence $\{c_{n,k}\}_{n\in\mathbb{N}}$ is Cauchy, that is, $(*)$ holds. Therefore (i) can be applied in each $D(0,r)$ to show that $\{f_n\}$ converges locally uniformly to the function $f_0(z) := \sum_{k=0}^{\infty} c_{0,k} z^k$. Since

$$\frac{1}{(1-z)^2} = \frac{d}{dz}\left(\frac{1}{1-z}\right) = \frac{d}{dz}\sum_{k=0}^{\infty} z^k = \sum_{k=1}^{\infty} k z^{k-1} \qquad \forall z \in \mathbb{D},$$

we can set $R := \frac{1}{2}(1+r)$ and employ the Cauchy estimate $|c_{n,k}| \leq R^{-k} M(R)$ to see that

$$\begin{aligned}\sum_{k=1}^{\infty} k|c_{n,k}|r^{k-1} &\leq \sum_{k=1}^{\infty} kR^{-k}M(R)r^{k-1} = \frac{M(R)}{R}\sum_{k=1}^{\infty} k\left(\frac{r}{R}\right)^{k-1} \\ &= \frac{M(R)}{R}\frac{1}{(1-r/R)^2} =: M^*(r)\end{aligned}$$

and so $|f'_n| \leq M^*(r)$ in $\bar{D}(0,r)$, for all $n \in \mathbb{N}$. Of course, $(*)$ trivially entails that

$$\lim_{n\to\infty} kc_{n,k} = kc_{0,k} \qquad \forall k \in \mathbb{N}.$$

Hence in each $D(0,r)$ the hypotheses of part (i) are met by the sequence $\{f'_n\}$.

(iii) *The union of* (i) *and* (ii) *is known as* Weierstrass' Double-Limit Theorem. *Comment on the significance of this designation and on the relevance of the example* $f_n(z) := 2^n z^n$ $(n \in \mathbb{N})$, $f_0 := \mathbf{0}$.

Hints: Succinctly put, what was shown is that

$$\lim_{n\to\infty}\lim_{m\to\infty}\sum_{k=0}^{m} c_{n,k}z^k = \lim_{m\to\infty}\lim_{n\to\infty}\sum_{k=0}^{m} c_{n,k}z^k \qquad \forall z \in \mathbb{D}.$$

Hypothesis $(*)$ is clearly satisfied by the given example, although the conclusion in (i) evidently fails, showing that without some boundedness hypothesis, even pointwise convergence does not follow from $(*)$.

(iv) (Montel's Compactness Theorem—cf. 7.6) *Show that there exist* $1 \leq n_1 < n_2 < \cdots$ *and a power-series function* f *such that the sequence* $\{f_{n_j}\}_{j\in\mathbb{N}}$ *converges locally uniformly in* $\mathbb{D}$ *to* f.

Hints: As noted earlier, $|c_{n,k}| \leq M$ for all n, k. The compactness of closed bounded sets in $\mathbb{C}$ therefore lets a diagonal argument (see chapter 0) be employed to produce a single sequence $\{n_j\}_{j\in\mathbb{N}}$ such that $\lim_{j\to\infty} c_{n_j,k} =: c_{0,k}$ exists for each $k \geq 0$. Apply (i) to the sequence $\{f_{n_j}\}_{j\in\mathbb{N}}$.

(v) (Vitali-Porter theorem on induced convergence—cf. 5.74(ii) and 7.5.) *Suppose* $z_j \in \mathbb{D}\setminus\{0\}$ *converge to* 0, *and hypothesis* $(*)$ *in* (i) *is replaced by hypothesis*

$$(**) \qquad \lim_{n\to\infty} f_n(z_j) \text{ exists in } \mathbb{C} \text{ for each } j \in \mathbb{N}.$$

Show that $(*)$ *follows from* $(**)$, *and consequently the sequence* $\{f_n\}$ *converges locally uniformly in* $\mathbb{D}$.

Hints: We again use Cauchy estimates, $|c_{n,k}| \leq M$ for all n, k, to see that for all $m, n, j \in \mathbb{N}$

$$\begin{aligned}|c_{m,0} - c_{n,0}| &= |f_m(0) - f_n(0)| \\ &\leq |f_m(0) - f_m(z_j)| + |f_m(z_j) - f_n(z_j)| + |f_n(z_j) - f_n(0)| \\ &\leq |f_m(z_j) - f_n(z_j)| + 2\sum_{k=1}^{\infty} M|z_j|^k = |f_m(z_j) - f_n(z_j)| + \frac{2M|z_j|}{1-|z_j|}.\end{aligned}$$

Given $\varepsilon > 0$, choose j so large that this quotient is less than $\varepsilon/2$. Then there exists $N_\varepsilon \in \mathbb{N}$ such that $|f_m(z_j) - f_n(z_j)| < \varepsilon/2$ for all $m, n \geq N_\varepsilon$. This argument shows that $\{c_{n,0}\}_{n\in\mathbb{N}}$ is a Cauchy, hence convergent sequence in $\mathbb{C}$.

Suppose the convergence of the sequence $\{c_{n,k}\}_{n\in\mathbb{N}}$ has been proved for all $k < K \in \mathbb{N}$. Again from the Cauchy estimates

$$\left|f_n(z) - \sum_{k=0}^{K-1} c_{n,k}z^k\right| \leq \sum_{k=K}^{\infty} M|z|^k = \frac{M|z|^K}{1-|z|} \leq 2M|z|^K \quad \text{for } z \in D(0,1/2).$$

Therefore the power series functions

$$F_n(z) := z^{-K}\left[f_n(z) - \sum_{k=0}^{K-1} c_{n,k}z^k\right] = \sum_{k=K}^{\infty} c_{n,k}z^{k-K}$$

are all bounded by $2M$ in $D(0,1/2)$. They obviously satisfy the other hypothesis (convergence at each z_j – here the induction hypothesis plays a role) which the f_n do, so by the result of the first paragraph the sequence $\{F_n(0)\}_{n\in\mathbb{N}} = \{c_{n,K}\}_{n\in\mathbb{N}}$ converges.

(vi) *Let the z_j be as in* (v) *but strengthen the hypothesis* (**) *to*

$$(**)_0 \qquad \lim_{n\to\infty} f_n(z_j) = 0 \quad \textit{for each } j \in \mathbb{N}.$$

Show that the stronger conclusion

$$(*)_0 \qquad \lim_{n\to\infty} c_{n,k} = 0 \quad \textit{for each } k$$

follows, and consequently the sequence $\{f_n\}$ converges locally uniformly in $\mathbb{D}$ to $\mathbf{0}$.

Hint: Only slight modifications are necessary in the argument suggested for (v).

(vii) *The power series $f_0(z) := \sum_{k=0}^{\infty} c_{0,k} z^k$ converges in $\mathbb{D}$ and $f_0(z_j) = 0$ for a sequence $\{z_j\}$ in $\mathbb{D}\setminus\{0\}$ which converges to* 0. *Show that $f_0(\mathbb{D}) = \{0\}$.*

Hint: Apply (vi) to the functions $f_n(z) := \sum_{k=0}^{n} c_{0,k} z^k$ $(n \in \mathbb{N})$. To get the needed bound M, work in $D(0, 1/2)$ instead of $\mathbb{D}$. [A more elementary argument appears in the proof of 5.62.]

(viii) (Uniqueness Theorem) *Ω is a region in $\mathbb{C}$ and $f : \Omega \mapsto \mathbb{C}$ has a power-series representation in a neighborhood of each point of Ω. Show that if $f^{-1}(0)$ has a limit point in Ω, then $f = \mathbf{0}$.*

Hints: Let C denote the set of limit points of $f^{-1}(0)$ *in* Ω. By continuity of f, C is relatively closed in Ω. Consider any $a \in C$. By hypothesis there is a power series $\sum c_n(z-a)^n$ representing f in some $D(a,r) \subset \Omega$. It follows from (vii) that $f(D(a,r)) = \{0\}$. Thus $D(a,r) \subset C$, showing that C is an open set. The non-void (by hypothesis) C must exhaust the connected set Ω of which it is a clopen subset, so $f^{-1}(0) \supset C = \Omega$, $f = \mathbf{0}$.

(ix) *The power series $\sum_{k=1}^{\infty} c_k z^k$ converges in $\mathbb{D}$ to a non-constant function $f(z)$. Show that $f(\mathbb{D})$ is a neighborhood of $f(0) = 0$.*

Hints: Let c_j be the first non-zero coefficient and select $r \in (0,1)$ so that

$$(1) \qquad \sum_{k=j+1}^{\infty} |c_k| r^{k-j} < |c_j|.$$

Since $f(z) = z^j \left[c_j + \sum_{k=j+1}^{\infty} c_k z^{k-j}\right]$, it follows from (1) that 0 does not lie in the compact set $f(C(0,r))$, consequently has positive distance d from it.

Let us show that $D(0,d/2) \subset f\big(\bar{D}(0,r)\big)$. Given $w \in D(0,d/2)$, there exists $a \in \bar{D}(0,r)$ with

$$|w - f(a)| = \inf\big\{|w - f(z)| : z \in \bar{D}(0,r)\big\}. \tag{2}$$

Notice that $|z| = r \Rightarrow d \le |f(z)| \le |w| + |f(z) - w| \Rightarrow |f(z) - w| \ge d - |w| > 2|w| - |w| = |w| = |f(0) - w|$. Therefore the minimizing point $a \notin C(0,r)$ and $r - |a| > 0$. We aim to prove that $f(a) = w$, and to this end, use an elementary argument due to Oswaldo Rio Branco de Oliveira (cf. Cater [2001/2002]). Consider the function $g(z) := w - f(z)$ for $z \in D(a, r - |a|)$. Use 3.12(iii) to write g as a power series centered at a and consequently having the form

$$g(z) = g(a) + (z-a)^k h(z-a) \tag{3}$$

for some $k \in \mathbb{N}$ and a power series h in $D(0, r - |a|)$ with $h(0) \ne 0$. From (2) we get, for all $z \in D(a, r - |a|) \subset D(0,r)$

$$|g(a)|^2 \le |g(z)|^2 = |g(a)|^2 + 2\operatorname{Re}[\overline{g(a)}(z-a)^k h(z-a)] + |(z-a)^k h(z-a)|^2.$$

Write $z := a + \rho u$ with $0 < \rho < r - |a|$, $u \in \mathbb{T}$, and this reads

$$0 \le 2\operatorname{Re}[\overline{g(a)}(\rho u)^k h(\rho u)] + \rho^{2k}|h(\rho u)|^2.$$

Divide by ρ^k and let $\rho \downarrow 0$ to get

$$0 \le 2\operatorname{Re}[\overline{g(a)} u^k h(0)], \quad \text{valid for all } u \in \mathbb{T}. \tag{4}$$

The validity of this inequality for $u = 1$ and for the three special u constructed in 2.21(i) [without the aid of the exponential function!] is already enough to entail

$$\overline{g(a)} h(0) = 0,$$

that is, since $h(0) \ne 0$, $g(a) = 0$. Which means that $f(a) = w$, as wanted.

(x) (Open-Mapping Theorem) *The power series $\sum_{k=1}^{\infty} C_k z^k$ converges in $\mathbb{D}$ to a non-constant function $F(z)$. Show that F is an open mapping.*

Hints: It suffices to show that for each $a \in \mathbb{D}$ and each $0 < r < 1 - |a|$, $F(a)$ is an interior point of $F(D(a,r))$. To this end, use 3.12(iii) to find $a_k \in \mathbb{C}$ such that

$$F(z) = \sum_{k=0}^{\infty} a_k (z-a)^k \qquad \forall z \in D(a,r).$$

Form the function

$$f(z) := F(rz + a) = \sum_{k=0}^{\infty} a_k r^k z^k, \quad z \in \mathbb{D}.$$

According to (viii), F non-constant in $\mathbb{D} \Rightarrow F$ non-constant in $D(a,r) \Rightarrow f$ non-constant in $\mathbb{D}$. Hence (ix) applies to assert that $f(\mathbb{D}) = F(D(a,r))$ is a neighborhood of $f(0) = F(a)$.

(xi) *From* (x) *infer that the range of every non-constant polynomial P is the whole complex plane. In particular, P has a zero in $\mathbb{C}$.*

Hints: As P eventually dominates a positive multiple of its highest-degree term, $|P(z)| \to \infty$ as $|z| \to \infty$. Consequently, if $P(z_n) \to w \in \mathbb{C}$, $\{z_n\}$ is bounded, a subsequence of it converges to some $z \in \mathbb{C}$, and by continuity $P(z) = w$. This shows that $P(\mathbb{C})$ is a closed set. Thanks to (x), $P(\mathbb{C})$ is therefore a clopen subset of $\mathbb{C}$.

(xii) (Maximum-Modulus Principle for power series). *Show that if the non-constant function F is represented by a power series in an open disk, then $|F|$ has no maximum in the disk.*

Hint: If $F(z_0) \neq 0$, the image set, being open by (x), will contain $rF(z_0)$ for some $r > 1$, so $|F(z_0)|$ is not the maximum of $|F|$. [Thanks to 3.12(iii) the present hypothesis implies that of 3.33(iii), from which the present conclusion then also follows.]

(xiii) (Reciprocal of a power series via Cauchy's Method of Majorants). *Suppose $f : U$ open $\subset \mathbb{C} \mapsto \mathbb{C} \setminus \{0\}$ is representable in a neighborhood of each point by a power series. Show that $1/f$ is so representable.*

Hints: Consider any $z_0 \in U$. Without loss of generality, $z_0 = 0$ and $f(z_0) = 1$. By hypothesis (and 3.12(iii)), for some positive $r \in \mathbb{R}$, $D(0,r) \subset U$ and $a_k \in \mathbb{C}$ exist such that

$$f(z) = \sum_{k=0}^{\infty} a_k z^k \qquad \forall z \in D(0,r) \qquad (\text{so} \quad a_0 = 1). \tag{1}$$

Suppose that

$$g(z) := \sum_{n=0}^{\infty} b_n z^n \tag{2}$$

converges for all z sufficiently near 0. According to 3.12(iv)

$$1 = f(z)g(z) \quad \text{for all } z \text{ sufficiently near } 0 \tag{3}$$

if and only if $b_0 = 1$ and

$$0 = \sum_{j=0}^{n} a_{n-j} b_j \qquad \forall n \in \mathbb{N},$$

that is, since $a_0 = 1$, for all $n \in \mathbb{N}$

(4.n) $$b_n := -\sum_{j=0}^{n-1} a_{n-j} b_j.$$

So, let us define $b_0 := 1$ and take (4.n) as recursively *defining* $b_n \in \mathbb{C}$, then try to show that series (2) has *positive* radius of convergence. To that end, recall (3.3(iii)) that $\overline{\lim} |a_k|^{1/k} \leq 1/r < \infty$. Hence for some $a \in \mathbb{R}$, $|a_k|^{1/k} < a$ for all k: For all $k \in \mathbb{N}$

(5.k) $$|a_k| < a^k.$$

This enables us to prove by induction that for $n = 0, 1, 2, \ldots$

(6.n) $$|b_n| \leq 2^n a^n.$$

Indeed, (6.0) holds, and if (6.j) holds for all $j < n$, then by (4.n) and (5.$n-j$)

$$|b_n| \leq \sum_{j=0}^{n-1} |a_{n-j}||b_j| \leq \sum_{j=0}^{n-1} a^{n-j} 2^j a^j = a^n \sum_{j=0}^{n-1} 2^j = a^n(2^n - 1).$$

Thus, $\overline{\lim} |b_n|^{1/n} \leq 2a$ and the radius of convergence of (2) is at least $\frac{1}{2a}$. [Later (5.30) we'll be able to easily show that this radius is not less than r. But to prove that using only the machinery available at this point is a challenge. See BROMWICH [1926], §89 or ULLRICH [1990], p. 98. The broader issue involved here is discussed in Remark (iii) following 5.30.]

(xiv) (Holomorphic n^{th}-roots without logarithms). *For every* $n \in \pm\mathbb{N}$, $w_0 \in \mathbb{C} \setminus \{0\}$ *there is an* $r_0 > 0$ *and a holomorphic function* φ *in* $D(w_0, r_0)$ *such that*

$$\varphi^n(w) = w \qquad \forall w \in D(w_0, r_0).$$

Hints: Let $P(z) := z^n$ denote the nth-power function. Use 1.26(ii) to find $z_0 \in \mathbb{C} \setminus \{0\}$ such that $P(z_0) = w_0$. By 3.7 there is an $0 < r \leq |z_0|$ such that P is one-to-one in $D(z_0, r)$. Then by (x) the set $W := P(D(z_0, r))$ is open and the function $\varphi := P^{-1}$ maps W continuously onto $D(z_0, r)$. Note that $W \supset D(w_0, r_0)$ for sufficiently small positive r_0. If $w, w_1 \in W$ are distinct and $w \to w_1$, then $z := \varphi(w)$, $z_1 := \varphi(w_1) \in D(z_0, r)$ are distinct and $z \to z_1$. Therefore $\frac{P(z)-P(z_1)}{z-z_1} \to P'(z_1) = n z_1^{n-1} = \frac{nP(z_1)}{z_1} \neq 0$, so

$$\frac{\varphi(w) - \varphi(w_1)}{w - w_1} = \frac{z - z_1}{P(z) - P(z_1)} \to \frac{1}{P'(z_1)} = \frac{z_1}{nP(z_1)} = \frac{\varphi(w_1)}{n w_1}.$$

Alternatively, the *binomial series* produces an explicit holomorphic n^{th}-root in $D(1,1)$—which can be (multiplicatively) transferred to $D(w_0, |w_0|)$ for each $w_0 \neq 0$: The ratio test 3.3(iv) shows that the series $\sum_{k=0}^{\infty} \binom{1/n}{k}(w-1)^k$ has radius of convergence 1 and so defines a holomorphic function $\varphi(w)$, $w \in D(1,1)$. The *generalized binomial coefficients* $\binom{x}{k}$ were defined in 3.12(vi), and one easily checks that $k\binom{x}{k} = x\binom{x-1}{k-1}$ and $\binom{x-1}{k} + \binom{x-1}{k-1} = \binom{x}{k}$, $k \in \mathbb{N}$. Using this and 3.6 we see that $nw\varphi'(w) = \varphi(w)$, and consequently the function $\varphi^n(w)/w$ has derivative 0, for all $w \in D(1,1)$.

Remarks: (i) For $\varnothing \neq U$ open $\subset \mathbb{C}$ denote by $A(U)$ the set of all functions $f : U \mapsto \mathbb{C}$ such that in a neighborhood of each point of U, f is representable by a power series. (Such an f is said to be *locally of power-series type*; Weierstrass called f *analytic*.) Then from 3.12(iv) and 3.35(xiii) it follows that $A(U)$ is a unital $\mathbb{C}$-algebra under the usual pointwise operations, and every zero-free function is invertible; from 3.35(viii) it further follows that $A(U)$ is an integral domain if U is connected, and the interested reader should have no difficulty extending 3.35(x) and (xii) from power series in disks to functions in $A(U)$, whenever U is connected. Moreover, with the help of remark (iii) following 5.30, 3.35(ii) asserts that $A(U)$ is closed under local uniform convergence. Of course (3.6), we have $A(U) \subset H(U)$. A deeper result, *the most important in this subject*, is that equality always holds here. We reach that "peak in Darien" at 5.30 (but cf. 3.39(ii)).

(ii) Two other basic facts about holomorphic functions that will be derived in Chapter V can be reached by pure power-series arguments: If f is defined by a power series centered at z_0 and g by one centered at $w_0 := f(z_0)$, then in some neighborhood of z_0, $g \circ f$ is represented by a power series centered at z_0. In words, a power series can be inserted into another and the result rearranged (using 3.12(ii)) into a power series. Part (xiii) above is the special case of this where $w_0 \neq 0$ and $g(w) := 1/w = \sum(-1)^n w_0^{-n-1}(w-w_0)^n$. If $f'(z_0) \neq 0$, then there is a function g defined by a power series centered at w_0 such that $g(w_0) = z_0$ and $f(g(w)) = w$ for all w in some neighborhood of w_0. For power-series proofs of these two facts, see pp. 180-181 and pp. 184-187 of KNOPP [1951], or §2.4 of HENRICI [1974]. Cauchy's majorant method is used in proving the second of these, known as the theorem on "reversion of power series." (Cf. 5.92(iii), (iv).)

Theorem 3.36 *Suppose that $\sum_{k=0}^{\infty} c_k z^k$ converges for all $z \in \mathbb{C}$. Call it $f(z)$ and set $M(r) := \sup\{|f(z)| : |z| = r\}$ for each $r \geq 0$. Let c be a non-negative real number. Then the following boundedness condition on the coefficients of f:*

$$\overline{\lim_{n\to\infty}}\, n|c_n|^{1/n} \leq ce \tag{1}$$

is equivalent to the following global boundedness condition on f itself:

(2) $\quad f(z)e^{-(c+\varepsilon)|z|}$ *is a bounded function of $z \in \mathbb{C}$ for each $\varepsilon > 0$.*

Moreover, if f is not the **0** *function, then*

$$\overline{\lim_{r\to\infty}} \frac{\log M(r)}{r} = \overline{\lim_{n\to\infty}} \left(\frac{n}{e}\right)|c_n|^{1/n} \tag{3}$$

$$= \overline{\lim_{n\to\infty}} \left|f^{(n)}(0)\right|^{1/n}. \tag{4}$$

Proof: (1) ⇒ (2). Consider any $\varepsilon > 0$ and define $\delta := \delta(\varepsilon) > 1$ by

$$\left(c + \frac{\varepsilon}{2}\right)\delta = c + \varepsilon. \tag{5}$$

By (1) there exists an $N = N(\varepsilon)$ such that

$$n|c_n|^{1/n} < \left(c + \frac{\varepsilon}{2}\right)e \qquad \forall n > N. \tag{6}$$

Moreover by 3.9 we have

$$\lim_{n\to\infty} \left[\frac{n!}{n^n}\right]^{1/n} = \lim_{n\to\infty} \left[\frac{(n+1)!}{(n+1)^{n+1}} \Big/ \frac{n!}{n^n}\right] = \lim_{n\to\infty} \left(1 + \frac{1}{n}\right)^{-n} \overset{(3.21)}{=} e^{-1} \tag{7}$$

(which, as has been previously noted, expresses e as twice the limit of the ratio of the arithmetic to the geometric means of $\{1, 2, \ldots, n\}$).

Consequently we can also suppose the N above so large that

$$\left[\frac{n!}{n^n}\right]^{1/n} < \delta e^{-1} \qquad \forall n > N. \tag{8}$$

Multiplying (6) and (8), we get from (5)

$$(n!|c_n|)^{1/n} < \left(c + \frac{\varepsilon}{2}\right)\delta = c + \varepsilon \qquad \forall n > N$$

and so

$$|f(z)| \le \left|\sum_{n=0}^{N} c_n z^n\right| + \sum_{n=N+1}^{\infty} \frac{(c+\varepsilon)^n}{n!}|z|^n$$

$$< \sum_{n=0}^{N} |c_n||z|^n + e^{(c+\varepsilon)|z|} \qquad \forall z \in \mathbb{C}. \tag{9}$$

For any $n \geq 0$, however, the power series for the exponential function yields

$$\frac{e^{\varepsilon|z|}}{|z|^n} > \frac{(\varepsilon|z|)^{n+1}/(n+1)!}{|z|^n} = \frac{\varepsilon^{n+1}|z|}{(n+1)!} \to \infty \quad \text{as } |z| \to \infty$$

and so there exists $K = K(\varepsilon)$ such that

$$\sum_{n=0}^{N} |c_n||z|^n < e^{\varepsilon|z|} \qquad \forall |z| \geq K. \tag{10}$$

From (9) and (10)

$$|f(z)| \leq 2e^{(c+\varepsilon)|z|} \qquad \forall |z| \geq K$$

and this proves that $f(z)e^{-(c+\varepsilon)|z|}$ is bounded in $\mathbb{C}$. As $\varepsilon > 0$ was arbitrary, (2) is established.

(2) $\Rightarrow$ (1). For each $\varepsilon > 0$ there exists $K(\varepsilon) < \infty$ such that

$$M(r)e^{-(c+\varepsilon)r} < K(\varepsilon) \qquad \forall r > 0. \tag{11}$$

Take $r := n/(c+\varepsilon)$ and use 3.34(i) and (11) to get

$$|c_n| \leq \left(\frac{n}{c+\varepsilon}\right)^{-n} e^n K(\varepsilon),$$

$$n|c_n|^{1/n} \leq (c+\varepsilon)e[K(\varepsilon)]^{1/n} \qquad \forall n.$$

Now cite 3.1.

For (3) note first that by 3.35(viii), f not the zero function means that $M(r)$ is not 0 for any $r > 0$, so $\log M(r)$ is defined. Moreover, 3.34(i) shows that $M(r)$ is bounded away from 0 for all large r, so the number $c := \overline{\lim}(\log M(r))/r$ lies in $[0, +\infty]$. It is easy to see that this number satisfies (2), hence (1) (whether c is finite or not). Thus "$\geq$" prevails in (3). On the other hand, since (1) holds for the number $c := \overline{\lim}(n/e)|c_n|^{1/n} \in [0, \infty]$, (2) holds for this c (whether it is finite or not); from which it follows that $\overline{\lim}(\log M(r))/r \leq c + \varepsilon$, for every $\varepsilon > 0$; so "$\leq$" prevails in (3). Finally, for (4) use 3.6 to write

$$\left|f^{(n)}(0)\right|^{1/n} = (n!|c_n|)^{1/n} = e\left[\frac{n!}{n^n}\right]^{1/n} \frac{n}{e}|c_n|^{1/n}$$

and then cite (7).

Remark: There are other relationships between the rate of growth of an entire function and the rate of growth of its coefficients. For example, if

$f(z) = \sum c_n z^n$ and $f(z)e^{-|z|^\rho}$ is bounded for some $0 < \rho < \infty$, then the infimum of such ρ is

$$\overline{\lim_{r\to\infty}} \frac{\log\log M(r)}{\log r} = \overline{\lim_{n\to\infty}} \frac{n \log n}{\log(1/|c_n|)}.$$

See BOAS [1954].

§ 4 Bernoulli Polynomials, Numbers and Functions

In this section we use the theory of the exponential and logarithm functions to compute $\sum_{k=1}^{\infty} k^{-2n}$ and some other interesting series. None of this material is used later.

Exercise 3.37 (i) *Show that* $\sum_{n=1}^{\infty} \frac{\sin(n\theta)}{n} = (\pi - \theta)/2$ *and* $\sum_{n=1}^{\infty} \frac{\cos(n\theta)}{n} = -\log(2\sin(\theta/2))$ *for all* $\theta \in (0, 2\pi)$, *and convergence is uniform on each compact subset of* $(0, 2\pi)$. *Infer the remarkable identity*

$$\sum_{k=1}^{\infty} \sin((2k-1)\theta)/(2k-1) = \pi/4 \quad \text{for all } \theta \in (0, \pi).$$

Hints: Consider the logarithm l constructed in the proof of 3.19:

$$(0) \qquad e^{-l(z)} = 1 - z,$$

$$(1) \qquad e^{-i\,\mathrm{Im}\, l(z)} = \frac{e^{-l(z)}}{|e^{-l(z)}|} = \frac{1-z}{|1-z|} \qquad \forall z \in \overline{\mathbb{D}} \setminus \{1\}.$$

For $z \in \overline{\mathbb{D}} \setminus \{1\}$ we have $1 - z \in \overline{D}(1,1) \setminus \{0\} \subset \mathbb{C} \setminus (-\infty, 0]$. It follows then from (1) that $-\mathrm{Im}\, l(z)$ is not an odd integral multiple of π, that is, $\mathrm{Im}\, l$ maps $\overline{\mathbb{D}} \setminus \{1\}$ into $\bigcup_{k=-\infty}^{\infty} ((2k-1)\pi, (2k+1)\pi)$. As $\overline{\mathbb{D}} \setminus \{1\}$ is connected [use 1.2 or 1.22], the range of $\mathrm{Im}\, l$ must lie in only one of the intervals $((2k-1)\pi, (2k+1)\pi)$. Since $l(0) = 0$, $k = 0$. For $\theta \in (0, 2\pi)$ we have $\sin\theta/2 > 0$ and so from

$$\begin{aligned} 1 - e^{i\theta} &= \left[e^{-i\theta/2} - e^{i\theta/2}\right] e^{i\theta/2} = -i(2\sin\theta/2)e^{i\theta/2} \\ &= (2\sin\theta/2)e^{i(\theta-\pi)/2}, \end{aligned}$$

follows that the *positive number* $2\sin(\theta/2)$ is $|1 - e^{i\theta}|$, whence

$$(2) \qquad \frac{1 - e^{i\theta}}{|1 - e^{i\theta}|} = e^{i(\theta-\pi)/2}.$$

Since $(\theta-\pi)/2 \in (-\pi,\pi)$, we get from (1),(2) and the above observation about $\operatorname{Im} l$, that $-\operatorname{Im} l(e^{i\theta}) = (\theta-\pi)/2$. Since $l(e^{i\theta}) = \sum_{k=1}^{\infty} e^{ik\theta}/k$, the first of the desired equalities follows. The second follows from taking logarithms of the moduli of the two sides of (0). The third follows by replacing θ with 2θ in the first equality and subtracting half the result from the original first equality.

(ii)
$$\sum_{n=1}^{\infty} (-1)^{n+1} \frac{\sin(nx)}{n} = \frac{x}{2} \qquad \forall x \in (-\pi,\pi)$$

with convergence uniform on compacta.

Hint: Replace θ by $\pi - x$ in the first identity in (i).

(iii)
$$\sum_{n=1}^{\infty} (-1)^{n+1} \frac{1-\cos(n\theta)}{n^2} = \frac{\theta^2}{4} \qquad \forall \theta \in [-\pi,\pi].$$

Hint: Both sides of the alleged equation are continuous functions of θ (since the series is uniformly convergent), so it suffices to confirm the equality for $\theta \in (-\pi,\pi)$. For such θ, integrate (ii) from $x = 0$ to $x = \theta$ and use the uniform convergence on $[-|\theta|,|\theta|]$ to interchange sum and integral. Alternatively, appeal to 2.12.

(iv)
$$\sum_{n=1}^{\infty} \frac{(-1)^{n+1}}{n^2} = \frac{\pi^2}{12}.$$

Hint: The series (iii), being uniformly convergent, can be integrated term by term over $[-\pi,\pi]$.

(v)
$$\sum_{n=1}^{\infty} \frac{\cos(n\theta)}{n^2} = \frac{3\theta^2 - 6\pi\theta + 2\pi^2}{12}, \qquad 0 \le \theta \le 2\pi.$$

Hint: Subtract (iv) from (iii) and change θ to $\pi - \theta$.

(vi)
$$\sum_{n=1}^{\infty} \frac{1}{n^2} = \frac{\pi^2}{6} \quad \textit{and} \quad \sum_{k=1}^{\infty} \frac{1}{(2k-1)^2} = \frac{\pi^2}{8}.$$

Hints: Take $\theta = 0$ in (v) to get the first sum, then add this to (iv) to get the second sum.

(vii) *Take $\theta = \pi/2$ in* (i) *to infer anew the* Gregory–Leibniz *formula in* 3.22. *Then use it and* (vi) *to confirm the striking identity*

$$\left(\sum_{n\in\mathbb{Z}} \frac{(-1)^n}{2n+1}\right)^2 = \sum_{n\in\mathbb{Z}} \frac{1}{(2n+1)^2}.$$

Exercise 3.38 (i) *There exists a sequence* $\{P_n\}$ $(n = 0, 1, 2, \dots)$ *of differentiable functions on* $[0, 1]$ *which satisfy*

(1) $$P_0 = \mathbf{1}$$

(2) $$P'_{n+1}(t) = P_n(t) \qquad \forall n \geq 0, \quad t \in [0, 1]$$

(3) $$\int_0^1 P_n = 0 \qquad \forall n \geq 1.$$

These conditions uniquely determine the P_n*; they are polynomials, with rational coefficients (confirm this) and are called the* Bernoulli polynomials.

(ii) *The polynomials* P_n *satisfy in addition*

(4) $P_1(z) = z - \frac{1}{2}$

(5) $P_2(z) = z^2/2 - z/2 + \frac{1}{12}$

(6) $P_n(z) = (-1)^n P_n(1 - z) \qquad \forall z \in \mathbb{C}, n \geq 0.$

(7) $P_n(0) = P_n(1/2) = P_n(1) = 0$ for all odd integers $n > 1$.

(8) $P_n(0) = P_n(1)$ for all integers $n > 1$.

(9) $$P_{n+1}(z + 1) - P_{n+1}(z) = \int_{[0,z]} P_n(\xi + 1)\, d\xi - \int_{[0,z]} P_n(\xi)\, d\xi \quad \forall z \in \mathbb{C}, n \in \mathbb{N}.$$

(10) $$P_n(z + 1) = P_n(z) + \frac{z^{n-1}}{(n-1)!} \qquad \forall z \in \mathbb{C}, n \in \mathbb{N}.$$

Hint: Show that the polynomials $p_n(z) := (-1)^n P_n(1 - z)$ have, *mutatis mutandis*, properties (1), (2), (3), and appeal to uniqueness. Go on to prove (7)–(10).

(iii) *Use* (4) *and* 3.37(i) *to deduce that*

(11) $$P_1(t) = -\frac{1}{\pi} \sum_{k=1}^{\infty} \frac{\sin(2\pi k t)}{k} \qquad \forall t \in (0, 1)$$

with convergence uniform on compacta. Similarly use (5) *and* 3.37(v) *to deduce that*

(12) $$P_2(t) = \frac{1}{2\pi^2} \sum_{k=1}^{\infty} \frac{\cos(2\pi kt)}{k^2} \qquad \forall t \in [0,1].$$

In general

(13) $$P_{2m}(t) = (-1)^{m-1} \frac{2}{(2\pi)^{2m}} \sum_{k=1}^{\infty} \frac{\cos(2\pi kt)}{k^{2m}} \qquad \forall t \in [0,1], m \in \mathbb{N},$$

(14) $$P_{2m+1}(t) = (-1)^{m-1} \frac{2}{(2\pi)^{2m+1}} \sum_{k=1}^{\infty} \frac{\sin(2\pi kt)}{k^{2m+1}} \quad \forall t \in [0,1], \ m \in \mathbb{N}.$$

Hints: (for (13), (14)) Since these series converge uniformly, the real-line version of 2.12 lets them be differentiated termwise, showing that the nth one is the derivative of the $(n+1)$st one $(n \geq 2)$. The claimed equalities in (13), (14) then follow from the uniqueness clause in (i). [The periodic functions defined on $\mathbb{R}$ by the convergent trigonometric series (11)–(14) are called the *Bernoulli functions.*]

(iv) *Define* $B_n := n!\,P_n(0)$ (*the so-called* Bernoulli numbers) *for each* $n \geq 0$ *and show via* (2) *that*

(15) $$P_n(z) = \frac{1}{n!} \sum_{k=0}^{n} \binom{n}{k} B_k z^{n-k} \qquad \forall z \in \mathbb{C}, n \geq 0;$$

then use (8) *to conclude that*

(16) $$B_n = -\frac{1}{n+1} \sum_{k=0}^{n-1} \binom{n+1}{k} B_k \qquad \forall n \in \mathbb{N}.$$

Deduce from (13) *that*

(17) $$\sum_{k=1}^{\infty} \frac{1}{k^{2m}} = \frac{(-1)^{m-1}(2\pi)^{2m} B_{2m}}{2(2m)!} \in \pi^{2m}\mathbb{Q} \qquad \forall m \in \mathbb{N}.$$

Using this and (16), *deduce the formulas*

(18) $$\sum_{k=1}^{\infty} \frac{1}{k^2} = \frac{\pi^2}{6}, \quad \sum_{k=1}^{\infty} \frac{1}{k^4} = \frac{\pi^4}{90}, \quad \sum_{k=1}^{\infty} \frac{1}{k^6} = \frac{\pi^6}{945}, \quad \sum_{k=1}^{\infty} \frac{1}{k^8} = \frac{\pi^8}{9450}.$$

By breaking (17) into sums over even and odd k, deduce that

(19) $$\sum_{k=1}^{\infty} \frac{1}{(2k-1)^{2m}} = (-1)^{m-1}\pi^{2m}\frac{2^{2m}-1}{2(2m)!}B_{2m} \qquad \forall m \in \mathbb{N}$$

(20) $$\sum_{k=1}^{\infty} \frac{(-1)^{k-1}}{k^{2m}} = (-1)^{m-1}\pi^{2m}\frac{2^{2m-1}-1}{(2m)!}B_{2m} \qquad \forall m \in \mathbb{N}.$$

Use (10) and (15) to deduce that

(21) $$\sum_{k=0}^{N-1} k^n = \frac{1}{n+1}\sum_{k=0}^{n}\binom{n+1}{k}N^{n+1-k}B_k \qquad \forall n, N \in \mathbb{N}.$$

Confirm then the following identities

(22) $$\begin{cases} \displaystyle\sum_{k=1}^{N} k^2 = \frac{N^3}{3}+\frac{N^2}{2}+\frac{N}{6}, & \displaystyle\sum_{k=1}^{N} k^3 = \frac{N^4}{4}+\frac{N^3}{2}+\frac{N^2}{4} = \left[\sum_{k=1}^{N} k\right]^2 \\ \displaystyle\sum_{k=1}^{N} k^4 = \frac{N^5}{5}+\frac{N^4}{2}+\frac{N^3}{3}-\frac{N}{30}, & \displaystyle\sum_{k=1}^{N} k^5 = \frac{N^6}{6}+\frac{N^5}{2}+\frac{5N^4}{12}-\frac{N^2}{12}. \end{cases}$$

With a little more ingenuity (see GOULD [1972]) (21) can be "inverted" as

(21)′ $$B_n = \sum_{j=1}^{n+1}(-1)^{j+1}\binom{n+1}{j}\frac{1}{j}\sum_{k=1}^{j}k^n.$$

(v) *For $t \in \mathbb{R}$, $z \in \mathbb{C}$ with $0 \le t \le 1$ and $0 < |z| < 2\pi$ we have* (Jacobi, 1834)

(23) $$\frac{ze^{tz}}{e^z-1} = \sum_{n=0}^{\infty} P_n(t)z^n.$$

Hints: Fix $z \in \mathbb{C}$ with $0 < |z| < 2\pi$. It follows easily from (13) and (14) that for any $n \in \mathbb{N}$ and $t \in [0,1]$

(24) $$|P_n(t)| \le \frac{2}{(2\pi)^n}\sum_{k=1}^{\infty}\frac{1}{k^2} \overset{(18)}{=} \frac{\pi^2}{3}\frac{1}{(2\pi)^n}.$$

Consequently the series in (23) converges uniformly for $t \in [0,1]$. Call the limit $f(t)$. Using (2),

$$\frac{f(t)-f(t_0)}{t-t_0} - zf(t_0) = \sum_{n=1}^{\infty}\left[\frac{P_n(t)-P_n(t_0)}{t-t_0} - P_{n-1}(t_0)\right]z^n$$

(25)
$$= \sum_{n=1}^{\infty}[P_{n-1}(t_n) - P_{n-1}(t_0)]z^n,$$

for some t_n between t and t_0 (Mean-Value Theorem, recalling that P_n has real coefficients). It follows from (24) and (25) that f' exists in $[0,1]$ and satisfies $f'(t) = zf(t)$. Consequently $e^{-tz}f(t)$ has derivative 0 with respect to t and so

(26)
$$f(t) = c(z)e^{tz} \qquad \forall t \in [0,1],$$

where $c(z) = f(0)$. Integrate (26) with respect to t. Computing $\int_0^1 f(t)\,dt$ termwise through the series $\sum z^n P_n$ for f and recalling (1) and (3), gives for this integral the value 1, whence (23).

(vi) *Take* $t = 0$ *in* (23) *and recall* (7) *to get*

$$\frac{z}{e^z-1} = 1 - \frac{1}{2}z + \sum_{m=1}^{\infty}\frac{B_{2m}}{(2m)!}z^{2m}, \qquad z \in \mathbb{C}, 0 < |z| < 2\pi.$$

Transpose $-\frac{1}{2}z$, *multiply and divide by* $e^{-z/2}$ *to get*

$$\frac{z}{2}\cdot\frac{e^{z/2}+e^{-z/2}}{e^{z/2}-e^{-z/2}} = \sum_{m=0}^{\infty}\frac{B_{2m}}{(2m)!}z^{2m}$$

and conclude by a change of variable that

(27)
$$z\cot z = \sum_{m=0}^{\infty}(-1)^m\frac{2^{2m}B_{2m}}{(2m)!}z^{2m}, \qquad z \in \mathbb{C}, 0 < |z| < \pi.$$

Use the identity $\tan z = \cot z - 2\cot 2z$ *to deduce from* (27) *the series*

(28)
$$\tan z = \sum_{m=1}^{\infty}(-1)^{m-1}\frac{2^{2m}(2^{2m}-1)B_{2m}}{(2m)!}z^{2m-1}, \qquad z \in \mathbb{C}, |z| < \frac{\pi}{2}.$$

Recalling from (iv) *that* $B_0 = 1$, *use* (17) *to write* (27) *as*

$$z\cot z - 1 = \sum_{m=1}^{\infty}\left[\sum_{k=1}^{\infty}\frac{-2}{(\pi k)^{2m}}\right]z^{2m}, \qquad 0 < |z| < \pi.$$

By 3.3 this series converges absolutely *in* $|z| < \pi$. *Use* 3.12(ii) *to conclude* (cf. 11.37(vi)) *that*

(29) $$z \cot z - 1 = \sum_{k=1}^{\infty} \frac{2z^2}{z^2 - \pi^2 k^2}, \qquad 0 < |z| < \pi.$$

(vii) (from Johnsonbaugh [1980]) *Notice that for* $t \in (0,1]$

$$2\sum_{k=1}^{\infty} \frac{\sin^2 k\pi t}{k^{2m}} = \sum_{k=1}^{\infty} \frac{1 - \cos 2k\pi t}{k^{2m}} = \sum_{k=1}^{\infty} \frac{1}{k^{2m}} - \sum_{k=1}^{\infty} \frac{\cos 2k\pi t}{k^{2m}}$$
$$\overset{(13)}{=} \frac{(-1)^{m-1}(2\pi)^{2m}}{2}[P_{2m}(0) - P_{2m}(t)].$$

Take $m = 1,2$, *compute* P_4 *and get the remarkable identities*

(30) $$\sum_{k=1}^{\infty} \left(\frac{\sin k\pi t}{k\pi t}\right)^2 = \frac{1-t}{2t} = \sum_{k=1}^{\infty} \frac{\sin k\pi t}{k\pi t} \qquad \forall t \in (0,1]$$

by 3.37(i) [cf. 5.4(v)],

(31) $$\sum_{k=1}^{\infty} \frac{\sin^2 k}{k^4} = \frac{(\pi-1)^2}{6}.$$

(viii) *By taking* $z := \pi/4$ *in* (29) *and recognizing* $\frac{2}{1-16k^2}$ *as* $\frac{1}{4k+1} - \frac{1}{4k-1}$ *deduce anew the Gregory-Leibniz formula in* 3.22.

§ 5 Cauchy's Integral Formula Adumbrated

Once before, in 2.20(iv), a special case of Cauchy's Integral Formula was derived (under heavier than necessary hypotheses). Here is a comparably brief treatment, using very different ideas, which produces a more computationally versatile form of the formula, the one that dominates most expositions of the subject.

Exercise 3.39 (i) *For any* $a \in \mathbb{C}$, $r > 0$

$$\int_{C(a,r)} \frac{d\xi}{\xi - z} = 2\pi i \qquad \forall z \in D(a,r).$$

Hint: 2.15(ii) reduces the question to $z = a$. Alternatively, use the geometric series expansion idea in the 3.8 hints (with $n = 0$).

(ii) *Let f be $\mathbb{C}$-differentiable and f' be continuous in a neighborhood of $\overline{D}(a,r)$. Show that*

$$(*)\qquad f(z) = \frac{1}{2\pi i}\int_{C(a,r)} \frac{f(\xi)}{\xi - z}\,d\xi \qquad \forall z \in D(a,r),$$

whence, via 3.8, f is represented by a power series $\sum c_k(z-a)^k$ in $D(a,r)$.

Hints: (Cf. 2.19(vi), (vii).) Fix such a z and note that for $0 < t \le 1$

$$\frac{d}{d\xi} t^{-1} f(z + t(\xi - z)) = f'(z + t(\xi - z)).$$

So by 2.10

$$0 = \int_{C(a,r)} f'(z + t(\xi - z))d\xi$$

for all $t \in (0,1]$, but then for $t = 0$ as well, by continuity. Integrate in t and use Fubini 2.19(ii) to get

$$\begin{aligned} 0 = \int_{C(a,r)} \int_0^1 f'(z + t(\xi - z))\,dtd\xi &= \int_{C(a,r)} \int_0^1 \frac{d}{dt}\left[\frac{f(z + t(\xi - z))}{\xi - z}\right] dtd\xi \\ &= \int_{C(a,r)} \frac{f(\xi) - f(z)}{\xi - z}\,d\xi. \end{aligned}$$

Now cite (i).

Remarks: The proof in (ii) is adapted from FALK [1883]. It is a precise version of a less rigorous calculus of variations argument used by Cauchy in 1825 (see p. 90 of SMITHIES [1997]). It works to show that

$$(**)\qquad f(z)\int_\gamma \frac{d\xi}{\xi - z} = \int_\gamma \frac{f(\xi)\,d\xi}{\xi - z} \qquad \forall z \in U \setminus \gamma$$

whenever U is open and convex, $f \in H(U)$, f' is continuous in U and γ is a piecewise smooth loop in U. [Convexity is needed so that the points $z+t(\xi-z)$ ($0 \le t \le 1$) remain in U.] As noted in 2.15(ii), the integral on the left in $(**)$ is constant in each component of $\mathbb{C}\setminus\gamma$. In the next chapter we will prove that it is always an integer multiple of $2\pi i$. Formula $(**)$ then looks like $(*)$. It is in fact valid for a much larger class of open sets than the convex ones and can be proved without the *a priori* assumption of the continuity of f'. [That continuity and more of course follow from $(**)$ via 3.8.] This is the theme of Chapters X and V, respectively. New techniques, not at all as straightforward and simple as those in 3.39, will be required. Their creation was a major advance in the foundations of the subject, which prior to 1900 had always hypothesized continuity of f', in addition to its existence.

§ 6 Holomorphic Logarithms Previewed

A $\mathbb{C}$-valued function L on a subset S of $\mathbb{C} \setminus \{0\}$ which satisfies $e^{L(z)} = z$ for all $z \in S$ is traditionally called a *logarithm function* on S. (The homomorphic property $L(z_1 z_2) = L(z_1) + L(z_2)$ is not required, although whenever meaningful it holds "modulo $2\pi i\mathbb{Z}$.") More generally, if X is any set and $f\colon X \mapsto \mathbb{C} \setminus \{0\}$, then any function $L_f\colon X \mapsto \mathbb{C}$ satisfying $e^{L_f} = f$ is called a *logarithm of* f. Thanks to 3.14(vi) such L and L_f always exist. But if continuity is demanded, the matter is much more subtle and will play a central role in all that follows. The existence of a continuous logarithm L on S depends on the shapes of S and $\mathbb{C} \setminus S$, and such an L encodes significant topological properties of S. Chapter IV will explore this in depth, but in this short section we will look at the most elementary situation. When X is a metric space and f is continuous, there need not be a continuous L_f, and when there is a continuous L_f it need not have the form $L \circ f$ for a continuous logarithm function L on $S := f(X)$. The foremost example of the first phenomenon is the identity function f on $X := \mathbb{T}$, and of the second phenomenon it is $f(x) := e^{ix}$ on $X := [0, 2\pi]$. See 3.42(iv). Although notation like $\log f$ and $\log(f)$ for an L_f is popular and mnemonic, it is hazardous because it evokes $\log \circ f$ and therewith the expectation that L_f is a function of f, in particular, that $f(x_1) = f(x_2)$ entails $L_f(x_1) = L_f(x_2)$—generally not the case, as the last example shows. So we will eschew such notation and just as quixotically, (usually) refrain from designating by $\sqrt{f}$ a function g satisfying $g^2 = f$. (A good case for this kind of notational and conceptual distinction is made by H. P. Boas [2016].)

The situation being thus, let us resolve to live within our means and accommodate to the non-existence of continuous logarithms in certain regions. A popular alternative is to devalue the currency, here, pervert the concept of function. In lieu of fulminating against this, I'll just cite scripture—p. 198 of Dieudonné [1969]—and reassure readers that they will encounter no schizophrenic "multiple-valued" functions in this book. However, in the "brave new world" of Riemann surfaces such exotic flora *can* take root. And there is another way to get a continuous inverse for the exponential function which is defined in the whole $\mathbb{C} \setminus \{0\}$: Consider such a function as mapping, not into $\mathbb{C}$, but into the (topological) group $\mathbb{C}/2\pi i\mathbb{Z}$ and as being the inverse of the isomorphism that the exponential function (whose kernel is $2\pi i\mathbb{Z}$) induces on this quotient group. It follows from 3.35(x) and the definition of the topology on $\mathbb{C}/2\pi i\mathbb{Z}$ that this inverse is continuous. Such a construction brings the theory to a satisfactory conclusion algebraically and topologically, if not analytically.

On the other hand, the problematic expression "multiple-valued function" can be given a perfectly proper meaning, that of set-valued function. One envisions "the logarithm" as the set of all vertical stacks over the points of $\mathbb{C}\setminus\{0\}$: For each non-zero z there is a unique $w \in \mathbb{R}\times[0,2\pi)$ with $z = e^w$ and the stack over z is the discrete set $\log(z) := w + 2\pi i\mathbb{Z}$. A continuous logarithm function in a region $\Omega \subset \mathbb{C}\setminus\{0\}$ is then envisioned as a continuous slice L through these stacks, meaning that $L(z) \in \log(z)$ for each $z \in \Omega$. This interpretation of L explains why it is often called a "continuous *branch*" (*Zweig*, in German) of "the" logarithm in or over Ω, although "continuous layer" might seem a more descriptive phrase here. This model is the historical prototype of continuous "liftings" into "covering spaces" so important in topology. With some effort one can perhaps imagine why, if it is possible for z to make in Ω a complete unbroken circuit around 0, starting and ending at z_0, then $\operatorname{Im} L(z)$ must undergo a change of $\pm 2\pi$ and be unable to approach $\operatorname{Im} L(z_0)$ near the end of the circuit. Consequently, for a continuous L to exist, z must find it impossible to complete such a circuit, which *suggests* Ω must omit a continuum from 0 to ∞ in order to block such circuits. This intuition is sound: see Remark 3.43(i).

Whenever $z \in \mathbb{C}\setminus i\mathbb{R}$ is written as $re^{i\theta}$ with $r > 0$, $\theta \in \mathbb{R}$, one has $\cos\theta = \operatorname{Re}(e^{i\theta}) = r^{-1}\operatorname{Re} z$, $\sin\theta = \operatorname{Im}(e^{i\theta}) = r^{-1}\operatorname{Im} z$, so $\tan\theta = \operatorname{Im} z/\operatorname{Re} z$. Since tan has period π and (as was shown in the proof of 3.22) maps $(-\pi/2, \pi/2)$ bijectively onto $\mathbb{R}$, the last equation means that θ differs from $\tan^{-1}(\operatorname{Im} z/\operatorname{Re} z)$ by an integer multiple of π. So all logarithms of $z \in \mathbb{C}\setminus\{0\}$ have the form, for appropriate $n(z) \in \mathbb{Z}$:

$$\log|z| + i\tan^{-1}(\operatorname{Im} z/\operatorname{Re} z) + n(z)\pi i \text{ if } \operatorname{Re} z \neq 0,\ \log|z| + \frac{\pi}{2}i + n(z)\pi i \text{ if } \operatorname{Re} z = 0.$$

Exercise 3.40 *Let U_1, U_2 be open subsets of $\mathbb{C}\setminus\{0\}$ such that $U_1 \cap U_2$ is connected. Show that if there exists a continuous logarithm L_j in U_j for each j, then there exists a continuous logarithm on $U_1 \cup U_2$.*

Hint: Show that $L_1 - L_2$ is constant in $U_1 \cap U_2$. (Cf. the hints to 4.73(ii).)

Exercise 3.41 (i) (Cf. 4.67.) *Prove that a continuous logarithm is necessarily holomorphic. More precisely, prove that if L is a continuous logarithm in an open subset U of $\mathbb{C}\setminus\{0\}$, then L is holomorphic and $L'(z) = 1/z$ for every $z \in U$.*

Hints: Imitate the argument in 3.18. Alternatively, use 3.19 to get local holomorphic logarithms and use continuity to show that locally L differs only by constants from such functions.

(ii) *From* (i) *infer that no continuous logarithm exists in any* $D(0,r)\setminus\{0\}$.

Hint: If L were such a function, $\int_{C(0,r/2)} \frac{1}{\xi}d\xi = \int_{C(0,r/2)} L'(\xi)d\xi = 0$ would follow from 2.10, contradicting 3.39(i).

Exercise 3.42 (i) *Prove that in every open half-plane which does not contain* 0 *there is a holomorphic logarithm.*

Hints: It suffices to consider $\mathbb{H} := \{z \in \mathbb{C} : \operatorname{Re} z > 0\}$. For each positive integer n there is a holomorphic logarithm L_n in $D(n,n)$, by 3.19. Then $[L_{n+1} - L_n]/2\pi i$ is continuous and integer-valued in $D(n+1,n+1) \cap D(n,n) = D(n,n)$. Therefore, by adding an appropriate (constant) integer multiple of $2\pi i$ to L_{n+1}, we can assume that L_{n+1} extends L_n. Do this *successively* for each $n = 1, 2, 3, \ldots$ It is then possible to define the desired logarithm L in $\mathbb{H}$ by the simple decree $L := L_n$ in $D(n,n)$, for every n, bearing in mind that $\bigcup_{n=1}^{\infty} D(n,n) = \mathbb{H}$. More simply, for $\mathbb{H}$ we may just define $L(z)$ by the first formula immediately preceding 3.40, with $n(z) := 0$, and follow by citing 3.41.

(ii) *Prove that there is a holomorphic logarithm in any slit-plane* $S := \mathbb{C} \setminus \{tu : t \le 0\}$, $u \in \mathbb{C} \setminus \{0\}$. (Cf. 4.28(i).) *Show moreover that any such function differs only by an additive constant from that defined by the integral* $\int_{[u,z]} d\xi/\xi$, $z \in S$.

Hints: $S = H_1 \cup H_2 \cup H_3$ where H_j are open half-planes such that $H_1 \cap H_2$ and $H_2 \cap H_3$ are connected while $H_1 \cap H_3$ is void. Part (i) provides holomorphic logarithms L_j in each H_j. Add appropriate constants to these to synthesize a single holomorphic logarithm L in $H_1 \cup H_2 \cup H_3$, as in 3.40. (Cf. with the Hints to 4.73(iii).) From $e^{L(\xi)} = \xi$ and the Chain Rule follows $L'(\xi) = 1/\xi$. Observing that S is starlike with respect to u, it follows then from 2.10 that $L(z) - L(u) = \int_{[u,z]} d\xi/\xi$, for all $z \in S$. [It is easy to see—as in the 2.22(ii) hint—that the *existence* of L can also be secured by means of this integral.]

(iii) *Show that in* $\mathbb{C}_+$ *there is a unique holomorphic logarithm* L *such that* $\operatorname{Im} L(z) \in (0, 2\pi)$ *for all* $z \in \mathbb{C}_+$. *Thus for any* $r > 0$ *and* $\theta \in (0, 2\pi)$, $re^{i\theta} \in \mathbb{C}_+$ *and* $L(re^{i\theta}) = \log r + i\theta$.

Hints: If l is any holomorphic logarithm in $\mathbb{C}_+$ (such exist by (ii)), then $e^{l(z)} = z \notin [0, \infty)$ implies that $\operatorname{Im} l(z) \ne 2\pi n$ for any integer n. Since $\operatorname{Im} l$ is a continuous real-valued function and $\mathbb{C}_+$ is connected (starlike), the set $\operatorname{Im} l(\mathbb{C}_+)$ is a connected subset of $\bigcup_{n=-\infty}^{\infty} (2\pi n, 2\pi(n+1))$ and so lies wholly in $(2\pi n, 2\pi(n+1))$ for some one integer n. Set $L := l - 2\pi n i$. Alternatively, note that $[l(re^{i\theta}) - \log r - i\theta]/2\pi i$ is a continuous, integer-valued function of (r, θ) in the connected set $(0, \infty) \times (0, 2\pi)$.

The most direct access to L is to simply make the desired equation the definition. The problem then comes down to proving that θ is a continuous function of $e^{i\theta}$, which can be done with the elementary compactness argument in the 4.7(i) hints. [In fact, this is the special case $\Gamma(t) := e^{it}$ of the result 4.7(i) itself.]

(iv) *L is holomorphic in the region $\Omega \subset \mathbb{C} \setminus \{0\}$. If L is a logarithm, the chain rule says $L'(z) = 1/z$. Formulate and prove the converse.*

Hints: The holomorphic function $F(z) := ze^{-L(z)}$ is constant ($\neq 0$) in Ω, thanks to 2.10(iv). Now recall 3.14(vi).

(v) *Show that there is no continuous logarithm in $\mathbb{T}$.*

Hints: If l is such a continuous logarithm, then $N(\theta) := l(e^{i\theta}) - i\theta$ is a continuous $2\pi i\mathbb{Z}$-valued function of $\theta \in \mathbb{R}$. Hence it is a constant, yet $N(2\pi) = l(1) - 2\pi i$ and $N(0) = l(1)$. (An alternative elementary proof is suggested in 4.21. Another consists in noting that a continuous logarithm l on $\mathbb{T}$ would map $\mathbb{T}$ homeomorphically to a compact connected subset of $i\mathbb{R}$, that is, to a non-degenerate interval I. Since connectedness of I is destroyed by removal of its midpoint, whereas removal of one point does not disconnect $\mathbb{T}$, no such l exists.)

Remarks 3.43: (i) The full story, of which 3.42 is only a fragment, is told in 8.22(iii): a continuous (holomorphic) logarithm exists in the open subset U of $\mathbb{C}$ if and only if 0 lies in an unbounded component of $\mathbb{C} \setminus U$. For compact subsets K of $\mathbb{C}$ the story (told in 4.28(ii) and 4.31) is the same: A continuous logarithm exists in K if and only if 0 lies in the unbounded component of $\mathbb{C} \setminus K$.

(ii) The function L in (iii) above is customarily called the *Principal Branch of the Logarithm in* $\mathbb{C}_+$. There is similarly a unique holomorphic logarithm in $\mathbb{C}_-$ which assigns to $re^{i\theta}$ the value $\log r + i\theta$ whenever $r > 0$ and $-\pi < \theta < \pi$. It is called the *Principal Branch of the Logarithm in* $\mathbb{C}_-$.

Note: There is no agreed-upon notation in the literature for these two important functions.

Exercise 3.44 *Let l denote the* Principal Branch *of the Logarithm in* $\mathbb{C}_-$.

(i) *Like* Vishnu *and* Loki, *this function manifests itself in numerous ways.*

Confirm the following instances:

(a) $l(z) = -\sum_{k=1}^{\infty} \frac{1}{k}(1-z)^k \forall z \in \bar{D}(1,1) \setminus \{0\},$

(b) $l(z) = 2\sum_{k=1}^{\infty} \frac{1}{2k-1}\left(\frac{z-1}{z+1}\right)^{2k-1} = \log|z| + i\arctan\left(\frac{\operatorname{Im} z}{\operatorname{Re} z}\right) \quad \forall z \in \mathbb{H},$

(c) $l(z) = \int_{[1,z]} \frac{d\xi}{\xi} = \log|z| + 2i\arctan\left(\frac{\operatorname{Im} z}{|z| + \operatorname{Re} z}\right) \quad \forall z \in \mathbb{C}_-.$

(d) *The exponential function maps the strip* $\mathbb{R} \times (-\pi, \pi)$ *conformally onto* $\mathbb{C}_-$ *and* l *is the inverse map.*

Hints: The function $L(z)$ defined by the first series satisfies $e^{L(z)} = z$ (cf. proof of 3.19). Since also $e^{l(z)} = z$, $L - l$ is a continuous map of $D(1,1)$ into $2\pi i\mathbb{Z}$ (3.14(iii)), hence it is constant on that connected set, equal to $L(1) - l(1) = 0$. Since $z \mapsto (z-1)/(z+1)$ maps $\mathbb{H}$ onto $\mathbb{D}$ (2.5), equation (2) in the proof of 3.20 establishes that the series in (b) exponentiates to z. Moreover, 3.20 itself says that this series maps onto $\mathbb{R} \times (-\pi/2, \pi/2)$. Hence it must be $l(z)$. The paragraph immediately preceding 3.40 explains the arctan. The first equality in (c) is from 3.42(ii). For $z \in \mathbb{C}_-$ let $\theta := \operatorname{Im} l(z)$, $w := |z|^{1/2}e^{i\theta/2}$. Then $w \in \mathbb{H}$ and $l(z) = 2l(w)$. Apply (b) to w and note that $\tan(\theta/2) = \sin\theta/(1+\cos\theta)$.

(ii) *Compute* l *as a limit of square-roots of positive real numbers.*

Hints: Define $S\colon \mathbb{C} \mapsto \mathbb{C}$ as follows:

$$S(w) := \sqrt{\frac{|w| + \operatorname{Re} w}{2}} + i\operatorname{sgn}(\operatorname{Im} w)\sqrt{\frac{|w| - \operatorname{Re} w}{2}},$$

where $\operatorname{sgn} y$ is 1 if $y \geq 0$ and is -1 if $y < 0$. Thus $S^2(w) = w$ and $S(\mathbb{C}_-) = \mathbb{H}$. Using 1.7, show that in $\mathbb{C}_-$, $S = e^{l/2}$. Letting $S^{[n]}$ denote the nth iterate of S, check that $S^{[n]} = e^{l/2^n}$ in $\mathbb{C}_-$. [Go by induction on n, first checking that $l(e^{\alpha l}) = \alpha l$ for every $0 < \alpha < 1$.] Consequently,

$$2^n\left[S^{[n]}(w) - 1\right] = \frac{e^{l(w)/2^n} - 1}{l(w)/2^n} \cdot l(w) \to E'(0)l(w) = l(w)$$

as $n \to \infty$, for each $w \in \mathbb{C}_- \setminus \{1\}$.

(iii) *Note that l lacks the expected homomorphic property by checking that $w_0 = z_0 := e^{3\pi i/4}$ and $w_0 z_0$ all lie in $\mathbb{C}_-$ but $l(w_0 z_0) \neq l(w_0) + l(z_0)$. Show that nevertheless*

$$w, z \in \mathbb{H} \Rightarrow wz,\ w/z \in \mathbb{C}_- \quad \textit{and}$$
$$l(wz) - l(z) = l(w) = l(w/z) + l(z).$$

Hints: Since $\mathbb{H}$ and $\mathbb{C}_-$ are closed under multiplication by positive real numbers, to prove $wz \in \mathbb{C}_-$ it is enough to consider $\operatorname{Re} w = \operatorname{Re} z = 1$. In that case, $\operatorname{Im}(wz) = \operatorname{Im}(w) + \operatorname{Im}(z)$, and if this is 0, then $\operatorname{Re}(wz) = 1 + (\operatorname{Im} z)^2 > 0$, showing that $wz \notin (-\infty, 0]$. Also, $\operatorname{Re}(1/z) = |z|^{-2} \operatorname{Re}(z)$, so $1/z \in \mathbb{H}$. Thus for fixed $w \in \mathbb{H}$, $f(z) := l(wz) - l(z)$ and $g(z) := l(w/z) + l(z)$ are well-defined holomorphic functions in $\mathbb{H}$. Since $f' = g' = \mathbf{0}$, they are each constant, equal to their values at 1, namely, $l(w)$.

Remarks: (i) It is possible to found the theory of the real logarithm function on the limit in 3.44(ii); see HURWITZ [1911] or E. LANDAU [1950]. The series in (b) is an effective way to compute $\log z$ for real $z > 0$. (Cf. the error estimate in (iii) below.) When the series (9) in 3.22 is brought into the picture, (c) too becomes computationally feasible.

(ii) The formulas themselves scarcely suggest it, but $S(w)$ in the hints to 3.44(ii) coincides, when $w \in \mathbb{C}_-$, with the square-root recipe in 1.26(i), and for $z \in D(1, 1)$ the square-root $q(z)$ of z constructed in 3.12(vi) coincides with $e^{\frac{1}{2} l(z)}$. Note too that since S maps $\mathbb{C}_-$ onto $\mathbb{H}$, it may be combined with (b) in 3.44(i) to give a series representation of l in its whole domain:

$$l(w) = 4 \sum_{k=1}^{\infty} \frac{1}{2k-1} \left(\frac{S(w) - 1}{S(w) + 1} \right)^{2k-1} \qquad \forall w \in \mathbb{C}_-.$$

(iii) The function $\log_{10}$ [defined as $(\log 10)^{-1} \log$] can easily be computed on $\mathbb{R}^+$ if one has available (via any of a number of simple algorithms) a table of successive square-roots of 10. For example, if $1 < x \leq 10$, use the table to determine the smallest $n_1 \in \mathbb{N}$ such that $10^{2^{-n_1}} < x$, and once minimal integers $n_1 < n_2 < \cdots < n_{N-1}$ satisfying $\prod_{j=1}^{N-1} 10^{2^{-n_j}} < x$ have been determined, find in the table the least integer n_N such that $\prod_{j=1}^{N} 10^{2^{-n_j}} < x$. A little reflection shows that if $n_N \leq n_{N-1}$, then some integer smaller than n_{N-1} could do the job for which n_{N-1} is minimal, a contradiction from which we infer that $n_N > n_{N-1}$. In turn, minimality of n_N means that

$$x \leq 10^{2^{-(n_N - 1)}} \prod_{j=1}^{N-1} 10^{2^{-n_j}} = 10^{2^{-n_N}} \prod_{j=1}^{N} 10^{2^{-n_j}}.$$

Since $n_j \geq j$ for all j, we thus have

$$\frac{1}{10^{2^{-N}}} \leq \frac{1}{10^{2^{-n_N}}} \leq \left(\prod_{j=1}^{N} 10^{2^{-n_j}}\right)/x < 1 \qquad \forall N \in \mathbb{N}$$

and then by 3.1

$$\lim_{N\to\infty} 10^{\sum_{j=1}^{N} 2^{-n_j}} = \lim_{N\to\infty} \prod_{j=1}^{N} 10^{2^{-n_j}} = x.$$

Since $y \mapsto 10^y$ is a continuous function on $\mathbb{R}$ and the series obviously converges, this entails that

$$\log_{10} x = \lim_{N\to\infty} \sum_{j=1}^{N} 2^{-n_j} = \sum_{j=1}^{\infty} 2^{-n_j}.$$

As to the rate of convergence, we have the crude estimate

$$0 < \log_{10} x - \sum_{j=1}^{N} 2^{-n_j} \leq 2^{-n_N} \leq 2^{-N} \qquad \forall N \in \mathbb{N}.$$

Now the most elementary maneuvers with geometric series show that in (b) of 3.44(i)

$$\left| l(z) - 2\sum_{k=1}^{N} \frac{1}{2k-1}\left(\frac{z-1}{z+1}\right)^{2k-1} \right| < \frac{1}{N}\left(1 - \left|\frac{z-1}{z+1}\right|\right)^{-1} \left|\frac{z-1}{z+1}\right|^{2N}$$

for all $N \in \mathbb{N}$, $z \in \mathbb{H}$ and in particular

$$\left| l(x) - 2\sum_{k=1}^{N} \frac{1}{2k-1}\left(\frac{x-1}{x+1}\right)^{2k-1} \right| < \frac{x+1}{2N}\left(1 - \frac{2}{x+1}\right)^{2N}$$

for all $N \in \mathbb{N}$, real $x > 1$, again a geometric rate of convergence, but now achieved without need of a table of roots of 10. According to GLAISHER [1916], around 1624 Henry Briggs constructed and used a table of successive square-roots of 10 (up to $10^{1/2^{54}}$) to compute logarithms, but in a less transparent way than above. Since $10^{1/2^2} < 2$, after the first two such square-root extractions, the rest can be computed via 3.12(vi). Briggs' calculations actually anticipated this result of Newton's — see pp. 16–19 of GOLDSTINE [1977]. Of course, the superior power-series method was not discovered until many years after the pioneering work of Briggs.

(iv) A simpler way to create logarithms on $\mathbb{R}^+$ is to select a *base* $b_k := 1+10^{-k}$ with $k \in \mathbb{N}$ and to compute the powers b_k^n, $n \in \mathbb{N}$, up to the largest n, call it N_k, such that $b_k^n \leq 10$. Even if k is only moderately large, these powers are fairly dense in the real interval $(1, 10]$. Hence via linear interpolation one gets a table of logarithms in base b_k. The larger k, the denser the powers b_k^n and the more accurate the table. In fact, if $x \in (1, 10]$ is given and n_k is the largest n for which $b_k^n \leq x$, then

$$0 \leq x - b_k^{n_k} < b_k^{1+n_k} - b_k^{n_k} = b_k^{n_k}(b_k - 1) \leq x10^{-k} \leq 10^{1-k},$$

so $b_k^{n_k} \to x$ as $k \to \infty$, uniformly in x. Consequently,

$$10^{-k} n_k \log \left(1 + \frac{1}{10^k}\right)^{10^k} = n_k \log \left(1 + \frac{1}{10^k}\right) = \log b_k^{n_k} \to \log x,$$

and from 3.21 we then learn that $10^{-k} n_k \to \log x$. Jost Bürgi (see **MR** 3233086) chose $k = 4$ and found $N_4 = 23027$. Thus he implicitly found that $\log 10 \approx 10^{-4} N_4 = 2.3027$ (accurate to 3 places). No square-root extractions, but 23027 multiplications (which, however, turn out to be merely additions!) are required, and precision can only be increased by increasing k and (necessarily) computing a whole new (and larger) table.

(v) A uniform cord hanging freely between two equally high points is called a *catenary* (*Kettenlinie* in German); evidently it's a curve symmetric about a vertical axis through its lowest point. From the physics of the matter one can derive a differential equation for such a curve, whose solution, if appropriate units are used, is (part of) the hyperbolic cosine function $K(x) := \frac{1}{2}(e^x + e^{-x})$, from $\mathbb{R}$ into $[1, \infty)$. Moreover, there is an elementary geometric construction on the catenary curve itself for determining those units, that is, for determining how to calibrate the horizontal and vertical axes. Any given catenary together with the coordinate system it thus determines provides a graph of K, from which we can ocularly find the x such that $K(x) = \frac{1}{2}(y + 1/y)$ for any $y > 0$. But those x are just $\pm \log(y)$. So nature has provided mechanical means for finding (natural) logarithms! For the fascinating details and some history, the reader is urged to consult the prize-winning paper BLÅSJÖ [2016].

Exercise 3.45 *Let $a, b \in \mathbb{R}$ with $a < b$ and consider $\phi(z) := (z - ia)/(z - ib)$ for $z \in \mathbb{C} \setminus \{ib\}$ and the Principal Branch l of the Logarithm in $\mathbb{C}_-$.*

(i) *Describe the mapping properties of ϕ. Show, in particular, that ϕ maps $\mathbb{H}$ onto $\mathbb{U}$.*

(ii) *If $\operatorname{Re} z > 0$, then*

$$\operatorname{Im} l(\phi(z)) = \int_a^b \frac{\operatorname{Re} z}{|z - it|^2}\, dt.$$

Hint: Since $l'(z) = 1/z$, 2.10 gives

$$l(z - ia) - l(z - ib) = \int_{[z-ib,z-ia]} \xi^{-1}\, d\xi = \int_a^b \frac{idt}{z - it}.$$

Consequently, using 3.44(iii)

$$\operatorname{Im} l\left(\frac{z - ia}{z - ib}\right) = \int_a^b \operatorname{Im}\left(\frac{i}{z - it}\right) dt.$$

(iii) *Show that* $\Phi := \operatorname{Im} l \circ \phi$ *is harmonic in* $\mathbb{H}$ *and as* $z \in \mathbb{H}$ *approaches* iy, $\Phi(z)$ *converges to* 0 *if* $y \in \mathbb{R} \setminus [a, b]$, *to* π *if* $a < y < b$. *Use the remarks following* 3.16 *to interpret* $\Phi(z)$ *as the angle subtended at* z *by* $[ia, ib]$.

(iv) *Check that* $(1 + iz)/(1 - iz)$ *maps* $z \in (\mathbb{C} \setminus i\mathbb{R}) \cup \mathbb{D}$ *onto* $\mathbb{C}_-$ *and then show that for all such* z *the number* $w := \frac{1}{2i} l\left(\frac{1+iz}{1-iz}\right)$ *satisfies* $\tan w = z$. *Finally, show that for* $z \in \overline{\mathbb{D}} \setminus \{-i, i\}$, $w = \sum_{n=0}^{\infty} (-1)^n (2n+1)^{-1} z^{2n+1}$.

Hint: By algebra one infers that for $z, w \in \mathbb{C}$

$$z = \tan w = \frac{1}{i} \frac{e^{iw} - e^{-iw}}{e^{iw} + e^{-iw}} = \frac{1}{i} \frac{e^{2iw} - 1}{e^{2iw} + 1} \Leftrightarrow e^{2iw} = \frac{1 + iz}{1 - iz},$$

and by 3.44(iii) we have

$$l\left(\frac{1 + iz}{1 - iz}\right) = l(1 + iz) - l(1 - iz) \qquad \forall z \in \mathbb{D}.$$

Now use 3.44(i)(a) and 3.13(ii) to prove the series claim.

(v) *Denote by* S *the strip* $(-\pi/2, \pi/2) \times \mathbb{R}$ *and by* Tan *the restriction of* tan *to* S. *Show that* Tan *maps* S *injectively into the doubly-slit plane* $(\mathbb{C} \setminus i\mathbb{R}) \cup \mathbb{D}$.

Hints: Suppose $\tan(w)$ is not in the doubly-slit plane, that is, $\tan(w) = iy$ with $y \in \mathbb{R}$ and $|y| \geq 1$. As shown in the preceding hint, such an equation means that $e^{2iw} = (1 - y)/(1 + y)$, which here is a negative real number. Consequently, $\operatorname{Im}(2iw)$ must be an odd-integer multiple of π. So $|\operatorname{Re}(2w)| = |\operatorname{Im}(2iw)| \geq \pi$, $|\operatorname{Re} w| \geq \pi/2$, and $w \notin S$. For injectivity, suppose $w_j \in S$ and $\tan(w_1) = \tan(w_2)$. Thus $0 = \sin(w_1)\cos(w_2) - \sin(w_2)\cos(w_1) = \sin(w_1 - w_2)$, so by 3.30(iii), $w_1 - w_2 = n\pi$ for some $n \in \mathbb{Z}$. Since $-\pi/2 < \operatorname{Re} w_1, \operatorname{Re} w_2 < \pi/2$, n must be 0.

(vi) *Show that the holomorphic injection* Tan *in* (v) *is bijective and has the holomorphic inverse*

$$\operatorname{Tan}^{-1}(z) = \frac{1}{2i} l\left(\frac{1 + iz}{1 - iz}\right), \quad \textit{with } \left(\operatorname{Tan}^{-1}\right)'(z) = (1 + z^2)^{-1}, \quad \textit{so}$$

$$\mathrm{Tan}^{-1}(z) = \int_{[0,z]} (1+\xi^2)^{-1} d\xi \textit{ for all } z \textit{ in the starlike region } (\mathbb{C}\backslash i\mathbb{R}) \cup \mathbb{D}.$$

Hint: Since l maps $\mathbb{C}_-$ onto $\mathbb{R} \times (-i\pi, i\pi) = 2iS$, it follows from (iv) that $A(z) := \frac{1}{2i} l\left(\frac{1+iz}{1-iz}\right)$, maps the doubly-slit plane holomorphically onto S and, moreover, satisfies $\tan A(z) = z$.

(vii) *With the aid of a diagram adapt the geometric interpretation in* (iii) *to the first formula in* (vi) *when z is real and positive.*

Hint: Take $a := -1$, $b := 1$ and consider $1/z$ in the role of z in (iii).

Exercise 3.46 (i) *Show that the sequence* $\log(n) - \sum_{k=1}^{n} \frac{1}{k}$ *increases to a finite limit.* The negative of this limit is known as *Euler's constant* or the *Euler-Mascheroni constant*, and is often denoted by γ or by C. It lies in $(0, 1]$.

Hints: $\log(n) - \sum_{k=2}^{n} \frac{1}{k} = \sum_{k=2}^{n} \int_{k-1}^{k} \left(\frac{1}{x} - \frac{1}{k}\right) dx$. The kth integral lies in $(0, \frac{1}{k-1} - \frac{1}{k}]$, so the sum is non-negative, increases with n, and never exceeds 1.

With a little more effort it can be proved that

$$\frac{1}{2(n+1)} < \sum_{k=1}^{n} \frac{1}{k} - \log(n) - \gamma < \frac{1}{2n} \qquad \forall n \in \mathbb{N}.$$

(ii) *Show that for every $b > 1$*

$$\lim_{n\to\infty} \sum_{k=n}^{nb} \frac{1}{k} = \log b.$$

Hint: Writing $[nb]$ for the greatest integer not greater than nb,

$$\sum_{k=n}^{nb} \frac{1}{k} - \log b = \left(\sum_{k=1}^{nb} \frac{1}{k} - \log[nb]\right) - \left(\sum_{k=1}^{n} \frac{1}{k} - \log(n)\right) + \frac{1}{n} + \log\left(\frac{[nb]}{nb}\right).$$

Note that $\log([nb]/nb) \to \log(1) = 0$, and the first two terms on the right each converge to γ as $n \to \infty$.

Remark: Grouping all even denominators and all odd denominators in the sum $\sum_{k=1}^{2n} (-1)^{k+1}/k$, shows that it equals $\sum_{k=n+1}^{2n} 1/k$, so (ii) furnishes alternative access to part of the 3.19 Scholium.

(iii) *Letting l denote the Principal Branch of the Logarithm in $\mathbb{C}_-$, show that for every $b > 1$, $z \in \mathbb{C}$*

$$\lim_{n\to\infty} \sum_{k=n}^{nb} l\left(1 + \frac{z}{k}\right) = z \log b.$$

Hints: Note that, strictly speaking, only $n > |z|$ should enter consideration. In view of (ii), it suffices to show that

$$\lim_{n\to\infty} \sum_{k=n}^{nb} \left[l\left(1 + \frac{z}{k}\right) - \frac{z}{k} \right] = 0.$$

To that end, use 3.44(i), according to which

$$|l(1+w) - w| \le \sum_{n=2}^{\infty} |w|^n = \frac{|w|^2}{1 - |w|} \qquad \forall w \in \mathbb{D}.$$

Consequently,

$$\left| l\left(1 + \frac{z}{k}\right) - \frac{z}{k} \right| \le \frac{2|z|^2}{k^2} \qquad \forall k \ge 2|z|.$$

(iv) *By exponentiating in* (iii) *infer that for every $b > 1$, $z \in \mathbb{C}$*

$$\lim_{n\to\infty} \prod_{k=n}^{nb} \left(1 + \frac{z}{k}\right) = b^z \qquad (:= e^{z \log b}).$$

(v) *What are the results corresponding to* (ii)–(iv) *for $0 < b < 1$?*

(vi) *Show that for $n, x \in \mathbb{R} \setminus \{1\}$*

$$\frac{1}{n^x} = \frac{1}{1-x}\left[\frac{1}{n^{x-1}} - \frac{1}{(n-1)^{x-1}}\right] + x \int_{n-1}^{n} \frac{n-1-t}{t^{x+1}}\, dt.$$

(vii) *Infer from* (vi) *that for real $x > 1$ and integer $N \ge 2$*

$$\sum_{n=2}^{N} \left(\frac{1}{n^x} - \frac{1}{x^n}\right) = \frac{1}{1-x}\left[\frac{1}{N^{x-1}} - \frac{1}{x^N}\right] + \frac{1}{x} + x \sum_{n=2}^{N} \int_{n-1}^{n} \frac{n-1-t}{t^{x+1}}\, dt.$$

Hint: Use the closed form of the geometric series $\sum_{n=2}^{N} \left(\frac{1}{x}\right)^n$.

(viii) *First let* $N \to \infty$*, then let* $x \downarrow 1$ *in* (vii) *to get*

$$\lim_{x\downarrow 1} \sum_{n=2}^{\infty} \left(\frac{1}{n^x} - \frac{1}{x^n} \right) = 1 + \lim_{N\to\infty} \sum_{n=2}^{N} \int_{n-1}^{n} \frac{n-1-t}{t^2}\, dt.$$

Hint: Since the nth integrand is majorized by $\frac{1}{t^{x+1}} \le \frac{1}{(n-1)^{x+1}} \le \frac{1}{(n-1)^2}$, the convergence in N is uniform with respect to $x \in [1, \infty)$.

(ix) *Do the integrations in* (viii) *to reveal its right-hand side as Euler's constant:*

$$\lim_{x\downarrow 1} \sum_{n=1}^{\infty} \left(\frac{1}{n^x} - \frac{1}{x^n} \right) = \gamma.$$

[For a *slightly* different proof see SONDOW [1998]. This equality appears in the literature most often in the visually less dramatic form $\lim_{z\to 1}[\zeta(z) - \frac{1}{z-1}] = \gamma$, when analyzing the singularity at $z = 1$ of Riemann's zeta function.]

Exercise 3.47 (a qualitative Stirling's formula) *Consider the positive numbers*

$$S_n := n! \left(\frac{e}{n} \right)^n \frac{1}{\sqrt{n}}, \qquad n \in \mathbb{N}.$$

(i) *Show that they satisfy*

$$\log \left(\frac{S_n}{S_{n+1}} \right) = \left(n + \frac{1}{2} \right) \log \left(\frac{n+1}{n} \right) - 1 \qquad \forall n \in \mathbb{N}.$$

(ii) *Deduce from the series in* (b) *of* 3.44 *that for* $0 < x < 1$

$$2x < \log \left(\frac{1+x}{1-x} \right) < 2x + \frac{2}{3} x^2 \sum_{n=1}^{\infty} x^{2n-1} = 2x + \frac{x^2}{3} \left(\frac{1}{1-x} - \frac{1}{1+x} \right).$$

(iii) *Apply* (ii) *with* $x := (2n+1)^{-1}$ *to infer from* (i) *that*

$$0 < \log \left(\frac{S_n}{S_{n+1}} \right) < \frac{1}{12} \left(\frac{1}{n} - \frac{1}{n+1} \right) \qquad \forall n \in \mathbb{N}.$$

(iv) *From* (iii) *infer that* S_n decreases *to a non-negative limit* S.

(v) *Refine* (iv) *by showing that for each* $n \in \mathbb{N}$ *there exists* $\theta_n \in (0, 1)$ *such that* $S_n = S e^{\theta_n / 12n}$*, that is,*

$$n! = S\sqrt{n} \left(\frac{n}{e} \right)^n e^{\theta_n / 12n} \qquad \forall n \in \mathbb{N}.$$

Infer from the case $n = 1$ that $e^{11/12} < S < e$.

Hint: Sum in (iii) to get

$$0 < \log(S_n) - \log(S_{n+k}) < \frac{1}{12}\left(\frac{1}{n} - \frac{1}{n+k}\right) \qquad \forall n, k \in \mathbb{N}$$

and then let $k \to \infty$.

Remark: In 11.37(xiv) the constant S will be identified. A derivation of Stirling's asymptotic formula that is somewhat more direct in this regard is given by RZADKOWSKI [1997]. Note that two trivial consequences of $0 < S < \infty$ are the non-obvious facts $\lim \sqrt[n]{n!}/n = 1/e$ and $\lim(n/e)^n/n! = 0$. Cf. 5.92(v).

Example 3.48 *Closing the chapter is one of several beautiful constructions of a holomorphic function* Γ *found by Euler in* 1729, *which satisfies the functional equation* (7) *of the factorial (whence* $\Gamma(m+1) = m!$ *for all* $m \in \mathbb{N}$*).* To motivate its definition (6), note that for $m, M \in \mathbb{N}$

$$\frac{M!M^m}{m(m+1)\dots(m+M)} = (m-1)!\,\frac{M!M^m}{(m+M)!} = \frac{(m-1)! \quad M^m}{(M+1)(M+2)\dots(M+m)}$$
$$= (m-1)!\left[\left(1+\frac{1}{M}\right)\left(1+\frac{2}{M}\right)\dots\left(1+\frac{m}{M}\right)\right]^{-1},$$

from which it follows trivially that

$$\lim_{M\to\infty} \frac{M!M^m}{m(m+1)\dots(m+M)} = (m-1)! \quad \forall m \in \mathbb{N}.$$

Now recall the definition $b^z := e^{z\log b}$ for real $b > 0$, $z \in \mathbb{C}$, and for integers $N \geq n \geq 0$ define

(1) $$f_{n,N}(z) := \frac{(z+n)(z+n+1)\dots(z+N)}{N!N^z}, \qquad z \in \mathbb{C},$$

which is zero-free in $D(0, n)$. A little calculation shows that

(2) $$f_{n,N}(z)/f_{n,N-1}(z) = (1 + z/N)(1 - 1/N)^z, \quad z \in D(0, n), N > n.$$

Let L denote the Principal Branch of the Logarithm in $\mathbb{C}_-$ and define

(3) $$L_N(z) := L(1 + z/N) + zL(1 - 1/N), \quad N > 1, \quad z \in D(0, N).$$

Note that $|z/N| < 1$, so by 3.44(i)

$$L_N(z) = \sum_{k=2}^{\infty} \frac{(-1)^{k-1}}{k} \left[\left(\frac{z}{N}\right)^k + z\left(-\frac{1}{N}\right)^k \right].$$

Consequently, for $z \in D(0,n)$ and $N \geq 2n$

$$\begin{aligned} |L_N(z)| &\leq \sum_{j=0}^{\infty} \left[\left|\frac{z}{N}\right|^{j+2} + |z| \left(\frac{1}{N}\right)^{j+2} \right] = \frac{|z/N|^2}{1-|z/N|} + |z| \frac{(1/N)^2}{1-1/N} \\ &\leq 2\left|\frac{z}{N}\right|^2 + 2|z|\left(\frac{1}{N}\right)^2 = 2\,\frac{|z|^2+|z|}{N^2}, \end{aligned}$$

and so $\lim_{M\to\infty} \sum_{N=n+1}^{M} L_N(z)$ exists uniformly for $z \in D(0,n)$. Exponentiating, we have, with convergence uniform in $z \in D(0,n)$ [cf. 7.1(ii)]:

$$\begin{aligned} \exp\left(\lim_{M\to\infty} \sum_{N=n+1}^{M} L_N(z) \right) &= \lim_{M\to\infty} \exp\left(\sum_{N=n+1}^{M} L_N(z) \right) \\ &= \lim_{M\to\infty} \prod_{N=n+1}^{M} \exp(L_N(z)) \\ &\overset{(3)}{=} \lim_{M\to\infty} \prod_{N=n+1}^{M} (1+z/N)(1-1/N)^z \\ &\overset{(2)}{=} \lim_{M\to\infty} \prod_{N=n+1}^{M} f_{n,N}(z)/f_{n,N-1}(z) \\ \text{(4)} \qquad &= \lim_{M\to\infty} f_{n,M}(z)/f_{n,n}(z). \end{aligned}$$

The limit function, being an exponential, is zero-free. Since the $f_{n,N}$ are zero-free in $D(0,n)$, it follows from (4) that $\lim_{M\to\infty} f_{n,M}(z)$ exists uniformly and is zero-free in $D(0,n)$, and therefore too the products

(5) $$f_{0,M}(z) = z(z+1)\ldots(z+n-1)f_{n,M}(z)$$

converge as $M \to \infty$ uniformly in $D(0,n)$, and the limit function has zeros there only at $z = 0, -1, \ldots, -n+1$. This holds for every $n \in \mathbb{N}$. Since each product $f_{0,M}$ is a power series in $\mathbb{C}$, 3.35(ii) insures that the limit as $M \to \infty$ defines an entire function $f(z)$, with zeros only in $\mathbb{Z} \setminus \mathbb{N}$. It satisfies

$$\begin{aligned} zf(z+1) &= \lim_{M\to\infty} \frac{z(z+1)(z+2)\ldots(z+1+M)}{M!M^{z+1}} \\ &= \lim_{M\to\infty} \left[\frac{z+1+M}{M} \cdot \frac{z(z+1)(z+2)\ldots(z+M)}{M!M^z} \right] \\ &= \lim_{M\to\infty} [1+(z+1)/M] \cdot f_{0,M}(z) = f(z) \qquad \forall z \in \mathbb{C}. \end{aligned}$$

The function

$$\Gamma(z) := \frac{1}{f(z)} = \lim_{M\to\infty} \frac{M!M^z}{z(z+1)\dots(z+M)} \tag{6}$$

is thus holomorphic in $\mathbb{C}\setminus(\mathbb{Z}\setminus\mathbb{N})$ and satisfies

$$\Gamma(z+1) = z\Gamma(z) \quad \text{for all such } z. \tag{7}$$

Hence by induction

$$\Gamma(m+1) = m! \qquad \forall m \in \mathbb{N}.$$

3.35(ii) also insures that $f'_{0,M} \to f'$, whence $f'_{0,M}/f_{0,M} \to f'/f = -\Gamma'/\Gamma$ in $\mathbb{C}\setminus(\mathbb{Z}\setminus\mathbb{N})$. Using "logarithmic differentiation" adroitly, the first quotient can be seen to be

$$\frac{1}{z} + \frac{1}{z+1} + \dots + \frac{1}{z+M} - \log M$$

and so

$$-\Gamma'(1)/\Gamma(1) = \lim_{M\to\infty}\left(1 + \frac{1}{2} + \dots + \frac{1}{1+M} - \log M\right) = \gamma,$$

that is, the two gammas are related by

$$\gamma = -\Gamma'(1)/\Gamma(1) = -\Gamma'(1). \tag{8}$$

Remarks: The elementary equalities $(n-1)! = \int_0^\infty t^{n-1}e^{-t}dt$ ($n \in \mathbb{N}$) suggest that the generalized factorial Γ might satisfy

$$\Gamma(z) = \int_0^\infty t^{z-1}e^{-t}dt \qquad \forall z \in \mathbb{H},$$

and this is indeed the case (REMMERT [1998], p. 51)! From this fact and (7), with an assist from 3.24(iii), follows the fantastic formula $\frac{1}{2}(-\frac{1}{2})! = (\frac{1}{2})! = \frac{1}{2}\sqrt{\pi}$. See 11.37(xvi) for another remarkable property of the function Γ.

Notes to Chapter III

The history and development of the ideas in this chapter are well detailed in the encyclopedia article of PRINGSHEIM, FABER and MOLK [1911]; see also REIFF [1889], PRINGSHEIM and MOLK [1907], PRINGSHEIM and FRÉCHET [1908], and (especially) FERRARO [2008], ROY [2011]. Like the limit processes in the theory of continuity and differentiability, the issue of convergence of series was still in limbo at the beginning of the 19th century and it too needed to be clarified (by Bolzano, Cauchy, Abel — chapter 4 of GRATTAN-GUINNESS

[1970]) for a rigorous grounding of analysis to proceed. Techniques other than the limit of partial sums had been devised for assigning numerical values to infinite series. The sound ones eventuated in the modern *regular summability methods* for series (TUCCIARONE [1973], HARDY [1944]). But many were bizarre, had metaphysical overtones, or led to contradictory results (see, e.g., chapters 9 and 19 of FERRARO [2008]), leaving us to marvel that Euler and others navigated the maze with so few fatal errors. In the "algebraic analysis" of the latter 18th century (which Cauchy was supplanting with "arithmetic analysis") the status of "=" between two formulae was never adequately clarified; quantification of variables and articulation of hypotheses were not recognized issues. See Chapter 4 of JAHNKE, *et al.* [2003] and SØRENSEN [2005].

3.3(iii) is due to CAUCHY [1821], pp. 151,286. It was rediscovered by HADAMARD [1888], p. 259 and [1892b], pp. 107–108. It is therefore customarily called the *Cauchy–Hadamard radius of convergence formula*, although Riemann may also have discovered it—see LAUGWITZ and NEUENSCHWANDER [1994].

The basic results 3.5 and 3.6 can be got just as easily without appealing to $\lim n^{1/n} = 1$, and that fact inferred as a corollary. The elegant details are in VÝBORNÝ [1987] and VAMANAMURTHY [1989]. (Cf. the argument in 3.12(iii).)

All sorts of behavior of a power series $\sum_{n=0}^{\infty} c_n z^n$ are possible *on* its circle of convergence: for example, the series may converge uniformly on this circle but not absolutely; an example due to FEJÉR [1917] is found on p. 122 of HILLE [1959] and one due to Hardy in E. LANDAU [1929a], §14; see also Gaier's notes thereon in the 3rd edition. In §3 Landau constructs a power series which converges in $\mathbb{D}$ to a function which extends continuously to $\overline{\mathbb{D}}$ while the partial sums of the series are unbounded when $z = 1$ (modification of an example in FEJÉR [1910]). The series may converge at every point of this circle but represent a function which is unbounded there (SIERPIŃSKI [1916]). The series may converge at every point of this circle but not uniformly on any arc: GAIER [1953]. We may have divergence at every point of this circle and yet have $c_n \to 0$ and it is also possible that the series converge at exactly one point of this circle. Examples of the latter two phenomena, due to Lusin and Sierpiński, respectively, can also be found in E. LANDAU [1929a]. See also FEJÉR [1910], NEDER [1920], MAZURKIEWICZ [1922b], DIENES [1931], HERZOG [1953–54], DVORETSKY and ERDÖS [1955], VIJAYARAGHAVAN [1947], and ŠTĚPÁNEK [1982]. The paper of KÖRNER [1983] is especially recommended; in addition to the examples above, it provides full details of Karl Zeller's 1955 proof (based on deep work of Andrei Kolmogorov) that for any $S \subset \mathbb{T}$ which is a countable union of closed sets there is a power series whose partial sums converge (in $\mathbb{C}$) at each point of S but are unbounded at each point of $\mathbb{T} \setminus S$. Boundary pathology of a different kind was realized by BAGEMIHL and PIRANIAN [1960/61]. They constructed a power series that converges absolutely in $\bar{\mathbb{D}}$ and maps $\partial\mathbb{D}$ onto a space-filling curve; their function f satisfies $f(\bar{\mathbb{D}}) = f(\partial\mathbb{D}) = [-1,1] \times [0,1]$.

Suppose the series $\sum_{n=0}^{\infty} c_n z^n =: f(z)$ has radius of convergence 1 and converges to 0 for each $z \in \mathbb{T}$. Then 3.13(v) asserts that $\lim_{r\uparrow 1} \sum c_n (ru)^n = 0$ for all $u \in \mathbb{T}$. If this convergence is uniform in u, it follows routinely that all c_n are 0. If, on the other hand, we know only that the partial sums of the series are bounded in $\mathbb{D}$, then an application of the elementary bounded convergence theorem for the Riemann integral (LEWIN [1986]) in 3.32(i) confirms that all $c_n = 0$. But that conclusion follows without any supplemental hypotheses, although the proof lies deeper. See MORTINI [2017] and the Chapter V Notes on 5.75.

Another boundary phenomenon, one which is perhaps unexpected in view of 3.6, is illustrated by 3.19: A power series can converge throughout $\overline{\mathbb{D}} \setminus \{1\}$ to a function f which is holomorphic in a whole neighborhood of $\overline{\mathbb{D}} \setminus \{1\}$, yet at every $u \in \mathbb{T} \setminus \{1\}$ the term-by-term derivative of this series can fail to converge to $f'(u)$, indeed can fail to converge at all.

In connection with the definition of radius of convergence contained in the proof of 3.3, there is the following curious result: If $\theta, \psi \in [0, 2\pi)$, $(\theta - \psi)/2\pi$ is irrational but algebraic and the two sets $\{\mathrm{Re}\,(c_n e^{in\theta})\}$, $\{\mathrm{Re}\,(c_n e^{in\psi})\}$ are bounded, then the series $\sum_{n=0}^{\infty} c_n z^n$ has radius of convergence at least 1. See TEISSIER DU CROS [1944] and VALIRON [1944].

The square-root series in 3.12(vi) involves neither exponential nor logarithm functions. But with the help of these functions, N. H. Abel, in 1826, extended the binomial expansion there to all complex exponents. Thus, let $\alpha \in \mathbb{C}$ be given. One easily checks (3.3(iv)) that the series $\sum \binom{\alpha}{n} z^n =: f_\alpha(z)$ converges throughout $\mathbb{D}$ and that $(1+z) f'_\alpha(z) \equiv \alpha f_\alpha(z)$. Also, the function $g_\alpha(z) := \exp(\alpha L(1+z))$, with L the Principal Branch of the Logarithm in $\mathbb{C}_-$, is holomorphic and zero-free in $\mathbb{D}$ and satisfies $(1+z) g'_\alpha(z) = \alpha g_\alpha(z)$. It follows that $(f_\alpha / g_\alpha)' = \mathbf{0}$, so $f_\alpha = [f_\alpha(0)/g_\alpha(0)] g_\alpha = g_\alpha$. The complex power $(1+z)^\alpha$ is usually *defined* to be [consistent with 1.47(vi) when $z, \alpha \in \mathbb{R}$] $g_\alpha(z)$, so this shows that $(1+z)^\alpha = \sum \binom{\alpha}{n} z^n$ for all $z \in \mathbb{D}$.

A good history of 3.13 plus a discussion of the uniform case is offered by HARDY [1907]. HOLZER [1939] constructs a series as in (v) for which f is continuous in $\mathbb{D} \cup \{b\}$ but at some points of $\mathbb{T}$ arbitrarily close to b the series fails to converge. When the function f is bounded, there are two quite different ways in which the more general angular continuity in (v) can be deduced from the radial continuity in (iv); see 5.56 and 12.33. The generality of those techniques suggests that a bounded example in (vi) is not so easy to come by (but such do exist). The natural converse of 3.13(v) is disproved by the simplest example: The series $\sum z^k$ converges in $\mathbb{D}$ to $f(z) := 1/(1-z)$ and therefore $\lim_{z\to -1, z\in\mathbb{D}} f(z)$ exists, but the series does not converge at the point $z = -1$. However, under appropriate decay conditions on the coefficients c_k a converse does hold. The first such is due to Alfred Tauber: If $kc_k \to 0$ and $\lim_{r\uparrow 1}\sum_{k=0}^{\infty} c_k r^k =: C \in \mathbb{C}$ exists, then the series $\sum_{k=0}^{\infty} c_k$ converges (and to C). Later eponymous *Tauberian* theorems weakened this hypothesis, for example, to the mere boundedness of the sequence $\{kc_k\}$. See STROMBERG [1981], p. 498.

I have followed PRINGSHEIM [1925] and RUDIN [1987] in the concise treatment of the exponential function. Other interesting approaches occur in WULF [1897], VOLPI [1903], HURWITZ [1911], WIGNER [1952], EBERLEIN [1954], [1966], ROBISON [1968] and chapter II of WEIL [1976]. The definition of π and the surjectivity of E are accomplished more abstractly (via topological group ideas) in REMMERT [1991], pp. 141–142.

As noted in the text, the power series $\sum z^n/n!$ was not just written down and its wonderful properties derived; it is still less imaginable that the logarithm was discovered by the forced march in §5 of chapter II, although on $\mathbb{R}$ it is certainly natural to consider the anti-derivative of $1/x$, and its homomorphic property follows rather naturally from the integral defining it. On the other hand too, b^y for $b > 0$ and arbitrary $y \in \mathbb{R}$, thought of as $\lim_k \left(b^{1/n_k}\right)^{m_k}$, where m_k, n_k are integers and $m_k/n_k \to y$, is natural and elementary from the perspective of calculus. (See 1.47.) Finally, the inverse relation these functions bear to each other is far from evident. [Around 1669 Newton integrated $\frac{1}{1+x} = \sum_{n=0}^{\infty}(-1)^n x^n$ termwise to get $y = \sum_{n=1}^{\infty}(-1)^{n-1}n^{-1}x^n$ (not yet recognized as $\log(1+x)$), then used his "method of reversion of series" to solve for x in terms of y, getting $\sum_{n=1}^{\infty} y^n/n!$ (not yet recognized as $e^y - 1$).] As a matter of fact, historically logarithms preceded the exponential function and the calculus. The founders, Jost (Joost, Jobst) Bürgi of Switzerland and John Napier (also Neper and numerous other spellings) of Scotland, *were* trying to reduce multiplication to addition; Napier's conceptual model was two moving points on two parallel lines, the speed of one being proportional to displacement, the speed of the other being constant. The displacement of the latter at each instant was called the logarithm of the displacement of the former at that instant. That this is so, up to a translation and multiplicative factor, is clear from 3.25(ii). How Napier, Henry Briggs and others computed large accurate tables without calculus and series, and how the relation of logarithms to areas under hyperbolas was discovered is very fascinating. Excellent, readable accounts will be found in chapter 6 of C. H. EDWARDS [1979], as well as the relevant sections of GOLDSTINE [1977]. (See also AYOUB [1993], KAUNZNER [1994], and pp. 119–148, volume I of LÜNEBURG [2008].) The reader desiring a more extensive account, including the controversies over logarithms of complex numbers, the evolution (due largely to Euler) of the complex exponential and the remarkable equation $e^{ix} = \cos x + i\sin x$ [which was *not*, as here, the definition of cos and sin!] can consult the two-volume work of NAUX [1966], [1971] (being wary of its deficiencies, as pointed out in the *Zentralblatt* reviews) and the numerous papers of J. W. L. Glaisher, Florian Cajori, and Joseph E. Hofmann.

3.22 inspired Leibniz to quote Virgil (on gambling!): "Numero Deus impari gaudet." Its interesting history and independent discovery by an Indian mathematician are detailed in the prize-winning paper ROY [1990]; also PARAMEŚWARAN [1992]. A beautiful elementary derivation is given by KAZARINOFF [1964], MATSUMOTO [1995] and ZHENG [2002]. Essentially the first equality in 3.22 was adopted by Weierstrass (in 1841) as his *definition* of π.) (This was also the definition adopted in the alternative treatment of the exponential function given in 2.22.) The present definition of π as "twice the least positive zero of the cosine function" dates at latest from an 1875 textbook of Richard Baltzer, and appears in PRINGSHEIM [1925] and E. LANDAU [1950]. When Landau used it in his Göttingen lectures in 1934, it occasioned a vicious attack by the Nazis—see chapter 6 of EBBINGHAUS, *et al.* [1990].

3.23 is from the Euler-Johann Bernoulli correspondence (see **MR** 88f:01011, **Zbl** 631.01011 for its history); different proofs may be found in HOPPE [1877] and SELBERG [1933], and the 3.23 method is pushed a little farther in GLAISTER [2001]. LORD [2003a] shows that many other continuous functions g satisfy $\int_0^1 g(x)\,dx = \sum_{n=1}^{\infty} g(n)$, including $g(x) := c^x$ for a unique $c > 0$.

A very nice alternative proof of 3.25 can be found on page 35 of GUICHARDET [1968] or page 235 of STROMBERG [1981]. Yet another proof runs thus: use an analog of 4.1 to find a continuous function $\phi\colon \mathbb{R} \mapsto \mathbb{C}$ such that $f = e^{\phi}$ and $\phi(0) = 0$. Deduce that ϕ is additive, whence $\phi(r) = r\phi(1)$ for all rational r, whence by continuity $\phi(x) = cx$ for all $x \in \mathbb{R}$ $(c := \phi(1))$. The integration trick in the text originates with SCHIMMACK [1908], rediscovered by KAC [1936/37]. It is further capable of showing that the only endomorphisms of $(\mathbb{R}, +)$ which are integrable over $[0, 1]$ are the constant multiples of the identity. Since the rate of change of many quantities in nature is (approximately) a constant multiple of their momentary magnitude, 3.25(i) is the reason for the ubiquity of the exponential function in science.

Though proofs of 3.32(i) appear in GUTZMER [1887], [1888], [1889], the result pre-dates this; he acknowledges the priority of A. Harnack, but in fact the formula goes back to an 1806 memoir of M. A. Parseval (see §23 of BURKHARDT [1914]) and is customarily referred to in the literature as Parseval's formula. See remark (iii) following 5.20 and Chapter XI. The Cauchy estimates date from 1831 and were discovered independently in 1841 by WEIERSTRASS (*Werke* I, pp. 67–68 and II, pp. 224–226 and ULLRICH [1990]), who gave an integral-free treatment. On the latter see also MÉRAY [1891b], pp. 278, 280, 606 of PRINGSHEIM [1925], A. KNESER [1920], MONTEL [1936], and pp.

85, 86 of HENRICI [1974]. For refinements of 3.32(ii) see TAMMI [1957], MACGREGOR [1969] and ROBERTSON [1978].

3.33(i) and 3.34(ii) are due, curiously, to CAUCHY [1844]—see BURKHARDT [1914], p. 1014 and PEIFFER [1983] concerning Liouville's role; its ultimate generalization is 12.18, Picard's little theorem. A proof avoiding integration is given by LENARD [1986] and a very geometric proof which generalizes to $\mathbb{R}^n$ by NELSON [1961]; see also p. 316 of OSGOOD [1928] and 5.34(iii). Since, as has been anticipated at several earlier points (and will be proved in Chapter V), every holomorphic function is represented by a power series, 3.33(i) is usually expressed by saying that the only bounded entire functions are the constants. This dramatic fact, like so many others in the subject, has no analog for differentiable functions on $\mathbb{R}$ (as the sine function illustrates). The generalization of 3.33(i) in 3.34(ii) was re-discovered by many: HERMITE [1885] (based on the interpolation technique in 5.73), VENEZIANI [1885], VON KOCH [1901]. A version of 3.34(ii) for harmonic functions is proved in 6.37 and 11.62. For material related to 3.34(v) see SIDON [1927], §4 of E. LANDAU [1929a], TOMIĆ [1962], and Remark (ii) following 6.39 below.

The proof of 3.35 rests in the final analysis on 3.34(i). In his approach to function theory, based solely on power series and eschewing integrals, Weierstrass had therefore to find another derivation of 3.34(i). For the details of how he did this, dispensing as well with the complex exponential function, see ULLRICH [1990] or REMMERT [1991a], p. 247. A different treatment was discussed in 3.32(v).

3.35(i) was proved by Weierstrass for power series in several variables (*Werke* I, 67–74 and II, 201–233). The proof suggested in the text is from CARATHÉODORY and LANDAU [1911]; they also get (v), by an indirect argument: Part (iv) and the uniqueness theorem 5.62 combine to show that there is no k for which the (bounded) sequence $\{c_{n,k}\}_{n\in\mathbb{N}}$ has *two* cluster points. The proof of (v) outlined in the text is almost verbatim from JENTZSCH [1918], which is a slight streamlining of LINDELÖF [1913]. For more on this important theorem see the Notes in chapters V and VII. For an unusual, linear-algebra proof of 3.35(xii), see SHALIT [2013].

Embryonic versions of 3.36 were discovered by HADAMARD [1893] and further elucidated by LINDELÖF [1903] and PRINGSHEIM [1902], [1904]; see §30 of BIEBERBACH [1921]. CHAUNDY [1937] extended 3.37(ii) thus: for any $r \in \mathbb{N}$ and for all real x with $|x| < \pi/r$, $\sum_{n=1}^{\infty} (-1)^{n+1} \frac{\sin^r(nx)}{n^r} = x^r/2$. This series ($r = 1$) is important historically, being perhaps the first Fourier series (Euler, 1744; see §7 of BURKHARDT [1914]). As shown in 3.37(ii), it converges pointwise on $\mathbb{R}$ to the 2π-periodic function f which satisfies $f(-\pi) = 0$, $f(x) = \frac{1}{2}x$ for $x \in (-\pi, \pi)$. It is thus an example of a series of continuous functions convergent to a discontinuous limit and was adduced by N. H. Abel in 1826 to show that a "theorem" to the contrary in CAUCHY [1821] "seems to admit exceptions"! (See SØRENSEN [2005].) Uniform convergence had yet to be conceptualized. See the Chapter VII notes. (One wonders why the simple example of the *sequence* $f_n(x) := 1/(1+x^{2n})$, $n \in \mathbb{N}$, $x \in \mathbb{R}$, had not bubbled to the surface even earlier.) Since the limit function f in 3.37(ii) is not continuous on $[-\pi, \pi]$, partial sums S_N of its Fourier series cannot converge uniformly on $[-\pi, \pi]$. The non-uniformity manifests itself in the so-called *Gibbs phenomenon*: $\lim_{N\to\infty} S_N((1-\frac{1}{N})\pi) = \int_0^\pi t^{-1} \sin t dt \approx 1.18\pi/2 = 1.18 \lim_{x\uparrow\pi} f(x)$. Pictorially the graph of S_N is crowned with a steeple of height about $.18\pi/2$ which looks more and more like an antenna at the point $(\pi, \pi/2)$ as $N \to \infty$. The graph of the limit of the S_N (i.e., that of f) is quite different from the limit of the graphs of the S_N, which includes the vertical segment $\{\pi\} \times [-1.18\pi/2, 1.18\pi/2]$. For details, see SHELUPSKY [1980]. Abel pointed out another feature of this series which confounded the *Zeitgeist*. Namely, although the convergence is uniform on each compact subset of $(-\pi, \pi)$ and the limit function is differentiable on $(-\pi, \pi)$, the series of termwise derivatives does not converge at any point! Indeed, the derivative of the nth term is $(-1)^{n+1}\cos(nx)$ and for every $x \in \mathbb{R}$ this does not even converge to 0. (See the 5.44(iv) hints.) On the other hand, integrating in 3.37(i) over $[\varepsilon, x]$, letting $\varepsilon \downarrow 0$, and utilizing 3.37(vi), shows that the series $\sum(2n-1)^{-2}\cos(2n-1)x$ converges (uniformly) on $[-\pi, \pi]$ to $\pi^2/8 - \pi|x|/4$, a function which is not differentiable at 0, although the termwise differentiated series converges throughout $[-\pi, \pi]$. Cf. 2.12. To complete this mini-museum of misbehavior, mention might be made of an example mined from DARBOUX [1875]: The functions $F_n(x) := e^{-n^2x^2}$ $(n \in \mathbb{N}, x \in \mathbb{R})$ have continuous derivatives which obviously (due to telescoping) satisfy $\sum_{n=1}^\infty \int_0^1 [F'_{n+1} - F'_n] \neq \int_0^1 \sum_{n=1}^\infty [F'_{n+1} - F'_n]$.

The notation in 3.38 is somewhat unorthodox: It is more usual to bestow the designation Bernoulli polynomials (functions) on the $n!P_n$ rather than the P_n themselves. Moreover, since $B_n = 0$ for all odd n (see (7) in 3.38), it is often our B_{2n}, or even $(-1)^{n-1}B_{2n}(>0)$, that get the label B_n.

For other elementary and *ab initio* treatments of the sums 3.38(17) the reader is invited to see K. S. WILLIAMS [1971a], APOSTOL [1973], STARK [1972], [1974] and BERNDT [1975]. The latter two papers contain extensive bibliographies, as does ELY [1882]. See also GOULD [1972] and pp. 350–364, vol. I of LEVEQUE [1974]. SAALSCHÜTZ [1893] and NIELSEN [1923] are exhaustive treatises. These formulae were first deduced by Euler. (See AYOUB [1974] and A. W. F. EDWARDS [1982].) The evaluation $\sum n^{-2} = \pi^2/6$ (known as 'the Basel problem') has considerable historical significance (see STÄCKEL [1907/08] and SANDIFER [2007]). Leibniz and the Bernoullis worked on it in vain, and its final resolution taxed even Euler. See **MR** 99c:11003 and 2008i:01006. There are numerous quite different proofs of it (11.37(viii), (xii) and Remark (iii) following 5.20), several by Euler; for a very nice exposition of six of them, see KALMAN [1993]. (One of these involves only a little algebra and an elementary, albeit recondite, cotangent identity; see also KORTRAM [1996], HOFBAUER [2002] and LORD [2016].) Interestingly, the closely related series $\sum_{n=2}^{\infty}(n^2-1)^{-1}$ and $\sum_{n=-\infty}^{\infty}(n^2+1)^{-1}$ have closed-form sums of quite different characters, namely, 3/4 [write $2(n^2-1)^{-1}$ as $(n-1)^{-1}-(n+1)^{-1}$] and $\pi\,\mathrm{cotanh}(\pi)$ [see 11.37(viii)], respectively. The telescoping sum phenomenon here has been employed by BENKO [2012] and DANERS [2012] to give a very elementary derivation of 3.37(vi) – using only integration by parts! For a beautiful and elementary connection of $\pi^2/6 = \sum 1/n^2$ with the trace of an integral operator and the eigenvalues of a certain differential equation see §1.7 of DYM and MCKEAN [1972], (also **MR** 2009a:58034), and for a connection with probability theory see PACE [2011]. It is a curious fact and not hard to prove, that the reciprocal of this sum, that is, $6/\pi^2$, is the probability that a random pair of positive integers is relatively prime, as well as the probability that a random integer is square-free. See G. A. JONES [1993]. These two facts and more generally the connection of the sums $\sum_k k^{-m}$ with factorization and prime numbers arise from another brilliant (but easy to confirm) insight of Euler's, that for each integer $m \geq 2$ this sum coincides with the *product* $\prod(1-p^{-m})^{-1}$ over all primes p. Later (in the 19th century) it was noted that, although this product does not converge when $m = 1$, it does generate a multiplicative analog of the definition of γ, namely the equality $\lim_{n\to\infty}\prod_{p\leq n}(1-p^{-1})^{-1}/\log n = e^{\gamma}$.

An *outline* of 22 different proofs of 3.37(vi), together with a purely combinatorial proof of 3.37(vii) that does not involve this identity is provided by ROGALSKI [1996]. For *all* the algebraic relations that obtain between different power sums, of which $\sum k^3 = \left(\sum k\right)^2$ is representative, see BEARDON [1996].

The sum $\sum_{n=1}^{\infty} n^{-k}$ $(k > 1)$ is the value at k of an important function which is holomorphic in the open half-plane $H := \{z \in \mathbb{C} : \mathrm{Re}\, z > 1\}$. This is Riemann's Zeta function, about which there is an enormous literature. Any of the basic texts in the bibliography discuss the elementary theory of this function and in addition several (e.g., BAK and NEWMAN [1982], ZAGIER [1997]) discuss its role in the proof of the Prime Number Theorem. One easily checks that the series $\zeta(z) := \sum_{n=1}^{\infty} n^{-z} = \sum_{n=1}^{\infty} e^{-z\log n}$ converges absolutely in H and uniformly on each compact subset. From the results of Chapter VII (or from 5.44) it follows that ζ is indeed holomorphic in H. (This series is a typical example of what are called *Dirichlet series*, another extensive chapter of function theory with which this book will not deal at all.)

For recent literature on the sums 3.38(21) see pp. 335–339, vol. I of LEVEQUE

[1974] and for some history, A. W. F. EDWARDS [1982]. The material of 3.38 can also be treated with the aid of Cauchy's Residue Theorem (Chapter XI). For such a treatment see §5, Chapter III of BEHNKE and SOMMER [1965] or pp. 32–37 of LINDELÖF [1905]. For a complete history of who did what and when, more relations and alternative developments, of the material of §4 see BURKHARDT [1914], §7 and LEHMER [1988]. An unusual alternative approach is that in GUINARD [1979]. How the great master Euler himself dealt with 3.37 and 3.38, as well as the infinite products in 11.37, is delightfully related in the prize-winning book DUNHAM [1999] (and LORD [2014]).

The Bernoulli polynomials and numbers come up naturally in an important asymptotic formula for $\sum_0^n f(k) - \int_0^n f(x)\,dx$ known as the Euler–Maclaurin summation formula. This difference is a Stieltjes integral, and repeated integration by parts on it leads to the Bernoulli polynomials. See §64 of KNOPP [1951] or §11.11 of HENRICI [1977]. A beautiful elementary account is found in APOSTOL [1999], where the connection with Euler's constant γ is emphasized.

A laborious calculational proof of 3.44(i)(a) was given by DANTSCHER [1895]. A more natural proof is via 5.30, i.e., from the knowledge that its Taylor series about 1 converges to the Principal Branch of the Logarithm in $D(1,1)$. Cf. also 2.21(xxi). In 1702 Johann Bernoulli observed that $(\tan^{-1})'(x) = (1+x^2)^{-1} = (1/2i)[i(1+ix)^{-1} - (-i)(1-ix)^{-1}]$ and felt that these summands had to be derivatives of "complex" logarithms, by analogy with $d/dx \log(1+cx) = c(1+cx)^{-1}$ for positive real c, x. This was evidently a big step toward 3.45(vi), although complex logarithms had yet to be clarified—by his student Euler!

The numbers e and π introduced in this chapter have many remarkable properties. For example, beyond their irrationality (3.28), each is *transcendental*, that is, satisfies no polynomial equation with rational coefficients. A. BAKER [1975], pp. 4, 5 gives nice short proofs of both these facts, as well as the relevant historical references. The appendix of MAHLER [1976] gives a unified treatment of many of the beautiful classical proofs, together with more history, and HESSENBERG [1912] is a completely self-contained account of the classical proofs. A compendium of the fascinating history and lore of π has been assembled by BERGGREN, BORWEIN, and BORWEIN [2004]. There you will find how close the state of Indiana came in 1897 to legislating a rational value (but they weren't sure *what* rational value!) for π, and, on a more constructive if equally quixotic note, a remarkable re-writing of Edgar Allan Poe's *The Raven*, retaining both its poetry and its story, while delivering the first 740 decimal digits of π. See also EYMARD and LAFON [2004], chapter 5 of EBBINGHAUS, *et al.* [1990], and CASTELLANOS [1988]. Each of the famous numbers π, i, e has had books written about its history and mystique (entertaining and informative, if occasionally unreliable in their mathematical details); see **MR** 56#8261, 2000a:01003, 95a:01002. Euler's constant γ too is the subject of a book HAVIL [2003]. The role of the logarithm function and the *harmonic series* $\sum 1/n$ in interesting diverse problems (desert crossings, cantilevered cards, optimal selection, first-digit frequency) is also explored in Havil's book. For short, elegant accounts of the Γ-function see also ARTIN [1964] and chapter 2 of REMMERT [1998].

Like the number on the uniform of a superstar athlete, the letters "i", "e" and "π" are now retired from general usage; however, "γ" and "Γ" will not be so honored.

What has been done with power series in this chapter is only one aspect of them, the analytic aspect. There is an algebraic (combinatorial) side which is also highly developed. This is the theory of *formal power series.* The set-up is a commutative ring R with identity I and an "indeterminate" X over R. For $\{r_n\}_{n\in\{0\}\cup\mathbb{N}} \subset R$ the formal power series $P(X) := \sum_{n=0}^{\infty} r_n X^n$ is the map of $\{0\}\cup\mathbb{N}$ into R which assigns value r_n to n, for each n. Thus X itself is the map of $\{0\}\cup\mathbb{N}$ into R which is I at 1 and 0 elsewhere. The set $R[[X]]$ of these is made into an R-algebra in the "obvious" way, that is, the way forced upon us by the distributive law and the coefficient formulas it leads to when two finitely-nonzero series (i.e., polynomials) are multiplied. This ring has a rich algebraic structure. For example, if $Q(X) := \sum_{n=0}^{\infty} a_n X^n$ with $a_0 = 0$, then sense can be made of $P(Q(X)) = \sum_{n=0}^{\infty} r_n [Q(X)]^n$ in $R[[X]]$, leading to "logarithms" and "exponentials", while if r_0 is invertible in R, then $P(X)$ is always invertible in the ring $R[[X]]$; also the "derivative" $P'(X) := \sum_{n=1}^{\infty} n r_n X^{n-1}$ can always be formed. There is even a useful metric in this ring: the distance from $P(X)$ to $Q(X) \neq P(X)$ is the reciprocal of the exponential of the least n such that $r_n \neq a_n$. Polynomials are dense, addition and multiplication are (jointly) continuous, and the metric is complete; moreover, a *series* $\sum P_k(X)$ in $R[[X]]$ converges in this metric if and only if the sequence $\big(P_k(X)\big)$ converges to $\mathbf{0}$. When R is $\mathbb{C}$ or a complex Banach algebra, most of the structural results in $R[[X]]$ hold in R upon replacing X everywhere by any element of R that makes the relevant series all converge. In this way the theory of $R[[X]]$ makes contact with (its historical roots) Lagrange's and Weierstrass' way of developing function theory. For an attractive glimpse of this theory see NIVEN [1969] or chapter I of HENRICI [1974]; for some history, FERRARO [2008]. One of its many uses is in proving (and discovering!) combinatorial identities via "generating functions". If we allow finitely many negative powers of X, we get the ring of *formal Laurent series*, and (as 11.46 might suggest) when R is a field, this ring is the field of quotients of the integral domain $R[[X]]$.

Chapter IV
The Index and Some Plane Topology

§ 1 Introduction

In this chapter we complete the preliminary investigations of Chapter I on the topology of the plane. The chief tools are the index and deformation and extension techniques. The notion of the index rests squarely on the theory of the exponential function developed in the previous chapter.

§ 2 Curves Winding around Points

The exponential function makes clear how to formalize the notion of the "the number of times the loop γ in $\mathbb{C} \setminus \{0\}$ winds counter-clockwise around 0": we have only to write $\gamma := |\gamma|e^{i\theta}$, where θ is continuous and real-valued. Evidently the function θ tells "how much angle γ sweeps out". Thus we have to produce such θ. It is convenient (following a suggestion of A. DiBucchianico) to base the existence of continuous logarithms on an elegant result from the theory of commutative Banach algebras (see 10.34 in RUDIN [1973]), whose proof in the limited generality we need is completely elementary.

Theorem 4.0 (EILENBERG [1935]) *Let K be a compact metric space, $C(K)$ the continuous complex-valued functions on K, $\mathcal{G}$ the set of zero-free functions in $C(K)$ and $\mathcal{E}$ the set of exponentials of functions in $C(K)$. Give $C(K)$ the uniform metric. Then $\mathcal{G}$ is an open multiplicative group in $C(K)$ and $\mathcal{E}$ is an open subgroup of $\mathcal{G}$ which coincides with the connected component of $\mathcal{G}$ that contains* $\mathbf{1}$.

Proof: *Clearly $\mathcal{E} \subset \mathcal{G}$ and each is a multiplicative group. The key thing to note for all the rest is that the open ball $B := \{f \in C(K) : \|f - \mathbf{1}\|_\infty < 1\} = \{f \in C(K) : f(K) \subset D(1,1)\}$ lies in $\mathcal{E}$. For if L is a continuous logarithm in*

R. B. Burckel, *Classical Analysis in the Complex Plane*, Cornerstones,
https://doi.org/10.1007/978-1-0716-1965-0_5

$D(1,1)$, (3.19), then $L \circ f \in C(K)$ and $f = e^{L\circ f}$ for each $f \in B$. If $f \in \mathcal{G}$ (resp., $\mathcal{E}$) and $g \in C(K)$ satisfies $\|f - g\|_\infty < \|1/f\|_\infty^{-1}$, then $g \in \mathcal{G}$ (resp., $\mathcal{E}$). Indeed, $\|\mathbf{1} - g/f\|_\infty \leq \|1/f\|_\infty \cdot \|g - f\|_\infty < 1$, so $g/f = e^h$ for some $h \in C(K)$, by what was just shown. Therefore $g = fe^h$ and g is also zero-free, i.e., $g \in \mathcal{G}$. If $f \in \mathcal{E}$, say $f = e^F$, then $g = e^{F+h} \in \mathcal{E}$. This proves that $\mathcal{G}$ and $\mathcal{E}$ are open sets. On the other hand, since $f = ge^{-h}$, this same argument shows that if $f \notin \mathcal{E}$, then $g \notin \mathcal{E}$; that is, $\mathcal{G} \setminus \mathcal{E}$ is open. If $f, g \in C(K)$, $t \mapsto e^{tf+(1-t)g}$ is a continuous map of the connected space $[0,1]$ into $\mathcal{E}$ which joins e^f to e^g. That is (1.4), any two points of $\mathcal{E}$ lie in a connected subset of $\mathcal{E}$. It follows (1.1) that $\mathcal{E}$ is connected. Letting $\mathcal{C}$ denote the connected component of $\mathcal{G}$ which contains $\mathbf{1}$, we have $\mathcal{E} \subset \mathcal{C}$ and both $\mathcal{E}$ and $\mathcal{C} \setminus \mathcal{E}$ are relatively open in $\mathcal{C}$. Hence $\mathcal{C} \setminus \mathcal{E} = \varnothing$ and $\mathcal{E} = \mathcal{C}$.

Corollary 4.1 *Let $K \subset \mathbb{R}^n$ be compact and starlike, $f\colon K \mapsto \mathbb{C} \setminus \{0\}$ continuous. Then f has a continuous logarithm. In particular, if $\gamma\colon [a,b] \mapsto \mathbb{C}$ is a curve and $z_0 \in \mathbb{C} \setminus \gamma$, then for any $D_0 := D(z_0, r)$ with $\bar{D}_0 \subset \mathbb{C} \setminus \gamma$ there is a continuous $\varphi\colon [a,b] \times \bar{D}_0 \mapsto \mathbb{C}$ such that $\gamma(t) - z = e^{\varphi(t,z)}$ for all $(t,z) \in [a,b] \times \bar{D}_0$.*

Proof: If c is a star-center for K, then for each $t \in [0,1]$ the function $f_t(x) := f((1-t)c + tx)$, $x \in K$, belongs to the group $\mathcal{G}$ of invertible elements of $C(K)$. In fact, $\varphi(t) := f_t$ clearly defines a continuous map of $[0,1]$ into $\mathcal{G}$. Since $\varphi(0)$ is the non-zero constant $f(c)$, it lies in $\mathcal{E}$ and therefore the maximal connected subset $\mathcal{E}$ of $\mathcal{G}$ contains the whole image $\varphi[0,1]$; in particular, $f = \varphi(1) \in \mathcal{E}$. The statement about $\gamma - z$ is the special case of the convex subset $K := [a,b] \times \bar{D}_0$ of $\mathbb{R}^3$. [In turn, this proof shows the corollary itself to be an instance of 4.24.]

Here is an amusing application of 4.1 (from TOTIK [1999]). Suppose $f, g\colon [0,1] \mapsto \mathbb{R}$ are Riemann integrable and $\int_0^1 f = \int_0^1 g = 1$. Must there be an interval $I \subset [0,1]$ such that $\int_I f = \int_I g = 1/2$? If $\int_I f$ represents the value one person attributes to the rectangular part over the base interval I of a square cake $[0,1] \times [0,1]$, and $\int_I g$ represents a second person's valuation, then we have a "fair-division": $\int_I f = \int_{[0,1]\setminus I} g = 1/2$, so by each person's criterion the cake is fairly divided. To prove that I exists, introduce the convex set $\Delta := \{(x,y) : 0 \leq x \leq y \leq 1\}$ and the function $F(x,y) := \int_x^y f - 1/2 + i\big(\int_x^y g - 1/2\big)$ on it. Our question is whether F is zero-free in Δ. Suppose that it is. Then 4.1 supplies continuous $\psi\colon \Delta \mapsto \mathbb{C}$ such that

$$F = e^{\psi}. \tag{1}$$

From $\int_0^1 f = \int_0^1 g = 1$ it follows that

$$F(0,t) = -F(t,1) \qquad \forall t \in [0,1].$$

From (1) therefore $\psi(0,t) - [\psi(t,1) + \pi i] \in 2\pi i\mathbb{Z}$ for each $t \in [0,1]$. So this continuous function is constant, say equal to $2\pi i N$. Taking first $t = 0$, then $t = 1$, and adding the resulting two equations gives

$$\psi(0,0) = \psi(1,1) + 2(2N+1)\pi i. \tag{2}$$

However, from $e^{\psi(t,t)} = F(t,t) = -1/2 - i/2$ for all $t \in [0,1]$, it follows that the continuous function $t \mapsto \psi(t,t)$ is constant on the connected set $[0,1]$. But then (2) says that $2N + 1 = 0$ —×—. Conclude from this contradiction that F must have a zero in Δ, so the fair-division problem is always solvable.

Definition 4.2 If $\gamma\colon [a,b] \mapsto \mathbb{C}$ is any loop and z any point not on γ, then the (circulation) *index* or *order* or *winding number* (=*Umlaufzahl*) of γ with respect to z, as well as that of z with respect to γ, noted $\operatorname{Ind}_\gamma(z)$, is defined to be $[\varphi(b) - \varphi(a)]/2\pi i$ for *any* continuous logarithm φ on $[a,b]$ of the function $\gamma - z$. (Notice that such φ exist by 4.1.) If $\gamma_1, \ldots, \gamma_n$ ($n \in \mathbb{N}$) are loops, not necessarily distinct, we introduce the n-tuple $(\gamma_1, \ldots, \gamma_n) =: \gamma$ and define $\operatorname{Ind}_\gamma(z) := \sum_{j=1}^{n} \operatorname{Ind}_{\gamma_j}(z)$ for all $z \in \mathbb{C}$ which lie on none of the γ_j.

Corollary 4.3 *If $\gamma\colon [a,b] \mapsto \mathbb{C}$ is a loop, then $\operatorname{Ind}_\gamma$ is a well-defined, continuous, integer-valued function on $\mathbb{C} \setminus \gamma[a,b]$ which is constant in each component of $\mathbb{C} \setminus \gamma[a,b]$. Moreover, the set $\mathbb{C} \setminus (\operatorname{Ind}_\gamma)^{-1}(0)$ is compact. In particular, $\operatorname{Ind}_\gamma$ is 0 throughout the unbounded component of $\mathbb{C} \setminus \gamma[a,b]$.*

Proof: To see that $\operatorname{Ind}_\gamma(z_0)$ is well defined for each $z_0 \in \mathbb{C} \setminus \gamma[a,b]$, we let $\varphi, \psi\colon [a,b] \mapsto \mathbb{C}$ be two continuous logarithms of the function $\gamma - z_0$, that is, $e^\varphi = \gamma - z_0 = e^\psi$. Then $[\varphi - \psi]/2\pi i$ is a continuous integer-valued (by 3.14(iii)) function on the connected set $[a,b]$, and so is constant. Thus $\varphi(b) - \varphi(a) = \psi(b) - \psi(a)$. To see that $\operatorname{Ind}_\gamma(z_0)$ is an integer, note that

$$e^{\varphi(b)-\varphi(a)} = e^{\varphi(b)}/e^{\varphi(a)} = (\gamma(b) - z_0)/(\gamma(a) - z_0) = 1$$

(since γ is closed), so $[\varphi(b) - \varphi(a)]/2\pi i$ is an integer. Next, given $z_0 \in \mathbb{C} \setminus \gamma[a,b]$, the preceding corollary provides a disk D_0 about z_0 and a continuous $\varphi\colon [a,b] \times D_0 \mapsto \mathbb{C}$ which satisfies $e^{\varphi(t,z)} = \gamma(t) - z$ for all $(t,z) \in [a,b] \times D_0$. Then for all $z \in D_0$ we have $\operatorname{Ind}_\gamma(z) = [\varphi(b,z) - \varphi(a,z)]/2\pi i$, showing that $\operatorname{Ind}_\gamma$ is continuous at z_0. As noted before, it follows that the continuous integer-valued function $\operatorname{Ind}_\gamma$ is constant on connected subsets of its domain.

Also, if we consider a fixed $|z| > \sup|\gamma[a,b]|$ and (3.19) a continuous logarithm L in the open disk $D(-1,1)$, then we may form $\psi(t) := L(z^{-1}\gamma(t) - 1)$. If we write $z = e^w$ for some complex number w (see 3.14(vi)) and set $\varphi := \psi + w$, then $e^{\varphi(t)} = e^w e^{\psi(t)} = e^w[z^{-1}\gamma(t) - 1] = \gamma(t) - z$, so by definition of the index, $\operatorname{Ind}_\gamma(z) = [\varphi(b) - \varphi(a)]/2\pi i = [\psi(b) - \psi(a)]/2\pi i = 0$ (since $\gamma(b) = \gamma(a)$). Thus

Ind_γ vanishes for sufficiently large $|z|$. Since $\{z \in \mathbb{C} : |z| > \sup|\gamma[a,b]|\}$ is a connected subset of $\mathbb{C} \setminus \gamma[a,b]$ (for example, it is the image of $\mathbb{D} \setminus \{0\}$ under $z \mapsto c/z$ for appropriate constant c), it lies in some component of $\mathbb{C} \setminus \gamma[a,b]$. All the other components, being disjoint from this one, lie in the disk of radius $\sup|\gamma[a,b]|$ and so are bounded. Thus $\mathbb{C} \setminus \gamma[a,b]$ has exactly one unbounded component and Ind_γ vanishes on it.

Since the components of the open set $\mathbb{C} \setminus \gamma[a,b]$ are open (1.30(i)′) and Ind_γ is constant in each, the subset $(\mathrm{Ind}_\gamma)^{-1}(0)$ is a union of such components and hence is an open set. Thus $\mathbb{C} \setminus (\mathrm{Ind}_\gamma)^{-1}(0)$ is a closed set.

For piecewise smooth curves continuous logarithms and indices are easy to compute. Suppose $\gamma\colon [a,b] \mapsto \mathbb{C}\setminus\{0\}$ is piecewise smooth: there exist $a =: t_0 < t_1 < \cdots < t_n := b$ such that γ is smooth in each $[t_{j-1}, t_j]$. Use 3.14(vi) to find $c \in \mathbb{C}$ such that $\gamma(a) = e^c$. Then define $\varphi(t) := \int_a^t \gamma'/\gamma, t \in [a,b]$. As noted in 2.7, φ exists and is continuous. The Fundamental Theorem of Calculus says that an indefinite Riemann integral has one-sided derivative equal to the integrand at each point where the latter has a one-sided limit. Therefore φ is differentiable in each $[t_{j-1}, t_j]$ with $\varphi' = \gamma'/\gamma$ there, whence $(\gamma e^{-\varphi})' = \mathbf{0}$ and $\gamma e^{-\varphi}$ is constant in $[t_{j-1}, t_j]$. Since adjacent pairs of these intervals share a point, all the constants are the same and therefore have the value that $\gamma e^{-\varphi}$ does at $t_0 = a$, that is, $\gamma(a)e^{\varphi(a)} = \gamma(a)$. Thus $\gamma = \gamma(a)e^{\varphi} = e^c e^{\varphi}$. Therefore $\varphi + c$ is a continuous logarithm for γ; and it is smooth in each interval $[t_{j-1}, t_j]$.

Corollary 4.4 *If $\gamma\colon [a,b] \mapsto \mathbb{C}$ is a piecewise smooth loop, then*

(1) $$\mathrm{Ind}_\gamma(z) = \frac{1}{2\pi i}\int_\gamma \frac{1}{\xi - z}\,d\xi \qquad \forall z \in \mathbb{C} \setminus \gamma[a,b].$$

Proof: Let $z_0 \in \mathbb{C} \setminus \gamma[a,b]$ and apply the preceding discussion with $\gamma - z_0$ in the role of the γ there. We get $\gamma - z_0 = e^{c+\varphi}$ and so by definition of the index

$$2\pi i \mathrm{Ind}_\gamma(z_0) = \varphi(b) - \varphi(a) = \int_a^b \frac{(\gamma - z_0)'}{\gamma - z_0} = \int_a^b \frac{\gamma'}{\gamma - z_0} = \int_\gamma \frac{d\xi}{\xi - z_0}.$$

Remark: Integral formula (1) is often adopted as the definition of the index. That has several disadvantages. For example, from 4.2 it is obvious that

(2) $$\mathrm{Ind}_\gamma(z) = \mathrm{Ind}_{\gamma - z}(0) = \mathrm{Ind}_{(\gamma - z)/|\gamma - z|}(0) \qquad \forall z \in \mathbb{C}\setminus\gamma[a,b].$$

But the second equality here is not apparent from (1), nor is fact 4.7(v). Also, (1) is limited to piecewise smooth γ, and hardly clarifies the "winding number" designation.

Example 4.5 *Let* $a \in \mathbb{C}$, $s > 0$. *Set* $b := \operatorname{Re} a$, $c := \operatorname{Im} a$, $Q := (b, b+s) \times (c, c+s)$, *and recall* 1.12(ii). *Then*

$$\operatorname{Ind}_{\partial Q}(z) = 1 \qquad \forall z \in Q.$$

Proof: The very definition of index shows that for any loop γ, any $w \in \mathbb{C}$ and any z not on γ we have

$$\operatorname{Ind}_{\gamma - w}(z - w) = \operatorname{Ind}_\gamma(z).$$

(In either case logarithms of $\gamma - z$ are involved.) Moreover, $\partial(Q-w) = \partial Q - w$ and so without loss of generality we may replace Q by any $Q - w$. Taking $w := a + \frac{1}{2}s(1+i)$, it therefore suffices to consider the case where $a = -\frac{1}{2}s(1+i)$, that is, with $r := \frac{1}{2}s$

$$Q = (-r, r) \times (-r, r)\,.$$

By 4.3 it then suffices to establish that

$$\operatorname{Ind}_{\partial Q}(0) = 1.$$

This follows from 4.4, 2.15(iii) and 3.22:

$$\operatorname{Ind}_{\partial Q}(0) = \frac{1}{2\pi i}\int_{\partial Q}\frac{d\xi}{\xi} = \frac{1}{2\pi i}\cdot 8i \cdot \int_0^1 \frac{1}{u^2+1}\,du = \frac{1}{2\pi i}\cdot 8i \cdot \frac{\pi}{4}.$$

Alternative Proofs: As before, reduce at once to $Q := (-r, r) \times (-r, r)$ and $z = 0$. Then let L denote the Principal Branch of the Logarithm in $\mathbb{C} \backslash (-\infty, 0]$. Set $\Gamma := \partial Q$, $\gamma_1 := [-r - ir, r - ir]$. Then (see 1.12)

$$\Gamma(t) = \begin{cases} \gamma_1(t), & 0 \le t \le 1 \\ i\gamma_1(t-1), & 1 \le t \le 2 \\ -\gamma_1(t-2), & 2 \le t \le 3 \\ -i\gamma_1(t-3), & 3 \le t \le 4. \end{cases} \tag{1}$$

Letting $\varphi_1(t) := L(\gamma_1(t))$ for $0 \le t \le 1$, we have (see 3.43(ii))

$$\varphi_1(0) = L(-r - ir) = L\big(\sqrt{2}re^{-i3\pi/4}\big) = \log\big(\sqrt{2}r\big) - \frac{3\pi i}{4} \tag{2}$$

$$\varphi_1(1) = L(r - ir) = L\big(\sqrt{2}re^{-i\pi/4}\big) = \log\big(\sqrt{2}r\big) - \frac{\pi i}{4}. \tag{3}$$

From these equations it follows that a function φ is well defined by

(4) $$\varphi(t) := \begin{cases} \varphi_1(t), & 0 \le t \le 1 \\ \varphi_1(t-1) + \dfrac{i\pi}{2}, & 1 \le t \le 2 \\ \varphi_1(t-2) + i\pi, & 2 \le t \le 3 \\ \varphi_1(t-3) + \dfrac{i3\pi}{2}, & 3 \le t \le 4. \end{cases}$$

Evidently φ is continuous on $[0,4]$. Since $\gamma_1 = e^{\varphi_1}$, it follows from (1) and (4) that $\Gamma(t) = e^{\varphi(t)}$, $0 \le t \le 4$ and consequently by definition 4.2

$$\mathrm{Ind}_\Gamma(0) = \frac{\varphi(4) - \varphi(0)}{2\pi i} \overset{(4)}{=} \frac{\varphi_1(1) + \frac{3\pi i}{2} - \varphi_1(0)}{2\pi i} = 1, \quad \text{by (2) and (3).}$$

In view of the discussion preceding 4.4, the integral of Γ'/Γ is also a logarithm of Γ, and so it differs from φ by a constant; but constructing the logarithm φ via (4) is easier than computing this integral. [Evidently a third proof of 4.5 could be based on 4.7(iv).]

According to 4.3, $\mathrm{Ind}_\gamma(z)$ is a continuous function of z. We next show that it is also a continuous function of γ.

Exercise 4.6 (i) *Let $\gamma_0\colon [a,b] \mapsto \mathbb{C}$ be a loop and $z \in \mathbb{C} \setminus \gamma_0$. Then there exists a positive $\delta := \delta(z, \gamma_0)$ with this property: for any loop $\gamma_1\colon [a,b] \mapsto \mathbb{C}$ with $|\gamma_0 - \gamma_1| < \delta$ on $[a,b]$, we have $z \in \mathbb{C} \setminus \gamma_1$ and*

(∗) $$\mathrm{Ind}_{\gamma_1}(z) = \mathrm{Ind}_{\gamma_0}(z).$$

Hints: As in the proof of 4.5, replacing γ_0 by $\gamma_0 - z$ and γ_1 by $\gamma_1 - z$, enables us to assume that $z = 0$. Then $\inf |\gamma_0[a,b]|$ is positive. This number has the properties sought for δ. For if $\gamma_1\colon [a,b] \mapsto \mathbb{C}$ is a loop satisfying $|\gamma_0 - \gamma_1| < \delta$, then $|\gamma_0 - \gamma_1| < |\gamma_0|$ on $[a,b]$. In particular, γ_1 never takes the value 0. Also the quotient $\gamma := \gamma_1/\gamma_0$ is a well-defined continuous function, in fact a loop, which maps $[a,b]$ into $D(1,1)$. Since $\mathbb{C} \setminus D(1,1)$ is connected and contains 0, it follows from 4.3 that $\mathrm{Ind}_\gamma(0) = 0$. But it is clear from the definition of index in terms of logarithms that $\mathrm{Ind}_{\gamma\gamma_0}(0) = \mathrm{Ind}_\gamma(0) + \mathrm{Ind}_{\gamma_0}(0)$. (Cf. 4.14.)

(ii) *Formulate and prove a theorem on the joint continuity of* $\mathrm{Ind}_\gamma(z)$ *as a function of the pair (γ, z).*

Exercise 4.7 (i) *If $\Gamma\colon [a,b] \mapsto \mathbb{C}$ is a simple loop, then Γ is bicontinuous on (a,b), and $\Gamma[a,b]$ is a Jordan-curve.*

Hints: We have to show that if $\Gamma(t_n) \to \Gamma(t)$ for some $t_n, t \in (a,b)$, then $t_n \to t$. It suffices to show that t is the only limit point of $\{t_n\}$ in $[a,b]$. If

$t_0 \in [a, b]$ and $t_{n_j} \to t_0$, then by continuity $\Gamma(t_0) = \Gamma(t)$. Since $\{t, t_0\} \subset [a, b)$ or $\{t, t_0\} \subset (a, b]$ and since Γ is one-to-one on $[a, b)$ and on $(a, b]$, it follows that $t_0 = t$. Now according to 3.26, $\varphi(e^{2\pi i t}) := \Gamma(a + t(b - a))$, $t \in [0, 1]$, well-defines a continuous function on $\mathbb{T}$. It is evidently one-to-one, hence a homeomorphism. Therefore $\Gamma[a, b] = \varphi(\mathbb{T})$ is a Jordan-curve (1.11(v)).

(ii) *$\gamma : [a, b] \mapsto \mathbb{C}$ is a loop, Γ its periodic extension to $\mathbb{R}$. Show that for any $c \in \mathbb{R}$ the map $t \mapsto \Gamma(c + t(b - a))$ defines a loop $\gamma_c : [0, 1] \mapsto \mathbb{C}$ with the same range and index function as γ.*

(iii) *If Γ_1, Γ_2 are simple loops with the same range J and $f : J \mapsto \mathbb{C}$ is continuous, then as functions in $\mathbb{C} \setminus f(J)$*

$$\textit{either} \quad \mathrm{Ind}_{f\circ\Gamma_1} = \mathrm{Ind}_{f\circ\Gamma_2} \quad \textit{or} \quad \mathrm{Ind}_{f\circ\Gamma_1} = -\mathrm{Ind}_{f\circ\Gamma_2}.$$

Hints: Thanks to (ii) we can re-parameterize, without changing any indices, to have both loops parameterized on $[0, 1]$ and both start (and end) at the same point of J. According to (i), Γ_1, Γ_2 map the open interval $(0, 1)$ homeomorphically onto $J \setminus \Gamma_1(0) = J \setminus \Gamma_2(0)$. Therefore, $\Gamma_1^{-1} \circ \Gamma_2$ is a homeomorphism of $(0, 1)$. This extends to a homeomorphism ψ of $[0, 1]$ by 1.46. Then $\Gamma_2 = \Gamma_1 \circ \psi$ holds in $[0, 1]$. Given $z \in \mathbb{C} \setminus f(J)$, let $\varphi : [0, 1] \mapsto \mathbb{C}$ be a continuous logarithm of $f \circ \Gamma_1 - z$ Then $\varphi \circ \psi$ is a logarithm for $f \circ \Gamma_2 - z$ and consequently

$$2\pi i \,\mathrm{Ind}_{f\circ\Gamma_2}(z) = \varphi(\psi(1)) - \varphi(\psi(0)) = \pm[\varphi(1) - \varphi(0)] = \pm 2\pi i \,\mathrm{Ind}_{f\circ\Gamma_1}(z),$$

since $\psi\{0, 1\} = \{0, 1\}$. Of course the same sign prevails for all z.

(iv) *Prove that the same sign prevails in* (iii) *for all f, and this is the plus sign if Γ_1 and Γ_2 start at the same point z_0 on J and visit some two points of $J \setminus \{z_0\}$ in the same order.*

Hint: The sign is determined by the function ψ in the preceding construction, and ψ is manufactured from Γ_1 and Γ_2. The supplemental hypothesis clearly precludes ψ being a *decreasing* function.

(v) *If $\Gamma : [a, b] \mapsto \mathbb{C} \setminus \{0\}$ is a loop which intersects each half-line through 0 at most once, then Γ is simple and $|\mathrm{Ind}_\Gamma(0)| = 1$.*

Hints: More exactly, the hypothesis means that each half-line starting at 0 contains $\Gamma(t)$ for at most one $t \in [a, b)$. This means that $\Gamma/|\Gamma|$ is injective on $[a, b)$; so Γ is too. Consider continuous φ such that $\Gamma = e^\varphi$. Thus

$$\mathrm{Ind}_\Gamma(0) = (2\pi i)^{-1}[\varphi(b) - \varphi(a)] = (2\pi)^{-1}[\mathrm{Im}\,\varphi(b) - \mathrm{Im}\,\varphi(a)].$$

If $|\mathrm{Ind}_\Gamma(0)| \geq 2$, then the real interval $(2\pi)^{-1}\,\mathrm{Im}\,\varphi[a, b)$ contains two different integers n_1, n_2; say $(2\pi)^{-1}\,\mathrm{Im}\,\varphi(t_j) = n_j$. Thus $t_1 \neq t_2$, yet $\Gamma(t_j)/|\Gamma|(t_j) =$

$\exp(i\,\mathrm{Im}\,\varphi(t_j)) = \exp(2\pi i n_j) = 1$, contradicting the hypothesis on Γ. On the other hand, if $\mathrm{Ind}_\Gamma(0) = 0$, then there must be a pair t_1, t_2 with $a < t_1 < t_2 < b$ for which $\mathrm{Im}\,\varphi(t_1) = \mathrm{Im}\,\varphi(t_2)$ (see 1.46), and the same contradiction arises.

(vi) *Γ_1, Γ_2 are simple loops with the same range J. Using the map ψ in the hints to* (iii), *and computing curve length via inscribed polygons as in* 2.13(ii), *show that $\ell(\Gamma_1) = \ell(\Gamma_2)$. This number could be denoted by $\ell(J)$ and called the length of the Jordan-curve J.*

Definition 4.8 (i) If $\gamma_0,\ \gamma_1\colon [a,b] \mapsto A \subset \mathbb{C}$ are loops, we say γ_0 and γ_1 are *A-homologous* or *homologous in A* if $\mathrm{Ind}_{\gamma_0}(z) = \mathrm{Ind}_{\gamma_1}(z)$ for every $z \in \mathbb{C} \setminus A$.

(ii) If $\mathrm{Ind}_{\gamma_0}(z) = 0$ for every $z \in \mathbb{C} \setminus A$, we say γ_0 is *A-nullhomologous* or *nullhomologous in A*. If $\gamma_1, \ldots, \gamma_k$ are loops in A, $\gamma := (\gamma_1, \ldots, \gamma_k)$ and $\mathrm{Ind}_\gamma = 0$ throughout $\mathbb{C} \setminus A$, then γ is said to be *A-nullhomologous.*

(iii) If every loop in A is nullhomologous in A, we call A *homologically-connected*, or *homologically trivial.*

Remark: When A is open, Definition 4.8(i) can be cast in an equivalent form that is entirely internal to A, namely

(i)′ $\exists$ compact $K \subset A$ such that $\gamma_0 \cup \gamma_1 \subset K$ & $\mathrm{Ind}_{\gamma_0} = \mathrm{Ind}_{\gamma_1}$ in $A \setminus K$.

That (i) implies (i)′ is an easy consequence of 1.31, according to which there is a compact $K \subset A$ which contains $\gamma_1 \cup \gamma_0$, such that each bounded component of $\mathbb{C} \setminus K$ contains a component of $\mathbb{C} \setminus A$; since the indexes are constant on connected sets and agree in $\mathbb{C} \setminus A$, they agree in $\mathbb{C} \setminus K$. Suppose (i)′ holds. Then by continuity (4.3), $\mathrm{Ind}_{\gamma_0} = \mathrm{Ind}_{\gamma_1}$ in $\bar{A} \setminus K \supset \partial A$. Since each component of $\mathbb{C} \setminus A$ meets ∂A (by 1.30(v)) and each index is constant in such a component, it follows from the preceding equality that $\mathrm{Ind}_{\gamma_0} = \mathrm{Ind}_{\gamma_1}$ throughout $\mathbb{C} \setminus A$.

We close this section with a technical result which we shall need later (in 5.59), and the reader should probably skip it now.

Exercise 4.9 *Let $0 < r < R < \infty$, $S := \{z \in \mathbb{H} : r < |z| < R\}$, $\gamma\colon [0,1] \mapsto \mathbb{H}$ a curve such that $|\gamma(0)| = r$, $|\gamma(1)| = R$. Show that if C is a component of $S \setminus \gamma$, then $\overline{C}$ does not meet both the negative and positive y-axis.*

Hints: Since γ is a compact subset of the open right half-plane, both the sets $S \cap [(0,\varepsilon) \times (-R, 0)]$ and $S \cap [(0,\varepsilon) \times (0,R)]$ lie in $S \setminus \gamma$ if $\varepsilon > 0$ is small. Each is connected (why?), so if $\overline{C}$ meets both $\pm[ir, iR]$, then C meets, hence contains, both these sets. It follows easily, via 1.28, that there exists a polygon

$\Gamma_1\colon [0,1] \mapsto \mathbb{C}$ with $\Gamma_1(0) = -ir$ and $\Gamma_1(1) = ir$, which lies, except for $\pm ir$, in C. We may then define

$$\begin{aligned}
\Gamma_2(t) &:= re^{i\pi t/2} && t \in [1,3],\\
\Gamma_3(t) &:= \begin{cases} \Gamma_1(t) \\ \Gamma_2(t) \end{cases} && \begin{matrix} t \in [0,1] \\ t \in [1,3], \end{matrix}\\
\Gamma_4(t) &:= \begin{cases} (2t-1)ir \\ \Gamma_2(t) \end{cases} && \begin{matrix} t \in [0,1] \\ t \in [1,3], \end{matrix}\\
\Gamma_5(t) &:= \begin{cases} (1-t)ir \\ \Gamma_1(t-2) \end{cases} && \begin{matrix} t \in [0,2] \\ t \in [2,3]. \end{matrix}
\end{aligned}$$

Since Γ_1 lies in $C \cup \{\pm ir\} \subset \bar{S} \setminus \gamma$, we see that $\Gamma_3 \subset \bar{A}(r,R) \setminus \gamma$. The set

$$\mathbb{C} \setminus \left[\bar{A}(r,R) \setminus \gamma\right] = \left[\mathbb{C} \setminus \bar{A}(r,R)\right] \cup \gamma = D(0,r) \cup \gamma \cup \left[\mathbb{C} \setminus \bar{D}(0,R)\right]$$

is connected (by 1.3), unbounded and lies in $\mathbb{C} \setminus \Gamma_3$. Consequently

$$\mathrm{Ind}_{\Gamma_3}(c) = 0 \qquad \forall c \in D(0,r).$$

For $c \in (-r,0)$ we compute via 4.4

$$\begin{aligned}
0 = 2\pi i \mathrm{Ind}_{\Gamma_3}(c) &= \int_{\Gamma_3} \frac{dz}{z-c}\\
&= \int_{\Gamma_1} \frac{dz}{z-c} + \int_{[ir,-ir]} \frac{dz}{z-c} + \int_{[-ir,ir]} \frac{dz}{z-c} + \int_{\Gamma_2} \frac{dz}{z-c}\\
&= \int_{\Gamma_5} \frac{dz}{z-c} + \int_{\Gamma_4} \frac{dz}{z-c}. \qquad (1)
\end{aligned}$$

However Γ_5 and $-\Gamma_4$ are each closed curves and c lies in an open connected unbounded subset of $\mathbb{C}$ disjoint from each, viz., the open left half-plane. Consequently

$$\int_{\Gamma_5} \frac{dz}{z-c} = 0 = \int_{-\Gamma_4} \frac{dz}{z-c}. \qquad (2)$$

Using this in (1) gives

$$\begin{aligned}
0 = \int_{\Gamma_4} \frac{dz}{z-c} + \int_{-\Gamma_4} \frac{dz}{z-c} = \int_{C(0,r)} \frac{dz}{z-c} &= 2\pi i \mathrm{Ind}_{C(0,r)}(c) \quad \text{by 4.4}\\
= 2\pi i \mathrm{Ind}_{C(0,r)}(0), \qquad &\text{by 4.3.}
\end{aligned}$$

This is an obvious contradiction.

§ 3 Homotopy and the Index

Definition 4.10 PartOne Let $S \subset \mathbb{R}^n$, f_0, $f_1 \colon S \mapsto A \subset \mathbb{C}$ continuous functions. An *A-homotopy* or a *homotopy in A* of f_0 with f_1 is a continuous function $h \colon [0,1] \times S \mapsto A$ such that

$$\left.\begin{aligned} h(0,x) &= f_0(x) \\ h(1,x) &= f_1(x) \end{aligned}\right\} \qquad \forall x \in S.$$

We sometimes write $f_t(x)$ for $h(t,x)$. We say f_0 is *A-homotopic* to f_1 or *homotopic in A* to f_1 if such an h exists. Occasionally the role of $[0,1]$ here is taken over by some other compact interval.

(ii) If f_1 is constant, we say f_0 is *A-nullhomotopic* or *nullhomotopic in A*. Sometimes f_0 is also said to be an *inessential* map in A.

Definition 4.11 PartOne Let $[a,b] \subset \mathbb{R}$, γ_0, $\gamma_1 \colon [a,b] \mapsto A \subset \mathbb{C}$ be loops. An *A-loophomotopy* or a *loophomotopy in A* of γ_0 with γ_1 is an A-homotopy h such that each curve $\gamma_t(x) := h(t,x)$ is a loop, i.e.,

$$h(t,a) = h(t,b) \qquad \forall t \in [0,1].$$

We say γ_0 is *A-loophomotopicto* γ_1 or *loophomotopic in A* to γ_1 if such an h exists.

(ii) If γ_1 is constant, we say γ_0 is *A-loopnullhomotopic* or *loopnullhomotopic in A*.

(iii) Call A *loophomotopically-connected* if every loop in A is loopnullhomotopic in A. The term *homotopically trivial* is also used here.

Remarks: (i) It is the work of a moment to confirm that both these notions define equivalence relations. Thus in particular the roles of f_0 and f_1 (resp., γ_0 and γ_1) can be interchanged above without affecting the definitions, and so symmetric terminology like "f_0, f_1 are A-homotopic", etc., is employed.

(ii) According to 3.26 every loop $\gamma : [0,1] \mapsto A$ can be "factored through the exponential" and realized as $\gamma(x) = \Gamma(e^{2\pi i x})$ for some continuous $\Gamma : \mathbb{T} \mapsto A$. Indeed, it is somewhat more natural to use $\mathbb{T}$ rather than $[0,1]$ to parameterize loops. With that perspective, and with $a = 0, b = 1$, 4.11(ii) is equivalent to the existence of a continuous $H : \overline{\mathbb{D}} \mapsto A$ such that $H(0) = \gamma_1(x)$ and $H(e^{2\pi i x}) = \gamma_0(x)$ for all $x \in [0,1]$. Such an H is well defined by $H(te^{2\pi i x}) := h(1-t, x)$, $(t,x) \in [0,1] \times [0,1]$.

Theorem 4.12 *Let S be a subset of $\mathbb{C}$, γ_0, $\gamma_1 \colon [a,b] \mapsto S$ loops which are loophomotopic in S. Then γ_0, γ_1 are S-homologous.*

Proof: Let H be a loophomotopy in S of γ_0 with γ_1. Then for each $\tau \in [0,1]$ the function $\gamma_\tau \colon [a,b] \mapsto S$ defined by $\gamma_\tau(t) := H(\tau,t)$ is a loop and $\tau \mapsto \gamma_\tau$ is a continuous map of $[0,1]$ into $C[a,b]$. It follows easily then from 4.6 that, for each $z \in \mathbb{C} \setminus S$, $\tau \mapsto \mathrm{Ind}_{\gamma_\tau}(z)$ is a continuous function. Since it is integer-valued and its domain is the connected set $[0,1]$, it is constant. Thus $\mathrm{Ind}_{\gamma_0}(z) = \mathrm{Ind}_{\gamma_1}(z)$. This holds for every $z \in \mathbb{C} \setminus S$ and therefore says that γ_0 and γ_1 are S-homologous.

The converse of 4.12 is true for some sets S (see, e.g., 4.15) but is generally false. An example to illustrate this is described in the next exercise. While intuitively plausible, a logical proof requires some serious work, and quite a few details will be left to the interested reader. This example is ubiquitous in algebraic topology texts to illustrate the non-commutativity of the fundamental group. See, e.g., 11.14 in FULTON [1995]. Their proofs use considerable machinery and usually require all the homotopies to be "based" at a single point, whereas here it is a question of "free" homotopies. A self-contained but intricate covering-space argument appears as a nine-part exercise on pp. 356–358 of GREENE and KRANTZ [2002]. At the other extreme, an analytic proof (using the holomorphic monodromy theorem) is suggested in FISCHER and LIEB [1988], p. 32 and ČERNÝ [1992], p. 313.

Exercise 4.13 (The "figure-eight"). *Set $\Omega := \mathbb{C}\setminus\{0,2\}$ and let L denote the Principal Branch of the Logarithm in $\mathbb{C} \setminus (-\infty, 0]$, $E\colon \mathbb{C} \mapsto \mathbb{C}$ the exponential function. Consider the functions $\Gamma_0, \varphi_0 : [0,4] \mapsto \mathbb{C}$ defined by*

$$\Gamma_0(t) := \begin{cases} E(2\pi i t) \\ 2 - E(2\pi i t) \\ E(2\pi i(3-t)) \\ 2 - E(2\pi i(4-t)), \end{cases} \qquad \varphi_0(t) := \begin{cases} 2\pi i t & t \in [0,1] \\ L(2 - E(2\pi i t)) + 2\pi i & t \in [1,2] \\ 2\pi i(3-t) & t \in [2,3] \\ L(2 - E(2\pi i(4-t))) & t \in [3,4]. \end{cases}$$

(i) *Show that these are well-defined, continuous functions, that Γ_0 is a loop in Ω, and that $\Gamma_0 = E \circ \varphi_0$. Show also that*

(1) $$\Gamma_0 \text{ is } \Omega\text{-nullhomologous},$$

that $\varphi_0|[3,4]$ is a loop, and that

(2) $$\mathrm{Ind}_{\varphi_0|[3,4]}(L(2)) = -1.$$

Hints: $\Gamma_0 = E \circ \varphi_0$ and 4.2 make $\mathrm{Ind}_{\Gamma_0}(0) = 0$ clear. After checking that $\Gamma_0(t) - 2 = -\Gamma_0(4-t) = E(\pi i + \varphi_0(4-t))$ for all $t \in [0,4]$, one sees as well that $\mathrm{Ind}_{\Gamma_0}(2) = 0$. Show that $\varphi_0 - L(2)$ maps the open interval $(3,4)$ into $\mathbb{C}_-$ and any continuous logarithm of it on $[3,4]$ differs on $(3,4)$ from $L \circ (\varphi_0 - L(2))$ only by a constant. $\mathrm{Ind}_{\varphi_0|[3,4]-L(2)}(0)$ therefore involves only the limiting behavior of L at $\varphi_0(3) - L(2) = -L(2) < 0$. More perspicuously, (2) is an instance of 4.49(ii).

Our goal is to show that in spite of (1), Γ_0 is not Ω-loopnullhomotopic. The argument is by *reductio ad absurdum:*

(ii) *Suppose that*

(∗) *$H\colon [0,1] \times [0,4] \mapsto \Omega$ is a loophomotopy of Γ_0 with a constant.*

Define

$$\widetilde{H}(\tau, t) := \begin{cases} H((t+1)\tau, 0) & (\tau, t) \in [0,1] \times [-1,0] \\ H(\tau, t) & (\tau, t) \in [0,1] \times [0,4] \\ H((5-t)\tau, 0) & (\tau, t) \in [0,1] \times [4,5], \end{cases}$$

$$\widetilde{\varphi}_0(t) := \begin{cases} 0 & t \in [-1,0] \\ \varphi_0(t) & t \in [0,4] \\ 0 & t \in [4,5]. \end{cases}$$

Prove that $\widetilde{H}$ is an Ω-loophomotopy in which all loops have initial point 1, and cite 4.1 to get a continuous function φ such that $\widetilde{H} = E \circ \varphi$. Show that φ may be chosen to satisfy in addition

$$\varphi(0, t) = \widetilde{\varphi}_0(t) \qquad \forall t \in [-1, 5]\,,$$

and that φ is then a loophomotopy of $\widetilde{\varphi}_0$ with the loop

$$\widetilde{\varphi}_1(t) := \varphi(1, t), \qquad t \in [-1, 5]\,.$$

(iii) *Since $\widetilde{H} = E \circ \varphi$ and $\widetilde{H}$ maps into Ω, φ maps into $E^{-1}(\Omega)$, that is, $\mathbb{C} \setminus (L(2) + 2\pi i\mathbb{Z})$. Infer from 4.12 that*

(3) $$\mathrm{Ind}_{\widetilde{\varphi}_0}(L(2)) = \mathrm{Ind}_{\widetilde{\varphi}_1}(L(2)).$$

(iv) *Show that $\widetilde{\varphi}_1$ is constant on $[0,4]$ and has the symmetry property*

$$\widetilde{\varphi}_1(t) = \widetilde{\varphi}_1(4-t) \qquad \forall t \in [4,5],$$

and infer from this that

(4) $$\mathrm{Ind}_{\widetilde{\varphi}_1}(L(2)) = 0.$$

(v) *Show that $\widetilde{\varphi}_0|[0,3]$ is a loop disjoint from the positive real axis, and from this and $\widetilde{\varphi}_0|\,[-1,0] = \widetilde{\varphi}_0|[4,5] = \mathbf{0}$ deduce that*

(5) $$\mathrm{Ind}_{\widetilde{\varphi}_0}(L(2)) = \mathrm{Ind}_{\varphi_0|[3,4]}(L(2)) \overset{(2)}{=} -1.$$

In (3), (4), and (5) we have the contradiction which proves (∗) untenable.

Hint: From 4.2 it is clear that if a loop visits its endpoints $k > 2$ times, then its Index is the sum of the Indexes of $k-1$ sub-loops.

Remark: The truly masochistic reader may wish to consider the loop $\Gamma : [0,6] \mapsto \Omega := \mathbb{C} \setminus \{0,2\}$ defined by

$$\Gamma(t) := \begin{cases} 2+3e^{\pi it} & , \quad 0 \le t \le 1 \\ -e^{2\pi it} & , \quad 1 \le t \le 2 \\ 2+3e^{\pi i(3-t)} & , \quad 2 \le t \le 3 \end{cases} \qquad \Gamma(t) := \overline{\Gamma}(t-3),\ 3 \le t \le 6.$$

A figure makes it appear that Γ is Ω-nullhomologous but not Ω-loopnullhomotopic. Can this be *proved*?

Exercise 4.14 *If γ is a loop and $z \in \mathbb{C} \setminus \gamma$, then for every integer n, $\mathrm{Ind}_{n*\gamma}(z) = n\mathrm{Ind}_\gamma(z)$. If γ_1, γ_2: $[a,b] \mapsto \mathbb{C} \setminus \{0\}$ are loops, then $\mathrm{Ind}_{\gamma_1\gamma_2}(0) = \mathrm{Ind}_{\gamma_1}(0) + \mathrm{Ind}_{\gamma_2}(0)$.*

Exercise 4.15 *Let U be $A(r,R)$ for some $0 \le r < R \le \infty$.*

(i) *Show that if γ_0, γ_1 are two loops in U with common parameter interval and $\mathrm{Ind}_{\gamma_0}(0) = \mathrm{Ind}_{\gamma_1}(0)$, then γ_0 and γ_1 are U-loophomotopic.*

Hints: Let the parameter interval of γ_0, γ_1 be $[a,b]$ and use 4.1 to write $\gamma_j = e^{h_j}$ for continuous h_j. Define $H\colon [0,1] \times [a,b] \mapsto \mathbb{C}$ by

$$H(\tau,t) := \exp\left(\tau h_1(t) + (1-\tau)h_0(t)\right).$$

Then $H(0,t) = \gamma_0(t)$, $H(1,t) = \gamma_1(t)$ for all $t \in [a,b]$, and for $\tau \in [0,1]$

$$\begin{aligned} &[\tau h_1(b) + (1-\tau)h_0(b)] - [\tau h_1(a) + (1-\tau)h_0(a)] \\ &\qquad = \tau[h_1(b) - h_1(a)] + (1-\tau)[h_0(b) - h_0(a)] \\ &\qquad = 2\pi i\tau\mathrm{Ind}_{\gamma_1}(0) + 2\pi i(1-\tau)\mathrm{Ind}_{\gamma_0}(0) = 2\pi i\mathrm{Ind}_{\gamma_0}(0) \in 2\pi i\mathbb{Z}. \end{aligned}$$

Therefore $t \mapsto H(\tau,t)$ is a *closed* curve for each τ. Why is H a homotopy *in* U?

(ii) *Show that if* γ_0, γ_1 *are two loops in* U, *then there exist integers* n_0, n_1, *not both* 0, *such that* $n_0 {*} \gamma_0$ *is* U*-loophomotopic to* $n_1 {*} \gamma_1$.

Hints: If $\mathrm{Ind}_{\gamma_0}(0)$ equals $\mathrm{Ind}_{\gamma_1}(0)$, then take $n_0 = n_1 = 1$ and cite (i). Otherwise, not both of the integers $n_0 := \mathrm{Ind}_{\gamma_1}(0)$ and $n_1 := \mathrm{Ind}_{\gamma_0}(0)$ are 0 and according to 4.14 we have $\mathrm{Ind}_{n_0 * \gamma_0}(0) = \mathrm{Ind}_{n_1 * \gamma_1}(0)$. So again (i) may be applied.

Exercise 4.16 *Every starlike subset of* $\mathbb{C}$ *is loophomotopically-connected.*

Hint: If S is starlike with respect to z_0 and $\gamma\colon [a,b] \mapsto S$ is a loop, then

$$H(t,\tau) := (1-t)\gamma(\tau) + t z_0, \quad (t,\tau) \in [0,1] \times [a,b]$$

is obviously a loophomotopy of γ with the constant z_0. By starlikeness of S the range of H is in S.

We close this section with a practical method for computing indices, which the reader will have no trouble varying and generalizing (getting thereby, e.g., yet a third proof of 4.5).

Exercise 4.17 $\Gamma\colon [0,1] \mapsto \mathbb{C}$ *is a loop which for some* $0 < t_0 < 1$ *satisfies*

(1) $$\Gamma[0,t_0] \subset \mathbb{C}_- := \mathbb{C} \setminus (-\infty, 0]$$

(2) $$\Gamma[t_0,1] \subset \mathbb{C}_+ := \mathbb{C} \setminus [0, +\infty)$$

(3) $$\mathrm{Im}\,\Gamma(0) < 0 < \mathrm{Im}\,\Gamma(t_0).$$

Show that $0 \in \mathbb{C} \setminus \Gamma$ *and* $\mathrm{Ind}_\Gamma(0) = 1$.

Hints: Let L_-, L_+ denote the Principal Branch of the Logarithm in $\mathbb{C}_-, \mathbb{C}_+$, resp. (See 3.43.) Notice that

(4) $$L_+(z) = L_-(z) \qquad \text{whenever } \mathrm{Im}\, z > 0$$

(5) $$L_+(z) = L_-(z) + 2\pi i \qquad \text{whenever } \mathrm{Im}\, z < 0,$$

and define

$$\varphi(t) := \begin{cases} L_-(\Gamma(t)), & t \in [0, t_0] \\ L_+(\Gamma(t)), & t \in [t_0, 1]. \end{cases}$$

Because of (3) and (4), φ is well defined at $t = t_0$; hence φ is continuous on $[0,1]$. Therefore by 4.2

$$\begin{aligned} 2\pi i \mathrm{Ind}_\Gamma(0) &= \varphi(1) - \varphi(0) = L_+(\Gamma(1)) - L_-(\Gamma(0)) = L_+(\Gamma(0)) - L_-(\Gamma(0)) \\ &= 2\pi i, \text{by (3) and (5)}. \end{aligned}$$

§ 4 Existence of Continuous Logarithms

Theorem 4.18 (Abstract Monodromy Theorem) *Let U be an open subset of $\mathbb{C}$ and $f\colon U \mapsto \mathbb{C}\setminus\{0\}$ a continuous function. Then the following are equivalent:*

(i) $\mathrm{Ind}_{f\circ\gamma}(0) = 0$ *for every (polygonal) loop γ in U.*

(ii) *f has a continuous logarithm in U.*

Proof: (ii) $\Rightarrow$ (i). Trivial.

(i) $\Rightarrow$ (ii). The components of U inherit property (i) and are open. Since we have only to manufacture the logarithm in each component, let us assume that U is connected. Fix $z_0 \in U$ and let $w_0 \in \mathbb{C}$ satisfy $e^{w_0} = f(z_0)$. For each $z \in U$ let $\gamma_z\colon [0,1] \mapsto U$ be a polygon joining z_0 to z. (See 1.28.) Let φ_z be a continuous logarithm for $f\circ\gamma_z$:

$$f\circ\gamma_z = e^{\varphi_z}, \qquad z\in U. \tag{1}$$

(See 4.1.) Evidently we may require that

$$\varphi_z(0) = w_0 \qquad \forall z \in U. \tag{2}$$

We define $\varphi(z)$ to be $\varphi_z(1)$ and have

$$e^{\varphi(z)} = e^{\varphi_z(1)} \overset{(1)}{=} f(\gamma_z(1)) = f(z), \quad \text{by definition of } \gamma_z.$$

So only continuity of φ has to be established. To this end, fix $z_1 \in U$ and pick $r > 0$ so small that $\bar{D}(z_1, r) \subset U$ and $f\big(\bar{D}(z_1,r)\big) \subset D(f(z_1), |f(z_1)|)$. Let L be a continuous logarithm in the latter disk. (See 3.19.) We may require that

$$L(f(z_1)) = \varphi_{z_1}(1), \tag{3}$$

since both these numbers are logarithms of $f(z_1)$. Now, given $z \in D(z_1, r)$, let $\gamma := [z_1, z]$ and consider the polygonal loop Γ defined by

$$\Gamma(t) := \begin{cases} \gamma_z(1-t) & 0 \le t \le 1 \\ \gamma_{z_1}(t-1) & 1 \le t \le 2 \\ \gamma(t-2) & 2 \le t \le 3 \end{cases} \tag{4}$$

and the function φ defined by

$$\varphi(t) := \begin{cases} \varphi_z(1-t) & 0 \le t \le 1 \\ \varphi_{z_1}(t-1) & 1 \le t \le 2 \\ L(f(\gamma(t-2))) & 2 \le t \le 3. \end{cases} \tag{5}$$

Conditions (2) and (3) ensure that φ is consistently defined at $t = 1$ and $t = 2$. Evidently φ is continuous on each of the intervals $[0, 1]$, $[1, 2]$, $[2, 3]$. Therefore φ is continuous. It is evidently a logarithm for $f \circ \Gamma$. By hypothesis therefore

$$0 = 2\pi i \mathrm{Ind}_{f\circ\Gamma}(0) = \varphi(3) - \varphi(0) = L(f(z)) - \varphi_z(1),$$
$$\varphi(z) = \varphi_z(1) = L(f(z)).$$

This shows that φ coincides with $L \circ f$ in $D(z_1, r)$ and hence is continuous in that disk.

Exercise 4.19 (i) *If U is an open subset of $\mathbb{C}$ and $f\colon U \mapsto \mathbb{C} \setminus \{0\}$ has a continuous n^{th} root ψ_n for every positive integer n (or even for just an infinite set of n), then f has a continuous logarithm. Show by an example that f may have a continuous square-root but lack a continuous logarithm.*

Hints: Let γ be any loop in U and φ_n a continuous logarithm for $\psi_n \circ \gamma$. Then $n\varphi_n$ is a continuous logarithm for $f \circ \gamma$ and so

$$\mathrm{Ind}_{f\circ\gamma}(0) = n\mathrm{Ind}_{\psi_n\circ\gamma}(0).$$

Since both indexes are integers and n does not feature on the left-hand side, the index on the right-hand side cannot be non-zero for large n. It follows that $\mathrm{Ind}_{f\circ\gamma}(0) = 0$. Now appeal to 4.18. [Could the ideas in 3.44(ii) be adapted to give an alternative proof?] For the example, consider $U := \mathbb{C}\setminus\{0\}$, $f(z) := z^2$, and glance ahead at 4.21 or back to 3.42(iv).

(ii) *Let k be a positive integer, U open $\subset \mathbb{C}$, $f\colon U \mapsto \mathbb{C}$ continuous. Show that the following are equivalent:*

(a) $\mathrm{Ind}_{f\circ\gamma}(0)$ *is divisible by k for every (polygonal) loop γ in $U\setminus f^{-1}(0)$.*

(b) $f = g^k$ *for some continuous function g on U.*

Hints: In each component of the open set $U \setminus f^{-1}(0)$ imitate the proof of 4.18, defining $g(z)$ to be $\exp(\varphi_z(1)/k)$. After (5) notice that $\frac{1}{2\pi i}[L(f(z)) - \varphi_z(1)] = \mathrm{Ind}_{f\circ\Gamma}(0)$ being a multiple of k, $\exp(\frac{1}{k}[L(f(z)) - \varphi_z(1)]) = 1$, and so $g(z) = \exp(L(f(z))/k)$ for all $z \in D(z_1, r)$. Finally, extend g from $U \setminus f^{-1}(0)$ to U by defining it to be 0 on $f^{-1}(0)$.

(iii) *If continuous functions f_n on U converge locally uniformly to f and, for some $k \in \mathbb{N}$, each f_n has a continuous k^{th} root, then f has a continuous k^{th} root.*

Hints: Condition (ii)(a) passes over from the f_n to f (by 4.6).

Corollary 4.20 *Let U be an open subset of $\mathbb{C}$. If U is loophomotopically-connected, in particular (see 4.16) if U is starlike, then every continuous, zero-free function on U has a continuous logarithm.*

Proof: Let $f\colon U \mapsto \mathbb{C}\setminus\{0\}$ be continuous and let $\gamma\colon [a,b] \mapsto \mathbb{C}$ be any loop in U. By hypothesis there is a loophomotopy $h\colon [0,1]\times[a,b] \mapsto U$ of γ with a constant. Then $H := f\circ h$ is a loophomotopy in $f(U)\subset \mathbb{C}\setminus\{0\}$ of $f\circ\gamma$ with a constant. Consequently by 4.12, $\mathrm{Ind}_{f\circ\gamma}(0)=0$. Now cite 4.18.

Exercise 4.21 *Let $\omega := e^{2\pi i/n}$ for some integer $n>1$, $\varepsilon \in \mathbb{C}\setminus\{1\}$, and $h : \mathbb{T} \mapsto \mathbb{C}\setminus\{0\}$ a continuous function which satisfies $h(\omega z) = \varepsilon h(z)$ for all $z\in\mathbb{T}$. Show that h does not have a continuous n^{th} root, hence no continuous logarithm. In particular, neither any non-zero integer power of the identity function nor any odd continuous function on $\mathbb{T}$ has a continuous logarithm.*

Hints: Suppose contrariwise that $g\colon \mathbb{T}\mapsto\mathbb{C}$ is continuous and $g^n = h$. Then $g^n(\omega z) = h(\omega z) = \varepsilon h(z) = \varepsilon g^n(z) \neq 0$,

$$\left[\frac{g(\omega z)}{g(z)}\right]^n = \varepsilon \qquad \forall z\in\mathbb{T}.$$

Thus the image of the connected set $\mathbb{T}$ under the continuous map $z\mapsto g(\omega z)/g(z)$ lies in the finite set of nth roots of ε, hence coincides with one of them, say ν. It follows that

$$1 = \frac{g(\omega)}{g(\omega)} = \frac{g(\omega^{n+1})}{g(\omega)} = \prod_{j=1}^{n} \frac{g(\omega^{j+1})}{g(\omega^j)} = \nu^n = \varepsilon \qquad \longrightarrow\!\!\!\!\!\!\times\!\!\!\longleftarrow .$$

The next exercise is a special case of a general and important fact (4.33) in whose proof it will play a role.

Exercise 4.22 (i) *If $f\colon \mathbb{T}\mapsto\mathbb{C}\setminus\{0\}$ is continuous, then there exists a unique integer n and a continuous $\varphi\colon\mathbb{T}\mapsto\mathbb{C}$ such that $f(z)\equiv z^n e^{\varphi(z)}$. (This integer is called the* Brouwer degree *of f.) Show further that $n = \mathrm{Ind}_{f\circ C(0,1)}(0)$, and that f has a zero-free continuous extension to $\overline{\mathbb{D}}$ if and only if this index is 0.*

Hints: Let $n = \mathrm{Ind}_\gamma(0)$, where γ is the loop $\gamma(t) := f\big(e^{2\pi i t}\big)$. Thus $\gamma(t) = e^{2\pi i\psi(t)}$, where ψ is continuous and $\psi(1)-\psi(0) = n$. Then

$$(*) \qquad e^{-2\pi i n t} f\big(e^{2\pi i t}\big) = e^{2\pi i[\psi(t)-nt]} \qquad \forall t\in[0,1].$$

The function $2\pi i[\psi(t)-nt]$ has the value $2\pi i\psi(0)$ at $t=0$ and also the value $2\pi i\psi(0)$ at $t=1$. According to 3.26 there is a continuous $\varphi\colon \mathbb{T}\mapsto\mathbb{C}$ such that

$$\varphi(e^{2\pi i t}) = 2\pi i[\psi(t)-nt] \qquad \forall t\in[0,1].$$

We get from (∗)

$$e^{-2\pi int} f\big(e^{2\pi it}\big) = e^{\varphi(e^{2\pi it})} \qquad \forall t \in [0,1],$$

that is,

$$z^{-n} f(z) = e^{\varphi(z)} \qquad \forall z \in \mathbb{T}.$$

For uniqueness of n, either cite 4.21 or just note (i.e., compute) that in *any* such representation of f, necessarily $n = \operatorname{Ind}_\gamma(0)$. If $n = 0$, $f(rz) := \exp(r\varphi(z))$, $(r, z) \in [0,1] \times \mathbb{T}$, defines a continuous zero-free extension. On the other hand, if such an extension F exists, then $H(r,t) := F(r\exp(2\pi it))$, $(r,t) \in [0,1] \times [0,1]$, is a loophomotopy in $\mathbb{C} \setminus \{0\}$ of $F \circ C(0,1) = f \circ C(0,1)$ with the constant function $F(0)$, and 4.12 may be cited.

(ii) *The set $\mathcal{G}$ of all continuous, zero-free, complex-valued functions on $\mathbb{T}$ is a commutative group under multiplication and the set $\mathcal{E}$ of all exponentials of continuous, complex-valued functions on $\mathbb{T}$ constitutes a subgroup of $\mathcal{G}$.*

(iii) *Let $\mathcal{G}_1$ be the subgroup of functions in $\mathcal{G}$ of modulus 1, $\mathcal{E}_1 := \mathcal{E} \cap \mathcal{G}_1$. Show that $\mathcal{G}_1$ is complete in the uniform metric and that $\mathcal{E}_1$ is both open and closed in $\mathcal{G}_1$ and connected.* (Cf. the proof of 4.0.) *Conclude that $\mathcal{E}_1$ is the component of the metric space $\mathcal{G}_1$ which contains* **1**. *Show also that $\mathcal{G}_1/\mathcal{E}_1 \cong \mathcal{G}/\mathcal{E}$.*

(iv) *Show that this quotient group is torsion-free, that is, if $f \colon \mathbb{T} \mapsto \mathbb{C} \setminus \{0\}$ is continuous and $f^n = e^\varphi$ for some non-zero integer n and some continuous φ, then $f = e^\psi$ for some continuous ψ. Conclude via* (i) *that $\mathcal{G}/\mathcal{E}$ is isomorphic to the (additive) group $\mathbb{Z}$.*

Hints: The sets $T_k := \big\{z \in \mathbb{T} : f(z) = e^{[\varphi(z)+2\pi ik]/n}\big\}$ $(k = 1, 2, \ldots, |n|)$ are disjoint, closed and their union is $\mathbb{T}$. Define $\psi := [\varphi + 2\pi ik]/n$ on each nonvoid T_k, getting a function continuous on $\mathbb{T}$ such that $f = e^\psi$. (Evidently the nature of the domain of f is irrelevant to the truth of (iv); this same proof shows that the quotient group $\mathcal{G}/\mathcal{E}$ in 4.0 is always torsion-free.) Alternatively, (iv) is a direct consequence of the uniqueness of the representation in (i).

Exercise 4.23 (i) *Prove the analog of* 4.22(i) *for squares: If $Q := (-1,1) \times (-1,1)$ and $f \colon \partial Q \mapsto \mathbb{C} \setminus \{0\}$ is continuous, then there exists an integer n and continuous $\varphi \colon \partial Q \mapsto \mathbb{C}$ such that $f(w) = w^n e^{\varphi(w)}$ for all $w \in \partial Q$.*

Hints: Define $\psi \colon \partial Q \mapsto \mathbb{T}$ by $\psi(w) := w/|w|$. Then ψ is a surjective homeomorphism. Apply 4.22(i) to $f \circ \psi^{-1}$ to come up with an integer n and a continuous φ on $\mathbb{T}$ such that

$$f\big(\psi^{-1}(z)\big) = z^n e^{\varphi(z)} \qquad \forall z \in \mathbb{T},$$

whence

$$\begin{aligned} f(w) &= \psi(w)^n e^{\varphi(\psi(w))} \\ &= w^n e^{\varphi(\psi(w)) - n\log|w|} \qquad \forall w \in \partial Q. \end{aligned}$$

(ii) *Use* 4.22(i) *to show that any continuous zero-free f on $\mathbb{T}$ which satisfies the functional equation $f(z^2) = f^2(z)$ for all z must be an integer power of z. Is there anything special about the exponent* 2 *here?*

Hints: Evidently $f(1) = 1$ and so for the φ supplied by 4.22(i) we can suppose that $\varphi(1) = 0$. Argue that $\varphi(z^2) - 2\varphi(z)$ is constant, hence 0. Therefore the sequence $\{2^k\varphi(z)\}_{k=1}^{\infty}$ lies in the compact set $\varphi(\mathbb{T})$, whence $\varphi(z) = 0$.

(iii) *Let $f\colon (\mathbb{R}, +) \mapsto (\mathbb{C} \setminus \{0\}, \cdot)$, $g\colon ((0,\infty), \cdot) \mapsto (\mathbb{C} \setminus \{0\}, \cdot)$, $h\colon (\mathbb{C} \setminus \{0\}, \cdot) \mapsto (\mathbb{C} \setminus \{0\}, \cdot)$ and $F\colon (\mathbb{C}, +) \mapsto (\mathbb{C} \setminus \{0\}, \cdot)$ be continuous group homomorphisms. Use* (ii) *above and* 3.25(ii) *to show successively that these maps have the forms*

$$f(x) = e^{\alpha x}, \quad g(t) = t^{\alpha}, \quad h(z) = |z|^{\alpha} z^n, \quad F(z) = e^{az + b\bar{z}}$$

for some $\alpha, a, b \in \mathbb{C}$, $n \in \mathbb{Z}$. [Analogous forms characterize the continuous automorphisms of the semigroup $(\overline{\mathbb{D}}, \cdot)$; see **MR** 2006b:30034.]

(iv) *f_1, $f_2\colon \mathbb{T} \mapsto \mathbb{T}$ are continuous, $n_j := \mathrm{Ind}_{f_j \circ C(0,1)}(0)$. Show that there exist at least $|n_1 - n_2|$ points at which f_1 and f_2 agree. Interpret this result in terms of a fast and a slow runner on a circular track.*

Hints: We may assume that $n := n_1 - n_2 > 0$. Set $F := f_1/f_2$. By 4.1 there exists continuous $\varphi\colon [0,1] \mapsto \mathbb{C}$ such that

$$F\left(e^{2\pi i t}\right) = e^{2\pi i \varphi(t)} \qquad \forall t \in [0,1].$$

Since $|F| = 1$, φ is real-valued. By 4.2 and 4.14 we have

$$\varphi(1) - \varphi(0) = \mathrm{Ind}_{F \circ C(0,1)}(0) = n \geq 1.$$

The range of the continuous function φ contains the closed interval $[\varphi(0), \varphi(1)]$ of length n, hence it contains n successive integers. Thus there exist $t_k \in [0,1]$ such that $\varphi(t_k)$, $k = 1, 2, \ldots, n$, are distinct integers. The numbers $u_k := e^{2\pi i t_k} \in \mathbb{T}$ are therefore distinct, and they satisfy $F(u_k) = e^{2\pi i \varphi(t_k)} = 1$.

(v) *Show that if $f\colon \mathbb{T} \mapsto \mathbb{T}$ is continuous and $\mathrm{Ind}_{f \circ C(0,1)}(0) \neq 0$, then f is surjective.*

Hints: If $u \in \mathbb{T} \setminus f(\mathbb{T})$, 3.42(ii) supplies a continuous logarithm L in $\mathbb{C} \setminus \{tu : t \geq 0\}$. Then $f = e^{L \circ f}$, so $\mathrm{Ind}_{f \circ C(0,1)}(0) = 0$.

(vi) *By combining* (iv) *and* (v), *show that every continuous, fixed-point-free* $f\colon \mathbb{T} \mapsto \mathbb{T}$ *is surjective.*

(vii) *Continuous* $f : \mathbb{T} \mapsto \mathbb{T}$ *and* $u \in \mathbb{T}$ *are given, and* $n := \mathrm{Ind}_{f \circ C(0,1)}(0)$. *Show that there exists* $z \in \mathbb{T}$ *such that* $f(uz) = u^n f(z)$.

Hints: 4.22(i) supplies continuous $\varphi : \mathbb{T} \mapsto \mathbb{C}$ such that $f(z) = z^n e^{\varphi(z)} \in \mathbb{T}$ for all $z \in \mathbb{T}$. Hence $\mathrm{Re}\,\varphi = \mathbf{0}$, so $\psi := -i\varphi/n$ is real-valued and satisfies $f(z) = [ze^{i\psi(z)}]^n$ for all $z \in \mathbb{T}$. It suffices to find $z \in \mathbb{T}$ such that $\psi(uz) = \psi(z)$. Write $u = e^{2\pi i\alpha}$ and $\Psi(t) := \psi(e^{2\pi it})$, $\alpha, t \in \mathbb{R}$. Because the function Ψ is 1-periodic, $\int_0^1 [\Psi(\alpha + t) - \Psi(t)]dt = 0$. Therefore the (continuous) integrand cannot be everywhere positive or everywhere negative, hence must have a zero.

(viii) $f : \mathbb{T} \mapsto \mathbb{R}$ *is continuous and* $u \in \mathbb{T}$. *Show that there exists* $z \in \mathbb{T}$ *such that* $f(uz) = f(z)$.

Hint: There exist $z_1, z_2 \in \mathbb{T}$ such that $f(z_1) \le f(z) \le f(z_2)$ for all $z \in \mathbb{T}$. The continuous function $F(z) := f(z) - f(uz)$ therefore satisfies $F(z_1) \le 0 \le F(z_2)$, hence must vanish somewhere on the connected set $\mathbb{T}$.

4.21 and 4.1 show that the domain of a continuous, zero-free function affects whether it has a continuous logarithm or not. In particular, if continuous, zero-free extension to a disk or to $\mathbb{C}$ is possible, continuous logarithms are assured. Thus the extension theorem 1.44 plays a big role.

Theorem 4.24 *Let K be a compact metric space, $f_0, f_1\colon K \mapsto \mathbb{C}\backslash\{0\}$ continuous functions which are homotopic in $\mathbb{C} \setminus \{0\}$. Then f_0/f_1 has a continuous logarithm on K. Hence f_0 has a continuous logarithm on K if and only if f_1 does.*

Proof: Let $h\colon [0,1] \times K \mapsto \mathbb{C} \setminus \{0\}$ be a homotopy of f_0 with f_1 and set $h_t(z) := h(t,z)/f_1(z)$ for each $t \in [0,1]$, $z \in K$. As noted before, $t \mapsto h_t$ is a continuous map φ of the connected space $[0,1]$ into the set $\mathcal{G}$ of invertible elements of the ring $C(K)$. Since $\varphi(1) = \mathbf{1} \in \mathcal{E}$, $\mathcal{E} \cup \varphi[0,1]$ is a connected (1.1) subset of $\mathcal{G}$ which contains $\mathbf{1}$, so by the maximality of the component $\mathcal{E}$ (4.0), $\mathcal{E} \cup \varphi[0,1] \subset \mathcal{E}$. Hence $f_0/f_1 = \varphi(0) \in \mathcal{E}$.

A more constructive proof is possible as follows: Compactness supplies $m, n \in \mathbb{N}$ such that

$$(1) \qquad |h(t,z)| \ge 1/m \qquad \forall (t,z) \in [0,1] \times K$$

and

$$(2) \qquad s,t \in [0,1] \ \& \ |s-t| \le 1/n \implies |h(t,z) - h(s,z)| < 1/m \quad \forall z \in K.$$

Set $g_j(z) := h(j/n, z)$ for $j \in \{0, 1, \ldots, n\}, z \in K$. Then (2) shows that

$$|g_j(z) - g_{j-1}(z)| < 1/m$$

from which and (1) follows

$$(3) \qquad \begin{aligned} &|g_j(z)/g_{j-1}(z) - 1| < 1, \text{ that is,} \\ &g_j(z)/g_{j-1}(z) \in D(1,1) \quad \forall j \in \{1, 2, \ldots, n\}, z \in K. \end{aligned}$$

Letting L denote a continuous logarithm in $D(1,1)$, we therefore have

$$g_j/g_{j-1} = \exp \circ L \circ (g_j/g_{j-1}), \quad 1 \leq j \leq n.$$

Multiplying these equations and noting that $g_0 = f_0$, $g_n = f_1$, gives

$$f_1/f_0 = g_n/g_0 = \exp \circ \sum_{j=1}^{n} L \circ (g_j/g_{j-1}).$$

Yet a third proof is to be found on p. 326 of KURATOWSKI [1945].

Theorem 4.25 (BORSUK [1932]) *For a compact subset C of $\mathbb{R}^n$ and a continuous function $f\colon C \mapsto \mathbb{C} \setminus \{0\}$ the following three conditions are equivalent:*

(i) *f is homotopic in $\mathbb{C} \setminus \{0\}$ to a constant,*

(ii) *f has an extension to a continuous function $F : \mathbb{R}^n \mapsto \mathbb{C} \setminus \{0\}$,*

(iii) *f has a continuous logarithm.*

Proof: (i) $\Rightarrow$ (iii). Immediate from 4.24.

(iii) $\Rightarrow$ (ii). If h is a continuous logarithm for f on C, then 1.45 furnishes a continuous extension of h to $H\colon \mathbb{R}^n \mapsto \mathbb{C}$. Then e^H is a continuous zero-free function which on C agrees with $e^h = f$.

(ii) $\Rightarrow$ (i). The continuous zero-free function F on $\mathbb{R}^n$ has a continuous logarithm h by 4.20. Then $H(t, z) := e^{th(z)}$ is a homotopy in $\mathbb{C} \setminus \{0\}$ of F with the constant function $\mathbf{1}$ and by restriction it furnishes such a homotopy between f and $\mathbf{1}$.

The next exercise is a special case of some important classical results about holomorphic functions. It uses only the index and the relation between zeros and factorizations. For holomorphic functions generally the latter must await subsequent developments, but for polynomials the relevant facts are elementary: If z_0 is a zero of the polynomial P, then there is a polynomial P_0 such that $P(z) = (z - z_0)P_0(z)$. (For the easy proof, glance ahead to 4.52(i).) The highest power of $z - z_0$ which thus divides $P(z)$ is called the *multiplicity* of z_0. Evidently then, if $z_0, \ldots, z_k$ are distinct zeros of P having multiplicities $m_0, \ldots, m_k$, there is a polynomial Q such that

$$(*) \qquad P(z) = (z - z_0)^{m_0} \cdots (z - z_k)^{m_k} Q(z).$$

Exercise 4.26 (i) (Argument Principle for Polynomials) *Let P be a polynomial, $z_0, \ldots, z_k$ all of its zeros in $\overline{D}(a, r)$, having multiplicities $m_0, \ldots, m_k$ and none lying on $\gamma := C(a, r)$. Show that*

$$\mathrm{Ind}_{P\circ\gamma}(0) = m_0 + m_1 + \cdots + m_k.$$

[This says that the change in the "argument" of P along γ, more precisely, the number of times $P \circ \gamma$ winds around 0, counts the zeros of P inside γ.]

Hints: For each j

$$\mathrm{Ind}_{\gamma - z_j}(0) = \mathrm{Ind}_{\gamma}(z_j) = 1,$$

so by $(*)$ above and 4.14

$$\mathrm{Ind}_{P\circ\gamma}(0) = \mathrm{Ind}_{Q\circ\gamma}(0) + \sum_{j=0}^{k} \mathrm{Ind}_{(\gamma - z_j)^{m_j}}(0) = \mathrm{Ind}_{Q\circ\gamma}(0) + \sum_{j=0}^{k} m_j.$$

Since Q is zero-free in $\overline{D}(a, r)$ (why?), $(\tau, t) \mapsto Q\big(a + \tau r e^{2\pi i t}\big)$ is a loophomotopy in $\mathbb{C} \setminus \{0\}$ of $Q \circ \gamma$ with the constant loop $Q(a)$. Hence $\mathrm{Ind}_{Q\circ\gamma}(0) = 0$, by 4.12.

(ii) (Rouché's Theorem) *Suppose f, g are polynomials which satisfy*

$$(**) \qquad |f + g| < |f| + |g| \qquad \textit{on } C(a, r).$$

Show that f and g have the same number of zeros in $\overline{D}(a, r)$, when each is counted according to its multiplicity.

Hints: Write γ for the loop $C(a, r)$. By $(**)$ g is zero-free on $C(a, r)$. Therefore f/g is continuous on $C(a, r)$ and $(f/g) \circ \gamma$ is a loop. Call it Γ. By $(**)$

$$|\mathbf{1} + \Gamma| < \mathbf{1} + |\Gamma|.$$

This shows that Γ never meets $[0, \infty)$, and so 0 lies in the unbounded component of $\mathbb{C} \setminus \Gamma$. Hence by 4.3, $\mathrm{Ind}_{\Gamma}(0) = 0$. It follows from 4.14 and the definition of Γ that

$$0 = \mathrm{Ind}_{f\circ\gamma}(0) + \mathrm{Ind}_{(1/g)\circ\gamma}(0) = \mathrm{Ind}_{f\circ\gamma}(0) - \mathrm{Ind}_{g\circ\gamma}(0).$$

Now invoke (i).

(iii) (FTA) *Show that every monic complex polynomial* $f(z) = \sum c_j z^j$ *of degree* $n > 0$ *has* n *zeros (counted according to multiplicity) in* $D(0, 1 + c)$*, where* $c := \max\{|c_j| : 0 \leq j \leq n-1\}$.

Hints: In (ii) take $a := 0$, $r := 1 + c$, $g(z) := -z^n$ and note that for all $z \in C(a, r) = C(0, 1 + c)$

$$|f(z) + g(z)| \leq \sum_{j=0}^{n-1} c|z|^j = \sum_{j=0}^{n-1} (|z| - 1)|z|^j = |z|^n - 1 < |g(z)|.$$

The essential idea underlying the proof of 4.26(ii) can be used to prove a more general result:

Exercise 4.27 *Let* X *be any topological space,* $f_0, f_1 \colon X \mapsto \mathbb{C}$ *continuous functions which satisfy*

$$|f_0 + f_1| < |f_0| + |f_1|.$$

Show that both functions are zero-free and one has a continuous logarithm if and only if the other does.

Hints: The strict inequality precludes zeros and implies that the quotient $f := f_0/f_1$ maps into $\mathbb{C} \setminus [0, \infty)$. Letting L denote a continuous logarithm in $\mathbb{C} \setminus [0, \infty)$ (cf. 3.43) and $\varphi := L \circ f$, we have $f_0/f_1 = e^{\varphi}$.

Exercise 4.28 (i) [generalizing 3.42(ii)] *If* U *is an open subset of* $\mathbb{C}$ *and* 0 *lies in an unbounded component of* $\mathbb{C} \setminus U$*, then there is a continuous logarithm in* U*. In particular, there is a continuous logarithm in every open, simply-connected subset of* $\mathbb{C} \setminus \{0\}$.

Hints: If γ is any loop in U, then $\mathrm{Ind}_\gamma(0) = 0$ by 4.3, since 0 lies in an unbounded component of $\mathbb{C} \setminus U$. From 4.18 it follows that the identity function on U has a continuous logarithm. Recall definition 1.37.

(ii) *Let* K *be a compact metric space and* $f \colon K \mapsto \mathbb{C} \setminus \{0\}$ *continuous. Show that if* 0 *belongs to the unbounded component of* $\mathbb{C} \setminus f(K)$*, then* f *is homotopic in* $\mathbb{C} \setminus \{0\}$ *to a constant, and then by* 4.24 f *has a continuous logarithm.*

Hints: Choose $r > \sup |f(K)|$ and use 1.28 and 1.30(i)′ to find a continuous $\gamma \colon [0, 1] \mapsto \mathbb{C} \setminus f(K)$ such that $\gamma(0) = 0$, $\gamma(1) = r$. Define h by

$$h(t, z) := \begin{cases} f(z) - \gamma(t), & (t, z) \in [0, 1] \times K \\ (2 - t)f(z) - r, & (t, z) \in [1, 2] \times K. \end{cases}$$

Then h is a homotopy in $\mathbb{C} \setminus \{0\}$ of f with the constant $-r$.

Alternatively, extend γ to $\Gamma\colon [0,\infty) \mapsto \mathbb{C}$ by setting $\Gamma(t) := rt$ for $t \geq 1$. Then (i) furnishes a continuous logarithm L for the identity function on $U := \mathbb{C} \setminus \Gamma \supset f(K)$. The composite $L \circ f$ is a continuous logarithm of f.

Theorem 4.29 *Let K be a compact subset of $\mathbb{C}$, $f\colon K \mapsto \mathbb{C} \setminus \{0\}$ be continuous. Then there exist finitely many points $p_1, \ldots, p_N \in \mathbb{C} \setminus K$ and integers $n_1, \ldots, n_N$ such that the function $F(z) := f(z)\prod_{j=1}^{N}(z-p_j)^{n_j}$ has a continuous logarithm on K.*

Proof: As a matter of convenience, subject $\mathbb{C}$ to a preliminary transformation $z \mapsto az+b$ in order to have

$$(1) \qquad K \subset (0,1) \times (0,1).$$

Apply 1.45 to get a continuous extension $f_0\colon Q := [0,1] \times [0,1] \mapsto \mathbb{C}$. Let $L := f_0^{-1}(\{0\})$. This is a closed subset of Q disjoint from K, so by compactness there exists an $r > 0$ such that

$$(2) \qquad |z - w| \geq r \qquad \forall z \in K,\ w \in L.$$

Let m be a positive integer greater than $\sqrt{2}/r$ and consider the squares

$$Q_{jk} := \left[\frac{j-1}{m}, \frac{j}{m}\right] \times \left[\frac{k-1}{m}, \frac{k}{m}\right], \quad \text{center } p_{jk} := \frac{j-\frac{1}{2}}{m} + i\frac{k-\frac{1}{2}}{m}$$

for $j,\ k = 1, 2, \ldots, m$. Set

$$\begin{aligned} \mathcal{K} &:= \{(j,k) : 1 \leq j,k \leq m \quad \text{and} \quad Q_{jk} \cap K \neq \varnothing\} \\ \mathcal{L} &:= \{(j,k) : 1 \leq j,k \leq m \quad \text{and} \quad Q_{jk} \cap K = \varnothing\}. \end{aligned}$$

It follows from the choice of m and r that

$$(3) \qquad K \subset \bigcup\{Q_{jk} : (j,k) \in \mathcal{K}\} =: K_1, \text{ closed } \subset Q \setminus L.$$

Let f_1 be the restriction of f_0 to K_1. By (3) and the definition of L, this function is zero-free. Since K_1 is comprised of some of the squares Q_{jk}, each interval $\{j/m\} \times [(k-1)/m, k/m]$ and each interval $[(j-1)/m, j/m] \times \{k/m\}$ either lies wholly in K_1 or meets K_1 (if at all) only at endpoints. At each endpoint where f_1 is not already defined give it the value 1. Then for any interval $[a,b]$ of the above kind which does not lie wholly in K_1, $f_1(a)$ and $f_1(b)$ are non-zero complex numbers and f_1 is not defined on (a,b). Since

$\mathbb{C} \setminus \{0\}$ is open and connected, 1.28 furnishes a continuous zero-free extension f_2 of f_1 to the closed subset

$$K_2 := K_1 \cup \bigcup_{j,k=1}^{m} \partial Q_{jk} \tag{4}$$

of Q. The definitions of $\mathcal{L}$ and K_1 then show that

$$K_2 \cap Q_{jk} = \partial Q_{jk} \qquad \forall\ (j,k) \in \mathcal{L}. \tag{5}$$

For each such (j,k) there exists an integer n_{jk} such that $(z-p_{jk})^{n_{jk}} f_2(z)$ has a zero-free, continuous extension F_{jk} to Q_{jk}. This follows from 4.23(i) and 4.25. We can then consistently define F_0 on Q by

$$F_0(z) := \begin{cases} f_2(z) \displaystyle\prod_{(j,k)\in\mathcal{L}} (z-p_{jk})^{n_{jk}}, & z \in K_2 \\ F_{j'k'}(z) \displaystyle\prod_{(j,k)\in\mathcal{L}\setminus(j',k')} (z-p_{jk})^{n_{jk}}, & z \in Q_{j'k'},\ (j',k') \in \mathcal{L}. \end{cases}$$

This function is continuous and zero-free on Q, hence has a continuous logarithm on Q by 4.1. The restriction of F_0 to K is governed by the first of the two formulas and therefore has the product form stated in the theorem.

Corollary 4.30 *Let K be a compact subset of $\mathbb{C}$, $\mathcal{C}$ the set of holes in K and for each $C \in \mathcal{C}$ let $p_C \in C$. Then given continuous $f\colon K \mapsto \mathbb{C} \setminus \{0\}$, there exist $C_1, \ldots, C_M \in \mathcal{C}$ and non-zero integers $n_1, \ldots, n_M$ such that the function $F(z) := f(z)\prod_{j=1}^{M}(z-p_{C_j})^{n_j}$ has a continuous logarithm on K. The case $\mathcal{C} = \varnothing$ is allowed, void products being interpreted as 1. [In 4.33(i) we shall see that the C_j and n_j are uniquely determined.]*

Proof: Let $p_1, \ldots, p_N$ and $n_1, \ldots, n_N$ be as provided by the preceding theorem. It suffices to show that if $p_j \in C_j \in \mathcal{C}$, then $g(z) := (z-p_{C_j})/(z-p_j)$ has a continuous logarithm on K and if p_j belongs to the unbounded component of $\mathbb{C} \setminus K$ then $z-p_j$ itself has such a logarithm. For then we just multiply the representation 4.29 by the appropriate powers of these factors. In the second case the existence of a logarithm for $z-p_j$ is guaranteed by 4.28(ii). In the first case, recalling that C_j is open (1.30(i)$'$), we use 1.28 to select a curve $\gamma\colon [0,1] \mapsto C_j$ joining p_{C_j} to p_j. Then

$$H(t,z) := \frac{z-\gamma(t)}{z-p_j}, \qquad (t,z) \in [0,1] \times K$$

is a homotopy in $\mathbb{C} \setminus \{0\}$ of g with $\mathbf{1}$, and so g has a continuous logarithm by 4.24.

Theorem 4.31 *Let K be a non-empty compact subset of $\mathbb{C}$, C a hole in K and p a point of C. Then the function f defined on K by $f(z) := z - p$ has no continuous, zero-free extension to $K \cup C$; neither does any non-zero power of f. In particular, f does not have a continuous logarithm.*

Proof: We may suppose $p = 0$. Suppose, contrariwise, that for some integer $n \neq 0$ the function f^n has a continuous extension $F\colon K \cup C \mapsto \mathbb{C} \setminus \{0\}$. Thus $F(z) = z^n$ for all $z \in K$. Choose $r > \sup |C|$ (recall, C is bounded) and define

$$g(z) := \begin{cases} z^n, & z \in \bar{D}(0,r) \setminus C \\ F(z), & z \in \bar{C}. \end{cases}$$

Notice that $\bar{C} \cap \left[\bar{D}(0,r) \setminus C\right] = \bar{C} \setminus C \subset K$ (C being relatively closed in $\mathbb{C} \setminus K$). This shows that the definition of g is consistent and that in fact g is continuous, C being open. Moreover, we see that g maps into $\mathbb{C} \setminus \{0\}$ (remember, $0 \in C$ and F is zero-free). Then 4.1 furnishes a continuous logarithm φ for g. In particular,

$$(*) \qquad z^n = g(z) = e^{\varphi(z)} \qquad \forall z \in C(0,r).$$

This contradicts 4.21. Alternatively, we can compute indexes at 0 on both sides of $(*)$ and reach the contradiction $n = 0$.

Exercise 4.32 *Let G be a torsion-free abelian group. Suppose that $\{f_1, \ldots, f_N\}$ is an independent set in G and for some $g_1, \ldots, g_n \in G$ we have $\{f_1, \ldots, f_N\} \subset$ $\mathrm{grp}\{g_1, \ldots, g_n\}$, the group generated by $g_1, \ldots, g_n$. Show that $n \geq N$.*

Hints: Writing G multiplicatively, let

$$(1) \qquad f_j := \prod_{k=1}^{n} g_k^{m_{kj}}, \quad m_{kj} \in \mathbb{Z}.$$

If $n < N$, the system of equations

$$(2) \qquad \sum_{j=1}^{N} c_j m_{kj} = 0, \quad k = 1, 2, \ldots, n$$

has a non-trivial rational solution $(c_1, \ldots, c_N)$ since the N vectors $(m_{1j}, \ldots, m_{nj})$ $(j = 1, 2, \ldots, N)$ are linearly dependent in the n-dimensional space $\mathbb{Q}^n$. Upon clearing denominators, there is then a non-trivial integer solution (c_j) to (2). From (1) and (2) together and the commutativity of the group we get

$$\prod_{j=1}^{N} f_j^{c_j} = \prod_{k=1}^{n} \prod_{j=1}^{N} g_k^{c_j m_{kj}} = \prod_{k=1}^{n} g_k^{\sum_{j=1}^{N} c_j m_{kj}} = \prod_{k=1}^{n} g_k^0 = \mathbf{1}.$$

Since not all c_j are 0, this violates the independence of $\{f_1, \ldots, f_N\}$.

Exercise 4.33 *Let K be a non-empty compact subset of $\mathbb{C}$. Form the groups $\mathcal{G} := \mathcal{G}(K)$ and $\mathcal{E} := \mathcal{E}(K)$ of all continuous, zero-free functions on K and all exponentials of continuous functions on K, respectively, as in 4.22. Let $\mathbb{Z}_0^{\mathbb{Z}}$ denote the additive group of all maps of $\mathbb{Z}$ into $\mathbb{Z}$ which vanish off finite sets.*

(i) *Let $C_1, \ldots, C_n$ be distinct holes in K and $p_j \in C_j$. Let $f_j(z) := z - p_j$, $z \in K$. Then $f_1, \ldots, f_n \in \mathcal{G}$. Show that their cosets are* independent *in the quotient group $\mathcal{G}/\mathcal{E}$. That is, show that $f_1^{k_1} \cdots f_n^{k_n} \in \mathcal{E}$ for integers $k_1, \ldots, k_n$ only if $k_1 = \cdots = k_n = 0$. Use this fact to formulate a uniqueness codicil to* 4.30.

Hints: Use 4.31: If $f_1^{k_1} \cdots f_n^{k_n} \in \mathcal{E}$ for some $k_j \in \mathbb{Z}$, we write

$$f_1^{k_1} = f_2^{-k_2} \cdots f_n^{-k_n} e^{\varphi}$$

for some continuous φ on K. That is,

$$f_1^{k_1}(z) = \frac{e^{\varphi(z)}}{(z - p_2)^{k_2} \cdots (z - p_n)^{k_n}} \qquad \forall z \in K.$$

The denominator on the right-hand side naturally extends to a continuous, zero-free function on C_1, since $p_2, \ldots, p_n \notin C_1$, and φ has by 1.45 a continuous extension to $\mathbb{C}$. Thus $f_1^{k_1}$ admits a continuous, zero-free extension to $K \cup C_1$, so $k_1 = 0$ by 4.31.

(ii) (Eilenberg [1936]) *Use* (i) *and* 4.30 *to show that the group $\mathcal{G}/\mathcal{E}$ is torsion-free. Then use* 4.30 *and* 4.32 *to show that this group is isomorphic to $\mathbb{Z}^n$ if K has exactly n holes and is isomorphic to $\mathbb{Z}_0^{\mathbb{Z}}$ if K has infinitely many holes.* (The group is said to have *rank n* or *infinite rank*, respectively.)

Hint: Notice that since components of $\mathbb{C} \setminus K$ are open (1.30(i)′), there are at most countably many of them, so in the second case they may be indexed by $\mathbb{Z}$.

(iii) *Noting that the groups $\mathcal{G}(K_j)$, as well as the groups $\mathcal{E}(K_j)$ ($j = 1, 2$), are naturally isomorphic whenever the compact sets K_j are homeomorphic, deduce that the number of components of the complement of a compactum is a homeomorphic invariant, i.e., is the same for any two homeomorphic compacta lying in $\mathbb{C}$.*

Corollary 4.34 (Borsuk [1932]). *For a compact subset K of $\mathbb{C}$ the following are equivalent:*

(i) $\mathbb{C} \setminus K$ *is connected.*

(ii) *Every continuous, zero-free function on K has a continuous logarithm.*

Proof: (i) $\Rightarrow$ (ii). Use 4.30.

(ii) $\Rightarrow$ (i). Use 4.31.

Corollary 4.35 *The property of having a connected complement is a homeomorphic invariant for compact subsets of* $\mathbb{C}$.

Exercise 4.36 *A compact subset of* $\mathbb{C}$ *is simply-connected if and only if each of its components is. The sequence of steps below (mainly from* EILENBERG [1936]) *will carry the reader through to this conclusion.*

(i) *Let* C *closed* $\subset K$ *closed* $\subset \mathbb{C}$, $f\colon K \mapsto \mathbb{C}\setminus\{0\}$ *continuous and suppose that* f *has a continuous logarithm on* C. *Show that there is a relatively open subset* U *of* K *containing* C *such that* f *has a continuous logarithm on* U.

Hints: Let $\varphi\colon K \mapsto \mathbb{C}$ be a continuous extension to K (after 1.45) of a continuous logarithm of f on C and consider $U := \{z \in K : |f(z)e^{-\varphi(z)} - 1| < 1\}$. This set is relatively open in K and contains C. It is mapped by $fe^{-\varphi}$ into $D(1,1)$ and there is (by 3.19) a continuous logarithm L in $D(1,1)$. Therefore we may form $\psi := L \circ (fe^{-\varphi})$ on U and have $f = e^{\varphi}e^{\psi} = e^{\varphi+\psi}$ on U.

(ii) *Let* K *be a compact subset of* $\mathbb{C}$, $f\colon K \mapsto \mathbb{C}\setminus\{0\}$ *continuous. Suppose that* f *has a continuous logarithm on each component of* K. *Show that* f *has a continuous logarithm on* K.

Hints: For each component C of K let U_C be a neighborhood of C as provided by (i). Use 1.33 to select a compact, relatively open (in K) neighborhood V_C of C lying in $K \cap U_C$. Cover compact K with finitely many of the sets V_C, say with $V_{C_1}, \ldots, V_{C_n}$. Then the sets $A_1 := V_{C_1}$, $A_k := V_{C_k} \setminus \bigcup_{j=1}^{k-1} V_{C_j}$ $(1 < k \le n)$ are disjoint, compact, they cover K and on each of them f has a continuous logarithm.

(iii) *If* K *is a compact subset of* $\mathbb{C}$ *and each component of* K *is simply-connected, then* K *is simply-connected.*

Hint: Use (ii) and 4.34.

(iv) *If* C_0 *is a component of a compact, simply-connected subset* K *of* $\mathbb{C}$, *then* C_0 *is simply-connected.*

Hints: It suffices to show that for each (other) component C of K, $\overline{\mathbb{C}\setminus K}$ meets C. For then $(\mathbb{C}\setminus K)\cup C$ is connected (by 1.3) and $\mathbb{C}\setminus C_0$ is a union of such sets. If $\overline{\mathbb{C}\setminus K} \cap C = \varnothing$, then for each $z \in C$ there exists an $r_z > 0$ such that $(\mathbb{C}\setminus K) \cap D(z, r_z) = \varnothing$. Then $U := \bigcup_{z\in C} D(z, r_z) = \bigcup_{z\in C} C \cup D(z, r_z)$ is connected, contains C and lies in K. Hence by maximality of components, $U = C$. But then C is clopen in $\mathbb{C}$.

Exercise 4.37 *Say that the set $S \subset \mathbb{C}$* separates *two points a, b if a, b lie in different components of $\mathbb{C} \setminus S$.*

(i) (EILENBERG [1936]) *Let K be a compact subset of $\mathbb{C}$, a, $b \in \mathbb{C} \setminus K$. Let $f\colon K \mapsto \mathbb{C}$ be defined by $f(z) := (z-a)/(z-b)$. Show that K does not separate a and b if and only if f has a continuous logarithm on K.*

Hints: ($\Rightarrow$) Examine the proof of 4.30.

($\Leftarrow$) According to 4.25, f has a continuous zero-free extension F to $\mathbb{C}$. If a, b are not in the same component of $\mathbb{C} \setminus K$, then one of them, say a, belongs to a bounded component C of $\mathbb{C} \setminus K$ which does not contain the other. Then the function $z - a = (z-b)f(z)$ $(z \in K)$ has the zero-free continuous extension $(z-b)F(z)$ $(z \in K \cup C)$, contradicting 4.31.

(ii) (JANISZEWSKI [1913]) *Suppose that K_1, K_2 are two compact subsets of $\mathbb{C}$ neither of which separates a, $b \in \mathbb{C} \setminus (K_1 \cup K_2)$. Show that if $K_1 \cap K_2$ is either void or connected, then $K_1 \cup K_2$ does not separate a, b either.*

Hint: Form $f(z) := (z-a)/(z-b)$ $(z \in K_1 \cup K_2)$. According to (i) this function has a continuous logarithm on K_1 and a continuous logarithm on K_2. Use connectedness of $K_1 \cap K_2$ to manufacture a single continuous logarithm on $K_1 \cup K_2$. Then cite the other direction of (i).

(iii) *Let U open $\subset \mathbb{C}$, $a, b \in \mathbb{C} \setminus U$. Use 4.18 to prove that if a, b lie in the same component of $\mathbb{C} \setminus U$, or if each lies in an unbounded component of $\mathbb{C} \setminus U$, then the function $f\colon U \mapsto \mathbb{C} \setminus \{0\}$ defined by $f(z) := (z-a)/(z-b)$ has a continuous logarithm.*

Hint: For every loop $\gamma \subset U$, $\mathrm{Ind}_{f\circ\gamma}(0) = \mathrm{Ind}_\gamma(a) - \mathrm{Ind}_\gamma(b)$. According to 4.3 this difference is 0 if a, b lie in a common connected subset of $\mathbb{C} \setminus \gamma$ or if each lies in an unbounded connected subset of $\mathbb{C} \setminus \gamma$.

(iv) *Use* (i) *and* 1.31 *to prove the converse of* (iii).

Hint: Let K_n be as supplied by 1.31. By (i), a and b lie in a common component C_n of $\mathbb{C} \setminus K_n$, for each n. 1.31(1) insures that the sets $X_n := \overline{C}_n$ satisfy $X := \bigcap X_n \subset \mathbb{C} \setminus U$. Since $a, b \in X$, it suffices to show that either X is connected or that each component C of X is unbounded. The argument for this is exactly that in the third paragraph of the 1.35 proof.

(v) *Use* (iii) *to show that every region Ω whose complement has at least one non-degenerate component is biholomorphic to a region lying in $\mathbb{D}$.*

Hints: Suppose $a \neq b$ lie in a common component of $\mathbb{C} \setminus \Omega$. (iii) provides a continuous logarithm φ in Ω for $f(z) := (z-a)/(z-b)$. By 3.41, φ is holomorphic, so $g := e^{\varphi/2}$ is holomorphic and $g^2 = f$. From injectivity of f follows that of g. If $z_1, z_2 \in \Omega$ and $-g(z_1) = g(z_2)$, then $f(z_1) = f(z_2)$, so $z_1 = z_2$, $g(z_1) = 0$ and the fact g is an exponential is contradicted. This proves that

$$-g(\Omega) \cap g(\Omega) = \varnothing. \tag{1}$$

Since $f(\Omega)$ is open (f being biholomorphic), the set $W := \{w \in \mathbb{C} : w^2 \in f(\Omega)\}$ is open. From $g^2 = f$ it is clear that

$$W = -g(\Omega) \cup g(\Omega). \tag{2}$$

It is thus apparent that the disjoint connected sets $-g(\Omega)$ and $g(\Omega)$ are the components of W, hence open sets. For all $w \in W$, $g^2(f^{-1}(w^2)) = w^2$ so from 1.7 it follows that

$$\begin{cases} \text{either} \quad g(f^{-1}(w^2)) = w & \text{for all } w \in g(\Omega) \\ \quad\text{or} \quad g(f^{-1}(w^2)) = -w & \text{for all } w \in g(\Omega). \end{cases}$$

Replacing g with $-g$ if necessary allows us to assume that the first alternative prevails. Then

$$g^{-1}(w) = f^{-1}(w^2) \qquad \forall w \in g(\Omega), \tag{3}$$

making g^{-1} holomorphic (since f^{-1} is). The open set $-g(\Omega)$ contains $\bar{D}(w_0, r)$ for some $w_0 \in \mathbb{C}$, $0 < r < \infty$, and $F(w) := r/(w - w_0)$ maps $\mathbb{C} \setminus \bar{D}(w_0, r) \supset g(\Omega)$ biholomorphically into $\mathbb{D}$. The desired biholomorphic map is $F \circ g$.

(vi) *Show that the conclusion in* (ii) *remains true for two disjoint closed subsets K_1, K_2 of $\mathbb{C}$, whether or not compact.*

Hints: The component C_j of $\mathbb{C} \setminus K_j$ to which both a, b belong is open, so there is a polygon $\gamma_j : [0, 1] \mapsto C_j$ joining a and b. Since the γ_j are polygons, it is clear that the open set $C_1 \cap C_2 \setminus (\gamma_1 \cup \gamma_2)$ is not void. Let c be a point in it and consider the homeomorphism $\varphi : \mathbb{C} \setminus \{c\} \to \mathbb{C} \setminus \{0\}$ given by $\varphi(z) := 1/(z - c)$. The points $\varphi(a)$, $\varphi(b)$ lie in the connected subset $\varphi(\gamma_j)$ of $\mathbb{C} \setminus \{0\} \setminus \varphi(K_j)$. Also $\{0\} \cup \varphi(K_1)$ and $\{0\} \cup \varphi(K_2)$ are compact sets (why?) with connected intersection $\{0\}$. By (ii), $\varphi(a), \varphi(b)$ lie in the same component C of

$$\mathbb{C} \setminus \{0\} \setminus [\varphi(K_1) \cup \varphi(K_2)] = \varphi(\mathbb{C} \setminus \{c\} \setminus (K_1 \cup K_2)).$$

Then $\varphi^{-1}(C)$ is a connected subset of $\mathbb{C} \setminus (K_1 \cup K_2)$ which contains both a and b.

Remark: An artifice used in the 4.37(v) hints proves effective in other situations (e.g., 5.76, 7.27, 9.7). It is the application of a square-root to transform a region in $\mathbb{C} \setminus \{0\}$ onto one that is disjoint from its negative, and whose complement consequently has a robust interior. We might call this *radical surgery*.

The special case in which $[a, b] \subset \mathbb{C} \setminus U$ is very geometrical: It is easy to check that f maps $\mathbb{C} \setminus [a, b]$ into $\mathbb{C}_-$ (cf. 2.5), so for φ we can take $L \circ f$, where L is the Principal Branch of the Logarithm in $\mathbb{C}_-$. Then $g := e^{\varphi/2}$ is built from a square-root that maps $\mathbb{C}_-$ onto $\mathbb{H}$ by halving angles. This example is also the simplest case of and possibly the inspiration for 4.37(iii). One finds it on p. 845 of KOEBE [1912], but it was in use much earlier.

Exercise 4.38 *This exercise is designed to culminate in a strengthened version of the non-trivial implication in* 4.18. *Let U be an open subset of $\mathbb{C}$ and $f\colon U \mapsto \mathbb{C} \setminus \{0\}$ a continuous function satisfying*

(*) $\quad \operatorname{Ind}_{f \circ \Gamma}(0) = 0$ for every simple polygonal loop Γ in U.

(i) *Consider two polygonal* arcs $\gamma_1, \gamma_2\colon [0, 1] \mapsto U$ *such that*

(1.0) $\quad \gamma_1(0) = \gamma_2(0)$

(1.1) $\quad \gamma_1(1) = \gamma_2(1)$

and let $\varphi_1, \varphi_2\colon [0, 1] \mapsto \mathbb{C}$ be continuous and satisfy

(2) $\quad f \circ \gamma_j = e^{\varphi_j} \quad (j = 1, 2)$

(3.0) $\quad \varphi_1(0) = \varphi_2(0)$.

Show that $\varphi_1(1) = \varphi_2(1)$.

Hints: Note that each γ_j is bicontinuous (cf. §6 of Chapter 0) and form $\psi_j := \varphi_j \circ \gamma_j^{-1} \quad (j = 1, 2)$. Set

$$p_0 := \gamma_1(0) - \gamma_2(0), \qquad p := \gamma_1(1) - \gamma_2(1).$$

Introduce the compact set

$$E := \gamma_1[0, 1] \cap \gamma_2[0, 1] \supset \{p_0, p\}.$$

Notice that for $z \in E$, we have

$$e^{\psi_j(z)} = \exp(\varphi_j(\gamma_j^{-1}(z))) \overset{(2)}{=} f \circ \gamma_j(\gamma_j^{-1}(z)) = f(z), \quad \text{so}$$

(4) $$\frac{\psi_2(z) - \psi_1(z)}{2\pi i} \text{ is an integer for each } z \in E.$$

We need to show that $\psi_1(p) = \psi_2(p)$ and so we suppose contrariwise that

(3.1) $$\psi_1(p) \neq \psi_2(p).$$

Then the set

$$E_1 := \left\{ z \in E : \left| \frac{\psi_2(z) = \psi_1(z)}{2\pi i} \right| \geq 1 \right\} \quad \text{contains } p = \gamma_1(1) - \gamma_2(1),$$

whereas by (3.0) the set

$$E_0 := \{ z \in E : \psi_2(z) - \psi_1(z) = 0 \} \quad \text{contains } p_0 = \gamma_1(0) = \gamma_2(0).$$

The sets E_j are evidently compact, disjoint and by (4) their union is E. Let

$$\begin{aligned} t_0 &:= \max \{ t \in [0,1] : \gamma_1(t) \in E_0 \}, \quad & a_0 &:= \gamma_1(t_0) \in E_0; \\ t_1 &:= \min \{ t \in [t_0,1] : \gamma_1(t) \in E_1 \}, \quad & a_1 &:= \gamma_1(t_1) \in E_1. \end{aligned}$$

Note that $t_0 < t_1$ and that $\gamma_1(t_0, t_1)$ is disjoint from $E_0 \cup E_1 = E$, hence by definition of E

$$\gamma_1(t_0, t_1) \cap \gamma_2[0,1] = \varnothing. \tag{5}$$

Since $a_j \in E \subset \gamma_2[0,1]$, we may form $s_j := \gamma_2^{-1}(a_j)$ $(j = 0, 1)$. Let

$$\tau_1 := [t_1, t_0], \quad \tau_2 := [s_1, s_0] \qquad \text{(cf. definition 1.12)}$$

and define

$$\Gamma(t) := \begin{cases} \gamma_1 \circ \tau_1(t) & 0 \leq t \leq 1 \\ \gamma_2 \circ \tau_2(2-t) & 1 \leq t \leq 2. \end{cases}$$

This is a well-defined polygonal loop and by (5) and the fact that γ_1, γ_2 are each arcs, it is simple. Next define

$$\varphi(t) := \begin{cases} \varphi_1 \circ \tau_1(t) & 0 \leq t \leq 1 \\ \varphi_2 \circ \tau_2(2-t) & 1 \leq t \leq 2. \end{cases}$$

This is a well-defined (because $a_0 \in E_0$) continuous function which by (2) satisfies $f \circ \Gamma = e^{\varphi}$. By hypothesis $(*)$ therefore

$$\begin{aligned} 0 = 2\pi i \operatorname{Ind}_{f \circ \Gamma}(0) &= \varphi(2) - \varphi(0) = \varphi_2(\tau_2(0)) - \varphi_1(\tau_1(0)) \\ &= \varphi_2(s_1) - \varphi_1(t_1) = \varphi_2\big(\gamma_2^{-1}(a_1)\big) - \varphi_1\big(\gamma_1^{-1}(a_1)\big) \\ &= \psi_2(a_1) - \psi_1(a_1). \end{aligned}$$

Since $a_1 \in E_1$, this is the sought-for contradiction which proves (3.1) untenable.

(ii) *Prove that under hypothesis $(*)$, f has a continuous logarithm in U.*

Hints: As in 4.18 we reduce at once to connected U. Fix a point $z_0 \in U$ and a $w_0 \in \mathbb{C}$ such that $e^{w_0} = f(z_0)$. Set $\gamma_{z_0} := z_0$ and for each $z \in U$ different from z_0 select [by 1.28 and 1.14] a polygonal *arc* $\gamma_z \colon [0,1] \mapsto U$ joining z_0 to z. Let φ_z and φ be defined as in 4.18, but with $\varphi_{z_0} := w_0$. To prove φ continuous at a point z_1, select r as in 4.18 and let t_1 be the smallest $t \in [0,1]$ such that $\gamma_{z_1}(t) \in \bar{D}(z_1, r)$. Then choose L as in 4.18 but with the normalization (3) replaced by

$$(3)' \qquad L(f(\gamma_{z_1}(t_1))) = \varphi_{z_1}(t_1).$$

For each $z \in D(z_1, r) \setminus \{z_0\}$ define

$$\Gamma_z(t) := \begin{cases} \gamma_{z_1}(t) & 0 \le t \le t_1 \\ \gamma_{z_1}(t_1) + \dfrac{t - t_1}{1 - t_1}(z - \gamma_{z_1}(t_1)), & t_1 \le t \le 1, \end{cases}$$

$$\tilde{\varphi}_z(t) := \begin{cases} \varphi_{z_1}(t) & 0 \le t \le t_1 \\ L(f(\Gamma_z(t))) & t_1 \le t \le 1. \end{cases}$$

We note that $\tilde{\varphi}_z$ is well defined at t_1 by $(3)'$ and so is continuous on [0,1]. Also clearly $f \circ \Gamma_z = e^{\tilde{\varphi}_z}$. If $z_1 \ne z_0$, then γ_{z_1} is one-to-one and if $z_1 = z_0$, then $t_1 = 0$. In either case therefore $\gamma_{z_1} = \Gamma_z$ is one-to-one on $[0, t_1]$. Now $z \ne \gamma_{z_1}(t_1)$ (why?), so Γ_z is one-to-one on $[t_1, 1]$. Moreover, by minimality of t_1, $\Gamma_z[0, t_1)$ is disjoint from $\bar{D}(z_1, r) \supset \Gamma_z[t_1, 1]$. Hence Γ_z is one-to-one. Finally, since $\varphi_z(0) = w_0 = \varphi_{z_1}(0) = \tilde{\varphi}_z(0)$, the conclusion $\varphi_z(1) = \tilde{\varphi}_z(1)$ follows from (i). This says that

$$(6) \quad \varphi(z) := \varphi_z(1) = L(f(\Gamma_z(1))) = L(f(z)) \qquad \forall z \in D(z_1, r) \setminus \{z_0\},$$

and proves continuity of φ at z_1 in case $z_1 \ne z_0$. But if $z_1 = z_0$, then $t_1 = 0$ and

$$\begin{aligned} \varphi(z_0) &= \varphi_{z_0}(1) = \varphi_{z_0}(t_1) \quad \text{by constancy of } \varphi_{z_0} \\ &= \varphi_{z_1}(t_1) = L(f(\gamma_{z_1}(t_1))) \quad \text{by } (3)' \\ &= L(f(\gamma_{z_1}(0))) = L(f(z_0)), \end{aligned}$$

so that now (6) is valid in the whole neighborhood $D(z_1, r)$ of z_1.

(iii) *Formulate and prove the analog of* 4.19(ii) *involving only simple loops.*

(iv) *Infer from* (iii) *the following remarkable relative of* 4.19(i)*: If U is open $\subset \mathbb{C}$, $k > 1$ an integer, $f \colon U \mapsto \mathbb{C} \setminus \{0\}$ continuous and one-to-one, then f has a continuous k^{th} root if and only if it has a continuous logarithm.*

Hints: We have to anticipate 4.43. If f has a continuous k^{th} root, then by (the trivial half of) (iii) the integer $\text{Ind}_{f \circ \gamma}(0)$ is divisible by k, for each simple loop γ in U. Since $f \circ \gamma$ is also a simple loop, this index is ± 1 or 0 by 4.43; hence, k being greater than 1, this index must be 0.

(v) *Let U be a non-void open subset of $\mathbb{C}$, I the identity function on U. Show that U is homologically connected if and only if for every $p \in \mathbb{C}\backslash U$, the function $I - p$ has a continuous square-root.* (Cf. 8.22(ii).)

Hint: Homologically connected means that

$$0 = \mathrm{Ind}_\gamma(p) = \mathrm{Ind}_{\gamma - p}(0) = \mathrm{Ind}_{(I-p)\circ\gamma}(0)$$

for all loops γ in U and all points $p \in \mathbb{C} \setminus U$. Therefore (iv) above and 4.18 settle the matter.

§ 5 The Jordan Curve Theorem

Corollary 4.39 *For any Jordan-curve J* [recall 1.11(v)] *the complement $\mathbb{C}\backslash J$ has exactly one bounded component.*

Proof: $J = \varphi(\mathbb{T})$ for some homeomorphism φ of $\mathbb{T}$. Since $\mathbb{C} \setminus \mathbb{T}$ has exactly one bounded component, so does $\mathbb{C}\backslash J$ by 4.33(iii). Alternatively, if one wishes to avoid the algebra, he can argue thus: If $\mathbb{C} \setminus J$ had no bounded component, then by 4.34 $\mathcal{G}(J) = \mathcal{E}(J)$, whence, via φ, $\mathcal{G}(\mathbb{T}) = \mathcal{E}(\mathbb{T})$. This contradicts 4.21. If $\mathbb{C}\setminus J$ has two distinct bounded components C_1, C_2, we select points $p_j \in C_j$ and form functions $f_j(z) := z - p_j$ $(j = 1, 2)$ on J. By 4.22 there exist integers n_j and continuous functions ψ_j on $\mathbb{T}$ such that

$$(*) \qquad \varphi(w) - p_j = w^{n_j} e^{\psi_j(w)} \qquad \forall w \in \mathbb{T}, \quad j = 1, 2.$$

Then

$$(\varphi(w) - p_1)^{n_2} (\varphi(w) - p_2)^{-n_1} = e^{n_2\psi_1(w) - n_1\psi_2(w)} \qquad \forall w \in \mathbb{T},$$

whence

$$(z - p_1)^{n_2} (z - p_2)^{-n_1} = e^{n_2\psi_1(\varphi^{-1}(z)) - n_1\psi_2(\varphi^{-1}(z))} \qquad \forall z \in J.$$

Then 4.33(i) implies that $n_1 = n_2 = 0$. Returning to $(*)$, we have then

$$\varphi(w) - p_1 = e^{\psi_1(w)} \qquad \forall w \in \mathbb{T},$$
$$f_1(z) = z - p_1 = e^{\psi_1(\varphi^{-1}(z))} \qquad \forall z \in J.$$

Thus f_1 extends to a continuous, zero-free function on $\mathbb{C}$ (since $\psi_1 \circ \varphi^{-1}$ is extendable), violating 4.31.

Exercise 4.40 *If K is a proper, closed subset of a Jordan-curve J, then $\mathbb{C} \setminus K$ is connected.*

Hints: By 4.35 it suffices to demonstrate this for the special case $J = \mathbb{T}$. However that case was handled in 1.25(i). [Using 4.45 and 1.2 one can easily prove that the conclusion remains valid if the hypothesis "closed" is dropped.]

Exercise 4.41 *If K is a compact subset of $\mathbb{C}$ homeomorphic to a subset of $\mathbb{R}$, then $\mathbb{C} \setminus K$ is connected.*

Hint: Treat the case $K \subset \mathbb{R}$ directly (via 1.2, say), then cite 4.35.

Theorem 4.42 *If J is a Jordan-curve and C a component of $\mathbb{C} \setminus J$, then $\partial C = J$.*

Proof: According to 1.30(i)′ and (v), C is an open set and

$$(*) \qquad \partial C \subset J.$$

By 4.39, $\mathbb{C} \setminus J$ is not connected, so there is another component K of $\mathbb{C} \setminus J$ disjoint from C. Thus K is disjoint from $\overline{C}$, since $\overline{C} \subset C \cup J$ (as was noted above). It follows that $\mathbb{C} \setminus \overline{C}$ is not void. Therefore the equality $\mathbb{C} \setminus \partial C = C \cup (\mathbb{C} \setminus \overline{C})$ displays $\mathbb{C} \setminus \partial C$ as a disjoint union of non-void, open sets, thus showing it to be disconnected. Consequently 4.40 tells us that the inclusion $(*)$ cannot be proper.

Theorem 4.43 *If $\gamma : [0,1] \mapsto \mathbb{C}$ is a simple loop, then $J := \gamma[0,1]$ is a Jordan-curve and in the unique bounded component of $\mathbb{C} \setminus J$, $\operatorname{Ind}_\gamma$ is either 1 or -1.*

Proof: That $\gamma[0,1]$ is a Jordan-curve was noted in 4.7(i). Thus there is a homeomorphism Γ from $\mathbb{T}$ onto J, and $f := \Gamma^{-1}$ maps J onto $\mathbb{T} \subset \mathbb{C} \setminus \{0\}$. Consequently by 4.30, for each p in the (unique, by 4.39) bounded component of $\mathbb{C} \setminus J$, there is $n \in \mathbb{Z}$ and continuous function φ on J such that

$$(z-p)^{-n} f(z) = e^{\varphi(z)} \qquad \forall z \in J,$$

that is,

$$f(\gamma(t)) = (\gamma(t) - p)^n e^{\varphi(\gamma(t))} \qquad \forall t \in [0,1].$$

Now $f \circ \gamma$ is a simple loop whose range is $\mathbb{T}$, the range of the simple loop $C(0,1)$. Hence by 4.7(iii) its index with respect to 0 is ± 1. Thus the preceding equation implies

$$\pm 1 = n \operatorname{Ind}_{\gamma - p}(0) + \operatorname{Ind}_{e^{\varphi \circ \gamma}}(0) = n \operatorname{Ind}_\gamma(p) + 0.$$

It follows that $\operatorname{Ind}_\gamma(p) = \pm 1$.

Definition 4.44 (i) For a Jordan-curve J the unique bounded component of $\mathbb{C} \setminus J$ is called the *inside* of J and is denoted by $\mathcal{I}(J)$. The unique unbounded component of $\mathbb{C} \setminus J$ is called the *outside* of J and is denoted by $\mathcal{O}(J)$.

(ii) A *Jordan region* is any set $\mathcal{I}(J)$.

(iii) If Γ is a homeomorphism of $\mathbb{T}$ onto J and $\gamma(t) := \Gamma(e^{2\pi i t})$, then Γ and γ are said to be *positively (negatively) oriented* if $\operatorname{Ind}_\gamma$ equals 1 (-1) throughout $\mathcal{I}(J)$.

Remark 4.45: The fundamental facts 4.39 and 4.42 can be summarized in the above language as follows:

(i) $\mathcal{I}(J)$, $\mathcal{O}(J)$ are disjoint, open, connected sets.

(ii) $\mathcal{I}(J)$ is bounded and simply-connected, $\mathcal{O}(J)$ is unbounded.

(iii) $\mathbb{C} \setminus J = \mathcal{I}(J) \cup \mathcal{O}(J)$.

(iv) $\overline{\mathcal{I}(J)} = J \cup \mathcal{I}(J)$, $\overline{\mathcal{O}(J)} = J \cup \mathcal{O}(J)$.

Exercise 4.46 (Generalization of the Jordan Curve Theorem).

(i) *Show that if J is a Jordan-curve lying in region Ω, then the components of $\Omega \setminus J$ are $\Omega \cap \mathcal{I}(J)$ and $\Omega \cap \mathcal{O}(J)$.*

Hint: Show that each of these intersections is a connected set by using 4.37(vi), with $K_1 := J$, $K_2 := \mathbb{C} \setminus \Omega$.

(ii) *Show that if $J_1, \ldots, J_n$ are mutually disjoint Jordan-curves lying in region Ω, then $\Omega \setminus (J_1 \cup \cdots \cup J_n)$ has exactly $n+1$ components. Can you describe them?*

Hint: Use (i) and induction on n, noting that J_n must lie wholly in a single component of $\Omega \setminus (J_1 \cup \cdots \cup J_{n-1})$.

Remark: Harder to prove than 4.46(ii) but accessible using only ideas developed here is the following useful (and "obvious") fact. If Ω is a *region* whose boundary consists of the mutually disjoint Jordan-curves $J_1, \ldots, J_n$, then there are only two possibilities: Either the sets $\overline{\mathcal{I}(J_k)}$ are mutually disjoint and Ω is the complement of their union, or there is $m \leq n$ such that $\overline{\mathcal{I}(J_k)}$ are mutually disjoint subsets of $\mathcal{I}(J_m)$ for $k \neq m$ and Ω is $\mathcal{I}(J_m)$ minus their union.

Exercise 4.47 *Let J be a Jordan-curve, a, b two distinct points on J and C a cross-cut in $\mathcal{I}(J)$ with endpoints a and b.* (Recall 1.15.)

(i) *Show that there are two arcs A_1, A_2 whose union is J and whose endpoints are each a and b.*

(ii) Show that $J_1 := A_1 \cup C$ and $J_2 := A_2 \cup C$ are Jordan-curves.

(iii) *Show that* $\mathcal{I}(J) \setminus C = \mathcal{I}(J_1) \cup \mathcal{I}(J_2)$.

Hints: The set $A_1' := A_1 \setminus \{a, b\}$ lies in J and J lies in the closure of the unbounded connected set $\mathcal{O}(J)$. (See 4.45.) Therefore $A_1' \cup \mathcal{O}(J)$ is unbounded and connected [by 1.2]. It is also disjoint from J_2. Hence it lies in $\mathcal{O}(J_2)$, whence

$$\mathcal{I}(J_2) \subset \mathbb{C} \setminus (A_1' \cup \mathcal{O}(J)) = \overline{\mathcal{I}(J)} \setminus A_1'. \tag{1}$$

Of course $\mathcal{I}(J_2)$ is disjoint from $J_2 = (J \setminus A_1') \cup C$ and so the inclusion (1) implies

$$\mathcal{I}(J_2) \subset \big(\overline{\mathcal{I}(J)} \setminus J\big) \setminus C = \mathcal{I}(J) \setminus C.$$

Similarly for $\mathcal{I}(J_1)$. Thus

$$\mathcal{I}(J_1) \cup \mathcal{I}(J_2) \subset \mathcal{I}(J) \setminus C. \tag{2}$$

To reverse the inclusion, consider any point $z \in \mathcal{I}(J) \setminus C$ and select

$$w \in \mathcal{O}(J). \tag{3}$$

If z belongs to neither $\mathcal{I}(J_1)$ nor $\mathcal{I}(J_2)$, then (since $z \notin J \cup C = J_1 \cup J_2$) z belongs to each of $\mathcal{O}(J_1)$, $\mathcal{O}(J_2)$, to which also the point w belongs by (2) and (3). Thus neither J_1 nor J_2 separates z and w. Since $J_1 \cap J_2 = C$ is connected, it follows from Janiszewski's theorem 4.37(ii) that $J_1 \cup J_2$ does not separate z and w. *A fortiori* then J, a subset of $J_1 \cup J_2$, does not separate z and w. That is false however, since $z \in \mathcal{I}(J)$ and $w \in \mathcal{O}(J)$.

(iv) *Why are* (i)–(iii) *known as "the theorem on θ-curves"?*

(v) *Show that if c, d are different from a, b and each lies in a different one of the arcs in* (i), *then c and d separate a and b in J.*

Hint: Use the exponential function to first settle the case $J = \mathbb{T}$.

Exercise 4.48 *Let J be a Jordan-curve, $f\colon \overline{\mathcal{I}(J)} \mapsto \mathbb{C}$ a continuous function.*

(i) (Schönflies)*Show that if f is one-to-one on J, then $J^* := f(J)$ is a Jordan-curve and $\mathcal{I}(J^*) \subset f\big(\mathcal{I}(J)\big)$.* (Cf. 9.16.)

Hints: Let J be the range of the homeomorphism $\Gamma\colon \mathbb{T} \mapsto \mathbb{C}$ and set $\gamma(t) := \Gamma(e^{2\pi i t})$, $0 \le t \le 1$. Thus J^* is the range of $f \circ \gamma$. Consider any $w \in \mathcal{I}(J^*)$. Then by 4.43 $\operatorname{Ind}_{f\circ\gamma}(w) \neq 0$. If $w \notin f\big(\overline{\mathcal{I}(J)}\big)$, then $f - w$ is zero-free in the compact simply-connected set $\overline{\mathcal{I}(J)}$. Therefore by 4.34 there exists continuous $L\colon \overline{\mathcal{I}(J)} \mapsto \mathbb{C}$ such that $f - w = e^L$ in $\overline{\mathcal{I}(J)}$. In particular,

$$f \circ \gamma - w = e^{L \circ \gamma},$$

whence

$$\operatorname{Ind}_{f\circ\gamma}(w) = \operatorname{Ind}_{f\circ\gamma - w}(0) = \frac{1}{2\pi i}[L(\gamma(1)) - L(\gamma(0))] = 0.$$

(ii) *Show that if f is one-to-one on $\overline{\mathcal{I}(J)}$, then $J^* := f(J)$ is a Jordan-curve and $\mathcal{I}(J^*) = f(\mathcal{I}(J))$.*

Hints: By univalence the connected set $f(\mathcal{I}(J))$ lies in $\mathbb{C} \setminus f(J) = \mathcal{I}(J^*) \cup \mathcal{O}(J^*)$, and so cannot meet both of these disjoint open sets. From (i) then $f(\mathcal{I}(J)) = \mathcal{I}(J^*)$.

(iii) (Open-Map Theorem; cf. 4.61.) *Deduce from* (ii) *that if $f\colon U \mapsto \mathbb{C}$ is continuous and one-to-one, and U is an open subset of $\mathbb{C}$, then $f(U)$ is open.*

Exercise 4.49 (Orientation of a Homeomorphism; see §62, XIII of Kuratowski [1968]).

(i) *Ω is a region in $\mathbb{C}$ and h is a homeomorphism of Ω into $\mathbb{C}$. Prove that there exists an integer N such that*

(1) $$\operatorname{Ind}_{h\circ C(a,r)}(h(a)) = (-1)^N \qquad \textit{whenever } \bar{D}(a,r) \subset \Omega.$$

Remark: h is called *orientation-preserving* if N is even, *orientation-reversing* if N is odd. Part (ii) below considerably generalizes equation (1). In 11.49(iii) it will be shown that this integer is even whenever h is a conformal mapping.

Hints: Define a function $d\colon \Omega \mapsto (0, \infty)$ as follows: If $\Omega = \mathbb{C}$, set $d := 1$. If $\Omega \neq \mathbb{C}$, define $2d(a)$ to be the distance from a to $\mathbb{C} \setminus \Omega$. According to 1.32 this is a continuous function. For each $a \in \Omega$

$$r \mapsto \operatorname{Ind}_{h\circ C(a,r)}(h(a))$$

is an integer-valued continuous function of r on the interval $(0, 2d(a))$, hence it is constant, say $I(a)$. (See 4.6(ii).) For similar reasons

$$a \mapsto \operatorname{Ind}_{h\circ C(a,d(a))}(h(a))$$

is a continuous integer-valued function on the connected set Ω. Hence it too is constant, say I. It follows that $I(a) = I$ for all $a \in \Omega$. It remains to note that this number is either -1 or 1 according to 4.43 and 4.48(ii).

(ii) (Preservation of Index under Homeomorphism.) *With the notation and hypotheses of* (i), *show that*

$$\mathrm{Ind}_{h\circ\gamma}(h(a)) = (-1)^N\,\mathrm{Ind}_\gamma(a)$$

for all Ω*-nullhomologous loops* γ *and all* $a \in \Omega\backslash\gamma$. (See also (iv) below.)

Hints: Let such γ and a be given. Pick $r > 0$ so that $\bar{D}(a,r) \subset \Omega \setminus \gamma$. Look at the special compact exhaustion of $\Omega \setminus \bar{D}(a, r/2)$ furnished by 1.31. It supplies compact $K \subset \Omega \setminus \bar{D}(a, r/2)$ such that

(2) $\gamma \cup C(a,r) \subset K$, and

(3) every hole in K contains a hole in $\Omega \setminus \bar{D}(a, r/2)$.

Apply 4.30 to the continuous function $f := h - h(a)$, which is zero-free in K. It supplies points $p_1, \ldots, p_M$ from holes in K, integers $n_1, \ldots, n_M$, and continuous $\varphi\colon K \mapsto \mathbb{C}$ such that

(4) $$h(z) - h(a) = e^{\varphi(z)} \prod_{j=1}^{M} (z - p_j)^{n_j} \qquad \forall z \in K.$$

In fact, the p_j can be arbitrarily chosen within their respective holes. We therefore take $p_M := a$ (with $n_M := 0$ if such a hole does not contribute to (4)) and the other p_j from the corresponding holes in Ω (cf. (3)). Then (4) and (2) yield

$$h\circ\gamma - h(a) = e^{\varphi\circ\gamma} \prod_{j=1}^{M} (\gamma - p_j)^{n_j}, \qquad \text{whence}$$

$$\mathrm{Ind}_{h\circ\gamma - h(a)}(0) = \sum_{j=1}^{M} n_j \mathrm{Ind}_{\gamma - p_j}(0)$$

$$\mathrm{Ind}_{h\circ\gamma}(h(a)) = \sum_{j=1}^{M} n_j \mathrm{Ind}_{\gamma}(p_j).$$

Since γ is Ω-nullhomologous and $p_j \in \mathbb{C} \setminus \Omega$ for $j < M$, this reads

(5) $$\mathrm{Ind}_{h\circ\gamma}(h(a)) = n_M \mathrm{Ind}_\gamma(a).$$

But the loop $\gamma_a := C(a,r)$ also lies in K and is Ω-nullhomologous (since $\bar{D}(a,r) \subset \Omega$), so the same argument that gave (5) also gives

(6) $$\mathrm{Ind}_{h\circ\gamma_a}(h(a)) = n_M \mathrm{Ind}_{\gamma_a}(a) = n_M.$$

According to (i) the left side of (6) equals $(-1)^N$. Therefore $n_M = (-1)^N$ and (5) proves what (ii) is claiming.

(iii) *Show that the conclusion in* (ii) *may fail for loops in* Ω *which are not* Ω*-nullhomologous.* (Cf. 11.49(iv).)

(iv) *Show that if* $h : \Omega \mapsto \mathbb{C}$ *is merely continuous and* Ω *merely open,* $h \circ \gamma$ *is an* $h(\Omega)$*-nullhomologous loop whenever* γ *is an* Ω*-nullhomologous loop. In other words, if* γ *is* Ω*-nullhomologous and* $w \in \mathbb{C} \setminus h \circ \gamma$ *satisfies* $\mathrm{Ind}_{h\circ\gamma}(w) \neq 0$, *then* $w \in h(\Omega)$.

Hints: In the hints to (ii) suppress all reference to a and replace $h(a)$ by a typical point in $\mathbb{C} \setminus h(\Omega)$.

For use later we record here a technical result.

Lemma 4.50 *Let* Ω *be a Jordan region,* $z \in \partial\Omega$.

(i) *For every* $\varepsilon > 0$ *there exists* $\delta > 0$ *such that any two points in* $\Omega \cap D(z, \delta)$ *lie on a polygon in* $\Omega \cap D(z, \varepsilon)$.

(ii) *Whenever* $z_n \in \Omega \to z$, *there is a curve* $\gamma : [0, 1] \mapsto \mathbb{C}$ *with* $\gamma(0, 1] \subset \Omega$ *and* $1 \geq t_n \downarrow 0$ *such that* $\gamma(t_n) = z_n$ *for all* $n \in \mathbb{N}$.

Proof: Let Ω be the inside of the Jordan-curve J. There is a homeomorphism φ of the unit circle onto J such that $\varphi(1) = z$. Let $\gamma(t) := \varphi(e^{it})$ $(t \in \mathbb{R})$. Choose $0 < t_0 < \pi$ sufficiently small that

(1) $$J_1 := \gamma[-t_0, t_0] \subset D(z, \varepsilon).$$

Now z is disjoint from the compact set $J_2 := J \setminus \gamma(-t_0, t_0) = \gamma[t_0, -t_0 + 2\pi]$, so for some $0 < \delta < \varepsilon$, $D(z, \delta)$ is also disjoint from this set. Consider any two points p, $q \in \Omega \cap D(z, \delta)$. Setting $C := \partial D(z, \varepsilon)$, it is obvious that

(2) $$p, q \text{ are not separated by } C \cup J_2,$$

since the latter set lies in the complement of $D(z, \delta)$. Of course

(3) $$p, q \text{ are not separated by } J \text{ either},$$

since they both belong to the component Ω of $\mathbb{C} \setminus J$. Moreover, since J_1 is disjoint from C (see (1)) and $J_1 \cup J_2 = J$, it follows that $C \cap J \subset J_2$ and so

(4) $$(C \cup J_2) \cap J = (C \cap J) \cup (J_2 \cap J) = J_2.$$

From (2), (3), (4) and Janiszewski's theorem 4.37(ii) we have that

(5) $$p, q \text{ are not separated by } (C \cup J_2) \cup J = C \cup J,$$

that is, p, q belong to the same component U of $\mathbb{C}\setminus(C\cup J) = (\mathbb{C}\setminus C)\cap(\mathbb{C}\setminus J)$. Since we know which of the two components of $\mathbb{C}\setminus C$ contains p and q (viz., $D(z,\varepsilon)$) and which of the two components of $\mathbb{C}\setminus J$ contains p and q (viz., Ω), it follows that $U \subset \Omega \cap D(z,\varepsilon)$. Since U is open there is, thanks to 1.28, a polygon in U joining p to q.

To prove (ii), use(i) to select $r_1 > r_2 > r_3 \cdots \to 0$ such that for every $j \in \mathbb{N}$

(6) any two points in $\Omega \cap D(z, r_{j+1})$ lie on a polygon in $\Omega \cap D(z, r_j)$.

We can assume all $z_n \in D(z, r_2)$. (Why?) Then select integers $1 = n_1 < n_2 < n_3 \cdots$ such that

$$z_n \in D(z, r_{j+1}) \qquad \forall n \geq n_j, j \in \mathbb{N}. \tag{7}$$

Using (6), for each j we construct, step by step, a polygon $P_j : [1/n_{j+1}, 1/n_j] \mapsto \Omega \cap D(z, r_j)$ which encounters all the z_n with $n_j \leq n \leq n_{j+1}$ in their natural order. That is, for an increasing sequence of $t_n \in [1/n_{j+1}, 1/n_j]$,

(8) beginning with $1/n_{j+1}$ and ending with $1/n_j$,

$$P_j(t_n) = z_n \quad \text{for every } n_j \leq n \leq n_{j+1}. \tag{9}$$

Clearly then $t_n \downarrow 0$. Thanks to (8) and (9) a (continuous) function $\gamma : (0,1] \mapsto \mathbb{C}$ is well defined by

$$\gamma(t) := P_j(t) \quad \text{for } 1/n_{j+1} \leq t \leq 1/n_j, j \in \mathbb{N}. \tag{10}$$

We have $\gamma(t_n) = z_n$ for all $n \in \mathbb{N}$, and γ maps into Ω, since every P_j does. Moreover, since the r_j are decreasing, for every k

$$\gamma(0, 1/n_k] \overset{(10)}{=} \bigcup_{j\geq k} P_j[1/n_{j+1}, 1/n_j] \subset \bigcup_{j \geq k} \Omega \cap D(z, r_j) = \Omega \cap D(z, r_k),$$

and this shows that $\gamma(t)$ converges to z as $t \downarrow 0$. Consequently, we have only to *define* $\gamma(0) := z$ in order to have the desired curve γ.

§ 6 Applications of the Foregoing Technology

The machinery developed so far leads to elegant proofs of some nontrivial topological properties of the plane and the sphere. This section is devoted to such a topological harvest, but nothing proved here is needed in the sequel, so it is certainly optional for the reader in a hurry.

Theorem 4.51 *If* $f\colon \mathbb{C} \mapsto \mathbb{C}$ *is continuous and for some non-zero integer* n *satisfies* $\lim_{|z|\to\infty} z^{-n} f(z) = c \in \mathbb{C} \setminus \{0\}$, *then* f *has a zero. If, in addition,* $n > 0$, *then* f *is surjective. In particular,* (cf. 4.26(iii)) *every non-constant polynomial has a zero.*

Proof: We may divide f by the hypothesized non-zero limit and so suppose from the start that it is 1. Choose then positive r sufficiently large that

$$\begin{aligned} \left|z^{-n} f(z) - 1\right| &< 1 \qquad \forall |z| \geq r \\ |f(z) - z^n| &< |z^n| \qquad \forall |z| \geq r. \end{aligned} \tag{1}$$

According to 4.21 the identity map I on $r\mathbb{T}$ has no continuous logarithm and neither does I^n. From (1) and 4.27 it follows that $f|_{r\mathbb{T}}$ has no continuous logarithm either. *A fortiori* then $f|_{r\overline{\mathbb{D}}}$ has no continuous logarithm. It follows from 4.1 that f must have a zero in $r\overline{\mathbb{D}}$. If $n > 0$, the functions $f - w$ ($w \in \mathbb{C}$) all fulfill the hypothesis that f does.

Finally, note that if p is a non-constant polynomial and $n \geq 1$ is its degree, then $z^{-n}p(z)$ approaches the leading coefficient of p as $|z| \to \infty$, so that p is an eligible f.

Exercise 4.52 (i) *For any polynomial* p *over* $\mathbb{C}$, $p(z) - p(z_0)$ *is divisible by* $z - z_0$, *that is,* $p(z) - p(z_0) = (z - z_0)q(z)$ *for some polynomial* q.

Hint: For any positive integer k, $z^k - z_0^k = (z - z_0)\sum_{j=0}^{k-1} z_0^{k-1-j} z^j$.

(ii) *Use* (i), *induction on* n, *and* 4.51 *to show that if* p *is any* nth *degree polynomial over* $\mathbb{C}$, *then there exist* $c, z_1, \ldots, z_n \in \mathbb{C}$ *such that* $p(z) \equiv c\prod_{j=1}^{n}(z - z_j)$.

(iii) *If* p *is a non-constant polynomial, then all the zeros of* p' *lie in* co $p^{-1}(0)$. *In particular, if all zeros of* p *are real, so are all zeros of* p'.

Hint: Let $z_1, \ldots, z_N$ be the distinct zeros of p and n_j the multiplicity of z_j. Then from (ii)

$$p(z) = c\prod_{j=1}^{N}(z - z_j)^{n_j} \qquad (\text{some } c \in \mathbb{C} \setminus \{0\}).$$

Then for $z \notin p^{-1}(0)$

$$\frac{p'(z)}{p(z)} = \sum_{j=1}^{N} \frac{n_j}{z - z_j} = \sum_{j=1}^{N} \frac{n_j}{|z - z_j|^2}(\bar{z} - \bar{z}_j).$$

If in addition $p'(z) = 0$, it follows that

$$\bar{z} \sum_{j=1}^{N} \frac{n_j}{|z - z_j|^2} = \sum_{j=1}^{N} \frac{n_j \bar{z}_j}{|z - z_j|^2}.$$

Take conjugates to see that $z = \sum_{j=1}^{N} \lambda_j z_j$, where

$$\lambda_j := \frac{n_j}{|z - z_j|^2} \left[\sum_{l=1}^{N} \frac{n_l}{|z - z_l|^2} \right]^{-1} \in (0, 1] \quad \text{and} \quad \sum_{j=1}^{N} \lambda_j = 1.$$

Part (iii) of the preceding exercise is known as the *Gauss-Lucas Theorem.* Its great power is ingeniously revealed by the next two results.

Exercise 4.53 (E. LANDAU [1906]) *Show that for any integer $n \geq 2$ and any complex number c the polynomial $P(z) := 1 + z + cz^n$ has a zero of modulus less-than-or-equal to 2.*

Hints: The result is clear if $c = 0$, so assume $c \neq 0$. Consider the polynomial

$$(*) \qquad Q(z) := z^n + z^{n-1} + c = z^n P(1/z)$$

The polynomial Q' has the zero $-(n-1)/n$ of modulus $(n-1)/n$. Since all zeros of Q' lie in the convex hull of those of Q [4.52(iii)], not all zeros of Q can have modulus *less* than $(n-1)/n$. Let w be a zero of Q with $|w| \geq (n-1)/n$. In particular, $w \neq 0$ so by $(*)$ $z := 1/w$ is a zero of P of modulus $|z| \leq n/(n-1) = 1 + 1/(n-1) \leq 1 + 1/(2-1) = 2$.

Theorem 4.54 (FEJÉR [1908], MONTEL [1923]) *For any integers $2 \leq n_1 < n_2 < \cdots < n_{k-1}$ and any complex numbers $c_1, c_2, \ldots, c_{k-1}$ the polynomial*

$$P(z) := 1 + z + c_1 z^{n_1} + \cdots + c_{k-1} z^{n_{k-1}}$$

has a zero of modulus less-than-or-equal-to k.

Proof: We go by induction on k and so suppose $c_{k-1} \neq 0$. The case $k = 2$ is 4.53, so we suppose $k > 2$. Thus n_{k-2} is defined. Form the polynomial

$$(1) \qquad Q(z) := z^{n_{k-1}} P(1/z)$$

$$(2) \qquad = z^{n_{k-1}} + z^{n_{k-1}-1} + c_1 z^{n_{k-1}-n_1} + \cdots + c_{k-2} z^{n_{k-1}-n_{k-2}} + c_{k-1}.$$

Then

$$
\begin{aligned}
Q'(z) &= n_{k-1}z^{n_{k-1}-1} + (n_{k-1}-1)z^{n_{k-1}-2} + (n_{k-1}-n_1)c_1 z^{n_{k-1}-n_1-1} + \\
&\quad \cdots + (n_{k-1}-n_{k-2})c_{k-2}z^{n_{k-1}-n_{k-2}-1} \\
&= n_{k-1}z^{n_{k-1}-n_{k-2}-1}\left[z^{n_{k-2}} + (1-1/n_{k-1})z^{n_{k-2}-1} + \right. \\
&\quad \left.(1-n_1/n_{k-1})c_1 z^{n_{k-2}-n_1} + \cdots + (1-n_{k-2}/n_{k-1})c_{k-2}\right] \\
&= n_{k-1}z^{n_{k-1}-n_{k-2}-1}(1-1/n_{k-1})^{n_{k-2}}R\big(n_{k-1}z/(n_{k-1}-1)\big), \qquad (3)
\end{aligned}
$$

where

$$
\begin{aligned}
R(z) := z^{n_{k-2}} + z^{n_{k-2}-1} + (1-1/n_{k-1})^{-n_1}(1-n_1/n_{k-1})c_1 z^{n_{k-2}-n_1} + \\
\cdots + (1-1/n_{k-1})^{-n_{k-2}}(1-n_{k-2}/n_{k-1})c_{k-2}. \qquad (4)
\end{aligned}
$$

We form the polynomial

$$
\begin{aligned}
S(z) &:= z^{n_{k-2}}R(1/z) \qquad (5) \\
&= 1 + z + (1-1/n_{k-1})^{-n_1}(1-n_1/n_{k-1})c_1 z^{n_1} + \\
&\quad \cdots + (1-1/n_{k-1})^{-n_{k-2}}(1-n_{k-2}/n_{k-1})c_{k-2}z^{n_{k-2}}.
\end{aligned}
$$

By the induction hypothesis, S has a zero w with $|w| \le k-1$. By (5), $S(0) = 1$, so $w \neq 0$. Therefore $z = 1/w$ is a zero of R, thanks to (5). Then from (3) the number $\dfrac{n_{k-1}-1}{n_{k-1}}z$ is a zero of Q'. Its modulus is

$$
\begin{aligned}
\frac{n_{k-1}-1}{n_{k-1}}|z| &= \frac{n_{k-1}-1}{n_{k-1}} \cdot \frac{1}{|w|} \ge \frac{n_{k-1}-1}{n_{k-1}} \cdot \frac{1}{k-1} \\
&= (1-1/n_{k-1})/(k-1) \\
&\ge (1-1/k)/(k-1), \quad \text{since } 2 \le n_1 < n_2 < \cdots < n_{k-1} \text{ implies } n_{k-1} \ge k \\
&= 1/k.
\end{aligned}
$$

By 4.52(iii) then not all zeros of Q can have modulus *less* than $1/k$. I.e., Q has a zero x of modulus $\ge 1/k$. In particular, $x \neq 0$. Then by (1), $1/x$ is a zero of P of modulus $\le k$.

Exercise 4.55 (i) *For each positive integer n there are multinomials $s_{n,0}, \ldots, s_{n,n-1}$ with integer coefficients such that the coefficient of z^j in the product $(z-\alpha_1)\cdots(z-\alpha_n)$ is $s_{n,j}(\alpha_1, \ldots, \alpha_n)$ for each $j = 0, 1, \ldots, n-1$ and all choices of complex numbers $\alpha_1, \ldots, \alpha_n$. In particular, the coefficients of a monic polynomial are continuous functions of its zeros.*

Hint: Induction on n.

(ii) *Let n be a positive integer and $\alpha_1, \ldots, \alpha_n \in \mathbb{C}$. Suppose that the polynomial $p(z) := (\alpha_1 - z) \cdots (\alpha_n - z)$ has integer coefficients. Show that then the polynomial $P(z) := \left(\alpha_1^2 - z\right) \cdots \left(\alpha_n^2 - z\right)$ also has integer coefficients.*

Hint: (Ostrowski) The polynomial $p(z)p(-z)$ has integer coefficients and equals $(\alpha_1 - z)(\alpha_1 + z) \cdots (\alpha_n - z)(\alpha_n + z) = \left(\alpha_1^2 - z^2\right) \cdots \left(\alpha_n^2 - z^2\right) = P\left(z^2\right)$.

(iii) *Let P be a monic polynomial with integer coefficients all of whose zeros lie on the unit circle. Show that each of these zeros is in fact a root of unity.*

Hints: Fix positive integer n and let $\mathcal{P}(n)$ denote the set of nth degree monic polynomials with integer coefficients whose zeros lie on the unit circle. It follows from part (i) that the coefficients of the polynomials in $\mathcal{P}(n)$ are bounded. As these coefficients are integers, it follows that $\mathcal{P}(n)$ is finite. Consequently the set $Z(n)$ of all the zeros of all the polynomials in $\mathcal{P}(n)$ is also finite. But if $\alpha \in Z(n)$, then so does α^2 by part (ii). Inductively, $\left\{\alpha^{2^k} : k \in \mathbb{N}\right\} \subset Z(n)$. Then for some $k_1 < k_2$ we must have $\alpha^{2^{k_2}} = \alpha^{2^{k_1}}$, $\alpha^{(2^{k_2} - 2^{k_1})} = 1$.

Exercise 4.56 (Cf. 11.42.) Let $\mathcal{R}$ denote the set of all polynomials with rational coefficients. Let P, Q be (non-zero) elements of $\mathcal{R}$ which have no common non-constant factor in $\mathcal{R}$. Show that there exist A, $B \in \mathcal{R}$ such that $AP + BQ = \mathbf{1}$.

Hint: All is trivial if P or Q is constant. For example, if Q is constant take $A := Q$ and $B := -P + 1/Q$. So suppose $\deg P > 0$ and $\deg Q > 0$. Use induction on the degree of PQ. If, say, $\deg Q \leq \deg P$, let B_0 be the quotient of the leading term of P by that of Q. Thus $B_0 \in \mathcal{R}$ and

$$P - B_0 Q =: P_1$$

has degree less than that of P. Therefore $\deg P_1 Q < \deg PQ$. Moreover P_1 and Q have no common factor in $\mathcal{R}$ (any such would be one for P and Q too). Therefore the induction hypothesis provides A, $B_1 \in \mathcal{R}$ such that $AP_1 + B_1 Q = \mathbf{1}$, that is,

$$A(P - B_0 Q) + B_1 Q = \mathbf{1}$$
$$AP + (B_1 - AB_0)Q = \mathbf{1}.$$

We now leave polynomials to investigate fixed-points and vector fields.

Theorem 4.57 *Suppose* $f, g\colon \overline{\mathbb{D}} \mapsto \mathbb{C}$ *continuous,* $g|\mathbb{T}$ *zero-free,* $\mathrm{Ind}_{g\circ C(0,1)}(0) \neq 0$, *and* $|f| \leq |g|$ *on* $\mathbb{T}$. *Then* $f(z) = g(z)$ *for some* $z \in \overline{\mathbb{D}}$. *In particular, if* $f(\mathbb{T}) \subset \overline{\mathbb{D}}$, *then* f *fixes some point in* $\overline{\mathbb{D}}$.

Proof: Suppose that $h := (f-g)/2$ has no zero. Then (4.1) h has a continuous logarithm in $\overline{\mathbb{D}}$. The inequality $|f| \leq |g|$ on $\mathbb{T}$ implies that

$$|\mathbf{1} + h/g| = \tfrac{1}{2}|f/g + \mathbf{1}| \leq \mathbf{1}, \text{ and so}$$
$$|\mathbf{1} + h/g| < \mathbf{1} + |h/g| \text{ on } \mathbb{T}.$$

Consequently (4.27), h/g has a continuous logarithm on $\mathbb{T}$. But then g/h and $g = h \cdot g/h$ also have continuous logarithms on $\mathbb{T}$. From the latter follows the contradiction $\mathrm{Ind}_{g\circ C(0,1)}(0) = 0$. The identity function on $\overline{\mathbb{D}}$ is an eligible g if $f(\mathbb{T}) \subset \overline{\mathbb{D}}$.

Exercise 4.58 (i) (BOHL [1904]) *Let* $F\colon \overline{\mathbb{D}} \mapsto \mathbb{T}$ *be continuous. Show that for each* $\lambda \in \mathbb{T}$ *there exists some* $z = z_\lambda$ *and* $w = w_\lambda \in \mathbb{T}$ *such that* $F(z) = \lambda z$ *and* $F(w) = F(\lambda w)$. *In particular, there is no continuous retraction of* $\overline{\mathbb{D}}$ *onto* $\mathbb{T}$ *fixing each point of* $\mathbb{T}$.

Hints: 4.57 supplies a $z \in \overline{\mathbb{D}}$ which $f := \lambda^{-1}F$ fixes. Of course, $z \in \lambda^{-1}F(\overline{\mathbb{D}}) \subset \mathbb{T}$. To get w, apply 4.23(viii) to the restriction to $\mathbb{T}$ of a continuous logarithm of F.

(ii) (BROUWER [1912b]) *Let* K *be a non-empty, convex and compact subset of* $\mathbb{C}$, $F\colon K \mapsto K$ *a continuous function. Show that* F *has a fixed-point.*

Hints: We can assume that $K \subset \overline{\mathbb{D}}$. According to 1.19(iii) there is a continuous retraction N of $\overline{\mathbb{D}}$ onto K. Now apply 4.57 to $f := F \circ N$. Since $f(\overline{\mathbb{D}}) = F(K) \subset K$, any point of $\overline{\mathbb{D}}$ fixed by f lies in K and is (therefore) fixed by N.

(iii) $f : \mathbb{C} \mapsto \mathbb{C}$ *is continuous and* $\overline{\lim}_{|z|\to\infty} |f(z)/z| < 1$. *Show that* f *has a fixed-point.*

Hints: In view of (ii), it suffices to prove that $f(\overline{D}(0,r)) \subset \overline{D}(0,r)$ for some real $r \geq 0$. If no such inclusion holds, then for every $n \in \mathbb{N}$ there exists z_n satisfying $|z_n| \leq n < |f(z_n)|$. Hence $|f(z_n)| \to \infty$, so the continuity of f entails that $\{z_n\}$ cannot be bounded. But then $|f(z_n)|/|z_n| > 1$ for all (sufficiently large) n contradicts the hypothesized inequality.

Exercise 4.59 *Let* $f\colon \overline{\mathbb{D}} \mapsto \mathbb{C}$ *be continuous,* C *a component of* $\mathbb{C} \setminus f(\mathbb{T})$, $w_0 \in C$.

(i) *Show that if* $\mathrm{Ind}_{f\circ\partial\mathbb{D}}(w_0) \neq 0$, *then* $C \subset f(\mathbb{D})$ (cf. 4.49(iv)); *infer that* w_0 *is an interior point of* $f(\mathbb{D})$. *In the special case where* $f(\overline{\mathbb{D}}) \subset \overline{\mathbb{D}}$ *and* f *fixes each point of* $\mathbb{T}$, *it follows that* $f(\overline{\mathbb{D}}) = \overline{\mathbb{D}}$.

Hints: Suppose, contrariwise, that for some $w \in C$, $f - w$ is zero-free in $\mathbb{D}$. By definition of w, $f - w$ is zero-free on $\mathbb{T}$. Hence $(r,t) \mapsto f(re^{2\pi it}) - w$ is a loophomotopy in $\mathbb{C} \setminus \{0\}$ of $(f - w) \circ \partial\mathbb{D}$ with a constant, making $\mathrm{Ind}_{(f-w)\circ\partial\mathbb{D}}(0) = 0$, by 4.12. On the other hand, by 4.3, $\mathrm{Ind}_{f\circ\partial\mathbb{D}}(w) = \mathrm{Ind}_{f\circ\partial\mathbb{D}}(w_0) \neq 0$. Finally, note that by 1.30(i)$'$, C is an open subset of $\mathbb{C}$.

(ii) *Show that the index hypothesis in* (i) *is equivalent to the function* $f|\mathbb{T} - w_0$ *having no continuous logarithm.*

Hint: Apply 4.22(i) to the function $f|\mathbb{T} - w_0$ in the role of its f.

(iii) (T. Radó[1936]) *Suppose that* f *is zero-free in* $\overline{\mathbb{D}} \setminus \{0\}$ *and that* $k := \mathrm{Ind}_{f\circ\partial\mathbb{D}}(0) \neq 0$. *Show that* $f(0) = 0$ *and that* f *has a continuous* $|k|^{\mathrm{th}}$*-root in* $\overline{\mathbb{D}}$. *Show moreover that for* $r := \inf|f(\partial\mathbb{D})|$ *every point in* $D(0,r) \setminus \{0\}$ *is the image of at least* $|k|$ *points in* $\mathbb{D}$.

Hints: According to (i), $0 \in f(\mathbb{D})$, and therefore necessarily $f(0) = 0$. If we replace $f(z)$ by $f(\bar{z})$, then k is replaced by $-k$ (why?). Since it suffices to prove the conclusion for the function $z \mapsto f(\bar{z})$, we may assume $k > 0$.

If we apply 4.30 to $K_n := \overline{\mathbb{D}} \setminus D(0,1/n)$, whose sole hole is $D(0,1/n)$, we find that the function $z^{-k}f(z)$ has a continuous logarithm in K_n. Harmonizing these logarithms in the usual way leads to a continuous φ on $\overline{\mathbb{D}} \setminus \{0\}$ such that

$$f(z) = z^k e^{\varphi(z)} \qquad \forall z \in \overline{\mathbb{D}} \setminus \{0\}.$$

The function $g(z) := ze^{\varphi(z)/k}$ satisfies $f = g^k$ in $\overline{\mathbb{D}} \setminus \{0\}$, but since $0 = |f(0)|^{1/k} = \lim_{|z|\downarrow 0}|g(z)|$, the definition $g(0) := 0$ extends it continuously to $\overline{\mathbb{D}}$, with $f = g^k$ now valid in the whole $\overline{\mathbb{D}}$. We have (4.14)

$$k = \mathrm{Ind}_{f\circ\partial\mathbb{D}}(0) = k\mathrm{Ind}_{g\circ\partial\mathbb{D}}(0),$$

so the latter index is non-zero. Let C denote the component of $\mathbb{C} \setminus g(\partial\mathbb{D})$ that contains 0. This is an open set and according to (i), $C \subset g(\mathbb{D})$. By definition of r, $D(0,r) \subset \mathbb{C} \setminus f(\partial\mathbb{D}) = \mathbb{C} \setminus g^k(\partial\mathbb{D})$. Therefore $D(0,r^{1/k}) \subset \mathbb{C} \setminus g(\partial\mathbb{D})$, whence $D(0,r^{1/k}) \subset C$.

Finally, consider any $w \in D(0,r) \setminus \{0\}$. The kth-roots of w, denote them $w_1, \ldots, w_k$, lie in $D(0,r^{1/k}) \subset C \subset g(\mathbb{D})$ and are distinct. Hence there exist distinct $z_j \in \mathbb{D}$ such that $g(z_j) = w_j$, whence

$$w = w_j^k = (g(z_j))^k = f(z_j), \quad \text{for } j = 1, \ldots, k.$$

(iv) *$f : \overline{\mathbb{D}} \mapsto \mathbb{C}$ is continuous and has only finitely many zeros, none on $\mathbb{T}$. Show that if these zeros are $p_1, \ldots, p_N$, $\varepsilon > 0$ is sufficiently small and for each $j \in \{1, 2, \ldots, N\}$ the loop γ_j in $D(p_j, \varepsilon) \setminus \{p_j\}$ satisfies $\operatorname{Ind}_{\gamma_j}(p_j) = 1$, then*

$$(0) \qquad \operatorname{Ind}_{f \circ C(0,1)}(0) = \sum_{j=1}^{N} \operatorname{Ind}_{f \circ \gamma_j}(0).$$

Hints: If $\varepsilon > 0$ is small enough, the disks $\bar{D}(p_j, \varepsilon)$ are disjoint and wholly contained in $\mathbb{D}$. Given the γ_j, choose $0 < \delta < \varepsilon$ so that γ_j is disjoint from $D(p_j, \delta)$ for each j. Then 4.30 supplies $n_j \in \mathbb{Z}$ and continuous function φ such that

$$(1) \qquad f(z) = e^{\varphi(z)} \prod_{j=1}^{N} (z - p_j)^{n_j} \qquad \forall z \in \overline{\mathbb{D}} \setminus \bigcup_{j=1}^{N} D(p_j, \delta).$$

Since $z - p_j$ has a continuous logarithm in $\bar{D}(p_k, \varepsilon)$ for each $k \neq j$ [by 4.1], it follows easily from (1) that

$$(2) \qquad \operatorname{Ind}_{f \circ \gamma_j}(0) = n_j \operatorname{Ind}_{\gamma_j}(p_j) = n_j, \qquad j = 1, 2, \ldots, N.$$

Also from (1) and 4.14 follows

$$(3) \qquad \operatorname{Ind}_{f \circ C(0,1)}(0) = \sum_{j=1}^{N} n_j \operatorname{Ind}_{C(0,1)}(p_j) = \sum_{j=1}^{N} n_j.$$

Remarks: If, in addition, f is holomorphic in $\mathbb{D}$, then in (1) the exponents n_j can all be taken positive, with φ holomorphic in $\mathbb{D}$. Equation (0) in the form (3) is then the important *argument principle* for holomorphic functions (5.86). Note too that if Γ is a simple loop and (iv) is formulated with $\mathcal{J}(\Gamma)$ in the role of $\mathbb{D}$, then the conclusion is still valid.

Exercise 4.60 (i) *Let $f : \overline{\mathbb{D}} \mapsto \mathbb{C}$ be continuous and one-to-one. Show that $f(0)$ is an interior point of $f(\mathbb{D})$.*

Hints: Define

$$\begin{aligned} f_0(z) &:= f(z) - f(0) && z \in \mathbb{T} \\ f_1(z) &:= f(z) - f(-z) && z \in \mathbb{T} \\ h(z,t) &:= f(z) - f(-tz), && z \in \mathbb{T}, t \in [0,1]. \end{aligned}$$

Since f is one-to-one, the function h never vanishes, i.e., maps into $\mathbb{C} \setminus \{0\}$. Evidently f_1 satisfies $f_1(-z) = -f_1(z)$ for all $z \in \mathbb{T}$, and so by 4.21 f_1 does not have a continuous logarithm. Therefore by 4.24 f_0 does not have a continuous logarithm, and so 4.59(ii) may be applied.

(ii) *Suppose $F\colon U$ open $\subset \mathbb{C} \mapsto \mathbb{C}$ is continuous, and is real-differentiable at the point $z_0 \in U$. That is, for some $\mathbb{R}$-linear map $L\colon \mathbb{C} \mapsto \mathbb{C}$*

$$(*) \qquad \lim_{z \to z_0} \frac{F(z) - F(z_0) - L(z - z_0)}{z - z_0} = 0.$$

(Cf. the discussion at the beginning of §4, chap. II).) *Suppose also that L is invertible. Show that $F(z_0)$ is an interior point of $F(U)$.*

Hints: Without loss of generality, $z_0 = 0 = F(z_0)$. Now L is a homeomorphism of $\mathbb{C}$ onto $\mathbb{C}$ which maps 0 to 0. The number $B := \sup\left|L^{-1}(\mathbb{T})\right|$ is therefore finite and positive, and so by $(*)$ there exists $\rho > 0$ such that

$$\left|\frac{F(z) - L(z)}{z}\right| \leq \frac{1}{2B} \qquad \forall z \in \bar{D}(0, \rho) \subset U, \ z \neq 0.$$

This, the definition of B, and the $\mathbb{R}$-linearity of L yield

$$\left|(L^{-1} \circ F)(z) - z\right| = \left|L^{-1}(F(z) - L(z))\right| \leq B|F(z) - L(z)| \leq \tfrac{1}{2}|z| \ \forall z \in \bar{D}(0, \rho).$$

In particular, if I denotes the identity map on $\mathbb{C}$,

$$\left|L^{-1} \circ F - I\right| \leq \tfrac{1}{2}|I| \quad \text{on } C(0, \rho)$$

and $L^{-1} \circ F$ is zero-free on $C(0, \rho)$. Since (4.21) I has no continuous logarithm on $C(0, \rho)$, it follows from 4.27 that $L^{-1} \circ F$ has no continuous logarithm on $C(0, \rho)$. Now apply 4.59(ii), with $w_0 = 0$ and $L^{-1}(F(\rho z))$ in the role of $f(z)$, to learn that 0 is an interior point of $L^{-1}(F(D(0, \rho)))$. Hence $0 = L(0)$ is an interior point of $F(D(0, \rho)) \subset F(U)$, since L is a homeomorphism of $\mathbb{C}$.

Exercise 4.61 (Open-Map Theorem) *If U is an open subset of $\mathbb{C}$ and $f\colon U \mapsto \mathbb{C}$ is continuous and one-to-one, then $f(U)$ is an open subset of $\mathbb{C}$ and f^{-1} is continuous.*

Hints: Given $a \in U$, choose $r > 0$ so that $\bar{D}(a, r) \subset U$ and apply 4.60(i). (A different (?) proof is contained in 4.48(iii) and another in A. Browder [2000].)

Exercise 4.62 *Prove the* Borsuk Antipoden Satz *in dimension two: If $\mathbb{S} := \{(x_1, x_2, x_3) \in \mathbb{R}^3 : x_1^2 + x_2^2 + x_3^2 = 1\}$ is the unit sphere in $\mathbb{R}^3$ and $f\colon \mathbb{S} \mapsto \mathbb{C}$ is continuous, then $f(p) = f(-p)$ for some $p \in \mathbb{S}$. What does this result say* *about the existence at any instant of antipodes on the earth which have the same temperature and the same barometric pressure?*

Hint: Consider the function $h\colon \overline{\mathbb{D}} \mapsto \mathbb{C}$ defined by

$$h(x,y) := f\big(x, y, \sqrt{1-x^2-y^2}\,\big) - f\big(-x,-y,-\sqrt{1-x^2-y^2}\,\big).$$

If no such point p exists, then h maps into $\mathbb{C}\setminus\{0\}$ and so has a continuous logarithm by 4.1. Since $h(-x,-y) = -h(x,y)$ for all $(x,y) \in \mathbb{T}$, a contradiction of 4.21 ensues.

Remark: It follows from 4.62 that no continuous complex-valued function on $\mathbb{S}$ can be one-to-one and so $\mathbb{R}^2$ contains no subset homeomorphic to $\mathbb{S}$. In particular, $\mathbb{R}^2$ is not homeomorphic to $\mathbb{R}^n$ for any $n > 2$.

Our next goal is to prove the "hairy billiard ball" theorem (also known as the "*Igelsatz*") of H. Poincaré. That is, we shall verify a sequence of four steps culminating in a proof (by *reductio ad absurdum*) that there exists no zero-free continuous tangent field on the 2-sphere $\mathbb{S}$ in $\mathbb{R}^3$.

One interpretation is that whenever a hairy sphere is combed flat, a cowlick must appear; another is that at any instant some point on the earth is wind-free.

We start with the relevant definitions. Let

$$\mathbb{S}_+ := \{(x_1,x_2,x_3) \in \mathbb{S} : x_3 \geq 0\}, \quad \mathbb{S}_- := \{(x_1,x_2,x_3) \in \mathbb{S} : x_3 \leq 0\}.$$

The *stereographic projection* of $\overline{\mathbb{D}}$ on $\mathbb{S}_+$ is the map

$$P_+(z) := \left(\frac{2\operatorname{Re} z}{1+|z|^2}, \frac{2\operatorname{Im} z}{1+|z|^2}, \frac{1-|z|^2}{1+|z|^2}\right), \qquad z \in \overline{\mathbb{D}}$$

and the stereographic projection of $\overline{\mathbb{D}}$ on $\mathbb{S}_-$ is the map

$$P_-(z) := \left(\frac{2\operatorname{Re} z}{|z|^2+1}, \frac{2\operatorname{Im} z}{|z|^2+1}, \frac{|z|^2-1}{|z|^2+1}\right), \qquad z \in \overline{\mathbb{D}}.$$

A *continuous tangent vector field* on $\mathbb{S}$ is a continuous map $V\colon \mathbb{S} \mapsto \mathbb{R}^3$ such that $V(x)$ is orthogonal to x for each $x \in \mathbb{S}$. It is convenient to denote by $\bullet$ the inner product in both $\mathbb{R}^2$ and $\mathbb{R}^3$ and to think of $\mathbb{R}^3$ as $\mathbb{C} \times \mathbb{R}$. Let then V_r denote the projection of V on $\mathbb{R}$, V_c the projection of V on $\mathbb{C}$ and define $W_+,\, W_- : \overline{\mathbb{D}} \mapsto \mathbb{C}$ by

$$W_+(z) := V_r(P_+(z))z - V_c(P_+(z))$$
$$W_-(z) := V_r(P_-(z))z + V_c(P_-(z)).$$

In what follows we assume

(∗) $\exists$ a zero-free, continuous tangent vector field V on $\mathbb{S}$

and we aim to deduce from this a contradiction.

Exercise 4.63 (i) *Prove that P_+ maps $\overline{\mathbb{D}}$ continuously onto $\mathbb{S}_+$ and P_- maps $\overline{\mathbb{D}}$ continuously onto $\mathbb{S}_-$. Actually, both maps are one-to-one as well; e.g.,*

$$P_+^{-1}(x_1, x_2, x_3) = \frac{x_1}{1+x_3} + \frac{ix_2}{1+x_3}.$$

(ii) (ii) *Show that W_+, W_- are each continuous and zero-free.*

Hint: Show that $W_+(z) = 0$, together with $P_+(z) \bullet V(P_+(z)) = 0$, forces the contradiction $V(P_+(z)) = 0$.

(iii) *Show that $W_+(u) = u^2 \overline{W}_-(u)$ for all $u \in \mathbb{T}$.*

Hints: For brevity we shall write the vector $(u, 0)$ as just u. From $u \bullet V(u) = 0$ we get

$$\begin{aligned} &0 = u \bullet V_c(u) = \operatorname{Re}\left[\bar{u} V_c(u)\right] \quad \text{and so} \\ &\bar{u} V_c(u) + \overline{\bar{u} V_c(u)} = 0, \\ &\bar{u} V_c(u) + u \overline{V}_c(u) = 0. \end{aligned}$$

Multiply by $u = 1/\bar{u}$ and get $-V_c(u) = u^2 \overline{V}_c(u)$. Add $V_r(u)u$ to both sides and have

$$V_r(u)u - V_c(u) = V_r(u)u + u^2 \overline{V}_c(u) = u^2 \left[V_r(u)\bar{u} + \overline{V}_c(u)\right].$$

Finally observe that for $u \in \mathbb{T}$ we have $P_+(u) = u = P_-(u)$, and so the left-hand side of the preceding chain is $W_+(u)$ and its right-hand side is $u^2 \overline{W}_-(u)$.

(iv) *Deduce from* (ii), (iii) *and* 4.31 *a contradiction.*

Hints: The function $W_+/\overline{W}_-$ is defined and continuous throughout $\overline{\mathbb{D}}$. According to (iii) it provides a continuous zero-free extension to $\overline{\mathbb{D}}$ of the function $f(u) := u^2$ on $\mathbb{T}$. But 4.31 affirms that no such extension exists. More elementarily, 4.1 and 4.21 also lead to contradiction in this situation.

Exercise 4.64 *Let $f \colon \mathbb{S} \mapsto \mathbb{S}$ be continuous.*

(i) *Show that either f fixes some point or f maps some point to its antipode.*

Hints: Suppose contrariwise that for every $p \in \mathbb{S}$, $f(p) \neq \pm p$. Then $f(p)$ is not parallel to p, so $V(p) := p \times f(p) \neq 0$. Here $\times$ denotes vector cross-product in $\mathbb{R}^3$. Since $p \times f(p)$ is orthogonal to p, we have in V a tangent vector field of the kind forbidden by Poincaré's Theorem.

(ii) *Show that if f is not surjective, then there exists $p \in \mathbb{S}$ with $f(p) = f(-p)$.*

Hint: If $(0,0,-1) \notin f(\mathbb{S})$, apply 4.62 to $P \circ f$, where P is the stereographic projection of $\mathbb{S} \setminus (0,0,-1)$ onto $\mathbb{C}$ $\big($cf. 4.63(i)$\big)$.

(iii) *Show that if f is injective, then it is also surjective.*

(iv) (Ljusternik-Schnirelmann and Borsuk [1933a]) *$\mathbb{S}$ is covered by the closed subsets S_1, S_2, S_3. Show that at least one of them must contain a pair of antipodes. Infer that whenever a closed ball of unit radius in $\mathbb{R}^3$ is a union of three closed sets, one of the three has diameter* 2.

Hints: The case $S_1 \cap S_2 \cap S_3$ non-void is easy. Define $d_j(x) := \inf\{\|x - y\| : y \in S_j\}$ for $x \in \mathbb{S}$, if $S_j \neq \varnothing$ (cf. 1.32) and set $d_j := 2$ if $S_j = \varnothing$. If $S_1 \cap S_2 \cap S_3 = \varnothing$, the function $D := \sqrt{d_1^2 + d_2^2 + d_3^2}$ is positive on $\mathbb{S}$ and we may form $f := (d_1/D, d_2/D, d_3/D) \colon \mathbb{S} \mapsto \mathbb{S}$. Evidently, f is not surjective, all its coordinates being non-negative.

(v) *$\mathbb{S}$ is covered by the closed subsets S_1, S_2, S_3, S_4 and none of them contains an antipodal pair. Show that then every three of them have non-empty intersection.*

Hint: If, say, $S_1 \cap S_2 \cap S_3 = \varnothing$, use the d_j of (iv). Get a p_0 with $f(p_0) = f(-p_0)$. Argue that $p_0 \in S_4$ and then that $-p_0 \in S_4$, contrary to hypothesis.

(vi) *Suppose $A, B, C \in \mathbb{S}$ are all different and form $N := (A-C) \times (B-C)$. Show that $(C - N) \times N \neq 0$, and $(C - N) \times N \perp C$.*

Hints: A diagram makes this geometrically clear, but here is a proof outline:

$0 = (C - N) \times N = C \times N \Rightarrow C \parallel N \Rightarrow C \perp (A - C) \Rightarrow 0 = C{\bullet}A - C{\bullet}C = C{\bullet}A - 1 = -\frac{1}{2}(A-C){\bullet}(A-C) \longrightarrow\!\!\!\!\!\!\times\;$. On the other hand, $\big[(C-N) \times N\big]{\bullet}C = [C \times N - N \times N]{\bullet}C = [C \times N]{\bullet}C = 0$.

(vii) *Show that if $f, g \colon \mathbb{S} \mapsto \mathbb{S}$ are continuous and injective, then either f, g or $f \circ g$ has a fixed-point.*

Hints: Write $h := f \circ g$. If the conclusion fails, then $f(p)$, $h(p)$, p will all be different, for each $p \in \mathbb{S}$. Form $N(p) := [f(p) - p] \times [h(p) - p]$ and $V(p) := [p - N(p)] \times N(p)$, for each $p \in \mathbb{S}$. According to (vi), $V(p)$ is never 0 and is always orthogonal to p. Since V is evidently a continuous function, Poincaré's theorem is contradicted.

(viii) *Show that for every continuous injection* $f\colon \mathbb{S} \mapsto \mathbb{S}$ *the composite* $f \circ f$ *has a fixed-point.*

Hint: If f itself has no fixed-point, (vii) may be applied with $g := f$.

(ix) F *maps the octant* $C := \{(x_1, x_2, x_3) \in \mathbb{R}^3 : x_1 \geq 0, x_2 \geq 0, x_3 \geq 0\}$ *continuously into itself. Show that there is a unit vector* $u \in C$ *and real* $\lambda \geq 0$ *such that* $F(u) = \lambda u$.

Hints: If F vanishes somewhere on $C \cap \mathbb{S}$, we may take $\lambda := 0$. Otherwise, $f(u) := \|F(u)\|^{-1} F(u)$ maps $C \cap \mathbb{S}$ into itself. Since this set is homeomorphic to the compact, convex set $C \cap \overline{\mathbb{D}}$, via the map $z \mapsto \left(\operatorname{Re} z, \operatorname{Im} z, \sqrt{1 - |z|^2}\right)$, f must fix some point, by Brouwer's theorem, 4.58(ii).

(x) (O. Perron and F. G. Frobenius) *If every entry in the* 3-by-3 *matrix* M *is a non-negative real number, then* M *has a non-negative eigenvalue with an eigenvector whose components are each non-negative.*

Hint: M defines a (linear) function F to which (ix) is applicable.

There is a classic application of the *Antipodensatz* which every reader should have the pleasure of experiencing. Unfortunately, one rather intuitively obvious step in it [(ii) below] is messy to prove rigorously without the machinery of Lebesgue integration; see §33 of CHINN and STEENROD [1966]. But here is the main idea:

For $v \in \mathbb{S}$ and $t \in \mathbb{R}$ consider the *open half-space* $H(v, t) := \{x \in \mathbb{R}^3 : x \bullet v < t\}$ and any (non-void) open, connected subset A of the unit ball in $\mathbb{R}^3$.

(i) *Prove that* $\varphi(t) := \operatorname{vol}(A \cap H(v, t))$ *is a continuous function of* t, *that* $\varphi(-1) = 0$ *and* $\varphi(1) = \operatorname{vol}(A)$. *Infer that there exists a* unique $t_A(v) \in [-1, 1]$ *such that* $\varphi(t_A(v)) = \frac{1}{2}\operatorname{vol}(A)$.

(ii) *Try to show that* $v \mapsto t_A(v)$ *is a continuous, odd function.*

Deduce the

Sandwich-slicing Theorem *If* A, B, C *are bounded, open, connected subsets of* $\mathbb{R}^3$, *then there is a plane which simultaneously bisects each of them.*

Hints: After scaling we can assume A, B, C all lie in the unit ball. Consider the functions t_A, t_B, t_C defined on $\mathbb{S}$ in (i) and the function $f := (t_A - t_B, t_A - t_C)\colon \mathbb{S} \mapsto \mathbb{R}^2$. By (ii) it is a continuous, odd function and by 4.62 $f(v) = f(-v)$ for some $v \in \mathbb{S}$. Hence $f(v) = (0, 0)$.

Remark: If the corresponding *Antipodensatz* in $\mathbb{R}^4$ is used instead, then only the continuity of $v \mapsto \operatorname{vol}(A \cap H(v, 0))$ is needed, and this is easier to prove than (ii). Moreover, A, B, C need only be measurable sets of finite volume. Cf. DUGUNDJI and GRANAS [1982], p. 79 (but also ARENS [1980]!). A more elementary, but perhaps more ingenious, application of 4.62 to a fair-division problem (the one discussed preceding 4.2) is given by TOTIK [1999]; see also the beautiful generalization of it in **MR** 54#13425.

We close this section with one more beautiful geometric result. If the dimension of the function in 4.62 is lowered, then the number of points collapsed together by it increases. More than two points cannot, of course, be mutually antipodal but we have instead the following

Theorem 4.65 (KAKUTANI [1942]) *For each continuous map $f\colon \mathbb{S} \mapsto \mathbb{R}$ there exist mutually orthogonal P_1, P_2, $P_3 \in \mathbb{S}$ such that $f(P_1) = f(P_2) = f(P_3)$.*

Proof: Let $k := (0,0,1)$ and regard $\mathbb{R}^3$ as $\mathbb{C} \oplus \mathbb{R}k$. Set $\omega := e^{2\pi i/3}$ and define

$$P_1 := \left(-\frac{1}{\sqrt{6}}, \frac{1}{\sqrt{2}}, -\frac{1}{\sqrt{3}}\right) = \sqrt{\frac{2}{3}}\left(-\frac{1}{2} + \frac{i\sqrt{3}}{2}\right) - \frac{1}{\sqrt{3}}k = \sqrt{\frac{2}{3}}\omega - \frac{1}{\sqrt{3}}k$$

$$P_2 := \left(-\frac{1}{\sqrt{6}}, -\frac{1}{\sqrt{2}}, -\frac{1}{\sqrt{3}}\right) = \sqrt{\frac{2}{3}}\left(-\frac{1}{2} - \frac{i\sqrt{3}}{2}\right) - \frac{1}{\sqrt{3}}k = \sqrt{\frac{2}{3}}\omega^2 - \frac{1}{\sqrt{3}}k$$

$$P_3 := \left(\frac{2}{\sqrt{6}}, 0, -\frac{1}{\sqrt{3}}\right) = \sqrt{\frac{2}{3}}(1 + i\cdot 0) - \frac{1}{\sqrt{3}}k = \sqrt{\frac{2}{3}}\omega^3 - \frac{1}{\sqrt{3}}k.$$

(The sine, cosine values above follow from $\sin(\pi/6) = 1/2$, proved in 3.30(ii) hints.)

Evidently this is an orthonormal set. Let $\mathcal{O}$ denote the set of all rotations of $\mathbb{S}$ and assume that no Kakutani triad exists. This means that

$$f(R(P_1)) = f(R(P_2)) = f(R(P_3))$$

does not occur for any $R \in \mathcal{O}$. Now one checks in a microsecond that

$$a, b, c \in \mathbb{R} \quad \& \quad a\omega + b\omega^2 + c\omega^3 = 0 \iff a = b = c.$$

Therefore the absence of Kakutani triads means that the function

$$(1) \qquad F(R) := f(R(P_1))\omega + f(R(P_2))\omega^2 + f(R(P_3))\omega^3, \quad R \in \mathcal{O}$$

is a zero-free map of $\mathcal{O}$ into $\mathbb{C}$. For each $u \in \mathbb{T}$ let $R_u \in \mathcal{O}$ be defined as rotation by u in the complex plane, that is,

$$R_u(x_1, x_2, x_3) := (x_1 + ix_2)u + x_3 k \qquad \text{for } (x_1, x_2, x_3) \in \mathbb{S}.$$

Evidently then

$$(2) \qquad (u, P) \mapsto R_u(P) \quad \text{is a continuous map of } \mathbb{T} \times \mathbb{S} \text{ into } \mathbb{S}.$$

Now note the following obvious identities for $j = 1, 2, 3$ (with P_4 defined to be P_1):

$$R_{\omega u}(P_j) = R_{\omega u}\left(\sqrt{\frac{2}{3}}\omega^j - \frac{1}{\sqrt{3}}k\right) = \sqrt{\frac{2}{3}}u\omega^{j+1} - \frac{1}{\sqrt{3}}k = R_u(P_{j+1}),$$

whence

$$\begin{aligned} F(R_{\omega u}) &= f(R_u(P_2))\omega + f(R_u(P_3))\omega^2 + f(R_u(P_1))\omega^3 \\ &= \bar{\omega}\big[f(R_u(P_2))\omega^2 + f(R_u(P_3))\omega^3 + f(R_u(P_1))\omega\big] \\ &= \bar{\omega}F(R_u) \qquad \forall u \in \mathbb{T}. \end{aligned} \tag{3}$$

From (1) and (2) it follows that the prescription

$$\varphi(u) := F(R_{u^2}), \quad u \in \mathbb{T}$$

yields a continuous map of $\mathbb{T}$ into $\mathbb{C} \setminus \{0\}$, and (3) says that it satisfies

$$\varphi(\omega u) = \bar{\omega}^2\varphi(u) \qquad \forall u \in \mathbb{T}. \tag{4}$$

It follows then from 4.21 that φ does not have a continuous logarithm.

Next, for each $t \in [0, 1]$ consider the rotation $T_t \in \mathcal{O}$ which is defined by the matrix

$$\begin{pmatrix} 1 & 0 & 0 \\ 0 & \cos \pi t & -\sin \pi t \\ 0 & \sin \pi t & \cos \pi t \end{pmatrix}. \tag{5}$$

Writing u in the form $e^{i\theta}$ and consequently the matrix of R_u as

$$\begin{pmatrix} \cos\theta & -\sin\theta & 0 \\ \sin\theta & \cos\theta & 0 \\ 0 & 0 & 1 \end{pmatrix},$$

enables us to compute at once that

$$R_u \circ T_1 \circ R_u = T_1 \qquad \forall u \in \mathbb{T}. \tag{6}$$

We then define a map $H\colon \mathbb{T} \times [0, 1] \mapsto \mathbb{C} \setminus \{0\}$ by

$$H(u, t) := F(R_u \circ T_t \circ R_u). \tag{7}$$

Since $(t, P) \mapsto T_t(P)$ is clearly a continuous map of $[0, 1] \times \mathbb{S}$ into $\mathbb{S}$, it follows from (2) that $(t, u) \mapsto T_t(R_u(P_j))$ is a continuous map of $[0, 1] \times \mathbb{S}$ into $\mathbb{S}$ for

$j = 1, 2, 3$. But then again by (2), $(v, t, u) \mapsto R_v(T_t(R_u(P_j)))$ is a continuous map of $\mathbb{S} \times [0, 1] \times \mathbb{S}$ into $\mathbb{S}$, for $j = 1, 2, 3$. Finally then continuity of f shows that $(t, u) \mapsto f(R_u(T_t(R_u(P_j))))$ is continuous, for $j = 1, 2, 3$. Hence (7) and (1) show that H is continuous. It follows from (5) and (6) that

$$\begin{cases} H(u, 0) = F(R_u \circ R_u) = F(R_{u^2}) = \varphi(u) \\ H(u, 1) = F(T_1) \neq 0. \end{cases} \tag{8}$$

Then 4.24 leads to the contradiction that φ has a continuous logarithm. Thus the absence of Kakutani triads is untenable.

§ 7 Continuous and Holomorphic Logarithms in Open Sets

The following variations on 4.30 are the first phase of a program to be completed in Chapter X of extending to open sets results of previous sections on existence of continuous logarithms and connectivity of the set.

Theorem 4.66 *Let U be an open subset of $\mathbb{C}$ which has only finitely many holes $C_1, \ldots, C_M$. Let $p_j \in C_j$. Then for any continuous, zero-free f on U there exist integers $k_1, \ldots, k_M$ and continuous function φ such that*

$$f(z) = e^{\varphi(z)} \prod_{j=1}^{M} (z - p_j)^{k_j} \qquad \forall z \in U.$$

If U has no holes, then $f = e^{\varphi}$ for such a φ. (See also 4.69(ii).)

Proof: Let $\{K_n\}$ be a compact exhaustion of U as provided by 1.31. Let K be a compact subset of U as furnished by 1.42. Since the sets $\overset{\circ}{K}_n$ increase and cover U, they eventually contain the compact set K. Ignoring early terms as necessary, we can suppose that $K \subset K_n$ for all n. Now according to 1.42, $C_1, \ldots, C_M$ lie in distinct bounded components $C_1(K), \ldots, C_M(K)$ of $\mathbb{C} \setminus K$. We have $C_j \subset \mathbb{C} \setminus U \subset \mathbb{C} \setminus K_n \subset \mathbb{C} \setminus K$. Therefore C_j lies in a component $C_j(K_n)$ of $\mathbb{C} \setminus K_n$ and in turn $C_j(K_n)$ lies in a (unique) component of $\mathbb{C} \setminus K$. The latter is perforce $C_j(K)$. Thus $C_1(K_n), \ldots, C_M(K_n)$ are distinct. On the other hand, by 1.31 the number of holes in K_n does not exceed M. It follows that $C_1(K_n), \ldots, C_M(K_n)$ account for them all and so $p_1, \ldots, p_M$ represent points from each of the distinct holes in K_n. Consequently, we may appeal to 4.30 to come up with integers $k_1(n), \ldots, k_M(n)$ and continuous functions φ_n on K_n such that

$$f(z) = e^{\varphi_n(z)} \prod_{j=1}^{M} (z - p_j)^{k_j(n)} \qquad \forall z \in K_n. \tag{1}$$

The uniqueness result 4.33(i) ensures that we have $k_j(1) = k_j(2) = k_j(3) = \cdots$ for each $j = 1, 2, \ldots, M$. Call the common value k_j. Then (1) says that the function

$$F(z) := f(z)\prod_{j=1}^{M}(z - p_j)^{-k_j}, \quad z \in U$$

has a continuous logarithm on each K_n. Since any loop γ in U lies in some K_n, it follows that $F \circ \gamma$ has index 0 at 0. By 4.18 therefore F has a continuous logarithm on U. (For holomorphic f a different proof is offered in 10.26(iii).)

It was conjectured by Kuratowski and later proved by HEILBRONN [1958] that the finite-connectivity hypothesis in 4.66 is unnecessary. All the machinery for a proof of this as KURATOWSKI [1968] later presented it, is at hand here, save for one significant tool from Chapter VIII. Reassuring readers that no logical circularity is involved, I think it appropriate to give the proof at this point in the book, rather than later:

Theorem 4.66* *For any open $U \subset \mathbb{C}$ and continuous $f : U \mapsto \mathbb{C} \setminus \{0\}$, there exist continuous $\psi : U \mapsto \mathbb{C}$ and holomorphic $h : U \mapsto \mathbb{C} \setminus \{0\}$ such that $f = he^{\psi}$.*

Proof: Let $\{K_n\}$ be a compact exhaustion of U as provided by 1.31. Apply 4.30 to each K_n, using 1.31(3) to take the points p_j to be, not merely in $\mathbb{C} \setminus K_n$, but in $\mathbb{C} \setminus U$. We get a rational function $r_n(z) = \prod_{j=1}^{N_n}(z - p_j)^{n_j}$ with poles and zeros only in $\mathbb{C} \setminus U$, and continuous function $u_n : K_n \mapsto \mathbb{C}$ such that

$$(1.n) \qquad f(z) = r_n(z)e^{u_n(z)} \quad \forall z \in K_n,\ n \in \mathbb{N}.$$

For convenience we may (1.44) and will assume u_n is defined in all of U. Together $(1.n)$ and $(1.n+1)$ entail

$$(2.n) \qquad r_n e^{u_n} = r_{n+1}e^{u_{n+1}} \quad \text{in } K_n \quad \forall n \in \mathbb{N}.$$

The rational function r_{n+1}/r_n has poles and zeros only in $\mathbb{C} \setminus U$, hence is holomorphic in $U \supset K_n$. A routine local argument (see 4.67(i)) shows that from $r_{n+1}/r_n \overset{(2.n)}{=} e^{u_n - u_{n+1}}$ in $\overset{\circ}{K}_n$ there follows that $u_n - u_{n+1}$ is holomorphic in $\overset{\circ}{K}_n$. We make an inductive construction of rational functions s_n with poles only in $\mathbb{C} \setminus U$ such that the (continuous in U) functions

$$(3.n) \qquad v_n := s_{n+1} - s_n + u_n - u_{n+1}, \quad n \in \mathbb{N}$$

satisfy

$$(4.n) \qquad |v_n| < 2^{-n} \quad \text{in } K_{n-1} \quad \forall n \in \mathbb{N} \quad (\text{taking } K_0 := \varnothing).$$

Start with $s_1 := \mathbf{0}$ and assume the rational functions $s_1, \dots, s_n$ have been constructed. Then $u_{n+1} - u_n + s_n$ is holomorphic in $\overset{\circ}{K}_n$. On the compact subset K_{n-1} of $\overset{\circ}{K}_n$ this holomorphic function can be approximated closer than 2^{-n} by a rational function s_{n+1} with poles only in A, for any set A which meets every bounded component of $\mathbb{C} \setminus K_{n-1}$. This is Rational Runge I, (8.11). For every $m \in \mathbb{N}$ each bounded component C of $\mathbb{C} \setminus K_m$ contains a point $p(C, m) \in \mathbb{C} \setminus U$, by 1.31(3). Hence the set of such points for all m and C is an eligible A and lies in $\mathbb{C} \setminus U$. So (8.11) supplies a rational function s_{n+1} with poles only in $\mathbb{C} \setminus U$ such that

$$|s_{n+1} - (u_{n+1} - u_n + s_n)| < 2^{-n} \quad \text{on } K_{n-1},$$

giving (4.n) and completing the induction. Thus

$$g_1 := r_1 = r_1 e^{s_1} \text{ and } g_{n+1} := \frac{r_{n+1} e^{s_{n+1}}}{r_n e^{s_n}}, \qquad n \in \mathbb{N} \tag{5}$$

are defined and holomorphic in U. From (1.n) and (3.n) we see that

$$g_{n+1} = e^{v_n} \quad \text{in } K_n, \quad n \in \mathbb{N}. \tag{6.n}$$

For $N > 1$ the inequalities (4.n) give

$$\sum_{n=N}^{\infty} |v_n| < \sum_{n=N}^{\infty} 2^{-n} = 2^{-N+1} \quad \text{in } K_{N-1}. \tag{7.N}$$

Thus the series $\sum_{n=1}^{\infty} v_n$ is uniformly absolutely convergent on K_{N-1} for each $N \in \mathbb{N}$. For all $M > N > 1$ the equations (6.n) give

$$g_1 \cdots g_N g_{N+1} \cdots g_{M+1} = g_1 \cdots g_N e^{v_N} \cdots e^{v_M} = g_1 \cdots g_N e^{v_N + \cdots + v_M} \quad \text{in } K_N.$$

So as $M \to \infty$, this product converges uniformly in K_N, for every $N \in \mathbb{N}$. That is,

$$h := \lim_{n \to \infty} g_1 \cdots g_n \text{ exists locally uniformly in } U. \tag{8}$$

Since each factor here is holomorphic in U, h is holomorphic in U. Next note that for $M > N > 1$

$$\begin{aligned} \big|[u_M - s_M] - [u_N - s_N]\big| &\le \sum_{n=N}^{M-1} \big|[u_{n+1} - s_{n+1}] - [u_n - s_n]\big| \\ &\overset{(3.n)}{=} \sum_{n=N}^{M-1} |v_n| < 2^{-N+1} \quad \text{in } K_{N-1} \end{aligned} \tag{9}$$

by (7.N). This shows that the sequence of continuous functions $\{u_n - s_n\}$ on U is locally uniformly Cauchy in U. Thus

(10) $$\psi := \lim_{n\to\infty} [u_n - s_n] \text{ exists locally uniformly in } U.$$

By (5) and (1.$n+1$)

$$g_1 \cdots g_{n+1} = r_{n+1} e^{s_{n+1}} = f e^{s_{n+1} - u_{n+1}} \quad \text{in } K_n \subset K_{n+1} \quad \forall n \geq N \geq 1.$$

Letting $n \to \infty$ here, and recalling (8) and (10), gives

$$h = f e^{-\psi} \quad \text{in } \bigcup_{N=1}^{\infty} K_N = U.$$ Q.E.D.

Exercise 4.67 (i) *Let U be an open subset of $\mathbb{C}$, $f\colon U \mapsto \mathbb{C} \setminus \{0\}$ holomorphic. Show that any continuous logarithm (resp., square-root) of f is holomorphic.* (Cf. 5.98.)

Hints: Let $z \in U$ be given. Choose $r > 0$ so small that $D(z,r) \subset U$ and $f(D(z,r)) \subset D(f(z), |f(z)|)$. Let L be a holomorphic logarithm in the latter disk (as furnished by 3.19). Then we can form $L \circ f$ in $D(z,r)$ and have a holomorphic function φ such that $e^\varphi = f$ in $D(z,r)$. If $g\colon U \mapsto \mathbb{C}$ is continuous and satisfies $e^g = f$ $\big($resp., $g^2 = f\big)$ throughout U, then $e^{g-\varphi} = \mathbf{1}$ $\big($resp., $g^2 - \big(e^{\varphi/2}\big)^2 = \mathbf{0}\big)$ in $D(z,r)$. It follows from 3.14(iii) (resp., 1.7) that $(g - \varphi)/2\pi i$ is a continuous integer-valued function on the connected set $D(z,r)$ $\big($resp., that either $g = e^{\varphi/2}$ or $g = -e^{\varphi/2}$ throughout $D(z,r)\big)$. Thus $(g - \varphi)/2\pi i$ is constant, say N, in $D(z,r)$ and $g = \varphi + 2\pi i N$ $\big($resp., either $g = e^{\varphi/2}$ or $g = -e^{\varphi/2}\big)$ exhibits g as a holomorphic function in $D(z,r)$.

A direct computational proof is also possible:

With g as before and $z_0 \in U$, write $w_0 := g(z_0)$, and define $F(w) := e^w$ (resp., w^2). Then

$$F(w) - F(w_0) = (w - w_0)h(w),$$

where h is continuous in $\mathbb{C}$ and $h(w_0) := F'(w_0) = e^{w_0}$ (resp., $2w_0$) $\neq 0$. Then $h(w)$ is non-zero for w at or near w_0 and

$$f(z) - f(z_0) = F(g(z)) - F(g(z_0)) = (g(z) - g(z_0))h(g(z)),$$

with $h(g(z))$ non-zero for z at or near z_0. Therefore as $z_0 \neq z \to z_0$, continuity of g and h entail that

$$\frac{g(z) - g(z_0)}{z - z_0} = \frac{1}{h(g(z))} \, \frac{f(z) - f(z_0)}{z - z_0} \to \frac{1}{h(g(z_0))} f'(z_0),$$

which gives $g'(z_0) = f'(z_0)/f(z_0)$ (resp., $f'(z_0)/2g(z_0)$).

(ii) *Adapt the argument in* (i) *to prove the holomorphy of any continuous n^{th}-root of a zero-free holomorphic function.* (Cf. 5.65.)

Hint: In the notation of the preceding hints, $ge^{-\varphi/n}$ is a continuous mapping of the connected set $D(z, r)$ into the finite set of nth-roots of 1.

Corollary 4.68 *If f in* 4.66 *is holomorphic, then so is φ.*

Proof: The function φ is a continuous logarithm for the holomorphic function $F(z) := f(z)\prod_{j=1}^{M}(z - p_j)^{-k_j}$ in U.

Exercise 4.69 (i) *Let U be an open subset of $\mathbb{C}$. Show that a necessary and sufficient condition that U have at most n holes is this: there exist zero-free holomorphic functions $\varphi_1, \ldots, \varphi_n$ such that every zero-free $f \in H(U)$ determines integers $k_1, \ldots, k_n$ and $\varphi \in H(U)$ for which $f = e^{\varphi}\prod_{j=1}^{n}\varphi_j^{k_j}$. [In fact, sufficiency requires only the continuity of the φ_j and the φ.]*

Hints: For sufficiency (with φ and the φ_j merely continuous), suppose $p_k \in C_k$ are distinct holes in U, and consider the functions $f_k(z) := z - p_k$, $z \in U$, $k = 1, \ldots, N$. By 1.42 there is a compact $K \subset U$ such that the p_k lie in different holes in K. Form $\mathcal{G}(K)$, $\mathcal{E}(K)$ as in 4.0. The Hints to 4.22(iv) indicated that the quotient group $\mathcal{G}(K)/\mathcal{E}(K)$ is torsion-free. The cosets $f_k\mathcal{E}(K)$ are independent in this group, by 4.33(i). Yet by hypothesis each f_k has a factorization which puts the coset $f_k\mathcal{E}(U)$ in the group generated by the $\varphi_j\mathcal{E}(U)$, hence (by restriction to K), the coset $f_k\mathcal{E}(K)$ is in the group generated by the $\varphi_j\mathcal{E}(K)$, $j = 1, \ldots, n$. Therefore by 4.32, $N \leq n$.

(ii) *Formulate and prove a uniqueness assertion to complement* 4.66.

(iii) (Kuratowski) [1945]) *Let $\mathcal{G} = \mathcal{G}(U)$ denote all continuous zero-free functions, $\mathcal{E} = \mathcal{E}(U)$ all exponentials of continuous functions on U. Show that the group $\mathcal{G}(U)/\mathcal{E}(U)$ is isomorphic to $\mathbb{Z}^n$ ($n \in \mathbb{N}$) if and only if U has exactly n holes.* (Cf. 10.30 and *Remark* (i) following it.)

Hints: If $\mathcal{G}/\mathcal{E}$ is isomorphic to $\mathbb{Z}^n$, then the images $\varphi_1\mathcal{E}, \ldots, \varphi_n\mathcal{E}$ of the natural generators of $\mathbb{Z}^n$, generate $\mathcal{G}/\mathcal{E}$, so by the sufficiency half of (i), U has at most n holes. Suppose that U has exactly $M \in \mathbb{N}$ holes, points p_k are chosen from each, and $f_k(z) := z - p_k$ ($z \in U$, $k = 1, \ldots, M$). Then by 4.66, $\mathcal{G}/\mathcal{E}$ is generated by the cosets $f_k\mathcal{E}$. Using maneuvers such as employed in (i) above, we can deduce from 4.33(i) that these cosets are independent. Hence the (torsion-free abelian) group $\mathcal{G}/\mathcal{E}$ is isomorphic to $\mathbb{Z}^M$. Notice that for $M \neq n$, the groups $\mathbb{Z}^M$ and $\mathbb{Z}^n$ are not isomorphic (4.32).

§ 8 Simple Connectivity for Open Sets

Definition 4.70 Let U be an open subset of $\mathbb{C}$. Say that U has
property (CL) if every zero-free continuous function in U has a continuous logarithm;
property (HL) if every zero-free holomorphic function in U has a holomorphic logarithm;
property (CS) if every zero-free continuous function in U has a continuous square root;
property (HS) [resp., (HS_1)] if every zero-free holomorphic [and one-to-one] function in U has a holomorphic square root.

These definitions and previous results enable us to state succinctly some important topological facts about $\mathbb{C}$:

Theorem 4.71 *For an open subset U of $\mathbb{C}$ the following eight conditions are equivalent:*

(i) *U is simply-connected*

(ii) *U has property (CL)*

(iii) *U has property (CS)*

(iv) *U has property (HS)*

(v) *U has property (HL)*

(vi) *U has property (HS_1)*

(vii) *U is homologically-connected*

(viii) *U contains the inside of each Jordan-curve in it.*

Proof: (i) $\Rightarrow$ (ii) from 4.66.

(ii) $\Rightarrow$ (iii) trivial.

(iii) $\Rightarrow$ (iv) 4.67.

(iv) $\Rightarrow$ (v). If $f\colon U \mapsto \mathbb{C} \setminus \{0\}$ is holomorphic, repeated application of the hypothesis produces (zero-free) $f_n \in H(U)$ such that $f_n^{2^n} = f$. It follows from 4.19(i) that f has a continuous logarithm in U, which by 4.67 must in fact be holomorphic.

(v) $\Rightarrow$ (vi) trivial.

(vi) $\Rightarrow$ (vii). For each $p \in \mathbb{C} \setminus U$ the function $f(z) := z - p$, being one-to-one, holomorphic and zero-free on U, has a holomorphic logarithm. The argument is the same as that in the proof (iv) $\Rightarrow$ (v). It follows that $\mathrm{Ind}_\gamma(p) = \mathrm{Ind}_{f\circ\gamma}(0) = 0$ for all loops γ in U. Now recall definition 4.8(iii).

(vii) $\Rightarrow$ (viii). If J is a Jordan-curve in U, let φ be a homeomorphism of the unit circle onto J and let $\gamma(t) := \varphi(e^{it})$. For any $z \in \mathcal{I}(J)$ we have $\mathrm{Ind}_\gamma(z) \neq 0$ by 4.43. Consequently by (vii), $z \notin \mathbb{C} \setminus U$.

(viii) $\Rightarrow$ (i). This is an obvious consequence of the following fact, which deserves to be highlighted:

$$(*) \quad \begin{cases} \textit{For each hole } C \textit{ in } U \textit{ there is a simple polygonal loop } \gamma \textit{ in } U \\ \textit{such that the (constant) value of } \mathrm{Ind}_\gamma \textit{ in } C \textit{ is non-zero.} \end{cases}$$

To prove $(*)$, fix $p \in C$ and suppose contrariwise that every simple polygonal loop γ in U satisfies $\mathrm{Ind}_{\gamma - p}\ (0) = \mathrm{Ind}_\gamma\ (p) = 0$. Then 4.38(ii) says that the function $f(z) := z - p$ has a continuous logarithm L in U. From 1.42 we get compact $K \subset U$ such that p lies in a hole in K, and then the existence of L contradicts 4.31. In particular, $(*)$ tells us that if U has a hole C, then there is a Jordan-curve J in U such that $\mathcal{I}(J) \supset C$, so $\mathcal{I}(J)$ meets $\mathbb{C} \setminus U$ and U lacks property (viii).

Remarks: In Chapter X these equivalences will be augmented by several more, the most dramatic of which is that each component of U is homeomorphic to a disk. The proof that (viii) $\Rightarrow$ (i) uses the non-trivial result 4.38. Alternatively, this implication is an immediate consequence of 8.21. It is also possible to close this chain with the Jordan Curve Theorem (and therewith the complicated 4.38) excised, by replacing 4.38 with 4.18 in the (viii) $\Rightarrow$ (i) argument to prove instead that (vii) $\Rightarrow$ (i).

Corollary 4.72 *An open subset U of $\mathbb{C}$ is simply-connected if and only if each of its components is.*

Proof: The components are open and disjoint, so a function on U is continuous if and only if it is continuous on each component. Therefore the corollary follows from the equivalence of (i) and (ii) in 4.71, the last sentence of the preceding remarks being relevant. There is an elementary way to see that every component U_0 of a simply-connected open set U is itself simply-connected. Each such component U_n is relatively closed in U, but not in $\mathbb{C}$, so $\overline{U}_n$ meets some (unbounded) component C_n of $\mathbb{C} \setminus U$, and consequently (1.3) $U_n \cup C_n$ is connected (and unbounded). Therefore, each component of U different from U_0 lies in an unbounded connected subset of $\mathbb{C} \setminus U_0$. The rest of $\mathbb{C} \setminus U_0$ consists of (unbounded) components of $\mathbb{C} \setminus U$.

Exercise 4.73 *Let U and V be open, simply-connected subsets of $\mathbb{C}$.*

(i) *Show that $U \cap V$ is simply-connected.*

Hint: Use 4.71(vii).

(ii) *Show that if* $U \cap V$ *is connected, then* $U \cup V$ *is simply-connected. Supply a counterexample to the conclusion when this intersection is not connected.* (Cf. 3.40.)

Hints: If $f\colon U \cup V \mapsto \mathbb{C} \setminus \{0\}$ is continuous, 4.71(ii) provides continuous logarithms φ_U, φ_V for f on U, V respectively. We have $e^{\varphi_U - \varphi_V} = 1$ on the connected set $U \cap V$, hence the integer-valued function $(\varphi_U - \varphi_V)/2\pi i$ is constant there, say N. Then the function φ defined by

$$\varphi := \begin{cases} \varphi_U - 2\pi i N & \text{on } U \\ \varphi_V & \text{on } V \end{cases}$$

is a (well-defined) continuous logarithm for f on $U \cup V$.

(iii) *Let* U_j $(j = 1, 2, 3)$ *be simply-connected regions such that* $U_1 \cap U_2$ *and* $U_2 \cap U_3$ *are each connected while* $U_1 \cap U_3 \subset U_2$*. Show that then* $U_1 \cup U_2 \cup U_3$ *is simply-connected.*

Hints: The sets $U_1 \cup U_2$ and $U_2 \cup U_3$ are each simply-connected by (ii). Their intersection is the connected set U_2, due to $U_1 \cap U_3 \subset U_2$. Now use (ii) again.

Exercise 4.74 *If* U, V *are open subsets of* $\mathbb{C}$ *which are homeomorphic, then one is simply-connected if and only if the other is. In fact, if* V *has exactly* n *(finite) holes, then so does* U.

Hints: If $p_1, \ldots, p_n$ are points from each of the holes in V, $f_j(z) := z - p_j$ $(j = 1, 2, \ldots, n;\ z \in V)$ and $h\colon U \mapsto V$ a homeomorphism, then consider $\varphi_j := f_j \circ h$. Using these functions, it follows from 4.66 and 4.69(i) that U has at most n holes. If the actual number is $m < n$, make the same argument all over with the roles of U, V reversed, to get a contradiction. A very quick alternative proof follows from 4.69(iii). Since property (*CS*) clearly passes from U to any homeomorph of U, 4.71 furnishes a third proof that simple-connectivity of U is a homeomorphic invariant.

Exercise 4.75 (i) *Use* 4.49(iv) *to give a direct proof that* (vii)$\Rightarrow$(ii) *in* 4.71*, that is, every continuous, zero-free function on an open, homologically-connected subset of* $\mathbb{C}$ *has a continuous logarithm.*

(ii) *Use* 4.49(iv) *and* 4.18 *to prove* (cf. 4.74) *that every homeomorphic image of an open, homologically-connected subset of* $\mathbb{C}$ *is itself homologically-connected.*

Hint: If the h in 4.49(iv) is a homeomorphism, then every loop Γ in $h(\Omega)$ is $h \circ \gamma$ for some loop γ in Ω (namely, $\gamma := h^{-1} \circ \Gamma$).

Notes to Chapter IV

The earliest idea of $\text{Ind}_\gamma(0)$ involved counting the sign changes in $\operatorname{Im}\gamma(t)/\operatorname{Re}\gamma(t)$ as t advances through the parameter interval. The ideas in the present approach to the index find their appropriate culmination in the theory of covering spaces in topology. There is an important higher-dimensional analog of index, called *degree* (due to Brouwer but anticipated by Kronecker). See, e.g., HEINZ [1959], chapters 4,5 of EISENACK and FENSKE [1978], LLOYD [1978] or ROTHE [1986]. For history see SIEGBERG [1981].

The use of the index to prove the Jordan Curve Theorem seems to have originated with AMES [1905]. Its use in the basic Cauchy theory was advocated by Emil Artin in his 1944 Notre Dame lectures (reproduced in ARTIN [1965]) and is now very popular. This point of view got its fullest articulation in BEARDON [1979].

An analysis of how the homotopy concept entered mathematics via complex function theory is given by VANDEN EYNDE [1992]. The concepts nullhomologous and nullhomotopic both seem to capture the idea that the loop does not surround a hole in the region. That these ideas are distinct took some time to emerge, which is not surprising in view of the complexity of the example to this effect in 4.13, which seems to be due to Camille Jordan but is known as *Pochhammer's curve.* (And there is good reason to believe that this is the simplest example possible.)

For elementary accounts of aspects of plane topology similar in spirit and method to this one see PÁL [1922/23], TUCKER [1945], EILENBERG [1936], [1948], FORT [1952], [1967], J. A. BAKER [1991], the last chapter of KURATOWSKI [1972], chapters VI and VII of ČECH [1969], R. F. BROWN [1974], SANDERSON [1980] and (especially) ČERNÝ [1992], A.BROWDER [2000], chapter 2 of SHEIL-SMALL [2002] and the first four chapters of FULTON [1995]. The definitive treatment of the topology of $\mathbb{R}^n$ is probably KURATOWSKI [1968]. His account (especially chapter X) clearly influenced the presentation here. A very readable, less encyclopedic account, with excellent historical notes, which also introduces the reader to algebraic methods, is AGOSTON [1976]. In chapter 7 of his book, among other things relevant to the material here, the reader will find a discussion of the degree of a map mentioned above. Another elegant introduction to algebraic topology is the little book of WALL [1972].

The Greek constituents of "monodromy" show what an apt description of 4.18 this word is. There is also an important holomorphic monodromy theorem, which relates simple-connectivity and analytic continuation. See 10.3(iv).

The important theorems 4.29, 4.30 are from EILENBERG [1936] and ČECH [1969], a generalization to zero-free $\mathbb{R}^n$-valued functions in BORSUK [1950]. Eilenberg (p. 90) gives an inductive proof of the independence result 4.33(i).

There is an interesting proof of 4.30 in A. Browder [2000] which uses 8.11 (proved via Green's Theorem) and 8.29(v). See also Kuratowski [1945] and [1958]. These results (indeed most material of this chapter) are valid, appropriately phrased, in $\mathbb{R}^n$ as well as in $\mathbb{C} = \mathbb{R}^2$. The important result 4.33(iii), that the number of components of the complement of a compactum in $\mathbb{R}^2$ is a positional invariant, goes back to Brouwer [1912c]. It has many extensions. Eilenberg [1941] showed that if A, B are homeomorphic subsets of the sphere S^n in $\mathbb{R}^{n+1}$ and $S^n \setminus A$ has finitely many components, then this number equals the number of components of $S^n \setminus B$. (If both the numbers are infinite cardinals, however, they need not coincide—he adduces an example.) Borsuk [1950] supplies an elementary proof (no algebraic topology) of this when A is compact (see also p. 47 of Lloyd [1978]) and Kuratowski [1950] (see also [1957], [1959]) modifies Borsuk's proof to cover all A. Fort [1961] (see also p. 180 of Černý [1992]) shows that even in case the number of components of $\mathbb{R}^n \setminus A$ is infinite, this cardinal equals that of the set of components of $\mathbb{R}^n \setminus B$ whenever A, B are bounded, connected open subsets of $\mathbb{R}^n$ which are homeomorphic to each other. (Cf. 4.74.) Finally I mention that if A, B are open, connected subsets of S^2 whose complements in S^2 have the same finite number N of components, then A and B are homeomorphic. (See Čech [1969], p. 264.)

The proof of 4.38 is adapted from T. Radó [1948] and Čech [1969]. For more on 4.37, see §62, II of Kuratowski [1968], and pp. 219 ff. of Čech [1969]. The version in $\mathbb{C}$ was discovered independently by Mullikin [1922].

The union of 4.39 and 4.42 is called the Jordan Curve Theorem even though Camille Jordan's original proof (1887) was incomplete (on which point see Hales [2007b]); he (and earlier Carl Neumann) at least recognized that something here needed proof in spite of how obvious it all seems. (Coincidentally, about this time the Dedekind dogma was promulgated: "Was beweisbar ist, soll in der Wissenschaft nicht ohne Beweis geglaubt werden.") The naive 19th-century intuitions about curves were further jolted in 1903 when W. F. Osgood produced a Jordan-curve having positive area. (See, e.g., Lance and Thomas [1991], and Stromberg and Tseng [1994].) For the history of the many proofs of the Jordan Curve Theorem which have appeared in the literature, see §13 of Borel and Rosenthal [1924], J. R. Kline [1942] and Guggenheimer [1977]. The first valid proof that covered all Jordan-curves was given by Oswald Veblen in 1905 (although some, like Hans Hahn, had reservations). By now we have not only unimpeachably rigorous proofs, but even a fully formalized computer proof (involving over 44,000 steps); see Hales [2007a]. The proofs of 4.39–4.43 in the text are essentially from Čech [1969]. Other civilized proofs are to be found in Kuratowski [1926], Sanderson [1980] and Armstrong [1983]. Černý [1992] derives the Jordan Curve Theorem directly from the results of Eilenberg and Janiszewski (4.37). Straszewicz

[1923] proved (and EILENBERG [1936] re-proved) the following generalization: if C_1, C_2 are compact, connected subsets of $\mathbb{C}$, neither separates $\mathbb{C}$, and $C_1 \cap C_2$ has exactly two components, then $C_1 \cup C_2$ separates $\mathbb{C}$ into exactly two components. There is a complete converse of the Jordan Curve Theorem due to SCHÖNFLIES [1904] (and subsequently elaborated by many others) which affirms that if C is a compact, connected subset of $\mathbb{C}$ and $\mathbb{C} \setminus C$ has exactly two components, for each of which every point of C is an accessible boundary point, then C is a Jordan-curve. See, e.g., M. H. A. NEWMAN [1951] or the Scholium to 9.14 for a proof and JOHNSON [1979] for history.

4.51 was proposed as a problem by D. J. Newman, with solutions by R. Moore and J. W. Gaddum in the *American Mathematical Monthly* 59 (1952), 110 and 60 (1953), 271. See also REICHBACH [1958] and p. 328 of KURATOWSKI [1945].

In connection with the Gauss-Lucas Theorem 4.52(iii) and its further applications, see the papers of HAYASHI [1914–15], SPECHT [1959], RUBEL [1966], M. MARDEN [1966], [1968], [1976], LABELLE [1971] and SCHMIEDER [1997].

The special case of a polynomial in 4.51 is called the *Fundamental Theorem of Algebra* (FTA), and we have already seen several proofs of it. If the complex numbers $\mathbb{C}$ are thought of as the minimal enlargement of $\mathbb{R}$ that permits all quadratic equations to be solved, who would dream that all polynomial equations whatsoever should be solvable in $\mathbb{C}$? The FTA is therefore a quite remarkable fact, and its evolution is an important (and well-studied) chapter in the history of mathematics. The number and variety of its proofs is astounding. Interestingly, it received a big impetus from calculus in the 18th century, as mathematicians tried to integrate the most general rational function in closed form (for which the partial-fraction decomposition 11.12 was needed); see GILAIN [1991]. Although the FTA looks like an algebraic result, since its conclusion is false if the coefficients and zeros sought are restricted to the field $\mathbb{Q} + i\mathbb{Q}$, the completeness (order, or metric) of $\mathbb{R}$ plays an essential role. In fact, this is the irreducible kernel of topology that is needed: given that every odd-degree polynomial with real coefficients has a real zero [a trivial consequence of the connectedness of the topological space $\mathbb{R}$ and the continuity of polynomials], the Fundamental Theorem of Algebra can be proved by several purely algebraic arguments. See, e.g., the standard algebra treatises of van der Waerden (**MR** 91h:00009) and Jacobson (**MR** 52#13719) or J. KÖNIG [1879], NETTO [1880], WALECKI [1883], BRISSE [1886], A. KNESER [1887], VAHLEN [1897], GIUDICE [1907], MERTENS [1911], STEINITZ [1914], HOROWITZ [1966] $\subset$ ARTIN and SCHREIER [1926] (exceedingly elegant), VAN DER CORPUT [1950], VITALI [1928], DÖRGE [1928], PUCCIO [1933], H. KNESER [1939], THOMAS [1940], MIGNOSI [1941], EATON [1960], ZASSENHAUS [1967], and KOWALSKY [1983] (a linear algebra *tour de force*). Interesting, more or less constructive proofs occur in MAN-

SION [1880] (which contains a historical appendix), BLUTEL [1911], MONTEL [1913], ROSENBLOOM [1945], E. BISHOP [1967], HIRSCH and SMALE [1979] and BRENNER and LYNDON [1981]. Unusual proofs occur in WEIERSTRASS [1891] (see also *Werke* I, pp. 247–256 and III, pp. 251–269), MÉRAY [1891a], JAMET [1894], KOEBE [1918b], H. KNESER [1940] (containing a good bibliography), IGLISCH [1940], MOHR [1942], [1953a], SCARF [1952], ISEKI [1954] (via Banach algebra; cf. also D.SINGH [2006]), KÔNO [1985] (via brownian motion!), HANTOUT [1991] (via Lie theory), BRÜCKNER [1991], VAGGIONE [1997], DERKSEN [2003] (exceptionally beautiful), SANTOS [2005], LEIBMAN [2005] (via non-standard analysis), and D. SINGH [2006] (via functional analysis). In addition there are elementary proofs that use only calculus (but usually involve the complex exponential function): SANTOS [2007], DE JONG [2009], and MATHEW and PATYI [2011].

An excellent history of the FTA, with details of some of its more historically significant proofs is given by R. Remmert in chapter 4 of EBBINGHAUS, *et al.* [1990]. See also §80–88 of NETTO and LEVAVASSEUR [1907] for references to over 100 proofs and "proofs" of the FTA which have occurred in the literature (up to that time!); LORIA [1891], AGOSTINI [1924], GASAPINA [1957], BAŠMAKOVA [1957], [1960], §5 of SPECHT [1958], PETROVA [1971], [1973], [1974] and GILAIN [1991], [1997]. The historical role of the FTA as a catalyst for the development of new techniques is the organizational principle of the book by FINE and ROSENBERGER [1997]. There each of analysis, algebra, and topology is developed in turn to the point where it can provide its own proof of the FTA. Another sweep across the centuries is the awesome 3-volume work of ALVAREZ and DHOMBRES [2011], [2012], which aims to get into the mind-set of the pioneers. Philosophical, psychological, as well as ontological (what kind of thing is a "root" of a polynomial equation?), epistemological and methodological issues are exhaustingly discussed. The most complete and accurate bibliography of proofs of the FTA was created in 2017 by HERBERT SCHROEDER, available online at https://www.researchgate.net/publication 321867034 Der Fundamentalsatz der Algebra — mit einer kommentierten Bibliographie.

A field is called *algebraically closed* if every non-constant polynomial with coefficients in the field has a zero in the field. The FTA affirms that $\mathbb{C}$ is algebraically closed. Interestingly, the repleteness of $\mathbb{C} = \mathbb{R}^2$ which this property proclaims, precludes every higher-dimensional euclidean space from being a field at all: If $n > 2$, it is not possible to define in $\mathbb{R}^n$ a multiplication which makes it a commutative unital $\mathbb{R}$-algebra without non-zero zero-divisors. See R. M. YOUNG [1988] or 4.3.6 of EBBINGHAUS, *et al.* [1990] for a beautiful, elementary proof that uses only the FTA and a little linear algebra. This fact was anticipated by Gauss, but first proved by Weierstrass. Without much additional effort it can be improved to a characterization of $\mathbb{C}$, the Gelfand-Mazur theorem: Every normed commutative division algebra over $\mathbb{R}$ is (isomorphic to) either $\mathbb{R}$ or $\mathbb{C}$. See §4, chapter 8 of EBBINGHAUS, *et al.* [1990], ULLRICH [1990], or REMMERT [1993]

It is difficult to assign credit for the first "rigorous" proof of the FTA. Some proofs regarded as rigorous in their day are not by present-day standards; some others that critical contemporaries (esp. Gauss) found inadequate were actually rigorous modulo the following points: (i) The complex numbers were not rigorously grounded or even understood; (ii) zeros (factorizations) were assumed to exist in some universe of symbols, but the construction of splitting fields to provide this universe was not available; (iii) the modulus of a polynomial was assumed to realize a minimum value, although the compactness concept and the Bolzano-Weierstrass Theorem had not evolved (and the definition of continuity was not in good shape). CAUCHY [1847] had the idea of constructing $\mathbb{C}$ as the space of cosets of the ideal generated by the polynomial $P(X) := X^2+1$ in the ring $\mathbb{R}[X]$ of polynomials over $\mathbb{R}$ in an indeterminate X, and this is the definition of $\mathbb{C}$ one finds in Bourbaki. (Since $P(X)$ represents 0 in the coset ring, which can be viewed as an enlargement of $\mathbb{R}$, X is, so to speak, an ideal zero of P. This may explain how such subsets of rings came to be called "ideals.") Gauss (earlier) and Kronecker (later) used the same idea to construct splitting fields. In 1837 W. R. Hamilton constructed $\mathbb{C}$ fully *algebraically* from $\mathbb{R}$ via ordered pairs, but a sound definition of $\mathbb{R}$ itself did not appear until later in the 19th century. Regarding logical lacunae (ii) and (iii), we may wonder how mathematicians before 1800 could have overlooked them. But later generations overlooked similar (if more subtle) logical points: Many arguments and constructions involving the apparently non-controversial selection of elements from non-empty sets, devised in the 19th century (and still used today) were assessed in the early 20th century to be deficient, absent a new foundational principle (the Axiom of Choice) to justify them.

Gauss' dissertation (1799) was devoted to a proof of the FTA, after a prolog in which he pinpointed the inadequacies of all previous proofs. Altogether Gauss gave 4 proofs of this theorem in the course of his life (see GAUSS [1890] and chapter 4 of BOTTAZZINI [1986]), the proofs in 2.20 and 4.51 being related to the third of them, 1816. The FTA is consequently often called "Gauss' Theorem". Ironically Gauss' first proof contains some unsubstantiated assertions about algebraic curves which were first proved as late as 1920, by A. Ostrowski while working on a supplement to vol. X of Gauss' *Werke*; see MOTZKIN and OSTROWSKI [1933]. The flavor of this first proof, with informal but transparent arguments that the level-curves $\{(x,y) : \operatorname{Re} p(x,y) = 0\}$ and $\{(x,y) : \operatorname{Im} p(x,y) = 0\}$ must intersect, can be gotten from GOMEZ–CALDERON and WELLS [1996]. The second proof (the most algebraic) is analyzed in detail by LÜNEBURG [2008], volume I, pp. 589-606.

Many French writers speak of "le théorème de d'Alembert". See PETROVA, and GILAIN, *op. cit.*, and BALTUS [2004] for details and assessment of d'Alembert's work.

Though finding d'Alembert's argument inadequate, Gauss opined that its key idea could probably become the basis of a valid proof. The argument was indeed refined by J. R. Argand (in 1806 and 1814): If P is a polynomial of degree $n > 1$, there exists $z_0 \in \mathbb{C}$ at which $|P|$ attains a minimum over $\mathbb{C}$. (Nowadays a routine compactness argument justifies this once dubious claim.) Suppose $P(z_0) \neq 0$. Write $P(z) - P(z_0)$ as a polynomial in $(z - z_0)$ (for example, by using 4.52(i) to set up a proof by induction on the degree of P) and let $b_k(z - z_0)^k$ be the lowest degree non-zero term which it contains. An easy argument shows that if u is a kth root of the complex number $-P(z_0)/b_k$, then for sufficiently small positive r the contradiction $|P(z_0 + ru)| < |P(z_0)|$ ensues. According to GILAIN [1997], Argand felt that this proof vindicated the nascent geometrical view of complex numbers, and variants of it were later repeated by Cauchy. What is the source of this k^{th} root, whose existence, after all, is a special case of the theorem being proved? The complex exponential (3.14) can of course be used, but that is arguably less elementary than the FTA itself. But this proof can proceed by induction on n and the induction hypothesis used in a clever way to get the needed u. With just a little more work we can still reach this contradiction if the number $b_k u^k/P(z_0)$, instead of equaling -1, merely has negative real part. The existence of such a u can be secured by clever but totally elementary means. One finds versions of these ideas in PRINGSHEIM [1925], pp. 184ff, LITTLEWOOD [1941], ESTERMANN [1956], SPIVAK [1967], p. 460, EBBINGHAUS, *et al.* [1990], p. 114, and KOCHOL [1999]. Its most simple expositions are to be found in SEARCÓID [2002] and in DE OLIVEIRA [2011] (who used iterated square-roots of -1 and $\pm i$ to this end, very efficiently, as in the hints to 2.21(ii)).

On the other hand, one can also get the desired k^{th} root by elementary real-variable arguments (MÉRAY [1885], §3 of MERTENS [1892], POMEY [1905], DIENES [1931], LITTLEWOOD [1941] and KÖRNER [2006]). Because every complex number has a square-root (see, e.g., the hint to 3.44(ii)), we only have to consider odd k. Thus $c \in \mathbb{C}\backslash\mathbb{R}$ is given and we seek $x,\, y \in \mathbb{R}$ so that $z := x + iy$ satisfies $z^k = c$. This equation is practically equivalent to the two equations $|z^{2k}| = |c|^2$ and $\operatorname{Im}(\bar{c}z^k) = 0$. The second of these is (after division by y^k) a polynomial in x/y of degree k with real coefficients. It consequently has a real zero r. The equation $|z|^{2k} = |c|^2$ then gives $y = \pm|c|^{1/k}/\sqrt{1 + r^2}$, and with the proper choice of sign $z := (r + i)y$ satisfies the original equation $z^k = c$. The elementary topological argument in 1.26 also secures k^{th} roots of complex numbers, as does DZEWAS [1968]. Interestingly, the above proof of the FTA simplifies for the special polynomials $P(z) := z^n - c$, $c \in \mathbb{C} \setminus i\mathbb{R}$, delivering n^{th} roots for such c. As noted, it suffices to deal with odd n. The minimizing

z_0 cannot be 0, since $z := (\operatorname{Re} c)^{1/n}$ satisfies $|P(z)| = |\operatorname{Im} c| < |c| = |P(0)|$. The k above is therefore 1, the b_k being $P'(z_0) = nz_0^{n-1}$, and the needed k^{th} root u is simply $-P(z_0)/b_k = (c - z_0^n)/nz_0^{n-1}$. For $c \in i\mathbb{R}$, say $c = iy$ with $y \in \mathbb{R}$, calculation confirms that $y^{1/n} i^n$ is an nth root of c because n odd entails $n^2 \in 4\mathbb{Z} + 1$.

Because it uses neither integration, differentiation, nor any transcendental function, the foregoing (perhaps along with 2.21) bids fair to be the *most* elementary (methodologically, but conceptually as well) of the numerous extant proofs of the FTA. A close competitor for most elementary (as well as most elegant) is the proof by DERKSEN [2003], which is essentially pure (linear) algebra; every reader should have a look at it. It is traditional to use FTA to prove that linear transformations on $\mathbb{C}$-vector spaces have eigenvalues. Derksen reverses this, proves from scratch that eigenvalues exist, and from this deduces FTA (see also D. SINGH [2006]). The issue of minimal assumptions is pushed further in DE OLIVEIRA [2012] (where roots of complex numbers are not needed at all, not even square-roots in $\mathbb{R}_+$, as the l_1-metric is employed in $\mathbb{C} = \mathbb{R}^2$), and in a quite different way in the companion piece EISERMANN [2012]. For example, he develops an algebraic theory of the index [free of the exponential function and integration], use of which renders 4.26 a wholly elementary proof of the FTA, and he employs only *first-order* predicate logic; a 71-item bibliography is included.

Two other rather elementary and economical proofs of the FTA which also belong philosophically and methodologically to the eighteenth century are REDHEFFER [1957] and LAZER [2006]. They note that when f' as well as f is complex-differentiable (cf. 5.30), a little computation with (C. R.) in its componentwise forms leads to the identities

$$(*) \quad \Delta|f|^2 := D_{11}|f|^2 + D_{22}|f|^2 = 4|f'|^2 \quad \& \quad (**) \quad |\nabla|f|^2|^2 = 4|f'|^2|f|^2.$$

(If the gradient vector is viewed as a complex number, then we actually have $\nabla|f|^2 = 2f\overline{f'}$.) Any non-constant polynomial p is clearly such a function f. Redheffer considers $\varphi_k(z) := |p(z)|^2 - |z|^2/k$ in a disk $\overline{D}(0,r)$ where r is chosen so large (thanks to $\lim_{|z|\to\infty} |p(z)| = \infty$) that $|p|^2 > 1 + |p(0)|^2$ on $C(0,r)$, and integer $k > r^2$. These choices mean that the minimum of φ_k on the compact set $\overline{D}(0,r)$ occurs (only) at some $z_k \in D(0,r)$. From elementary calculus therefore $\nabla\varphi_k(z_k) = 0$ and $\Delta\varphi_k(z_k) \geq 0$, whence via $(*)$ and $(**)$ follows $|p(z_k)|^2 \leq r^2/k$. Any limit point of the z_k in $\overline{D}(0,r)$ is therefore a zero of p. Lazer, on the other hand, seeks a contradiction from the assumption that p is zero-free, in which case $1/p$ is also an eligible f. His innovation is $\varphi(z) := \sum_{j=1}^n |f(z+j)|^2$, where $n > 1$ is the degree of p. This continuous function vanishes at ∞, hence (compactness again) experiences an absolute maximum at some $w \in \mathbb{C}$. Again, calculus shows that at such a point $\Delta\varphi(w) \leq 0$. That is, in view of $(*)$,

$$0 \geq \sum_{j=1}^{n} \Delta|f|^2(w+j) = \sum_{j=1}^{n} 4|f'(w+j)|^2 = 4\sum_{j=1}^{n} |p'(w+j)/p^2(w+j)|^2,$$

which says that the polynomial p', whose degree is $n-1>0$, has the n distinct zeros $w+1,\ldots,w+n$, the sought-for contradiction (cf. 4.52(i))—*kurz und bündig*. For the record, note that compactness was used (inevitable, given that it distinguishes $\mathbb{Q}$ from $\mathbb{R}$), but neither the exponential nor any integration.

One of the purely algebraic proofs (due to P. S. Laplace and L. Euler) goes by induction on the highest power of 2 which divides the degree. It begins by factoring the polynomial (in an unspecified extension of $\mathbb{C}$!). A clever argument via the Fundamental Theorem on Symmetric Polynomials (see FINE and ROSENBERGER [1997]) is used to show that for some pair X_1, X_2 of these zeros and distinct real numbers r_1, r_2 both $X_1+X_2+r_1X_1X_2$ and $X_1+X_2+r_2X_1X_2$ necessarily lie in $\mathbb{C}$, whence X_1, X_2 themselves lie in $\mathbb{C}$. STEINITZ [1914] is a proof similar to this, as was Gauss' second proof (1816). (See H. KNESER [1939] and BAŠMAKOVA [1960].) Another beautiful algebraic proof, due to ARTIN and SCHREIER [1926] (pp. 258–272 of ARTIN [1965]), uses Galois theory and the Sylow theorems from group theory. A tortuous algebraic proof (modulo existence of splitting fields) which even Gauss admired was advanced by J. L. Lagrange in 1772; the reader can reach his own conclusion about its merits by consulting SUZUKI [2006].

As noted, the naive 17th- and 18th-century view was that every polynomial factors ("formally") into linear ones, the problem being to show that these zeros have the form $a+ib$ ($a,b \in \mathbb{R}$). This suggests another way to attack the FTA: Show that the factored (in $\mathbb{C}$) polynomials indeed comprise all polynomials over $\mathbb{C}$. That is, let functions S_j be defined by $(X-z_1)\cdots(X-z_n) = X^n + \sum_{j=0}^{n-1} S_j(z_1,\ldots,z_n)X^j$ [cf. 4.55(i)] and try to show that $(S_0,\ldots,S_{n-1})$ maps $\mathbb{C}^n$ *onto* $\mathbb{C}^n$ (by showing that the image set is both open and closed). For this see WEIERSTRASS [1891], KOEBE [1918b], and ALMIRA, *et al.* [2002].

REDHEFFER [1957], [1964] uses very elementary ideas about harmonic functions and maximum principles (see next chapter); both these papers are recommended reading. SCHMIDT [1936] and WIGERT [1936] are other interesting proofs using harmonic functions, without mentioning $\mathbb{C}$ at all: they establish that every real polynomial is a product of real linear and quadratic factors. (The proofs in PHRAGMÉN [1891], PERRON [1949], MAČYS [2002] and MOHR [1942], [1951/52], [1953] also involve only $\mathbb{R}$.) Of course, using results about holomorphic functions, many proofs are possible: the Fundamental Theorem of Algebra is an immediate consequence of (1) Liouville's theorem (3.33) applied to the reciprocal of a zero-free polynomial, (2) the Maximum-Modulus Principle for rational functions (as in 2.21(iv)), (3) Picard's Little Theorem (12.18 and its remark), (4) the mean-value theorem 5.5(ii) (see the remarks following it), (5) Rouché's Theorem 4.26(iii) or (6) Hadamard's generalized Liouville Theorem (6.37). In addition there are many contour-integral proofs, the most elementary of which are ANKENY [1947], BOAS [1964b], ZALCMAN [1978], p. 20 of LEBORGNE [1991], and 2.20(vi) (the latter two do not involve the complex exponential function).

As noted in 3.35(xi), the Open-Map Theorem gives a short proof of the FTA, since (by a simple compactness argument, detailed in the hints to 3.35(xi)) $P(\mathbb{C})$ is a closed set for any non-constant polynomial P. The iterative procedure in the proof of 5.79 is totally elementary and leads almost as quickly to a proof (MONTEL [1913], WOLFENSTEIN [1967]). Indeed, if Z denotes the zero-set of P', then Z is finite (possibly empty) and the iterative procedure shows that $P(\mathbb{C} \setminus Z)$ is open. For the finite set $F := P(\mathbb{C}) \setminus P(\mathbb{C} \setminus Z)$, the difference set $\mathbb{C} \setminus F$ is connected (1.24) and equals the disjoint union of the open sets $\mathbb{C} \setminus P(\mathbb{C})$ and $P(\mathbb{C} \setminus Z)$. It follows that $\mathbb{C} \setminus P(\mathbb{C}) = \varnothing$. See also SEN [2000], and cf. 6.15.

Finally, it should be noted that there are genuinely constructive proofs of the FTA. One of the nicest is that of M. KNESER [1981]. For each $n \in \mathbb{N}$ he defines $q_n := 1 - \frac{1}{2} \cdot 3^{1-n^2} < 1$ and then shows how to construct for each monic polynomial P of degree n a sequence $\{z_k\} \subset \mathbb{C}$ which satisfies $|z_{k+1} - z_k| \le |P(0)|^{1/n} (q_n^{1/n})^k$ and $|P(z_k)| \le |P(0)| q_n^k$ for all $k \in \mathbb{N}$. The argument is intricate, but completely elementary. Likewise constructive and elementary, but intricate, is ROSENBLOOM [1945]. There are even algorithms which can be computer implemented; see §6 of HIRSCH and SMALE [1979] and their bibliography.

The result 4.55(iii) is due to KRONECKER [1857]. The suggested proof is from BIEBERBACH [1953] and PÓLYA and SZEGÖ [1976], p. 346, and is quite close to the original. For a generalization see KLOOSTERMAN [1927] and for another elementary proof GREITER [1978]. Part (ii) is a trivial consequence of the Fundamental Theorem on Symmetric Polynomials, but the suggested proof is self-contained and avoids appeal to this basic result from algebra.

4.57 was pointed out to me by S. Saeki. It is from SCHIRMER [1966] and a later version of it appeared in HARTE and KEOGH [2003]. The fixed-point theorem that it contains is due to WAVRE and BRUTTIN [1926] and was rediscovered by ABIAN [1961], with other proofs in ABIAN and BROWN [1962] and A. B. BROWN [1962]. These authors get the fixed-point result with $\mathbb{D}$ replaced by the inside of any Jordan-curve, a corollary of 4.57 and 9.19 taken together. The n-dimensional version of the fixed-point result was secured by KNASTER, KURATOWSKI and MAZURKIEWICZ [1929]. An infinite-dimensional version dates from 1937. (See, e.g., p. 51 of GRANAS [1962].)

For much more on the topological argument principle in 4.59(iv) see SHEIL-SMALL [2002].

4.61 first occurs in JÜRGENS [1879] but is often attributed to SCHÖNFLIES [1899]. Proofs were also supplied by OSGOOD [1900], F. BERNSTEIN [1900], SCHÖNFLIES [1908] (pp. 162–164), PÁL [1922/23], WINTERNITZ [1927], and NÖBELING [1950]. The n-dimensional version was secured by BROUWER [1912a] and another proof occurs in EILENBERG [1936]. (See also OSTROWSKI [1946].) The proof in the text is from GRANAS [1962], chapter VII. This result is called the "invariance of domain" and is closely related to the problem of the (topological) invariance of the dimension number n in $\mathbb{R}^n$. (Cf. the Remark following 4.62.) The latter problem was treated by J. Lüroth, G. Cantor, E. Netto and others in the late 19th century (see DAUBEN [1975]) but only definitively resolved by BROUWER in [1911], amid an interesting controversy. In a note following Brouwer's in the 1911 *Mathematische Annalen* Henri Lebesgue claimed to give a simple proof of the invariance of the dimension number. This proof was in error at several points and totally inadequate from the point of view of logical completeness. Though Lebesgue never recanted (as Brouwer intensely hoped and campaigned that he would), in subsequent patch-up attempts (none completely successful) he gave birth to the important concept that lies at the basis of modern (covering) dimension theory. (See SPERNER [1928] and HUREWICZ [1929].) For the fascinating details of the controversy, including the names of the other protagonists, the reader should see FREUDENTHAL [1975] and his notes on pp. 435–452 of BROUWER [1976], as well as **MR** 2007i:03001. For comprehensive history see JOHNSON [1979], [1981].

Exercise 4.62 was conjectured by Ulam and proved by BORSUK [1933a]. Its meteorological interpretation seems to be due to H. Steinhaus. For history and extensions see §59.V of KURATOWSKI [1968], as well as HADWIGER [1959], [1960]. E.g., in 1944 Heinz Hopf proved the following: with f, $\mathbb{S}$ as in 4.62, for every $d \in [0, 2]$ there exist points x, $y \in \mathbb{S}$ with $\|x - y\| = d$ and $f(x) = f(y)$. (For an elegant, elementary proof, see **MR** 95e:55004.) STEINLEIN [1985] is an extensive and [1993] is a shorter survey of generalizations and applications of the Borsuk–Ulam and Hopf theorems. A "colorful" proof of 4.64(iv) is offered in WOLFF and DENJOY [1933]. There is an analogous conclusion in which the role of the antipodal mapping $x \mapsto -x$ is taken over by an arbitrary continuous self-map f of $\mathbb{S}$: At least one of the sets S_1, S_2, S_3 contains a point x *and* its image $f(x)$. See **MR** 95b:55003.

For a somewhat different proof of Kakutani's theorem 4.65 see HADWIGER

[1959]. For references to the many generalizations that this theorem has experienced see the bibliography of HADWIGER [1959] and §59.V of KURATOWSKI [1968]. A planar version of the theorem is offered by DAIA [1986–87], and DYSON [1951] proved that there exist orthogonal $P_1, P_2 \in \mathbb{S}$ such that $f(P_1) = f(-P_1) = f(P_2) = f(-P_2)$. The model for the proof of Poincaré's Theorem 4.63 is EISENBERG and GUY [1979] and another very pretty elementary proof is to be found in GAMELIN and GREENE [1983]. Less successful elementary treatments are FENCHEL [1932], JESSEN [1933] and HOLMANN [1964].

On the theme of 4.33 and 4.69 see RUDIN [1955b] and 10.30 below. A quite different proof of 4.74 will be found in MUELLER and RUDIN [1991] and in ČERNÝ [1992] (p. 180), while a proof of the "only if" half of 4.72 direct from the definition of simple-connectivity is given in SAKS and ZYGMUND [1971] (p. 31). 4.73 is a special case of the important Seifert-van Kampen theorem in algebraic topology (FULTON [1995], p. 199).

For extensions of many of the results of §6 to the infinite dimensional (i.e., Banach space) setting see the beautiful little monograph of GRANAS [1962] or the books of EISENACK and FENSKE [1978] and DUGUNDJI and GRANAS[1982].

Chapter V

Consequences of the Cauchy–Goursat Theorem—Maximum Principles and the Local Theory

§ 1 Goursat's Lemma and Cauchy's Theorem for Starlike Regions

Of great utility in the differential calculus of functions on $\mathbb{R}$ are *primitives* or antiderivatives. The Fundamental Theorem here affirms that integration always produces such primitives. In particular, every continuous function on $\mathbb{R}$ has a primitive. [This is not so for continuous functions on regions in $\mathbb{C}$ because in $\mathbb{C}$, as it turns out, having a primitive is equivalent to having a derivative!] Naturally we look to integration to produce primitives in the plane too. But now we must face the fact that integration can be carried out over many paths from z_0 to z, whereas if the integrand f has a primitive F, then the integral over any such path must equal $F(z) - F(z_0)$ (see 2.10) and be consequently independent of the path chosen. This independence of path is equivalent to the integral over any *closed* path being 0. Conversely, when the integral is path independent, a primitive can be manufactured by using an "indefinite" integral (see the proof of 2.11). Thus the corresponding fundamental theorem(s) in the plane affirm that certain integrals over closed paths are 0. Everything flows from

Theorem 5.1 (Cauchy-Goursat) *Let U be an open subset of $\mathbb{C}$, f holomorphic in U, a, b, $c \in \mathbb{C}$ and* $\operatorname{co}\{a,b,c\} \subset U$. *Then* $\int_{[a,b,c,a]} f = 0$. (Cf. 2.19(xiv).)

Proof: Write $\triangle$ for $\operatorname{co}\{a,b,c\}$ and d for the diameter of $\triangle$. Consider

$$a' := \frac{b+c}{2}, \qquad b' := \frac{a+c}{2}, \qquad c' := \frac{a+b}{2}.$$

R. B. Burckel, *Classical Analysis in the Complex Plane*, Cornerstones,
https://doi.org/10.1007/978-1-0716-1965-0_6

By 2.9

$$\int_{[a,b,c,a]} f = \int_{[a,b]} f + \int_{[b,c]} f + \int_{[c,a]} f$$

$$= \int_{[a,c']} f + \int_{[c',b]} f + \int_{[b,a']} f + \int_{[a',c]} f + \int_{[c,b']} f + \int_{[b',a]} f$$

$$= \left(\int_{[a,c']} f + \int_{[c',b']} f + \int_{[b',a]} f\right) + \left(\int_{[c',b]} f + \int_{[b,a']} f + \int_{[a',c']} f\right)$$

$$+ \left(\int_{[a',c]} f + \int_{[c,b']} f + \int_{[b',a']} f\right) + \left(\int_{[a',b']} f + \int_{[b',c']} f + \int_{[c',a']} f\right)$$

(1)

$$= \int_{[a,c',b',a]} f + \int_{[c',b,a',c']} f + \int_{[a',c,b',a']} f + \int_{[a',b',c',a']} f.$$

Note that the sets co $\{a, c', b'\}$, co $\{c', b, a'\}$, co $\{a', c, b'\}$ and co $\{a', b', c'\}$ each lie in co $\{a, b, c\}$ and each has diameter equal to $\frac{1}{2}$ the diameter of co $\{a, b, c\}$, by 1.19(i). It follows from (1) that at least one of the integrals on its right-hand side has modulus not less than $\frac{1}{4}\left|\int_{[a,b,c,a]} f\right|$; that is, there are points a_1, b_1, c_1 with co $\{a_1, b_1, c_1\} =: \triangle_1$ such that

(1.1)
$$\begin{cases} \triangle_1 \subset \triangle, \ \operatorname{diam} \triangle_1 = 2^{-1} \operatorname{diam} \triangle \quad \text{and} \\ \left|\displaystyle\int_{[a_1,b_1,c_1,a_1]} f\right| \geq 4^{-1}\left|\displaystyle\int_{[a,b,c,a]} f\right|. \end{cases}$$

Proceed inductively to construct, for every positive integer n, points a_n, b_n, c_n with $\triangle_n :=$ co $\{a_n, b_n, c_n\}$ satisfying

(1.n)
$$\begin{cases} \triangle_n \subset \triangle_{n-1}, \ \operatorname{diam} \triangle_n = 2^{-1} \operatorname{diam} \triangle_{n-1} \quad \text{and} \\ \left|\displaystyle\int_{[a_n,b_n,c_n,a_n]} f\right| \geq 4^{-1}\left|\displaystyle\int_{[a_{n-1},b_{n-1},c_{n-1},a_{n-1}]} f\right|. \end{cases}$$

Now by induction on this we have for all positive integers n

(2.n) $$\triangle_n \subset \triangle_{n-1} \subset \cdots \subset \triangle.$$

(3.n) $$\operatorname{diam} \triangle_n = 2^{-n} d.$$

(4.n) $$\left|\int_{[a_n,b_n,c_n,a_n]} f\right| \geq 4^{-n}\left|\int_{[a,b,c,a]} f\right|.$$

According to (1.19)(ii) $\triangle_n$ is compact. From $(2.n)$ therefore it follows that there exists $z_0 \in \bigcap_{n\in\mathbb{N}} \triangle_n \subset U$. (See §4 of chapter 0.) Alternatively, $(2.n)$ and $(3.n)$ insure that $\{a_n\}_{n\in\mathbb{N}}$ is a Cauchy sequence; its limit z_0 lies in every $\triangle_n$ and is the unique such point. As noted before 2.3, $\mathbb{C}$-differentiability of f at z_0 just means that there is a continuous function ε on U such that

$$f(z) = f(z_0) + \big(f'(z_0) + \varepsilon(z)\big)(z - z_0) \qquad \forall z \in U, \text{ and} \tag{5}$$

$$\lim_{z\to z_0} \varepsilon(z) = 0. \tag{6}$$

Now

$$\int_{[a_n,b_n,c_n,a_n]} \big(f(z_0) + f'(z_0)(z - z_0)\big)\, dz = 0 \qquad \forall n \in \mathbb{N} \tag{7}$$

by 2.10(ii), since the integrand is a polynomial and since every polynomial is a derivative. By 2.9(ii) the length of the curve $[a_n, b_n, c_n, a_n]$ is not greater than $3\,\mathrm{diam}\,(\triangle_n)$, which is $3\cdot 2^{-n}d$ according to $(3.n)$. It follows from (5), (7) and 2.9(i) that

$$\begin{aligned}\left|\int_{[a_n,b_n,c_n,a_n]} f\right| &= \left|\int_{[a_n,b_n,c_n,a_n]} \varepsilon(z)(z - z_0)\, dz\right| \\ &\le 3\cdot 2^{-n}d\cdot \max\big\{|\varepsilon(z)(z - z_0)| : z \in [a_n, b_n, c_n, a_n]\big\} \\ &\le 3\cdot 2^{-n}d\cdot \max\big\{|\varepsilon(z)||z - z_0| : z \in \triangle_n\big\} \\ &\le 3\big(2^{-n}d\big)^2\cdot \max\big\{|\varepsilon(z)| : z \in \triangle_n\big\}\end{aligned} \tag{8}$$

by $(3.n)$ and the fact that $z_0 \in \triangle_n$. From $(4.n)$ and (8)

$$\left|\int_{[a,b,c,a]} f\right| \le 3d^2 \max\big|\varepsilon(\triangle_n)\big| \qquad \forall n \in \mathbb{N}. \tag{9}$$

Since $z_0 \in \triangle_n$, it follows from $(3.n)$ and (6) that the right-hand side of (9) converges to 0 as $n \to \infty$. Thus, $\int_{[a,b,c,a]} f = 0$, and the proof is complete.

Corollary 5.2 (Cauchy's Theorem for Starlike and Convex Regions) *If Ω is an open starlike subset of $\mathbb{C}$ and f is holomorphic in Ω, then there is a holomorphic function F in Ω such that $F' = f$. Therefore $\int_\gamma f = 0$ for every piecewise smooth loop γ in Ω. If additionally Ω is convex and f is continuous on $\overline{\Omega}$, then $\int_\gamma f = 0$ for every piecewise smooth loop γ in $\overline{\Omega}$.* (Cf. 9.27 and the remark following it.)

Proof: 5.1 and the Scholium to 2.11 combine to confirm the first claim. Since $F' = f$ is continuous, the second claim follows from 2.10(ii).

Finally, suppose in addition that Ω is convex and that f is continuous on $\overline{\Omega}$. Without loss of generality, suppose $0 \in \Omega$. Then for every $z \in \overline{\Omega}$ the half-open segment $[0, z)$ lies in Ω, by 1.19(iv). Hence if $\gamma\colon [0,1] \mapsto \mathbb{C}$ is a piecewise smooth loop in $\overline{\Omega}$, $\tau\gamma$ is one in Ω for every $0 \le \tau < 1$ and so by the foregoing

$$0 = \int_{\tau\gamma} f = \tau \int_0^1 f\big(\tau\gamma(t)\big)\gamma'(t)\,dt.$$

On the compact subset $\{\tau\gamma(t) : 0 \le \tau \le 1,\ 0 \le t \le 1\}$ of $\overline{\Omega}$, f is uniformly continuous, so $\int_{\tau\gamma} f$ converges to $\int_\gamma f$ as $\tau \uparrow 1$.

Remarks: (i) In view of 5.2 the conclusion of 5.1 remains valid if there is a point z_0 such that f is continuous in U but only known to be holomorphic in $U \setminus \{z_0\}$. To prove this, only $z_0 \in \operatorname{co}\{a, b, c\}$ need be considered. For such z_0 the integrals of f over $[a, b, z_0, a]$, $[b, c, z_0, b]$ and $[c, a, z_0, c]$ each vanish by 5.2, and they add up to the integral of f over $[a, b, c, a]$. (By using all the triangles in all the subdivision stages of Goursat's proof, HERVÉ [1982] is able to accommodate an arbitrary set of isolated exceptional points.) Using this stronger 5.1, the proof of 5.2 goes through with its hypotheses weakened in the same way. This modest extension of 5.2 proves useful (if not indispensable) later.

(ii) The subdivision idea in 5.1 and the use of primitives in 5.2 can be combined to yield a generalization of 5.1 corresponding to 2.19(v)*; see also 7.23(iii). Namely, instead of triangles, consider $S := [0,1] \times [0,1]$ and a continuous $\phi\colon S \mapsto U$ for which $\phi \circ \partial S$ is piecewise smooth. If $n \in \mathbb{N}$ is sufficiently large, ϕ maps each square $S_{j,k} := \big[(j-1)/n, j/n\big] \times \big[(k-1)/n, k/n\big]$ into a disk $D_{j,k} \subset U$. Set $z_{j,k} := \phi\,(j/n, k/n)$. According to 5.2 f has a primitive $F_{j,k}$ in each $D_{j,k}$. Since the intersection of two disks is a convex set, the difference

(1) $F_{j,k} - F_{j-1,k}$ is constant in $D_{j,k} \cap D_{j-1,k}$, and

(2) $F_{j,k-1} - F_{j,k}$ is constant in $D_{j,k-1} \cap D_{j,k}$, thanks to 2.10(iv).

There is no reason to suppose that $\phi \circ \partial S_{j,k}$ is piecewise smooth, or even rectifiable, but if it were, then the integral of f over it, computed along each side of $S_{j,k}$ via 2.10, would be the number

$$\begin{aligned} I_{j,k} :=& [F_{j,k}(z_{j,k-1}) - F_{j,k}(z_{j-1,k-1})] + [F_{j,k}(z_{j,k}) - F_{j,k}(z_{j,k-1})] + \\ & [F_{j,k}(z_{j-1,k}) - F_{j,k}(z_{j,k})] + [F_{j,k}(z_{j-1,k-1}) - F_{j,k}(z_{j-1,k})], \end{aligned}$$

namely, 0. If we add up the $I_{j,k}$, (1) and (2) see to it that terms interior to S cancel and all that remains is the sum over those sides of the various $S_{j,k}$

which lie in ∂S, a sum which equals $\int_{\phi\circ\partial S} f$. Consequently, this integral is 0. Careful confirmation of the combinatorial details of this argument will be found on pp. 109–111 of HEINS [1968]; see also EBERLEIN [1975] and pp. 173–174 of REMMERT [1998]. HANCHE-OLSEN [2008] applies Goursat's descent directly to $\phi(S)$.

Exercise 5.3 (Sneak preview) (i) *Deduce from* 5.2 *and* 3.39(ii) *that if f is holomorphic in an open set U, then in fact f is infinitely differentiable and even locally of power series type in U.*

Hints: The problem is a local one, so assume U is a disk. Then 5.2 furnishes a primitive F for f. Since $F' = f$, we know F' is continuous. Consequently by 3.39(ii) F has a power series expansion near each point of U. By 3.6 so does f.

(ii) *From* (i) *and* 2.16(i) *deduce, without appeal to any other theorems of analysis* (*like* 2.19(iii)), *that all partial derivatives of all orders of a holomorphic* (*resp., harmonic*) *function exist, are independent of the order in which they are carried out and are themselves holomorphic* (*resp., harmonic*) *functions.*

Hints: In view of (i) above, 2.16(i) is applicable to all higher derivatives of f. Consequently, for any $n \in \mathbb{N}$ and $(j, k, \ldots, l) \in \{1, 2\}^n$ with m components equal 2, $D_j D_k \ldots D_l f = i^m f^{(n)}$; and $D_j D_k \ldots D_l(\operatorname{Re} f) = \operatorname{Re}(i^m f^{(n)})$.

The next exercise (due to Euler) is an application of 5.2, but not a convincing demonstration of its power, since there are so many other ways to get this result, some quite elementary (HARDY [1909–11], EGAN [1915–16] and K. S. WILLIAMS [1971b]); there will be many more convincing applications in the sequel. But because of this variety of approaches and the obstacles faced and (eventually) overcome in each, FISCHER [2007] makes the case that the integral $\int_{-\infty}^{\infty} \sin x/x \, dx$ represents a "history of analysis in a nutshell."

Exercise 5.4 (i) *Prove that*

$$\left| \int_{-r}^{r} \frac{\sin x}{x} \, dx - \pi \right| < \frac{\pi}{r} \qquad \forall r > 0.$$

Hints: The function

$$h(z) := \frac{e^{iz} - 1}{z} = \sum_{n=1}^{\infty} \frac{i^n}{n!} z^{n-1}$$

is entire. Therefore for the piecewise smooth loop

$$\Gamma_r(t) := \begin{cases} re^{\pi i t}, & 0 \le t \le 1 \\ (2t - 3)r, & 1 \le t \le 2 \end{cases}$$

we have $\int_{\Gamma_r} h = 0$. This follows either from 5.2 or by simply integrating termwise the locally uniformly convergent series representation of h. Splitting apart this integral,

$$\int_{-r}^{r} h(x)\,dx - \pi i = -i \int_0^{\pi} \exp\left(ire^{ix}\right) dx$$

$$\left|\int_{-r}^{r} h(x)\,dx - \pi i\right| \leq \int_0^{\pi} \left|\exp\left(ire^{ix}\right)\right| dx = \int_0^{\pi} \exp\left(\operatorname{Re}\left(ire^{ix}\right)\right) dx$$

$$= \int_0^{\pi} e^{-r\sin x}\,dx = 2\int_0^{\pi/2} e^{-r\sin x}\,dx$$

$$\leq 2\int_0^{\pi/2} e^{-2rx/\pi}\,dx \quad \text{by 3.30(i)}$$

$$= \left[-\frac{\pi}{r}e^{-2rx/\pi}\right]_{x=0}^{x=\pi/2} < \frac{\pi}{r}.$$

Since

$$\int_{-r}^{r} \frac{\sin x}{x}\,dx - \pi = \operatorname{Im}\left[\int_{-r}^{r} h(x)\,dx - \pi i\right],$$

the result follows.

(ii) *Show that*

$$\left|\int_{-r}^{r} \left(\frac{\sin x}{x}\right)^2 dx - \pi\right| < \frac{4}{r} \qquad \forall r > 0.$$

Hints: Integrating by parts, we see that

$$\int_{-2r}^{2r} \frac{\sin y}{y}\,dy = \int_{-r}^{r} \frac{\sin 2x}{x}\,dx = 2\int_{-r}^{r} \frac{\sin x \cos x}{x}\,dx$$

$$= \left[\frac{\sin^2 x}{x}\right]_{x=-r}^{x=r} + \int_{-r}^{r} \frac{\sin^2 x}{x^2}\,dx = \frac{2\sin^2 r}{r} + \int_{-r}^{r} \left(\frac{\sin x}{x}\right)^2 dx.$$

Therefore from part (i) follows

$$\left|\int_{-r}^{r} \left(\frac{\sin x}{x}\right)^2 dx - \pi\right| = \left|-\frac{2\sin^2 r}{r} + \int_{-2r}^{2r} \frac{\sin y}{y}\,dy - \pi\right|$$

$$\leq \frac{2\sin^2 r}{r} + \frac{\pi}{2r} \leq \frac{2}{r} + \frac{\pi}{2r} < \frac{4}{r}.$$

(iii) *Show that*

$$\left|\int_{-r}^{r} \frac{\sin^4 x}{x^2}\,dx - \frac{\pi}{2}\right| < \frac{5}{r} \qquad \forall r > 0.$$

Hints:

$$\begin{aligned}\int_{-r}^{r}\left(\frac{\sin x}{x}\right)^2 dx &= \int_{-r}^{r}\frac{\sin^2 x(\sin^2 x+\cos^2 x)}{x^2}\,dx\\ &= \int_{-r}^{r}\frac{\sin^4 x}{x^2}\,dx + \int_{-r}^{r}\left(\frac{\sin x\cos x}{x}\right)^2 dx\\ &= \int_{-r}^{r}\frac{\sin^4 x}{x^2}\,dx + \int_{-r}^{r}\left(\frac{\sin 2x}{2x}\right)^2 dx\\ &= \int_{-r}^{r}\frac{\sin^4 x}{x^2}\,dx + \frac{1}{2}\int_{-2r}^{2r}\left(\frac{\sin y}{y}\right)^2 dy.\end{aligned}$$

Therefore

$$\begin{aligned}\left|\int_{-r}^{r}\frac{\sin^4 x}{x^2}\,dx - \frac{\pi}{2}\right| &\le \left|\int_{-r}^{r}\left(\frac{\sin x}{x}\right)^2 dx - \pi\right| + \left|\frac{\pi}{2} - \frac{1}{2}\int_{-2r}^{2r}\left(\frac{\sin y}{y}\right)^2 dy\right|\\ &< \frac{4}{r}+\frac{1}{r} \quad \text{by part (ii).}\end{aligned}$$

(iv) *Show that*

$$\left|\int_{-r}^{r}\left(\frac{\sin x}{x}\right)^4 dx - \frac{2\pi}{3}\right| < \frac{2}{3r^3} \qquad \forall r>0.$$

Hints:

$$\begin{aligned}\int_{-2r}^{2r}\left(\frac{\sin y}{y}\right)^2 dy &= 2\int_{-r}^{r}\left(\frac{\sin 2x}{2x}\right)^2 dx = 2\int_{-r}^{r}\left(\frac{\sin x\cos x}{x}\right)^2 dx\\ &= 2\int_{-r}^{r}\left[\sin^2 x\cos x\right]\left[\frac{\cos x}{x^2}\right] dx.\end{aligned}$$

Integrate by parts twice to get

$$\begin{aligned}\int_{-2r}^{2r}\left(\frac{\sin y}{y}\right)^2 dy &= \left[2\frac{\sin^3 x}{3}\frac{\cos x}{x^2}\right]_{x=-r}^{x=r} - 2\int_{-r}^{r}\frac{\sin^3 x}{3}\left[-\frac{\sin x}{x^2} - \frac{2\cos x}{x^3}\right] dx\\ &= \frac{4}{3}\frac{\sin^3 r\cos r}{r^2} + \frac{2}{3}\int_{-r}^{r}\frac{\sin^4 x}{x^2}\,dx + \frac{4}{3}\int_{-r}^{r}\frac{\sin^3 x\cos x}{x^3}\,dx\\ &= \frac{4}{3}\frac{\sin^3 r\cos r}{r^2} + \frac{2}{3}\int_{-r}^{r}\frac{\sin^4 x}{x^2}\,dx\\ &\quad + \frac{4}{3}\left[\frac{\sin^4 x}{4x^3}\right]_{x=-r}^{x=r} + \int_{-r}^{r}\frac{\sin^4 x}{x^4}\,dx.\end{aligned}$$

It follows then from parts (ii) and (iii) that

$$\lim_{r\to\infty}\int_{-r}^{r}\frac{\sin^4 x}{x^4}\,dx \quad \text{exists and equals } \frac{2\pi}{3}.$$

Therefore

$$\begin{aligned}\frac{2\pi}{3}-\int_{-r}^{r}\left(\frac{\sin x}{x}\right)^4 dx &= 2\lim_{R\to\infty}\int_0^R\frac{\sin^4 x}{x^4}\,dx-2\int_0^r\frac{\sin^4 x}{x^4}\,dx\\ &= 2\lim_{R\to\infty}\int_r^R\frac{\sin^4 x}{x^4}\,dx < 2\lim_{R\to\infty}\int_r^R\frac{1}{x^4}\,dx\\ &\le 2\lim_{R\to\infty}\left[\frac{1}{3r^3}-\frac{1}{3R^3}\right]=\frac{2}{3r^3}.\end{aligned}$$

(v) *With* $\sin(0)/0$ *interpreted as* 1, *prove the striking equalities*

$$\sum_{k=-\infty}^{\infty}\frac{\sin k}{k}=\int_{-\infty}^{\infty}\frac{\sin x}{x}\,dx=\int_{-\infty}^{\infty}\left(\frac{\sin x}{x}\right)^2 dx=\sum_{k=-\infty}^{\infty}\left(\frac{\sin k}{k}\right)^2=\pi.$$

Hint: Look back at (i) and (ii), and at (30) in 3.38.

(vi) *Show that for every integer* $m \ge 2$

$$\int_{-\infty}^{\infty}\left(\frac{\sin x}{x}\right)^m dx=\frac{\pi}{(m-1)!\,2^{m-1}}\sum_{0\le k<m/2}(-1)^k\binom{m}{k}(m-2k)^{m-1}.$$

Hints: Repeated integration-by-parts, as before, together with appropriate trigonometric identities. For details see Apostol and Young [1980], or Mitrinović and Kečkić [1993], p. 36.

Of central significance are the following circumferential-mean-value properties of holomorphic functions:

Theorem 5.5 *Let* $c\in\mathbb{C}$, $0\le a<b\le\infty$, $a\le\rho\le b, g$ *be continuous in the annulus* $A:=\{z\in\mathbb{C}:a<|z-c|<b\}$ *and holomorphic in* $A\setminus C(c,\rho)$. *Then*

(i) $\displaystyle\int_0^1 g(c+re^{2\pi it})\,dt$ *is a constant function of* $r\in(a,b)$.

If g *is continuous in* $D(c,b)$ *and holomorphic in* $D(c,b)\setminus C(c,\rho)$, *for some* $0\le\rho<b$, *then*

(ii) $\quad g(c) = \int_0^1 g(c + re^{2\pi it})\,dt \qquad \forall 0 < r < b.$

If f is holomorphic in $D(c,b)$ and $0 < r < b$, then

(iii) $\quad f(z_0) = \frac{1}{2\pi i}\int_{C(c,r)} \frac{f(z)}{z - z_0}\,dz \qquad \forall z_0 \in D(c,r).$

Proof: We may assume $c = 0$. Consider first the case in which g is holomorphic in the whole of A. Define $E(z) := e^z$ for $z \in \mathbb{C}$. Then the open set $E^{-1}(A)$ has the form $(\alpha, \beta) \times \mathbb{R}$ and so is convex. Fix $r, R \in (a, b)$ and set

$$\gamma := [\log r, \log R, \log R + 2\pi i, \log r + 2\pi i, \log r].$$

Evidently γ maps into $E^{-1}(A)$, is piecewise smooth and is closed. Apply 5.2 (or even 5.1 itself) to the function $f := g \circ E$, which is holomorphic in $E^{-1}(A)$. We get

$$\begin{aligned} 0 = \int_\gamma f = \int_0^1 & f(t \log R + (1-t)\log r)(\log R - \log r)\,dt \\ & + \int_1^2 f(\log R + 2\pi i(t-1))2\pi i\,dt \\ & + \int_2^3 f((t-2)\log r + (3-t)\log R + 2\pi i)(\log r - \log R)\,dt \\ & - \int_3^4 f(\log r + 2\pi i(4-t))2\pi i\,dt \\ = (\log R & - \log r)\int_0^1 g \circ E(t\log R + (1-t)\log r)\,dt \\ & + 2\pi i \int_0^1 g \circ E(\log R + 2\pi i t)\,dt \\ & + (\log r - \log R)\int_0^1 g \circ E(t \log r + (1-t)\log R)\,dt \\ & - 2\pi i \int_0^1 g \circ E(\log r + 2\pi i t)\,dt, \end{aligned}$$

by changing variables and using the $2\pi i$-periodicity of E. By the change of variable $\tau := 1 - t$ the first integral on the right of the last equality is found to be equal to the third. Thus we get

$$\int_0^1 g \circ E(\log r + 2\pi i t)\,dt = \int_0^1 g \circ E(\log R + 2\pi i t)\,dt,$$

that is,

$$\int_0^1 g\big(re^{2\pi it}\big)\,dt = \int_0^1 g\big(Re^{2\pi it}\big)\,dt.$$

Since r, R are arbitrary in (a,b), part (i) is established.

In the general case, where $a<\rho<b$, g is uniformly continuous in the compact subset $\bar{A}(\frac{1}{2}(a+\rho),\frac{1}{2}(b+\rho))$ of A. Therefore

$$\lim_{r\uparrow\rho}\int_0^1 g(re^{2\pi it})dt = \int_0^1 g(\rho e^{2\pi it})dt = \lim_{R\downarrow\rho}\int_0^1 g(Re^{2\pi it})dt,$$

while the previously proven result is available in $A(a,\rho)$ and in $A(\rho,b)$.

For (ii), notice that $D(0,b)\backslash\{0\}$ is the A of part (i) with $a=0$. As $r\in(a,b)=(0,b)$ decreases to 0, the (constant!) integral converges to $g(0)$, by continuity of g at 0. For (iii), fix $z_0\in D(0,r)$ and define

$$g(z) := z\frac{f(z)-f(z_0)}{z-z_0}$$

if $z_0\neq z\in D(0,b)$, and $g(z_0):=z_0f'(z_0)$. Then g is continuous in $D(0,b)$, holomorphic in $D(0,b)\setminus\{z_0\}\supset D(0,b)\setminus C(0,\rho)$ with $\rho:=|z_0|$. So according to (ii)

$$0=g(0)=\int_0^1 g(re^{2\pi it})dt = \frac{1}{2\pi i}\int_{C(0,r)}\frac{f(z)-f(z_0)}{z-z_0}dz.$$

Now split this integral at the minus sign in the numerator and cite 3.39(i).

Remarks: After examining the special case in which g is a polynomial, Cauchy discerned in 1831 the following route to 5.5(i) [$c:=0$ and no exceptional circle]; similar ideas were used in 2.19(v) to get a weak form of 5.1. By the chain rule $D_\rho g(\rho e^{i\theta}) = e^{i\theta}g'(\rho e^{i\theta})$ and $D_\theta g(\rho e^{i\theta}) = \rho ie^{i\theta}g'(\rho e^{i\theta})$. Hence $D_\rho g(\rho e^{i\theta}) = (\rho i)^{-1}D_\theta g(\rho e^{i\theta})$, the polar form of (C. R.). Now, on one hand, for $a<R<r<b$

$$\int_{-\pi}^{\pi}\int_R^r D_\rho g(\rho e^{i\theta})\,d\rho\,d\theta = \int_{-\pi}^{\pi}\big[g(re^{i\theta})-g(Re^{i\theta})\big]\,d\theta$$

and on the other hand

$$\int_R^r\int_{-\pi}^{\pi}\frac{1}{\rho i}D_\theta g(\rho e^{i\theta})\,d\theta\,d\rho = \int_R^r\frac{1}{\rho i}\big[g(\rho e^{i\pi})-g(\rho e^{-i\pi})\big]\,d\rho = 0.$$

The polar (C. R.) then combines with 2.19(ii) to give

$$\int_{-\pi}^{\pi} g(re^{i\theta})\,d\theta = \int_{-\pi}^{\pi} g(Re^{i\theta})\,d\theta,$$

and part (ii) follows as before. Alternatively, define $G(\rho) := \int_{-\pi}^{\pi} g(\rho e^{i\theta})d\theta$ and use 2.19(i) to write $G'(\rho) = \int_{-\pi}^{\pi} D_\rho g(\rho e^{i\theta})d\theta = \int_{-\pi}^{\pi}(i\rho)^{-1}D_\theta g(\rho e^{i\theta})d\theta = 0$ (by FTC) for $a < \rho < b$, thus learning that G is constant. These arguments use the continuity of $D_\rho g$, $D_\theta g$. The proof in the text, which is a specialization of a general homotopy argument (see 7.23), does not use continuity of g' (although after 5.3 that fact is available).

Operating, as usual, under less than totally explicit hypotheses, Cauchy got his famous integral representation (iii) from (ii), as above (see p. 159 of SMITHIES [1997]). Independently Poisson found 5.5(ii), for power series functions in 1823. But even earlier G. Frullani had a version of it. See the *Nachtrag* to STÄCKEL [1901] and pp. 1024ff. of BURKHARDT [1914].

Still earlier, in 1816, Gauss had used this technique of interchanging the order of integration in a polar integral in his third proof of the fundamental theorem of algebra, although he did not carry things through to the mean-value equality 5.5(ii): If the non-constant polynomial P is zero-free in $\mathbb{C}$, then $g := 1/P$ is holomorphic, and in fact $g' = -P'/P^2$ is continuous throughout $\mathbb{C}$. Since $g(re^{i\theta}) \to 0$ uniformly in θ as $r \to \infty$, the fact that $0 \neq g(0) = \frac{1}{2\pi}\int_{-\pi}^{\pi} g(re^{i\theta})\,d\theta$ holds for all $r > 0$ leads to a contradiction. (Cf. 2.20(vii) hints, as well as SANTOS [2007].) The same idea proves Liouville's theorem 3.33(i): If $f \in H(\mathbb{C})$ is bounded and $w \in \mathbb{C}$, then the function $g(z) := \left\{ \begin{array}{ll} \frac{f(w+z)-f(w)}{z}, & z \in \mathbb{C}\setminus\{0\} \\ f'(w), & z = 0 \end{array} \right\}$ is continuous on $\mathbb{C}$, holomorphic in $\mathbb{C}\setminus\{0\}$ and satisfies $\lim_{|z|\to\infty} g(z) = 0$. Consequently, $f'(w) = g(0) = \int_0^1 g(re^{2\pi it})\,dt \to 0$ as $r \to \infty$. That is, $f'(w) = 0$ for every $w \in \mathbb{C}$, f is constant.

§ 2 Maximum Principles

With a view toward the later applications it is convenient to atomize the theory and proceed a bit more generally here than our immediate goals dictate. For the moment the next definition serves merely as a convenient abbreviation; the rationale for the terminology will emerge later—in 5.22.

Definition 5.6 Let U be an open subset of $\mathbb{C}$, $f\colon U \mapsto \mathbb{R}$ a continuous function.

(i) Say that f is *subharmonic in* U if

$$(*) \qquad f(a) \le \frac{1}{2\pi}\int_0^{2\pi} f\left(a + re^{i\theta}\right)d\theta$$

whenever $\bar{D}(a, r) \subset U$. (Cf. 5.23(i).) Say that f is *superharmonic in* U if $-f$ is subharmonic in U; i.e., if the reverse inequality always obtains in $(*)$. The integrals in $(*)$ are called *circumferential means* of f.

(ii) f is said to have the *circumferential mean-value property in* U if equality always holds in $(*)$, that is, if

$$\text{(CM)} \qquad f(a) = \frac{1}{2\pi} \int_0^{2\pi} f\left(a + re^{i\theta}\right) d\theta \quad \text{whenever } \bar{D}(a, r) \subset U.$$

If for each $a \in U$ there is *some* $r = r(a) > 0$ such that $\bar{D}(a, r) \subset U$ and the equality above holds, we say that f has the *restricted circumferential mean-value property in* U, or f has the *one-radius property in* U.

Exercise 5.7 Let U be an open subset of $\mathbb{C}$, $F: U \mapsto \mathbb{C}$ holomorphic. Show that $f := \operatorname{Re} F$ is both subharmonic and superharmonic in U and that $|F|$ is subharmonic in U. Show that $\log|F|$ is harmonic if F is zero-free. Cf. 9.33(ii).

Hint: Whenever $\bar{D}(a, r) \subset U$ use 5.5(ii) to show that the equality holds in $(*)$ above. As to $|F|$, pass $|\ \ |$ through the integral in 5.5(ii). If $F(z_0) \neq 0$, $D := D(F(z_0), |F(z_0)|)$ is a neighborhood of $F(z_0)$, so $V := F^{-1}(D)$ is a neighborhood of z_0. 3.19 supplies a holomorphic logarithm L in D, so $F = e^{L \circ F}$ and $\log|F| = \operatorname{Re}(L \circ F)$ in V.

Remark: In 5.21(iii) we shall see that for harmonic functions (CM) holds. The proof suggested for 5.7 then shows that the modulus of a harmonic function is subharmonic. 10.8(vii) describes when a harmonic function has the form $\log|F|$.

Theorem 5.8 (Maximum Principle) *Let Ω be a region, f a subharmonic function in Ω. If f attains a maximum in Ω, then f is a constant.*

Proof: Suppose f attains a maximum over Ω at some point $a \in \Omega$. Then whenever $R > 0$ is such that $D(a, R) \subset \Omega$, we have from $(*)$ above that

$$0 \leq \int_0^{2\pi} \left[f\left(a + re^{i\theta}\right) - f(a)\right] d\theta, \qquad 0 \leq r < R.$$

On the other hand, $f - f(a) \leq 0$ throughout Ω, by definition of a, so the integrand above is non-positive. As it is continuous, its integral can only be non-negative if it is identically 0. We infer that

$$f\left(a + re^{i\theta}\right) - f(a) = 0 \qquad \forall \theta \in [0, 2\pi].$$

This holds for all $r \in [0, R)$, so we conclude that

$$f(z) = f(a) \qquad \forall z \in D(a, R).$$

Let Ω_0 be the set of all points at which f attains its maximum value, M say, over Ω. Thus $\Omega_0 = f^{-1}(M)$ and by hypothesis Ω_0 is not empty. Since f is continuous, Ω_0 is relatively closed in Ω. The result of the first paragraph says that Ω_0 is open. Since Ω is connected, it follows that $\Omega_0 = \Omega$.

Scholium: This proof does not use the full force of the definition 5.6. In fact, an examination of the proof shows that the desired conclusion follows if f is continuous and for each $a \in \Omega$ there is *some* $r_a > 0$ such that inequality $(*)$ of 5.6 holds for all $r < r_a$. Another improvement of this kind is contained in

Exercise 5.9 *Let U be a bounded open subset of $\mathbb{C}$, $f\colon \overline{U} \mapsto \mathbb{R}$ a continuous function such that for each $a \in U$ there exists an $r = r(a) > 0$ with $D(a, r) \subset U$ and $f(a) \le \frac{1}{2\pi}\int_0^{2\pi} f\big(a + re^{i\theta}\big)\, d\theta$. Show that* $\max f\big(\overline{U}\,\big) = \max f(\partial U)$.

Hints: We are to show that for $M := \max f\big(\overline{U}\,\big)$, the closed subset $f^{-1}(M)$ of $\overline{U}$ meets ∂U. If this is not the case, then there exists a point $a \in f^{-1}(M) \subset U$ nearest to ∂U. From $M = f(a) \le (2\pi)^{-1}\int_0^{2\pi} f(a + r(a)e^{i\theta})\, d\theta$ infer as in the preceding proof that the whole circle $C(a, r(a))$ lies in $f^{-1}(M)$, thereby getting a point of $f^{-1}(M)$ closer to ∂U than a.

Corollary 5.10 (Maximum-Modulus Principle) *Let Ω be a region, $F \in H(\Omega)$ and $f\colon \Omega \mapsto \mathbb{R}$ harmonic. If either f or $|f|$ attains a maximum in Ω, then f is a constant. If either $|\operatorname{Re} F|$ or $|F|$ attains a maximum in Ω, then F is a constant.*

Proof: If $|f|$ attains a maximum in Ω, this is either the maximum of f or of $-f$; say the former. Since locally f is the real part of a holomorphic function g, 5.5(ii) shows that the conditions of the preceding scholium are met, so the conclusion of 5.8 may be inferred. Since $\operatorname{Re} F$ is a harmonic function, if $|\operatorname{Re} F|$ attains a maximum, then $\operatorname{Re} F$ is constant, so according to 2.16(ii), F itself is a constant. If $|F|$ attains a maximum in Ω at z_0, write $|F(z_0)| = uF(z_0)$ for some $u \in \mathbb{T}$ and observe that the maximum of $\operatorname{Re}(uF)$ in Ω is attained at z_0 and is $|F(z_0)|$. From the preceding result it follows that uF, whence F itself, is constant.

Exercise 5.11 *U is an open subset of $\mathbb{C}$, Γ a loop in U, and C is a component of $\mathbb{C} \setminus \Gamma$ in which $\operatorname{Ind}_\Gamma$ does not vanish. Show that if $C \subset U$, then*

$$\sup |f(C)| \le \sup |f(\Gamma)| \qquad \forall f \in H(U).$$

Infer that if U is simply-connected, then

$$|f(z)| \le \sup |f(\Gamma)| \qquad \forall f \in H(U)$$

holds at each $z \in U \setminus \Gamma$ for which $\mathrm{Ind}_\Gamma(z) \neq 0$.

Hints: Since C is open and relatively closed in $\mathbb{C} \setminus \Gamma$ (1.30), we have $\partial C \subset \Gamma$. Moreover, since Ind_Γ vanishes in the unbounded component of $\mathbb{C} \setminus \Gamma$, C must be a bounded component. Thus $\overline{C}$ is a compact subset of U, so by 5.10 $\sup |f(C)| = \sup |f(\partial C)| \le \sup |f(\Gamma)|$. If U is simply-connected, every component of $\mathbb{C} \setminus U$ is unbounded, so Ind_Γ vanishes in each, and therefore the component C of z in $\mathbb{C} \setminus \Gamma$ does not meet $\mathbb{C} \setminus U$.

Theorem 5.12 (Boundary Maximum Principle) *Let U be a bounded open subset of $\mathbb{C}$, h subharmonic in U and suppose that $\overline{\lim}_{z \to z_0} h(z) \le M$ for every $z_0 \in \partial U$. Then h is bounded above by M in U. If in addition h is continuous on $\overline{U}$, then* $\max h(\overline{U}) = \max h(\partial U)$. *In particular, this is applicable to $h := |f|$ with $f \in H(U)$.*

Proof: Since, according to 1.30(v), the boundary of each component of U lies in that of U, the hypothesis is also satisfied in each component of U. Therefore it is enough to treat the case of connected U. Let $M_0 := \sup\{h(z) : z \in U\}$. This is possibly infinite. In any case, there exists a sequence $\{z_n\} \subset U$ such that $h(z_n) \to M_0$. As $\overline{U}$ is compact, a subsequence of $\{z_n\}$, say without loss of generality $\{z_n\}$ itself, converges to an element $z_0 \in \overline{U}$. If $z_0 \in U$, then M_0 is finite and $h(z_0) = M_0$ by continuity of h at z_0. Then by the Maximum Principle and the connectedness of U, h is constant in U. Of course that constant is not greater than M. If $z_0 \in \overline{U} \setminus U = \partial U$, then by hypothesis $M_0 = \lim_{n\to\infty} h(z_n) \le M$.

Remarks: (i) The conclusion $h \le M$ throughout U still obtains if the hypothesis that U be bounded is dropped, provided we interpret the hypotheses on h to include the condition that

$$(*) \qquad \overline{\lim_{\substack{|z|\to\infty \\ z \in U}}}\ h(z) \le M.$$

The above proof needs only trifling changes. See 5.46 and 7.17 for other extensions. The delicacy of matters here is well illustrated by two examples. Without $(*)$ the conclusion may fail for unbounded regions ($U := \mathbb{U}$, $h(z) := \mathrm{Im}\, z$, $M := 0$) and for bounded regions even one $z_0 \in \partial U$ failing to meet the hypothesis can invalidate the conclusion ($U := D(\frac{1}{2}, \frac{1}{2})$, $h(z) := \mathrm{Re}(1/z) - 1$, $h(z_0) = 0 =: M$ for all $z_0 \in \partial U$ except $z_0 = 0$). Surprisingly, there is a simple topological condition on unbounded U which characterizes when the first conclusion in 5.12 holds for all subharmonic functions h on U; namely, that ∞

not be *pathwise accessible* from U: there exists no continuous $\gamma\colon [0,\infty) \mapsto U$ satisfying $\lim_{t\to\infty} |\gamma(t)| = \infty$. See ARMITAGE and GAUTHIER [1996]. Earlier IVERSEN [1914] had shown that if U is an unbounded region and h satisfies the hypotheses of 5.12 (but not $(*)$), then such a γ exists for which $\lim_{t\to\infty} |h(\gamma(t))| = \infty$.

(ii) Another seemingly innocuous weakening of the hypothesis in 5.12 is also fatal to its conclusion: There is a non-zero function h which is harmonic in $U := \mathbb{D}$ and satisfies $\lim_{r\uparrow 1} h(re^{i\theta}) = 0$ for every $\theta \in \mathbb{R}$. In fact, the limit exists in the stronger (non-tangential) sense of 3.13(v). See ASH and BROWN [1991].

Theorem 5.13 *Let U be an open proper subset of $\mathbb{C}$, f a bounded holomorphic function in U and suppose that $\overline{\lim}_{z\to\xi} |f(z)| \le B$ for every $\xi \in \partial U$. Then f is bounded by B throughout U.*

Proof: Replacing B by $B + \varepsilon$, $\varepsilon > 0$, it is enough to suppose that

$$\overline{\lim_{z\to\xi}}\, |f(z)| < B \qquad \forall \xi \in \partial U \tag{1}$$

and to show that this implies $|f| \le B$ throughout U. (Then let $\varepsilon \downarrow 0$.) Now U is open but not closed (since $\mathbb{C}$ is connected), so $\partial U = \overline{U} \setminus U \neq \varnothing$. Let $b \in \partial U$. Then by (1) there is an open neighborhood V of b such that

$$|f| < B \quad \text{in } U \cap V. \tag{2}$$

Since V is a neighborhood of a boundary point, this intersection is not empty and so it contains the closure of some non-empty open disk D. Now

$$\begin{aligned}\partial(U \setminus \overline{D}) &= \overline{U \setminus \overline{D}} \setminus (U \setminus \overline{D}) \subset (\overline{U} \setminus D) \setminus (U \setminus \overline{D}) \subset (\overline{U} \setminus U) \cup \overline{D} \\ &= (\partial U) \cup \overline{D} \subset (\partial U) \cup (V \cap U).\end{aligned}$$

Therefore by (1) and (2) we have

$$\overline{\lim_{z\to\xi}}\, |f(z)| < B \qquad \forall \xi \in \partial(U \setminus \overline{D}). \tag{1$'$}$$

Since we already have $|f| < B$ in $\overline{D}$ and $(1)'$ says that our original hypothesis is fulfilled by $U \setminus \overline{D}$ in the role of U, it suffices to prove the whole theorem for $U \setminus \overline{D}$, i.e., to suppose that there is a non-empty open disk D whose closure is disjoint from U. This we now assume, as well as (1). Consideration of $F(z) := f(c + Rz)$ for appropriate $c \in \mathbb{C}$ and $R > 0$ shows that we may further assume D to be $\mathbb{D}$.

Let M be a finite bound for f in U and consider a fixed $z_1 \in U$ and $n \in \mathbb{N}$. For $r > |z_1|$, $D_r := D(0,r)$ contains z_1, so by 5.12

$$|f^n(z_1)/z_1| \le M_n, \tag{3}$$

where we may take for M_n any number satisfying

$$\overline{\lim_{z\to\xi}} \left|\frac{f^n(z)}{z}\right| \le M_n \qquad \forall \xi \in \partial(U \cap D_r). \tag{4}$$

(Notice that $0 \notin U$, so $f^n(z)/z$ is holomorphic in $U \cap D_r$.) Now we have

$$\begin{aligned}\partial(U \cap D_r) &= \overline{U \cap D_r} \setminus (U \cap D_r) = \left(\overline{U \cap D_r} \setminus U\right) \cup \left(\overline{U \cap D_r} \setminus D_r\right)\\ &\subset \left(\bar{U} \setminus U\right) \cup \left(\bar{D}_r \setminus D_r\right) = \partial U \cup \partial D_r.\end{aligned} \tag{5}$$

Consider any $\xi \in \partial(U \cap D_r)$ and any $z \in U \cap D_r$ convergent to ξ. We have $|z| > 1$, since U is disjoint from the closed unit disk. Glancing back at (1) we see therefore that if $\xi \in \partial U$,

$$\overline{\lim_{z\to\xi}} \left|\frac{f^n(z)}{z}\right| \le B^n.$$

But if ξ is not in ∂U, then by (5) it is in ∂D_r, $|\xi| = r$ and we get, since $|f| \le M$

$$\overline{\lim_{z\to\xi}} \left|\frac{f^n(z)}{z}\right| \le \frac{M^n}{r}.$$

It follows that $\max\{B^n, M^n/r\}$ is an eligible M_n in (4). We have then from (3)

$$|f(z_1)| \le \left[|z_1| \max\{B^n, M^n/r\}\right]^{1/n} = |z_1|^{1/n} \max\{B, Mr^{-1/n}\},$$

valid for every positive integer n and all $r > |z_1|$. Let $r \to \infty$ and get

$$|f(z_1)| \le B|z_1|^{1/n}$$

and then let $n \to \infty$ and get (3.1)

$$|f(z_1)| \le B.$$

Theorem 5.14 (Lindelöf [1915]) *Let U be an open subset of $\mathbb{C}$, $z_0 \in U$, $r > 0$ such that $\left(\mathbb{C} \setminus \overline{U}\right) \cap C(z_0, r)$ contains an arc of length $2\pi r/k$ for some positive integer k. Let f be holomorphic and bounded by M in U and for some $\varepsilon \ge 0$ satisfy*

$$\overline{\lim_{z\to\xi}} |f(z)| \le \varepsilon \qquad \forall \xi \in D(z_0, r) \cap \partial U. \tag{$*$}$$

Then

$$|f(z_0)| \le \left(\varepsilon M^{k-1}\right)^{1/k}.$$

Proof: We may suppose that $z_0 = 0$ and that

(1) $$\mathbb{C} \setminus \overline{U} \supset \left\{ re^{i\theta} : 0 \le \theta \le \frac{2\pi}{k} \right\} =: A.$$

For each $j = 1, 2, \ldots, k$ let

$$U_j := e^{2\pi ij/k} U, \quad A_j := e^{2\pi ij/k} A, \quad U_0 := \bigcap_{j=1}^{k} U_j.$$

The arc A_j lies outside $\overline{U}_j$ and so

$$C(0, r) = \bigcup_{j=1}^{k} A_j \quad \text{lies outside} \quad \bigcap_{j=1}^{k} \overline{U}_j,$$

that is,

$$\overline{U}_0 \subset \mathbb{C} \setminus C(0, r).$$

Consequently, if Ω_0 is the component of U_0 which contains 0, we have

(2) $$\overline{\Omega}_0 \subset D(0, r).$$

Now

(3) $$\partial\Omega_0 \subset \bigcup_{j=1}^{k} \partial U_j,$$

because (by 1.30(v)) $\partial\Omega_0 \subset \partial U_0$, and $\partial U_0 \subset \bigcup_{j=1}^{k} \partial U_j$. The latter is so since if every neighborhood N of a point z meets both U_0 and $\mathbb{C} \setminus U_0 = \bigcup_{j=1}^{k} (\mathbb{C} \setminus U_j)$, then every such N meets some one $\mathbb{C} \setminus U_j$, as well, of course, as U_j. From (2) and (3) we have

(4) $$\partial\Omega_0 \subset \bigcup_{j=1}^{k} D(0, r) \cap \partial U_j = \bigcup_{j=1}^{k} e^{2\pi ij/k} [D(0, r) \cap \partial U].$$

We form

(5) $$F(z) := \prod_{j=1}^{k} f\left(ze^{-2\pi ij/k}\right), \quad z \in \Omega_0.$$

By (4) any $\xi \in \partial\Omega_0$ satisfies $\xi e^{-2\pi ij/k} \in D(0,r) \cap \partial U$ for some j. Consequently, since $e^{-2\pi ij/k}\Omega_0 \subset U$,

$$\varlimsup_{\substack{z\to\xi \\ z\in\Omega_0}} \left| f\left(ze^{-2\pi ij/k}\right)\right| = \varlimsup_{\substack{ze^{-2\pi ij/k}\to\xi e^{-2\pi ij/k} \\ z\in\Omega_0}} \left| f\left(ze^{-2\pi ij/k}\right)\right| \leq \varlimsup_{\substack{w\to\xi e^{-2\pi ij/k} \\ w\in U}} |f(w)|,$$

which, according to hypothesis (∗), does not exceed ε. Therefore

$$\varlimsup_{\substack{z\to\xi \\ z\in\Omega_0}} |F(z)| \leq \varepsilon M^{k-1} \qquad \forall \xi \in \partial\Omega_0. \tag{6}$$

Since Ω_0 is bounded (see (2)), it follows from (6) and 5.12 that

$$|F| \leq \varepsilon M^{k-1} \quad \text{throughout } \Omega_0.$$

In particular,

$$|f(0)|^k = |F(0)| \leq \varepsilon M^{k-1}.$$

Theorem 5.15 (A "Subharmonic Liouville Theorem") *If the subharmonic function $u\colon \mathbb{C} \mapsto \mathbb{R}$ satisfies*

$$\varliminf_{r\to\infty} \frac{1}{\log r} \sup_{|z|=r} u(z) \leq 0, \tag{$*$}$$

then u is constant. In particular, every non-negative function which is superharmonic in the whole plane is constant. (Cf. 5.97(ii), 7.17, 11.57(iv), and 11.60(iv).)

Proof: Let ε, $r > 0$ be given. Then $h(z) := u(z) - \varepsilon \log|z/r|$ is subharmonic in $\mathbb{C} \setminus \{0\}$ (why?) and for $R > r$

$$\sup_{|z|=R} h(z) = \sup_{|z|=R} \left[\frac{u(z)}{\log R} - \left(1 - \frac{\log r}{\log R}\right)\varepsilon\right] \cdot \log R.$$

It follows from (∗) then that

$$\varliminf_{R\to\infty} \sup_{|z|=R} h(z) = -\infty.$$

If $|w| > r$ is given, this equation insures that there exists an $R > |w|$ such that $\sup_{|z|=R} h(z) < h(w)$. Apply 5.12 to h in $U := A(r,R)$ and this inequality entails that

$$h(w) \leq \sup_{|z|=r} h(z).$$

Because of the definition of h this says that

$$u(w) - \varepsilon \log|w/r| \leq \sup_{|z|=r} u(z).$$

Here $\varepsilon > 0$ is arbitrary, so it follows that

$$u(w) \leq \sup_{|z|=r} u(z) \qquad \forall w \in \mathbb{C} \setminus D(0,r).$$

Fix $w \in \mathbb{C} \setminus \{0\}$ and let $r \downarrow 0$ to get

$$u(w) \leq u(0)$$

now valid for all $w \in \mathbb{C}$. From 5.8 infer that u is a constant.

Finally, if $v \colon \mathbb{C} \mapsto [0, \infty)$ is superharmonic, then $u := -v$ is subharmonic and trivially satisfies $(*)$.

Exercise 5.16 (Lindelöf [1909] and [1915]) *Let* $\theta_1 < \theta_2$ *with* $\theta_2 - \theta_1 \leq \pi$, $S := \{re^{i\theta} : r > 0, \theta_1 < \theta < \theta_2\}$, $f \colon \overline{S} \mapsto \mathbb{C}$ *bounded and continuous, holomorphic in* S.

(i) *Suppose that for some finite* m *we have*

$$\varlimsup_{r\to\infty} |f(re^{i\theta_1})| \leq m \quad \text{and} \quad \varlimsup_{r\to\infty} |f(re^{i\theta_2})| \leq m.$$

Show that then $\varlimsup_{\substack{|z|\to\infty \\ z\in S}} |f(z)| \leq m$.

Hints: After a rotation we may assume that $\theta_2 = -\theta_1 > 0$. Let M be a bound for f in S. Given $0 < \varepsilon < M/(m+1)$, there exists by hypothesis positive $r_1 = r_1(\varepsilon)$ such that

$$(1) \qquad |f(re^{\pm i\theta_1})| \leq m + \varepsilon \qquad \forall r \geq r_1.$$

Set $\lambda := \lambda(\varepsilon) := r_1 M/(m+\varepsilon)$ and form

$$(2) \qquad F_\varepsilon(z) := \frac{z}{z+\lambda} f(z) \qquad \forall z \in \overline{S}.$$

Since $\theta_2 - \theta_1 = 2\theta_2 \leq \pi$, we have $(-\infty, 0) \cap \overline{S} = \varnothing$, so F_ε is well defined and (so) continuous on $\overline{S}$, holomorphic in S. We have

$$(3) \quad |F_\varepsilon(re^{i\theta})| = \frac{r}{\sqrt{r^2 + 2r\lambda\cos\theta + \lambda^2}} |f(re^{i\theta})| \leq \frac{r}{\sqrt{r^2+\lambda^2}} |f(re^{i\theta})|$$

for $r \geq 0$ and $|\theta| \leq \theta_2 \leq \pi/2$. In particular,

$$(4) \qquad |F_\varepsilon| \leq M \quad \text{in } \overline{S}$$

and from (1) and (3)

$$(5) \qquad \left|F_\varepsilon\left(re^{\pm i\theta_1}\right)\right| \leq m + \varepsilon \qquad \forall r \geq r_1.$$

Furthermore, from the definition of λ

$$\frac{r_1}{\sqrt{r_1^2 + \lambda^2}} \leq \frac{r_1}{\lambda} = \frac{m+\varepsilon}{M},$$

so (3) and the definition of M give

$$(6) \qquad \left|F_\varepsilon\left(r_1 e^{i\theta}\right)\right| \leq m + \varepsilon \qquad \forall |\theta| \leq \theta_2.$$

With (4), (5) and (6) in hand we cite 5.13 to conclude that

$$|F_\varepsilon| \leq m + \varepsilon \quad \text{throughout } S \setminus \overline{D}(0, r_1),$$

and so

$$|f(z)| = \left|1 + \frac{\lambda}{z}\right| |F_\varepsilon(z)| \leq \left(1 + \frac{\lambda}{|z|}\right)(m+\varepsilon) < (1+\varepsilon)(m+\varepsilon)$$

for all $z \in S$ with $|z| > \lambda/\varepsilon = r_1(\varepsilon)M/\varepsilon(m+\varepsilon)$.

(ii) *Suppose that for some a, $b \in \mathbb{C}$ we have*

$$(*) \qquad \lim_{r\to\infty} f(re^{i\theta_1}) = a, \quad \lim_{r\to\infty} f(re^{i\theta_2}) = b.$$

Show that $a = b$ and that $\displaystyle\lim_{\substack{|z|\to\infty \\ z\in S}} f(z) = a.$

Hints: Apply (i) with the function $[f-a][f-b]$ in the role of the f there to learn that

$$(7) \qquad \lim_{\substack{|z|\to\infty \\ z\in S}} |f(z) - a||f(z) - b| = 0.$$

Given $\varepsilon > 0$, there exists by $(*)$ and (7) an $R > 0$ such that

$$(8) \qquad \left|f(re^{i\theta}) - a\right|\left|f(re^{i\theta}) - b\right| < \varepsilon^2 \qquad \forall r \geq R, \theta \in [\theta_1, \theta_2], \text{ and}$$

$$(9) \qquad \left|f(re^{i\theta_1}) - a\right| < \varepsilon, \quad \left|f(re^{i\theta_2}) - b\right| < \varepsilon \qquad \forall r \geq R.$$

Therefore the sets

$$A := \{\theta \in [\theta_1, \theta_2] : |f(Re^{i\theta}) - a| \le \varepsilon\}$$
$$B := \{\theta \in [\theta_1, \theta_2] : |f(Re^{i\theta}) - b| \le \varepsilon\}$$

are each closed (continuity of f) and non-void (by (9)) and by (8)

$$A \cup B = [\theta_1, \theta_2].$$

Since the interval is connected, we cannot therefore have $A \cap B = \varnothing$. So let $\theta \in A \cap B$. Then

$$|a - b| \le |f(Re^{i\theta}) - a| + |b - f(Re^{i\theta})| \le \varepsilon + \varepsilon.$$

As $\varepsilon > 0$ is arbitrary, we conclude that $a = b$.

Remarks: The restriction $\theta_2 - \theta_1 \le \pi$ is only a convenience. For if $0 < \theta_2 - \theta_1 \le 2\pi$, then we can work with the function $f(z^2)$ in the appropriate sector of aperture $\frac{1}{2}(\theta_2 - \theta_1) \le \pi$. The result (ii) also follows without restriction from two citations of 12.34. See 9.20 for a more drastic permissible change in the geometry of the problem and 5.56, 12.33, 12.37, and 12.44(ii) for related results. Evidently any continuous function on $\bar{S}$ for which the conclusion in (i) holds must be bounded. Nevertheless, unbounded entire functions f exist which satisfy $\lim_{r\to\infty} f(re^{i\theta}) = 0$ for every $\theta \in \mathbb{R}$. See D. J. Newman [1976] and Beardon and Minda [2003].

Even without hypothesis $(*)$ one has $f^{(k)}(re^{i\theta}) \to 0$ as $r \to \infty$ for each $\theta \in (\theta_1, \theta_2)$, $k \ge 1$. (See 5.37(iii) and 6.27(iii) below, Rogosinski [1945] and Eggleston [1951].)

The idea of using an *auxiliary function* containing an ε parameter [the summand $\varepsilon \log |z/r|$ in 5.15, the factor $z/(z + \lambda(\varepsilon))$ in 5.16], getting a global inequality and letting $\varepsilon \downarrow 0$ at the end originates with E. Phragmén and E. Lindelöf and is very versatile and powerful. We will use it often (e.g. in 12.41(ii)). Here is another elegant example of it, shown me by Sadahiro Saeki. In 5.26(ii) we will see that $h := \log(\max\{|f|, M\})$ is subharmonic whenever f is holomorphic and M is a positive finite constant. Using this fact 5.16(i), under even the (formally) weaker hypothesis that $e^{-|z|^\gamma} f(z)$ be bounded for some $\gamma < \pi/(\theta_2 - \theta_1)$, is a few-line deduction from

Theorem 5.16′ *Let $S := \{re^{i\theta} : r > 0, -\alpha < \theta < \alpha\}$, where $0 < \alpha < \pi$. Suppose U open, unbounded $\subset S$ and $h : \overline{U} \mapsto \mathbb{R}$ a continuous function which is subharmonic in U and satisfies*

(i) $$\varlimsup_{\substack{|z|\to\infty \\ z\in U}} |z|^{-\beta} h(z) \le 0$$

for some $\beta \in (0, \pi/2\alpha)$. *Then*

(ii) $$\varlimsup_{\substack{|z|\to\infty \\ z\in U}} h(z) = \varlimsup_{\substack{|z|\to\infty \\ z\in\partial U}} h(z).$$

Proof: Since the left-hand side of (ii) majorizes the right, it suffices to treat the case where

(1) $$\varlimsup_{\substack{|z|\to\infty \\ z\in\partial U}} h(z) =: m < \infty.$$

Consider any finite $m' > m$. There is a finite positive r' such that

(2) $$h(z) < m' \qquad \forall z \in \partial U \setminus D(0, r').$$

Let L denote the Principal Branch of the Logarithm in $\mathbb{C}_-$ and write $z^\beta := \exp(\beta L(z))$ for $z \in \mathbb{C}_-$. This function is holomorphic in $\mathbb{C}_- \supset \overline{U} \setminus \{0\}$. For $-\alpha \le \theta \le \alpha$, $r > 0$ and $z = re^{i\theta}$, we have

$$z^\beta = r^\beta e^{i\beta\theta}$$

(3) $$\operatorname{Re} z^\beta = r^\beta \cos(\beta\theta) \ge |z|^\beta \cos(\beta\alpha) > 0, \quad \text{and}$$

(4) $$\operatorname{Re}(1/z^\beta) = r^{-\beta} \cos(-\beta\theta) \ge |z|^{-\beta} \cos(\beta\alpha) > 0.$$

Since h is bounded on the compact set $\overline{U} \cap C(0, r')$, (4) shows that for M' sufficiently large

(5) $$h(z) < m' + M' \operatorname{Re}(1/z^\beta) \qquad \forall z \in \overline{U} \cap C(0, r').$$

Now consider any $\varepsilon > 0$ and form

$$h_\varepsilon(z) := h(z) - \big[m' + M' \operatorname{Re}(1/z^\beta) + \varepsilon \operatorname{Re}(z^\beta)\big], \quad z \in \overline{U} \setminus \{0\}.$$

This function is continuous, and it is subharmonic in U. It is trivial topology to check that for any open sets V, W

$$\partial(V \cap W) \subset \overline{V \cap W} \cap (\partial V \cup \partial W) \subset (\overline{V} \cap \partial W) \cup (\overline{W} \cap \partial V),$$

so

$$\partial(U \setminus \overline{D}(0, r')) \subset (\partial U \setminus D(0, r')) \cup (\overline{U} \cap C(0, r')).$$

Consequently, (2) and (5) together with (3) and (4) assert that

(6) $$h_\varepsilon < 0 \qquad \text{on } \partial(U \setminus \overline{D}(0, r')).$$

Moreover,

$$\begin{aligned} h_\varepsilon(z) &\leq h(z) - m' - \varepsilon \operatorname{Re} z^\beta \overset{(3)}{\leq} h(z) - m' - \varepsilon |z|^\beta \cos(\beta\alpha) \\ &= |z|^\beta \left[|z|^{-\beta} h(z) - \varepsilon \cos(\beta\alpha) \right] - m' \\ &< |z|^\beta \left[-\frac{\varepsilon}{2} \cos(\beta\alpha) \right] - m' \quad \text{for all sufficiently large } z \in U, \end{aligned}$$

by (i). Therefore, since U is unbounded,

$$\lim_{\substack{|z|\to\infty \\ z\in U}} h_\varepsilon(z) = -\infty. \tag{7}$$

From (6), (7) and the Remarks following 5.12 we infer that $h_\varepsilon < 0$ throughout $U \setminus \bar{D}(0, r')$. Let $\varepsilon \downarrow 0$ to conclude that

$$h(z) - \left[m' + M' \operatorname{Re}(1/z^\beta) \right] \leq 0 \qquad \forall z \in U \setminus \bar{D}(0, r').$$

From this and the inequalities in (4) it follows that

$$\overline{\lim_{\substack{|z|\to\infty \\ z\in U}}}\; h(z) \leq m'.$$

Since $m' > m$ is arbitrary, the desired conclusion is obtained, namely

$$\overline{\lim_{\substack{|z|\to\infty \\ z\in U}}}\; h(z) \leq m.$$

[Theorem 7.6 in MARKUSCHEWITSCH [1965] further strengthens this result.]

There will be more maximum-modulus results but they must await further developments.

§ 3 The Dirichlet Problem for Disks

The utility of the mean-value formula 5.5(ii) was well confirmed by the results of the previous section. Such an integral representation formula is likely to be even more useful if it can be made to yield the value of the function at *any* point a inside the disk as an integral around the boundary circle. Such a representation was found in 5.5(iii). But there is another, gotten by moving 0 to a via a conformal automorphism T of $\mathbb{D}$ and applying 5.5(ii) to the composite (a technique which generalizes — see 9.52):

Theorem 5.17 (Poisson Formula for Holomorphic Functions) *Let f be holomorphic in $\mathbb{D}$ and continuous on $\overline{\mathbb{D}}$. Then*

$$f(a) = \frac{1}{2\pi}\int_0^{2\pi} \operatorname{Re}\left[\frac{e^{i\theta}+a}{e^{i\theta}-a}\right] f\left(e^{i\theta}\right)d\theta = \frac{1}{2\pi}\int_0^{2\pi} \frac{1-|a|^2}{\left|e^{i\theta}-a\right|^2} f\left(e^{i\theta}\right)d\theta \quad \forall a \in \mathbb{D}.$$

Proof: Let such an f and such an a be given. The case $a = 0$ is 5.5(ii). For $a \neq 0$ consider

$$T(z) := \frac{z+a}{1+\bar{a}z}, \quad z \in \mathbb{C} \setminus \{-1/\bar{a}\}.$$

As noted in 2.6, T maps $\mathbb{D}$ conformally onto $\mathbb{D}$ and $\overline{\mathbb{D}}$ homeomorphically onto $\overline{\mathbb{D}}$. Therefore 5.5(ii) is applicable to $g := f \circ T$ and it yields

$$g(0) = \frac{1}{2\pi}\int_0^{2\pi} g\left(re^{it}\right)dt \qquad \forall\, 0 < r < 1.$$

Use uniform continuity of g on $\overline{\mathbb{D}}$ and let $r \uparrow 1$ to get

$$(1) \qquad f(a) = g(0) = \frac{1}{2\pi}\int_0^{2\pi} g\left(e^{it}\right)dt = \frac{1}{2\pi}\int_0^{2\pi} f\left(T\left(e^{it}\right)\right)dt.$$

According to 2.6

$$T^{-1}(w) = \frac{w-a}{1-\bar{a}w}, \quad \text{so } \left(T^{-1}\right)'(w) = \frac{1-|a|^2}{(1-\bar{a}w)^2} \qquad \forall w \in \mathbb{C} \setminus \{1/\bar{a}\}.$$

Now the discussion preceding 4.4 supplies us with a smooth $\phi\colon [0, 2\pi] \mapsto \mathbb{R}$ such that

$$(2) \qquad T^{-1}\left(e^{is}\right) = e^{i\phi(s)}, \quad 0 \le s \le 2\pi.$$

Therefore the chain rule gives

$$i\phi'(s)e^{i\phi(s)} = \left(T^{-1}\right)'\left(e^{is}\right)ie^{is}$$

$$(3) \qquad \phi'(s) = e^{is}\frac{\left(T^{-1}\right)'\left(e^{is}\right)}{e^{i\phi(s)}} = \frac{1-|a|^2}{\left|e^{is}-a\right|^2} \qquad \forall s \in [0, 2\pi].$$

In particular, ϕ is strictly increasing, so

$$\frac{\phi(2\pi)-\phi(0)}{2\pi} > 0.$$

On the other hand, by 4.7(iii) [with $\Gamma_1 = C(0,1)$, $\Gamma_2 = T^{-1} \circ \Gamma_1$ and f the identity function there], this quotient has modulus 1. We infer that

$$\phi(2\pi) - \phi(0) = 2\pi. \tag{4}$$

Since the integrand in (1) is 2π-periodic, that integral can just as well be evaluated over $[\phi(0), \phi(2\pi)]$ in view of (4). Hence (1) says

$$\begin{aligned} f(a) &= \frac{1}{2\pi} \int_{\phi(0)}^{\phi(2\pi)} f\left(T\left(e^{it}\right)\right) dt \\ &= \frac{1}{2\pi} \int_0^{2\pi} f\left(T(e^{i\phi(s)})\right) \phi'(s)\, ds \overset{(2)}{=} \frac{1}{2\pi} \int_0^{2\pi} f\left(e^{is}\right) \phi'(s)\, ds \\ &= \frac{1}{2\pi} \int_0^{2\pi} f\left(e^{is}\right) \frac{1-|a|^2}{\left|e^{is}-a\right|^2}\, ds \quad \text{by (3).} \end{aligned} \tag{5}$$

But direct computation confirms that for $z \in \mathbb{T}$

$$\operatorname{Re}\left[\frac{z+a}{z-a}\right] = \frac{1}{2}\left[\frac{z+a}{z-a} + \frac{\bar{z}+\bar{a}}{\bar{z}-\bar{a}}\right] = \frac{1-|a|^2}{|z-a|^2}, \qquad \text{since } z\bar{z} = 1.$$

(This function is known as the *Poisson kernel* in the unit disk.) And so (5) can also be written as

$$f(a) = \frac{1}{2\pi} \int_0^{2\pi} f\left(e^{is}\right) \operatorname{Re}\left[\frac{e^{is}+a}{e^{is}-a}\right] ds. \tag{6}$$

Remarks: The geometric significance of the argument function (logarithm) ϕ in the preceding proof suggests another way to access it. Introduce $\gamma(s) := e^{is}$, $s \in [0, 2\pi]$, and $\Gamma := T^{-1} \circ \gamma$, that is,

$$\Gamma(s) := T^{-1}(e^{is}), \qquad s \in [0, 2\pi],$$

a smooth loop. Since T maps $\mathbb{T}$ homeomorphically onto itself, Γ is a bijection of $[0, 2\pi)$ with $\mathbb{T}$, so its length is 2π, by 4.7(vi). Its arclength function

$$A(s) := \int_0^s |\Gamma'|, \qquad s \in [0, 2\pi]$$

is strictly increasing, since $A' = |\Gamma'| > 0$, and is therefore a homeomorphism of $[0, 2\pi]$ with an interval in $\mathbb{R}$ of length 2π. The initial point of Γ is $\Gamma(0) = T^{-1}(1) = \frac{1-a}{1-\bar{a}}$, which is $e^{i\alpha}$ for a unique $\alpha \in [0, 2\pi)$. After elapsed time s, the moving point Γ is describing will have *advanced* counterclockwise (why?) a length $A(s)$ along its trajectory $\mathbb{T}$ from this initial point, and will consequently be $e^{i(\alpha+A(s))}$, so that if we let ϕ denote the (smooth) function $\alpha + A$, we will have $\Gamma(s) = e^{i\phi(s)}$. In summary, we recover the representation

$$(2) \qquad T^{-1}(e^{is}) = \Gamma(s) = e^{i\phi(s)}, \qquad s \in [0, 2\pi].$$

In the literature various notational legerdemain is employed to circumvent the transitory function ϕ and move directly from (1) to (5). For example, by writing equation (1) as a contour integral (with $\gamma(t) := e^{it}$)

$$(1)^* \qquad f(a) = (2\pi i)^{-1} \int_\gamma \zeta^{-1} f(T(\zeta)) d\zeta,$$

introducing $\xi := T(\zeta)$, and computing carefully, one gets

$$(5)^* \qquad f(a) = (2\pi i)^{-1} \int_{T\circ\gamma} (1 - |a|^2)|\xi - a|^{-2} \xi^{-1} f(\xi) d\xi.$$

But of course (?), $\int_{T\circ\gamma} = \int_\gamma$, so in fact $(5)^*$ is the contour-integral version of (5). Short and sweet? Well, the only parameterization-invariance for integrals proved in most texts is the *elementary* one in the Remarks following Definition 2.8. To infer the asserted equality of integrals from it, one needs to get $T \circ \gamma$ into the form $\gamma \circ \tau$, for some smooth increasing function τ. That is what the digression above with $\phi(=: \tau^{-1})$ accomplished. When (as here) γ is a closed curve, hence not an injective function, the desired τ is apparently not simply accessible as $\gamma^{-1} \circ T \circ \gamma$. In 5.21 (iv) a proof of 5.17 not involving this issue is shown to follow from (independently proven) 5.20.

Corollary 5.18 (Cauchy-Schwarz Formulae) *If f is holomorphic in $\mathbb{D}$ and continuous on $\overline{\mathbb{D}}$, then besides* 5.17, *f has three other integral representations:*

$$(*) \qquad f(z) = i \operatorname{Im} f(0) + \frac{1}{2\pi} \int_0^{2\pi} \left[\frac{e^{it} + z}{e^{it} - z}\right] \operatorname{Re} f(e^{it})\, dt \qquad \forall z \in \mathbb{D};$$

$$(**) \qquad f(z) = -\overline{f(0)} + \frac{1}{\pi i} \int_{C(0,1)} \frac{\operatorname{Re} f(\xi)}{\xi - z} d\xi \qquad \forall z \in \mathbb{D}; \text{ and}$$

$$(***) \qquad f(z) = \frac{1}{2\pi i} \int_{C(0,1)} \frac{f(\xi)}{\xi - z} d\xi \qquad \forall z \in \mathbb{D}.$$

Proof: According to 5.17 we have

$$(1) \qquad f(z) = \frac{1}{2\pi} \int_0^{2\pi} \operatorname{Re} \left[\frac{e^{it} + z}{e^{it} - z}\right] f(e^{it})\, dt \qquad \forall z \in \mathbb{D}.$$

Upon taking real parts of both sides

$$\text{(2)} \qquad \operatorname{Re} f(z) = \operatorname{Re}\left[\frac{1}{2\pi}\int_0^{2\pi} \frac{e^{it}+z}{e^{it}-z}\operatorname{Re} f(e^{it})\,dt\right].$$

Now consider the function F defined by

$$\text{(3)} \qquad F(z) := \frac{1}{2\pi}\int_0^{2\pi} \frac{e^{it}+z}{e^{it}-z}\operatorname{Re} f(e^{it})\,dt, \quad z \in \mathbb{D}.$$

The function $(e^{it}+z)/(e^{it}-z)$ here is sometimes called the *Herglotz kernel.*

We have

$$\text{(4)} \qquad F(0) = \frac{1}{2\pi}\int_0^{2\pi} \operatorname{Re} f(e^{it})\,dt$$

$$\text{(5)} \qquad = \operatorname{Re} f(0), \quad \text{by (2)}.$$

Consequently,

$$F(z) + F(0) \overset{(4)}{=} \frac{1}{2\pi}\int_0^{2\pi}\left[\frac{e^{it}+z}{e^{it}-z}+1\right]\operatorname{Re} f(e^{it})\,dt = \frac{1}{\pi}\int_0^{2\pi}\frac{e^{it}\operatorname{Re} f(e^{it})}{e^{it}-z}\,dt$$

$$\text{(6)} \qquad = \frac{1}{\pi i}\int_{C(0,1)}\frac{\operatorname{Re} f(\xi)}{\xi - z}\,d\xi.$$

In particular, from (6) and 2.14 we have that F is holomorphic in $\mathbb{D}$. By (2) and (3) we have $\operatorname{Re} f = \operatorname{Re} F$, so from 2.16(ii) we infer that

$$f - F = f(0) - F(0) \quad \text{throughout } \mathbb{D}$$

$$\text{(7)} \qquad = i\operatorname{Im} f(0) \quad \text{by (5)}.$$

This and (3) give $(*)$. On the other hand, from (7) and (6)

$$f(z) = F(z) + i\operatorname{Im} f(0) = \frac{1}{\pi i}\int_{C(0,1)}\frac{\operatorname{Re} f(\xi)}{\xi - z}\,d\xi - F(0) + i\operatorname{Im} f(0)$$

and together with (5) this proves $(**)$. Write $f(z) = \frac{1}{2}[f(z) + i(-if(z))]$ and apply $(**)$ to the function $-if$, as well as to f, to get $(***)$.

Scholium: It follows from $(***)$ and (the proof of) 3.8 that f equals a power series about 0 in $\mathbb{D}$; see also the beginning of the proof of 5.20.

Remarks: (i) A curiosity emerges from replacing $\operatorname{Re} f$ by $\frac{1}{2}(f+\bar{f})$ in (**), splitting the integral at the $+$, and evaluating the first integral via (***):

$$\bar{f}(0) = \frac{1}{2\pi i}\int_{C(0,1)} \frac{\bar{f}(\xi)}{\xi - z}\,d\xi \qquad \forall z \in \mathbb{D}.$$

But this is de-mystified by replacing ξ with $e^{i\theta}$ and conjugating both sides: One thereupon recognizes the mean-value formula 5.5(ii) for the holomorphic function $g(\xi) := (1-\bar{z}\xi)^{-1}f(\xi)$. This is, in fact, the "exterior" Cauchy formula 11.49 (vii) applied to the function $z \mapsto \bar{f}(1/\bar{z})$, which is holomorphic in $U := \mathbb{C} \setminus \overline{\mathbb{D}}$ (by 2.4). The n there will be 1, F the constant function $\bar{f}(0)$ and the integral there is $\int_{C(0,1)} \bar{f}(1/\bar{\xi})[\xi - z]^{-1}\,d\xi = \int_{C(0,1)} \bar{f}(\xi)[\xi - z]^{-1}\,d\xi$ for $z \in \mathbb{D}$.

(ii) The proof of 5.18 deduces the Cauchy integral formula (***) from Poisson's formula 5.17. The role of the function g in (i) suggests a proof of the reverse implication: Apply (***), instead of merely the mean-value formula, to get (cf. ÅKERBERG [1961]):

$$\frac{f(z)}{1-|z|^2} = g(z) = \frac{1}{2\pi i}\int_{C(0,1)} \frac{g(\xi)}{\xi - z}\,d\xi = \frac{1}{2\pi}\int_0^{2\pi} \frac{f(e^{i\theta})e^{i\theta}}{\left(1 - e^{i\theta}\bar{z}\right)\left(e^{i\theta} - z\right)}\,d\theta$$
$$= \frac{1}{2\pi}\int_0^{2\pi} \frac{f(e^{i\theta})}{\left|e^{i\theta} - z\right|^2}\,d\theta.$$

Definition 5.19 Let U be an open subset of $\mathbb{C}$. The *Dirichlet problem for U* (also called the *first boundary-value problem of potential theory*) is the problem of finding for any continuous $h\colon \partial U \mapsto \mathbb{R}$ a continuous extension $H\colon \overline{U} \mapsto \mathbb{R}$ of h which is harmonic in U.

Solving the Dirichlet problem for a disk is easy. For 5.17 tells us that if h is harmonic in $\mathbb{D}$ and is given by $h = \operatorname{Re} f$ on $\overline{\mathbb{D}}$, where f is holomorphic in $\mathbb{D}$ and continuous on $\overline{\mathbb{D}}$, then

$$h(z) = \frac{1}{2\pi}\int_0^{2\pi} \operatorname{Re}\left[\frac{e^{it} + z}{e^{it} - z}\right] h\left(e^{it}\right) dt \qquad \forall z \in \mathbb{D}.$$

Naturally, we will use this integral to create a harmonic function $h(z)$ when all that is given is a continuous $h : \partial\mathbb{D} \mapsto \mathbb{R}$. In that case, the function on $\mathbb{D}$ defined by the above integral is designated $P[h]$ and called the *Poisson integral* of h. The factor of h in its integrand (see also (2) below) is called the *Poisson kernel*. It will be obvious that the harmonic extension into $\mathbb{D}$ of h which 5.20 produces is indeed $P[h]$. Moreover, the solution of the Dirichlet problem to follow does not actually use 5.17, so it furnishes an alternative proof of 5.17, when it is applied to $h := \operatorname{Re} f \mid \partial\mathbb{D}$ and $h := \operatorname{Im} f \mid \partial\mathbb{D}$ and 5.12 is invoked.

Since harmonic functions are determined by their boundary values (maximum principle), the Poisson integral has an interesting commutation property:

$$(0) \qquad P[h] \circ \psi = P[h \circ \psi] \qquad \text{(more precisely, } P[h \circ \psi \mid \partial\mathbb{D}]\text{)}$$

whenever $\psi : \overline{\mathbb{D}} \mapsto \overline{\mathbb{D}}$ is continuous, maps $\partial\mathbb{D}$ into $\partial\mathbb{D}$ and is holomorphic in $\mathbb{D}$. (Such functions ψ are characterized in 6.16(i) and (ii).) This is just because both sides of (0) are harmonic in $\mathbb{D}$ and have continuous extensions to $\overline{\mathbb{D}}$ equal to $h \circ \psi$ on $\partial\mathbb{D}$. The integrals in (0) give no hint of this equality. But one of them involves $h(\psi(e^{it}))dt$ and the other $h(e^{is})ds$, which might suggest, as in 5.17, the "change of variable" $e^{is} := \psi(e^{it})$. The proof of 5.17 succeeded in making such an argument rigorous. In the case of a conformal automorphism $\psi(z) := (z+a)/(1+\bar{a}z)$ (that is, ψ being the T in the 5.17 proof), an arduous elaboration of that argument *is* successful in proving (0). For the more general ψ described above it probably fails.

Theorem 5.20 (Dirichlet Problem for the Disk) *Let $h\colon \partial\mathbb{D} \mapsto \mathbb{R}$ be continuous and define*

$$f(z) := \begin{cases} \dfrac{1}{2\pi}\displaystyle\int_0^{2\pi} \frac{e^{it}+z}{e^{it}-z} h(e^{it})\,dt, & z \in \mathbb{D} \\ h(z) & z \in \partial\mathbb{D}. \end{cases}$$

Then f equals a power series about 0 *in $\mathbb{D}$ and* $\operatorname{Re} f$ *is continuous on $\overline{\mathbb{D}}$, equal to h on $\partial\mathbb{D}$.*

Proof: For fixed $z \in \mathbb{D}$ the series $\sum_{k=0}^{\infty}(ze^{-it})^k$ converges uniformly in $t \in [0, 2\pi]$ to $1/(1 - ze^{-it})$. Therefore the series

$$(*) \qquad 1 + \sum_{k=1}^{\infty} 2(ze^{-it})^k \quad \text{converges uniformly in } t \in [0, 2\pi] \text{ to } \frac{e^{it}+z}{e^{it}-z}.$$

It follows that

$$(1) \quad f(z) = \frac{1}{2\pi}\int_0^{2\pi} h(e^{it})\,dt + \sum_{k=1}^{\infty} \frac{1}{2\pi}\int_0^{2\pi} 2(ze^{-it})^k h(e^{it})\,dt = \sum_{k=0}^{\infty} c_k z^k,$$

where

$$c_0 := \frac{1}{2\pi}\int_0^{2\pi} h(e^{it})\,dt \quad \text{and} \quad c_k := \frac{1}{\pi}\int_0^{2\pi} e^{-ikt} h(e^{it})\,dt \quad \text{for } k \in \mathbb{N}.$$

Now

$$2\,\mathrm{Re}\left[\frac{e^{it}+z}{e^{it}-z}\right]=\frac{e^{it}+z}{e^{it}-z}+\frac{e^{-it}+\bar z}{e^{-it}-\bar z}=2\frac{1-|z|^2}{\left|e^{it}-z\right|^2}. \tag{2}$$

So, recalling that h is real-valued,

$$\mathrm{Re}\, f(z)=\frac{1}{2\pi}\int_0^{2\pi}\mathrm{Re}\left[\frac{e^{it}+z}{e^{it}-z}\right]h\big(e^{it}\big)\,dt=\frac{1}{2\pi}\int_0^{2\pi}\frac{1-|z|^2}{\left|e^{it}-z\right|^2}h\big(e^{it}\big)\,dt. \tag{3}$$

It remains only to show that for each $u\in\partial\mathbb{D}$

$$\lim_{\substack{z\to u\\ z\in\overline{\mathbb{D}}}}\mathrm{Re}\, f(z)\ \text{exists and equals}\ h(u).$$

Moreover, because of the definition of f on the boundary of $\overline{\mathbb{D}}$ and the continuity of h, it is enough to establish this limit relation for approach from within $\mathbb{D}$. We return to $(*)$ and notice via (2) that for $z\in\mathbb{D}$

$$\begin{aligned}\int_0^{2\pi}\frac{1-|z|^2}{\left|e^{it}-z\right|^2}\,dt&=\mathrm{Re}\int_0^{2\pi}\frac{e^{it}+z}{e^{it}-z}\,dt\\ &=\mathrm{Re}\left[\int_0^{2\pi}1\,dt+\sum_{k=1}^{\infty}\int_0^{2\pi}2\big(ze^{-it}\big)^k\,dt\right]=2\pi,\end{aligned} \tag{4}$$

facts which also follow by applying 5.17 to the constant function **1**. Therefore, writing $u=e^{is}$ with $s\in[0,2\pi)$ and recalling (3), we have for $z\in\mathbb{D}$

$$\mathrm{Re}\, f(z)-h(u)=\frac{1}{2\pi}\int_0^{2\pi}\frac{1-|z|^2}{\left|e^{it}-z\right|^2}\left[h\big(e^{it}\big)-h(u)\right]dt$$

$$|\mathrm{Re}\, f(z)-h(u)|\le\frac{1}{2\pi}\int_0^{2\pi}\frac{1-|z|^2}{\left|e^{it}-z\right|^2}\left|h\big(e^{it}\big)-h\big(e^{is}\big)\right|dt. \tag{5}$$

Set $M:=\max|h(\partial\,\mathbb{D})|$. From (3) and (4) follows $|\,\mathrm{Re}\, f|\le M$. If $\varepsilon>0$ is given, there exists $\delta\in(0,\pi)$ small enough that

$$\left|h\big(e^{it}\big)-h\big(e^{is}\big)\right|\le\varepsilon\quad\text{whenever}\quad|t-s|\le\delta. \tag{6}$$

Periodicity allows integration to be done over $[s-\delta,2\pi+s-\delta]$ in place of $[0,2\pi]$. For any $z\in\mathbb{D}$, (5) and (6) therefore give

$$\begin{aligned}|\,\mathrm{Re}\, f(z)-h(u)|&\le\frac{1}{2\pi}\int_{s-\delta}^{s+\delta}\frac{1-|z|^2}{\left|e^{it}-z\right|^2}\varepsilon\,dt+\frac{1}{2\pi}\int_{s+\delta}^{2\pi+s-\delta}\frac{1-|z|^2}{\left|e^{it}-z\right|^2}2M\,dt\\ &\overset{(4)}{\le}\varepsilon+\frac{(1-|z|^2)M}{\pi}\int_{s+\delta}^{2\pi+s-\delta}\frac{dt}{\left|e^{it}-z\right|^2}\\ &=\varepsilon+\frac{(1-|z|^2)M}{\pi}\int_{\delta}^{2\pi-\delta}\frac{d\tau}{\left|ue^{i\tau}-z\right|^2}.\end{aligned} \tag{7}$$

The point 1 is at a positive distance ρ from the compact set $\{e^{i\tau} : \delta \le \tau \le 2\pi - \delta\}$. Therefore for $z \in D(u, \rho/2)$ the integrand in (7) is majorized by $(2/\rho)^2$ and so it follows from (7) that

$$\limsup_{\substack{z \to u \\ z \in \mathbb{D}}} |\operatorname{Re} f(z) - h(u)| \le \varepsilon,$$

holding for every $\varepsilon > 0$. This limit therefore exists and is 0, which completes the proof.

Remarks and contact with Fourier analysis.

(i) Analogous arguments with the less tractable

$$\textit{conjugate Poisson kernel} := \operatorname{Im}\left(\frac{e^{it}+z}{e^{it}-z}\right) = \frac{2\operatorname{Im}(ze^{-it})}{|e^{it}-z|^2}$$

succeed in proving that in 5.20 $\operatorname{Im} f$ is also continuous on $\overline{\mathbb{D}}$ if $h(e^{i\theta})$ is a continuously differentiable function of θ (cf. also (iv) below), but not in general, with good reason: This conclusion fails for some h. Let g be a continuous, odd, real-valued function on $[-\pi, \pi]$ with $g(-\pi) = g(\pi) = 0$, $g \le 0$ on $[0, \pi]$ and $\int_0^1 t^{-1}|g(t)|\, dt = \infty$. An example of such a function is the odd extension of

$$g(t) := \frac{1}{\log\left(\frac{1}{2}\sin t\right)}, \qquad t \in [0, \pi].$$

Use 3.26 to find continuous $h\colon \partial\,\mathbb{D} \mapsto \mathbb{R}$ with $h(e^{it}) = g(t)$. Then for $0 \le x < 1$

$$\begin{aligned}\frac{1}{2\pi}\int_{-\pi}^{\pi} \frac{2\operatorname{Im}(xe^{-it})h(e^{it})}{|e^{it}-x|^2}\, dt &= \frac{2x}{\pi}\int_0^{\pi} \frac{\sin t|g(t)|}{(1-x)^2 + 2x(1-\cos t)}\, dt \\ &\ge \frac{2x}{\pi}\int_0^1 \frac{2\pi^{-1}t|g(t)|}{(1-x)^2 + xt^2}\, dt,\end{aligned}$$

since $\sin t \ge 2t/\pi$ and $1 - \cos t < t^2/2$ on $[0, \pi/2] \supset [0, 1]$; that is,

$$\frac{1}{2\pi}\int_{-\pi}^{\pi} \frac{2\operatorname{Im}(xe^{-it})h(e^{it})}{|e^{it}-x|^2}\, dt \ge \frac{4x}{\pi^2}\int_0^1 \frac{t|g(t)|}{(1-x)^2 + t^2}\, dt.$$

As $x \uparrow 1$ the last integrand increases to $\frac{t|g(t)|}{t^2}$, whose integral is ∞.

A geometric perspective on this problem is suggested by rephrasing it: There exist $F \in H(\mathbb{D})$ such that $\operatorname{Re} F$ but not $\operatorname{Im} F$ extends continuously to $\overline{\mathbb{D}}$. A conformal map of F of $\mathbb{D}$ onto an infinitely attenuated tear-drop would seem to be a candidate. A specific F that works is

$$F := \sqrt{i \sin^{-1} \circ \phi}$$

where $\phi(z) := i\frac{1+z}{1-z}$ maps $\mathbb{D}$ conformally onto $\mathbb{U}$ (2.5), sin maps the semi-finite strip $S := \{w \in \mathbb{C} : -\pi/2 < \operatorname{Re} w < \pi/2\} \cap \mathbb{U}$ conformally onto $\mathbb{U}$ (cf. 5.81), and $\surd$ denotes the square-root defined by the Principal Branch of the Logarithm in $\mathbb{C}_+$. Continuous extendability of F to $\overline{\mathbb{D}} \setminus \{1\}$ involves some work. That $\operatorname{Re} F$ but not $\operatorname{Im} F$ extends continuously to 1 depends on the fact that $\operatorname{Re} \sqrt{iw} \to 0$ and $\operatorname{Im} \sqrt{iw} \to \infty$ as $|w| \to \infty$, $w \in S$. The examples were shown to me by Sadahiro Saeki and Pietro Poggi-Corradini; another is given by Novinger [1975b]. A less explicit example in which $\operatorname{Im} F$, while not continuous in $\overline{\mathbb{D}}$, nevertheless is bounded, is given in Zalcman [1974].

(ii) For any continuous $h \colon \mathbb{T} \mapsto \mathbb{C}$ the number

$$\hat{h}(n) := \frac{1}{2\pi} \int_0^{2\pi} h(e^{it}) e^{-int}\, dt, \qquad n \in \mathbb{Z}$$

is called the nth *Fourier coefficient* of h. If h is real-valued, we clearly have

$$\hat{h}(0) \in \mathbb{R} \quad \text{and} \quad \hat{h}(-n) = \overline{\hat{h}}(n) \qquad \forall n \in \mathbb{Z}. \tag{8}$$

A glance back at (1) in the 5.20 proof shows that if f solves the Dirichlet problem for h, then

$$f(z) = \hat{h}(0) + \sum_{n=1}^{\infty} 2\hat{h}(n) z^n \qquad \forall z \in \mathbb{D} \tag{9}$$

and therefore

$$\begin{aligned} \operatorname{Re} f(re^{i\theta}) &= \hat{h}(0) + \sum_{n=1}^{\infty} \left[\hat{h}(n) r^n e^{in\theta} + \overline{\hat{h}}(n) r^n e^{-in\theta}\right] \\ &= \sum_{n\in\mathbb{Z}} \hat{h}(n) r^{|n|} e^{in\theta} \qquad \forall r \in [0,1)\,,\ \theta \in \mathbb{R}. \end{aligned} \tag{10}$$

This has two important consequences. First, given $\varepsilon > 0$, 5.20 lets us choose $r < 1$ so that

$$\left|\operatorname{Re} f(re^{i\theta}) - h(e^{i\theta})\right| < \varepsilon \qquad \forall \theta \in \mathbb{R}$$

and then (10) lets us choose $N \in \mathbb{N}$ so that

$$\left|\operatorname{Re} f(re^{i\theta}) - \sum_{n=-N}^{N} \hat{h}(n) r^{|n|} e^{in\theta}\right| < \varepsilon \qquad \forall \theta \in \mathbb{R}.$$

Together these say that

$$(11) \qquad \left| h(e^{i\theta}) - \sum_{n=-N}^{N} \hat{h}(n) r^{|n|} e^{in\theta} \right| < 2\varepsilon \qquad \forall \theta \in \mathbb{R}.$$

The sum here, a finite linear combination of the exponentials $e^{in\theta}$, is called an *exponential polynomial* and (11) is a version of Weierstrass' approximation theorem [8.29(iii)]: *Every continuous 2π-periodic function can be uniformly approximated by exponential polynomials.* The proof of that reduces at once to real-valued functions, and each such can be put in the form $h(e^{i\theta})$ [by 3.26], at which point (11) furnishes explicit approximating exponential polynomials.

As $r \uparrow 1$ the left-hand side of (10) converges (uniformly in θ) to $\operatorname{Re} f(e^{i\theta}) = h(e^{i\theta})$, but this limit does not generally commute with the summation on the right-hand side of (10), and indeed the bilateral series

$$(12) \qquad \sum_{n \in \mathbb{Z}} \hat{h}(n) e^{in\theta}, \quad \theta \in \mathbb{R},$$

called the *Fourier Series* of h, may diverge for a dense set of θ. This actually happens for many continuous functions h on $\mathbb{T}$ (§5.11, RUDIN [1987]). But the *Fourier uniqueness theorem* for continuous functions on $\mathbb{T}$ does follow:

$$(13) \qquad \hat{h}(n) = 0 \quad \forall n \in \mathbb{Z} \;\Rightarrow\; h = \mathbf{0}.$$

If (12) does converge, i.e., $\lim_{N\to\infty} \sum_{n=-N}^{N} \widehat{h}(n) e^{in\theta}$ exists in $\mathbb{C}$, for some $\theta \in \mathbb{R}$, then its value is $h(e^{i\theta})$. To see this (for real-valued h), given such θ, it is convenient to introduce $c_0 := \widehat{h}(0)$, $c_k := \widehat{h}(k)e^{ik\theta} + \widehat{h}(-k)e^{-ik\theta}$ for $k \in \mathbb{N}$. The (convergent) bilateral series (12) is just $\sum_{k=0}^{\infty} c_k$. Therefore $c_k \to 0$ and the series $\sum c_k z^k$ converges for all $z \in \mathbb{D}$. For $0 \le r < 1$, $\sum c_k r^k \overset{(10)}{=} \operatorname{Re} f(re^{i\theta})$ and consequently by 5.20 converges to $h(e^{i\theta})$ as $r \uparrow 1$. According to 3.13(iv) with $B := \{1\}$, $\sum c_k r^k$ is a continuous function of $r \in [0,1]$, so this limit $h(e^{i\theta})$ is $\sum c_k$, namely (12), as claimed.

As an attractive application, consider the continuous (why? – cf. 3.26) function h on $\mathbb{T}$ defined by $h(e^{i\theta}) := \theta^2$ for $\theta \in [-\pi, \pi]$. Then

$$\theta^2 = h(e^{i\theta}) = \pi^2/3 + \sum_{|n|=1}^{\infty} 2(-1)^n n^{-2} e^{in\theta}, \quad \theta \in [-\pi, \pi].$$

Evaluation at $\theta = 0$ recovers 3.37(iv), using which in the preceding series recovers 3.37(iii) as well.

(iii) The second important consequence of (9) is a more satisfactory version of *Parseval's formula*: Every continuous $h\colon \mathbb{T} \mapsto \mathbb{C}$ satisfies

$$\frac{1}{2\pi}\int_0^{2\pi} \left|h(e^{it})\right|^2 dt = \sum_{n\in\mathbb{Z}} \left|\hat{h}(n)\right|^2. \tag{14}$$

Again, suppose h is real-valued. From (9) above and 3.32(ii)

$$\frac{1}{2\pi}\int_0^{2\pi} \left[\operatorname{Re} f(re^{it})\right]^2 dt = \left[\hat{h}(0)\right]^2 + \frac{1}{2}\sum_{n=1}^{\infty} \left|2\hat{h}(n)\right|^2 r^{2n} \qquad \forall r \in [0,1)\,.$$

Letting $r \uparrow 1$ gives (why?)

$$\frac{1}{2\pi}\int_0^{2\pi} \left[\operatorname{Re} f(e^{it})\right]^2 dt = \left[\hat{h}(0)\right]^2 + 2\sum_{n=1}^{\infty} \left|\hat{h}(n)\right|^2.$$

Since $\operatorname{Re} f = h$ on $\mathbb{T}$ and the Fourier coefficients satisfy (8), the equality (14) follows. Complex-valued h reduces to $h_1 := \operatorname{Re} h$ and $h_2 := \operatorname{Im} h$ via the observation (i.e., the *calculation*, using $|z|^2 = \operatorname{Re}(z^2)$) that

$$\left|\widehat{h_1}(n)\right|^2 + \left|\widehat{h_2}(n)\right|^2 + \left|\widehat{h_1}(-n)\right|^2 + \left|\widehat{h_2}(-n)\right|^2 = \left|\hat{h}(n)\right|^2 + \left|\hat{h}(-n)\right|^2 \qquad \forall n \in \mathbb{Z}.$$

The validity of (14) is not confined to continuous $h\colon \mathbb{T} \mapsto \mathbb{C}$. A modest but quite useful extension involves arbitrary continuous functions $h_0\colon [0,2\pi] \mapsto \mathbb{C}$: If Fourier coefficients $\widehat{h_0}(n)$ are defined by using $h_0(t)$ in place of $h(e^{it})$ in (ii), then

$$\frac{1}{2\pi}\int_0^{2\pi} \left|h_0(s)\right|^2 ds = \sum_{n\in\mathbb{Z}} \left|\widehat{h_0}(n)\right|^2. \tag{14_0}$$

To confirm this, first for $\mathbb{R}$-valued h_0, return to the proof of 5.20 and replace $h(e^{it})$ with $h_0(t)$ and $h(u) = h(e^{is})$ with $h_0(s)$. Given $0 < \eta < \pi$, $\epsilon > 0$, there exists $0 < \delta < \eta$ such that

$$|h_0(t) - h_0(s)| \le \epsilon \text{ whenever } s,t \in [0,2\pi] \text{ and } |t-s| \le \delta. \tag{6_0}$$

Redefine h_0 at 2π so as to be 2π-periodic. No integrals are affected. Because $[s-\delta, s+\delta] \subset (0,2\pi)$ whenever $s \in [\eta, 2\pi-\eta]$, the concluding argument in 5.20, *mutatis mutandis*, proves that

$$\limsup_{z\to e^{is}} |\operatorname{Re} f(z) - h_0(s)| \le \epsilon \qquad \forall \epsilon > 0, s \in [\eta, 2\pi-\eta], \text{ that is}$$

(15) $$\lim_{z\to e^{is}} |\operatorname{Re} f(z) - h_0(s)| = 0 \text{ uniformly in } s \in [\eta, 2\pi - \eta].$$

Now from (3) and (4) we learn that

$$|\operatorname{Re} f(z)| \le \max\big|h_0[0,2\pi]\big| =: M \qquad \forall z \in \mathbb{D}.$$

From this inequality and (15)

$$\limsup_{r\uparrow 1} \int_0^{2\pi} \big|\big[\operatorname{Re} f(re^{is})\big]^2 - h_0^2(s)\big|\, ds \le \int_0^{\eta} 2M^2 + \int_{2\pi-\eta}^{2\pi} 2M^2 = 4\eta M^2,$$

valid for every $0 < \eta < \pi$. Hence

$$\lim_{r\uparrow 1} \int_0^{2\pi} \big[\operatorname{Re} f(re^{is})\big]^2\, ds = \int_0^{2\pi} h_0^2(s)\, ds.$$

The desired $(14)_0$ then follows from 3.32(ii), just as (14) did. And the complex case also follows from the real one as before.

A weak conclusion from $(14)_0$ that is useful [see (iv) below] and not *a priori* obvious is that $\widehat{h_0}(n) \to 0$ as $|n| \to \infty$. That is (*Riemann-Lebesgue lemma*), for every continuous $h_0\colon [0, 2\pi] \mapsto \mathbb{C}$

(16) $$\lim_{|n|\to\infty} \widehat{h_0}(n) = \lim_{|n|\to\infty} \frac{1}{2\pi} \int_0^{2\pi} h_0(t) e^{-int}\, dt = 0.$$

Finally, there is a standard artifice for generalizing $(14)_0$ into

(17) $$\frac{1}{2\pi} \int_0^{2\pi} f(s)\bar{g}(s)\, ds = \sum_{n\in\mathbb{Z}} \hat{f}(n)\bar{\hat{g}}(n)$$

for all continuous $f, g\colon [0, 2\pi] \mapsto \mathbb{C}$, the series being always absolutely convergent. We have only to note the *polarization identity*

$$4z\bar{w} = |z+w|^2 - |z-w|^2 + i|z+iw|^2 - i|z-iw|^2 \qquad \forall z, w \in \mathbb{C},$$

apply $(14)_0$ successively to the functions $h_0 = f \pm g$, $f \pm ig$, and add the four resulting equations.

Often $(14)_0$ and (17) are seen with all functions, integrals and Fourier coefficients based on $[-\pi, \pi]$ instead of $[0, 2\pi]$. A striking application of $(14)_0$ in this form will be found in 11.37(v). A more modest one consists of applying it to $h_0(t) := t/2$ on $[-\pi, \pi]$ to get another proof that $\sum_{n=1}^{\infty} 1/n^2 = \pi^2/6$.

(iv) A moderate hypothesis on a 2π-periodic function $h : \mathbb{R} \mapsto \mathbb{C}$ which insures uniform, even absolute, convergence of its Fourier series (12) is that it be continuously differentiable on $[0, 2\pi]$. For this enables the Fourier coefficients $\hat{h}(n)$ to be expressed via partial integration in terms of those of $h_0 := h'$ as

$$\widehat{h_0}(n) = in\hat{h}(n) \qquad \forall n \in \mathbb{Z},$$

whence, from the Cauchy–Schwarz inequality in $\mathbb{C}^{2N}$

$$\sum_{|n|=1}^{N} |\hat{h}(n)| = \sum_{|n|=1}^{N} \frac{1}{|n|}|\widehat{h_0}(n)| \leq \left[\sum_{|n|=1}^{N} \frac{1}{n^2}\right]^{1/2} \left[\sum_{|n|=1}^{N} |\widehat{h_0}(n)|^2\right]^{1/2} \qquad \forall N \in \mathbb{N}.$$

Since [by (iii)] both series on the right-hand side converge, the series (12) is absolutely convergent.

Back to 5.20, equation (1) in its proof can be written as (9) above. If now $t \mapsto h(e^{it}) \in \mathbb{R}$ is continuously differentiable, then $\sum_{n=1}^{\infty} |\hat{h}(n)| < \infty$, and so this equation shows that f itself, not just $\operatorname{Re} f$, extends continuously to $\overline{\mathbb{D}}$, as asserted in remark (i) above.

With no more effort (but some ingenuity supplied by CHERNOFF [1980]) pointwise convergence follows from a weaker hypothesis:

If $h\colon \mathbb{R} \mapsto \mathbb{C}$ is continuous and 2π-periodic and is differentiable at a point θ_0, then the series (12) *converges at $\theta = \theta_0$ to $h(\theta_0)$.*

After a trivial reduction to the case $\theta_0 = 0 = h(\theta_0)$, one considers $h_0(\theta) := \frac{h(\theta)}{e^{i\theta}-1}$ on $[-\pi, \pi]$. This function is continuous, because

$$\lim_{\theta \to 0} h_0(\theta) = \lim_{\theta \to 0} \frac{\frac{h(\theta)-h(0)}{\theta - 0}}{\frac{e^{i\theta}-e^{i0}}{\theta - 0}} = -ih'(0).$$

Since $h(\theta) = e^{i\theta}h_0(\theta) - h_0(\theta)$, it follows directly from the defining integrals that $\hat{h}(n) = \widehat{h_0}(n-1) - \widehat{h_0}(n)$ for all n, so for $\theta = \theta_0 = 0$ series (12) telescopes. Consequently (16) insures its convergence to $0 = h(\theta_0)$.

(v) In view of 5.20 it is natural to inquire whether every continuous $h\colon \mathbb{T} \mapsto \mathbb{C}$ has a *holomorphic* extension into $\mathbb{D}$. Setting $h_1 := \operatorname{Re} h$, $h_2 := \operatorname{Im} h$, the harmonic extension of h_j was seen in (10) to be

$$H_j(re^{i\theta}) = \sum_{n \in \mathbb{Z}} \widehat{h_j}(n) r^{|n|} e^{in\theta} \qquad \forall r \in [0, 1)\,,\, \theta \in \mathbb{R}.$$

The function $H := H_1 + iH_2$ is then continuous on $\overline{\mathbb{D}}$ and equal to h on $\mathbb{T}$. Is there any reason to believe that H is holomorphic in $\mathbb{D}$? Evidently $\hat{h}(n) = \widehat{h_1}(n) + i\widehat{h_2}(n)$ for every n, so

$$H(re^{i\theta}) = H_1(re^{i\theta}) + iH_2(re^{i\theta}) = \sum_{n\in\mathbb{Z}} \hat{h}(n)r^{|n|}e^{in\theta}, \quad r \in [0,1)\,,\, \theta \in \mathbb{R}.$$

This equation shows that under the hypothesis

$$\hat{h}(n) = 0 \qquad \forall n < 0 \tag{18}$$

we have

$$H(z) = \sum_{n=0}^{\infty} \hat{h}(n)z^n \qquad \forall z \in \mathbb{D}, \tag{19}$$

and so H is indeed holomorphic. This proves half of the assertion:

If $h\colon \mathbb{T} \mapsto \mathbb{C}$ is continuous, then $h = H|\mathbb{T}$ for an H which is continuous on $\overline{\mathbb{D}}$ and holomorphic in $\mathbb{D}$ if and only if (18) *holds. In that case the unique such H is specified in terms of h by* (19).

The "only if" direction is immediate from Cauchy's theorem on the boundary: If H is continuous on $\overline{\mathbb{D}}$ and holomorphic on $\mathbb{D}$, then so is $z^{n-1}H(z)$ for every $n \in \mathbb{N}$, and 5.2 gives

$$0 = \int_{C(0,1)} z^{n-1}H(z)\,dz = i\int_0^{2\pi} H(e^{it})e^{int}\,dt = 2\pi i\widehat{(H|\mathbb{T})}\,(-n)\,.$$

(vi) Putting aside holomorphic functions, thinking of harmonic functions instead as solutions of Laplace's equation [see 9.29, 9.31] and expressing the relevant partial derivatives in polar coordinates, led Joseph Fourier and others to create the harmonic extension of h from $\mathbb{T}$ into $\mathbb{D}$ directly from the Fourier coefficients of h via (10). That is, the series (10) are taken as *defining* a function $h(re^{i\theta})$ in $\mathbb{D}$, which one then proves is harmonic and converges to the given $h(e^{i\theta})$ as $r \uparrow 1$. In this sense (now called *Abel summation*) every continuous function h is "represented" by its Fourier series, a controversial thesis of Fourier's, who maintained that his series "represented" its function by actually converging pointwise to it.

This is a fruitful idea of broad applicability; e.g., it is also one way to solve the Dirichlet problem in an annulus [cf. 11.54 and the notes on it].

(vii) A second, even more important way in which the Fourier series of a continuous function f represents it is latent in Parseval's formula; this is the so-called L^2- or *mean-square convergence*: Apply Parseval to the functions $R_N(\theta) := f(\theta) - \sum_{k=-N}^{N} \hat{f}(k)e^{ik\theta}$ to get

$$\frac{1}{2\pi}\int_0^{2\pi}\left|f(\theta) - \sum_{k=-N}^{N}\hat{f}(k)e^{ik\theta}\right|^2 d\theta = \sum_{k\in\mathbb{Z}}\left|\widehat{R}_N(k)\right|^2 = \sum_{|k|>N}\left|\hat{f}(k)\right|^2$$

which converges to 0 as $N \to \infty$.

Exercise 5.21 (i) *Show that if* $h : \overline{\mathbb{D}} \mapsto \mathbb{R}$ *is continuous and is harmonic in* $\mathbb{D}$, *then* (Poisson Formula for Harmonic Functions)

$$h(z) = \frac{1}{2\pi}\int_0^{2\pi}\operatorname{Re}\left(\frac{e^{it}+z}{e^{it}-z}\right)h(e^{it})\,dt = \frac{1}{2\pi}\int_0^{2\pi}\frac{1-|z|^2}{|e^{it}-z|^2}h(e^{it})\,dt \quad \forall z \in \mathbb{D}.$$

Infer that if h *is any (real) harmonic function in* $\mathbb{D}$, *then* $h = \operatorname{Re} F$ *for some* $F \in H(\mathbb{D})$.

Hints: Define F on $\overline{\mathbb{D}}$ as h on $\partial\mathbb{D}$ and as

$$F(z) := \frac{1}{2\pi}\int_0^{2\pi}\frac{e^{it}+z}{e^{it}-z}h(e^{it})\,dt \quad \forall z \in \mathbb{D}.$$

Then $\operatorname{Re} F$ is continuous on $\overline{\mathbb{D}}$ and F is holomorphic in $\mathbb{D}$ (by 5.20). Since the harmonic functions $\operatorname{Re} F$ and h are equal on $\partial\mathbb{D}$, they must be equal throughout $\mathbb{D}$, by the maximum principle (5.10).

If h is only defined and harmonic in $\mathbb{D}$, there are, by what was just shown, holomorphic functions F_n in $D(0, 1-1/n)$ such that $h = \operatorname{Re} F_n$ in $D(0, 1-1/n)$. 2.16(ii) insures that by adding appropriate imaginary constants to successive F_n we may suppose that $F_{n+1} = F_n$ in $D(0, 1-1/n)$. The desired F is then well defined by the decree $F := F_n$ in $D(0, 1-1/n)$, $n = 2, 3, \ldots$. See 10.9 for a different proof.

(ii) *Show that* (i) *fails for a punctured disk or an annulus. Specifically, show that for the harmonic function* $f(z) := \log|z|$ (*see* 5.7) *there is no annulus centered at* 0 *in which* f *is the real part of a single holomorphic function.*

Hints: If $0 \le r < R \le \infty$, F is holomorphic in $A := A(r, R)$ and $f = \operatorname{Re} F$ in A, then $\left|ze^{-F(z)}\right| = 1$ in A, so by 5.10 or 2.16(iii) $ze^{-F(z)} = u$, for some $u \in \mathbb{T}$ and all $z \in A$. This contradicts 4.21 (or 3.42(iv)).

(iii) (Gauss Mean-Value Theorem) *Show that every harmonic function is both subharmonic and superharmonic. In other words, harmonic functions have the circumferential mean-value property.*

Hint: Use (i) and take real parts in the equation of 5.5(ii).

(iv) *Though* 5.17 *served to motivate* 5.20, *the proof of the latter is independent of the former. Show that in fact* 5.17 *is a consequence of* 5.20.

Hints: Given continuous $f : \overline{\mathbb{D}} \mapsto \mathbb{C}$ which is holomorphic in $\mathbb{D}$, apply the Poisson formula in (i) with $h := \operatorname{Re} f$ and with $h := \operatorname{Im} f$. Then to the first equation add i times the second, to get 5.17.

Exercise 5.22 (i) *Let U be an open subset of $\mathbb{C}$, $f \colon U \mapsto \mathbb{R}$ a continuous function. Say that f has the* harmonic-majorant *property* (HM) *if $f \leq F$ in V whenever V is a bounded open set with $\overline{V} \subset U$ and F is a continuous function on $\overline{V}$ which is harmonic in V and coincides with f on ∂V. Show that this condition is equivalent to the subharmonicity of f; in fact, subharmonicity of f follows if this condition is only known to hold for (small) disks V.*

Hints: If f is subharmonic and F is as above, then $f - F$ is subharmonic in V and assumes a maximum in the compact set $\overline{V}$. Now cite 5.12 and remember that $f - F = 0$ on ∂V.

Suppose f has the property (HM). If $\overline{D}(a, r) \subset U$, let F be the solution of the Dirichlet problem in $D(a, r)$ with boundary data f. Then by (HM)

$$f(a) \leq F(a) \overset{5.21(\text{iii})}{=} \frac{1}{2\pi} \int_0^{2\pi} F\left(a + re^{i\theta}\right) d\theta = \frac{1}{2\pi} \int_0^{2\pi} f\left(a + re^{i\theta}\right) d\theta.$$

(ii) *Use* (i) *to show that if U, W are open subsets of $\mathbb{C}$, ϕ maps W biholomorphically onto U and u is subharmonic in U, then $u \circ \phi$ is subharmonic in W.* (Cf. 5.97(i).)

Exercise 5.23 (Littlewood [1927]) (i) *Suppose that U is an open subset of $\mathbb{C}$ and $f \colon U \mapsto \mathbb{R}$ is a continuous function with this property: for each $a \in U$ there is a sequence $\{r_n(a)\}_{n=1}^{\infty}$ of positive numbers such that $\overline{D}(a, r_n(a)) \subset U$, $r_n(a) \to 0$ and*

$$(**) \qquad f(a) \leq \frac{1}{2\pi} \int_0^{2\pi} f\left(a + r_n(a)e^{i\theta}\right) d\theta, \quad n \in \mathbb{N}.$$

Show that f is subharmonic in U and draw the following elementary but important conclusions: In spite of the quantification "whenever $\overline{D}(a, r) \subset U$" in definition 5.6,

(a) *subharmonicity is a local property;*

and

(b) *a non-negative continuous function f is subharmonic in U if it is subharmonic in $U \setminus f^{-1}(0)$.*

Hints: Use the harmonic-majorant property: let V be any open disk with $\overline{V} \subset U$ and let F be continuous on $\overline{V}$, harmonic in V and equal to f on ∂V. Since F has the circumferential mean-value property (5.21(iii)), the present hypothesis on f is also fulfilled by $f - F$. Consequently by 5.9, $f - F$ satisfies

$$\max_{\overline{V}}(f - F) = \max_{\partial V}(f - F) = 0.$$

That is, $f \leq F$ in V. Recall 5.22.

(ii) (Extension of (i) due to SAKS [1930–32].) *Suppose that U is an open subset of $\mathbb{C}$ and $f\colon U \mapsto \mathbb{R}$ is a continuous function with this property:*

$$\overline{\lim_{r \to 0}}\, r^{-2}\left[\frac{1}{2\pi}\int_0^{2\pi} f\left(a + re^{i\theta}\right) d\theta - f(a)\right] \geq 0$$

for each $a \in U$. Show that f is subharmonic in U.

Hint: For every $\varepsilon > 0$ the function $f_\varepsilon(z) := f(z) + \varepsilon|z|^2$ fulfills the hypothesis of (i), and $f_\varepsilon \to f$ locally uniformly in U as $\varepsilon \downarrow 0$.

(iii) (Vitali-Volterra strong converse of Gauss' mean-value theorem.) *Let U be a bounded open subset of $\mathbb{C}$ for which the Dirichlet problem is solvable. Show that any continuous function $f\colon \overline{U} \mapsto \mathbb{R}$ which has the* restricted *circumferential mean-value property in U is harmonic in U.*

Hints: Let h solve the Dirichlet problem in $\overline{U}$ with boundary data $f|_{\partial U}$. Then h has the circumferential mean-value property, so both $\pm(h - f)$ have the restricted circumferential mean-value property. Apply 5.9 to each to learn that throughout U, $\pm(h - f) \leq \max_{\partial U} \pm(h - f) = 0$.

(iv) (Converse of Gauss' mean-value theorem.) *Let U be an open subset of $\mathbb{C}$, $f\colon U \mapsto \mathbb{R}$ any continuous function which has the circumferential mean-value property in U. Show that f is harmonic in U; that is, a function which is both subharmonic and superharmonic is harmonic.*

Hints: Apply (iii) in every open disk whose closure lies in U.

Remarks: The strong converse in (iii) above is insensitive to what may seem its major secondary hypothesis, that U be a Dirichlet domain, but is surprisingly sensitive to the other two secondary hypotheses: that U be bounded and that f be continuous on ∂U. KELLOGG [1934] proved that the hypothesis U be a Dirichlet domain can be dropped and FENTON [1976a] gave a delicate example with $U := \mathbb{D}$, where f is not continuous on ∂U and (iii) fails. (A simpler example is suggested in p. 22 of LITTLEWOOD [1968].) There is even an example like this with f bounded (HANSEN and NADIRASHVILI [1994]). If we anticipate a result to be proved later, we can (following FENTON [1979]) give a very instructive counter-example to (iii) when U is unbounded.

Exercise 5.24 (i) *Show that* $J_0(z) := \sum_{n=0}^{\infty} \frac{(-1)^n}{(n!)^2} \left(\frac{z}{2}\right)^{2n}$ *defines an entire function which satisfies*

$$J_0(z) = \frac{1}{2\pi} \int_0^{2\pi} e^{iz\cos\theta} \, d\theta \qquad \forall z \in \mathbb{C}.$$

(This is the important 0th-order *Bessel function.*)

Hint: Term-by-term integration of the power series appearing under the integral. It is convenient to write $(\cos\theta)^n = \left((e^{i\theta} + e^{-i\theta})/2\right)^n$ and use the binomial expansion before integrating it.

(ii) *Show that* $|J_0(z)| \leq e^{|z|} \qquad \forall z \in \mathbb{C}.$

Hint: Majorize the integrand in (i).

(iii) *Show that* $J_0(z) = 1$ *has a root* $\lambda \in \mathbb{C} \setminus \{0\}$.

Hints: $g(z) := (J_0(z) - 1)/z^2$ is entire, has value $-\frac{1}{4}$ at the origin, and is an even function. If $J_0 - 1$ has no zeros in $\mathbb{C} \setminus \{0\}$, then g is zero-free, hence of the form e^f for some entire f (4.66, 67 or 5.35). Using (ii) and Hadamard's theorem 6.37 infer that $f(z) = \alpha z + \beta$ for some constants α, β. From g even, infer that $e^{\alpha z}$ is even, so $\alpha = 0$. But then $J_0(z) - 1 = e^{\beta} z^2$, which is absurd.

(iv) *Show that the continuous function* $f(z) := e^{i\lambda \operatorname{Re} z}$ *has the restricted mean-value property (with radius* 1*) at every point of* $\mathbb{C}$*, that is,*

$$\frac{1}{2\pi} \int_0^{2\pi} f\left(z + e^{i\theta}\right) d\theta = J_0(\lambda) f(z) = f(z) \qquad \forall z \in \mathbb{C}.$$

Hint: Use (i).

(v) (Failure of the strong converse of Gauss' mean-value theorem.) *Show that the function* $u := \operatorname{Re} f$ *is not harmonic in* $\mathbb{C}$*. Contrast with the Vitali–Volterra theorem.*

Hints: If in some disk there is a continuous real-valued function v such that $u + iv$ is holomorphic, then from the Cauchy-Riemann equation and 5.3(ii) we would have $\mathbf{0} = D_{22}u = D_2(-D_1 v) = -D_1(D_2 v) = -D_{11}u$. That is, upon writing $a := \operatorname{Re} \lambda$, $b := \operatorname{Im} \lambda$, $x := \operatorname{Re} z$, $0 = e^{bx} D_{11}u(z) = (b^2 - a^2)\cos ax + 2ab \sin ax$ would hold for a whole interval of x's. This entails $a = b = 0$, a contradiction.

Exercise 5.25 (i) *Let U be an open subset of $\mathbb{C}$, $f\colon U \mapsto \mathbb{R}$ superharmonic. Suppose V is a bounded open set with $\overline{V} \subset U$, $g\colon \overline{V} \mapsto \mathbb{R}$ is continuous, harmonic in V and equal to f on ∂V. Show that the function*

$$F := \begin{cases} g & \text{in } \overline{V} \\ f & \text{in } U \setminus \overline{V}, \end{cases}$$

called a "Poisson modification" of f, is superharmonic and satisfies $f \geq F$ in U.

Hints: Evidently F is continuous on U. Also if $a \in V$ $\left[\text{respectively, } a \in U \setminus \overline{V}\right]$, then for some $r_a > 0$ we have $\overline{D}(a, r_a) \subset V$ $\left[\text{respectively, } \overline{D}(a, r_a) \subset U \setminus \overline{V}\right]$. Since g is harmonic [respectively, f is superharmonic], it follows that

$$F(a) = g(a) = \frac{1}{2\pi} \int_0^{2\pi} g\left(a + re^{i\theta}\right) d\theta = \frac{1}{2\pi} \int_0^{2\pi} F\left(a + re^{i\theta}\right) d\theta$$

[respectively,

$$F(a) = f(a) \geq \frac{1}{2\pi} \int_0^{2\pi} f\left(a + re^{i\theta}\right) d\theta = \frac{1}{2\pi} \int_0^{2\pi} F\left(a + re^{i\theta}\right) d\theta]$$

for every $r < r_a$. On the other hand, $f \geq g$ in $\overline{V}$ by the harmonic-majorant property (5.22). Therefore, $f \geq F$ in U and if $a \in \partial V$, $\overline{D}(a, r) \subset U$, then

$$F(a) = g(a) = f(a) \geq \frac{1}{2\pi} \int_0^{2\pi} f\left(a + re^{i\theta}\right) d\theta \geq \frac{1}{2\pi} \int_0^{2\pi} F\left(a + re^{i\theta}\right) d\theta.$$

We conclude from 5.23(i) that F is superharmonic in U.

(ii) *By imitating the above procedure establish the following: V, U open $\subset \mathbb{C}$, $f\colon \overline{V} \cap U \mapsto (-\infty, 1]$ continuous, superharmonic in $V \cap U$ and $f(U \cap \partial V) = 1$. Then the extension of f to U given by $f\left(U \setminus \overline{V}\right) := 1$ is superharmonic in U.*

Exercise 5.26 (i) *If f is subharmonic in $\mathbb{D}$ and $\phi\colon \mathbb{D} \mapsto \mathbb{D}$ is holomorphic with $\phi(0) = 0$, then*

$$\int_0^{2\pi} f(\phi(re^{it}))\, dt \leq \int_0^{2\pi} f\left(re^{it}\right) dt \qquad \forall r \in [0, 1)\,.$$

In particular, $r \mapsto \int_0^{2\pi} f\left(re^{it}\right) dt$ is non-decreasing. (Cf. 11.56(iii).)

Hints: Fix $r \in [0, 1)$. We need to anticipate the elementary fact (6.1) that $|\phi(z)| \le |z|$ for all $z \in \mathbb{D}$, so

$$\phi\big(D(0,r)\big) \subset D(0,r). \tag{1}$$

Let h be the solution of the Dirichlet problem in $D(0,r)$ with boundary data f. By the harmonic-majorant property

$$f \le h \quad \text{in } \bar{D}(0,r). \tag{2}$$

By (1) we may form $h \circ \phi$. This is also a harmonic function, which according to (2) satisfies

$$f\big(\phi(re^{it})\big) \le h\big(\phi(re^{it})\big) \qquad \forall t \in \mathbb{R}. \tag{3}$$

Integrate (3) and use the mean-value property of harmonic functions (5.21(iii)) on both $h \circ \phi$ and h:

$$\int_0^{2\pi} f\big(\phi(re^{it})\big)\,dt \le \int_0^{2\pi} h\big(\phi(re^{it})\big)\,dt = h\big(\phi(0)\big) = h(0) = \int_0^{2\pi} h\big(re^{it}\big)\,dt$$
$$= \int_0^{2\pi} f\big(re^{it}\big)\,dt, \quad \text{since } f = h \text{ on } \partial D(0,r).$$

If $0 \le R < r < 1$, we can use the map $\phi(z) := Rz/r$, which trivially satisfies (1), and the inequality just proved says

$$\int_0^{2\pi} f\big(Re^{it}\big)\,dt \le \int_0^{2\pi} f\big(re^{it}\big)\,dt.$$

(ii) *If F, f, g are holomorphic in the open subset U of $\mathbb{C}$, and α, β are positive real numbers, then $|f|^\alpha |g|^\beta$ and $\log^+|F| := \log\left(\max\{|F|, \mathbf{1}\}\right)$ are each subharmonic in U.*

Hints: Let $h = |f|^\alpha |g|^\beta$ or $\log^+|F|$. By 5.23(i) it suffices to show that h is subharmonic in each disk D_0 lying in $U \setminus h^{-1}(0)$. In D_0, $h > 0$, which means that $|fg| > 0$ or $|F| > 1$ in D_0, as the case may be. Hence f, g and F have holomorphic logarithms f_0, g_0, F_0 in D_0 [4.66 and 4.67]. Thus in D_0, $h = \left|e^{\alpha f_0 + \beta g_0}\right|$ or $\operatorname{Re} F_0$, as the case may be; each of these is subharmonic in D_0 by 5.7.

Here is a nice application:

(iii) *Let f be entire and for some non-constant polynomial P and some positive real α and K satisfy*

$$|f(z)|\,|\operatorname{Re} P(z)|^\alpha \le K \qquad \textit{for all } z \in \mathbb{C}. \tag{$*$}$$

Show that $f = \mathbf{0}$.

Hints: (Saeki) We may suppose $K = 1$. Write $P(z) := \sum_{j=0}^{n} c_j z^j$ with $n \geq 1$ and $c_n \neq 0$, and introduce $Q(w, z) := \sum_{j=0}^{n} \bar{c}_j w^{2j} z^{n-j}$. Since $\bar{z} = |z|^2 z^{-1}$, we see that $\overline{P(z)} = z^{-n} Q(|z|, z)$, so

$$2|\mathrm{Re}\, P(z)| = \left|P(z) + \overline{P(z)}\right| = \left|z^{-n}[z^n P(z) + Q(|z|, z)]\right|$$

and $(*)$ reads

$$\begin{aligned} &|f(z)| \leq 2^\alpha |z|^{n\alpha} |z^n P(z) + Q(|z|, z)|^{-\alpha}, && \text{that is,} \\ &|z^n P(z) + Q(|z|, z)|^\alpha |f(z)| \leq 2^\alpha |z|^{n\alpha} && \forall z \in \mathbb{C}. \end{aligned}$$

Therefore, for z on the circle $C(0, r)$, $r > 0$, we have

$$|z^n P(z) + Q(r, z)|^\alpha |f(z)| \leq 2^\alpha r^{n\alpha}.$$

Since the sum on the left is a polynomial, (ii) and the maximum principle for subharmonic functions insure that this inequality persists for all $z \in D(0, r)$. In particular it holds for $z = 0$: $|Q(r, 0)|^\alpha |f(0)| \leq 2^\alpha r^{n\alpha}$. That is, since $Q(r, 0) = \bar{c}_n r^{2n}$,

$$|f(0)| \leq \left(2|c_n|^{-1} r^{-n}\right)^\alpha \qquad \forall r > 0.$$

It follows that $f(0) = 0$. For any $a \in \mathbb{C}$, the entire function $f_a(z) := f(z + a)$ and the polynomial $P_a(z) := P(z + a)$ satisfy the hypotheses which f and P do, so $0 = f_a(0) = f(a)$ follows. [This proof also works if, instead of $(*)$, $f(z)$ is majorized by $\sum_{k=1}^{N} |z|^{\beta_k} |\mathrm{Re}\, P_k(z)|^{-\alpha_k}$ for polynomials P_k and non-negative β_k, α_k, the conclusion then being that f is either constant or a polynomial of degree not exceeding $\max\{\beta_k - \alpha_k \deg P_k : 1 \leq k \leq N\}$. Cf. 3.34(iv).]

(iv) (Erdös & Radó [1948]) *P and Q are monic polynomials of positive degree n and m, respectively, and $|Q(z)| \leq 1$ whenever $|P(z)| = 1$. Show that $P^m = Q^n$. Infer that P, Q are powers of a common polynomial.*

Hints: Replacing P, Q by P^m, Q^n, shows that it suffices to prove that $P = Q$ if additionally $\deg P = \deg Q$. Being monic then

$$\lim_{|z| \to \infty} Q(z)/P(z) = 1. \tag{1}$$

Let Ω denote the unbounded component of $U := \{z \in \mathbb{C} : |Q(z)| > 1\}$. Ω is open, contains $\mathbb{C} \setminus D(0, r_0)$ for some finite r_0, and satisfies

$$\partial\Omega \subset \partial U \subset \{z \in \mathbb{C} : |Q(z)| = 1\}. \tag{2}$$

Since the sets given by $\{z \in \mathbb{C} : |P(z)| > 1\}$, $\{z \in \mathbb{C} : |P(z)| < 1\}$, and $\{z \in \mathbb{C} : |P(z)| = 1\}$ cover $\mathbb{C}$ and the latter is disjoint from Ω, thanks to the hypothesis on Q, the first two sets cover Ω and effect a disconnection of Ω, if each of them meets it. Since the first set evidently does meet Ω, the second cannot. That is, $|P| > 1$ throughout Ω. Therefore $F := Q/P$ is holomorphic in Ω and, in view of (2),

$$|F| = 1/|P| \le 1 \text{ on } \partial\Omega. \tag{3}$$

For each $r > r_0$, (1) furnishes $r' \ge r$ such that

$$1 - 1/r < \operatorname{Re} F(z) \le |F(z)| < 1 + 1/r \qquad \forall |z| \ge r'. \tag{4}$$

The boundary of $D(0, r') \cap \Omega$ lies in $C(0, r') \cup \partial\Omega$, so (3), (4) and the maximum-modulus principle insure that $|F| < 1 + 1/r$ in $D(0, r') \cap \Omega$. Letting $r \uparrow \infty$ shows that

$$|F| \le 1 \text{ throughout } \Omega. \tag{5}$$

For all $r > r_1 > r_0$ we have from (4) and 5.5(i)

$$1 - \frac{1}{r} \le \frac{1}{2\pi} \int_0^{2\pi} \operatorname{Re} F(r' e^{i\theta})\, d\theta = \frac{1}{2\pi} \int_0^{2\pi} \operatorname{Re} F(r_1 e^{i\theta})\, d\theta.$$

Letting $r \uparrow \infty$, it follows that $\frac{1}{2\pi} \int_0^{2\pi} \operatorname{Re} F(r_1 e^{i\theta})\, d\theta \ge 1$ and then on account of (5) we are forced to conclude that

$$\operatorname{Re} F(r_1 e^{i\theta}) = 1 = \left|F(r_1 e^{i\theta})\right| \qquad \forall \theta \in [0, 2\pi].$$

As this holds for all $r_1 > r_0$ we have

$$F = 1 \text{ throughout } \mathbb{C} \setminus \bar{D}(0, r_0),$$

which means that $Q = P$ throughout $\mathbb{C} \setminus \bar{D}(0, r_0)$. Therefore the polynomial $Q - P$ is identically 0. [As soon as 5.62 (or even 5.30) is available, the interested reader will see how to generalize this result to any entire functions in the roles of P, Q which satisfy (1).]

For the last claim, factor P, Q, noting that they have the same zeros.

Exercise 5.27 (Cf. 6.34) *Let $f \in H(\mathbb{D})$ satisfy $f(0) = 0$, $|\operatorname{Re} f| \le A$. Show that*

$$|\operatorname{Im} f| \le \frac{2A}{\pi} \log\left(\frac{1+r}{1-r}\right) \quad \text{on } \bar{D}(0, r) \qquad \forall 0 \le r < 1.$$

Hint: From the Cauchy-Schwarz formula 5.18(∗) applied to $f(Rz)$, with $z \in D(0, 1/R) \supset \overline{\mathbb{D}}$, deduce that

$$\operatorname{Im} f\left(Rre^{i\psi}\right) = \frac{1}{2\pi}\int_{-\pi}^{\pi} \frac{2r\sin t}{1 - 2r\cos t + r^2} \operatorname{Re} f\left(Re^{i(\psi - t)}\right) dt,$$

$$0 \leq r, R < 1, \psi \in \mathbb{R}.$$

Write $h(t) := 1 - 2r\cos t + r^2$ and have

$$\left|\operatorname{Im} f\left(Rre^{i\psi}\right)\right| \leq \frac{1}{\pi}\int_0^{\pi} \frac{h'}{h} A = \frac{A}{\pi}[\log(h(\pi)) - \log(h(0))].$$

Exercise 5.28 (Harnack [1886]) (i) *If h is continuous on $\overline{D}(c, R)$, harmonic and non-negative in $D(c, R)$, then*

$$(*) \qquad \frac{R - |z - c|}{R + |z - c|} h(c) \leq h(z) \leq \frac{R + |z - c|}{R - |z - c|} h(c) \qquad \forall z \in D(c, R)$$

Hint: Routine changes of variables suffice to infer (∗) from its special case $c = 0$, $R = 1$. In that case, look at the integral representation of h in $\overline{\mathbb{D}}$ given by 5.21(i) and use obvious lower and upper bounds for its Poisson kernel

$$\operatorname{Re}[(e^{it} + z)/(e^{it} - z)] = (1 - |z|^2)/|e^{it} - z|^2.$$

(ii) *For every region Ω and compact $K \subset \Omega$ there exists a finite positive constant $c = c(K, \Omega)$ with this property:*

$$(**) \qquad \frac{1}{c} \leq \frac{h(z)}{h(w)} \leq c$$

holds for all z, $w \in K$ and all positive harmonic functions h on Ω.

Hints: Let $\mathcal{P}$ denote the set of positive harmonic functions in Ω. Given z_0, $w_0 \in \Omega$, there is a (polygonal) curve $\gamma : [0, 1] \mapsto \Omega$ with $\gamma(0) = z_0$, $\gamma(1) = w_0$. By compactness there is a real $r > 0$ such that $\overline{D}(\gamma(t), 2r) \subset \Omega$ for all $t \in [0, 1]$. Uniform continuity of γ means there exists $n \in \mathbb{N}$ such that $|\gamma(t) - \gamma(t')| < r$ whenever $|t - t'| \leq 1/n$. Set $t_j := j/n$ ($j = 0, 1, \ldots, n$). Then $\gamma(t_j) \in D(\gamma(t_{j-1}), r)$ ($j = 1, 2, \ldots, n$). Apply (i) with $R = 2r$, $c = \gamma(t_j)$, $z = \gamma(t_{j-1})$ to see that for every $h \in \mathcal{P}$

$$\frac{h(\gamma(t_{j-1}))}{h(\gamma(t_j))} \leq \frac{2r + |\gamma(t_{j-1}) - \gamma(t_j)|}{2r - |\gamma(t_{j-1}) - \gamma(t_j)|} \leq \frac{2r + r}{2r - r} = 3 \qquad (j = 1, 2, \ldots, n).$$

From $n+2$ applications of $(*)$ we learn that for any $h \in \mathcal{P}$, $z \in D(\gamma(0), r)$, $w \in D(\gamma(1), r)$

$$\frac{h(z)}{h(w)} = \frac{h(z)}{h(\gamma(t_0))} \cdot \frac{h(\gamma(t_0))}{h(\gamma(t_1))} \cdots \frac{h(\gamma(t_{n-1}))}{h(\gamma(t_n))} \cdot \frac{h(\gamma(t_n))}{h(w)} \leq 3^{n+2}.$$

This means that to every $(z_0, w_0) \in \Omega \times \Omega$ is associated a neighborhood $N(z_0, w_0)$ and a power of 3, $C(z_0, w_0)$, such that

$$\frac{h(z)}{h(w)} \leq C(z_0, w_0) \qquad \forall (z, w) \in N(z_0, w_0), h \in \mathcal{P}.$$

For any compact $K \subset \Omega$, there exist $(z_k, w_k) \in \Omega \times \Omega$ $(k = 1, \ldots, m)$ such that the neighborhoods $N(z_k, w_k)$ cover $K \times K$. The number

$$c(K, \Omega) := \max\{C(z_k, w_k) : k = 1, \ldots, m\}$$

does what's wanted.

(iii) *h_n are positive harmonic functions in the region Ω and the sequence $\{z_n\}$ lies in a compact subset of Ω. Use* (ii) *to show that*

(a) *if $\{h_n(z_n)\}$ is bounded, then $\{h_n\}$ is bounded on each compact subset of Ω;*

(b) *if $\lim_{n\to\infty} h_n(z_n) = \infty$, then $h_n \to \infty$ uniformly in each compact subset of Ω;*

(c) *if $\lim_{n\to\infty} h_n(z_n) = 0$, then $h_n \to \mathbf{0}$ uniformly in each compact subset of Ω.*

Exercise 5.29 (Tsuji [1955]) *Let $\gamma\colon (0,1] \mapsto \mathbb{H}$ be a continuous function such that $\operatorname{Im}\gamma / \operatorname{Re}\gamma$ is bounded and $\lim_{t\to 0}\gamma(t) = 0$. Let h be a positive harmonic function in $\mathbb{H}$. For each positive real c, define the triangular set*

$$T_c := \{x + iy : 0 < x \leq c, |y| \leq cx\}.$$

Show that

(i) *if h is bounded on γ, then h is bounded on T_c for every $c > 0$;*

(ii) *if $\lim_{t\to 0} h(\gamma(t)) = \infty$, then $\lim_{\substack{z\to 0\\ z\in T_c}} h(z) = \infty$ for every $c > 0$;*

(iii) *if $\lim_{t\to 0} h(\gamma(t)) = 0$, then $\lim_{\substack{z\to 0\\ z\in T_c}} h(z) = 0$ for every $c > 0$.*

Hints: It suffices to consider only c satisfying $c \geq 1 + 1/\operatorname{Re}\gamma(1)$ and $c \geq |\operatorname{Im}\gamma(t)|/\operatorname{Re}\gamma(t)$ for all $t \in (0,1]$. Then the range of $\operatorname{Re}\gamma$ contains $(0, \operatorname{Re}\gamma(1)] \supset (0, c^{-1}]$, so for each $n \in \mathbb{N}$ there exists t_n such that $\operatorname{Re}\gamma(t_n) = c^{-n}$. Since for each $0 < \varepsilon < 1$, $\gamma[\varepsilon, 1]$ is a compact subset of $\mathbb{H}$, hence a subset of $\{z : \operatorname{Re} z > c^{-n}\}$ for all large n, it follows that

$$(1) \qquad t_n \to 0 \quad \text{as} \quad n \to \infty.$$

The functions $h_n(z) := h(c^{-n}z)$ are harmonic and positive in $\mathbb{H}$ and the points $z_n := c^n\gamma(t_n)$ lie in the compact subset $K := \{x + iy : 1 \le x \le c^2, |y| \le cx\}$ of $\mathbb{H}$, because $c \ge |\operatorname{Im}\gamma|/\operatorname{Re}\gamma$. Thus

$$(2) \qquad h(\gamma(t_n)) = h_n(z_n) \quad \text{and} \quad z_n \in K \qquad \forall n \in \mathbb{N}.$$

Notice that for any $k \in \mathbb{N}$

$$(3) \quad \bigcup_{n \ge k} h_n(K) = \bigcup_{n \ge k} h(c^{-n}K) = h\{x + iy : 0 < x \le c^{-k+2}, |y| \le cx\}.$$

From (2), (3) [with $k = 1$] and 5.28(iii)(a) it follows that boundedness of h on γ implies boundedness of h on T_c. On the other hand, (1), (2), and (3) show that (ii) follows from 5.28(iii)(b) and (iii) follows from 5.28(iii)(c).

§ 4 Existence of Power Series Expansions, Cauchy's Formula, Removable Singularities

Returning to the fundamental 5.18 we derive at once (cf. 5.3):

Corollary 5.30 *If U is an open subset of $\mathbb{C}$ and f is $\mathbb{C}$-differentiable in U, then f is infinitely $\mathbb{C}$-differentiable in U and for each $z_0 \in U$ the series*

$$\sum_{n=0}^{\infty} \frac{f^{(n)}(z_0)}{n!}(z - z_0)^n,$$

called the Taylor series *of f at z_0, converges to $f(z)$ in every open disk centered at z_0 which lies in U.*

Proof: Given such a z_0 and such a disk $D(z_0, r)$, we just take an $0 < R < r$ and consider the function $h(z) := f(z_0 + Rz)$. This function is holomorphic in $\mathbb{D}$ and continuous in $\overline{\mathbb{D}}$. By the scholium to 5.18 we have then $h(z) = \sum_{n=0}^{\infty} a_n z^n$ for all $z \in \mathbb{D}$ and some $a_n \in \mathbb{C}$. Therefore

$$(1) \qquad f(z) = \sum_{n=0}^{\infty} c_n(z - z_0)^n \qquad \forall z \in D(z_0, R)$$

with $c_n := a_n R^{-n}$. By 3.6 we get from (1) that f is infinitely differentiable in $D(z_0, R)$ and that $f^{(n)}(z_0) = n!c_n$. In particular, the c_n do not depend on R and so, as R in (1) was arbitrary in $(0, r)$, we get from (1)

$$(2) \qquad f(z) = \sum_{n=0}^{\infty} \frac{f^{(n)}(z_0)}{n!}(z - z_0)^n \qquad \forall z \in D(z_0, r).$$

Remarks: (i) As Cauchy himself stressed, the mere fact 5.30 ought to be supplemented with error estimates, and 3.34(i) provides reasonable ones: If $\bar{D}(z_0, r) \subset U$ and $M := \max\{|f(z)| : |z - z_0| = r\}$, then for all $N \in \mathbb{N}$, $z \in D(z_0, r)$

$$\left| f(z) - \sum_{n=0}^{N} \frac{f^{(n)}(z_0)}{n!}(z - z_0)^n \right| \leq M\left(1 - \frac{|z - z_0|}{r}\right)^{-1} \left(\frac{|z - z_0|}{r}\right)^{N+1}.$$

(ii) If U is $\mathbb{D}$, f is continuous in $\bar{\mathbb{D}}$, and $S_N(z) := \sum_{n=0}^{N} f^{(n)}(0)z^n/n!$, the sequence $\{S_N\}_{N \in \mathbb{N}}$ need not converge uniformly in $\mathbb{D}$ or converge at all on $\mathbb{T}$ or be (uniformly) bounded in $\mathbb{D}$. See the notes on bizarre boundary behavior in chapter III.

(iii) In Weierstrass' theory an *analytic* function is *defined* as one that has a local power series representation at each point. The reader may suspect that 5.30 is not trivial even for such functions. How Weierstrass proved convergence in the *largest* disk lying in the domain, without recourse to integrals, is detailed in Ullrich [1990] and Remmert [1991], p. 234 references such a proof by Adolf Hurwitz. Here is another way, from G. T. Whyburn [1962b]. It uses only power-series results, those recounted in the Remarks following 3.35. Suppose f is analytic in $\mathbb{D}$ and $f(z) = \sum_{n=1}^{\infty} c_n z^n$ in $D(0, r_0)$, for some $0 < r_0 < 1$. Fix $k \in \mathbb{N}$ and form $F(z) := \sum_{j=0}^{2k-1} f(ze^{\pi ij/k})$. This function is also analytic in $\mathbb{D}$ and in $D(0, r_0)$ it is represented by a power series whose lowest-degree term is $2kc_k z^k$. The function g, defined as $2kc_k$ at 0 and as $z^{-k}F(z)$ elsewhere, is evidently analytic in $\mathbb{D}$. (At any $p \neq 0$, $1/z = \sum p^{-j-1}(z - p)^j$ whenever $|z - p| < |p|$, so g is a product of $k+1$ power series centered at p and 3.12(iv) can be repeatedly cited.) We have $F(z) = z^k g(z)$ for all $z \in \mathbb{D}$. For each $0 < r < 1$ let $M(r) := \max\{|f(z)| : |z| = r\}$. By the maximum-modulus principle (for analytic functions, 3.33(iii)) and the definitions of g and F

$$|g(0)| \leq \max_{|z|=r} |g(z)| = r^{-k} \max_{|z|=r} |F(z)| \leq r^{-k} 2kM(r).$$

That is, since $g(0) = 2kc_k$,

$$(3.k) \qquad |c_k| r^k \leq M(r), \quad \text{whenever } 0 < r < 1, \quad k \in \mathbb{N}.$$

For each $N \in \mathbb{N}$ consider $R_N(z) := f(z) - \sum_{n=1}^{N} c_n z^n$. The inequalities (3.$k$) give the crude bound

$$(4) \qquad |R_N(\xi)| \le (N+1)M(r) \qquad \forall \xi \in C(0,r).$$

Just as before, there exists an analytic function h in $\mathbb{D}$ such that $R_N(z) = z^N h(z)$ for all $z \in \mathbb{D}$. Apply the maximum-modulus principle 3.33(iii) to get

$$|h(z)| \le \max_{|\xi|=r} |h(\xi)| = r^{-N} \max_{|\xi|=r} |R_N(\xi)| \overset{(4)}{\le} r^{-N}(N+1)M(r) \quad \forall z \in D(0,r).$$

Given $0 < \rho < 1$, take $r := (1+\rho)/2$. Then

$$(5) \qquad |R_N(z)| = |z|^N |h(z)| \le \rho^N r^{-N}(N+1)M(r) \qquad \forall z \in D(0,\rho).$$

Since $\rho/r < 1$, we know $N(\rho/r)^N \to 0$ as $N \to \infty$. Consequently, (5) says $\lim_{N\to\infty} R_N(z) = 0$ uniformly in $D(0,\rho)$, for each $\rho < 1$. Alternatively, (3.k) shows that $\sum c_n z^n$ converges in $D(0,r)$ for every $r < 1$ and the uniqueness theorem for *analytic* functions (3.35(viii)) can be applied to $f(z) - \sum c_n z^n$.

(iv) The series (2) in 5.30 could have radius of convergence $R > r$. In such a case its value at points $z \in U \cap A(z_0; r, R)$ need not equal $f(z)$. Trivial examples with disconnected U are immediate. The most natural example with a connected U is $U := \mathbb{C}_+$, $z_0 := 1 - i$, $r := 1$, and $f := L$, the Principal Branch of the Logarithm in $\mathbb{C}_+$ [3.24(iii)]. Then (2) has radius of convergence $R := \sqrt{2}$ [see 3.19]. If series (2) were equal to $L(z)$ in the whole $\mathbb{C}_+ \cap D(z_0, \sqrt{2})$, then L would thereby be continuously extendible to points of $\mathbb{R} \cap D(z_0, \sqrt{2})$, which it clearly is not. However, in general, the series does represent f in every *region* $\Omega \subset U \cap D(z_0, R)$ which meets $D(z_0, r)$, by 5.62.

Exercise 5.31 (i) (Cauchy Product of Series) *Show that if* $f(z) := \sum_{n=0}^{\infty} a_n z^n$ *and* $g(z) := \sum_{n=0}^{\infty} b_n z^n$ *have radii of convergence* $R_1 > 0$ *and* $R_2 > 0$, *then* $h(z) := \sum_{n=0}^{\infty} c_n z^n$ *converges for all* $|z| < R := \min\{R_1, R_2\}$ *and* $h(z) = f(z)g(z)$ *for all such* z, *where* $c_n := \sum_{k=0}^{n} a_k b_{n-k}$ *for* $n = 0, 1, 2, \ldots$.

Hint: 3.12(iv) is one proof; here is another. The functions f and g are both holomorphic in $D(0,R)$ and $f^{(n)}(0)/n! = a_n$, $g^{(n)}(0)/n! = b_n$ by 3.6. Therefore $f \cdot g$ is holomorphic in $D(0,R)$ and

$$f(z)g(z) = (f \cdot g)(z) = \sum_{n=0}^{\infty} \frac{(f \cdot g)^{(n)}(0)}{n!} z^n \qquad \text{in } D(0,R)$$

by the preceding result. Finally, verify by induction on n the so-called *Leibniz formula:*

$$(f \cdot g)^{(n)} = \sum_{k=0}^{n} \binom{n}{k} f^{(k)} \cdot g^{(n-k)}.$$

(ii) *Part* (i) *suggests (by analogy) the question how to construct the Taylor series of $f \circ g$ at* 0 *from those of g at* 0 *and f at $g(0)$. Can the two methods used in* (i) *be harnessed to do this job?*

Hints: If $g(z) = \sum_{k=0}^{\infty} b_k z^k$ and $f(w) = \sum_{m=0}^{\infty} a_m (w - b_0)^m$, then the coefficient c_n of z^n in the Taylor series of $f \circ g$ at 0 involves only a_m and the series $(g(z) - b_0)^m = (\sum_{k=1}^{\infty} b_k z^k)^m$ for $m \le n$. Actually, only the *polynomials* $(\sum_{k=1}^{n} b_k z^k)^m$, and their multinomial expansions, enter the computations. Careful bookkeeping results, for $n \in \mathbb{N}$, in

(1)

$$c_n = \frac{(f \circ g)^{(n)}(0)}{n!} = \sum_{m=1}^{n} a_m \sum_{S^*_{n,m}} b_{k_1} \dots b_{k_m} = \sum_{k=1}^{n} a_k \sum_{S_{k,n}} \frac{k!}{k_1! \dots k_n!} b_1^{k_1} \dots b_n^{k_n},$$

where $S^*_{n,m} := \{(k_1, \dots, k_m) \in \mathbb{N}^m : k_1 + k_2 + \dots + k_m = n\}$ and $S_{k,n} := \{(k_1, \dots, k_n) \in \mathbb{Z}_+^n : k_1 + k_2 + \dots + k_n = k$ and $k_1 + 2k_2 + \dots + nk_n = n\}$. Naturally, after the a's and b's in (1) are replaced by derivatives divided by factorials, one gets a recipe for $(f \circ g)^{(n)}$ somewhat analogous to Leibniz' for $(f \cdot g)^{(n)}$, namely, *Faà di Bruno's formula*

$$(2)\quad (f \circ g)^{(n)} = \sum_{k=1}^{n} (f^{(k)} \circ g) \cdot \sum_{S_{k,n}} \frac{n!}{k_1! \dots k_n!} \left(\frac{g^{(1)}}{1!}\right)^{k_1} \dots \left(\frac{g^{(n)}}{n!}\right)^{k_n}, \qquad n \in \mathbb{N}.$$

It is valid for n-times differentiable functions in $\mathbb{R}$, as well as holomorphic functions in $\mathbb{C}$. A proof in this greater generality is accomplished by reduction to Taylor polynomials, as outlined here, by SPINDLER [2005], who gives more details and also points out that Francesco Faà di Bruno (1825–1888) was canonized a saint in 1988!

Exercise 5.32 (i) *Let Ω be a region in $\mathbb{C}$, $z_0 \in \Omega$, f a zero-free holomorphic function in Ω. For each $z \in \Omega$ let γ_z be a piecewise smooth curve in Ω joining z_0 to z. Show that*

$$\exp\left(\int_{\gamma_z} \frac{f'}{f}\right) = \frac{f(z)}{f(z_0)} \qquad \forall z \in \Omega.$$

Hints: First note that by 5.30 f' is holomorphic, so these integrals exist. Define

$$L(z) := \int_{\gamma_z} \frac{f'}{f}, \qquad z \in \Omega,$$

$$(1)\qquad F := e^L.$$

Consider a fixed disc $\overline{D}(z_1, r) \subset \Omega$ and set

$$L_1(z) := \int_{[z_1,z]} \frac{f'}{f} \qquad \forall z \in D(z_1, r).$$

For $z \in D(z_1, r)$ the curves γ_{z_1}, γ_z, $[z_1, z]$ can be assembled into a loop Γ in Ω in such a way that

$$L(z) - L(z_1) - L_1(z) = \int_\Gamma \frac{f'}{f} = 2\pi i \mathrm{Ind}_{f\circ\Gamma}(0) \in 2\pi i\mathbb{Z}. \tag{2}$$

Similarly, since γ_{z_0} is a loop, we have

$$L(z_0) \in 2\pi i\mathbb{Z}. \tag{3}$$

It follows from (2) that

$$F(z) = e^{L(z_1)+L_1(z)} \qquad \forall z \in D(z_1, r). \tag{4}$$

Since L_1 is holomorphic with $L_1' = f'/f$ [cf. the proof of 2.11], it then follows from (4) that F is holomorphic with $F' = F \cdot L_1' = F \cdot f'/f$, that is, $\big(F/f\big)' = 0$ in $D(z_1, r)$. Since z_1 is arbitrary, we have shown that F/f is holomorphic in Ω and has derivative $\mathbf{0}$. Hence it is constant. Since $F(z_0) = 1$ [by (1) and (3)], the constant is $1/f(z_0)$.

(ii) *In case Ω is the annulus $A(0, 2)$, $z_0 = 1$ and f is the identity function, why does this result not contradict* 4.21 *and* 3.42(v)*?*

(iii) *Let Ω be a region in $\mathbb{C}$, $f \in H(\Omega)$. Show that $f = g'/g$ for some zero-free $g \in H(\Omega)$ if and only if $\frac{1}{2\pi i}\int_\gamma f \in \mathbb{Z}$ for every piecewise smooth loop γ in Ω. How is this related to* 2.11*?*

Hints: ($\Rightarrow$) Since $\frac{1}{2\pi i}\int_\gamma g'/g = \mathrm{Ind}_{g\circ\gamma}(0)$, this implication is trivial.

($\Leftarrow$) Fix $z_0 \in \Omega$ and for each $z \in \Omega$ let γ_z be a polygon in Ω joining z_0 to z. Form

$$L(z) := \int_{\gamma_z} f, \qquad z \in \Omega,$$

and $g := e^L$. Argue locally, as in (i): If $D(z_1, r) \subset \Omega$ and $L_1(z) := \int_{[z_1,z]} f$ for $z \in D(z_1, r)$, then $g = g(z_1)e^{L_1}$ in that disk. Infer that g is holomorphic and satisfies $g'/g = L_1' = f$ in $D(z_1, r)$. This disk being arbitrary, the results hold throughout Ω.

For the next result, a pillar of the theory, which will undergo numerous generalizations, we have seen ample antecedents, in 2.20(iv), 3.39(ii), 5.5(iii) and 5.18. Four quite different proofs were available due to the simple geometry of the regions involved, rectangles and disks.

Theorem 5.33 (Cauchy's Formulae for Starlike Regions) *Let Ω be an open, starlike subset of $\mathbb{C}$, f holomorphic in Ω, γ a piecewise smooth loop in Ω. Then for every non-negative integer k and every $z \in \Omega \setminus \gamma$ we have*

$$(*) \qquad \frac{\operatorname{Ind}_\gamma(z)}{k!} f^{(k)}(z) = \frac{1}{2\pi i} \int_\gamma \frac{f(\xi)}{(\xi - z)^{k+1}}\, d\xi.$$

If, in addition, Ω is convex and f is defined and continuous in $\overline{\Omega}$, then γ may be anywhere in $\overline{\Omega}$.

Proof: For fixed $z_0 \in \Omega \setminus \gamma$ the function $F(z) := (f(z) - f(z_0))/(z - z_0)$ for $z \in \Omega \setminus \{z_0\}$, $F(z_0) := f'(z_0)$ is continuous in Ω, holomorphic in $\Omega \setminus \{z_0\}$. We want to apply 5.2 to it. To that end we can cite 5.30 to see that $F(z)$ is represented by the convergent power series $\sum_{n=1}^{\infty} f^{(n)}(z_0)/n!(z - z_0)^{n-1}$ in a neighborhood of z_0, thus insuring that F is holomorphic throughout Ω, as the hypotheses of 5.2 ostensibly require. But even without resort to power series 5.2 is applicable to F, thanks to Remark (i) following it. So

$$0 = \int_\gamma F(\xi)\, d\xi = \int_\gamma \frac{f(\xi) - f(z_0)}{\xi - z_0}\, d\xi = \int_\gamma \frac{f(\xi)}{\xi - z_0}\, d\xi - f(z_0) \int_\gamma \frac{1}{\xi - z_0}\, d\xi.$$

This is the case $k = 0$ of $(*)$ (recall 4.4). Using this, 2.14 and induction [recall that the function $\operatorname{Ind}_\gamma$ is locally constant], we get $(*)$ for all k. The fact that we already know all higher derivatives exist (5.30) makes possible an alternative endgame here: Apply the proven case $k = 0$ to $f^{(n)}$ in the role of f, and then carry out n integrations by parts to get $(*)$ for $k = n$.

A useful fact suggested by this proof is

Exercise 5.34 Let U be an open subset of $\mathbb{C}$, f holomorphic in U. Define $g \colon U \times U \mapsto \mathbb{C}$ by

$$g(z, w) := \begin{cases} \dfrac{f(z) - f(w)}{z - w}, & z, w \in U,\ z \neq w \\ f'(w), & z = w \in U. \end{cases}$$

(i) *Show that g is continuous and that for each $w \in U$, $g(z, w)$ is a holomorphic function of $z \in U$.*

Hints: Let $(z_0, w_0) \in U \times U$ be given. If $z_0 \neq w_0$, then in a whole neighborhood of (z_0, w_0) the first formula in the definition of g prevails, so the continuity of g at (z_0, w_0) and the differentiability of $g(z, w_0)$ at z_0 are immediate. If $z_0 = w_0 \in U$, pick $r > 0$ so that $\overline{D}(w_0, r) \subset U$ and use 5.18$(***)$ to write

$$(1) \qquad g(z, w) = \frac{1}{2\pi i} \int_{C(w_0, r)} \frac{f(\xi)}{(\xi - z)(\xi - w)}\, d\xi \qquad \forall z, w \in D(z_0, r).$$

Differentiability of $g(z, w_0)$ at z_0 then follows from 2.14 and continuity of g at (z_0, w_0) is an easy exercise in the uniform continuity of the integrand. Inequality (4) in 3.6 readily yields a second proof. Here is a third: For $z, w \in D(z_0, r)$

$$(z-w)g(z,w) = f(z) - f(w) = \int_{[w,z]} f' = \int_0^1 f'(w + t(z-w))(z-w)\,dt.$$

Consequently, whether or not $z = w$,

$$g(z,w) = \int_0^1 f'(w + t(z-w))\,dt, \text{ whence}$$

$$g(z,w) - g(z_0, z_0) = \int_0^1 [f'(w + t(z-w)) - f'(z_0)]\,dt,$$

$$|g(z,w) - g(z_0,z_0)| \le \sup\{|f'(\xi) - f'(z_0)| : \xi \in D(z_0, r)\}.$$

(ii) *Show that for each compact $K \subset U$ there is a finite positive r such that $K_r := \bigcup\{\bar{D}(z,r) : z \in K\}$ is a compact subset of U, and that for any such r*

$$|f(z) - f(w)| \le \frac{3}{r} \sup |f(K_r)| \cdot |z - w| \qquad \forall z, w \in K.$$

Hints: Write $M := \sup |f(K_r)|$ and use (1) to show that for all $z \in K$ and $w \in D(z, 2r/3)$

$$|f(z) - f(w)| \le \frac{3M}{r}|z - w|. \tag{2}$$

On the other hand, if $z, w \in K$ and $|z - w| \ge 2r/3$, then

$$|f(z) - f(w)| \le 2M \le 2M \cdot \frac{3}{2r} \cdot |z - w|.$$

(iii) *Prove* Liouville's Theorem: *every bounded entire function is constant.*

Hints: 3.33(i) ∪ 5.30 is one approach. Alternatively, if $f \in H(\mathbb{C})$ is bounded by $M \in \mathbb{N}$, and $z, w \in \mathbb{C}$ are given, then (2) holds for all finite, positive r. For a third proof, see the Remarks following 5.5.

(iv) *It is interesting that* (iii) *can bootstrap itself up to the ostensibly stronger assertion that the range of every non-constant entire function f is dense in* $\mathbb{C}$. (Cf. 12.18 and 7.29(v).)

Hint: If $D(a, r) \subset \mathbb{C} \setminus f(\mathbb{C})$, consider the entire function $1/(f - a)$.

As a further consequence of 5.30 we have

Corollary 5.35 *Let Ω be a region in which every holomorphic function has a primitive. Then every zero-free holomorphic function in Ω has a holomorphic logarithm. In particular, every zero-free holomorphic function in a starlike region has a holomorphic logarithm there.*

Proof: Let $f \in H(\Omega)$ be zero-free. Since f' is also holomorphic (5.30), the hypothesis furnishes $F \in H(\Omega)$ with $F' = f'/f$. From 2.10 it follows that $\int_{\gamma_z} f'/f = F(z) - F(z_0)$ for any piecewise smooth curve γ_z in Ω joining z_0 and z. Writing $f(z_0) = e^{c_0}$, 5.32 yields $\exp\big(F(z) - F(z_0) + c_0\big) = f(z)$ for all $z \in \Omega$. Alternatively, simply note that $(fe^{-F})' = \mathbf{0}$. Starlike Ω fulfill the hypothesis about primitives, thanks to 5.2. (Alternatively, the result for starlike Ω follows from conjoining 4.20 and 4.67(i).)

Exercise 5.36 (i) *Let F be holomorphic and zero-free in $\mathbb{D}$ and satisfy $|F| \leq 1$. Show that*

$$\sup_{|z|\leq 1/5} |F(z)|^2 \leq \inf_{|z|\leq 1/7} |F(z)|.$$

Hints: There is a holomorphic function f in $\mathbb{D}$ such that $F = e^{-f}$. From $e^{-\operatorname{Re} f} = |F| \leq 1$ we see $\operatorname{Re} f \geq 0$. From 5.28(i)

$$\begin{aligned} \operatorname{Re} f(z) &\geq \tfrac{2}{3} \operatorname{Re} f(0) \qquad \forall |z| \leq \tfrac{1}{5}, \\ \operatorname{Re} f(z) &\leq \tfrac{4}{3} \operatorname{Re} f(0) \qquad \forall |z| \leq \tfrac{1}{7}. \end{aligned}$$

Therefore

$$\sup_{|z|\leq 1/5} [-2 \operatorname{Re} f(z)] \leq -\tfrac{4}{3} \operatorname{Re} f(0) \leq \inf_{|z|\leq 1/7} [-\operatorname{Re} f(z)].$$

(ii) *For each region Ω and each compact subset K of Ω there is a positive integer $n = n(K, \Omega)$, dependent only on K and Ω, such that*

$$(*) \qquad |f(z)|^{1/n} \leq |f(w)| \leq |f(z)|^n \qquad \forall z, w \in K$$

holds for all $f \in H(\Omega)$ which satisfy $|f| \geq 1$ in Ω.

Hints: From (i) we see that

$$(1) \qquad |F(z)|^2 \leq |F(w)| \leq |F(z)|^{1/2} \qquad \forall z, w \in \overline{D}\big(0, \tfrac{1}{7}\big)$$

holds for all F which are holomorphic in $\mathbb{D}$ and satisfy $0 < |F| \leq 1$. Consequently

$$(2) \qquad |f(z)|^{1/2} \leq |f(w)| \leq |f(z)|^2 \qquad \forall z, w \in \overline{D}\big(a, \tfrac{1}{7}r\big)$$

holds for all f which are holomorphic in $D(a,r)$ and satisfy $|f| \geq 1$. Use (2) and a covering compactness argument to get $(*)$ [for an n of the form 2^m]. For the (easy) details see pp. 25 ff. of OSTROWSKI [1931] or the hints to 5.28(ii). (The result $(*)$ is also an easy consequence of 5.28(ii) itself, applied to $h := \log|f|$.)

Exercise 5.37 (i) *$\mathcal{F}$ is a family of functions holomorphic in the open set U, K is a compact subset of U and M is a constant such that $|f| \leq M$ on U for every $f \in \mathcal{F}$. Show that there is a constant $C = C(M,K)$ such that $|f'| \leq C$ on K for every $f \in \mathcal{F}$.*

Hint: There is a positive number $r = r(K)$ such that $\bar{D}(z,r) \subset U$ for every $z \in K$. According to 3.34(i), $|f'(z)| \leq M/r$ for every such z and $f \in \mathcal{F}$.

(ii) *Let f be entire and suppose that $f(z)e^{-c|z|}$ is bounded by M for some $c > 0$. Show that $f'(z)e^{-c|z|}$ is bounded by ceM.*

Hints: By Cauchy's formula for disks 5.33, for all $z \in \mathbb{C}$ and all $r > 0$

$$\begin{aligned} f'(z) &= \frac{1}{2\pi i}\int_{C(z,r)} \frac{f(\xi)}{(\xi - z)^2}\,d\xi, \\ |f'(z)| &\leq r \sup_{\xi \in C(z,r)} \frac{|f(\xi)|}{|\xi - z|^2} = \frac{1}{r}\sup_{|\xi - z| = r} |f(\xi)| \\ &\leq \frac{1}{r}\sup_{|\xi - z| = r} Me^{c|\xi|} = \frac{1}{r}Me^{c(|z|+r)} = \frac{e^{cr}}{r}Me^{c|z|}. \end{aligned}$$

The function $\phi(r) := e^{cr}/r$ has its minimum over $(0,\infty)$ at the point $r = 1/c$.

(iii) (PAINLEVÉ [1887], p. 18) *Let f be holomorphic and bounded in $S := \{re^{i\theta} : r > 0,\ \alpha < \theta < \beta\}$. Show that $\lim_{r\to\infty} f'(re^{i\theta}) = 0$ for each $\theta \in (\alpha, \beta)$ and uniformly in every compact subset of such θ.*

Hints: Let $\alpha < \alpha' < \beta' < \beta$ be given, as well as $\varepsilon > 0$, and set $S' := \{re^{i\theta} : r > 0,\ \alpha' < \theta < \beta'\}$. Let M be a (finite) bound for $|f|$ in S. Evidently if r_ε is sufficiently large, $\bar{D}(z, \varepsilon^{-1}) \subset S$ for each $z \in S'$ with $|z| > r_\varepsilon$. For such z the Cauchy estimates 3.34 then yield

$$|f'(z)| \leq \frac{M}{\varepsilon^{-1}} = M\varepsilon.$$

[A different proof is suggested in 5.44(vi).]

Exercise 5.38 (i) *Let F be holomorphic in $S := (0,1) \times \mathbb{R}$ and continuous in $\bar{S}$. Suppose that F is bounded in $\bar{S}$ and set*

$$M(x) := \sup\{|F(x+iy)| : y \in \mathbb{R}\} \quad \textit{for each } x \in [0,1].$$

Show that

$$|F(z)| \le M(0)^{1-\operatorname{Re} z} M(1)^{\operatorname{Re} z} \qquad \forall z \in S,$$

with strict inequality for all z unless $F(z) \equiv ae^{bz}$ for constants $a \in \mathbb{C}$, $b \in \mathbb{R}$, in which case equality prevails for all z.

Hints: Consider any $M_0 > M(0)$, $M_1 > M(1)$ and form $G(z) := M_0^{1-z} M_1^z$, $f := F/G$. [For $b > 0$, b^z means $e^{z \log b}$.] Evidently f is holomorphic in S and bounded and continuous in $\bar{S}$, since $|G(z)| = M_0^{1-\operatorname{Re} z} M_1^{\operatorname{Re} z} \ge \min\{M_0, M_1\} > 0$ when $0 \le \operatorname{Re} z \le 1$. Also $|f| < 1$ on ∂S. It follows from 5.13 that $|f| < 1$ throughout S, i.e.,

$$|F(z)| < M_0^{1-\operatorname{Re} z} M_1^{\operatorname{Re} z} \qquad \forall z \in S.$$

Let $M_0 \downarrow M(0)$, $M_1 \downarrow M(1)$. If F is not the function $\mathbf{0}$, the inequality shows both $M(0)$, $M(1)$ are positive, which allows it to be written as

$$|F(z)| \le M(0) \left[\frac{M(1)}{M(0)}\right]^{\operatorname{Re} z} = \left| M(0) \left[\frac{M(1)}{M(0)}\right]^{z} \right| \qquad \forall z \in S.$$

Hence the occurrence of an equality in S implies the constancy of the holomorphic function $F(z)[M(1)/M(0)]^{-z}$ (5.10).

(ii) (Hadamard Three Circles Theorem) *Let $0 < r_1 < r_2 < \infty$, f holomorphic in $A(r_1, r_2)$ and continuous in $\bar{A}(r_1, r_2)$. For each $r_1 \le r \le r_2$ set $M(r) := \sup\{|f(z)| : |z| = r\}$. Then*

$$(*) \quad M(r)^{\log(r_2/r_1)} \le M(r_1)^{\log(r_2/r)} M(r_2)^{\log(r/r_1)} \qquad \forall r \in [r_1, r_2].$$

Can the functions that experience equality for some $r \in (r_1, r_2)$ be specified?

Hints: Consider the map

$$\phi(z) := \exp(\log r_1 + z \log(r_2/r_1)) = \exp((1 - z) \log r_1 + z \log r_2)$$

of $[0, 1] \times \mathbb{R}$ onto $\bar{A}(r_1, r_2)$ and the composite $F := f \circ \phi$. Evidently F fulfills the hypothesis of (i) and $F(x + iy) = f\big(r_1^{1-x} r_2^x e^{iy \log(r_2/r_1)}\big)$, so

$$\sup\{|F(x + iy)| : y \in \mathbb{R}\} = \sup\big\{\big|f\big(r_1^{1-x} r_2^x e^{i\theta}\big)\big| : \theta \in \mathbb{R}\big\} = M\big(r_1^{1-x} r_2^x\big).$$

For any $x \in (0, 1)$ we get then from (i)

$$M\big(r_1^{1-x} r_2^x\big) \le M(r_1)^{1-x} M(r_2)^x.$$

If we take $x := [\log(r/r_1)]/[\log(r_2/r_1)]$, we have $r_1^{1-x}r_2^x = r$ and $(*)$ ensues when both sides of the preceding inequality are raised to power $\log(r_2/r_1)$. Suppose equality holds in the inequality for some $r = r_0 \in (r_1, r_2)$. Compactness furnishes $\theta_0 \in \mathbb{R}$ such that $|f(r_0e^{i\theta_0})| = M(r_0)$. Because $F = f \circ \phi$, equality then holds in (i) at $z := \phi(r_0e^{i\theta_0})$, which entails that there exist $a \in \mathbb{C}$, $b \in \mathbb{R}$ such that $F(z) = ae^{bz}$ for all $z \in S$. Given $r \in (r_1, r_2)$, $\theta \in \mathbb{R}$, a brutal calculation confirms that

$$x := \log(r/r_1)/\log(r_2/r_1) \quad \text{and} \quad y := \theta/\log(r_2/r_1)$$

satisfy $\phi(x + iy) = re^{i\theta}$, so

$$|f(re^{i\theta})| = |F(x+iy)| = |a|\left(\frac{r}{r_1}\right)^{b/\log(r_2/r_1)} = cr^\alpha$$

for constants $c > 0$, $\alpha \in \mathbb{R}$. A glance ahead to 6.18(iii) reveals that such an equality for all $r \in (r_1, r_2)$ forces α to be an integer, whence $z^{-\alpha}f(z)$ is constant.

(iii) *Let Ω be a region, U a non-empty open subset of Ω, K a non-empty compact subset of Ω. Show that there exists a positive constant $c = c(\Omega, U, K) < 1$ such that*

$$(*) \qquad \sup|f(K)| \le [\sup|f(\Omega)|]^{1-c}[\sup|f(U)|]^c \qquad \forall f \in H(\Omega).$$

Hints: Since $(*)$ holds for any $c \in (0,1)$ if $\sup|f(\Omega)|$ is 0 or ∞, we can exclude such f and then, upon dividing f by this number, $(*)$ is seen to be equivalent to

$$(**) \qquad \sup|f(K)| \le [\sup|f(U)|]^c \qquad \forall f \in H(\Omega) \text{ with } \sup|f(\Omega)| = 1.$$

If the open disks $D_1, \ldots, D_n$ with $\bar{D}_j \subset \Omega$ cover K and the constants $c(\Omega, U, \bar{D}_j)$ have been produced satisfying $(**)$ with $K := \bar{D}_j$, then the choice $c(\Omega, U, K) := \min\{c(\Omega, U, \bar{D}_j) : 1 \le j \le n\}$ will satisfy $(**)$. Therefore it suffices to prove that for each point $a \in \Omega$ there is an $r > 0$ such that $\bar{D}(a, r) \subset \Omega$ and an appropriate constant $c(\Omega, U, \bar{D}(a, r))$ exists. To this end, set $m := \sup|f(U)|$, select any point $a_0 \in U$ and for the given a let $\gamma : [0,1] \mapsto \Omega$ be a curve with initial point a_0, terminal point a. (Ω is connected!) By compactness there is an $r > 0$ such that $\bar{D}(a_0, r) \subset U$ and $\bar{D}(\gamma(t), 4r) \subset \Omega$ for all t. By uniform continuity of γ there is $k \in \mathbb{N}$ such that $|t - t'| \le 1/k \Rightarrow |\gamma(t) - \gamma(t')| < r$, whence $D(\gamma(t), r) \subset D(\gamma(t'), 2r)$. Defining $a_j := \gamma(j/k)$, the preceding inclusions entail that for $j = 1, 2, \ldots, k$

$$(1.j) \qquad D(a_j, r) \subset D(a_{j-1}, 2r).$$

Define $M_j(\rho) := \sup|f(C(a_j,\rho))|$ for $j = 0, \ldots, k$, $\rho \in \{r, 2r, 4r\}$. We want to apply (ii) to the annuli $A_j := \{z \in \mathbb{C} : r < |z - a_j| < 4r\}$ and the intermediate radius $2r$. Notice that $\bar{A}_j \subset \overline{D}(a_j, 4r) \subset \Omega$. Then initially (ii)(∗) asserts that

$$M_0(2r)^{\log 4} \leq M_0(r)^{\log 2} M_0(4r)^{\log 2}$$

or, setting $\lambda := \log 2/\log 4 \in (0,1)$ and recalling that $|f| \leq 1$ throughout Ω,

$$\text{(2.0)} \qquad M_0(2r) \leq M_0(r)^{\lambda} \leq m^{\lambda}, \quad \text{since } C(a_0, r) \subset U.$$

Via the Maximum Principle this entails

$$\text{(3.0)} \qquad \sup|f(\overline{D}(a_0, 2r))| \leq m^{\lambda}.$$

Apply (ii) again to get

$$M_1(2r) \leq M_1(r)^{\lambda}.$$

Since $C(a_1, r) \subset \overline{D}(a_1, r) \subset \overline{D}(a_0, 2r)$ by (1.1), $M_1(r) \leq \sup|f(\overline{D}(a_0, 2r)| \overset{(3.0)}{\leq} m^{\lambda}$, so inequality (2.1) yields

$$M_1(2r) \leq (m^{\lambda})^{\lambda} = m^{\lambda^2},$$

whence by the Maximum Principle

$$\text{(3.1)} \qquad \sup|f(\overline{D}(a_1, 2r))| \leq m^{\lambda^2}.$$

Further iterating this bootstrap procedure results in

$$\text{(3.k)} \qquad \sup|f(\overline{D}(a_k, 2r))| \leq m^{\lambda^{k+1}}.$$

Since $a_k = a$, this says that λ^{k+1} is an eligible exponent $c(\Omega, U, \overline{D}(a, r))$.

(iv) (Stieltjes [1894]) *Let V be a region, U a non-empty open subset of V, $\{f_n\} \subset H(V)$. If $\{f_n\}$ is uniformly convergent on U and uniformly bounded on each compact subset of V, then $\{f_n\}$ is in fact uniformly convergent on each compact subset of V.* Cf. 7.5.

Hints: Let D be a non-empty open disk such that $\overline{D} \subset U$. Given compact $K \subset V$, it is not hard to construct a bounded region Ω such that $\overline{D} \cup K \subset \Omega \subset \overline{\Omega} \subset V$. According to (iii) there is then a constant $c = c(\Omega, D, K) \in (0,1)$ such that

$$(*) \qquad \sup|f_n - f_m|(K) \leq [\sup|f_n - f_m|(\Omega)]^{1-c}[\sup|f_n - f_m|(D)]^{c}.$$

By hypothesis there is a finite constant $M = M(\overline{\Omega})$ such that $\sup|f_n(\Omega)| \leq M$ for all n. Therefore $(*)$ yields

$$\sup|f_n - f_m|(K) \leq (2M)^{1-c}[\sup|f_n - f_m|(D)]^c.$$

Exercise 5.39 (i) (MORERA [1886]) *Prove the converse of the Cauchy-Goursat theorem: Let f be continuous in the open subset U of $\mathbb{C}$. Suppose that f satisfies $\int_{[a,b,c,a]} f = 0$ whenever* $\operatorname{co}\{a,b,c\} \subset U$. *Then f is holomorphic in U.*

Hint: The problem is a local one: if Ω is any open disk in U, then f has a primitive in Ω by the Scholium to 2.11 and so is holomorphic in Ω by 5.30.

(ii) *Let $f\colon \mathbb{D} \mapsto \mathbb{C}$ be a continuous function such that $\int_{\partial R} f = 0$ for every $R := [0,x]\times[0,y]$ (x, y real) which lies in $\mathbb{D}$. Show that f is holomorphic in $\mathbb{D}$.* (This significantly generalizes 2.19(ix).)

Hint: Form F in $\mathbb{D}$ by $F(z) := \int_{[0,\operatorname{Re} z, z]} f$ and use the hypothesis on f to show that also $F(z) = \int_{[0, i \operatorname{Im} z, z]} f$. From the first representation of F infer that D_2F exists and $D_2F = if$. From the second representation infer that D_1F exists and equals f. Conclude from 2.17 that F is holomorphic. Then $F' = D_1F = f$ is too, by 5.30.

(iii) (Extension of 2.11) *Suppose $f\colon U$ open $\subset \mathbb{C} \mapsto \mathbb{C}$ is continuous and the sequence $\{\int_{\Gamma_n} f\}$ is bounded whenever the piecewise smooth loops Γ_n uniformly converge to a point in U. Show that f is holomorphic in U.*

Hints: (PERRY and YOUNGS [1947]). Suppose that f is not $\mathbb{C}$-differentiable at some $a \in U$. Then by (i) every disk $D(a,r) \subset U$ must contain a triangle $\triangle$ such that $\int_{\partial\triangle} f \neq 0$. So we get a sequence $\gamma_n\colon [0,1] \mapsto D(a, r/n)$ of piecewise smooth loops with $\int_{\gamma_n} f \neq 0$. Let $N = N_n$ be a positive integer such that $N\left|\int_{\gamma_n} f\right| > n$. One checks easily that $\Gamma_n := N * \gamma_n$ (see 1.11) is a piecewise smooth loop in $D(a, r/n)$ which satisfies $\int_{\Gamma_n} f = N\int_{\gamma_n} f$, and therefore by choice of N the sequence $\{\int_{\Gamma_n} f\}$ is not bounded.

Exercise 5.40 (i) *Suppose $f\colon \mathbb{D} \mapsto \mathbb{C}$ is continuous and is holomorphic in $\mathbb{D}\setminus\mathbb{R}$. Use* 5.39(ii) *to show that f is holomorphic throughout $\mathbb{D}$.* (See 9.28 for an extension of this result and Remark (c) after 5.41 concerning a harmonic-function analog.)

Hints: (LEVI-CIVITA [1912]) Whenever $R := [0,x] \times [0,y]$ lies in $\mathbb{D}$, the convex region $\Omega := (0,x) \times (0,y)$ lies in $\mathbb{D}\setminus\mathbb{R}$. Therefore f is continuous on $\overline{\Omega} = R$ and holomorphic in Ω, and so $\int_{\partial R} f = 0$ by 5.2. Cite 5.39(ii).

(ii) *$f \in H(\mathbb{D} \cap \mathbb{U})$ has limit* 0 *at each point in the interval* $(-1,1)$. *Show that* $f = \mathbf{0}$. (Cf. 5.76.)

Hints: Defining F to be f in $\mathbb{D} \cap \mathbb{U}$, to be 0 in $(-1,1)$ and $F(z) := \bar{f}(\bar{z})$ for $z \in \mathbb{D} \setminus \overline{\mathbb{U}}$, gives a function which is evidently continuous in $\mathbb{D}$ and, by 2.4, holomorphic in $\mathbb{D} \setminus \mathbb{R}$. Thanks to (i), F is therefore holomorphic in the whole disk $\mathbb{D}$. The coefficients of its Taylor series at 0 can be computed via $(F \mid \mathbb{R})^{(n)}(0) = 0$. [Another proof can be based on 8.36(ii) and 5.62.]

(iii) *Suppose f is continuous in the open set U, $I_1, \ldots, I_n$ are closed intervals lying in U, and f is holomorphic in $U \setminus (I_1 \cup \cdots \cup I_n)$. Use* (i) *to show that f is holomorphic in U.*

Hints: By subdividing the intervals appropriately, we can assume they intersect only at endpoints; call the set of these E. Each $z \in I_j \setminus E$ is then the center of a disk D lying in U and disjoint from the other intervals. Hence (i) is applicable to f in D and shows it to be holomorphic there. Thus f is holomorphic in $U \setminus E$. Consequently, each point of E being the center of a disk lying in U minus the other points of E, (i) is now applicable to f in such a disk.

Theorem 5.41 (Removable Singularities) *Let U be an open subset of $\mathbb{C}$, f a holomorphic, h a harmonic function in $U \setminus \{a\}$ for some $a \in U$.*

(i) *If f satisfies the* (*weaker than boundedness near a*) *hypothesis*

$$\lim_{z \to a} (z-a) f(z) = 0,$$

then f is the restriction to $U \setminus \{a\}$ of a (*unique*) *F which is holomorphic in U.* (Cf. 11.60(i).)

(ii) *If h satisfies the* (*weaker than boundedness near a*) *hypothesis*

$$\lim_{z \to a} \frac{h(z)}{\log |z-a|} = 0,$$

then h is the restriction to $U \setminus \{a\}$ of a (*unique*) *H which is harmonic in U.* (Cf. 11.57(i).)

Proof: We may suppose $a = 0$ and $\overline{\mathbb{D}} \subset U$.

(i) Define g in $\mathbb{D}$ by

$$g(z) := \begin{cases} 0, & z = 0 \\ z^2 f(z), & z \in \mathbb{D} \setminus \{0\}. \end{cases} \tag{1}$$

Then g is holomorphic in $\mathbb{D} \setminus \{0\}$. Moreover $(g(z) - g(0))/z = z f(z)$ ($z \in \mathbb{D} \setminus \{0\}$) so the hypothesis in (i) shows that g is differentiable at 0, with

$g'(0) = 0$. Therefore by 5.30 there is a series which represents g throughout $\mathbb{D}$:

$$g(z) = \sum_{n=0}^{\infty} c_n z^n, \qquad z \in \mathbb{D}.$$

And $c_0 = g(0) = 0$, $c_1 = g'(0) = 0$, so we have

(2) $$g(z) = \sum_{n=2}^{\infty} c_n z^n = z^2 \sum_{k=0}^{\infty} c_{k+2} z^k, \qquad z \in \mathbb{D}.$$

It follows from (1) and (2) that

$$f(z) = \sum_{k=0}^{\infty} c_{k+2} z^k \qquad \forall z \in \mathbb{D} \setminus \{0\}.$$

Therefore the desired F is given by

$$F(z) := \begin{cases} f(z), & z \in U \setminus \{0\} \\ \displaystyle\sum_{k=0}^{\infty} c_{k+2} z^k, & z \in \mathbb{D}. \end{cases}$$

A different proof is suggested in Remark (ii) following 11.7. Both of these proofs rely, in the final analysis, on the representation of a holomorphic function as a series of powers of z. But there is an interesting maneuver which circumvents that and derives the result directly from the Cauchy integral formula. It amounts to using an integral remainder formula in the Taylor expansion (cf. 5.73(iv)): From the identity

$$\frac{1}{1-w} = \frac{1-w+w}{1-w} = 1 + \frac{w}{1-w} = 1 + w\left[1 + \frac{w}{1-w}\right]$$

with $w := z/\xi$ and a little algebra one sees that

$$\frac{1}{\xi - z} = \frac{1}{\xi} + \frac{z}{\xi^2} + \frac{z^2}{(\xi - z)\xi^2} \qquad \forall \xi \in \mathbb{T}, \ z \in \mathbb{D}.$$

Multiply by $g(\xi)$ and integrate to get, for all $z \in \mathbb{D}$

$$\int_{C(0,1)} \frac{g(\xi)}{\xi - z}\, d\xi = \int_{C(0,1)} \frac{g(\xi)}{\xi}\, d\xi + z \int_{C(0,1)} \frac{g(\xi)}{\xi^2}\, d\xi + z^2 \int_{C(0,1)} \frac{g(\xi)}{(\xi - z)\xi^2}\, d\xi.$$

From 5.18(∗∗∗) and 2.14 we see that the first three integrals are $2\pi i g(z)$, $2\pi i g(0)$ and $2\pi i g'(0)$, respectively; and, as earlier noted, $g(0) = g'(0) = 0$. Thus we have

$$2\pi i g(z) = z^2 \int_{C(0,1)} \frac{g(\xi)}{(\xi - z)\xi^2}\, d\xi \qquad \forall z \in \mathbb{D}.$$

That is,

$$f(z) = \frac{1}{2\pi i} \int_{C(0,1)} \frac{f(\xi)}{\xi - z}\, d\xi \qquad \forall z \in \mathbb{D} \setminus \{0\}. \tag{3}$$

This integral, being by 2.14 a holomorphic function of z throughout $\mathbb{D}$, furnishes the claimed holomorphic extension of f.

An alternative approach used by OSGOOD [1895/96] and PRINGSHEIM [1896a] is based on 5.5. As above, assume $a = 0$ and $\overline{\mathbb{D}} \subset U$. With $z \in \mathbb{D} \setminus \{0\}$ fixed, consider the function F defined in $U \setminus \{0\}$ by $F(\xi) := [f(\xi) - f(z)]/[\xi - z]$ if $\xi \neq z$, $F(z) := f'(z)$. It is holomorphic (5.34), so by 5.5(i)

$$\int_{C(0,1)} F = \int_{C(0,r)} F \qquad \forall 0 < r < 1. \tag{4}$$

Actually, Pringsheim does not mention 5.34 explicitly but proves (as 5.5(i) does) that (4) is valid for any F which is continuous in $U \setminus \{0\}$ and holomorphic there save for finitely many exceptional points (see also the chapter notes on this matter). Routine estimates using the hypothesis $\lim_{\xi \to 0} \xi f(\xi) = 0$ show that the right side of (4) converges to 0 as $r \downarrow 0$. Hence $\int_{C(0,1)} F = 0$, an equality which routinely unpacks into (3). It should be noted that (4) effectively contains Cauchy's integral representation of f in the annulus $A(r, 1)$; see 11.1 and its short proof.

(ii) Let H solve the Dirichlet problem in $\mathbb{D}$ with boundary data $h|_{\partial\mathbb{D}}$. For each $\varepsilon > 0$, look at the harmonic function $u_\varepsilon(z) := H(z) - h(z) + \varepsilon \log|z|$ in $\overline{\mathbb{D}} \setminus \{0\}$. We have $u_\varepsilon = 0$ on $\partial\mathbb{D}$ and the continuity of H at 0 plus the hypothesized limiting behavior of h at 0 yield that

$$u_\varepsilon(z) = H(z) + \log|z| \left[\varepsilon - \frac{h(z)}{\log|z|}\right] \to -\infty \text{ as } z \to 0.$$

It follows from 5.12 that $u_\varepsilon \leq 0$ throughout $\mathbb{D} \setminus \{0\}$:

$$H(z) - h(z) \leq -\varepsilon \log|z| \qquad \forall z \in \mathbb{D} \setminus \{0\}, \quad \forall \varepsilon > 0.$$

Let $\varepsilon \downarrow 0$ to conclude $H - h \leq 0$ in $\mathbb{D} \setminus \{0\}$. The same argument applied to $h(z) - H(z) + \varepsilon \log |z|$ gives the reverse inequality. Thus $H = h$ in $\mathbb{D} \setminus \{0\}$ and furnishes the desired extension of h.

Remarks: (a) Of course, the case of continuous f is already contained in 5.40(i) and in 5.5(iii). It is also a routine exercise to deduce (i) from (ii), but the given proof of (i) is too elegant to withhold. On the other hand, in the *special case that* $h = \operatorname{Re} g$ for some $g \in H(U \setminus \{a\})$, it is also easy to deduce (ii) by applying (i) to $f := e^g$. Notice also that, in conjunction with 5.22, the method of (ii) above establishes a removable singularity theorem for subharmonic functions: If v is continuous in U and subharmonic in $U \setminus \{a\}$, then in fact v is subharmonic in U. It is a good exercise for the reader to supply the details of these arguments.

(b) From (ii) it follows that the Dirichlet problem is not solvable in a punctured disk: If the (continuous) boundary data ϕ are 1 at 0 and 0 on $\partial\mathbb{D}$, then no function h can be continuous on $\overline{\mathbb{D}}$, harmonic in $\mathbb{D} \setminus \{0\}$ and equal to ϕ on $\{0\} \cup \partial\mathbb{D}$. This is because (by(ii)) h would be harmonic throughout $\mathbb{D}$ and the maximum principle would then be violated. There is not even a *subharmonic* extension u of ϕ, as an easy argument using 5.46 shows.

(c) In contrast to (b), the Dirichlet problem *is* solvable for a disk with a whole interval deleted, which fact generates an example showing that there is no analog of 5.40(i) for *harmonic* functions. Indeed, for any $0 < b < 1$, 9.44 supplies continuous $h : \overline{\mathbb{D}} \mapsto \mathbb{R}$ which is harmonic in $\mathbb{D} \setminus [-b, b]$ and satisfies $h(\mathbb{T}) = \{0\}$, $h[-b, b] = \{1\}$. By 5.12, $h \leq 1$ throughout $\overline{\mathbb{D}}$. Therefore, if function h were harmonic in a neighborhood of some point $x_0 \in [-b, b]$, then $h(x_0) = 1$ would violate 5.8.

(d) It is even possible to holomorphically extend over a whole countable closed set at once (cf. 5.40(i)) if we invoke an elementary result from topology, namely, that every complete metric space X with no isolated points is uncountable. The proof that no sequence $\{x_n\}_{n\in\mathbb{N}} \subset X$ can exhaust X starts with the construction of closed balls B_n of positive radius $r_n < 1/n$ such that $B_{n+1} \subset B_n \subset X \setminus \{x_1, \ldots, x_n\}$ for all n. This is done inductively: If B_k has been constructed, its center z_k is not isolated and so some ball with that center and radius less than r_k contains a point $z_{k+1} \in X \setminus \{x_{k+1}\}$. We may evidently take B_{k+1} to be the closed ball of radius r_{k+1} centered at z_{k+1} for any sufficiently small positive r_{k+1}. The completeness of X guarantees there is a point in $\bigcap_{n\in\mathbb{N}} B_n \subset X \setminus \{x_n\}_{n\in\mathbb{N}}$.

With this fact we can prove

Theorem 5.42 *Let U be a non-empty open subset of $\mathbb{C}$, K a countable subset of U which is closed in $\mathbb{C}$, f a bounded holomorphic function in $U \setminus K$.*

Then f has a (unique, bounded) holomorphic extension to U. Cf. 9.28(ii) and 10.10(iii).

Proof: Let, say, $|f| \leq M < \infty$ in $U \setminus K$. If $U \setminus K \subset U_j$ open $\subset U$ and $f_j \in H(U_j)$ extends f $(j = 1, 2)$, then $f_1 = f_2$ in $U_1 \cap U_2$. This is because $U_1 \cap U_2 \setminus K$ is dense in $U_1 \cap U_2$ (why?) and the continuous functions f_1, f_2 agree with f, hence with each other, in this dense set. For the same reason $|f_j| \leq M$ holds in each U_j.

Now let $\mathcal{U}$ denote the class of all open sets U_1 such that $U \setminus K \subset U_1 \subset U$ and there exists $f_1 \in H(U_1)$ with $f_1 = f$ on $U_1 \setminus K$. Set $U_0 := \bigcup \mathcal{U}$. By what was noted in the first paragraph, a function f_0 is well defined in the open set U_0 by the simple decree $f_0 := f_1$ in U_1 whenever $U_1 \in \mathcal{U}$ and $f_1 \in H(U_1)$ extends f. Of course f_0 is holomorphic and bounded by M in U_0. Let $K_0 := U \setminus U_0$. This is a relatively closed subset of K, hence is closed. Moreover, K_0 has no isolated points. For if z_0 is an isolated point of K_0, there exists $r_0 > 0$ such that $D(z_0, r_0) \subset U$ and $D(z_0, r_0) \cap K_0 = \{z_0\}$. Therefore $D(z_0, r_0) \setminus \{z_0\} \subset U \setminus K_0 = U_0$ and by the removable singularity theorem 5.41(i) f_0 would extend holomorphically to the open set $D(z_0, r_0) \cup U_0 = \{z_0\} \cup U_0$. This would put $\{z_0\} \cup U_0$ in the class $\mathcal{U}$, so $\{z_0\} \subset \bigcup \mathcal{U} = U_0$, a contradiction.

In summary, if $K_0 \neq \varnothing$, then it is a complete metric space (being a closed subset of $\mathbb{C}$) which is countable (being a subset of K). This would contradict Remark (d) above. So we must conclude that K_0 is indeed empty and hence f_0 is the desired extension of f to $U_0 = U \setminus K_0 = U$.

Exercise 5.43 (i) *Let V be an open subset of $\mathbb{C}$, $f\colon [0,1] \times V \mapsto \mathbb{C}$ a locally bounded function. Suppose $f(t,z)$ is a Riemann integrable (resp., continuous) function of $t \in [0,1]$ for each $z \in V$ and a holomorphic function of $z \in V$ for each $t \in [0,1]$. Show that then $D_2 f(t,z)$ is a Riemann integrable (resp., continuous) function of $t \in [0,1]$ for each z and is also locally bounded on $[0,1] \times V$.* (See 7.21 for more on this theme.)

Hints: Given $z \in V$, let $r > 0$ be such that $\overline{D}(z,r) \subset V$ and for each integer $n > 2/r$ define a Riemann integrable (resp., continuous) function f_n on $[0,1]$ by

$$f_n(t) := n\big[f\big(t, z + n^{-1}\big) - f(t,z)\big].$$

Then the Cauchy Integral Formulae 5.33 give for all $t \in [0,1]$

$$\begin{aligned} f_n(t) - D_2 f(t,z) &= \frac{1}{2\pi i} \int_{C(z,r)} \left[\frac{nf(t,\xi)}{\xi - z - n^{-1}} - \frac{nf(t,\xi)}{\xi - z} - \frac{f(t,\xi)}{(\xi - z)^2}\right] d\xi \\ &= \frac{1}{2\pi i n} \int_{C(z,r)} \frac{f(t,\xi)}{(\xi - z - n^{-1})(\xi - z)^2}\, d\xi. \end{aligned}$$

It follows from this, 2.9(i), and the fact $n \geq 2/r$ that

$$|f_n(t) - D_2 f(t,z)| \leq \frac{2M}{nr^2},$$

holding for all $t \in [0,1]$. Here M is a bound for f on $[0,1] \times C(z,r)$. Thus $D_2 f(t,z)$ is the limit, uniformly in t, of the Riemann integrable (resp., continuous) functions $f_n(t)$. Finally, recall 5.37(i).

(ii) (Bôcher [1910–1911]) *Let T be a compact, U an open subset of $\mathbb{C}$. Let $F\colon T \times U \mapsto \mathbb{C}$ be continuous and $z \mapsto F(t,z)$ be holomorphic in U for each $t \in T$. Show that then $D_2^{(j)} F$ is also continuous throughout $T \times U$ for each $j \in \mathbb{N}$.*

Hints: Fix $z_0 \in U$ and $r > 0$ such that $\bar{D}(z_0, r) \subset U$. Then Cauchy's Integral Formula (5.33) gives

$$D_2 F(t,z) = \frac{1}{2\pi i} \int_{C(z_0,r)} \frac{F(t,\xi)}{(\xi - z)^2}\, d\xi \qquad \forall t \in T,\ z \in D(z_0, r).$$

All claims follow from this integral representation via 2.14 and the fact that F is *uniformly* continuous on $T \times C(z_0, r)$.

(iii) *With notation and hypotheses as in* (ii) *and $T := [0,1]$, show that the function $F^*(z) := \int_0^1 F(t,z)dt$ is holomorphic in U and has derivative $\int_0^1 D_2 F(t,z)dt$.*

Hints: By (ii), $D_2 F$ is continuous, so an idea in the 2.19(i) hints can be exploited: If $D(z_0, r) \subset U$, then for z in this disk

$$F^*(z) - F^*(z_0) = \int_0^1 [F(t,z) - F(t,z_0)]dt = \int_0^1 \int_{[z_0,z]} D_2 F(t,\xi) d\xi dt.$$

Consequently, for such $z \neq z_0$

$$\frac{F^*(z) - F^*(z_0)}{z - z_0} - \int_0^1 D_2 F(t, z_0)dt = \frac{1}{z - z_0} \int_0^1 \int_{[z_0,z]} [D_2 F(t,\xi) - D_2 F(t, z_0)] d\xi dt.$$

Given $\varepsilon > 0$, there exists $0 < r_\varepsilon < r$ such that

$$|D_2 F(t,\xi) - D_2 F(t,z_0)| < \varepsilon \qquad \forall (t,\xi) \in [0,1] \times D(z_0, r_\varepsilon)$$

and so by 2.9(i) the inner integral above is majorized by $\varepsilon|z - z_0|$, for every z with $0 < |z - z_0| < r_\varepsilon$. [See 7.21(iii) for another proof.]

We are overdue to highlight another pillar of our subject, of which power series are the simplest examples and 5.74 will be a dramatic generalization:

Exercise 5.44 (i) (WEIERSTRASS *Werke I*, 67–74; *II*, 201–233.) *A uniform limit of holomorphic functions is holomorphic.* (7.5 and 7.8 are a considerable strengthening of this result.)

Hint: Since uniform convergence commutes with integration, 5.1 combines with its converse Morera 5.39(i) to give a quick proof. Alternatively, 2.12 and 5.2 can be used to show that locally the limit function is a derivative, hence holomorphic by 5.30. Together 3.35(ii) and 5.30 constitute a third proof.

(ii) *If f_n are holomorphic in the open set U and $f_n \to f$ uniformly in each compact subset of U, then not only is f holomorphic in U but for each positive integer j, $f_n^{(j)} \to f^{(j)}$ uniformly in each compact subset of U.*

Hint: Go to the limit in the Cauchy formulae 5.33. More precisely, as the problem is a local one, whenever $\bar{D}(a, 2r) \subset U$, use these formulae to majorize $|f_n^{(j)} - f^{(j)}|$ on $D(a, r)$ by $2j!r^{-j}$ times the supremum of $|f_n - f|$ over $C(a, 2r)$.

(iii) *If harmonic functions h_n converge uniformly on compacta to a (finite) limit function h, then h is harmonic. Moreover, all the partial derivatives of h_n converge uniformly on compacta to the corresponding h partial derivatives.*

Hints: In analogy with (i), use the integral characterization of harmonicity in 5.23(iv) to prove the first assertion. To prove both assertions at once, reduce to (ii) via 5.20 and the idea in the hint to 5.3(ii).

(iv) *There exist differentiable functions $f_n \colon \mathbb{R} \mapsto \mathbb{R}$ which converge uniformly to $\mathbf{0}$ on $\mathbb{R}$ but, in contrast to* (ii), *their derivatives do not converge. It is even possible that the sequence $\{f_n'(x)\}_{n\in\mathbb{N}}$ be unbounded for every $x \in \mathbb{R}$.*

Hints: Consider $f_n(x) := n^{-1/2}\sin(nx)$. We have $f_n'(x) = n^{1/2}\cos(nx)$. From the identity $\cos(2\theta) = 2\cos^2\theta - 1$ therefore follows that

$$\frac{2}{\sqrt{k}}[f_k'(x)]^2 - \frac{1}{\sqrt{2}}f_{2k}'(x) = \sqrt{k} \qquad \forall x \in \mathbb{R},\ k \in \mathbb{N}.$$

This shows that the sequence $\{f_n'(x)\}_{n\in\mathbb{N}}$ cannot be bounded for any $x \in \mathbb{R}$.

(v) *In contrast to* (i) *there exist differentiable functions $f_n : [-1, 1] \mapsto \mathbb{R}$ which converge uniformly to the absolute-value function. It is even possible to find polynomial f_n.*

Hints: Evidently the functions $f_n(x) := (x^2 + n^{-2})^{1/2}$ do the job. To get polynomials, cite Weierstrass' approximation theorem (in Remark (ii) following 5.20, or 8.29(iii)). Explicitly, the power series $q(z)$ in 3.12(vi), which has real coefficients, converges uniformly in $\bar{D}(1,1)$, maps into $\bar{D}(1,1)$, and satisfies $(q(z))^2 = z$ for all z. It follows (how?) that $q(x^2) = |x|$ for all $x \in [-1,1]$. The reader is invited to resolve an apparent paradox here: whereas power series are differentiable functions, the function $|x|$ is not $\mathbb{R}$-differentiable at $x = 0$.

(vi) *Use* (ii) *to give an alternative proof of* 5.37(iii).

Hints: Let notation and hypotheses be as in 5.37(iii). Set $K := \{re^{i\theta} : 1 \leq r \leq 2,\ \alpha' \leq \theta \leq \beta'\}$ and $f_n(z) := n^{-1}f(nz)$, $z \in S$, $n \in \mathbb{N}$. Since f is bounded, $f_n \to 0$ uniformly in S, whence by (ii) $f_n' \to 0$ locally uniformly in S. Given $\varepsilon > 0$, there exists n_ε such that

$$n \geq n_\varepsilon \Rightarrow |f_n'(z)| = |f'(nz)| \leq \varepsilon \qquad \forall z \in K.$$

Given $w \in S'$ with $|w| \geq n_\varepsilon$, choose $n \in \mathbb{N}$ via $n \leq |w| < n+1$. Then $n \geq n_\varepsilon$ and $z := (1/n)w \in K$. Consequently, $|f'(w)| = |f'(nz)| \leq \varepsilon$.

(vii) *Show that if f_n are holomorphic in the open set U, $f_n \to f$ uniformly on each compact subset of U and each f_n has a primitive, then f has a primitive.*

Hints: Note that for any piecewise smooth loop γ in U, $\int_\gamma f_n = 0$ by 2.10(ii), and $\int_\gamma f_n \to \int_\gamma f$. Then invoke 2.11. Alternatively, if F_n is a primitive of f_n, then 2.12 even shows that for any $a \in U$ the sequence $\{F_n - F_n(a)\}$ converges to a primitive of f throughout the component of U in which a lies.

§ 5 Harmonic Majorization

A simple form of the theme of this section, which continues that of 5.38, is illustrated in the following exercise.

Exercise 5.45 (Ostrowski [1922/23]) (i) *Let f be holomorphic in $\mathbb{D}$ and continuous on $\bar{\mathbb{D}}$. Let A be a subarc of $\mathbb{T}$ of length $\alpha > 0$. Define*

$$\lambda(z) := \frac{\alpha}{2\pi}\frac{1-|z|}{1+|z|}, \qquad z \in \mathbb{D}.$$

Show that

$$(*) \qquad |f(z)| \leq [\sup|f(A)|]^{\lambda(z)}[\sup|f(\mathbb{D})|]^{1-\lambda(z)} \qquad \forall z \in \mathbb{D}.$$

Hints: Consider any $m > \sup|f(A)|$, $M > \sup|f(\mathbb{D})|$ and $M \geq m$. Define

$$(1) \qquad \phi(z) := \log(\max\{|f(z)|, m\}) = \log m + \log^+\left|\frac{f(z)}{m}\right|, \qquad z \in \bar{\mathbb{D}}.$$

By 5.26(ii) ϕ is subharmonic in $\mathbb{D}$. Therefore if we define

$$(2) \qquad h(z) := \frac{1}{2\pi}\int_0^{2\pi} \operatorname{Re}\left[\frac{e^{i\theta}+z}{e^{i\theta}-z}\right]\phi(e^{i\theta})\,d\theta, \quad z \in \mathbb{D},$$

and $h := \phi$ on $\partial\mathbb{D}$, we have a continuous function on $\overline{\mathbb{D}}$ which is harmonic in $\mathbb{D}$ (5.20) and from 5.12 (applied to the difference $\phi - h$)

$$(3) \qquad \phi \le h \quad \text{in } \overline{\mathbb{D}}.$$

After a rotation we can assume that

$$A := \{e^{i\theta} : 0 \le \theta \le \alpha\}.$$

On A we have $|f| \le m$, so $\phi = \log m$. Therefore

$$h(z) \le \frac{1}{2\pi}\int_0^{\alpha} \operatorname{Re}\left[\frac{e^{i\theta}+z}{e^{i\theta}-z}\right]\log m\,d\theta + \frac{1}{2\pi}\int_{\alpha}^{2\pi} \operatorname{Re}\left[\frac{e^{i\theta}+z}{e^{i\theta}-z}\right]\log M\,d\theta$$

(since, by equation (2) in the proof of 5.20, the kernel

$$\operatorname{Re}\left[\frac{e^{i\theta}+z}{e^{i\theta}-z}\right] = \frac{1-|z|^2}{\left|e^{i\theta}-z\right|^2} \text{ is nonnegative)}$$

$$= \psi(z)\log m + (1-\psi(z))\log M$$

$$(4) \qquad = \log M - \psi(z)[\log M - \log m],$$

where

$$(5) \qquad \psi(z) := \frac{1}{2\pi}\int_0^{\alpha} \operatorname{Re}\left[\frac{e^{i\theta}+z}{e^{i\theta}-z}\right]d\theta \ge \frac{1}{2\pi}\int_0^{\alpha} \frac{1-|z|}{1+|z|}\,d\theta = \lambda(z)$$

and (taking $h = \mathbf{1}$ in 5.20)

$$\frac{1}{2\pi}\int_0^{2\pi} \operatorname{Re}\left[\frac{e^{i\theta}+z}{e^{i\theta}-z}\right]d\theta = 1.$$

From (5) and (4) we get

$$(6) \quad h(z) \le \log M - \lambda(z)[\log M - \log m] = \lambda(z)\log m + (1-\lambda(z))\log M.$$

Since $|f| \le \max\{|f|, m\} = e^{\phi}$, (6) and (3) yield

$$|f(z)| \le m^{\lambda(z)}M^{1-\lambda(z)}$$

and $(*)$ follows from this by letting $m \downarrow \sup|f(A)|$, $M \downarrow \sup|f(\mathbb{D})|$.

(ii) *Let A, B be complementary, non-degenerate arcs of the unit circle. Show that for each $r \in (0,1)$, there exists $\lambda(r) = \lambda_A(r) \in (0,1)$ such that*

$$(*) \qquad \sup_{|z|\leq r} |f(z)| \leq M_A^{\lambda(r)} M_B^{1-\lambda(r)} \qquad \forall 0 < r < 1$$

holds for all numbers $0 \leq M_A \leq M_B < \infty$ and all $f \in H(\mathbb{D})$ which satisfy

$$(**) \qquad \overline{\lim_{z\to u}}\,|f(z)| \leq \begin{cases} M_A & \forall u \in \bar{A} \\ M_B & \forall u \in \bar{B}. \end{cases}$$

Hints: Define $\lambda_A(r) := \alpha(1-r)/2\pi(1+r)$, α the length of A. Consider any f of the described kind. Since $M_A \leq M_B$, it follows from $(**)$ and 5.12 that

$$(7) \qquad |f| \leq M_B \quad \text{throughout } \mathbb{D}.$$

If we define f_n on $\overline{\mathbb{D}}$ by $f_n(z) := f((n-1)z/n)$, then (i) applied to f_n yields

$$\left|f\left(\frac{n-1}{n}z\right)\right| = |f_n(z)| \leq \left[\sup\left|f_n\left(\bar{A}\right)\right|\right]^{\lambda(r)} \left[\sup|f_n(\mathbb{D})|\right]^{1-\lambda(r)}$$
$$\leq \left[\sup\left|f_n\left(\bar{A}\right)\right|\right]^{\lambda(r)} M_B^{1-\lambda(r)} \qquad \forall |z| = r.$$

So by the Maximum-Modulus Principle

$$(8) \qquad \left|f\left(\frac{n-1}{n}z\right)\right| \leq \left[\sup\left|f_n\left(\bar{A}\right)\right|\right]^{\lambda(r)} M_B^{1-\lambda(r)} \qquad \forall z \in \bar{D}(0,r).$$

Choose $u_n \in \bar{A}$ such that $|f((n-1)u_n/n)| = \sup\left|f_n\left(\bar{A}\right)\right|$. Then choose $1 \leq n_1 < n_2 \cdots$ such that $\sup\left|f_{n_j}\left(\bar{A}\right)\right| \to \overline{\lim}_{n\to\infty} \sup\left|f_n\left(\bar{A}\right)\right|$ and also such that $u_{n_j} \to u \in \bar{A}$. By $(**)$ we have then

$$(9) \qquad \overline{\lim_{n\to\infty}} \sup\left|f_n\left(\bar{A}\right)\right| = \lim_{j\to\infty} \left|f\left(\frac{n_j - 1}{n_j} u_{n_j}\right)\right| \leq M_A.$$

From (8) and (9) we get

$$|f(z)| \leq M_A^{\lambda(r)} M_B^{1-\lambda(r)} \qquad \forall z \in \bar{D}(0,r).$$

(iii) (Cf. §90, MONTEL [1927]) *Let A be as in* (ii), *F_n holomorphic functions in $\mathbb{D}$ which are continuous on $\mathbb{D} \cup A$. Suppose that $\{F_n\}$ is uniformly bounded on $\mathbb{D}$ and uniformly convergent on A. Show that $\{F_n\}$ is then uniformly convergent on each compact subset of $\mathbb{D}$.*

Hints: It suffices to show that for each $r < 1$ and each subsequence $1 \leq n_1 < n_2 < \cdots$ the functions $f_j := F_{n_{j+1}} - F_{n_j}$ converge to 0 uniformly in $D(0, r)$. If we set $M_j := \sup |f_j(A)|$, then $M_j \to 0$ by hypothesis, and by (ii)

$$|f_j(z)| \leq M_j^{\lambda(r)} (2M)^{1-\lambda(r)} \qquad \forall z \in D(0, r),$$

where M is some bound for all the F_n in $\mathbb{D}$. Since $\lambda(r) > 0$, the result follows.

To exploit the full potential of the idea in 5.45(i) we need to have available for various standard sets (like slit disks, half-disks, etc.) harmonic functions with very tightly specified boundary behavior and for which some appraisals of the interior values can be got. These are used as comparison functions to provide upper bounds for holomorphic functions via the various maximum principles. In many important applications we can manufacture the relevant harmonic function by composing elementary transformations but usually a few "corner" points on the boundary will be points of discontinuity, and therefore the first order of business is a technical result: we extend 5.12 to allow some singular points on the boundary. Notice that hypothesis (2) below is weaker than

$$(2)' \qquad \varlimsup_{\substack{z \to \zeta \\ z \in U}} u(z) < \infty \qquad \forall \zeta \in E,$$

which is often fulfilled in applications of 5.46.

Theorem 5.46 (Cf. 7.17) *Let U be a bounded open subset of $\mathbb{C}$, E a countable subset of ∂U, u a subharmonic function in U which for some finite M satisfies*

$$(1) \qquad \varlimsup_{\substack{z \to \zeta \\ z \in U}} u(z) \leq M \qquad \forall \zeta \in \partial U \setminus E,$$

and

$$(2) \qquad \varlimsup_{\substack{z \to \zeta \\ z \in U}} \frac{u(z)}{-\log|z - \zeta|} \leq 0 \qquad \forall \zeta \in E.$$

Then $u \leq M$ throughout U.

Proof: Let $d :=$ diameter U and $z_1, z_2, \ldots$ be an enumeration of E. Each function $\log(|z - z_n|/d)$ is harmonic in U, by 5.7 and

$$h(z) := \sum_{n=1}^{\infty} 2^{-n} \log(|z - z_n|/d)$$

converges uniformly in a neighborhood of each $z_0 \in U$: If $D(z_0, 2r) \subset U$, we have

$$r \le |z - \zeta| \le d \qquad \forall z \in D(z_0, r),\ \zeta \in \partial U,$$

so that

$$\log(r/d) \le \log\bigl(|z - z_n|/d\bigr) \le 0 \qquad \text{for all such } z \text{ and all } n \in \mathbb{N}.$$

It follows from 5.44(iii) that h is harmonic and non-positive in U. Given $\varepsilon > 0$, form $u_\varepsilon := u + \varepsilon h$. From (1) and the non-positivity of h we get

$$(1)' \qquad \varlimsup_{\substack{z \to \zeta \\ z \in U}} u_\varepsilon(z) \le M \qquad \forall \zeta \in \partial U \setminus E.$$

By hypothesis (2) there exist $r_n > 0$ such that

$$(3) \qquad \frac{u(z)}{-\log|z - z_n|} \le \varepsilon 2^{-n-1} \qquad \forall z \in U \cap D(z_n, r_n).$$

We can suppose $r_n < 1$ and so small that in addition

$$(4) \qquad \log|z - z_n| \le \varepsilon^{-1} 2^{n+1} M + 2\log(d) \qquad \forall z \in U \cap D(z_n, r_n).$$

Then (3) yields

$$u(z) \le -\varepsilon 2^{-n-1} \log|z - z_n|,$$
$$u(z) + \varepsilon 2^{-n} \log|z - z_n| \le \varepsilon 2^{-n-1} \log|z - z_n| \overset{(4)}{\le} M + \varepsilon 2^{-n} \log(d),$$
$$u(z) + \varepsilon 2^{-n} \log\bigl(|z - z_n|/d\bigr) \le M \qquad \forall z \in U \cap D(z_n, r_n).$$

Adding the other (non-positive) terms in $\varepsilon h(z)$ gives

$$u_\varepsilon(z) \le M \qquad \forall z \in U \cap D(z_n, r_n),$$

so that

$$(5) \qquad \varlimsup_{\substack{z \to z_n \\ z \in U}} u_\varepsilon(z) \le M, \quad \text{holding for each } z_n \in E.$$

In view of $(1)'$ and (5), 5.12 insures that $u_\varepsilon \le M$ throughout U, that is,

$$u(z) + \varepsilon h(z) \le M \qquad \forall z \in U.$$

And this holds for every $\varepsilon > 0$. So fix $z \in U$ and let $\varepsilon \downarrow 0$.

Lemma 5.47 *There is a non-negative harmonic function h in $\mathbb{D} \setminus (-1,0]$ which is bounded by* 1 *and satisfies*

$$\text{(i)} \qquad \lim_{z\to\zeta} h(z) = \begin{cases} 0, & \zeta \in \mathbb{T}\setminus\{-1\} \\ 1, & \zeta \in (-1,0) \end{cases}$$

and

$$\text{(ii)} \qquad h(z) \geq 1 - \frac{4}{\pi}\tan^{-1}\sqrt{|z|} \geq \frac{1}{\pi}(1-|z|) \qquad \forall z \in \mathbb{D}\setminus(-1,0]\,.$$

Proof: Let L be the holomorphic logarithm in $\mathbb{C}\setminus(-\infty,0]$ which satisfies

$$\text{(1)} \qquad L\big(re^{i\theta}\big) = \log r + i\theta, \quad r>0, |\theta|<\pi,$$

and consider

$$\phi(z) := \frac{1+z}{1-z}, \qquad z \in \mathbb{C}\setminus\{1\}.$$

We know (cf. 2.5) that ϕ is conformal in $\mathbb{D}$ and that

$$\text{(2)} \qquad \phi(\mathbb{D}\cap\mathbb{U}) = \mathbb{H}\cap\mathbb{U},$$

$$\text{(3)} \qquad \phi(\mathbb{T}\cap\mathbb{U}) = i(0,\infty),$$

$$\text{(4)} \qquad \phi(-1,1) = (0,\infty).$$

Therefore the composite

$$\text{(5)} \qquad v := \frac{2}{\pi}\,\mathrm{Im}\, L\circ\phi$$

is defined in $\overline{\mathbb{D}}\cap\mathbb{U}$, is harmonic in $\mathbb{D}\cap\mathbb{U}$ and satisfies

$$\text{(2)}' \qquad v(\mathbb{D}\cap\mathbb{U}) = (0,1) \quad \text{by (1), (5) and (2)}$$

$$\text{(3)}' \qquad v(\mathbb{T}\cap\mathbb{U}) = \{1\} \quad \text{by (1), (5) and (3)}$$

$$\text{(4)}' \qquad v(-1,1) = \{0\} \quad \text{by (1), (5) and (4).}$$

Now look at the rotated square-root map $z \mapsto ie^{L(z)/2}$. Because of (1) it evidently maps $\mathbb{D}\setminus(-1,0]$ onto $\mathbb{D}\cap\mathbb{U}$, $\mathbb{T}\setminus\{-1\}$ onto $\mathbb{T}\cap\mathbb{U}$ and (0,1) onto $i(0,1)$. Therefore the composite

$$\text{(6)} \qquad h(z) := 1 - v\Big(ie^{L(z)/2}\Big), \quad z \in \overline{\mathbb{D}}\setminus[-1,0]$$

is continuous, harmonic in $\mathbb{D}\setminus(-1,0]$, and we have from $(2)'$ and $(3)'$ that

$$\text{(7)} \qquad h(\mathbb{D}\setminus(-1,0]) \subset (0,1)$$

$$\text{(8)} \qquad h(\mathbb{T}\setminus\{-1\}) = \{0\}.$$

But from (1) it is clear that for any $x \in (-\infty, 0)$

$$\operatorname{Im} L(z) \to \begin{cases} \pi & \text{if } \operatorname{Im} z > 0 \text{ and } z \to x \\ -\pi & \text{if } \operatorname{Im} z < 0 \text{ and } z \to x. \end{cases}$$

Consequently, for any $x \in (-\infty, 0)$

$$ie^{L(z)/2} \to \begin{cases} -\sqrt{|x|} & \text{if } \operatorname{Im} z > 0 \text{ and } z \to x \\ \sqrt{|x|} & \text{if } \operatorname{Im} z < 0 \text{ and } z \to x. \end{cases}$$

It follows from (6) and (4)′ that

$$\lim_{z \to x} h(z) = 1 \qquad \forall x \in (-1, 0)\,. \tag{9}$$

All that remains is to check (ii). To this end, note that by direct computation, for any $w \in \mathbb{C} \setminus \{-1\}$

$$\frac{\operatorname{Im}\left(\dfrac{1+w}{1-w}\right)}{\operatorname{Re}\left(\dfrac{1+w}{1-w}\right)} = \frac{2 \operatorname{Im} w}{1 - |w|^2}. \tag{10}$$

We look at $z = re^{i\theta} \in \mathbb{D} \setminus (-1, 0]$ $(0 < r < 1,\ |\theta| < \pi)$. Then

$$w := ie^{L(z)/2} = i\sqrt{r}e^{i\theta/2} \in \mathbb{D} \cap \mathbb{U}$$

and so by (2)

$$\frac{1+w}{1-w} = \phi\left(ie^{L(z)/2}\right) = Re^{i\tilde{\theta}} \tag{11}$$

with $R > 0$ and $0 < \tilde{\theta} < \pi/2$. According to (10) we have

$$\tan \tilde{\theta} = \frac{2\sqrt{r}\cos\theta/2}{1-r}. \tag{12}$$

From (1) and (11)

$$\begin{aligned} \operatorname{Im} L\left(\phi\left(ie^{L(z)/2}\right)\right) &= \tilde{\theta} \\ &\overset{(12)}{=} \tan^{-1}\left(\frac{2\sqrt{r}\cos\theta/2}{1-r}\right) \\ &\leq \tan^{-1}\left(\frac{2\sqrt{r}}{1-r}\right). \end{aligned}$$

It follows from (5), (6) and the preceding inequality that

$$h(z) \geq 1 - \frac{2}{\pi}\tan^{-1}\left(\frac{2\sqrt{r}}{1-r}\right) = 1 - \frac{2}{\pi}\tan^{-1}\left(\frac{2\sqrt{|z|}}{1-|z|}\right).$$

If we use the "double angle" formula for the function tan, the preceding inequality becomes

$$\begin{aligned} h(z) &\geq 1 - \frac{4}{\pi}\tan^{-1}\sqrt{|z|} \\ &= \frac{4}{\pi}\left[\frac{\pi}{4} - \tan^{-1}\sqrt{|z|}\right], \qquad z \in \mathbb{D} \setminus (-1,0]\,. \end{aligned}$$

Recalling (3) and (3)′ in 3.22, we have then

$$\begin{aligned} h(z) \geq \frac{4}{\pi}\int_{\sqrt{|z|}}^{1}\frac{dt}{1+t^2} \geq \frac{4}{\pi}\int_{\sqrt{|z|}}^{1}\frac{dt}{2} &= \frac{2}{\pi}\left(1-\sqrt{|z|}\right) \\ &= \frac{2}{\pi}\frac{1-|z|}{1+\sqrt{|z|}} > \frac{1}{\pi}(1-|z|). \end{aligned}$$

Corollary 5.48 *Let f be holomorphic and bounded by 1 in $\mathbb{D} \setminus (-1,0]$ and for some $0 < \delta < 1$ satisfy*

$$\varlimsup_{z\to\zeta}|f(z)| \leq \delta \qquad \forall \zeta \in (-1,0)\,.$$

Then

$$|f(z)| \leq \delta^{(1-|z|)/\pi} \qquad \forall z \in \mathbb{D} \setminus (-1,0]\,.$$

Proof: The function

$$\phi := \log\left(\max\{|f|,\delta\}\right) = \log\delta + \log^{+}\left|\frac{f}{\delta}\right|$$

is subharmonic in $\mathbb{D} \setminus (-1,0]$, by 5.26(ii). It is bounded and satisfies

$$\varlimsup_{z\to\zeta}\phi(z) \leq \begin{cases} 0 & \text{if } \zeta \in \mathbb{T} \\ \log\delta & \text{if } \zeta \in (-1,0). \end{cases} \tag{1}$$

Let h be the harmonic function of 5.47. According to 5.47(i) and (1) we have

$$\varlimsup_{z\to\zeta}\,[\phi - (\log\delta)h](z) \leq 0$$

for all points $\zeta \in (\mathbb{T} \setminus \{-1\}) \cup (-1,0) = \partial(\mathbb{D} \setminus (-1,0]) \setminus \{-1,0\}$. Therefore by 5.46 the (bounded) subharmonic function $\phi - (\log \delta)h$ is non-positive throughout $\mathbb{D} \setminus (-1,0]$:

$$\phi \le (\log \delta)h,$$
$$|f| \le e^{\phi} \le e^{(\log \delta)h} = \delta^h \quad \text{in } \mathbb{D} \setminus (-1,0]. \tag{2}$$

Since $0 < \delta < 1$ and $h(z) \ge (1 - |z|)/\pi$ for all $z \in \mathbb{D} \setminus (-1,0]$ [5.47(ii)], the desired upper bound on $|f(z)|$ follows from (2).

Exercise 5.49 *Let* $\gamma\colon [0,1] \mapsto \mathbb{C}$ *be a curve such that* $\gamma(0), \gamma(1) \in \mathbb{R}$ *and* $|\operatorname{Im} \gamma| > 0$ *on* $(0,1)$. *Form*

$$\Gamma(t) := \begin{cases} \gamma(t), & 0 \le t \le 1 \\ \overline{\gamma}(2-t), & 1 \le t \le 2. \end{cases}$$

Notice that Γ *is a loop. Show that* $|\operatorname{Ind}_\Gamma(x)| = 1$ *for every real* x *strictly between* $\gamma(0)$ *and* $\gamma(1)$.

Hints: Replacing γ by $\overline{\gamma}$ if necessary, we can suppose that $\operatorname{Im} \gamma > 0$ in $(0,1)$. By reversing the parameter direction if necessary, we can suppose that $\gamma(0) \le \gamma(1)$. Therefore we are to consider points x such that

$$\gamma(0) < x < \gamma(1). \tag{1}$$

Let L be the holomorphic logarithm in $\mathbb{C} \setminus i(-\infty, 0]$ which satisfies

$$L\big(re^{i\theta}\big) = \log r + i\theta \quad \text{when } r > 0,\ -\frac{\pi}{2} < \theta < \frac{3\pi}{2}. \tag{2}$$

(Cf. 3.43.) Form

$$\phi(t) := \begin{cases} L(\gamma(t) - x), & 0 \le t \le 1 \\ \overline{L}(\gamma(2-t) - x), & 1 < t \le 2. \end{cases}$$

Notice that ϕ is continuous at $t = 1$ (and trivially elsewhere) because by (1) and (2), $L(\gamma(1) - x)$ is real. Since $\overline{e^w} = e^{\overline{w}}$ for any $w \in \mathbb{C}$, we see that

$$e^{\phi} = \Gamma - x. \tag{3}$$

It follows that

$$\begin{aligned} \operatorname{Ind}_{\Gamma - x}(0) = \frac{1}{2\pi i}[\phi(2) - \phi(0)] &= \frac{1}{2\pi i}\Big[\overline{L}(\gamma(0) - x) - L(\gamma(0) - x)\Big] \\ &= -\frac{1}{\pi} \operatorname{Im} L(\gamma(0) - x). \end{aligned}$$

Since $\gamma(0) - x$ is a negative real number by (1), (2) shows that $\operatorname{Im} L(\gamma(0) - x) = \pi$.

Theorem 5.50 (MILLOUX [1924]) *Let γ be a curve with initial point* 0, *terminal point on $\mathbb{T}$ and all other points in $\mathbb{D}$. Let $f \in H(\mathbb{D})$ be bounded by* 1 *and for some $0 < \delta < 1$ satisfy*

$$|f(\zeta)| \leq \delta \qquad \forall \zeta \in \gamma \cap \mathbb{D}.$$

Then

$$|f(z)| \leq \delta^{(1-|z|)/2\pi} \qquad \forall z \in \mathbb{D}.$$

Proof: (PÓLYA [1929], p. 634) After a rotation we may suppose that the terminal point of γ is -1. Consider the function

$$(1) \qquad F(z) := f(z)\bar{f}(\bar{z}), \quad z \in \mathbb{D}.$$

By 2.4 it is holomorphic. Of course it is bounded by 1 in $\mathbb{D}$ and it satisfies

$$(2) \qquad |F(\zeta)| \leq \delta \qquad \forall \zeta \in (\gamma \cup \overline{\gamma}) \cap \mathbb{D}.$$

Consider any $x \in (-1, 0)$. If $x \in \gamma$, then $|F(x)| \leq \delta$ by (2). Suppose $x \notin \gamma$. Set $t_0 := \max \gamma^{-1}[x, 1]$, $t_1 := \min[t_0, \infty) \cap \gamma^{-1}[-1, x]$. Then $\gamma(t_0), \gamma(t_1) \in \mathbb{R}$, x lies between them, and $|\operatorname{Im} \gamma| > 0$ on the interval (t_0, t_1). Hence the curve $\gamma_1 := \gamma \mid [t_0, t_1]$, except for having a different parameter interval, answers to the description of γ in 5.49. Therefore $x \in C$, a bounded component of $\mathbb{C} \setminus \gamma_1 \cup \overline{\gamma}_1$. Necessarily $C \subset \mathbb{D}$, since $\mathbb{C} \setminus (\mathbb{D} \cup \{-1\})$ is connected and disjoint from $\gamma \cup \overline{\gamma}$. Since C is a component of $\mathbb{C} \setminus \gamma_1 \cup \overline{\gamma}_1$, we have $\partial C \subset \gamma_1 \cup \overline{\gamma}_1 \subset \gamma \cup \overline{\gamma}$. From (2) then

$$|F| \leq \delta \quad \text{on } (\partial C) \setminus \{-1\}.$$

Therefore by the Maximum-Modulus Principle (in the form 5.46) $|F| \leq \delta$ holds throughout C, in particular at the point x. We have therefore shown that

$$|F(x)| \leq \delta \qquad \forall x \in (-1, 0)\,,$$

$$|f(x)|^2 \leq \delta \qquad \forall x \in (-1, 0)\,.$$

We are therefore in a position to apply 5.48 with $\delta^{1/2}$ in the role of the δ there.

Here are two more examples of harmonic majorization very similar to 5.50. The first is a particularly unencumbered illustration of the method, its hypotheses being just a little too weak for it to be an instance of 5.16′. In these examples we will use the abbreviations $\mathbb{D}^+ := \mathbb{D} \cap \mathbb{U}$ and $\mathbb{T}^+ := \mathbb{T} \cap \mathbb{U}$.

Example 5.51 (Phragmén-Lindelöf) *If f is continuous in $\overline{\mathbb{U}}$, holomorphic in $\mathbb{U}$, and satisfies*

(i) $f(z)e^{-\varepsilon|z|}$ *is bounded in $\mathbb{U}$ for each $\varepsilon > 0$,*

(ii) $|f| \le 1$ *on* $\mathbb{R}$,

then $|f| \le 1$ *throughout* $\mathbb{U}$.

Proof: Consider the functions ϕ and v constructed in the proof of 5.47. It was (essentially) shown there that v maps $\overline{\mathbb{D}^+} \setminus \{\pm 1\}$ continuously onto $[0,1]$, is harmonic in $\mathbb{D}^+$, and satisfies

$$v(\mathbb{T}^+) = \{1\} \tag{1}$$

$$v(-1,1) = \{0\}. \tag{2}$$

Since $L(\phi(z))/z = [L(\phi(z)) - L(\phi(0))]/z \to (L \circ \phi)'(0) = 2$ as $z \to 0$, there exists $0 < \delta < 1$ such that

$$|L(\phi(z))/z| < \pi \quad \text{whenever} \quad 0 < |z| < \delta.$$

Hence $v := 2\pi^{-1} \operatorname{Im} L \circ \phi$ satisfies

$$|v(z)| \le 2|z| \qquad \forall z \in D(0,\delta) \cap \mathbb{U}. \tag{3}$$

Suppose now that

$$|f(z_0)| > 1 \quad \text{for some} \quad z_0 \in \mathbb{U}. \tag{4}$$

Consider only $r > \delta^{-1}|z_0| > |z_0|$, and set $M(r) := \max\{|f(z)| : z \in \overline{\mathbb{U}} \cap \overline{\mathbb{D}}(0,r)\}$. Then

$$M(r) \ge |f(z_0)| > 1. \tag{5}$$

Introduce the function

$$u := \log \max\{|f|, 1\}.$$

It is subharmonic in $\mathbb{U}$ (5.26(ii)) and it satisfies

$$u(z) \le \log M(r) = v(z/r) \log M(r) \qquad \forall z \in \mathbb{U} \cap C(0,r)$$

thanks to (5) and (1), and

$$u(x) = \log \max\{|f(x)|, 1\} = 0 = v(x/r) \log M(r) \qquad \forall x \in (-r, r)$$

thanks to (ii) and (2). In $\mathbb{U} \cap D(0,r)$ the function $u(z) - v(z/r) \log M(r)$ is subharmonic and bounded. In view of the preceding two inequalities, 5.46 is therefore applicable, E being $\{-r, r\}$ and gives

$$u(z) \le v(z/r) \log M(r) \qquad \forall z \in \mathbb{U} \cap D(0,r).$$

Hence from (5) and (3)

$$u(z) \le 2r^{-1}|z| \log M(r) \qquad \forall z \in \mathbb{U} \cap D(0, \delta r).$$

In particular, since $z_0 \in D(0, \delta r)$,

$$\log |f(z_0)| = u(z_0) \le 2r^{-1}|z_0| \log M(r).$$

If we denote the positive number $\frac{1}{2}|z_0|^{-1} \log |f(z_0)|$ by 2ε, the preceding inequality says that $2\varepsilon \le r^{-1} \log M(r)$,

$$e^{-\varepsilon r} M(r) \ge e^{\varepsilon r}, \quad \text{valid for all} \quad r > \delta^{-1}|z_0|,$$

clearly contradicting hypothesis (i), thereby proving that (4) is untenable.

Remarks: (i) As its last step shows, this proof works with the boundedness hypothesis

(i)′ $\quad \underline{\lim}_{r \to \infty} \frac{1}{r} M(r) = 0$

in lieu of (i). Ostensibly this is weaker than (i), but either hypothesis implies $|f| \le 1$ and consequently implies the other, for the relevant f. The example $f(z) := e^{-i\lambda z}$ $(\lambda > 0)$ shows that some hypothesis like this is needed.

(ii) Actually 5.51 *can* be deduced from 5.16′, by applying it separately in $\mathbb{H} \cap \mathbb{U}$ and $(-\mathbb{H}) \cap \mathbb{U}$, along the lines of the hints to 11.36(viii).

Lemma 5.52 *There exists a non-negative harmonic function h_+ in $\mathbb{D}^+$ which is bounded by* 1 *and satisfies*

(i)
$$\lim_{z \to \zeta} h_+(z) = \begin{cases} 0 & \text{if } \zeta \in \mathbb{T}^+ \\ 0 & \text{if } \zeta \in (-1, 0) \\ 1 & \text{if } \zeta \in (0, 1), \end{cases}$$

(ii)
$$\lim_{r \downarrow 0} \left[1 - \frac{\theta}{\pi} - h_+\left(re^{i\theta}\right) \right] = 0 \quad \text{uniformly for } \theta \in (0, \pi).$$

Proof: Consider first the function

$$\phi_1(z) := \frac{1+z}{1-z}, \qquad z \in \overline{\mathbb{D}} \setminus \{1\}$$

which was examined in 5.47. From the results of 5.47 and by direct inspection it is easy to see that

$$\begin{aligned} \phi_1(-1, 0) &= (0, 1) \\ \phi_1(0, 1) &= (1, \infty) \\ \phi_1\left(\mathbb{T}^+\right) &= i(0, \infty) \\ \phi_1\left(\mathbb{D}^+\right) &= (0, \infty) \times (0, \infty). \end{aligned}$$

Therefore the function

$$\phi_2 := \phi_1^2$$

satisfies

$$\begin{aligned}
\phi_2\,(-1,0) &= (0,1)\\
\phi_2(0,1) &= (1,\infty)\\
\phi_2\big(\mathbb{T}^+\big) &= (-\infty,0)\\
\phi_2\big(\mathbb{D}^+\big) &= \mathbb{U}.
\end{aligned}$$

Finally, the (Koebe) function

$$\phi_3(z) := \phi_2(z) - 1 = \frac{4z}{(1-z)^2}$$

satisfies

$$\begin{aligned}
\phi_3\,(-1,0) &= (-1,0)\\
\phi_3(0,1) &= (0,\infty)\\
\phi_3\big(\mathbb{T}^+\big) &= (-\infty,-1)\\
\phi_3\big(\mathbb{D}^+\big) &= \mathbb{U}.
\end{aligned}$$

Now let L be the holomorphic logarithm in $\mathbb{C}\setminus i\,(-\infty,0]$ which satisfies

$$L\big(re^{i\theta}\big) = \log r + i\theta \qquad \forall r>0,\ \theta\in(-\pi/2, 3\pi/2)\,.$$

(Cf. 3.43.) Look at

$$h(z) := 1 - \frac{1}{\pi}\operatorname{Im} L(z), \qquad z\in\mathbb{C}\setminus i\,(-\infty,0]\,.$$

It is a harmonic function, is non-negative and bounded by 1 in $\mathbb{U}$, and satisfies

$$\begin{aligned}
h(-\infty,0) &= \{0\}\\
h(0,\infty) &= \{1\}.
\end{aligned}$$

If we set $h_+ := h\circ\phi_3$ on $\mathbb{D}^+$, property (i) is clear from all the foregoing. Now for $r>0$ and $\theta\in(0,\pi)$ we have

$$\begin{aligned}
-1+\frac{\theta}{\pi}+h_+\big(re^{i\theta}\big) &= \frac{\theta}{\pi} - \frac{1}{\pi}\operatorname{Im} L\left[\frac{4re^{i\theta}}{\big(1-re^{i\theta}\big)^2}\right]\\
&= \frac{1}{\pi}\operatorname{Im} L\big(e^{i\theta}\big) - \frac{1}{\pi}\operatorname{Im} L\left[\frac{e^{i\theta}}{\big(1-re^{i\theta}\big)^2}\right].
\end{aligned}$$

Since L is uniformly continuous on a neighborhood of the closure of $\mathbb{T}^+$, the assertion (ii) is now obvious.

Exercise 5.53 *With h_+ as in 5.52, show that $h_-(z) := h_+(-\bar{z})$ defines a non-negative harmonic function on $\mathbb{D}^+$ which is bounded by 1 and satisfies*

$$\text{(i)} \qquad \lim_{z\to\zeta} h_-(z) = \begin{cases} 0 & \text{if } \zeta \in \mathbb{T}^+ \\ 1 & \text{if } \zeta \in (-1,0) \\ 0 & \text{if } \zeta \in (0,1), \end{cases}$$

$$\text{(ii)} \qquad \lim_{r\downarrow 0} \left[\frac{\theta}{\pi} - h_-\left(re^{i\theta}\right)\right] = 0 \quad \text{uniformly for } \theta \in (0,\pi).$$

Lemma 5.54 *Let $\gamma\colon [0,1] \mapsto \overline{\mathbb{D}^+}$ be a curve with $\gamma(0) = 0$, $\gamma(1) \in \mathbb{T}^+$ and $\gamma(0,1) \subset \mathbb{D}^+$. If C is a component of $\mathbb{D}^+ \setminus \gamma$, then $\overline{C}$ cannot meet both the intervals $(-1,0)$ and $(0,1)$.*

Proof: Suppose contrariwise that there exist points

$$c_{-1} \in (-1,0) \cap \overline{C} \quad \text{and} \quad c_1 \in (0,1) \cap \overline{C}.$$

We can choose $r > 0$ so small that $\overline{D}(c_j, r)$ are disjoint from each other and from the compact set $\mathbb{T} \cup \gamma$. Then there exist $z_j \in D(c_j, r) \cap C$ and (by (1.30)(i)$'$ + (1.28)) a polygon P in C joining z_{-1} to z_1. The polygon $[c_{-1}, z_{-1}] \cup P \cup [z_1, c_1]$ lies in $\overline{\mathbb{D}^+} \setminus \gamma$. Form the loop Γ consisting of this polygon and its reflection, as in 5.49. This curve is disjoint from γ. According to 5.49, $\operatorname{Ind}_\Gamma(0) \neq 0$, so (4.3)

(1) $\gamma(0) = 0$ lies in a bounded component of $\mathbb{C} \setminus \Gamma$.

Since $\Gamma \subset \mathbb{D}$, the unbounded connected set $\mathbb{C} \setminus \mathbb{D}$ lies wholly in $\mathbb{C} \setminus \Gamma$, hence in the unbounded component of $\mathbb{C} \setminus \Gamma$. By hypothesis $\gamma(1)$ belongs to $\mathbb{T} \subset \mathbb{C} \setminus \mathbb{D}$. Thus

(2) $$\partial C \subset \mathbb{T}^+ \cup (0,1] \cup \gamma$$

The components of $\mathbb{C} \setminus \Gamma$ are open (1.30(i)$'$) and they cover the compact subset γ of $\mathbb{C} \setminus \Gamma$, hence finitely many of them , say, $C_1, C_2, \ldots, C_n$ cover γ. According to (1) and (2) we have $n \geq 2$. Hence this cover constitutes a disconnection of γ —×—.

Lemma 5.55 *Let $\gamma\colon [0,1] \mapsto \overline{\mathbb{D}^+}$ be a curve with $\gamma(0) = 0$, $\gamma(1) \in \mathbb{T}^+$ and $\gamma(0,1) \subset \mathbb{D}^+$. Let w be a non-negative superharmonic function on $\mathbb{D}^+$ such that $w \geq 1$ on $\gamma(0,1)$. Then for each $\varepsilon > 0$ there exists an $0 < r_\varepsilon < 1$ such that*

$$w\big(re^{i\theta}\big) \geq \min\left\{\frac{\theta}{\pi}, 1 - \frac{\theta}{\pi}\right\} - \varepsilon \qquad \forall 0 < r < r_\varepsilon,\ 0 < \theta < \pi.$$

Proof: Consider any component C of $\mathbb{D}^+ \setminus \gamma$. Since it is relatively closed, its boundary lies wholly in that of $\mathbb{D}^+ \setminus \gamma$, hence in $\mathbb{T}^+ \cup [-1,0) \cup (0,1] \cup \gamma$. The preceding lemma shows that we must then actually have

either

(1) $$\partial C \subset \mathbb{T}^+ \cup [-1,0) \cup \gamma$$

or

(2) $$\partial C \subset \mathbb{T}^+ \cup [0,1) \cup \gamma.$$

Since $w \geq 1 \geq h_+$ on $\gamma(0,1)$ and $w \geq 0$ throughout $\mathbb{D}^+$, it follows from 5.52(i) that

$$\varlimsup_{z\to\zeta}\,(h_+ - w)(z) \leq 0 \qquad \forall \zeta \in \mathbb{T}^+ \cup (-1,0) \cup \gamma(0,1)$$

and so from the maximum principle in the form 5.46 we will have

$$w \geq h_+ \quad \text{in } C \text{ if (1) holds.}$$

Similar reasoning using 5.53(i) gives

$$w \geq h_- \quad \text{in } C \text{ if (2) holds.}$$

Therefore in either case

(3) $$w \geq \min\{h_+, h_-\}$$

holds in C, and this for each component C of $\mathbb{D}^+ \setminus \gamma$. That is, (3) holds throughout $\mathbb{D}^+ \setminus \gamma$. Since $h_+ \leq 1$, and $w \geq 1$ on $\gamma(0,1)$, we actually have (3) holding throughout $\mathbb{D}^+$. The assertion of the lemma follows easily from this fact and parts (ii) of 5.52 and 5.53.

Corollary 5.56 (LINDELÖF [1915]) *Let f be holomorphic and bounded in $\mathbb{D}^+ := \{z \in \mathbb{C} : |z| < 1 < \operatorname{Im} z + 1\}$ and $\Gamma\colon [0,1] \mapsto \mathbb{C}$ be a curve such that $\Gamma(0) = 0$, $\Gamma(0,1) \subset \mathbb{D}^+$ and $\lim_{t\downarrow 0} f(\Gamma(t)) = a$, for some $a \in \mathbb{C}$. Then for each $\varepsilon > 0$*

(∗) $$\lim_{r\downarrow 0} f\big(re^{i\theta}\big) = a \quad \textit{uniformly for } \theta \in (\varepsilon, \pi - \varepsilon).$$

If $\widetilde{\Gamma}$ and $\tilde{a}$ are another such curve and limit, then from (∗) and its tilde analog it follows that $\tilde{a} = a$.

Proof: We can assume $a = 0$ and $|f| \le 1$. Let $0 < \varepsilon < 1$ be given and set $\delta(\varepsilon) := e^{(2/\varepsilon)\log\varepsilon}$. Then choose $0 < T < 1$ sufficiently small that

$$|f(\Gamma(t))| \le \delta(\varepsilon) \qquad \forall 0 < t \le T. \tag{1}$$

The function

$$h := \log\left(\max\{|f|, \delta(\varepsilon)\}\right) = \log\delta(\varepsilon) + \log^+\left|\frac{f}{\delta(\varepsilon)}\right|$$

is subharmonic in $\mathbb{D}^+$ by 5.26(ii) and is non-positive. Since $\Gamma(0) = 0$ and Γ is continuous, there is a least $t \in (0, T]$, call it t^*, such that $|\Gamma(t^*)| = |\Gamma(T)| =: R$, say. Note that $R < 1$. By the Intermediate-value Theorem we must have

$$|\Gamma(t)| < R \quad \text{for all } 0 \le t < t^*. \tag{2}$$

From (1) and $t^* \le T$ it follows that $|f(\Gamma(t^*t))| \le \delta(\varepsilon)$ for all $0 < t \le 1$, whence

$$h(\Gamma(t^*t)) = \log\delta(\varepsilon) + \log^+\left|\frac{f(\Gamma(t^*t))}{\delta(\varepsilon)}\right| = \log\delta(\varepsilon) = (2/\varepsilon)\log\varepsilon. \tag{3}$$

The function

$$w(z) := \frac{\varepsilon h(Rz)}{2\log\varepsilon}, \qquad z \in \mathbb{D}^+ \tag{4}$$

is non-negative and superharmonic (recall, $\log\varepsilon < 0$), and the curve $\gamma(t) := R^{-1}\Gamma(t^*t)$, $0 \le t \le 1$, has $\gamma(0) = 0$, $\gamma(0,1) \subset \mathbb{D}^+$, by (2), and $\gamma(1) \in \mathbb{T}^+$. These two functions satisfy

$$w(\gamma(t)) \overset{(4)}{=} \frac{\varepsilon h(\Gamma(t^*t))}{2\log\varepsilon} \overset{(3)}{=} 1 \qquad \forall 0 < t < 1,$$

and consequently fulfill the hypotheses of 5.55. Therefore there exists an $0 < r_\varepsilon < 1$ such that

$$\begin{aligned} w\big(re^{i\theta}\big) &\ge \min\left\{\frac{\theta}{\pi}, 1 - \frac{\theta}{\pi}\right\} - \frac{\varepsilon}{2} && \forall 0 < r < r_\varepsilon,\ \theta \in (0,\pi) \\ &> \varepsilon - \frac{\varepsilon}{2} && \forall 0 < r < r_\varepsilon,\ \theta \in (\varepsilon\pi, (1-\varepsilon)\pi) \\ &= \frac{\varepsilon}{2}. \end{aligned} \tag{5}$$

From (4), (5) and the fact that $\log\varepsilon < 0$

$$h\big(re^{i\theta}\big) \le \log\varepsilon \qquad \forall 0 < r < Rr_\varepsilon,\ \theta \in (\varepsilon\pi, (1-\varepsilon)\pi),$$

whence

$$|f(re^{i\theta})| \le e^{h(re^{i\theta})} \le \varepsilon \qquad \forall\, 0 < r < Rr_\varepsilon,\, \theta \in (\varepsilon\pi, (1-\varepsilon)\pi).$$

Remarks 5.57: For a beautiful geometric proof of 5.56 see §§307–308 of CARA-THÉODORY [1960], or pp. 205–208 of HERVÉ [1982]. Lindelöf proved 5.56 under the weaker hypothesis that there be at least two complex numbers absent from the range of f. (See also TANAKA [1964], [1976].) We shall secure a version of this in Chapter XII. (See 12.33 and 12.38.) But without any sort of boundedness hypothesis the conclusion fails; e.g., $f(z) := e^{i/z^2}$, $\Gamma(t) := te^{3\pi i/4}$, $\theta := \pi/2$. The hypotheses can also be weakened in another way: HALL [1937] shows that if f is a bounded holomorphic function in $\operatorname{Im} z > 0$ and $\inf\{|f(re^{i\theta})| : 0 < \theta < \pi\} \to 0$ as $r \to \infty$ through an appropriately dense set of r values, then $\lim_{r\to\infty} f(re^{i\theta}) = 0$ uniformly in $\varepsilon \le \theta \le \pi - \varepsilon$, for each $\varepsilon > 0$. (Cf. 12.38 and CARTWRIGHT [1935].) An account of Hall's work may also be found in FUCHS [1967], pp. 81–87 and an extension in KAWAKAMI [1956]. For a generalization of 5.56 of a different kind see GEHRING and LOHWATER [1958] (or p. 20 of COLLINGWOOD and LOHWATER [1966]).

We offer one final interesting application of harmonic majorization.

Exercise 5.58 (i) *Let* $a > 0$, $R := \left(-\frac{\pi}{2}, \frac{\pi}{2}\right) \times (0, a)$, $B := \left(\left\{\pm\frac{\pi}{2}\right\} \times [0, a]\right) \cup \left[-\frac{\pi}{2}, \frac{\pi}{2}\right]$. *Prove that there exists a continuous function* $g \colon \overline{R} \mapsto [0, 1]$ *which is harmonic and positive in* R *and* 0 *on* B.

Hints: Let $\psi(w) := (w - i)/(w + i)$ $\left(w \in \overline{\mathbb{U}}\right)$. Then ψ maps $\overline{\mathbb{U}}$ onto $\overline{\mathbb{D}} \setminus \{1\}$ and $\mathbb{R}$ onto $\mathbb{T} \setminus \{1\}$. The function sin maps $\overline{R}$ into $\overline{\mathbb{U}}$ and B into a compact subset of $\mathbb{R}$. Therefore the composite $\psi \circ \sin$ maps $\overline{R}$ into $\overline{\mathbb{D}} \setminus \{1\}$ and B onto a compact subset K of $\mathbb{T} \setminus \{1\}$. Let $f \colon \mathbb{T} \mapsto [0, 1]$ be any continuous function satisfying $f(K) = 0$, $f(1) = 1$ and let F solve the Dirichlet problem in $\mathbb{D}$ for f. Since $F = f \ge 0$ on $\mathbb{T}$, it follows from 5.10 that $F > 0$ in $\mathbb{D}$. Let $g := F \circ \psi \circ \sin$.

(ii) *Let* $r > 1$, $S_r := \{z \in \mathbb{C} : \operatorname{Re} z > 0, 1/r < |z| < r\}$. *Prove that there is a continuous function* $f_r \colon \overline{S}_r \mapsto [0, 1]$ *which is harmonic and positive in* S_r *and* 0 *on* $\partial S_r \setminus [i/r, ir]$.

Hint: Let L denote the Principal Branch of the Logarithm in $\mathbb{C}_-$ and take $a := \pi^2/(2\log r)$ in (i). It follows easily that $\pi/(2\log r) \cdot L + i\pi^2/(4\log r)$ maps S_r conformally onto R, $\overline{S}_r$ homeomorphically onto $\overline{R}$, and $\partial S_r \setminus (i/r, ir)$ onto B.

Exercise 5.59 *Let* $r > 1$, $S_r := \{z \in \mathbb{H} : 1/r < |z| < r\}$. *Prove that there is a positive constant* $c(r)$, *dependent only on* r, *such that* $u(1) \ge c(r)$ *holds for* *all continuous functions u on $\overline{S}_r \cap \mathbb{H}$ which are harmonic and non-negative in S_r and for which there exists a curve $\gamma\colon [0,1] \mapsto \overline{S}_r \cap \mathbb{H}$ with $|\gamma(0)| = 1/r$, $|\gamma(1)| = r$ and $u \geq 1$ on γ.*

Hints: Let such a function u and such a curve γ be given. It suffices to deal with the case where $1 \notin \gamma$. Let Ω be the component of $S_r \setminus \gamma$ which contains 1. According to 4.9 $\overline{\Omega}$ cannot meet both the positive and negative y-axis. Therefore either $\partial\Omega \subset (\gamma \cap S_r) \cup \partial S_r \setminus [i/r, ir]$, or $\partial\Omega \subset (\gamma \cap S_r) \cup \partial S_r \setminus [-i/r, -ir]$. Say, the former holds. If f_r is as provided by 5.58(ii), then $u - f_r \geq 1 - f_r \geq 0$ on $\gamma \cap S_r$ and $u - f_r = u \geq 0$ on $\partial S_r \setminus [i/r, ir]$. Therefore $u - f_r \geq 0$ on $\partial\Omega$. It follows from the Maximum Principle that $u - f_r \geq 0$ throughout Ω. In particular, $u(1) \geq f_r(1)$.

Theorem 5.60 (JØRGENSEN [1939]) *Let $f\colon \mathbb{H} \mapsto \overline{\mathbb{H}}$ be holomorphic. Suppose that*

(i) $$\lim_{\substack{x\to\infty \\ x\in\mathbb{R}}} \frac{\operatorname{Re} f(x)}{x} = 0$$

and that for some finite constant M

(ii) $$\overline{\lim_{z\to iy}} \operatorname{Re} f(z) \leq M \qquad \forall y \in \mathbb{R}.$$

Then $\operatorname{Re} f \leq M$ throughout $\mathbb{H}$.

Proof: From (i) and (ii) it follows easily that for each $\delta > 0$ and each $\varepsilon > 0$ there is a finite constant $M(\varepsilon) \geq M$ such that

(1) $$\overline{\lim_{z\to\zeta}} \,[\operatorname{Re} f(z) - \varepsilon \operatorname{Re} z - \delta |\operatorname{Im} z|] \leq M(\varepsilon) \qquad \forall \zeta \in i\mathbb{R} \cup (0,\infty).$$

We want first to establish the inequality

(2) $$\operatorname{Re} f(z) - \varepsilon \operatorname{Re} z - \delta |\operatorname{Im} z| \leq M(\varepsilon) \qquad \forall z \in \mathbb{H}.$$

The arguments for $\mathbb{H}_+ := \{z \in \mathbb{H} : \operatorname{Im} z > 0\}$ and $\mathbb{H}_- := \{z \in \mathbb{H} : \operatorname{Im} z < 0\}$ are the same, so we (only) look at $\mathbb{H}_+$ and argue by contradiction, i.e., we assume that the (open) set

$$U := \{z \in \mathbb{H}_+ : \operatorname{Re} f(z) - \varepsilon \operatorname{Re} z - \delta \operatorname{Im} z > M(\varepsilon)\}$$

is not void. Let C be any component of U. We have $\partial C \subset \partial U \subset \overline{\mathbb{H}}_+ \setminus U$. At points z of $\mathbb{H}_+ \setminus U$ we have (2). The other points ζ of $\overline{\mathbb{H}}_+ \setminus U$ lie in $\overline{\mathbb{H}}_+ \setminus \mathbb{H}_+ \subset i\mathbb{R} \cup (0,\infty)$ and for them we have (1). Therefore

$$\overline{\lim_{z\to\zeta}} \,[\operatorname{Re} f(z) - \varepsilon \operatorname{Re} z - \delta \operatorname{Im} z] \leq M(\varepsilon) \qquad \forall \zeta \in \partial C.$$

If C were bounded, then this inequality would imply (by 5.12) the validity of the inequality in (2) for all $z \in C \subset U$, in contradiction with the definition of U. Therefore C must be unbounded. Since C is open (1.30(i)′) and connected, it follows readily from 1.28 that there exists a continuous $\gamma\colon [0,1) \mapsto C$ such that

$$\overline{\lim_{t\uparrow 1}}\, |\gamma(t)| = \infty. \tag{3}$$

We have then

$$\operatorname{Re} f(z) - \varepsilon \operatorname{Re} z - \delta \operatorname{Im} z > M(\varepsilon) \qquad \forall z \in \gamma$$

and so *a fortiori*

$$\operatorname{Re} f(z) > \eta\{\operatorname{Re} z + \operatorname{Im} z\} \geq \eta|z| \qquad \forall z \in \gamma, \tag{4}$$

where $\eta := \min\{\varepsilon, \delta\}$. For each $r > r_0 := \inf|\gamma|[0,1)$, the curve γ meets $C(0,r)$. Set $S_2 := \left\{z \in \mathbb{C} : \operatorname{Re} z > 0,\ \frac{1}{2} < |z| < 2\right\}$ and consider $x > 2r_0$. (4) implies that

$$\operatorname{Re} f(xz) > \eta|xz| > \tfrac{1}{2}\eta x \quad \text{whenever } xz \in \gamma \cap xS_2.$$

Therefore if we set $u(z) := 2\operatorname{Re} f(xz)/\eta x$, then

$$u(z) > 1 \quad \text{whenever } z \in \frac{1}{x}\gamma \cap S_2. \tag{5}$$

We cite 5.59 to conclude therefore that $u(1) \geq c(2)$, that is,

$$\frac{2\operatorname{Re} f(x)}{\eta x} \geq c(2),$$
$$\frac{\operatorname{Re} f(x)}{x} \geq \frac{\eta c(2)}{2}.$$

This holds for all real $x > 2r_0$, in violation of (i). It follows that U is void and (2) holds throughout $\mathbb{H}_+$. Similarly (2) holds throughout $\mathbb{H}_-$, thus throughout $\mathbb{H}$. Let $\delta \downarrow 0$ in (2) to conclude that

$$\operatorname{Re} f(z) - \varepsilon \operatorname{Re} z \leq M(\varepsilon) \qquad \forall z \in \mathbb{H}. \tag{6}$$

To the function $e^{f(z)-\varepsilon z}$ apply 5.13 and conclude from (6) and (ii) that

$$\operatorname{Re} f(z) - \varepsilon \operatorname{Re} z \leq M \qquad \forall z \in \mathbb{H}, \quad \forall \varepsilon > 0. \tag{7}$$

Let $\varepsilon \downarrow 0$ in (7) and the proof is complete.

Remarks 5.61: Of course 5.60 is a theorem about positive harmonic functions in the right half-plane $\mathbb{H}$. To secure this (apparent) generalization, the reader can either repeat the above proof, citing 7.17 instead of 5.13, or note that any positive harmonic function in $\mathbb{H}$ is the real part of a single holomorphic function there (5.21(i)+2.5) and apply 5.60 as proven. [Another proof of 5.60 occurs in 8.39(iv).]

§ 6 Uniqueness Theorems

The general theme of this section, well illustrated by the first theorem, is the complete determination of a holomorphic function from its behavior on a small subset of its domain. The Cauchy integral formula already illustrates this, since it re-captures all the function values inside a disk in terms of those on the boundary.

Theorem 5.62 (Fundamental Uniqueness Theorem of Weierstrass) *If f is holomorphic in the region Ω and $f^{-1}(0)$ has a limit point in Ω, then $f = \mathbf{0}$.*

Proof: (Cf.3.35(viii).) Let Ω_0 be the set of limit points *in* Ω of $f^{-1}(0)$. Thus $\Omega_0 \neq \varnothing$ by hypothesis. We shall show that the hypothesis of the Basic Connectedness Lemma 1.27 is fulfilled with $\theta = 1$ and conclude therefrom that $\Omega_0 = \Omega$. Since f is continuous, $\Omega_0 \subset f^{-1}(0)$ and we shall be finished. So suppose $z_0 \in \Omega_0$ and $D(z_0, r) \subset \Omega$. Use 5.30 to write

$$f(z) = \sum_{n=0}^{\infty} c_n (z - z_0)^n \qquad \forall z \in D(z_0, r).$$

We claim that $D(z_0, r) \subset f^{-1}(0)$, whence $D(z_0, r) \subset \Omega_0$. For if not, there is a first non-zero coefficient in this series, say c_k:

$$f(z) = (z - z_0)^k \sum_{n=0}^{\infty} c_{n+k} (z - z_0)^n = (z - z_0)^k h(z),$$

where h is defined in $D(z_0, r)$ by the obvious power series and $h(z_0) = c_k \neq 0$. Since h is continuous, it is non-zero sufficiently near z_0 and so $f(z) = (z - z_0)^k h(z)$ shows that f has no zeros near z_0 except z_0 itself, contrary to z_0 being a limit point of $f^{-1}(0)$.

Remarks 5.63: (i) This result is best possible in the sense that if A is any subset of Ω with no limit points in Ω, then there is a holomorphic function on Ω whose zero-set is exactly A. This will be proved in 7.35 and 11.40. On the other hand, RUDIN [1992] constructs a non-zero power series in $\mathbb{D}$ which converges to 0 at countably many points of $\partial\mathbb{D}$. In 5.73 the reader will find an alternative, *constructive* proof of 5.62. An obvious consequence of 5.62 is that a non-constant $f \in H(\Omega)$ has only finitely many zeros in any compact subset of Ω, hence by 1.31 only countably many zeros altogether. Thus $H(\Omega)$ is an *integral domain* (see also 11.46).

(ii) Exactly the opposite of 5.62 is true of *harmonic* functions—their zeros are never isolated. Indeed if $h\colon \Omega \mapsto \mathbb{R}$ is harmonic, $D(a,r) \subset \Omega$, $h(a) = 0$ and h is zero-free in $D(a,r) \setminus \{a\}$, then on this latter connected set h is either everywhere positive or everywhere negative, and consequently either $-h$ or h is a harmonic function which violates the maximum principle in $D(a,r)$. Much more can be said: see 10.9(iii). A partial analog of 5.62 for harmonic functions will nevertheless be found in 5.67(iv).

(iii) *A fortiori* from 5.62 every f which is holomorphic in a region Ω is determined throughout Ω by its behavior in any neighborhood of any point. How can its values elsewhere in Ω be recovered from some such local data? One beautiful way is laid out in 5.73. A simpler way, Weierstrass' *analytic continuation along curves*, uses neither integration nor 5.62, but only 5.30 and 3.6: Given $z_0, z \in \Omega$, there is a curve $\gamma : [0,1] \mapsto \Omega$ with $\gamma(0) = z_0$, $\gamma(1) = z$. Compactness of $\gamma[0,1]$ ensures that there is an $r > 0$ such that $D(\gamma(t), r) \subset \Omega$ for all t. Uniform continuity of γ ensures that for some $n \in \mathbb{N}$, $|\gamma(t) - \gamma(t')| < r$ whenever $|t - t'| \leq 1/n$. Let $z_j := \gamma(j/n)$ for $1 \leq j \leq n$. The disk $D(z_j, r)$ lies in Ω, the Taylor series of f at z_j converges in $D(z_j, r)$ to f (5.30), and $z_j \in D(z_{j-1}, r)$ for $1 \leq j \leq n$. Because of the latter, the Taylor coefficients of f at z_j, namely the numbers $f^{(k)}(z_j)/k!$, can be computed from those of f at z_{j-1} via (2.k) in 3.6. Thus, in principle, after n steps the Taylor coefficients of f at $z_n = z$ are expressed in terms of those at z_0. This technique can sometimes lead to a holomorphic extension of f into a region larger than Ω. See 10.3(iv).

Example 5.64 (i) *If f is holomorphic in the region Ω and for each $z \in \Omega$ there is a non-negative integer $n = n(z)$ such that $f^{(n)}(z) = 0$, then f is a polynomial.*

Proof: Setting $\Omega_n := \{z \in \Omega : f^{(n)}(z) = 0\}$ $(n = 0, 1, 2, \dots)$, the zero-set of the holomorphic function $f^{(n)}$, the hypothesis is that $\Omega = \bigcup_{n=0}^{\infty} \Omega_n$. We may suppose that $\overline{\mathbb{D}} \subset \Omega$, and then we have $\overline{\mathbb{D}} = \cup \overline{\mathbb{D}} \cap \Omega_n$. According to Remark (d) following 5.41, the complete metric space $\overline{\mathbb{D}}$ is uncountable. But a countable union of finite sets is only countable, so $\overline{\mathbb{D}} \cap \Omega_n$ must be infinite for some n, and hence it must have a limit point in the compact set $\overline{\mathbb{D}} \subset \Omega$. From 5.62 then $f^{(n)} = 0$ identically in Ω and consequently $f(z)$ coincides with the polynomial $\sum_{k=0}^{n-1} \left(f^{(k)}(0)/k!\right) z^k$ in $\mathbb{D}$, hence throughout Ω.

(ii) *f, g are holomorphic in the region Ω and for each $z \in \Omega$, $f(z)$ is a positive integer power of $g(z)$. Then $f = g^n$ for some $n \in \mathbb{N}$.*

Proof: For each $n \in \mathbb{N}$ let $\Omega_n := (f - g^n)^{-1}(0)$. Argue as in (i) that at least one of these sets has a limit point in Ω and then invoke 5.62.

Remark: The proof of (i) evidently works if it is only hypothesized that there is a countable set S such that for each $z \in \Omega$ some derivative of f maps z into S. COROMINAS and SUNYER I BALAGUER [1954] used an intricate Baire-category argument to prove the analog of this for an infinitely differentiable function on an open interval in $\mathbb{R}$. Expositions (albeit, somewhat inadequate) of their result also occur in DONOGHUE [1969] and WERMUTH [1991]. For a meticulous account see ROSS [2013], p. 353 ff.

Exercise 5.65 (i) (Extension of 4.67) *Let U be an open subset of $\mathbb{C}$, $f \in H(U)$, $g\colon U \mapsto \mathbb{C}$ a continuous function such that $g^n = f$ for some positive integer n. Show that g is holomorphic in U.*

Hints: We may assume U is connected. All is trivial if $f = \mathbf{0}$. Otherwise $A := f^{-1}(0)$ consists of isolated points, by 5.62, so $U \setminus A$ is open. g is holomorphic there by 4.67(ii) and, being continuous, it is $\mathbb{C}$-differentiable at each point of A as well by 5.41. [Note that this result is a trivial corollary of 5.98.]

(ii) *Show that $f \in H(U)$ has a holomorphic n^{th} root if and only if* $\operatorname{Ind}_{f\circ\gamma}(0)$ *is divisible by n for every (polygonal) loop γ in $U \setminus f^{-1}(0)$.* Cf. 10.6.

Hint: Use 4.19(ii) together with (i) above. An alternative (but essentially very similar) proof can be based on a variant of the technique used in 5.32(i); for details see WALLEN [1992].

(iii) *Ω is a region and $f \in H(\Omega) \setminus \{\mathbf{0}\}$ has a continuous n^{th}-root g_n for infinitely many $n \in \mathbb{N}$. Show that f is zero-free and has a holomorphic logarithm.*

Hints: It suffices to show that f is zero-free; 4.19(i) and 4.67(i) take over after that. So suppose $f(z_0) = 0$. Then each $g_n(z_0) = 0$. By (i), g_n is holomorphic. As shown in the proof of 5.33, there is then $G_n \in H(\Omega)$ such that $g_n(z) = g_n(z) - g_n(z_0) = (z - z_0)G_n(z)$. Hence $f(z) = g_n^n(z) = (z - z_0)^n G_n^n(z)$ shows that the first n terms in the Taylor series of f at z_0 are 0, making f the $\mathbf{0}$ function —⨯—.

Exercise 5.66 (i) *Let f be holomorphic in $\mathbb{D}$ and suppose that $\left|f\left(re^{i\theta}\right)\right|$ is constant for $r \in [0, 1)$ and also that $\left|f\left(re^{i\psi}\right)\right|$ is constant for such r. If $(\theta - \psi)/\pi$ is irrational, then f is a constant.*

Hints: If the constant value of $|f|$ on (either of) the rays is 0, then f is 0 by 5.62. Otherwise $f(0) \neq 0$ and by considering $f\left(\rho e^{i\psi} z\right)/f(0)$ for sufficiently small positive ρ, we may suppose that f has no zeros in $\mathbb{D}$, that $\psi = 0$ and that $f(0) = 1$. 5.35 provides a $g \in H(\mathbb{D})$ such that $f = e^{ig}$. Write $g(z) = \sum_{n=0}^{\infty} c_n z^n$. Since $|f(t)| = |f(0)| = 1$ for all $0 \le t < 1$, we see that g is real-valued in $[0, 1)$ and so

$$\text{(1)} \qquad c_n = \frac{1}{n!}\big[g\,|_{\mathbb{R}}\big]^{(n)}(0) \in \mathbb{R}, \quad n = 0, 1, 2, \ldots.$$

Similarly $g\big(e^{i\theta}t\big)$ is real for all $0 \le t < 1$, so we conclude that

$$\text{(2)} \qquad c_n e^{in\theta} \in \mathbb{R}, \quad n = 0, 1, 2, \ldots.$$

If for some $n > 0$, $c_n \ne 0$ we conclude from dividing (2) by (1) that $e^{in\theta} \in \mathbb{R}$. This means $e^{in\theta} = \pm 1$ and so there is an integer m such that either $n\theta - \pi = 2\pi m$ or $n\theta = 2\pi m$. In either case θ is a rational multiple of π (since $n \ne 0$).

(ii) *Let h be harmonic in $\mathbb{D}$ and vanish on two sequences $\{r_n e^{i\theta}\}$, $\{\rho_n e^{i\psi}\}$, where $\{r_n\}$ and $\{\rho_n\}$ each have a limit point in $[0, 1)$. If $(\theta - \psi)/\pi$ is irrational, then $h = \mathbf{0}$.*

Hints: Look ahead to 5.67(iv) to see that h is identically 0 on the radii terminating at $e^{i\theta}$ and $e^{i\psi}$. Then use 5.21(i) to find $F \in H(\mathbb{D})$ with $\operatorname{Re} F = h$. Finally, apply (i) to $f := e^F$, or just repeat the argument in (i) for $g := iF$.

(iii) *Show that the irrationality hypothesis in* (i) *and* (ii) *is essential.*

Hints: Given $m, n \in \mathbb{N}$, set $\theta := m\pi/n$ and $F(z) := z^n$. The entire function $f := e^{iF}$ is not constant in $\mathbb{D}$ but satisfies $|f(z)| = 1$ whenever $z = re^{ik\theta}$, $r \ge 0$, $k \in \mathbb{Z}$, since $F(z)$ is real for such a z.

Remark: The reader knowledgeable of measure theory and the generalized Fubini theorem can derive from 5.66(ii) the following attractive uniqueness theorem for harmonic functions: If h is harmonic in the region Ω and $h^{-1}(0)$ has positive 2-dimensional Lebesgue measure, then $h = 0$ throughout Ω. Cf. the weaker version in 10.9(ii).

Exercise 5.67 *$a_n \in \mathbb{R}$ are distinct, $a_n \to 0$, f is holomorphic in $D(0, r)$ for some $r > 0$ and $f(a_n)$ is real for all large n.*

(i) *Show that $f(\bar{z}) = \bar{f}(z)$ for all $z \in D(0, r)$.*

Hint: Form $F(z) := \bar{f}(\bar{z})$, which is holomorphic in $D(0, r)$ [by 2.4], and consider $g := F - f$. It is holomorphic and vanishes at a_n for all large n.

(ii) *If all a_n are positive and $f(a_{2n}) = f(a_{2n+1})$ for all large n, then f is constant.*

Hint: Since by (i), f is real (and differentiable) on $(-r, r)$, the Mean-Value Theorem provides (for each $n \geq N$, say) b_n between a_{2n} and a_{2n+1} such that

$$0 = \frac{f(a_{2n+1}) - f(a_{2n})}{a_{2n+1} - a_{2n}} = f'(b_n).$$

Since $a_n \to 0$, the set $\{b_N, b_{N+1}, \ldots\}$ is not finite and accumulates at 0. Conclude that $f' = \mathbf{0}$.

(iii) *$f \in H(\mathbb{D})$, $f(0) = 0$ but f is not constant. Show that there exists $\varepsilon > 0$ such that for every $\theta \in \mathbb{R}$ the function $r \mapsto \left|f(re^{i\theta})\right|^2$ is convex and strictly increasing on $[0, \varepsilon]$.*

Hints: As in the proof of 5.62 there exists $g \in H(\mathbb{D})$ with $g(0) \neq 0$ and $n \in \mathbb{N}$ such that $f(z) = z^n g(z)$. For each $u \in \mathbb{T}$ form $f_u(z) := f(uz)$, $F_u(z) := f_u(z)\overline{f_u(\bar{z})}$, $z \in \mathbb{D}$. They are holomorphic (see 2.4), and $F_u(r) = |f(ru)|^2$ for $r \in [0, 1)$. Compute that for such r

$$F_u'(r) = r^{2n-1}\left[2n|g(ru)|^2 + 2r \operatorname{Re}\left(u\overline{g'(ru)}g(ru)\right)\right]$$

and infer that for an appropriate continuous function

$$G\colon [0,1) \times \mathbb{T} \mapsto \mathbb{R}$$
$$F_u''(r) = r^{2n-2}\left[2n(2n-1)|g(ru)|^2 + rG(r,u)\right].$$

Since $g(0) \neq 0$ and all functions are continuous, there exists $0 < \varepsilon < 1$ such that the bracketed expressions are strictly positive for all $(r, u) \in [0, \varepsilon] \times \mathbb{T}$. [Note that (iii) subsumes (ii).]

(iv) *Suppose $[a, b] \subset U$ open $\subset \mathbb{C}$, $h\colon U \mapsto \mathbb{R}$ is harmonic, and $h^{-1}(0) \cap [a, b]$ is infinite. Show that then $h[a, b] = 0$.*

Hints: We may suppose that $[a, b] \subset \mathbb{R}$. Let x_0 be a limit point of $h^{-1}(0) \cap [a, b]$. Choose $r > 0$ so that $D := D(x_0, r) \subset U$ and then choose $f \in H(D)$ so that $h = \operatorname{Im} f$ in D. From (i) infer that $f(\bar{z}) = \bar{f}(z)$ for all $z \in D$. In particular, for $z = x \in \mathbb{R} \cap D$, $0 = \operatorname{Im} f(x) = h(x)$. This shows that the (closed) set of all limit points of $h^{-1}(0) \cap [a, b]$ is relatively open in $[a, b]$.

(v) *Ω is a region properly contained in $\mathbb{C}$, and for every $z \in \mathbb{C}$, $\rho(z) := \inf\{|z - w| : w \in \mathbb{C} \setminus \Omega\}$ denotes the distance from z to $\mathbb{C} \setminus \Omega$. Show that if ρ is harmonic in Ω, then Ω is a half-plane.*

Hints: By definition of ρ

(1) $$\Omega = \rho^{-1}(0, \infty) \quad \text{and} \quad D(z, \rho(z)) \subset \Omega \quad \forall\, z \in \Omega.$$

Fix $x_0 \in \Omega$. There exists $b_0 \in \partial\Omega$ with $\rho(x_0) = |x_0 - b_0|$. We may assume that $b_0 = 0$ and that x_0 is real and positive. Then an application of the triangle inequality shows that $\rho(x) = x$ for all $x \in (0, x_0]$:

$$(2) \qquad D(x_0, x_0) \subset \Omega \quad \& \quad \rho(x) = x \quad \forall x \in (0, x_0]\,.$$

But then by (iv), $\rho(x) = x$ for all $x \in (0, 2x_0)$, whence by continuity of ρ on $\mathbb{C}$ (1.32), $\rho(2x_0) = 2x_0 > 0$. It follows from (1) that

$$(3) \qquad D(2x_0, 2x_0) \subset \Omega \quad \& \quad \rho(x) = x \quad \forall x \in (0, 2x_0]\,.$$

Iterating the argument which led from (2) to (3) shows that

$$(4) \qquad \mathbb{H} = \bigcup_{n\in\mathbb{N}} D(2^n x_0, 2^n x_0) \subset \Omega \quad \& \quad \rho(x) = x \quad \forall x \in (0, \infty).$$

Consequently, for $z \in \mathbb{H}$

$$(5) \qquad \rho(z) = \operatorname{dist}\{z, \mathbb{C}\setminus\Omega\} \geq \operatorname{dist}\{z, \mathbb{C}\setminus\mathbb{H}\} = \operatorname{Re} z.$$

From (5), (4) and the Maximum Principle (5.8) infer that

$$(6) \qquad \rho(z) = \operatorname{Re} z \qquad \forall z \in \mathbb{H} \subset \Omega.$$

Since ρ is continuous on $\mathbb{C}$, this shows $\rho(i\mathbb{R}) = 0$, so by (1) $i\mathbb{R} \cap \Omega = \varnothing$. The fact that $\mathbb{H} \subset \Omega$ and that Ω is connected then leads to $\Omega = \mathbb{H}$.

Definition 5.68 If f is holomorphic and not constant in a connected neighborhood of a, it follows from 5.30 and 5.62 that $f^{(n)}(a) \neq 0$ for some non-negative integer n. The non-negative integer $n(a, f) := \min\{n : f^{(n)}(a) \neq 0\}$ is called the *multiplicity* (or *order*) *of f at a* or the *multiplicity* (or *order*) *of a as a zero of f*, or the *local degree* of f at a. The zero is called *simple* if the multiplicity is 1. If $f^{(n)}(a) = 0$ for all n (equivalently, $f = 0$ throughout some disk centered at a), the point a is sometimes called a zero of *infinite multiplicity.* Zeros can also be quantified without reference to power series: Evidently a is a zero of f of multiplicity n if and only if there is a function g holomorphic in a neighborhood V of a, with $g(a) \neq 0$ and $f(z) = (z-a)^n g(z)$ for all $z \in V$. From the other direction, the size of the set $f^{-1}(w)$ is called the *valence* of w, for each $w \in \mathbb{C}$. More precisely, the valence of w is the cardinality of $(f-w)^{-1}(0)$ when each zero is counted according to its multiplicity. (Cf. 6.12(iv).)

Exercise 5.69 *Let f be holomorphic and not constant in the region Ω. Let $a_1, \ldots, a_k \in \Omega$. Using 5.68 and induction on k show that there is a holomorphic function F in Ω such that*

$$f(z) = F(z) \prod_{j=1}^{k} (z - a_j)^{n(a_j, f)} \qquad \forall z \in \Omega$$

and that $F(a_j) \neq 0$ *for each* $j = 1, \ldots, k$. (Cf. 4.66 + 4.68.)

Exercise 5.70 (i) (CAUCHY [1844]) *An entire function* F *such that* $|F(z)| \geq \delta$ *for some* $\delta > 0$ *and all sufficiently large* z *is a polynomial.*

Hints: There is an $r_1 > 0$ such that $|F(z)| \geq \delta$ for all $|z| \geq r_1$. Therefore the zeros of F are all in the compact set $\overline{D}(0, r_1)$, hence there are only finitely many, say $a_1, \ldots, a_l$ with multiplicities $m_1, \ldots, m_l$. Let $P(z) := \prod_{j=1}^{l}(z - a_j)^{m_j}$, a monic polynomial of degree $n := m_1 + \cdots + m_l$. Evidently $\delta P/F$ is an entire function f and satisfies $|f(z)| \leq |P(z)| \leq |z|^n$ for all $|z| \geq R$ and some $R \geq r_1$. (See (1) in the proof of 4.51.) Then 3.34(ii) says that f is a polynomial. But by the definition of P, the function P/F has no zeros. Thus the polynomial f has no zeros and so is constant, say $f = c \neq 0$. Then $F = (\delta/c)P$.

(ii) *If* f *and* g *are non-constant entire functions and* $f \circ g$ *is a polynomial, then* f *and* g *are each polynomials.*

Hints: First note that $f \circ g$ is not constant. For, g being non-constant, $g[0, 1]$ is an infinite compact set on which f would be constant. If g is not a polynomial, (i) furnishes $z_n \in \mathbb{C}$ with $|z_n| \to \infty$ and $|g(z_n)| \leq 1$. We can suppose $g(z_n)$ converge to some $w_0 \in \overline{\mathbb{D}}$. Then $(f \circ g)(z_n) = f(g(z_n)) \to f(w_0)$, contrary to the fact that the non-constant polynomial $f \circ g$ must satisfy $\lim_{|z| \to \infty} |(f \circ g)(z)| = \infty$. Thus g is a (non-constant) polynomial. Therefore, given $w_n \in \mathbb{C}$ with $|w_n| \to \infty$, $g - w_n$ has a zero z_n, and necessarily $|z_n| \to \infty$. Therefore, $f \circ g$ being a non-constant polynomial, $|f(w_n)| = |(f \circ g)(z_n)| \to \infty$. That is, $|f(w_n)| \to \infty$ whenever $|w_n| \to \infty$. Hence from (i) it follows that f is a polynomial.

Exercise 5.71 *Let* U *be a non-empty open subset of* $\mathbb{C}$, $f \in H(U)$. *Suppose that for each* $b \in \partial U$, f *has a holomorphic extension to a neighborhood of* b. *Show that there is a holomorphic extension of* f *to a neighborhood of* $\overline{U}$.

Hints: (MINDA [1973]) For each $b \in \overline{U}$ there exists finite $r_b > 0$ and f_b holomorphic in $D(b, 2r_b)$ such that $f_b = f$ in $U \cap D(b, 2r_b)$. Set $V := \bigcup\{D(b, r_b) : b \in \overline{U}\}$, an open neighborhood of $\overline{U}$. If $a, b \in \overline{U}$ and $D(a, r_a) \cap D(b, r_b) \neq \varnothing$, then $|a - b| < r_a + r_b$. Let, say, $r_a \leq r_b$. Then $a \in D(b, 2r_b)$. The neighborhood $D(a, 2r_a) \cap D(b, 2r_b)$ of a meets U (since $a \in \overline{U}$) in a non-empty open subset W. In W we have $f_a = f = f_b$. Hence by 5.62, $f_a = f_b$ throughout the convex set $D(a, 2r_a) \cap D(b, 2r_b)$. This fact enables us to well define an extension of f to V by decreeing $F := f_b$ in $D(b, r_b)$ for each $b \in \overline{U}$. Evidently F is holomorphic in V and coincides with f in U.

Exercise 5.72 (i) *Let* $f_1, \ldots, f_n$ *be holomorphic in the region* Ω. *Show that if* $\phi := |f_1| + \cdots + |f_n|$ *assumes a maximum value in* Ω, *then each* f_j

is constant.

Hints: If $\sup\phi(\Omega) = \phi(z_0)$, $z_0 \in \Omega$, then by replacing each f_j by the appropriate scalar multiple of itself, we can suppose that $f_j(z_0) = |f_j(z_0)|$. Then the maximum of $|f_1 + \cdots + f_n|$ over Ω is achieved at z_0 and is $\phi(z_0)$. It follows from 5.10 that

$$f_1 + \cdots + f_n = f_1(z_0) + \cdots + f_n(z_0) = \phi(z_0) \geq \phi.$$

Taking real parts, we have $\sum \operatorname{Re} f_j \geq \phi = \sum |f_j|$. Since $\operatorname{Re} f_j \leq |f_j|$ for each j, we infer that $\operatorname{Re} f_j = |f_j|$ for each j. Thus $\operatorname{Im} f_j = 0$ and f_j is constant by 2.16(ii).

(ii) *Let $F_1, \ldots, F_N$ be holomorphic in the region Ω, $p_1, \ldots, p_N$ positive real numbers and suppose that $\phi := |F_1|^{p_1} + \cdots + |F_N|^{p_N}$ assumes a maximum value in Ω. Show that each F_j is constant.*

Hints: Let $z_0 \in \Omega$ and $\phi \leq \phi(z_0)$. If $\phi(z_0) = 0$, all is trivial. Otherwise, after re-naming, we can suppose that $F_j(z_0) \neq 0$ for $1 \leq j \leq n \leq N$ and $F_k(z_0) = 0$ for $n < k \leq N$. Then for some $r > 0$, $D := D(z_0, r) \subset \Omega$ and $|F_k| > 0$ in D for $j = 1, 2, \ldots, n$. By 5.35 there exist $g_j \in H(D)$ such that $F_j = e^{g_j}$. It follows that

$$|F_j|^{p_j} = \left[e^{\operatorname{Re} g_j}\right]^{p_j} = e^{\operatorname{Re} p_j g_j} = \left|e^{p_j g_j}\right|.$$

Setting $f_j := e^{p_j g_j}$, we therefore have $\sum_{j=1}^{n} |f_j| \leq \phi \leq \phi(z_0) = \sum_{j=1}^{n} |f_j(z_0)|$ in D. According to (i), each f_j is constant in D. So then is each $|F_j| = |f_j|^{1/p_j}$ and then by 5.10 or 2.16(iii) each F_j is constant in D, thence (5.62) constant in Ω. It follows that

$$\phi(z_0) = \sum_{j=1}^{n} |F_j(z_0)|^{p_j} = \sum_{j=1}^{n} |F_j|^{p_j} \leq \sum_{j=1}^{N} |F_j|^{p_j} = \phi \leq \phi(z_0),$$

so $\sum_{j=n+1}^{N} |F_j|^{p_j} = \mathbf{0}$.

(iii) *Let $F_1, \ldots, F_n$ be holomorphic in the region Ω and suppose that $|F_1| + \cdots + |F_n| = |F|$ for some $F \in H(\Omega)$. Show that each F_j is a constant multiple of F. What if $\Omega = \mathbb{C}$ and the hypothesis "$=$" is relaxed to "$\leq$"?*

Hints: All is trivial if $F = \mathbf{0}$. Otherwise, there is a non-void open disk $\Omega_0 \subset \Omega$ in which F is zero-free, and we may apply (i) in Ω_0 with $f_j := F_j/F$.

According to 5.62 a holomorphic function in a region is uniquely determined by its values on any sequence with a limit point in the region. But how specifically can these values of the function at an arbitrary point be constructed from its values on the sequence? The next exercise answers this nicely, at the same time furnishing an independent proof of 5.62.

Suppose that f is holomorphic in a neighborhood Ω of z_0 and that z_n are distinct points converging to z_0. For every $n \in \mathbb{N}$ there are polynomials whose values at $z_1, z_2, \cdots, z_n$ coincide with those of f. For example, the *Lagrange polynomial*

$$L_n(f)(z) := \sum_{k=1}^{n} f(z_k) \prod_{j=1, j\neq k}^{n} \frac{z - z_j}{z_k - z_j}$$

(void products meaning 1) has this property and thanks to 4.52(ii) it is the only polynomial of degree less than n that does. It is determined by the sequences $z_1, z_2, \ldots$ and $f(z_1), f(z_2), \ldots$. Is it too much to hope that $L_n(f)$ converge to f throughout Ω as $n \to \infty$? The defining equation for $L_n(f)$, involving a sum of $(n-1)$-degree polynomials, is not suitable for investigating this question. Rather, in analogy with Taylor series, we introduce the polynomials $\psi_0 := \mathbf{1}$, $\psi_k(z) := \prod_{j=1}^{k}(z - z_j)$ $(k \in \mathbb{N})$ and determine $A_k \in \mathbb{C}$ so that

$$L_n(f) = \sum_{k=0}^{n-1} A_k \psi_k,$$

in which form it is also known as a *Newton polynomial.* This is always possible (uniquely) due to the obvious linear independence of the ψ_k, though one should allow for the circumstance that the coefficients depend on n as well as k. Such abstract considerations are, however, unnecessary because the demand that $L_n(f)(z_j) = f(z_j)$ is equivalent to

$$f(z_j) = \sum_{k=0}^{j-1} A_k \psi_k(z_j),$$

an equation which allows the coefficients to be recursively defined. When this recursion is unwound, a quite simple expression for the coefficients emerges which involves only the sequences $\{z_n\}$ and $\{f(z_n)\}$; this is the left-hand side of 5.73(iii). The right-hand side of 5.73(iii) shows, when compared with 5.33(∗), that the kth coefficient A_k would be $f^{(k)}(z_0)/k!$ if $z_1, z_2, \ldots, z_k$ were all coincident with z_0. In general it can be shown that $A_k = \lambda_k f^{(k)}(\zeta_k)/k!$ for some $\lambda_k \in \overline{\mathbb{D}}$, $\zeta_k \in \text{co}\{z_1, z_2, \ldots, z_k\}$. These observations further enhance the analogy of $L_n(f)$ with a Taylor polynomial. The remainder $f - L_n(f)$ is realized in 5.73(iv) by a Cauchy integral of such simplicity that not only can its convergence to 0 as $n \to \infty$ be easily proved, but this convergence is clearly uniform over any uniformly bounded family of functions f, affording alternative access to the important convergence theorem 3.35(v). Appearance of the functions $\psi_k(z)$ in the role of the monomials z^k is part of a larger theory of *polynomial bases*: a sequence $\{p_k\}_{k\geq 0}$ of polynomials is given and one investigates which functions have representations $\sum_{k=0}^{\infty} c_k p_k$, the nature of the coefficients, the existence of a radius of convergence and a formula for it, etc. See BOAS and BUCK [1964].

After this glimpse of the golden fleece, the reader is invited to help build the *Argo*:

Exercise 5.73 *Let f be holomorphic in a neighborhood of $\overline{D}(z_0, r)$, $z_j \in D(z_0, r)$ and $\lim_{j\to\infty} z_j = z_0$. Suppose the z_j are all distinct. Define $\psi_0 := \mathbf{1}$,*

$$\psi_k(z) := \prod_{j=1}^{k}(z - z_j), \qquad z \in \mathbb{C},\ k \in \mathbb{N},$$

$$M_k(z) := \sup_{\xi \in C(z_0,r)} \left|\frac{z - z_k}{\xi - z_k}\right|, \qquad z \in \mathbb{C},\ k \in \mathbb{N},$$

$$M := \sup_{\xi \in C(z_0,r)} |f(\xi)|.$$

(i) *Show that (if a void product is interpreted to be* 1*)*

$$\psi_k'(z_l) = \prod_{\substack{j=1\\ j\neq l}}^{k}(z_l - z_j) \neq 0, \qquad l = 1, \ldots, k;\ k \in \mathbb{N}.$$

(ii) *Show that*

$$\frac{1}{\psi_k(\xi)} = \sum_{j=1}^{k} \frac{1}{(\xi - z_j)\psi_k'(z_j)}, \qquad \xi \in C(z_0, r),\ k \in \mathbb{N}.$$

Hint: After a few algebraic maneuvers (using (i)), the desired equation is equivalent to

$$1 - \sum_{j=1}^{k} \prod_{\substack{l=1\\ l\neq j}}^{k} \frac{\xi - z_l}{z_j - z_l} = 0.$$

Since this is a polynomial of degree less than k with the k distinct zeros $z_1, \ldots, z_k$, it is identically zero (4.52 (ii)).

(iii) *Deduce from* (ii) *that*

$$A_k := \sum_{j=1}^{k+1} \frac{f(z_j)}{\psi'_{k+1}(z_j)} = \frac{1}{2\pi i} \int_{C(z_0,r)} \frac{f(\xi)}{\psi_{k+1}(\xi)}\, d\xi, \quad k = 0, 1, \ldots.$$

(iv) (FROBENIUS [1871], HERMITE [1877]) *Verify that*

$$f(z) - \sum_{k=0}^{n-1} A_k \psi_k(z) = \frac{1}{2\pi i} \int_{C(z_0,r)} \frac{f(\xi)}{\xi - z} \cdot \frac{\psi_n(z)}{\psi_n(\xi)}\, d\xi,\ z \in D(z_0, r),\ n \in \mathbb{N}.$$

Hint: (iii) and induction on n.

(v) *Deduce from* (iv) *that*

$$\begin{aligned} \left| f(z) - \sum_{k=0}^{n-1} A_k \psi_k(z) \right| &\leq \sup_{\xi \in C(z_0,r)} \left| \frac{\psi_n(z)}{\psi_n(\xi)} \right| \frac{rM}{r - |z - z_0|} \\ &\leq \frac{rM}{r - |z - z_0|} \prod_{k=1}^{n} M_k(z),\ z \in D(z_0, r),\ n \in \mathbb{N}, \end{aligned}$$

and from the vanishing of this product at all $z \in \{z_1, \ldots, z_n\}$ *infer that the polynomial* $\sum_{k=0}^{n-1} A_k \psi_k$ *is the interpolatory polynomial* $L_n(f)$.

(vi) *For each* $z \in D(z_0, r)$, $\xi \in C(z_0, r)$, $k \in \mathbb{N}$

$$\left| \frac{z - z_k}{\xi - z_k} \right| \leq \frac{|z - z_0| + |z_k - z_0|}{||\xi - z_0| - |z_k - z_0||} = \frac{|z - z_0| + |z_k - z_0|}{r - |z_k - z_0|}.$$

(vii) *Deduce from* (v) *and* (vi) *that*

$$\lim_{n \to \infty} \left| f(z) - \sum_{k=0}^{n-1} A_k \psi_k(z) \right| = 0, \qquad z \in D(z_0, r).$$

Therefore, recalling the definition of A_k *from* (iii), *we have*

$$f(z) = \sum_{k=0}^{\infty} \left[\sum_{j=1}^{k+1} \frac{f(z_j)}{\psi'_{k+1}(z_j)} \right] \psi_k(z), \qquad z \in D(z_0, r).$$

Show that the convergence is uniform for z *in any compact subset of* $D(z_0, r)$.

Hints: If $R < r$ and $R/r < \rho < 1$, (vi) shows that

$$\left|\frac{z - z_k}{\xi - z_k}\right| \le \frac{R + |z_k - z_0|}{r - |z_k - z_0|} \qquad \forall z \in \overline{D}(z_0, R),\ \xi \in C(z_0, r),\ k \in \mathbb{N}.$$

Hence for all large k

$$\sup_{z \in \overline{D}(z_0,R)} M_k(z) \le \frac{R + |z_k - z_0|}{r - |z_k - z_0|} < \rho.$$

The right-hand side of (v) therefore converges to 0 uniformly in $z \in \overline{D}(z_0, R)$.

Exercise 5.74 *Let* $\{f_m\}_{m=1}^{\infty}$ *be holomorphic and uniformly bounded in a neighborhood of* $\overline{D}(z_0, r)$ *and let* $z_1, z_2, \ldots$ *be distinct elements of* $D(z_0, r)$ *convergent to* z_0.

(i) *Suppose that* $\lim_{m\to\infty} f_m(z_j) = 0$ *for each* $j = 1, 2, \ldots$. *Show that then* $\lim_{m\to\infty} f_m = \mathbf{0}$ *uniformly in each closed subdisk of* $D(z_0, r)$.

Hints: Let M be a (finite) bound for the f_m in $\overline{D}(z_0, r)$. Given $R < r$ and $\varepsilon > 0$, use the 5.73(vii) hints to choose n so large that

$$\frac{rM}{r - |z - z_0|} \prod_{k=1}^{n} M_k(z) \le \frac{\varepsilon}{2} \qquad \forall z \in \overline{D}(z_0, R).$$

Deduce from 5.73(v) that then

$$\left| f_m(z) - \sum_{k=0}^{n-1} \left[\sum_{j=1}^{k+1} \frac{f_m(z_j)}{\psi'_{k+1}(z_j)} \right] \psi_k(z) \right| \le \frac{\varepsilon}{2} \qquad \forall z \in \overline{D}(z_0, R),\ m \in \mathbb{N}.$$

Finally, choose m_ε so large that this double sum has modulus less than $\varepsilon/2$ for every $m \ge m_\varepsilon$ and conclude that

$$|f_m(z)| \le \varepsilon \qquad \forall z \in \overline{D}(z_0, R),\ \forall m \ge m_\varepsilon.$$

(ii) *Suppose that* $\lim_{m\to\infty} f_m(z_j)$ *exists in* $\mathbb{C}$ *for each* $j \in \mathbb{N}$. *Show that then there exists an* $f \in H(D(z_0, r))$ *such that* $f_m \to f$ *uniformly on each closed subdisk of* $D(z_0, r)$. (Cf. 7.5 and its notes, as well as 3.35(v) and 7.4.)

Hints: If $\{f_m\}$ is not uniformly Cauchy in $\overline{D}(z_0, R)$ for some $0 < R < r$, there will be $\delta > 0$ and $1 \le m_1 < m_2 < \cdots$ such that $\sup\{|f_{m_{k+1}}(z) - f_{m_k}(z)| : z \in \overline{D}(z_0, R)\} \ge \delta$ for all k. Deduce a contradiction by applying (i) to the sequence $\{f_{m_{k+1}} - f_{m_k}\}_{k=1}^{\infty}$. Finally, use 5.44.

(iii) *Suppose that $\{P_m\}$ is a sequence of polynomials, each of degree less than d, and that for d distinct points z_k the sequence $\{P_m(z_k)\}$ converges to some $w_k \in \mathbb{C}$. Show that $\{P_m\}$ converges locally uniformly in $\mathbb{C}$ and express its limit in terms of the numbers w_k.*

Hints: Use the Lagrange polynomials from the discussion preceding 5.73. We saw there that

$$P_m(z) = \sum_{k=1}^{d} P_m(z_k) \prod_{\substack{j=1 \\ j\neq k}}^{d} \frac{z - z_j}{z_k - z_j} \qquad \forall z \in \mathbb{C}, m \in \mathbb{N}.$$

Each of the d products here is a polynomial (independent of m), and so is bounded on any compact subset of $\mathbb{C}$.

The simple idea in the next exercise will experience several profound extensions in the sequel. 5.76 is an outright generalization. A weak version of 5.76 may be found in E. H. TAYLOR [1913].

Exercise 5.75 (i) *Let f be holomorphic in $\mathbb{D}$ and continuous on $\overline{\mathbb{D}}$. Suppose that $0 \le \theta_0 < \theta_1 \le 2\pi$ and $f(e^{i\theta}) = 0$ for all $\theta_0 \le \theta \le \theta_1$. Show that $f = \mathbf{0}$.*

Hint: (CHERNOFF [1967]): Without loss of generality, $\theta_0 = 0$. Consider then the function

$$F(z) := f(z) f\left(ze^{i\theta_1}\right) \cdots f\left(ze^{in\theta_1}\right)$$

where $n\theta_1 \ge 2\pi$. Show that $F = 0$ on $\mathbb{T}$. Alternatively, cite 5.45(i).

(ii) *f is holomorphic in the interior of a regular polygon having n sides and center c, and is continuous on its closure. On the respective sides it is bounded by constants $M_1, \ldots, M_n$. Use the preceding rotation device to infer that $|f(c)|^n \le M_1 \ldots M_n$.*

(iii) *The continuity hypothesis in* (i) *can be dropped: If $f \in H(\mathbb{D})$ satisfies $\lim_{z \to e^{i\theta}} f(z) = 0$ for all $\theta \in [\theta_1, \theta_2]$, then $f = \mathbf{0}$.*

Hint: 2.5 provides a continuous function φ^* on $\overline{\mathbb{U}}$ which maps $\mathbb{U}$ conformally onto $\mathbb{D}$ and maps some interval in $\mathbb{R}$ onto the arc $\{e^{i\theta} : \theta_1 < \theta < \theta_2\}$. Apply 5.40(ii) to $f \circ \varphi^*$.

Theorem 5.76 (T. RADÓ [1924]) *Let Ω be a region, f a bounded holomorphic function on Ω which satisfies*

$$\lim_{z \to z'} f(z) = 0 \qquad \forall z' \in V \cap \partial\Omega$$

for some neighborhood V of some $b \in \partial\Omega$. If b is not a one-point component of $\mathbb{C} \setminus \Omega$, then $f = \mathbf{0}$. [Notice that this topological hypothesis is satisfied if Ω is simply-connected.]

Proof: We may assume $b = 0$. We first deal with the case in which Ω satisfies the stronger hypothesis that

$$(*) \qquad \text{every neighborhood of } 0 \text{ meets } \mathbb{C} \setminus \overline{\Omega}.$$

Pick $r_0 > 0$ so that $D(0, 2r_0) \subset V$ and consider any $z_0 \in \Omega \cap D(0, r_0)$. Then $D(z_0, r_0)$ is a neighborhood of 0 and so by $(*)$ contains a point $z_1 \in \mathbb{C} \setminus \overline{\Omega}$. Therefore, for a sufficiently large k, the points $z_0 + (z_1 - z_0)e^{i\theta}$ lie in $D(z_0, r_0)$ and in $\mathbb{C} \setminus \overline{\Omega}$ whenever $|\theta| \le 2\pi/k$. We can therefore apply Lindelöf's theorem 5.14, with $r := |z_1 - z_0| < r_0$ and $\varepsilon := 0$, since $D(z_0, r) \subset D(0, 2r_0) \subset V$. The conclusion is that $f(z_0) = 0$. This is true for every z_0 in the non-empty open subset $\Omega \cap D(0, r_0)$ of Ω, and so by 5.62 and the connectedness of Ω the desired conclusion $f = \mathbf{0}$ is obtained.

The role of the hypothesis on $b = 0$, not yet used, is to enable us to reduce the general case to $(*)$, via the radical surgery device introduced following 4.37. The component C of 0 in $\mathbb{C} \setminus \Omega$ contains a point $c \ne 0$. We may reduce the neighborhood V and thereby suppose that it is open and

$$(1) \qquad c \notin V.$$

Consider

$$\psi(z) := \frac{z}{z - c} \qquad (z \in \mathbb{C} \setminus \{c\}).$$

This function and its inverse $\psi^{-1}(w) = cw/(w-1)$ are holomorphic. According to 4.37(iii) and 4.67(i) it has a holomorphic logarithm, hence a holomorphic square-root ϕ in Ω:

$$(2) \qquad \phi^2(z) = \psi(z) = \frac{z}{z - c} \qquad \forall z \in \Omega.$$

The function ϕ is consequently one-to-one, $\phi(\Omega)$ is open, and we may suppose that $\phi^{-1}(w) = \psi^{-1}(w^2)$ for all $w \in \phi(\Omega)$. These assertions are substantiated in the hints to 4.37(v). Moreover, 0 is a boundary point of $\phi(\Omega)$, since $z_n \in \Omega$ and $z_n \to 0$ imply $\phi^2(z_n) = \psi(z_n) \to \psi(0) = 0$. Of course, 0 is also a boundary point of $-\phi(\Omega)$. Now the open set $-\phi(\Omega)$ is disjoint from $\phi(\Omega)$. For, if $z, w \in \Omega$, $-\phi(z) = \phi(w) \Rightarrow \psi(z) = [-\phi(z)]^2 = [\phi(w)]^2 = \psi(w) \Rightarrow z = w$, and then $-\phi(z) = \phi(w)$ says $2\phi(z) = 0$, so $0 = \phi^2(z) = \frac{z}{z-c}$, contrary to $0 \notin \Omega$. Therefore every neighborhood of 0 meets the open set $-\phi(\Omega)$, which is disjoint from $\overline{\phi(\Omega)}$. This shows that the condition $(*)$ is satisfied by $\Omega_0 := \phi(\Omega)$ in the role of Ω. We shall apply the conclusion of the first paragraph to $f_0 := f \circ \phi^{-1}$ on Ω_0 and this will finish the proof. The relevant boundary neighborhood for f_0 will be (recall (1), and the fact $\psi(V)$ is open)

$$V_0 := \{z \in \mathbb{C} : z^2 \in \psi(V)\}. \tag{3}$$

To see this, what we have to show is that

$$\text{if } z_n \in \Omega_0 \text{ and } z_n \to z \in V_0 \cap \partial\Omega_0, \text{ then} \tag{4}$$

$$\{\phi^{-1}(z_n)\} \subset \Omega \text{ converges to some boundary point of } \Omega \text{ lying in } V. \tag{5}$$

We may suppose in (4) that all $z_n \in V_0$. Then z^2 and all z_n^2 lie in $\psi(V)$. Writing $w_n := \phi^{-1}(z_n) \in \Omega$, we have $z_n^2 = \phi^2(w_n) = \psi(w_n)$, so $w_n = \psi^{-1}(z_n^2) \to \psi^{-1}(z^2) \in V$. Write $w := \psi^{-1}(z^2)$. Thus

$$w_n \to w \in V. \tag{6}$$

Since all $w_n \in \Omega$, $w \in \overline{\Omega}$. If $w \in \Omega$, then $z_n = \phi(w_n) \to \phi(w) \in \phi(\Omega) = \Omega_0$ would contradict the fact that $z_n \to z \in \partial\Omega_0$. Hence

$$w_n \to w \in \partial\Omega. \tag{7}$$

With (6) and (7), (5) is established and the proof is complete.

Remark: Evidently the weakest possible hypothesis on b under which the conclusion of 5.76 could be true is that b not be an isolated point in $\mathbb{C} \setminus \Omega$. This version (without the boundedness hypothesis on f) is proved in 10.10(iv) by very different means.

§ 7 Local Theory

The first two theorems, being very fundamental results, are susceptible to proof from many directions and the reader will experience several in this section. (Cf. also 3.35(x).) In the first theorem the injectivity hypothesis of the open-map theorem 4.61 is simply replaced by a holomorphy hypothesis.

Theorem 5.77 (Holomorphic Open-Map Theorem) *Let f be holomorphic and not constant in the region Ω. Then $f(\Omega)$ is an open set in $\mathbb{C}$ and f is an open mapping.*

Proof: Both claims follow if we prove that for each $z_0 \in \Omega$

(0) $$D(z_0, r_0) \subset \Omega \Rightarrow f(z_0) \text{ is an interior point of } f(D(z_0, r_0)).$$

Set $w_0 := f(z_0)$. By hypothesis $f - w_0$ is not the function $\mathbf{0}$, o by 5.62 its zeros do not accumulate at z_0, and for some $0 < r < r_0$ the function $f - w_0$ has no zeros in $\bar{D}(z_0, r) \setminus \{z_0\}$. Therefore the number $m := \min\{|f(z) - w_0| : z \in C(z_0, r)\}$ is positive. The proof will be completed by showing that $D(w_0, m/2) \subset f\big(D(z_0, r)\big)$. To this end, consider any $w \in \mathbb{C}$ satisfying

$$|w - w_0| < m/2.$$

Then the function $f - w$ satisfies

(1) $$|f(z) - w| \geq |f(z) - w_0| - |w_0 - w| > m - m/2 = m/2 \quad \forall z \in C(z_0, r)$$

while

(2) $$\big|f(z_0) - w\big| = |w_0 - w| < m/2.$$

We want to show that $f - w$ has a zero in $D(z_0, r)$. If this were not the case, then $1/(f - w)$ would be continuous on $\bar{D}(z_0, r)$ and holomorphic in $D(z_0, r)$, and (1) and (2) would violate the boundary maximum-modulus principle for holomorphic functions.

Elegant and transparent as this argument is, it is rivalled by one that is more topological and less analytical, based on the index, a tool which will be exploited later to get much more detailed local information about the function. Here, as noted above, all that we have to do is confirm (0), for each $z_0 \in \Omega$. To that end, use 5.69 to get $m \in \mathbb{N}$, $F \in H(\Omega)$ with $F(z_0) \neq 0$ and

(3) $$f(z) = f(z_0) + (z - z_0)^m F(z) \qquad \forall z \in \Omega.$$

Consider any $0 < r < r_0$ such that F is a zero-free in $\bar{D}(z_0, r)$. Write γ_r for $C(z_0, r)$. According to 4.14 and (3)

(4) $$\mathrm{Ind}_{f \circ \gamma_r}\big(f(z_0)\big) = \mathrm{Ind}_{[f - f(z_0)] \circ \gamma_r}(0) = m + \mathrm{Ind}_{F \circ \gamma_r}(0).$$

Thanks to 4.12 we have $\mathrm{Ind}_{F \circ \gamma_r}(0) = \mathrm{Ind}_{F \circ \gamma_0}(0) = 0$, so (4) yields

(5) $$\mathrm{Ind}_{f \circ \gamma_r}\big(f(z_0)\big) = m > 0.$$

According to 4.59(i) fact (5) insures that $f(z_0)$ is an interior point of $f\big(D(z_0, r)\big)$. Moreover, if f' is zero-free, we may take $m = 1$ and use for F the (Carathéodory) function in the Remarks following 2.2, so no properties of holomorphic functions beyond the definition are needed!

Corollary 5.78 (Differentiability of the Inverse) *Let U be an open subset of $\mathbb{C}$, f holomorphic and one-to-one in U. Then f' is zero-free, $f(U)$ is open, f^{-1} is holomorphic therein and $\left(f^{-1}\right)' = 1/f' \circ f^{-1}$.*

Proof: U is a union of open disks and to f on each such disk we may apply the Open-Map Theorem (5.77 or 4.48(iii)) to learn that f is an open map: $f(U)$ is open and f^{-1} is continuous. Let g denote f^{-1}. At every point $w_0 \in f(U)$ such that $f'(g(w_0)) \neq 0$ we use the fact that $g(w) \neq g(w_0)$ for $w \neq w_0$ (g being one-to-one) to see that

$$\frac{g(w) - g(w_0)}{w - w_0} = \frac{1}{\dfrac{w - w_0}{g(w) - g(w_0)}} = \frac{1}{\dfrac{f(g(w)) - f(g(w_0))}{g(w) - g(w_0)}}.$$

Since g is continuous at w_0 and f is differentiable at $g(w_0)$, we see from this that g is differentiable at w_0 and that $g'(w_0) = 1/f'(g(w_0))$. That is,

$$g'\big(f(z_0)\big)f'(z_0) = 1 \quad \text{for every } z_0 \ \big(= g(w_0)\big) \text{ such that } f'(z_0) \neq 0.$$

It remains only to show that f' never vanishes. Since f is not constant on any open disk in U, f' is not identically zero on any such disk so (5.62) the zeros of the holomorphic (5.30) function f' cannot accumulate in U. Since f is a homeomorphism, the f-images of these zeros do not accumulate in $f(U)$. Therefore by the above, the set of points in $f(U)$ where g does not have a derivative does not accumulate in $f(U)$, i.e., is composed of isolated points. Since g is continuous on $f(U)$, we can conclude from 5.41 (or 5.40) that g is differentiable at these points as well. Therefore by 5.30 g' is continuous. In summary, $(g' \circ f) \cdot f'$ is a continuous function on U which equals 1 except perhaps at the points of an isolated set. By continuity then $(g' \circ f) \cdot f' = 1$ throughout U, so f' is zero-free.

We have the following local converse to the preceding result. 3.7 (together with 5.30), proves part of it, as does 4.60(ii). The argument to follow, which is more constructive, proves a little more. A fourth proof occurs in 5.89, and local univalence is also a rather obvious consequence of 5.34(i).

Theorem 5.79 *Let U be an open subset of $\mathbb{C}$, f a holomorphic function in U, $z_0 \in U$ and $f'(z_0) \neq 0$. Then there exist open sets V, W with $z_0 \in V \subset U$ such that f maps V bijectively onto W. Moreover, $(f|V)^{-1}$ is holomorphic on W.*

Proof: We lose no generality by supposing that $z_0 = f(z_0) = 1 - f'(z_0) = 0$. Let $\rho \in (0,1)$ and set $\phi(z) := f(z) - z$. Use continuity of ϕ' at 0 (5.30) to select $r > 0$ such that

(1) $$|\phi'| < \rho \quad \text{in } \overline{D}(0,r) \subset U.$$

Notice that if $w \in D(0, r(1-\rho))$ and $z \in D(0, r)$, then

$$|w - \phi(z)| = \left| w - \int_{[0,z]} \phi' \right| \leq |w| + |z| \sup |\phi'[0, z]| < r(1-\rho) + r\rho,$$

using (1), and so $w - \phi(z) \in D(0, r)$. Therefore the recursive definition $g_0 := \mathbf{0}$, $g_{n+1}(w) := w - \phi(g_n(w))$ $(n \geq 0)$ always makes sense for $w \in D(0, r(1-\rho))$. For such w these functions map into $D(0, r)$ and satisfy $g_n(0) = 0$ and

$$\begin{aligned} |g_{n+1}(w) - g_n(w)| &= |\phi(g_n(w)) - \phi(g_{n-1}(w))| = \left| \int_{[g_n(w), g_{n-1}(w)]} \phi' \right| \\ &\leq \rho |g_n(w) - g_{n-1}(w)| \leq \cdots \\ &\leq \rho^n |g_1(w) - g_0(w)| = \rho^n |w|, \\ |g_{N+k}(w) - g_N(w)| &\leq \sum_{j=1}^{k} |g_{N+j}(w) - g_{N+j-1}(w)| \\ &\leq \sum_{j=1}^{k} \rho^{N+j-1} |w| \leq |w| \rho^N \sum_{j=1}^{\infty} \rho^{j-1} \\ &= |w| \frac{\rho^N}{1-\rho} < r\rho^N. \end{aligned}$$

Therefore $\{g_N\}$ converges uniformly in $D(0, r(1-\rho))$ and the limit function g maps into $\bar{D}(0, r) \subset U$ and satisfies $g(0) = 0$ and

$$\begin{aligned} g(w) = \lim_{n\to\infty} g_{n+1}(w) &= \lim_{n\to\infty} [w - \phi(g_n(w))] \\ &= w - \phi\Big(\lim_{n\to\infty} g_n(w)\Big) = w - \phi(g(w)), \end{aligned}$$

that is,

$$(2) \qquad w = g(w) + \phi(g(w)) = f(g(w)) \in f(U) \qquad \forall w \in D(0, r(1-\rho)).$$

Evidently (2) implies that f is one-to-one in the set $g\big(D(0, r(1-\rho))\big)$ and that $f(U)$ is a neighborhood of $f(0) = 0$. Now g is holomorphic in $D(0, r(1-\rho))$ [by 5.44(i)] and using the chain rule we get from (2) that $g'(0) = 1/f'(0) = 1$. Therefore the function g fulfills all the hypotheses which f does and we can attribute to it the conclusions so far gained for f, to learn that $g\big(D(0, r(1-\rho))\big)$ is a neighborhood of $g(0) = 0$. Let V be an open subset of $g\big(D(0, r(1-\rho))\big)$ which contains 0. Then $W := g^{-1}(V)$ is an open subset of $D(0, r(1-\rho))$, the domain of the holomorphic function g, and $g(W) = V$. On the latter set (as indeed on the whole range of g) f is one-to-one, and for each $w \in W$, $f(g(w)) = w$, by (2), so $f(V) = f(g(W)) = W$ and $g = (f|V)^{-1}$.

Remarks: (i) The full open-map theorem 5.77 can be inferred from 5.79 via an interesting purely topological argument. With the notation and hypotheses of 5.77, the zero-set A of f' has no limit point in Ω, and 5.79 says that on $U := \Omega \setminus A$ the function f is an open mapping. Hence the following topological fact yields 5.77:

If $A \subset \Omega$ open $\subset \mathbb{C}$, A has no limit point in Ω, $f\colon \Omega \mapsto \mathbb{C}$ is continuous and $f|(\Omega \setminus A)$ is an open mapping, then f is an open mapping.

To prove this it suffices to show that for each $a \in A$ and all sufficiently small $R > 0$ the set $f\big(D(a,R)\big)$ is open. Consider any $R > 0$ such that $D(a,R) \subset \Omega$ and $D(a,R) \cap A = \{a\}$. It clearly then suffices to show that $f(a)$ is an interior point of $f\big(D(a,R)\big)$, and that is trivial if $f(a)$ lies in the open subset $f\big(D(a,R)\setminus\{a\}\big)$ of $f\big(D(a,R)\big)$. So suppose that $(*)$: $f(a) \notin f\big(D(a,R)\setminus\{a\}\big)$. Set $r := R/2$ and pick $\rho > 0$ so that $D(f(a),\rho) \cap f(C(a,r)) = \varnothing$. Consider the component C of $\mathbb{C} \setminus \big[f\big(C(a,r)\big) \cup \{f(a)\}\big]$ which contains the connected subset $D(f(a),\rho)\setminus\{f(a)\}$. The closed sets $f\big(\bar{D}(a,r)\big)$ and $\mathbb{C}\setminus f\big(D(a,r)\setminus\{a\}\big) \subset \big[\mathbb{C}\setminus f\big(D(a,r)\big)\big] \cup \{f(a)\}$ cover $\mathbb{C}$, and their intersection lies in $f\big(C(a,r)\big) \cup \{f(a)\}$ which is disjoint from C. Hence not both of the former closed sets can meet C (that would disconnect C). Since $f\big(\bar{D}(a,r)\big)$ meets $D(f(a),\rho) \setminus \{f(a)\} \subset C$, by continuity of f at a together with the assumption $(*)$, it must be that $C \subset f\big(\bar{D}(a,r)\big)$. From $D(f(a),\rho) \setminus \{f(a)\} \subset C \subset f\big(\bar{D}(a,r)\big)$ and the compactness of this image set it follows that $\bar{D}(f(a),\rho) \subset f\big(\bar{D}(a,r)\big) \subset f\big(D(a,R)\big)$, which shows that $f(a)$ is an interior point of $f\big(D(a,R)\big)$, as wanted.

The idea of getting 5.77 from 5.79 this way was suggested by Crowe and Samperi [1989], but the preceding proof goes back to work of E. Connell and P. Porcelli [**MR** 32#5899], and something like it may be found on pp. 122–123 of Stoïlow [1956].

Outright reduction of 5.77 to 5.79 is also possible and involves a useful technique: near any zero z_0 of f' we can express $f - f(z_0)$ as the nth-power of a holomorphic function g with zero-free derivative. The details will be found in the proof of 5.83.

(ii) 5.78 and 5.79 are two more examples of properties enjoyed by all differentiable functions in $\mathbb{C}$ but lacked by some differentiable functions in $\mathbb{R}$. The function $f(x) := x^3$ is one-to-one on $\mathbb{R}$, but $f'(0) = 0$. If $f\colon \mathbb{R} \mapsto \mathbb{R}$ is differentiable and f' is continuous at 0, then $f'(0) \neq 0$ easily entails that f is one-to-one near 0. On the other hand, with a little effort a differentiable $f\colon \mathbb{R} \mapsto \mathbb{R}$ can be constructed which is constant in small neighborhoods of each $1/n$ and yet satisfies $f'(0) \neq 0$.

(iii) The global version of 5.79 fails, as $f(z) := z^2$ is not one-to-one in the region $\mathbb{C} \setminus \{0\}$ although its derivative is zero-free there. In this respect things are simpler in $\mathbb{R}$: If I is an open interval in $\mathbb{R}$, $f : I \mapsto \mathbb{R}$ is differentiable and f' is zero-free, then f is one-to-one in I (by the Mean-Value Theorem).

Exercise 5.80 (i) (Cf. 9.17(iii).)*Prove the following result from* POMPEIU [1911]: *If f is holomorphic in the open subset U of $\mathbb{C}$ and $z_0 \in U$, then in every neighborhood of z_0 there is a pair of distinct points z_1, z_2 such that*

$$\frac{f(z_2) - f(z_1)}{z_2 - z_1} = f'(z_0).$$

Hint: $g(z) := f(z) - zf'(z_0)$ has $g'(z_0) = 0$ and so by 5.78 is not one-to-one in any neighborhood of z_0.

(ii) (RUBINSTEIN [1969]) *f is holomorphic in the region Ω. Show that*

$$\text{(MVP):} \qquad \frac{f(z) - f(w)}{z - w} \in f'[z, w] \quad \textit{whenever } [z, w] \subset \Omega \textit{ and } z \neq w$$

if and only if f is a polynomial of degree at most 2.

Hints: ($\Rightarrow$) The conclusion is that $f''' = \mathbf{0}$. Because of the uniqueness theorem 5.62 it suffices to show this in some disk in Ω. So we suppose there exists $a \in \Omega$ with $f''(a) \neq 0$. According to 5.70, f' maps some open neighborhood V of a in Ω conformally onto an open $W \subset \mathbb{C}$. Take $r > 0$ so that $D(a, r) \subset V$. (MVP) says there is $\eta(z) \in [0, 1]$ such that

$$[f(z) - f(a)]/[z - a] = f'(a + \eta(z)(z - a)) \in f'(V) = W \qquad \forall z \in D(a, r) \setminus \{a\}.$$

Letting g denote $(f'|V)^{-1}$, this equation yields

$$\eta(z) = (-a + g([f(z) - f(a)]/[z - a]))/(z - a)$$

which shows that η is holomorphic in $D(a, r) \setminus \{a\}$. From 2.16(ii) then follows the constancy of the $\mathbb{R}$-valued function η, say, $\eta = c$. The power series for f at a now satisfies

$$\sum_{n=1}^{\infty} \frac{f^{(n)}(a)}{n!}(z - a)^{n-1} = \frac{f(z) - f(a)}{z - a} = f'(a + c(z - a))$$

$$= \sum_{n=1}^{\infty} \frac{f^{(n)}(a)}{(n-1)!}(c(z - a))^{n-1} \qquad \forall z \in D(a, r).$$

Equating coefficients,

$$\frac{f^{(n)}(a)}{n!} = \frac{f^{(n)}(a)}{(n-1)!}c^{n-1}, \qquad \text{that is,}$$

$$f^{(n)}(a)[1 - nc^{n-1}] = 0 \qquad \forall n \in \mathbb{N}.$$

From $f''(a) \neq 0$ we learn $c = 1/2$. Since $n(1/2)^{n-1} < 1$ for all integers $n \geq 3$, it now follows that $f^{(n)}(a) = 0$ for all $n \geq 3$.

The next example illustrates a devious device for rendering the inverse function quite explicit (cf. 5.92).

Example 5.81 *Let* $\Omega := \mathbb{C} \setminus \{x \in \mathbb{R} : |x| \geq 1\}$, $V := \{z \in \mathbb{C} : -\frac{\pi}{2} < \operatorname{Re} z < \frac{\pi}{2}\}$. *The function* sin *maps the vertical strip* V *conformally onto* Ω *and the inverse map* A *has the power series*

$$A(w) = \sum_{k=0}^{\infty} \frac{(2k)!}{2^{2k}(k!)^2} \frac{1}{2k+1} w^{2k+1} = w + \sum_{k=1}^{\infty} \frac{1 \cdot 3 \cdot \cdots \cdot (2k-1)}{2 \cdot 4 \cdot \cdots \cdot 2k} \frac{w^{2k+1}}{2k+1}, \quad w \in \mathbb{D}.$$

Moreover, this series converges (absolutely) throughout $\overline{\mathbb{D}}$. Cf. 3.45(v), (vi).

Proof: As is easily confirmed, the function $E(z) := -ie^{iz}$ maps V conformally onto $-\mathbb{U}$. And, according to 2.5(v), the Joukowski function $\psi(z) := \frac{1}{2}(z + \frac{1}{z})$ maps $-\mathbb{U}$ conformally onto Ω. Checking that $\sin z = \psi(E(z))$, we then see that sin maps V conformally onto Ω. Denote by A the inverse of $\sin|V$,

$$(1) \qquad \sin A(w) = w \qquad \forall w \in \Omega,$$

and by L the inverse of $E|V$. Note that

$$(2) \qquad L(-i) = 0.$$

In 2.5(vi) we *essentially* computed the inverse of $\psi|-\mathbb{U}$ to be the map $w \mapsto w - is^{-1}(1-w^2)$, where $s^{-1} : \mathbb{C}_- \mapsto \mathbb{H}$ is the square-root constructed in 1.26(i). [Note that $s^{-1} = \exp \circ \frac{1}{2}l$, where l is the principal branch of the logarithm in $\mathbb{C}_-$, thanks to (1.7).] Thus

$$(3) \qquad A(w) = L(w - is^{-1}(1 - w^2)), \qquad w \in \Omega.$$

It's worthwhile unraveling the notation when $-1 \leq w \leq 1$, to see that for such w equation (3) reads simply: $\arcsin(w) = \operatorname{Im} l(\sqrt{1 - w^2} + iw)$, a diagram of which then reveals the geometric origin of (3). An adroit application of the chain rule to (3) gives

$$(4) \qquad A'(w) = \frac{1}{s^{-1}(1-w^2)}, \qquad w \in \Omega.$$

This shows that A' is an even function, so the coefficients of all odd powers in its Taylor expansion around 0 are 0 :

$$A'(w) = \sum_{k=0}^{\infty} a_k w^{2k}, \qquad w \in \mathbb{D}.$$

Therefore,

$$S(w) := \sum_{k=0}^{\infty} a_k w^k$$

defines a holomorphic function in $\mathbb{D}$ such that

$$(5) \qquad A'(w) = S(w^2) \qquad \forall w \in \mathbb{D}.$$

From this and (4)

$$S(w^2) = \frac{1}{s^{-1}(1-w^2)} \qquad \forall w \in \mathbb{D}, \quad \text{that is,}$$

$$S(w) = \frac{1}{s^{-1}(1-w)} \qquad \forall w \in \mathbb{D}.$$

The square-root here, from 1.26(i), is the usual positive one for $w \in (-1,1)$. From this it follows, by computing derivatives along the real axis only, that

$$S^{(k)}(0) = \frac{1}{2}\left(\frac{1}{2}+1\right)\cdots\left(\frac{1}{2}+k-1\right) = \frac{1\cdot 3\cdot\cdots\cdot(2k-1)}{2^k},$$

$$(6) \qquad \frac{S^{(k)}(0)}{k!} = \frac{1\cdot 3\cdot\cdots\cdot(2k-1)}{2\cdot 4\cdot\cdots\cdot(2k)}, \qquad k \in \mathbb{N}.$$

(The power series found for $S(w)$ this way is, *mutatis mutandis*, the binomial expansion discussed in the 3.35(xiv) hints.) Now from (5) and (6)

$$A'(w) = \sum_{k=0}^{\infty} \frac{S^{(k)}(0)}{k!} w^{2k} = 1 + \sum_{k=1}^{\infty} \frac{1\cdot 3\cdot\cdots\cdot(2k-1)}{2\cdot 4\cdot\cdots\cdot(2k)} w^{2k}$$

whence, since $A(0) = 0$ [see (2) and (3)],

$$A(w) = w + \sum_{k=1}^{\infty} \frac{1 \cdot 3 \cdot \dots \cdot (2k-1)}{2 \cdot 4 \cdot \dots \cdot (2k)} \frac{w^{2k+1}}{2k+1} \qquad \forall w \in \mathbb{D}. \tag{7}$$

Thus A maps $(-1, 1)$ into $\mathbb{R} \cap V = \left(-\frac{\pi}{2}, \frac{\pi}{2}\right)$. Therefore, for any $0 < x < 1$

$$x + \sum_{k=1}^{N} \frac{1 \cdot 3 \cdot \dots \cdot (2k-1)}{2 \cdot 4 \cdot \dots \cdot (2k)} \frac{x^{2k+1}}{2k+1} < A(x) < \frac{\pi}{2} \qquad \forall N \in \mathbb{N}.$$

Let $x \uparrow 1$ and conclude that

$$1 + \sum_{k=1}^{N} \frac{1 \cdot 3 \cdot \dots \cdot (2k-1)}{2 \cdot 4 \cdot \dots \cdot (2k)} \frac{1}{2k+1} \leq \frac{\pi}{2} \qquad \forall N \in \mathbb{N}.$$

This shows that series (7) is absolutely convergent throughout $\overline{\mathbb{D}}$. Calling the limit $A(w)$ for $w \in \{-1, 1\}$ as well, the identity (1) persists for such w by continuity.

In particular, $\sin A(1) = 1$, while we also know that $A(1) \in \left[-\frac{\pi}{2}, \frac{\pi}{2}\right]$. It follows that $A(1) = \frac{\pi}{2}$, giving

Scholium 5.82: (i) $\displaystyle \frac{\pi}{2} = 1 + \sum_{k=1}^{\infty} \frac{1 \cdot 3 \cdots (2k-1)}{2 \cdot 4 \cdots (2k)} \cdot \frac{1}{2k+1}.$

(ii) *Factoring the polynomial $X^3 - i = (X+i)(X^2 - iX - 1)$ produces the cube roots of i and confirms that $e^{\pi i/6} = (\sqrt{3}+i)/2$. Hence $w = 1/2$ in (1) and (7) above yields*

$$\frac{\pi}{6} = \frac{1}{2} + \frac{1}{2} \cdot \frac{1}{3 \cdot 2^3} + \frac{1}{2} \cdot \frac{3}{4} \cdot \frac{1}{5 \cdot 2^5} + \cdots.$$

(This series converges more rapidly than that in (i).)

(iii) *Use integration by parts and induction to confirm that $\int_0^{\pi/2} \sin^{2k+1} = \frac{2 \cdot 4 \cdots (2k)}{1 \cdot 3 \cdot 5 \cdots (2k+1)}$ for each integer $k \geq 0$. Then take $w = \sin t$ in (7), remember that $A(\sin t) = t$ for $t \in [0, \pi/2]$, and integrate. You will rediscover the value of the sum $\sum_{k=0}^{\infty} (2k+1)^{-2}$.*

(iv) *For every real x, $\theta := \tan^{-1}(x)$ lies in $\left(-\frac{\pi}{2}, \frac{\pi}{2}\right)$ and $\sin\theta = x/\sqrt{1+x^2}$, that is,*

$$\tan^{-1}(x) = \sin^{-1}\left(x/\sqrt{1+x^2}\right) \qquad \forall x \in \mathbb{R}.$$

It follows from this and 5.81 *that*

$$\tan^{-1}(x) = \frac{x}{\sqrt{1+x^2}} \sum_{k=0}^{\infty} \frac{(2k)!}{2^{2k}(k!)^2} \frac{1}{2k+1} \left(\frac{x^2}{1+x^2} \right)^k \quad \forall x \in \mathbb{R}.$$

Cf.(9) in 3.22. [To which $x \in \mathbb{C}$ can this formula be extended?]

(v) Newton (1664) originally derived (7) above for $w = y \in [-1, 1]$, by geometric arguments based on the formula for arclength on $\mathbb{T} \cap \mathbb{H}$ parameterized by $\gamma(t) := \sqrt{1-t^2} + it$, $-1 \le t \le 1$. For $0 \le y < 1$ the arc on $\mathbb{T} \cap \mathbb{H}$ from $\gamma(0) = 1$ to $\gamma(y)$ has length

$$(8) \qquad \theta = \int_0^y |\gamma'(t)| dt = \int_0^y \frac{1}{\sqrt{1-t^2}} dt = -y\sqrt{1-y^2} + 2\int_0^y \sqrt{1-t^2}\, dt.$$

[The last equality comes from a non-obvious integration by parts, but can be simply checked by differentiation; it also represents twice the area of the circular sector formed by the arc.] Thus $\gamma(y) = e^{i\theta}$, so $y = \operatorname{Im} e^{i\theta} = \sin\theta$, and $\sin^{-1}(y) = \theta$, which he computed, using on the right side of (8) his binomial series (3.12(vi)) — discovered in the course of this work. The reader can check that this does result in the right side of (7), with $w = y$. This shows that the power series in (7) converges for all $w \in [0, 1)$ and that equation (7) holds for such w. Therefore the series converges for all $w \in \mathbb{D}$ and by the uniqueness theorem for holomorphic functions equation (7) holds as well for all $w \in \mathbb{D}$. As noted before 3.14, using (8) to *define* the inverse-sine function is a way to rigorously ground the theory of the cyclometric functions within real analysis. In fact, Newton went on to discover from (7) the sine series by his method of "reversion of power series." (See FERRARO [2008], pp. 76–78.)

Theorem 5.83 *Let f be holomorphic and not constant in a connected neighborhood Ω of z_0. Then there is a positive R, a positive integer n, an open neighborhood V of z_0 in Ω which f maps onto $D(f(z_0), R)$, and a one-to-one holomorphic function F in V such that $f = f(z_0) + F^n$ in V and $f^{-1}(w) \cap V$ consists of n distinct points for each $w \in D(f(z_0), R) \setminus \{f(z_0)\}$.* Cf. 4.59(iii).

Proof: We may suppose that $f(z_0) = 0$, for otherwise we can work with $f - f(z_0)$. According to 5.69 there is a positive integer n and a holomorphic function h in Ω such that

$$(1) \qquad f(z) = (z - z_0)^n h(z) \quad \text{and} \quad h(z_0) \neq 0, \quad \forall z \in \Omega.$$

There is therefore an $r_0 > 0$ such that $D(z_0, r_0) \subset \Omega$ and $0 \notin h(D(z_0, r_0))$ and then (5.35) a holomorphic function g in $D(z_0, r_0)$ such that

$$(2) \qquad h = e^g \quad \text{in } D(z_0, r_0).$$

Form

$$(3) \qquad F(z) := (z - z_0)e^{g(z)/n}, \qquad z \in D(z_0, r_0).$$

Since $F'(z_0) = e^{g(z_0)/n} \neq 0$, it follows from 5.79 that there is an $0 < r < r_0$ such that F is one-to-one in $D(z_0, r)$ and $F(D(z_0, r))$ is an open neighborhood of $F(z_0) = 0$, and so there is an $R > 0$ such that $D\big(0, R^{1/n}\big) \subset F(D(z_0, r))$. Let $V := D(z_0, r) \cap F^{-1}\big(D\big(0, R^{1/n}\big)\big)$. Then V is an open neighborhood of z_0, F is one-to-one in V and $f = F^n$ in V [from (1), (2) and (3)]. Finally, given non-zero $w \in D(0, R)$, let $w_1, \ldots, w_n$ be its n distinct n^{th}-roots. These points lie in $D\big(0, R^{1/n}\big)$ so they are F-images of exactly n points $z_1, \ldots, z_n$ in V. Then $w = w_j^n = [F(z_j)]^n = f(z_j)$. Conversely, if $z \in V \setminus \{z_0\}$ and $f(z) = w$, then $w = [F(z)]^n$ so $F(z) \in \{w_1, \ldots, w_n\}$, whence $z \in F^{-1}\{w_1, \ldots, w_n\} \cap V = \{z_1, \ldots, z_n\}$.

Remarks: (i) The Open-Map Theorem 5.77 is a genuine corollary of 5.83 because the preceding argument involved only 5.79, whose proof is independent of 5.77. Note too that if $f'(z_0) = 0$, then the integer n in the proof of 5.83 exceeds 1, so (part of) 5.83 shows that f is not one-to-one near z_0, proving 5.78 anew.

(ii) If in the proof of 5.83 we excise from $D(f(z_0), R) = D(0, R)$ not just its center $f(z_0) = 0$, but the whole radius $[0, R)$, then the $z_1, \ldots, z_n$ become holomorphic functions of w. More precisely, let $\sqrt[n]{w}$ denote the holomorphic n^{th}-root function of $w \in \mathbb{C} \setminus [0, \infty)$ which maps onto $\{\rho e^{i\theta} : \rho > 0, 0 < \theta < 2\pi/n\}$. Then for $w \in D(0, R) \setminus [0, R)$ the $w_1, \ldots, w_n$ in the proof of 5.83 are $\sqrt[n]{w}e^{2\pi ij/n}$, $1 \leq j \leq n$, and the $z_1, \ldots, z_n$ are the points $F^{-1}\big(\sqrt[n]{w}e^{2\pi ij/n}\big)$, holomorphic functions of w.

The following vivid picture emerges: The radii $\big[0, R^{1/n}\big)\, e^{2\pi ij/n}$ partition the disk $D(0, R^{1/n})$ into n disjoint open sectors, each of which is mapped conformally by $w \mapsto w^n$ onto $D(0, R) \setminus [0, R)$. As $(F \mid V)^{-1}$ maps $D(0, R^{1/n})$ conformally onto V, we envision it mapping these radii to curves Γ_j with only the point z_0 in common, and these curves effecting a partition of V into n disjoint open "sectors." Then f, which is F^n, maps each of these "sectors" conformally onto $D(0, R) \setminus [0, R)$ and each separating curve Γ_j onto $[0, R)$. The smaller r is taken, the closer $F(z)$ is approximated by the first term $F'(0)(z - z_0)$ of its Taylor series, and the closer the partition of V resembles that of $D(0, R^{1/n})$ into congruent angular sectors. In fact, this whole discussion has merely been an attempt to elaborate the superficial observation that near z_0 the original function $f(z)$ behaves like the first term $h(z_0)(z - z_0)^n$ of its Taylor series.

But there is more to be seen here. For $0 < r < r_0$ and $0 \leq t \leq 2\pi$ let $\gamma_r(t) := z_0 + re^{it}$ and $\Gamma_r := f \circ \gamma_r$. From the representation $f(z) = (z - z_0)^n e^{g(z)}$,

$z \in D(z_0, r_0)$, we see

$$f(\gamma_r(t)) = f(z_0 + re^{it}) = r^n \exp[int + g(\gamma_r(t))], t \in [0, 2\pi].$$

Consequently, if we set $\theta_r(t) := nt + \operatorname{Im} g(\gamma_r(t))$, then

$$\Gamma_r(t)/|\Gamma_r(t)| = \exp(i\theta_r(t)) \tag{4}$$

and, by the chain rule,

$$\theta_r'(t) = n + \operatorname{Im}[\gamma_r'(t)g'(\gamma_r(t))] = n + r\operatorname{Re}[e^{it}g'(\gamma_r(t))] \geq n - r|g'(\gamma_r(t))|.$$

Now if $0 < r < r_0$ is sufficiently small, $|g'| < 1/r$ on $\bar{D}(z_0, r)$, and consequently,

$$\theta_r'(t) > n - r \cdot \frac{1}{r} \geq 0 \qquad \forall t \in [0, 2\pi],$$

showing that θ_r is strictly increasing. Thus (4) says that, like the circle $t \mapsto z_0 + re^{it} = \gamma_r(t)$, its image curve $t \mapsto f(z_0 + re^{it}) = \Gamma_r(t) = |\Gamma_r(t)|e^{i\theta_r(t)}$ is traversing its trajectory counterclockwise *at all times.* This is a *local sense* in which non-constant holomorphic f *preserve orientation.*

To "round out" the preceding local description of f, recall 5.67(iii).

There is another approach (due to Cauchy) to the local theory which we shall now explore. It utilizes the index and leads to other interesting and useful facts, as well as different proofs of some results already secured. Several items (e.g., 5.84–5.86) are special cases of the Residue Theorem 11.29, and it is worth returning later to view them in that light. In deriving them from 11.29, for example, the integration path $C(a, r)$ can be replaced y much more general curves.

Theorem 5.84 *If f and h are holomorphic in $D(a, R)$, then for each $0 < r < R$ and each $w \in \mathbb{C} \setminus f(C(a, r))$ the set $f^{-1}(w) \cap D(a, r)$ is finite and*

$$\frac{1}{2\pi i}\int_{C(a,r)} \frac{hf'}{f - w} = \sum_{z \in f^{-1}(w) \cap D(a,r)} n(z, f - w)h(z).$$

Proof: First note that since $w \in \mathbb{C} \setminus f(C(a, r))$,

$$f^{-1}(w) \cap D(a, r) = f^{-1}(w) \cap \bar{D}(a, r). \tag{1}$$

If this set were infinite, then it would have a limit point in $\bar{D}(a, r)$ and by 5.62 $f = w$ in $D(a, R)$. In particular, $f(C(a, r)) = \{w\}$, a contradiction. Let then

$a_1, \ldots, a_m$ be the (distinct) points of $f^{-1}(w) \cap D(a,r)$, if any. 5.69 provides a holomorphic function F in $D(a,R)$ such that (void products being 1)

$$f(z) - w = F(z) \prod_{j=1}^{m} (z - a_j)^{n_j}, \qquad z \in D(a,R) \tag{2}$$

and

$$F(a_j) \neq 0, \quad j = 1, \ldots, m. \tag{3}$$

Here (recall 5.68)

$$n_j := n(a_j, f - w), \quad j = 1, \ldots, m. \tag{4}$$

Since $a_1, \ldots, a_m$ are all the zeros of $f - w$ in $\bar{D}(a,r)$, it follows from (2) and (3) that F is zero-free in $\bar{D}(a,r)$, hence by continuity zero-free in some (convex) open neighborhood V of $\bar{D}(a,r)$ in $D(a,R)$. From (2) then

$$\frac{f'(z)}{f(z) - w} = \frac{F'(z)}{F(z)} + \sum_{j=1}^{m} \frac{n_j}{z - a_j} \qquad \forall z \in V \setminus \{a_1, \ldots, a_m\},$$

whence

$$\int_{C(a,r)} \frac{hf'}{f - w} = \int_{C(a,r)} \frac{hF'}{F} + \sum_{j=1}^{m} n_j \int_{C(a,r)} \frac{h(z)}{z - a_j}\, dz.$$

By Cauchy's Theorem and Integral Formula for Convex Regions this gives

$$\int_{C(a,r)} \frac{hf'}{f - w} = \sum_{j=1}^{m} 2\pi i n_j h(a_j) \mathrm{Ind}_{C(a,r)}(a_j). \tag{5}$$

But $a_j \in D(a,r)$, which is a connected subset of $\mathbb{C} \setminus C(a,r)$, so

$$\mathrm{Ind}_{C(a,r)}(a_j) = \mathrm{Ind}_{C(a,r)}(a).$$

The latter is trivially 1, so (5) and (4) give

$$\frac{1}{2\pi i} \int_{C(a,r)} \frac{hf'}{f - w} = \sum_{j=1}^{m} n(a_j, f - w) h(a_j).$$

When we recall that $\{a_1, \ldots, a_m\} = f^{-1}(w) \cap D(a,r)$, the desired formula is at hand.

Corollary 5.85 (Cf. 11.48) *If f is holomorphic in a neighborhood V of a, then for each $r > 0$ such that $\overline{D}(a,r) \subset V$ and such that f has no zero on $C(a,r)$, the number of zeros of f in $D(a,r)$ and the sum of these zeros, counted according to multiplicity, are respectively*

$$\frac{1}{2\pi i}\int_{C(a,r)} \frac{f'(\xi)}{f(\xi)}\,d\xi \quad \textit{and} \quad \frac{1}{2\pi i}\int_{C(a,r)} \frac{\xi f'(\xi)}{f(\xi)}\,d\xi.$$

Proof: Take $w = 0$, and h in the preceding theorem successively equal to the constant function $\mathbf{1}$ and the identity function.

Corollary 5.86 (Argument Principle; cf. 9.15 and 4.26(i), as well as 4.59(iv)) *Let f be continuous in $\overline{\mathbb{D}}$ and holomorphic in $\mathbb{D}$. Set $\gamma := C(0,1)$. Then for every $w \in \mathbb{C} \setminus f(\gamma)$ the number of zeros of $f - w$ in $\mathbb{D}$, counted according to multiplicity, is* $\mathrm{Ind}_{f\circ\gamma}(w)$.

Proof: For each $0 < r \le 1$ let $\gamma_r(t) := re^{2\pi i t}$, $t \in [0,1]$. If $r < 1$ and $w \in \mathbb{C}\setminus f(\gamma_r)$, then the number of zeros of $f - w$ in $D(0,r)$ is, by the preceding corollary (with $f - w$ in the role of the f there)

$$\begin{aligned}\frac{1}{2\pi i}\int_{\gamma_r} \frac{f'}{f-w} &= \frac{1}{2\pi i}\int_0^1 \frac{f'\circ\gamma_r}{f\circ\gamma_r - w}\cdot\gamma_r' \\ &= \frac{1}{2\pi i}\int_0^1 \frac{(f\circ\gamma_r)'}{f\circ\gamma_r - w} = \frac{1}{2\pi i}\int_{f\circ\gamma_r} \frac{1}{\xi - w}\,d\xi,\end{aligned}$$

which is $\mathrm{Ind}_{f\circ\gamma_r}(w)$ (4.4).

If $w \in \mathbb{C}\setminus f(\gamma_1)$, then by continuity $w \in \mathbb{C}\setminus f(\gamma_r)$ and $\mathrm{Ind}_{f\circ\gamma_r}(w) = \mathrm{Ind}_{f\circ\gamma_1}(w)$ for all $r \in (0,1)$ sufficiently close to 1. So for such r the set of zeros of $f - w$ in $D(0,r)$ coincides with the set of zeros of $f - w$ in $\mathbb{D}$ and their number, counted according to multiplicity, is $\mathrm{Ind}_{f\circ\gamma_r}(w) = \mathrm{Ind}_{f\circ\gamma_1}(w)$.

Alternatively, notice that in view of 5.69 and the fact that $f - w$ has only finitely many zeros in $\overline{\mathbb{D}}$ (why?), the proof of 4.26(i) can be repeated verbatim, with $f - w$ in the role of the polynomial P there, to secure the present corollary.

Exercise 5.87 (Cf. remark following 9.15.) *Let f and g be continuous on $\overline{\mathbb{D}}$, holomorphic in $\mathbb{D}$, and let γ denote the loop $C(0,1)$.*

(i) *Prove that if f and g are zero-free on γ, then the following are equivalent:*

(a) *f and g have the same number (multiplicities counted) of zeros in $\mathbb{D}$.*

(b) $\mathrm{Ind}_{f\circ\gamma}(0) = \mathrm{Ind}_{g\circ\gamma}(0)$.

(c) *$f \circ \gamma$ and $g \circ \gamma$ are loophomotopic in $\mathbb{C} \setminus \{0\}$.*

Hints: (a) $\Leftrightarrow$ (b) is 5.86, (b) $\Rightarrow$ (c) is 4.15(i), and (c) $\Rightarrow$ (b) is 4.12.

(ii) (Rouché's Theorem; cf. 4.26(ii) and 8.18) *An oft-realized hypothesis that implies* (a), (b), (c) *above is*

$$|f + g| < |f| + |g| \quad \textit{on } \partial\mathbb{D}.$$

Hint: In the hints to 4.27 it is shown that the quotient f/g has a continuous logarithm on $\gamma := C(0, 1)$, and so $\text{Ind}_{(f/g)\circ\gamma}(0) = 0$. It follows from 4.14 that $\text{Ind}_{f\circ\gamma}(0) = \text{Ind}_{(f/g)\circ\gamma}(0) + \text{Ind}_{g\circ\gamma}(0) = \text{Ind}_{g\circ\gamma}(0)$.

(iii) *Show that if in* (ii) *the holomorphy of f is dropped, then it is still possible to infer that g is zero-free in $\mathbb{D}$ if f is.*

Hints: The preceding hint gives $\text{Ind}_{f\circ\gamma}(0) = \text{Ind}_{g\circ\gamma}(0)$. Since f is zero-free in $\overline{\mathbb{D}}$, it has a continuous logarithm there (4.1), so $\text{Ind}_{f\circ\gamma}(0) = 0$. Hence $\text{Ind}_{g\circ\gamma}(0) = 0$ and by 5.86 g is zero-free in $\mathbb{D}$.

(iv) *f is continuous in $\overline{\mathbb{D}}$, holomorphic in $\mathbb{D}$, and C is a component of $\mathbb{C} \setminus f(\mathbb{T})$ which meets $f(\mathbb{D})$. Show that in fact $C \subset f(\mathbb{D})$.*

Hints: If $w_0 \in C \cap f(\mathbb{D})$, then by 5.86, $\text{Ind}_{f\circ\partial\mathbb{D}}(w_0) \neq 0$. Hence 4.59(i) is applicable. Alternatively, $\text{Ind}_{f\circ\partial\mathbb{D}}(w) = \text{Ind}_{f\circ\partial\mathbb{D}}(w_0) \neq 0$ for every $w \in C$, so by 5.86 again such w lie in $f(\mathbb{D})$. There is also a purely topological argument which uses only the fact that f is a continuous, open map, and which is valid with any region in the role of $\mathbb{D}$: The intersection of the closures of the complementary sets $f(\mathbb{D})$ and $\mathbb{C} \setminus f(\mathbb{D})$ is $f(\overline{\mathbb{D}}) \cap [\mathbb{C} \setminus f(\mathbb{D})] \subset f(\mathbb{T})$, which does not meet C. Hence these two sets cover C and would disconnect C if each of them meets C. Since $f(\mathbb{D})$ does meet C, $\mathbb{C} \setminus f(\mathbb{D})$ therefore cannot.

Corollary 5.88 *Let f be holomorphic and not constant in the region Ω. If $z_0 \in \Omega$, $w_0 := f(z_0)$ and m is the multiplicity of z_0 as a zero of $f - w_0$, then there exist open sets V and W such that $z_0 \in V \subset \Omega$, $W = f(V)$ and for each $w \in W \setminus \{w_0\}$ the set $f^{-1}(w) \cap V$ consists of exactly m distinct points.* (Cf. 5.83.)

Proof: Since $f - w_0$ is not identically 0, f' is not either (2.10(iv)) and so (5.62) neither the zeros of $f - w_0$ nor those of f' accumulate at z_0, i.e., there exists $r > 0$ such that

(1) $$|f - w_0| > 0 \quad \text{and} \quad |f'| > 0 \quad \text{in } \overline{D}(z_0, r) \setminus \{z_0\} \subset \Omega.$$

In particular, $w_0 \in \mathbb{C} \setminus f(C(z_0, r))$ and it follows from the preceding corollary that $\text{Ind}_{f\circ\gamma_r}(w_0)$ ($\gamma_r := C(z_0, r)$) is the number of zeros of $f - w_0$ in $D(z_0, r)$, counted according to multiplicities. By (1) this is just the multiplicity of z_0 as a zero of $f - w_0$, viz., m: $\mathrm{Ind}_{f\circ\gamma_r}(w_0) = m$. Since $\mathrm{Ind}_{f\circ\gamma_r}$ is a continuous integer-valued function on the open set $\mathbb{C} \setminus f(\gamma_r)$, there is an $R > 0$ such that $W := D(w_0, R) \subset \mathbb{C} \setminus f(\gamma_r)$ and

$$(2) \qquad \mathrm{Ind}_{f\circ\gamma_r}(w) = m \qquad \forall w \in W.$$

Set $V := f^{-1}(W) \cap D(z_0, r)$. It follows from (2) and the preceding corollary that for each $w \in W$, $f - w$ has exactly m zeros in $D(z_0, r)$, hence in V. In particular, since $m > 0$, it follows that $f(V) = W$. Moreover if $w \in W \setminus \{w_0\}$ and $z \in f^{-1}(w) \cap D(z_0, r)$, then $z \neq z_0$ and so $(f - w)'(z) = f'(z) \neq 0$ by (1). Therefore the multiplicity of each zero of $f - w$ in V is 1 and the m zeros are therefore distinct.

Corollary 5.89 *If f is holomorphic in a neighborhood of z_0, then f is one-to-one near z_0 if and only if $f'(z_0) \neq 0$.*

Proof: f is one-to-one near z_0 if and only if the integer m of the preceding corollary is 1. Consulting the definition of multiplicity, we see that this is equivalent to $f'(z_0) \neq 0$.

Remarks 5.90: (i) The Open-Map Theorem (5.77) is an immediate consequence of 5.88, and the differentiability of the inverse of a one-to-one holomorphic function follows at once from this and from the preceding corollary. For the calculation at the beginning of the proof of 5.78 needed only the facts that f^{-1} is continuous (f is open) and f' never vanishes. Alternatively, yet another proof (5.92) follows from the next corollary.

(ii) Some of the several proofs so far presented for 5.89 reveal that the radius of univalence predicated there is to a surprising extent independent of the function. For example, suppose $f \in H(\mathbb{D})$ is bounded by $M < \infty, f(0) = 0$ and $|f'(0)| = 1$. If $0 < r < 1$ is sufficiently small, $\sum_{n=2}^{\infty} nr^{n-1} \leq (2M)^{-1}$. For any such r, f is one-to-one in $D(0, r)$. Indeed, writing $\sum_{n=0}^{\infty} c_n z^n$ for the Taylor series of f, we have $|c_n| \leq M$ by 3.34(i), whence

$$\sum_{n=2}^{\infty} n|c_n|r^{n-1} \leq M \sum_{n=2}^{\infty} nr^{n-1} \leq 1/2 \leq |c_1|/2$$

and consequently, as shown in the hints to 3.7,

$$|f(z) - f(w)| \geq \big(|c_1|/2\big)|z - w| \geq |z - w|/2 \qquad \forall z, w \in D(0, r).$$

Since $\sum_{n=2}^{\infty} nr^{n-1} = \frac{d}{dr}\left[\frac{r^2}{1-r}\right] = \frac{2r-r^2}{(1-r)^2}$, the largest r that can be found by refining this method is $1 - \sqrt{M/(M+1)}$. However, E. Landau [1926] determined that the largest r which works for all such functions f is actually $M - \sqrt{M^2 - 1}$. Heins [1962b], p. 37 and Nehari [1952], pp. 171-172 give a proof somewhat different from Landau's.

Corollary 5.91 *If f is holomorphic and one-to-one in the open set U, then for every $a \in U$ and every $r > 0$ such that $\overline{D}(a,r) \subset U$ we have*

$$f^{-1}(w) = \frac{1}{2\pi i}\int_{C(a,r)} \frac{zf'(z)}{f(z)-w}\,dz \qquad \forall w \in f(D(a,r)).$$

Proof: There exists $R > r$ such that $D(a,R) \subset U$. Apply 5.84 with $h(z) \equiv z$ and get

$$\frac{1}{2\pi i}\int_{C(a,r)} \frac{zf'(z)}{f(z)-w}\,dz = \sum_{z \in f^{-1}(w)\cap D(a,r)} n(z, f-w)z \qquad \forall w \in \mathbb{C}\setminus f(C(a,r)).$$

Now if $w \in f(D(a,r))$, then $w \in \mathbb{C} \setminus f(C(a,r))$ since f is one-to-one. Also because f is one-to-one, $f^{-1}(w) \cap D(a,r)$ consists just of the unique $z \in U$ such that $f(z) = w$ and for this z we have $n(z, f-w) = 1$ because $f'(z) \neq 0$ (5.89). The result follows. Here is a somewhat more illuminating and direct proof: Fix $z_0 \in D(a,r)$. Then $g(z) := zf'(z)(z-z_0)/(f(z)-f(z_0))$ is well defined for $z \in U \setminus \{z_0\}$ and converges to z_0 as z does, since f is $1-1$ and $f'(z_0) \neq 0$. Defining $g(z_0) := z_0$ thus puts g in $H(U)$, and Cauchy's Integral Formula for $g(z_0)$ becomes

$$z_0 = \frac{1}{2\pi i}\int_{C(a,r)} \frac{g(z)}{z-z_0}dz = \frac{1}{2\pi i}\int_{C(a,r)} \frac{zf'(z)}{f(z)-f(z_0)}dz,$$

which is the claimed equality for $w = f(z_0)$, the typical point in $f(D(a,r))$.

Exercise 5.92 In what follows notation and hypotheses are as in 5.91.

(i) *Deduce from the preceding corollary that f^{-1} is holomorphic and exhibit integral formulae for $\left(f^{-1}\right)^{(k)}$ analogous to the one there for f^{-1}.*

Hint: Use a complex version of 2.19(i), or 2.19(i) itself together with (C. R.), or make appropriate modifications in the proof of 2.14. (Cf. also 7.21.) Actually, once it is shown that f^{-1} is continuous, 2.14 can simply be *applied* with $g := f^{-1}$, $\gamma := f \circ C(a,r)$.

(ii) *Using the formula*

$$\left(f^{-1}\right)'(w) = \frac{1}{2\pi i}\int_{C(a,r)} \frac{zf'(z)}{(f(z)-w)^2}\,dz, \qquad w \in f(D(a,r))$$

from (i), *and the identity*

$$\frac{zf'(z)}{(f(z)-w)^2} = \frac{1}{f(z)-w} - \frac{d}{dz}\left(\frac{z}{f(z)-w}\right),$$

get

$$(f^{-1})'(w) = \frac{1}{2\pi i}\int_{C(a,r)} \frac{1}{f(z)-w}\,dz, \qquad w \in f(D(a,r)).$$

(iii) *For simplicity let* $a = 0 = f(a)$. *Then* $0 \notin f(C(0,r))$ *and consequently* $m := \inf|f(C(0,r))| > 0$. *If* $r > R > 0$ *is sufficiently small,* f *maps* $D(0,R)$ *into* $D(0,m)$. *Consider* w *in the neighborhood* $W := f(D(0,R))$ *of* 0 *and use* (ii) *to show that*

$$\begin{aligned}(f^{-1})'(w) &= \frac{1}{2\pi i}\int_{C(0,r)} \frac{1}{f(z)}\left[\frac{1}{1-w/f(z)}\right] dz \\ &= \sum_{n=1}^{\infty} w^{n-1}\frac{1}{2\pi i}\int_{C(0,r)} \frac{1}{f^n(z)}\,dz, \\ f^{-1}(w) &= \sum_{n=1}^{\infty} w^n \frac{1}{2n\pi i}\int_{C(0,r)} \frac{1}{f^n}.\end{aligned}$$

(iv) *Use* 5.33 *to identify the coefficient of* w^n *in* (iii) *as* $\frac{1}{n!}$ *times the* $(n-1)^{\text{st}}$ *derivative of* $(z/f(z))^n$ *evaluated at* $z = 0$. [Why is $z/f(z)$ holomorphic near 0?]

(v) *Apply* (iii) *and* (iv) *to* $f(z) := ze^{-z}$ *to infer* (Eisenstein, 1844) *that the series* $g(w) := \sum_{n=1}^{\infty} n^{n-1}w^n/n!$ *converges in* $D(0,1/e)$ *and satisfies* $f(g(w)) = w$ *for all such* w. *Show further that* $g(w) \uparrow 1$ *as* $w \uparrow 1/e$, *hence*

$$1 = \sum_{n=1}^{\infty} \frac{n^{n-1}}{n!\,e^n}.$$

Hint: Use (iv), 3.3(iv), and the Remark following 3.47 to find the radius of convergence. For the last claim use the fact that $f(z)$ achieves its maximum value of $1/e$ for $z \in [0,\infty)$ only at $z = 1$. [The function g here is not just a curiosity: $w(z) = -g(-z)$, which satisfies $w(z)e^{w(z)} = z$, is called the *Lambert W-function* and has numerous applications; see **MR** 98j:33015.]

We can now present an elegant proof of a useful *Implicit-Function Theorem* for functions of two complex variables. The reader will observe that the preceding corollary is (essentially) the special case $F(w,z) := f(z) - w$ of this theorem. Moreover, as SOKAL [2009] has recently shown, the function $z = G(w)$ implicitly defined by the equation $F(w,z) = 0$ in 5.93 can be represented by a mere notational modification of the series for $f^{-1}(w)$ in 5.92(iv).

Theorem 5.93 *Let V and W be neighborhoods in $\mathbb{C}$ of z_0, w_0 and let $F\colon W \times V \mapsto \mathbb{C}$ be continuous with $z \mapsto F(w,z)$ holomorphic in V for each $w \in W$. Then D_2F is continuous on $W \times V$. Suppose $F(w_0, z_0) = 0$ and $D_2F(w_0, z_0) \neq 0$. Then there exist $r > 0$ and a neighborhood U of w_0 such that F is zero-free in $U \times C(z_0, r) \subset W \times V$ and the function*

$$G(w) := \frac{1}{2\pi i} \int_{C(z_0,r)} \frac{\xi D_2F(w,\xi)}{F(w,\xi)}\, d\xi, \qquad w \in U$$

maps U into $D(z_0, r) \subset V$, is continuous and satisfies

$$F(w,z) = 0 \iff z = G(w) \qquad \forall (w,z) \in U \times D(z_0, r).$$

Proof: For each $w \in W$ let $F_w\colon V \mapsto \mathbb{C}$ be the holomorphic function $F_w(z) := F(w,z)$. We have

$$(1) \qquad F'_{w_0}(z_0) = D_2F(w_0, z_0) \neq 0,$$

so there is an $r > 0$ such that

$$(2) \qquad F_{w_0} \text{ has no zero in } \bar{D}(z_0, r) \setminus \{z_0\} \subset V.$$

The continuity of D_2F in $W \times V$ follows routinely from the Cauchy integral formula and is outlined in Exercise 5.43(ii). Because of this continuity, it follows from (1) that $|D_2F| > 0$ on some neighborhood of (w_0, z_0), so we may also take r small enough that

$$(3) \qquad |D_2F| > 0 \quad \text{on } U \times \bar{D}(z_0, r)$$

for some compact, connected neighborhood U of w_0 in W. From (2) it follows that $\{w_0\} \times C(z_0, r)$ is a compact subset of $W \times V$ disjoint from the (relatively) closed subset $F^{-1}(0)$, and so there is a neighborhood of this set which is disjoint from $F^{-1}(0)$. In particular, we may take the U of (3) sufficiently small that

$$(4) \qquad |F| > 0 \quad \text{on } U \times C(z_0, r).$$

Then according to 5.85

$$(5) \quad N(w) := \frac{1}{2\pi i} \int_{C(z_0,r)} \frac{F'_w(\xi)}{F_w(\xi)}\, d\xi = \frac{1}{2\pi i} \int_{C(z_0,r)} \frac{D_2F(w,\xi)}{F(w,\xi)}\, d\xi, \qquad w \in U$$

is the number of zeros of F_w in $D(z_0, r)$, counted according to multiplicity. But $F'_w(z) = D_2F(w,z) \neq 0$ by (3) for each $z \in D(z_0, r)$, so each zero of F_w has multiplicity 1. Then from 5.85 also follows that

$$(6) \quad G(w) := \frac{1}{2\pi i} \int_{C(z_0,r)} \frac{\xi F'_w(\xi)}{F_w(\xi)}\, d\xi = \frac{1}{2\pi i} \int_{C(z_0,r)} \frac{\xi D_2F(w,\xi)}{F(w,\xi)}\, d\xi, \qquad w \in U$$

is the sum of the zeros of F_w in $D(z_0, r)$. Now by (1) and (2) we have

$$N(w_0) = 1. \tag{7}$$

Also it is clear that

$$\frac{D_2F(w,\gamma(t))\gamma'(t)}{F(w,\gamma(t))} \quad \text{and} \quad \frac{\gamma(t)D_2F(w,\gamma(t))\gamma'(t)}{F(w,\gamma(t))}$$

are uniformly continuous on $U \times [0,1]$, where $\gamma := C(z_0, r)$, and so from (5) and (6) it follows that N and G are continuous. Being integer-valued, N is therefore constantly 1 in the connected neighborhood U of w_0. Thus for $w \in U$ the equation $F(w,z) = F_w(z) = 0$ has a unique solution $z \in D(z_0, r)$ and that z is $G(w)$.

Remark 5.94: In the above argument W can be any topological space, e.g., any subset of $\mathbb{C}$ or any interval in $\mathbb{R}$. If however W is an open subset of $\mathbb{C}$ and $w \mapsto F(w,z)$ is holomorphic in W for each fixed $z \in V$, then 7.21 will show that G is actually holomorphic in U.

A typical application of 5.93 (and 5.94) is the following fact about the continuity of the (simple) zeros of a polynomial as a function of its coefficients:

Corollary 5.95 *Let the functions $c_0, \ldots, c_n$ be continuous (holomorphic) in a neighborhood of w_0 and let z_0 be a simple zero of the non-constant polynomial $\sum_{j=0}^n c_j(w_0)z^j$. Then there is a neighborhood U of w_0 and a function G which is continuous (holomorphic) in U, such that $G(w_0) = z_0$ and $\sum_{j=0}^n c_j(w)G^j(w) = 0$ for all $w \in U$.*

The idea in the proof of 5.78 can be further usefully exploited; its essence is captured in the next theorem, whose proof is a nice application of many of the results of this section.

Factorization Lemma 5.96 *Let G, Ω be regions, $g\colon G \mapsto \mathbb{C}$, $f\colon G \mapsto \mathbb{C}$ be holomorphic functions with $g(G) = \Omega$. Suppose that f is constant on $g^{-1}(w)$ for each $w \in \Omega$. Then there exists a unique holomorphic function F on Ω such that $f = F \circ g$. Viewed differently, if Ω is a region in $\mathbb{C}$, $h\colon \Omega \mapsto \mathbb{C}$ a function such that $h \circ g$ is holomorphic for some holomorphic function g whose range is Ω, then h itself is holomorphic.*

Proof: Since $g(G) = \Omega$, at most one F can satisfy the equation $F \circ g = f$. In fact, such an F is well defined on Ω by the decree

$$F(g(z)) := f(z) \qquad \forall\, z \in G, \tag{1}$$

since by hypothesis $f(z) = f(z')$ whenever $g(z) = g(z')$. So it remains only to show that F defined this way is holomorphic in Ω. To this end, first note that since g is an open map (5.77), the set $F^{-1}(U) = g(f^{-1}(U))$ is open for every open subset U of $\mathbb{C}$. This shows that at least

$$\text{(2)} \qquad F \text{ is continuous in } \Omega.$$

Since g is non-constant, the zero-set N' of g' has no limit point in G (5.62). Then from the continuity and openness of g it follows that the set

$$\text{(3)} \qquad N := \{w \in \Omega : g^{-1}(w) \subset N'\} \quad \text{has no limit point in } \Omega.$$

Indeed, suppose w is a limit point of N in Ω. Take any $z \in g^{-1}(w)$. For each $n \in \mathbb{N}$ there exists $r_n > 0$ such that $g(D(z, r_n)) \subset D(w, 1/n)$ and we may suppose $r_n \downarrow 0$. Since $g(D(z, r_n))$ is an open neighborhood of $g(z) = w$, there is a point $w_n \in N \cap [g(D(z, r_n)) \setminus \{w\}]$ say, $w_n = g(z_n)$, $z_n \in D(z, r_n) \setminus \{z\}$. Thus $0 < |z - z_n| < r_n \to 0$. Since $z_n \in g^{-1}(w_n)$ and $w_n \in N$, we have $z_n \in N'$. Thus we have the contradiction that z is a limit point of N' in G. This confirms (3).

As a result of (2), (3) and the removable singularity theorem (5.41), it now suffices to prove that F is holomorphic in $\Omega \setminus N$. So consider any point $w_0 \in \Omega \setminus N$. Since $w_0 \notin N$, the set $g^{-1}(w_0)$ contains a point $z_0 \notin N'$. Thus $g'(z_0) \neq 0$ and so there is an open disk D centered at z_0 and lying in G on which g is one-to-one (5.79). The set $W := g(D)$ is an open neighborhood of $g(z_0) = w_0$ and the restriction $\tilde{g}$ of g to D maps D biholomorphically onto W (5.78). From (1) we see that

$$F(w) = F(g(\tilde{g}^{-1}(w))) = f(\tilde{g}^{-1}(w))$$

for all $w \in W$, that is, in the neighborhood W of w_0, F coincides with the holomorphic function $f \circ \tilde{g}^{-1}$.

In the second scenario, *define* f to be $h \circ g$; then h evidently coincides with the F constructed for this f.

Here are two more illustrations of the local-to-global technique used in 5.78 and 5.96.

Exercise 5.97 (i) *Show that if U, W are open subsets of $\mathbb{C}$, $u\colon U \mapsto \mathbb{R}$ is subharmonic and $\phi\colon W \mapsto U$ is holomorphic, then $u \circ \phi$ is subharmonic in W.*

Hints: We may assume that W is connected and that ϕ is not constant. Then the zero-set Z of the holomorphic function ϕ' consists of isolated points. In a small neighborhood V of the typical $W \setminus Z$ point ϕ is biholomorphic

(5.79). That the function $v := u \circ \phi$ is subharmonic in V can then be inferred from 5.22(ii). Thus v is continuous in W and subharmonic except for the isolated points Z. Now recall remark (a) after 5.41.

(ii) *Deduce from* (i) *and* 5.15 *that a function which is superharmonic in* $\mathbb{C} \setminus \{0\}$ *and bounded from below is constant.*

Hints: If $v \geq m > -\infty$ is superharmonic in $\mathbb{C} \setminus \{0\}$, then $u(z) := v(e^z) - m$ is non-negative and by (i) is superharmonic in $\mathbb{C}$, so by 5.15 it is constant.

Exercise 5.98 *Let U be an open subset of $\mathbb{C}$, $g\colon U \mapsto \mathbb{C}$ continuous, f holomorphic and non-constant in a region $\Omega \supset g(U)$. Show that if $f \circ g$ is holomorphic in U, then g is holomorphic in U.* (This is a sort of "dual" to 5.96.)

Hints: Let $a \in U$ be given and pick $r > 0$ so that $\bar{D}(a,r) \subset U$. We need to show that g is holomorphic in $D := D(a,r)$, and to that end we can assume it's not constant there. Since f is not constant, $N' := (f')^{-1}(0)$ has no limit point in Ω (5.62). We claim that

$$(*) \qquad N := g^{-1}(N') \cap D \quad \text{has no limit point in } D.$$

For suppose $z_n \in N$, $z_0 \in D$ and $0 < |z_0 - z_n| \to 0$. Then $g(z_n) \in N' \to g(z_0) \in \Omega$, so the set $\{g(z_n)\}$ must be finite; in particular, $g(z_n) = g(z_0)$ for infinitely many n. The holomorphic function $f \circ g$ has then by 5.62 the constant value $f(g(z_0))$ throughout D. The subset $g(\bar{D})$ of Ω, on which f is constant, is not a singleton but it is compact and connected. Hence it is infinite, so by 5.62 again f is constant. This contradiction proves $(*)$. Finally, consider any point $z \in D \setminus N$. We have $g(z) \in \Omega \setminus N'$, so $f'(g(z)) \neq 0$ and therefore f is one-to-one in some open neighborhood W of $g(z)$ (5.79). The open set $g^{-1}(W)$ is a neighborhood V of z and $g = (f \,|\, W)^{-1} \circ (f \circ g)$ in V. Hence g is holomorphic in V (5.78). In summary, the function g is holomorphic in $D \setminus N$ and continuous in D, hence by $(*)$ and 5.41, holomorphic in D.

Notes to Chapter V

In a 1743 book Alexis-Claude Clairaut claimed that integrals of the form $\int (P\,dx + Q\,dy)$ were path independent if and only if $\frac{\partial P}{\partial y} = \frac{\partial Q}{\partial x}$ (see Stäckel [1901]); this is a precursor of 5.1. In a famous 1811 letter to F. W. Bessel, Gauss pointed out that the integrals of a function ϕ over two paths in $\mathbb{C}$ would be equal if and only if "ϕ did not become infinite at any point in the region enclosed by the two curves." Cauchy's own work started with an important 200–page memoir in 1814 (published only 13 years later) and another shorter one in 1825. See David [1882], [1885], Jourdain [1905a], Cauchy [1900], Ettlinger [1921–22], chapter 4 of Bottazzini [1986] and chapters 2 and 4 of Smithies [1997].

The integral theorem was discovered while trying to evaluate certain integrals over intervals of $\mathbb{R}$ and investigating when orders of integration in iterated integrals can be reversed. The very natural arguments in 2.19 were the first to be used; they are rather real-variable in nature and dominated the early work, before complex numbers and functions became fully enfranchized. A clearer articulation in terms of contour integrals appears in the 1825 memoir, and even more in his 1846 *Comptes Rendus* notes. Cauchy discovered the integral formula as a special case of his residue theorem (see the remark preceding 11.29). In turn, that was discovered when orders of integration were reversed in iterated integrals of functions which were "allowed to become infinite" at certain interior points of the rectangle and, instead of 0, a "residue" was found as the difference of the two iterated integrals, which in their turn reduce to contour integrals as in 2.19. See §35 of BURKHARDT [1914]. Some of these ideas, in still more inchoate form, were used by Gauss (cf. his third proof of the FTA following 5.5). Bear in mind that the exact hypotheses on the functions f to which the theory should apply were never clearly set down and their discovery was part of the evolution of the theory. Moreover, the role of "holes" in the region and their being "surrounded" or not by the integration paths γ was only very dimly perceived at first. The grandeur of the edifice that arose from Cauchy's pioneering work can be seen in 11.29 and in the multi-part 10.2.

Continuity of f' seems necessary in the proof of 2.19(v), although Cauchy vaccilated as to whether or not it needed to be hypothesized; see pp. 1010–1012 of BURKHARDT [1914] and p. 473 of PRINGSHEIM [1900]. GOURSAT [1900] set out a method of descent (subsequently known as "Goursat's lemma" – cf. 2.19(xi)) that did not involve continuity of f' (cf. also DE LA VALLÉE POUSSIN [1914]), and he remarked that it enabled a proof of Cauchy's theorem he'd given in 1884 to run without that hypothesis. But there are non-trivial technical difficulties in implementing the latter, as PRINGSHEIM [1901] noted. A slight modification of Goursat's idea by E. H. MOORE [1900] partially overcomes this problem, but his proof is still not very transparent regarding how various integrals created by a subdivision add up to the original one, and is further complicated by trying to have that initial curve of integration as general as possible. The idea of restricting to a rectangle or triangle and proceeding as we did here is the end-product of PRINGSHEIM [1901], [1903] and HEFFTER [1902], [1903]. J. J. GRAY [2000] analyzes the original papers and this evolution in some detail. It was gradually recognized that determination of the most general curve for which Cauchy's theorem holds is a different issue, best resolved by more sophisticated topological and approximation arguments (to be given in chapters VIII and IX). For the whole rich local theory Cauchy's theorem for a triangle (thence for a circle, as in 5.2) suffices. *It is consequently the most central and important single result of the theory.* For history and references to original memoirs and many subsequent proofs and extensions, see the (confused and confusing) monograph of HEFFTER [1960] and the bibliography and notes of FALLIN and GOULD [1965]. See also corollary 9.27 and the notes thereon at the end of Chapter IX, STÄCKEL [1900], section II of BRILL and NOETHER [1892–93], FABBRI [1900], pp. 27–36 of MITTAG-LEFFLER [1923] and §35 of BURKHARDT [1914]. The latter is a good history of Cauchy's many fundamental contributions; less satisfactory scientifically but useful as biography is BELHOSTE [1991]. The best short scientific biography and critical appraisal is FREUDENTHAL [1971a] and the best short perspective on Cauchy's founding role in complex analysis is NEUENSCHWANDER [1979]. For the latter see also §4.6 of BOTTAZZINI [1986] and §3.c of his 1992 Introduction to CAUCHY [1821]. ETTLINGER [1921/22] is an invaluable "explication de texte", and the survey in chapter 10 of MITRINOVIĆ and KEČKIĆ [1984] is also very helpful. The monograph of SMITHIES [1997] is a detailed analysis and exegesis of Cauchy's work in function theory from 1814 up to the important 1831 Turin memoirs; it also contains a valuable 18-page summary chapter.

It is sometimes maintained that the use of integration theory to secure 5.3 is a methodological adulteration; see the discussion in PRINGSHEIM [1920]. The renunciation of integration theory here is debilitating but not fatal; circuitous alternative paths to 5.3 have been hewn out. See STOÏLOW [1956], CONNELL [1965], G. T. WHYBURN [1962], [1964], [1968] for one approach and PRINGSHEIM's book [1925], [1932] for another. In the latter, the limits of Riemann sums based on $(2^n)^{\text{th}}$ roots of 1 replace integrals over circles. These roots can easily be constructed inductively (cf. 1.26(i) and the hints to 3.44(ii)) and so this procedure has the added advantage of freeing the local theory from dependence on the exponential function (which the use of rectangles, as in 2.20, also accomplishes). A completely different approach is offered in ADEL'SON-VEL'SKIĬ and KRONROD [1945] and another in LELAND [1964–65], [1971]. It should be noted that the Stoïlow–Whyburn program is much broader than merely eliminating integration. It aims to derive the whole local theory of the non-constant holomorphic function f topologically from just two properties: *lightness* (f is not constant on any non-singleton continuum) and *openness*; these in turn are derived using no other analytic hypothesis than the $\mathbb{C}$-differentiability of f. A substantial sample of this program is elegantly condensed into a three-page appendix in BEARDON [1979].

Each of the (improper) integrals in 5.4 can be computed by various contour integrals or by the Residue Theorem of Chapter XI, but the development presented has a certain unity and is perhaps minimally arduous. For more on these kinds of integrals see SCHLÖMILCH [1860], HAYASHI [1922/24], [1923], A. N. SINGH [1925], DEAUX and DELCOURTE [1957], LAGRANGE [1957], BUTLER [1960], SPIEGEL [1971], CLARKE [1998] and the prize-winning paper TRAININ [2010]. In connection with 5.4(i) see 10.44 of RUDIN [1987] and the amusing essay of HARDY [1909–11]. The identities 5.4(v) are very special cases of relations between certain Fourier series and integrals, involving the famous *Poisson summation formula* (11.33(vi)). For a unified treatment of these identities and others from this point of view see BOAS and POLLARD [1973] and the fascinating later paper BAILLIE, *et al.* [2008]. In particular, the last equality in 5.4(v) is true when the exponent 2 there is replaced by $3, 4, 5, 6$, but *not* 7.

The concept of subharmonicity was introduced by F. Hartogs and greatly elaborated by F. Riesz. See the little monograph of T. RADÓ [1937] for history and relevant papers. These functions have many analogies with convex functions on open subintervals of $\mathbb{R}$, and it's not hard to prove that every convex function on an open convex subset of $\mathbb{R}^2$ is subharmonic. They certainly enter function theory naturally and provide just the appropriate level of generality in which to prove many results for the most efficient application to function theory, as the results of §2 and §3 illustrate. See MONTEL [1928], [1940], BRELOT [1939a] and HAYMAN and KENNEDY [1976] for more applications of them to function theory. By contrast to the integral (mean-value) properties of harmonic and subharmonic functions which occupy center stage in §2 and §3, we will look at differential properties (e.g., Laplace's equation) in §7 of Chapter IX and we will show that the Dirichlet problem is solvable for bounded simply-connected regions.

It is customary to require only upper semi-continuity in the definition of subharmonicity and to allow the function to take the value $-\infty$. While a bit more delicacy is then called for in the early stages of the theory, there are many advantages. For example, now $\log |f|$ is subharmonic whenever f is holomorphic and artifices like that of considering $\log(\max\{|f|, \delta\})$ and letting $\delta \downarrow 0$, used in this chapter, can be eliminated.

The maximum principle for harmonic functions was proved in RIEMANN's dissertation [pp. 22–23 of his *Werke*, 1953 ed.] by brief reference to the mean-value property 5.21(iii) (the heart of the matter, to be sure). Subsequently more extensive treatments were given by NEUMANN [1871] (pp. 340–344) and H. A. SCHWARZ (pp. 203–204, vol. II of his *Gesammelte Mathematische Abhandlungen*).

The device of taking powers in the proof of 5.13 is due to E. Landau (in footnote 1, p. 340 of M. RIESZ [1916]). The need to *hypothesize* in 5.13 that f be bounded in U is illustrated by $U := \mathbb{H}$, $f(z) := e^z$. Though the proof of 5.12 in the text is taken from LINDELÖF [1915], the result itself pre-dates this; e.g., it is mentioned as a known result in PHRAGMÉN and LINDELÖF [1908].

The rotation trick used in 5.14 (and also in 9.10) can be seen in its most pristine form in 5.75(i) and is due to PAINLEVÉ [1887], p. 29. For an extension of 5.14

see exercise IV.134 of PÓLYA and SZEGÖ [1976] and for another proof SAKS [1928/29]. For improvements of 5.16(ii) see SEIDEL and LITTAUER [1932], GEHRING and LOHWATER [1958] and SINDALOVSKIĬ [1987]. In the latter paper the boundedness of f is replaced by the hypothesis that $|f|$ be integrable over S. Gehring and Lohwater prove that if $\operatorname{Re} f$ is bounded and f is continuous in $\bar{S}\backslash\{0\}$, holomorphic in S, $\lim_{r\downarrow 0} \operatorname{Re} f\left(re^{i\theta_1}\right) = a_1$ and $\lim_{r\downarrow 0} \operatorname{Im} f\left(re^{i\theta_2}\right) = a_2$, then for any open sector S^* with $\overline{S^*} \subset S \cup \{0\}$ the limit of $f(z)$ exists and equals $a_1 + ia_2$ for $z \to 0$, $z \in S^*$.

The proof of 5.17 in the text is adapted from H. A. SCHWARZ (pp. 181–185, vol. II, *op. cit.*, where earlier work of Carl Neumann is mentioned), to whom is also due the first rigorous proof (that in the text—see pp. 186–188, and 360–361, *op. cit.*) of the important continuity of $\operatorname{Re} f$ on $\overline{\mathbb{D}}$ in 5.20. See §34 of BURKHARDT [1914] for early history. On 5.18 see also BOGGIO [1911–12].

Geometric proofs of 5.20 appear in BÔCHER [1906a], DARBOUX [1910], PERKINS [1928] and DUFFIN [1957]. A detailed analysis of all the geometry behind the Poisson formula and the Dirichlet problem in $\mathbb{D}$, including Schwarz' own geometric interpretation of 5.17, will be found in NEEDHAM [1994]; see also VIGNATI [1992]. A clever but "natural" analytic proof is offered by VILLAT [1911]. FEJÉR [1904] solved the Dirichlet problem in the disk as follows: His *Hauptsatz* of this paper affirms that the arithmetic means of the partial sums of the Fourier series of any continuous, 2π-periodic function h converge uniformly to it. These sums have natural harmonic extensions in $\mathbb{D}$ and they converge there too, by the maximum principle, to the desired harmonic extension f. This method is quite versatile and can be adapted to an annulus, as was done by GRONWALL [1912/13b] and PICONE [1926a]. See also S. N. BERNSTEIN [1909] and the notes to Chapter XI. Finally, one can get the Poisson formula (and indeed the Cauchy integral formula too) from Green's Theorem. For this see OSGOOD [1928], pp. 658–666.

The proof of 5.20 really shows more than was claimed: If h is merely Lebesgue integrable on $\partial\mathbb{D}$, then $\operatorname{Re} f$ is continuous at every point of $\partial\mathbb{D}$ where h is. In particular, for any finite $S \subset \partial\mathbb{D}$, if h is bounded and continuous on $\partial\mathbb{D} \setminus S$ (so Riemann integration suffices), then $\operatorname{Re} f$ is continuous on $\overline{\mathbb{D}} \setminus S$. (Cf. 8.35(i).) This fact is very useful in constructing harmonic-majorants—one takes for h the indicator function of an appropriate set. The use of such a function occurs implicitly in the hints to 5.45(i), but its explicit use was circumvented with the ϕ constructed there. A short proof of 5.20 in this generality, using an elementary version of Lebesgue's Dominated Convergence Theorem for the Riemann integral, is given by MINDA [1990].

5.22 and 5.97(i) are due to F. RIESZ [1922/23b] and [1926] (p. 340), respectively.

There is a remarkable holomorphic-function version of 5.23(i)(b), due to T. RADÓ [1924]: If f is continuous in the open set U and holomorphic in $U \setminus f^{-1}(0)$, then f is holomorphic in U. We prove this in 10.10 and there is an elegant proof in KAUFMAN [1967] (somewhat elaborated in ABIAN [1990]). The harmonic-function analog of 5.23(i)(b) is false, as the function $f(z) := |\operatorname{Re} z|$ in $\mathbb{C}$ shows. But if $f\colon U \mapsto \mathbb{R}$ is continuous, has continuous partial derivatives and is harmonic in $U \setminus f^{-1}(0)$, then it is harmonic in U (GREEN [1960]). For a further strengthening of 5.23 see READE [1943] and RUDIN [1951]. That harmonic functions are characterized by equality in 5.23(ii) goes back to ZAREMBA [1905] (see also E. HOPF [1930]) and PLANCHEREL [1914]. In fact, if f is twice continuously differentiable, then the limit in (ii) exists and coincides with $\frac{1}{4}\triangle f(a)$; see BLASCHKE [1916], [1918] (or 7.15 of FEYEL and PRADELLE [1973]) and glance ahead at 9.29, 9.31 below. For 5.23(iii) and (iv) see KOEBE [1906b], LEVI [1909], TONELLI [1909], VOLTERRA [1909], VITALI [1912], LEBESGUE [1920], KELLOGG [1934], GLOBEVNIK [1990], and GLOBEVNIK and RUDIN [1988]. On the issue of one-radius and two-radius theorems see ZALCMAN [1972], [1980], FENTON [1976], [1979], BERENSTEIN and GAY [1986], the beautiful work of HANSEN and NADIRASHVILI [1993], [1994], HANSEN [1996] and the numerous references given later in discussing Morera's theorem, which is a kind of holomorphic-function analog of 5.23(iv). An attractive and thorough survey of mean-value theorems and harmonic functions is presented by NETUKA and VESELÝ [1994]. There is an obvious areal mean-value property (see Chapter XVIII) which also characterizes harmonicity. See NICOLESCO [1945]. These theorems also have interesting connections with probability theory—see, e.g., HANSEN and NADIRASHVILI, *op. cit.* and the references there for more on this.

In a sense, the circumferential and the areal mean-value properties also characterize circles and disks, respectively. See GREEN [1956], KURAN [1972] and KOSMODEM'YANSKI [1981]. In Chapter XVIII Kuran's beautiful result will be presented, together with a complete survey of the relevant literature, which can also be found in NETUKA and VESELÝ [1994].

5.26(i) is due to LITTLEWOOD [1925] and F. RIESZ [1925] (see also PRIVALOV [1935]). That the integral is an increasing function of r was proved by F. RIESZ [1922/23b]. The further corollary (upon taking, $f := |F|^p$) that $\int_0^{2\pi} |F(re^{i\theta})|^p \, d\theta$ is an increasing function of r for $F \in H(\mathbb{D})$, $p > 0$ is due to HARDY [1915]. See also E. LANDAU [1917], HILLE [1916–17], JULIA [1927d], MONTEL [1928] (p. 41), PRAWITZ [1927–1928] and A. E. TAYLOR [1950b]; a proof based on 5.72 is described in exercise III.310 of PÓLYA and SZEGÖ [1972]. For more references on this matter see T3, pp. 118, 225 of the bibliography BERNARDI [1982].

If u is a continuous real-valued function on $[0, 1]$ and ϕ is convex, increasing (whence continuous) function on $\mathbb{R}$, then a well known inequality due to JENSEN [1906] asserts that $\phi(\int_0^1 u) \leq \int_0^1 (\phi \circ u)$. It follows at once that $\phi \circ u$ is subharmonic if u is. This subsumes parts of 5.26. MONTEL [1928] shows that if $f > 0$, then f^p is subharmonic for all positive p if and only if $\log \circ f$ is subharmonic.

5.26(iv) should be compared with the following fact, for which RUDIN [1993] offers an elementary, purely algebraic proof: If P, Q are non-constant polynomials and the sets $P^{-1}(c), Q^{-1}(c)$ coincide for two different $c \in \mathbb{C}$, then $P = Q$.

5.27 is due to LINDELÖF [1909] (p. 10) (but see footnote 5 in CARATHÉODORY [1914] and pp. 190–191 of KOEBE [1915]). For an extension of 5.28 see p. 16 of LINDELÖF [1909], and for a proof via normal families, lemma 1, p. 174 of MANDELBROJT [1929]. For other proofs of 5.29 see OU [1957], lemma 1 of GEHRING and LOHWATER [1958] and also theorem 7 of GOLDBERG [1962]. For the holomorphic function version of it see 5.56 and 5.57 (also 12.33–12.39).

The particular series in 5.30 is called the *Taylor series* of f at z_0. (See PRINGSHEIM [1900].) The all-important fact 5.30 (due to Cauchy, 1831) constitutes the essential difference between differentiable (complex-valued) functions on an open subset of $\mathbb{R}$ and those on an open subset of $\mathbb{C}$. The demand that the difference quotients have a limit is so much more restrictive when the approach can be made through a disk than merely along one line, that the existence of such limits entails the existence of all higher derivatives as well as the convergence to the function of its Taylor series. No such conclusions hold on the line. Most readers know differentiable functions on $\mathbb{R}$ whose derivatives are not differentiable. (A primitive of Weierstrass' everywhere continuous and nowhere differentiable function on $\mathbb{R}$—see BANACH [1931] and MAZURKIEWICZ [1931]—even provides a function which is everywhere differentiable but whose derivative is nowhere differentiable.) Furthermore there exist infinitely differentiable functions on $(0, 1)$ (a plenitude of them—see MORGENSTERN [1954], SALZMANN and ZELLER [1955], DARST [1973], and CATER [1984]) which do not coincide with their Taylor series in any subinterval of $(0, 1)$, and given *any* $c_n \in \mathbb{C}$, there is an infinitely differentiable function f on $\mathbb{R}$ having $f^{(n)}(0) = c_n$ for all n—see REMMERT [1998], p. 300, or **MR** 82h:26005.

The derivation of 5.30 (thence 5.33) from the Cauchy–Schwarz formula 5.18($*$), "the road less travelled by" of the poet Robert Frost, is the path followed by H. A. SCHWARZ on p. 189, vol. II of his *Gesammelte Mathematische Abhandlungen*. See also footnotes 889, 888 of BURKHARDT [1914].

5.30 can be secured without the intervention of an integral representation; see chapter VI of G. T. WHYBURN [1964], and CONNELL and PORCELLI [1962]. The key is to get (without integration) a maximum-modulus theorem in which the function is continuous everywhere and complex-differentiable at all but possibly one point, and then to use this to prove 5.34(i). See CONNELL [1965], G. T. WHYBURN [1968] and the Appendix of BEARDON [1979]. From this perspective the maximum-modulus principle rather than the Cauchy integral formula is the essential reason why complex-differentiable functions are of power-series type. And in a sense the maximum-modulus property entails differentiability itself! See RUDIN [1953] or [1987], p. 262.

One can also deduce 5.30, under the additional assumption that f' is locally bounded, directly from the Cauchy-Riemann equation by Fourier analysis. See W. H. YOUNG [1911]. Similarly, 5.30 and 11.2, under the additional assumption f' is continuous, were derived by Fourier analysis in HARNACK [1883] and LELAND [1971].

I took 5.36(i) from HOFFMAN [1962]; see also 7.3 in FEYEL and PRADELLE [1973]. A generalization of 5.36(ii) in which $0, 1 \notin f(\Omega)$ but $|f| \geq 1$ is required only on K, is proved in I. N. BAKER [1981].

The Three Circles Theorem (5.38(ii)) [so named by E. Landau] occurs without proof in HADAMARD [1896], but in a footnote on p. 49 of HADAMARD [1912] he supplied a proof. See footnote 13 of OSTROWSKI [1922] and ROBINSON [1944] for history. For an interesting alternative proof and more history see COLLINGWOOD [1932]. See also p. 14 of F. and R. NEVANLINNA [1922], CARLSON [1938], ROBINSON [1943], SCHNEIDER [1967] and §3.4 of AHLFORS [1973] for sharper versions. MILNER [1968] addresses a gap in many textbook treatments—the use of $z^\lambda F(z)$ for non-integer λ (such a gap could also be closed using 5.26(ii)). Highly technical extensions of the theorem will be found in CLUNIE [1986] and an abstract version on pp. 20–21 of FUCHSSTEINER and LUSKY [1981]. There is even a "three ellipses" theorem [in Chinese—see **Zbl** 599#30014]. For the "three lines" theorem in 5.38(i) see DOETSCH [1920], WALTHER [1921], HARDY and ROGOSINSKI [1946] and BONSALL [1949]. The boundedness requirement here can be weakened with Phragmén-Lindelöf techniques (see, e.g., theorem 12.9 of RUDIN [1987]) but cannot be entirely abandoned, as shown by the example $F(z) := \exp(\exp z)$, $z \in S := \mathbb{R}\times(-\pi/2, \pi/2)$. It is interesting to note that 5.38(i) is a quantitative extension (for strip regions) of the result 5.13. For a proof of 5.38(i) and (ii) involving harmonic and subharmonic functions (cf. the following remarks) see MONTEL [1928] (pp. 45, 55), (HERVÉ [1970] (pp. 11, 12), BEAR [1974] and RANSFORD [1995]. For a beautiful application of 5.38(i) to Fourier series (the Hausdorff-Young inequality, generalizing Parseval's equality), see RUDIN [1987]. Modulo problems with 0, the Three Circles Theorem asserts that $\log M(r) := \max_{|z|=r} \log |f(z)|$ is a convex function of $\log r$, $r \in (0, 1)$. In this form it generalizes: if $h\colon A(R_1, R_2) \mapsto \mathbb{R}$ is subharmonic, then $M(r) := \max_{|z|=r} h(z)$ is a convex function of $\log r$ for $r \in (R_1, R_2)$. The proof is effected by comparing $h(z)$ to

$\alpha \log |z| + \beta$ via 5.22(i), for appropriate α and β. (Cf. the proof of 11.56(iii).) The case $h := \log |f|$ is 5.38(ii). This theorem leads to another proof and generalization of the Subharmonic Liouville Theorem 5.15; see RAMANKUTTY [1982].

The result 5.38(iii) is from OSTROWSKI [1922]. When applied to uniformly convergent sequences of holomorphic functions [as was done in (iv)], it implies that the "rates of convergence on any two compacta with interiors are comparable". (See also MILLOUX [1924] and MANDELBROJT [1929].) Even more: convergence in one region induces convergence in another, and so the phenomenon in 5.38(iv) [and that in 5.45(iii) and 5.74 too] is called "induced convergence". Many more striking examples of it occur in Chapters VII and XII. The harmonic function version of 5.38(iii) was proved by KOREVAAR and MYERS [1994].

The attribution of 5.39(i) to MORERA [1886] is a bit gratuitous: what he really proved is 2.11; the further inference from existence of a primitive to the holomorphy of f itself was common knowledge by then. 2.11 seems too natural not to have been noted earlier, and this suspicion is reinforced by the fact that subsequent writers prove 2.11 without acknowledging any priority. See, e.g., DE LA VALLÉE POUSSIN [1892–93], OSGOOD [1895/96] and POMPEIU [1905a]. All these writers point out how easily Weierstrass' theorem 5.44(i) follows from this result and in addition de la Vallée Poussin shows how, using Cauchy estimates, 5.44(ii) can be deduced from 2.11 too. Weierstrass proved 5.44(ii) in 1841, as in 3.35, and the proof in the text from Cauchy's Integral Formula was given by PAINLEVÉ [1887]. The stronger form of 2.11 in 5.39(ii), involving only rectangles, seems to have first occurred in the second edition (1912) of the book by OSGOOD [1928] and under weaker hypotheses on f in RADEMACHER [1919b]. Much smaller classes of loops suffice in the role of the triangles and rectangles of 5.39 to imply holomorphy of f. See WOLFF [1922-1924], RIDDER [1930a], BOCHNER [1955a], DELSARTE [1958], [1961], ROYDEN [1962], SMITH [1972], ZALCMAN [1972] (especially its notes and bibliography), [1974], [1980]; BROWN, SCHREIBER and TAYLOR [1973], BERENSTEIN and TAYLOR [1977], AGRANOVSKIĬ [1978], BERENSTEIN and GAY [1986], and GLOBEVNIK [1989], [1990]. See C. MÜLLER [1954] and G. SPRINGER [1957], exercise III.171 of PÓLYA and SZEGÖ [1972] and p. 333 of RICHARDS [1970] for proofs that circles suffice. The proof consists of convolving f with a smooth "approximate identity" or "mollifier" (cf. 8.29(ii)) to get smooth f_n which converge locally uniformly to f and which (by Fubini) inherit the property of integrating to 0 over circles. Each f_n is therefore holomorphic by 2.19(ix) or by the corresponding arguments for disks (in the remarks preceding 3.17). Smith and Zalcman show that if the ratio of two positive numbers r_1, r_2 does not coincide with that of any two zeros of the Bessel function J_1, then any continuous function f in $\mathbb{C}$ that satisfies $\int_{C(a,r_1)} f = \int_{C(a,r_2)} f = 0$ for all $a \in \mathbb{C}$, is necessarily entire. VOLCHKOV [1994], [1993] gets definitive 1-radius and 2-radius results in $\overline{\mathbb{D}}$. Wolff, Müller, and Springer show that the continuous function f in U is holomorphic if $(\text{area}\, R)^{-1} \int_{\partial R} f$ converges to 0 whenever rectangles or disks R in U decrease to a point in U. Further generalizations of Morera's theorem and a survey of others will be found in GLOBEVNIK and QUINTO [1996].

A different proof of 5.40 is offered in 13.29. For a survey of the vast field of "removable sets" touched on in 5.40, see **MR** 3429163.

5.41(i) appears in Riemann's dissertation [pp. 23–24 of his *Werke*, 1953 ed.] but was known earlier to Weierstrass; see ULLRICH [1994c] for the tangled history and faulty proofs. HÖLDER [1882] did two integrations to get a holomorphic function Φ with $\Phi'' = f$ and then differentiated the Taylor series of Φ. Independently, CHESSIN [1895–96], OSGOOD [1895/96] and SCHOTTKY [1916] first treat the case where f is continuous at a, then reduce 5.41(i) to that case applied to a primitive of $h(z) := (z-a)f(z)$. Use of Cauchy's Integral Theorem to produce such a primitive (as in 2.11) required, before Goursat's intervention, not just the differentiability of h throughout U, but the continuity of h', and some writers overlooked this issue. E. LANDAU [1905–06] showed how the matter could have been resolved with pre-Goursat reasoning. The first two proofs in the text are from CURTISS [1905–06], taking full advantage of Goursat. BÔCHER [1906b] showed that $\int_{\partial R} f = 0$ for all rectangles R in U and then invoked Morera 5.39. I do not discuss Riemann's original proof (*q.v.*), because I do not understand it. However, in his 1861 lecture notes (NEUENSCHWANDER [1987]) he gave a proof similar to the third one in the text, that is, he used Cauchy's Integral Formula in the annulus $A(r,1)$ and let $r \downarrow 0$ to get (3). Weierstrass' proof (p. 63 of *Werke* I) is described in the Remarks following 11.7 below. The special case of 5.41(ii) in which h is bounded around a was proved in 1872 by H. A. SCHWARZ (see p. 209, vol. II of his *Gesammelte Mathematische Abhandlungen*). BESICOVITCH [1931] is a significant extension of 5.42. See also theorem 1.4 of GARNETT [1972] and the Notes to Chapter II.

For material on removable singularities of harmonic functions see HILBERT [1904] §10, ZAREMBA [1909], PICARD [1923], [1924], LEBESGUE [1923], STOŻEK [1926], RAYNOR [1926], PICONE [1926b], [1929], KELLOGG [1926a], NOAILLON [1927], ASCOLI [1928a], BRELOT [1932] (good survey) and the last chapter of CARLESON [1967]. Isolated singularities of subharmonic functions are treated in BRELOT [1934] and in chapter VII of T. RADÓ [1937].

5.46 (for E finite and u harmonic) and the proof given are due to ZAREMBA [1909]. There is a version in which $u = |f|$, $f \in H(U)$ and hypothesis (2) is replaced by f univalent near each point of E. See MINAMI [1937].

For the history of the important theorem 5.50 (which derives from anterior work of Carleman) see the notes in E. LANDAU [1930] and R. NEVANLINNA [1933]. Landau proves that the curve γ can be abandoned in favor of the weaker hypothesis

$$(*) \qquad \sup_{0\le r<1}\inf_{\theta\in\mathbb{R}}\left|f\left(re^{i\theta}\right)\right|\le\delta,$$

that is, on each circle centered at 0 there be at least one point at which $|f|$ does not exceed δ. Landau's clever proof of this does not use the harmonic-majorant technique. V. I. LEVIN [1934] replaced $\sup_{0\le r<1}$ in $(*)$ with $\sup_{\rho<r<R}$ and found a similar inequality as Milloux but with the constant $1/2\pi$ replaced by one dependent on ρ and R. Though he seems to rely on Landau's result, his proof is independent and different. See also R. NEVANLINNA [1978]. The question of the largest admissible constant in the role of $1/2\pi$ is also discussed in Landau's paper. Also, in the curve version of the theorem, we might (as in 5.48) relax $f\in H(\mathbb{D})$ to $f\in H(\mathbb{D}\setminus\gamma)$. [The inequalities about $|f|$ on γ become $\overline{\lim}\,|f|$ as γ is approached through $\mathbb{D}\setminus\gamma$.] In addition to Landau's and Nevanlinna's references, see MILLOUX [1930], [1932], AHLFORS [1932b], BEURLING [1933], MAITLAND [1939], HEINS [1945], C. J. LEWIS [1958], the treatment in CARATHÉODORY's book [1960] (pp. 107–109) and the definitive treatment in NEVANLINNA's book [1970] (pp. 102–113). MILLOUX [1948a] is an elegant qualitative version of Landau's result above with a very elementary *reductio ad absurdum* proof based on 7.5 or 7.6. FANG [1992] gets the inequality with the exponent $1-|z|^{1/\pi}$, which is larger than $(1-|z|)/2\pi$.

The method of harmonic-majorization is very powerful and of broad applicability. It goes back to Lindelöf, Carleman, Ostrowski and the Nevanlinnas. The general setting, of which the examples in the text are special cases, is this: a region Ω and a portion B of $\partial\Omega$ are given. For any $f\in H(\Omega)$ bounded by M and satisfying $\overline{\lim}\,|f|\le m\,(\le M)$ on B we have $|f|\le m^\omega M^{1-\omega}$, where ω is the harmonic function with values in $[0,1]$ which approaches 1 (resp.,0) as the variable tends to B (resp., $\partial\Omega\backslash B$). This is known as the *two-constants theorem* of Ostrowski and F. and R. NEVANLINNA [1922]. (See also R. NEVANLINNA [1923], SAKS [1928/29], and SCHNEIDER [1967].) The problem is to ensure that such an ω exists and to find useful bounds for it in Ω. The function ω is traditionally called (after R. Nevanlinna) the *harmonic measure* of B at z with respect to Ω, and one writes $\omega(z,B,\Omega)$ for its value at $z\in\Omega$. This language derives from the fact that for each fixed z, $\omega(z,B,\Omega)$ is countably-additive over disjoint (Borel) sets B, i.e., is a measure. (For the geometric significance when $\Omega=\mathbb{D}$, see §158 of CARATHÉODORY [1958], and when $\Omega=\mathbb{H}$, see 3.45. This function is intimately involved in the mathematical theory of brownian motion.) For more applications of the method see R. NEVANLINNA [1935], BRELOT [1939], §18.3 of HILLE [1962], FUCHS [1967], R. NEVANLINNA [1970],

AHLFORS [1973], and SCHARM [1987]; the definitive treatment is GARNETT and MARSHALL [2005].

On 5.62 see BURDICK and LESLEY [1975] and the historical notes on pp. 231–232 of REMMERT [1991a]. 5.62, also known as The Identity Theorem, is a much stronger version of the "permanence of functional identities," widely and uncritically used by eighteenth-century mathematicians: an identity, like $\sin^2 x + \cos^2 x = 1$, that is valid for all $x \in \mathbb{R}$, is necessarily valid for all $x \in \mathbb{C}$.

Some subsets of $\mathbb{C}$ are uniqueness sets in a very strong sense. For example, $S := \{1/n : n \in \mathbb{N}\}$ has the remarkable property that $f = g$ for any entire functions f, g which satisfy $f(S) = g(S)$. Many such sets are constructed, and by elementary means, in DIAMOND, POMERANCE and RUBEL [1981], which is also the source of 5.67(iii). A somewhat different kind of uniqueness has been the subject of research since about 1990: There exists a subset A of $\mathbb{C}$ having 7 elements and the remarkable property that $f - a$ and $g - a$ possessing the same zeros, each with the same multiplicity, for every $a \in A$ ensures the equality of the entire functions f and g. See the literature survey in **MR** 2001e:30051.

5.66(i) is from ROSENBLATT [1945]; the proof given was suggested by S. Saeki. The folkloric result 5.66(ii) may be found in ABIAN [1984] and can also be proved via a reflection argument. 5.67(i) is from p. 66, vol. II of H. A. SCHWARZ' *Gesammelte Mathematische Abhandlungen*, and 5.67(v) appears as exercises in FUCHS [1967] and PALKA [1991]. ARMITAGE and KURAN [1985] proved the following relative of 5.67(v): For a proper subregion Ω of $\mathbb{R}^n$, the function $d(z) := \operatorname{dist}(z, \partial\Omega)$ is subharmonic in Ω if and only if $\mathbb{R}^n \setminus \Omega$ is convex. Using this PARKER [1987] gave a proof (different from that in the text) for the $\mathbb{R}^n$-version of 5.67(v).

In contrast to 5.70 the entire function $f(z) := z + e^z$ satisfies $\lim_{r\to\infty} \left|f\left(re^{i\theta}\right)\right| = \infty$ for each fixed θ, yet f is not a polynomial. (See VON KOCH [1903–04].) 5.72 is exercise III.299 and III.302 of PÓLYA and SZEGÖ [1972]; for harmonic function versions see PÓLYA [1930].

5.73(vii) is from PEANO [1882], BENDIXSON [1886/87], H. LAURENT [1902], JOURDAIN [1905b]. Related series representations for regions other than disks are developed by ANGHELUTZA [1929], MORANT [1938a], [1938b] and DEMIN [1946] (the latter two for strips). A "recovery" formula like that in 5.73(vii) in which the points z_j are merely required not to approach the boundary too rapidly ($\sum(r - |z_j|) = \infty$) is proved by TOTIK [1984] and a quite different one in SCHIFF and WALKER [1987]. A weaker version of 5.73 appears in MÉRAY [1884]. A different treatment of the problem is given by FRANKLIN [1927]. For related matter see ABIAN [1974] and SMITH [1975]. A more general problem (see, e.g., IZUMI [1929]), that of "interpolation", is to find necessary and sufficient conditions on complex numbers w_n which will insure the existence of a holomorphic function f satisfying $f(z_n) = w_n$ for all $n \geq 1$ and, as in 5.73, to find formulas for the values of $f(z)$ in terms of z, z_n and the w_n. The literature on this and allied problems in which supplemental requirements are imposed on the interpolating function f (e.g., that it be bounded, or univalent) is too enormous to attempt to survey, but the monographs J. L. WALSH [1969] and PARTINGTON [1997] may be mentioned. Isolated special cases will be treated from time to time in the sequel. (See e.g., 6.4(vi), 6.9(iii), 7.36, 11.38 and Chapter XIV, where VALIRON [1925c] will be discussed.) Note however that on a subset with *no* limit points in the region, arbitrary interpolation is always possible (Mittag-Leffler's theorem 11.41). The idea for 5.74 came from BOWEN and MACINTYRE [1954].

5.75 has many generalizations. Here is one (FATOU [1906], p. 395): If f is bounded and holomorphic in $\mathbb{D}$ and $\lim_{r\uparrow 1} f\bigl(re^{i\theta}\bigr) = 0$ for each θ in $[\theta_0, \theta_1]$, then $f = \mathbf{0}$. See also Chapter XIII, WOLF [1940/1941] and RIDDER [1967]. The ultimate theorem of this type is due to Lusin and Privalov: If f is holomorphic in $\mathbb{D}$ and for some Lebesgue measurable subset S of positive measure in $[0, 2\pi)$ and for each $\theta \in S$ we have $f(z) \to 0$ as z converges to $e^{i\theta}$ within any angle (less than π) in $\mathbb{D}$ whose vertex is $e^{i\theta}$ (so-called *non-tangential* approach), then $f = \mathbf{0}$. Moreover there is such a set S and an f holomorphic in $\mathbb{D}$ such that S has measure 2π, f is not the zero function and yet $\lim_{r\uparrow 1} f\bigl(re^{i\theta}\bigr) = 0$ for each $\theta \in S$. For the (complicated) details see chapter 4 of PRIVALOV [1956], pp. 222 ff. or the simpler construction via Runge's theorem in SCHNEIDER [1980]. For the last quoted result see also WOLFF [1928] and BURGERS [1929]. Schneider (*op. cit.*) also constructs $\mathbf{0} \neq f \in H(\mathbb{D})$ and for each θ a curve P_θ in $\mathbb{D}$ and asymptotic to the radius $\bigl\{re^{i\theta} : 0 \leq r < 1\bigr\}$ such that $f(z) \to 0$ as $z \in P_\theta \to e^{i\theta}$. In 9.10 we will see another kind of generalization of 5.75.

There are several explicit formulas which recover the function values inside $\mathbb{D}$ from those on a *part* of the boundary. For example, each non-degenerate arc A of $\mathbb{T}$ determines a "quenching" function $\phi = \phi_A \in H(\mathbb{D})$ (cf. 5.45 and the above remarks about harmonic measure) such that

$$f(z) = \lim_{k\to\infty} \frac{1}{2\pi i} \int_A f(\xi) \left[\frac{\phi(\xi)}{\phi(z)}\right]^k \frac{d\xi}{\xi - z} \qquad \forall z \in \mathbb{D}$$

whenever f is bounded and continuous on $\mathbb{D} \cup A$ and holomorphic in $\mathbb{D}$. See CARLEMAN [1926], GOLUZIN and KRYLOW [1933], ZIN [1953], PICONE [1954], PATIL [1972], SCHEMPP [1980], ZAYED [1983] and, especially, AIZENBERG [1993]; other "recovery" techniques in CIMA and STRESSIN [1996]. The idea behind this limit and the term "quenching" function is illuminated in the Remark following 11.1 below.

For an extension of 5.76 to subharmonic functions, see for instance SAKS [1928/29] and BRELOT [1939b], p. 469.

5.77 remains true for real-differentiable maps with zero-free Jacobians; see 4.60(ii) or J. A. BAKER [1991]. The maximum-modulus principle underlies the proof of 5.77, but is obviously a trivial corollary of 5.77. However, 5.77 leads to stronger forms of that principle. For example, if f, g are continuous on $\overline{\mathbb{D}}$ and holomorphic in $\mathbb{D}$, then the maximum of $|f(z) - g(w)|$ over $\overline{\mathbb{D}} \times \overline{\mathbb{D}}$ is attained on $\mathbb{T} \times \mathbb{T}$, and the diameter of $f(\overline{\mathbb{D}})$ equals that of $f(\mathbb{T})$.

For an interesting direct proof of the fact (5.78) that univalent functions have zero-free derivatives, see POMPEIU [1933]. The converse of 5.78, i.e., f' zero-free implies f univalent, is obviously false. (Take $f(z) := e^z$.) When U is a region, U or $f(U)$ is simply-connected, and f has the additional property that $f^{-1}(K)$ is compact for each compact subset K of $f(U)$, then the non-vanishing of f' does imply the univalence of f. See 10.12 and the notes on it, and 5.17 in FEYEL and PRADELLE [1973].

On 5.79 see GOURSAT [1903a] and KRAFFT [1932].

The question of a holomorphic function version of the Mean-Value Theorem in real analysis would seem to be settled by 5.80(ii). But there is more on this theme: POMPEIU [1905b], D. R. CURTISS [1906–07]; pp. 1021, 1022, vol. 59 of the *Jahrbuch*, HUBER [1948], SAMUELSSON [1973], JOHNSTON [1983], M. MARDEN [1985], GEVIRTZ [1986], ROSENHOLTZ [1991], JIANG [1991], EVARD and JAFARI [1992].

For further elaboration of 5.83 and 5.88 (versions of the Weierstrass *Vorbereitungssatz*) see p. 267 of HILLE [1959], chapter VII, §6 of BIEBERBACH [1934], or §4.4.3 of MITRINOVIĆ and KEČKIĆ [1984]. A version of 5.88 for continuous functions is proved by T. RADÓ [1936]. Cf. our 4.59(iii).

The popular designation "argument principle" for 5.86 is readily appreciated when one realizes that $\mathrm{Ind}_{f\circ\gamma}(w)$ is the "number of times $f(z)$ winds around the point w when z runs (once, counterclockwise) over a circle centered at 0", and 2π times it is the "increase in the argument of $f(z) - w$ as $z \ldots$". 9.18(i) and (ii) are two nice applications.

5.87(ii) is a version of Rouché's Theorem, to be discussed in detail in Chapter VIII. (See 8.18 and 11.49.) This theorem can be elevated to one of the main workhorses of the theory and used to prove, *inter alia* 5.77, the fact that univalent functions have zero-free derivatives (HERVÉ [1982], p. 104), and that non-constant limits of univalent or zero-free functions inherit these properties (7.11, 7.12, 8.19). The proof suggested is from GLICKSBERG [1976]; there it is pointed out (cf. also MONTEL [1932b] and **MR** 2001f:30038) that the weakest hypothesis about f and g on $\partial\mathbb{D}$ under which this technique works is (in view of

4.28) that 0 lie in the unbounded component of $\mathbb{C}\setminus\{f(z)/g(z) : z \in \partial\mathbb{D}\}$. This proof yields a geometrically very satisfying version of the theorem for Jordan regions once the Argument Principle 9.15, generalizing 5.86, is available. But such a result is subsumed by 8.18, in which no special assumptions about the boundary occur. The reader with more background may be titillated by the exotic proof in VAN DULST [1971] or that in ABIAN [1977b]. The strong form of Rouché's theorem in 5.87(ii) has an interesting converse; see CHALLENER and RUBEL [1982], and especially Theorem 18 in MORTINI and RUPP [2014].

5.87(iii) was taken from TSARPALIAS [1989].

The residue theorem can also be made the basis of 5.91 and 5.92. For this approach and a slight generalization see p. 87–89 of MARKUSCHEWITSCH [1965], vol. II, or §4.4.2 of MITRINOVIĆ and KEČKIĆ [1984]. 5.92(iii) is from HAMEL [1940]. The names of H. Bürmann and J. L. Lagrange are associated with these expansions and the reader will find an exhaustive account in §1.9 and §2.4 of HENRICI [1974], and some history in OSGOOD [1901], §15. A recent novel treatment is GROSSMAN [2005].

For the history of 5.93 (Cauchy 1831, Weierstrass 1861, Poincaré 1879) see p. 27 of LINDELÖF [1905].

5.95 is a manifestation of the much more general fact that "the zeros of a polynomial are continuous functions of its coefficients." An elementary interpretation of this statement can be found in 8.19 and in UHERKA and SERGOTT [1977], and a topologically more sophisticated one in BHATIA and MUKHERJEA [1983], HARRIS and MARTIN [1987], CUCKER and CORBALAN [1989] and ĆURGUS and MASCIONI [2006]. Other references on this theme are BONNET [1871], OSTROWSKI [1939], [1940], IGLISCH [1942], SPECHT [1958], BURGHELEA [1977] and MURAKAMI [1993]. For the sense in which the zeros of $z^n + c_{n-1}z^{n-1} + \cdots + c_0$ are *holomorphic* functions of $(c_0, \ldots, c_{n-1}) \in \mathbb{C}^n$, see BRILLINGER [1966].

In 5.97(ii) much larger exceptional sets can be allowed and a restricted mean-value property substituted for the superharmonicity hypothesis; the constancy of the function still follows. See W. HANSEN [1998] and [2008]. The theme of 5.98 is explored in great generality in **MR** 87i:26027.

Chapter VI
Schwarz' Lemma and its Many Applications

§ 1 Schwarz' Lemma and the Conformal Automorphisms of Disks

This chapter is fairly tightly unified around the following simple result:

Theorem 6.1 (Schwarz' Lemma) *Let f be holomorphic and bounded by* 1 *in* $\mathbb{D}$ *and* $f(0) = 0$. *Then*

(i) $|f'(0)| \le 1$

(ii) $|f(z)| \le |z| \qquad \forall z \in \mathbb{D} \qquad$ (cf. 6.7(ii))

and equality in (i) *or in* (ii) *for some non-zero z occurs if and only if*

(iii) $f(z) \equiv uz$ *for some* $u \in \mathbb{T}$.

Proof: Write the power series expansion about 0 of f in $\mathbb{D}$: with $c_n := f^{(n)}(0)/n!$

$$f(z) = \sum_{n=0}^{\infty} c_n z^n \qquad \forall z \in \mathbb{D}. \tag{1}$$

Set

$$g(z) := \sum_{n=1}^{\infty} c_n z^{n-1}, \quad z \in \mathbb{D}. \tag{2}$$

This is a holomorphic function in $\mathbb{D}$, which, since $c_0 = f(0) = 0$, satisfies

$$zg(z) = f(z) \qquad \forall z \in \mathbb{D}. \tag{3}$$

R. B. Burckel, *Classical Analysis in the Complex Plane*, Cornerstones,
https://doi.org/10.1007/978-1-0716-1965-0_7

If $z \in \mathbb{D}$ and we pick $1 > r > |z|$, then by the Maximum-Modulus Principle

$$|g(z)| \le \sup_{|w|=r} |g(w)| \overset{(3)}{=} \sup_{|w|=r} \frac{|f(w)|}{r} \le \frac{1}{r}.$$

Let $r \uparrow 1$ to get

$$(4) \qquad |g(z)| \le 1 \qquad \forall z \in \mathbb{D}.$$

From (4) and (3) follows (ii). But also, taking $z = 0$ in (4) and noting that $g(0) = c_1 = f'(0)$, gives (i).

If equality holds in (i), then (4) shows that $|g|$ attains a maximum in $\mathbb{D}$ at 0, and so by the Maximum-Modulus Principle g is constant in $\mathbb{D}$. If u is that constant, then $|u| = |g(0)| = |f'(0)| = 1$ and (iii) holds: $f(z) \equiv zg(z) \equiv uz$.

If equality occurs in (ii) for some non-zero z, then $|g(z)| = 1$ and so (4) says that $|g|$ attains a maximum in $\mathbb{D}$ at z, making g constant and (iii) follows as before.

Scholia: Part (iii) is an extreme case. More generally this method provides a good quantitative estimate of how close $f(z)$ is to $f'(0)z$ in terms of how close $|f'(0)|$ is to 1, namely

$$6.1(\text{iv}) \qquad |f(z) - f'(0)z| \le \frac{1 - |f'(0)|^2}{1 - |f'(0)|\,|z|}|z|^2 \qquad \forall z \in \mathbb{D}.$$

In proving this we can assume, in view of (iii), that $f'(0) = g(0)$ lies in $\mathbb{D}$. Thanks to 2.6, $f^*(z) := [g(z) - g(0)]/[1 - \bar{g}(0)g(z)]$ is then a holomorphic self-map of $\mathbb{D}$ with $f^*(0) = 0$. The arguments in the preceding proof furnish a $g^* \in H(\mathbb{D})$ such that $|g^*| \le 1$ and $f^*(z) \equiv zg^*(z)$. Unwinding back to f, we get

$$f(z) - f'(0)z = \frac{\left(1 - |f'(0)|^2\right)z^2 g^*(z)}{1 + \overline{f'(0)}zg^*(z)} \qquad \forall z \in \mathbb{D}$$

from which (iv) follows, thanks to $|g^*| \le 1$. An easy and useful consequence of inequality (iv) is

$$6.1(\text{v}) \qquad \begin{cases} f_n \in H(\mathbb{D}),\ |f_n| \le 1,\ f_n(0) \to 0\ \&\ f_n'(0) \to u \in \mathbb{T} \Rightarrow \\ f_n(z) \to uz \textit{ locally uniformly in } \mathbb{D}. \end{cases}$$

A variant of (v) is used to establish the Cauchyness of the approximating sequence in the constructive proof of the Riemann mapping theorem 9.7.

Remarks: A less general version of 6.1 (for univalent f only) will be found on pp. 109–111 of H. A. SCHWARZ' *Gesammelte Mathematische Abhandlungen*, vol. II. The full scope of the result was recognized by CARATHÉODORY [1907], [1912a], who gave the above proof (based on a verbal suggestion of Erhard Schmidt) and instituted the designation. However, a less transparent variant of this proof occurs already in an 1884 paper of H. Poincaré. See footnote 427 of LICHTENSTEIN [1919] for history and KOEBE [1920] for refinements. Another elementary proof of Schmidt's establishes (i) and a weaker form of inequality (iv) [from which (iii) still follows] by exploiting Parseval's formula, 3.32(i), and the Cauchy-Schwarz inequality; CARATHÉODORY [1960] records the details in a footnote on p. 21. The reader will soon appreciate the keystone character of this deceptively simple result. Trivial modifications in the proof of 6.1 establish that if 0 is a zero of multiplicity at least m, then $|f(z)| \le |z|^m$. An analog of 6.1 with diam $f(\mathbb{D})$ replacing the maximum modulus of f is the other bookend of this chapter.

Theorem 6.2 *For each $a \in \mathbb{D} \setminus \{0\}$ define f_a by*

$$f_a(z) := \frac{z-a}{1-\bar{a}z}, \qquad z \in \mathbb{C} \setminus \{1/\bar{a}\},$$

and let f_0 denote the identity function on $\mathbb{C}$. Then for all $a \in \mathbb{D}$

(i) *f_a is one-to-one, continuous on $\overline{\mathbb{D}}$, holomorphic in $\mathbb{D}$ and $f_a(\mathbb{D}) = \mathbb{D}$, $f_a(\overline{\mathbb{D}}) = \overline{\mathbb{D}}$. Furthermore, the inverse map to f_a on $\overline{\mathbb{D}}$ is f_{-a}.*

Conversely:

(ii) *If $f \in \operatorname{Aut}(\mathbb{D})$, then $f = uf_a$ for unique $a \in \mathbb{D}$ and $u \in \mathbb{T}$. In particular, f extends to be holomorphic in a neighborhood of $\overline{\mathbb{D}}$ and to a homeomorphism of $\overline{\mathbb{D}}$ onto $\overline{\mathbb{D}}$.*

Proof: Part (i) is quite elementary and was completely attended to in 2.6. In proving (ii), consider first the case $f(0) = 0$. Schwarz' lemma may then be applied to each of the holomorphic functions f and f^{-1} to get

$$|f(z)| \le |z| = \left|f^{-1}\big(f(z)\big)\right| \le |f(z)| \qquad \forall z \in \mathbb{D}.$$

It follows that $|f(z)| = |z|$, so again from Schwarz, $f(z) \equiv uz$ for some unimodular constant u.

A little calculation reveals that in any representation $f = uf_a$, $a = f^{-1}(0)$ and $u = f'(0)/(1-|a|^2)$, which shows that u and a are uniquely determined by f, if they exist at all. To settle existence, given $f \in \operatorname{Aut}(\mathbb{D})$, set $a := f^{-1}(0)$ and form $F := f \circ f_{-a}$. By part (i), $F \in \operatorname{Aut}(\mathbb{D})$; moreover $F(0) = f\big(f_{-a}(0)\big) = f(a) = 0$. By what was shown above, there is $u \in \mathbb{T}$ such that $F(z) = uz$, whence $f(z) = F\big(f_{-a}^{-1}(z)\big) = F\big(f_a(z)\big) = uf_a(z)$, for all $z \in \mathbb{D}$.

Remark: Every $f \in \mathrm{Aut}\,(\mathbb{D})$ also has a more symmetric (but not quite unique) representation

$$f(z) = \frac{vz + b}{\bar{b}z + \bar{v}}, \qquad v \in \mathbb{T}, b \in \mathbb{D}.$$

[Take for v one of the square-roots of u, then $b := -va$.]

Exercise 6.3 (i) *If f is a holomorphic function in $\mathbb{D}$ which satisfies $|f| < 1$, then* (Schwarz-Pick Inequality)

$$\left|\frac{f(z) - f(w)}{1 - \bar{f}(w)f(z)}\right| \leq \left|\frac{z - w}{1 - \bar{w}z}\right| \qquad \forall z, w \in \mathbb{D}$$

and

$$\frac{|f'(z)|}{1 - |f(z)|^2} \leq \frac{1}{1 - |z|^2} \qquad \forall z \in \mathbb{D}.$$

Moreover, if $f \in \mathrm{Aut}\,(\mathbb{D})$, then equality prevails in both these inequalities.

Hints: Let

$$\phi(z) := \frac{z - w}{1 - \bar{w}z}, \quad \psi(\xi) := \frac{\xi - f(w)}{1 - \bar{f}(w)\xi}, \qquad z, \xi \in \mathbb{D}$$

and apply 6.1 to the composite $\psi \circ f \circ \phi^{-1}$. For the second inequality, divide the first one by $|z - w|$ and let $w \to z$.

(ii) *If F is a holomorphic function in $D(0, R)$ bounded by M, then*

$$\left|\frac{F(z) - F(w)}{z - w}\right| \leq \frac{2MR}{|R^2 - \bar{w}z|} \qquad \forall z, w \in D(0, R),\ z \neq w.$$

Hint: Take $f(z) := F(Rz)/M$ in (i). [*Notice that letting $R \to \infty$ in this inequality gives yet another proof of Liouville's Theorem* (5.34)(iii).]

(iii) *Show that* (i) *is equivalent to*

(a) $$\frac{|1 - \bar{f}(w)f(z)|^2}{1 - \bar{w}z|^2} \leq \frac{1 - |f(z)|^2}{1 - |z|^2} \cdot \frac{1 - |f(w)|^2}{1 - |w|^2} \qquad \forall z, w \in \mathbb{D}$$

and implies

(b) $$\left|\frac{f(z) - f(w)}{z - w}\right|^2 \leq \frac{1 - |f(z)|^2}{1 - |z|^2} \cdot \frac{1 - |f(w)|^2}{1 - |w|^2} \qquad \forall z, w \in \mathbb{D}, z \neq w.$$

Hints: After confirming the identity

$$(*) \qquad |1-\bar{u}v|^2 = (1-|u|^2)(1-|v|^2) + |u-v|^2 \qquad \forall u, v \in C,$$

square the first inequality in (i) and transform both numerators via (∗) to see (i) equivalent to

$$-\frac{(1-|f(z)|^2)(1-|f(w)|^2)}{|1-\bar{f}(w)f(z)|^2} \le -\frac{(1-|z|^2)(1-|w|^2)}{|1-\bar{w}z|^2}.$$

Multiplying by −1 and inverting the resulting positive quotients here shows this inequality equivalent to (a). Inequality (b) follows, since by (i) its left side is majorized by the left side of (a).

(iv) *Show that*

$$\left|\frac{a-b}{1-\bar{a}b}\right| \ge \frac{|a|-|b|}{1-|a||b|} \qquad \forall a, b \in \mathbb{D}.$$

Hint: As in the Hints to 2.6,

$$\begin{aligned}\left|\frac{a-b}{1-\bar{a}b}\right|^2 &= 1 - \frac{\left(1-|a|^2\right)\left(1-|b|^2\right)}{|1-\bar{a}b|^2} \ge 1 - \frac{\left(1-|a|^2\right)\left(1-|b|^2\right)}{\left(1-|a||b|\right)^2}\\ &= \frac{\left(|a|-|b|\right)^2}{\left(1-|a||b|\right)^2}.\end{aligned}$$

(v) (Lindelöf [1909], p. 11.) *Let f be holomorphic and bounded by* 1 *in* $\mathbb{D}$. *Then*

$$\frac{|f(0)|-|z|}{1-|f(0)||z|} \le |f(z)| \le \frac{|f(0)|+|z|}{1+|f(0)||z|} \qquad \forall z \in \mathbb{D}.$$

Hint: If $|f(z)| < 1$, then by (iv) and (i)

$$\frac{|f(0)|-|f(z)|}{1-|f(0)||f(z)|}, \frac{|f(z)|-|f(0)|}{1-|f(0)||f(z)|} \le \left|\frac{f(z)-f(0)}{1-\bar{f}(0)f(z)}\right| \le |z|.$$

Solve the extreme inequalities algebraically for $|f(z)|$.

(vi) *Suppose that equality holds for some $z \ne w$ in the first inequality of* (i) *or for some $z \ne 0$ in either* (v) *or the second inequality of* (i). *Show that* $f \in \text{Aut}(\mathbb{D})$.

(vii) (Invariant formulation of Schwarz' lemma, after PICK [1916]). *Show that* $d(z,w) := |z-w||1-\overline{w}z|^{-1}$ *defines a metric in* $\mathbb{D}$ *which generates the usual Euclidean topology. Interpret* (i) *and* (vi) *as saying that every holomorphic* $f\colon \mathbb{D} \mapsto \mathbb{D}$ *is d-nonexpansive, and is d-isometric if and only if it is a conformal automorphism of* $\mathbb{D}$. See also 6.5(iv).

Hints: From (v) we get

$$d(z,w) \le \frac{d(z,0)+d(w,0)}{1+d(z,0)d(w,0)} \le d(z,0)+d(0,w) \qquad \forall z,w \in \mathbb{D}.$$

Given $a \in \mathbb{D}$, apply this with $f_a(z)$, $f_a(w)$ in the roles of z, w and use the fact (i) that f_a^{-1} and f_a are d-isometric to infer that

$$d(z,w) \le d(z,a)+d(a,w) \qquad z,w,a \in \mathbb{D}.$$

In 6.5 we will see that the open d-balls are just the Euclidean open disks lying $\mathbb{D}$.

(viii) $F:\mathbb{D} \mapsto \mathbb{C}$ *maps* 0 *to* 0 *and for any* $a,b \in \mathbb{D}$ *there is a holomorphic* $f:\mathbb{D} \mapsto \mathbb{D}$ *mapping* 0 *to* 0 *and satisfying* $f(a) = F(a)$, $f(b) = F(b)$. *Use* (i) *to show that* F *is differentiable at* 0.

Hints: (Don Marshall)**:** Consider any $a,b \in D(0,1/2) \setminus \{0\}$. The proof of 6.1 shows that there is a holomorphic $g:\mathbb{D} \mapsto \mathbb{D}$ such that $f(z) = zg(z)$. From (i) [5.34(ii) could serve as well] follows

$$\left|\frac{F(a)}{a} - \frac{F(b)}{b}\right| = |g(a)-g(b)| \le |1-\bar{g}(a)g(b)|\left|\frac{a-b}{1-\bar{a}b}\right| \le 2\left|\frac{a-b}{1-\bar{a}b}\right| \le \frac{8}{3}|a-b|.$$

Consequently, if $\{z_n\}$ is *any* sequence in $\mathbb{D}\setminus\{0\}$ that converges to 0, the sequence $\{F(z_n)/z_n\}$ of difference quotients is Cauchy.

(ix) (A three-point characterization of holomorphy) $f:\mathbb{D} \mapsto \mathbb{C}$ *has the property that for any three points in* $\mathbb{D}$ *there is a holomorphic self-map of* $\mathbb{D}$ *that agrees with* f *at these points. Infer from* (viii) *that* f *is holomorphic in* $\mathbb{D}$.

Hints: Note: $f(\mathbb{D}) \subset \mathbb{D}$. Given $z_0 \in \mathbb{D}$, 6.2 furnishes conformal automorphisms φ, ψ of $\mathbb{D}$ satisfying $\varphi(0) = z_0$ and $\psi(f(z_0)) = 0$. Clearly $F := \psi \circ f \circ \varphi :$ $\mathbb{D} \mapsto \mathbb{D}$ inherits from f the three-point property, and $F(0) = 0$. Consequently, by (viii), F is differentiable at 0, whence $f = \psi^{-1} \circ F \circ \varphi^{-1}$ is differentiable at $\varphi(0) = z_0$.

Exercise 6.4 (i) *If* $0 \le \theta_0 < \theta_1 < \theta_2 < 2\pi$ *and* $0 \le \psi_0 < \psi_1 < \psi_2 < 2\pi$, *then there exists* $f \in \operatorname{Aut}(\mathbb{D})$ *such that* $f(e^{i\theta_j}) = e^{i\psi_j}$ *for* $j = 0,1,2$.

Hints: It suffices to treat the special case

$$\theta_0 = \psi_0 = 0, \quad \theta_1 = \pi/2, \quad \theta_2 = \pi. \tag{1}$$

For this special case it is convenient to go over to the half-plane $\mathbb{U}$. According to 2.5 the function $\phi(z) := -i\frac{z+1}{z-1}$, $z \in \mathbb{C} \setminus \{1\}$, maps $\mathbb{D}$ conformally onto $\mathbb{U}$. Notice that

$$\phi(i) = -1, \quad \phi(-1) = 0 \tag{2}$$

and that for $\psi \in (0, 2\pi)$

$$\phi(e^{i\psi}) = -i\frac{e^{i\psi/2} + e^{-i\psi/2}}{e^{i\psi/2} - e^{-i\psi/2}} = -\cot(\psi/2).$$

Since $-\cot$ is a strictly increasing map of $(0, \pi)$ onto $\mathbb{R}$, the numbers

$$y_j := -\cot(\psi_j/2) = \phi(e^{i\psi_j}) \quad (j = 1, 2) \tag{3}$$

satisfy $y_1 < y_2$. Hence the function $F(w) := (y_2 - y_1)w + y_2$ maps $\mathbb{U}$ conformally onto $\mathbb{U}$ and satisfies

$$F(-1) = y_1, \quad F(0) = y_2. \tag{4}$$

In view of (2), (4) and (3) the composite $f := \phi^{-1} \circ F \circ \phi$ does the required job. Notice that without unraveling the complicated formula for f, we know that it is a conformal automorphism of $\mathbb{D}$ that maps $e^{i\theta_1}$, $e^{i\theta_2}$ as desired and that $\lim_{z\to 1} f(z) = 1$, since $\lim_{z\to 1} |\phi(z)| = \infty$, $\lim_{|w|\to\infty} |F(w)| = \infty$ and $\lim_{|w|\to\infty} \phi^{-1}(w) = 1$.

(ii) *f is a non-identity automorphism of $\mathbb{D}$, g a non-identity holomorphic self-map of $\mathbb{D}$. Show that either f fixes exactly one point in $\mathbb{D}$ and none on $\mathbb{T}$, or f fixes no points in $\mathbb{D}$ and one or two points on $\mathbb{T}$. Show that g fixes at most one point in $\mathbb{D}$.*

Hints: 6.2 supplies $u \in \mathbb{T}$, $a \in \mathbb{D}$ such that $f(z) = u(z-a)/(1-\bar{a}z)$ for all $z \in \overline{\mathbb{D}}$. The point $z_0 \in \overline{\mathbb{D}}$ is fixed by f if and only if it is a zero of the polynomial $(1-\bar{a}z)(f(z) - z) = \bar{a}z^2 + (u-1)z - ua$. Since f is not the identity function, either $a \neq 0$ or $u \neq 1$, so this polynomial has degree 2 or 1. If $a = 0$, it has the unique zero $z_0 = 0$. If $a \neq 0$, the factorization 4.52(ii) supplies $z_1, z_2 \in \mathbb{C}$ such that this polynomial is $\bar{a}(z-z_1)(z-z_2)$. Consequently $\bar{a}z_1z_2 = -ua$ and $|z_1z_2| = 1$. Thus either exactly one of z_1, z_2 lies in $\mathbb{D} \setminus \{0\}$ or both lie on $\mathbb{T}$. If g fixes $a \in \mathbb{D}$, then $f_a \circ g \circ f_a^{-1}$ fixes 0, but is not the identity function. Hence by 6.1(iii) it fixes no other point of $\mathbb{D}$, so g fixes no other point than a.

(iii) *Use* (ii) *to show that the function f in* (i) *is unique.*

(iv) *Prove the converse of* (i). *That is, if θ_j, ψ_j are as in* (i), *then there is no $g \in \operatorname{Aut}(\mathbb{D})$ satisfying $g(e^{i\theta_0}) = e^{i\psi_0}$, $g(e^{i\theta_1}) = e^{i\psi_2}$, $g(e^{i\theta_2}) = e^{i\psi_1}$.*

Hints: Suppose such a g exists and let f be as supplied by (i). Then the conformal automorphism $h := g^{-1} \circ f$ fixes $e^{i\theta_0}$ and interchanges $e^{i\theta_1}$ with $e^{i\theta_2}$. Consequently, $h \circ h$ fixes three points of $\mathbb{T}$ and so by (ii) it must be the identity map. Write $h(z) = u \cdot \frac{z-\alpha}{1-\bar{\alpha}z}$, $u \in \mathbb{T}$, $\alpha \in \mathbb{D}$. Then $h(0) = -\alpha u$, so

$$0 = h\big(h(0)\big) = -\alpha u \cdot \frac{1+u}{1+|\alpha|^2 u}.$$

Hence, either $\alpha = 0$ or $u = -1$. If $\alpha = 0$, then $h(z) \equiv uz$, so from $h(e^{i\theta_0}) = e^{i\theta_0}$ we get $u = 1$, whence h is the identity map. The contradiction $g = f$ would follow. On the other hand, if $u = -1$, then the fact $h(e^{i\theta_0}) = e^{i\theta_0}$ leads to the equality $2e^{i\theta_0} = \alpha + \bar{\alpha}e^{2i\theta_0}$, which is untenable since the modulus of this sum is majorized by $|\alpha| + |\bar{\alpha}| < 2$.

(v) *Show that for any distinct points a_0, $a_1 \in \mathbb{T}$ and any distinct points b_0, $b_1 \in \mathbb{T}$ there exists $f \in \operatorname{Aut}(\mathbb{D})$ with $f(a_0) = b_0$, $f(a_1) = b_1$, but that such a map is usually not unique.*

Hints: It suffices to consider the special case $a_0 = b_0 = 1$ and then compose the function found with appropriate rotations on its left and right. In case $a_0 = b_0 = 1$, a_1 and b_1 may be written as $e^{i\theta_1}$ and $e^{i\psi_1}$ with $0 < \theta_1, \psi_1 < 2\pi$. Therefore the conclusion is a special case of (i), upon picking any $\theta_2 \in (\theta_1, 2\pi)$ and any $\psi_2 \in (\psi_1, 2\pi)$.

(vi) *Suppose $\alpha_1, \ldots, \alpha_n$ distinct points of $\mathbb{T}$, $\beta_1, \ldots, \beta_n$ arbitrary points of $\mathbb{T}$. Show that there is a rational function B mapping $\mathbb{D}$ into $\mathbb{D}$ such that $B(\alpha_k) = \beta_k$ for all $k = 1, \ldots, n$.*

Hints: For the shortest treatment (Younis [1980]) we need to anticipate a later result. First, by rotations we reduce to the case that 1 is not among the α's and β's. The map $\phi(z) := i\frac{1+z}{1-z}$ of $\overline{\mathbb{D}}\backslash\{1\}$ onto $\overline{\mathbb{U}}$ sends $\mathbb{D}$ conformally onto $\mathbb{U}$. Set $z_j := \phi(\alpha_j)$, $w_j := \phi(\beta_j)$ and let f be the rational function supplied by 11.52(v). It maps z_j onto w_j, $\mathbb{U}$ onto $\mathbb{U}$ and its poles are in $\mathbb{R}$. The rational function $B := \phi^{-1} \circ f \circ \phi$ does the desired interpolation.

(vii) *For any $a, b \in \mathbb{D}, u, v \in \mathbb{T}$ there exist $f, F \in \operatorname{Aut}(\mathbb{D})$ with $f(a) = b$, $f = f^{-1}$, $F(a) = b$, $F(u) = v$. Such an F is unique and, if $a \neq b$, f is also unique* (cf. part (ii)).

Hint: Consider (2.6) $\phi := f_a, \psi := -f_{\phi(b)}, f := \phi^{-1} \circ \psi \circ \phi$. For F it suffices to deal with the case $a = 0$. (Why?)

Exercise 6.5 (i) *Show that every conformal automorphism of* $\mathbb{D}$ *maps each subdisk of* $\mathbb{D}$ *onto another.* (Cf. 6.32 and 11.20(vii).)

Hints: Given $b \in \mathbb{D} \setminus \{0\}$, the map $f(z) := (z+b)/(1+\bar{b}z)$ is the composite $f_1 \circ f_3 \circ f_2 \circ f_1$, where $f_1(z) := z + 1/\bar{b}$, $f_2(z) := 1/z$ and $f_3(z) := z\big(|b|^2 - 1\big)/\bar{b}^2$. Note that $r > 0$ and $D(c,r) \subset \mathbb{D} \Rightarrow |c| + r \leq 1$. Consequently, $c_1 := f_1(c)$ satisfies

$$|c_1| = \left|\frac{\bar{b}c + 1}{\bar{b}}\right| \geq \frac{1 - |b||c|}{|b|} > \frac{|b| - |b||c|}{|b|} = 1 - |c| \geq r.$$

Thus $0 \notin D(c_1, r) = f_1(D(c,r))$, and calculation reveals that f_2 maps this disk onto $D(c_2, R)$, where $c_2 := \frac{\bar{c}_1}{|c_1|^2 - r^2}$, $R := \frac{r}{|c_1|^2 - r^2}$.

(ii) *For the conformal automorphism* $f_a(z) := (z-a)/(1-\bar{a}z)$ *of* $\mathbb{D}$ *and* $0 < r \leq 1$, *find the center and radius of the disk* $f_a^{-1}\big(D(0,r)\big)$. [*In terms of the metric* d *of* 6.3(vii) *this set is the open* d*-ball of radius* r *and center* a.]

Hint: We may assume $a \neq 0$. Direct multiplication shows that for all $z \in \mathbb{C} \setminus \{1/\bar{a}\}$

$$\big|f_a(z)\big|^2 < r^2 \iff \big(1 - |a|^2 r^2\big)|z|^2 + 2(r^2 - 1)\operatorname{Re}(\bar{a}z) < r^2 - |a|^2$$

and "completion of the square" shows that the last inequality is equivalent to

$$\left|z - \frac{(1-r^2)a}{1 - |a|^2 r^2}\right|^2 < \left[\frac{\big(1 - |a|^2\big)r}{1 - |a|^2 r^2}\right]^2.$$

Calculation confirms that the center c and radius R of the Euclidean disk defined by this last inequality satisfy $|c| + R \leq 1$, proving that this disk lies wholly in $\mathbb{D}$. (Is such a calculation really necessary?)

(iii) *Show that any open disk whose closure lies in* $\mathbb{D}$ *can be mapped onto one centered at* 0 *by a conformal automorphism of* $\mathbb{D}$.

Hints: It suffices to look at $0 < c < 1$ and $0 < r < 1 - c$ and to find $0 < a < 1$ such that $f(z) := (z-a)/(1-\bar{a}z)$ maps $D(c,r)$ onto a disk (in $\mathbb{D}$) centered at 0. For any such a, the function f satisfies $f(\bar{z}) = \overline{f(z)}$ for all z and maps $D(c,r)$ onto a disk in $\mathbb{D}$ (by (i)). Therefore this disk has its center on $\mathbb{R}$. We have only to see that by proper choice of a, this real number can be made 0. The points $f(c \pm r)$ are real and lie on the circle $\partial\big(f\big(D(c,r)\big)\big) = f\big(C(c,r)\big)$. Its center is therefore

$$\frac{1}{2}\big[f(c-r) + f(c+r)\big] = \frac{ca^2 - \big(1 + c^2 - r^2\big)a + c}{(1-ac)^2 - (ar)^2}.$$

It remains to see that the real quadratic $P(a) := ca^2 - \big(1 + c^2 - r^2\big)a + c$ vanishes for some $0 < a < 1$. Since $P(0) = c > 0 > r^2 - (1 - c)^2 = P(1)$, that is assured by continuity.

(iv) *Show that the mapping $f\colon \mathbb{D} \mapsto \mathbb{D}$ is d-isometric if and only if either f or $\bar{f}$ is a conformal automorphism.*

Hints: ($\Leftarrow$) was part of 6.3(vii), and will be used in dealing with ($\Rightarrow$). Suppose $f\colon \mathbb{D} \mapsto \mathbb{D}$ is d-isometric. By following f with the conformal automorphism $\phi(z) := (z - f(0))/\big(1 - \overline{f(0)}z\big)$ we get a d-isometry $\phi \circ f$ that fixes 0. Since $d(0, w) = |w|$ for all $w \in \mathbb{D}$, we have $|\phi \circ f(z)| = d(0, \phi \circ f(z)) = d(\phi \circ f(0), \phi \circ f(z)) = d(0, z) = |z|$ for all z. Fix $0 < a < 1$ (say, $a := 1/2$). Then $u := a/\phi \circ f(a) \in \mathbb{T}$ and $u\phi \circ f$ is a d-isometry F (why?) which fixes both 0 and a. For any $z \in \mathbb{D} \setminus \{0, a\}$, $F(z)$ lies on $C(0, |z|)$ and on the d-circle $\{w \in \mathbb{C} : d(a, w) = r\}$, where $r := d(a, z)$. The hints to (ii) show that the latter set is a Euclidean circle with center on the positive real axis. These two circles therefore have in common just the two points z and $\bar{z}$. Thus either $F(z) = z$ or $F(z) = \bar{z}$. Since the d-metric topology is the Euclidean topology and F is d-continuous, the sets $\{z \in \mathbb{U} \cap \mathbb{D} : F(z) = z\}$ and $\{z \in \mathbb{U} \cap \mathbb{D} : F(z) = \bar{z}\}$ are relatively closed, disjoint, and cover $\mathbb{U} \cap \mathbb{D}$. So one of them is the whole connected $\mathbb{U} \cap \mathbb{D}$. Say, $F(z) = z$ for all $z \in \mathbb{U} \cap \mathbb{D}$. In the same way, one alternative must prevail throughout the lower half-disk $\mathbb{D} \setminus \overline{\mathbb{U}}$. But if there exists $z_0 \in \mathbb{D} \setminus \overline{\mathbb{U}}$ with $F(z_0) = \overline{z_0}$, then $\overline{z_0} \in \mathbb{U} \cap \mathbb{D}$ and consequently $F\big(\overline{z_0}\big) = \overline{z_0}$, which violates the injectivity of the isometry F. Thus $F(z) = z$ for all $z \in \mathbb{D}$, and $f(z) = \phi^{-1}(\bar{u}z) = \bar{u}(z + uf(0))(1 + \overline{uf(0)}z)^{-1}$. In the other scenario, $F(z) = \bar{z}$ for all $z \in \mathbb{D}$, so $\bar{f}(z) = \overline{\phi^{-1}(\bar{u}\bar{z})} = u(z + \overline{uf(0)})(1 + uf(0)z)^{-1}$.

Next we look at a remarkable property of holomorphic self-maps of $\mathbb{D}$, which in the next chapter will lead to a complete description of the limiting behavior of the iterates of such maps. Here the terminology "oricycle" or "horocycle", for a circle in $\overline{\mathbb{D}}$ which is tangent to $\mathbb{T}$, is often encountered. The theorem is due to J. Wolff [1926b] and the unique "fixed-point" it produces has been called the Denjoy-Wolff point or the "Wolff trap", imagery which is somewhat clarified by the second paragraph of the proof of 7.56.

Theorem 6.6 *Let $f\colon \mathbb{D} \mapsto \mathbb{D}$ be holomorphic and fixed-point-free. Then there exists a unique $u \in \mathbb{T}$ such that $f(D) \subset D$ for every subdisk D of $\mathbb{D}$ whose closure is tangent to $\mathbb{T}$ at u.* (See 6.28(x), (xii) and 7.56.)

Proof: If u_1, u_2 are two different points of $\mathbb{T}$ with this property, consider disks D_j internally tangent to $\mathbb{T}$ at u_j and tangent to each other, say at $z_0 \in \mathbb{D}$. From $f(D_j) \subset D_j$ for $j = 1, 2$ it follows that $f(z_0) \in \bar{D}_j$ for $j = 1, 2$, so $f(z_0) \in \bar{D}_1 \cap \bar{D}_2 = \{z_0\}$, contrary to f being fixed-point-free in $\mathbb{D}$.

Let I denote the identity map of $\mathbb{D}$ into $\mathbb{D}$, pick $r_k \in (0,1)$ with $r_k \uparrow 1$, and set $f_k := r_k f$. Since

$$\big|I - (I - f_k)\big| < r_k = |I| \quad \text{on } \partial D(0, r_k),$$

it follows from Rouché's Theorem (5.87(ii)) that $I - f_k$ has the same number of zeros in $D(0, r_k)$ as does I. Thus $I - f_k$ has exactly one zero in $D(0, r_k)$. Let a_k denote this zero:

$$(1) \qquad r_k f(a_k) = a_k \qquad \forall k \in \mathbb{N}.$$

[Since the continuous function f_k maps $\overline{D}(0, r_k)$ into itself, Brouwer's theorem 4.58(ii) is another way to see that it has a fixed-point; see also 7.41(iii).] Pass to a subsequence of $\{a_k\}$ and so assume with no loss of generality that

$$(2) \qquad \lim_{k\to\infty} a_k =: u$$

exists in $\overline{\mathbb{D}}$. If $u \in \mathbb{D}$, then letting $k \to \infty$ in (1) shows that $f(u) = u$. Since f is supposed to be fixed-point-free, this is not possible. Hence

$$(3) \qquad |u| = 1.$$

Since $f_k(a_k) = a_k$, 6.3(iii)(a) says that

$$(4) \qquad \frac{\big|1 - \bar{a}_k f_k(z)\big|^2}{1 - \big|f_k(z)\big|^2} \le \frac{|1 - \bar{a}_k z|^2}{1 - |z|^2} \qquad \forall z \in \mathbb{D},\ k \in \mathbb{N}.$$

Fixing z and letting $k \to \infty$, we get

$$(5) \qquad \frac{|1 - \bar{u} f(z)|^2}{1 - |f(z)|^2} \le \frac{|1 - \bar{u} z|^2}{1 - |z|^2} \qquad \forall z \in \mathbb{D}.$$

But by simple algebra one sees that for any $r > 0$ and any $w \in \mathbb{D}$

$$(6) \qquad \frac{|1 - w|^2}{1 - |w|^2} < r \iff \left| w - \frac{1}{1+r} \right| < \frac{r}{1+r}.$$

Hence, remembering (3), we have the implications, for all $z \in \mathbb{D}$:

$$\left| z - \frac{u}{1+r} \right| < \frac{r}{1+r} \iff \left| \bar{u} z - \frac{1}{1+r} \right| < \frac{r}{1+r} \overset{(6)}{\iff} \frac{|1 - \bar{u} z|^2}{1 - |\bar{u} z|^2} < r \overset{(5)}{\Rightarrow}$$

$$\frac{|1 - \bar{u} f(z)|^2}{1 - |\bar{u} f(z)|^2} < r \overset{(6)}{\iff} \left| \bar{u} f(z) - \frac{1}{1+r} \right| < \frac{r}{1+r} \iff \left| f(z) - \frac{u}{1+r} \right| < \frac{r}{1+r}.$$

This shows that

$$f\Big(D\Big(\frac{u}{1+r},\frac{r}{1+r}\Big)\Big) \subset D\Big(\frac{u}{1+r},\frac{r}{1+r}\Big) \qquad \forall r > 0$$

or equivalently

$$f\big(D(Ru, 1-R)\big) \subset D\big(Ru, 1-R\big) \qquad \forall 0 < R < 1,$$

proving the theorem.

Remark: Simple algebra as in (6) shows that the circle $C\big(\frac{u}{1+r},\frac{r}{1+r}\big)$ coincides with the set $\big\{z \in \mathbb{D} : \frac{|1-\bar{u}z|^2}{1-|z|^2} = r\big\} = \big\{z \in \mathbb{D} : \frac{1-|z|^2}{|u-z|^2} = \frac{1}{r}\big\}$, and the latter is a level-set for the *Poisson kernel* $P_u(z) := \frac{1-|z|^2}{|u-z|^2}$ at $u \in \mathbb{T}$. The graph, in $\mathbb{D} \times \mathbb{R}_+$, of P_u is therefore an easily visualizable stalagmite with circular cross-sections each tangent to the spine $\{u\} \times \mathbb{R}_+$.

An extremely important class of functions in the disk $\mathbb{D}$ is that comprised of unimodular constants times products of the automorphisms f_a. These are called *Blaschke products*, though we find them being used already in the 1897 first edition of PICARD [1905], and in POINCARÉ [1907]. For a survey of their many remarkable properties and uses see COLWELL [1985]. The finite Blaschke product $f_{a_1} \cdot f_{a_2} \cdot \dots \cdot f_{a_n}$ has modulus 1 on $\partial\mathbb{D}$ and zeros at $a_1, a_2, \dots, a_n$. Dividing by such a product removes these zeros from an f without increasing its upper bound in $\mathbb{D}$. This is often more advantageous than removing the zeros by merely dividing off the factors $z - a_j$. Cf. 6.7, 6.9 and 6.16 below. Our first use of them will be to refine the uniqueness theorem 5.62 for *bounded* functions f in $H(\mathbb{D}) \setminus \{\mathbf{0}\}$: Not only can the zeros a_j of f not accumulate in $\mathbb{D}$, but they must rush toward $\partial\mathbb{D}$ so fast that $\sum(1 - |a_j|) < \infty$.

Exercise 6.7 (JENSEN [1898–99] and PETERSEN [1900])

(i) *f is holomorphic in $D(0, r)$, bounded by $M < \infty$, and $z_1, \dots, z_k$ are zeros of f, each appearing no more often than its multiplicity. Show that*

$$|f(0)| \le r^{-k} M |z_1 \dots z_k|.$$

Hints: (CARATHÉODORY and FEJÉR [1907]) The function $f_j(z) := r(z - z_j)/\big(r^2 - \bar{z}_j z\big)$ is holomorphic in a neighborhood of $\bar{D}(0, r)$ and has exactly one zero at z_j, and of order 1. Consequently, $F := f/(f_1 \dots f_k)$ is holomorphic in $D(0, r)$. Since $|f_j| = 1$ on $C(0, r)$ and f_j is continuous on $\bar{D}(0, r)$, we see that $\overline{\lim}_{\rho\uparrow r} \max\{|F(z)| : |z| = \rho\} \le M$, and so by the Maximum-Modulus Principle $|F(z)| \le M$ for all $z \in D(0, r)$; that is, $|f| \le M|f_1 \dots f_k|$.

(ii) *From* (i) *deduce the following refinement of* 6.1(ii)*: If $F : \mathbb{D} \mapsto \mathbb{D}$ is holomorphic, $F(0) = 0$, $w \in F(\mathbb{D})$ and $z, z_1, \ldots, z_n$ are zeros of $F - w$, each appearing no more often than its multiplicity, then*

$$|F(z)| \leq |z||z_1| \ldots |z_n|.$$

Hint: Apply (i) to $f := (F - w)/(1 - \bar{w}F)$, $r := 1$.

(iii) *If f is bounded, holomorphic and not identically* 0 *in $\mathbb{D}$ and its zeros repeated according to multiplicity are $z_1, z_2, \ldots$, then $\sum(1 - |z_j|) < \infty$.*

Hints: We can suppose $f(0) \neq 0$. Since $\log x = \int_1^x t^{-1}\,dt \geq \int_1^x x^{-1}\,dt = 1 - x^{-1}$ for $x \geq 1$, we have $\log |z_j|^{-1} \geq 1 - |z_j|$ for all j. Hence (i) yields for all k

$$0 < \sum_{j=1}^{k}(1 - |z_j|) \leq \sum_{j=1}^{k} \log |z_j|^{-1} = \log \left|\prod_{j=1}^{k} z_j\right|^{-1} \leq \log\left(\frac{M}{|f(0)|}\right) < \infty.$$

Exercise 6.8 (i) *Let f be an entire function with $|f(0)| = 1$. For each $\rho > 0$ let $M(\rho)$ denote* $\sup\left|f\left(\bar{D}(0, \rho)\right)\right|$ *and $n(\rho)$ the number of zeros, counted according to multiplicity, of f in $\bar{D}(0, \rho)$. Show that*

$$2^{n(\rho)} \leq M(2\rho) \qquad \forall \rho > 0.$$

Hints: Let $z_1, z_2, \ldots$ be an enumeration of the zeros of f repeated according to multiplicity and ordered by increasing modulus. Thus for each $R > 0$

$$(*) \qquad |z_j| \leq R \quad \text{for } 1 \leq j \leq n(R).$$

With $\rho > 0$ fixed, consider $r = 2\rho$, $k = n(2\rho)$, $M = M(2\rho)$ in 6.7. Then

$$2^{n(\rho)} \overset{(*)}{\leq} \prod_{j=1}^{n(\rho)} \left|\frac{2\rho}{z_j}\right| \overset{(*)}{\leq} \prod_{j=1}^{n(2\rho)} \left|\frac{2\rho}{z_j}\right| = \frac{r^k |f(0)|}{|z_1 \ldots z_k|} \overset{(6.7)}{\leq} M = M(2\rho).$$

(ii) *Suppose that for some $0 < \rho < \infty$ the entire function $f \neq \mathbf{0}$ satisfies the growth condition*

$$(**) \qquad f(z) \cdot e^{-|z|^\rho} \textit{ is bounded.}$$

Let $z_1, z_2, \ldots$ be an enumeration of the non-zero zeros of f repeated according to multiplicity and ordered by increasing modulus. Show that for every $\rho' > \rho$

$$\sum |z_j|^{-\rho'} < \infty.$$

Hints: Dividing away the appropriate power of z does not imperil (**) but lets us assume that $|f(0)| = 1$. From (i) and (**), for some $0 < B < \infty$ and all $r > 0$

$$n(r)\log 2 \le \log M(2r) \le B + (2r)^\rho.$$

Therefore, for $C := (B + 2^\rho)/\log 2$ we will have

$$n(r) \le Cr^\rho \qquad \forall r \ge 1.$$

As the zeros do not accumulate, this means that

$$j \le n\big(|z_j|\big) \le C|z_j|^\rho \text{ for all large } j, \text{ so}$$

$$\frac{1}{|z_j|^{\rho'}} \le \left(\frac{C}{j}\right)^{\rho'/\rho} \quad \text{for all large } j.$$

Since $\rho' > \rho$, $\sum j^{-\rho'/\rho} < \infty$.

Exercise 6.9 (*Following* LÖWNER *and* RADÓ [1923]; *see also* M. M. LAVRENTIEFF [1956].)

(i) *For all a, $z \in \mathbb{D}$*

$$\left|\frac{z-a}{1-\bar{a}z}\right| < e^{(1-|a|)(|z|-1)/(|z|+1)}.$$

Hints: According to 6.3(v)

$$\left|\frac{z-a}{1-\bar{a}z}\right| \le \frac{|z|+|a|}{1+|a||z|} = 1 - \big(1-|z|\big)\frac{1-|a|}{1+|a||z|} \le 1 - \big(1-|z|\big)\frac{1-|a|}{1+|z|}.$$

The desired inequality follows from this when we recall that $1 - t < e^{-t}$ for positive t.

(ii) *For each set of k distinct numbers $z_1, \ldots, z_k \in \mathbb{D}$ there is a positive finite constant $C_k := C(z_1, \ldots, z_k)$ such that*

$$|f(z)| \le 2mC_k + Me^{[(|z|-1)/(|z|+1)]\sum_{j=1}^k (1-|z_j|)}, \quad z \in \mathbb{D}$$

for all f which are holomorphic in $\mathbb{D}$, bounded by M there and by m on $\{z_1, \ldots, z_k\}$.

Hints: Set $C_k := \sum_{j=1}^k \prod_{l=1, l\ne j}^k 2|z_l - z_j|^{-1}$. Let such an f be given and form the (Lagrange) polynomial

$$L(z) := \sum_{j=1}^k f(z_j) \prod_{\substack{l=1\\ l\ne j}}^k (z - z_l)(z_j - z_l)^{-1}.$$

Then we evidently have

$$(1) \qquad |L(z)| \le C_k \max_{1\le j\le k} |f(z_j)| \le C_k m, \quad z \in \mathbb{D}$$

and

$$(2) \qquad L(z_j) = f(z_j), \quad j = 1, 2, \ldots, k.$$

Form the Blaschke product $B(z) := \prod_{j=1}^{k}(z - z_j)/(1 - \bar{z}_j z)$. Then because of (2), $(L - f)/B$ is holomorphic in $\mathbb{D}$. Compute its maximum on $C(0, r)$, cite the Maximum-Modulus Principle, let $r \uparrow 1$ and notice that $|B(z)| \to 1$ as $|z| \to 1$. We learn that

$$\begin{aligned} \left|\frac{L - f}{B}\right| &\le \sup_{\mathbb{D}} |L - f| \\ &\le mC_k + M, \qquad \text{by (1) and the definition of } M. \end{aligned}$$

It follows that

$$\begin{aligned} |f| &\le (mC_k + M)|B| + |L| \le (mC_k + M)|B| + mC_k, \quad \text{by (1) again,} \\ &\le 2mC_k + M|B|, \quad \text{since } |B| \le 1. \end{aligned}$$

Finally, use (i) to majorize the Blaschke product in this last inequality.

(iii) *Let f be holomorphic and bounded by M in $\mathbb{D}$ and let $z_1, z_2, \ldots$ be zeros of f repeated according to multiplicity. Show that*

$$|f(z)| \le Me^{[(|z|-1)/(|z|+1)]\sum_{j=1}^{k}(1-|z_j|)}, \qquad \forall z \in \mathbb{D}, \forall k \in \mathbb{N}.$$

Deduce that if $\sum_{j=1}^{\infty}(1 - |z_j|) = \infty$, then $f = \mathbf{0}$. (Cf. 6.7(iii) and 12.6.)

Hint: Let B be the Blaschke product on $z_1, \ldots, z_k$ and argue as in the hints to (ii): form f/B, apply the Maximum-Modulus Principle and use (i).

(iv) *The bounded holomorphic function F in $\mathbb{H}$ vanishes on the sequence $\{w_n\}$ of distinct points in $\mathbb{H}$. If the sequence has a limit point in $\mathbb{H}$, then $F = \mathbf{0}$* (5.62). *If the sequence has no limit point in $\mathbb{H}$ but is bounded, then* $\operatorname{Re} w_n \to 0$. *Show that if this convergence is so slow that $\sum \operatorname{Re} w_n = \infty$, then $F = \mathbf{0}$ again follows. Draw the same conclusion if instead $\{w_n\}$ is bounded away from 0 and $\sum \operatorname{Re}(1/w_n) = \infty$.*

Hints: In the first scenario there exists $M \in \mathbb{N}$ such that $|w_n + 1| \le |w_n| + 1 \le M$ for all n. The function $\phi(w) := (w - 1)/(w + 1)$ maps $\mathbb{H}$ conformally onto $\mathbb{D}$ (2.5) and the points $z_n := \phi(w_n)$ satisfy

$$2(1 - |z_n|) \ge 1 - |z_n|^2 = \frac{2 \operatorname{Re} w_n}{|w_n + 1|^2} \ge \frac{2}{M^2} \operatorname{Re} w_n \qquad \forall n \in \mathbb{N}$$

and are zeros of the bounded function $f := F \circ \phi^{-1} \in H(\mathbb{D})$. In the second scenario apply the conclusion of the first to the function $F(1/w)$, whose sequence of zeros $\{1/w_k\}$ is bounded.

Exercise 6.10 *Let f be holomorphic and satisfy $|f| < 1$ in $\mathbb{D}$. Define g in $\mathbb{D}$ by $g(z) := \big(f(0) - f(z)\big)/\big(z - z\bar{f}(0)f(z)\big)$. Suppose that g^* is holomorphic and bounded by 1 in $\mathbb{D}$ and that $g^* - g$ has a zero of multiplicity at least n at 0 $(n \geq 1)$. Show that for the function $f^*(z) := \big(f(0) - zg^*(z)\big)/\big(1 - \bar{f}(0)zg^*(z)\big)$ the difference $f^* - f$ has a zero of multiplicity at least $n + 1$ at 0.*

Hint: Direct computation reveals that for all $z \in \mathbb{D}$

$$f(z) - f^*(z) = z\big[g^*(z) - g(z)\big]\left[\frac{1 - \bar{f}(0)f(z)}{1 - \bar{f}(0)zg^*(z)}\right].$$

The third factor on the right-hand side is holomorphic in $\mathbb{D}$, since $|\bar{f}(0)| < 1$ and $|g^*| \leq 1$.

Exercise 6.11 *Let f, g be entire functions which for some $c < 1$ satisfy*

(i) $|f(z)| \leq e^{c|z|}$ & $|g(z)| \leq e^{c|z|}$ *for all (sufficiently large) z,*

(ii) $f(n) = g(n)$ *for all positive integers n.*

Show that then $f = g$. (See 11.38 for an enhancement.)

Hints: (Pólya [1915]) Suppose not. Then (5.62) there exists at least one $a \in (0, 1)$ such that $f(a) \neq g(a)$. Thus

$$h(z) := f(z + a) - g(z + a)$$

is entire, not identically 0 and satisfies

$$h(a_n) = 0, \quad n = 1, 2, 3, \ldots,$$

where $a_n := n - a$. For any $c < C < 1$ we have

$$|h(z)| \leq e^{C|z|} \quad \text{for all sufficiently large } z. \tag{1}$$

Notice that $0 < a_j < j$ for every $j = 1, 2, \ldots$. Therefore for large n we may apply 6.7(i) and 5.12 to h with $r := n$ and, because of (1), $M := e^{Cn}$. There results

$$\frac{n^n|h(0)|}{n!} < \frac{n^n|h(0)|}{|a_1 \cdots a_n|} \leq e^{Cn},$$

$$\frac{n}{(n!)^{1/n}}|h(0)|^{1/n} \leq e^{C}. \tag{2}$$

Notice that if we set $c_n := n^n/n!$, then $\frac{c_{n+1}}{c_n} = (1+1/n)^n$ and so $\lim_{n\to\infty} \frac{c_{n+1}}{c_n} = e$ by 3.21. It follows then from 3.9 that $\lim_{n\to\infty}(c_n)^{1/n}$ exists and equals e. Insert this fact into (2), recall that $h(0) \neq 0$ and cite 3.1. We conclude that $e \leq e^C$, whence $C \geq 1$, contrary to the definition of C.

§ 2 Boundary-preserving Maps

The (finite) Blaschke products introduced in the preceding section have an interesting characterization due to FATOU [1923]: they are just those holomorphic functions $f\colon \mathbb{D} \mapsto \mathbb{D}$ which satisfy $\lim_{|z|\to 1} |f(z)| = 1$ (see 6.16 below). The generalization of this phenomenon to other regions leads to deep results affirming that certain locally univalent maps are in fact univalent. This fruit will be picked in Chapter X but the initial spade-work, involving some results which are quite attractive in their own right, will be done now. These present a beautiful interplay of very elementary topology (ideas from the theory of covering spaces) with the local theory of holomorphic functions. But first, the relevant definitions and the mentioned generalization of Fatou's theorem.

Definition 6.12 Let U, V be open subsets of $\mathbb{C}$, $f\colon U \mapsto V$ a continuous map.

(i) Call f a *proper map* if $f^{-1}(K)$ is a compact subset of U for every compact $K \subset V$.

(ii) Call f a *boundary-preserving map* if whenever $\{z_n\} \subset U$ and either z_n converges to a boundary point of U or $|z_n| \to \infty$, every (finite) subsequential limit of $\{f(z_n)\}$ is on the boundary of V. It comes to the same thing to call a sequence $\{z_n\} \subset U$ *compactafugal in* U or *a boundary sequence in* U, if it is eventually outside each compact $K \subset U$ (i.e., has no limit point *in* U) and to require that f map every such sequence in U onto one in V.

(iii) Call f a *closed map* if $f(C)$ is a relatively closed subset of V for every relatively closed subset C of U.

(iv) If f is holomorphic and $m \in \mathbb{N}$, say that f is *exactly* m*-to-*1, or is m*-valent*, if each $w \in V$ has valence m, that is, counting multiplicities, exactly m pre-images in U. Generally a function that is m-valent for an $m > 1$ is called *multivalent*.

The reader may already have anticipated that (i) and (ii) are the same; nevertheless, both terms are prevalent in the literature. The beautiful and remarkable fact is that for non-constant holomorphic functions on connected domains all four properties are equivalent.

Lemma 6.13 *Let U, V be open subsets of $\mathbb{C}$, $f\colon U \mapsto V$ continuous. Then* (i) *and* (ii) *of* 6.12 *are equivalent and imply* (iii). *If U is connected and f is non-constant and holomorphic, then* (iii) *implies* (i) *and* (ii).

Proof: (i) $\Rightarrow$ (ii)&(iii). Both conclusions follow if we prove: whenever $\{z_n\}$ is a sequence in a relatively closed $C \subset U$ and $f(z_n) \to w \in V$, then a subsequence of $\{z_n\}$ converges to a $z \in C$ with $f(z) = w$. To this end, note that $\{f(z_n) : n \in \mathbb{N}\} \cup \{w\}$ is a compact subset K of V. By (i) $f^{-1}(K)$ is therefore compact. Of course, $\{z_n\} \subset f^{-1}(K)$. A subsequence consequently converges to some $z \in f^{-1}(K) \subset U$. We have $z \in C$ because C contains all its limit points which lie in U, and by continuity $w = f(z) \in f(C)$.

(ii) $\Rightarrow$ (i). Consider K compact $\subset V$ and $C := f^{-1}(K)$. First note that C is closed in $\mathbb{C}$. For if not, then there exist $z_n \in C$ with $z_n \to z_0 \in \overline{C} \setminus C$. Since C is relatively closed in U, $z_0 \in \overline{C} \setminus U \subset \overline{U} \setminus U = \partial U$. By compactness of K, some subsequence $\{f(z_{n_j})\}$ converges to some $w \in K \subset V$, which contradicts (ii). The same argument shows that there is no sequence $\{z_n\} \subset C$ with $|z_n| \to \infty$, i.e., C is bounded.

Finally, suppose U is connected, f is holomorphic, non-constant and a closed map. Consider $\{z_n\}$ as in 6.12(ii) and suppose $\{f(z_n)\}$ has a limit point w in V. We want to deduce a contradiction. Without loss of generality then suppose $|f(z_n) - w| < 1/n$ for all n. Since f is not constant, $f^{-1}(w)$ has no limit point in the connected set U (5.62). In particular, z_n is not such a limit point, so there exist points $\xi_n \in D(z_n, 1/n) \cap f^{-1}\big(D(f(z_n), 1/n)\big) \cap \big[U \setminus f^{-1}(w)\big]$. Thus on one hand $\{\xi_n\}$, like $\{z_n\}$, behaves as in 6.12(ii), which makes $\{\xi_n : n \in \mathbb{N}\}$ a relatively closed subset of U, and on the other hand $\{f(\xi_n)\}$, like $\{f(z_n)\}$, converges to $w \notin f\{\xi_n : n \in \mathbb{N}\}$. This contradicts f being a closed map.

Theorem 6.14 *Let U, V be open subsets of $\mathbb{C}$, $f\colon U \mapsto V$ holomorphic and exactly m-to-1 for some positive integer m. Then f is a boundary-preserving map.*

Proof: (T. Radó [1922/23c]) Given $w_0 \in V$, let $z_1, \ldots, z_k$ be the distinct zeros of $f - w_0$, with respective multiplicities $m_1, \ldots, m_k$. Thus

$$(*) \qquad m_1 + \cdots + m_k = m.$$

Applying 5.83 to each z_j we find $\varepsilon = \varepsilon(w_0) > 0$ and disjoint $D_j := D(z_j, r_j)$ with $\overline{D}_j \subset U$ such that each $w \in D(w_0, \varepsilon) \setminus \{w_0\}$ has m_j distinct pre-images in D_j ($j = 1, \ldots, k$). Therefore by $(*)$ w has m distinct pre-images in the disjoint union $D_1 \cup \cdots \cup D_k$, and since f is exactly m-to-1, this means that $f^{-1}\big(D(w_0, \varepsilon)\big) \subset D_1 \cup \cdots \cup D_k$. In summary, $f^{-1}\big(D(w_0, \varepsilon(w_0)\big)$ lies in the compact subset $\bar{D}_1 \cup \cdots \cup \bar{D}_k$ of U. Any compact $K \subset V$ can be covered by finitely many $D(w_0, \varepsilon(w_0))$ and so $f^{-1}(K)$ lies in a compact subset of U. Thus f is a proper map, hence boundary-preserving by 6.13.

Theorem 6.15 (CARATHÉODORY and RADEMACHER [1917]; T. RADÓ *op. cit.*) *Let U, V be regions in $\mathbb{C}$ and $f\colon U \mapsto V$ a holomorphic function which is a boundary-preserving mapping. Then f is surjective and in fact an exactly m-to-1 mapping for some positive integer m (which is called the* degree *of f).*

Proof: It is clear from 6.12(ii) that if $f(U)$ is a single number ($\in V$), then f is not boundary-preserving. Hence f is not constant. Therefore by 5.77 f is an open map. In conjunction with 6.13 this makes $f(U)$ a clopen subset of V. As V is connected, this entails that $f(U) = V$.

Next we show that for each $w \in V$ the set $f^{-1}(w)$ is finite. For by 6.13 $f^{-1}(w)$ is a compact subset of U. If it were infinite, then by the uniqueness theorem and the connectedness of U it would follow that f is constant ($= w$), contrary to $f^{-1}(w)$ being compact! Similarly, no point $z \in f^{-1}(w)$ can have infinite multiplicity, since that would make $f - w$ constantly 0 in some disk centered at z, hence throughout the connected set U. We can therefore define $N\colon V \mapsto \mathbb{N}$ by letting $N(w)$ denote the number of zeros, each counted according to multiplicity, of $f - w$ in U. That is, $N(w)$ is the valence of w.

We shall show that N is continuous at each point of V. Let $w \in V$ be given, and set $m := N(w)$. Let the distinct zeros of $f-w$ be $a_1, \ldots, a_k$ with respective multiplicities $m_1, \ldots, m_k$. Thus $m_1 + \cdots + m_k = m$. Use 5.83 to select $\varepsilon > 0$ sufficiently small and for each j an open disk D_j centered at a_j whose closure lies in U such that $\bar{D}(w, \varepsilon) \subset V$ and

(∗) each point of $D(w, \varepsilon) \setminus \{w\}$ has m_j distinct pre-images in D_j.

It suffices to show that for n sufficiently large no point of $D(w, \varepsilon/n)$ has any pre-images outside $D_1 \cup \cdots \cup D_k$. For then (∗) shows that $N = m$ identically in $D(w, \varepsilon/n)$. To prove this, note that $f^{-1}\big(\bar{D}(w, \varepsilon)\big)$ is compact (6.13). If for each n there exists $z_n \in U \setminus (D_1 \cup \cdots \cup D_k)$ such that $f(z_n) \in D(w, \varepsilon/n)$, then the z_n all belong to the compact subset $f^{-1}\big(\bar{D}(w, \varepsilon)\big)$ of U, so some subsequence $\{z_{n_j}\}$ converges to some $z \in U$. But $f(z_{n_j}) \in D(w, \varepsilon/n_j)$. Therefore $f(z) = w$. But like all the z_n, z lies in the closed set $U \setminus (D_1 \cup \cdots \cup D_k) \subset U \setminus \{a_1, \ldots, a_k\} = U \setminus f^{-1}(w)$, so in $f(z) = w$ we have a contradiction. This proves continuity of N at the point w.

The continuous integer-valued function N must be constantly m on the connected set V, which means that f is exactly m-to-1 in U.

Exercise 6.16 (i) *If f is a boundary-preserving holomorphic map of $\mathbb{D}$ into itself, then f is a finite Blaschke product* (FATOU [1923]).

Hints: f can have only finitely many zeros. If these are $a_1, \ldots, a_n$ repeated according to multiplicity, then let $B(z) := \prod_{j=1}^{n}(z - a_j)/(1 - \bar{a}_j z)$ [void products being 1] and note that f/B and B/f are each holomorphic in $\mathbb{D}$ and their moduli are continuous on $\overline{\mathbb{D}}$. Apply the Maximum-Modulus Principle to each to conclude that f/B is a unimodular constant.

(ii) *Show that every finite Blaschke product* $B(z) := \prod_{j=1}^{m}(z - c_j)/(1 - \bar{c}_j z)$ *is an exactly* m*-to-1 map of* $\mathbb{D}$ *into* $\mathbb{D}$*; in particular, the converse of* (i) *holds.* (Cf. TAHA [1991].)

Hints: Note that such a product is clearly a boundary-preserving map and cite 6.15. Alternatively, given $b \in \mathbb{D}$, apply 5.87(ii) with $f := B$ and $g := b - B$.

(iii) *Deduce from* (i) *that an entire function of constant modulus on the unit circle is a monomial.* (Cf. 11.14.)

(iv) *Suppose* f *is holomorphic and non-constant in the bounded region* Ω *and that* $\lim_{z \to \xi} |f(z)| = 1$ *for each* $\xi \in \partial\Omega$*. Show that* $f(\Omega) = \mathbb{D}$*.*

Hints: By the maximum principles 5.12 and 5.10, $|f| < 1$ in Ω. The same would be true of $1/f$ if f were zero-free. Hence $0 \in f(\Omega) \subset \mathbb{D}$. For any $c \in \mathbb{D}$ the function $F := (f - c)/(1 - \bar{c}f)$ satisfies the same hypotheses as f, so by what has been shown, there exists $z \in \mathbb{D}$ with $0 = F(z)$, i.e., with $c = f(z)$.

(v) *Suppose* f *is holomorphic and non-constant in the region* Ω*,* J *a Jordan-curve such that* dist $(f(z_n), J) \to 0$ *whenever* $\{z_n\} \subset \Omega$ *converges to any point in* $\{\infty\} \cup \partial\Omega$*. Show that* $f(\Omega) = \mathcal{I}(J)$.(See 4.44.)

Hints: From the given convergence condition (and 5.77) it follows that $f(\Omega) \setminus J$ is relatively closed (and open) in $\mathbb{C} \setminus J$. Hence if $f(\Omega) \setminus J$ meets the connected set $\mathcal{O}(J)$, it must contain $\mathcal{O}(J)$. But then there would be a sequence $w_n \in f(\Omega) \setminus J$ with $|w_n| \to \infty$, which contradicts the given convergence condition. Hence $f(\Omega) \setminus J$ does not meet $\mathcal{O}(J)$ and therefore lies in $\mathcal{I}(J)$. The open set $f(\Omega)$ cannot meet $J = \partial(\mathcal{O}(J))$, else it would meet $\mathcal{O}(J)$ too. Thus $f(\Omega) \subset \mathcal{I}(J)$. As $f(\Omega) = f(\Omega) \setminus J$ is relatively clopen in $\mathbb{C} \setminus J$ and $\mathcal{I}(J)$ is connected, $f(\Omega) = \mathcal{I}(J)$ follows. (This is a generalization and alternative treatment of (iv).)

(vi) *Let* $\mathbb{D}^+ := \mathbb{D} \cap \mathbb{U}$ *and for each* $a \in \mathbb{D}$ *form* f_a *as in* 6.2*. Show that if* $a \in \mathbb{D}^+$*, then* $\phi := f_a/f_{\bar{a}}$ *maps* $\mathbb{D}^+$ *conformally onto* $\mathbb{D}$*.*

Hints: The sole zero of $f_{\bar{a}}$ is at $\bar{a} \notin \mathbb{D}^+$, so the rational function ϕ is holomorphic in $\mathbb{D}^+$. On $\mathbb{T}$ it satisfies $|\phi| = 1$ (see hints to 2.6), and for $x \in (-1, 1)$, $f_{\bar{a}}(x) = \bar{f}_a(x)$, so $|\phi(x)| = 1$. From (iv) it follows that $\phi(\mathbb{D}^+) = \mathbb{D}$. Since ϕ is boundary-preserving, it is exactly m-to-1 for some $m \in \mathbb{N}$ (6.15). Since obviously a is the unique pre-image of 0 and $\phi'(a) \neq 0$, m must be 1. [For the choice $a := (\sqrt{2} - 1) \cdot i$, $\phi(z) = -(z^2 + 2iz + 1) \div (z^2 - 2iz + 1)$, a mapping that can be realized in a very geometrically revealing way: According to 2.5, $\psi(z) := (1+z)/(1-z)$ maps $\mathbb{D}$ conformally onto $\mathbb{H}$. Since $\operatorname{Im}\psi(z) = 2\operatorname{Im} z/|1-z|^2$, ψ therefore maps $\mathbb{D}^+$ onto $\mathbb{H}\cap\mathbb{U}$, which $w \mapsto -iw^2$ maps back to $\mathbb{H}$, so that $\psi^{-1}\circ\left(-i\psi^2\right)$ is a conformal map of $\mathbb{D}^+$ onto $\mathbb{D}$. It is the function $i\phi$.]

Remarks: The next result about proper mappings has a somewhat intimidating multi-part proof. But the techniques involved are important enough to justify its inclusion. They come from the theory of covering spaces in topology and feature in the classical proof of the holomorphic monodromy theorem (10.3(iv)); in fact they are considerably more transparent in the present context because of the simple geometry of the disk. The appropriate setting for this result has an arbitrary topological space X in the role of U and an arbitrary simply-connected region Ω in the role of $D(0,R)$. The latter generalization follows from 6.17 via 10.2 (according to which every such Ω is homeomorphic to $\mathbb{D}$; cf. also 10.12), whereas the proof here essentially applies in X as well. I am indebted to Sadahiro Saeki for showing me how to dispense with all holomorphy hypotheses. One interesting, non-trivial classical consequence of 6.17 is that a local homeomorphism f of $\mathbb{C}$ into $\mathbb{C}$ satisfying $\lim_{|z|\to\infty}|f(z)| = \infty$ is necessarily bijective.

Theorem 6.17 *Let U be an open subset of $\mathbb{C}$, $0 < R \le \infty$, $f\colon U \mapsto D(0,R)$ a proper local homeomorphism. Then U has only finitely many components, each of which f maps homeomorphically onto $D(0,R)$.*

Proof: Being locally homeomorphic, f is (by 4.61) an open mapping. The proof of 6.13 shows that f being proper implies f is a closed mapping. Therefore $f(U)$ is a relatively open-and-closed subset of the connected $D(0,R)$, hence must be all of $D(0,R)$: f is surjective. For each $w \in D(0,R)$ the set $f^{-1}(w)$ is compact, but it can have no accumulation point, for that would violate the local injectivity of f. Hence $f^{-1}(w)$ is finite.

Let us designate as a *partial right-inverse* (of f) any continuous open mapping g from an open subset W of $D(0,R)$ into U such that $f\circ g$ coincides on W with the identity map. Note that such a g is necessarily one-to-one.

I: Let $g_j\colon W_j \mapsto U$ be a partial right-inverse ($j = 1,2$). Then

$$W_0 := \{w \in W_1 \cap W_2 : g_1(w) = g_2(w)\}$$

is open and relatively closed in $W_1 \cap W_2$.

Continuity of the g_j of course insures that W_0 is relatively closed. Fix $w_0 \in W_0$ and set $z_0 := g_j(w_0)$. Since f is locally univalent, z_0 has a neighborhood V_0 on which f is one-to-one. By continuity of g_j, w_0 has a neighborhood

$N_0 \subset W_1 \cap W_2$ such that $g_1(N_0) \cup g_2(N_0) \subset V_0$. For $w \in N_0$, $f(g_1(w)) = w = f(g_2(w))$, whence $g_1(w) = g_2(w)$, which means $w \in W_0$. That is, W_0 contains the whole neighborhood N_0 of its typical point w_0.

II: Choose and fix $a \in f^{-1}(0)$. For some $0 < r \leq R$ the local homeomorphism f has a partial right-inverse $g_r \colon D(0,r) \mapsto U$ such that $g_r(0) = a$. If $0 < s \leq r$, it follows from **I** that $g_r|D(0,s)$ is the only partial right-inverse g_s on $D(0,s)$ satisfying $g_s(0) = a$. Thus there is a largest $\rho \leq R$ for which there exists a partial right-inverse $g = g_\rho \colon D(0,\rho) \mapsto U$ satisfying $g(0) = a$.

For notational ease in the sequel, write $D_r := D(0,r)$, $0 < r \leq \infty$. Let $w_0 \in D_R \cap \overline{D}_\rho = f(U) \cap \overline{D}_\rho$ be given.

III: There exists $z_0 \in f^{-1}(w_0) \cap \overline{g(D_\rho)}$.

Indeed, consider any sequence $\{w_n\}_{n=1}^{\infty}$ in D_ρ which converges to w_0. Since f is a proper map, $f^{-1}(\{w_n\}_{n=0}^{\infty})$ is a compact subset of U, and it contains the sequence $\{g(w_n)\}_{n=1}^{\infty}$. So there exist $1 \leq n_1 < n_2 \cdots$ and $z_0 \in U$ such that $g(w_{n_j}) \to z_0$. Clearly $z_0 \in \overline{g(D_\rho)}$ and $w_{n_j} = f(g(w_{n_j})) \to f(z_0)$. Hence $w_0 = f(z_0)$.

IV: For *any* $z_0 \in f^{-1}(w_0) \cap \overline{g(D_\rho)}$ there is an $\varepsilon > 0$ such that g extends to a partial right-inverse $\tilde{g}$ on $D(w_0,\varepsilon) \cup D_\rho \subset D_R$ with $\tilde{g}(w_0) = z_0$.

To effect this, choose an $\varepsilon > 0$ and an open neighborhood V_0 of z_0 such that f maps V_0 homeomorphically onto $D(w_0,\varepsilon)$. Denote by h the function $(f|V_0)^{-1}$. We have $V_0 \cap g(D_\rho) \neq \varnothing$. If $z \in V_0 \cap g(D_\rho)$, then $f(z) \in D_\rho$ and $g(f(z))$ can be formed. Since f is one-to-one on the range of g, from $f(g(f(z))) = f(z)$ follows that $g(f(z)) = z$. On the other hand, $h(f(z)) = z$. Therefore $g(f(z)) = h(f(z))$. That is

$$h = g \text{ on } f(V_0 \cap g(D_\rho)) \subset f(V_0) \cap D_\rho = D(w_0,\varepsilon) \cap D_\rho.$$

Hence **I** insures that $h = g$ throughout the connected (convex) superset $D(w_0,\varepsilon) \cap D_\rho$. Thus an extension $\tilde{g}$ of g to $D(w_0,\varepsilon) \cup D_\rho$ is well defined by $\tilde{g} := h$ on $D(w_0,\varepsilon)$. Clearly $\tilde{g}$ is a partial right-inverse with $\tilde{g}(w_0) = h(w_0) = (f|V_0)^{-1}(w_0) = z_0$.

V: $\rho = R$.

Suppose contrariwise that $\rho < R$. According to **IV**, for each $w_0 \in \partial D_\rho$ there exists $\varepsilon = \varepsilon(w_0) > 0$ and a partial right-inverse g_{w_0} on $G_{w_0} := D(w_0,\varepsilon) \cup D_\rho$ which extends g. If also $w_1 \in \partial D_\rho$, then $g_{w_0} = g_{w_1}$ on the connected set $G_{w_0} \cap G_{w_1}$, by **I**. Thus from the family $\{g_w : w \in \partial D_\rho\}$ we get a single extension of g to a partial right-inverse $\tilde{g}$ on $\widetilde{G} := \bigcup\{G_w : w \in \partial D_\rho\}$ (cf. 5.71). As this open set contains $\overline{D}_\rho$, it also contains $D_{\tilde{\rho}}$ for some $\rho < \tilde{\rho} < R$.

But this violates the maximality of ρ. Hence $\rho = R$, and g is defined on the whole D_R.

VI: $g(D_R)$ is relatively closed in U.

This follows from the fact that g is a maximal partial right-inverse. Indeed, consider any $z_0 \in U \cap \overline{g(D_R)}$. Then $w_0 := f(z_0) \in D_R$. Therefore by **IV**, there is an extension $\tilde{g}$ of g which satisfies $\tilde{g}(w_0) = z_0$. Since however $\rho = R$, this $\tilde{g}$ is necessarily just g, so $g(w_0) = z_0$. That is, $z_0 \in g(D_R)$, for every $z_0 \in U \cap \overline{g(D_R)}$.

VII: The open and relatively closed subset $g(D_R)$ of U must contain any component that it meets, and being itself connected, it must therefore be a component, the component that contains a (see **II**). Of course, f maps it homeomorphically onto D_R.

VIII: If C is a component of U, then C is open (1.30) and $f|C$ is also a proper mapping into D_R. For if K is a compact subset of D_R, then $f^{-1}(K)$ is a compact subset of U, and C is a relatively closed (1.30) subset of U; so $(f|C)^{-1}(K) = C \cap f^{-1}(K)$ is compact. As noted in the first paragraph of this proof, $f|C$ is therefore surjective. Hence C meets $f^{-1}(0)$. As the latter is finite and components are disjoint, there can be only finitely many C. Taking the a in **II** to be a point in $C \cap f^{-1}(0)$, **VII** shows that f maps C homeomorphically onto D_R, thereby completing this long proof.

A complete classification of proper holomorphic self-maps of annuli is also elementary, but the results are surprising and contrast with 6.16. We shall consider two annuli $A(r_j, R_j)$ with $0 < r_j < R_j < \infty$ $(j = 1, 2)$. By scaling in the domain and range there is no loss in supposing $r_1 = r_2 = 1$.

Exercise 6.18 *Let* $A_j := \{z \in \mathbb{C} : 1 < |z| < R_j\}$, $1 < R_j < \infty$ $(j = 1, 2)$ *be two annuli and* $f\colon A_1 \mapsto A_2$ *a holomorphic boundary-preserving map.*

(i) *Prove that the following dichotomy prevails:*
either

(a) $$\lim_{|z|\to 1} |f(z)| = 1 \quad \textit{and} \quad \lim_{|z|\to R_1} |f(z)| = R_2$$

or

(b) $$\lim_{|z|\to 1} |f(z)| = R_2 \quad \textit{and} \quad \lim_{|z|\to R_1} |f(z)| = 1.$$

Hints: According to 6.13, the compact set $C(0, \sqrt{R_2})$ has a compact pre-image, which must therefore lie in $A(1 + \varepsilon_1, R_1 - \varepsilon_1)$ for some $0 < \varepsilon_1 < \frac{R_1 - 1}{2}$.

The f-image of the compact set $\bar{A}(1+\varepsilon_1, R_1-\varepsilon_1)$ is compact, so we have

$$(0)\ f\big(\bar{A}(1+\varepsilon_1, R_1-\varepsilon_1)\big) \subset A(1+\varepsilon_2, R_2-\varepsilon_2), \text{ for some } 0 < \varepsilon_2 < \frac{1}{2}(R_2-1).$$

The connected sets $A(1, 1+\varepsilon_1)$ and $A(R_1-\varepsilon_1, R_1)$ map into the disconnected union $A\big(1, \sqrt{R_2}\big) \cup A\big(\sqrt{R_2}, R_2\big)$. Hence one of four alternatives prevails:

$$(1)\quad f\big(A(1,1+\varepsilon_1)\big) \subset A\Big(1, \sqrt{R_2}\Big)\ \&\ f\big(A(R_1-\varepsilon_1, R_1)\big) \subset A\Big(\sqrt{R_2}, R_2\Big)$$

$$(2)\quad f\big(A(1,1+\varepsilon_1)\big) \subset A\Big(\sqrt{R_2}, R_2\Big)\ \&\ f\big(A(R_1-\varepsilon_1, R_1)\big) \subset A\Big(1, \sqrt{R_2}\Big)$$

$$(3)\quad f\big(A(1,1+\varepsilon_1)\big) \subset A\Big(1, \sqrt{R_2}\Big)\ \&\ f\big(A(R_1-\varepsilon_1, R_1)\big) \subset A\Big(1, \sqrt{R_2}\Big)$$

$$(4)\quad f\big(A(1,1+\varepsilon_1)\big) \subset A\Big(\sqrt{R_2}, R_2\Big)\ \&\ f\big(A(R_1-\varepsilon_1, R_1)\big) \subset A\Big(\sqrt{R_2}, R_2\Big).$$

Since f is boundary-preserving, $f(z)$ approaches ∂A_2 whenever z approaches ∂A_1. Hence (1) implies (a), and (2) implies (b). According to 6.15, f is surjective and this fact makes both (3) and (4) untenable. For example, if (3) holds, then

$$f(A_1) \subset A\Big(1, \sqrt{R_2}\Big) \cup A(1+\varepsilon_2, R_2-\varepsilon_2) \subset A(1, R_2-\varepsilon_2), \quad \text{by (0).}$$

(ii) *Write* $\alpha := \log R_2 / \log R_1$. *Show that if* (a) *prevails, then*

$$|f(z)| = |z|^\alpha \qquad \forall z \in A_1.$$

Hints: Notice that (a) says

$$\lim_{z \to \partial A_1} \big[\log|f(z)| - \alpha \log|z|\big] = 0,$$

so by the maximum principle the bounded harmonic function $\log|f(z)| - \alpha \log|z|$ vanishes identically in A_1.

(iii) *Show that if* F *is holomorphic in* A_1, $\alpha \in \mathbb{R}$, *and*

$$|F(z)| = |z|^\alpha \qquad \forall z \in A_1,$$

then α *is an integer and* $z^{-\alpha} F(z)$ *is constant.*

Hints: Let L denote the Principal Branch of the Logarithm in $\mathbb{C} \setminus [0, \infty)$. Then the holomorphic function $e^{-\alpha L(z)} F(z)$ has constant modulus 1 throughout the connected set $A_1 \setminus [0, \infty)$. Therefore 2.16(iii) says that for some $u \in \mathbb{T}$

$$(5) \qquad e^{-\alpha L(z)} F(z) = u \qquad \forall z \in A_1 \setminus [0, \infty).$$

Fix $1 < r < R_1$ and recall that $L(re^{it}) = \log r + it$ for $0 < t < 2\pi$. Using this and (5)

$$\frac{F\left(re^{i(2\pi-t)}\right)}{F(re^{it})} = \frac{e^{\alpha i(2\pi-t)}}{e^{\alpha it}} = e^{2\alpha i(\pi-t)}, \quad 0 < t < 2\pi.$$

Let $t \downarrow 0$ and cite continuity of F at r to infer that

$$1 = e^{2\alpha i\pi},$$

showing that α is an integer. [Alternatively, one can infer from (5) that $F'(z)/F(z) = \alpha/z$ for all $z \in A_1$ and using that realize α as $\mathrm{Ind}_{F\circ C(0,r)}(0)$.] Now (5) reads $z^{-\alpha}F(z) = u$, holding for all $z \in A_1 \setminus [0,\infty)$, hence for all $z \in A_1$, by continuity.

(iv) *Show that $R_2 = R_1^n$ for some positive integer n and that f has either the form $F(z) \equiv uz^n$ or the form $f(z) \equiv uR_2z^{-n}$, for some $u \in \mathbb{T}$. Infer that every boundary-preserving, holomorphic self-map of an annulus is in fact a conformal automorphism.*

Hints: If (a) prevails, combine (ii) and (iii). If (b) prevails for f, then (a) prevails for R_2/f.

(v) *Suppose F maps A_1 conformally onto A_2. Show that then $R_2 = R_1$ and F has either the form $F(z) \equiv uz$ or the form $F(z) \equiv uR_2/z$, for some $u \in \mathbb{T}$.*

Hint: Since F^{-1} exists and is continuous, F is a proper map. The integer n supplied by (iv) can only be 1, since F is one-to-one.

§ 3 Applications to Half-planes, Strips and Annuli

Exercise 6.19 (i) *Let $f\colon \mathbb{U} \mapsto \mathbb{U}$ be holomorphic. Show that*

$$\left|\frac{f(z) - f(z_0)}{f(z) - \bar{f}(z_0)}\right| \le \left|\frac{z - z_0}{z - \bar{z}_0}\right| \qquad \forall z,\ z_0 \in \mathbb{U},$$

and that equality holds for some distinct pair if and only if $f \in \mathrm{Aut}\,(\mathbb{U})$, in which case equality holds for all pairs.

Hints: For each $w_0 = x_0 + iy_0 \in \mathbb{U}$ define $\phi = \phi_{w_0}\colon \mathbb{U} \mapsto \mathbb{C}$ by

$$\begin{aligned}\phi(w) &:= \frac{w - w_0}{w - \overline{w}_0}, \quad w \in \mathbb{U}\\ &= \frac{w' - i}{w' + i}, \quad w' := \frac{w - x_0}{y_0} \in \mathbb{U}.\end{aligned}$$

Show (as in 2.5) that ϕ is a conformal map of $\mathbb{U}$ onto $\mathbb{D}$. Then form $\phi_{f(z_0)} \circ f \circ \phi_{z_0}^{-1}$ and apply Schwarz' lemma to it.

(ii) *For the f in* (i) *suppose that there is a positive $y \neq 1$ such that*

$$f(yi) = yf(i). \tag{1}$$

Show that $f(z) = |f(i)|z$ for all $z \in \mathbb{U}$.

Hints: (Erwin Mues and Helmut Köditz in REMMERT [1998].) There is no loss of generality in requiring that

$$|f(i)| = 1. \tag{2}$$

Apply (i) with $z_0 := i, z := yi$ to see that

$$\left|\frac{f(yi) - f(i)}{f(yi) - \bar{f}(i)}\right| \overset{(1)}{=} \left|\frac{yf(i) - f(i)}{yf(i) - \bar{f}(i)}\right| \leq \left|\frac{yi - i}{yi + i}\right| = \frac{|y-1|}{y+1}. \tag{3}$$

Due to (2) this is equivalent to $|yf(i) - \bar{f}(i)| \geq y + 1$, that is

$$y^2|f(i)|^2 - 2\,\mathrm{Re}[yf^2(i)] + |\bar{f}(i)|^2 \geq y^2 + 2y + 1,$$

so, again recalling (2), $\mathrm{Re}\, f^2(i) \leq -1$. Since $f^2(i) \in \mathbb{T}$, this tells us $f^2(i) = -1$, and then $f(i) \in \mathbb{U}$ yields

$$f(i) = i, \tag{4}$$

which means that equality prevails throughout (3). Consequently by (i), $f \in \mathrm{Aut}\,(\mathbb{U})$. Since f fixes the distinct points yi and i (by (1) and (4)), f must be the identity function on $\mathbb{U}$ (cf. 6.4(iii)).

Exercise 6.20 *Show that the conformal automorphisms of $\mathbb{U}$ are exactly all maps of the form*

$$\phi(z) = \frac{az+b}{cz+d}, \quad \text{where } a,\ b,\ c,\ d \in \mathbb{R} \text{ and } ad - bc > 0. \tag{$*$}$$

Hints: $f(z) := (z-i)/(z+i)$ maps $\mathbb{U}$ conformally onto $\mathbb{D}$. (See the hints to 6.19.) Therefore, the typical element of $\mathrm{Aut}\,(\mathbb{U})$ is $\phi := f^{-1} \circ \psi \circ f$, $\psi \in \mathrm{Aut}\,(\mathbb{D})$. Representing ψ as in the remark following 6.2 leads to the representation $(*)$ for ϕ. Why, conversely, is every map of form $(*)$ in $\mathrm{Aut}\,(\mathbb{U})$?

Exercise 6.21 *Suppose $\phi \in \mathrm{Aut}\,(\mathbb{U})$ has no fixed-point in $\mathbb{U}$.*

(i) *Show that there is* $\psi \in \operatorname{Aut}(\mathbb{U})$ *such that*

$$\psi\big(\phi(\psi^{-1}(z))\big) = \alpha z + \beta \qquad \forall z \in \mathbb{U}$$

for some $\alpha > 0$, $\beta \in \mathbb{R}$.

Hints: Let ϕ have the form $(*)$ in 6.20. If $c = d - a = 0$, then ϕ already has the desired form. Otherwise the equation $(**)$ $cx^2 + (d-a)x - b = 0$ has a solution x in $\mathbb{C}$, which says that ϕ fixes some $x \in \mathbb{C}$. By hypothesis ϕ fixes no $z \in \mathbb{U}$ and so by $(*)$ fixes no $\bar{z}$ with $z \in \mathbb{U}$. Hence $x \in \mathbb{R}$. Take any real $B < -x^2$ and form $\psi(z) := (xz + B)/(z - x)$. A little calculation and $(**)$ reveal that $\psi \circ \phi \circ \psi^{-1}$ has the form $z \mapsto \alpha z + \beta$ with $\alpha, \beta \in \mathbb{R}$, and since it maps $\mathbb{U}$ into $\mathbb{U}$, necessarily $\alpha > 0$.

(ii) *Show that if* α, $\beta \in \mathbb{R}$, $\alpha > 0$ *and* $\Phi(z) := \alpha z + \beta$, *then there is a conformal automorphism* Ψ *of* $\mathbb{U}$ *such that*

$$\begin{cases} \textit{either} & \Psi\big(\Phi(\Psi^{-1}(z))\big) = z + \beta \qquad \forall z \in \mathbb{U} \\ \textit{or} & \Psi\big(\Phi(\Psi^{-1}(z))\big) = \alpha z \qquad \forall z \in \mathbb{U}. \end{cases}$$

Hint: If $\alpha = 1$ or $\beta = 0$, there is nothing to prove. Otherwise, choose $\beta' := \pm 1$ so that $(\alpha - 1)\beta'/\beta > 0$, set $\alpha' := (\alpha - 1)\beta'/\beta$ and take $\Psi(z) := \alpha' z + \beta'$.

(iii) *There is a conformal automorphism* θ *of* $\mathbb{U}$ *such that*

$$\begin{cases} \textit{either} & \theta\big(\phi(\theta^{-1}(z))\big) = z + \beta \qquad \forall z \in \mathbb{U} \\ \textit{or} & \theta\big(\phi(\theta^{-1}(z))\big) = \alpha z \qquad \forall z \in \mathbb{U} \end{cases}$$

for some $\alpha > 0$, $\beta \in \mathbb{R}$.

Hint: Combine (i) and (ii).

Theorem 6.22 *Let* $S := \mathbb{R} \times (-\pi/2, \pi/2)$ *and* $\phi : S \mapsto S$ *be holomorphic. Then*

(i) $|\phi'(x)| \leq 1 \qquad \forall x \in \mathbb{R}$,

and

(ii) $|\phi'(x_0)| = 1$ *for some* $x_0 \in \mathbb{R}$ *implies that* $\phi(x_0)$ *is real.*

Proof: Fix $x_0 \in \mathbb{R}$ and let $x_1 := \operatorname{Re}\phi(x_0)$, $y_1 := \operatorname{Im}\phi(x_0) \in (-\pi/2, \pi/2)$. Define

$$\phi_1(z) := \frac{1+z}{1-z}, \qquad z \in \mathbb{D}$$
$$\phi_3(z) := e^{z - x_1}, \qquad z \in \mathbb{C}$$
$$\phi_4(z) := \frac{z - e^{iy_1}}{z + e^{-iy_1}}, \qquad z \in \mathbb{H}$$

As in hints to 2.6, computation confirms that

$$1 - |\phi_4(z)|^2 = \frac{4\cos y_1 \operatorname{Re} z}{|z + e^{-iy_1}|^2} > 0,$$

so ϕ_4 maps $\mathbb{H}$ into $\mathbb{D}$. Let l be the Principal Branch of the Logarithm in $\mathbb{C}_-$, so $l(\mathbb{H}) \subset S$ (see 3.43(ii)); finally, set $\phi_2 := x_0 + l$. One checks that $\phi_5 := \phi_4 \circ \phi_3 \circ \phi \circ \phi_2 \circ \phi_1$ is well defined, maps $\mathbb{D}$ into itself, and $\phi_5(0) = 0$. From Schwarz' lemma, $|\phi_5'(0)| \leq 1$. Computing via the Chain Rule,

$$\phi_5'(0) = \frac{e^{iy_1}}{\cos y_1}\phi'(x_0).$$

It follows that

$$|\phi'(x_0)| \leq \cos y_1 \leq 1.$$

Moreover, we see that $|\phi'(x_0)| = 1$ entails $\cos y_1 = 1$, whence $y_1 = 0$.

Theorem 6.23 *Let* $S\colon = \mathbb{R} \times (-\pi/2, \pi/2)$ *and* $\phi\colon S \mapsto S$ *be holomorphic. Then*

(i) $|\phi(x_2) - \phi(x_1)| \leq |x_2 - x_1| \qquad \forall x_1, x_2 \in \mathbb{R}.$

(ii) *Equality in* (i) *for some pair* $x_1 \neq x_2$ *implies that either* $\phi(z) \equiv z + c$ *or* $\phi(z) \equiv -z + c$, *for some real constant* c.

Proof: To get (i), integrate the inequality of the preceding theorem: if $x_2 \geq x_1$,

$$|\phi(x_2) - \phi(x_1)| = \left|\int_{x_1}^{x_2} \phi'\right| \leq \int_{x_1}^{x_2} |\phi'| \leq \int_{x_1}^{x_2} 1 = |x_2 - x_1|.$$

Equality here for some $x_1 < x_2$ implies that $|\phi'| \equiv 1$ in $[x_1, x_2]$. Then by (ii) of the preceding theorem ϕ, whence also ϕ', is real in $[x_1, x_2]$ and so $|\phi'| \equiv 1$ and connectedness mean (via 1.7) that either $\phi' \equiv 1$ or $\phi' \equiv -1$ in $[x_1, x_2]$. By the Fundamental Uniqueness Theorem (5.62), either $\phi' \equiv 1$ or $\phi' \equiv -1$ holds throughout the connected set S. It follows that for some constant c, either $\phi(z) \equiv z + c$ or $\phi(z) \equiv -z + c$. In either case, the fact that the range of ϕ lies in $\mathbb{R} \times (-\pi/2, \pi/2)$ clearly forces c to be real.

Theorem 6.24 *Let* $0 < r_j < R_j < \infty$, $A_j := \{z \in \mathbb{C} : r_j < |z| < R_j\}$ $(j = 1, 2)$, *and* $\gamma := C(0, r)$ *for some* $r_1 < r < R_1$. *Suppose* $f\colon A_1 \mapsto A_2$ *is holomorphic and set* $n := \operatorname{Ind}_{f\circ\gamma}(0)$. *Then*

(i) $|n| \leq \dfrac{\log(R_2/r_2)}{\log(R_1/r_1)}.$

(ii) *If equality holds in* (i), *then* $f(z) \equiv u\sqrt{R_2 r_2}(z/\sqrt{R_1 r_1})^n$ *for some* $u \in \mathbb{T}$,

and f *is surjective.* (Cf. 10.13(iii).)

Proof: Since the curves $f \circ C(0,r)$ are evidently loophomotopic in $\mathbb{C} \setminus \{0\}$ for $r_1 < r < R_1$, the number n is the same for all these r (4.12). The most convenient r to use in the proof is $r := \sqrt{R_1 r_1}$. We may also assume that $n \neq 0$. Let $M_j := \log(R_j/r_j)$, $F(z) := f(re^{iM_1 z/\pi})$. Then F is a holomorphic map of the strip $S := \mathbb{R} \times (-\pi/2, \pi/2)$ into A_2. By 5.35 it has a holomorphic logarithm L:

$$L \in H(S) \ \ \& \ \ e^L = F. \tag{1}$$

Since $F(S) = f(A_1) \subset A_2$, $\operatorname{Re} L(S) \subset (\log r_2, \log R_2)$. Therefore, if we set

$$\phi := \frac{i\pi}{M_2}\left[-L + \log\sqrt{R_2 r_2}\right], \tag{2}$$

then ϕ is a holomorphic map of S into S. Now for $t \in [0,1]$

$$(f \circ \gamma)(t) = f\left(re^{2\pi i t}\right) = F\left(\frac{2\pi^2}{M_1}t\right) \overset{(1)}{=} e^{L((2\pi^2/M_1)t)}.$$

It follows from this and the definition (4.2) of the index that

$$\begin{aligned} n &= \operatorname{Ind}_{f\circ\gamma}(0) \\ &= \frac{1}{2\pi i}\left[L\left(\frac{2\pi^2}{M_1}\right) + L(0)\right] \\ &= \frac{M_2}{2\pi^2}\left[\phi\left(\frac{2\pi^2}{M_1}\right) - \phi(0)\right], \quad \text{by (2).} \end{aligned} \tag{3}$$

Since ϕ meets all the hypotheses of the preceding theorem, we conclude that

$$|n| = \frac{M_2}{2\pi^2}\left|\phi\left(\frac{2\pi^2}{M_1}\right) - \phi(0)\right| \leq \frac{M_2}{2\pi^2}\left|\frac{2\pi^2}{M_1} - 0\right| = \frac{M_2}{M_1}. \tag{4}$$

This proves inequality (i). Moreover, if equality holds in (i), then from (4) $\left|\phi(2\pi^2/M_1) - \phi(0)\right| = 2\pi^2/M_1$, so from part (ii) of the preceding theorem we infer that for some real constant a

$$\text{either} \quad \phi(z) \equiv z + a \quad \text{or} \quad \phi(z) \equiv -z + a. \tag{5}$$

In fact, from (3) we see that the sign in (5) is that of n, i.e.,

$$\phi(z) \equiv \frac{n}{|n|}z + a. \tag{5$'$}$$

Insert $(5)'$ into (2) and (1) to see that

$$F(z) \equiv e^{M_2 i\phi(z)/\pi + \log\sqrt{R_2 r_2}} = \sqrt{R_2 r_2} e^{M_2 i(nz/|n|+a)/\pi}.$$

For any $\rho \in (r_1, R_1)$ and any real θ then

$$\begin{aligned} f\left(\rho e^{i\theta}\right) &= F\left(\frac{\pi}{M_1}\theta + \frac{i\pi}{M_1}\log\frac{r}{\rho}\right) \\ &= \sqrt{R_2 r_2} e^{M_2 ia/\pi} e^{(n/|n|)(M_2/M_1)(\log(\rho/r)+i\theta)}. \end{aligned}$$

Remembering that $|n| = M_2/M_1$, this gives finally

$$\begin{aligned} f\left(\rho e^{i\theta}\right) &= \sqrt{R_2 r_2} e^{M_2 ia/\pi} e^{n(\log(\rho/r)+i\theta)} \\ &= \frac{\sqrt{R_2 r_2}}{r^n} e^{M_2 ia/\pi}\left(\rho e^{i\theta}\right)^n. \end{aligned}$$

Recalling that $r = \sqrt{R_1 r_1}$ shows that this equation has the form claimed in (ii). Surjectivity is confirmed via routine calculation.

Remarks 6.25: This result will be quite important to us later on. Here is an alternative proof of part of it. (Cf. E. REICH [1966] and the review thereof in *Zentralblatt*; also p. 207 of MARDEN, RICHARDS and RODIN [1967].) Use 4.68 to write

$$f(z) = z^m e^{\phi(z)}, \qquad \forall z \in A_1$$

for some integer m and holomorphic function ϕ in A_1. Evidently then $m = n$. Take logarithms of the moduli of both sides of this equation and have, since $f(A_1) \subset A_2$,

$$\log r_2 \le \log|f(z)| = n\log|z| + \operatorname{Re}\phi(z) \le \log R_2.$$

For each $r_1 < r < R_1$, set $z = re^{i\theta}$ and integrate in θ:

$$\text{(1.}r\text{)} \qquad \log r_2 \le n\log r + \frac{1}{2\pi}\int_0^{2\pi} \operatorname{Re}\phi\left(re^{i\theta}\right)d\theta \le \log R_2.$$

The integral here is independent of r, by 5.5(i). Therefore if we subtract $(1.r)$ from $(1.R)$ and let $r \downarrow r_1$, $R \uparrow R_1$, we get $\log r_2 - \log R_2 \le n\log R_1 - n\log r_1 \le \log R_2 - \log r_2$.

Exercise 6.26 *Use an argument similar to that in the remarks above to prove the following* (A. J. MACINTYRE [1939b])*: If* $f\colon \bar{A}(r,R) \mapsto \mathbb{C}\setminus\{0\}$ *is continuous and holomorphic in* $A(r,R)$*, then*

$$\sup_{\theta\in\mathbb{R}}\left|\frac{f\left(Re^{i\theta}\right)}{f\left(re^{i\theta}\right)}\right| \ge \left(\frac{R}{r}\right)^n,$$

where $n := \operatorname{Ind}_{f\circ C(0,\rho)}(0)$ *for any* $r \le \rho \le R$.

§ 4 The Theorem of Carathéodory, Julia, Wolff, *et al.*

This section is about the forms Schwarz' Lemma takes when the interior fixed-point is allowed to migrate to the boundary (cf. 6.6). There are both disk and half-plane versions. For an excellent overview, highlighting the (non-euclidean) geometry and history involved, see the ingeniously entitled paper by H. P. Boas [2010].

Theorem 6.27 *Let* $f\colon \mathbb{H} \mapsto \mathbb{H}$ *be holomorphic and consider the non-negative real number*

$$c := \inf\left\{\frac{\operatorname{Re} f(z)}{\operatorname{Re} z} : z \in \mathbb{H}\right\}.$$

It satisfies

(i) $\operatorname{Re}[f(z) - cz] \ge 0 \qquad \forall z \in \mathbb{H}$,

and

(ii) $$\lim_{\substack{|z|\to\infty\\ z\in S_\varepsilon}} \frac{f(z)}{z} = \lim_{\substack{|z|\to\infty\\ z\in S_\varepsilon}} \frac{\operatorname{Re} f(z)}{\operatorname{Re} z} = c \qquad \forall 0<\varepsilon<1,$$

where $S_\varepsilon := \{z \in \mathbb{H} : \operatorname{Re} z > \varepsilon|z|\}$.

Moreover,

(iii) $$\lim_{\substack{|z|\to\infty\\ z\in S_\varepsilon}} f'(z) = c \qquad \forall 0<\varepsilon<1, \text{ and}$$

(iv) $$c \ge \varlimsup_{n\to\infty}\left[\frac{\operatorname{Re} z_n}{\operatorname{Re} f(z_n)}\cdot\left|\frac{f(z_n)}{z_n}\right|^2\right]$$

whenever $z_n \in \mathbb{H}$ *satisfy* $|z_n| \to \infty$ *and* $|f(z_n)| \to \infty$.

(v) $$|f'(z)| \le \frac{\operatorname{Re} f(z)}{\operatorname{Re} z} \qquad \forall z \in \mathbb{H}$$

and along every horizontal line

$$\frac{\operatorname{Re} f(z)}{\operatorname{Re} z} \quad \text{is a decreasing function of } \operatorname{Re} z.$$

(vi) *Finally,* f *also satisifies*

$$\frac{(\operatorname{Re} f(z))(\operatorname{Re} f(w))}{|f(z)+\bar f(w)|^2} \ge \frac{(\operatorname{Re} z)(\operatorname{Re} w)}{|w+\bar w|^2} \qquad \forall z, w \in \mathbb{H}.$$

Proof: Introduce the function

$$g(z) := f(z) - cz \quad (z \in \mathbb{H}). \tag{1}$$

Then $\operatorname{Re} g \geq 0$ in $\mathbb{H}$ and by the Maximum Principle we can assume strict inequality

$$\operatorname{Re} g > 0 \quad \text{in } \mathbb{H}. \tag{2}$$

(Otherwise $g(z) \equiv ia$, for some $a \in \mathbb{R}$, and all is clear.) Using 6.19 *mutatis mutandis*, we have

$$\left|\frac{g(z) - g(z_0)}{g(z) + \bar{g}(z_0)}\right| \leq \left|\frac{z - z_0}{z + \bar{z}_0}\right| \qquad \forall z, z_0 \in \mathbb{H}. \tag{3}$$

For the moment fix z and z_0 and denote the right-hand side of (3) by r. Evidently $r < 1$. Set

$$w := \frac{g(z) - i \operatorname{Im} g(z_0)}{\operatorname{Re} g(z_0)} \quad \text{[recalling (2)]}.$$

Then (3) says that

$$\frac{|w| - 1}{|w| + 1} \leq \left|\frac{w - 1}{w + 1}\right| \leq r.$$

Solve the extreme inequality for $|w|$ remembering that $r < 1$:

$$\begin{aligned} |w| &\leq \frac{1 + r}{1 - r} = \frac{(1 + r)^2}{1 - r^2} = \frac{\big(|z + \bar{z}_0| + |z - z_0|\big)^2}{|z + \bar{z}_0|^2 - |z - z_0|^2} \\ &= \frac{\big(|z + \bar{z}_0| + |z - z_0|\big)^2}{4(\operatorname{Re} z)(\operatorname{Re} z_0)} \leq \frac{\big(|z| + |z_0|\big)^2}{(\operatorname{Re} z)(\operatorname{Re} z_0)}. \end{aligned} \tag{4}$$

Then

$$\begin{aligned} \left|\frac{g(z)}{z}\right| &\leq \left|\frac{i \operatorname{Im} g(z_0)}{z}\right| + \left|\frac{g(z) - i \operatorname{Im} g(z_0)}{z}\right| = \left|\frac{\operatorname{Im} g(z_0)}{z}\right| + |w| \left|\frac{\operatorname{Re} g(z_0)}{z}\right| \\ &\overset{(4)}{\leq} \left|\frac{\operatorname{Im} g(z_0)}{z}\right| + \frac{\big(|z| + |z_0|\big)^2}{(\operatorname{Re} z)(\operatorname{Re} z_0)} \cdot \frac{\operatorname{Re} g(z_0)}{|z|}. \end{aligned}$$

If now $z \in S_\varepsilon$, this inequality yields

$$\begin{aligned} \left|\frac{g(z)}{z}\right| &\leq \left|\frac{\operatorname{Im} g(z_0)}{z}\right| + \frac{\big(|z| + |z_0|\big)^2}{\varepsilon |z|^2} \cdot \frac{\operatorname{Re} g(z_0)}{\operatorname{Re} z_0} \\ &= \left|\frac{\operatorname{Im} g(z_0)}{z}\right| + \frac{1}{\varepsilon}\left(1 + \left|\frac{z_0}{z}\right|\right)^2 \frac{\operatorname{Re} g(z_0)}{\operatorname{Re} z_0}. \end{aligned} \tag{5}$$

Given $\delta > 0$, the definition of c insures that there exists a $z_0 \in \mathbb{H}$ such that $\operatorname{Re} g(z_0)/\operatorname{Re} z_0 < \varepsilon\delta$. It is then clear that if ($z \in S_\varepsilon$ and) $|z|$ is sufficiently large, the right-hand side of (5) is less than δ. This proves half of the claim (ii). For the other half, just note that

$$\left|\frac{\operatorname{Re} g(z)}{\operatorname{Re} z}\right| \le \left|\frac{g(z)}{z}\right|\left|\frac{z}{\operatorname{Re} z}\right| < \frac{1}{\varepsilon}\left|\frac{g(z)}{z}\right| \qquad \forall z \in S_\varepsilon.$$

For (iii), re-write (3) as

$$\left|\frac{g(z)-g(z_0)}{z-z_0}\right| \le \left|\frac{g(z)+\bar{g}(z_0)}{z+\bar{z}_0}\right|$$

and let $z_0 \to z$ to get

$$\text{(6)} \qquad |f'(z)-c| = |g'(z)| \le \left|\frac{\operatorname{Re} g(z)}{\operatorname{Re} z}\right| = \frac{\operatorname{Re} g(z)}{\operatorname{Re} z}, \quad \text{by (ii)}$$
$$\to 0 \quad \text{as } |z| \to \infty,\ z \in S_\varepsilon.$$

Finally, apply (3) to the original function f, square both sides, cross-multiply and simplify. The calculations are formidable, but result in

$$\begin{aligned}\frac{\operatorname{Re} f(z)}{\operatorname{Re} z}\cdot\frac{\operatorname{Re} f(z_0)}{\operatorname{Re} z_0} &\ge \frac{|f(z)|^2+|f(z_0)|^2-2\operatorname{Im} f(z)\operatorname{Im} f(z_0)}{|z|^2+|z_0|^2-2\operatorname{Im} z\operatorname{Im} z_0}\\ &\ge \frac{|f(z)|^2+|f(z_0)|^2-2|f(z)||f(z_0)|}{|z|^2+|z_0|^2+2|z||z_0|}\\ &= \left(\frac{|f(z)|-|f(z_0)|}{|z|+|z_0|}\right)^2.\end{aligned}$$

Thus

$$\frac{\operatorname{Re} f(z_0)}{\operatorname{Re} z_0} \ge \frac{\operatorname{Re} z}{\operatorname{Re} f(z)}\left|\frac{f(z)}{z}\right|^2\left(1-\left|\frac{f(z_0)}{f(z)}\right|\right)^2\left(1+\left|\frac{z_0}{z}\right|\right)^{-2} \qquad \forall z, z_0 \in \mathbb{H}.$$

Take $z = z_n$, let $n \to \infty$ and use the facts that $|z_n| \to \infty$ and $|f(z_n)| \to \infty$. We get

$$\text{(7)} \qquad \frac{\operatorname{Re} f(z_0)}{\operatorname{Re} z_0} \ge \varlimsup_{n\to\infty}\left[\frac{\operatorname{Re} z_n}{\operatorname{Re} f(z_n)}\left|\frac{f(z_n)}{z_n}\right|^2\right] \qquad \forall z_0 \in \mathbb{H}.$$

Then (iv) follows from (7) and the definition of c.

The inequality in (v) follows from (6), the non-negativity of c, and the triangle-inequality.

Writing $h := \operatorname{Re} f$ and $x := \operatorname{Re} z$, routine calculation shows that

$$D_1\left(\frac{h(z)}{x}\right) = \frac{1}{x}\left(D_1 h(z) - \frac{1}{x}h(z)\right) = \frac{1}{x}\left(\operatorname{Re} f'(z) - \frac{\operatorname{Re} f(z)}{\operatorname{Re} z}\right)$$

which is non-positive according to the inequality in (v). It follows that $h(z)/x$ is a non-increasing function of x.

The inequality in (vi) is best accessed with the help of a useful identity:

$$\left|\frac{z-w}{z+w}\right|^2 = 1 - \frac{(4\operatorname{Re} z)(\operatorname{Re} w)}{|z+\bar{w}|^2} \qquad \forall z, w \in \mathbb{H}, \tag{8}$$

which is confirmed by a routine calculation, made more transparent by assuming (with no loss of generality) that $\operatorname{Re} z = 1$. Again the $\mathbb{H}$-version of 6.19 comes into play, supplying the inequality

$$\left|\frac{f(z) - f(w)}{f(z) + \bar{f}(w)}\right|^2 \leq \left|\frac{z-w}{z+\bar{w}}\right|^2 \qquad \forall z, w \in \mathbb{H}.$$

Use (8) to transform each side of this inequality, resulting in (vi).

Interesting and non-trivial facts emerge when we transfer the preceding via a conformal map to the unit disk. The transferring function is fractional-linear, so it maps the boundary rays of the approach-regions S_ε in $\mathbb{H}$ onto arcs of circles in $\mathbb{D}$. The appropriate approach-regions in $\mathbb{D}$ are therefore lens-shaped. However, as the reader can easily check, for any $v \in \mathbb{D}$ the triangle $\operatorname{co}\{1, v, \bar{v}\} \setminus \{1\}$ lies in such a lens, and the part of any lens near 1 lies in such a triangle. So here too it is really a question of sectorial or angular approach (to 1), which is often also called *non-tangential* approach.

Exercise 6.28 (Angular Derivative) *Let F be a holomorphic self-map of $\mathbb{D}$. Recall* (2.5) *that $\phi(z) := (1+z)/(1-z)$ maps $\mathbb{D}$ conformally onto $\mathbb{H}$, and for each $0 < \varepsilon < 1$ define*

$$S_\varepsilon := \{w \in \mathbb{C} : \operatorname{Re} w > \varepsilon|w|\} \subset \mathbb{H}$$
$$C_\varepsilon := \left\{z \in \mathbb{C} : 1 - |z|^2 > \varepsilon|1 - z^2|\right\} \subset \mathbb{D}.$$

(i) *Show that*

$$C_\varepsilon = \phi^{-1}(S_\varepsilon) = D\left(\frac{i\varepsilon}{\sqrt{1-\varepsilon^2}}, \frac{1}{\sqrt{1-\varepsilon^2}}\right) \bigcap D\left(\frac{-i\varepsilon}{\sqrt{1-\varepsilon^2}}, \frac{1}{\sqrt{1-\varepsilon^2}}\right),$$

then sketch these convex regions.

Hints: The first equality is a straightforward calculation. The second equality is also a calculation, but it is perhaps easier if one first introduces $\delta := \varepsilon/\sqrt{1-\varepsilon^2}$ and notes that the condition for $w = \phi(z)$ ($z \in \mathbb{D}$) to be in S_ε translates into $\pm\delta \operatorname{Im} w < \operatorname{Re} w$.

(ii) *Define* $f := \phi \circ F \circ \phi^{-1} : \mathbb{H} \mapsto \mathbb{H}$ *and show that*

$$\frac{f(\phi(z))}{\phi(z)} = \frac{1-z}{1-F(z)} \cdot \frac{1+F(z)}{1+z} \qquad \forall z \in \mathbb{D}. \tag{1}$$

(iii) *Let* $c \in \mathbb{R}^+$ *be as in the preceding theorem. Show that for each* $0 < \varepsilon < 1$

$$\lim_{\substack{z\to 1\\ z\in C_\varepsilon}} \frac{1-F(z)}{1-z} = \frac{1}{c} \quad \textit{if } c > 0, \tag{2}$$

whereas

$$\lim_{\substack{z\to 1\\ z\in C_\varepsilon}} \left|\frac{1-F(z)}{1-z}\right| = \infty \quad \textit{if } c = 0. \tag{3}$$

Hints: Fix $0 < \varepsilon < 1$. Evidently $|\phi(z)| \to \infty$ as $z \to 1$. Therefore (i), (1) and 6.27(ii) yield

$$\lim_{\substack{z\to 1\\ z\in C_\varepsilon}} \frac{1-z}{1-F(z)} \cdot \frac{1+F(z)}{1+z} = c. \tag{4}$$

Consider two cases.

Case I: $\lim_{z\to 1, z\in C_\varepsilon} F(z) = 1$. In this case, (2) and (3) follow immediately from (4).

Case II: There is a sequence in C_ε convergent to 1 on which $|F-1|$ remains bounded away from 0. In this case, it follows from (4) (and the boundedness of F) that $c = 0$. Now consider any $z_n \in C_\varepsilon$ such that $z_n \to 1$. We have for each n

$$\text{either} \quad |1-F(z_n)| \geq 1 \quad \text{or} \quad |1+F(z_n)| \geq 1. \tag{5}$$

And by (1)

$$\left|\frac{1-F(z_n)}{1-z_n}\right| = \left|\frac{\phi(z_n)}{f(\phi(z_n))}\right| \left|\frac{1+F(z_n)}{1+z_n}\right| \geq \left|\frac{\phi(z_n)}{f(\phi(z_n))}\right| \frac{|1+F(z_n)|}{2}. \tag{6}$$

From (5) and (6) above we have for each n

$$(7)\qquad \begin{cases} \text{either} & \left|\dfrac{1-F(z_n)}{1-z_n}\right| \geq \dfrac{1}{|1-z_n|} \\ \text{or} & \left|\dfrac{1-F(z_n)}{1-z_n}\right| \geq \dfrac{1}{2}\cdot\left|\dfrac{\phi(z_n)}{f\big(\phi(z_n)\big)}\right|. \end{cases}$$

Since $z_n \to 1$, we have $|\phi(z_n)| \to \infty$ and therefore (recalling (i) and 6.27(ii))

$$(8)\qquad \left|\frac{\phi(z_n)}{f\big(\phi(z_n)\big)}\right| \to \left|\frac{1}{c}\right| = \infty.$$

From (7) and (8) we learn that

$$\left|\frac{1-F(z_n)}{1-z_n}\right| \to \infty.$$

(iv) *Show that if $c > 0$, then for each $0 < \varepsilon < 1$*

$$\lim_{\substack{z\to 1\\ z\in C_\varepsilon}} F'(z) = \frac{1}{c}.$$

How do these derivatives behave when $c = 0$?

Hints: By the Chain Rule

$$F'(z) = \left[\frac{1-F(z)}{1-z}\right]^2 \cdot f'\big(\phi(z)\big).$$

Now cite 6.27(iii) and (2) above. As to the query, let α be *any* complex number. Choose integer $n > |\alpha|$ and form $F(z) := \alpha z^n/n$. For this F the associated c is 0, since $|F-1|$ is bounded away from 0. However, $F'(z) \to \alpha$ as $z \to 1$. Therefore if $c = 0$, any number is a possible value for the limit in (iv).

(v) *Show that if $F(0) = 0$, then $c \leq 1$.* (See UNKELBACH [1938], [1940], HERZIG [1940] and OSSERMAN [2000] for refinements of this inequality.)

Hint: By 6.1 we have

$$(9)\qquad |F(z)| \leq |z| \qquad \forall z \in \mathbb{D}.$$

For $0 < x < 1$ then

$$\left|\frac{1-F(x)}{1-x}\right| \geq \frac{1-|F(x)|}{|1-x|} \geq \frac{1-|x|}{|1-x|} = \frac{1-x}{1-x} = 1.$$

It follows from (2) that either $c = 0$ or $1/c \geq 1$.

(vi) *Suppose there exist $z_n \in \mathbb{D}$ such that*

$$z_n \to 1, \quad F(z_n) \to 1 \quad \textit{and} \quad \lim_{n\to\infty} \frac{1-|F(z_n)|}{1-|z_n|} =: \alpha \quad \textit{is finite.}$$

Show that then $\alpha > 0$ and $c \geq 1/\alpha$; consequently the limits in (2) *and* (iv) *exist.*

Hint: Set $w_n := \phi(z_n)$ and note that $|w_n| \to \infty$, $|f(w_n)| = |\phi(F(z_n))| \to \infty$. Then check by (routine) calculation that

$$\frac{\operatorname{Re} w_n}{\operatorname{Re} f(w_n)} \left|\frac{f(w_n)}{w_n}\right|^2 = \frac{1-|z_n|}{1-|F(z_n)|} \cdot \frac{1+|z_n|}{1+|F(z_n)|} \cdot \left|\frac{1+F(z_n)}{1+z_n}\right|^2.$$

Finally, cite 6.27(iv).

(vii) *Recalling how $c \in [0,\infty)$ is defined in the statement of* 6.27, *show that*

$$\frac{1}{c} = \sup_{z\in\mathbb{D}} \frac{1-|z|^2}{|1-z|^2} \cdot \frac{|1-F(z)|^2}{1-|F(z)|^2} = \sup_{z\in\mathbb{D}} \operatorname{Re}\left(\frac{1+z}{1-z}\right) \Big/ \operatorname{Re}\left(\frac{1+F(z)}{1-F(z)}\right)$$

and infer that

$$\frac{|1-F(z)|^2}{1-|F(z)|^2} \leq \frac{1}{c}\frac{|1-z|^2}{1-|z|^2} \qquad \forall z \in \mathbb{D} \tag{10}$$

$$F\left(D\left(\frac{1}{1+a}, \frac{a}{1+a}\right)\right) \subset D\left(\frac{c}{c+a}, \frac{a}{c+a}\right) \qquad \forall a > 0. \tag{11}$$

Hint: (11) is trivial if $c = 0$. To get (11) from (10) when $c > 0$, "complete the square" [as was done in the derivation of (6) in the proof of 6.6] to confirm that for all $w \in \mathbb{D}$, $r > 0$

$$\frac{|1-w|^2}{1-|w|^2} < r \iff \left|w - \frac{1}{1+r}\right| < \frac{r}{1+r}. \tag{12}$$

Alternatively, (ii) can be gotten from (12) via the identity $F = \phi^{-1} \circ f \circ \phi$ and the fact [6.27(i)] that f maps each $r + \mathbb{H}$ into $cr + \mathbb{H}$.

(viii) *By way of a converse to* (vi), *show that if $c > 0$, then there exist $z_n \in \mathbb{D}$ such that*

$$z_n \to 1, F(z_n) \to 1 \quad \textit{and} \quad \lim_{n\to\infty} \frac{1-|F(z_n)|}{1-|z_n|} = \lim_{z\to 1} \frac{1-|F(z)|}{1-|z|} = \frac{1}{c}.$$

Hints (using ideas from ABATE [1989], which also lead to an alternative treatment of this whole exercise)**:** Call this limit inferior α. If we take $z = x \in [0,1)$, then

$$0 < \frac{1-|F(x)|}{1-|x|} \leq \frac{|1-F(x)|}{1-|x|} = \left|\frac{1-F(x)}{1-x}\right| \to \frac{1}{c} \quad \text{as } x \uparrow 1$$

by (iii). Hence $0 \leq \alpha \leq 1/c$. In particular, α is finite. Choose $w_n \in \mathbb{D}$ with $w_n \to 1$ and $\frac{1-|F(w_n)|}{1-|w_n|} \to \alpha$. Since α is finite, this entails that $|F(w_n)| \to 1$, and so by factoring we see that

$$(13) \qquad \frac{1-|F(w_n)|^2}{1-|w_n|^2} \to \alpha.$$

A subsequence of $\{F(w_n)\}$ converges to some $u \in \mathbb{T}$. Letting w run through this subsequence, it follows from (13) and 6.3(v) that

$$(14) \qquad \frac{|1-\bar{u}F(z)|^2}{1-|F(z)|^2} \leq \alpha \frac{|1-z|^2}{1-|z|^2} \qquad \forall z \in \mathbb{D}.$$

A consequence of (14) is that $\alpha \neq 0$. Use (12) [as in the proof of 6.6] to see that inequality (14) is equivalent to

$$(15) \qquad F\left(D\left(\frac{1}{1+r}, \frac{r}{1+r}\right)\right) \subset D\left(\frac{u}{1+\alpha r}, \frac{\alpha r}{1+\alpha r}\right) \qquad \forall r > 0.$$

From (15) and (11) we infer that $u = 1$. Indeed, if $u \neq 1$ and $r > 0$ is sufficiently small, then the disks $D(\frac{u}{1+\alpha r}, \frac{\alpha r}{1+\alpha r})$ and $D(\frac{c}{c+r}, \frac{r}{c+r})$, which are internally tangent to $\mathbb{T}$ at u and 1, respectively, will be disjoint, so that not both (15) and (11) can hold. With $u = 1$, inequality (14) reads

$$\frac{|1-F(z)|^2}{1-|F(z)|^2} \leq \alpha \frac{|1-z|^2}{1-|z|^2} \qquad \forall z \in \mathbb{D}$$

and with (vii) this implies that $1/c \leq \alpha$. Since we already have $\alpha \leq 1/c$, we are done. [Alternatively, $1/c \leq \alpha$ follows from (vi), since we have a subsequence of $\{F(w_n)\}$ converging to $u = 1$.]

(ix) (Generalization of (v)) *Show that if F has a fixed-point in $\mathbb{D}$, then $c \leq 1$.*

Hint: In this case the equation in (vii) says that $1/c \geq 1$.

(x) *Show that if F has no fixed-point in $\mathbb{D}$ and $u = 1$ is its Denjoy-Wolff point from 6.6, then $c \geq 1$. Defining $F(1) := 1$, we thus have*

$$(16) \qquad \lim_{\substack{z\to 1\\ z\in C_\varepsilon}} F(z) = F(1) \text{ and } \lim_{\substack{z\to 1\\ z\in C_\varepsilon}} \frac{F(z)-F(1)}{z-1} \text{ exists in } (0,1] \qquad \forall\, 0 < \varepsilon < 1.$$

Hint: Combine (5) in the proof of 6.6 with the equality in (vii) above to see that $1/c \leq 1$. Therefore by (iii) the limits exist.

(xi) *Show that if F has no fixed-point in $\mathbb{D}$ and 1 is not its Denjoy-Wolff point, then $c < 1$.*

Hints: Suppose contrariwise that $c \geq 1$. Then from (11) it follows that

$$F\left(D\left(\frac{1}{1+a}, \frac{a}{1+a}\right)\right) \subset D\left(\frac{1}{1+a}, \frac{a}{1+a}\right) \qquad \forall a > 0.$$

The very definition of the Denjoy-Wolff point u of F in 6.6 entails that

$$F\left(D\left(\frac{u}{1+a}, \frac{a}{1+a}\right)\right) \subset D\left(\frac{u}{1+a}, \frac{a}{1+a}\right) \qquad \forall a > 0.$$

Since by hypothesis $u \neq 1$, the argument at the beginning of the 6.6 proof produces from these two families of inclusions a fixed-point in $\mathbb{D}$.

(xii) (Partial converse of (x)) *Show that if $c > 1$ (that is, by* (iii)*, F has at 1 an angular derivative less than 1), then F has no fixed-point in $\mathbb{D}$ and $u = 1$ is its Denjoy-Wolff point.*

Hints: Setting $\rho := 1/c \in (0, 1)$, inclusion (11) reads

$$\begin{aligned} F\left(D\left(\frac{1}{1+a}, \frac{a}{1+a}\right)\right) &\subset D\left(\frac{1}{1+\rho a}, \frac{\rho a}{1+\rho a}\right) \\ &\subset D\left(\frac{1}{1+a}, \frac{a}{1+a}\right) \qquad \forall a > 0. \end{aligned} \tag{11$_+$}$$

Any two of the circles $C\left(\frac{1}{1+b}, \frac{b}{1+b}\right)$, $b > 0$ have only the point 1 in common, and the union of all of them is $\mathbb{D} \cup \{1\}$. Therefore the first inclusion in $(11)_+$ implies that every such circle is disjoint from its F-image. Hence F has no fixed-point in $\mathbb{D}$. Consequently, $(11)_+$ and the uniqueness clause in 6.6 insure that 1 is the Denjoy-Wolff point of F.

Summary: Angular Limits and Angular Derivatives: The exceptional role of the point 1 here is misleading. The theory really concerns an arbitrary holomorphic self-map ψ of $\mathbb{D}$ and the limiting behavior of the "difference quotient" $\frac{\psi(z)-w}{z-u}$ for points $u, w \in \mathbb{T}$ when z approaches $\mathbb{T}$ through a lens uC_ε. This behavior is described by applying 6.28 to $F(z) := \bar{w}\psi(uz)$. First, some overdue notation: For $g : \mathbb{D} \mapsto \mathbb{C}$ and $u \in \mathbb{T}$, $v \in \mathbb{C}$, say that g has *angular limit* v at u and write $\angle \lim_{z \to u} g(z) = v$, if $\lim_{\substack{z\to u \\ z \in uC_\varepsilon}} g(z) = v$ for all $0 < \varepsilon < 1$. We are concerned with when

$$\angle \lim_{z \to u} \frac{w - \psi(z)}{u - z} \text{ exists in } \mathbb{C} \text{ for some } w \in \mathbb{T}. \tag{17}$$

and when

$$\angle \lim_{z \to u} \psi'(z) \text{ exists in } \mathbb{C}. \tag{18}$$

Clearly (17) implies that ψ has angular limit w at u and in this case we naturally *define* $\psi(u) := w$, call the limit in (17) the *angular derivative* of ψ at u and denote it $\psi'(u)$. Contact with 6.28 is provided by the three obvious identities

$$\begin{aligned} \frac{w - \psi(z)}{u - z} &= \frac{1 - F(\bar{u}z)}{1 - \bar{u}z} \cdot \frac{w}{u}, \\ \psi'(z) &= F'(\bar{u}z) \cdot \frac{w}{u}, \\ \frac{1 - |\psi(z)|}{1 - |z|} &= \frac{1 - |F(\bar{u}z)|}{1 - |\bar{u}z|}. \end{aligned} \tag{19}$$

In view of these, (2), (3) and (iv) tell us (17) implies that $c > 0$, that (18) holds, and that these limits coincide and equal $\frac{1}{c}\frac{w}{u}$. In fact, (17) and (18) are equivalent. For if the angular limit in (18) is called L and used to extend the function ψ' to $\mathbb{D} \cup \{u\}$, then ψ' is continuous in each $u\overline{C}_\varepsilon$, and so continuous in co$\{0, z, u\}$ for each $z \in \mathbb{D}$. (Why?) Consequently, the boundary version (5.2) of Cauchy's Integral Theorem in triangles is valid for ψ' and yields

$$\psi(z) - \psi(0) - \int_{[0,u]} \psi' = \int_{[0,z]} \psi' - \int_{[0,u]} \psi' = \int_{[u,z]} \psi' \qquad \forall z \in \mathbb{D}. \tag{20}$$

Given $\varepsilon, \delta > 0$, $\exists r > 0$ such that $|\psi'(z) - L| < \delta$ for all $z \in D(u, r) \cap \overline{C}_\varepsilon$. For such z the whole interval $[u, z]$ lies in this convex set, so $|\psi' - L| < \delta$ throughout this interval, showing that

$$\left| L - \frac{1}{z - u} \int_{[u,z]} \psi' \right| = \frac{1}{|u - z|} \left| \int_{[u,z]} (L - \psi') \right| \leq \delta.$$

Thus,

$$L = \lim_{\substack{z \to u \\ z \in uC_\varepsilon}} \frac{1}{z - u} \int_{[u,z]} \psi' \qquad \forall \varepsilon > 0. \tag{21}$$

Then, as z converges to u inside any uC_ε

$$\frac{1}{z - u} \left[\psi(z) - \psi(0) - \int_{[0,u]} \psi' \right] \overset{(20)}{=} \frac{1}{z - u} \int_{[u,z]} \psi' \overset{(21)}{\to} L. \tag{22}$$

Multiplying (22) by $z - u$ shows that

$$\angle \lim_{z \to u} \psi(z) \text{ exists and equals } \psi(0) + \int_{[0,u]} \psi'.$$

Since whenever it exists this angular limit serves to *define* $\psi(u)$, (22) now affirms that

$$\angle \lim_{z\to u} \frac{\psi(z)-\psi(u)}{z-u} \text{ exists and equals } L,$$

and confirms that (17) follows from (18), completing the proof of the claimed equivalence of the two.

Note that for $0 < r < 1$

$$\frac{1-|\psi(ru)|}{1-|ru|} = \frac{|w|-|\psi(ru)|}{|u-ru|} \leq \frac{|w-\psi(ru)|}{|u-ru|}.$$

Consequently, (17) implies that

$$\alpha := \varliminf_{z\to u} \frac{1-|\psi(z)|}{1-|z|} < \infty. \tag{23}$$

Conversely, if (23) holds, then a routine compactness argument produces $\zeta_n \in \mathbb{D}$ and $w \in \mathbb{T}$ such that $\zeta_n \to u$, $(1-|\psi(\zeta_n)|)/(1-|\zeta_n|) \to \alpha$ and $\psi(\zeta_n) \to w$. In view of (19), F therefore satisfies the hypotheses of (vi), with $z_n := \bar{u}\zeta_n$. It follows from (vi) that

$$\angle \lim_{\zeta\to 1} \frac{1-F(\zeta)}{1-\zeta} = \frac{1}{c} < \infty,$$

that is, via (19), (17) holds. In summary, (17) is equivalent to (18) and to (23). Note too that (buried in the hints to (viii)) $\alpha = 1/c$.

But there is one more useful equivalent to these, namely

$$\varlimsup_{r\uparrow 1} |\psi(ru)| = 1 \quad \text{and} \quad \varlimsup_{r\uparrow 1} |\psi'(ru)| < \infty. \tag{24}$$

Clearly (17) and (18) imply (24). Suppose conversely that (24) holds. The equality there and a compactness argument supply $w \in \mathbb{T}$ and $r_n \uparrow 1$ such that

$$\lim_{n\to\infty} \psi(r_n u) = w.$$

The inequality in (24) means that for some $N \in \mathbb{N}$, $|\psi'(tu)| \leq N$ for all $0 \leq t < 1$ whence for all $0 \leq r < 1$, $n \in \mathbb{N}$

$$|\psi(r_n u) - \psi(ru)| = \left| \int_r^{r_n} \psi'(tu)dt \right| \leq N|r_n - r|.$$

Letting $n \to \infty$,

$$|w - \psi(ru)| \leq N|1-r| \qquad \forall 0 \leq r < 1.$$

From this, by a now familiar march

$$\varliminf_{z\to u} \frac{1-|\psi(z)|}{1-|z|} \le \varliminf_{r\uparrow 1} \frac{1-|\psi(ru)|}{1-|ru|} \le \varliminf_{r\uparrow 1} \frac{|w-\psi(ru)|}{1-r} \le N < \infty.$$

That is, (23) holds.

Gathering everything together, we have proved that:

For any holomorphic $\psi : \mathbb{D} \mapsto \mathbb{D}$ and any $u \in \mathbb{T}$

$$(17) \Leftrightarrow (18) \Leftrightarrow (23) \Leftrightarrow (24)$$

and when these hold, $w = \angle \lim_{z\to u} \psi(z) =: \psi(u)$, each of the limits in (17)*and* (18) *equals $\alpha w/u = \alpha\psi(u)/u =: \psi'(u)$, and*

$$|\psi'(u)| = \sup_{z\in\mathbb{D}} \frac{1-|z|^2}{|u-z|^2} \cdot \frac{|\psi(u)-\psi(z)|^2}{1-|\psi(z)|^2} = \sup_{z\in\mathbb{D}} \operatorname{Re}\left(\frac{u+z}{u-z}\right) \Big/ \operatorname{Re}\left(\frac{\psi(u)+\psi(z)}{\psi(u)-\psi(z)}\right),$$

as routine calculation with the equalities in 6.28(vii) reveals.

Especially interesting is the case $w = u$, that is, a "boundary fixed-point" is at hand. Suppose ψ has no fixed-point in $\mathbb{D}$ and $u \in \mathbb{T}$ is its Denjoy-Wolff point given by 6.6. Then (17) holds with $w = u$, and $0 < \psi'(u) \le 1$. For any *other* $u \in \mathbb{T}$ where (17) holds with $w = u$, $\psi'(u) > 1$. Conversely, if (17) holds with $w = u$ and $|\psi'(u)| < 1$, then ψ has no fixed-point in $\mathbb{D}$ and u is its Denjoy-Wolff (fixed-)point. On the other hand, if ψ fixes some $z \in \mathbb{D}$, then *every* $u \in \mathbb{T}$ at which (17) holds with $w = u$, satisfies $|\psi'(u)| = \alpha \ge 1$. This all follows from 6.28(ix), (x), (xi).

The prototype of all of this is $\psi(z) := e^{i\theta} \cdot \frac{z-a}{1-\bar{a}z}$ $(\theta \in \mathbb{R}, a \in \mathbb{D})$, for which a routine calculation confirms (using the first equality in the 2.6 hints) that

$$\psi'(u) = \frac{1-|a|^2}{|1-\bar{a}u|^2} \cdot \frac{\psi(u)}{u} \quad \text{and} \quad \lim_{z\to u} \frac{1-|\psi(z)|}{1-|z|} = |\psi'(u)| \qquad \forall u \in \mathbb{T}.$$

An example in VISSER [1931-1932], when transferred from $\mathbb{H}$ to $\mathbb{D}$ via ϕ, fulfills (17) and (18) but the limits do not exist if z approaches u *arbitrarily* from inside $\mathbb{D}$.

Remark: It is a deeper fact that each bounded $\psi \in H(\mathbb{D}) \setminus \{\mathbf{0}\}$ has a non-zero angular limit *almost everywhere* on $\mathbb{T}$, that is, at $u = e^{2\pi i t}$ for every t in some subset of $[0, 1]$ having Lebesgue measure 1. If $\Psi \in H(\mathbb{D})$ is *conformal*, then $\psi(z) := z/(\Psi(z) - \Psi(0))$ is a bounded, zero-free function in $H(\mathbb{D})$, and

$\Psi(z) = \Psi(0) + z/\psi(z)$. So from the preceding it follows that Ψ too has an angular limit at almost every point of $\mathbb{T}$.

Exercise 6.29 *Let $f\colon \mathbb{H} \mapsto \overline{\mathbb{H}}$ be holomorphic and S_ε be as in 6.27. Show that if f is not* $\mathbf{0}$*, then*

either there exists a positive finite constant a such that

(i) $$\lim_{\substack{|z|\to\infty \\ z\in S_\varepsilon}} zf(z) = a \qquad \forall\varepsilon > 0$$

or

(ii) $$\lim_{\substack{|z|\to\infty \\ z\in S_\varepsilon}} |zf(z)| = \infty \qquad \forall\varepsilon > 0.$$

(iii) *Deduce from* (i) *and* (ii) *that if $\underline{\lim}_{r\to\infty} \left|rf\left(re^{i\theta}\right)\right| = 0$ for some $\theta \in (-\pi/2, \pi/2)$, then f is* $\mathbf{0}$.

Hint: If f is not constant, then it has no zeros (Maximum Principle applied to $-\operatorname{Re} f$), so $1/f$ is a candidate for 6.27.

We close this section with an application of 6.28 to a technical result which will be needed in the analysis of convex functions in Chapter XVIII.

Exercise 6.30 *Let h be non-constant and holomorphic in a neighborhood of $\overline{D}(0, r)$ and let $z_0 \in C(0, r)$ satisfy*

$$|h(z_0)| = \sup\{|h(z)| : z \in D(0, r)\}.$$

(i) *Show that $h'(z_0) \neq 0$.*

(ii) *Show that if $h(0) = 0$, then in fact*

$$\frac{z_0 h'(z_0)}{h(z_0)} \quad (\textit{is real and}) \geq 1.$$

Hints: (Ruscheweyh [1981]) Since h is non-constant, $h(z_0) \neq 0$ and we can form

(1) $$F(z) := \frac{h(z_0 z)}{h(z_0)}.$$

Then F is holomorphic and non-constant in a neighborhood of $\overline{\mathbb{D}}$ and

(2) $$F(1) = 1 = \sup\left\{|F(z)| : z \in \overline{\mathbb{D}}\right\}.$$

Since F is not constant in $\mathbb{D}$, from (2) and the Maximum-Modulus Principle we infer that $F(\mathbb{D}) \subset \mathbb{D}$. Since F is differentiable at 1, the (real) constant c in 6.28 is positive (see 6.28(iii)) and in fact

$$F'(1) = \lim_{x \uparrow 1} \frac{1 - F(x)}{1 - x} = \frac{1}{c}. \tag{3}$$

Since a trivial calculation confirms that

$$\frac{z_0 h'(z_0)}{h(z_0)} = F'(1), \tag{4}$$

we get from (3) that $h'(z_0) \neq 0$. Moreover, if $h(0) = 0$, then $F(0) = 0$ and $c \leq 1$ by 6.28(v). Consequently, (3) and (4) yield (ii).

§ 5 Subordination

If $f, g \in H(\mathbb{D})$ and there is a holomorphic self-map ϕ of $\mathbb{D}$ with $\phi(0) = 0$ such that $g = f \circ \phi$, then g is said to be *subordinate to* f. If f is univalent, then every $g \in H(\mathbb{D})$ satisfying $g(0) = f(0)$ and $g(\mathbb{D}) \subset f(\mathbb{D})$ is subordinate to f, via the map $\phi := f^{-1} \circ g$. Now Schwarz (6.1) gives $\phi\big(D(0,r)\big) \subset D(0,r)$ and so $g\big(D(0,r)\big) \subset f\big(D(0,r)\big)$, for all $0 < r < 1$. Moreover, $|\phi'(0)| \leq 1$, so $|g'(0)| = |f'(0)\phi'(0)| \leq |f'(0)|$. We codify this in the useful

Principle of Subordination 6.31: *If $f, g \in H(\mathbb{D})$ and g is subordinate to f, then*

$$|g'(0)| \leq |f'(0)| \;\;\&\;\; g\big(D(0,r)\big) \subset f\big(D(0,r)\big) \qquad \forall\, 0 < r \leq 1.$$

Variations on this simple idea lead to many useful results:

Theorem 6.32 *If f is a one-to-one holomorphic function in $\mathbb{D}$ and $f(\mathbb{D})$ is convex, then $f(D)$ is also convex for each open disk D lying in $\mathbb{D}$.*

Proof: Consider any $0 < r < 1$ and any pair of points $w_1, w_2 \in f\big(D(0,r)\big)$, say $w_j = f(z_j)$ and $|z_1| \leq |z_2| < r$. We shall show that for each fixed $0 < t_0 < 1$ the point

$$w_0 := t_0 w_1 + (1 - t_0) w_2$$

belongs to $f\big(D(0,r)\big)$. This is trivial if $z_2 = 0$ (for then $z_1 = 0$ too and all $w_j = f(0)$.) If $z_2 \neq 0$, define

$$g(z) = t_0 f(z z_1 / z_2) + (1 - t_0) f(z), \quad z \in \mathbb{D}.$$

Since $zz_1/z_2 \in \mathbb{D}$ for every $z \in \mathbb{D}$, g is well defined and holomorphic in $\mathbb{D}$. Moreover $g(\mathbb{D}) \subset f(\mathbb{D})$ by convexity of the latter, and $f(0) = g(0)$. Thus by subordination

$$w_0 = g(z_2) \in g\big(D(0,r)\big) \subset f\big(D(0,r)\big).$$

The proof thus far has established that each image set $f\big(D(0,r)\big)$, $0 < r \leq 1$ is convex. But given any open disk D with $\bar{D} \subset \mathbb{D}$, 6.5(iii) furnishes a conformal automorphism ϕ of $\mathbb{D}$ and an $r \in (0,1)$ such that $\phi\big(D(0,r)\big) = D$. Apply to the function $f^* := f \circ \phi$ what was learned about f to conclude that $f(D) = f^*\big(D(0,r)\big)$ is convex. If only $D \subset \mathbb{D}$ holds, choose $D_n \uparrow D$ with $\bar{D}_n \subset \mathbb{D}$. Then $f(D) = \bigcup_{n\in\mathbb{N}} f(D_n)$ is convex. (Why?)

Exercise 6.33 *Deduce this starlike analog of the preceding result: If f is a one-to-one holomorphic function in $\mathbb{D}$ and $f(\mathbb{D})$ is starlike with respect to $f(0)$, then $f\big(D(0,r)\big)$ is also starlike with respect to $f(0)$ for each $0 < r \leq 1$.*

Hint: Without loss of generality, $f(0) = 0$. Given $0 < r < 1$ and $|z_0| < r$ and $0 < t < 1$, we are to see that $tf(z_0) \in f\big(D(0,r)\big)$. Form

$$g(z) := tf(z), \quad z \in \mathbb{D},$$

and apply the Subordination Principle, as was done in 6.32.

Exercise 6.34 *Use 3.20 and Schwarz' lemma to improve 5.27 thus: If $f \in H(\mathbb{D})$, $f(0) = 0$ and $|\operatorname{Re} f| < A$, then*

(i) $$|f| \leq \frac{2A}{\pi} \log\left(\frac{1+r}{1-r}\right) \quad \text{on } \bar{D}(0,r), \quad \forall\, 0 \leq r < 1.$$

Hints: Consider the map

(1) $$L(z) := \frac{4A}{\pi i} \sum_{k=1}^{\infty} \frac{z^{2k-1}}{2k-1}, \qquad z \in \mathbb{D}.$$

According to 3.20 it maps $\mathbb{D}$ conformally onto $(-A, A) \times \mathbb{R} \supset f(\mathbb{D})$. Since $L(0) = 0$, all the conditions are met for an application of the Principle of Subordination. This supplies us with the inclusions

(2) $$f\big(\bar{D}(0,r)\big) \subset L\big(\bar{D}(0,r)\big) \qquad \forall 0 \leq r < 1.$$

Now it was further shown in 3.20 that

(3) $$e^{(\pi i/2A)L(z)} = \frac{1+z}{1-z} \qquad \forall z \in \mathbb{D},$$

whence

$$(4) \qquad -\frac{\pi}{2A}\operatorname{Im} L(z) = \log\left|\frac{1+z}{1-z}\right| \qquad \forall z \in \mathbb{D}.$$

For $z \in \overline{D}(0,r)$ two applications of (1) yield

$$\begin{aligned} |L(z)| &\le \frac{4A}{\pi}\sum_{k=1}^{\infty}\frac{|z|^{2k-1}}{2k-1} = -\operatorname{Im} L(|z|) \\ &\overset{(4)}{=} \frac{2A}{\pi}\log\left|\frac{1+|z|}{1-|z|}\right| \le \frac{2A}{\pi}\log\left(\frac{1+r}{1-r}\right). \end{aligned}$$

From the last inequality and (2) follows the claim (i).

Using the same method, deduce that

$$\text{(ii)} \qquad |\operatorname{Re} f(z)| \le \frac{4A}{\pi}\tan^{-1}\big(|z|\big) \qquad \forall z \in \mathbb{D}.$$

Hints: Since $\operatorname{Re} L$ maps into $(-A, A)$, $\operatorname{Im}\big[(\pi i/2A)L\big]$ maps into $(-\pi/2, \pi/2)$. It follows therefore from (3) that for $z \in \mathbb{D}$

$$\operatorname{Im}\left[\frac{\pi i}{2A}L(z)\right] = \tan^{-1}\left[\frac{\operatorname{Im}\left(\dfrac{1+z}{1-z}\right)}{\operatorname{Re}\left(\dfrac{1+z}{1-z}\right)}\right] = \tan^{-1}\left[\frac{2\operatorname{Im} z}{1-|z|^2}\right]$$

$$\begin{aligned} |\operatorname{Re} L(z)| &= \frac{2A}{\pi}\tan^{-1}\left[\frac{2|\operatorname{Im} z|}{1-|z|^2}\right] \le \frac{2A}{\pi}\tan^{-1}\left[\frac{2|z|}{1-|z|^2}\right] \\ &= \frac{4A}{\pi}\tan^{-1}\big(|z|\big). \end{aligned}$$

[The "double angle" formula for the tangent function intervened in the last step.] Consequently

$$|\operatorname{Re} L| \le \frac{4A}{\pi}\tan^{-1}(r) \quad \text{in } \overline{D}(0,r),$$

and so from (2) the same inequality can be asserted for $|\operatorname{Re} f|$.

If only $\operatorname{Re} f$ and not $|\operatorname{Re} f|$ is bounded above, useful bounds on $|f|$ can still be found:

Theorem 6.35 (Hadamard-Borel-Carathéodory Inequalities) *Let f be holomorphic in $D(0,R)$ and continuous on $\bar{D}(0,R)$. For each $0 \le r \le R$ let $M(r) := \sup\left|f\big(C(0,r)\big)\right|$ and $A(r) := \sup \operatorname{Re} f\big(C(0,r)\big)$. Then*

$$\text{(i)} \qquad A(r) \le \frac{R-r}{R+r}\operatorname{Re} f(0) + \frac{2r}{R+r}A(R),$$

$$\text{(ii)} \qquad M(r) \le \frac{R+r}{R-r}|f(0)| + \frac{2r}{R-r}A(R).$$

Proof: We have $A(R) \ge \operatorname{Re} f(0)$ by the Maximum Principle and equality holds only if f is constant, in which case (i) and (ii) are obvious. So assume in what follows that

$$(1) \qquad A(R) > \operatorname{Re} f(0).$$

We are going to use the Principle of Subordination. Recall that by 2.5 $z \mapsto (1+z)/(1-z)$ maps $\mathbb{D}$ conformally onto $(0,\infty)\times\mathbb{R}$, and so $z \mapsto (1+z)/(1-z)-1 = 2z/(1-z)$ maps $\mathbb{D}$ conformally onto $(-1,\infty)\times\mathbb{R}$. Therefore, because of (1),

$$(2) \qquad \psi(z) := f(0) - [A(R) - \operatorname{Re} f(0)]\frac{2z}{1-z}$$

maps $\mathbb{D}$ conformally onto $(-\infty, A(R))\times\mathbb{R}$. By definition of $A(R)$ [plus the Maximum Principle] the latter set contains $f\big(D(0,R)\big)$. After re-scaling the variable under f, the stage is set to invoke the Principle of Subordination to conclude that $f(Rz) \in \psi\big(D(0,\rho)\big)$ for all $z \in D(0,\rho)$ and all $0 \le \rho < 1$. That is,

$$(3) \qquad f(z) \in \psi\big(D(0,r/R)\big) \qquad \forall z \in D(0,r),\ 0 \le r < R.$$

From (3) we learn that

$$\begin{aligned}
M(r) &\le \sup\left|\psi\big(D(0,r/R)\big)\right| \\
&\overset{(2)}{=} \sup\left\{\left|f(0) - \big[A(R) - \operatorname{Re} f(0)\big]\frac{2z}{1-z}\right| : |z| < r/R\right\} \\
&\le |f(0)| + \big[A(R) - \operatorname{Re} f(0)\big]\frac{2r/R}{1-r/R} \\
&\le |f(0)| + \big[A(R) + |f(0)|\big]\frac{2r/R}{1-r/R} \\
&= \frac{R+r}{R-r}|f(0)| + A(R)\frac{2r}{R-r}.
\end{aligned}$$

A similar deduction of (i) from (3) is possible. Only one step will be non-trivial, viz., checking that

$$\operatorname{Re}\left(\frac{-z}{1-z}\right) \leq \frac{\rho}{1+\rho} \qquad \forall z \in D(0,\rho),\ 0 \leq \rho < 1.$$

This can be accomplished by computing that

$$\frac{|z|}{1+|z|} + \operatorname{Re}\left(\frac{z}{1-z}\right) = \frac{\big(1-|z|\big)\big(|z| + \operatorname{Re} z\big)}{\big(1+|z|\big)|1-z|^2} \geq 0$$

and noting that $|z|/\big(1+|z|\big) = 1/\big(1+1/|z|\big)$ is an increasing function of $|z|$. Alternatively, (i) is an almost immediate consequence of the lower Harnack inequality (5.28(i)) applied to the function $A(R) - \operatorname{Re} f$.

Remark: By defining $A(R) := \sup \operatorname{Re} f\big(D(0,R)\big)$, the conclusion and the proof remain valid if continuity of f on $\overline{D}(0,R)$ is relaxed to $\operatorname{Re} f$ being bounded above.

Corollary 6.36 *If f is holomorphic and zero-free in a neighborhood of $\overline{D}(0,R)$, then for every $0 \leq r \leq R$*

$$M(r) \leq |f(0)|^{(R-r)/(R+r)} M(R)^{2r/(R+r)}.$$

Proof: Choose $\rho > R$ such that f is holomorphic and zero-free in $D(0,\rho)$ and then use 5.35 to select an F holomorphic in $D(0,\rho)$ such that $f = e^F$. Apply the preceding theorem to F:

$$\operatorname{Re} F(z) \leq \frac{2r}{R+r} \sup_t \operatorname{Re} F\big(Re^{it}\big) + \frac{R-r}{R+r} \operatorname{Re} F(0) \quad \text{for } |z| = r.$$

Since the real exponential is monotone, we may exponentiate this inequality and permute sup and exp to get

$$\begin{aligned}
|f(z)| = e^{\operatorname{Re} F(z)} &\leq \left[\exp \sup_t \operatorname{Re} F\big(Re^{it}\big)\right]^{2r/(R+r)} \big[\exp \operatorname{Re} F(0)\big]^{(R-r)/(R+r)} \\
&= \left[\sup_t \exp \operatorname{Re} F\big(Re^{it}\big)\right]^{2r/(R+r)} \big[\exp \operatorname{Re} F(0)\big]^{(R-r)/(R+r)} \\
&= \left[\sup_t \big|f\big(Re^{it}\big)\big|\right]^{2r/(R+r)} |f(0)|^{(R-r)/(R+r)} \\
&= M(R)^{2r/(R+r)} |f(0)|^{(R-r)/(R+r)} \qquad \forall |z| = r.
\end{aligned}$$

Corollary 6.37 (Hadamard [1892a]) *If f is an entire function which for some real R, B and K satisfies*

$$(*) \qquad \operatorname{Re} f(z) \leq B|z|^K \qquad \forall |z| \geq R > 0,$$

then f is a polynomial of degree not greater than K.

Proof: In the notation of 6.35 we have from $(*)$

$$A(r) \leq Br^K \qquad \forall r \geq R,$$

whence

$$\begin{aligned} M(r) &\leq \frac{2r}{2r-r}A(2r) + \frac{2r+r}{2r-r}|f(0)| = 2A(2r) + 3|f(0)| \\ &\leq 2B(2r)^K + 3|f(0)| \qquad \forall r \geq R/2. \end{aligned} \tag{1}$$

From the Cauchy estimates (3.34(i)) and (1) we get

$$\frac{|f^{(n)}(0)|}{n!} \leq 2^{K+1}Br^{K-n} + 3|f(0)|r^{-n}, \tag{2}$$

holding for all positive integers n and all $r \geq R/2$. For each fixed $n > K$ let $r \uparrow \infty$ to see from (2) that $f^{(n)}(0) = 0$. Since (5.30) we have

$$f(z) = \sum_{n=0}^{\infty} \frac{f^{(n)}(0)}{n!} z^n \qquad \forall z \in \mathbb{C},$$

the vanishing of $f^{(n)}(0)$ for all $n > K$ proves the assertion. (Cf. also 6.45.)

Remarks: (i) The proof shows that inequality $(*)$ need only be hypothesized to hold on a set of concentric circles whose radii are unbounded, and the conclusion still follows. For three other proofs see 3.34(iii), 6.45 and 11.62.

(ii) If $F \in H(\mathbb{C})$ is zero-free and for some real B and all large z satisfies $|F(z)| \leq e^{B|z|}$, then $F = e^f$ for an entire f (5.35), which satisfies $\operatorname{Re} f(z) \leq B|z|$ for all large z. By 6.37, $f(z) = az + b$. If in addition F' has a zero, then $a = 0$ and F is constant. This fact can be used to give a quick proof of the FTA, by induction on the degree, as follows (**MR** 87d:12004): Suppose for some integer $n(> 2)$ every complex polynomial of degree less than n has a zero in $\mathbb{C}$ and that F is a complex polynomial of degree n. Since for each integer $k \geq 0$, $e^{|z|} \geq |z|^{k+1}/(k+1)!$ entails $|z|^k e^{-|z|} \to 0$ as $|z| \to \infty$, $|F(z)| \leq e^{|z|}$ for all large z. As F' is a polynomial of degree less than n, it has by assumption a zero in $\mathbb{C}$. According to the above consequence of 6.37, F cannot be zero-free, otherwise it would be constant, contrary to its having degree $n > 2$.

Theorem 6.38 *Let $\phi(z) = \sum_{n=1}^{\infty} a_n z^n$ and $\psi(z) = \sum_{n=1}^{\infty} b_n z^n$ be holomorphic in $\mathbb{D}$. Suppose that ϕ is one-to-one, $\phi(\mathbb{D})$ is convex and $\psi(\mathbb{D}) \subset \phi(\mathbb{D})$. Then*

$$|b_k| \leq |a_1| \qquad \forall k \in \mathbb{N}.$$

The conclusion applies in particular to $\psi := \phi$.

Proof: According to the Principle of Subordination 6.31, $|\psi'(0)| \le |\phi'(0)|$. That is,

$$(1) \qquad |b_1| \le |a_1|.$$

Next we create a new function to play the role of ψ in (1) and thereby parlay (1) into the final conclusion.

Given $k \in \mathbb{N}$, we have, by convexity of $\phi(\mathbb{D}) \supset \psi(\mathbb{D})$, that

$$(2) \qquad \frac{1}{k}\sum_{j=1}^{k} \psi\left(e^{2\pi ij/k} r^{1/k} e^{i\theta/k}\right) \in \phi(\mathbb{D}) \qquad \forall (r,\theta) \in [0,1) \times [0,2\pi).$$

Using the power series for ψ, this convex sum is

$$(3) \qquad \sum_{n=1}^{\infty} b_n \left[\frac{1}{k}\sum_{j=1}^{k} e^{2\pi ijn/k}\right] r^{n/k} e^{in\theta/k}.$$

Now if $e^{2\pi in/k} \neq 1$, that is, if k does not divide n, then

$$\sum_{j=1}^{k} \left(e^{2\pi in/k}\right)^j = \frac{e^{2\pi in/k} - \left(e^{2\pi in/k}\right)^{k+1}}{1 - e^{2\pi in/k}} = 0,$$

while if k divides n, this sum is clearly k. Therefore the (outer) sum in (3) need only to be extended over the n which are multiples of k, and so that sum is $\sum_{m=1}^{\infty} b_{mk} r^m e^{im\theta}$ and the assertion (2) is therefore that

$$\Psi(z) := \sum_{m=1}^{\infty} b_{mk} z^m \in \phi(\mathbb{D}) \qquad \forall z \in \mathbb{D}.$$

The function Ψ so defined satisfies all the hypotheses which ψ does, so from (1) we are entitled to conclude that $|b_k| \le |a_1|$.

Corollary 6.39 *If $f(z) = 1 + \sum_{n=1}^{\infty} c_n z^n$ maps $\mathbb{D}$ into $\mathbb{H}$, then $|c_n| \le 2$ for all n. It follows* (cf. 5.28(i)) *that*

$$\frac{1-|z|}{1+|z|} \le |f(z)| \le \frac{1+|z|}{1-|z|} \qquad \forall z \in \mathbb{D}.$$

Proof: The function $z \mapsto (1+z)/(1-z)$ maps $\mathbb{D}$ conformally onto $\mathbb{H}$, according to 2.5. Therefore the function

$$(*) \qquad \phi(z) := \frac{1+z}{1-z} - 1 = \frac{2z}{1-z} = 2z\sum_{n=0}^{\infty} z^n = \sum_{n=1}^{\infty} 2z^n, \quad z \in \mathbb{D}$$

maps $\mathbb{D}$ conformally onto $-1+\mathbb{H}$. By hypothesis $\psi(z) := f(z)-1 = \sum_{n=1}^{\infty} c_n z^n$ also maps $\mathbb{D}$ into this half-plane. Apply 6.38 to conclude from $(*)$ that $|c_n| \le 2$ for all n. Then $|f(z)| \le 1 + \sum 2|z|^n = (1+|z|)/(1-|z|)$. Since $1/f$ satisfies the same hypotheses as f, the lower inequality follows as well.

Remarks: (i) Using the Principle of Subordination directly, the lower inequality in 6.39 can be improved to

$$\frac{1-|z|}{1+|z|} \le \operatorname{Re} f(z) \qquad \forall z \in \mathbb{D}.$$

For $(1+z)/(1-z)$ maps $D(0,r)$ onto the disk D_r in $\mathbb{H}$ centered in $\mathbb{R}$ and having antipodes $(1-r)/(1+r)$ and $(1+r)/(1-r)$. Consequently, f also maps $D(0,r)$ into D_r, hence $\bar{D}(0,r)$ into $\overline{D_r} \subset (1-r)/(1+r) + \overline{\mathbb{H}}$. Given $z \in \mathbb{D}$, take $r := |z|$ and get $f(z) \in (1-|z|)/(1+|z|) + \overline{\mathbb{H}}$. As noted, 5.28(i) furnishes an alternative proof.

(ii) A curious result of RAJAGOPAL [1957] is any easy consequence of 6.39: If $g(z) := \sum_{n=0}^{\infty} a_n z^n$ converges in $\mathbb{D}$, a_0 is positive and $\operatorname{Re} g < 1$, then we have $\sum_{n=0}^{\infty} |a_n| 3^{-n} \le 1$. To see this, apply 6.39 to $f := (1-a_0)^{-1}(1-g)$ to learn that $(1-a_0)^{-1}|a_n| \le 2$ for all $n \in \mathbb{N}$, whence $\sum_{n=1}^{\infty} |a_n| 3^{-n} \le 1 - a_0$. No smaller positive real number can serve in the role of "3" in this inequality, as the function $g(z) := 1/3 + 4z/(3+3z)$ illustrates.

Exercise 6.40 *If f is holomorphic in $\mathbb{D}$ and satisfies $0 < |f| \le 1$ there, then $|f'(0)| \le 2/e$.*

Hints: (ROBERTSON [1953]) We may assume that $|f| < 1$ throughout $\mathbb{D}$; for if equality is attained, f is constant by the Maximum-Modulus Principle, and then $f'(0) = 0$. We may also assume that $f(0) = |f(0)|$, multiplying f by a unimodular constant if necessary. Let (5.35) L be a holomorphic logarithm for $1/f$. We may assume $L(0) = -\log|f(0)| = -\log f(0)$. Since $|f| < 1$, $\operatorname{Re} L > 0$. Thus $L/L(0)$ maps into the right half-plane and the case $n=1$ of 6.39 implies that $|L'(0)/L(0)| \le 2$,

$$|L'(0)| \le 2|L(0)|. \tag{1}$$

However, $e^{-L} = f$, so

$$\begin{aligned} -L'(0)e^{-L(0)} &= f'(0) \\ -L'(0)f(0) &= f'(0) \\ \left|\frac{f'(0)}{f(0)}\right| &= |L'(0)| \overset{(1)}{\le} 2|L(0)| = 2\log\frac{1}{|f(0)|}, \end{aligned}$$

$$|f'(0)| \le -2|f(0)|\log|f(0)|. \tag{2}$$

Now compute the maximum of the function $-x\log x$ for $0 < x \le 1$, remembering that $x \log x \to 0$ as $x \downarrow 0$ (by 3.18(4)).

Exercise 6.41 *Find all functions* ϕ *for which equality at* $k = 1$ *holds in* 6.38, *all functions for which equality at* $n = 1$ *holds in* 6.39, *and all functions for which equality holds in* 6.40.

Hint: Return to 6.31 and see what happens there if $|g'(0)| = |f'(0)|$.

Theorem 6.42 *If* $g \in H(\mathbb{D})$ *has convex image* $g(\mathbb{D})$ *and if* $g(0) = 0$ *and* $|g'(0)| \ge 1$, *then* $g(\mathbb{D}) \supset D\big(0, \frac{1}{2}\big)$.

Proof: We can suppose $g(\mathbb{D}) \ne \mathbb{C}$. The set $C := g(\mathbb{D})$ is open (5.77) and convex; $\mathbb{C}\setminus C$ is closed, non-void and does not contain 0. By a trivial compactness argument there is an $a \in \mathbb{C} \setminus C$ closest to 0. According to 1.19(v)

$$\operatorname{Re}\left[\bar{a}(w-a)\right] < 0 \qquad \forall w \in C.$$

Therefore if we let u denote the unimodular complex number $-a/|a|$, we have

(1) $$\operatorname{Re}(\bar{u}w) > -|a| \qquad \forall w \in C.$$

Consider now the function $\phi(z) := \frac{1+z}{1-z} - 1 = \frac{2z}{1-z}$ used in the proof of 6.39. It maps $\mathbb{D}$ conformally onto $\{v \in \mathbb{C} : \operatorname{Re} v > -1\}$. Hence the function

$$f(z) := u|a| \cdot \frac{2z}{1-z}$$

maps $\mathbb{D}$ conformally onto the open half-plane described by the inequality (1). Therefore (1) says that

(2) $$g(\mathbb{D}) = C \subset f(\mathbb{D}).$$

In view of (2) the Principle of Subordination 6.31 gives

(3) $$|g'(0)| \le |f'(0)|.$$

But a trivial computation reveals that $f'(0) = 2u|a|$, while $1 \le |g'(0)|$ by hypothesis. Hence (3) says that $1 \le 2|a|$. By definition of a, $D\big(0, |a|\big) \subset C = g(\mathbb{D})$. Thus $D\big(0, \frac{1}{2}\big) \subset g(\mathbb{D})$.

Exercise 6.43 *Use the Poisson formula* (5.17) *to establish the following variant of* 6.35(ii)*:*

$$M(r) \le \frac{R-r}{R+r}|f(0)| + \frac{2r}{R+r}M(R) \qquad \forall 0 \le r \le R,$$

whenever f is holomorphic in $D(0,R)$, continuous on $\bar{D}(0,R)$ and $M(r) :=$ $\sup\bigl|f\bigl(C(0,r)\bigr)\bigr|$ for each $0 \le r \le R$.

Hints: Without loss of generality, $R = 1$. Set $P(\theta, a) := \bigl(1-|a|^2\bigr)/\bigl|e^{i\theta}-a\bigr|^2$, $P(a) := \bigl(1-|a|\bigr)/\bigl(1+|a|\bigr)$ for $\theta \in \mathbb{R}$, $a \in \mathbb{D}$. Then

$$P(\theta,a) \ge \frac{1-|a|^2}{\bigl(1+|a|\bigr)^2} = \frac{1-|a|}{1+|a|} = P(a).$$

Consequently, from 5.17

$$\begin{aligned}
|f(a)| &= \left|\frac{1}{2\pi}\int_0^{2\pi} P(\theta,a) f\bigl(e^{i\theta}\bigr)\,d\theta\right| \\
&\le \left|\frac{1}{2\pi}\int_0^{2\pi} P(a) f\bigl(e^{i\theta}\bigr)\,d\theta\right| + \left|\frac{1}{2\pi}\int_0^{2\pi} \bigl[P(\theta,a)-P(a)\bigr] f\bigl(e^{i\theta}\bigr)\,d\theta\right| \\
&\le P(a)\left|\frac{1}{2\pi}\int_0^{2\pi} f\bigl(e^{i\theta}\bigr)\,d\theta\right| + \frac{1}{2\pi}\int_0^{2\pi} \bigl[P(\theta,a)-P(a)\bigr]\bigl|f\bigl(e^{i\theta}\bigr)\bigr|\,d\theta \\
&\le P(a)\left|\frac{1}{2\pi}\int_0^{2\pi} f\bigl(e^{i\theta}\bigr)\,d\theta\right| + \frac{M(1)}{2\pi}\int_0^{2\pi} \bigl[P(\theta,a)-P(a)\bigr]\,d\theta \\
&= P(a)|f(0)| + M(1)[1-P(a)] \\
&= \frac{1-|a|}{1+|a|}|f(0)| + M(1)\frac{2|a|}{1+|a|}, \qquad \text{by definition of } P(a).
\end{aligned}$$

Take the supremum over $a \in C(0,r)$.

Exercise 6.44 (Cf. 5.38(iii).) *Let Ω be a region, $a \in \Omega$, K compact $\subset \Omega$ and $0 \le r < 1$. Show that there exists a constant $C := C(\Omega, K, a, r) < 1$ such that $\sup|f(K)| \le C$ for all $f \in H(\Omega)$ which satisfy $|f(a)| \le r$, $|f| \le 1$.*

Hints: Pick $R > 0$ so that $\bar{D}(a,R) \subset \Omega$ and let $U := D(a, R/2)$. According to 6.43 and the Maximum-Modulus Principle we have for any $f \in H(\Omega)$ with $|f| \le 1$

$$|f(z)| \le \frac{R-R/2}{R+R/2}|f(a)| + \frac{2R/2}{R+R/2} \qquad \forall z \in U,$$

so

$$\sup|f(U)| \le \frac{1}{3}|f(a)| + \frac{2}{3} \le \frac{1}{3}r + \frac{2}{3}$$

whenever $|f(a)| \le r$. Now use 5.38(iii).

Exercise 6.45 *Supply details in the following outline of an interesting alternative proof of* 6.37 *(due to* E. Landau*).*

(i) *If K is a non-negative integer, f minus its Kth-degree Taylor polynomial is another entire function f^* satisfying $(*)$, say with constant $B^* > 0$.*

(ii) *For $r \geq R > 0$ infer that* $\text{Re}[2B^* r^K - f^*] \geq B^* r^K$ (> 0) *and that* $(2B^* r^K - \text{Re}\, f^*)^2 \geq (\text{Re}\, f^*)^2$, *whence* $|2B^* r^K - f^*|^2 \geq |f^*|^2$ *in* $\bar{D}(0, r)$.

(iii) *The function $r^{K+1} f^*(z)/[2B^* r^K - f^*(z)] z^{K+1}$ is holomorphic in $D(0, r)$. From its behavior on $C(0, r)$ infer that it is bounded by 1 in $D(0, r)$.*

(iv) *In the inequality $|f^*(z)| \leq r^{-K-1} |2B^* r^K - f^*(z)||z^{K+1}|$ let $r \to \infty$.*

It is appropriate to close this chapter as it began, with Schwarz' Lemma, more precisely, with a variant of it in which the diameter of the image set takes over the role of the maximum modulus of the function. Namely, every (bounded) $f \in H(\mathbb{D})$ satisfies

$$\left|f^{(n)}(0)\right|/n! \leq \tfrac{1}{2}\text{diam}\, f(\mathbb{D}) := \tfrac{1}{2} \sup\{|f(z) - f(w)| : z, w \in \mathbb{D}\} \qquad \forall n \in \mathbb{N}$$

and equality holds for some n if and only if f has the form

$$f(z) \equiv f(0) + cz^n, \quad c \in \mathbb{C}, \quad |c| = \tfrac{1}{2}\text{diam}\, f(\mathbb{D}).$$

The inequality follows easily from applying Parseval's equality (3.32(i)) to the function $f(z) - f(ze^{\pi i/n})$, which is bounded by $\text{diam}\, f(\mathbb{D})$. The problem of equality is more subtle. An ingenious but elementary argument of Friedrich Hartogs (in LANDAU and TOEPLITZ [1907]) settles the case $n = 1$ and a further argument of POUKKA [1907] reduces the general case to that one. Details and extensions may be found in BURCKEL, MARSHALL, MINDA, POGGI-CORRADINI, and RANSFORD [2008].

Notes to Chapter VI

For more references to literature on Schwarz' lemma see pp. 126 and 263 of the bibliography BERNARDI [1982]. Some generalizations are given by JULIA [1920], KOEBE [1920], and R. NEVANLINNA [1929a]. Much of the Schwarz-lemma lore and more is captured by BEARDON and MINDA [2004] in terms of hyperbolic geometry. OSSERMAN [2000] noted that by applying 6.3(iv) to $f(z)/z$ when $f(0) = 0$, the general inequality 6.1(ii) could be refined to $|f(z)| \leq \big(|z| + |f'(0)|\big)\big(1 + |f'(0)||z|\big)^{-1}|z|$. If additionally the f in 6.1 is normalized by the requirement $f'(0) \in \mathbb{R}$, then not only is $|f(z)| \leq |z|$ for all $z \in \mathbb{D}$, but for every such z, $f(z)$ lies in the lens-shaped region centered at 0 having vertices $\pm z$ and thickness $2|z|^2$ (ROGOSINSKI [1934]). There is also the following quite general version due to N. Sibony: If $u\colon \mathbb{D} \mapsto [0, 1]$ has continuous second-order partial derivatives, $u(0) = 0$ and $\log u$ is subharmonic, then $u(z) \le |z|^2$ and $\Delta u(0) \le 4$, with equality (for some $z \ne 0$) if and only if $u(z) \equiv |z|^2$. See DINEEN [1989], where in addition the several-variable and Banach-algebra versions of Schwarz' lemma are exhaustively treated.

The derivative inequality 6.1(i) is generalized to other regions in 7.46 and for $\mathbb{D}$ can be extended: $|f'| \le 1$ throughout $D(0, \sqrt{2} - 1)$; see CARATHÉODORY [1960], p. 19. Much of 6.3 is also to be found in CARATHÉODORY [1912a]. For a strengthening of 6.3(i) see BEARDON and CARNE [1992]. Generalizing the derivative inequality in 6.3(i), RUSCHEWEYH [1985] proved that for all such f and z and all $n \in \mathbb{N}$

$$\frac{|f^{(n)}(z)|}{1 - |f(z)|^2} \le \frac{n!}{(1 - |z|)^{n-1}} \cdot \frac{1}{1 - |z|^2}.$$

For many more inequalities of this kind, and in regions other than $\mathbb{D}$, see AVKHADIEV and WIRTHS [2009].

The best general result of the kind 6.4(ii) is the following: For *any* region Ω, the only conformal automorphism of Ω which fixes more than two points of Ω is the identity. See SUITA [1981] for the shortest proof and his bibliography for original credits. The following remarkable result of ZINTERHOF [1972] should be contrasted to 6.4(ii): For every $\varepsilon > 0$ there exists a conformal $f \in H(\mathbb{D})$, not the identity function, such that $f(\mathbb{D})$ is a convex subset of $D(0, 1 + \varepsilon)$ and f fixes infinitely many points in $\mathbb{D}$. Moreover, for any closed $K \subset \mathbb{T}$ having Lebesgue measure 0, there is a continuous function on $\overline{\mathbb{D}}$, which is holomorphic in $\mathbb{D}$ and fixes each point of $\{0\} \cup K$ but is not the identity function. See Example 2.2 of COWEN and POMMERENKE [1982].

6.4(vi) is due to CANTOR and PHELPS [1965]. They produce B as a Blaschke product, by induction on n, based on a simple result about transitive families of maps on semigroups; very elegant and ingenious. From 6.16 it follows that the B produced in our 6.4(vi) hints is indeed a finite Blaschke product. In fact, JONES and RUSCHEWEYH [1987] show that this Blaschke product can be chosen to have only $n - 1$ factors. (The one produced in the above proof may have as many as $n(n-1)$ factors.) There have been other dramatic extensions and refinements of 6.4(vi). CLUNIE, HALLENBECK and MACGREGOR [1985] proved that if $0 < \theta_1 < \theta_2 < \cdots < \theta_n \le 2\pi$ and $0 < \psi_1 < \psi_2 < \cdots < \psi_n \le 2\pi$, then there exists a polynomial f with $|f| < 1$ on $\overline{\mathbb{D}} \setminus \{e^{i\theta_1}, \ldots, e^{i\theta_n}\}$, $f(e^{i\theta_j}) = e^{i\psi_j}$ for all j and f one-to-one in $\mathbb{D}$. An improved proof of this was given by MACGREGOR and TEPPER [1988]. If some of the β_k lie in $\mathbb{D}$, the interpolation can be accomplished with infinite Blaschke products (CARGO [1979]). When the α_j and the β_j are ordered counterclockwise and the interpolating rational function B is allowed to have poles anywhere in $\mathbb{C} \setminus \mathbb{T}$ but required to map *injectively* $\mathbb{T}$ onto $\mathbb{T}$, see LANGER [2007] and 11.14.

An important relative of 6.4(vi) is the interpolation theorem of Pick and Nevanlinna which gives necessary and sufficient conditions on $\{z_n\}_{n=1}^{\infty}$, $\{w_n\}_{n=1}^{\infty}$ $\subset \mathbb{D}$ that there exist an $f \in H(\mathbb{D})$ with $|f| \leq 1$ and $f(z_n) = w_n$ for all n. It has innumerable generalizations, proofs and applications to operator theory. See Pick [1916a], [1920], R. Nevanlinna [1919], [1922b], [1929c] and Marshall [1974].

The mappings f_a belong to a larger family of the form $M(z) := \frac{az+b}{cz+d}$, $ad-bc \neq 0$, called *Möbius transformations* or *fractional linear transformations*, which each map circles and lines to circles and lines. This family comes up naturally as Aut $(\mathbb{C}_\infty)$; cf. 11.20(ii). Their theory is very geometric and is especially revealingly presented in Needham [1997] and in the film by Arnold and Rogness [2008]; see also Palka [1991]. It culminates in the hyperbolic geometry of $\mathbb{D}$ and $\mathbb{U}$.

According to 6.2(ii) and 6.18(v), for Ω a disk or an annulus the group Aut (Ω) is a set of Möbius transformations. Minda [1979c] proved an attractive converse of this: If Ω is either a simply-connected or a doubly-connected region in $\mathbb{C}$ and Aut (Ω) consists only of Möbius transformations, then Ω is either a disk or an annulus.

Carathéodory [1937b] (see also Hua [2008]) proved a strong converse of 6.5(i): If $f\colon \mathbb{D} \mapsto \mathbb{C}$ is a one-to-one map which transforms every circle in $\mathbb{D}$ onto a circle in $\mathbb{C}$ (f is not hypothesized to be holomorphic or even continuous), then either f or $\bar{f}$ is a Möbius map. Cf. 11.20(vii).

On 6.7 see E. Landau [1914] and §61 of Bieberbach's encyclopedia article. An equality refining inequality 6.7(i) may be obtained by taking $z = 0$ in equation (1) of the proof of 12.4. The conclusion in 6.7(iii) follows under weaker hypotheses on $f \in H(\mathbb{D})$; e.g., (12.6) it is enough that the set $\left\{\int_0^{2\pi} \log^+\left|f(re^{i\theta})\right| d\theta : 0 \leq r < 1\right\}$ be bounded.

6.8 is implicit in Jensen [1898–99] and in Petersen [1900]; a slightly weaker version is explicit in Schou [1897]. A refinement of 6.8(ii) asserts that $\sum |z_j|^{-\rho}$ $< \infty$ for any $\rho > 0$ such that $\int_1^\infty r^{-1-\rho} \log M(r)dr < \infty$, $M(r)$ being $\max\{|f(z)| : |z| \leq r\}$. The uniqueness part of 6.9(iii) is due to Blaschke [1915]. For a sectorial version of it see Tsuji [1926] and Ou [1957]. Boundedness of f is not necessary here: Hayman and Korenblum [1980] discovered continuous, increasing functions $k\colon [0,1) \mapsto \mathbb{R}$ having the property that $f \in H(\mathbb{D}) \setminus \{\mathbf{0}\}$, $|f(z)| \leqq e^{k(|z|)}$ and $f(z_j) = 0 \Rightarrow \sum(1-|z_j|) < \infty$. In fact, k has this property if and only if $\int_0^1 \sqrt{k(r)/(1-r)}\, dr < \infty$. On the other hand, in 6.9(iii) the function need not be hypothesized to vanish at the z_j; it need only be small there, as 6.9(ii) suggests: Each sequence of distinct numbers $z_j \in \mathbb{D}$ satisfying $\sum(1-|z_j|) = \infty$ determines positive numbers ε_j such that the only bounded $f \in H(\mathbb{D})$ satisfying $|f(z_j)| \le \varepsilon_j$ for all j is the function $\mathbf{0}$. See DANIKAS [1994].

The idea for 6.10 comes from GRONWALL [1914–15], SCHUR [1917] and R. NEVANLINNA [1919]. 6.10 itself is only ancillary to 8.26. A closely related application of it is to produce the power series expansion of a holomorphic function without using any integral representation; see CONNELL and PORCELLI [1962].

WIGERT [1916] weakened the hypothesis $c < 1$ in 6.11 to $c < \pi$, but the function $f(z) := \sin \pi z$ attests that no further weakening of this kind is possible. On the other hand, with $c < \pi$ the conclusion still follows if f is only holomorphic in $\mathbb{H}$. See p. 34 of RANSFORD [1995], as well as chapters 9, 10 of BOAS [1954]. In BOAS [1972] there is a generalization to harmonic functions.

For results related to 6.16(i) and (iii) see 11.14 and 11.47. Other characterizations of finite Blaschke products will be found in HEINS [1986]. The reader should formulate the half-plane analog of 6.16(i) and compare it with 11.52. [It is also instructive to compare 6.20 with 11.52.] A proof of (i) occurs in CARATHÉODORY [1936] without any mention of Fatou. A considerable extension of (i) was achieved by LOHWATER [1952b]. For extensions of (iii) see SHIMIZU [1931b] and HEJHAL [1969] and its references. The situation in (iii) when the unit disk is replaced by a rectangle is striking: An entire function that maps some rectangle onto itself is a polynomial of degree 1. For this and much more, see BAK and DING [2009].

For a different proof and extension of 6.15 see GORDON [1972], MUELLER and RUDIN [1991]. Cf. also STOÏLOW [1933], ROBERTS [1940] and MIODUSZEWSKI [1961]. For regions of *finite* connectivity greater than 2 with no 1-point holes, and for $m > 1$, no holomorphic self-maps exist which are exactly m-to-1. More positively, every boundary-preserving holomorphic self-map of such a region is a conformal automorphism; moreover, the set (group) of all such maps is finite. See HEINS [1941c]. Roughly speaking, this is because a proper map must "permute" the finitely many components of the boundary (as 6.18 illustrates). See MUELLER and RUDIN [1991] for a detailed explication of this idea; they also construct, for each $m > 1$, a region having an exactly m-to-1 holomorphic self-map. For the literature on 6.17 see the notes on 10.12.

A more cavalier version of the topological argument in 6.18(i) occurs already in KOEBE [1906a], p. 143. 6.22–6.24 are from HUBER [1951]. Cf. also KHAJALIA [1937], SCHIFFER [1946], TSUJI [1960] and MINDA [1979b] (where 6.27 is utilized). RENGGLI [1979] is a differential geometry treatment. These are the key to a more elegant proof of 6.18(v) which will be given in 10.14; the chapter X notes will discuss still other proofs. For another proof of 6.24 see pp. 427–428 of HEINS [1941b] and LANDAU and OSSERMAN [1959].

6.27 and 6.28 are the work of many hands: JULIA [1920], WOLFF [1926c] (direct derivation from 6.6), CARATHÉODORY [1929a], LANDAU and VALIRON [1929], WOLFF and DE KOK [1932]. See also R. NEVANLINNA [1929c], AHLFORS [1930] (chapter III), [1973] (pp. 7–10), M. RIESZ [1931], DINGHAS [1938], TSUJI [1939], HERZIG [1940] and VERBLUNSKY [1949]. For unified treatments see SERRIN [1956], Herzig, KOMATU [1961], GOLDBERG [1962], MERCER [2000], and SHOIKHET [2001]. Much of the development of 6.27 in the text is straight out of LANDAU and VALIRON [1929]; a similar account appears in VALIRON [1931]. Wolff and de Kok, Verblunsky, Komatu, and Serrin proceed from an integral representation formula. We shall explore this in §6 of Chapter VIII. (Cf. 8.39(iii).) CARATHÉODORY's treatment may also be found in volume II of his book [1960] and in chapter II of JULIA [1958]; it is quite geometric. SARASON [1988] treats the angular derivative by Hilbert-space methods. Extensions to operator-valued holomorphic functions are to be found in PFLUGER [1966], ANDÔ and FAN [1979], FAN [1979], [1986] and TAO [1984]. JØRGENSEN [1946] relaxes the hypothesis $\operatorname{Re} f \geq 0$ in 6.27. In 6.28(v), if $F^{(j)}(0) = 0$ for all $j \leq n$, then $c \leq 1/n$, and if equality holds in (10) for at least one z, then it holds for all z and $F \in \operatorname{Aut}(\mathbb{D})$. Further refinements of 6.27, 6.28 are in POMMERENKE [1979], BAKER and POMMERENKE [1979], and COWEN and POMMERENKE [1982]. In 6.27 it important that $\operatorname{Im} z/\operatorname{Re} z$ remain bounded as $|z| \to \infty$ (see VISSER [1931–1932] and WOLFF [1930b]). HEINS [1989] constructs an example where $c = 0$ and the limit in 6.28(iv) does not exist, although F does have sectorial limit 1 at 1. 6.27 and 6.28 have been generalized in many directions, e.g., to holomorphic functions in Banach spaces and (as already mentioned) to operator-valued holomorphic functions; see GOEBEL and REICH [1984], pp. 124 ff. and WŁODARCZYK [1987], together with its extensive bibliography. 6.28 is related to the question of when the composition operator $g \mapsto g \circ F$ on certain Banach subspaces of $H(\mathbb{D})$ is compact, and it is proved from this perspective by J. H. SHAPIRO [1993]; his very revealing treatment is geometric and emphasizes the pseudo-hyperbolic metric d of 6.3(vi), as does the account in BEAR [1992] and BEARDON [1997]. Since $\mathbb{H}$ is a "covering space" for the annulus, 6.27 will see interesting service in Chapter VII when we analyze the iterates of self-maps of annuli.

6.29(iii) was originally proved in R. NEVANLINNA [1922a] by a somewhat different (though basically the same!) method.

6.30 is from PÓLYA and SZEGÖ [1972], exercise III.291; they attribute it to K. Löwner and the argument there is more elementary (free of 6.28) than here. See also their exercise III.144 and pp. 87-90 of BAK and NEWMAN [2010]. 6.30 was re-discovered by JACK [1971], who is usually cited in the many applications of this fact in the study of univalent functions, although his proof is criticized by MILLER and MOCANU [1978], who give another elementary proof (using ideas of L. Brickman; cf. the proof of our 18.1 and 18.8). For generalizations and discussions of these applications see PARVATHAM [1989]. In 1994 a whole conference was devoted to this theme; see **MR** 95m:30001.

The idea of subordination originates with LINDELÖF [1909]. For a revealing discussion of it and several more applications the reader is strongly encouraged to look at the expository paper of MACGREGOR [1972] (source of 6.41). See also 18.5 of HILLE [1962] and the important basic papers of LITTLEWOOD [1925] (who coined the term) and ROGOSINSKI [1939], [1943]; also ROBINSON [1947]. On pp. 118 and 254 of the bibliography BERNARDI [1982] the reader will find more references to the literature on subordination. DUREN [1977] is a short survey.

The part of 6.32 dealing with $D := D(0,r)$ is due to STUDY [1913], p. 109 (though only after the analytic characterization of range-convexity in Chapter XVIII will the reader recognize it there); the extension to all D was noted by CARATHÉODORY [1919] and ROBERTSON [1936]. The proofs given in the text are due to T. RADÓ [1930], for 6.32, and TAKAHASHI [1930], for 6.33. See also SEIDEL [1931] and NABETANI [1935–36]. For a half-plane analog, see STANKIEWICZ and STANKIEWICZ [1991]. Interestingly the stronger conclusion about all D fails for 6.33—see J. E. BROWN [1989]. FORD [1935] shows that if T is holomorphic in $\mathbb{C}^n$ and $T(w_1,\ldots,w_n) \in f(\mathbb{D})$ whenever $w_1,\ldots,w_n \in f(\mathbb{D})$ then $T(w_1,\ldots,w_n) \in f\big(D(0,r)\big)$ whenever $w_1,\ldots,w_n \in D(0,r)$, for each $0 < r < 1$ and this is generalized by BECKENBACH and GRAHAM [1952] and ALEKSANDROV [1958]. POMMERENKE [1962] and HEINS [1970], [1973] find necessary and sufficient conditions that $\Omega \subset \mathbb{D}$ have $f(\Omega)$ convex for every conformal map f of $\mathbb{D}$ onto a convex region and FENTON [1976b] carries out a similar program for starlikeness.

The result 6.34(ii) is due to H. A. SCHWARZ (p. 190 and pp. 361–362, vol. II of his *Gesammelte Mathematische Abhandlungen*). For more on it see KOEBE [1920]. A related inequality is proved in VALIRON [1928b]. For other proofs of the Hadamard-Borel-Carathéodory inequality 6.35 see pp. 275–277 of E. LANDAU [1906], pp. 191–194 of E. LANDAU [1908b] (this is the method used in 6.45), p. 15 of LINDELÖF [1909] (which actually contains a generalization), ELKINS [1971] and ZALCMAN [1978]. A short proof can be based on the Poisson integral formula. See, e.g., HOLLAND [1973] or E. LANDAU (*loc. cit.*). Such derivations lead to useful bounds on $f^{(n)}$ as well. Cf. also RAJAGOPAL [1953]. If $\sum c_n z^n$ is the power series representation of $f(z)$, then 6.35(ii) can be improved to

$$\sum |c_n| r^n \le |f(0)| + \frac{2r}{R-r}\big(A(R) - \operatorname{Re} f(0)\big);$$

see p. 18 of TAUVEL [1994]. For an encyclopedic account of 6.35- and 6.43-type inequalities, see **MR** 2008d:30001.

For history and extensive generalizations of 6.1, 6.3, 6.7, 6.19 and 6.35 see JENSEN [1919–20] and for material of a related kind, MACINTYRE and ROGOSINSKI [1945].

The weaker version of 6.37 with the hypothesis on $|\operatorname{Re} f|$ instead of $\operatorname{Re} f$ was proved using Poisson's formula by H. A. SCHWARZ (p. 207, vol. II of his *Gesammelte Mathematatische Abhandlungen*) and also follows easily from 3.32(ii). Likewise, COWLING [1956] gives a straightforward deduction of 6.39 from Poisson. (See also CARATHÉODORY [1912].) Compare the hints to 6.43. Using similar techniques (a Poisson-Stieltjes, instead of a simple Poisson formula) HERZIG [1940] gives a complete description of the f for which equality holds in 6.39 for at least one n or for a relatively prime pair of n. A third kind of proof of 6.39, echoing ideas in the hints to 3.34(iv), may be found in 3.7.4 of MITRINOVIĆ and KEČKIĆ [1993]. For more on functions with positive real part see the literature references on pp. 118 and 258 of BERNARDI [1982]. The weak form (due to Cauchy—(3.34)(ii)) of 6.37 in which $M_f(r) := \max\{|f(z)| : |z| = r\} \leq Br^K$ has the following interesting generalization (see I. N. BAKER [1958]): If there are positive constants a, b and positive integer n such that $M_f(ar^n) \leq b[M_f(r)]^n$ for all large r, then f is a polynomial of degree no higher than n. A version of 6.37 for harmonic functions will be found in OTHMAN [1996].

6.38 is from ROGOSINSKI [1943]. The special case of it in which $\psi = \phi$ is due to LÖWNER [1917].

6.40(2) appears in ROGOSINSKI [1945] but dates from V. I. Levin and W. Fenchel in 1933. For its history see HUMMEL, SCHEINBERG and ZALCMAN [1977]; they prove that also $|a_2| \leq 2/e$, $|a_3| \leq 2/e$ (where $a_n := f^{(n)}(0)/n!$), and $|a_4| \leq 2/e$, $|a_5| \leq 2/e$ are known too (see **MR** 96m: 30021 and 2004g: 30027). It is conjectured that $|a_n| \leq 2/e$ for all n. For progress on this and the role of the principle of subordination in the problem see PERETZ [1992]. That the inequality 6.40 cannot be improved is witnessed by the function $f(z) := \exp\big((z-1)/(z+1)\big)$; see also 6.41.

Chapter VII
Convergent Sequences of Holomorphic Functions

§ 1 Convergence in $H(U)$

We have already seen four amazing results on the convergence of sequences of holomorphic functions [5.38(iv), 5.44(ii), 5.45(iii) and 5.74] which have no analog for differentiable functions on the line. In this chapter we want to explore this theme more thoroughly. We start with a convenient definition, which, it must be acknowledged, was anticipated at several points in prior chapters.

Definition 7.0 A sequence (set) of complex-valued functions on an open subset U of $\mathbb{C}$ is said to converge or be Cauchy (be bounded) *locally uniformly on* U, if U is a union of open sets in each of which the sequence (set) is uniformly convergent or uniformly Cauchy (uniformly bounded).

Remarks 7.1: (i) A routine covering-compactness argument shows that "locally uniformly on U" is equivalent to "uniformly on each compact subset of U" for each of the phenomena above. We use this equivalence frequently in the sequel without further comment.

(ii) Suppose U, V open, $f_n : U \mapsto V$, $F: V \mapsto \mathbb{C}$ are continuous and $f_n \to f_0$ locally uniformly in U. Then $F \circ f_n \to F \circ f_0$ locally uniformly in U. We will also use (and perhaps already have used!) this fact in the sequel without further comment. To prove it, check that for any compact $K \subset U$ the set $\bigcup_{n=0}^{\infty} f_n(K)$ must lie in a compact subset C of V, then exploit the fact that F is uniformly continuous on C. Another easy proof follows from 7.2. Slight elaboration of the argument even proves a bit more: If $F_n : V \mapsto \mathbb{C}$ are continuous and $F_n \to F_0$ locally uniformly in V, then $F_n \circ f_n \to F_0 \circ f_0$ locally uniformly in U.

R. B. Burckel, *Classical Analysis in the Complex Plane*, Cornerstones,
https://doi.org/10.1007/978-1-0716-1965-0_8

There is another way to look at convergence in $H(U)$; CARATHÉODORY [1929b] even thought it was the most natural way. Say that functions $f_n\colon U \mapsto \mathbb{C}$ *converge continuously* to f if $\lim_{n\to\infty} f_n(z_n)$ exists and equals $f(z_0)$ whenever $z_n \to z_0 \in U$.

Exercise 7.2 *Show that for functions $f_n\colon U \mapsto \mathbb{C}$ the following are equivalent:*

(i) $\{f_n\}$ *converges locally uniformly to the* continuous *function f.*

(ii) $\{f_n\}$ *converges continuously to the function f.*

Hints: (i) $\Rightarrow$ (ii). If z_n, $z_0 \in U$ and $z_n \to z_0$, then $K := \{z_n : n \in \{0\} \cup \mathbb{N}\}$ is a compact subset of U, so $f_n(z) \to f(z)$ uniformly for $z \in K$, while $f(z_n) \to f(z_0)$.

(ii) $\Rightarrow$ (i). If $f_n \to f$ does not happen uniformly on the compact set K, then there exist $\varepsilon > 0$, $1 \leq m_1 < m_2 < \cdots$ and $z_j \in K$ such that

$$\left|f_{m_j}(z_j) - f(z_j)\right| \geq \varepsilon \qquad \forall j \in \mathbb{N}.$$

By passing to a subsequence we can suppose that for some $z_0 \in K$

$$z_j \to z_0 \;\;\&\;\; \mathbb{N} \setminus \{m_1, m_2, \ldots\} \quad \text{is infinite.}$$

Since $\lim_{n\to\infty} f_n(z_j) = f(z_j)$ for each fixed j (why?), there then exist $1 < n_1 < n_2 < \cdots \in \mathbb{N} \setminus \{m_1, m_2, \ldots\}$ such that

$$\left|f_{n_j}(z_j) - f(z_j)\right| \leq \varepsilon/2 \qquad \forall j \in \mathbb{N}$$

whence

$$\left|f_{m_j}(z_j) - f_{n_j}(z_j)\right| \geq \varepsilon/2 \qquad \forall j \in \mathbb{N}.$$

Because of the disjointness of the m-sequence from the n-sequence, we can define

$$z^*_{m_j} := z_j,\; z^*_{n_j} := z_j \qquad \forall j \in \mathbb{N} \text{ and } z^*_n := z_0 \qquad \forall n \in \mathbb{N}\setminus\{n_1, m_1, n_2, m_2, \ldots\}$$

and have $z^*_n \to z_0$. Consequently $f_n(z^*_n) \to f(z_0)$, so that in particular $f_{m_j}(z_j) = f_{m_j}(z^*_{m_j}) \to f(z_0)$ and $f_{n_j}(z_j) = f_{n_j}(z^*_{n_j}) \to f(z_0)$, contradicting the third inequality above. Local uniform convergence and continuous convergence together easily entail continuity of the limit function ($\varepsilon/2$-argument).

Exercise 7.3 *Let U be an open subset of $\mathbb{C}$. According to* 1.31 *there exists a sequence $\mathcal{K}$ of compact sets $K_1 \subset K_2 \subset \cdots \subset U$ such that every compact subset of U lies in some K_n. For f, $g \in H(U)$ define*

$$\|f\|_{K_n} := \sup|f(K_n)|$$

$$d_{\mathcal{K}}(f,g) := \sum_{n=1}^{\infty} 2^{-n} \min\{1, \|f-g\|_{K_n}\}.$$

(i) *Show that $d_{\mathcal{K}}$ is a metric in $H(U)$.*

(ii) *Show that $d_{\mathcal{K}}$ is translation invariant, i.e.,*

$$d_{\mathcal{K}}(f-h, g-h) = d_{\mathcal{K}}(f,g) \qquad \forall f, g, h \in H(U).$$

(iii) *Show that for $f, f_1, f_2, \cdots \in H(U)$ the following are equivalent:*

(a) $\lim_{k\to\infty} f_k = f$ *locally uniformly in U,*

(b) $\lim_{k\to\infty} d_{\mathcal{K}}(f, f_k) = 0$.

(iv) *Show that any translation-invariant metric d in $H(U)$ which generates the topology of local uniform convergence is necessarily complete.*

Hints: Suppose $\{f_n\}$ is d-Cauchy but not locally uniformly Cauchy. Then there is a compact $K \subset U$, $\varepsilon > 0$ such that associated with every $n \in \mathbb{N}$ is an $m > n$ for which $\|f_m - f_n\|_K \geq \varepsilon$. This enables us to successively choose $n_1 < n_2 < n_3 < \cdots$ so that $\left\|f_{n_{j+1}} - f_{n_j}\right\|_K \geq \varepsilon$ for all j. But $d(f_{n_{j+1}} - f_{n_j}, \mathbf{0}) = d(f_{n_{j+1}}, f_{n_j}) \to 0$, which means that $f_{n_{j+1}} - f_{n_j} \to \mathbf{0}$ locally uniformly in U $\longrightarrow\!\!\!\!\!\!\times$.

(v) *Does the conclusion* (iv) *follow without the translation-invariance hypothesis?*

Hints: (Saeki) There is a metric d_0 in $\mathbb{R}$ which generates the usual order-topology of $\mathbb{R}$ but is incomplete. For example, $\phi(x) := \tan(x)$ maps the set $\{x \in \mathbb{R} : |x| < \pi/2\}$ homeomorphically onto $\mathbb{R}$, so

$$d_0(x_1, x_2) := |\phi^{-1}(x_1) - \phi^{-1}(x_2)|$$

defines a metric which generates the usual topology of $\mathbb{R}$. The sequence $\phi(\pi/2 - 1/n)$ is d_0-Cauchy but unbounded in the Euclidean metric. Fix $u_0 \in U$, and for $f, g \in H(U)$ define

$$d(f,g) := d_0(\operatorname{Re} f(u_0), \operatorname{Re} g(u_0)) + d_{\mathcal{K}}(f - \operatorname{Re} f(u_0), g - \operatorname{Re} g(u_0)).$$

Show that d is an incomplete metric which generates the topology of local uniform convergence in $H(U)$.

(vi) *Show that $D(f) := f'$ is a continuous map of $H(U)$ onto a closed subset (vector subspace) of itself.*

Hint: 5.44(ii) and (vii).

Remarks: (i) The convergence in 7.1 is also called "almost uniform convergence" and the topology defined by any of the equivalent metrics in 7.3 is called the "c.c."(compact convergence) or the "compact-open" topology. Admittedly most people carry on the business of function theory without ever mentioning these metrics. The main reason for injecting 7.3 is that it makes the Baire Category Theorem applicable to $H(U)$. In turn, that device leads to the conclusion that many exotic and pathological properties are not just possessed by certain holomorphic functions, but "most" of the functions in $H(U)$ have the property in question [i.e., the set of functions lacking the property is merely of the first Baire category]. Many examples will be produced in Chapters XV and XVI.

(ii) There is yet a third natural and useful convergence concept, inspired by Weierstrass' M (=majorant)-test. It has already come up, but has not been isolated and named. That was first done in 1908 by René Baire. (See REMMERT [1991a], p. 107.) To appreciate the terminology, consider a vector space $\mathcal{V}$ with a norm $\| \ \|$. For $v_n \in \mathcal{V}$ the series $\sum v_n$ is called *normally convergent* if the numerical series $\sum \|v_n\|$ converges in $\mathbb{R}$. It is rather easy to show that the norm $\| \ \|$ is complete (meaning that every Cauchy sequence converges to something) if and only if every normally convergent series converges in $\mathcal{V}$ (i.e., a $v \in \mathcal{V}$ exists such that $\left\|v - \sum_{n=1}^{N} v_n\right\| \to 0$). If U is connected and K is an infinite compact subset of U, then $\|f\|_K := \max |f(K)|$ defines a norm in $H(U)$, thanks to the uniqueness theorem 5.62. The series $\sum f_n$ is called *normally convergent* in $H(U)$ in $H(U)$ if $\sum \|f_n\|_K < +\infty$ for each such K. Using Cauchy's Integral Formula (in a circle) adroitly, it is not too hard to see that this is equivalent to the local uniform convergence in U of the series $\sum |f_n|$. Of course, then there is indeed an $f \in H(U)$ to which the series $\sum f_n$ converges locally uniformly.

Notice that all the metrics $d = d_{\mathcal{K}}$ are bounded by 1, so boundedness in the *metric space* $(H(U), d)$ is not a useful notion. None of the boundedness conditions in the sequel concerns d-boundedness.

Theorem 7.4 (BLASCHKE [1915]) *Let $f_1, f_2, \ldots$ be holomorphic and uniformly bounded in $\mathbb{D}$, let $z_1, z_2, \ldots$ be* distinct *elements of $\mathbb{D}$ satisfying*

(i) $$\sum_{j=1}^{\infty}(1 - |z_j|) = \infty$$

and suppose that

(ii) $$\lim_{n\to\infty} f_n(z_j) \quad \textit{exists for each } j.$$

Then $\{f_n\}$ is locally uniformly convergent on $\mathbb{D}$. (Cf. 12.7(ii).)

Proof: (LÖWNER and RADÓ [1923]) Let $M/2$ be a uniform bound for the f_n in $\mathbb{D}$ and consider any $0 < r < 1$ and any $\varepsilon > 0$. Use hypothesis (i) to select

k so large that

$$M \exp\left[\frac{r-1}{r+1}\sum_{j=1}^{k}(1-|z_j|)\right] < \frac{\varepsilon}{2}. \tag{1}$$

With this k in hand, let C_k be the constant furnished for $\{z_1, \ldots, z_k\}$ by 6.9(ii). Then use hypothesis (ii) to select N so large that

$$\max_{1\le j\le k} |f_n(z_j) - f_l(z_j)| < \frac{\varepsilon}{4C_k} \qquad \forall n, l \ge N. \tag{2}$$

For each pair of n, $l \ge N$ apply the inequality from 6.9(ii) to the function $f := f_n - f_l$, with $m := \varepsilon/4C_k$. It follows, using (1), that

$$|f_n(z) - f_l(z)| \le \varepsilon \qquad \forall z \in \overline{D}(0, r).$$

This establishes the local uniform Cauchyness of $\{f_n\}$ and finishes the proof.

Remark: Notice that hypothesis (i) is (dramatically) fulfilled if $\{z_j\}$ has a limit point in $\mathbb{D}$. That case, however, was handled already in 5.74(ii). It suffices for most of what follows.

Easy connectedness arguments free us from the disk:

Corollary 7.5 (Vitali-Porter Theorem on Induced Convergence) *Let Ω be a region, $\{F_n\} \subset H(\Omega)$ be locally uniformly bounded in Ω. If $\{F_n(z)\}$ converges for a set of z having a limit point in Ω, then $\{F_n\}$ converges locally uniformly on Ω to some $F \in H(\Omega)$.*

Proof: Let C denote the set of $z \in \Omega$ such that $\{F_n(z)\}$ converges and let Ω_0 denote the set of limit points of C *in* Ω. Note that $\Omega_0 \ne \varnothing$ by hypothesis. 7.4 (or 5.74(ii)) shows that

$$z_0 \in \Omega_0 \;\;\&\;\; \overline{D}(z_0, r) \subset \Omega \Rightarrow D(z_0, r) \subset \Omega_0 \cap C. \tag{$*$}$$

It follows from $(*)$ and the Basic Connectedness Lemma (1.27) that $\Omega_0 = \Omega$ and then a second citation of $(*)$ shows that $C = \Omega$. A return to 7.4 (or 5.74) shows that in fact convergence is locally uniform in Ω. The limit function is holomorphic by 5.44(i).

Alternatively, a complete one-line proof is provided by citing 5.74(ii) and 5.38(iv). See also 3.35(v).

Corollary 7.6 (Montel [1907]) *If Ω is a region, $\{F_n\}$ a locally uniformly bounded sequence in $H(\Omega)$, then there exists a subsequence of $\{F_n\}$ which is locally uniformly convergent on Ω.*

Proof: (LINDELÖF [1913]). Let K be a countable subset of Ω with a limit point in Ω. [For example, select $z_0 \in \Omega$, $r > 0$ with $D(z_0, r) \subset \Omega$ and let $K := \{z_0 + r/m : m = 2, 3, \dots\}$.] Now $\{F_n(z)\}$ is a bounded set for each $z \in \Omega$. Therefore an application of the diagonal process (see chapter 0) to the countable set K produces a subsequence $\{F_{n_k}\}$ such that $\{F_{n_k}(z)\}$ converges for each $z \in K$. The asserted local uniform convergence follows upon applying 7.5 to the sequence $\{F_{n_k}\}$.

Exercise 7.7 (i) *Show that the connectedness hypothesis in the preceding corollary is unnecessary.*

Hint: The components of Ω are open and disjoint, hence only countable in number. Apply 7.6 successively to each and take the diagonal subsequence (see chapter 0).

(ii) *Show that if $f_n \in H(U)$ and $f_n \to f_0$ locally uniformly in U, then the set $\{f_n\}$ is locally uniformly bounded in U.*

(iii) *Show that for $U := \mathbb{D}$ the conclusion in* (ii) *can be strengthened: If $\mathcal{F}$ is a locally uniformly bounded subset of $H(\mathbb{D})$ and $\sum_{k=0}^{\infty} c_k(f) z^k$ is the Taylor series of $f \in \mathcal{F}$, then*

$$\sum_{k=0}^{\infty} \sup_{f \in \mathcal{F}} |c_k(f)| r^k < \infty \qquad \forall r \in [0, 1)\,.$$

Hint: Use the Cauchy estimates 3.34(i).

Corollary 7.8 (MONTEL [1910], p. 21) *Let U be an open subset of $\mathbb{C}$, $\{f_n\}$ a locally uniformly bounded sequence in $H(U)$ with the property that every one of its convergent subsequences converges to g. Then the sequence $\{f_n\}$ itself converges to g, locally uniformly on U.*

Proof: Suppose there exists compact $K \subset U$ such that $\{f_n\}$ does not converge to g uniformly on K: $\sup |g - f_n|(K) \not\to 0$ as $n \to \infty$. This means that there exist $\varepsilon > 0$ and integers $1 \le n_1 < n_2 < \cdots$ such that

$$\sup |g - f_{n_j}|(K) \ge \varepsilon \qquad \forall j \in \mathbb{N}.$$

By 7.7 $\{f_{n_j}\}_{j \in \mathbb{N}}$ contains a locally uniformly convergent subsequence. Its limit cannot be g.

Corollary 7.9 *Let $f_n(z) := \sum_{k=0}^{\infty} c_{n,k} z^k$ $(n = 0, 1, 2, \dots)$ be holomorphic in $\mathbb{D}$ and locally uniformly bounded there. Then $f_n \to f_0$ locally uniformly if and only if $c_{n,k} \to c_{0,k}$ for each $k = 0, 1, 2, \dots$.*

Proof: ($\Rightarrow$) Since $c_{n,k} = f_n^{(k)}(0)/k!$, this assertion follows from 5.44(ii). Alternatively, it follows from 3.34(i).

($\Leftarrow$) If g is a subsequential limit of $\{f_n\}$, then by 5.44(ii) and the condition $c_{n,k} \to c_{0,k}$, we see that $g^{(k)}(0)/k! = c_{0,k}$, i.e., by 5.30 $g = f_0$. Now cite 7.8. A direct proof using the Cauchy estimates 3.34(i) was outlined in 3.35.

Theorem 7.10 *Let Ω be a region, f_n zero-free, holomorphic, and locally uniformly bounded on Ω. If $\lim_{n\to\infty} f_n(z_0) = 0$ for* some $z_0 \in \Omega$, *then $\lim_{n\to\infty} f_n(z) = 0$ for all $z \in \Omega$ and the convergence is locally uniform.*

Proof: Let C denote the set of $z \in \Omega$ such that $\{f_n(z)\}$ converges to 0 and let U denote the set of $z \in \Omega$ such that $\{f_n\}$ converges uniformly to 0 in some neighborhood of z. Note that $C \neq \varnothing$ by hypothesis and that $U \subset C$. We shall show that

$$(*) \qquad z_1 \in C \;\&\; \bar{D}(z_1, 2r) \subset \Omega \Rightarrow D(z_1, r) \subset U \subset C.$$

Indeed let $B := B(z_1, r)$ be a bound for $\{f_n\}$ in the compact subset $\bar{D}(z_1, 2r)$ and take $R := 2r$ in 6.36, getting for each n

$$\begin{aligned}\sup_{|z-z_1|=r} |f_n(z)| &\le |f_n(z_1)|^{(R-r)/(R+r)} \left[\sup_t \left|f_n\left(z_1 + Re^{it}\right)\right|\right]^{2r/(R+r)} \\ &\le |f_n(z_1)|^{1/3} B^{2/3}.\end{aligned}$$

This clearly shows that if $\{f_n(z_1)\}$ converges to 0, then $\{f_n\}$ converges to 0 uniformly in $D(z_1, r)$, whence $(*)$. In the light of $(*)$ the Basic Connectedness Lemma 1.27 is applicable and asserts that $C = \Omega$. But then $(*)$ says that $U = \Omega$.

Corollary 7.11 *Ω is a region, $F_n \in H(\Omega)$ converge locally uniformly to the* non-constant *function F.*

(i) *If all $F_n(\Omega)$ are subsets of a set S, then $F(\Omega) \subset S$.*

(ii) *If all F_n are one-to-one, then F is one-to-one.*

Proof: (i) If $F(w) \in \mathbb{C} \setminus S$ for some $w \in \Omega$,then $F_n - F(w)$ is zero-free for all n, so by 7.10 and 7.7(ii), $F - F(w)$ is constant. —⨯—

(ii) Fix $w \in \Omega$ and consider the functions $f_n := F_n - F_n(w)$ in the connected (by 1.24) open set $\Omega \setminus \{w\}$. They are zero-free and their limit is the non-constant function $F - F(w)$, which is also zero-free in $\Omega \setminus \{w\}$ (by (i)); that is, $F(z) = F(w)$ for no $z \neq w$.

Notice that this theorem too has no analog on the line. For example, if F_n is defined by $F_n(x) := e^{-1/x^2}$ for $x \in (-\infty, 0)$, $F_n(0) := 0$ and $F_n(x) := -\frac{1}{n}e^{-1/x^2}$ for $x \in (0, \infty)$, then F_n is a one-to-one C^∞-function on $\mathbb{R}$. The (uniform) limit function is also C^∞ but is neither constant nor one-to-one.

In 7.10 it is clear that the conclusion follows under the weaker hypothesis that locally all but finitely many of the f_n are zero-free. Rouché's theorem can be used to deduce the range conclusion in 7.11 without the holomorphy hypothesis on the F_n (cf. TSARPALIAS [1989]):

Exercise 7.12 *Let Ω be a region in $\mathbb{C}$, $f_n\colon \Omega \mapsto S \subset \mathbb{C}$ continuous functions which converge locally uniformly in Ω to an $f \in H(\Omega)$. Show that either f is constant or $f(\Omega) \subset S$.*

Hints: Suppose f is not constant. We have to show that $f(\Omega) \subset \mathbb{C} \setminus \{a\}$ for each $a \in \mathbb{C} \setminus S$, and it suffices [via a translation] to consider $a = 0$. Since f is not identically 0, given $z_0 \in \Omega$, there exists $0 < r < \infty$ such that $\overline{D}(z_0, r) \subset \Omega$ and f is zero-free on $C(z_0, r)$. The number $m := \inf|f(C(z_0, r))|$ is consequently positive. Therefore there exists n such that $|f_n - f| < m$ on the compact set $C(z_0, r)$. In particular then,

$$|f_n - f| < |f| \le |f_n| + |-f| \quad \text{on } C(z_0, r).$$

Apply 5.87(iii) [with g there equal $-f$ and f there equal f_n] to infer that f is zero-free in $\overline{D}(z_0, r)$.

Though 7.6 appears here as a trivial glissade from the Vitali–Porter theorem, in fact the two are easily seen to be equivalent via the fundamental uniqueness theorem (5.62). An *ab initio* proof of 7.6 is then desirable. Montel himself followed this path in [1907] and in his [1910] monograph, and prior to LINDELÖF [1913] (see p. 178) it was customary to deduce 7.5 from 7.6. Here's how it goes:

Exercise 7.13 *Let Ω be a region in $\mathbb{C}$, S a countable dense subset of Ω (e.g., S might be $(\mathbb{Q} + i\mathbb{Q}) \cap \Omega$), $\mathcal{F} \subset H(\Omega)$ a locally uniformly bounded family.*

(i) *Show that for each compact $K \subset \Omega$ there is a finite constant $k := k(\mathcal{F}, K)$ such that*

$$|f(z) - f(w)| \le k|z - w| \qquad \forall z, w \in K, f \in \mathcal{F}.$$

Hint: Use 5.34(ii).

(ii) *Using only* (i), *show that if $\{F_j\} \subset \mathcal{F}$ and $\{F_j(z)\}$ converges for each $z \in S$, then $\{F_j\}$ converges locally uniformly in Ω.*

Hint: We need to show that $\{F_n\}$ is uniformly Cauchy in any open $U \subset \overline{U}$ compact $\subset \Omega$. Given $\varepsilon > 0$, pick $r_\varepsilon > 0$ such that $r_\varepsilon k(\overline{U}) < \varepsilon/3$. Denseness of S ensures existence of $w_1, \dots, w_m \in U \cap S$ such that the disks $D(w_j, r_\varepsilon)$ cover U. There exists $N(\varepsilon) \in \mathbb{N}$ such that

$$|F_n(w_j) - F_m(w_j)| < \varepsilon/3 \qquad \forall n, m \geq N(\varepsilon), \quad j = 1, 2, \dots, m.$$

Given $z \in U$, there is j with $z \in D(w_j, r_\varepsilon)$. For $n, m \geq N(\varepsilon)$

$$\begin{aligned} |F_n(z) - F_m(z) &\leq |F_n(z) - F_m(w_j)| + |F_n(w_j) - F_m(w_j)| + |F_m(w_j) - F_m(z)| \\ &\leq k(\overline{U})|z - w_j| + \varepsilon/3 + k(\overline{U})|w_j - z|, \text{ since } z, w_j \in \overline{U} \\ &\leq k(\overline{U})r_\varepsilon + \varepsilon/3 + k(\overline{U})r_\varepsilon, \text{ since } z \in D(w_j, r_\varepsilon) \\ &\leq \varepsilon/3 + \varepsilon/3 + \varepsilon/3, \text{ by definition of } r_\varepsilon. \end{aligned}$$

(iii) *Show that if $\{f_n\} \subset \mathcal{F}$, then there exist $1 \leq n_1 < n_2 < \cdots$ such that $\lim_{j\to\infty} f_{n_j}(z)$ exists in $\mathbb{C}$ for every $z \in S$.*

Hint: Diagonal argument (see chapter 0); recall each sequence $\{f_n(z)\}_{n\in\mathbb{N}}$ is bounded.

(iv) (Montel's Compactness Theorem) *Show that every locally uniformly bounded sequence in $H(\Omega)$ contains a locally uniformly convergent subsequence.*

Hint: Combine (ii) and (iii).

(v) *From* (iv) *and* 5.62 *alone deduce* 7.5.

Hints: Let notation and hypotheses be as in 7.5. Consider any subsequential limits h, g of $\{F_n\}$. They agree on the convergence-set of $\{F_n\}$, which by hypothesis has a limit point in Ω. Hence by 5.62, $h = g$. But now we can invoke 7.8 (whose simple proof used only (iv)).

Theorem 7.14 *Let Ω be a region, f_n holomorphic in Ω such that $\{f_n(z_0)\}$ converges for some $z_0 \in \Omega$. Suppose that either*

(i) *$\{\operatorname{Re} f_n\}$ converges locally uniformly on Ω*

or

(ii) *$\operatorname{Re} f_n(z) \leq \operatorname{Re} f_{n+1}(z)$ for all $n \in \mathbb{N}$ and all $z \in \Omega$*

or

(iii) *$|f_n(z)| \leq |f_{n+1}(z)|$ for all $n \in \mathbb{N}$ and all $z \in \Omega$ and $\lim_{n\to\infty} f_n(z_0) \neq 0$.*

Then $\{f_n\}$ converges locally uniformly in Ω.

Proof: Let C denote the set of $z \in \Omega$ such that $\{f_n(z)\}$ converges and let U denote the set of $z \in \Omega$ such that $\{f_n\}$ converges uniformly in some neighborhood of z. Note that $z_0 \in C$, so $C \neq \varnothing$. Also $U \subset C$. For each $z_1 \in \Omega$ and each $r > 0$ such that

$$\bar{D}(z_1, 2r) \subset \Omega \tag{1}$$

we may apply the second Hadamard–Borel–Carathéodory Inequality 6.35(ii) to $f_m - f_n$ for every pair (m, n) with $n \geq m$ to get

$$\sup_{|z-z_1|=r} |f_n - f_m|(z) \leq \frac{2r}{2r-r} \sup_{|z-z_1|=2r} \operatorname{Re}(f_m - f_n)(z) + \frac{2r+r}{2r-r}|f_n - f_m|(z_1)$$

and, applying the Maximum Principle,

$$\sup_{z \in \bar{D}(z_1,r)} |f_n(z) - f_m(z)| \leq 2 \sup_{|z-z_1|=2r} \operatorname{Re}(f_m - f_n)(z) + 3|f_n(z_1) - f_m(z_1)|. \tag{2}$$

If hypothesis (i) prevails, then it follows at once from (2) that

$$z_1 \in C \;\&\; \bar{D}(z_1, 2r) \subset \Omega \Rightarrow D(z_1, r) \subset U \subset C. \tag{3}$$

On the other hand, if hypothesis (ii) prevails, then (2) yields

$$\sup_{z \in \bar{D}(z_1,r)} |f_n(z) - f_m(z)| \leq 3|f_n(z_1) - f_m(z_1)| \qquad \forall n \geq m,$$

from which (3) again follows. From (3) and the Basic Connectedness Lemma it follows that $C = \Omega$. But then (3) says that $U = \Omega$.

Now consider hypothesis (iii). Because $\lim_{n\to\infty} f_n(z_0) \neq 0$, we can discard finitely many f_n and simply suppose that $f_1(z_0) \neq 0$. Then the set $A := f_1^{-1}(0)$ has no limit points in Ω (5.62) and $\Omega_0 = \Omega \setminus A$ is a region (1.24) in which all the f_n are zero-free. As before, form the pointwise convergence set C and the locally uniform convergence set U, in Ω_0. We have $z_0 \in C$, so $C \neq \varnothing$. If $z_1 \in C$ and $D(z_1, r) \subset \Omega_0$, then by 5.35 each f_n has a holomorphic logarithm g_n in $D(z_1, r)$:

$$f_n = e^{g_n} \quad \text{in } D(z_1, r). \tag{4}$$

Set $w_1 := \lim f_n(z_1)$, which is non-zero by (iii) and the definition of Ω_0. Let L be a continuous logarithm in $D(w_1, |w_1|)$. For some $N_1 \in \mathbb{N}$, $f_n(z_1) \in D(w_1, |w_1|)$, so $e^{L(f_n(z_1))} = f_n(z_1) = e^{g_n(z_1)}$, for all $n \geq N_1$. Thus for these n

there exists $N(n) \in \mathbb{Z}$ such that $g_n(z_1) = L(f_n(z_1)) - 2\pi i N(n)$. By replacing g_n with $g_n + 2\pi i N(n)$, we may have (4) and also

$$\text{(5)} \qquad \{g_n(z_1)\} \quad \text{converges.}$$

Because of (4), (5) and (iii) the hypotheses of (ii) are met by the g_n in $D(z_1, r)$, so $\{g_n\}$ converges locally uniformly in $D(z_1, r)$. It follows that $D(z_1, r) \subset U$ and by the Basic Connectedness Lemma 1.27 we again get $C = U = \Omega_0$.

Now consider any $a \in A$. Since the points of A are isolated, there exists an $r > 0$ such that $\bar{D}(a, r) \subset \Omega$ and $\bar{D}(a, r) \cap A = \{a\}$. In particular, $C(a, r) \subset \Omega \setminus A = \Omega_0$ and so by the result of the preceding paragraph $\{f_n\}$ is uniformly convergent on the compact set $C(a, r)$. By the Maximum-Modulus Principle then, $\{f_n\}$ is uniformly convergent throughout $\bar{D}(a, r)$.

We have shown that each point of Ω_0 and each point of $\Omega \setminus \Omega_0 = A$ has a neighborhood in which $\{f_n\}$ is uniformly convergent; that is, $\{f_n\}$ is locally uniformly convergent in Ω.

Corollary 7.15 *If h_n are harmonic in a region Ω, $h_n \le h_{n+1}$ for all n and $\{h_n(z_0)\}$ converges (in $\mathbb{R}$) for some $z_0 \in \Omega$, then $\{h_n\}$ is locally uniformly convergent on Ω (to a harmonic function, by* 5.44(iii)*).* Cf. 5.28(iii)

Proof: 7.14 and 5.21(i) combine to show that the local-uniform-convergence set A fulfills the hypothesis of 1.27 with $\theta = 1$.

Remark: The scope of all these theorems will be greatly extended when we prove in Chapter XII the remarkable fact that a family $\mathcal{F}$ of holomorphic functions is locally uniformly bounded if there are two distinct complex numbers not in the range of any function in $\mathcal{F}$ and the set of function values at some one point is bounded.

Exercise 7.16 (i) *Let U be an open subset of $\mathbb{C}$, $h_n \colon U \mapsto \mathbb{R}$ harmonic functions such that $\{h_n\}$ is locally uniformly bounded in U. Show that $\{h_n\}$ contains a locally uniformly convergent subsequence.*

Hints: It suffices to consider the case in which U is a disk and $\{h_n\}$ is uniformly bounded in U. This is because U is a countable union of such disks and we can apply the weaker conclusion *successively* in each disk and take a diagonal subsequence. Thus, suppose h_n are harmonic in the disk D and $|h_n| < M < \infty$ in D for all n. According to 5.21(i) there exist $f_n \in H(D)$ such that $h_n = \operatorname{Re} f_n$. Then the functions $F_n := e^{f_n} \in H(D)$ satisfy $e^{-M} \le e^{h_n} = |F_n| \le e^M$. According to 7.6 there then exist $1 \le n_1 < n_2 < \cdots$ and $F \in H(D)$ such that $\lim_{j\to\infty} F_{n_j} = F$ locally uniformly in D. Then $|F_{n_j}| \to |F|$ and $e^{-M} \le |F| \le e^M$. It follows from this and the uniform continuity of log on $\left[e^{-M}, e^{M}\right]$ that $h_{n_j} = \log|F_{n_j}| \to \log|F|$ locally uniformly in D.

(ii) *Suppose that in* (i) *the sequence of harmonic functions is only locally uniformly bounded from above. What conclusion can be drawn?*

Hints: Denote by U_b the set of $p \in U$ for which $\{h_n(p)\}$ has a bounded subsequence. By hypothesis each $p \in U$ lies in a connected open set V_p in which $M_p - h_n > 0$ for some $M_p \in \mathbb{R}$ and all n. If $\{h_{n_j}(p)\}$ is bounded, 5.28(ii) can be applied to $\{M_p - h_{n_j}\}$ to learn that $\{h_{n_j}\}$ is uniformly bounded in a subneighborhood W_p. Thus U_b is an open set. The complementary set consists of those p for which $h_n(p) \to -\infty$, that is, $M_p - h_n(p) \to \infty$. For such p, 5.28(ii) produces a subneighborhood of V_p in which $\{h_n\}$ diverges uniformly to $-\infty$, so $U \setminus U_b$ is also open. Each component Ω of U must lie wholly in one of these clopen subsets of U. In the first case, (i) insures that $\{h_n\}$ has a subsequence locally uniformly convergent in Ω. In the second case, $\{h_n\}$ diverges to $-\infty$ locally uniformly in Ω.

(iii) *Let $h\colon \mathbb{D} \mapsto \mathbb{R}$ be subharmonic. Show that the following two properties of h are equivalent:*

$$(1.h) \qquad \sup_{r<1} \int_0^{2\pi} h\left(re^{i\theta}\right) d\theta < \infty,$$

$$(2.h) \qquad h \le H \quad \textit{for some harmonic function } H \textit{ in } \mathbb{D}.$$

Hints: If (2.h) holds, then the integral in (1.h) is majorized by $\int_0^{2\pi} H\left(re^{i\theta}\right) d\theta$, which equals $2\pi H(0)$.

Suppose (1.h) holds. Take a sequence $0 < r_1 < r_2 < \cdots \to 1$ and let (5.17) h_n be the function which is harmonic in $D(0, r_n)$ and satisfies

$$(3) \qquad h_n = h \quad \text{on } C(0, r_n), \quad n = 1, 2, \ldots.$$

Then by 5.22 we have

$$(4) \qquad h \le h_{n+1} \quad \text{in } D(0, r_{n+1}), \quad n = 1, 2, \ldots.$$

From (3), (4) and $r_n < r_{n+1}$ we have

$$h_n = h \le h_{n+1} \quad \text{on } C(0, r_n)$$

and so, application of the Maximum Principle to $h_n - h_{n+1}$ in $D(0, r_n)$ gives

$$(5) \qquad h_n \le h_{n+1} \quad \text{in } D(0, r_n),\ n = 1, 2, \ldots.$$

By the mean-value property of harmonic functions

$$2\pi h_n(0) = \int_0^{2\pi} h_n\left(r_n e^{i\theta}\right) d\theta \overset{(3)}{=} \int_0^{2\pi} h\left(r_n e^{i\theta}\right) d\theta, \quad n = 1, 2, \ldots .$$

From this equation and (1.h) we see that the sequence $\{h_n(0)\}$ is bounded above, hence, by (5), convergent. Apply 7.15 in $D(0, r_k)$ to the sequence $\{h_n\}_{n>k}$ for each k to come up with a harmonic function H in $\mathbb{D}$ such that $h_n \to H$ throughout $\mathbb{D}$. Go to the limit in (4) to see that $h \leq H$ throughout $\mathbb{D}$.

(iv) *Show that if h is harmonic in $\mathbb{D}$, then*

(3.h) $h = h_1 - h_2$, *for some non-negative harmonic functions h_j in $\mathbb{D}$*

is a third condition equivalent to (1.$|h|$) *and* (2.$|h|$).

Hint: Given h_j, set $H := h_1 + h_2$. Given H, set $h_j := H - (-1)^j h/2$.

The next result is an extension of 5.12 and 5.46 which we shall need later. An interesting but rather complicated proof of it can be based on 7.15 (as was done in my earlier book; see also HEINS [1962b]) and this explains its appearance at this juncture. However, there is a more perspicuous arrangement of the proof which just uses ideas from Chapter V that I want to present instead.

Theorem 7.17 *Let U be an open subset of $\mathbb{C}$, E a countable* proper *subset of ∂U, $u\colon U \mapsto \mathbb{R}$ a subharmonic function which for some finite constant M satisfies*

(1) $$\varlimsup_{\substack{z \to \zeta \\ z \in U}} u(z) \leq M \qquad \forall \zeta \in \partial U \setminus E,$$

(2) $$\varlimsup_{\substack{z \to \zeta \\ z \in U}} \frac{u(z)}{-\log|z - \zeta|} \leq 0 \qquad \forall \zeta \in E$$

and, in case U is unbounded,

(2)′ $$\varlimsup_{\substack{|z| \to \infty \\ z \in U}} \frac{u(z)}{\log|z|} \leq 0.$$

Then $u \leq M$ throughout U.

Proof: Things become more transparent if we divide the proof into three phases:

Step I. U is bounded. This is just 5.46.

Step II. U is unbounded but not dense in $\mathbb{C}$. Here, after a translation, we may suppose that $0 \in \mathbb{C}\backslash\overline{U}$. Thus for some $0 < r < \infty$, $D(0,r) \subset \mathbb{C}\backslash U$ and the map $\phi(z) := 1/z$ sends U conformally onto an open subset U_0 of $D(0, 1/r)$. As noted in 5.22(ii), the function $u_0 := u \circ \phi^{-1} = u \circ \phi$ is subharmonic in U_0. Since ϕ is a conformal map of a whole neighborhood of $\overline{U}$ into $\mathbb{C} \setminus \{0\}$, $\phi(\partial U) = (\partial U_0) \cap (\mathbb{C} \setminus \{0\}) = (\partial U_0) \setminus \{0\}$. On the other hand, if $w \in U_0$ and $w \to 0$, then $\left|\phi^{-1}(w)\right| \to \infty$. Hence by $(2)'$

$$(2)'_0 \qquad \varlimsup_{\substack{w \to 0 \\ w \in U_0}} \frac{u_0(w)}{-\log|w|} = \varlimsup_{\substack{|z| \to \infty \\ z \in U}} \frac{u(z)}{\log|z|} \leq 0,$$

and, if we set $E_0 := \phi(E)$, for all $w_0 \in (\partial U_0) \setminus E_0 \setminus \{0\} = \phi\big((\partial U) \setminus E\big)$

$$(1)_0 \qquad \varlimsup_{\substack{w \to w_0 \\ w \in U_0}} u_0(w) \leq M.$$

If $w \in U_0$ and $w \to w_0 \in E_0$, say $w_0 = \phi(\zeta)$, then $z := \phi^{-1}(w)\ (\in U) \to \zeta \in E$ and

$$\begin{aligned} \frac{u_0(w)}{-\log|w - w_0|} &= \frac{u(z)}{-\log\left|1/z - 1/\zeta\right|} = \frac{u(z)}{-\log|z - \zeta| + \log|z\zeta|} \\ (3) \qquad &= \frac{u(z)}{-\log|z - \zeta|}\left[1 - \frac{\log|z\zeta|}{\log|z - \zeta|}\right]^{-1}. \end{aligned}$$

Clearly the square-bracketed expression converges to 1 as z converges to the non-zero number ζ. Hence from (3), (2), and $(2)'_0$ we have

$$(2)_0 \qquad \varlimsup_{\substack{w \to w_0 \\ w \in U_0}} \frac{u_0(w)}{-\log|w - w_0|} \leq 0 \qquad \forall w_0 \in E_0 \cup \{0\}.$$

In view of $(1)_0$ and $(2)_0$ Step I applies to u_0 in U_0 (with exceptional set $E_0 \cup \{0\}$) and affirms that $u_0 \leq M$ throughout U_0. It follows that $u \leq M$ throughout U.

Step III. For an arbitrary unbounded open set U, consider any $z_0 \in U$ and any $\varepsilon > 0$. Since u is continuous at z_0, there exists $0 < r < \infty$ such that $\overline{D}(z_0, r) \subset U$ and

$$(4) \qquad u < u(z_0) + \varepsilon \quad \text{on } \overline{D}(z_0, r).$$

The open set $U_\varepsilon := U \setminus \bar{D}(z_0, r)$ comes under Step II and because of (4), u satisfies in U_ε the original hypotheses with $M_\varepsilon := \max\{M, u(z_0) + \varepsilon\}$ in the role of M. It follows that $u \le M_\varepsilon$ throughout U_ε and so by (4)

$$u \le M_\varepsilon \quad \text{throughout } U. \tag{5}$$

In (5) let $\varepsilon \downarrow 0$ and get

$$u(z) \le \max\{M, u(z_0)\} \qquad \forall z, z_0 \in U. \tag{6}$$

If

$$u(z_0) > M \quad \text{for some } z_0 \in U, \tag{7}$$

then (6) gives

$$u(z) \le u(z_0) \qquad \forall z \in U. \tag{8}$$

But, upon reversing the roles of z, z_0 in (6), we also have

$$u(z_0) \le \max\{M, u(z)\} \qquad \forall z \in U \tag{9}$$

and (7) shows that this maximum cannot be M, so

$$u(z_0) \le u(z) \qquad \forall z \in U.$$

Combined with (8) and (7) this says that

$$u(z) = u(z_0) > M \qquad \forall z \in U. \tag{10}$$

Since by hypothesis $E \neq \partial U$, there exist ζ which satisfy (1), rendering (10) untenable. Hence (7) is also untenable.

Remark: The argument in step III above covers as well the case that $U := \mathbb{C}$, in which hypotheses (1) and (2) evaporate and only (2)′ is relevant. Now there is no M, and (6) simply proves that $u(z) \le u(z_0)$ for all $z, z_0 \in U = \mathbb{C}$, that is, u is constant. Cf. 5.15.

As every reader probably suspects, pointwise convergence alone does not entail local uniform convergence. (But because of Vitali–Porter, examples are not so easy to construct! See 8.15 for one, SCHMIEDER [1994], and BEARDON and MINDA [2003] for others.) Surprisingly though, pointwise convergence does lead to uniform convergence on a large subset:

Exercise 7.18 (OSGOOD [1901–02]) *Let U be an open subset of $\mathbb{C}$, f_n holomorphic in U. Suppose that $\lim_{n\to\infty} f_n(z)$ exists (as a complex number) for* *each $z \in U$. Call it $f(z)$. Show that there is a dense open subset U_0 of U such that the convergence is locally uniform on U_0 and (so) f is holomorphic in U_0.*

Hints: Thanks to 7.5 we need only prove that $\{f_n\}$ is locally uniformly bounded in some dense open subset U_0 of U. The argument for this (MONTEL [1910], p. 109; ULLRICH [1994a]; BEARDON and MINDA [2003]) uses only the continuity of the f_n, and it works as well if U is any metric space which is either complete or locally compact. The claim is: Every non-void open $V \subset U$ contains a non-void open disk $D(V)$ in which $\{f_n\}$ is bounded. Were this false for some V, we could inductively find $g_k \in \{f_n\}$ and non-void open disks $D_{k+1} \subset \bar{D}_k \subset V$ such that $|g_k| > k$ in D_{k+1}. At each point z of the non-void set $\bigcap \bar{D}_k$ the set $\{g_k(z)\} \subset \{f_n(z)\}$ would be unbounded $\longrightarrow\!\!\!\!\!\!\times$. Let $U_0 := \bigcup\{D(V) : \varnothing \neq V \text{ open } \subset U\}$. Evidently U_0 is open and dense in U. Each compact $K \subset U_0$ is covered by finitely many of the disks $D(V)$, making $\{f_n\}$ bounded in K.

By now many readers will have wondered about what operations preserve local uniform convergence and in particular whether "inversion" does. This important issue comes up in Chapter IX, §6. The question is rather delicate, because, while the different functions may have the same domain, their inverses do not. Thus the "local" result [in part (ii) below] is rather convoluted, but the global result [in part (iii)] is quite clean.

Exercise 7.19 (i) (converse of 7.11): *Suppose regions $\Omega_n \uparrow \Omega$ and $f_n \in H(\Omega_n)$ converge locally uniformly in Ω to f. (Note that every compact $K \subset \Omega$ lies in all but finitely many Ω_n.) Show that if f is not constant, then any compact $L \subset f(\Omega)$ is a subset of $f_n(\Omega_n)$ for all but finitely many n. Show further that if f is one-to-one in Ω, then in any compact $K \subset \Omega$ all but finitely many f_n are one-to-one.*

Hints: If L lacks this property, there exist $1 \leq n_1 < n_2 < \cdots$ and $w_j \in L \setminus f_{n_j}(\Omega_{n_j})$. We can suppose $w_j \to w_0 \in L$. Then $w_0 = f(z_0)$, for some $z_0 \in \Omega$, say $z_0 \in \Omega_{n_0}$. Then there exists $r > 0$ such that $\bar{D}(z_0, r) \subset \Omega_{n_0}$. We can suppose all $n_j \geq n_0$. The functions $f_{n_j} - w_j$ are zero-free in $\bar{D}(z_0, r)$ and converge uniformly there to the non-constant function $f - w_0$, which does have a zero in $D(z_0, r)$. This contradicts 7.10.

We have $K \subset \Omega_{n_0}$ for some n_0. If K lacks the property claimed, there exist $n_0 \leq n_1 < n_2 < \cdots$, $z_j \in K$, $z_j' \in K \setminus \{z_j\}$ with $f_{n_j}(z_j) = f_{n_j}(z_j')$ $(j \in \mathbb{N})$. We can suppose $z_j \to z_0 \in K$, $z_j' \to z_0' \in K$. Then (7.2) $f(z_0) = \lim f_{n_j}(z_j) = \lim f_{n_j}(z_j') = f(z_0')$, so by univalence of f, $z_0 = z_0'$. Pick $r > 0$ such that $\bar{D}(z_0, 2r) \subset \Omega_{n_0}$. The compact sets $f\big(C(z_0, 2r)\big)$ and $f\big(\bar{D}(z_0, r)\big)$ are disjoint (f being one-to-one), hence at a positive distance d from one another:

$$(1) \qquad |f(\zeta) - f(z)| \geq d \qquad \forall \zeta \in C(z_0, 2r), z \in \bar{D}(z_0, r).$$

Since convergence is uniform in $\bar{D}(z_0, 2r)$,

$$|f_n(\zeta) - f(z)| \geq d/2 \qquad \forall \zeta \in C(z_0, 2r), z \in \bar{D}(z_0, r)$$

holds for all sufficiently large n, without loss of generality for all $n \geq n_0$. That is,

$$(2) \qquad |f_n(\zeta) - w| \geq d/2 \qquad \forall \zeta \in C(z_0, 2r), w \in f\big(\bar{D}(z_0, r)\big), n \geq n_0.$$

Then according to 5.85 the number of zeros of $f_n - w$ in $D(z_0, 2r)$ is

$$(3) \qquad N_n(w) := \frac{1}{2\pi i} \int_{C(z_0,2r)} \frac{f_n'}{f_n - w} \qquad \forall w \in f\big(\bar{D}(z_0, r)\big), n \geq n_0.$$

Now $f_n \to f$ and $f_n' \to f'$ uniformly on $C(z_0, 2r)$. Keeping (2) in mind, it follows then from (3) that

$$(4) \qquad N_n(w) \to N(w) := \frac{1}{2\pi i} \int_{C(z_0,2r)} \frac{f'}{f - w} \qquad \forall w \in f\big(\bar{D}(z_0, r)\big).$$

But according to 5.85 and (1) the integral $N(w)$ is the number of zeros of $f - w$ in $D(z_0, 2r)$, and that is 1, since f is one-to-one. The convergence (4), which is uniform for w in the compact set $f\big(\bar{D}(z_0, r)\big)$, therefore means that

$$N_n(w) = 1 \qquad \forall w \in f\big(\bar{D}(z_0, r)\big)$$

and all large n; without loss of generality, for all $n \geq n_0$. That is, for each $n \geq n_0$

(5) every value $w \in f\big(\bar{D}(z_0, r)\big)$ is assumed by f_n exactly once in $D(z_0, 2r)$.

Since $f_{n_j}(z_j) = f_{n_j}(z_j') \to f(z_0)$, these numbers are eventually in the open set $f\big(D(z_0, r)\big)$, while z_j, z_j' are eventually in $D(z_0, 2r)$. The facts $f_{n_j}(z_j) = f_{n_j}(z_j')$ and $z_j \neq z_j'$ then contradict (5).

(ii) (Cartan [1932] and Carathéodory [1932], *Satz* 2). *Let notation and hypotheses be as in* (i) *and suppose that f is one-to-one in Ω and L is compact $\subset f(\Omega)$. Show that there exist $k \in \mathbb{N}$ and compact $K \subset \Omega$ such that $K \subset \Omega_n$, f_n is one-to-one on K, $L \subset f_n(K)$ for all $n \geq k$, and $\big\{(f_n|K)^{-1}\big\}_{n \geq k}$ converges uniformly in L to f^{-1}.*

Hints: Pick open W such that $L \subset W \subset \overline{W}$ compact $\subset f(\Omega)$. Then $K := f^{-1}\big(\overline{W}\big)$ is compact (since f^{-1} is continuous). According to (i) there is $n_0 \in \mathbb{N}$

such that $K \subset \Omega_n$, f_n is one-to-one on K and $\overline{W} \subset f_n(\Omega_n)$ for all $n \geq n_0$. It suffices now to prove that each $w_0 \in L$ has a neighborhood W_0 which lies in all but finitely many $f_n(K)$ and on which $(f_n|K)^{-1}$ converges uniformly to f^{-1}.

We have $w_0 = f(z_0)$ for some $z_0 \in f^{-1}(L) \subset f^{-1}(W) \subset K \subset \Omega_{n_0}$. Notice that $f^{-1}(L)$ lies in the open set $f^{-1}(W)$, enabling us to get $r > 0$ so that $\bar{D}(z_0, 2r) \subset f^{-1}(W) \subset K$. Now pick up the thread of the preceding hints at (5). The open set $f(D(z_0, r))$ is our candidate for W_0.

In particular, $W_0 \subset f_n\big(D(z_0, 2r)\big) \subset f_n(K)$. Since $D(z_0, 2r) \subset f^{-1}(W) \subset K$ and each f_n is one-to-one on K, the unique point in $D(z_0, 2r)$ where (5) says that f_n assumes value $w \in W_0$ must be $(f_n|K)^{-1}(w)$. Finally then 5.91 gives

$$\text{(6)}\quad (f_n|K)^{-1}(w) = \frac{1}{2\pi i}\int_{C(z_0,2r)} \frac{\zeta f_n'(\zeta)}{f_n(\zeta) - w}\, d\zeta \qquad \forall w \in W_0, n \geq n_0, \text{ and}$$

$$\text{(7)}\qquad f^{-1}(w) = \frac{1}{2\pi i}\int_{C(z_0,2r)} \frac{\zeta f'(\zeta)}{f(\zeta) - w}\, d\zeta \qquad \forall w \in W_0.$$

From $f_n \to f$, $f_n' \to f'$ uniformly on $\bar{D}(z_0, 2r)$, and (2) we see that

$$\frac{\zeta f_n'(\zeta)}{f_n(\zeta) - w} \longrightarrow \frac{\zeta f'(\zeta)}{f(\zeta) - w}$$

uniformly in $\zeta \in C(z_0, 2r)$, $w \in W_0$. This together with (6) and (7) gives

$$(f_n|K)^{-1}(w) \to f^{-1}(w) \qquad \text{uniformly for } w \in W_0.$$

(iii) *Suppose Ω_1, Ω_2 are regions in $\mathbb{C}$, f_n conformal maps of Ω_1 onto Ω_2 which converge locally uniformly in Ω_1 to the non-constant limit f. Show that if Ω_1 is bounded, then f is a conformal map of Ω_1 onto Ω_2 and f_n^{-1} converges to f^{-1} locally uniformly in Ω_2.*

Hints: Thanks to 7.11, f is one-to-one and $f(\Omega_1) \subset \Omega_2$. The bounded sequence $\{f_n^{-1}\}$ has a subsequence which is locally uniformly convergent in Ω_2, by 7.6. Say, $f_{n_j}^{-1} \to g \in H(\Omega_2)$. For each $z \in \Omega_1$ the set $\{f(z)\} \cup \{f_{n_j}(z) : j \in \mathbb{N}\}$ is a compactum in Ω_2, so the convergence of $f_{n_j}^{-1}$ to g is uniform on it. It follows easily (cf. 7.2) that

$$f_{n_j}^{-1}(f_{n_j}(z)) \to g(f(z)).$$

Thus $g \circ f$ is the identity map on Ω_1. Similarly (after citing 7.11 to see that $g(\Omega_2) \subset \Omega_1$), $f \circ g$ is the identity map on Ω_2. Thus f is bijective and $g = f^{-1}$.

It further follows from 7.8 that the whole sequence $\{f_n^{-1}\}$ converges locally uniformly to $g = f^{-1}$.

(iv) *Replace the boundedness hypothesis in* (iii) *by the hypothesis that the limit function f be surjective. Show that f is injective and f_n^{-1} converges to f^{-1}, locally uniformly in Ω_2.*

Hints: Injectivity of f follows from 7.11 and its hypothesized non-constancy. According to 7.2 we need to confirm that whenever $w_n \in \Omega_2$ $(n \geq 0)$ and $w_n \to w_0$, then $z_n := f_n^{-1}(w_n)$ converge to $f^{-1}(w_0) =: z_0$. If this fails, then by passing to a subsequence we can have an $r > 0$ such that $D(z_0, r) \subset \Omega$, and $z_n \in \Omega_1 \backslash D(z_0, r)$ for all $n \geq 1$. Since each function $g_n := f_n - w_n = f_n - f_n(z_n)$ is $1-1$ in Ω_1 and has a zero at $z_n \in \Omega_1 \setminus D(z_0, r)$, it must be zero-free in $D(z_0, r)$. The assumption $w_n \to w_0$ means that $g_n \to f - w_0 = f - f(z_0)$, a function which has a zero at $z_0 \in D(z_0, r)$. According to 7.11, this makes f constant in $D(z_0, r)$, a contradiction.

Remarks 7.20: Despite their intricacy, the arguments in 7.19(i), (ii) are actually rather routine. The boundedness hypothesis in (iii) is unnecessary; the argument requires only that $\{f_n^{-1}\}$ be locally uniformly bounded. That follows from 12.27 if $\mathbb{C} \setminus \Omega_1$ merely contains more than one point. [For 12.27 ensures that $\{f_n^{-1}\}_{n \in \mathbb{N}}$ is a normal family (12.24), but no subsequence of it can diverge to infinity *uniformly* on any compact subset of Ω_2 of the form $\{f(z), f_n(z) : n \in \mathbb{N}\}$. Thus the other alternative in the definition of normality must prevail, that is, $\{f_n^{-1}\}$ is a locally uniformly bounded family in $H(\Omega_2)$. Therefore, as in the (iii) hints, 7.6 ensures the existence of a locally uniformly convergent subsequence $\{f_{n_j}^{-1}\}$. Matters then proceed as before.] The cases $\Omega_1 = \mathbb{C}$ and $\Omega_1 = \mathbb{C} \backslash \{0\}$ then need to be dealt with *ab ovo*. But in these cases we know exactly the form which the f_n must have (by 11.19 and 11.20(ii), respectively) and it is so simple that the asserted conclusion in 7.19(iii) is evident for them.

An attractive and useful way to phrase (iv) above, in case $\Omega_1 = \Omega_2 = \Omega$, is that the set Aut (Ω) with the local-uniform-convergence topology is a *topological group*: the maps $(f, g) \mapsto f \circ g$ and $f \mapsto f^{-1}$ of Aut $(\Omega) \times$ Aut (Ω) into Aut (Ω), respectively, Aut (Ω) into Aut (Ω), are continuous. (Recall 7.1(ii).)

§ 2 Applications of the Convergence Theorems; Boundedness Criteria and Covering Results

Exercise 7.21 (i) *Let U be an open subset of $\mathbb{C}$, let $f \colon U \times [0, 1] \mapsto \mathbb{C}$ be bounded on each compact subset of $U \times [0, 1]$ and let $f(z, t)$ be a* *holomorphic function of z for each t and a Riemann integrable function of t for each z. Show that the function*

$$F(z) := \int_0^1 f(z,t)\,dt$$

is holomorphic in U, that for each positive integer j, $D_1^{(j)} f(z,t)$ is a Riemann integrable function for every $z \in U$ and that

$$F^{(j)}(z) = \int_0^1 D_1^{(j)} f(z,t)\,dt.$$

Hints: Let $D \neq \varnothing$ be any open disk with $\bar{D} \subset \mathbb{U}$. Thus for some $M \in \mathbb{N}$, $|f| \leq M$ on $\bar{D} \times [0,1]$. Now consider any sequence of partitions

$$\mathcal{P}_n := \{0 =: t_{n,0} < \cdots < t_{n,N_n} := 1\}$$

whose meshes converge to 0, and, for each $z \in \bar{D}$, Riemann sums

$$S_n(z) := \sum_{k=1}^{N_n} (t_{n,k} - t_{n,k-1}) f(z, \tau_{n,k}), \quad S_n'(z) := \sum_{k=1}^{N_n} (t_{n,k} - t_{n,k-1}) D_1 f(z, \tau_{n,k}),$$

$\tau_{n,k} \in [t_{n,k-1}, t_{n,k}]$. The S_n are bounded by M in $\bar{D}$ and $S_n(z) \to \int_0^1 f(z,t)dt = F(z)$ for each z. Vitali-Porter says that this convergence is locally uniform, so F is holomorphic in D, and [by 5.44(ii)], $S_n' \to F'$ in D. As this is true for all such sequences of partitions, it follows that $D_1 f(z,t)$ is a Riemann integrable function of t for each z, and $S_n'(z) \to \int_0^1 D_1 f(z,t)dt = F'(z)$. Finally, thanks to 5.37(i) the given local uniform boundedness of the family $\{f(\cdot,t) : 0 \leq t \leq 1\}$ entails that of the family $\{D_1 f(\cdot,t) : 0 \leq t \leq 1\}$, so $D_1 f$ fulfills all the hypotheses that f does and the proof can be finished by induction on j.

(ii) *X is a metric space, U an open subset of $\mathbb{C}$ and the function $f : U \times X \mapsto \mathbb{C}$ is bounded on each compact subset. Suppose that for each $z_0 \in U$, $x_0 \in X$*

$$x \mapsto f(z_0, x) \text{ is continuous and } z \mapsto f(z, x_0) \text{ is holomorphic.}$$

Show that f is continuous.

Hints: Let $z_0 \in U$, $x_0 \in X$ be given and consider any sequence $\{x_n\}_{n\in\mathbb{N}} \subset X$ convergent to x_0 and any $\{z_n\}_{n\in\mathbb{N}} \subset U$ convergent to z_0. Define $f_n(z) := f(z, x_n)$, $n \geq 0$, $z \in U$. For any compact $K \subset U$, f is bounded on the compact set $K \times \{x_n : n \geq 0\}$. That is, the sequence $\{f_n\}_{n\geq 0}$ is uniformly bounded on K. We have $f_n(z) \to f_0(z)$ for each $z \in U$, so thanks to 7.5, $f_n \to f_0$ locally uniformly in U. But then the (i) $\Rightarrow$ (ii) part of 7.2 insures that f_n converges continuously to f_0, which means that $f(z_n, x_n) = f_n(z_n) \to f_0(z_0) = f(z_0, x_0)$.

(iii) *Strengthen the hypotheses of* (i) *slightly to require that $f(z,t)$ be a con-* tinuous *function of t for each z. Then use Morera and Fubini to get a quick proof of the holomorphy of F.*

Hints: By (ii) f is continuous on $U \times [0,1]$, from which follows at once that F is continuous. For any open disk $D \subset U$ and any piecewise smooth loop $\gamma : [0,1] \mapsto D$, 2.19(ii) gives

$$\int_\gamma F = \int_0^1 \int_0^1 f(\gamma(s), t)dt\gamma'(s)ds = \int_0^1 \int_0^1 f(\gamma(s), t)\gamma'(s)dsdt = \int_0^1 \int_\gamma f(\xi, t)d\xi dt.$$

According to Cauchy's Theorem in disks, each integral $\int_\gamma f(\xi, t)d\xi$ is 0, so $\int_\gamma F = 0$ and Morera 5.39(i) steps up to proclaim F holomorphic in D. [Cf. 5.43(iii).]

Remarks: An elegant proof of 7.21(i) based on replacing the integrands $f(z,t)$, $D_1 f(z,t)$ by Cauchy integrals (over small circles) representing them is given on pp. 66,67 of Sansone and Gerretsen [1960]. Using Lebesgue's Dominated Convergence Theorem the boundedness hypothesis can be weakened to the requirement that there exist an (Lebesgue) integrable function ϕ on $[0,1]$ such that $|f(z,t)| \leq \phi(t)$ for all z and t. See Hayman [1979] and Lewin [1987]. It can also be weakened to the requirement that $\int_0^1 |f(z,t)|\, dt$ be bounded on each compact set of $z \in U$. See Świątkowski [1967] and Mattner [2001]. However, without *some* such hypothesis the conclusion may fail. Hayman (*op. cit.*) shows that if g is any non-constant entire function which is bounded on every line through 0 (see D. J. Newman [1976] for the existence of such functions), then the function $F(z) := \int_0^\infty e^{-t} g(tz)\, dt$ is not holomorphic in any neighborhood of 0.

The next result will recur in a more natural habitat in Chapter VIII where a proof based on rational approximation will be given. But it is hard to resist presenting the elegant and wholly different proof discovered by Dixon [1971], and later slightly simplified by Loeb [1991].

Theorem 7.22 (Homology Form of Cauchy's Theorem and Formula. Cf. 8.14.) *Let U be an open subset of $\mathbb{C}$, $\gamma_1, \ldots, \gamma_k : [0,1] \mapsto U$ be piecewise smooth loops, and $\gamma := (\gamma_1, \ldots, \gamma_k)$. Then the following are equivalent:*

(i) *γ is U-nullhomologous, that is,* $\mathrm{Ind}_\gamma := \sum_{j=1}^k \mathrm{Ind}_{\gamma_j} = 0$ *in $\mathbb{C} \setminus U$.*

(ii) $\displaystyle\int_\gamma F = 0 \qquad \forall F \in H(U).$

(iii) $\displaystyle f(z)\,\mathrm{Ind}_\gamma(z) = \frac{1}{2\pi i}\int_\gamma \frac{f(\xi)}{\xi - z}d\xi \qquad \forall z \in U \setminus \gamma_1 \cup \cdots \cup \gamma_k, f \in H(U).$

Proof: (iii) $\Rightarrow$ (ii) By selecting any $z \in U \setminus \gamma_1 \cup \cdots \cup \gamma_k$ and using $f(\xi) := (\xi - z)F(\xi)$. (ii) $\Rightarrow$ (i) because for each $z \in \mathbb{C}\setminus U$ the function $F(\xi) := 1/(\xi - z)$ featuring in the integral form of $\mathrm{Ind}_\gamma(z)$ lies in $H(U)$. So it remains to prove (i) $\Rightarrow$ (iii). To that end, define $g : U \times U \mapsto \mathbb{C}$ by

$$g(z,\xi) := \begin{cases} \dfrac{f(z) - f(\xi)}{z - \xi} & z, \xi \in U, z \neq \xi \\ f'(z) & z = \xi \in U. \end{cases}$$

According to 5.34(i), $g(z, \xi)$ is a holomorphic function of z for each fixed ξ and (so) a holomorphic function of ξ for each fixed z. If $z \in U \setminus \gamma_1 \cup \cdots \cup \gamma_k$, then the first formula for g prevails for all $\xi \in \gamma_1 \cup \cdots \cup \gamma_k$ and we have

$$\int_\gamma g(z,\xi)\,d\xi = -f(z)\int_\gamma \frac{d\xi}{\xi - z} - \int_\gamma \frac{f(\xi)}{z - \xi}\,d\xi$$

(1) $$= -f(z)2\pi i\ \mathrm{Ind}_\gamma(z) + \int_\gamma \frac{f(\xi)}{\xi - z}\,d\xi \quad \text{(by 4.4 and 4.2).}$$

Now define $V := \{z \in \mathbb{C} \setminus \gamma_1 \cup \cdots \cup \gamma_k : \sum_{j=1}^k \mathrm{Ind}_{\gamma_j}(z) = 0\}$. Since each Ind_{γ_j} is locally constant in $\mathbb{C} \setminus \gamma_j$, the function $\sum_{j=1}^k \mathrm{Ind}_{\gamma_j} = \mathrm{Ind}_\gamma$ is locally constant in $\mathbb{C} \setminus \gamma_1 \cup \cdots \cup \gamma_k$, so V is open. According to 5.34 g is continuous on $U \times U$, so $(z,t) \mapsto g(z, \gamma_j(t))\gamma_j'(t)$ is continuous on $U \times I$ for each closed interval I where γ_j is smooth. Thus we may form

(2) $$G(z) := \begin{cases} \displaystyle\int_\gamma g(z,\xi)\,d\xi, & z \in U \\ \displaystyle\int_\gamma \frac{f(\xi)}{\xi - z}\,d\xi, & z \in V. \end{cases}$$

This is consistent. For if $z \in U \cap V$, then on the one hand $z \in U \setminus \gamma_1 \cup \cdots \cup \gamma_k$, so (1) holds, and on the other hand by the definition of V, $\mathrm{Ind}_\gamma(z) = \sum_{j=1}^k \mathrm{Ind}_{\gamma_j}(z) = 0$, so from (1) it follows that the two integrals in (2) are equal. Thanks to 7.21(iii) [or 5.43(iii)], G is holomorphic in U. And 2.14 says that G is holomorphic in V. Now hypothesis (i) means that $C \setminus U \subset V$ so $U \cup V = \mathbb{C}$ and G is entire. Since $V \supset \mathbb{C}\setminus D(0,r)$ for all large r (4.3), use of the second formula for G and obvious estimates show that $\lim_{|z|\to\infty} G(z) = 0$. By Liouville G is therefore identically 0 in $\mathbb{C}$. Equation (1) and the first formula in (2) therefore show that (iii) holds.

Exercise 7.23 (Homotopy Versions of Cauchy's Theorem) *Let U be an open subset of $\mathbb{C}$, $f \in H(U)$.*

(i) *Show that if γ_1, γ_2 are piecewise smooth loops which are U-loophomotopic, then $\int_{\gamma_1} f = \int_{\gamma_2} f$.*

Hints: By 4.12, γ_1 and γ_2 are U-homologous, so the assertion follows from (i) of the preceding theorem, with $(-1) * \gamma_2$ in the role of γ_2 there.

(ii) *Deduce from* (i) *the following fixed-endpoint homotopy version of Cauchy's theorem: If $h\colon [0,1] \times [0,1] \mapsto U$ is continuous, $h(0,\tau) = h(0,0)$, $h(1,\tau) = h(1,0)$ for all $\tau \in [0,1]$ and the curves $\gamma_0(t) := h(t,0)$, $\gamma_1(t) := h(t,1)$ $(0 \le t \le 1)$ are piecewise smooth, then $\int_{\gamma_0} f = \int_{\gamma_1} f$.*

Hints: Define $H\colon [0,2] \times [0,1] \mapsto \mathbb{C}$ by

$$H(t,\tau) := \begin{cases} \gamma_0(t) & 0 \le t \le 1,\ 0 \le \tau \le 1 \\ h(2-t,\tau), & 1,\ 02\tau 21 \end{cases}$$

and let $\Gamma_0(t) := H(t,0)$, $\Gamma_1(t) := H(t,1)$ for $t \in [0,2]$. Check that Γ_0, Γ_1 are piecewise smooth loops and that H is a loophomotopy between them in U. Apply part (i) and observe that $\int_{\Gamma_0} f = 0$ and $\int_{\Gamma_1} f = \int_{\gamma_0} f - \int_{\gamma_1} f$.

(iii) *Prove the so-called* "Cauchy-Goursat lemma for a singular cell" (*cf.* Remarks following 5.2): *If $S := [0,1] \times [0,1]$, $h\colon S \mapsto U$ is continuous and $h \circ \partial S$ is piecewise smooth, then $\int_{h\circ\partial S} f = 0$.* [This generalizes (ii).]

Hints: Define $H\colon [0,1] \times [0,4] \mapsto h(S) \subset U$ by $H(r,t) := h(r\partial S(t))$. Then H is a loophomotopy in U of $h \circ \partial S$ with the constant $h(0,0)$.

(iv) *Show that every continuous $\gamma\colon [0,1] \mapsto U$ determines an $r > 0$ such that $\int_{\gamma_0} f = \int_{\gamma_1} f$ for every $f \in H(U)$ and every pair $\gamma_0, \gamma_1\colon [0,1] \mapsto U$ of piecewise smooth curves which have the same initial points, the same terminal points, and lie uniformly within r of γ.*

Hints: By compactness there is an $r > 0$ such that $\bar{D}(\gamma(t), r) \subset U$ for every $t \in [0,1]$. If γ_0, γ_1 are as described,

$$\left|\tau\gamma_1 + (1-\tau)\gamma_0 - \gamma\right| \le \tau|\gamma_1 - \gamma| + (1-\tau)|\gamma_0 - \gamma| \le r \qquad \forall \tau \in [0,1],$$

so the point $h(\tau,t) := \tau\gamma_1(t) + (1-\tau)\gamma_0(t)$ lies in U for every $(\tau,t) \in [0,1]\times[0,1]$ and we have a fixed-endpoint homotopy in U of γ_0 with γ_1.

(v) *Suppose $\gamma_1^*, \ldots, \gamma_k^*$ are loops in U such that $\gamma^* := (\gamma_1^*, \ldots, \gamma_k^*)$ is U-null-homologous and that $r > 0$ satisfies $D(\xi, r) \subset U$ for all $\xi \in \gamma_1^* \cup \cdots \cup \gamma_k^*$. Show that if $\gamma := (\gamma_1, \ldots, \gamma_k)$, in which γ_j is a* piecewise smooth *loop satisfying $|\gamma_j - \gamma_j^*| < r$ for each j, then conclusion* (ii) *of* 7.22 *holds.*

Hints: It suffices to show that γ is U-nullhomologous. The hint to (iv) shows that for each j, γ_j and γ_j^* are U-loophomotopic, so according to 4.12 they are U-homologous.

Remarks: Part (iv) would enable us to define $\int_\gamma f$ for *arbitrary* (continuous) curves γ and holomorphic functions f, namely, as $\lim \int_{\gamma_n} f$ for any sequence of piecewise smooth curves γ_n having the same endpoints as γ (a condition readily dispensed with) which converges uniformly to γ. The limit of course equals $\int_{\gamma_n} f$ itself as soon as n is sufficiently large. Perhaps the simplest way to accomplish this is to subdivide the domain of γ finely enough that γ maps each subinterval $[t_{j-1}, t_j]$ into a disk $D_j \subset U$. Then select a primitive F_j for f in D_j and let γ_n be the polygon with vertices $\gamma(t_j)$. We then have $\int_{\gamma_n} f = \sum \int_{[\gamma(t_{j-1}),\gamma(t_j)]} f = \sum \bigl[F_j(\gamma(t_j)) - F_j(\gamma(t_{j-1}))\bigr]$. The limit $\int_\gamma f$ would have most of the expected properties of an integral; and, for example, 7.22 would be true without the smoothness hypothesis (thanks to (v)). However, without further hypotheses on γ, not all choices of intermediate points $\tau_j \in [t_{j-1}, t_j]$ will produce Riemann sums $\sum f(\gamma(\tau_j))[\gamma(t_j) - \gamma(t_{j-1})]$ which converge to this integral. And what will replace the useful bound 2.9(i)? See DENJOY [1934], ZORN [1947], PERRY and YOUNGS [1947] and SCHATTE [1967]. [There's an analogy here with *defining* $\int_0^1 f'$ to be $f(1) - f(0)$ whenever $f : [0,1] \mapsto \mathbb{R}$ is a differentiable function for which f' is not Riemann integrable, $f(x) := x^2 \sin(x^{-2})$, for example.] So, often this "ideal" integral needs to be distinguished from the "genuine" Riemann one in calculations. Like the ideal point "∞" adjoined to create $\mathbb{C}_\infty$, this more general integral streamlines the *formulations* of results, at the expense of sometimes requiring case considerations in proofs. I elect not to pay this price, and so forego formally incorporating these objects into the theory.

In 5.39(iii) we saw a kind of converse of (iv): if f is a continuous function in U such that the sequence $\{\int_{\gamma_n} f\}$ converges whenever the piecewise smooth curves γ_n converge uniformly to a curve in U, then necessarily f is holomorphic in U.

In the next clutch of theorems we are going to use Montel's compactness criterion to prove some important covering theorems. The constants that emerge are by no means best possible; the reader should concentrate on the interesting methodology.

The next definition is convenient in what follows; the designation is traditional.

Definition 7.24 The class $\mathcal{S}$ (for *schlicht*) consists of all holomorphic and univalent functions f on $\mathbb{D}$ with the normalizations $f(0) = 0$ and $f'(0) = 1$.

Exercise 7.25 *Show that $\mathcal{S}$ is closed in $H(\mathbb{D})$ in the metric of* 7.3. *Consequently, by* 7.6 *and* 7.27 *below, $\mathcal{S}$ is compact.*

Hint: 7.11.

Exercise 7.26 *For each $f \in \mathcal{S}$ the number $\alpha := \inf|\mathbb{C} \setminus f(\mathbb{D})|$ satisfies $0 < \alpha \leq 1$.*

Hints: Since $f(\mathbb{D})$ is open and contains 0 and (Liouville) is not all of $\mathbb{C}$, the number α is clearly positive and finite, and $D(0, \alpha) \subset f(\mathbb{D})$. Therefore f^{-1} is defined in $D(0, \alpha)$ and the function $F(w) := f^{-1}(\alpha w)$ is a holomorphic map of $\mathbb{D}$ into $\mathbb{D}$ which fixes 0. Apply the Schwarz lemma to F, using 5.78.

Theorem 7.27 *The family $\mathcal{S}$ is locally uniformly bounded on $\mathbb{D}$.*

Proof: Let an $f \in \mathcal{S}$ be given and let $\alpha_f := \alpha$ be as defined in the preceding exercise. Evidently there exists an $a_f \in \mathbb{C}$ with $|a_f| = \alpha_f$ and $a_f \in \partial f(\mathbb{D})$. Form

$$(1) \qquad g_f := \frac{1}{a_f} f - \mathbf{1}$$

and have

$$(2) \qquad D(-1, 1) \subset g_f(\mathbb{D}), \quad 0 \notin g_f(\mathbb{D}), \quad g_f(0) = -1.$$

As g_f is a conformal map, it follows from 5.35 that there exists on $g_f(\mathbb{D})$ a holomorphic square root h_f of the identity. Since $h_f^2(-1) = h_f^2(g_f(0)) = g_f(0) = -1$, we have $h_f(-1) = \pm i$. Replacing h_f with $-h_f$ if necessary, we can suppose $h_f(-1) = i$. Let I denote the identity function on $\mathbb{D}$. Then $I \in \mathcal{S}$ and $h_I^2(z) = z = h_f^2(z)$ for all $z \in D(-1, 1)$ and $h_f(-1) = i = h_I(-1)$. It follows from 1.7 that $h_f = h_I$ throughout $D(-1, 1)$. Now $i = h_I(-1) \in h_I(D(-1, 1))$, so there is an $0 < r < 1$ such that $D(i, r) \subset h_I(D(-1, 1)) = h_f(D(-1, 1))$. Therefore from (2)

$$(3) \qquad D(i, r) \subset h_f(g_f(\mathbb{D})).$$

But then $D(-i, r) \cap h_f(g_f(\mathbb{D})) = \varnothing$. Indeed, if $z \in \mathbb{D}$ and $h_f(g_f(z)) =: w \in D(-i, r)$, then use (3) to choose $z' \in \mathbb{D}$ such that $h_f(g_f(z')) = -w$. Squaring, we get $g_f(z') = [h_f(g_f(z'))]^2 = (-w)^2 = w^2 = [h_f(g_f(z))]^2 = g_f(z)$. Since g_f is one-to-one, it follows that $z' = z$ and then $w = -w$, $w = 0$, a contradiction: $r < 1$, so $0 \notin D(-i, r)$. Thus $h_f(g_f(\mathbb{D})) \subset \mathbb{C} \setminus D(-i, r)$, that is,

$$(4) \qquad |h_f \circ g_f + i| \geq r \quad \text{on } \mathbb{D}.$$

Now for each $f \in \mathcal{S}$ we have $0 < \alpha_f \leq 1$ by the preceding exercise, so by (1) it suffices to show that the family $\{g_f : f \in \mathcal{S}\}$ is uniformly bounded on each compact subset of $\mathbb{D}$. If this fails for some compact $K \subset \mathbb{D}$, then there exist $f_n \in \mathcal{S}$, $g_n := g_{f_n}$ and $h_n := h_{f_n}$ such that

$$\sup |g_n(K)| > n \qquad \forall n \in \mathbb{N}. \tag{5}$$

The sequence of functions $F_n := 1/(h_n \circ g_n + i)$ is uniformly bounded in $\mathbb{D}$ by (4) and so by 7.6 some subsequence of it, say without loss of generality $\{F_n\}$ itself, converges locally uniformly on $\mathbb{D}$ to a holomorphic limit F. We have

$$F_n(0) = \frac{1}{h_n(g_n(0)) + i} = \frac{1}{h_n(-1) + i} = \frac{1}{2i},$$

so $F(0) = 1/2i$ and F is not identically zero. Since none of the functions F_n is ever zero on $\mathbb{D}$, it follows from 7.10 that F is zero-free. Therefore $1/F$ is holomorphic in $\mathbb{D}$ and $1/F_n$ converges to $1/F$ uniformly on K (cf. 7.1(ii)). Since

$$\left[\frac{1}{F_n} - i\right]^2 = (h_n \circ g_n)^2 = g_n,$$

we have a contradiction to (5).

Besides this *reductio* argument, there is a more direct one from (4) via Harnack's inequality. The function

$$F := \log\bigl|h_f \circ g_f + i\bigr| - \log(r/2)$$

is harmonic in $\mathbb{D}$ and strictly positive, by (4). Therefore, for each compact $K \subset \mathbb{D}$, 5.28(ii) supplies a finite constant $b(K)$ (there denoted $c(K \cup \{0\}, \mathbb{D})$), which depends only on K, such that

$$F \le b(K)F(0) \qquad \text{on } K.$$

Since $F(0) = \log 2 - \log(r/2) = \log(4/r)$, this implies that

$$\bigl|h_f \circ g_f + i\bigr| \le \frac{r}{2}\left(\frac{4}{r}\right)^{b(K)} < \left(\frac{4}{r}\right)^{b(K)},$$

$$|g_f| = \bigl|h_f \circ g_f\bigr|^2 \le \left[1 + \left(\frac{4}{r}\right)^{b(K)}\right]^2 \qquad \text{on } K.$$

Finally, since $|f| \overset{(1)}{=} |g_f + \mathbf{1}|\alpha_f \le |g_f + \mathbf{1}|$, the inequality for $|g_f|$ implies that

$$|f| \le 1 + \left[1 + \left(\frac{4}{r}\right)^{b(K)}\right]^2 \qquad \text{on } K,$$

valid for all $f \in \mathcal{S}$.

Corollary 7.28 (Koebe [1907]) *There exists a positive constant α such that $D(0,\alpha) \subset f(\mathbb{D})$ for each $f \in \mathcal{S}$.* (Cf. 12.17.)

Proof: For $f \in \mathcal{S}$ let $\alpha_f := \inf |\mathbb{C} \setminus f(\mathbb{D})|$. The claim is that $\inf\{\alpha_f : f \in \mathcal{S}\}$ is positive. If this is not the case, pick $f_j \in \mathcal{S}$ such that $\alpha_j := \alpha_{f_j} < 1/j$ and set $g_j := f_j/\alpha_j$. In the proof of 7.27 it was shown that $\{g_j\}$ is locally uniformly bounded on $\mathbb{D}$. It follows that $f_j = \alpha_j g_j$ converges to 0 locally uniformly on $\mathbb{D}$. But then $f_j'(0)$ converges to 0 by 5.44(ii), contrary to $f_j \in \mathcal{S}$.

Latent in the proof of 7.27 is a lower bound of $1/24$ for α. This can be extricated as follows: (4) says that the function

$$F := 1/(h_f \circ g_f + i) - 1/2i$$

is bounded in $\mathbb{D}$ by $1/r + 1/2$. Since $h_f \circ g_f(0) = h_f(-1) = i$, it therefore follows from 6.1 that $|F'(0)| \leq 1/r + 1/2$. Using the Chain Rule and the identity $(h_f \circ g_f)^2 = g_f$, we compute that $2(h_f \circ g_f) \cdot (h_f \circ g_f)' = g_f'$. Evaluating at 0 gives $(h_f \circ g_f)'(0) = 1/2ia_f$. Another calculation with the Chain Rule then reveals that $F'(0) = 1/8ia_f$. Consequently, $1/r + 1/2 \geq 1/|8ia_f|$, so $\alpha_f \geq |a_f| \geq r/4(r+2)$. In the 7.27 proof, r is subject only to $D(i,r) \subset h_I(D(-1,1))$. Recall that on $D(-1,1)$, h_I agrees with the holomorphic square root that maps $\mathbb{C}_+$ onto $\mathbb{U}$. Also, $|w - i| < \sqrt{2} - 1 \Rightarrow w \in \mathbb{U}$ & $|w + i| < \sqrt{2} + 1$. Consequently, $|w - i| < \sqrt{2} - 1 \Rightarrow |w^2 + 1| = |w - i||w + i| < 1 \Rightarrow w^2 \in D(-1,1)$. Therefore, every such w is $h_I(w^2) \in h_I(D(-1,1))$. This says that $\sqrt{2} - 1$ is an eligible r. Using it, the inequality for α_f becomes $\alpha_f \geq \frac{1}{4}(3 - 2\sqrt{2}) > 1/24$, valid for all $f \in \mathcal{S}$.

Remarks: The largest α for which the conclusion above is true is $\frac{1}{4}$. This was conjectured by Koebe and proved by Bieberbach [1916] and by Faber [1916]. The proof is based on an area method and will be developed in Chapter XVIII. It is possible to parlay 7.28 into a much broader covering theorem (Bloch [1925]): if f is holomorphic in a neighborhood of $\overline{\mathbb{D}}$ and $|f'(0)| = 1$, then $f(\mathbb{D})$ contains "schlichtly" a disk of radius $\alpha/6$, for any α which satisfies the conclusion of 7.28. (See, e.g., Veech [1967], pp. 149–150.) In particular, after $\alpha = \frac{1}{4}$ is shown to be admissible, we get that some open subset U_f of $\mathbb{D}$ is mapped conformally by f onto some open disk D_f of radius $\frac{1}{24}$. However, one can proceed *ab initio* with similar elementary means and get a better constant than $\frac{1}{24}$. E. Landau [1929b] got .39 and below we will get $1/3\sqrt{3} > .19$.

Exercise 7.29 (i) *Let $f \in H(\mathbb{D})$ satisfy $f(0) = 0$, $f'(0) = 1$ and $|f'| < M < \infty$ (so that $M > |f'(0)| = 1$). Show that f maps $D(0, 1/M)$ conformally onto a set which contains $D(0, 1/2M)$.*

Hints: Let

$$g := \frac{M(f' - \mathbf{1})}{M^2 - f'} = \frac{\dfrac{f'}{M} - \dfrac{1}{M}}{\mathbf{1} - \dfrac{f'}{M} \cdot \dfrac{1}{M}}. \tag{1}$$

Apply Schwarz' lemma (6.1) to learn that

$$|g(z)| \leq |z| \qquad \forall z \in \mathbb{D}. \tag{2}$$

By the usual algebraic maneuvers involving "completing the square" show that for positive $r < M$ we have

$$\left|\frac{M(w-1)}{M^2 - w}\right| \leq r \iff \left|w - \frac{M^2(1-r^2)}{M^2 - r^2}\right| \leq \frac{rM(M^2-1)}{M^2 - r^2} \tag{3}$$

and consequently, since $M^2(1-r^2)/(M^2-r^2) \leq 1$, the point w in the disk (3) which is farthest from 1 is

$$\frac{M^2(1-r^2)}{M^2 - r^2} - \frac{rM(M^2-1)}{M^2 - r^2} = \frac{M(1-rM)}{M - r}.$$

If $r < 1$ and $|z| = r$, then $z \in \mathbb{D}$ and by (2), $|g(z)| \leq r$ so that by (1), $f'(z)$ lies in the disk (3). Thus by the preceding observation

$$|f'(z) - 1| \leq 1 - \frac{M(1-rM)}{M-r} = \frac{r(M^2-1)}{M-r} \quad \text{whenever } |z| = r < 1. \tag{4}$$

If now $r \leq 1/M (< 1)$, then the right-hand side of inequality (4) is majorized by $rM = |z|M$, and so we have

$$|f'(z) - 1| \leq |z|M \qquad \forall z \in D(0, 1/M). \tag{5}$$

It follows by integration that for $z = |z|e^{i\theta} \in D(0, 1/M)$

$$\begin{aligned}|f(z) - z| &= \left|\int_{[0,z]} (f' - \mathbf{1})\right| = \left|\int_0^{|z|} (f' - \mathbf{1})\left(\rho e^{i\theta}\right) d\rho\right| \\ &\leq \int_0^{|z|} |f' - \mathbf{1}|\left(\rho e^{i\theta}\right) d\rho \leq \int_0^{|z|} \rho M \, d\rho = \frac{|z|^2}{2} M < \frac{1}{2M},\end{aligned}$$

(6)

$$|f(z) - z| < \frac{1}{2M} \qquad \forall z \in D(0, 1/M).$$

Let I denote the identity function on $\mathbb{D}$. If $w \in D(0, 1/2M)$, then (6) implies that

$$|I - w| > \frac{1}{2M} \geq |f - I| \quad \text{on } C(0, 1/M).$$

Therefore by Rouché's theorem (5.87), $I - w$ and $(f - I) + (I - w) = f - w$ have the same number of zeros in $D(0, 1/M)$. Thus $f - w$ has exactly one zero in $D(0, 1/M)$ for every $w \in D(0, 1/2M)$. Furthermore, (5) shows that f' maps $D(0, 1/M)$ into $D(1, 1)$, so that $\operatorname{Re} f' > 0$. For z_1, $z_2 \in D(0, 1/M)$ with $z_1 \neq z_2$ we therefore have

$$\frac{f(z_2) - f(z_1)}{z_2 - z_1} = \int_0^1 \frac{d}{dt}\left[\frac{f(z_1 + t(z_2 - z_1))}{z_2 - z_1}\right] dt = \int_0^1 f'(z_1 + t(z_2 - z_1))\, dt.$$

Since all $z_1 + t(z_2 - z_1) \in D(0, 1/M)$, the last integrand lies in $D(1, 1)$, thanks to (5). Thus it and its integral have positive real part and it follows, in particular, from the preceding equality that $f(z_2) \neq f(z_1)$.

(ii) *Let F be holomorphic in $D(0, r)$ and satisfy $F'(0) \neq 0$, $|F'| < K < \infty$. Show that for some $\rho \leq r$, F maps $D(0, \rho)$ conformally onto a set which contains $D\big(F(0), r|F'(0)|^2/2K\big)$.*

Hint: Apply (i) to $f(z) := (F(rz) - F(0))/rF'(0)$ with $M := K/|F'(0)|$.

(iii) *Let f be holomorphic in a neighborhood of $\overline{\mathbb{D}}$ with $|f'(0)| \geq 1$. Show that f maps some open subdisk D_f of $\mathbb{D}$ conformally onto a set which contains a disk of radius $1/3\sqrt{3}$.*

Hints: The continuous function $\phi(z) := (1 - |z|^2)|f'(z)|$ attains its maximum k over $\overline{\mathbb{D}}$ at some point z_0 of $\mathbb{D}$, since $\phi(0) \geq 1 > \phi(\partial\mathbb{D}) = 0$. Let $\psi(z) := (z + z_0)/(1 + \overline{z_0}z)$ and $F := f \circ \psi$, a function on $\mathbb{D}$. Check (cf. hints to 2.6) that

$$\big(1 - |z|^2\big)|F'(z)| = \big(1 - |\psi(z)|^2\big)|f'(\psi(z))| \qquad \forall z \in \mathbb{D}.$$

Consequently,

$$\big(1 - |z|^2\big)|F'(z)| \leq k,$$

$$|F'(z)| \leq \frac{k}{1 - |z|^2} \qquad \forall z \in \mathbb{D}.$$

For each $0 \leq r < 1$ we can therefore apply (ii) with $K := k/(1 - r^2)$ to learn that for some $\rho \leq r$, F maps $D(0, \rho)$ conformally onto a set which contains $D(f(z_0), kr(1 - r^2)/2)$. But $k \geq \phi(0) \geq 1$, so this implies that

(7)

$$F \text{ maps } D(0, \rho) \text{ conformally onto a set which contains } D\Big(f(z_0), \tfrac{r(1-r^2)}{2}\Big).$$

This holds for any $0 < r < 1$, so choose r to maximize the radius $r(1-r^2)/2$, viz., $r := 1/\sqrt{3}$, $r(1-r^2)/2 = 1/3\sqrt{3}$. Since ψ maps $\mathbb{D}$ conformally onto $\mathbb{D}$ and $D(0,\rho)$ onto some subdisk of $\mathbb{D}$ (see 6.5(i)) and since $f = F \circ \psi^{-1}$, the assertion in (iii) follows from (7) with this choice of r.

(iv) *F is a non-constant entire function. Deduce from* (iii) *that $F(\mathbb{C})$ contains disks of arbitrarily large radius. Show in fact that arbitrarily large open disks are bijective F-images of open sets.*

Hint: Pick a with $F'(a) \neq 0$ and, given $0 < R < \infty$, set $\rho := 3R\sqrt{3}/|F'(a)|$. Then (iii) is applicable to $f(z) := F(a+\rho z)/3R\sqrt{3}$.

(v) (Picard's Little Theorem) *Use* (iv) *to derive what might seem the ultimate refinement of Liouville's Theorem: For each non-constant entire function f, $\mathbb{C} \setminus f(\mathbb{C})$ contains at most one point.* (Actually, 12.29 and 12.30 refine this even further.)

Hints: (E. LANDAU [1926], H. KÖNIG [1957]): Suppose $f \in H(\mathbb{C})$ omits two values from its range, say 0 and 1. According to 5.35, f has a logarithm $2\pi i L \in H(\mathbb{C})$. Necessarily $\mathbb{Z} \cap L(\mathbb{C}) = \varnothing$. Therefore (5.35 again) square-root functions $\sqrt{L}, \sqrt{L-\mathbf{1}} \in H(\mathbb{C})$ exist. The entire function $G := \sqrt{L} - \sqrt{L-\mathbf{1}}$ is evidently zero-free, and so it too has a logarithm $F \in H(\mathbb{C})$. For $n \in \mathbb{N}$ define $x_n := \log\left(\sqrt{n} + \sqrt{n-1}\right)$, so that $-x_n = \log\left(\sqrt{n} - \sqrt{n-1}\right)$. Note that $F(z) = \pm x_n + \frac{1}{2} m\pi i$ holds for no $z \in \mathbb{C}, n \in \mathbb{N}, m \in \mathbb{Z}$. Indeed since

$$(*) \quad 4L = (2\sqrt{L})^2 = \left(G + 1/G\right)^2 = 2 + G^2 + 1/G^2 = 2 + e^{2F} + e^{-2F},$$

such an equality would entail the contradiction

$$\begin{aligned} 4L(z) &= 2 \pm \left[e^{\pm 2x_n} + e^{\mp 2x_n}\right] = 2 \pm \left[\left(\sqrt{n} \pm \sqrt{n-1}\right)^2 + \left(\sqrt{n} \mp \sqrt{n-1}\right)^2\right] \\ &= 2 \pm [4n - 2] \in 4\mathbb{Z}. \end{aligned}$$

Now $0 = x_1 \le x_n \uparrow \infty$, and

$$0 < x_{n+1} - x_n = \log\left(\left(\sqrt{1+1/n} + 1\right)/\left(\sqrt{1-1/n} + 1\right)\right) \le \log\left(\sqrt{2} + 1\right) < 1$$

for all $n \in \mathbb{N}$, since this is a decreasing function of n. (For the last inequality, exponentiate and recall the inequalities following 3.15.) Consequently, given $x, y \in \mathbb{R}$, there exists $m \in \mathbb{Z}$ with $|y - m| \le 1/2$, and $n \in \mathbb{N}$ with (for some choice of sign) $|x \mp x_n| \le 1/2$; which means that $\left|\left(x + \frac{1}{2}\pi i y\right) - \left(\pm x_n + \frac{1}{2} m\pi i\right)\right| \le \sqrt{(1/2)^2 + (\pi/4)^2} < 1$ [by (8) in the proof of 3.14]. That is, every point in $\mathbb{C}$ lies within 1 of some point $\pm x_n + \frac{1}{2} m\pi i \in \mathbb{C} \setminus F(\mathbb{C})$. In view of (iv), this means that F is constant, but then by $(*)$ so are L and $e^{2\pi i L} = f$.

Theorem 7.30 *The sets $\mathcal{S}_1 := \{f' : f \in \mathcal{S}\}$ and $\mathcal{S}_2 := \{1/f' : f \in \mathcal{S}\}$ are locally uniformly bounded on $\mathbb{D}$.*

Proof: For $\mathcal{S}_1$ use 7.27 and 5.37(i). If $\mathcal{S}_2$ fails to be uniformly bounded on some compact $K \subset \mathbb{D}$, then there exist $f_n \in \mathcal{S}$ and $z_n \in K$ such that

$$(*) \qquad |f_n'(z_n)| < 1/n.$$

A subsequence of $\{z_n\}$ converges and by 7.6 and 7.27 a subsequence of $\{f_n\}$ converges locally uniformly on $\mathbb{D}$. Without loss of generality then, $z_n \to z_0 \in K$ and $f_n \to f_0$ locally uniformly on $\mathbb{D}$. Then $f_n' \to f_0'$ locally uniformly on $\mathbb{D}$, by 5.44(ii), whence $f_0'(0) = 1$ and f_0 is not constant. From 7.11 f_0 is therefore one-to-one. However, $f_0'(z_0) = \lim_{n\to\infty} f_n'(z_n)$ follows easily from the uniform convergence of f_n' to f_0' on K. (Cf. 7.2.) Therefore from $(*)$ we have $f_0'(z_0) = 0$. This contradicts the fact that f_0 is one-to-one (5.78).

In more modern dress our proof goes thus: $(z, f) \mapsto f'(z)$ is a continuous function Ψ on the metric space $\mathbb{D} \times H(\mathbb{D})$ (easy to check using the Cauchy integral formula) and for each compact $K \subset \mathbb{D}$ the set $K \times \mathcal{S}$ is a compact subset of $\mathbb{D} \times H(\mathbb{D})$ (7.25). Ψ maps this compact set to a compact subset of $\mathbb{C} \setminus \{0\}$.

Exercise 7.31 (i) *Let Ω be a region, $a \in \Omega$ and $\mathfrak{F}$ the set of all univalent $f \in H(\Omega)$ which satisfy $|f(a)| \le 1$ and $|f'(a)| \le 1$. Show that both $\mathfrak{F}$ and $\mathfrak{F}' := \{f' : f \in \mathfrak{F}\}$ are locally uniformly bounded in Ω.*

Hints: Let Ω_0 denote the subset of points $z \in \Omega$ such that $\mathfrak{F}$ and $\mathfrak{F}'$ are uniformly bounded in some neighborhood of z. Show that

$$(*) \quad \sup\{|f(z_0)| + |f'(z_0)| : f \in \mathfrak{F}\} < \infty \ \& \ D(z_0, r) \subset \Omega \Rightarrow D(z_0, r/4) \subset \Omega_0$$

and conclude with 1.27. To this end, denote the above supremum by $A < \infty$. The functions

$$\tilde{f}(z) := \frac{f(z_0 + rz) - f(z_0)}{rf'(z_0)}, \quad z \in \mathbb{D},\ f \in \mathfrak{F}$$

all belong to $\mathcal{S}$. Therefore, by 7.27 there is a constant B such that

$$|\tilde{f}| \le B \quad \text{in } D(0, \tfrac{1}{2}) \text{ for each } f \in \mathfrak{F}.$$

This says that

$$|f(z)| \le r|f'(z_0)|B + |f(z_0)| \le rAB + A \qquad \forall z \in D(z_0, r/2),\ f \in \mathfrak{F}.$$

From this inequality and Cauchy's integral (around $C(z_0, r/2)$) for f' we infer that

$$|f'(z)| \le (rAB + A)8/r \qquad \forall z \in D(z_0, r/4),\ f \in \mathcal{F}$$

and $(*)$ is established.

(ii) *Let Ω be a region, $a \in \Omega$ and $\mathcal{F}$ the set of all univalent $f \in H(\Omega)$ which satisfy $|f(a)| \le 1$ and $|f'(a)| \ge 1$. Show that $\{1/f' : f \in \mathcal{F}\}$ is locally uniformly bounded in Ω.*

Hint: Use 7.30 and an argument like that in (i) above.

(iii) (Montel [1925], p. 253) *Let Ω be a region, $a \in \Omega$ and $\{F_n\}$ univalent holomorphic functions in Ω such that $\{F_n(a)\}$ is convergent and $\{F_n'(a)\}$ converges to 0. Show that $\{F_n\}$ converges to a constant, locally uniformly in Ω.*

Hints: Let $L := \lim_{n\to\infty} F_n(a)$. Given $\varepsilon > 0$, pick $n(\varepsilon)$ so that $|F_n(a) - L| \le \varepsilon$ and $|F_n'(a)| \le \varepsilon$ for all $n \ge n(\varepsilon)$. Then the functions $f_n := (F_n - L)/\varepsilon$, $n \ge n(\varepsilon)$, belong to the family $\mathcal{F}$ of (i) and so for each compact $K \subset \Omega$ there is a constant $c(K) < \infty$ such that $|f_n| \le c(K)$ on K, for all $n \ge n(\varepsilon)$.

(iv) (Koebe's Distortion Theorem) *Let Ω be a region. Show that for each compact $K \subset \Omega$ there is a finite positive constant $C := C(\Omega, K)$, dependent only on Ω and K, such that*

$$\frac{1}{C} \le \left|\frac{f'(z)}{f'(w)}\right| \le C \qquad \forall z, w \in K$$

holds for every univalent $f \in H(\Omega)$.

Hints: Let $\mathcal{F}$ denote the set of all univalent $f \in H(\Omega)$ and fix an $a \in \Omega$. For $f \in \mathcal{F}$ define $\tilde{f} := (f - f(a))/f'(a)$. According to (ii) and (i) there is a positive finite constant $c = c(\Omega, K)$ such that

$$\frac{1}{c} \le |\tilde{f}'| \le c \quad \text{on } K \text{ for each } f \in \mathcal{F},$$

that is,

$$\frac{1}{c} \le \left|\frac{f'(z)}{f'(a)}\right| \le c \qquad \forall z \in K,\ f \in \mathcal{F}.$$

Multiplying a pair of such inequalities gives

$$\frac{1}{c^2} \le \left|\frac{f'(z)}{f'(w)}\right| \le c^2 \qquad \forall z, w \in K,\ f \in \mathcal{F}.$$

Therefore we may take $C(\Omega, K) := c^2(\Omega, K)$.

Exercise 7.32 (Cf. 12.31(v).) *Let Ω be a region, A a subset of Ω with a limit point in Ω, $\mathcal{F}$ a set of univalent holomorphic functions in Ω.*

(i) *Show that if $\{f(a) : f \in \mathcal{F}\}$ is bounded for each $a \in A$, then $\mathcal{F}$ is locally uniformly bounded on Ω.*

Hints: For each $a \in A$ let $M(a) := \sup\{|f(a)| : f \in \mathcal{F}\} < \infty$. Fix an $a_0 \in A$ and consider the functions

$$\tilde{f} := \frac{f - f(a_0)}{f'(a_0)}, \quad f \in \mathcal{F}.$$

The family $\widetilde{\mathcal{F}} := \{\tilde{f} : f \in \mathcal{F}\}$ is locally uniformly bounded on Ω by 7.31(i). Also $\{f(a_0) : f \in \mathcal{F}\}$ is bounded by $M(a_0)$. Since $f = f(a_0) + f'(a_0)\tilde{f}$, the local uniform boundedness of $\mathcal{F}$ will follow if we prove that the set $\{f'(a_0) : f \in \mathcal{F}\}$ is bounded. If this set is not bounded, pick $f_n \in \mathcal{F}$ such that $|f_n'(a_0)| > n$. Then we have

$$|\tilde{f}_n(a)| \leq \frac{M(a) + M(a_0)}{n} \qquad \forall a \in A,\ n \in \mathbb{N}$$

and so $\tilde{f}_n(a) \to 0$ for each $a \in A$. From the Vitali–Porter theorem we then have $\tilde{f}_n \to 0$ locally uniformly on Ω and so from 5.44(ii), $\tilde{f}_n'(a_0) \to 0$, a manifest absurdity since $\tilde{f}'(a_0) = 1$ for every $f \in \mathcal{F}$.

(ii) *If $\{f_1, f_2, \ldots\} \subset \mathcal{F}$ and $\{f_n(a)\}$ converges for each $a \in A$, then $\{f_n\}$ converges locally uniformly on Ω.*

(iii) *If $\Omega = \mathbb{D}$ and $A = \{a_1, a_2, \ldots\}$ are* distinct *points of Ω satisfying $\sum_{j=1}^{\infty}(1 - |a_j|) = \infty$, formulate and try to prove versions of* (i) *and* (ii).

Hint: Recall 7.4 and look at the chapter-notes on it and on this exercise.

Exercise 7.33 *Let Ω be a region, $a_n \in \Omega$ and $a_n \to a \in \Omega$. Let $f_n : \Omega \mapsto \mathbb{D}$ be holomorphic and satisfy $|f_n(a_n)| \to 1$. Show that $|f_n| \to \mathbf{1}$ locally uniformly on Ω.*

Hint: Assume not. Then, by passing to a subsequence, we may assume that $\sup_n |f_n(z_n)| \leq 1 - \delta < 1$ for some $z_1, z_2, \ldots$ in a compact subset K of Ω. We may suppose a and all a_n lie in K. Use 7.6 to select $1 = n_1 < n_2 < \ldots$ such that $\{f_{n_j}\}$ converges to some holomorphic function f locally uniformly on Ω. Then (cf. 7.2) $f_{n_j}(a_{n_j}) \to f(a)$, so $|f(a)| = 1$ and by the Maximum-Modulus Principle f is constant. In particular, $|f| = |f(a)| = 1$. Then

$$\delta \leq 1 - |f_{n_j}(z_{n_j})| \leq |f(z_{n_j}) - f_{n_j}(z_{n_j})| \leq \sup |f - f_{n_j}|(K) \qquad \forall j \in \mathbb{N}$$

contradicts the uniform-on-K convergence.

Exercise 7.34 *Let* $a, b, c \in \mathbb{R}$ *with* $a < b$ *and* $f\colon S := (a, b) \times (c, \infty) \mapsto \mathbb{D}$ *a holomorphic function. Suppose that the numbers* $x_n + iy_n \in S$ *satisfy*

(1) $$x_n \to d \in (a, b) \;\;\&\;\; y_n \to \infty$$

(2) $$\overline{\lim_{n\to\infty}}\,(y_{n+1} - y_n) < \infty$$

(3) $$\lim_{n\to\infty} |f(x_n + iy_n)| = 1.$$

Show that $\lim_{y\to\infty} |f(x + iy)| = 1$ *uniformly for* x *in any compact subset of* (a, b).

Hints: There exists $\delta > 0$ such that $\{y_n\} \subset (c+\delta, \infty)$. Choose $0 < M < \infty$ such that $y_{n+1} - y_n < M$ for all n. Then form the functions

$$f_n(z) := f(z + i(y_n - c)), \quad z \in \Omega := (a, b) \times (c - \delta, c + \delta + M).$$

They satisfy

$$|f_n(x_n + ic)| = |f(x_n + iy_n)| \to 1.$$

Since $x_n + ic \to d + ic \in \Omega$, the preceding exercise gives

$$\lim_{n\to\infty} |f_n| = 1 \quad \text{uniformly on compact subsets of } \Omega.$$

In particular, given $\varepsilon > 0$ and a compact $K \subset (a, b)$, there exists N such that

$$|f_n(x + iy)| > 1 - \varepsilon \qquad \forall n \geq N,\ (x, y) \in K \times [c, c + M].$$

If $x \in K$ and $y > y_N$, let n be maximal such that $y_n < y$. Then $y_{n+1} \geq y > y_n$ and $n \geq N$. Consequently $y = y_n + t$ for some $t \in (0, y_{n+1} - y_n] \subset (0, M)$ and

$$f(x + iy) = f_n(x + i(t + c)) \in f_n(K \times [c, c + M]).$$

§ 3 Prescribing Zeros

I shall now deliver on a promise made in 5.63 and show the reader how to construct a holomorphic function with arbitrary prescribed limit-point-free zero-set. The theorem is very useful in investigations of the ideal structure of the algebra $H(U)$ (see Chapter XI) and in constructing functions with natural boundaries (see Chapter XVI and the notes to the present chapter). Since a factor $(z - a_n)$ is wanted for each zero a_n, some sort of infinite product $\prod(z - a_n)$ suggests itself. This will probably not converge, so convergence-factors $g_n(z)$ are introduced that are zero-free and bring each factor close enough to 1 that the product does converge. $g_n(z)$ should be close enough to $\frac{1}{z-a_n}$, say the exponential of a partial sum of a logarithm of $\frac{1}{z-a_n}$. This idea occurs already in Riemann's 1861 Göttingen lectures (see pp. 62–65 of NEUENSCHWANDER [1996]). The exposition below follows HEINS [1968]. Another proof, based on the (independently proved) Mittag-Leffler theorem, is offered in 11.40.

Theorem 7.35 (WEIERSTRASS [1876]) *Let U be an open subset of $\mathbb{C}$, A a subset of U with no limit point in U. With each $a \in A$ associate a positive integer $m(a)$. Then there exists a holomorphic function on U which has a zero of multiplicity $m(a)$ at each $a \in A$ and no others.*

Proof: Since A has no limit point in U, it meets each compact subset of U in only a finite set. Since (1.31) U is a countable union of compact subsets, it follows that A is at most countable. If A is finite, the polynomial $\prod_{a\in A}(z - a)^{m(a)}$ is evidently a function of the desired kind. So we shall assume that A is not finite. Let $a_1, a_2, \ldots$ be an enumeration of A with each a appearing $m(a)$ times. Consider first the case in which

$$U \neq \mathbb{C} \text{ and } A \text{ is bounded.} \tag{1}$$

For each $a_n \in A$ there is, by a routine compactness argument, a point $b_n \in \mathbb{C} \setminus U$ such that

$$|a_n - b_n| \leq |a_n - b| \qquad \forall b \in \mathbb{C} \setminus U. \tag{2}$$

Next notice that

$$\lim_{n\to\infty} |a_n - b_n| = 0. \tag{3}$$

For if not, then $n_1 < n_2 < \ldots.$ and $\varepsilon > 0$ exist such that $|a_{n_k} - b_{n_k}| \geq \varepsilon$. Using the boundedness of A, we can assume in addition, by passing to a subsequence of $\{n_k\}$ if necessary, that $\lim_{k\to\infty} a_{n_k} = a$ exists. We have for any $b \in \mathbb{C} \setminus U$ and all k

$$\begin{aligned} |a - b| \geq |a_{n_k} - b| - |a - a_{n_k}| &\overset{(2)}{\geq} |a_{n_k} - b_{n_k}| - |a - a_{n_k}| \\ &\geq \varepsilon - |a - a_{n_k}|. \end{aligned}$$

Therefore $|a - b| \geq \varepsilon$, $a \neq b$ and $a \notin \mathbb{C} \setminus U$. That is, the limit point a of $\{a_n\}$ lies in U, a contradiction. This establishes (3).

For each positive integer n define

$$\phi_n(z) := \frac{a_n - b_n}{z - b_n}, \qquad z \in \mathbb{C} \setminus \{b_n\} \supset U \tag{4}$$

$$\lambda_n(w) := -\sum_{k=n+1}^{\infty} \frac{w^k}{k}, \qquad w \in \mathbb{D} \tag{5}$$

$$E_n(z) := (1 - z)\exp\left[\sum_{k=1}^{n} \frac{z^k}{k}\right], \qquad z \in \mathbb{C}. \tag{6}$$

Now equation (2) in the proof of 3.19 says that

$$1 - w = \exp\left[-\sum_{k=1}^{\infty} \frac{w^k}{k}\right]$$
$$= \exp\left[-\sum_{k=1}^{n} \frac{w^k}{k}\right] \exp \lambda_n(w) \qquad \forall w \in \mathbb{D}.$$

Therefore

$$(7) \qquad E_n(w) = \exp \lambda_n(w) \qquad \forall w \in \mathbb{D}.$$

Now let K be any compact subset of U. The compact set K and the closed set $\mathbb{C} \setminus U$ are disjoint, so by a routine compactness argument they are a positive distance apart. That is, there exists an $\varepsilon = \varepsilon_K > 0$ such that

$$(8) \qquad |z - b| \geq \varepsilon \qquad \forall z \in K,\ b \in \mathbb{C} \setminus U.$$

There exists by (3) an n_K such that

$$|a_n - b_n| \leq \varepsilon/2, \qquad \forall n \geq n_K$$

and together with (8) this gives

$$(9) \qquad |\phi_n(z)| = \left|\frac{a_n - b_n}{z - b_n}\right| \leq \frac{1}{2} \qquad \forall n \geq n_K,\ z \in K.$$

We therefore have for all $n \geq n_K$ and all $z \in K$

$$\left|\lambda_n(\phi_n(z))\right| \leq \sum_{k=n+1}^{\infty} \frac{1}{k} |\phi_n(z)|^k \leq \sum_{k=n+1}^{\infty} |\phi_n(z)|^k \leq \sum_{k=n+1}^{\infty} 2^{-k} = 2^{-n},$$

that is,

$$(10) \qquad \left|\lambda_n(\phi_n(z))\right| \leq 2^{-n} \qquad \forall z \in K,\ n \geq n_K.$$

It follows that for every positive integer $M \geq n_K$ the series $\sum_{n=M+1}^{\infty} \lambda_n \circ \phi_n$ converges uniformly on K. The limit function, call it $g_{M,K}$, is holomorphic in the interior of K by 5.44. Define

$$(11) \qquad F_N(z) := \prod_{n=1}^{N} E_n(\phi_n(z)), \quad z \in \mathbb{C} \setminus \{b_1, \ldots, b_N\},\ N \in \mathbb{N}.$$

The definitions make clear that the only zero of $E_n \circ \phi_n$ is a_n and it has multiplicity 1. Hence the zero-set of the product F_N in (11) is $\{a_1, \ldots, a_N\}$. More precisely,

(12) the N-tuple $(a_1, ..., a_N)$ specifies the zeros of F_N with their multiplicities.

For $N > M \geq n_K$ and $z \in K$ we have

$$\begin{aligned} F_N(z) &= F_M(z) \prod_{n=M+1}^{N} E_n(\phi_n(z)) \\ &= F_M(z) \prod_{n=M+1}^{N} \exp(\lambda_n(\phi_n(z)) \quad \text{by (7) and (9)} \\ &= F_M(z) \exp\left(\sum_{n=M+1}^{N} \lambda_n \circ \phi_n \right)(z). \qquad (13) \end{aligned}$$

Therefore we see that the sequence $\{F_N\}$ converges at each point of U and uniformly on each compact subset K of U, with

$$\lim_{N\to\infty} F_N(z) = F_M(z) e^{g_{M,K}(z)} \qquad \forall z \in K,\ M \geq n_K.$$

Call this limit F. Thus, F is holomorphic in U (5.44) and

$$F(z) = F_M(z) e^{g_{M,K}(z)} \qquad \forall z \in K,\ M \geq n_K. \qquad (14)$$

This shows first of all that the zeros of F are those of the F_M $(M = 1, 2, \ldots)$. That is, by (12), $F^{-1}(0) = \{a_n\}_{n\in\mathbb{N}} = A$. Given $a \in A$, consider any compact $K \subset U$ with a in its interior. If $n > M \geq n_K$, then by (9) $|(a_n - b_n)/(a - b_n)| < 1/2$, so a_n cannot be a; that is, all $m(a)$ occurrences of a in the sequence $\{a_n\}_{n\in\mathbb{N}}$ are in the M-tuple $(a_1, \ldots, a_M)$, so by (12) the multiplicity of a as a zero of F_M is $m(a)$, and consequently by (14) the multiplicity of a as a zero of F is also $m(a)$, as wanted.

Because of (3) and the boundedness of A there exists an $R \in \mathbb{R}$ such that

$$|a_n|,\ |b_n| < R \qquad \forall n \in \mathbb{N}.$$

Then for $|z| \geq 5R$ and all n

$$\begin{aligned} &|z - b_n| \geq |z| - |b_n| > 4R > 2|a_n| + 2|b_n| \geq 2|a_n - b_n| \\ &|\phi_n(z)| = \left| \frac{a_n - b_n}{z - b_n} \right| < \frac{1}{2} \qquad \forall |z| \geq 5R,\ \forall n \in \mathbb{N}. \qquad (15) \end{aligned}$$

As in the derivation of (10) it follows from (15) that

$$|\lambda_n(\phi_n(z))| < 2^{-n} \qquad \forall |z| \geq 5R,\ \forall n \in \mathbb{N}. \tag{16}$$

From (7) and (11)

$$F_N(z) = \exp\left[\sum_{n=1}^{N} \lambda_n \circ \phi_n(z)\right] \qquad \forall |z| \geq 5R,\ \forall N \in \mathbb{N}.$$

Therefore (16) shows that

$$|F_N(z)| \leq \exp \sum_{n=1}^{N} 2^{-n} < e \qquad \forall |z| \geq 5R,\ \forall N \in \mathbb{N}.$$

Let $N \to \infty$ and conclude that

$$|F(z)| \leq e \qquad \forall z \in U,\ |z| \geq 5R. \tag{17}$$

Now in the general case we may suppose, after a translation, that

$$0 \in U \setminus A. \tag{18}$$

Let $\phi(z) := 1/z$ and consider

$$U_0 := \phi(U \setminus \{0\}),\ A_0 := \phi(A).$$

Since $0 \notin A$ and, being in $U, 0$ is not a limit point of A either, the set A is bounded away from 0. Therefore A_0 is bounded and U_0, A_0 satisfy the conditions in (1). Therefore there is a function F which is holomorphic in U_0 with zeros of multiplicity $m(a)$ at $\phi(a)$, for each $a \in A$, and no others. Form $f := F \circ \phi$ on $U \setminus \{0\}$. Then f has zeros only at the $a \in A$. In fact if we write

$$F(w) - F(\phi(a)) = [w - \phi(a)]^{m(a)} G_a(w), \qquad \forall w \in U_0,$$

where G_a is holomorphic in U_0 and $G_a(\phi(a)) \neq 0$, then

$$\begin{aligned} f(z) - f(a) &= [\phi(z) - \phi(a)]^{m(a)} G_a(\phi(z)) \\ &= (z-a)^{m(a)} \frac{G_a(1/z)}{(-az)^{m(a)}}, \qquad \forall z \in U \setminus \{0\}, \end{aligned}$$

so a is a zero of f of multiplicity $m(a)$. Since $0 \in U$ and U is open, there is an $r > 0$ such that $D(0, r) \subset U$. We may suppose that $r < 1/(5R)$. Then for $0 < |z| < r$ we have by (17)

$$|f(z)| = |F(1/z)| \leq e.$$

Therefore according to 5.41, f extends to be differentiable at 0, i.e., holomorphic throughout U. If 0 is a zero of (the extension of) f, say of multiplicity m_0, then $f(z)/z^{m_0}$ is the desired function.

In specific cases it is often possible to exhibit the above function rather explicitly. We shall look at just one especially important instance of this.

Exercise 7.36 *If $a_n \in \mathbb{D} \setminus \{0\}$ and $\sum_{n=1}^{\infty}(1-|a_n|) < \infty$, then*

$$\lim_{N\to\infty} \prod_{n=1}^{N} \frac{|a_n|}{a_n} \frac{z-a_n}{\overline{a_n}z-1}$$

exists throughout $\mathbb{D}$ and defines a holomorphic function which is bounded by 1 and whose zero-set is $\{a_1, a_2, \dots\}$. (See 12.6, 6.7(iii) and 6.9(iii) for converses of this.) *The limit exists as well for all $z \in \mathbb{C} \setminus \overline{\mathbb{D}} \setminus \{1/\overline{a_n} : n \in \mathbb{N}\}$ and defines a holomorphic function in that open set. What happens at points $z \in \mathbb{T}$?*

Hints: Suppose first that $|a_n| > r$ for all n and some $r > 0$. Then the functions

$$B_N(z) := \prod_{n=1}^{N} \frac{|a_n|}{a_n} \frac{z-a_n}{\overline{a_n}z-1}, \qquad N \in \mathbb{N}$$

are holomorphic in $\mathbb{D}$ and zero-free in $D(0,r)$. There exist then (see 4.20 and 4.67(ii)) holomorphic L_N in $D(0,r)$ such that $e^{L_N} = B_N$ in $D(0,r)$. Since

$$B_N(0) = \prod_{n=1}^{N} |a_n| \in (0,\infty),$$

we can suppose that $L_N(0) = \log B_N(0)$. Notice that $\operatorname{Re} L_N = \log|B_N|$ and $|B_{N+1}| \le |B_N| < 1$, so $0 > \operatorname{Re} L_1 \ge \operatorname{Re} L_2 \ge \cdots$ in $D(0,r)$. Since $\log x = \int_1^x (1/t)\,dt < x - 1$ for $x > 1$, we have

$$\log\frac{1}{|a_n|} < \frac{1}{|a_n|} - 1 = \frac{1-|a_n|}{|a_n|} < \frac{1-|a_n|}{r},$$

$$L_N(0) = \log B_N(0) > -\sum_{n=1}^{N} \frac{1-|a_n|}{r} = -\frac{1}{r}\sum_{n=1}^{N}(1-|a_n|)$$

$$\ge -\frac{1}{r}\sum_{n=1}^{\infty}(1-|a_n|) > -\infty.$$

Therefore the decreasing sequence $\{L_N(0)\}$ converges in $\mathbb{R}$. It follows then from 7.14(ii) that $\{L_N\}$ converges throughout $D(0,r)$, then from 7.5 that $\{B_N\}$ converges locally uniformly throughout $\mathbb{D}$. Call the latter limit B.

Since, as just noted, $B(0) > 0$, we have from 7.10 that B is zero-free in $D(0, r)$.

In general, since $|a_n| \to 1$, given $r < 1$ we can select n_r such that $|a_n| > r$ for all $n > n_r$ and then conclude from the above that

$$\lim_{N\to\infty} \prod_{n=n_r+1}^{N} \frac{|a_n|}{a_n} \frac{z - a_n}{\overline{a_n}z - 1}$$

is zero-free in $D(0, r)$, i.e., B/B_{n_r} is zero-free in $D(0, r)$. It follows that the zeros of B are precisely the points $\{a_1, a_2, \dots\}$. The limit points, if any, of $\{a_n\}$ all lie in $\mathbb{T}$; consequently, $U := \mathbb{C} \setminus \overline{\mathbb{D}} \setminus \{1/\overline{a_n} : n \in \mathbb{N}\}$ is an open set. Since $z \mapsto 1/\bar{z}$ maps U into $\mathbb{D} \setminus \{a_n : n \in \mathbb{N}\}$, the functions $\overline{B}_N(1/\bar{z})$, $\overline{B}(1/\bar{z})$, are holomorphic in U (2.4) *and* zero-free. Therefore $B_N(z) := 1/\overline{B}_N(1/\bar{z})$ converges, for each $z \in U$, to the holomorphic limit $1/\overline{B}(1/\bar{z})$.

Any function of the form $e^{i\theta} z^m B_N(z)$ $(\theta \in \mathbb{R}, m \in \mathbb{Z}^+)$ is called a *finite Blaschke product*; the limit $e^{i\theta} z^m B(z)$ is called an *infinite Blaschke product.* Perhaps the most important application of these products is the following factorization:

Exercise 7.37 *For each bounded $F \in H(\mathbb{D}) \setminus \{\mathbf{0}\}$ there exists a Blaschke product B and a zero-free $f \in H(\mathbb{D})$ such that $F = fB$. Prove this and show also that* $\sup_{\mathbb{D}} |F| = \sup_{\mathbb{D}} |f|$.

Hints: Without loss of generality, $F(0) \neq 0$. The zeros of F are at most countable in number and if $z_1, z_2, \dots$ is any enumeration of them (repeated according to multiplicity), we know from 6.9(iii) that $\sum_{n=1}^{\infty}(1 - |z_n|) < \infty$. Hence by 7.36 the products

$$B_N(z) := \prod_{n=1}^{N} \frac{|z_n|}{z_n} \frac{z - z_n}{\overline{z_n}z - 1}, \qquad z \in \mathbb{D},\ N \in \mathbb{N}$$

converge locally uniformly in $\mathbb{D}$ to a holomorphic function B whose zeros (counted according to multiplicity) are the z_n. If a is a zero of F of multiplicity k, and $n_1, \dots, n_k$ are the distinct indices such that $z_{n_1} = \cdots = z_{n_k} = a$, then $B(z)$ is the product of $\prod_{j=1}^{k} \frac{|z_{n_j}|}{z_{n_j}} \frac{z - z_{n_j}}{\overline{z_{n_j}}z - 1}$ and a holomorphic function which has no zero at a. Hence F/B is holomorphic near a. It follows that F/B is holomorphic in $\mathbb{D}$. Call this function f. Routine arguments with the maximum-modulus principle (cf. the proof of 6.1) show that F/B_N is bounded in $\mathbb{D}$ by $M := \sup |F(\mathbb{D})|$, for each N. Hence in the limit the function $f = F/B$ satisfies $|f| \leq M$ in $\mathbb{D}$. On the other hand, since $|B| \leq 1$ throughout $\mathbb{D}$, we have $|F| \leq |f|$, so $\sup_{\mathbb{D}} |f| \geq \sup_{\mathbb{D}} |F|$.

The (infinite) Blaschke products introduced in 7.36 have a beautiful characterization analogous to that provided by 6.16 (i) for the finite Blaschke products. The development of this characterization (due to FROSTMAN [1935]) is the closing topic of this section.

Exercise 7.38 *Let $z_j \in \mathbb{D}$ with $|z_1| \le |z_2| \le \cdots$ and for each $0 \le r < 1$ let $n(r)$ denote the number of j such that $|z_j| \le r$. Show that if*

(1) $$\sum_{j=1}^{\infty}(1 - |z_j|) < \infty,$$

then

(2) $$\lim_{r \uparrow 1} n(r)\log r = 0.$$

Hints: Use the following elementary fact:

(3) $$a_n \downarrow 0 \ \ \& \ \sum_{n=1}^{\infty} a_n < \infty \Rightarrow na_n \to 0.$$

The idea behind (3), which the reader can easily expand into a proof, is that $0 \le na_{2n} \le \sum_{k=n+1}^{2n} a_k$ for all $n \in \mathbb{N}$. From (1) we have $|z_j| \to 1$, so to prove (2) it suffices, after discarding some finite number N_0 of the initial z_j (a maneuver which changes $n(r)$ to $n(r) - N_0$) to assume $d := |z_1| > 0$. Then

$$\log \frac{1}{|z_n|} < \frac{1}{|z_n|} - 1 = \frac{1 - |z_n|}{|z_n|} \le \frac{1 - |z_n|}{d},$$

$$0 \le \sum_{n=1}^{\infty} \log \frac{1}{|z_n|} \le \frac{1}{d}\sum_{n=1}^{\infty}(1 - |z_n|) < \infty, \quad \text{by (1).}$$

Citing (3) we then get

(4) $$N \log |z_N| \to 0 \quad \text{as } N \to \infty.$$

Since $|z_n| \le r$ for all $n \le n(r)$, we have

$$\log |z_{n(r)}| \le \log r < 0$$

(5) $$n(r)\log |z_{n(r)}| \le n(r)\log r \le 0.$$

Now $n(r) \to \infty$ as $r \uparrow 1$, by (1), so the left-hand side of (5) converges to 0 as $r \uparrow 1$, by (4). This proves (2).

Now let F be a holomorphic function in $\mathbb{D}$ with $F(0) \neq 0$ and $\sup|F(\mathbb{D})| = 1$. Since F is not identically 0, its zeros do not have a limit point in $\mathbb{D}$, hence are at most countable in number. Let them be, in increasing modulus and repeated according to multiplicity, $z_1, z_2, \ldots$. We shall assume that in fact there are infinitely many, since otherwise the goals below are trivially realized with finite Blaschke products. According to 6.7(iii) then

$$\sum_{n=1}^{\infty}(1 - |z_n|) < \infty.$$

Define

$$B_N(z) := \prod_{n=1}^{N} \frac{|z_n|}{z_n} \frac{z - z_n}{\overline{z_n}z - 1}, \qquad z \in \mathbb{D}, N \in \mathbb{N}.$$

According to 7.36

$$B(z) := \lim_{N\to\infty} B_N(z)$$

exists for each $z \in \mathbb{D}$ and defines a holomorphic function which is bounded by 1 and whose zeros, with appropriate multiplicity, are just the z_n. Let C denote the countable subset $\{|z_n| : n \in \mathbb{N}\}$ of $(0, 1)$. Notice that $\theta \mapsto \left|F\left(re^{i\theta}\right)\right|$ and $\theta \mapsto \left|B\left(re^{i\theta}\right)\right|$ are zero-free continuous functions on $[0, 2\pi]$ for each $r \in (0, 1) \setminus C$. Hence $\int_0^{2\pi} \log\left|F\left(re^{i\theta}\right)\right| d\theta$ and $\int_0^{2\pi} \log\left|B\left(re^{i\theta}\right)\right| d\theta$ make sense for such r.

Exercise 7.39 *With the notation and hypotheses above, show that*

$$(*) \qquad \lim_{\substack{r\uparrow 1 \\ r\notin C}} \frac{1}{2\pi} \int_0^{2\pi} \log\left|F\left(re^{i\theta}\right)\right| d\theta = 0$$

if and only if F is a unimodular constant multiple of B.

Hints: For each $r \in (0, 1) \setminus C$ let $n(r)$ denote the number of k such that $|z_k| \leq r$. Fix r and for brevity write $N := n(r)$. Define

$$B_r^*(z) := \prod_{n=1}^{N} \frac{r(z - z_n)}{r^2 - \overline{z_n}z}, \qquad z \in \mathbb{C} \setminus \left\{r^2/\overline{z_1}, \cdots, r^2/\overline{z_N}\right\}.$$

Then F/B_r^* is holomorphic and zero-free in a neighborhood of $\bar{D}(0, r)$. [Its smallest zero has modulus $\min\{|z_{N+1}|, |r^2/\bar{z}_N|\}$, which exceeds r, by definition of $N = n(r)$ and the fact that $r \notin C$.] Hence $\log|F/B_r^*|$ is harmonic and so enjoys the mean-value property:

$$(1) \qquad \log|F(0)/B_r^*(0)| = \frac{1}{2\pi} \int_0^{2\pi} \log\left|F\left(re^{i\theta}\right)/B_r^*\left(re^{i\theta}\right)\right| d\theta.$$

But each factor of B_r^* has modulus 1 on the circle $C(0,r)$, so (1) reads

$$\log|F(0)/B_r^*(0)| = \frac{1}{2\pi}\int_0^{2\pi} \log\left|F\left(re^{i\theta}\right)\right| d\theta. \tag{2}$$

Now notice that $|B_r^*(0)| = \prod_{n=1}^N \frac{|z_n|}{r} = r^{-N}B_N(0)$. Therefore (2) reads (*Jacobi-Jensen* formula)

$$\log|F(0)/B_N(0)| + N\log r = \frac{1}{2\pi}\int_0^{2\pi} \log\left|F\left(re^{i\theta}\right)\right| d\theta,$$

$$\log|F(0)/B_N(0)| + n(r)\log r = \frac{1}{2\pi}\int_0^{2\pi} \log\left|F\left(re^{i\theta}\right)\right| d\theta. \tag{3}$$

Now as $r \uparrow 1$, $B_N(0) \to B(0) \neq 0$ (why?), so in view of (3) and 7.38 the condition $(*)$ is equivalent to

$$\log|F(0)/B(0)| = 0, \quad |F(0)| = |B(0)|.$$

That is, the function $f := F/B$ of 7.37 satisfies

$$|f(0)| = 1. \tag{4}$$

Since, as shown in 7.37, $|f| \leq \sup|F| = 1$, (4) happens if and only if f is a unimodular constant.

§ 4 Elementary Iteration Theory

Definition 7.40 If $S \subset \mathbb{C}$ and $f\colon S \mapsto S$, then we define $f^{[0]}$ to be the identity map of S onto S and inductively $f^{[n+1]} = f \circ f^{[n]}$ for each non-negative integer n. We call $f^{[n]}$ the nth *iterate* of f. If f is a bijection, $f^{-[n]}$ will mean $(f^{-1})^{[n]}$.

We investigate in this section elementary results, derivable from the convergence criteria of previous sections, about the sequence $\left\{f^{[n]}\right\}$ when f is holomorphic. Points fixed by f and the convergence of $f^{[n]}$ to such a point are on center stage, as the first example illustrates:

Exercise 7.41 (i) (Cf. 7.44, 7.46) *Let U be an open subset of $\mathbb{C}$, $f \in H(U)$, $f(U) \subset U$. Suppose $z_0 \in U$, $f(z_0) = z_0$ and $|f'(z_0)| < 1$. Show that there is an open disk about z_0 in which the iterates of f converge uniformly to z_0. (Such a point z_0 is consequently called an* attractive fixed-point.) *Hence, if in addition U is bounded and connected, the iterates converge everywhere in U to z_0.*

Hints: Let $R := \frac{1}{2}[1 + |f'(z_0)|]$. By definition of $f'(z_0)$ then there exists $r > 0$ such that $D(z_0, r) \subset U$ and

$$\left|\frac{f(z) - f(z_0)}{z - z_0}\right| \le R \qquad \forall z \in D(z_0, r) \setminus \{z_0\}.$$

Since $f(z_0) = z_0$, this says that

$$(*) \qquad |f(z) - z_0| \le R|z - z_0| \le Rr < r \qquad \forall z \in D(z_0, r).$$

The extreme inequality make $f(z)$ eligible for the role of z in the first inequality, giving

$$|f(f(z)) - z_0| \le R|f(z) - z_0| \le R^2|z - z_0|.$$

By induction then

$$|f^{[n]}(z) - z_0| \le R^n|z - z_0| \qquad \forall z \in D(z_0, r), n \in \mathbb{N}.$$

If U is bounded and connected, cite 7.5 (and 5.62) to conclude that $\{f^{[n]}\}$ converges locally uniformly on U to the constant z_0.

(ii) *Prove the following converse of* (i)*: If $f\colon U \mapsto U$ is holomorphic, $z_0 \in U$ and $f^{[n]}$ converges locally uniformly in U to z_0, then $f(z_0) = z_0$ and $|f'(z_0)| < 1$.*

Hints: From $f^{[n]}(z_0) \to z_0$ and continuity of f at z_0 follows $f\big(f^{[n]}(z_0)\big) \to f(z_0)$. On the other hand, $f\big(f^{[n]}(z_0)\big) = f^{[n+1]}(z_0) \to z_0$. We have $f^{[n]\prime} \to \mathbf{0}$ by 7.3(vi) and $f^{[n]\prime}(z_0) = (f'(z_0))^n$ by the Chain Rule and the fact that $f(z_0) = z_0$. It must therefore be that $|f'(z_0)| < 1$.

(iii) (Ritt [1920–21]) *Let Ω be a region, f a holomorphic map of Ω into a compact subset of Ω. Show that the iterates of f converge uniformly in Ω to a constant, and this number is the unique fixed-point of f.*

Hints: For $n \in \mathbb{N}$ let $\Omega_n := f^{[n]}(\Omega)$, $K := \bigcap_{n=1}^{\infty} \Omega_n$. By induction we establish that $\overline{\Omega}_{n+1} \subset \Omega_n \subset \overline{\Omega}_n$ compact $\subset \Omega$, for all $n \in \mathbb{N}$. Hence $K = \bigcap_{n=2}^{\infty} \overline{\Omega}_n$ is compact and not empty. Consider any locally uniformly convergent subsequence $\{f^{[n_j]}\}$ of iterates, [such exist, by 7.6] and its limit function g. Evidently $g(\Omega) \subset K$. On the other hand, if $w \in K$, then $w \in \Omega_{n+1} = f^{[n+1]}(\Omega) = f^{[n]}(f(\Omega)) = f^{[n]}(\Omega_1)$ for all n, so $w = f^{[n]}(w_n)$ for some $w_n \in \Omega_1$. A subsequence $\{w_{n_{j_k}}\}$ of $\{w_{n_j}\}$ converges to some w_0 in the compact set $\overline{\Omega}_1$. Since $\{f^{[n_j]}\}$ converges to g uniformly on this compact set, it follows easily (cf. 7.2) that $w = f^{[n_{j_k}]}(w_{n_{j_k}})$ converges to $g(w_0)$, i.e., $w \in g(\overline{\Omega}_1) \subset g(\Omega)$. Thus $g(\Omega) = K$. By the open-map theorem (5.77) g is therefore constant and the compact set K is a single point. By 7.8 then the whole sequence $\{f^{[n]}\}$ converges to this point.

(iv) *Let Ω be a region, $F_n \in H(\Omega)$ map into a common compact subset of Ω for $n \in \mathbb{N}$. Show that as $n \to \infty$ the composite $F_1 \circ \cdots \circ F_n$ converges locally uniformly in Ω to a constant.*

Hints: Set $f_n := F_1 \circ \cdots \circ F_n$, $\Omega_n := f_n(\Omega)$. Show that $\overline{\Omega}_{n+1} = \overline{f_n(F_n(\Omega))} \subset f_n\big(\overline{F_n(\Omega)}\big) \subset f_n(\Omega) = \Omega_n$, and then argue as in the preceding hints.

Exercise 7.42 *Let f be holomorphic in the open set U and satisfy $f(U) \subset U$. Show that if some subsequence of $\{f^{[n]}\}$ converges locally uniformly to a conformal automorphism of U, then f itself is a conformal automorphism of U. If U is connected and the whole sequence $\{f^{[n]}\}$ converges locally uniformly to a non-constant limit, then f is the identity map on U.*

Hints: Suppose $\{f^{[n_j]}\}$ converges locally uniformly to $\phi \in \operatorname{Aut}(U)$. If z, $z' \in U$ and $f(z) = f(z')$, then $f^{[n]}(z) = f^{[n]}(z')$ for all n, so $\phi(z) = \phi(z')$. It follows that $z = z'$. Since every $f^{[n]}(U)$ $(n \geq 1)$ lies in the open set $f(U)$, 7.11 (applied in each component of U) assures that the limit function ϕ satisfies $\phi(U) \subset f(U)$; that is, since ϕ is surjective, $U \subset f(U)$. If $\{f^{[n]}\} \to g$, non-constant, then (7.12) $g(U) \subset U$, so $g = \lim f^{[n+1]} = \lim f \circ f^{[n]} = f \circ g$. Thus $f = \mathrm{id}_U$ on the open set $g(U)$.

Exercise 7.43 (Carathéodory [1932], Cartan [1932]) *Let Ω be a bounded region, f a holomorphic self-map of Ω, $\mathcal{G}$ the set of non-constant subsequential limits of $\{f^{[n]}\}_{n\in\mathbb{N}}$. Show that if it is not empty, then $\mathcal{G}$ is a commutative subgroup of* $\operatorname{Aut}(\Omega)$ *which contains f.*

Hints: Suppose $f^{[n_j]} \to g \in \mathcal{G}$, $f^{[p_j]} \to h \in \mathcal{G}$. These limits are holomorphic functions and by 7.11 they map Ω into itself. Set $m_j := n_{j+1} - n_j$, which we may suppose strictly increasing. From boundedness of $\{f^{[n]}\}$ and 7.6 we get $j_1 < j_2 < \cdots$ such that $\{f^{[m_{j_k}]}\}$ converges locally uniformly in Ω to some $\phi \in H(\Omega)$. From 7.1(ii) it follows that $f^{[m_{j_k}]} \circ f^{[n_{j_k}]} \to \phi \circ g$. Since $f^{[m_j]} \circ f^{[n_j]} = f^{[m_j+n_j]} = f^{[n_{j+1}]} \to g$, we see that $\phi \circ g = g$. This says that on the non-void open subset $g(\Omega)$ of Ω, ϕ coincides with id_Ω, the identity map on Ω. Consequently, $\phi = \mathrm{id}_\Omega$, showing that $\mathrm{id}_\Omega \in \mathcal{G}$ and that $f \in \operatorname{Aut}(\Omega)$, by 7.42. From 7.19(iii) it then follows that $g, h \in \operatorname{Aut}(\Omega)$ and moreover $(f^{[p_j]})^{-1} \to h^{-1}$. From 7.1(ii) or the final remark in 7.20 we know that $f^{[n_j+p_j]} = f^{[n_j]} \circ f^{[p_j]} = f^{[p_j]} \circ f^{[n_j]} \to g \circ h = h \circ g$, putting this composite in $\mathcal{G}$ and proving $\mathcal{G}$ to be a commutative subsemigroup of $\operatorname{Aut}(\Omega)$. At the very beginning of the argument we could have first passed to a subsequence of $\{n_j\}$

in order to have, for all j, $n_{j+1} > 2n_j + p_j$. Then on the one hand

$$f^{[n_{1+j_k}-n_{j_k}-p_{j_k}]} = f^{[m_{j_k}]} \circ (f^{[p_j]})^{-1} \longrightarrow \phi \circ h^{-1} = h^{-1}$$

and on the other hand this is a subsequential limit of $\{f^{[n]}\}$, and so lies in $\mathcal{G}$.

Exercise 7.44 *Let Ω be a bounded region, $f\colon \Omega \mapsto \Omega$ holomorphic but $f \notin$* Aut (Ω).

(i) *Show that if f fixes some point $a \in \Omega$, then $\{f^{[n]}\}$ converges to a, locally uniformly in Ω.*

Hints: Any convergent subsequence of the iterates of f converges to the constant a. For the limit is a constant function by 7.43 and it fixes the point a. By 7.8 then the whole sequence $\{f^{[n]}\}$ converges to a.

(ii) *Show that if f fixes no point of Ω, then each subsequential limit of $\{f^{[n]}\}$ is a constant and the totality of these constants comprises a closed, connected subset C of $\partial\Omega$.*

Hints: Since $f \notin \mathrm{Aut}(\Omega)$, it follows from 7.43 that every subsequential limit of $\{f^{[n]}\}$ is constant. Evidently $\varnothing \neq C \subset \overline{\Omega}$. If some $a \in C$ lay in Ω, say $f^{[n_j]}(w) \to a$ for all $w \in \Omega$ [in particular, for all $f(w)$ too], then by continuity of f at a, $f(a) = \lim f\big(f^{[n_j]}(w)\big) = \lim f^{[n_j]}(f(w)) = a$, contrary to f being fixed-point-free; so $C \subset \overline{\Omega} \setminus \Omega = \partial\Omega$. It is also obvious that if K is any non-void compact subset of Ω and $N \in \mathbb{N}$, then

$$C \subset \overline{\bigcup_{n \geq N} f^{[n]}(K)} =: K_N, \quad \text{and so}$$

$$(*) \qquad C \subset \bigcap_{N \in \mathbb{N}} K_N.$$

Actually, the last inclusion is equality. For if c lies in this intersection, we can successively find $1 \leq n_1 < n_2 < \cdots$ such that c is within $1/k$ of the set $f^{[n_k]}(K)$ for each $k \in \mathbb{N}$. Since (7.6) a subsequence of $\{f^{[n_k]}\}$ converges locally uniformly to a constant, c must be that constant. Pick any $a \in \Omega$ and let K be a curve in Ω joining a to $f(a)$. Then $f(a) \in K \cap f(K)$, so by 1.4 and 1.1, $K \cup f(K)$ is connected. Repetitions of this argument show that each K_N is connected. From 1.5(i) and the fact of equality in $(*)$ above then follows the connectedness of C.

(iii) *Show that if f fixes no point of Ω, then every orbit $\{f^{[n]}(z)\}_{n\in\mathbb{N}}$ is compactafugal.*

Hints: Suppose contrariwise that for some compact $K \subset \Omega$, $z \in \Omega$, and $1 \leq n_1 < n_2 < \ldots$ we have $\{f^{[n_j]}(z)\}_{j\in\mathbb{N}} \subset K$. Thanks to Montel (7.6) we can further suppose that $\{f^{[n_j]}\}$ converges locally uniformly in Ω. By (ii) the limit function is a constant $c \in \partial\Omega$, whereas $f^{[n_j]}(z) \in K$ for all j implies $c \in K \subset \Omega$.

Exercise 7.45 (Cf. 12.51(i).) *If Ω is a bounded region, f a holomorphic map of Ω into Ω and f has at least two fixed-points in Ω, then $f \in \operatorname{Aut}(\Omega)$.* (It is much harder to prove, but true, that if f has more than two fixed-points, then it is the identity map—see SUITA [1981], FISHER and FRANKS [1987], and KRANTZ [2005].)

Hint: If $f \notin \operatorname{Aut}(\Omega)$, then $\{f^{[n]}\}$ would, by 7.44(i), converge to any fixed-point. Hence there could not be two fixed-points.

Complementing 7.41 we have

Theorem 7.46 *Let Ω be a bounded region, f a holomorphic map of Ω into Ω with $f(a) = a$ for some $a \in \Omega$. Then*

(i) $|f'(a)| \leq 1$;

(ii) $f'(a) = 1 \Leftrightarrow f$ *is the identity map*;

(iii) $|f'(a)| = 1 \Leftrightarrow f$ *is a conformal automorphism of* Ω.

In particular, if $f \in \operatorname{Aut}(\Omega)$, $f(a) = a$ and $f'(a) > 0$, then f is the identity map. [This latter is true of *every* region $\Omega \neq \mathbb{C}$—see 12.51(ii).]

Proof: (i) We may translate Ω by $-a$ and so suppose without loss of generality that $a = 0$. It is notationally convenient to set $f_n := f^{[n]}$. Then $f_n'(0) = f'(f_{n-1}(0)) \cdot f_{n-1}'(0)$. But $f(0) = 0$ implies $f_n(0) = 0$ for all n, so $f_n'(0) = f'(0)f_{n-1}'(0)$. This is the inductive step in the proof that

(1)
$$f_n'(0) = (f'(0))^n \qquad \forall n \in \mathbb{N}.$$

Now there exist $0 < r < M < \infty$ such that $\overline{D}(0, r) \subset \Omega \subset D(0, M)$. Thus $|f_n| < M$ for all n, so the Cauchy estimates 3.34(i) give

(2.k)
$$\frac{|f_n^{(k)}(0)|}{k!} \leq \frac{M}{r^k}, \quad n \in \mathbb{N}.$$

From (1) and (2.1) follows $|f'(0)| \leq 1$.

(ii) Now assume $f'(0) = 1$. If there are any $k \geq 2$ such that $f^{(k)}(0) \neq 0$, let m be the smallest and set $c_k := f^{(k)}(0)/k!$ for $k \geq m$.

$$f(z) = z + \sum_{k=m}^{\infty} c_k z^k$$

(3)
$$= z + c_m z^m g(z), \quad \text{where } g(0) = 1.$$

Let us show that for each n there is an $r_n > 0$ and a holomorphic function g_n in $D(0, r_n)$ such that for all $z \in D(0, r_n)$

$$(3.n) \qquad f_n(z) = z + c_m z^m g_n(z) \quad \text{and} \quad g_n(0) = n.$$

The equation (3) is just (3.1). If (3.n) holds, then since $f_n(0) = 0$ we may choose $r_n > r_{n+1} > 0$ so that $f_n(D(0, r_{n+1})) \subset D(0, r_n)$. Then from (3) and (3.$n$) will follow for $z \in D(0, r_{n+1})$

$$\begin{aligned} f_{n+1}(z) &= f(f_n(z)) = f_n(z) + c_m f_n^m(z) g(f_n(z)) \\ &= z + c_m z^m g_n(z) + c_m [z + c_m z^m g_n(z)]^m g(f_n(z)) \\ &= z + c_m z^m [g_n(z) + (1 + c_m z^{m-1} g_n(z))^m g(f_n(z))]. \end{aligned}$$

Let $g_{n+1}(z)$ denote the square-bracketed term and note that since $m - 1 \geq 1$, the term z^{m-1} is 0 when $z = 0$, so

$$\begin{aligned} g_{n+1}(0) &= g_n(0) + g(f_n(0)) \\ &= g_n(0) + 1 \quad \text{by (3)} \\ &= n + 1 \quad \text{by (3.}n\text{)}, \end{aligned}$$

which establishes (3.$n+1$).

It follows from (3.n) that

$$f_n^{(m)}(0) = m! c_m g_n(0) = n m! c_m, \qquad n \in \mathbb{N}$$

and since $c_m \neq 0$ this contradicts inequality (2.m) for sufficiently large n.

The conclusion is that $f^{(k)}(0) = 0$ for all $k \geq 2$, so $f(z) = z$ in $D(0, r)$, hence throughout the connected superset Ω.

(iii) Suppose $|f'(0)| = 1$. The sequence $\{f_n\}$ of iterates of f is bounded on Ω and so has a convergent subsequence $\{f_{n_j}\}$ (by 7.6). If the limit is ϕ, then by 5.44(ii)$(f_{n_j})'(0) \to \phi'(0)$. But, by (1), $f_n'(0) = (f'(0))^n$ for all n, so $|f_{n_j}'(0)| = 1$, whence $|\phi'(0)| = 1$. Therefore ϕ is not constant and it follows from 7.43 that f is a conformal automorphism of Ω.

For the converse direction in (iii), apply (i) to both f and f^{-1}, noting that, by 5.78, $(f^{-1})'(0) = 1/f'(f^{-1}(0)) = 1/f'(0)$.

Exercise 7.47 *Let Ω be a bounded region, $a \in \Omega$, $\mathcal{G} = \mathcal{G}_a$ the set of all conformal automorphisms of Ω which fix a.*

(i) *$\mathcal{G}$ is a group under composition and is isomorphic to a subgroup of the unit circle. Hence $\mathcal{G}$ is commutative.*

Hint: The isomorphism is $f \mapsto f'(a)$. Compare parts (ii) and (iii) of 7.46.

(ii) *If $\{f_n\} \subset \mathcal{G}$ and $\{f_n\}$ converges on Ω to f, then $f \in \mathcal{G}$.*

Hints: By 7.5 the convergence is locally uniform on Ω and f is holomorphic, since $\{f_n\}$ is bounded. Then $f_n'(a) \to f'(a)$ by 5.44(ii). As $|f_n'(a)| = 1$, it follows that $|f'(a)| = 1$, so f is not constant. Consequently, by 7.11, $f(\Omega) \subset \Omega$ and then by 7.46(iii) f is a conformal automorphism of Ω.

(iii) *If $\{f_n\} \subset \mathcal{G}$ and $\{f_n'(a)\}$ converges, then $\{f_n\}$ converges to an $f \in \mathcal{G}$.*

Hints: According to 7.8 it suffices to demonstrate that all convergent subsequences of $\{f_n\}$ have the same limit. (This limit will be in $\mathcal{G}$ by (ii).) If $\{f_{n_j}\}$, $\{f_{m_j}\}$ are two such convergent subsequences with limits ϕ and ψ, say, then $f_{n_j}'(a) \to \phi'(a)$ and $f_{m_j}'(a) \to \psi'(a)$. Therefore $\phi'(a) = \psi'(a)$. Since ϕ, $\psi \in \mathcal{G}$ by (ii), it follows from (i) that $\phi = \psi$.

(iv) *The map $f \mapsto f'(a)$ of $\mathcal{G}$ into the circle is bicontinuous and the image of $\mathcal{G}$ is a compact subgroup of the circle. Hence $\mathcal{G}$ is either finite and cyclic or isomorphic to the circle.*

Hints: Identify $\mathcal{G}$ with its image in $\mathbb{T}$ and let G be the (closed) subgroup $\{x \in \mathbb{R} : e^{ix} \in \mathcal{G}\}$. If G is dense in $\mathbb{R}$, then its exponential image $\mathcal{G}$ is both dense and compact in $\mathbb{T}$, hence coincides with $\mathbb{T}$. If G is not dense, it contains a minimum positive element x_0 (why?) and is consequently the group generated by x_0. Indeed, any $x \in G$ determines an $n \in \mathbb{Z}$ with $nx_0 \leq x < (n+1)x_0$, and then $x - nx_0 \in G$ satisfies $0 \leq x - nx_0 < x_0$, so by minimality $x - nx_0 = 0$. Since $2\pi \in G = \mathbb{Z}x_0$, $2\pi x_0^{-1} =: n_0 \in \mathbb{Z}$. The image $e^{2\pi i/n_0}$ of generator x_0 of G is a generator of $\mathcal{G}$.

Remarks 7.48: The group $\mathcal{G}$ is the whole circle if Ω is simply-connected. This follows easily (see 10.7) from the Riemann Mapping Theorem. AUMANN and CARATHÉODORY [1934] (see also BEHNKE and ZUMBUSCH [1936]) demonstrated the converse: If Ω is not simply-connected, then every element of $\mathcal{G}$ has finite order, hence by the above, $\mathcal{G}$ is a finite (cyclic) group. The proof has points of similarity with 10.7 and 10.20; one needs a generalization of the Riemann map: a locally univalent map f of $\mathbb{D}$ onto Ω with a special property—analytic continuability of the local inverses of f along curves in Ω. The construction of this function is very similar to the Koebe–Carathéodory construction of the mapping function for Riemann's theorem which will be presented in Chapter IX. The reader will find it carried out in §§315–321 of CARATHÉODORY's book [1960] and in §7, chapter 3 of VEECH [1967]. In exercises 12–19 of the latter he will also find the elegant details of Aumann and Carathéodory's proof. In §4 of Chapter X we shall present a quite different proof of the Aumann-Carathéodory theorem. A less constructive existence proof for f along the lines of 9.9(ii) appears in §5, chapter 16 of CONWAY [1995].

For more about groups of conformal automorphisms of a region see T. RADÓ [1924–26], BRUNÉ [1932], HEINS [1946b], MARDEN, RICHARDS and RODIN [1967], MUELLER and RUDIN [1991], BEARDON and MINDA [2007] and their bibliographies, as well as the notes on the Aumann-Carathéodory theorem in Chapter X below.

In [1932] CARATHÉODORY discovered a remarkable refinement of 7.46(iii): for a holomorphic $f\colon \Omega \mapsto \Omega$ which fixes a but is not a conformal automorphism, not only can $|f'(a)|$ not be 1, but in fact $|f'(a)|$ cannot be too close to 1. Theorem 7.51 below presents the beautiful proof H. CARTAN [1932] found a short while later. This is known as the *Starrheitssatz*, i.e., the rigidity theorem: A holomorphic self-map of Ω which fixes a point and is close to a conformal automorphism *is* a conformal automorphism. The proof is purely existential. To actually find the rigidity constant in 7.51 or give good estimates for it is a difficult task, though some work has been done on this; see, e.g., HERVÉ [1951] and MINDA [1979]. There exist regions Ω for which this constant is 0. In fact, every holomorphic self-map of $\mathbb{C}\backslash\{-1,1\}$ which is not constant turns out to be a conformal automorphism, as will be shown in the Remarks following 12.50. The extension of 7.51 to unbounded regions was achieved (using covering-space arguments) by AUMANN and CARATHÉODORY [1934]; we'll get this in 12.48 by other means. No such constant exists for $\mathbb{D}$ and its point 0, as the maps $z \mapsto cz$ ($c \in \mathbb{D}$) attest. After the Riemann mapping theorem 9.7 has been proved, it will then be evident that such constants do not exist for any simply-connected proper subregion of $\mathbb{C}$. Hence Theorem 7.51 may be viewed as yet another (cf. 10.2) characterization of simple-connectivity.

We need some preliminary results, due to CARTAN [1932], which will be formulated as

Exercise 7.49 (i) *Let Ω be a bounded region, $a \in \Omega$, f a holomorphic self-map of Ω which fixes a. Suppose that γ is a loop in Ω which is not Ω-loopnullhomotopic, but is Ω-loophomotopic to $f \circ \gamma$. Show that $f \in \operatorname{Aut}(\Omega)$.*

Hints: After 7.44(i) the conclusion follows if we can show that the sequence $\{f^{[n]}\}$ does not converge to the constant a. Suppose contrariwise that $f^{[n]} \to a$ locally uniformly in Ω. Choosing $r > 0$ so that $D(a,r) \subset \Omega$, we will have $f^{[n]}(\gamma) \subset D(a,r)$ for a large enough $n > 1$. Via $(\tau, t) \mapsto \tau f^{[n]}(\gamma(t)) + (1-\tau)a \in D(a,r)$ the loop $f^{[n]} \circ \gamma$ is rendered Ω-loophomotopic to the constant a. If H is a loophomotopy in Ω of γ with $f \circ \gamma$, then $f^{[k]} \circ H$ is a loophomotopy in Ω of $f^{[k]} \circ \gamma$ with $f^{[k+1]} \circ \gamma$, for any $k \in \mathbb{N}$. Since homotopy is transitive, it follows that γ is Ω-loophomotopic to $f^{[n]} \circ \gamma$ and thence to the constant a. This is the contradiction we were seeking.

(ii) *Suppose the bounded region Ω has at least one, but only finitely many* *holes, none a singleton, and that $f : \Omega \mapsto \Omega$ is holomorphic and maps every loop $\gamma \subset \Omega$ which is not Ω-nullhomologous onto another such loop. Deduce the amazing conclusion $f \in \mathrm{Aut}\,(\Omega)$.* [Cf. 10.13 (iii).]

Hints: In view of 7.44, it suffices to show that no convergent sequence of iterates of f has a constant limit. Suppose $\{g_k\}_{k\in\mathbb{N}}$ is a locally uniformly convergent subsequence of $\{f^{[n]}\}_{n\in\mathbb{N}}$ with constant limit g. According to the remarks following 4.71, there is a loop γ in Ω which is not Ω-nullhomologous. Enumerate the holes in Ω as $K_1, \ldots, K_m$. By hypothesis, for each $n \in \mathbb{N}$ the loop $f^{[n]} \circ \gamma$ is not Ω-nullhomologous, so it has non-zero index throughout some K_j. There is at least one j such that $\mathrm{Ind}_{g_k\circ\gamma}$ is non-zero throughout K_j for infinitely many k. At least one point in K_j, say p, differs from the constant g. That is, p is not on the (constant) loop $g \circ \gamma$, and therefore by 4.6(i) for all sufficiently large k, $\mathrm{Ind}_{g_k\circ\gamma}(p) = \mathrm{Ind}_{g\circ\gamma}(p) = 0$ —⨯—.

(iii) *Prove the converse of* (ii).

Hint: 4.49(iv) [or 11.48].

Exercise 7.50 *Let Ω be a region, $F_n : \Omega \mapsto \Omega$ holomorphic functions which converge locally uniformly in Ω to the identity function. Show that for any loop γ in Ω, there is an $n \in \mathbb{N}$ such that $F_n \circ \gamma$ is loophomotopic in Ω to γ.*

Hints: By compactness, there exists $\varepsilon > 0$ such that $D(\gamma(t), \varepsilon) \subset \Omega$ for all $t \in [a, b]$, the parameter interval of γ. Choosing n large enough that $|F_n(z) - z| < \varepsilon$ for all z in the compact set γ, we have $F_n(\gamma(t)) \in D(\gamma(t), \varepsilon)$ for all $t \in [a, b]$. Therefore

$$(\tau, t) \mapsto \tau\gamma(t) + (1 - \tau)F_n(\gamma(t)) \in D(\gamma(t), \varepsilon) \subset \Omega$$

is a loophomotopy in Ω of $F_n \circ \gamma$ with γ.

Theorem 7.51 *Let Ω be a bounded region, $a \in \Omega$. If Ω is not simply-connected, then there is a constant $c < 1$, which depends only on Ω and a, such that any holomorphic $f : \Omega \mapsto \Omega$ fixing a and satisfying $|f'(a)| > c$ already satisfies $|f'(a)| = 1$ and is therefore a conformal automorphism of Ω.*

Proof: Suppose for a given pair a, Ω no such constant exists. This means that there exist holomorphic $f_n : \Omega \mapsto \Omega$ which fix a, have $|f_n'(a)| < 1$ and satisfy $|f_n'(a)| \to 1$. By passing (via 7.6) to a subsequence, we can suppose the sequence $\{f_n\}$ converges locally uniformly in Ω to a holomorphic function f. Of course then $|f'(a)| = \lim |f_n'(a)| = 1$, so f is not constant. Hence by 7.12 $f(\Omega) \subset \Omega$. From 7.46(iii) therefore f is in fact a conformal automorphism of Ω. Evidently then $f^{-1} \circ f_n$ converges locally uniformly to the identity function on Ω. Since Ω is not simply-connected, there is a loop γ in Ω which is not Ω-loopnullhomotopic (see 4.75). According to 7.50 there is an n such that $f^{-1} \circ f_n \circ \gamma$ is Ω-loophomotopic to γ. But then by 7.49(i) $f^{-1} \circ f_n$ is a conformal automorphism of Ω, and so therefore is $f \circ f^{-1} \circ f_n = f_n$. Since $|f_n'(a)| < 1$, this contradicts 7.46(iii).

Exercise 7.52 *Let* $\Omega := \{z \in \mathbb{C} : \operatorname{Re} z > 0,\ \operatorname{Im} z > 0,\ |z| < 1\}$, $f(z) := e^{\frac{1}{2}\pi i z}$. *Show that* $f(\Omega) \subsetneq \Omega$ *and that* f *has a fixed-point in* Ω.

Hints: If $0 < \theta < \pi/2$ and $0 < r < 1$, then $f(re^{i\theta}) = e^{-\frac{1}{2}\pi r \sin\theta} e^{\frac{1}{2}\pi i r \cos\theta}$. Since $\frac{1}{2}\pi r \cos\theta \in (0, \pi/2)$ and $\sin\theta > 0$, it is therefore clear that $f(re^{i\theta}) \in \Omega$. Also we see that this point is fixed by f if and only if $r = e^{-\frac{1}{2}\pi r \sin\theta}$ and $\theta = \frac{1}{2}\pi r \cos\theta$, that is, if and only if the function $F(\theta) := -\frac{2\theta}{\pi \cos\theta} + e^{-\theta \tan\theta}$ has a zero in $(0, \pi/2)$. The existence of such a zero follows from the Intermediate-Value Theorem for continuous functions, since $F(0) = 1$ and $F(\pi/4) = e^{-\pi/4} - 1/\sqrt{2} < 0$, because of the trivial inequalities $e^{\pi/4} = \sqrt{e^{\pi/2}} > \sqrt{e} > \sqrt{2}$. Finally, note that $\frac{1}{2}i \in \overline{\Omega}$ but $\frac{1}{2}i \notin f(\overline{\Omega})$. Indeed if $re^{i\theta} \in \overline{\Omega}$ and $\frac{1}{2}e^{\frac{1}{2}\pi i} = \frac{1}{2}i = f(re^{i\theta}) = e^{-\frac{1}{2}\pi r \sin\theta} e^{\frac{1}{2}\pi i r \cos\theta}$, then $\frac{1}{2}\pi = \frac{1}{2}\pi r \cos\theta$ and it follows that $r = 1$, $\cos\theta = 1$, whence $re^{i\theta} = 1$ and $f(re^{i\theta}) = i$, a contradiction. Since $\frac{1}{2}i \in \overline{\Omega} \setminus f(\overline{\Omega}) = \overline{\Omega} \setminus \overline{f(\Omega)}$, certainly $f(\Omega) \neq \Omega$.

Exercise 7.53 (i) *Show that* $i^{i^{\cdot^{\cdot^{\cdot}}}}$ *converges, that is, the sequence* i_n *defined inductively by* $i_1 := i$, $i_{n+1} := i^{i_n} := e^{\frac{1}{2}i\pi i_n}$, *converges.*

Hints: Let Ω and f be as in the preceding exercise. The conclusion of 7.44(i) is valid for Ω, since by the preceding exercise f is a holomorphic map of Ω into itself which is not onto but which fixes some point $\alpha \in \Omega$. It follows that the sequence $\{f^{[n]}\}$ of iterates of f converges throughout Ω to α. Now check that $\beta := f^{[4]}(1) \in \Omega$ and that $i_n = f^{[n]}(1) = f^{[n-4]}(\beta)$ for all $n > 4$ and we are done.

(ii) *Let* $a > 0$, $b > 0$, $a + b = 1$ *and* $f(z) := az + b/z$ $(z \in \mathbb{C} \setminus \{0\})$. *Then the iterates of* f *converge to* 1 *in the open right half-plane and to* -1 *in the open left half-plane. What happens on the imaginary axis?*

Hints: Note that each of $\mathbb{H}$ and $\mathbb{C} \setminus \overline{\mathbb{H}}$ is conformal to a disk [see 2.5] and is invariant under f. Since $f(1) = 1$, $f(-1) = -1$, f has a fixed-point in each half-plane, but since $f'(\sqrt{b/a}) = f'(-\sqrt{b/a}) = 0$, f is not one-to-one in either half-plane (cf. 5.78). The assertion follows then from 7.44(i), applied to the appropriate composite $\phi^{-1} \circ f \circ \phi$.

§ 5 Iteration in Disks and Annuli

In this section we make a deeper study of the convergence behavior of iterates of holomorphic self-maps of disks and annuli. The results are satisfyingly complete, and in view of 10.2 and 10.20 they (when properly interpreted) are results about iterates in arbitrary simply-connected or doubly-connected (proper) subregions of $\mathbb{C}$, not just disks and annuli. We first look at conformal automorphisms of $\mathbb{D}$, and find it easier to work in the upper half-plane $\mathbb{U}$.

Theorem 7.54 *Let $f \in \operatorname{Aut}(\mathbb{D})$ be fixed-point-free in $\mathbb{D}$. Then $\{f^{[n]}\}$ converges locally uniformly in $\mathbb{D}$ to a unimodular constant which is a fixed-point of f. It is the Denjoy-Wolff point u from 6.6 and satisfies $0 < f'(u) \leq 1$.*

Proof: Consider the map $\phi(w) := \frac{w-i}{w+i}$, which we know (2.5) sends $\mathbb{U}$ conformally onto $\mathbb{D}$. If ψ is any other conformal map of $\mathbb{U}$ onto $\mathbb{D}$, then $\psi \circ \phi^{-1}$ is a conformal automorphism T_ψ of $\mathbb{D}$ and $\psi = T_\psi \circ \phi$. Now T_ψ is continuous on $\overline{\mathbb{D}}$ by 6.2(ii). Since

$$\lim_{\substack{w \to 0 \\ w \in \mathbb{U}}} \phi(w) = -1 \quad \text{and} \quad \lim_{\substack{|w| \to \infty \\ w \in \mathbb{U}}} \phi(w) = 1,$$

we therefore infer that

$$\text{(1)} \qquad \lim_{\substack{w \to 0 \\ w \in \mathbb{U}}} \psi(w) = T_\psi(-1) \quad \text{and} \quad \lim_{\substack{|w| \to \infty \\ w \in \mathbb{U}}} \psi(w) = T_\psi(1).$$

Let $F := \phi^{-1} \circ f \circ \phi$. It is a fixed-point-free conformal automorphism of $\mathbb{U}$. According to 6.21(iii) there is $\Psi \in \operatorname{Aut}(\mathbb{U})$ and an $\alpha > 0$, $\beta \in \mathbb{R}$ such that

either

$$\text{(2)} \qquad \Psi \circ F \circ \Psi^{-1}(w) = \alpha w \qquad \forall w \in \mathbb{U},$$

or

$$\text{(3)} \qquad \Psi \circ F \circ \Psi^{-1}(w) = \beta + w \qquad \forall w \in \mathbb{U}.$$

Consequently, $\psi := \phi \circ \Psi^{-1}$ is a conformal map of $\mathbb{U}$ onto $\mathbb{D}$ and we have

either

$$\text{(2)}' \qquad f^{[n]}(z) = \psi\big(\alpha^n \psi^{-1}(z)\big) \qquad \forall z \in \mathbb{D},\ n \in \mathbb{N}$$

or

$$\text{(3)}' \qquad f^{[n]}(z) = \psi\big(n\beta + \psi^{-1}(z)\big) \qquad \forall z \in \mathbb{D},\ n \in \mathbb{N}.$$

Since F has no fixed-point in $\mathbb{U}$, it follows from (2) and (3) that $\alpha \neq 1$ and $\beta \neq 0$. Now $n\beta + \psi^{-1}(z) \to \infty$ in $\mathbb{U}$ as $n \to \infty$, and either $\alpha^n \psi^{-1}(z) \to 0$ in

$\mathbb{U}$ or $\alpha^n \psi^{-1}(z) \to \infty$ in $\mathbb{U}$ as $n \to \infty$ (according as $\alpha < 1$ or $\alpha > 1$). It follows then from $(2)'$, $(3)'$, and (1) that

either

$$\lim_{n\to\infty} f^{[n]}(z) = T_\psi(-1) \qquad \forall z \in \mathbb{D}$$

or

$$\lim_{n\to\infty} f^{[n]}(z) = T_\psi(1) \qquad \forall z \in \mathbb{D}.$$

It is easy to see directly from $(2)'$, $(3)'$ [or one can cite 7.5] that this convergence is locally uniform in $\mathbb{D}$. Denoting the limit by a, we have $a \in \overline{\mathbb{D}}$ and since (by 6.2(ii)) f is continuous on $\overline{\mathbb{D}}$, we get

$$a \leftarrow f^{[n+1]}(z) = f\big(f^{[n]}(z)\big) \to f(a).$$

Thus $f(a) = a$. As f fixes no points of $\mathbb{D}$ (hypothesis), $a \in \overline{\mathbb{D}} \setminus \mathbb{D} = \mathbb{T}$.

Since the Denjoy-Wolff point u of 6.6 satisfies $f(D(Ru, 1-R)) \subset D(Ru, 1-R)$ for each $R \in [0,1]$, it is clear that $f^{[n]} \to a$ entails $a = u$, while 6.28(x) applied to the function $F(z) := \bar{u} f(uz)$ insures that $f'(u) \in (0,1]$.

Here is an alternative, more computational attack. After 6.2(ii) our function has the form $f(z) = e^{i\theta}(z-c)/(1-\bar{c}z)$, with $\theta \in \mathbb{R}$, $c \in \mathbb{D} \setminus \{0\}$. Without loss of generality let us suppose one of the points of $\mathbb{T}$ which f fixes (6.4(ii)) is 1. Then $e^{i\theta} = (1-\bar{c})/(1-c)$, and after a little algebra one finds that the point

$$u := -\frac{c}{\bar{c}} \cdot \frac{\bar{c}-1}{c-1} \in \mathbb{T}$$

is also fixed by f. Again, routine calculations show that

$$f'(1) = \frac{1-|c|^2}{|1-c|^2} > 0 \quad \text{and} \quad f'(u) = \frac{|1-c|^2}{1-|c|^2} > 0.$$

A simple calculation shows that since $c \in \mathbb{D}$, u cannot be 1, and this entails $\operatorname{Re} c \neq |c|^2$. It follows that exactly one of these derivatives is less than 1. In fact, it is the derivative at the point farther from $f^{-1}(0)$, since

$$|f^{-1}(0) - u| = \frac{1-|c|^2}{|1-c|} = \frac{1-|c|^2}{|1-c|^2}\,|f^{-1}(0) - 1|.$$

Convergence in $\mathbb{D}$ of the sequence $\{f^{[n]}\}$ to this fixed-point then follows by a slight modification of the 7.41(i) hints. [More is true: If we think of f

as mapping $\mathbb{C}_\infty$ into $\mathbb{C}_\infty$, with $f(\infty) := -e^{i\theta}/\bar{c}$ and $f(1/\bar{c}) := \infty$, then f is continuous and $\{f^{[n]}(z)\}$ converges to the attracting fixed-point for all $z \in \mathbb{C}_\infty$ except the other fixed-point.]

Exercise 7.55 *$f \in \mathrm{Aut}(\mathbb{D})$ and f fixes $z_0 \in \mathbb{D}$. Show that either f is periodic (that is, $f^{[n]}$ is the identity map for some n) or the semigroup $\{f^{[n]} : n \in \mathbb{N}\}$ generated by f is dense in the compact group $\mathcal{G}$ of all conformal automorphisms of $\mathbb{D}$ which fix z_0.*

Hints: Use the bicontinuous isomorphism $g \mapsto g'(z_0)$ of $\mathcal{G}$ onto a compact subgroup of $\mathbb{T}$ furnished by 7.47(iv). If for some $n \in \mathbb{N}$, $\omega := f'(z_0)$ is an nth root of 1, then $f^{[n]}$ is the identity automorphism. Otherwise the group $\{\omega^k : k \in \mathbb{Z}\}$ is dense in $\mathbb{T}$, as was shown in the 7.47(iv) hints. Evidently then $\{\omega^k : k \in \mathbb{Z}, |k| > |m|\}$ is also dense for any $m \in \mathbb{Z}$, and so given $\varepsilon > 0$, there is $k \in \mathbb{Z}$ with $|\omega^k - 1| < \varepsilon$ and $|k| > |m|$. Since $|\omega^k - 1| = |1 - \omega^{-k}|$, we can assume $k > 0$. Thus $k > |m|$ and $n := k + m \in \mathbb{N}$ satisfies $|\omega^n - \omega^m| < \varepsilon$. This puts the whole group $\{\omega^m : m \in \mathbb{Z}\}$ in the closure of the semigroup $\{\omega^n : n \in \mathbb{N}\}$.

Finally, the Denjoy-Wolff fixed-point theorem 6.6 lets us handle the iterates of non-automorphisms:

Theorem 7.56 *Let $f\colon \mathbb{D} \mapsto \mathbb{D}$ be holomorphic but $f \notin \mathrm{Aut}\,(\mathbb{D})$. Then $\{f^{[n]}\}$ converges locally uniformly in $\mathbb{D}$ to a constant α. Either α is the unique fixed-point of f in $\mathbb{D}$, or f has no fixed-point in $\mathbb{D}$, α is the "fixed-point" $u \in \mathbb{T}$ supplied by 6.6, and f has at α an angular derivative lying in $(0,1]$.*

Proof: Suppose f has a fixed-point $\alpha \in \mathbb{D}$. Then by 7.44(i) the sequence $\{f^{[n]}\}$ of iterates converges to α locally uniformly in $\mathbb{D}$ and α is the unique fixed-point of f (7.45). For an estimate of the rate of convergence, see Beardon and Carne [1992].

Suppose f has no fixed-point in $\mathbb{D}$. Then 6.6 furnishes a point $u \in \mathbb{T}$ such that

$$(*) \qquad f\big(D(ru, 1-r)\big) \subset D(ru, 1-r) \qquad \forall 0 < r < 1.$$

Consider any subsequence $\{f^{[n_j]}\}$ of the iterates which converges locally uniformly in $\mathbb{D}$. By 7.43 its limit is a constant α. By induction on $(*)$

$$f^{[n]}\big(D(ru, 1-r)\big) \subset D(ru, 1-r) \qquad \forall n \in \mathbb{N}, \quad \forall 0 < r < 1.$$

Therefore $\alpha = \lim_{j\to\infty} f^{[n_j]}(ru)$ must lie in $\overline{D}(ru, 1-r)$. The only number in all these disks is u, so $\alpha = u$. It now follows from 7.8 that the whole sequence $\{f^{[n]}\}$ converges locally uniformly in $\mathbb{D}$ to u. The statement about angular derivative follows from applying 6.28(x) to the function $F(z) := \bar{u}f(uz)$.

Next, we use these results to learn something about common fixed-points of commuting mappings.

Theorem 7.57 (MITCHELL [1979]) *Let $\mathcal{G}$ be a group (composition) of holomorphic self-maps of $\mathbb{D}$. Suppose $\mathcal{G}$ is compact (with respect to locally uniform convergence). Then either $\mathcal{G}$ consists of a single constant function, or $\mathcal{G}$ consists of all powers of a single conformal automorphism of $\mathbb{D}$ which has finite order, or $\mathcal{G}$ consists of all conformal automorphisms of $\mathbb{D}$ which fix some one point of $\mathbb{D}$. In any case, $\mathcal{G}$ is commutative and a unique point in $\mathbb{D}$ is fixed by every function in $\mathcal{G}$.*

Proof: Let E denote the identity element of $\mathcal{G}$. If E is a constant function, say $E(z) = z_0 \in \mathbb{D}$ for every $z \in \mathbb{D}$, then $f(z) = (E \circ f)(z) = E(f(z)) = z_0$ for all $f \in \mathcal{G}$. Thus $\mathcal{G} = \{E\} = \{z_0\}$.

If E is not constant, then $E(\mathbb{D})$ is an open subset of $\mathbb{D}$ (5.77) and from $E(w) = (E \circ E)(w) = E(E(w))$ we infer that in the open set $E(\mathbb{D})$, E coincides with the identity map I of $\mathbb{D}$. By the uniqueness theorem then $E = I$ throughout $\mathbb{D}$. It follows that the functions in the group $\mathcal{G}$ are all bijective, i.e., are conformal automorphisms of $\mathbb{D}$.

Consider any $I \neq f_0 \in \mathcal{G}$. If f_0 is fixed-point-free in $\mathbb{D}$, then by 7.54 the iterates $f_0^{[n]}$ converge to a constant, in particular, to a function not in the group $\mathcal{G}$. This contradicts the compactness of $\mathcal{G}$. Hence f_0 must have a fixed-point, a_0, in $\mathbb{D}$. Set $T(z) := \frac{z+a_0}{1+\bar{a}_0 z}$. Then set $F_0 := T^{-1} \circ f_0 \circ T$ and $\mathcal{G}_0 := T^{-1} \circ \mathcal{G} \circ T$, another compact group of conformal automorphisms of $\mathbb{D}$. By 6.2(ii) the function $F_0 \in \mathcal{G}_0$ has the form

$$(1) \qquad F_0(z) = e^{i\theta} z, \quad z \in \mathbb{D}$$

for some $0 < \theta < 2\pi$. Since also $F_0^{-1}(z) = e^{-i\theta} z = e^{(2\pi-\theta)i} z$ is in $\mathcal{G}_0$, we can replace F_0 by F_0^{-1} if necessary in order to have the θ in (1) satisfy

$$(2) \qquad 0 < \theta \leq \pi.$$

Since $f \mapsto |f(0)|$ is a continuous function on the compact set $\mathcal{G}_0$, there is an $f \in \mathcal{G}_0$ which maximizes it, say

$$f(z) = u\frac{z+a}{1+\bar{a}z}, \quad |u| = 1 > |a|.$$

Write $u := e^{i\psi}$. Since $0 < \frac{\theta}{\pi} \leq 1$, there is an integer n such that

$$n \cdot \frac{\theta}{\pi} \in \left[-\frac{1}{2} - \frac{\psi}{\pi}, \frac{1}{2} - \frac{\psi}{\pi}\right].$$

Thus

$$\frac{n\theta + \psi}{\pi} \in \left[-\frac{1}{2}, \frac{1}{2}\right],$$

so

$$\operatorname{Re}\left(e^{in\theta}u\right) = \operatorname{Re}\left(e^{i(n\theta+\psi)}\right) \geq 0.$$

Set $w := e^{in\theta}u$ and note that then $\operatorname{Re} w \geq 0$ and so

$$|a|^4 + 2\operatorname{Re}\left[w|a|^2\right] < 2\operatorname{Re} w + 1$$

$$|w|^2|a|^4 + 2\operatorname{Re}\left[w|a|^2\right] + 1 < |w|^2 + 2\operatorname{Re} w + 1, \quad \text{since } |w| = 1,$$

$$\left|1 + w|a|^2\right|^2 < |1 + w|^2. \tag{3}$$

Now consider $f_1 := F_0^{[n]} \circ f = e^{in\theta} f$, an element of $\mathcal{G}_0$. It satisfies

$$f_1^{[2]}(0) = f_1\big(f_1(0)\big) = f_1(wa) = w \cdot \frac{wa + a}{1 + \bar{a}wa} = wa \cdot \frac{1 + w}{1 + w|a|^2}.$$

Therefore

$$\left|f_1^{[2]}(0)\right| = |a| \left|\frac{1 + w}{1 + w|a|^2}\right| = |f(0)| \cdot \left|\frac{1 + w}{1 + w|a|^2}\right|. \tag{4}$$

From (3) and (4) we infer that $|f(0)| = 0$, for otherwise the maximality of $|f(0)|$ would be violated. It follows that $g(0) = 0$ for *every* $g \in \mathcal{G}_0$. From 6.2(ii) then $\mathcal{G}_0$ is a (compact) subgroup of $\{\omega I : \omega \in \mathbb{T}\}$. Hence (see the hints to 7.47(iv)) it is either the whole rotation group, or it is finite and cyclic $= \left\{\omega^k I : k = 1, 2, \ldots, n\right\}$ for some n^{th} root ω of 1. It follows that $\mathcal{G} = T \circ \mathcal{G}_0 \circ T^{-1}$ is commutative and is either the totality of conformal automorphisms which fix $T(0)$, or the cyclic subgroup $\left\{T \circ (\omega^k I) \circ T^{-1} = (T \circ (\omega T^{-1}))^{[k]} : k = 1, 2, \ldots, n\right\}$ thereof.

If inverses are lacking (i.e., the group hypothesis is dropped) but commutativity is available by fiat, then a common fixed-point can be found by other arguments:

Theorem 7.58 (Shields [1964]) *Let $\mathcal{F}$ be a family of continuous self-maps of $\overline{\mathbb{D}}$ which are holomorphic in $\mathbb{D}$ and commute with one another under composition. Then the functions in $\mathcal{F}$ have a common fixed-point.*

Proof: If $\mathcal{F}$ contains a constant function, then this is a common fixed-point. Otherwise, by the maximum-modulus principle $f(\mathbb{D}) \subset \mathbb{D}$ for each $f \in \mathcal{F}$. If some $f \in \mathcal{F}$ is not a conformal automorphism of $\mathbb{D}$, then some subsequence of the iterates has a limit (7.6) and by 7.43 the limit is a constant (lying in $\overline{\mathbb{D}}$). Say $f^{[n_k]} \to z_0$ (on $\mathbb{D}$). Then for any $g \in \mathcal{F}$, $f^{[n_k]}(g(0)) \to z_0$. Since $f^{[n_k]}(g(0)) = g\big(f^{[n_k]}(0)\big) \to g(z_0)$ $\big[$continuity of g on $\overline{\mathbb{D}}\big]$, $g(z_0) = z_0$, for every $g \in \mathcal{F}$.

Finally, suppose $\mathcal{F} \subset \operatorname{Aut}(\mathbb{D})$. If some $f \in \mathcal{F}$ has no fixed-points in $\mathbb{D}$, then by 7.54, $\{f^{[n]}\}$ converges to some $z_0 \in \mathbb{T}$ and the preceding argument shows that z_0 is a common fixed-point. There remains the case that $\mathcal{F}$ contains a non-identity map f which fixes some $z_0 \in \mathbb{D}$. Then $g(z_0) = g(f(z_0)) = f(g(z_0))$ shows that $g(z_0)$ is also a fixed-point of f in $\mathbb{D}$, for each $g \in \mathcal{F}$. By 6.4(ii) then $g(z_0) = z_0$, for each $g \in \mathcal{F}$.

Remarks: There is an important perspective on the material of this section which needs to be mentioned, however cursorily. Let

$$h(t) := \frac{1}{2}\log\left(\frac{1+t}{1-t}\right), \quad 0 \le t < 1$$

and, as in 6.3(vii), define

$$d(z,w) := \left|\frac{z-w}{1-\bar{w}z}\right|, \quad z, w \in \mathbb{D},$$
$$\rho := h \circ d.$$

Then it can be verified that ρ is a complete metric in $\mathbb{D}$ (although a direct check of the $\triangle$-inequality is arduous). The topology it generates in $\mathbb{D}$ is the usual Euclidean topology. In fact, the ρ-ball of center z and radius r is the d-ball of center z and radius $\frac{e^{2r}-1}{e^{2r}+1}$, which by 6.5(ii) is a Euclidean disk. In the remarks following 2.8 we reviewed the usual procedure for "rectifying" a smooth curve γ in $\mathbb{C}$ which results in its being assigned the length $\int |\gamma'|$. If in that subdivision procedure the absolute value is replaced by the d-metric, it is not hard to see that the length assigned to γ is $\int \frac{|\gamma'|}{1-|\gamma|^2}$. It turns out that for points $z \neq w$ in $\mathbb{D}$ the minimum of the lengths so measured of smooth curves joining z and w in $\mathbb{D}$ equals $\rho(z,w)$, and the unique curve having this length is the arc of the circle through z and w which intersects $\mathbb{T}$ orthogonally.

d is called the *pseudo-hyperbolic* metric, while ρ is called variously the *hyperbolic*, *Poincaré* or *Harnack* metric. The latter name derives from the fact that $e^{2\rho(z_1,z_2)}$ is the smallest constant $c(\{z_1,z_2\},\mathbb{D})$ that works in the Harnack inequality 5.28(ii). (See, e.g., BEAR [1991] or §1.3 of RANSFORD [1995].) Since h is an increasing function, 6.3(i) really says that any holomorphic self-map f of $\mathbb{D}$ is ρ-nonexpansive: $\rho(f(z_1), f(z_2)) \le \rho(z_1,z_2)$ for all z_1, $z_2 \in \mathbb{D}$. (Cf. also 6.3(vii) and 6.5(iv).) The contraction-mapping theorem of S. Banach and its very elementary and transparent proof support the view that it is this property of f which accounts for the convergence behavior of its iterates and the existence of fixed-points in $\mathbb{D}$ (or of "generalized" Denjoy–Wolff fixed-points on $\partial\mathbb{D}$). Many books (e.g., AHLFORS [1973], KRANTZ [1990] and J. H. SHAPIRO [1993]) put this very geometric point of view in the foreground. In addition, SINE [1989], BEARDON [1990], BEAR [1992], and SHOIKHET [2001] explore how much of the above theory remains valid, or how its results get modified, if the holomorphy of f is dispensed with and it is only hypothesized that f is a ρ-nonexpansive self-map of $\mathbb{D}$. In fact, at this level of generality it is just as easy to leave $\mathbb{D}$ for the unit ball of a Hilbert space; see GOEBEL and REICH [1984].

The next exercise, involving (without mentioning the word) covering-space ideas, studies the iterates of a self-map of an annulus. There is no loss of generality in assuming that the annulus is symmetric with respect to the unit circle. We follow the prize-winning paper of HEINS [1941b].

Exercise 7.59 *Let $r > 0$, $R_1 := e^{-r\pi/2}$, $R_2 := e^{r\pi/2}$, $\lambda := e^{2\pi/r}$, $A := \{w \in \mathbb{C} : R_1 < |w| < R_2\}$, $g\colon A \mapsto A$ holomorphic. Let L denote the Principal Branch of the Logarithm in $\mathbb{C} \setminus (-\infty, 0]$.*

(i) *Show that the function E defined on $\overline{\mathbb{H}} \setminus \{0\}$ as e^{irL} maps $\mathbb{H}$ onto A and $\overline{\mathbb{H}} \setminus \{0\}$ onto $\bar{A}$.*

(ii) *Show that there exists holomorphic $g^*: \mathbb{H} \mapsto \mathbb{H}$ satisfying $g \circ E = E \circ g^*$.*

Hints: $g \circ E$ is a zero-free holomorphic function on $\mathbb{H}$ and so (see 5.35) it has the form e^{irh} for some holomorphic function h on $\mathbb{H}$. Since $e^{irh(\mathbb{H})} = g(E(\mathbb{H})) = g(A) \subset A$, we must have $R_1 < \left|e^{irh}\right| < R_2$, so $\log R_1 < \operatorname{Re}(irh) < \log R_2$, that is, $-\frac{\pi}{2} < -\operatorname{Im} h < \frac{\pi}{2}$. Therefore the function $g^* := e^h$ maps $\mathbb{H}$ into $\mathbb{H}$ and we can form $L \circ g^*$. Since $e^{L \circ g^*} = g^* = e^h$ and $L \circ g^*$, h both have imaginary part in $\left(-\frac{\pi}{2}, \frac{\pi}{2}\right)$, we infer that $L \circ g^* = h$. Thus finally

$$E \circ g^* = e^{irL} \circ g^* = e^{irL \circ g^*} = e^{irh} = g \circ E.$$

(iii) *Show that either g has the form $g(z) \equiv \alpha z$ or $g(z) \equiv \alpha/z$ for some unimodular constant α, or else g^* satisfies*

$$g^*(\lambda z) = g^*(z) \qquad \forall z \in \mathbb{H}.$$

Hints: For any $R_1 < \rho < R_2$ and $0 \le t \le 1$ we have $-\frac{\pi}{2} < \frac{\log \rho}{r} < \frac{\pi}{2}$ and so

$$\begin{aligned} g(\rho e^{2\pi i t}) &= g\left(\exp\left(ir\left[\frac{2\pi t}{r} - \frac{i}{r}\log\rho\right]\right)\right) = g\left(E\left(\exp\left(\frac{2\pi t}{r} - \frac{i}{r}\log\rho\right)\right)\right) \\ &\overset{\text{(ii)}}{=} E\left(g^*\left(\exp\left(\frac{2\pi t}{r} - \frac{i}{r}\log\rho\right)\right)\right). \end{aligned}$$

Consequently, by definition 4.2 of index

$$\mathrm{Ind}_{g\circ C(0,\rho)}(0)=\frac{1}{2\pi i}\left[irL\left(g^*\left(\exp\left(\frac{2\pi}{r}-\frac{i}{r}\log\rho\right)\right)\right)-irL\left(g^*\left(\exp\left(-\frac{i}{r}\log\rho\right)\right)\right)\right].$$

This is a continuous integer-valued function of ρ, hence is constant, say n. Thus

$$L\left(g^*\left(\exp\left(\frac{2\pi}{r}-\frac{i}{r}\log\rho\right)\right)\right)=\frac{2\pi n}{r}+L\left(g^*\left(\exp\left(-\frac{i}{r}\log\rho\right)\right)\right),\qquad R_1<\rho<R_2.$$

In other words, setting $\theta := -\frac{\log\rho}{r}$ and exponentiating,

$$g^*\left(e^{\frac{2\pi}{r}}e^{i\theta}\right)=e^{\frac{2\pi n}{r}}g^*\left(e^{i\theta}\right),\qquad -\frac{\pi}{2}<\theta<\frac{\pi}{2}.$$

From this, the definition of λ, and the Uniqueness Theorem 5.62 we infer that

$$g^*(\lambda z)=\lambda^n g^*(z)\qquad \forall z\in\mathbb{H}. \tag{1}$$

Now according to 6.24(i) the integer n must be 0 or 1 or -1. If we assume that g does not have either of the special forms indicated, then by 6.24(ii) $n=0$ and equation (1) is the desired conclusion about g^*.

(iv) *Show that if g has neither of the forms $g(z)\equiv\alpha z$ or $g(z)\equiv\alpha/z$ (α a unimodular constant), then the iterates $(g^*)^{[n]}$ converge locally uniformly in $\mathbb{H}$ to a constant $b^*\in\overline{\mathbb{H}}\setminus\{0\}$.*

Hints: From (iii) we have $g^*(\lambda^k z)=g^*(z)$ for all $z\in\mathbb{H}$ and $k\in\mathbb{N}$. In particular, g^* is *not* a conformal automorphism of $\mathbb{H}$. Moreover, for the constant c of 6.27(ii) we have

$$c=\lim_{k\to\infty}\frac{g^*(\lambda^k)}{\lambda^k}=\lim_{k\to\infty}\frac{g^*(1)}{\lambda^k}=0. \tag{2}$$

Let $\phi(w):=\frac{1+w}{1-w}$ ($w\in\mathbb{D}$) and form $G^*:=\phi^{-1}\circ g^*\circ\phi$ as in 6.28. Then G^* is not a conformal automorphism of $\mathbb{D}$. If G^* has a fixed-point a^* in $\mathbb{D}$, then by 7.44(i) the sequence $(G^*)^{[n]}$ converges locally uniformly in $\mathbb{D}$ to a^*. If G^* has no fixed-point in $\mathbb{D}$, then by 7.56 the sequence $(G^*)^{[n]}$ converges locally uniformly in $\mathbb{D}$ to the Denjoy–Wolff fixed-point a^* of G^* in $\partial\mathbb{D}$. According to 6.28(x), if this point is 1, then $c\geq 1$. Therefore, by (2), $a^*\neq 1$. Now the function $g^{**}(z):=\frac{1}{g^*(1/z)}$ also maps $\mathbb{H}$ holomorphically into $\mathbb{H}$ and satisfies $g^{**}(\lambda z)=g^{**}(z)$ for all $z\in\mathbb{H}$, and so the associated constant c for g^{**} is also 0. The function $G^{**}:=\phi^{-1}\circ g^{**}\circ\phi$ is just $G^{**}(w)=-G^*(-w)$. [Verification simply involves repeated use of the identities $\phi(\omega)=1/\phi(-\omega)$

and $\phi^{-1}(w) = -1/\phi(\omega)$,valid for all ω at which both ϕ and ϕ^{-1} are defined, namely, all $\omega \in \mathbb{C} \setminus \{-1, 1\}$.] Hence $(G^{**})^{[n]}$ converges locally uniformly in $\mathbb{D}$ to $-a^*$ and by the above argument, $c = 0$ implies $-a^* \neq 1$. Since ϕ maps $\overline{\mathbb{D}} \setminus \{-1, 1\}$ homeomorphically onto $\overline{\mathbb{H}} \setminus \{0\}$, it follows in either case that the sequence $(g^*)^{[n]}$ converges locally uniformly in $\mathbb{H}$ to $\phi(a^*) \in \overline{\mathbb{H}} \setminus \{0\}$.

(v) *Show that if g has neither of the forms $g(z) \equiv \alpha z$ or $g(z) \equiv \alpha/z$ (α a unimodular constant), then the iterates $g^{[n]}$ converge locally uniformly in A to a constant.*

Hints: L is continuous on $\overline{\mathbb{H}} \setminus \{0\}$ and (ii) says that $g^{[n]} \circ E = E \circ (g^*)^{[n]}$ for every $n \in \mathbb{N}$. Therefore from (iv) it follows that

$$g^{[n]}(E(z)) = \exp\Big(irL\Big((g^*)^{[n]}(z)\Big)\Big) \to e^{irL(b^*)} = E(b^*) \in \bar{A}$$

locally uniformly in $z \in \mathbb{H}$. Because E is a continuous and open mapping, each point of $A = E(\mathbb{H})$ has a neighborhood V which is the E-image of a compact subset of $\mathbb{H}$. It follows that $g^{[n]} \to E(b^*)$ uniformly in V.

Remarks: HEINS [1941b] uses (1) and 6.27 to infer directly that $|n| \leq 1$ and that if equality holds, then g has one of the special forms indicated. Thus the results of 6.24 are not needed. The behavior of the iterates $\{g^{[n]}\}$ when g has one of the special forms (i) $g(z) \equiv \alpha z$ or (ii) $g(z) \equiv \alpha/z$ is, of course, trivial: In case (i) the set $\{\alpha^n : n \in \mathbb{N}\}$ is either a finite subgroup of $\mathbb{T}$ or is dense in $\mathbb{T}$ (see the hints to 7.55). In case (ii) $g^{[2n]}$ is the identity map and $g^{[2n+1]} = g$ for all $n \in \mathbb{N}$. It should be noted that these special forms are just exactly all the conformal automorphisms of A, as was proved in 6.18(v).

Notes to Chapter VII

The convergence concept appropriate to the theory of holomorphic functions took some time to be born. It was absent from Cauchy's great corpus of work (but see his [1853]) and a case can be made that he never appreciated the failure of pointwise convergence to preserve the class of functions his theorems were all about. In view of how non-trivial it is to construct pointwise convergent sequences of holomorphic functions which are not also locally uniformly convergent (cf. 7.18 and 8.15), this is not surprising. It is true that in the context of non-standard analysis, which has legitimized infinitesimals and infinitely large numbers, many convergence and continuity questions evaporate. [This is because hypotheses operate over this larger class and hence may be more or less restrictive than these same expressions in $\mathbb{R}$ or $\mathbb{C}$, and in addition more candidates are now available to substantiate existence claims; the situation is analogous to the different consequences that the single concept of differentiability has in $\mathbb{R}$ and in $\mathbb{C}$.] Despite Cauchy's significant role in rigorizing analysis, he had no formal definition of $\mathbb{R}$, and his definitions of continuity and convergence are sufficiently vague that they can be construed as proto-non-standard analysis, thereby disqualifying "counterexamples" like that of N. H. Abel (cited in the chapter III notes). Other Cauchy apologists find uniformity (possibly in the form of continuous convergence) encoded in his vague definitions. On both points see pp. LXXXV, ff. of Bottazzini's introduction to CAUCHY [1821] and Lützen's chapter 6 in JAHNKE, *et al.* [1999]. A considerable literature, sometimes bombastic and tendentious, excoriating or exculpating Cauchy exists, including a 285-page polemic (whose coordinates I will mercifully withhold from the reader). Non-belligerents in this "war of the frogs and the mice" may dismiss the matter by paraphrasing Horace on Homer: even good old Cauchy sometimes nods; but curious readers can glimpse its forensic fervor and exegetical excesses in ARANTES SAD, *et al.* [2001] and SPALT [2002]. And readers wanting a deeper analysis of the controversy can look at chapter VI of SCHUBRING [2005], a scholarly examination of the numerous subtly different and time-dependent definitions of "infinitely small", "convergence", and "continuity" used by Cauchy and his colleagues who mutually influenced one another. It aims to clarify, without anticipating non-standard analysis, what Cauchy meant in 1821 when he asserted that a limit of continuous functions is a continuous function, his path to the 1853 revision, and the dawning recognition of a larger issue, the legitimacy of commuting two limit operations.

Weierstrass is generally credited with having introduced the concept of uniform convergence, around 1840. For more history see pp. 96–98 of REMMERT [1991a], pp. 202–208 of BOTTAZZINI [1986], and especially GRATTAN-GUINNESS [1986].

There is another way to look at uniform convergence in $H(U)$: We can map $\mathbb{C}$ to the Riemann sphere $\mathbb{C}_\infty$ by stereographic projection and for a, $b \in \mathbb{C}$ let $\chi(a, b)$ denote the (chordal) distance between their images on this sphere. This is arduously computable to be $2|a-b|(1+|a|^2)^{-\frac{1}{2}}(1+|b|^2)^{-\frac{1}{2}}$ from the P_- formula preceding 4.63. Then set $\chi(f, g) := \sup\{\chi(f(z), g(z)) : z \in U\}$ for f, $g \in H(U)$. The great advantage of this is that it handles *meromorphic* functions (which at poles take the value "∞"), as well as holomorphic ones, with equal ease.

E. LANDAU [1918] points out that 7.4 is a simple consequence of well-known earlier results of Mittag-Leffler and Jensen and generally downplays BLASCHKE [1915] by giving elementary proofs of the other results there. (History, however, has dealt more kindly with Blaschke on this point.) The uniqueness condition $\sum(1-|z_j|) = \infty$ also insures that a formula like that in 5.73(vii) for recovering f from the sequence $\{f(z_j)\}$ is available; see TOTIK [1984]. The use of Blaschke products to get bounds as in 6.9(ii) and to secure convergence as in 7.4 has been discovered independently by other writers. See especially the formula on p. 57 of WHITTAKER [1935] and the way this is used by BOWEN and MACINTYRE [1954] and WHITTAKER [1970b]. This device is also used very effectively by V. I. LEVIN [1934]. A related technique, expansion of a function into a series of finite Blaschke products with remainder to prove 7.5, occurs in LAMMEL [1958].

The important convergence theorem 7.5 is due independently to VITALI [1903], [1904] and PORTER [1904–1905b]. There is a beautiful proof based on 7.9 (itself proved *en route*) in LINDELÖF [1913] and, essentially the same proof, in JENTZSCH [1918]; the reader will find it outlined in the 3.35 hints. These proofs were probably inspired by MONTEL [1907], pp. 300–301 and [1910], pp. 21 ff. For a similar and comparably elegant proof see CARATHÉODORY and LANDAU [1911] (who point out gaps in the treatments by Vitali and Montel). For other proofs see JACOBSTHAL [1927], BOWEN and MACINTYRE [1954], LAMMEL [1958] and for interesting related results LINDWART and PÓLYA [1914] and SHISHA [1965]. For history and references to the precursors of Vitali and Porter (like STIELTJES [1894], OSGOOD [1901–02]) see §58 of BIEBERBACH's encyclopedia article. A good historical discussion may also be found in OSTROWSKI [1923].

The terminology associated with the 7.5 phenomenon is colorful and suggestive. It is said that convergence on a set S with a limit point in Ω "induces" convergence throughout Ω, or "propagates" the convergence into the whole of Ω, or "infects" the whole of Ω with convergence. A particularly vivid metaphor is that of seeding a crystal.

There are extensions of the Vitali–Porter theorem in which the set on which convergence is hypothesized, instead of having a limit point in the region, is allowed to accumulate on the boundary (e.g., 7.4) or even lie wholly on the boundary (e.g., 5.45(iii)): See MONTEL [1907], p. 309, [1917], KHINTCHINE [1922–24], [1923], OSTROWSKI [1922/23], F. RIESZ [1922/23a], KUNUGUI [1941–43]. The conclusion also follows if convergence (in a certain sense) is hypothesized on a subset of the boundary having positive measure. See KOVANKO [1924], HARTOGS [1928], pp. 157 ff. of BIEBERBACH's text [1931] or p. 456 of DIENES [1931]; also pp. 397 ff. of GOLUZIN [1969]. For related matter see WHITTAKER [1970b].

In Chapter XIV we shall see that the positive integers are a uniqueness set for entire functions with certain growth restrictions. (Cf. 6.11 and 11.38.) When these conditions are fulfilled uniformly for a sequence, then the convergence of the sequence on the positive integers implies its local uniform convergence in the whole plane. (See, e.g., 15.3.5 of HILLE [1962].)

From 12.27 and 12.26 follows a dramatic strengthening of the Vitali–Porter convergence theorem: It suffices for the conclusion in 7.5 that there be two distinct complex numbers which are not in the range of any of the functions F_n. See *Satz* VI of CARATHÉODORY and LANDAU [1911].

7.6 is due to MONTEL [1907] (pp. 298–302), with an independent proof by KOEBE [1908], p. 349. The proof outlined in 7.13 is based on earlier ideas of C. Arzelà and G. Ascoli from a larger context. They focused on the equicontinuity phenomenon [which 7.13(i) shows to be a consequence of local uniform boundedness for *holomorphic* functions] and showed that for a compact metric space X, a subset $\mathcal{F}$ of $C(X)$ is pre-compact (in the uniform norm) if and only if it is equicontinuous on X and for each $x \in X$, $\{f(x) : f \in \mathcal{F}\}$ is a bounded subset of $\mathbb{C}$. See STROMBERG [1981] or KLEPPNER [1967], where a more general version is proved. Generally, 7.8 can be coupled with any uniqueness theorem to yield a convergence theorem. DENJOY [1929] indicated how, given distinct $z_n \in \mathbb{D}$ and arbitrary $w_n \in \mathbb{D}$, to construct auxiliary parameters $w_n^* \in \overline{\mathbb{D}}$ such that the equality

$$(*) \qquad \sum_{n=1}^{\infty} \frac{1-|z_n|}{1-|w_n^*|} = +\infty$$

is necessary and sufficient that there be *exactly one* holomorphic $f\colon \mathbb{D} \mapsto \mathbb{D}$ satisfying $f(z_n) = w_n$ for all $n \in \mathbb{N}$. R. NEVANLINNA [1929d] pointed out that this entails the following generalization of 7.4: The condition $(*)$ is necessary and sufficient that every sequence of holomorphic self-maps of $\mathbb{D}$, which at each z_n converge to w_n, actually converge locally uniformly throughout $\mathbb{D}$. For details see pp. 37 ff. of R. NEVANLINNA [1929c].

7.10 is from MONTEL [1912] (see also [1925], p. 247). Alternative proofs of it can be based on 5.36 or 5.86. Earlier HURWITZ [1889] had proved (by the Rouché argument in 7.12) the result under the stronger hypothesis that $\{f_n\}$ be locally uniformly convergent; see also TONELLI [1910b]. Notice that the positive integer multiples of a fixed holomorphic function illustrate the need for the hypothesis that the limit at z_0 be *non-zero* in 7.14(iii). In proving 7.14(i) the Cauchy-Schwarz integral formula 5.18(∗) could as easily and naturally be employed as 6.35. The result 7.15 is due to HARNACK [1886]. 7.16(i) is asserted in KOEBE [1909], IV. *Mitteilung*, p. 329 as being a consequence of 7.6. A proof occurs in OSGOOD [1928], p. 697. See footnote 524 of LICHTENSTEIN [1919] for the geneology of this result. In contrast to (i), there is no harmonic-function analog of the Vitali–Porter induced-convergence (7.5), as the example $h_{2n}(z) := 0, h_{2n-1}(z) := \operatorname{Im} z$ $(z \in \mathbb{C})$ illustrates; but see 10.9(iv). 7.16(iii) is due to PLESSNER [1923]; see also F. and R. NEVANLINNA [1922], G. C. EVANS [1923], EVANS and BRAY [1923], LITTLEWOOD [1928] and FICHTENHOLZ [1929]. These authors show that for a harmonic function h in

$\mathbb{D}$ the boundedness of $\{\int_0^{2\pi} |h(re^{i\theta})|\, d\theta : 0 \leq r < 1\}$ is necessary and sufficient for h to have a Poisson integral representation (using, generally, a Riemann-Stieltjes integral). See also the last three paragraphs of the Chapter VIII notes. PRIVALOV [1935] contains a subharmonic function version of 7.16(iv) and in [1936] he shows that a family $\mathcal{F}$ of harmonic functions is locally uniformly bounded above in $\mathbb{D}$ if and only if the set $\{\int_0^{2\pi} u^+(re^{i\theta})\, d\theta : 0 \leq r < 1,\ u \in \mathcal{F}\}$ is bounded. A beautiful half-plane version of 7.16(iii) and one for strips may be found in KURAN [1970], BRAWN [1973].

For generalizations of 7.17 see pp. 15 ff. of F. and R. NEVANLINNA [1922], ASCOLI [1928b], PRIVALOV [1935] and BRUCKNER, LOHWATER and RYAN [1969]. The Phragmén–Lindelöf device in 5.46 and 7.17 will be explored further in Chapter XIV. HARTOGS and ROSENTHAL [1928], [1931] make an exhaustive topological study of the set U_0 of 7.18. It is, e.g., simply-connected if U is. (To see this, use the Maximum-Modulus Principle and the criterion 4.71(viii) for simple-connectivity.) See also ULLRICH [1994a].

For another proof of 7.18 see MONTEL [1932a] and pp. 315 ff. of MONTEL [1907]. On 7.21 and 5.43 see DE LA VALLÉE POUSSIN [1892–93], SEVERINI [1901], OSGOOD [1901–02], W. H. YOUNG [1910], BÔCHER [1910–11] and HARDY [1914]. The statement and proof of 7.21 are from MONTEL [1910], pp. 28, 29. For a nice related result see CUNNINGHAM [1967]. The weakest known hypotheses which yield the conclusion in 7.21, together with clever counterexamples when they are further weakened, are elucidated by MATTNER [2001].

The formulation of Cauchy's Theorem in 7.22, emphasizing the role of the index, can be traced to lectures Emil Artin gave at Notre Dame University in 1944. See ARTIN [1965]. A proof using the grid technique employed in 8.1 was given by WYLER [1965] and by PALKA [1991], pp. 188–191. There is corresponding to 7.22 a cohomology version of Cauchy's theorem; for that see p. 124 of NARASIMHAN [1985].

For an interesting alternative proof of 7.23 based on approximation by Bernstein polynomials in two variables, see FLATTO and SHISHA [1973] and for yet another proof EBERLEIN [1975] (see remarks following 5.2). A nice self-contained treatment occurs in REDHEFFER [1969] and on pp. 87–89 of CONWAY [1978]. See also GONZÁLEZ–VELASCO [1980]. KOPFERMANN [1981] implements the hint to 7.23(i) with a proof somewhat different from ours for 4.12. LAX [2007] proves the special case γ_2 constant of 7.23(i) by "mollifying" the homotopy to make it twice continuously differentiable in both its coordinates (cf. then 2.19(v)*).

7.26, 7.31 and 7.42, 7.43 and 7.47 were culled from the pages of BEHNKE and SOMMER [1965]. For 7.27 and 7.28 see also pp. 29–31 of MONTEL [1933b].

A sharp quantitative version of 7.27 and sharp, explicit bounds in 7.30 and 7.31(iv) will be produced in Chapter XVIII. (See p. 111 of E. LANDAU [1929a].) Though 7.28 occurs in KOEBE [1907] (pp. 204–205) and almost in JOHANSSON [1910] (Satz C′), it is contained in an earlier stronger result of HURWITZ [1904] (namely, 12.17 below) in which moreover a lower bound for the constant α occurs. With $\alpha = \frac{1}{10}$ the result occurs in PLEMELJ [1912]. KOEBE [1909b], p. 215 conjectured that the largest value for α is $\frac{1}{4}$ and that that is attained (only) when $f(\mathbb{D})$ is the plane slit along $\left(-\infty, -\frac{1}{4}\right]$ or a rotation thereof. The result 7.29(iii) (with an unspecified universal constant replacing $1/3\sqrt{3}$) is from BLOCH [1924b] and [1925]. Many proofs and refinements followed: FEKETE [1925], [1927b], VALIRON [1926], [1930], E. LANDAU [1926], [1929b], [1940], GRÜNWALD and TURÁN [1936–1937], A. J. MACINTYRE [1938] (whose bibliography mentions other references), ESTERMANN [1971], and MINDA [1992]. The proof in the text (essentially that on pp. 616–618 of E. LANDAU [1929b]) was taken from MILLOUX [1953]. AHLFORS [1938] and HEINS [1962a] proved that if $f \in H(\mathbb{D})$ and $|f'(0)| = 1$, then for some $\varepsilon_f > 0$, some region $\Omega_f \subset \mathbb{D}$ is mapped conformally onto a disk of radius $\varepsilon_f + \frac{\sqrt{3}}{4}$. Their proofs are not elementary but BONK [1990] gave a stunning, elementary proof (wholly accessible to any reader who followed 7.29) of the (stronger) result that $\varepsilon_f = 10^{-14}$ always works. Aiming only to get $\varepsilon_f \geq 0$, REMMERT [1998] simplifies this proof a bit more.

Other covering theorems will be presented and discussed in Chapter XVIII and its notes. See also MONTEL [1929] and chapter VII of [1933b] for covering theorems and the paper of E. LANDAU [1929b] and its detailed historical notes.

Buried in the hints to 7.29(i) is an elementary and oft-used result known as the Noshiro–Warschawski Theorem, after two independent discoverers: If f is holomorphic in the convex region Ω and $\operatorname{Re} f'$ is zero-free in Ω, then f is one-to-one in Ω. Other proofs and historical references will be found in AVKHADIEV and AKSENT'EV [1975]. JANIEC [1989] gave this generalization: If $\phi\colon \mathbb{R} \mapsto \mathbb{R}$ is continuous and $\operatorname{Re} f' + \phi(\operatorname{Im} f) \cdot \operatorname{Im} f > 0$, then f is one-to-one. A more elementary generalization is R. K. WILLIAMS [1980].

7.31(iv), proved via 7.6, occurs first in KOEBE [1909a], p. 73, then later with a different proof in KOEBE [1910], pp. 46–52. See also JOHANSSON [1910], PICK [1916b] and LÖWNER [1917]. For related harmonic function results see OSGOOD [1923–24], PERKINS [1925–26], MANDELBROJT [1929] and PLEMELJ [1933]; for locally univalent functions see MONTEL [1937].

7.32(i) is from MONTEL [1912], p. 508 (see also p. 70 of his book [1927]) and 7.32(ii) is from MONTEL [1912], p. 533. Both results were rediscovered by PRIVALOV [1924e], whose proofs are those suggested in the text. 7.32(iii) was proposed as a problem in vol. 32 (1923), p. *16* of *Jahresbericht der Deutschen*

Mathematiker-Vereinigung by LÖWNER and RADÓ.

7.33 is from BOWEN [1964–65a], though his proof is different; 7.34 is from the same paper and represents an extension of a result of CARTWRIGHT [1962], who took the hypothesis $\lim_{y\to\infty}|f(x_0+iy)| = 1$ for some $x_0 \in (a,b)$. See 12.42 and the Chapter XII notes for related material. Cartwright's proof is a slight variation on that of 12.41(i). For nice extensions of 7.33 and 7.34 to subharmonic functions see BEARDON [1971].

Actually Weierstrass only did the case of 7.35 where $A = \{a_1, a_2, \ldots\}$ and $|a_n| \to \infty$; this case, treated via infinite products, was also considered by PICARD [1881]. A considerable generalization of it was achieved by BARTH and SCHNEIDER [1972]. They showed that if $S_k \subset \mathbb{C}$ has no limit point in $\mathbb{C}$ (for each $k \in \mathbb{N}$), then there exist $n_k \in \mathbb{N}$ and a non-polynomial entire function f such that $f^{(n_k)}(S_k) = 0$ for each $k \in \mathbb{N}$ (a less general version by more elementary means in RUBEL [1990]). HURWITZ [1891] showed that given $a_n \in \mathbb{C}$ with $|a_n| \to \infty$, there exist *rational* complex c_n such that $f(z) := \sum_{n=0}^{\infty} c_n z^n$ is entire and has zero-set $\{a_1, a_2, \ldots\}$.

The general case of 7.35 occurs in MITTAG-LEFFLER [1884a] and it is essentially the latter's proof which is given in the text. 7.36 is from BLASCHKE [1915]; for another proof see p. 133 of SCHUR [1918] and FATOU [1923]. See the fourth from the last paragraph of the Chapter XII notes. One of the first applications of 7.35 was to deciding which regions Ω in $\mathbb{C}$ support a holomorphic function which cannot be extended to a holomorphic function in any strictly larger region. Mittag-Leffler confirmed an 1880 conjecture of Weierstrass that every $\Omega \neq \varnothing$ has this property. Then RUNGE [1885] gave the following simple and elegant proof: An elementary construction produces a subset A of Ω whose set of limit points is exactly $\partial\Omega$: For each $n \in \mathbb{N}$ cover the compact set $\bar{D}(0,n) \cap \partial\Omega$ with finitely many disks centered in it and having radii $1/n$. In each disk pick a point of Ω and denote by A_n the set of points chosen. Then it is not hard to see that $A := \bigcup_{n\in\mathbb{N}} A_n$ has the property wanted. Consider an $f \in H(\Omega)$ having zero-set A. If region $U \supsetneq \Omega$, then U contains some point $z_0 \in \partial\Omega$ (why?). Hence if $F \in H(U)$ extends f, the zeros of F have a limit point at z_0, so by 5.62 $F = 0$ throughout the region U, hence throughout Ω. But the only zeros of $f = F$ in Ω are the points of A and $A \neq \Omega$. For more on this theme of "domains of holomorphy" see §2, chapter 5 of REMMERT [1998].

In 7.39 the restriction $r \notin C$ is necessitated only by our insisting on Riemann integration. The integral there exists as a Lebesgue integral (and even as an improper Riemann integral) for $r \in C$ and $(*)$ holds with the condition $r \notin C$ deleted. See, for example, pp. 151–152 of SANSONE and GERRETSEN [1960].

7.41(i) is due to SCHRÖDER [1870] and the proofs of (i) and (ii) are from pp. 344–345 of KOENIGS [1883] and from RITT [1920–21]. 7.41(iii) is stated in

FARKAS [1884] but the proof there is inadequate. A multi-dimensional version occurs in WAVRE [1926]. An interesting proof of 7.41(iii) which does not use Montel's compactness theorem is sketched on p. 109 of HERVÉ [1982]. Closely related to 7.41(iii) are results of EDELSTEIN [1961] and SHIH and YEH [1981], while an operator-valued function version of 7.41(iii) occurs in TAO [1985b] and a semigroup version in LORENTZEN [1990], [1996]. The hypothesis in (iv) that the F_n map into a common compact subset has been significantly relaxed by BEARDON, CARNE, MINDA & NG [2004].

The special case in the last sentence of 7.46 is due to BIEBERBACH [1913a]. The proof given for (i) and (ii) is from T. RADÓ [1924–26]. See also T. RADÓ [1922/23a] and SAKS and ZYGMUND [1971], p. 232 (also 12.48 below) for a generalization to unbounded regions. Generalizing (i), BEARDON and MINDA [2008b] investigate the existence of neighborhoods N of a, depending on a and Ω, such that $|f'| \leq 1$ throughout N. For other approaches to and extensions of 7.46(iii) see CARATHÉODORY [1932], pp. 779–786, H. CARTAN [1932], AUMANN and CARATHÉODORY [1934], HEINS [1941a], and HERVÉ [1951]. Proofs of 7.45 and 7.46 also occur in KAMOWITZ [1976]. Notice that when $\Omega = \mathbb{D}$, 7.44 and 7.46 follow very easily from Schwarz' lemma, without using compactness of the sequence of the iterates (JULIA [1918] and BEARDON and CARNE [1992]).

The hypothesis in 7.43–7.51 that Ω be bounded can be dramatically weakened; see 12.48.

7.53(i) is due to SHELL [1962]; the proof in the text is from A. J. MACINTYRE [1966]. The real-variable analog of this "stacked exponent" or "tower of powers" problem is rather elementary, was considered already by Euler and was settled long ago: For positive real a the sequence $a^{a^{\cdot^{\cdot^{\cdot}}}}$, meaning $a_1 := a$, $a_{n+1} := a^{a_n}$, converges to a real number if and only if $e^{-e} \leq a \leq e^{1/e}$. The interested reader will find this presented as a well-hinted problem sequence on p. 185 of STROMBERG [1981]. See also ANDREWS and LACHER [1977], CREUTZ and STERNHEIMER [1980], DE VILLIERS and ROBINSON [1986], ELSTRODT [1994], LÄNGER [1996], WOLTERMANN [1997], and J. ANDERSON [2004]. An extensive treatment with graphs and history is that of KNOEBEL [1981]. Another excellent expository account, covering the complex case as well, is RIPPON [1983]. The most definitive results to date for the complex case are those of BAKER and RIPPON [1983], [1984], [1985] (especially accessible to students) and [1989]. To make sense of the tower of exponents for complex a, they introduce the function $f_c(z) := e^{cz}$ ($c \in \mathbb{C}$ thought of as a logarithm of a) and consider the sequence $\{f_c^{[n]}(0)\}$. They show that if this sequence converges to a number in $\mathbb{C} \setminus \{f_c^{[n]}(0)\}$, then c lies in the closure of the cardioid $C := \{we^{-w} : w \in \mathbb{D}\}$ and that for every $c \in C$, the sequence does converge to a complex number, but for most $c \in \partial C$ the sequence does not converge!

For an alternative treatment of 7.53(ii), which also settles the case $z \in i\mathbb{R}$, see III.262 in PÓLYA and SZEGÖ [1972]. This result is a very special case of FATOU [1917], who treats arbitrary rational functions mapping $\mathbb{H}$ into $\mathbb{H}$.

For iterated square-roots see SCHUSKE and THRON [1961], [1962], GERBER [1973], ROHDE [1974] and WALKER [1983].

There is a classification scheme for conformal automorphisms f of $\mathbb{D}$ based on their fixed-points in $\overline{\mathbb{D}}$. In each case there is a canonical form for f involving these fixed-points and from it the behavior of $\{f^{[n]}\}$ is easily read off. This approach (glimpsed in the second proof of 7.54) requires some preliminary work and so overall is perhaps not as efficient as the first treatment in 7.54. On the other hand, it treats Möbius mappings generally, on the whole Riemann sphere $\mathbb{C}_\infty$. See, e.g., pp. 402–407 of PALKA [1991] and LIEBECK [1961]. (On the question of locating the limit point in 7.56, some information may be found in CRAIG and MACINTYRE [1967].) Although 7.54, 7.55 are well known folklore, proofs recently appeared in a Chinese journal; see **MR** 98c: 30010.

7.56 is due independently to DENJOY [1926] and WOLFF [1926]; a version for *rational* functions mapping $\mathbb{U}$ into $\mathbb{U}$ had been done by FATOU [1917]. Wolff gave two quite different proofs, the one in the text being his second. Here is an outline of a modification of his first proof by VALIRON [1931], [1954]. Suppose f has no fixed-point in $\mathbb{D}$. Then each subsequential limit of $\{f^{[n]}\}$ is constant (7.43), and (7.44) the set of these limits is connected, i.e., is an arc A of $\partial\mathbb{D}$. In case f is continuous on $\overline{\mathbb{D}}$, it is further easy to see that each point of A is a fixed-point of f. Consequently, by 5.75, since f is not the identity map, A must be a single point. In the absence of continuity on $\overline{\mathbb{D}}$ we proceed a little differently. A rather deep theorem of FATOU [1906], pp. 364–367 (mentioned also in the Chapter IX notes) asserts that for almost all (in the sense of Lebesgue measure) $\theta \in [0, 2\pi]$, $f(z)$ has a limit as z approaches $e^{i\theta}$ inside any triangle in $\mathbb{D}$ with vertex at $e^{i\theta}$. Calling this limit suggestively $f(e^{i\theta})$, Fatou further showed that if for two such functions f and g, $f(e^{i\theta}) = g(e^{i\theta})$, at almost every $e^{i\theta}$ in some non-degenerate arc of $\partial\mathbb{D}$, then $f = g$. Now set $L := [0, f(0)]$. A routine connectedness argument shows that for every $u \in A$ with at most two exceptions the interval $[ru, u]$ and the connected set $\bigcup_{n=N}^{\infty} f^{[n]}(L)$ have non-void intersection for every $r < 1$ and every $N \in \mathbb{N}$. This fact enables one to select $r_k = r_k(u) \uparrow 1$ such that $f(r_k u) \to u$ for each $u \in A$. It follows, since f is not the identity function, that the arc A must be a single point. A third proof, due to E. Vesentini, based on 6.28, will be found on pp. 80–81 of ABATE [1989]. These theorems of Fatou are also exploited by BAKER and RIPPON [1990] to give a very elegant proof of the following: If Ω is a bounded convex region, $z_0 \in \Omega$, $b < 1$ and

$\mathcal{F}$ is a family of continuous self-maps f of $\overline{\Omega}$ which are holomorphic in Ω and satisfy $|f'| \leq 1$, $|f'(z_0)| \leq b$, then for every sequence $\{f_n\} \subset \mathcal{F}$, the sequence $\phi_n := f_1 \circ f_2 \circ \cdots \circ f_n$ converges uniformly in $\overline{\Omega}$ to a constant. When all f map into a single compact subset K of Ω, but continuity on $\overline{\Omega}$ is dispensed with, we have a similar conclusion (analog of 7.41(iii)—see LORENTZEN [1990]).

7.56 and its generalizations to higher dimensions and beyond (e.g., to operator-valued holomorphic functions) are extensively treated in the literature. See especially VALIRON [1931], [1954], MONTEL [1933c] (final section), CRAIG and MACINTYRE [1967], FAN [1982], [1983], POMMERENKE [1979], [1981], TAO [1985a], §30 of GOEBEL and REICH [1984], ABATE [1989], BEARDON [1990] and SHOIKHET [2001]. The relation of angular derivatives to convergence of iterates in 7.56 is explored in depth by J. H. SHAPIRO [1993], in his chapter 5.

The proof of 7.58 in SHIELDS [1964] and that of 7.57 in MITCHELL [1979] uses fewer results from function theory but instead relies in an interesting way on elementary facts about compact topological semigroups. Theorem 1.3.22 of ABATE [1989] extends 7.58 to regions other than $\mathbb{D}$. An extension of 7.58 to two complex dimensions is given by EUSTICE [1972] and SUFFRIDGE [1974]. The fixed-point claim in 7.57 (generalized to $\mathbb{C}^n$) is proved by an elegant use of the Haar integral on $\mathcal{G}$ in 3.2 of RUDIN [1982]. COWEN [1984] classifies the pairs f, g of holomorphic self-maps of $\mathbb{D}$ which commute under composition. [A classical and elementary result of this kind states that $f, g \in \operatorname{Aut}(\mathbb{D})$, not the identity map, commute if and only if they have the same fixed-points.]

Let f be a non-identity holomorphic self-map of $\mathbb{D}$. According to 6.4(ii), 7.54 and 7.56, either f has a unique fixed-point u_f in $\mathbb{D}$, or else there exists a unique point $u_f \in \partial\mathbb{D}$ to which the sequence $\left\{f^{[n]}\right\}$ converges. From 6.28(x) it follows that in the latter case $f(z) \to u_f$ whenever z converges to u_f inside a triangle in $\mathbb{D}$ with vertex at u_f. This means that the extension of f to $\mathbb{D} \cup \{u_f\}$ which fixes u_f is almost continuous. Hence the following result of BEHAN [1973] generalizes 7.58: If f, $g\colon \mathbb{D} \mapsto \mathbb{D}$ are holomorphic functions, neither a conformal automorphism of $\mathbb{D}$ having two fixed-points on $\mathbb{T}$, and $f \circ g = g \circ f$, then $u_f = u_g$. [The exceptional case is not illusory: $f(z) := (z-1/2)/(1-z/2)$ and $g(z) := -f(-z)$ each fix 1 and -1 and they commute, yet $u_f = -1$ while $u_g = 1$.]

In [1941b] HEINS goes well beyond the beautiful result 7.59(v) and proves a definitive result about arbitrary multiply-connected regions, which appears in 12.50 below. See also HERVÉ [1950] and chapter 1.3 of ABATE [1989].

For other interesting techniques and problems in iteration theory see RÅDSTRÖM [1953], the papers of I. N. Baker (only a few of which are listed in the bibliography), VALIRON [1931], JULIA [1918], [1919b], FATOU [1919], [1920], [1926],

CREMER [1925] (survey), WOLFF [1927], [1929], LIEBECK [1961], ELJOSEPH [1968], SHORT [1993] (the latter five being relevant to 7.51), the book of MONTEL [1957] and chapter 8 of GROSS [1972]. Since $f^{[n+m]} = f^{[n]} \circ f^{[m]}$, a natural generalization of iteration is a family $\{f_t : t > 0\}$ of holomorphic self-maps of $\mathbb{D}$ having the *semigroup* property $f_{s+t} = f_s \circ f_t$. Many results concerning $\lim_{n\to\infty} f^{[n]}$ remain valid for $\lim_{t\to\infty} f_t$. The question of embedding a given f as f_1 in such a family is particularly interesting. For this extensive theory see SHOIKHET [2001]. In the era of the computer, iteration has acquired a new vogue and new viewpoints, those of dynamical systems and fractals. The literature is already too vast to survey, so I shall mention only BEARDON [1991], CARLESON and GAMELIN [1993], STEINMETZ [1993], and MILNOR [1999].

D. S. ALEXANDER [1994] traces the development of iteration theory from the papers of Schröder and Koenigs in the 1870's and 1880's up to the monumental works of FATOU [1919], [1920] and JULIA [1918] which are the basis of the whole modern theory. Impetus for them was a 3000-franc prize offered by the French Academy in 1915 for a work on iteration; it was awarded to the war hero Julia in 1918. A minor priority issue arose (see chapter 9, *op. cit.* and **MR** 2011k:01008) because the work of the two, though independent, overlapped significantly.

Chapter VIII

Polynomial and Rational Approximation—Runge Theory

§ 1 The Basic Integral Representation Theorem

The power of "integral representations" was demonstrated in 3.8: they lead to local power series expansions. If the (Riemann) integrals involved are approximated by Riemann sums, we get global rational approximations to the function. (See 8.8 below.) If further the domains are properly disposed in $\mathbb{C}$, the "poles" in these rational functions can be "shoved to infinity" and the rational functions thereby approximated by polynomials. Global polynomial approximants are thus produced. They are a very powerful tool for investigating holomorphic functions: often a theorem is easy for polynomials and remains valid under local uniform convergence, hence its validity passes over to holomorphic functions generally.

So our first goal is to get a global integral representation for a holomorphic function, to replace the local ones in 5.18. The development is routine but is almost completely combinatorial and so rather unpleasant to write out: certain integrals occur in oppositely oriented pairs and cancel each other (as in 2.20(i) and Remarks (ii) following 5.2); allegedly all that remains are integrals of such-and-such a kind. Our main task is to keep track of the cancellations and confirm this.

Theorem 8.1 (The Basic Integral Representation Theorem) *Let U be an open subset of $\mathbb{C}$, K a non-void compact subset of U. Then there exist non-degenerate intervals $\gamma_1, \gamma_2, \ldots, \gamma_k$ lying in $U \setminus K$, intersecting pairwise only in endpoints and parallel to the coordinate axes such that for $\gamma := (\gamma_1, \ldots, \gamma_k)$ and all $z \in \mathbb{C} \setminus \gamma$ the integral $(2\pi i)^{-1} \int_\gamma d\xi/(\xi - z)$ is either 1 or 0; it is 1 for*

R. B. Burckel, *Classical Analysis in the Complex Plane*, Cornerstones,
https://doi.org/10.1007/978-1-0716-1965-0_9

all $z \in K$, 0 *for all* $z \in \mathbb{C} \setminus U$. *In addition*

(1) $$f(z) = \frac{1}{2\pi i} \int_\gamma \frac{f(\xi)}{\xi - z}\, d\xi \qquad \forall f \in H(U),\ z \in K.$$

Proof: We can suppose $U \neq \mathbb{C}$. Let

$$0 < \delta := \frac{1}{2}\text{dist } (K, \mathbb{C} \setminus U)$$

and for every pair of integers n and m set

$$\begin{aligned} Q(n,m) &:= [n\delta, (n+1)\delta] \times [m\delta, (m+1)\delta] \\ v(n,m) &:= n\delta + im\delta. \end{aligned}$$

[The notation is meant to suggest *Quadrat* = square (German) and vertex.] Let $\mathcal{Q}$ denote the set of all $Q(n,m)$ which meet K. This is a finite set, by compactness of K. On account of the size of δ we then have

(2) $$K \subset \bigcup \mathcal{Q} \subset U.$$

Let $\mathcal{S}$ denote the set of intervals (sides)

$$\begin{aligned} &[v(n,m), v(n+1,m)]\,, \quad [v(n+1,m), v(n+1,m+1)]\,, \\ &[v(n+1,m+1), v(n,m+1)]\,, \quad [v(n,m+1), v(n,m)] \end{aligned}$$

such that $Q(n,m) \in \mathcal{Q}$. Delete from $\mathcal{S}$ each pair of the form $[a,b]$, $[b,a]$ that it contains, and let $\mathcal{S}_0$ denote the family of intervals which remain. If g is continuous on $\bigcup\{\partial Q : Q \in \mathcal{Q}\}$, then it is clear from the definition 1.12 of ∂Q that

(3) $$\sum_{Q \in \mathcal{Q}} \int_{\partial Q} g = \sum_{S \in \mathcal{S}} \int_S g.$$

Furthermore, since $\int_{[a,b]} g = -\int_{[b,a]} g$, it is clear that

(4) $$\sum_{S \in \mathcal{S}} \int_S g = \sum_{S \in \mathcal{S}_0} \int_S g.$$

Notice that if an interval $[a,b]$ in $\mathcal{S}$ meets K, then the square of which $[a,b]$ is a side and the square of which $[b,a]$ is a side both lie in $\mathcal{Q}$. Therefore, $[b,a]$ is also in $\mathcal{S}$, hence the pair $[a,b]$, $[b,a]$ is absent from $\mathcal{S}_0$. That is,

(5) the intervals in $\mathcal{S}_0$ lie in $U \setminus K$.

These intervals are the $\gamma_1, \ldots, \gamma_k$ sought. For $z \in \mathbb{C} \setminus \cup \mathcal{S}_0 = \mathbb{C} \setminus \gamma_1 \cup \cdots \cup \gamma_k$, define

$$(6) \qquad W(z) := \sum_{S \in \mathcal{S}_0} \frac{1}{2\pi i} \int_S \frac{d\xi}{\xi - z} = \sum_{j=1}^{k} \frac{1}{2\pi i} \int_{\gamma_j} \frac{d\xi}{\xi - z}.$$

For now we aim to show that the range of this function is $\{0, 1\}$; later it will prove to be a winding number, as the notation is meant to suggest. For any $Q \in \mathcal{Q}$, $(2\pi i)^{-1} \int_{\partial Q} d\xi/(\xi - z)$ is 1 if $z \in \overset{\circ}{Q}$ and is 0 if $z \in \mathbb{C} \setminus Q$, according to 4.5. Therefore

$$(7) \qquad \sum_{Q \in \mathcal{Q}} \frac{1}{2\pi i} \int_{\partial Q} \frac{d\xi}{\xi - z} = \begin{cases} 1 & \text{if } z \in \bigcup\{\overset{\circ}{Q} : Q \in \mathcal{Q}\} \\ 0 & \text{if } z \in \mathbb{C} \setminus \cup \mathcal{Q}. \end{cases}$$

Each z in (7) lies outside $\bigcup\{\partial Q : Q \in \mathcal{Q}\}$, and so $\xi \mapsto 1/(\xi - z)$ is an eligible g for (3) and (4), which therefore transform (6) and (7) into

$$(8) \qquad W(z) = \sum_{j=1}^{k} \frac{1}{2\pi i} \int_{\gamma_j} \frac{d\xi}{\xi - z} = \begin{cases} 1 & \text{if } z \in \bigcup\{\overset{\circ}{Q} : Q \in \mathcal{Q}\} \\ 0 & \text{if } z \in \mathbb{C} \setminus \cup \mathcal{Q}. \end{cases}$$

The union of the two sets in (8) is $\mathbb{C} \setminus \cup\{\partial Q : Q \in \mathcal{Q}\}$, which is dense in $\mathbb{C}$ and according to (8) is mapped by W into $\{0, 1\}$. Since W is continuous (2.14), this means that it takes only the values 0 and 1. Specifically, $\mathbb{C} \setminus \cup \mathcal{Q} \supset \mathbb{C} \setminus U$ entails that $W = 0$ in $\mathbb{C} \setminus U$, and $\bigcup\{\overset{\circ}{Q} : Q \in \mathcal{Q}\}$ dense in $\cup \mathcal{Q} \supset K$ means that $W = 1$ in K. We have therefore proved all the claims about γ except (1).

Now Cauchy's theorem for rectangles says that

$$0 = \int_{\partial Q} g \qquad \forall g \in H(U),\ Q \in \mathcal{Q}.$$

If we add these equalities and again use (3) and (4), we get

$$0 = \sum_{S \in \mathcal{S}_0} \int_S g = \sum_{j=1}^{k} \int_{\gamma_j} g = \int_\gamma g \qquad \forall g \in H(U).$$

Given $f \in H(U)$ and $z \in U$, apply this conclusion to $g(\xi) := [f(\xi) - f(z)]/[\xi - z]$ ($:= f'(z)$ at $\xi := z$) and proceed as in 5.33 to split g apart and get

$$(9) \quad f(z) \cdot \frac{1}{2\pi i} \int_\gamma \frac{d\xi}{\xi - z} = \frac{1}{2\pi i} \int_\gamma \frac{f(\xi)}{\xi - z}\, d\xi \qquad \forall z \in U \setminus \gamma_1 \cup \cdots \cup \gamma_k.$$

By definition the factor multiplying $f(z)$ is $W(z)$ and, as noted above, this is 1 for every $z \in K$, so (9) implies (1).

One senses that the intervals γ_j can be organized into closed polygons; a more useful form of the representation (1) would emerge. To prove this, we need a little combinatorial fact. (Interestingly, the version $k = \infty$ of this is false—see LENARD [1990] and PRIESTLEY [1992].)

Lemma 8.2 *Let $a_1, \ldots, a_k$, $b_1, \ldots, b_k \in \mathbb{C}$ satisfy*

$$(*) \qquad \sum_{j=1}^{k} p(a_j) = \sum_{j=1}^{k} p(b_j) \quad \textit{for all polynomials } p.$$

Then there exists a permutation ϕ of $\{1, 2, \ldots, k\}$ such that $b_j = a_{\phi(j)}$ for each $j = 1, 2, \ldots, k$.

Proof: To construct a proof by induction on k, it suffices to show that a_k lies in the set $\{b_1, b_2, \ldots, b_k\}$. For if $a_k = b_m$, then the induction hypothesis would supply a bijection ϕ of $\{1, 2, \ldots, k-1\}$ onto $\{1, 2, \ldots, k\} \setminus \{m\}$ such that $b_j = a_{\phi(j)}$ for $j \leq k-1$, and $\phi(k) := m$ would extend it to the desired permutation. Choose the notation so that $a_l = a_{l+1} = \cdots = a_k$ but $a_j \neq a_k$ if $j < l$, and suppose, in order to reach a contradiction, that $a_k \notin \{b_1, b_2, \ldots, b_k\}$. There is then a polynomial p which has zeros at $\{a_1, \ldots, a_{l-1}\} \cup \{b_1, b_2, \ldots, b_k\}$ but $p(a_k) \neq 0$. For example,

$$p(z) := \prod_{j=1}^{l-1} (z - a_j) \prod_{n=1}^{k} (z - b_n).$$

For this p hypothesis $(*)$ reads $(k - l + 1)p(a_k) = 0$. As $k \geq l$ and $p(a_k) \neq 0$, this is a contradiction.

Corollary 8.3 *Let K be non-empty and compact, U open, $K \subset U \subset \mathbb{C}$. Then there exist closed polygons $\Gamma_1, \ldots, \Gamma_n$ lying in $U \setminus K$ with edges parallel to the coordinate axes such that for $\Gamma := (\Gamma_1, \ldots, \Gamma_n)$ and all $z \in \mathbb{C} \setminus \Gamma$, $\mathrm{Ind}_\Gamma(z)$ is either 1 or 0; it is 1 for all $z \in K$, 0 for all $z \in \mathbb{C} \setminus U$. Consequently*

$$f(z) = \frac{1}{2\pi i} \int_\Gamma \frac{f(\xi)}{\xi - z} \, d\xi \qquad \forall f \in H(U),\, z \in K.$$

Proof: Let $\gamma_1, \ldots, \gamma_k$ be the intervals provided by 8.1, say $\gamma_j = [a_j, b_j]$. Fix $z \in K$. For any polynomial p we apply 8.1 to the functions $f(\xi) := (\xi - z)p'(\xi)$ and get $\sum_{j=1}^{k} \frac{1}{2\pi i} \int_{\gamma_j} p'(\xi) \, d\xi = 0$, that is,

$$\sum_{j=1}^{k} [p(b_j) - p(a_j)] = 0.$$

Then according to lemma 8.2 there exists a permutation ϕ of $\{1, 2, \ldots, k\}$ such that $b_j = a_{\phi(j)}$ for each $j = 1, 2, \ldots, k$. Write ϕ as a product of disjoint cycles $\phi_1, \phi_2, \ldots, \phi_n$ where, say, ϕ_l is the cycle $\left(j_l, \phi(j_l), \ldots, \phi^{[N_l]}(j_l), j_l\right)$. Then

$$\begin{aligned}\{\gamma_1, \gamma_2, \ldots, \gamma_k\} &= \left\{[a_1, a_{\phi(1)}], \ldots, [a_k, a_{\phi(k)}]\right\}\\ &= \bigcup_{l=1}^{n} \left\{[a_{j_l}, a_{\phi(j_l)}], \ldots, [a_{\phi^{[N_l]}(j_l)}, a_{j_l}]\right\},\end{aligned}$$

a disjoint union, since the k intervals $\gamma_j = [a_j, a_{\phi(j)}]$ provided by 8.1 intersect only in endpoints, hence are distinct. Set $\Gamma_l := \left[a_{j_l}, a_{\phi(j_l)}, \ldots, a_{\phi^{[N_l]}(j_l)}, a_{j_l}\right]$.

Remark: In the circumstances of Corollary 8.3 it is convenient to say that "Γ surrounds K in U." In 8.20 we will see that after further dissection the loops $\Gamma_1, \ldots, \Gamma_n$ can even be assumed to be simple.

§ 2 Applications to Approximation

Definition 8.4 (i) A *rational* function is a quotient of two polynomials. When the common factors are removed from these polynomials (via 4.52), the zeros of the denominator are called the *poles* of the rational function. The domain of the rational function is the plane minus these poles.

(ii) For $w \in \mathbb{C}$ we shall, in this section at least, use the special designation F_w for the rational function $F_w(z) := 1/(z - w)$ $(z \in \mathbb{C} \setminus \{w\})$.

Remark: The reader can easily check that poles are well defined by this procedure (cf. 11.13). However this definition will be subsumed in Chapter XI by the more general notion of a pole of a meromorphic function.

Theorem 8.5 *Let K be a non-void compact subset of $\mathbb{C}$, Ω a component of $\mathbb{C} \setminus K$ and $a \in \Omega$. Then for each $w \in \Omega$, the function F_w can be uniformly approximated on K by polynomials in F_a.*

Proof: Let $C(K)$ denote the set of all continuous complex functions on K and $A(K)$ the uniform closure in $C(K)$ of the polynomials in F_a. Let S be the set of $w \in \Omega$ for which $F_w \in A(K)$. Note that Ω is open (by 1.30(i)$'$). If

$w_0 \in S$ and $D(w_0, r) \subset \Omega$, then for any $w \in D(w_0, r)$ and all $z \in K$

$$\begin{aligned} F_w(z) &= (z-w)^{-1} = \big[(z-w_0)-(w-w_0)\big]^{-1} \\ &= (z-w_0)^{-1}\big[1-(w-w_0)(z-w_0)^{-1}\big]^{-1} \\ &= (z-w_0)^{-1}\sum_{k=0}^{\infty}\big[(w-w_0)(z-w_0)^{-1}\big]^k \\ &= \sum_{k=0}^{\infty}(w-w_0)^k(z-w_0)^{-k-1} \\ &= \sum_{k=0}^{\infty}(w-w_0)^k F_{w_0}^{k+1}(z) \end{aligned}$$

and the convergence is uniform in K because for $z \in K$, $\big|(w-w_0)(z-w_0)^{-1}\big| \le |w-w_0|r^{-1} < 1$, by definition of r. Therefore the functions

$$R_n := \sum_{k=0}^{n}(w-w_0)^k F_{w_0}^{k+1}$$

approximate F_w uniformly on K. Since $w_0 \in S$, we have $F_{w_0} \in A(K)$, which is a uniformly closed algebra (why?), so $R_n \in A(K)$ and then $F_w \in A(K)$. That is, $w \in S$. We have shown that the non-void (a belongs) subset S of Ω satisfies

$$w_0 \in S \;\;\&\;\; D(w_0, r) \subset \Omega \Rightarrow D(w_0, r) \subset S,$$

so by the Basic Connectedness Lemma 1.27, $S = \Omega$.

Scholium: A closer look at this proof reveals that at each stage (finite) sums having the special form $F_a + \sum_{j\ge 2} c_j(F_a)^j$ always suffice for the approximation. This will be useful later.

Theorem 8.6 *If K is a non-void compact subset of $\mathbb{C}$, then the set S of w in $\mathbb{C}\setminus K$ for which F_w is uniformly approximable on K by polynomials is the unbounded component Ω of $\mathbb{C}\setminus K$.*

Proof: We first show that $S \subset \Omega$, that is, S does not meet any bounded component C of $\mathbb{C}\setminus K$. Suppose otherwise, say $w \in S \cap C$. Let

$$M := \max\big\{|z-w| : z \in K\big\} < \infty$$

and use the fact $w \in S$ to choose a polynomial p such that

$$|p(z) - F_w(z)| < \frac{1}{M} \qquad \forall z \in K,$$

whence, recalling the definition of F_w,

(1) $$|(z-w)p(z)-1| < \frac{|z-w|}{M} \leq 1 \qquad \forall z \in K.$$

Next note that by 1.30(v)the boundary of C lies entirely in K. The first inequality in (1) now implies that

(2) $$|(z-w)p(z)-1| < 1 \qquad \forall z \in \partial C.$$

Now the polynomial $1-(z-w)p(z)$ is bounded on $\overline{C}$ because C is a bounded set. Its maximum modulus on $\overline{C}$ must occur (at least once) on ∂C and so is less than 1 by (2). In particular, this polynomial has modulus less than 1 at $z=w$, an evident contradiction. This contradiction establishes the claim $S \subset \Omega$.

Let $P(K)$ be the uniform closure of the polynomials in $C(K)$ and set $M^* := \sup|K|$. If $|a| \geq M^*+1$, then $a \in \Omega$ and

$$F_a(z) = -a^{-1}\left(1-za^{-1}\right)^{-1} = -a^{-1}\sum_{k=0}^{\infty}\left(za^{-1}\right)^k,$$

with convergence uniform in $z \in K$, since for such z we have $|za^{-1}| \leq M^*/(M^*+1)$. The partial sums of this series are polynomials, so $F_a \in P(K)$ and $a \in S$. As $P(K)$ is a uniformly closed algebra, it follows that the uniform closure of the polynomials in F_a lies in $P(K)$. But by the preceding theorem F_w belongs to the closure of the algebra of polynomials in F_a, for every $w \in \Omega$, so $F_w \in P(K)$ for all $w \in \Omega$. That is, $\Omega \subset S$; so from the result of the first paragraph, $S = \Omega$, as claimed.

Exercise 8.7 (i) *Modify the proof of* 8.6 *to show that if* a *and* w *belong to different components of* $\mathbb{C}\setminus K$, *then* F_w *is not uniformly approximable on* K *by polynomials in* F_a. *Show too that the proof of* 8.5 *works if* K *is merely closed.*

Hints: Let Ω denote the component of $\mathbb{C}\setminus K$ which contains w, so $a \in \mathbb{C}\setminus(K\cup\Omega)$. Set $M := \max\left\{|z-w|\cdot|z-a|^{-1} : z \in K\right\}$, $\varepsilon := \left(2|w-a|M\right)^{-1}$ and suppose there exists a polynomial P such that $|F_w - P\circ F_a| \leq \varepsilon$ on K. Multiply by F_a/F_w to get $|F| \leq \left(2|w-a|\right)^{-1}$ on K, where $F := F_a - (P\circ F_a)F_a/F_w$. This function is holomorphic in $\mathbb{C}\setminus\{a\}$ and is bounded outside any neighborhood of a, in particular in Ω. Use 5.13 and the fact $\partial\Omega \subset K$ to infer that $|F| \leq \left(2|w-a|\right)^{-1}$ holds throughout Ω. Evaluate at w to reach a contradiction.

(ii) (Filling holes in compacta) *Associated to every non-void compact* $K \subset \mathbb{C}$ *is its* polynomially convex hull

$$\widehat{K} := \{w \in \mathbb{C} : |P(w)| \leq \max|P(K)| \text{ for all polynomials } P\}.$$

Use ideas from the proof of 8.6 *to show that*

$$\widehat{K} = \mathbb{C} \setminus \Omega,$$

where Ω *is the unbounded component of* $\mathbb{C} \setminus K$, *and infer that* K *is simply-connected if and only if it is* polynomially convex, *i.e., if and only if* $K = \widehat{K}$.

Hints: If $w \in \Omega$, then according to 8.6, F_w is uniformly approximable on K by polynomials. Inequality (1) in the proof says that the polynomial $P(z) := (z-w)p(z) - 1$ satisfies $\max|P(K)| < 1 = |P(w)|$, showing that $w \notin \widehat{K}$. If $w \in \mathbb{C} \setminus \Omega$, then $w \in C$ for some bounded component C of $\mathbb{C} \setminus K$. Now C is open and, according to 1.30(v), $\partial C \subset K$. Therefore (maximum modulus principle) every polynomial P satisfies

$$|P(w)| \leq \max|P(\partial C)| \leq \max|P(K)|,$$

showing that $w \in \widehat{K}$.

(iii) *Prove that if* Ω *is a bounded region and* C *the (unique) unbounded component of* $\mathbb{C} \setminus \Omega$, *then the set* $\widehat{\Omega} := \mathbb{C} \setminus C$ *is a bounded, simply-connected region containing* Ω *such that*

(∗) $$|P(z)| \leq \sup|P(\Omega)| \quad \forall \text{ polynomials } P, \forall z \in \widehat{\Omega}.$$

Consequently, any sequence of polynomials which converges uniformly on Ω *also converges uniformly on* $\widehat{\Omega}$.

Hints: The set $\widehat{\Omega} := \mathbb{C} \setminus C$ is the union of Ω with all the compact components of $\mathbb{C} \setminus \Omega$, that is, the union of Ω with all its holes. According to 1.39 it is open, connected, and simply-connected. By 1.30(v) we have

$$\partial\widehat{\Omega} = \partial(\mathbb{C} \setminus C) = \partial C \subset \partial(\mathbb{C} \setminus \Omega) = \partial\Omega.$$

Therefore the maximum modulus principle insures that every $z \in \widehat{\Omega}$ and every polynomial P satisfy

$$|P(z)| \leq \sup|P(\widehat{\Omega})| = \max|P(\partial\widehat{\Omega})| \leq \max|P(\partial\Omega)| = \max|P(\overline{\Omega})|,$$

which is (∗).

(iv) *Inequality* (∗) *and the definition of* $\widehat{\overline{\Omega}}$ *in* (ii) *show that* $\widehat{\Omega} \subset \widehat{\overline{\Omega}}$, *whence* $\overline{\widehat{\Omega}} \subset \widehat{\overline{\Omega}}$. *Is this inclusion an equality?*

Hint: (Don Marshall): Examine $\Omega := \mathbb{D} \setminus \overline{D}(1/2, 1/2)$.

Theorem 8.8 *If K is a non-void compact subset of $\mathbb{C}$ and f is holomorphic in a neighborhood U of K, then f is uniformly approximable on K by linear combinations of the functions $F_w(z) = 1/(z - w)$ with $w \in \mathbb{C} \setminus K$.*

Proof: Because of 8.1 it suffices to consider f of the form

$$f(z) := \int_{[a,b]} \frac{\phi(\xi)}{\xi - z}\, d\xi = \int_0^1 \frac{\phi\big(tb + (1-t)a\big)(b-a)}{tb + (1-t)a - z}\, dt$$

for some $[a, b] \subset U \setminus K$ and some continuous function ϕ on $[a, b]$. As in the hints to 7.21(i), the Riemann sums here converge uniformly for $z \in K$. That is, given $\varepsilon > 0$, for sufficiently large $n \in \mathbb{N}$ the points $t_j := j/n$ satisfy

$$\left| f(z) - \sum_{j=1}^{n} \frac{\phi\big(t_j b + (1-t_j)a\big)(b-a)(t_j - t_{j-1})}{t_j b + (1-t_j)a - z} \right| < \varepsilon \qquad \forall z \in K.$$

(Alternatively, one can get this by using 2.13(iii) and the uniform continuity of the integrand in $(\xi, z) \in [a, b] \times K$.) Setting $w_j := t_j b + (1 - t_j)a$, $c_j := \phi(w_j)(a - b)(t_j - t_{j-1})$, this inequality reads

$$\left| f(z) - \sum_{j=1}^{n} c_j (z - w_j)^{-1} \right| < \varepsilon \qquad \forall z \in K, \quad \text{i.e.,}$$

$$\left| f(z) - \sum_{j=1}^{n} c_j F_{w_j}(z) \right| < \varepsilon \qquad \forall z \in K.$$

Note that $w_j \in [a, b] \subset \mathbb{C} \setminus K$ completes the proof.

Remark: Approximation by linear combinations of positive *powers* F_w^n (which suffices for the applications of 8.8 that follow) can be gotten from the integral representation in a somewhat different way via the idea in the proof of 8.5: By partitioning $[a, b]$ into short subintervals and approximating separately the integral over each we reduce to the case where $[a, b]$ lies wholly in some disk $\overline{D}(w_0, r) \subset \mathbb{C} \setminus K$. As one easily checks, the convergence of

$$F_w(z) = \sum_{k=0}^{\infty} (w - w_0)^k F_{w_0}^{k+1}(z)$$

is then uniform in $(w, z) \in \overline{D}(w_0, r) \times K$. Consequently

$$\begin{aligned} f(z) &= \int_{[a,b]} \frac{\phi(w)}{w - z}\, dw = \int_{[b,a]} \phi(w) F_w(z)\, dw \\ &= \sum_{k=0}^{\infty} \left[\int_{[b,a]} \phi(w)(w - w_0)^k\, dw \right] F_{w_0}^{k+1}(z), \end{aligned}$$

uniformly for $z \in K$.

Corollary 8.9 (The Polynomial Runge Theorem I) *For compact non-void $K \subset \mathbb{C}$, every f holomorphic in a neighborhood of K is uniformly approximable on K by polynomials if and only if $\mathbb{C} \setminus K$ is connected.*

Proof: Due to 8.8, the claim can be equivalently formulated mentioning only f of the form $F_w, w \in \mathbb{C} \setminus K$, and 8.6 confirms that formulation.

Corollary 8.10 (The Polynomial Runge Theorem II) *Let U be an open, simply-connected subset of $\mathbb{C}$. If f is holomorphic in U, then f is uniformly approximable by polynomials on each compact subset of U.*

Proof: Given compact $C \subset U$, 1.31 furnishes a compact K such that $C \subset K \subset U$ and $\mathbb{C} \setminus K$ is connected. Now apply the preceding corollary.

Corollary 8.11 (The Rational Runge Theorem I) *Let K be a compact subset of $\mathbb{C}$ and $A \subset \mathbb{C} \setminus K$. Suppose that A meets each bounded component of $\mathbb{C} \setminus K$. If f is holomorphic in an open neighborhood of K, then f is uniformly approximable on K by rational functions with poles only in A.*

Proof: First approximate f uniformly on K via 8.8 by linear combinations of the functions F_w, $w \in \mathbb{C} \setminus K$. For a given w let C be the component of $\mathbb{C} \setminus K$ which contains w. If C is unbounded, then F_w is uniformly approximable on K by polynomials (8.6), while if C is bounded, there is by hypothesis an $a \in C \cap A$ and then (8.5) F_w is uniformly approximable on K by polynomials in F_a.

Corollary 8.12 (The Rational Runge Theorem II) *Let U be an open subset of $\mathbb{C}$ and let $A \subset \mathbb{C} \setminus U$. Suppose that A meets each bounded component of $\mathbb{C} \setminus U$. If f is holomorphic in U, then f is uniformly approximable on each compact subset of U by rational functions with poles only in A.*

Proof: This follows from 8.11 via 1.31 just as 8.10 was deduced from 8.9 via 1.31. (Notice that ∂U is such an A, thanks to 1.30(v).)

Remark: The Riemann Mapping Theorem, which will be developed in the next chapter, can be used to give an interesting alternative proof of Polynomial Runge II. It will be shown there that for any simply-connected region $\Omega \neq \mathbb{C}$ there is a conformal map ϕ of $\mathbb{D}$ onto Ω. Apply the proof of 8.8 to the representation 5.91 of ϕ^{-1} to get ϕ^{-1} approximated by linear combinations of the functions F_w, $w \in \phi(C(0,r))$. Then use the simple-connectivity of $\phi\big(\bar{D}(0,r)\big)$ [see 4.35] and 8.6 to approximate each F_w by polynomials. Given $f \in H(\Omega)$, approximate $f \circ \phi$ by partial sums of its power series, and then finally have $f = (f \circ \phi) \circ \phi^{-1}$ approximated by polynomials. (The last part of this argument will feature in the proof of 9.26, where the reader will find more details.)

Exercise 8.13 *Deduce directly from the Polynomial Runge Theorem* Cauchy's Theorem for Simply-Connected Regions *in the following form: Let U be an open, simply-connected subset of $\mathbb{C}$. Then $\int_\gamma f = 0$ for every f which is holomorphic in U and every piecewise smooth loop γ in U.*

Hint: The result is clear for any f which is a derivative, in particular for all polynomials, whence for any f which is a uniform limit of polynomials on the range of γ. That range being a compact subset of U, all $f \in H(U)$ qualify by 8.10.

Exercise 8.14 (Cf. 7.22) *Consider the following*

General Cauchy Theorem: *Let U be an open subset of $\mathbb{C}$, $\gamma_1, \ldots, \gamma_k$ piecewise smooth loops in U, and set $\gamma := (\gamma_1, \ldots, \gamma_k)$. Then $\int_\gamma F = 0$ for every $F \in H(U)$ if and only if $\int_\gamma F_w = 0$ for all $w \in \mathbb{C} \setminus U$, that is, if and only if γ is U-nullhomologous.*

(i) *Deduce this from the Rational Runge Theorem.*

Hints: Since $F_w \in H(U)$ for every $w \in \mathbb{C} \setminus U$, one implication is trivial. Suppose $\sum_{j=1}^{k} \int_{\gamma_j} F_w =: \int_\gamma F_w = 0$ for all $w \in \mathbb{C} \setminus U$, and let $F \in H(U)$ be given. On the compact subset $K := \gamma_1 \cup \cdots \cup \gamma_k$ of U, F is uniformly approximable by sums of polynomials and linear combinations of powers of the functions F_w with $w \in \mathbb{C} \setminus U$. This follows from the *proof* of 8.12. We have $\int_{\gamma_j} g = 0$ for every g which is a polynomial or a power F_w^n with $n \neq 1$ because such g are derivatives and each γ_j is a closed curve. On the other hand, $\sum_{j=1}^{k} \int_{\gamma_j} g =: \int_\gamma g = 0$ by hypothesis if $g = F_w$ for any $w \in \mathbb{C} \setminus U$.

(ii) *Deduce this Cauchy Theorem directly from the representation* 8.1 *and an elementary Fubini argument* (cf. 2.19(ii) *and* 8.29(vi)).

Hints: The intervals that 8.1 furnishes need to be differently designated because in the notation here the K to which 8.1 will be applied is $\gamma_1 \cup \cdots \cup \gamma_k \cup \{z \in \mathbb{C} \setminus \gamma : \mathrm{Ind}_\gamma(z) \neq 0\}$. That it is indeed a compact subset of U if γ is U-nullhomologous, is confirmed by the simple argument in the last paragraph of the 4.3 proof, which works just as well for the $\gamma = (\gamma_1, \ldots, \gamma_k)$ here as for the loop γ in 4.3. Details can be found in Černý [1976], Ahlfors [1979], pp. 143–144, Beardon [1979], pp. 163–165, Luecking and Rubel [1984], p. 84, or Sarason [2007], p. 118.

(iii) General Cauchy Formula: *Show that*

$$f(z)\mathrm{Ind}_\gamma(z) = \frac{1}{2\pi i} \int_\gamma \frac{f(\xi)}{\xi - z}\, d\xi$$

holds for all $f \in H(U)$ and all $z \in U \setminus (\gamma_1 \cup \cdots \cup \gamma_k)$ if and only if $\gamma := (\gamma_1, \ldots, \gamma_k)$ is U-nullhomologous.

Hint: Apply the Cauchy theorem just proved to $F(\xi) := \frac{f(\xi)-f(z)}{\xi - z}$ and split the integral in two, as was done in the proof of 5.33.

Example 8.15 For $n \in \mathbb{N}$ consider

$$K_n := [-n, n] \times [0, n] \subset \overline{\mathbb{U}}, \quad K_n^* := [-n, n] \times [-n, -n^{-1}] \subset \mathbb{C} \setminus \overline{\mathbb{U}}.$$

Their union is compact and has connected complement. They lie in disjoint open sets U_n, U_n^* and the indicator function of U_n is holomorphic in $U_n \cup U_n^*$. Consequently 8.9 furnishes polynomials P_n such that

$$|P_n - 1| < 1/n \ \text{ in } K_n \ \text{ and } \ |P_n| < 1/n \ \text{ in } \ K_n^*.$$

It follows that

$$\lim_{n\to\infty} P_n = \begin{cases} 1 & \text{uniformly on compact subsets of } \overline{\mathbb{U}} \\ 0 & \text{uniformly on compact subsets of } \mathbb{C} \setminus \overline{\mathbb{U}}. \end{cases}$$

and (so) the polynomials $Q_n(z) := P_n(z)P_n(-z)$ satisfy

$$\lim_{n\to\infty} Q_n = \begin{cases} 1 & \text{uniformly on compact subsets of } \mathbb{R} \\ 0 & \text{uniformly on compact subsets of } \mathbb{C} \setminus \mathbb{R}. \end{cases}$$

This example shows that the local uniform boundedness hypotheses of Chapter VII cannot (in general) be weakened to pointwise boundedness. (See also **MR** 87i:30043.)

Exercise 8.16 *Produce a sequence of polynomials p_n such that* $\lim_{n\to\infty} p_n(z) = 0$ *for every $z \in \mathbb{C}$ but such that this convergence is not locally uniform on* $\mathbb{C}$.

Hint: Set $p_n := P_n^2 - P_n$, where P_n is the polynomial of 8.15. Then by 7.5 and the result of 8.15 it follows that there is a compact subset K of $\mathbb{C}$ such that $\{P_n\}$ is not uniformly bounded on K. Since $|p_n| \geq |P_n|^2 - |P_n| = |P_n|(|P_n| - 1)$, the sequence $\{p_n\}$ is not uniformly bounded on K either.

Exercise 8.17 (i) *Show that in* 8.12 *it is enough that the closure of A meet each bounded component of $\mathbb{C} \setminus U$.*

(ii) *Show that in the various Runge theorems interpolation at any finite set of points can be achieved by the approximating functions.* (Cf. WALSH [1928] and SINCLAIR [1958].)

Hints: Consider 8.9 and a finite subset A of K together with a positive integer $n(a)$ for each $a \in A$. Let p be any polynomial such that $f - p$ has a zero of order at least $n(a)$ at each $a \in A$. Then the function $[f(z) - p(z)] \prod_{a\in A}(z -$

$a)^{-n(a)} =: F(z)$ is holomorphic in U and given $\varepsilon > 0$, 8.9 provides a polynomial P_ε such that $|F - P_\varepsilon| \le \varepsilon'$ in K, where $\varepsilon' := \varepsilon / \max_{z\in K} \prod_{a\in A} |z-a|^{n(a)}$. The polynomial $p_\varepsilon(z) := p(z) + P_\varepsilon(z) \prod_{a\in A}(z-a)^{n(a)}$ satisfies $|f - p_\varepsilon| \le \varepsilon$ on K and $f - p_\varepsilon$ has a zero of order at least $n(a)$ at each $a \in A$.

(iii) *The role of* $\mathbb{C}$ *in* 8.9 *and* 8.11 *can be taken over by an arbitrary open superset* U *of* K. *The appropriate conclusion is then the equivalence of* (a) *and* (b)*:*

(a) *Every function which is holomorphic in a neighborhood of* K *is uniformly approximable on* K *by* (*rational*) *functions in* $H(U)$.

(b) *The closure of every bounded component of* $U \setminus K$ *meets* $\mathbb{C} \setminus U$.

Hints: (a)⇒(b): If C is any bounded component of $U \setminus K$, then C is open (1.30) and $\partial C \subset \partial(U \setminus K) \subset K \cup \partial U$. Hence if $\overline{C} \cap (\mathbb{C} \setminus U) = \varnothing$, that is, if $\overline{C} \subset U$, then $\partial C \subset K$. As in the proof of 8.6, upon selecting any point $w \in C$, (a) furnishes $f \in H(U)$ such that $|(z-w)f(z) - 1| < 1$ for all $z \in K \supset \partial C$, hence for $z := w$, by the maximum-modulus principle —⨯—.

(b)⇒(a): Proceed as in the proof of 8.11 and consider the case where C is bounded. If $w \in \mathbb{C} \setminus U$, then already $F_w \in H(U)$. Otherwise $w \in U$, and the component C_0 of $U \setminus K$ in which it lies is bounded (being a subset of C). Therefore (b) supplies $w_0 \in \overline{C}_0 \cap (\mathbb{C} \setminus U) \subset \overline{C} \cap (\mathbb{C} \setminus K)$. Since the component C is relatively closed in $\mathbb{C} \setminus K$, this means that $w_0 \in C$. Therefore 8.5 insures that F_w is uniformly approximable on K by polynomials in F_{w_0}. The latter functions lie in $H(U)$ because $w_0 \notin U$.

Remark: A "relativized" version of Runge somewhat more subtle than 8.17(iii) is offered in NARASIMHAN [1985], Chapter 5, §2: If $V \subset U \subset \mathbb{C}$ are open, then $H(U)|V$ is dense in $H(V)$ (for the local uniform convergence topology) if and only if $U \setminus V$ has no bounded component.

§ 3 Other Applications of the Integral Representation

Theorem 8.18 (ROUCHÉ [1862]) *Let* U *be a non-empty, bounded open subset of* $\mathbb{C}$, Φ, Ψ *continuous on* $\overline{U}$ *and holomorphic in* U. *Suppose that*

$$(*) \qquad |\Phi + \Psi| < |\Phi| + |\Psi| \quad \text{on } \partial U.$$

Then, counting multiplicities, the functions Φ *and* Ψ *have the same number* [*which is finite*] *of zeros in* U. [An extension to meromorphic functions is outlined in 11.49.]

Proof: Consider first compact $K \subset U$ and use 8.3 to select polygonal loops $\Gamma_1, \ldots, \Gamma_n \subset U \setminus K$ having the property that, with $\Gamma := (\Gamma_1, \ldots, \Gamma_n)$,

$$(1) \qquad f(z) = \frac{1}{2\pi i} \int_\Gamma \frac{f(\xi)}{\xi - z}\, d\xi \qquad \forall z \in K$$

holds for every $f \in H(U)$. If $F \in H(U)$ and z is any point in K, we may apply (1) to the holomorphic function $f(\xi) := (\xi - z)F(\xi)$, which is 0 at z, to get

$$(2) \qquad \frac{1}{2\pi i} \int_\Gamma F(\xi)\, d\xi = 0.$$

Next, suppose $f \in H(U)$ has no zeros in $U \setminus K$ and only finitely many in K, say at $a_1, \ldots, a_N$ with respective multiplicities $m_1, \ldots, m_N$. Then the function

$$(3) \qquad F(z) := \frac{f'(z)}{f(z)} - \sum_{j=1}^{N} \frac{m_j}{z - a_j}$$

is (extendable to be) holomorphic in U. Indeed, according to 5.69 there is a zero-free holomorphic function h in U such that

$$f(z) = h(z) \prod_{j=1}^{N} (z - a_j)^{m_j} \qquad \forall z \in U$$

and so the right-hand side of (3) is $h'(z)/h(z)$. Apply (2) to this F to conclude that

$$(4) \qquad \frac{1}{2\pi i} \int_\Gamma \frac{f'(\xi)}{f(\xi)}\, d\xi = \sum_{j=1}^{N} m_j \frac{1}{2\pi i} \int_\Gamma \frac{1}{\xi - a_j}\, d\xi.$$

But N applications of (1) with the constant function $\mathbf{1}$ in the role of f there (and $z = a_1, a_2, \ldots, a_N$) show that

$$1 = \frac{1}{2\pi i} \int_\Gamma \frac{1}{\xi - a_j}\, d\xi, \qquad j = 1, 2, \ldots, N,$$

and so (4) becomes (cf. 5.85)

$$(5) \qquad \frac{1}{2\pi i} \int_\Gamma \frac{f'(\xi)}{f(\xi)}\, d\xi = \sum_{j=1}^{N} m_j.$$

Now to prove the theorem, let $K := \{z \in \overline{U} : |\Phi(z) + \Psi(z)| = |\Phi(z)| + |\Psi(z)|\}$. This is a closed, hence compact, subset of $\overline{U}$ and by (∗) it does not meet ∂U, hence lies in U. By definition we have

$$(**) \qquad |\Phi + \Psi| < |\Phi| + |\Psi| \quad \text{in } \overline{U} \setminus K.$$

If Φ has infinitely many zeros in K, then they cluster at some $z \in K \subset U$, and so $\Phi = 0$ throughout the component C of U containing z, by 5.62. But then $\Phi = 0$ in $\bar{C} \supset \partial C$. Since, by 1.30(v), $\partial C \subset \partial U$, this contradicts (∗). Thus $\Phi^{-1}(0)$ is finite (as is $\Psi^{-1}(0)$), and the argument of the first paragraph of the proof applies with either Φ or Ψ in the role of f. Consequently, to prove the theorem it suffices by (5) to confirm that

$$(6) \qquad \frac{1}{2\pi i}\int_\Gamma \frac{\Phi'}{\Phi} = \frac{1}{2\pi i}\int_\Gamma \frac{\Psi'}{\Psi}.$$

But (∗∗) shows that $\Phi(z)$ is not a non-negative multiple of $\Psi(z)$ for any $z \in U \setminus K$, so Φ/Ψ maps $U \setminus K$ into $\mathbb{C} \setminus [0, \infty)$. Letting L be the Principal Branch of the Logarithm in $\mathbb{C} \setminus [0, \infty)$ (see 3.43), we can therefore form $g := L \circ \left(\frac{\Phi}{\Psi}\right) \in H(U \setminus K)$. A little calculation gives

$$(7) \qquad g' = \frac{\Phi'}{\Phi} - \frac{\Psi'}{\Psi} \quad \text{in } U \setminus K.$$

Since each Γ_l is a piecewise smooth loop in $U \setminus K$, we have from 2.10(ii)

$$(8) \qquad \int_{\Gamma_l} g' = 0, \quad l = 1, 2, \ldots, n.$$

If we sum over l in (8) and use (7), the equation (6) is confirmed, and the proof is complete.

Remarks: Upon squaring, one sees that (∗) in 8.18 is equivalent to

$$\mathrm{Re}(\bar{\Phi}\Psi) < |\Phi\Psi| \quad \text{on } \partial U.$$

Consequently,

$$(9) \qquad \mathrm{Re}(\bar{\Phi}\Psi) < 0 \quad \text{on } \partial U$$

is a stronger hypothesis than (∗). From it Shih [1982] deduced a complex version of the intermediate-value theorem. The latter can be expressed thus: If $f\colon [a, b] \mapsto \mathbb{R}$ is continuous and $f(b) < 0 < f(a)$, then f has a zero in the open interval (a, b). In case $a < 0 < b$ the condition on f reads $xf(x) < 0$

for all $x \in \{a, b\}$, the boundary of (a, b). Hence the following immediate consequence of 8.18 is one possible complex generalization of this theorem:

If U is a bounded, open neighborhood of 0 in $\mathbb{C}$, Ψ is continuous on $\overline{U}$, holomorphic in U and satisfies $\mathrm{Re}(\bar{z}^n \Psi(z)) < 0$ for all $z \in \partial U$ and some $n \in \mathbb{N}$, then Ψ has exactly n zeros in U.

Notice that Φ, Ψ need only be functions in $H(U)$ if hypothesis (9) is reformulated as

$$\limsup_{z \to \xi} \mathrm{Re}(\bar{\Phi}(z)\Psi(z)) < 0 \qquad \forall \xi \in \partial U.$$

Exercise 8.19 *Use Rouché's theorem to prove the following relatives of* 7.10:

(i) *If U is a bounded open subset of $\mathbb{C}$, f_n, f_0 are continuous on $\overline{U}$, holomorphic in U, $0 \notin f_0(\partial U)$ and $f_n \to f_0$ uniformly on ∂U, then there is an n_0 such that for each $n \geq n_0$ the functions f_n and f_0 have exactly the same number of zeros (counted according to multiplicity) in $\overline{U}$.*

(ii) *If f_n are holomorphic in the open set U, $f_n \to f_0$ locally uniformly in U and f_0 has a zero of multiplicity $m < \infty$ at $z_0 \in U$, then for all sufficiently small positive r there exists n_r such that for each $n \geq n_r$ the function f_n has exactly m zeros (counted according to multiplicity) in $D(z_0, r)$.*

Hint: Since the zero of f_0 at z_0 is of finite order, f_0 is not identically 0 in any neighborhood of z_0. Therefore for $r > 0$ sufficiently small we have $\bar{D}(z_0, r) \subset U$ and f_0 is zero-free on $\bar{D}(z_0, r) \setminus \{z_0\}$. Therefore (i) is applicable.

The next exercise prepares us for the second important application of the integral representation 8.3, that of lassoing holes in multiply-connected regions. This will be very useful in chapter X.

Exercise 8.20 *Suppose $v_1, v_2, \ldots, v_n \in \mathbb{C}$ and the intervals*

$$[v_1, v_2], [v_2, v_3], \ldots, [v_{n-1}, v_n], [v_n, v_1]$$

intersect pairwise only in endpoints. Show that there exist simple closed polygons $L_1, \ldots, L_N$ such that

(1) $[v_1, \ldots, v_{n-1}, v_n, v_1]$ *and* $L_1 \cup \cdots \cup L_N$ *coincide as sets, and*

(2) $\displaystyle\int_{[v_1,\ldots,v_{n-1},v_n,v_1]} f = \int_{L_1} f + \cdots + \int_{L_N} f$

for all continuous functions f on $[v_1, \ldots, v_{n-1}, v_n, v_1]$.

Hints: Go by induction on n. If $[v_1, \ldots, v_{n-1}, v_n, v_1]$ is simple, there is nothing to prove. Otherwise, for some $1 \le \nu < \mu \le n$ we have $v_\nu = v_\mu$. Consider then the polygonal loops

$$L := [v_1, \ldots, v_\nu, v_{\mu+1}, \ldots, v_n, v_1] \quad \text{and} \quad K := [v_\nu, v_{\nu+1}, \ldots, v_\mu].$$

Each displays fewer than n vertices and we clearly have:

$$[v_1, \ldots, v_{n-1}, v_n, v_1] \quad \text{and} \quad L \cup K \quad \text{coincide as sets, and}$$
$$\int_{[v_1,\ldots,v_{n-1},v_n,v_1]} f = \int_L f + \int_K f$$

for all continuous functions f on $[v_1, \ldots, v_{n-1}, v_n, v_1]$. Now refer each of L and K to the induction hypothesis.

Theorem 8.21 *Let U be an open subset of $\mathbb{C}$, $C_0, C_1, \ldots, C_m$ distinct holes in U; if $m > 0$, suppose in addition that U is connected. Then there exists a simple* (polygonal) *loop Γ in U such that* $\mathrm{Ind}_\Gamma(C_0) = 1$ *and* $\mathrm{Ind}_\Gamma(C_j) = 0$ *for $j = 1, 2, \ldots, m$.*

Proof: In case $m = 0$, 1.34 furnishes a compact set V which contains C_0 and is a relatively open subset of $\mathbb{C} \setminus U$. Thus $\mathbb{C} \setminus (U \cup V) = (\mathbb{C} \setminus U) \setminus V$ is relatively closed in the closed set $\mathbb{C} \setminus U$, so it is closed in $\mathbb{C}$, and $U \cup V$ is open. The rest of the proof in this case will be an obvious simplification of what follows, in which $m > 0$ and we assume (only in order to deal with the other holes) that U is connected. Connectedness of U lets us use 1.40 to produce unbounded connected closed subsets $K_1, \ldots, K_m$ of $\mathbb{C} \setminus C_0$ which contain $C_1, \ldots, C_m$, respectively. Then $\mathbb{C} \setminus K_1 \cup \cdots \cup K_m$ is an open neighborhood of the compact set C_0, so 1.34 provides a compact set V which is relatively open in $\mathbb{C} \setminus U$, contains C_0 and lies in $\mathbb{C} \setminus K_1 \cup \cdots \cup K_m$. It follows (as in the second sentence of the proof) that $U \cup V$ is open. Apply 8.3 with the U there equal to the set $(U \cup V) \setminus (K_1 \cup \cdots \cup K_m)$, $K := V$ to come up with polygonal loops $\Gamma_1, \ldots, \Gamma_n$ in $U \setminus (V \cup K_1 \cup \cdots \cup K_m)$ such that

$$1 = \sum_{l=1}^{n} \frac{1}{2\pi i} \int_{\Gamma_l} \frac{1}{\xi - z} \, d\xi = \sum_{l=1}^{n} \mathrm{Ind}_{\Gamma_l}(z) \qquad \forall z \in V.$$

For any fixed $z_0 \in C_0 \subset V$ there is then some l such that

$$\mathrm{Ind}_{\Gamma_l}(z_0) > 0. \tag{1}$$

Now it is clear from the genesis of Γ_l in the proof of 8.3 (via 8.1) that it is of the form $[v_1, v_2, \ldots, v_{r-1}, v_r, v_1]$, where any two of the intervals

$$[v_1, v_2], [v_2, v_3], \ldots, [v_{r-1}, v_r], [v_r, v_1]$$

meet only in endpoints. Hence 8.20 furnishes simple polygonal loops $L_1, \ldots, L_N$ such that

$$(2)\quad \begin{cases} \Gamma_l \text{ and } L_1 \cup \cdots \cup L_N \text{ coincide as sets, and} \\ \operatorname{Ind}_{\Gamma_l}(z_0) = \dfrac{1}{2\pi i}\displaystyle\int_{\Gamma_l} \frac{d\xi}{\xi - z_0} = \sum_{j=1}^{N} \frac{1}{2\pi i}\int_{L_j} \frac{d\xi}{\xi - z_0} = \sum_{j=1}^{N} \operatorname{Ind}_{L_j}(z_0). \end{cases}$$

From this last equation and (1), after a change of notation if necessary, we can conclude that

$$(3)\qquad \operatorname{Ind}_{L_N}(z_0) > 0.$$

Now $L_N \subset \Gamma_l \subset \mathbb{C} \setminus K_1 \cup \cdots \cup K_m$ and the K_j are unbounded and connected, so (4.3)

$$\operatorname{Ind}_{L_N}(K_j) = 0 \quad \text{for } j = 1, 2, \ldots, m.$$

Since $C_j \subset K_j$ we thus have

$$(4)\qquad \operatorname{Ind}_{L_N}(C_j) = 0 \quad \text{for } j = 1, 2, \ldots, m.$$

Now 4.43 says that the only values available to the index of a simple loop are 0, 1, -1. Therefore, we can take $\Gamma := L_N$ in order to convert (3) and (4) into the assertion of the theorem. (For simple *polygonal* loops this index fact gets a direct proof on pp. 285–286 of REMMERT [1998].)

Although 8.21 looks rather special, it turns out to be extremely basic, as will be seen in the proofs of the important theorems 10.20, 10.26, and 10.27. Therefore, it is worth looking at it from other points of view. Adequate for these applications is the case in which *U is connected and has only finitely many (say, m) holes*, a hypothesis we make for what follows.

Alternative proofs of 8.21: Use 1.41 to get open $U_0 \subset U$ such that C_0 is the unique hole in U_0. According to ($*$) in the proof of 4.71 there is a simple polygonal loop $\gamma \subset U_0$ such that $\operatorname{Ind}_\gamma(C_0) \neq 0$, that is (4.43), $\operatorname{Ind}_\gamma$ is either 1 throughout C_0 or -1 throughout C_0. Since $C_1, \ldots, C_m$ lie in unbounded components of $\mathbb{C} \setminus U_0$, $\operatorname{Ind}_\gamma$ is 0 throughout each.

The Riemann Mapping Theorem, from the next chapter, can be used to give a proof that does not involve the Jordan Curve Theorem. Consider U_0 as in the preceding paragraph. 1.39 says that $U_0 \cup C_0$ is open and simply-connected. Let Ω_0 denote the component of it which contains C_0. Then Ω_0 is open (1.30(i)$'$) and simply-connected (by the easy half of 4.72). If $\Omega_0 = \mathbb{C}$, let ϕ be the identity map. If $\Omega_0 \neq \mathbb{C}$, there is a conformal map ϕ of Ω_0 onto $\mathbb{D}$ ((i)$\Rightarrow$(vi) in 4.71 + 9.7). We may assume that $0 \in C_0$ and $\phi(0) = 0$. Since $\phi(\Omega_0)$ is either $\mathbb{C}$ or $\mathbb{D}$, there is an $r > 0$ such that the compact set $\phi(C_0)$ lies in $D(0, r)$ and $\bar{D}(0, r) \subset \phi(\Omega_0)$. Set $\gamma := C(0, r)$, $\Gamma := \phi^{-1} \circ \gamma$, a smooth, simple loop which lies in Ω_0. According to 5.86 $\text{Ind}_\Gamma(0)$ equals the number of zeros of ϕ^{-1} in $D(0, r)$, namely 1. Since Ω_0 is simply-connected, Ind_Γ vanishes in its complement, which contains $\mathbb{C} \setminus (U \cup C_0) \supset C_1 \cup \cdots \cup C_m$.

Remark: For some applications it suffices to effect the separation of holes in 8.21 by a tuple $\Gamma = (\Gamma_1, \ldots, \Gamma_n)$ of loops, and the proof of this weaker version is considerably easier: As before, cite 1.34 to find compact V which is relatively open in $\mathbb{C} \setminus U$, contains C_0 and lies in $\mathbb{C} \setminus C_1 \cup \cdots \cup C_m$. The $\Gamma = (\Gamma_1, \ldots, \Gamma_n)$ which 8.3 supplies for the open set $U \cup V$ and its compact subset $K := V$ does the job.

Exercise 8.22 (i) *Show that an open subset U of $\mathbb{C}$ is homologically-connected if and only if it is simply-connected.* (Cf. 4.71.)

Hints: ($\Rightarrow$) If $\mathbb{C} \setminus U$ has a bounded component C, then 8.21 provides a loop Γ in U such that $\text{Ind}_\Gamma(C) \neq 0$. In particular, Γ is not U-nullhomologous. Note that for this argument only the relatively elementary first part of the 8.21 proof is needed; in particular, the Jordan Curve Theorem is *not* needed.

($\Leftarrow$) If Γ is a loop in U, then Ind_Γ is constant in each component C of $\mathbb{C} \setminus U$ and vanishes at infinity. Since each C is unbounded, it follows that $\text{Ind}_\Gamma(C) = 0$.

(ii) *Denoting by I the identity function on the open set U, show that U is simply-connected if and only if for each $p \in \mathbb{C} \setminus U$ the function $I - p$ has a continuous square-root.* (Cf. 4.38(v).)

Hints: If $p \in C_0$ a bounded component of $\mathbb{C} \setminus U$, let Γ be as in 8.21. Then $I - p$ cannot have a continuous square-root s, else $(s \circ \Gamma)^2 = \Gamma - p$ and by 4.14 the loop $s \circ \Gamma$ would satisfy $2\text{Ind}_{s\circ\Gamma}(0) = \text{Ind}_{\Gamma - p}(0) = 1$, contrary to $\text{Ind}_{s\circ\Gamma}(0) \in \mathbb{Z}$. Conversely, if U has no holes, $\text{Ind}_{\gamma - p}(0) = \text{Ind}_\gamma(p) = 0$ for all loops $\gamma \subset U, p \in \mathbb{C} \setminus U$. Hence by 4.18, $I - p$ has a continuous logarithm.

(iii) *Show that a continuous logarithm exists in the open subset U of $\mathbb{C} \setminus \{0\}$ if and only if 0 does not lie in a hole in U.*

Hints: ($\Rightarrow$) If $0 \in C_0$ a hole in U, the proof of 8.21 down to (1) furnishes a loop $\Gamma \colon [0, 1] \mapsto U$ such that $\text{Ind}_\Gamma(C_0) \neq 0$. But if L is a continuous logarithm in U, then $\Gamma = e^{L\circ\Gamma}$, so $2\pi i \text{Ind}_\Gamma(0) = L(\Gamma(1)) - L(\Gamma(0)) = 0$.

($\Leftarrow$) This is 4.28(i).

Exercise 8.23 *Let U be an open subset of $\mathbb{C}$ with the property:*

$$(*) \quad \begin{cases} \textit{For any two loops } \gamma_0, \gamma_1 \textit{ in } U \textit{ which are defined on the same} \\ \textit{parameter interval } \exists\ n_0, n_1 \in \mathbb{Z} \textit{ not both } 0 \textit{ such that } n_0 {*} \gamma_0 \\ \textit{is loophomotopic to } n_1 {*} \gamma_1 \textit{ in } U. \end{cases}$$

Show that U is connected and has at most one hole.

Hints: If c_0, c_1 are two points of U, we may apply $(*)$ with $\gamma_0\colon [0,1] \mapsto c_0$ and $\gamma_1\colon [0,1] \mapsto c_1$ constant curves. If $H\colon [0,1]\times[0,1] \mapsto U$ is the loophomotopy, then $H(\tau,0)$ $(0 \le \tau \le 1)$ is a curve in U from c_0 to c_1. Thus U is connected. If C_0, C_1 are two different bounded components of $\mathbb{C}\setminus U$, choose loops γ_0, γ_1 in U after 8.21 so that

$$\operatorname{Ind}_{\gamma_0}(C_0) = \operatorname{Ind}_{\gamma_1}(C_1) = 1 \ \ \& \ \ \operatorname{Ind}_{\gamma_0}(C_1) = \operatorname{Ind}_{\gamma_1}(C_0) = 0.$$

It follows easily from 4.14 and 4.12 that $(*)$ fails for these loops.

Our third application of the integral representation foreshadows the familiar Laurent decomposition 11.2, but in considerably greater generality. We follow BEHNKE and SOMMER [1965], pp. 194–5. See also SAAKIAN [1989].

Theorem 8.24 *Let U be an open subset of $\mathbb{C}$, K a compact subset of U, $f \in H(U\setminus K)$. Then there exists a unique pair (g,h) with $g \in H(U)$, $h \in H(\mathbb{C}\setminus K)$, $\lim_{|z|\to\infty} h(z) = 0$ and $f = g+h$ in $U\setminus K$.*

Proof: It is convenient to prove uniqueness first. Let (g^*, h^*) be another pair with the same properties as (g,h). Then $g+h = g^*+h^*$, so $g-g^* = h^*-h$, in $U\setminus K$. This last equation says that a function ϕ is well defined in $U\cup(\mathbb{C}\setminus K) = \mathbb{C}$ by the decree that $\phi := g-g^*$ in U and $\phi := h^*-h$ in $\mathbb{C}\setminus K$. Since

$$0 = \lim_{|z|\to\infty} h^*(z) - \lim_{|z|\to\infty} h(z) = \lim_{|z|\to\infty} \phi(z),$$

it follows from Liouville (3.33) that $\phi = \mathbf{0}$. Thus $g-g^* = 0$ in U and $h^*-h = 0$ in $\mathbb{C}\setminus K$.

To prove existence, choose (1.31) open sets U_n such that, $\overline{U}_n$ compact $\subset U_{n+1} \subset U$ and $\{U_n\}$ covers U. Since compact K is covered by $\{U_n\}$, we have $K \subset U_{n_1}$ for some n_1. It is no loss of generality to assume that $n_1 = 1$. The set $\bigcup\{D(z,1/n)\cap U_1 : z \in K\}$ is open and its closure K_n lies in $\overline{U}_1$. 8.1 provides a finite set $\mathcal{I}_n$ of intervals γ lying in $(U\setminus K)\setminus(\overline{U}_n\setminus \mathring{K}_n)$ such that

$$(1.n) \qquad f(z) = \sum_{\gamma\in\mathcal{I}_n} \frac{1}{2\pi i}\int_\gamma \frac{f(\xi)}{\xi - z}\,d\xi \qquad \forall z \in \overline{U}_n \setminus \mathring{K}_n.$$

Since $K \subset \mathring{K}_n \subset \overline{U}_n \subset U$ for all n, we have $(U\setminus K)\setminus(\overline{U}_n\setminus\mathring{K}_n) = (U\setminus \overline{U}_n)\cup(\mathring{K}_n\setminus K)$. The latter two sets are open and disjoint. Therefore each connected subset γ must lie wholly in one or the other of them. Let $\mathcal{K}_n$ denote the $\gamma \in \mathcal{I}_n$ which lie in $\overset{\circ}{K}_n \setminus K$, $\mathcal{U}_n$ those that lie in $U \setminus \overline{U}_n$. Then with the convention that a sum over the empty index set means 0, define

$$(2) \qquad g_n(z) := \sum_{\gamma \in \mathcal{U}_n} \frac{1}{2\pi i} \int_\gamma \frac{f(\xi)}{\xi - z}\, d\xi, \qquad z \in U_n$$

$$(3) \qquad h_n(z) := \sum_{\gamma \in \mathcal{K}_n} \frac{1}{2\pi i} \int_\gamma \frac{f(\xi)}{\xi - z}\, d\xi, \qquad z \in \mathbb{C} \setminus K_n.$$

These functions are holomorphic where defined (by 2.14) and since $\mathcal{K}_n$, $\mathcal{U}_n$ are disjoint and cover $\mathcal{I}_n$, we have from (1.n)

$$(4.n) \qquad f = g_n + h_n \quad \text{in } U_n \setminus K_n.$$

It is, moreover, clear from (3) that

$$(5.n) \qquad \lim_{|z| \to \infty} h_n(z) = 0.$$

Since $U_n \setminus K_n \subset U_{n+1} \setminus K_{n+1}$, it follows from (4.$n$), (4.$n+1$), (5.$n$), (5.$n+1$) and the uniqueness already proven (applied to f restricted to $U_n \setminus K_n$) that

$$(6) \qquad g_{n+1} = g_n \quad \text{in } U_n \qquad \text{and} \qquad h_{n+1} = h_n \quad \text{in } \mathbb{C} \setminus K_n.$$

Since $U_n \uparrow U$ and $K_n \downarrow K$, we can therefore define g in U, h in $\mathbb{C} \setminus K$ by decreeing that $g := g_n$ in U_n, $h := h_n$ in $\mathbb{C} \setminus K_n$ and the equations (6) insure that g and h are well defined, hence holomorphic. From (4.n) we then have $f = g + h$ in $U \setminus K$. Finally, since $h = h_1$ outside the compact set K_1, it follows from (5.1) that $\lim_{|z| \to \infty} h(z) = 0$.

One of the problems that historically was treated before Runge's theorem and adumbrated it was the following: If f is holomorphic in a region Ω, $z_0 \in \Omega$ and the Taylor series of f at z_0 is known, i.e., all the numbers $f^{(l)}(z_0)$, $l = 0, 1, 2, \ldots$, are known, then f is completely determined in Ω (5.62). One would like a constructive method of recovering the other values $f(z)$ from these data (cf. 5.73). The following example illustrates a beautiful and successful instance that brings the integral representations and the Polynomial Runge Theorem into play. Moreover, the monomials $c_{ln}(z - z_0)^l$ that enter the expansion, like the $(z - z_0)^l$ in the Taylor expansion, do not depend on the function f. The key idea embodied in (2) below admits much generalization—see footnote 166 in Bieberbach [1921].

Example 8.25 (from Borel [1899]) *Let Ω be an open set which is starlike with respect to some point z_0. Then complex numbers $\{c_{ln} : 0 \le l \le k_n, n \in \mathbb{N}\}$ exist with the following property: For every $f \in H(\Omega)$ the polynomials*

$$f_n(z) := \sum_{l=0}^{k_n} \frac{f^{(l)}(z_0)}{l!} c_{ln}(z - z_0)^l$$

converge to f locally uniformly throughout Ω.

Proof: There is clearly no loss of generality in assuming $z_0 = 0$, so we do. For the simply-connected region $\mathbb{C} \setminus [1, \infty)$ Polynomial Runge II furnishes polynomials

(1) $$P_n(w) := \sum_{l=0}^{k_n} c_{ln} w^l, \quad n \in \mathbb{N}$$

such that

(2) $$\lim_{n\to\infty} \left| \frac{1}{1-w} - P_n(w) \right| = 0 \quad \text{locally uniformly in } \mathbb{C} \setminus [1, \infty).$$

Now let $f \in H(\Omega)$ and compact $K \subset \Omega$ be given. Ω contains the set $K^* := \bigcup_{z \in K} [0, z]$, which is evidently starlike with respect to 0. Since it is the image of the compact set $[0, 1] \times K$ under the continuous map $(t, z) \mapsto tz$, it is also compact. The Basic Integral Representation 8.1 supplies intervals $\gamma_1, \ldots, \gamma_m \subset \Omega \setminus K^*$ such that for $\gamma := (\gamma_1, \ldots, \gamma_m)$

(3) $$f(z) = \frac{1}{2\pi i} \int_\gamma \frac{f(\xi)}{\xi - z} \, d\xi$$

(4) $$= \frac{1}{2\pi i} \int_\gamma \frac{1}{1 - \dfrac{z}{\xi}} \cdot \frac{f(\xi)}{\xi} \, d\xi \qquad \forall z \in K^*.$$

Consider the set $W := \{z/\xi : z \in K^*, \, \xi \in \gamma_1 \cup \cdots \cup \gamma_m\}$. As a continuous image of a compact set, it is compact. Furthermore, it lies in $\mathbb{C} \setminus [1, \infty)$. For if $z \in K^*$ and $\xi \in \mathbb{C}$ satisfy $z/\xi =: t \in [1, \infty)$, then $\xi = \frac{1}{t} z \in [0, z] \subset K^* \subset \Omega \setminus (\gamma_1 \cup \cdots \cup \gamma_m)$. It follows from (2) that

$$\lim_{n\to\infty} \left| \frac{1}{1 - \dfrac{z}{\xi}} - P_n\left(\frac{z}{\xi}\right) \right| = 0 \quad \text{uniformly for } (z, \xi) \in K^* \times (\gamma_1 \cup \cdots \cup \gamma_m).$$

Using this in (4) we see that

(5) $$f(z) = \lim_{n\to\infty} \frac{1}{2\pi i} \int_\gamma P_n\left(\frac{z}{\xi}\right) \frac{f(\xi)}{\xi} \, d\xi \quad \text{uniformly for } z \in K^*.$$

Putting (1) into (5) yields

(6) $$f(z) = \lim_{n\to\infty} \sum_{l=0}^{k_n} \left[\frac{1}{2\pi i} \int_\gamma \frac{f(\xi)}{\xi^{l+1}} \, d\xi \right] c_{ln} z^l,$$

uniformly for $z \in K^*$. If we assume, as we may, that K is a neighborhood of 0, then from (3) and 2.14 we learn that the square-bracketed term in (6) is just $f^{(l)}(0)/l!$ and so (6) says that

$$f(z) = \lim_{n\to\infty} \sum_{l=0}^{k_n} \frac{f^{(l)}(0)}{l!} c_{ln} z^l$$

uniformly for $z \in K^* \supset K$.

Remark: To get (2) we do not actually need Runge, because explicit polynomial approximants to $\frac{1}{1-z}$ are not hard to find. E.g., we can define $p_1(z) := 1 + z$ and $p_n(z) := p_{n-1}(z) + zp_{n-1}^2(z)$ for $n > 1$. Then it can be shown that $p_n(z/n) \to 1/(1-z)$ locally uniformly in $\mathbb{C} \setminus [1, \infty)$. See GOURSAT [1903b] or §138 of PRINGSHEIM [1932]. An even more explicit and dramatic instance of this is worked out in 11.34.

§ 4 Some Special Kinds of Approximation

Exercise 8.26 *Let F be holomorphic in $\mathbb{D}$ and $|F| \le 1$. Then there exists a sequence of rational functions E_n such that*

(i) *E_n is continuous on $\overline{\mathbb{D}}$ and holomorphic in $\mathbb{D}$,*

(ii) *$|E_n| = 1$ on $\mathbb{T}$* (cf. 6.16(i)),

(iii) *$E_n \to F$ locally uniformly on $\mathbb{D}$.*

Hints: (CARATHÉODORY [1960]) The usual device of approximating F locally uniformly on $\mathbb{D}$ by the functions $F_r(z) := F(rz)$ reduces the problem to the case where F is continuous on $\overline{\mathbb{D}}$. Let us call any function which has properties (i) and (ii) a *unit function.* (We know from 6.16(i) that these are precisely the finite Blaschke products.) Take as inductive hypothesis

$$(H_n) \qquad \begin{cases} \text{If } f \text{ is continuous on } \overline{\mathbb{D}}, \text{ holomorphic in } \mathbb{D} \text{ and} \\ |f| \le 1, \text{ then there is a unit function } E \text{ such that} \\ f - E \text{ has a zero at 0 of multiplicity at least } n. \end{cases}$$

If $|f(0)| = 1$, then we may take $E := f(0)$ and if $|f(0)| < 1$, we may take $E(z) := (f(0) - z)/(1 - \bar{f}(0)z)$, in order to establish that (H_1) holds. If (H_n) holds and f is given which is continuous on $\overline{\mathbb{D}}$, holomorphic in $\mathbb{D}$ and $|f| < 1$ in $\mathbb{D}$, then the function

$$g(z) := \frac{1}{z} \frac{f(0) - f(z)}{1 - \bar{f}(0)f(z)}$$

is also continuous on $\overline{\mathbb{D}}$, holomorphic in $\mathbb{D}$ and $|g| \le 1$ (by 6.1). Apply (H_n) to g and come up with a unit function g^* such that $g^* - g$ has a zero at 0 of multiplicity at least n. Evidently the unit functions are a semigroup under composition so

$$f^*(z) := \frac{f(0) - zg^*(z)}{1 - \bar{f}(0)zg^*(z)}$$

is a unit function, and according to 6.10 for it we know that $f^* - f$ has a zero at 0 of multiplicity at least $n+1$. With (H_n) in hand for all n, and noting that $|E_n - f| \le 2$ for all n, we can cite 7.9 to get (iii).

Remarks: The preceding proof can be concluded more elementarily by noting that because the zero of $E_n - F$ at 0 has multiplicity at least n, we have $|E_n - F|(z) \le 2|z|^n$ for all $z \in \mathbb{D}$ (see Remarks following 6.1), whence $|E_n - F| \le 2r^{2n}$ in $\overline{D}(0,r)$ for each $r < 1$ (Maximum Modulus Principle). The alternative definition of multiplicity in 5.68 suggests that in getting (H_n) above it is perhaps not necessary to involve power series. But, of course the *rational* functions E_n have power-series representations, for elementary reasons (partial fractions). So, does this all lead via 3.35(iv) to the Weierstrassian holy grail, an *integration-free* proof that (every) holomorphic F has a power-series representation? An answer, involving unexpected intricacies, is provided by Connell and Porcelli [1962].

Exercise 8.27 (i) (Fisher [1968]) *Let f be continuous on $\overline{\mathbb{D}}$, holomorphic in $\mathbb{D}$ with $|f| \le 1$. Then f is a uniform limit on $\overline{\mathbb{D}}$ of convex combinations of unit functions.*

Hints: f is uniformly continuous on $\overline{\mathbb{D}}$ so, setting $f_r(z) := f(rz)$ $\big(z \in \overline{\mathbb{D}}, r \in [0,1]\big)$, we have that $\lim_{r\uparrow 1} f_r = f$ uniformly on $\overline{\mathbb{D}}$. Given $\varepsilon > 0$, choose $r < 1$ so that $|f - f_r| < \varepsilon/2$ in $\overline{\mathbb{D}}$ and then use the preceding exercise to choose a unit function E such that $|f - E| < \varepsilon/2$ in $\overline{D}(0,r)$. It follows that $|f_r - E_r| < \varepsilon/2$, whence $|f - E_r| < \varepsilon$, in $\overline{\mathbb{D}}$. Therefore it suffices that we show that E_r is a convex combination of unit functions. As $(gh)_r = g_r h_r$ for any functions g, h on $\overline{\mathbb{D}}$ and as the convex combinations of unit functions are clearly a multiplicative semigroup, it suffices in view of 6.16(i) to restrict attention to E of the form $E(z) := (z-a)/(1-\bar{a}z)$, $|a| < 1$. Then we have for all $z \in \overline{\mathbb{D}}$

$$E_r(z) = \frac{r\left(1 - |a|^2\right)}{1 - |a|^2 r^2}\left[\frac{z - ar}{1 - \bar{a}rz}\right] + \frac{|a|\left(1 - r^2\right)}{1 - |a|^2 r^2}\left[\frac{-a}{|a|}\right] + \frac{(1-r)(1-|a|)}{2(1+|a|r)}[1] + \frac{(1-r)(1-|a|)}{2(1+|a|r)}[-1].$$

On the right-hand side the square-bracketed terms are unit functions and their coefficients are non-negative and add up to 1.

There is also an elementary constructive proof (BERNARD, GARNETT and MARSHALL [1977]) which is too elegant to forego: It suffices to approximate each polynomial $p(z) := \sum_{j=0}^{n} c_j z^j$ which is bounded in $\mathbb{D}$ by an $r < 1$. The related polynomial $p^*(z) := \sum_{j=0}^{n} \overline{c_j} z^{n-j}$ equals $z^n \overline{p(z)}$ for $z \in \mathbb{T}$, and so $|p^*(z)| = |p(z)| \leq r$ for $z \in \mathbb{T}$, hence for $z \in \overline{\mathbb{D}}$. The function

$$F(\xi, z) := \frac{\xi z^n + p(z)}{1 + p^*(z)\xi}, \qquad (\xi, z) \in \overline{\mathbb{D}} \times \overline{\mathbb{D}}$$

is consequently well-defined, and for each $(\xi, z) \in \mathbb{T} \times \mathbb{T}$

$$|F(\xi, z)| = \left| \frac{\xi z^n + p(z)}{1 + \xi z^n \overline{p(z)}} \right| = \left| \frac{\xi + z^{-n} p(z)}{\bar{\xi} + \overline{z^{-n} p(z)}} \right| = 1.$$

Since for each $z \in \overline{\mathbb{D}}$, $\xi \mapsto F(\xi, z)$ is continuous in $\overline{\mathbb{D}}$, 5.5(ii) gives

$$p(z) = F(0, z) = \int_0^1 F(e^{2\pi i t}, z) dt \qquad \forall z \in \overline{\mathbb{D}}.$$

The Riemann sums $N^{-1} \sum_{k=1}^{N} F(e^{2\pi i k/N}, z)$ are convex combinations of unit functions, and as $N \to \infty$ they converge to this integral uniformly for $z \in \overline{\mathbb{D}}$, via 2.13(iii) and the uniform continuity of F on $\mathbb{T} \times \overline{\mathbb{D}}$.

(ii) *By iterating the approximation in part* (i), *show that for each function f which is continuous on $\overline{\mathbb{D}}$, holomorphic in $\mathbb{D}$ and satisfies* $\sup |f(\mathbb{D})| < 1$, *there exist unit functions E_n and positive numbers c_n such that $\sum_{n=1}^{\infty} c_n = 1$ and $\sum_{n=1}^{\infty} c_n E_n = f$.*

Hints: Pick $\delta > 0$ so that $|f| \leq (1 + \delta)^{-1}$, set $f_0 := (1 + \delta) f$ and choose successively $C_0, C_1, \ldots$, each a convex combination of unit functions, so that for each $n \in \mathbb{N}$

$$(*) \qquad \left| f_0 - C_0 - \sum_{k=1}^{n} \delta 2^{-k} C_k \right| \leq \delta 2^{-n-1} \quad \text{on } \overline{\mathbb{D}}.$$

For example, select C_{n+1} via $(*)$ and (i) so that

$$\left| 2^{n+1} \delta^{-1} \left(f_0 - C_0 - \sum_{k=1}^{n} \delta 2^{-k} C_k \right) - C_{n+1} \right| \leq \frac{1}{2}$$

on $\overline{\mathbb{D}}$. Then $f = (1+\delta)^{-1}C_0 + \sum_{k=1}^{\infty}(1+\delta)^{-1}\delta 2^{-k}C_k$.

Exercise 8.28 (HELSON and SARASON [1967]) *Show that every continuous* $f\colon \mathbb{T} \mapsto \mathbb{T}$ *can be uniformly approximated by functions of the form* $z^{-m}B(z)$*,* B *a finite Blaschke product and* m *a non-negative integer.*

Hints: 4.22(i) supplies $n \in \mathbb{Z}$ and continuous $h : \mathbb{T} \mapsto \mathbb{R}$ such that $f(z) = z^n e^{ih(z)}$ for all $z \in \mathbb{T}$. Hence it suffices to approximate e^{ih}. Let $\varepsilon > 0$ be given. 5.20 provides a holomorphic function F whose real part extends continuously to $\overline{\mathbb{D}}$ and coincides on $\mathbb{T}$ with $H := -h/2$. Denote by S_N the Nth partial sum of the Taylor series of H. If $0 < r < 1$ is close to 1, $\operatorname{Re} F(rz)$ is close to $\operatorname{Re} F(z) = H(z)$ uniformly for $z \in \mathbb{T}$ (uniform continuity of $\operatorname{Re} F$ on $\overline{\mathbb{D}}$), and if N is large enough, $\operatorname{Re} S_N(rz)$ will be close to $\operatorname{Re} F(rz)$, hence close to $H(z)$, uniformly for $z \in \mathbb{T}$. In short, this argument supplies polynomials P such that $\operatorname{Re} P$ approximates $-h/2$ uniformly well on $\mathbb{T}$. According to 7.1(ii) the approximation can be made so good that

$$\text{(1)} \qquad \left|e^{2i\operatorname{Re} P} - e^{-ih}\right| \le \frac{\varepsilon}{2} \quad \text{on } \mathbb{T}.$$

The partial sums of the Taylor series of the entire function e^{iP} are polynomials Q_k which approximate e^{iP} uniformly well on $\overline{\mathbb{D}}$. Consequently for all sufficiently large k

$$\text{(2)} \qquad Q_k \text{ is zero-free in } \overline{\mathbb{D}}.$$

Consider the continuous function $\phi(w) := \overline{w}/w$ on $\mathbb{C}\setminus\{0\}$. According to 7.1(ii) the Q_k may be chosen to approximate e^{iP} on $\mathbb{T}$ so well that

$$\left|\phi \circ Q_k - \phi \circ e^{iP}\right| \le \varepsilon/2 \quad \text{on } \mathbb{T}.$$

That is, $Q := Q_k$ satisfies

$$\left|\overline{Q}/Q - e^{-2i\operatorname{Re} P}\right| = \left|\frac{\overline{Q}}{Q} - \frac{e^{-i\overline{P}}}{e^{iP}}\right| \le \frac{\varepsilon}{2} \quad \text{on } \mathbb{T}.$$

Combined with (1) this says that

$$\text{(3)} \qquad \left|\overline{Q}/Q - e^{ih}\right| \le \varepsilon \quad \text{on } \mathbb{T}.$$

Let the degree of Q be m. Since the zeros of Q lie in $\mathbb{C} \setminus \overline{\mathbb{D}}$ (by (2)), a calculation shows that the rational function

$$\text{(4)} \qquad \frac{\overline{Q(1/\overline{z})}}{Q(z)} \quad \text{has the form } uz^{-m}B(z),$$

where $u \in \mathbb{T}$ and B is the Blaschke product based on the conjugate reciprocals of the zeros of Q. Since (4) coincides with $\overline{Q(z)}/Q(z)$ for $z \in \mathbb{T}$, inequality (3) proves the claim.

The next exercise will be useful later on. It represents Weierstrass' own *Ansatz* to his famous polynomial approximation theorem. The proof is another "convolution approximate identity" argument similar to that in the solution of the Dirichlet problem for a disk. That idea can be ultimately traced back to the work of Poisson and Fourier on the solution of the heat equation. (See **MR** 90b:01034 and 93m:01036.) The actual numerical value of the integral in part (i) is clearly of no relevance. The literature abounds with elementary computations of it (e.g., 3.24) and we shall deduce its value as a by-product of some residue calculations in Chapter XI. It is of special significance in probability theory.

Exercise 8.29 (i) *Prove that* $\int_{-\infty}^{\infty} e^{-x^2}\,dx := \lim_{R\to\infty} \int_{-R}^{R} e^{-x^2}\,dx$ *exists. Calling it* I, *show that* $0 < I < 1 + 2e^{-1/4}$.

Hint: For $x \geq \frac{1}{2}$, $e^{-x^2} \leq 2xe^{-x^2} = \frac{d}{dx}\left[-e^{-x^2}\right]$. Therefore for $R \geq \frac{1}{2}$

$$\begin{aligned}\phi(R) &:= \int_{-R}^{R} e^{-x^2}\,dx \\ &= \int_{-1/2}^{1/2} e^{-x^2}\,dx + 2\int_{1/2}^{R} e^{-x^2}\,dx < 1 + 2\left[-e^{-x^2}\right]_{1/2}^{R} < 1 + 2e^{-1/4}.\end{aligned}$$

The function ϕ is therefore bounded. As it is evidently increasing, the desired limit exists.

(ii) *Let* $a, b \in \mathbb{R}$, $a < b$ *and* $f\colon [a,b] \mapsto \mathbb{C}$ *a continuous function. For each positive integer* k *form*

$$f_k(x) := \frac{k}{I}\int_a^b e^{-k^2(x-t)^2} f(t)\,dt, \quad x \in \mathbb{R}.$$

Show that for each $\varepsilon > 0$

$$\lim_{k\to\infty} f_k(x) = \begin{cases} f(x) & \textit{uniformly for } x \in [a+\varepsilon, b-\varepsilon] \\ 0 & \textit{uniformly for } x \in \mathbb{R} \setminus [a-\varepsilon, b+\varepsilon]. \end{cases}$$

Hints: Fix $\varepsilon > 0$. For $x \in \mathbb{R}\setminus[a-\varepsilon, b+\varepsilon]$ and $t \in [a,b]$ we have $(x-t)^2 \geq \varepsilon^2$ so, upon setting $M := \sup|f[a,b]|$,

$$\begin{aligned}|f_k(x)| &\leq \frac{k}{I}\int_a^b Me^{-k^2\varepsilon^2}\,dt = \frac{(b-a)M}{I}ke^{-k^2\varepsilon^2} \\ &< \frac{(b-a)M}{I}k\frac{1}{k^2\varepsilon^2} = \frac{(b-a)M}{\varepsilon^2 Ik} \quad (\text{since } e^{k^2\varepsilon^2} > k^2\varepsilon^2).\end{aligned}$$

Now extend f to $\mathbb{R}$ by making $f(-\infty, a-1] = f[b+1, \infty) := 0$ and f linear on $[a-1, a]$ and on $[b, b+1]$. Note that $\sup|f(\mathbb{R})|$ is still M. Changing the variable of integration, we see that

$$f_k(x) = \frac{1}{I}\int_{k(a-x)}^{k(b-x)} e^{-\tau^2} f\left(x + \frac{\tau}{k}\right) d\tau.$$

Remember that

$$\int_{-\infty}^{\infty} e^{-\tau^2}\, d\tau =: I. \tag{1}$$

Next write

$$\begin{aligned} f_k(x) - f(x) &= \frac{1}{I}\int_{\sqrt{k}(a-x)}^{\sqrt{k}(b-x)} e^{-\tau^2}\left[f\left(x+\frac{\tau}{k}\right) - f(x)\right] d\tau \\ &+ \frac{1}{I}\int_{k(a-x)}^{\sqrt{k}(a-x)} e^{-\tau^2}\left[f\left(x+\frac{\tau}{k}\right) - f(x)\right] d\tau \\ &+ \frac{1}{I}\int_{\sqrt{k}(b-x)}^{k(b-x)} e^{-\tau^2}\left[f\left(x+\frac{\tau}{k}\right) - f(x)\right] d\tau \\ &+ f(x)\left[\frac{1}{I}\int_{k(a-x)}^{k(b-x)} e^{-\tau^2}\, d\tau - 1\right]. \end{aligned} \tag{2}$$

Since f is continuous and compactly supported, it is uniformly continuous and so, given $\delta > 0$, there exists k_δ such that

(3)

$$\left|f\left(x+\frac{\tau}{k}\right) - f(x)\right| \le \frac{1}{2}\delta \quad \forall k \ge k_\delta,\ x \in [a,b],\ \tau \in \left[\sqrt{k}(a-x), \sqrt{k}(b-x)\right].$$

In view of (1), we may also require of k_δ that

$$1 - \frac{1}{I}\int_{-\varepsilon\sqrt{k}}^{\varepsilon\sqrt{k}} e^{-\tau^2}\, d\tau < \frac{1}{8M}\delta \qquad \forall k \ge k_\delta. \tag{4}$$

For $x \in [a+\varepsilon, b-\varepsilon]$ and $k \geq k_\delta$ we have from (2) and (3)

$$
\begin{aligned}
|f_k(x) - f(x)| \leq{} & \frac{1}{I}\int_{\sqrt{k}(a-x)}^{\sqrt{k}(b-x)} \frac{\delta}{2} e^{-\tau^2}\, d\tau \\
& + \left[\frac{1}{I}\int_{k(a-x)}^{\sqrt{k}(a-x)} 2Me^{-\tau^2}\, d\tau + \frac{1}{I}\int_{\sqrt{k}(b-x)}^{k(b-x)} 2Me^{-\tau^2}\, d\tau\right] \\
& + M\left|\frac{1}{I}\int_{k(a-x)}^{k(b-x)} e^{-\tau^2}\, d\tau - 1\right| \\
\overset{(1)}{\leq}{} & \frac{\delta}{2} + 2M\left[\frac{1}{I}\int_{-\infty}^{\sqrt{k}(a-x)} e^{-\tau^2}\, d\tau + \frac{1}{I}\int_{\sqrt{k}(b-x)}^{\infty} e^{-\tau^2}\, d\tau\right] \\
& + M\left|\frac{1}{I}\int_{k(a-x)}^{k(b-x)} e^{-\tau^2}\, d\tau - 1\right| \\
\overset{(4)}{\leq}{} & \frac{\delta}{2} + 2M\left[\frac{1}{I}\int_{-\infty}^{-\varepsilon\sqrt{k}} e^{-\tau^2}\, d\tau + \frac{1}{I}\int_{\varepsilon\sqrt{k}}^{\infty} e^{-\tau^2}\, d\tau\right] + \frac{\delta}{8}, \\
& \text{since } x \in [a+\varepsilon, b-\varepsilon], \\
\overset{(1)}{=}{} & \frac{\delta}{2} + 2M\left[1 - \frac{1}{I}\int_{-\varepsilon\sqrt{k}}^{\varepsilon\sqrt{k}} e^{-\tau^2}\, d\tau\right] + \frac{\delta}{8} \\
\overset{(4)}{\leq}{} & \frac{\delta}{2} + \frac{\delta}{4} + \frac{\delta}{8} < \delta.
\end{aligned}
$$

This holds for all $k \geq k_\delta$.

(iii) (Weierstrass [1885]) *If $A, B \in \mathbb{R}$, $A < B$ and $f\colon [A,B] \mapsto \mathbb{C}$ is continuous, then for each $\varepsilon > 0$ there is a polynomial p such that $|f - p| < \varepsilon$ on $[A,B]$. Given finitely many points in $[A,B]$, there is such a p which agrees with f at these points.*

Hints: Let $a := A-1$, $b := B+1$ and extend f continuously to $[a,b]$ (as was done in (ii)). Then pick k so that $|f - f_k| < \varepsilon/2$ on $[A,B]$. Finally, choose N so large that for all $x \in [a,b]$

$$
(*) \quad \left| f_k(x) - \int_a^b \frac{k}{I}\sum_{n=0}^{N} \frac{\left[-k^2(x-t)^2\right]^n}{n!} f(t)\, dt \right|
$$

$$
\leq (b-a)\frac{k}{I}\sup\bigl|f[a,b]\bigr| \cdot \sup_{(x,t)\in[a,b]\times[a,b]} \left| e^{-k^2(x-t)^2} - \sum_{n=0}^{N} \frac{\left[-k^2(x-t)^2\right]^n}{n!} \right| < \frac{\varepsilon}{2}.
$$

This is possible since the series converges to the exponential uniformly on compact subsets of $\mathbb{R}$. The integral on the left-hand side of $(*)$ is a linear combination of terms of the form $x^\ell \int_a^b t^j f(t)\,dt$, which one sees by multiplying out the various terms $(x-t)^{2n}$.

A quite different proof, using Fourier series ideas, is outlined in the discussion around inequality (11) in Remark (ii) following 5.20.

To achieve simultaneous approximation and interpolation in a finite set S, first choose a polynomial P with $|f-P| \le \varepsilon$ on $[A,B]$, then let L be the Lagrange polynomial that matches $f-P$ at each point of S. An examination of its definition (in the discussion preceding 5.73) shows that on $[A,B]$ it satisfies $|L| \le C \max\{|f(x)-P(x)| : x \in S\}$ for a (finite) constant C which depends only on $[A,B]$ and the set S. Thus $|f-(P+L)| < \varepsilon + C\varepsilon$. [Why is the method outlined in 8.17(ii) inapplicable?]

(iv) *Formulate and prove the two-dimensional version of* (ii).

(v) *Formulate and prove the two-dimensional analog of* (iii). (*Approximation will be by linear combinations of* $x^\ell y^m$, $\ell, m \ge 0$.)

Hints: Given continuous $F : [A,B] \times [c,d] \mapsto \mathbb{C}$, set $a := A-1$, $b := B+1$ and extend each function $x \mapsto F(x,y)$ to a continuous function on $[a,b]$ by making it 0 on $\{a,b\}$, linear on $[a,A]$, and linear on $[B,b]$. The result is an extension of F to $[a,b] \times [c,d]$. This extension, also to be denoted by F, is continuous. (Why?) For $k \in \mathbb{N}$ form

$$F_k(x,y) := \frac{k}{I} \int_a^b e^{-k^2(x-t)^2} F(t,y)dt, \qquad (x,y) \in [a,b] \times [c,d].$$

Examine the proof of (ii) to see that, because F is uniformly continuous on $[a,b] \times [c,d]$, all convergence there is uniform in the parameter y. Apply the procedure of (iii) to F_k to come up with a polynomial P such that

$$(*) \qquad \left| F_k(x,y) - \frac{k}{I} \int_a^b P(x-t)F(t,y)dt \right| < \frac{\varepsilon}{2} \qquad \forall (x,y) \in [A,B] \times [c,d].$$

This integral is a linear combination of terms $x^\ell \int_a^b t^j F(t,y)dt$ and each of the latter integrals, being a continuous function of y, is approximable by polynomials in y, by (iii). The Fourier series alternative proof mentioned in (iii) also works here because the Fourier coefficients of $x \mapsto F(x,y)$ are obviously continuous functions of y.

(vi) *Using* (v) *prove* Fubini's theorem (cf. 2.19(ii)) *in the form*

$$\int_a^b \int_c^d f(x,y)\,dy\,dx = \int_c^d \int_a^b f(x,y)\,dx\,dy$$

whenever $a, b, c, d \in \mathbb{R}$ *and* $f\colon [a,b] \times [c,d] \mapsto \mathbb{C}$ *is continuous.*

Hints: The claim is evidently true when f is a linear combination of multinomials $x^\ell y^m$ (ℓ, m non-negative integers) and if f_n converges to f uniformly on $[a,b] \times [c,d]$, each iterated integral of f_n converges to the corresponding integral of f.

§ 5 Carleman's Approximation Theorem

Carleman [1927] discovered the remarkable fact that a continuous function on $\mathbb{R}$ can be approximated by an entire function *uniformly on* $\mathbb{R}$. This generalizes the Weierstrass Theorem 8.29(iii), since on any *compact* portion of $\mathbb{R}$ the entire function can in turn be approximated uniformly by polynomials (partial sums of its power series). The proof given below is based on Hoischen [1967] and Scheinberg [1976]. Carleman's original proof is also quite beautiful and the reader is urged to look at it. It uses an integral representation like that used to prove the Runge theorems. Unfortunately the Runge theorem itself is not strong enough because the compacta on which approximation is to take place touch the boundary. This necessitates some delicate maneuvering in Carleman's proof. The techniques below circumvent this problem.

Exercise 8.30 (Poincaré [1892]) *Let* $f\colon \mathbb{R} \mapsto \mathbb{C}$ *be continuous. Construct a zero-free entire function* g *which is real-valued on* $\mathbb{R}$ *and satisfies*

$$g(x) > |f(x)| \qquad \forall x \in \mathbb{R}.$$

Hints: (from Scheinberg [1976]; cf. also Newman and Barlaz [1953]) For each $n \in \mathbb{N}$ let

$$M_n := \max\{|f(x)| : |x| \le n+1\} \tag{1}$$

and for $n \ge 2$ choose integer k_n to satisfy both

$$k_n \ge n \tag{2}$$

and

$$\left[\frac{n^2}{n+1}\right]^{k_n} > M_n. \tag{3}$$

If $|z| \le N$, then $\left|z^2/(n+1)\right| < \frac{1}{2}$ for all $n \ge 2N^2$, so from (2) we see that the series

$$M_1 + \sum_{n=2}^{\infty} \left[\frac{z^2}{n+1}\right]^{k_n} \tag{4}$$

converges uniformly for $z \in \bar{D}(0, N)$. This is the case for every N, so (4) defines an entire function h. Evidently for $|x| < 2$, $h(x) \geq M_1 \geq |f(x)|$; while for $2 \leq n \leq |x| < n + 1$, we have

$$h(x) \geq \left[\frac{x^2}{n+1}\right]^{k_n} \geq \left[\frac{n^2}{n+1}\right]^{k_n} > M_n \geq |f(x)|.$$

Therefore $h(x) \geq |f(x)|$ for all real x. It remains only to set $g := e^h$ to get a strict majorant that is zero-free in $\mathbb{C}$.

Exercise 8.31 (i) *Let $f\colon \mathbb{R} \mapsto \mathbb{C}$ be continuous. Then for each integer n there exists a continuous function $f_n\colon \mathbb{R} \mapsto \mathbb{C}$ which is supported in $[-1, 1]$ such that*

$$f(x) = \sum_{n=-\infty}^{\infty} f_n(x - n) \qquad \forall x \in \mathbb{R}.$$

Hints: Let ϕ be the function on $\mathbb{R}$ which is 0 outside $(-1, 1)$, is 1 on $\left[-\frac{1}{2}, \frac{1}{2}\right]$ and is linear on the adjacent intervals. Then set

$$\Phi(x) := \sum_{n=-\infty}^{\infty} \phi(x - n) \qquad \forall x \in \mathbb{R}.$$

We have

$$\Phi(x) = \sum_{n=-N-1}^{N+1} \phi(x - n) \qquad \forall x \in [-N, N],$$

so Φ is continuous at every point of $\mathbb{R}$. Also, given t, we can select an integer n so that $n - \frac{1}{2} \leq t \leq n + \frac{1}{2}$. Then $\phi(t - n) = 1$, so $\Phi(t) \geq 1$. Finally set

$$f_n(x) := \frac{\phi(x) f(x + n)}{\Phi(x + n)}, \qquad x \in \mathbb{R}.$$

(ii) *Let $f\colon \mathbb{R} \mapsto \mathbb{C}$ be continuous and have support in $[-1, 1]$. Define the set $S := \{z \in \mathbb{C} : |\mathrm{Re}\, z| > 3 \;\&\; |\mathrm{Re}\, z| > 2|\mathrm{Im}\, z|\}$. Given $\varepsilon > 0$, there exists an entire function F such that*

(1) $$|f(x) - F(x)| < \varepsilon \qquad \forall x \in \mathbb{R}$$

(2) $$|F(z)| < \varepsilon \qquad \forall z \in S.$$

Hints: For each positive integer k form

$$f_k(z) := \frac{k}{I} \int_{-1}^{1} e^{-k^2(z-t)^2} f(t)\, dt, \quad z \in \mathbb{C}, \qquad I := \int_{-\infty}^{\infty} e^{-x^2}\, dx.$$

Upon expanding the exponential here into a series and interchanging that sum with the integral, we have a power series for f_k. Thus f_k is entire. By 8.29(ii) with $a = -2$, $b = 2$ we have $f_k \to f$ uniformly in $\left[-\frac{3}{2}, \frac{3}{2}\right]$ and with $a = -1$, $b = 1$ we have $f_k \to 0$ uniformly in $\mathbb{R} \setminus \left[-\frac{3}{2}, \frac{3}{2}\right]$. Thus

$$f_k \to f \quad \text{uniformly on } \mathbb{R}. \tag{3}$$

For $z = x + iy \in S$ and $t \in [-1, 1]$ we have

$$\begin{aligned}\operatorname{Re} k^2(z-t)^2 = k^2\big((x-t)^2 - y^2\big) &= k^2x^2\left(1 - \frac{2t}{x} + \frac{t^2}{x^2} - \frac{y^2}{x^2}\right)\\ &\geq k^2x^2\left(1 - \frac{2}{|x|} - \left|\frac{y}{x}\right|^2\right)\\ &\geq k^2x^2\left(1 - \frac{2}{3} - \left(\frac{1}{2}\right)^2\right) = \frac{k^2x^2}{12} > \frac{3k^2}{4}\end{aligned}$$

and so

$$\begin{aligned}|f_k(z)| &\leq \frac{k}{I}\int_{-1}^{1}\left|e^{-k^2(z-t)^2}f(t)\right|dt = \frac{k}{I}\int_{-1}^{1} e^{-\operatorname{Re} k^2(z-t)^2}|f(t)|\,dt\\ &\leq \frac{k}{I}e^{-3k^2/4}\int_{-1}^{1}|f(t)|\,dt.\end{aligned}$$

Since $e^{3k^2/4} > 3k^2/4$, this gives

$$|f_k(z)| \leq \frac{k}{I}\cdot\frac{4}{3k^2}\cdot\int_{-1}^{1}|f| = \frac{1}{k}\cdot\frac{4}{3I}\cdot\int_{-1}^{1}|f| \qquad \forall z \in S. \tag{4}$$

We get (1) and (2) from (3) and (4) by taking $F := f_k$ for sufficiently large k.

(iii) *Let $f \colon \mathbb{R} \mapsto \mathbb{C}$ be continuous. Then there exists an entire function F such that*

$$|f(x) - F(x)| < 1 \qquad \forall x \in \mathbb{R}.$$

Hints: Let f_n be as furnished by (i). For each n, part (ii) provides an entire function F_n such that

$$|f_n(t) - F_n(t)| < 2^{-|n|-2} \qquad \forall t \in \mathbb{R} \tag{1.n}$$

$$|F_n(z)| < 2^{-|n|} \qquad \forall z \in S. \tag{2.n}$$

If $|z| \leq N$ and $|n| > 3N + 3$, then

$$|\operatorname{Re}(z - n)| \geq |n| - |\operatorname{Re} z| \geq |n| - N > 3$$

and

$$|\mathrm{Im}(z-n)| = |\mathrm{Im}\, z| \le N < \tfrac{1}{2}\big(|n|-N\big) \le \tfrac{1}{2}|\mathrm{Re}(z-n)|.$$

It follows that $z-n \in S$, and so from $(2.n)$

$$|F_n(z-n)| < 2^{-|n|}, \quad \text{if } |z| \le N \ \ \& \ \ |n| > 3N+3.$$

Therefore the series $\sum_{n=-\infty}^{\infty} F_n(z-n)$ converges uniformly for $z \in \overline{D}(0,N)$, for each positive integer N, and consequently defines an entire function F. For this F and all real x we have

$$\begin{aligned}
|F(x) - f(x)| &= \left| \sum_{n=-\infty}^{\infty} F_n(x-n) - \sum_{n=-\infty}^{\infty} f_n(x-n) \right| \\
&\le \sum_{n=-\infty}^{\infty} |F_n(x-n) - f_n(x-n)| \\
&< \sum_{n=-\infty}^{\infty} 2^{-|n|-2} \quad \text{by } (1.n) \\
&= \frac{3}{4}.
\end{aligned}$$

Theorem 8.32 (CARLEMAN [1927]) *Let $f\colon \mathbb{R} \mapsto \mathbb{C}$, $\varepsilon\colon \mathbb{R} \mapsto (0,\infty)$ be continuous. Then there exists an entire function g such that*

$$|f(x) - g(x)| < \varepsilon(x) \qquad \forall x \in \mathbb{R}.$$

Proof: 8.30 provides a zero-free entire function h such that

$$h(x) > \frac{1}{\varepsilon(x)} \qquad \forall x \in \mathbb{R}$$

and the preceding exercise provides an entire function F such that

$$\big|(hf)(x) - F(x)\big| < 1 \qquad \forall x \in \mathbb{R}.$$

Consequently

$$\left| f(x) - \frac{F(x)}{h(x)} \right| < \frac{1}{h(x)} < \varepsilon(x) \qquad \forall x \in \mathbb{R}$$

and we take for g the entire function F/h.

§ 6 Harmonic Functions in a Half-Plane

Corollary 8.33 (Dirichlet Problem in a Half-Plane) *If $g\colon \mathbb{R} \mapsto \mathbb{R}$ is continuous, then there exists f holomorphic in $\mathbb{U}$ such that the function G defined by*

$$G(z) := \begin{cases} \operatorname{Re} f(z), & z \in \mathbb{U} \\ g(z), & z \in \mathbb{R}. \end{cases}$$

is continuous on $\overline{\mathbb{U}} = \mathbb{U} \cup \mathbb{R}$.

Proof: First set $\varepsilon(x) := 1/(1+|x|)$ and apply Carleman's theorem to come up with an entire function G_1 such that

$$|G_1(x) - g(x)| < \varepsilon(x) \qquad \forall x \in \mathbb{R}.$$

Therefore, by replacing g with the function $g_1(x) := [g - \operatorname{Re} G_1](x)$, it suffices to suppose in addition that the given function g satisfies

$$(*) \qquad \lim_{|x|\to\infty} g(x) = 0.$$

The function

$$\phi(z) := i\frac{1+z}{1-z}$$

maps $\mathbb{D}$ onto $\mathbb{U}$, $\mathbb{T} \setminus \{1\}$ onto $\mathbb{R}$ and has inverse

$$\psi(w) := \frac{w-i}{w+i}.$$

Since $\lim_{z\to 1} |\phi(z)| = \infty$, the continuous function $g \circ \phi$ on $\mathbb{T} \setminus \{1\}$ has by $(*)$ an extension to a continuous function h on $\mathbb{T}$. Solve the Dirichlet Problem in the Disk for h [5.20] and then compose this solution with ψ.

Remark: Using ϕ, ψ above we can carry the Poisson formula for $\mathbb{D}$ into one for $\mathbb{U}$ (see pp. 121–123 of Hoffman [1962] for details), motivating and providing an alternative development of the following; Lange and Walsh [1985] and Needham [1994] give a valuable heuristic derivation.

Exercise 8.34 (Poisson Formula for a Half-Plane) *Let $f\colon \mathbb{R} \mapsto \mathbb{R}$ be continuous and bounded. For each positive integer n define*

$$F_n(z) := \frac{1}{\pi}\int_{-n}^{n} \frac{\operatorname{Im} z}{|z-t|^2} f(t)\,dt, \qquad z \in \mathbb{U}.$$

(i) *Show that each F_n is harmonic in $\mathbb{U}$.*

Hint: Use 2.14, noting that $\frac{\operatorname{Im} z}{|z-t|^2} = \operatorname{Re}\left(\frac{i}{z-t}\right)$, a function known as the *Poisson kernel in $\mathbb{U}$.*

(ii) *Show that for all* $x \in \mathbb{R}$, $y \in (0,\infty)$, $\frac{1}{\pi}\int_{-\infty}^{\infty} \frac{y}{(x-t)^2+y^2}\,dt = 1$.

Hint: Make an appropriate change of variable as in the proof of 3.22.

(iii) *Show that the sequence* $\{F_n\}$ *converges uniformly on each compact subset of* $\mathbb{U}$ *as* $n \to \infty$.

Hint: Note that $|z-t|^2 = (\operatorname{Re} z - t)^2 + (\operatorname{Im} z)^2$ and use (ii).

(iv) *Letting* F *denote the limit in* (iii), *show that* F *is harmonic in* $\mathbb{U}$.

Hint: 5.44(iii).

(v) *Show that* $\lim_{z\to x, z\in\mathbb{U}} F(z) = f(x)$ *for each* $x \in \mathbb{R}$, *and so* F *is (extendable to be) continuous on* $\overline{\mathbb{U}}$.

Hint: Modify the techniques used in the proof of 5.20, using now (ii). (See the hints to 8.35(vi) below.)

(vi) *Show that* F *is bounded: in fact* $\sup_{z\in\mathbb{U}} |F(z)| = \sup_{x\in\mathbb{R}} |f(x)|$.

Hint: Use (ii) again.

(vii) *Let* $g\colon \overline{\mathbb{U}} \mapsto \mathbb{R}$ *be bounded and continuous and be harmonic in* $\mathbb{U}$. *Form*

$$G(z) := \frac{1}{\pi}\int_{-\infty}^{\infty} \frac{\operatorname{Im} z}{|z-t|^2} g(t)\,dt = \lim_{n\to\infty} \frac{1}{\pi}\int_{-n}^{n} \frac{\operatorname{Im} z}{|z-t|^2} g(t)\,dt, \qquad z \in \mathbb{U}.$$

Show that $G - g \le 0$ *in* $\mathbb{U}$.

Hint: According to (iv), (v) and (vi) the function $u := G - g$ satisfies the hypothesis of 7.17, with $E := \varnothing$, $M := 0$.

(viii) *Show that* $G = g$ *in* $\mathbb{U}$.

Hint: Apply (vii) with $-g$ in the role of g to secure the opposite inequality.

(ix) *Explore integrability hypotheses on the continuous function* f, *weaker than the boundedness assumption above, under which all the conclusions* (i)–(v) *remain valid. One such is* $\int_{-\infty}^{\infty} |f(t)| \,/\, \left(1+t^2\right) dt < \infty$. (For other conditions see, e.g., §2 chapter V of B. Levin [1980].) We shall see in the next exercise that a sufficient condition for (i)–(v) to hold is that f be the restriction of a non-negative harmonic function in the upper-half plane.

Exercise 8.35 (i) *Let* $f\colon \mathbb{R} \mapsto \mathbb{R}$ *be bounded and continuous save at the points of a finite set* S. *Show that* (i)–(vi) *of* 8.34 *remain valid if in* (v) *and* (vi) *only* $x \in \mathbb{R}\setminus S$ *are considered and* F *is extended only to* $\overline{\mathbb{U}}\setminus S$. *An important special case is* f *the indicator function of a compact interval, the right-half-plane version of which was treated in* 3.45(iii).

(ii) *Let $g\colon \overline{\mathbb{U}} \mapsto [0,\infty)$ be continuous and harmonic in $\mathbb{U}$. Show that for positive a and b the function*

$$(*) \qquad G_{a,b}(z) := \frac{1}{\pi}\int_{-a}^{a} \frac{\operatorname{Im} z}{|z-t|^2} g(t+ib)\,dt, \quad z \in \mathbb{U},$$

satisfies

(1) $G_{a,b}$ *is non-negative, bounded and harmonic in* $\mathbb{U}$,

(2) $$\lim_{\substack{z\to x\\ z\in\mathbb{U}}} G_{a,b}(z) = \begin{cases} g(x+ib), & x\in(-a,a)\\ 0 & x\in\mathbb{R}\setminus[-a,a], \end{cases}$$

(3) $$\lim_{\substack{|z|\to\infty\\ z\in\mathbb{U}}} G_{a,b}(z) = 0.$$

Hints: (1) and (2) are from (i) with $f(t) := g(t+ib)$ for $t\in[-a,a]$ and $f(t) := 0$ for $t\in\mathbb{R}\setminus[-a,a]$. For the proof of (3), consider $z\in\mathbb{U}$ with $|z|\geq 2a$. Then $|z-t|\geq|z|-|t|\geq|z|-a\geq|z|/2$, so $|z-t|^2\geq|z|^2/4$. It follows that

$$G_{a,b}(z) \leq \frac{1}{\pi}\cdot\frac{4}{|z|^2}\cdot\operatorname{Im} z\int_{-a}^{a} g(t+ib)\,dt \leq \frac{4}{\pi|z|}\int_{-a}^{a} g(t+ib)\,dt.$$

(iii) *With g and $G_{a,b}$ as in* (ii) *show that*

$$G_{a,b}(z) \leq g(z+ib) \qquad \forall z\in\mathbb{U}.$$

Hints: Let $z_0\in\mathbb{U}$ and $\varepsilon>0$ be given. Use (3) to select $r>|z_0|$ such that $G_{a,b}(z)\leq\varepsilon$ on $C(0,r)\cap\mathbb{U}$. From this, the non-negativity of g and (2) above we have

$$\lim_{\substack{z\to\zeta\\ z\in\mathbb{U}}}\big[G_{a,b}(z)-g(z+ib)\big]\leq\varepsilon \qquad \forall\zeta\in\partial[D(0,r)\cap\mathbb{U}]\setminus\{\pm a\}.$$

Infer via 5.46 that

$$G_{a,b}(z)-g(z+ib)\leq\varepsilon \qquad \forall z\in D(0,r)\cap\mathbb{U}.$$

Note that hypothesis (2) of 5.46 is satisfied because both participating functions are bounded near the exceptional set $E := \{-a,a\}$. Since $z_0\in D(0,r)\cap\mathbb{U}$, this yields

$$G_{a,b}(z_0)\leq g(z_0+ib)+\varepsilon$$

and as $\varepsilon>0$ is arbitrary here, the desired conclusion follows.

(iv) *Show that*

$$G_a(z) := \frac{1}{\pi}\int_{-a}^{a} \frac{\operatorname{Im} z}{|z-t|^2} g(t)\,dt \le g(z) \qquad \forall z \in \mathbb{U},\ a > 0.$$

Hint: Let $b \downarrow 0$ in (iii).

(v) *Show that*

$$G(z) := \lim_{a\to\infty} \frac{1}{\pi}\int_{-a}^{a} \frac{\operatorname{Im} z}{|z-t|^2} g(t)\,dt =: \frac{1}{\pi}\int_{-\infty}^{\infty} \frac{\operatorname{Im} z}{|z-t|^2} g(t)\,dt$$

exists for each $z \in \mathbb{U}$ and defines a harmonic function in $\mathbb{U}$.

Hint: The integrand being non-negative, the integral increases with a. Cite 7.15 and (iv) above.

(vi) *Show that*

$$\lim_{\substack{z\to x\\ z\in\mathbb{U}}} G(z) = g(x) \qquad \forall x \in \mathbb{R}.$$

Hints: Let $x \in \mathbb{R}$ and $\varepsilon > 0$ be given. Set $g^* := g + g(x)$. This is a non-negative harmonic function in $\mathbb{U}$ for which the conclusions (ii)–(v) above hold. Choose $\delta > 0$ so that $|g(t) - g(x)| < \varepsilon$ whenever $t \in \mathbb{R}$ and $|t - x| \le 8\delta$. Then, using 8.34(ii), we have for $z \in \mathbb{U}$

$$\begin{aligned}
|G(z) - g(x)| &= \frac{1}{\pi}\left|\int_{-\infty}^{\infty} \frac{\operatorname{Im} z}{|z-t|^2}\,[g(t) - g(x)]\,dt\right| \\
&\le \frac{1}{\pi}\int_{x-8\delta}^{x+8\delta} \frac{\operatorname{Im} z}{|z-t|^2}|g(t) - g(x)|\,dt + \frac{1}{\pi}\int_{|t-x|\ge 8\delta} \frac{\operatorname{Im} z}{|z-t|^2} g^*(t)\,dt \\
&\le \frac{\varepsilon}{\pi}\int_{x-8\delta}^{x+8\delta} \frac{\operatorname{Im} z}{|z-t|^2}\,dt + \frac{\operatorname{Im} z}{\pi}\int_{|t-x|\ge 8\delta} \frac{g^*(t)}{|z-t|^2}\,dt \\
&\le \varepsilon + \frac{\operatorname{Im} z}{\pi}\int_{|t-x|\ge 8\delta} \frac{g^*(t)}{|z-t|^2}\,dt \quad \forall z \in \mathbb{U}. \qquad (5)
\end{aligned}$$

Now consider $z \in D(x, \delta) \cap \mathbb{U}$ and $t \in \mathbb{R} \setminus [x - 8\delta, x + 8\delta]$. We have

$$|x + 2i\delta - t| \ge |x - t| - 2\delta \ge 6\delta \qquad (6)$$

so

$$\begin{aligned}
|z - t| &\ge |x + 2i\delta - t| - |z - x - 2i\delta| \\
&\ge |x + 2i\delta - t| - |z - x| - 2\delta \\
&\ge |x + 2i\delta - t| - 3\delta \\
&\ge \tfrac{1}{2}|x + 2i\delta - t|, \quad \text{by (6).} \qquad (7)
\end{aligned}$$

From (7) and (5) we get

$$\begin{aligned}|G(z)-g(x)| &\le \varepsilon + \frac{\operatorname{Im} z}{\pi}\int_{|t-x|\ge 8\delta} \frac{4g^*(t)}{|x+2i\delta-t|^2}\,dt \\ &\le \varepsilon + \frac{\operatorname{Im} z}{\delta}\cdot\frac{2}{\pi}\int_{-\infty}^{\infty} \frac{2\delta g^*(t)}{|x+2i\delta-t|^2}\,dt \\ &=: \varepsilon + \frac{\operatorname{Im} z}{\delta} 2G^*(x+2i\delta) \\ &\le \varepsilon + \frac{\operatorname{Im} z}{\delta} 2g^*(x+2i\delta), \quad \text{by (iv) and (v) applied to } g^*.\end{aligned}$$

Since x is real, it follows that

$$\varlimsup_{\substack{z\to x \\ z\in\mathbb{U}}} |G(z)-g(x)| \le \varepsilon.$$

The next exercise is one of many *reflection principles* in function theory.

Exercise 8.36 (i) *Let* $a\in\mathbb{R}$, $r>0$, $D:=D(a,r)$, $D^+ := \{z\in D : \operatorname{Im} z > 0\}$. *Let* $h\colon D^+ \mapsto \mathbb{R}$ *be harmonic and satisfy* $\lim_{z\to x} h(z) = 0$ *for each* $x\in(a-r,a+r)$. *Define*

$$H(z) := \begin{cases} h(z) & z\in D^+ \\ 0 & z\in(a-r,a+r) \\ -h(\bar z) & \bar z \in D^+ \end{cases}$$

and show that H *is harmonic in* D.

Hints: Evidently H is continuous in D and harmonic in each of the open subsets D^+ and $\{z : \bar z\in D^+\}$. Consequently, for any z in either of these sets we have $2\pi H(z) = \int_0^{2\pi} H(z+\rho e^{it})\,dt$ for all sufficiently small $\rho>0$. (See 5.21(iii).) On the other hand, if $x\in(a-r,a+r)$ and $\rho>0$ is small enough that $D(x,\rho)\subset D$, then

$$\begin{aligned}\int_0^{2\pi} H(x+\rho e^{it})\,dt &= \int_0^{\pi} H(x+\rho e^{it})\,dt + \int_{\pi}^{2\pi} H(x+\rho e^{i\tau})\,d\tau \\ &= \int_0^{\pi} H(x+\rho e^{it})\,dt - \int_{\pi}^{0} H\left(x+\rho e^{i(2\pi-t)}\right)dt \\ &\quad (\tau := 2\pi - t) \\ &= \int_0^{\pi} \left[H(x+\rho e^{it}) + H(x+\rho e^{-it})\right]dt \\ &= 0, \quad \text{since } H(\bar z) = -H(z) \text{ for all } z\in D, \\ &= H(x).\end{aligned}$$

It now follows either from 5.23(i) that H is harmonic in D. (For another elegant, simple proof, see PHRAGMÉN [1890–91].)

(ii) *Notation as in* (i), *suppose* $f \in H(D^+)$ *and* $\operatorname{Re} f$ *tends to* 0 *at each point in* $(a-r, a+r)$. *Show that* f *has a holomorphic extension to* D.

Hint: Part (i) above and 5.21(i) furnish an $F \in H(D)$ such that $\operatorname{Re} F = \operatorname{Re} f$ in D^+. It follows from 2.16(ii) that $F - f$ is constant in the connected set D^+.

Theorem 8.37 *If* $h\colon \mathbb{U} \mapsto [0, \infty)$ *is a harmonic function such that* $\lim_{z\to x} h(z) = 0$ *for every real* x, *then* $h(z) \equiv c \operatorname{Im} z$ *for some non-negative constant* c.

Proof: (LOOMIS and WIDDER [1942]) From 8.36(ii) we get an $F \in H(\mathbb{C})$ with $\operatorname{Im} F = h$ in $\mathbb{U}$. Since F is thus real-valued on $\mathbb{R}$, the holomorphic function $F(z) - \bar{F}(\bar{z})$ is 0 for all $z \in \mathbb{R}$. Consequently

$$(1) \qquad F(z) = \bar{F}(\bar{z}) \qquad \forall z \in \mathbb{C}.$$

If we write the power series $F(z) = \sum_{n=0}^{\infty} c_n z^n$, then (1) and the uniqueness of the c_n show that $c_n = \bar{c}_n$ for all n, so all c_n are real and we have

$$h(re^{i\theta}) = \operatorname{Im} F(re^{i\theta}) = \sum_{n=0}^{\infty} c_n \operatorname{Im}(re^{i\theta})^n$$

$$(2) \qquad = \sum_{n=0}^{\infty} c_n r^n \sin(n\theta) \qquad \forall r > 0,\ \theta \in \mathbb{R}.$$

Using the orthogonality relations (from $2\sin u \sin v = \cos(u-v) - \cos(u+v)$)

$$\int_0^{\pi} \sin(n\theta)\sin(k\theta)\, d\theta = \begin{cases} \pi/2 & \text{if } n = k \neq 0 \\ 0 & \text{if } n \neq k \end{cases}$$

and integrating term by term the uniformly (for each fixed r) convergent series arising from (2), we get for $k \in \mathbb{N}$

$$(3) \qquad \int_0^{\pi} h(re^{i\theta}) \sin(k\theta)\, d\theta = \sum_{n=0}^{\infty} c_n r^n \int_0^{\pi} \sin(n\theta)\sin(k\theta)\, d\theta = \frac{\pi}{2} c_k r^k,$$

$$|c_k| \le \frac{2}{\pi} r^{-k} \int_0^{\pi} \left| h(re^{i\theta}) \sin(k\theta) \right| d\theta.$$

Since $h \ge 0$, this reads

$$(4) \qquad |c_k| \le \frac{2}{\pi} r^{-k} \int_0^{\pi} h(re^{i\theta}) |\sin(k\theta)|\, d\theta \qquad \forall r > 0,\ k \in \mathbb{N}.$$

But an easy induction confirms that

$$(5) \qquad |\sin(k\theta)| \leq k|\sin\theta| \qquad \forall\theta \in \mathbb{R},\ k \in \mathbb{N}.$$

Indeed, equality holds in (5) when $k = 1$ and if (5) holds for $k = m$, then

$$\begin{aligned}|\sin(m+1)\theta| &= |\sin(m\theta)\cos\theta + \cos(m\theta)\sin\theta| \\ &\leq |\sin(m\theta)| + |\sin\theta| \leq m|\sin\theta| + |\sin\theta|,\end{aligned}$$

giving (5) for $k = m + 1$. Use of (5) in (4) gives

$$\begin{aligned}|c_k| &\leq \frac{2}{\pi} r^{-k} \int_0^\pi kh\big(re^{i\theta}\big)\sin\theta\, d\theta = \frac{2}{\pi} r^{-k} k\Big(\frac{\pi}{2} c_1 r\Big) \quad \text{by (3)} \\ &= \frac{kc_1}{r^{k-1}}.\end{aligned}$$

This inequality holds for all $k \in \mathbb{N}$ and all $r > 0$. For $k \geq 2$ we let $r \to \infty$ and conclude that $c_k = 0$. It follows that $F(z) \equiv c_0 + c_1 z$ with real coefficients c_0 and c_1. Consequently $h(z) = \operatorname{Im} F(z) = c_1 \operatorname{Im} z$, for all $z \in \mathbb{U}$.

Remarks: (i) An amusing consequence of this theorem is the following: *If g is an entire function which is real-valued only on $\mathbb{R}$, then g is linear.* For by a connectedness argument, $h := \operatorname{Im} g$ is of constant sign in $\mathbb{U}$, and in $\mathbb{C} \setminus \overline{\mathbb{U}}$. If it is of the same sign in both, then one of $e^{\pm ig}$ is a bounded entire function and g is constant. If they are of different signs, then continuity causes $h(\mathbb{R})$ to be 0. By the theorem, $h(z) = c \operatorname{Im} z$ for some $c \in \mathbb{R}$. It follows that $\operatorname{Im}[g(z) - cz] = 0$ for all $z \in \mathbb{U}$ and then $g(z) - cz$ is a real constant. Of course, the conclusion follows whenever there is a pair of lines L_1, L_2 such that $g^{-1}(L_2) \subset L_1$, as noted by Evard, Jafari and Moorhouse [1994]. For an alternative, short and clever derivation based on 12.30(i), see D. J. Newman [1960].

(ii) There are naturally equivalent versions of 8.37 in $\mathbb{D}$, realized by conformal mapping, which characterize the Poisson kernel in $\mathbb{D}$, that is, the family of harmonic functions $P_u(z) := \frac{1-|z|^2}{|u-z|^2} = \operatorname{Re}\big(\frac{u+z}{u-z}\big)$, $z \in \mathbb{D}$, $u \in \mathbb{T}$. For each such u, $\phi_u(z) := i\big(\frac{u+z}{u-z}\big)$ maps $\mathbb{D}$ conformally onto $\mathbb{U}$ and $\overline{\mathbb{D}} \setminus \{u\}$ homeomorphically onto $\overline{\mathbb{U}}$ (cf. 2.5). Consequently, if H is a positive harmonic function on $\mathbb{D}$, $H(0) = 1$ and H converges to 0 at each point of $\mathbb{T} \setminus \{u\}$, then $H \circ \phi_u^{-1}$ is an eligible h in 8.37, according to which $H(\phi_u^{-1}(w)) = \operatorname{Im} w$ for all $w \in \mathbb{U}$, that is, $H(z) = \operatorname{Im}\phi_u(z) = P_u(z)$ for all $z \in \mathbb{D}$.

Corollary 8.38 *Let g be a continuous function on $\overline{\mathbb{U}}$ which is harmonic and non-negative in $\mathbb{U}$. Then for some $c \geq 0$*

$$g(z) = c \operatorname{Im} z + \frac{1}{\pi} \int_{-\infty}^{\infty} \frac{\operatorname{Im} z}{|z - t|^2} g(t)\, dt \qquad \forall z \in \mathbb{U}.$$

Proof: According to 8.35(iv)–(vi) the integral

$$G(z) := \frac{1}{\pi}\int_{-\infty}^{\infty} \frac{\operatorname{Im} z}{|z-t|^2} g(t)\,dt$$

is finite for every $z \in \mathbb{U}$ and defines a harmonic function which satisfies

$$\begin{aligned} G &\le g \quad \text{in } \mathbb{U}, \\ \lim_{\substack{z\to x\\ z\in\mathbb{U}}} G(z) &= g(x) \qquad \forall x \in \mathbb{R}. \end{aligned}$$

Therefore $h := g - G$ fulfills all the hypotheses of 8.37.

Exercise 8.39 *Let g, G, G_a be as* 8.35(ii)–(v). *For each $\varepsilon > 0$ set*

$$S_\varepsilon := \{z \in \mathbb{C} : \operatorname{Im} z > \varepsilon |z|\}.$$

(i) *Show that*

$$\lim_{\substack{|z|\to\infty\\ z\in S_\varepsilon}} \frac{G(z)}{\operatorname{Im} z} = 0 \qquad \forall\, 0 < \varepsilon < 1.$$

Hint: For $z := x + iy$ with $y > \varepsilon|z| \ge \varepsilon|x|$ and $|t| \ge 1$ we have

$$\begin{aligned} |z-t|^2 &= (x-t)^2 + y^2 > t^2 - 2tx + (1+\varepsilon^2)x^2 \\ &= t^2 - \frac{t^2}{1+\varepsilon^2} + (1+\varepsilon^2)\left[x - \frac{t}{1+\varepsilon^2}\right]^2 \\ &\ge \frac{\varepsilon^2 t^2}{1+\varepsilon^2} \ge \frac{\varepsilon^2(1+t^2)}{2(1+\varepsilon^2)}. \end{aligned}$$

Consequently, for all $a \ge 1$

$$\begin{aligned} \frac{G(z)}{\operatorname{Im} z} &= \frac{G_a(z)}{\operatorname{Im} z} + \frac{1}{\pi}\int_{|t|>a} \frac{g(t)}{|z-t|^2}\,dt \le \frac{G_a(z)}{\operatorname{Im} z} + \frac{2(1+\varepsilon^2)}{\pi\varepsilon^2}\int_{|t|>a} \frac{g(t)}{1+t^2}\,dt \\ &= \frac{G_a(z)}{\operatorname{Im} z} + \frac{2(1+\varepsilon^2)}{\varepsilon^2}[G(i) - G_a(i)]. \end{aligned}$$

For large a the second summand is small, and the integral defining G_a makes clear that for fixed a the first summand tends to 0 as $|z| \to \infty$.

(ii) *Use* (i) *and the representation* 8.38 *to prove,* mutatis mutandis, *the special case of the* Carathéodory-Julia-Wolff *theorem* 6.27 *in which the function is continuous on the closed half-plane. I.e., with the notation and hypotheses of* 8.38 *use* (i) *above to show that for each* $0 < \varepsilon < 1$

(∗) $$\lim_{\substack{|z|\to\infty \\ z\in S_\varepsilon}} \frac{g(z)}{\operatorname{Im} z} = \inf\left\{\frac{g(z)}{\operatorname{Im} z} : \operatorname{Im} z > 0\right\}.$$

(iii) *Conserving all the notation and hypotheses of* (ii) *except the continuity of* g *on the closed half-plane, prove that* (∗) *still holds.*

Hint: For each fixed (small) positive b, apply (ii) to the function $g_b(z) := g(z+ib)$. For any L greater than the infimum in (∗) there exists $z_L \in \mathbb{U}$ with $g(z_L)/\operatorname{Im}(z_L) < L$ and for $b > 0$ sufficiently small $g_b(z_L)/\operatorname{Im}(z_L) < L$. Routine maneuvers with the triangle inequality confirm that $z \in S_{2\varepsilon}$ and $|z| > (1+\varepsilon^{-1})b$ imply that $w := z - ib \in S_\varepsilon$. Therefore

$$\varlimsup_{\substack{|z|\to\infty \\ z\in S_{2\varepsilon}}} \frac{g(z)}{\operatorname{Im} z} = \varlimsup_{\substack{|z|\to\infty \\ z\in S_{2\varepsilon}}} \frac{g_b(z-ib)}{\operatorname{Im}(z-ib)}, \quad \text{since} \quad \frac{\operatorname{Im} z - b}{\operatorname{Im} z} \to 1,$$

$$\le \varlimsup_{\substack{|w|\to\infty \\ w\in S_\varepsilon}} \frac{g_b(w)}{\operatorname{Im} w} = \inf\left\{\frac{g_b(w)}{\operatorname{Im} w} : w \in \mathbb{U}\right\} \le \frac{g_b(z_L)}{\operatorname{Im}(z_L)} < L.$$

(iv) *Use the representation* 8.38, *adapted to* $\mathbb{H}$, *to secure the version of* 5.60 *in which* $\operatorname{Re} f$ *is continuous on the closed half-plane but the hypothesis* 5.60(i) *is weakened to*

(∗∗) $$\lim_{\substack{x\to\infty \\ x\in\mathbb{R}}} \frac{\operatorname{Re} f(x)}{x} = 0.$$

Hint: (∗∗) and the non-negativity of everything in sight insure that the c of 8.38 is 0. Now recall 8.34(v) and (vi).

Exercise 8.40 *Let* L *denote the Principal Branch of the Logarithm in* $\mathbb{C} \setminus (-\infty, 0]$, *and*

$$\phi(z) := \frac{1+z}{1-z}, \quad z \in \mathbb{C} \setminus \{1\}.$$

(i) *Show that* ϕ *maps* $\overline{\mathbb{D}} \setminus \{\pm 1\}$ *onto* $\overline{\mathbb{H}} \setminus \{0\}$ *and that the function* $g := \operatorname{Im} L \circ \phi$ *satisfies*

(1)

g is bounded and continuous in $\overline{\mathbb{D}} \setminus \{\pm 1\}$ and harmonic in $\mathbb{D}$,

(2)

$$g\,(-1, 1) = 0,$$

(3)

$$g(\mathbb{U} \cap \mathbb{T}) = \pi/2.$$

Hint: Examine the proof of 5.47, up to the introduction of the square-root transformation used there.

(ii) *Show that*

$$g(z) = \tan^{-1}\left(\frac{2\,\mathrm{Im}\, z}{1-|z|^2}\right) \qquad \forall z \in \mathbb{D}.$$

Hints: If $-\pi/2 < \theta < \pi/2$, $r > 0$ and $w = re^{i\theta}$, then

$$\tan\theta = \frac{\sin\theta}{\cos\theta} = \frac{\mathrm{Im}\, w}{\mathrm{Re}\, w}.$$

Since by definition of L, $\mathrm{Im}\, L(w) = \theta$, it follows that $\mathrm{Im}\, L(w) = \tan^{-1}\left(\frac{\mathrm{Im}\, w}{\mathrm{Re}\, w}\right)$. Finally, note that by direct computation

$$\frac{\mathrm{Im}\,\phi(z)}{\mathrm{Re}\,\phi(z)} = \frac{2\,\mathrm{Im}\, z}{1-|z|^2} \qquad \forall z \in \mathbb{D}.$$

(iii) *Show that*

$$\lim_{r\to\infty} rg(z/r) = 2\,\mathrm{Im}\, z \qquad \forall z \in \mathbb{C}.$$

Hints: Fix $z \in \mathbb{C}$. Of course, only $r > |z|$ will then be considered. By (ii)

$$\begin{aligned} rg(z/r) &= r\tan^{-1}\left(\frac{2\,\mathrm{Im}\, z/r}{1-|z/r|^2}\right) \\ &= \frac{\tan^{-1}\left(\frac{2\,\mathrm{Im}\, z/r}{1-|z/r|^2}\right) - \tan^{-1}(0)}{\frac{2\,\mathrm{Im}\, z/r}{1-|z/r|^2}} \cdot \frac{2\,\mathrm{Im}\, z}{1-|z/r|^2}. \end{aligned}$$

As r grows, the expression $2(\mathrm{Im}\, z/r)\,/\,\left(1-|z/r|^2\right)$ approaches 0, so the first factor on the right-hand side approaches $(\tan^{-1})'(0) = 1/\tan'(0) = \cos^2(0) = 1$, while the second factor on the right-hand side obviously approaches $2\,\mathrm{Im}\, z$.

Exercise 8.41 (F. and R. Nevanlinna [1922]) *Let h be a harmonic function in $\mathbb{U}$ which satisfies*

$$(*) \qquad \varliminf_{\substack{z\to x \\ z\in\mathbb{U}}} h(z) \geq 0 \qquad \forall x \in \mathbb{R}.$$

For each $r > 0$ set $m(r) := \inf\{h(z) : z \in \mathbb{U} \cap C(0,r)\}$ and suppose that

$$(**) \qquad b := \varlimsup_{r\to\infty} \frac{m(r)}{r} > -\infty.$$

(i) *Use the preceding exercise and harmonic majorization to show that*

$$h(z) \geq \frac{4b}{\pi} \operatorname{Im} z \qquad \forall z \in \mathbb{U}.$$

In particular, b is finite.

Hints: Notice first that on account of $(*)$, $m(r)$ is finite for every r. Fix $r > 0$ and in $\mathbb{U} \cap D(0,r)$ consider the function

$$u(z) := \frac{2}{\pi} m(r) g(z/r) - h(z),$$

where g is the function constructed in 8.40. Thanks to 8.40(i), the definition of $m(r)$ and $(*)$ above, we have

$$\varlimsup_{\substack{z \to \zeta \\ z \in \mathbb{U}}} u(z) \leq 0 \qquad \forall \zeta \in [\mathbb{U} \cap C(0,r)] \cup (-r,r) = \partial[\mathbb{U} \cap D(0,r)] \setminus \{\pm r\}.$$

Because of $(*)$ and the boundedness of g we are therefore entitled to cite 5.46 in order to conclude that

$$u(z) \leq 0 \qquad \forall z \in \mathbb{U} \cap D(0,r),$$

that is,

$$h(z) \geq \frac{2}{\pi} m(r) g(z/r) \qquad \forall z \in \mathbb{U} \cap D(0,r).$$

By definition of b there are positive $r_j \to \infty$ such that

$$\frac{m(r_j)}{r_j} \to b.$$

Any $z \in \mathbb{U}$ lies in $D(0, r_j)$ for all large j. Therefore the last inequality gives

$$h(z) \geq \frac{2}{\pi} \cdot \frac{m(r_j)}{r_j} \cdot r_j g(z/r_j)$$

and from 8.40(iii) it follows that

$$h(z) \geq \frac{2}{\pi} \cdot b \cdot 2 \operatorname{Im} z.$$

(ii) *Assuming that in addition to the hypotheses above, h is continuous on $\overline{\mathbb{U}}$, show that for some real constant a*

$$h(z) = a \operatorname{Im} z + \frac{1}{\pi} \int_{-\infty}^{\infty} \frac{\operatorname{Im} z}{|z-t|^2} h(t)\, dt \qquad \forall z \in \mathbb{U}.$$

Hints: Because of (i) the function $g(z) := h(z) - (4b/\pi)\operatorname{Im} z$ fulfills the hypotheses of 8.38.

(iii) *Under the hypotheses of* (ii) *show that*

$$\lim_{r\to\infty} \frac{m(r)}{r} \quad \textit{exists and equals } \min\{0, a\}.$$

Hints: From (ii) above and 8.39(i) we have that

$$a = \lim_{\substack{|z|\to\infty \\ z\in S_\varepsilon}} \frac{h(z)}{\operatorname{Im} z} \qquad \forall 0 < \varepsilon < 1,$$

where $S_\varepsilon := \{z \in \mathbb{C} : \operatorname{Im} z > \varepsilon|z|\}$. In particular,

$$a = \lim_{r\to\infty} \frac{h(re^{i\theta})}{r\sin\theta} \qquad \forall\theta \in (0,\pi).$$

Since by definition of m we have

$$\frac{m(r)}{r} \le \frac{h(re^{i\theta})}{r} = \sin\theta \cdot \frac{h(re^{i\theta})}{r\sin\theta},$$

it follows that

$$\varlimsup_{r\to\infty} \frac{m(r)}{r} \le \sin\theta \cdot \varlimsup_{r\to\infty} \frac{h(re^{i\theta})}{r\sin\theta} = \sin\theta\cdot a \qquad \forall\theta \in (0,\pi),$$

whence

(1) $$\varlimsup_{r\to\infty} \frac{m(r)}{r} \le \inf\{a\sin\theta : 0 < \theta < \pi\}.$$

On the other hand, since $h(t) \ge 0$ for all $t \in \mathbb{R}$ (by hypothesis (∗) and continuity of h on $\overline{\mathbb{U}}$), it follows from (ii) that

$$h(z) \ge a\operatorname{Im} z \qquad \forall z \in \mathbb{U},$$

that is,

$$ar\sin\theta \le h(re^{i\theta}) \qquad \forall r > 0, \theta \in (0,\pi)$$

and so

$$r\cdot\inf\{a\sin\theta : 0 < \theta < \pi\} \le \inf\{h(re^{i\theta}) : 0 < \theta < \pi\} = m(r).$$

It follows that

(2) $$\inf\{a\sin\theta : 0<\theta<\pi\} \le \varliminf_{r\to\infty} \frac{m(r)}{r}.$$

Finally, note that the infimum featuring in (1) and (2) equals 0 if $a \ge 0$ and equals a if $a < 0$.

Exercise 8.42 *Let h be harmonic and non-negative in $\mathbb{U}$.*

(i) *Show that if h is continuous in $\overline{\mathbb{U}}$, then*

$$\int_{-\infty}^{\infty} h(x+iy)\,dx \ge \int_{-\infty}^{\infty} h(t)\,dt \qquad \forall y>0.$$

[*The finiteness of these integrals is not being claimed.*]

Hints: Fix $y > 0$. Then for any n, $m \in \mathbb{N}$, $t \in \mathbb{R}$

$$\frac{1}{\pi}\int_{-n}^{n} \frac{y\,dx}{|x+iy-t|^2} = \frac{1}{\pi}\int_{t-n}^{t+n} \frac{y\,d\tau}{|iy-\tau|^2}$$

and so it follows from 8.34(ii) that

(1) $$\lim_{n\to\infty} \frac{1}{\pi}\int_{-n}^{n} \frac{y\,dx}{|x+iy-t|^2} = 1 \quad \text{uniformly for } t \in [-m,m].$$

The non-negative constant c supplied us by 8.38 can be assumed to be 0, because $h(x+iy) \ge cy$ and if $c > 0$ the left-hand integral in the desired inequality is ∞. With $c = 0$ it follows from 8.38 that for all n, $m \in \mathbb{N}$

$$\begin{aligned}\int_{-n}^{n} h(x+iy)\,dx &= \int_{-n}^{n}\left(\frac{1}{\pi}\int_{-\infty}^{\infty}\frac{yh(t)\,dt}{|x+iy-t|^2}\right)dx \\ &\ge \int_{-n}^{n}\left(\frac{1}{\pi}\int_{-m}^{m}\frac{yh(t)\,dt}{|x+iy-t|^2}\right)dx \\ &= \int_{-m}^{m} h(t)\left(\frac{1}{\pi}\int_{-n}^{n}\frac{y\,dx}{|x+iy-t|^2}\right)dt, \quad \text{by 8.29(vi) or 2.19(ii).}\end{aligned}$$

From this inequality and (1) infer that

$$\int_{-\infty}^{\infty} h(x+iy)\,dx = \lim_{n\to\infty}\int_{-n}^{n} h(x+iy)\,dx \ge \int_{-m}^{m} h(t)\,dt,$$

holding for all $m \in \mathbb{N}$.

(ii) *Show that if h is continuous on $\overline{\mathbb{U}}$ and $\int_{-\infty}^{\infty} h(x+iy_0)\,dx$ is finite for some $y_0 > 0$, then*

$$\int_{-\infty}^{\infty} h(x+iy)\,dx = \int_{-\infty}^{\infty} h(t)\,dt \qquad \forall y > 0.$$

Hints: Taking $y = y_0$ in (i) shows that $\int_{-\infty}^{\infty} h(t)\,dt < \infty$. Consequently, if we fix $y > 0$ and note that for any $x \in \mathbb{R}$, $0 < m < M$

$$\int_{m<|t|<M} \frac{h(t)\,dt}{|x+iy-t|^2} \leq \frac{1}{y^2} \int_{m<|t|<M} h(t)\,dt,$$

then it follows that the left-hand side approaches 0 uniformly in x as $m \to \infty$. Therefore

$$\lim_{m\to\infty} \frac{1}{\pi} \int_{-m}^{m} \frac{yh(t)\,dt}{|x+iy-t|^2} \quad \text{exists uniformly for } x \in \mathbb{R}$$

and consequently for any $n \in \mathbb{N}$

$$\begin{aligned}
\int_{-n}^{n} \left(\frac{1}{\pi} \int_{-\infty}^{\infty} \frac{yh(t)\,dt}{|x+iy-t|^2} \right) dx &= \int_{-n}^{n} \lim_{m\to\infty} \left(\frac{1}{\pi} \int_{-m}^{m} \frac{yh(t)\,dt}{|x+iy-t|^2} \right) dx \\
&= \lim_{m\to\infty} \int_{-n}^{n} \left(\frac{1}{\pi} \int_{-m}^{m} \frac{yh(t)\,dt}{|x+iy-t|^2} \right) dx \\
&= \lim_{m\to\infty} \int_{-m}^{m} h(t) \left(\frac{1}{\pi} \int_{-n}^{n} \frac{y\,dx}{|x+iy-t|^2} \right) dt, \\
&\qquad \text{by 8.29(vi) or 2.19(ii)}, \\
&\leq \lim_{m\to\infty} \int_{-m}^{m} h(t)\,dt, \quad \text{by 8.34(ii)}. \qquad (2)
\end{aligned}$$

As in (i), the constant c supplied by 8.38 is 0 and so according to the representation 8.38 the left-hand side of (2) is just $\int_{-n}^{n} h(x+iy)\,dx$.

(iii) (Rectilinear analog of 5.5). *Show that*

$$\int_{-\infty}^{\infty} h(x+iy_1)\,dx = \int_{-\infty}^{\infty} h(x+iy_0)\,dx \qquad \forall y_1,\, y_0 > 0.$$

Hints: We may assume that the integral is finite for some $y_0 > 0$. That being the case, (ii) may be applied to the function $h(z+ib)$ in the role of $h(z)$, for any $0 < b < y_0$. We get

$$\int_{-\infty}^{\infty} h(x+iy)\,dx = \int_{-\infty}^{\infty} h(t+ib)\,dt \qquad \forall y > b.$$

(iv) *Show that if* $\int_{-\infty}^{\infty} h(x+iy_0)\,dx =: M$ *is finite for some* $y_0 > 0$, *then*

$$(*) \qquad h(z) \le \frac{M}{\pi \operatorname{Im} z} \qquad \forall z \in \mathbb{U}.$$

Hints: Consider any $0 < y < y_0$ and the function $g(z) := h(z+iy)$. It is harmonic in $\mathbb{U}$, continuous on $\overline{\mathbb{U}}$ and the integral condition implies that the constant c which 8.38 furnishes for g is 0. Thus for $z \in \mathbb{U}$

$$\begin{aligned} g(z) &= \frac{1}{\pi}\int_{-\infty}^{\infty} \frac{\operatorname{Im} z}{|z-t|^2} g(t)\,dt = \frac{1}{\pi}\int_{-\infty}^{\infty} \frac{\operatorname{Im} z}{(\operatorname{Re} z - t)^2 + (\operatorname{Im} z)^2} g(t)\,dt \\ &\le \frac{1}{\pi \operatorname{Im} z}\int_{-\infty}^{\infty} g(t)\,dt = \frac{M}{\pi \operatorname{Im} z}, \quad \text{by (iii); that is,} \\ h(z+iy) &\le \frac{M}{\pi \operatorname{Im} z} \qquad \forall z \in \mathbb{U}. \end{aligned}$$

Now let $y \downarrow 0$.

Remarks: (i) A beautiful elementary argument involving none of the machinery developed here, but merely the areal mean-value inequality and passage back and forth between $\mathbb{D}$ and $\mathbb{U}$, leads to an inequality like $(*)$ for all non-negative *subharmonic* functions h in $\mathbb{U}$ satisfying $\sup_{y>0}\int_{-\infty}^{\infty} h(x+iy)\,dx =: M < \infty$, with, unfortunately, the larger majorant $\frac{4M}{3\pi \operatorname{Im} z}$. See KRYLOV [1939]. Less elementary arguments nevertheless successfully reduce the majorant back to (the best possible) $\frac{M}{\pi \operatorname{Im} z}$; see MCCALL [1983].

(ii) According to 8.33 every continuous $g\colon \mathbb{R} \mapsto \mathbb{R}$ is the restriction to $\mathbb{R}$ of a function which is continuous on $\overline{\mathbb{U}}$ and harmonic in $\mathbb{U}$. In view of the maximum principle it is perhaps surprising that for certain positive g there is no such extension which is non-negative throughout $\mathbb{U}$. For whenever there is a non-negative harmonic extension of g, 8.38 entails that $\int_{-\infty}^{\infty} g(t)|z-t|^{-2}\,dt$ is finite for every $z \in \mathbb{U}$, whereas $g(t) := |i-t|^2 = 1+t^2$, for example, lacks this property.

An attractive application of the Poisson formula in a half-plane is the derivation from it, via conformal mapping, of a Poisson formula for a strip, and this is the content of the closing exercise of the chapter. KOEBE [1913b] indicates how the Dirichlet problem in ellipses, annuli and rectangles can be solved by similar techniques. The formula we get, 8.43(v), could be derived directly, without mention of half-planes, and this is done by WIDDER [1961], under even somewhat weaker hypotheses on the boundary function.

Exercise 8.43 (Dirichlet problem in a strip). *Let* $S := \left(-\frac{\pi}{2}, \frac{\pi}{2}\right) \times \mathbb{R}$, E *the exponential function in* $\mathbb{C}$, *and* L *the Principal Branch of the Logarithm in* $\mathbb{C}_-$. *A bounded continuous function* $g\colon \partial S \mapsto \mathbb{R}$ *is given.*

(i) *Show that* $L_1 := iL$ *maps* $\overline{\mathbb{H}} \setminus \{0\}$ *continuously onto* $\overline{S}$, $\mathbb{H}$ *conformally onto* S, *and that* $E_1(z) := E(-iz)$ *is the inverse map.*

(ii) *Show that* $f := g \circ L_1$ *is a continuous bounded function on* $\partial\mathbb{H} \setminus \{0\} = i\mathbb{R} \setminus \{0\}$.

(iii) *Define* $F := f$ *on* $\partial\mathbb{H} \setminus \{0\}$ *and*

$$F(w) := \frac{1}{\pi} \int_{-\infty}^{\infty} \frac{\operatorname{Re} w}{|w + it|^2} f(it)\, dt, \quad w \in \mathbb{H}. \tag{1}$$

Show that F *is a continuous extension of* f *to* $\overline{\mathbb{H}} \setminus \{0\}$ *which is harmonic in* $\mathbb{H}$.

Hint: 8.35(i).

(iv) *Show that* $G := F \circ E_1$ *is continuous on* $\overline{S}$, *harmonic in* S *and agrees with* g *on* ∂S.

(v) *From* (iv) *deduce that for all* $(x, y) \in S$

$$G(x+iy) = \frac{1}{2\pi} \sum_{k=1}^{2} \int_{-\infty}^{\infty} \frac{\cos x}{\cosh(y - t) + (-1)^k \sin x} g\left((-1)^k \pi/2 + it\right) dt.$$

(vi) *From* (v) *deduce that if* g *is* $2\pi i$*-periodic, then so is* G.

Remark: By deploying E and L slightly differently we can in turn map S onto an annulus and thereby reduce the Dirichlet problem on the latter to the one just solved for S. See the introductory discussion in §7 of chapter XI.

Notes to Chapter VIII

Besides the several uses we made of the Basic Integral Representation 8.1, it is also essential elsewhere; e.g., in constructing the holomorphic functional calculus in Banach algebras and in describing the dual space of $H(\Omega)$. This will all be done in chapter XIX.

Our development of Runge theory follows SAKS and ZYGMUND [1971] and RUDIN [1966], with some modifications, especially in the matter of "pushing the poles to infinity" (see pp. 177 ff. of SAKS and ZYGMUND); it is similar to GRABINER [1976]. Actually RUNGE [1885] only proved rational approximation, but the pole-shoving technique occurs in this paper and, as 8.9 shows, the step from rational functions to polynomials via this technique is pretty small. (See pp. 218–220 of MITTAG-LEFFLER [1900] on this point, which also gives Runge some claim to the Weierstrass Approximation Theorem 8.29(v).) Polynomial approximation occurs explicitly (proved by other means) in HILBERT

[1897]. APPELL [1883] is an important precursor of Runge. A proof of Polynomial Runge II for Jordan regions follows from BIEBERBACH [1914] (pp. 107–112) and JULIA [1927a].

For more on rational approximation see the treatises of WALSH [1935], [1969] and GAIER [1987], and the surveys of KOREVAAR [1980], GAIER [1984a] and SAFF [1986]. In connection with 8.8, WOLFF [1921] shows that if K is the closure of a Jordan region, then one can choose $a_k \in \mathbb{C} \setminus K$ and $A_k \in \mathbb{C}$ such that $f = \sum_{k=1}^{\infty} A_k F_{a_k}$ and $\sum_{k=1}^{\infty} |A_k| < \infty$. See also DENJOY [1922]. For a discussion of the pre-1920 progeny of Runge's approximation theorem (e.g., PAINLEVÉ [1886], [1887], [1898], HILBERT [1897], FABER [1903], FEJÉR [1918]), see §59 of BIEBERBACH's encyclopedia article and for details of some of the methods, MONTEL's monograph [1910]. The problems of polynomial and rational approximation and those of analytic continuation are closely connected. E.g., both ends are served by summability methods which alter a power series to make it converge in regions other than disks. Also, from the Basic Integral Representation 8.1 [or from the various general Cauchy Integral Formulas like 7.22] follow, by the method of 8.8, other polynomial approximations to the function, if polynomial approximations to $1/(1-z)$ other than $\sum_{n=0}^{\infty} z^n$ are used (8.25 and 11.34 are examples). These may converge in larger regions and provide holomorphic extensions of the given function. See §§37–44 of BIEBERBACH's encyclopedia article, chapters 10, 11 of HILLE [1962] and BUHL [1925]. Yet another source of polynomial approximants is subsequences of the full sequence of partial sums of the power series of the function. Such subsequences may converge outside the disk of convergence of the full series and thus lead to polynomial approximations in and holomorphic extensions to larger regions. This is the phenomenon of overconvergence to be discussed in Chapter XVI and its notes. There is also the phenomenon of expansion in series of Faber polynomials, a direct analog of power series expansions in disks. See the notes to Chapter IX. Finally, polynomial and rational approximants arise in interpolation theory. Here one is attempting to find explicit expressions for a function in terms of its values at certain points. 5.73 and 11.38 are good examples. I refer the reader to the work of Fejér, Fekete and Leja. See also chapters 1 and 2 of SMIRNOV and LEBEDEV [1968], chapter 16 of HILLE [1962], and PARTINGTON [1997].

Runge's theorem is a wonderful tool for producing holomorphic functions with exotic properties. This will be well illustrated in chapter XV. See also RUBEL [1976] and SCHNEIDER [1980].

WALSH [1933] was the first to discern the elegant path to the general Cauchy theorem (8.14) via Runge's theorem (see 9.27) and shortly thereafter the idea appeared in the text of SAKS and ZYGMUND [1971] (1st ed. 1938). Many subsequent expositors, however, inexplicably chose not to follow this *camino real*, until HEINS [1962b] and RUDIN [1966] took up and extended the theme. (See also HEINS [1968] and CONWAY [1978].)

There is a remarkable strengthening of Polynomial Runge Theorem I (8.9) due to MERGELYAN [1952]: if K is a compact, simply-connected set, then every f which is continuous on K and holomorphic in the interior of K is uniformly approximable on K by polynomials. Note that this includes every continuous function on K if K has a void interior (M. A. LAVRENTIEFF [1936], pp. 25–35 and HARTOGS and ROSENTHAL [1931]). Thus this result subsumes Weierstrass' classical polynomial approximation theorem, 8.29(iii). For a readable account of this profound theorem, see the last chapter of RUDIN [1987]. A good survey of the precursors of this theorem (one of which is proved in 9.26 below) is offered in the first few pages of MERGELYAN [1952]. (The reader who looks there for Mergelyan's own proof should also take advantage of the simplification wrought in it by F. E. BROWDER [1960].) With such a theorem in hand, general Cauchy theorems can be proved in which the curve of integration lies on the boundary of the domain of the holomorphic function. Cf. 9.27. In [1952] MERGELYAN also proved a corresponding strengthening (i.e., f continuous on K but holomorphic only in the interior of K) of 8.11 for the case where $\mathbb{C} \setminus K$ has only finitely many components; see GARNETT [1968] or pp. 110–119 of GAIER [1987], which also describes a K for which such approximation fails.

The preoccupation with *compact* sets in this whole theory is explained by the following elementary observation: If polynomials P_n converge uniformly to F on an *unbounded* subset of $\mathbb{C}$, then F is a polynomial and for each sufficiently large n, $P_n - F$ is constant. Indeed, for all sufficiently large n the polynomial $P_{n+1} - P_n$ has modulus less than 1 on an unbounded set and consequently cannot have positive degree.

Mergelyan's theorem, together with the functional analytic proof of Glicksberg and Wermer (see CARLESON [1965]), has spawned a big industry, wedding functional analysis techniques with hard classical analysis. See chapters II and VII of GAMELIN [1969], chapters five and six of STOUT [1971], the notes of ZALCMAN [1968a], and SCHWINGENDORF [1980]; for the classical approach see SMIRNOV and LEBEDEV [1968]. Actually, these functional analysis techniques can be used to build the Runge theory and other parts of classical function theory as well, a program carried out in LUECKING and RUBEL [1984].

There are more dramatic examples than 8.15 and 8.16: For any $a_n \in \mathbb{C}$ there are polynomials P_k such that for every integer $n \geq 0$, $\sum_{k=1}^{\infty} |P_k^{(n)}(z)| < \infty$ and $\sum_{k=1}^{\infty} P_k^{(n)}(z) = \begin{cases} a_n & \text{if } z = 0 \\ 0 & \text{if } z \neq 0 \end{cases}$. See MATTNER [2001].

The technique in the 8.17(ii) hints is specific to polynomial approximation, but generally it is easy to refine approximation from a *convex* set into simultaneous interpolation and approximation. For an elegant treatment see YAMABE [1950].

The proof of 8.18 is adapted from those in BIEBERBACH [1934], SAKS and ZYGMUND [1971] and ESTERMANN [1962]. A somewhat different treatment is given by NIEVERGELT [1998]. For related material see BANK and ORLAND [1968] and Theorem 8, p. 576 of KURATOWSKI [1968] (a Rouché theorem for continuous functions). Actually Rouché only proved the 5.87 version of 8.18 and under the stronger hypothesis $|f+g|<|f|$ on the boundary; even in this he was, despite the time-honored attribution, anticipated by Cauchy in 1831 (see p. 198 of SMITHIES [1997]). For other extensions of Rouché's theorem and a different (area-based) proof of 8.18 see MORTINI and RUPP [2014].

An obvious generalization of 8.24 would be: For every pair of open subsets U_1, U_2 of $\mathbb{C}$ and every $f_{12} \in H(U_1 \cap U_2)$ there exist $f_j \in H(U_j)$ such that $f_{12} = f_1 + f_2$. This is true, but its proof is more sophisticated and requires one to leave the realm of holomorphic functions for the larger class of C^∞-functions on $\mathbb{R}^2$. See p. 123 of NARASIMHAN [1985] and 4.9 in FEYEL and PRADELLE [1973], and, for a different approach, MÜLLER and WENGENROTH [1998]. For the harmonic function version see p. 195 of AXLER, BOURDON and RAMEY [2001]. Historically see HAAR [1914], SCHOTTKY [1916] and ARONSZAJN [1935]. The partial-fraction decomposition of a rational function is another example of the 8.24 phenomenon: If $a_1, a_2, \ldots, a_n$ are the zeros of the polynomial Q, none is a zero of the polynomial P, and we set $U_j := \mathbb{C} \setminus \{a_j\}$, then 11.12 represents $P/Q \in H(U_1 \cap U_2 \cap \cdots \cap U_n)$ as a sum of functions in the $H(U_j)$. Mittag-Leffler's theorem performs the same service for entire functions P and Q: See the chapter Notes on 11.39–11.41.

For the last word on 8.27 and 8.28 see MARSHALL [1976] and P. W. JONES [1981], and for more on approximation by finite Blaschke products, chapter 9, §C of COLWELL [1985]. FISHER [1971] extends the themes in 8.26–8.28. CHALLENER and RUBEL [1982] apply 8.28 to prove an interesting converse of Rouché's theorem.

Many of the numerous proofs of Weierstrass' important theorem 8.29(iii) by succeeding generations are reprised in the wonderful, long paper of PINKUS [2000], which contains a 110-item bibliography. One of its most beautiful generalizations, the theorem of C. H. Müntz and O. Szász, says that for the linear span of the powers x^{r_n}, $0 = r_0 < r_1 < r_2 < \ldots$, to be uniformly dense in the space of all continuous functions on $[0,1]$ it is necessary and sufficient that $\sum_{n=1}^{\infty} 1/r_n = \infty$. [For example, 1 and prime powers of x suffice in Weierstrass' theorem.] For a proof, see RUDIN [1987], p. 313 (and **MR** 85b:41005 for a marvelously elementary proof of sufficiency when additionally $r_n \to \infty$). 8.29(v) was asserted on p. 797 of WEIERSTRASS [1885], with a proof supplied on pp. 27 ff., vol. 3 of his *Werke*. Proofs were also supplied by LEBESGUE [1898], [1908], MITTAG-LEFFLER [1900], PICARD [1905], BOREL [1928] (see chapter 4), E. LANDAU [1908a], TONELLI [1910a], TAKENAKA [1919] and FRANKLIN [1925b]. For references to other proofs see pp. 1186–1187 of BOREL and ROSENTHAL [1924]. The method in 8.29(v) is similar to that in VALIRON [1948].

The most elegant, explicit, and elementary proof of the original Weierstrass theorem ($f : [0,1] \mapsto \mathbb{C}$ continuous) is that due to Sergei N. Bernstein (in 1912). He showed that the polynomials $B_n(f)(x) := \sum_{k=0}^{n} f\left(\frac{k}{n}\right)\binom{n}{k}x^k(1-x)^{n-k}$ (now named for him) converge to $f(x)$, uniformly for $x \in [0,1]$. All readers should savor the proof, e.g., in ROSS [2013]!

In 2.19 we deduced Fubini's theorem from Leibniz' rule about differentiation under the integral sign; with Fubini's theorem proved independently in 8.29(vi) we can conversely deduce Leibniz' result from it—see SEELEY [1961] and FISHER and SHILLETO [1986]. When (as here) f is continuous on $[a,b] \times [c,d]$, the most straightforward and elementary proof of Fubini consists of exploiting the fact that f is *uniformly* continuous to show that both iterated integrals are approximable by sums of the form $\sum f(x_j, y_k)(x_j - x_{j-1})(y_k - y_{k-1})$. But continuity is not essential. 8.29(vi) is also valid for bounded functions f such that $f(x,y)$ is Riemann integrable in x for each fixed y and Riemann integrable in y for each fixed x. In this case $\int_a^b f(x,y)\,dx$ is a Riemann integrable function of y and $\int_c^d f(x,y)\,dy$ is a Riemann integrable function of x and the equality in 8.29(vi) holds. See LICHTENSTEIN [1910], [1911], FICHTENHOLZ [1913], GILLESPIE [1918–19], BESICOVITCH [1918–1919], ETTLINGER [1926–27], RIDDER [1930b,c], RENNIE [1974] and especially LEWIN [1987] (for a clean, elementary proof). But the function $f(x,y) := (x-y)(x+y)^{-3} = D_1\left[-x(x+y)^{-2}\right] = D_2\left[y(x+y)^{-2}\right]$ suggests the necessity of the boundedness hypothesis.

Independently of Carleman, in 1934 Hassler Whitney proved a version of 8.32 for continuous functions f on open subsets of $\mathbb{R}^n$, with approximation by real-analytic functions g. See PINKUS [2000], p. 54 and **MR** 82j:31001. As noted in the text, Runge's theorems are inadequate to Carleman's proof of his theorem 8.32. However, with the aid of the more potent theorem of Mergelyan mentioned above, his proof simplifies greatly; compare, for example, the proof of KAPLAN [1955–56]. It is from the latter paper that 8.33 is taken. Earlier, RUNGE [1885/86] had used a clever but very different approximation argument to settle the special case of 8.33 in which g vanishes at infinity. (This case is also easily reducible to $\mathbb{D}$ via the Cayley transform and is *constructively* solved in 8.34.) The satisfyingly constructive approach in 8.34 is not available in the more general case 8.33 but FINKELSTEIN and SCHEINBERG [1975] and GARDINER [1981] have found a constructive proof of a somewhat different nature for 8.33, and a similar solution of 8.33 was given by R. NEVANLINNA [1925]. The Dirichlet problem for a wide class of unbounded regions and unbounded boundary data has been solved using Carleman-like approximation as in 8.33 by ARMITAGE and GOLDSTEIN [1990]—see their theorem 13.1. A. SINCLAIR [1958], [1985], showed that the g in 8.32 could be chosen to satisfy in addition $g(n) = f(n)$ for all non-zero integers n. Finally HOISCHEN [1975] has shown that if $0 \le c_n \le c_{n+1} \to \infty$, $\{x_k\} \subset \mathbb{R}$ has no limit points and f is infinitely differentiable on $\mathbb{R}$, then an entire function g exists such that $|f^{(n)} - g^{(n)}| < \varepsilon$ in $\mathbb{R} \backslash [-c_n, c_n]$ for all $n = 0, 1, 2, \ldots$ and $g(x_k) = f(x_k)$ for all $k = 1, 2, \ldots$. See also RUBEL and VENKATESWARAN [1976] and GAUTHIER and HENGARTNER [1977]. In other directions Carleman's theorem has been extended by ROTH [1938–39], [1976], KELDYCH and LAVRENTIEFF [1939], MERGELYAN [1952], ARAKÉLIAN [1964], [1982], NERSESJAN [1971] and BROWN, GAUTHIER and SEIDEL [1975]. For a closed $C \subset \mathbb{C}$, Arakélian gave topological conditions on C and $\mathbb{C} \setminus C$ which ensure that every continuous complex-valued function on C is uniformly approximable on C by entire functions. See FUCHS [1968] for a detailed exposition of Arakélian's work, ARAKÉLIAN [1971], SCHEINBERG [1980] for surveys and ROSAY and RUDIN [1989] for a reduction to Mergelyan's theorem. The best treatment of this whole circle of ideas may be chapter IV of GAIER [1987] together with GAUTHIER [1994].

The suggested proof of 8.35 is adapted from LOOMIS and WIDDER [1942], as is the proof of 8.37. The idea in the 8.37 proof was used earlier to a somewhat different end by SZÁSZ [1933]. The result 8.37 itself goes back at least to TSCHEBOTAREFF [1928] (p. 675), who credits Krein and Grandjot, and may also be found in BOULIGAND [1932]. A proof based on the argument principle appears in B. LEVIN's book [1980]. Another beautiful proof based on the reflection principle is given by TIDEMAN [1954]. See also BALK [1957]. The reflection principle originates with H. A. SCHWARZ; see p. 12 of vol. I and pp. 66–67 of vol. II of his *Gesammelte Mathematische Abhandlungen.* There will be more on it in Chapter XIII. KURAN [1970] contains the following result, to which 8.37 is an easy corollary: If $h\colon \overline{\mathbb{U}} \mapsto \mathbb{R}$ is continuous, harmonic in $\mathbb{U}$ and $h(\mathbb{R}) = 0$, then

$$\sup_{y>0} \int_{-\infty}^{\infty} \frac{h^+(x+iy)}{x^2+y^2}\, dx < \infty \quad \text{is equivalent to} \quad h(x+iy) \equiv cy$$

for some real constant c. Bearing in mind the extension to $\mathbb{C}$ provided by 8.36, the main result of KURAN [1966b] is also a generalization of 8.37, as is theorem 8 of KURAN [1980]. Our 11.52(iv) below is a generalization of 8.37 in a different direction.

The importance of the (Poisson) integral representation 8.38 is clear from the consequences we derived from it. Such a representation can be obtained under less restrictive hypotheses than those of 8.38 or 8.41(ii). In particular, continuity on the closed half-plane can be dispensed with if the Riemann integral is replaced by a Riemann-Stieltjes integral. A compactness argument under the boundedness provided by 8.35(iii) and (iv), in the form of Helly's Selection Theorem, is used to get the integrator function. See R. NEVANLINNA [1925], FICHTENHOLZ [1929], WOLFF and DE KOK [1932], CAUER [1932], VERBLUNSKY [1935], TSUJI [1939], [1959] (pp. 149–152) and, especially, LOOMIS and WIDDER [1942] and NEVANLINNA and NIEMINEN [1955].

The basic fact that positive harmonic functions admit a representation as a (Poisson) kernel integrated against a positive measure, which is so central to the modern theory, goes back to HERGLOTZ [1911], F. RIESZ [1911], and R. NEVANLINNA [1922a]. SERRIN [1956] is the source of the proof in 8.39(i) and of the ideas in 8.41(ii) and (iii). The proof of 8.41(i) is from p. 19 of AHLFORS [1930] and the result 8.41(iii) is due to HEINS [1946a]. See also AHLFORS [1937], [1966], DINGHAS [1938], HEINS [1948], AHLFORS and HEINS [1949], TSUJI [1956] and pp. 412–420 of HILLE's book [1962]. In the last two papers it is shown that not only does $\lim m(r)/r$ exist, but even $\lim h\big(re^{i\theta}\big)/r$ exists for "most" θ. HEINS [1948] shows that if h is only superharmonic and satisfies 8.41($*$), then either $h \geq 0$ in $\mathbb{U}$ or $\lim m(r)/r^{1/2}$ exists in $[-\infty, 0)$. ESSÉN [1975] is a longer treatment of these themes, which are related to the ideas discussed in 5.51. Finally, KURAN [1966a] was my source for 8.42; for related results see HARDY, INGHAM and PÓLYA [1927], [1928], CARLSON [1943] and YÛJÔBÔ [1952].

As remarked before 8.34, some of the results of §6 can be proved a little more easily by transforming them into statements about the unit disk. This is a good exercise program for the reader. Notice that the disk analog of 8.37 is contained in LOHWATER [1952]. (The actual result proved there suggests that the boundary continuity hypothesis in 8.37 may be weakened to $\lim_{y\downarrow 0} h(x + iy) = 0$ for all real x. The interested reader might try to confirm this. On this point compare with WOLF [1947].) See also BRUCKNER, LOHWATER and RYAN [1969].

Chapter IX
The Riemann Mapping Theorem

§ 1 Introduction

In his Göttingen dissertation of 1851 Bernhard Riemann enunciated (p. 40 of his *Werke*, 1953 ed.) and attempted to prove the famous theorem that now bears his name:

Theorem 9.1 *Every simply-connected region other than $\mathbb{C}$ itself is conformal to an open disk.*

The gap in Riemann's proof consisted in the assumption that a certain extremal problem in the calculus of variations necessarily had a solution. (See Monna [1975], pp. 32–34.) Hilbert and others subsequently filled this gap (Monna [1975], chapter IV). In essence the matter comes down to the compactness criterion of Montel (7.6). A remnant of this method of proof persists in what has become the canonical (for concision and simplicity) proof, that of L. Fejér and F. Riesz, which is sketched in exercise 9.9 below. Here we examine a proof which, by contrast, is constructive, in that it produces more or less explicit sequences of maps that converge to the desired conformal equivalence. Moreover, the maps are elementary: repeated compositions of square-roots and conformal automorphisms of the unit disk. Explicit estimates are obtained for the rate of convergence and even the induced convergence and compactness theorems of Chapter VII can be avoided, so this proof is truly elementary (in its ingredients, if not in its genesis and execution!). It evolved at the hands of Paul Koebe and Constantin Carathéodory around 1910 and will be presented in all its glorious detail in section 2 below. The reader is reminded of 4.71, according to which property (HS_1) is equivalent to simple-connectivity. However, no use is made of this equivalence in the proof: we simply take (HS_1) as hypothesis, since only that property of the region is

R. B. Burckel, *Classical Analysis in the Complex Plane*, Cornerstones,
https://doi.org/10.1007/978-1-0716-1965-0_10

used. (A proof directly from the simple-connectivity hypothesis via 4.37(iii) can be found in ČERNÝ [1992], p. 186.) The grand synthesis occurs in 10.2.

If U is an open subset of $\mathbb{C}$ which contains 0, it is convenient to refer to

$$R := \sup\{r : D(0,r) \subset U\} \in (0,\infty]$$

as the *inner radius* of U. Naturally, $D(0,R) \subset U$, and if $U \neq \mathbb{C}$, then

$$R = \inf\{|w| : w \in \mathbb{C} \setminus U\} < \infty$$

and there exists at least one w in the closed set $\mathbb{C} \setminus U$ with $|w| = R$. Of course, $w \in (\mathbb{C} \setminus U) \cap \overline{U} = \partial U$.

The regions and maps which feature in the remarkable iterative scheme of Koebe and Carathéodory are described in

Definition 9.2 (i) A *Koebe region* is an open, connected, proper subset of the open unit disk $\mathbb{D}$ which contains 0 and has property (HS_1).

(ii) If Ω is a Koebe region and R is its inner radius, then a *Koebe map* on Ω is any conformal map k of Ω with the following properties

(a) $$k(\Omega) \subsetneq \mathbb{D}$$

(b) $$k(0) = 0$$

(c) $$k'(0) = \frac{1+R}{2\sqrt{R}}$$

(d) $$|k(z)| \geq |z| \qquad \forall z \in \Omega.$$

Note that (a) implies $R < 1$, so by (c), $k'(0) > 1$, which means that $z/k(z)$ is holomorphic in Ω with value at 0 equal to $1/k'(0) < 1$. If equality were to hold in (d) for some $z \in \Omega \setminus \{0\}$, this ratio would be constant, of modulus 1, by 5.10. Consequently, we must have

$(d)_0$ $$|k(z)| > |z| \qquad \forall z \in \Omega \setminus \{0\}.$$

Remark: The reader is cautioned not to confuse the novel (and temporary) terminology in (ii) with the standard terminology *Koebe function.* The latter traditionally designates the family of functions $k_\alpha(z) := z/(1-\alpha z)^2 = \sum n\alpha^{n-1}z^n$ $(\alpha \in \mathbb{T}, z \in \mathbb{D})$ which lie in the schlicht class $\mathcal{S}$ (7.24) and are significant because they are extremal for numerous mapping problems.

Lemma 9.3 *Let $0 \in U$ be an open subset of $\mathbb{C}$. Suppose $f,\ g \in H(U)$, $f(0) = 0$, $|f| \geq |g|$, f is non-constant and g is one-to-one. Then*

(i) *the inner radius of $f(U)$ is not less than that of $g(U)$.*

In particular, if Ω is a Koebe region and k a Koebe map on it, then

(ii) *$k(\Omega)$ is also a Koebe region, and its inner radius is not less than that of Ω.*

Proof: Note that $g(0) = 0$, that g is a homeomorphism, and that $f(U)$, $g(U)$ are open sets (5.77). Let R denote the inner radius of $g(U)$, so that $D(0,R) \subset g(U)$. It suffices to show that $D(0,R) \subset f\big(g^{-1}(D(0,R))\big) \subset f(U)$. So let $h := f \circ g^{-1}$, R^* be the inner radius of $h\big(D(0,R)\big)$ and $w \in \partial h\big(D(0,R)\big)$ with $|w| = R^*$. (If $R^* = \infty$, there is nothing to prove.) Now there exist $z_n \in D(0,R)$ such that $h(z_n) \to w$. The sequence $\{z_n\}$ cannot have an accumulation point z in $D(0,R)$, for then $\{h(z_n)\}$ would accumulate at $h(z)$ in the open set $h\big(D(0,R)\big)$, contrary to $h(z_n) \to w \in \partial h\big(D(0,R)\big)$. Hence $|z_n| \to R$, so from $|z_n| = \big|g\big(g^{-1}(z_n)\big)\big| \leq \big|f\big(g^{-1}(z_n)\big)\big| = |h(z_n)| \to |w| = R^*$ we infer that $R \leq R^*$. This establishes (i).

Evidently the conformal map k transmits property (HS_1) from Ω to $k(\Omega)$. Property (d) of the Koebe map k makes (i) applicable (with $f = k$ and g the identity map), from which it follows that the inner radius of $k(\Omega)$ is at least that of Ω.

Lemma 9.4 *Every Koebe region possesses a Koebe map.*

Proof: Let Ω be a Koebe region, with inner radius R. As noted above, there is a boundary point a of Ω with

$$|a| = R. \tag{1}$$

Since $R < 1$, $a \in \mathbb{D}$ and we may form

$$f_a(z) := \frac{z-a}{1-\bar{a}z}, \qquad z \in \mathbb{D}, \tag{2}$$

a conformal map of $\mathbb{D}$ onto $\mathbb{D}$ (see 2.6) which is zero-free in Ω. By property (HS_1) of Ω, f_a has a holomorphic square-root g *on* Ω. Evidently, $\Omega \subsetneq \mathbb{D}$ and f_a a one-to-one map of $\mathbb{D}$ onto $\mathbb{D}$ mean that $f_a(\Omega) \subsetneq \mathbb{D}$, that is, $g^2(\Omega) \subsetneq \mathbb{D}$. It follows that $g(\Omega) \subsetneq \mathbb{D}$. So we may set

$$b := g(0), \tag{3}$$

form

(4) $$G := f_b \circ g$$

and have

(5) $$G(\Omega) \subsetneq \mathbb{D}$$

and

(6) $$G(0) = 0.$$

Notice that $g^2 = f_a$ and (3) yield

(7) $$b^2 = -a.$$

By a direct calculation on (2)

$$2g(0)g'(0) = \left[g^2\right]'(0) = f_a'(0) = 1 - |a|^2$$

(8) $$g'(0) = \frac{1-|a|^2}{2g(0)} = \frac{1-|a|^2}{2b} \quad \text{by (3)}.$$

Therefore an application of the Chain Rule gives

$$G'(0) \overset{(4)}{=} f_b'(g(0))g'(0) \overset{(3)}{=} f_b'(b)g'(0) = \frac{1-|b|^2}{\left(1-\bar{b}b\right)^2} \cdot g'(0)$$

$$\overset{(8)}{=} \frac{1}{1-|b|^2} \cdot \frac{1-|a|^2}{2b} \overset{(7)}{=} \frac{1-|a|^2}{1-|a|} \cdot \frac{1}{2b}$$

(9) $$\overset{(1)}{=} \frac{1+R}{2b}.$$

We finally set

$$k := \frac{|G'(0)|}{G'(0)} \cdot G$$

and we see from (5) and (6) that k has properties (a) and (b) of 9.2(ii), while (9) shows that

$$k'(0) = |G'(0)| = \frac{1+R}{2|b|} = \frac{1+R}{2\sqrt{R}},$$

thus establishing property (c). Let s denote the square map on $\mathbb{C}$: $s(z) = z^2$. Then the above definitions show that

$$f_b^{-1} \circ G = g \qquad \text{on } \Omega$$

(10) $$s \circ f_b^{-1} \circ G = g^2 = f_a \quad \text{on } \Omega.$$

Now $B := f_a^{-1} \circ s \circ f_b^{-1}$ is a holomorphic map of $\mathbb{D}$ into $\mathbb{D}$ and since $f_a^{-1} = f_{-a}$, $f_b^{-1} = f_{-b}$ (see 2.6), we see that

$$B(0) = f_{-a}\Big(\big(f_{-b}(0)\big)^2\Big) = f_{-a}\left(b^2\right) \overset{(7)}{=} f_{-a}(-a) = 0.$$

Therefore by Schwarz

$$|B(w)| \le |w| \qquad \forall w \in \mathbb{D}$$

and combining this with (10) gives

$$|z| = \big|B\big(G(z)\big)\big| \le |G(z)| = |k(z)| \qquad \forall z \in \Omega,$$

which is property (d) of a Koebe map.

Remark: Because of (d) and the way it is used in the proof of Riemann's theorem to spread a region into the whole of $\mathbb{D}$, there is a temptation to think that $k(\Omega) \supset \Omega$. Remmert [1998], p. 193 offers the following dramatic example to illustrate that this need not be so: If $0 < t < 1$, then $\Omega := \mathbb{D}\setminus[t^2, 1]$ is a Koebe region and $k(\Omega)$ is the crescent-shaped region $\mathbb{D} \setminus \overline{D}(c, r)$, where $c := (1 + t^2)/2t$, $r := (1 - t^2)/2t$, q is the holomorphic square-root mapping $\mathbb{C}_-$ onto $\mathbb{H}$, and $k := -f_t \circ q \circ (-f_{t^2})$.

Lemma 9.5 *Let $r > 0$, $1 > \varepsilon > 0$ and F be holomorphic in a neighborhood of $\overline{D}(0, r)$ with $|F| \le 1$ and $F(0) \ge 1 - \varepsilon$. Then*

$$|F(z) - 1| \le \frac{\sqrt{2\varepsilon}}{1 - |z|/r} \qquad \forall z \in D(0, r).$$

Proof: We may assume $r = 1$. For $|w| \le 1$ we have

$$|w - 1|^2 = 1 - w - \overline{w} + w\overline{w} \le 2 - w - \overline{w}.$$

Therefore

$$\begin{aligned}\int_0^{2\pi} \left|F\left(e^{i\theta}\right) - 1\right|^2 d\theta &\le \int_0^{2\pi} \left[2 - F\left(e^{i\theta}\right) - \overline{F}\left(e^{i\theta}\right)\right] d\theta \\ &= 2\pi\big[2 - F(0) - \overline{F}(0)\big] \quad \text{by 5.5(ii)} \\ &\le 4\pi\varepsilon, \quad \text{since } F(0) \ge 1 - \varepsilon.\end{aligned}$$

The Cauchy-Schwarz inequality for integrals then yields

$$\int_0^{2\pi} \left|F\left(e^{i\theta}\right) - 1\right| d\theta \le \left[\int_0^{2\pi} \left|F\left(e^{i\theta}\right) - 1\right|^2 d\theta \int_0^{2\pi} 1\, d\theta\right]^{1/2} \le 2\pi\sqrt{2\varepsilon}.$$

It follows from the last inequality and Cauchy's formula that

$$\begin{aligned}|F(z) - 1| &= \frac{1}{2\pi}\left|\int_0^{2\pi} \frac{F\left(e^{i\theta}\right) - 1}{e^{i\theta} - z} ie^{i\theta}\, d\theta\right| \qquad \forall z \in \mathbb{D} \\ &\le \frac{1}{2\pi} \cdot \frac{1}{1 - |z|} \int_0^{2\pi} \left|F\left(e^{i\theta}\right) - 1\right| d\theta \\ &\le \frac{\sqrt{2\varepsilon}}{1 - |z|}.\end{aligned}$$

Corollary 9.6 *Let $0 < R < 1$, Ω a region containing 0 and f a conformal map of Ω such that $f(0) = 0$, $f'(0) > 0$. Suppose that*

$$D(0,R) \subset \Omega \cap f(\Omega) \subset \Omega \cup f(\Omega) \subset \mathbb{D}.$$

Then

$$|f(z) - z| \le \frac{3\sqrt{1-R}}{R^2 - |z|} \qquad \forall z \in D\big(0, R^2\big)\,.$$

Proof: Let $g := f^{-1}$ and apply Schwarz' lemma to both $f(Rz)$ and $g(Rz)$, $z \in \mathbb{D}$, to get $(w = Rz)$

$$(1) \qquad |f(w)| \le |w|/R, \quad |g(w)| \le |w|/R \qquad \forall w \in D(0,R).$$

Consider $z \in D\big(0,R^2\big)$. Then $|f(z)| < R$ by (1), so

$$(2) \qquad |z| = \big|g\big(f(z)\big)\big| \le |f(z)|/R \quad \text{by (1).}$$

Combining (1) and (2) we have

$$R^2 \le \big|Rf(z)/z\big| \le 1 \qquad \forall z \in D\big(0, R^2\big)\,.$$

If we define $r := R^2$, $\varepsilon := 1 - R^2$ and $F(z) := Rf(z)/z$, for $z \in \Omega$, then $F \in H(\Omega)$ (recall, $f(0) = 0$) and the last inequality becomes

$$(3) \qquad 1 - \varepsilon = R^2 \le |F(z)| \le 1 \qquad \forall z \in D(0,r).$$

Therefore F fulfills the hypotheses of the last lemma. Consequently, for $z \in D(0,r)$

$$\begin{aligned}
|f(z) - z| &\le \big|f(z) - z/R\big| + \big|z/R - z\big| \\
&= \frac{|z|}{R}\left|\frac{Rf(z)}{z} - 1\right| + \frac{|z|}{R}(1-R) \\
&\le \left|\frac{Rf(z)}{z} - 1\right| + (1-R), \quad \text{since } |z| < r = R^2 < R, \\
&= |F(z) - 1| + 1 - R \\
&\le \frac{\sqrt{2\varepsilon}}{1 - |z|/r} + 1 - R, \quad \text{by the preceding lemma,} \\
&= \frac{R^2\sqrt{2(1-R)(1+R)}}{R^2 - |z|} + 1 - R, \quad \text{by definition of } \varepsilon \text{ and } r, \\
&< \frac{2R^2\sqrt{1-R}}{R^2 - |z|} + 1 - R, \quad \text{since } 1 + R < 2, \\
&= \frac{2R^2 + \sqrt{1-R}\big(R^2 - |z|\big)}{R^2 - |z|}\sqrt{1-R} \\
&< \frac{3}{R^2 - |z|}\sqrt{1-R}.
\end{aligned}$$

This holds for all $z \in D(0,r) = D\big(0,R^2\big)$ and completes the proof.

§ 2 The Proof of Carathéodory and Koebe

The proof below is synthesized from the ideas of Koebe, Carathéodory and Ostrowski and follows somewhat the exposition in BISHOP [1967]; some alternative shorter versions are outlined in §3.

Theorem 9.7 (Riemann Mapping Theorem) *Every region having property* (HS_1) *other than* $\mathbb{C}$ *itself is conformal to a disk.*

Proof: We first show that it is enough to treat Koebe regions, that is, every region $\Omega \neq \mathbb{C}$ having property (HS_1) is conformally equivalent to a Koebe region. Pick $w_0 \in \mathbb{C}\backslash\Omega$ and use property (HS_1) of Ω to produce a holomorphic function ϕ on Ω such that

$$\phi^2(z) = z - w_0 \qquad \forall z \in \Omega.$$

Then ϕ is one-to-one and

$$\phi(\Omega) \cap -\phi(\Omega) = \varnothing. \tag{1}$$

This is just the radical surgery technique discussed following 4.37. (See the proof of corresponding facts in 5.76.) Since ϕ is an open map, $-\phi(\Omega)$ contains $\bar{D}(a,r)$ for some $a \in \mathbb{C}$ and some $r > 0$, so by (1)

$$\bar{D}(a,r) \subset \mathbb{C} \setminus \phi(\Omega). \tag{2}$$

Therefore, we may form $\psi := r/(\phi - a)$ and have a conformal map of Ω into $\mathbb{D}$. Indeed, by (2), $|\phi(z) - a| > r$ for all $z \in \Omega$. Note that $\psi(\Omega) \neq \mathbb{D}$, since by its very form the function ψ is never 0. We look at $f_c\big(\psi(\Omega)\big)$ for some $c \in \psi(\Omega)$, where f_c is the conformal map of $\mathbb{D}$ onto $\mathbb{D}$ given by

$$f_c(z) := \frac{z - c}{1 - \bar{c}z}, \qquad z \in \mathbb{D}.$$

Since $f_c(c) = 0$, we have $0 \in f_c\big(\psi(\Omega)\big)$ and since $\psi(\Omega) \subsetneq \mathbb{D}$, we have $f_c\big(\psi(\Omega)\big) \subsetneq \mathbb{D}$. Clearly property (HS_1) passes over to this set from Ω under the conformal map $f_c \circ \psi$. Hence $f_c\big(\psi(\Omega)\big)$ is a Koebe region.

So we are reduced to the case of a Koebe region Ω. We make an inductive construction thus: let $\Omega_0 := \Omega$ and let f_1 be a Koebe map of Ω_0. Set $\Omega_1 := f_1(\Omega_0)$. Suppose defined Koebe regions $\Omega_0, \ldots, \Omega_n$ and functions $f_1, \ldots, f_n$ such that f_j is a Koebe map of Ω_{j-1} and $\Omega_j = f_j(\Omega_{j-1})$ for each $j = 1, 2, \ldots, n$. Then let f_{n+1} be a Koebe map on Ω_n (such exist by 9.4) and let Ω_{n+1} be the Koebe region $f_{n+1}(\Omega_n)$. This completes the inductive definition.

Let R_n be the inner radius of Ω_n $(n = 0, 1, 2, \dots)$. Notice that by 9.3 we have then

$$(3) \qquad 0 < R_0 \leq R_1 \leq R_2 \leq \cdots < 1$$

and by definition of the inner radius

$$(4) \qquad D(0, R_n) \subset \Omega_n \qquad \forall n \geq 0.$$

For each $n \geq 0$ and each $p \geq 1$ set

$$(5) \qquad \psi_{n,p} := f_{n+p} \circ f_{n+p-1} \circ \cdots \circ f_{n+1} \quad \text{from } \Omega_n \text{ onto } \Omega_{n+p}.$$

By the Chain Rule, the fact that $f_k(0) = 0$ for all k, and induction we have

$$\psi'_{n,p}(0) = f'_{n+p}(0) \cdots f'_{n+1}(0) = \prod_{j=n}^{n+p-1} \left(\frac{1 + R_j}{2\sqrt{R_j}} \right),$$

by property (c) of Koebe maps. Now $R \mapsto \frac{1+R}{2\sqrt{R}}$ is a decreasing function on $(0, 1]$. Hence the preceding equality implies that

$$(6) \qquad \psi'_{n,p}(0) \geq \left(\frac{1 + R_{n+p}}{2\sqrt{R_{n+p}}} \right)^p \qquad \forall n \geq 0,\ p \geq 1.$$

Now all Ω_{n+p} lie in $\mathbb{D}$, so

$$(7) \qquad |\psi_{n,p}(z)| < 1 \qquad \forall z \in \Omega_n,\ \forall n \geq 0,\ p \geq 1.$$

If we apply Schwarz' lemma to each $\psi_{n,p}$ in the disk $D(0, R_0)$, which lies in each Ω_n by (3) and (4), we get from (7)

$$\left|\psi'_{n,p}(0)\right| \leq 1/R_0 \qquad \forall n \geq 0,\ p \geq 1,$$

which together with (6) yields

$$\frac{1}{R_0} \geq \left(\frac{1 + R_{n+p}}{2\sqrt{R_{n+p}}} \right)^p = \left[1 + \frac{\left(1 - \sqrt{R_{n+p}}\right)^2}{2\sqrt{R_{n+p}}} \right]^p > \left[1 + \frac{(1 - \sqrt{R_{n+p}})^2}{4} \right]^p.$$

Taking $n = 0$ and solving for $\sqrt{R_p}$ yields

$$\sqrt{R_p} \geq 1 - 2\sqrt{R_0^{-1/p} - 1},$$

holding for all $p \in \mathbb{N}$. According to 3.1, $R_0^{-1/p} \to 1$. Hence the preceding inequality and those in (3) show that

$$\lim_{p\to\infty} R_p = 1. \tag{8}$$

Now let K compact $\subset \Omega_0$ be given. Since $\psi_{0,n}(0) = 0$ and $|\psi_{0,n}| < 1$ for each n, it follows from 6.44 that for some $c = c(K) < 1$ we have

$$|\psi_{0,n}(z)| \le c \qquad \forall z \in K,\ n \in \mathbb{N}. \tag{9}$$

Since $c < 1$, (8) insures that there exists N such that

$$c < R_n^2 \qquad \forall n \ge N. \tag{10}$$

From (9) and (10) we have

$$\psi_{0,n}(K) \subset D(0, R_n^2) \qquad \forall n \ge N. \tag{11}$$

Now by (3), (4) and (5) we have

$$\begin{aligned} D(0,R_n) \subset D(0,R_{n+p}) \subset \Omega_{n+p} &= f_{n+p}(\Omega_{n+p-1}) = \cdots \\ &= f_{n+p} \circ f_{n+p-1} \circ \cdots \circ f_{n+1}(\Omega_n) \\ &= \psi_{n,p}(\Omega_n). \end{aligned} \tag{12}$$

From (4), (7) and (12) we thus have

$$D(0,R_n) \subset \Omega_n \cap \psi_{n,p}(\Omega_n) \subset \Omega_n \cup \psi_{n,p}(\Omega_n) \subset \mathbb{D}.$$

These inclusions together with (6) show that the hypotheses of 9.6 are fulfilled by $f := \psi_{n,p}$ and we can therefore assert that

$$|\psi_{n,p}(w) - w| \le \frac{3\sqrt{1-R_n}}{R_n^2 - |w|} \qquad \forall w \in D(0, R_n^2)\,.$$

In conjunction with (11) this yields

$$\left|\psi_{n,p}(\psi_{0,n}(z)) - \psi_{0,n}(z)\right| \le \frac{3\sqrt{1-R_n}}{R_n^2 - |\psi_{0,n}(z)|} \qquad \forall z \in K,\ n \ge N,\ p \ge 1.$$

From (5) however we see that $\psi_{n,p} \circ \psi_{0,n} = \psi_{0,n+p}$. Using this and (9) in the last inequality yields

$$|\psi_{0,n+p}(z) - \psi_{0,n}(z)| \le \frac{3\sqrt{1-R_n}}{R_n^2 - c} \qquad \forall z \in K,\ n \ge N,\ p \ge 1.$$

In view of (8) this last inequality asserts that the sequence $\{\psi_{0,k}\}$ is uniformly Cauchy on K. Therefore we have shown that

$$\psi := \lim_{k\to\infty} \psi_{0,k} \quad \text{exists locally uniformly in } \Omega_0. \tag{13}$$

This limit is a holomorphic function (5.44) and it further follows from (9) that $|\psi(z)| \le c(K) < 1$ for each $z \in K$ and each compact $K \subset \Omega_0$. That is,

$$\psi(\Omega_0) \subset \mathbb{D}. \tag{14}$$

Next we show that ψ is surjective. Let $w \in \mathbb{D}$ be given. Use (8) to select N so that

$$R_n > |w| \qquad \forall n \ge N. \tag{15}$$

Then by (4) we have for $p \ge 1$

$$\begin{aligned} w \in D\left(0, R_{N+p}\right) \subset \Omega_{N+p} &= f_{N+p}\left(\Omega_{N+p-1}\right) = \cdots \\ &= f_{N+p} \circ f_{N+p-1} \circ \cdots \circ f_{N+1}\left(\Omega_N\right) \\ &= f_{N+p} \circ f_{N+p-1} \circ \cdots \circ f_{N+1} \circ \psi_{0,N}\left(\Omega_0\right). \end{aligned}$$

Therefore there exists $z_p \in \Omega_0$ such that

$$w = f_{N+p} \circ f_{N+p-1} \circ \cdots \circ f_{N+1} \circ \psi_{0,N}(z_p) \tag{16}$$
$$= \psi_{0,N+p}(z_p). \tag{17}$$

By (16) and successive applications of property (d) of Koebe maps we have

$$|w| \ge \left|\psi_{0,N}(z_p)\right| \qquad \forall p \in \mathbb{N}. \tag{18}$$

From (4) and (15) we have that $\bar{D}\left(0,|w|\right)$ is a compa ctsubset of Ω_N. Therefore $\psi_{0,N}^{-1}\left(\bar{D}\left(0,|w|\right)\right)$ is a compact subset C of Ω_0. It follows from (18) that z_p belongs to this compact subset of Ω_0 for all $p \ge 1$. Therefore some subsequence $\left\{z_{p_j}\right\}$ converges to a $z \in C$. Since $\{\psi_{0,k}\}$ converges to ψ *uniformly* on C, it follows easily (cf. 7.2) that $\psi_{0,N+p_j}(z_{p_j}) \to \psi(z)$. That is, recalling (17), $w = \psi(z)$. As w is any point of $\mathbb{D}$, this establishes the surjective character of ψ.

The fact that ψ is one-to-one, which completes the proof, is now a consequence of 7.11. For ψ, being surjective, is not constant but it is the limit of one-to-one functions.

We can however prove the univalence of ψ directly without appeal to 7.11 thus. Define

$$(5)' \qquad \phi_{n,p} := \psi_{n,p}^{-1} = f_{n+1}^{-1} \circ \cdots \circ f_{n+p-1}^{-1} \circ f_{n+p}^{-1} \quad \text{from } \Omega_{n+p} \text{ onto } \Omega_n.$$

Since all maps fix 0 and are one-to-one, 5.78 gives

$$(6)' \qquad \phi'_{n,p}(0) = \frac{1}{\psi'_{n,p}(0)} \overset{(6)}{>} 0.$$

Let compact $L \subset \mathbb{D}$ be given. Choose $R' < R < 1$ so that $L \subset D(0, R')$ and then, recalling (18), choose N so large that

$$(19) \qquad R < R_n^2 \qquad \forall n \geq N.$$

By (4) then

$$(20) \qquad L \subset D(0, R') \subset D(0, R) \subset D\big(0, R_n^2\big) \subset \Omega_n \qquad \forall n \geq N.$$

We have

$$D(0, R_n) \subset D(0, R_{n+p}) \subset \Omega_{n+p}$$

and

$$D(0, R_n) \subset \Omega_n = \phi_{n,p}(\Omega_{n+p}),$$

whence

$$D(0, R_n) \subset \Omega_{n+p} \cap \phi_{n,p}(\Omega_{n+p}) \subset \Omega_{n+p} \cup \phi_{n,p}(\Omega_{n+p}) \subset \mathbb{D} \quad \forall n \geq 0,\ p \geq 1.$$

From these inclusions and (6)′ we infer via 9.6 that

$$(21) \qquad |\phi_{n,p}(w) - w| \leq \frac{3\sqrt{1 - R_n}}{R_n^2 - |w|} \qquad \forall w \in D\big(0, R_n^2\big),\ n \geq 0,\ p \geq 1.$$

Then (20) teams up with (21) to give

$$|\phi_{n,p}(w) - w| \leq \frac{3\sqrt{1 - R_n}}{R_n^2 - R'} \qquad \forall w \in L,\ n \geq N,\ p \geq 1,$$

$$(22) \qquad \overset{(19)}{\leq} \frac{3\sqrt{1 - R_n}}{R - R'} \qquad \forall w \in L,\ n \geq N,\ p \geq 1.$$

Property (d) of a Koebe map k says that $\left|k^{-1}(w)\right| \leq |w|$ for all w in the range of k. It follows by induction from (5)′ that

$$(23) \qquad |\phi_{n,p}(w)| \leq |w| \qquad \forall w \in \Omega_{n+p},\ n \geq 0,\ p \geq 1,$$

hence from (20) follows also

$$\phi_{n,p}(L) \subset D(0,R) \qquad \forall n \geq N,\ p \geq 1.$$

For $w \in L$, $n \geq N$ and $p \geq 1$ the point $z := \phi_{n,p}(w)$ thus lies in $D(0,R)$ which also contains w and which lies in the domain Ω_n of $\phi_{0,n}$, by (20). We are therefore in a position to apply 6.3(ii) with the F there equal to $\phi_{0,n}$ and the M there equal to 1. We get

$$\begin{aligned} \bigl|\phi_{0,n+p}(w) - \phi_{0,n}(w)\bigr| &= \bigl|\phi_{0,n}\bigl(\phi_{n,p}(w)\bigr) - \phi_{0,n}(w)\bigr| \\ &\leq \frac{2R}{\bigl|R^2 - \bar{w}\phi_{n,p}(w)\bigr|}\bigl|\phi_{n,p}(w) - w\bigr| \\ &\leq \frac{2R}{R^2 - |w|\bigl|\phi_{n,p}(w)\bigr|} \cdot \frac{3\sqrt{1-R_n}}{R-R'} \qquad \text{by (22)}, \\ &\leq \frac{2R}{R^2 - (R')^2} \cdot \frac{3\sqrt{1-R_n}}{R-R'} \qquad \text{by (23) and (20)}, \end{aligned} \tag{24}$$

an inequality which holds for all $w \in L$, all $n \geq N$ and all $p \geq 1$. Since the right-hand side of (24) converges to 0 as $n \to \infty$, we see that the sequence $\{\phi_{0,n}\}_{n=N}^{\infty}$ is uniformly Cauchy on L. Therefore

$$\phi := \lim_{k\to\infty} \phi_{0,k} \quad \text{exists locally uniformly in } \mathbb{D}. \tag{13$'$}$$

Now consider any $z_0 \in \Omega_0$. If in (9) we take for K the set $\{z_0\}$, we shall have

$$\bigl\{\psi_{0,n}(z_0) : n \in \mathbb{N}\bigr\} \subset \bar{D}(0,c). \tag{25}$$

As the latter is a compact subset of $\mathbb{D}$, the sequence $\{\phi_{0,k}\}$ converges to ϕ *uniformly* on it. It follows easily from (13) and the inclusion (25) (cf. 7.2) that therefore

$$\phi_{0,k}\bigl(\psi_{0,k}(z_0)\bigr) \to \phi\bigl(\psi(z_0)\bigr)$$

and so, since $\phi_{0,k}$ is the inverse of $\psi_{0,k}$,

$$z_0 = \phi\bigl(\psi(z_0)\bigr). \tag{26}$$

This holds for all $z_0 \in \Omega_0$. It establishes at once the univalence of ψ.

Remark: CARATHÉODORY [1914] proves the univalence and the surjectivity of ϕ with the zero-counting integral of 5.85 (see the hints to 9.22 and 7.19(i)). Since Koebe maps merely call for the creation of square-roots of the identity on certain simply-connected subsets of $\mathbb{D}$ [the set $f_a(\Omega)$ in the proof of 9.4] and since this can be done more or less constructively, the last vestige of non-constructivity (I use the term informally, not in the strict sense of E. Bishop, *et al.*) can thereby be eliminated from the proof of the mapping theorem. On the other hand, the *corroboration* that the preceding construction works can be shortened by non-constructive arguments (Montel's Compactness Theorem), and a pure existence proof can be fashioned very expediously, as shown in the next section.

§ 3 Fejér and Riesz' Proof and the Uniqueness Theorem

Exercise 9.8 (i) (Poincaré's Uniqueness Theorem) *Let Ω be a region in $\mathbb{C}$ which is conformal to an open disk, $w_0 \in \Omega$, $z_0 \in \mathbb{D}$, $\theta \in \mathbb{R}$. Show that there is exactly one conformal map f of $\mathbb{D}$ onto Ω which satisfies $f(z_0) = w_0$ and $f'(z_0) = e^{i\theta}\left|f'(z_0)\right|$. Formulate a uniqueness property involving two points of Ω analogous to that in* 6.4(ii).

Hints: (p. 111 of Carathéodory [1912a]) For existence, just move the disk $\mathbb{D}$ around by an appropriate conformal automorphism under any Riemann map of $\mathbb{D}$ onto Ω. For uniqueness, consider any other function g answering to this same description and form $h := \phi^{-1} \circ f^{-1} \circ g \circ \phi$, where $\phi(z) := (z+z_0)/(1+\bar{z}_0 z)$. Then h is a conformal automorphism of $\mathbb{D}$ which fixes 0, so by (6.2)(ii) it is a rotation of the identity map by some $u \in \mathbb{T}$. But a routine calculation reveals that $h'(0) = g'(z_0)/f'(z_0)$, which is positive. Hence $u = 1$.

(ii) *Let $\mathbb{E} := \left\{x + iy : \frac{x^2}{a^2} + \frac{y^2}{b^2} < 1\right\}$, where a, b are positive real numbers. Show that if ϕ is the conformal map of $\mathbb{D}$ onto $\mathbb{E}$ satisfying $\phi(0) = 0 < \phi'(0)$, then ϕ maps each quadrant of $\mathbb{D}$ onto the corresponding quadrant of $\mathbb{E}$.*

Hint: Uniqueness of ϕ (from (i)) shows that $\phi(z) = \bar{\phi}(\bar{z}) = -\phi(-z)$ for all $z \in \mathbb{D}$. It follows that $\phi((-1,1)) = (-a,a)$ and $\phi((-i,i)) = (-ib,ib)$. [For a construction of ϕ see Szegö [1950]; of related interest is Chernoff [1977].]

(iii) *Let Ω_n be a simply-connected region containing 0, $R_n < \infty$ its inner radius. Suppose $\Omega_n \subset \Omega_{n+1}$ for all n and set $\Omega := \bigcup_{n\in\mathbb{N}} \Omega_n$.*

(a) *Show that Ω is a simply-connected region.*

Hints: Use 1.1 and, say, 4.71(vii): any loop in Ω lies wholly in some Ω_n.

(b) *According to* (i) *there is a unique conformal map f_n of $\mathbb{D}$ onto Ω_n satisfying $f_n(0) = 0 < f_n'(0)$. Show that for each n*

$$|f_n(z)| \geq R_n|z| \qquad \forall z \in \mathbb{D}.$$

Hints: Applying 6.1 to the function $w \mapsto f_n^{-1}(R_n w)$ on $\mathbb{D}$ gives $\left|f_n^{-1}(w)\right| \leq |w|/R_n$ for all $w \in D(0, R_n)$. Of course, this inequality is trivially true for any $w \in \Omega_n \setminus D(0, R_n)$, since $\left|f_n^{-1}\right| \leq 1$.

(c) *Show that if* $\Omega = \mathbb{D}$, *then the sequence* $\{f_n\}$ *converges locally uniformly in* $\mathbb{D}$ *to the identity function.*

Hints: Since all $f_n'(0)$ are positive, any subsequential limit g satisfies $g'(0) \geq 0$. From (b), $|g(z)| \geq |z|$ for all $z \in \mathbb{D}$. These two facts and 6.1 show that g is the identity map on $\mathbb{D}$. Now invoke 7.8.

(d) *Show that if* $\Omega \neq \mathbb{C}$, *then the sequence* $\{f_n\}$ *converges locally uniformly in* $\mathbb{D}$ *to the unique conformal map* f *of* $\mathbb{D}$ *onto* Ω *satisfying* $f(0) = 0 < f'(0)$, *which* (i) *supplies, and* $\{f_n^{-1}\}$ *converges to* f^{-1}.

Hints: Apply (c) to $f^{-1}(\Omega_n)$ in the role of Ω_n, with correspondingly $f^{-1} \circ f_n$ in the role of f_n. Cf. 7.1(ii). For convergence of $\{f_n^{-1}\}$ see 7.19(ii).

Exercise 9.9 *Let* Ω *be a Koebe region and* $\mathcal{F}$ *the family of all conformal maps* f *of* Ω *into* $\mathbb{D}$ *with* $f(0) = 0$.

(i) *Show that* $\mathcal{F}$ *contains a function* f_0 *which maximizes* $|f'(0)|$ *among all* f *in* $\mathcal{F}$.

Hint: If R denotes the inner radius of Ω, $|f'(0)| \leq 1/R$ for all $f \in \mathcal{F}$, by 6.1(i). Use 7.6 and 7.11.

(ii) *Show that any such* f_0 *provides a conformal map of* Ω *onto* $\mathbb{D}$.

Hint: If f_0 is not surjective, it can be followed by a Koebe map to produce an $f_1 \in \mathcal{F}$ with $|f_1'(0)| > |f_0'(0)|$, by the Chain Rule and property (c) of Koebe maps.

(iii) *Use Montel's theorem* 7.8 *and* (8) *from the proof of* 9.7 *to give an alternative proof* (*avoiding* 9.5, 9.6 *and* 6.44) *that the whole sequence* $\{\psi_{0,p}\}_{p=1}^{\infty}$ *converges to a conformal map of* Ω *onto* $\mathbb{D}$.

Hints: Suppose $\{\psi_{0,p_j}\}_{j=1}^{\infty}$ is a locally uniformly convergent subsequence of $\{\psi_{0,p}\}$; let ψ denote its limit. We have $\psi(\Omega) \subset \overline{\mathbb{D}}$ and $\psi'(0) \geq 1$ (cf. (6) and (8) in the proof of 9.7), so ψ is not constant. Hence $\psi(\Omega) \subset \mathbb{D}$ and ψ is a conformal map (7.11). From $|\psi_{0,k}| \leq |\psi_{0,k+1}|$ (a consequence of property (d) of Koebe maps) for all k get $|\psi_{0,p_j}| \leq |\psi|$ for all j. Infer from 9.3(i) that the inner radius of $\psi(\Omega)$ is not less than that of $\psi_{0,p_j}(\Omega)$, which is $R_{p_j} \to 1$ as $j \to \infty$. Conclude that $\psi(\Omega) = \mathbb{D}$. Any other subsequential limit function ψ^* has these same properties, so by Poincaré's uniqueness theorem $\psi^* = \psi$. Now invoke 7.8 to see that the whole sequence $\{\psi_{0,p}\}_{p=1}^{\infty}$ converges.

(iv) *Show how property* (d) *of Koebe maps can be combined with* 7.14(iii) *to secure the convergence of the sequence* $\{\psi_{0,p}\}_{p=1}^{\infty}$ *in the proof of* 9.7.

Hints: Since $\psi_{0,p}(0) = 0$, each function $h_p(z) := \frac{\psi_{0,p}(z)}{z}$ is holomorphic in Ω. By property (d) of Koebe maps and the definition of $\psi_{0,p}$ we have $|\psi_{0,p}| \leq |\psi_{0,p+1}|$ for all p. Moreover $h_p(0) = \psi'_{0,p}(0) > 0$ and this increasing sequence is bounded above by $1/R_0$ (cf. the proof of 9.7), hence converges to a positive real number. Therefore 7.14(iii) may be applied to $\{h_p\}_{p=1}^{\infty}$.

(v) *Show how* 6.15 *can be used to prove that the limit function* ψ *in the proof of* 9.7 *maps onto* $\mathbb{D}$.

Hints: We need to see that ψ is a boundary-preserving map of Ω into $\mathbb{D}$. So consider a sequence $z_n \in \Omega$ with $z_n \to z_0 \in \partial\Omega$ and any $0 < R < 1$. Using (8) in the proof of 9.7, pick p so that $R_p > R$. Then $\bar{D}(0, R) \subset \psi_{0,p}(\Omega)$. Since $\psi_{0,p}$ is conformal, it is a proper map of Ω onto $\psi_{0,p}(\Omega)$ and so for some N we have $|\psi_{0,p}(z_n)| > R$ for all $n \geq N$. Since (cf. hints to (iii)) $|\psi| \geq |\psi_{0,p}|$, we therefore have $|\psi(z_n)| > R$ for all $n \geq N$.

§ 4 Boundary Behavior for Jordan Regions

A Jordan region is simply-connected (see 4.45) hence has the property (HS_1), by 4.71. Osgood conjectured in the encyclopedia article [1901] (p. 56) that any conformal map of a Jordan region onto $\mathbb{D}$ extends to a homeomorphism of its closure. This was confirmed more or less simultaneously by Osgood and Taylor [1913] and Carathéodory [1913b]. We present here one solution of the problem and later discuss some other approaches. All are quite geometrical, presenting formidable problems of exposition. The reason for singling out Jordan regions is simple: if $f\colon \mathbb{D} \mapsto \Omega$ is conformal and extends to a homeomorphism of $\overline{\mathbb{D}}$, then Ω is *necessarily* a Jordan region, the inside of the Jordan-curve $f(\mathbb{T})$.

Lemma 9.10 (Koebe [1913a]) *Let a, b be two distinct numbers of modulus 1, γ_n curves in $\mathbb{D}$ with endpoints a_n, b_n satisfying*

(i) $$a_n \to a,\ b_n \to b.$$

Let f be a bounded holomorphic function in $\mathbb{D}$ such that

(ii) $$\sup |f(\gamma_n)| \to 0 \quad \textit{as } n \to \infty.$$

Then $f = \mathbf{0}$.

Proof: There is a conformal automorphism T of $\mathbb{D}$ which maps a to i and b to $-i$ (see 6.4(v)) and is, of course, a homeomorphism of $\overline{\mathbb{D}}$. So $T(a_n) \to T(a) = i$ and $T(b_n) \to T(b) = -i$. Since $f = \mathbf{0}$ if $f \circ T^{-1} = \mathbf{0}$, we can replace f by

$f \circ T^{-1}$ and each γ_n by $T \circ \gamma_n$ and so simply assume (i) has the more specific form

(i)′ $$a_n \to i,\ b_n \to -i.$$

Next note that it suffices to prove the lemma under the additional hypothesis that

(iii) $$\inf |\gamma_n| \to 1 \text{ as } n \to \infty \quad \text{and} \quad |\gamma_n| > 0 \text{ for all } n.$$

For if (iii) fails, there is an $r_1 < 1$ and $1 < n_1 < n_2 < \ldots$ such that γ_{n_j} meets $D(0, r_1)$ for each j. By (i) we can suppose n_1 so large that $|a_{n_j}|$, $|b_{n_j}| \geq (1 + r_1)/2 =: r_2$ for all j. Connectedness considerations then show that γ_{n_j} meets $C(0, r)$ for each j and each $r \in (r_1, r_2)$. From compactness considerations and (ii) it follows that f has a zero on $C(0, r)$ for each $r \in (r_1, r_2)$, hence from 5.62, $f = \mathbf{0}$.

Now (i)′ says that γ_n visits both the upper half-plane and the lower half-plane for all large n, so such γ_n must cut the real axis. Replacing $f(z)$ by $f(-z)$ and γ_n by $-\gamma_n$ if necessary, and going to a subsequence again, we can simply suppose

(1) $$\gamma_n \text{ meets } (0, 1) \text{ for every } n.$$

We may also suppose, from (i)′, that

(2) $$\operatorname{Im} a_n > |\operatorname{Re} a_n| \quad \text{for all } n.$$

We form
$$S := \{z \in \mathbb{D} : \operatorname{Re} z > |\operatorname{Im} z|\}.$$

Let the parameter interval for each γ_n be $[0, 1]$. We have $\gamma_n(0) = a_n$ (or this can be arranged by replacing $\gamma_n(t)$ with $\gamma_n(1 - t)$) and $\gamma_n(\tau_n) \in (0, 1)$ for some $\tau_n > 0$. Thus $\operatorname{Re} \gamma_n - |\operatorname{Im} \gamma_n|$ is (by (2)) negative at 0 and positive at τ_n. Hence there is a $t \in (0, \tau_n)$ such that $\operatorname{Re} \gamma_n(t) = |\operatorname{Im} \gamma_n(t)|$. Let t_n denote the largest such t. Then

(3) $$\operatorname{Re} \gamma_n(t_n) = |\operatorname{Im} \gamma_n(t_n)|, \quad \gamma_n\,(t_n, \tau_n] \subset S.$$

Finally, let s_n be the least $t \in [t_n, \tau_n]$ such that $\gamma_n(t) \in [0, 1]$. By (iii) and (3), $s_n \neq t_n$. Thus $(t_n, s_n) \neq \varnothing$ and γ_n maps this interval into S [by (3)] and misses $[0, 1]$. Consequently, γ_n either maps (t_n, s_n) into the open upper half-plane or into the open lower half-plane. One of these alternatives prevails for infinitely many n and by discarding the others and renumbering, we can have

$$\text{(4)} \qquad \operatorname{Re}\gamma_n(t_n) = |\operatorname{Im}\gamma_n(t_n)|, \ \gamma_n(s_n) \in (0,1), \ \gamma_n(t_n, s_n) \subset S_0 \qquad \forall n,$$

where S_0 is the intersection of S with one of the upper or lower open half-planes. Let $\Gamma_n \colon [0,1] \mapsto \mathbb{C}$ be a curve with

$$\text{(5)} \qquad \Gamma_n(0) = \gamma_n(t_n), \quad \Gamma_n(1) = \overline{\gamma}_n(t_n)$$

$$\text{(6)} \qquad \Gamma_n(0,1) = \gamma_n\,(t_n, s_n] \cup \overline{\gamma}_n\,(t_n, s_n] \subset S.$$

By (iii) Γ_n is zero-free, so by 4.1 we can find continuous $\phi_n \colon [0,1] \mapsto \mathbb{C}$ such that

$$\text{(7)} \qquad \Gamma_n = e^{2\pi i \phi_n}.$$

According to (4) we have

$$\Gamma_n(0) = \gamma_n(t_n) = |\gamma_n(t_n)| e^{\mp i\pi/4}, \quad \Gamma_n(1) = \overline{\gamma}_n(t_n) = |\gamma_n(t_n)| e^{\pm i\pi/4}.$$

Therefore (7) shows that

$$\text{(8)} \qquad \phi_n(1) - \phi_n(0) = \pm\tfrac{1}{4} + k_n, \quad \text{for some integer } k_n.$$

If we define

$$\text{(9)} \qquad \psi_n(t) := \begin{cases} \phi_n(t) & 0 \le t \le 1 \\ \phi_n(t-1) + \phi_n(1) - \phi_n(0) & 1 \le t \le 2 \\ \phi_n(t-2) + 2[\phi_n(1) - \phi_n(0)] & 2 \le t \le 3 \\ \phi_n(t-3) + 3[\phi_n(1) - \phi_n(0)] & 3 \le t \le 4, \end{cases}$$

then ψ_n is (well defined and) continuous and by (7) and (8) it satisfies

$$\text{(10)} \qquad J_n(t) := e^{2\pi i \psi_n(t)} = \begin{cases} \Gamma_n(t) & 0 \le t \le 1 \\ \pm i\Gamma_n(t-1) & 1 \le t \le 2 \\ -\Gamma_n(t-2) & 2 \le t \le 3 \\ \mp i\Gamma_n(t-3) & 3 \le t \le 4. \end{cases}$$

We have

$$\psi_n(4) - \psi_n(0) \overset{(9)}{=} 4[\phi_n(1) - \phi_n(0)] \overset{(8)}{=} \pm 1 + 4k_n.$$

Therefore J_n is a *closed* curve and

$$\text{(11)} \qquad \operatorname{Ind}_{J_n}(0) = \pm 1 + 4k_n \neq 0.$$

Now (6) and (10) show that the range of J_n is contained in $\pm\big(\gamma_n \cup \overline{\gamma}_n \cup i\gamma_n \cup i\overline{\gamma}_n\big)$, hence by (iii)

$$\inf|J_n| \to 1 \quad \text{as } n \to \infty.$$

Therefore, given $0 < r < 1$, we have $D(0,r) \subset \mathbb{C} \setminus J_n$ for all large n. Fix such an n and let C_n denote the component of $\mathbb{C} \setminus J_n$ which contains 0. Then $D(0,r) \subset C_n$. Moreover, since $J_n \subset \mathbb{D}$, $\mathbb{C} \setminus \mathbb{D}$ is a connected unbounded subset of $\mathbb{C} \setminus J_n$, so Ind_{J_n} vanishes in $\mathbb{C} \setminus \mathbb{D}$ (by 4.3). Consequently, by (11) the component C_n does not meet $\mathbb{C} \setminus \mathbb{D}$:

$$D(0,r) \subset C_n \subset \mathbb{D}. \tag{12}$$

Since C_n is open in $\mathbb{C}$ and relatively closed in $\mathbb{C} \setminus J_n$, we have

$$\partial C_n = \overline{C}_n \setminus C_n \subset J_n. \tag{13}$$

Form the function

$$F(z) := f(z)\bar{f}(\bar{z})f(iz)\bar{f}(-i\bar{z})f(-z)\bar{f}(-\bar{z})f(-iz)\bar{f}(i\bar{z}), \qquad z \in \mathbb{D}.$$

It is holomorphic by 2.4. For any $z \in J_n$, (10) shows that one of the numbers $\pm z$, $\pm iz$ belongs to the range of Γ_n. Therefore, glancing back to (6), we see that one of the factors in $F(z)$ is majorized by $\varepsilon_n := \sup|f(\gamma_n)|$. If M is a majorant for $|f|$ on $\mathbb{D}$, this means that

$$|F| \le \varepsilon_n M^7 \qquad \text{on } J_n. \tag{14}$$

From (12), (13), (14) and the Maximum-Modulus Principle we infer that

$$|F| \le \varepsilon_n M^7 \qquad \text{on } D(0,r).$$

This holds for all sufficiently large n and by (ii), $\varepsilon_n \to 0$. We conclude that $F = 0$ in $D(0,r)$. Since $0 < r < 1$ is arbitrary, $F = \mathbf{0}$. But then from the definition of F and 5.62 follows easily that $f = \mathbf{0}$.

Lemma 9.11 *Let Ω be a Jordan region, g a conformal map of Ω onto $\mathbb{D}$. If $w_n \in \Omega$ and $w_n \to w \in \partial\Omega$, then $z_n := g(w_n)$ converges to some point of $\partial\mathbb{D}$.*

Proof: Let $f := g^{-1}$. We have $|z_n| \to 1$. For otherwise $\{z_n\}$ has a limit point z in $\mathbb{D}$ and then $f(z)$ is a limit point of $f\big(\{z_n\}\big) = \{w_n\}$ in Ω, contrary to $w_n \to w \in \partial\Omega$. By compactness of $\overline{\mathbb{D}}$, it suffices to show that $\{z_n\}$ has only one limit point. (Cf. 7.8.) So we suppose a, b two distinct limit points of $\{z_n\}$ and try to deduce a contradiction.

Thus we have two sequences $\{a_n\}$, $\{b_n\} \subset \mathbb{D}$ such that

$$(1) \qquad a_n \to a, \quad b_n \to b \neq a, \quad |a| = |b| = 1,$$

$$(2) \qquad f(a_n) \to w, \quad f(b_n) \to w.$$

4.50(ii) provides a curve $\gamma : [0,1] \mapsto \mathbb{C}$ with $\gamma(0,1] \subset \Omega$ and points $1 \geq t_n \downarrow 0$ such that $\gamma(t_{2n}) = f(a_n)$ and $\gamma(t_{2n+1}) = f(b_n)$ for every $n \in \mathbb{N}$. Denote by Γ_n the restriction of γ to $[t_{2n}, t_{2n+1}]$, a curve joining $f(a_n)$ to $f(b_n)$ in Ω such that $\Gamma_n \to w$:

for any $\varepsilon > 0$, $\Gamma_n \subset D(w, \varepsilon)$ for all sufficiently large n.

Now form $\gamma_n := g \circ \Gamma_n$. These are curves in $\mathbb{D}$ and their endpoints are a_n, b_n. Since $f(\gamma_n) = \Gamma_n \to w$, all the hypotheses of Koebe's lemma above are met (with $f - w$ in the role of the function f there). Conclude that $f - w = \mathbf{0}$, a contradiction.

Lemma 9.12 *Let Ω be a Jordan region, g a conformal map of Ω onto $\mathbb{D}$. Then g extends to a continuous map of $\overline{\Omega}$ onto $\overline{\mathbb{D}}$.*

Proof: For each $w \in \partial\Omega = \overline{\Omega} \setminus \Omega$ the preceding lemma provides a unique (why?) $z \in \partial\mathbb{D}$ such that

$$(1) \qquad g(w_n) \to z \quad \text{whenever } w_n \in \Omega \text{ and } w_n \to w.$$

Define $g(w)$ to be this z. This extends g to $\overline{\Omega}$. To check continuity of the extended g on $\overline{\Omega}$, we only have to see that

$$w_n' \in \overline{\Omega} \ \& \ w_n' \to w \in \partial\Omega \quad \text{imply } g(w_n') \to g(w).$$

To this end, select $w_n \in \Omega$ such that

$$(2) \qquad |w_n - w_n'| < 1/n$$

and

$$(3) \qquad |g(w_n) - g(w_n')| < 1/n.$$

If $w_n' \in \Omega$ we can just take $w_n := w_n'$. Otherwise, $w_n' \in \partial\Omega$ and the choice of such w_n is possible because of the way $g(w_n')$ is defined. From (2) and $w_n' \to w$, we have $w_n \to w$. Then from (1) we get $g(w_n) \to g(w)$. Finally then, from (3) follows $g(w_n') \to g(w)$.

Since $\overline{\Omega}$ is compact, the set $g(\overline{\Omega})$ is compact and contains $g(\Omega) = \mathbb{D}$. Hence $g(\overline{\Omega}) \supset \overline{\mathbb{D}}$, and we have $g(\overline{\Omega}) = \overline{\mathbb{D}}$.

Lemma 9.13 *Let Ω be a Jordan region, g a conformal map of Ω onto $\mathbb{D}$. Then the extension of the preceding lemma is one-to-one on $\overline{\Omega}$.*

First Proof: The proof of the preceding lemma shows that g maps $\partial\Omega$ into $\partial\mathbb{D}$, so only univalence on $\partial\Omega$ has to be checked. (Here we are using the name g for the extension too.) We can argue by contradiction and suppose that

$$a, b \in \partial\Omega, \quad a \neq b \quad \text{and} \quad g(a) = g(b).$$

We may rotate $\mathbb{D}$ and assume that $g(a) = 1$. Let ϕ be a homeomorphism of $\partial\mathbb{D}$ onto $\partial\Omega$ (which is a Jordan-curve). By preceding ϕ with an appropriate homeomorphism of $\overline{\mathbb{D}}$ onto $\overline{\mathbb{D}}$ (see 6.4) we can suppose that $\phi(+i) = a$, $\phi(-i) = b$. Let

$$\Gamma(t) := \phi\left(e^{it}\right), \qquad t \in \mathbb{R}.$$

Then

$$\partial\Omega = \Gamma[0, 2\pi], \quad \Gamma\left(\frac{\pi}{2}\right) = a, \quad \Gamma\left(\frac{3\pi}{2}\right) = b.$$

Since the accessible points of $\partial\Omega$ are dense in $\partial\Omega$ (1.16) and Γ is a local homeomorphism, we can also select a_n, b_n accessible points of $\partial\Omega$ sufficiently near a and b, respectively, that

$$\text{(1)} \qquad \begin{cases} a_n \in \Gamma\left(\dfrac{\pi}{4}, \dfrac{3\pi}{4}\right) & \& \ |g(a_n) - 1| < \dfrac{1}{n} \\[2ex] b_n \in \Gamma\left(\dfrac{5\pi}{4}, \dfrac{7\pi}{4}\right) & \& \ |g(b_n) - 1| < \dfrac{1}{n}. \end{cases}$$

Now 5.76 insures that $g|_{\partial\Omega}$ is not constant in any neighborhood of b. Therefore we may select a_n first, and in choosing b_n require in addition

$$(*) \qquad g(b_n) \neq g(a_n).$$

Let, say,

$$\text{(2)} \qquad \begin{cases} a_n = \Gamma(\alpha_n), & \dfrac{\pi}{4} < \alpha_n < \dfrac{3\pi}{4} \\[2ex] b_n = \Gamma(\beta_n), & \dfrac{5\pi}{4} < \beta_n < \dfrac{7\pi}{4}. \end{cases}$$

Since a_n, b_n are distinct accessible points of $\partial\Omega$, there is by 1.29 a cross-cut C_n in Ω joining a_n to b_n. Form

$$\text{(3)} \qquad \begin{cases} \Gamma_n' := \Gamma[\alpha_n, \beta_n] \supset \Gamma\left[\dfrac{3\pi}{4}, \pi\right] =: A' \\[2ex] \Gamma_n'' := \Gamma[\beta_n, \alpha_n + 2\pi] \supset \Gamma\left[\dfrac{7\pi}{4}, 2\pi\right] =: A''. \end{cases}$$

According to 4.47, $C_n \cup \Gamma'_n$ and $C_n \cup \Gamma''_n$ are Jordan-curves which satisfy

$$(4) \qquad \Omega \setminus C_n = \mathfrak{I}\,(C_n \cup \Gamma'_n) \cup \mathfrak{I}\,(C_n \cup \Gamma''_n)\,, \text{ disjoint union.}$$

We write $\gamma(t) := e^{it}$, $-\pi \le t \le \pi$, and for each n select s_n, $t_n \in [-\pi, \pi]$ such that $\gamma(s_n) = g(a_n)$, $\gamma(t_n) = g(b_n)$. According to (1) we then have

$$(5) \qquad s_n, t_n \in \left(-\frac{\pi}{2}, \frac{\pi}{2}\right) \quad \text{and} \quad s_n \to 0,\, t_n \to 0.$$

Let

$$(6) \qquad \gamma'_n := \gamma[s_n, t_n], \quad \gamma''_n, \text{ the complementary arc in } \partial\mathbb{D}.$$

Notice that, since g maps Ω univalently onto $\mathbb{D}$, $g(C_n)$ is a cross-cut c_n in $g(\Omega) = \mathbb{D}$ with endpoints $g(a_n) = \gamma(s_n)$ and $g(b_n) = \gamma(t_n)$, which are distinct by $(*)$. Therefore, citing 4.47 once again, we have from (6) that $c_n \cup \gamma'_n$ and $c_n \cup \gamma''_n$ are Jordan-curves which satisfy

$$(7) \qquad \mathbb{D} \setminus c_n = \mathfrak{I}\,(c_n \cup \gamma'_n) \cup \mathfrak{I}\,(c_n \cup \gamma''_n)\,, \text{ disjoint union.}$$

Since g is a homeomorphism of $\Omega \setminus C_n$ onto $g(\Omega \setminus C_n) = g(\Omega) \setminus g(C_n) = \mathbb{D} \setminus c_n$, connectedness considerations show that each of the two sets on the right-hand side of (4) is mapped by g onto one of those on the right-hand side of (7). In particular, for every $n \in \mathbb{N}$

$$(8) \quad \text{either} \quad g\big(\mathfrak{I}(C_n \cup \Gamma'_n)\big) = \mathfrak{I}(c_n \cup \gamma'_n) \quad \text{or} \quad g\big(\mathfrak{I}(C_n \cup \Gamma''_n)\big) = \mathfrak{I}(c_n \cup \gamma'_n).$$

Consider an n for which the first half of the alternative prevails:

$$(8)' \qquad g\big(\mathfrak{I}(C_n \cup \Gamma'_n)\big) = \mathfrak{I}(c_n \cup \gamma'_n).$$

For any $w \in \Gamma'_n$ there is (by 4.42) a sequence $\{w_k\} \subset \mathfrak{I}\,(C_n \cup \Gamma'_n)$ such that $w_k \to w$. Then from $(8)'$ we have

$$g(w_k) \in \mathfrak{I}(c_n \cup \gamma'_n) \quad \& \quad g(w_k) \to g(w) \in \partial\mathbb{D}.$$

Therefore

$$g(w) \in \overline{\mathfrak{I}(c_n \cup \gamma'_n)} \cap \partial\mathbb{D} = \big[\mathfrak{I}(c_n \cup \gamma'_n) \cup (c_n \cup \gamma'_n)\big] \cap \partial\mathbb{D} \subset (\mathbb{D} \cup \gamma'_n) \cap \partial\mathbb{D} = \gamma'_n.$$

This all shows that $g\,(\Gamma'_n) \subset \gamma'_n$. In particular, from (3)

$$(9)' \qquad g(A') \subset \gamma'_n.$$

We show in the same way that if for a given n the second half of the alternative (8) prevails, then

$$(9)'' \qquad g(A'') \subset \gamma_n'.$$

Now one of (9)′, (9)″ must prevail for infinitely many n. Let $n \to \infty$ through these values and remember (5) and (6). It follows that

$$\text{either} \quad g(A') \subset \{1\} \quad \text{or} \quad g(A'') \subset \{1\}.$$

Consequently, by 5.76, g is constant. Contradiction

Second Proof: This one is a bit more analytic and a bit less topological, and it proves a little more (see the Scholium after 9.14). As before, we have only to show that g is univalent on $\partial\Omega$. To this end, consider any $a, \tilde{a} \in \partial\Omega$ for which $g(a) = g(\tilde{a}) = z$, say. According to 4.50 there exist continuous $\gamma, \tilde{\gamma}\colon [0,1] \mapsto \mathbb{C}$ which map $(0,1]$ into Ω and 0 onto $a, \tilde{a}$, respectively. Then $\Gamma := g \circ \gamma$, $\widetilde{\Gamma} := g \circ \tilde{\gamma}$ are curves mapping $(0,1]$ into $\mathbb{D}$ and 0 onto $z \in \mathbb{T}$. We have $f \to a$ along Γ and $f \to \tilde{a}$ along $\widetilde{\Gamma}$, so by 5.56 $a = \tilde{a}$. This proves the univalence of g on $\partial\Omega$. Actually what was needed here is the version of 5.56 with $\mathbb{D}$ in the role of $\mathbb{D}^+$. It is obvious (by scaling) that 5.56 holds with $\mathbb{U}$ in the role of $\mathbb{D}^+$. Now $\psi(w) := -z \cdot \frac{w-i}{w+i}$ maps $\mathbb{U}$ conformally onto $\mathbb{D}$, with $\partial\mathbb{U} = \mathbb{R}$ mapping onto $\mathbb{T} \setminus \{-z\}$ (cf. 2.5). Consider $f \circ \psi$ in $\mathbb{U}$ along the curves $\psi^{-1} \circ \Gamma$, $\psi^{-1} \circ \widetilde{\Gamma}$ running into $\mathbb{U}$ from the boundary point $\psi^{-1}(z) = 0$, and apply 5.56.

Theorem 9.14 (OSGOOD and TAYLOR [1913], CARATHÉODORY [1913b]) *Let Ω be a Jordan region and f a conformal map of $\mathbb{D}$ onto Ω. Then f extends to a homeomorphism of $\overline{\mathbb{D}}$ onto $\overline{\Omega}$. For any distinct points $z_0, z_1 \in \Omega$, $b_0, b_1, b_2 \in \partial\Omega$ the map f is uniquely determined by the images of z_0, z_1; by the images of z_0, b_0; and by the images of b_0, b_1, b_2.*

Proof: Combine the preceding two lemmas. The uniqueness assertions follow from 6.4(ii).

Scholium: A review of the 9.11–9.13 proofs (the second one being employed for 9.13) reveals that beyond boundedness and simple-connectivity, the Ω there needed to have only one other property of a Jordan region, namely the boundary accessibility property 4.50(ii). Thus we have in fact proved that:

> *For a bounded simply-connected region Ω which has property 4.50(ii) any conformal map g of Ω onto $\mathbb{D}$ extends to a homeomorphism of $\overline{\Omega}$ onto $\overline{\mathbb{D}}$, whence (since by 9.1 such g exist) Ω is a Jordan region.*

(See NOVINGER [1975](a) and 4.4.13 of RANSFORD [1995] for other proofs.)

Every convex region clearly enjoys property 4.50(i), since $D(z,r) \cap \Omega$ is convex for every $r > 0$, $z \in \partial\Omega$. Thus, as the proof of 4.50 shows, Ω enjoys property 4.50(ii) as well. Of course, Ω is simply-connected, since $\mathbb{C} \setminus \Omega$ consists of half-lines. So we have the amazing fact:

Every bounded convex region is a Jordan region.

Theorem 9.16 below presents a sort of converse to 9.14. Another way to phrase 9.14 itself is

If f maps $\mathbb{D}$ conformally onto Ω, then f extends to a homeomorphism of $\overline{\mathbb{D}}$ onto $\overline{\Omega}$ if and only if $\partial\Omega$ is a Jordan-curve.

For this formulation there is a very useful parallel characterization of continuous extendability, for whose proof we have to refer the reader to ARSOVE [1968a], SCHÖNHAGE [1969] or POMMERENKE [1992]:

If f maps $\mathbb{D}$ conformally onto Ω, then f extends to a continuous map of $\overline{\mathbb{D}}$ onto $\overline{\Omega}$ if and only if $\partial\Omega$ is (the image of) a loop. In this case the Taylor series of f at 0 converges uniformly in $\mathbb{D}$.

Remarks: The crescent-shaped region in 10.25(iv) [and in Remark (i) following 9.27] illustrates the preceding. Another illuminating and canonical example is the slit disk $\Omega := \mathbb{D} \setminus [-1, 0]$. A conformal map g of this starlike region onto $\mathbb{D}$ is easily constructed with the help of the 6.16(vi) hints and the square-root function $\sqrt{\ }$ built from the Principal Branch of the Logarithm in $\mathbb{C}_-$, namely

$$g(z) := (z + 2\sqrt{z} - 1) \div (z - 2\sqrt{z} - 1), \quad z \in \Omega.$$

Although $\partial\Omega$ can be parameterized as a piecewise smooth loop, 4.50 shows it is not a Jordan-curve. By the theorem cited above, g^{-1} extends to a continuous function f of $\overline{\mathbb{D}}$ onto $\overline{\Omega}$, with $\partial\mathbb{D}$ evidently mapping to $\partial\Omega$, and f not 1–1 on $\partial\mathbb{D}$. It follows (how?) that g cannot be continuously extended to $\overline{\Omega}$. It is an instructive exercise to derive this formula for g and one for g^{-1}, and then use them to directly check all the assertions here.

§ 5 A Few Applications of the Osgood-Taylor-Carathéodory Theorem

Corollary 9.15 (Argument Principle) *Let Γ be a positively oriented simple loop, f a continuous function on $\overline{\mathfrak{I}(\Gamma)}$ which is holomorphic in $\mathfrak{I}(\Gamma)$ and zero-free on Γ. Then the number of zeros of f in $\mathfrak{I}(\Gamma)$, counted according to multiplicity, equals* $\text{Ind}_{f \circ \Gamma}(0)$. [This generalizes 5.86.]

Proof: Recall (4.7(i)) that the range J of Γ is a Jordan-curve; here $\mathfrak{I}(\Gamma)$ is abuse of language for $\mathfrak{I}(J)$. Let ϕ be a conformal map of $\mathbb{D}$ onto $\mathfrak{I}(\Gamma)$ and extend it to a homeomorphism of $\overline{\mathbb{D}}$ onto $\overline{\mathfrak{I}(J)}$ with $\partial\mathbb{D}$ mapping to J. Define $\Gamma_r(t) := re^{2\pi it}$ for $0 \leq r, t \leq 1$. Then $\phi \circ \Gamma_1$ and Γ have the same range J, so

$$(1) \qquad \pm\mathrm{Ind}_{f\circ\Gamma}(0) = \mathrm{Ind}_{f\circ\phi\circ\Gamma_1}(0)$$

by 4.7(iii). Consider $F := f \circ \phi$. It is easy to see that

$$(2) \qquad \begin{cases} F \text{ has a zero of order } n \text{ at } z \in \mathbb{D} \text{ if and only if} \\ f \text{ has a zero of order } n \text{ at } \phi(z). \end{cases}$$

We essentially checked this near the end of the proof of 7.35. Therefore the number of zeros of f in $\mathfrak{I}(\Gamma)$, counted according to multiplicity, equals the number of zeros of F in $\mathbb{D}$, counted according to multiplicity. Call this number N. Since the zero-set of F in $\overline{\mathbb{D}}$ is closed and lies wholly in $\mathbb{D}$ [the zero-set of f lies wholly in $\mathfrak{I}(\Gamma) = \phi(\mathbb{D})$ by hypothesis], this set lies in $D(0, r)$ for all r sufficiently near 1. Consequently we have from 5.86

$$(3) \qquad N = \mathrm{Ind}_{F\circ\Gamma_r}(0)$$

for all $r \in (0, 1)$ sufficiently close to 1. Let $r \uparrow 1$ and cite 4.6 to assert that

$$(4) \qquad N = \mathrm{Ind}_{F\circ\Gamma_1}(0) = \mathrm{Ind}_{f\circ\phi\circ\Gamma_1}(0).$$

Next take any $a \in \mathfrak{I}(\Gamma)$ and consider the special case of the foregoing where $f(z) := z - a$. Then $F = \phi - a$ and so is conformal, whence $N = 1$ and (4) reads

$$(5) \qquad 1 = \mathrm{Ind}_{F\circ\Gamma_1}(0) = \mathrm{Ind}_{(\phi-a)\circ\Gamma_1} = \mathrm{Ind}_{f\circ\phi\circ\Gamma_1}(0).$$

Because Γ is positively oriented we also have

$$(6) \qquad 1 = \mathrm{Ind}_{\Gamma}(a) = \mathrm{Ind}_{\Gamma-a}(0) = \mathrm{Ind}_{f\circ\Gamma}(0).$$

Together (5), (6) proclaim that the plus sign prevails in (1) for this special f. But then, thanks to 4.7(iv), the plus sign prevails for *all* eligible f:

$$(1)_+ \qquad \mathrm{Ind}_{f\circ\Gamma}(0) = \mathrm{Ind}_{f\circ\phi\circ\Gamma_1}(0).$$

The original equation (4) and $(1)_+$ now tell us that

$$N = \mathrm{Ind}_{f\circ\Gamma}(0),$$

which, in view of the definition of N, is what we wished to show.

Remark: This corollary and its proof technique, involving transfer via ϕ, also enable one to easily prove the appropriate generalization of 5.87, which the reader is encouraged to do.

Theorem 9.16 (Darboux–Picard) *Let D be an open disk, $f\colon \bar{D} \mapsto \mathbb{C}$ be continuous and satisfy*

(i) f *is holomorphic in* D.

(ii) f *is one-to-one on* ∂D.

Then f is one-to-one throughout $\bar{D}$ and $f(D)$ is the inside of the Jordan-curve $f(\partial D)$. [Cf. 4.48(i), 5.87(iv) and 6.16(v).]

Proof: If $w \in \mathbb{C} \setminus f(\partial D)$, then $f - w$ is zero-free on ∂D and so according to 5.86

(1) the number of zeros of $f - w$ in D equals $\mathrm{Ind}_{f\circ\partial D}(w)$.

But by hypothesis $f \circ \partial D$ is a simple loop, so by 4.43 this index is always ± 1 or 0, according as w belongs to the inside of the Jordan-curve $J := f(\partial D)$ or to its outside. We learn therefore from (1) that

(2) $f - w$ has a unique zero in D for all $w \in \mathcal{I}(J)$

(3) $f - w$ has no zero in D for all $w \in \mathcal{O}(J)$.

In particular

(4) $$\mathcal{I}(J) \subset f(D) \subset \mathbb{C} \setminus \mathcal{O}(J) = \mathcal{I}(J) \cup J.$$

Since (5.77) $f(D)$ is an open set, it cannot meet J without meeting $\mathcal{O}(J)$ [see 4.42] and therefore (4) entails

(5) $$\mathcal{I}(J) = f(D).$$

Then $\varnothing \overset{(5)}{=} J \cap f(D) = f(\partial D) \cap f(D)$ shows that f takes no value in D which it also takes on ∂D. By (2) and (5) f is one-to-one in D and by hypothesis f is one-to-one on ∂D. It follows that f is one-to-one throughout $\bar{D}$ and the proof is complete.

Exercise 9.17 (i) (Satz III of CARATHÉODORY and RADEMACHER [1917]) *Formulate and prove a version of the preceding in which D is any Jordan region.*

Hint: In the given proof, cite 9.15 instead of 5.86.

(ii) (Satz 3 of T. RADÓ [1922/23 c]) *Suppose Ω_1, Ω_2 are Jordan regions, $f\colon \Omega_1 \mapsto \Omega_2$ a boundary-preserving holomorphic map. Show that f extends to a continuous map of $\overline{\Omega}_1$ onto $\overline{\Omega}_2$.*

Hints: By 4.45 and 4.71 each Ω_j has property (HS_1). Hence 9.7 and 9.14 furnish homeomorphisms $\phi_j : \overline{\mathbb{D}} \mapsto \overline{\Omega}_j$ which map $\mathbb{D}$ conformally onto Ω_j ($j = 1, 2$). These reduce the problem to the special case where $\Omega_1 = \Omega_2 = \mathbb{D}$. But this case is settled by 6.16, which explicitly exhibits the boundary-preserving holomorphic self-maps of $\mathbb{D}$ as the finite Blaschke products, which are certainly continuous on $\overline{\mathbb{D}}$.

(iii) (Global version of 5.80 from POMPEIU [1927] and MONTEL [1932d].) *Let Γ be a Jordan-curve, $F : \overline{\mathfrak{I}(\Gamma)} \mapsto \mathbb{C}$ continuous and holomorphic in $\mathfrak{I}(\Gamma)$. Show that for each $z_0 \in \mathfrak{I}(\Gamma)$ there exist two distinct points z_1, $z_2 \in \Gamma$ such that*

$$\frac{F(z_2) - F(z_1)}{z_2 - z_1} = F'(z_0).$$

Hint: The function $f(z) := F(z) - F'(z_0)z$ has $f'(z_0) = 0$ and so (5.78) is not univalent in $\mathfrak{I}(\Gamma)$. By (i) it cannot be univalent on Γ either.

(iv) *Show that every point inside of a Jordan-curve is the midpoint of some chord of the curve.*

Hint: Consider $F(z) := z^2$ in (iii).

Typical applications of the Argument Principle are offered in the next exercise. For more complicated examples see KRZYŻ [1971], §3.9 and BEARDON [1979], Example 8.1.1.

Exercise 9.18 (i) *Show that for any positive numbers a, b, c the polynomial $P(z) := z^8 + az^3 + bz + c$ has exactly two zeros in the (open) first quadrant, exactly one of which lies in $S := \{z \in \mathbb{U} : \operatorname{Im} z < \operatorname{Re} z\}$.*

Hints: Evidently there exists an $r > 0$ such that $|z^{-8}P(z) - 1| < 1$ for all $|z| \geq r$, so that if we set $g(z) := z^{-8}P(z)$, then

$$g(z) \in D(1,1) \qquad \forall |z| \geq r, \tag{1}$$

$$P(z) = z^8 g(z) \qquad \forall z \neq 0. \tag{2}$$

Now $P(it) = (t^8 + c) + i\left(-at^3 + bt\right)$ for real t and so

$$P[0, r] \subset [c, \infty) \subset (0, \infty) \tag{3}$$

$$P[0, ir] \subset [c, \infty) \times \mathbb{R} \subset (0, \infty) \times \mathbb{R}. \tag{4}$$

Let γ be the positively oriented piecewise smooth simple loop defined by

$$\gamma(t) := \begin{cases} -it & -r \leq t \leq 0 \\ t & 0 \leq t \leq r \\ re^{\pi i(t-r)/2} & r \leq t \leq r+1 \end{cases} \tag{5}$$

and set $\Gamma := P \circ \gamma$. Thus by (2)

$$\Gamma(t) = \begin{cases} P(\gamma(t)) & -r \le t \le r \\ \gamma^8(t) g(\gamma(t)) & r \le t \le r+1. \end{cases} \tag{6}$$

Let L be the holomorphic logarithm in $\mathbb{C} \setminus (-\infty, 0]$ which satisfies

$$-\pi < \operatorname{Im} L < \pi. \tag{7}$$

According to (3), (4) and (1) we may form $L(P[0,r])$, $L(P[0,ir])$ and $L(g(z))$ for all $|z| \ge r$. Therefore we may define a continuous function $\phi\colon [-r, r+1] \mapsto \mathbb{C}$ by

$$\phi(t) := \begin{cases} L(P(\gamma(t))) & -r \le t \le r \\ L(r^8) + 4\pi i(t-r) + L(g(\gamma(t))) & r \le t \le r+1. \end{cases}$$

We must check that this is well defined; then continuity is obvious. So we must check that

$$L(P(\gamma(r))) = L(r^8) + L(g(\gamma(r))),$$

that is,

$$L(P(r)) = L(r^8) + L(g(r)).$$

Since the exponentials of both sides of this putative equation are equal (by (2)), we know that the real parts of both sides are equal and the imaginary parts differ by an integral multiple of 2π, say

$$\operatorname{Im}\left[L(g(r)) + L(r^8) - L(P(r))\right] = 2\pi n.$$

Since L is real on $(0, \infty)$, this means

$$\operatorname{Im}\left[L(g(r)) - L(P(r))\right] = 2\pi n$$

and from (7) it follows that $n = 0$.

Because of (6) and (5)

$$\Gamma = e^{\phi}. \tag{8}$$

Now we have

$$e^{L(r^8) + L(g(ir))} = r^8 g(ir) = (ir)^8 g(ir) \overset{(2)}{=} P(ir) = e^{L(P(ir))}$$

and therefore the reasoning above shows that

$$L(r^8) + L(g(ir)) = L(P(ir)). \tag{9}$$

From (8) and the definition of the index we have

$$\begin{aligned}\operatorname{Ind}_{P\circ\gamma}(0) = \operatorname{Ind}_{\Gamma}(0) &= \frac{1}{2\pi i}\left[\phi(r+1) - \phi(-r)\right] \\ &= \frac{1}{2\pi i}\Big[L\left(r^8\right) + 4\pi i + L\big(g(\gamma(r+1))\big) - L\big(P(\gamma(-r))\big)\Big] \\ &= \frac{1}{2\pi i}\Big[L\left(r^8\right) + 4\pi i + L\big(g(ir)\big) - L\big(P(ir)\big)\Big] \\ &= 2, \quad \text{by (9).}\end{aligned}$$

Consequently, by 9.15 P has exactly two zeros in $\mathfrak{I}(\gamma)$. Of course, by (1) and (2) P has no zeros of modulus greater-than-or-equal-to r. Thus P has exactly two zeros in the open first quadrant, counting multiplicity. Checking that P maps the ray $\{\rho e^{i\pi/4} : \rho > 0\}$ into $\mathbb{U}$, letting $re^{i\pi/4}$ take over the role of ir, and modifying γ and ϕ accordingly, a similar argument will show that P has exactly one zero in S.

(ii) *Show that if* $0 < c_0 < c_1 < \cdots < c_n$ $(n \geq 1)$, *then the function*

$$f(\theta) := \sum_{j=0}^{n} c_j \cos(j\theta)$$

has $2n$ *distinct zeros in* $(0, 2\pi)$ *and no non-real zeros.*

Hints: Introduce the polynomial $P(z) := \sum_{j=0}^{n} c_j z^j$ and note that

$$(z-1)P(z) = c_n z^{n+1} - c_0 - \sum_{j=1}^{n}(c_j - c_{j-1})z^j,$$

so

$$\left|(z-1)P(z)\right| \geq \left|c_n z^{n+1}\right| - \left|c_0 + \sum_{j=1}^{n}(c_j - c_{j-1})z^j\right|. \tag{1}$$

If $|z| \geq 1$

$$\begin{aligned}\left|c_0 + \sum_{j=1}^{n}(c_j - c_{j-1})z^j\right| &\leq c_0 + \sum_{j=1}^{n}(c_j - c_{j-1})\left|z^j\right| \\ &\leq c_0|z|^n + \sum_{j=1}^{n}(c_j - c_{j-1})|z|^n = c_n|z|^n.\end{aligned} \tag{2}$$

Hence from (1)

$$\left|(z-1)P(z)\right| \geq \left|c_n z^{n+1}\right| - c_n|z|^n = c_n|z|^n\big(|z|-1\big) > 0 \quad \text{if } |z| > 1,$$

so P has no zeros in $\mathbb{C} \setminus \overline{\mathbb{D}}$. But in fact, if $z \in \mathbb{T} \setminus \{1\}$, the first inequality in (2) is strict (since for such z, $|c_0 + (c_1 - c_0)z| < c_0 + (c_1 - c_0)$, which entails that strict inequality persists through the $n-1$ subsequent invocations of the triangle-inequality), so the same argument shows that $P(z) \neq 0$ for such z. Of course $P(1) > 0$. In summary (Eneström-Kakeya Theorem)

(3) $$P \text{ has no zeros in } \mathbb{C} \setminus \mathbb{D}.$$

Let $\gamma(t) := e^{it}$, $0 \leq t \leq 2\pi$, and ϕ be a continuous logarithm for $P \circ \gamma$ (cf. 4.1). Since $P\big(\gamma(0)\big) = P(1)$ is positive, we can suppose $\operatorname{Im}\phi(0) = 0$. According to the Fundamental Theorem of Algebra the polynomial P has n zeros and according to (3) they all lie in $\mathbb{D}$. Hence the Argument Principle (already in its simple 5.86 form) says that

$$\frac{\phi(2\pi) - \phi(0)}{2\pi i} = \operatorname{Ind}_{P\circ\gamma}(0) = n, \quad \text{that is,}$$
$$\operatorname{Im}\phi(2\pi) - \operatorname{Im}\phi(0) = 2\pi n.$$

Since $\operatorname{Im}\phi$ is a continuous function on $[0, 2\pi]$ and is 0 at 0, the preceding equation insures the existence of

(4) $$\text{distinct points } t_1, t_2, \ldots, t_{2n} \in (0, 2\pi)$$

such that

$$\operatorname{Im}\phi(t_k) = \left(k - \frac{1}{2}\right)\pi, \quad 1 \leq k \leq 2n$$

and so $P\big(\gamma(t_k)\big) = e^{\phi(t_k)}$ is purely imaginary:

(5) $$0 = \operatorname{Re} P\big(\gamma(t_k)\big) = \sum_{j=0}^{n} c_j \cos(jt_k) = f(t_k), \quad 1 \leq k \leq 2n.$$

Notice that if $z = e^{i\theta}$, then $2\cos(j\theta) = z^j + z^{-j}$, so

(6) $$2f(\theta) = \sum_{j=0}^{n} c_j \left(z^j + z^{-j}\right) = z^{-n} Q(z),$$

where $Q(z) := c_n + c_{n-1}z + \cdots + c_1 z^{n-1} + 2c_0 z^n + c_1 z^{n+1} + \cdots + c_n z^{2n}$. According to (5), (6) and (4) the numbers $z_k := e^{it_k}$ $(1 \leq k \leq 2n)$ are distinct zeros of the $2n$-degree polynomial Q. Hence there are no others. In particular, if $\theta \in \mathbb{C} \setminus \mathbb{R}$, then $z = e^{i\theta} \notin \{z_1, \ldots, z_{2n}\}$, so $Q(z) \neq 0$ and (6) shows that $f(\theta) \neq 0$.

Exercise 9.19 (Schönflies Extension Theorem) *Let J be a Jordan-curve, h any homeomorphism of the unit circle $\mathbb{T}$ onto J.*

(i) *If Ω is the inside of J, then h extends to a homeomorphism of $\overline{\mathbb{D}}$ onto $\overline{\Omega}$. If $0 \in \Omega$, then it can be arranged that the extension map 0 to 0.*

Hints: There is by the Osgood–Taylor-Carathéodory theorem a homeomorphism F of $\overline{\mathbb{D}}$ onto $\overline{\Omega}$ which maps $\mathbb{D}$ onto Ω and (hence) maps $\mathbb{T}$ onto J. Furthermore, we can have $F(0) = 0$ if $0 \in \Omega$. Evidently the equations

$$H(z) := \begin{cases} 0, & z = 0 \\ |z|F^{-1}(h(z/|z|)) & z \in \overline{\mathbb{D}} \setminus \{0\} \end{cases}$$

extend $F^{-1} \circ h$ to a homeomorphism of $\overline{\mathbb{D}}$ onto $\overline{\mathbb{D}}$, and $F \circ H$ is an extension of h of the desired kind.

(ii) *If U is the outside of J, then h extends to a homeomorphism of $\mathbb{C} \setminus \mathbb{D}$ onto $\overline{U}$.*

Hints: We may assume that $0 \in \Omega$. Let ϕ be the homeomorphism $\phi(z) := 1/z$ of $\mathbb{C} \setminus \{0\}$ onto $\mathbb{C} \setminus \{0\}$. It is easy to see that $\phi(J)$ is a Jordan-curve $\widetilde{J}$ and that $\widetilde{\Omega} := \phi(U) \cup \{0\}$ is its inside. Consequently by part (i), the homeomorphism $\widetilde{h} := \phi \circ h \circ \phi^{-1}$ of $\mathbb{T}$ onto $\widetilde{J}$ extends to a homeomorphism $\widetilde{H}$ of $\overline{\mathbb{D}}$ onto $\overline{\widetilde{\Omega}} = \phi(\overline{U}) \cup \{0\}$ which satisfies $\widetilde{H}(0) = 0$. Then $\phi^{-1} \circ \widetilde{H} \circ \phi$ maps $\phi^{-1}(\overline{\mathbb{D}} \setminus \{0\}) = \mathbb{C} \setminus \mathbb{D}$ homeomorphically onto $\phi^{-1}(\phi(\overline{U})) = \overline{U}$ and extends h.

(iii) *h extends to a homeomorphism of $\mathbb{C}$ onto $\mathbb{C}$.*

Hint: Put (i) and (ii) together.

Remarks: If $f(z) = \sum_{n=0}^{\infty} c_n z^n$ is a conformal map of $\mathbb{D}$ onto the inside of some Jordan-curve J, then not only does f extend continuously to $\overline{\mathbb{D}}$, but the series converges uniformly in $\overline{\mathbb{D}}$. This beautiful result of FEJÉR [1913], [1914], proved by an area method, can be found in VEECH [1967]. One consequence of it is that every Jordan-curve J is the range of a homeomorphism h of $\mathbb{T}$ which has the form $h(z) = \sum_{n=0}^{\infty} c_n z^n$, convergence uniform in $z \in \mathbb{T}$. Moreover, if J has a rectifiable parametrization, then the series $\sum_{n=0}^{\infty} c_n z^n$ converges absolutely in $\overline{\mathbb{D}}$, that is, $\sum_{n=0}^{\infty} |c_n| < \infty$. (See KOOSIS [1980], chapter II, D. 3, p. 72.) A topological characterization of the regions $f(\mathbb{D})$ for which the Taylor series of the Riemann map f converges uniformly in $\mathbb{D}$ is given by DENJOY [1942].

Exercise 9.20 (Extension of 5.16) *Let Ω be a Jordan region, $z_0 \in \partial\Omega$, $F\colon \overline{\Omega} \setminus \{z_0\} \mapsto \mathbb{C}$ bounded and continuous, holomorphic in Ω. Pick any* other *point $z_1 \in \partial\Omega$ and let A_1, A_2 be the two arcs of $\partial\Omega$ having endpoints z_0 and z_1.*

(i) *Suppose that*

$$\varlimsup_{\substack{z \to z_0 \\ z \in A_j}} |F(z)| \le m \quad (j = 1, 2).$$

Show that then

$$\varlimsup_{\substack{z \to z_0 \\ z \in \Omega}} |F(z)| \le m.$$

(ii) *Suppose that* $\lim_{z \to z_0, z \in A_j} F(z)$ *each exist* $(j = 1, 2)$. *Show that* F *is (extendable to be) continuous on* $\overline{\Omega}$.

Hints: There is a conformal map ϕ of Ω onto $\mathbb{D}$. This map extends to a homeomorphism of $\overline{\Omega}$ onto $\overline{\mathbb{D}}$. We may suppose (after a rotation) that $\phi(z_0) = 1$. The map $\psi(z) := (z+1)/(z-1)$ sends $\mathbb{D}$ conformally onto $S := (-\infty, 0) \times \mathbb{R}$ (see 2.5), the upper unit semi-circle onto the negative imaginary axis, the lower unit semi-circle onto the positive imaginary axis and $\overline{\mathbb{D}} \setminus \{1\}$ homeomorphically onto $\overline{S}$. Apply 5.16 (with $\theta_1 = \frac{\pi}{2}, \theta_2 = \frac{3\pi}{2}$) to the composite $f := F \circ \phi^{-1} \circ \psi^{-1}$.

§ 6 More on Jordan Regions and Boundary Behavior

The object here is to use the Osgood–Taylor–Carathéodory theorem to get uniform polynomial approximation on the whole closure of a Jordan region, thus improving the Polynomial Runge Theorems. Along the way we look at some (further—cf. 9.8(iii)) elementary instances of the following phenomenon which Carathéodory [1912a] explored in depth: if the simply-connected regions Ω_n converge in some geometric sense to Ω, do the associated (appropriately normalized) Riemann maps of Ω_n onto $\mathbb{D}$ converge to that of Ω onto $\mathbb{D}$? None of the material of this admittedly rather heavy section is needed anywhere else in the book and it may be skipped by the less ardent reader.

Exercise 9.21 *Let* $\Gamma_n, \Gamma \colon \mathbb{T} \mapsto \mathbb{C}$ *be homeomorphisms such that*

(i) $|\Gamma_n - \Gamma| \to 0$ *uniformly on* $\mathbb{T}$,

(ii) $\overline{\mathcal{J}(\Gamma)} \subset \mathcal{J}(\Gamma_{n+1}) \subset \mathcal{J}(\Gamma_n)$ *for all* $n \in \mathbb{N}$.

Show that

$$\overline{\mathcal{J}(\Gamma)} = \bigcap_{n \in \mathbb{N}} \overline{\mathcal{J}(\Gamma_n)} = \bigcap_{n \in \mathbb{N}} \mathcal{J}(\Gamma_n).$$

Hints: It suffices to show that $\bigcap_{n\in\mathbb{N}} \overline{\mathcal{J}(\Gamma_n)} \subset \overline{\mathcal{J}(\Gamma)}$. Let $w \in \bigcap_{n\in\mathbb{N}} \overline{\mathcal{J}(\Gamma_n)}$. If $w \in \Gamma$, there is nothing further to prove, since $\Gamma \subset \overline{\mathcal{J}(\Gamma)}$. If $w \notin \Gamma$, then for all large n, $w \notin \Gamma_n$ by (i). Consequently $w \in \overline{\mathcal{J}(\Gamma_n)} \setminus \Gamma_n = \mathcal{J}(\Gamma_n)$ and [by 4.43] $|\mathrm{Ind}_{\Gamma_n}(w)| = 1$. From (i) and 4.6 we know that $\mathrm{Ind}_{\Gamma_n}(w) \to$

$\mathrm{Ind}_\Gamma(w)$. Consequently, $|\mathrm{Ind}_\Gamma(w)| = 1$. This means that w is not in the (unique) unbounded component of $\mathbb{C} \setminus \Gamma$, so $w \in \mathcal{I}(\Gamma)$, the unique bounded component of $\mathbb{C} \setminus \Gamma$.

Exercise 9.22 *Let Ω_n be bounded regions in $\mathbb{C}$, each conformal to an open disk, such that $\Omega_{n+1} \subset \Omega_n$ for all $n \in \mathbb{N}$. Let $\Omega := \bigcap_{n\in\mathbb{N}} \Omega_n$ and suppose that 0 is an interior point of Ω. Let Ω_0 be the component of the interior of Ω which contains 0. Let f_n be the unique [9.8(i)] conformal map of $\mathbb{D}$ onto Ω_n which satisfies $f_n(0) = 0$, $f_n'(0) > 0$. Show that*

(i) $f = \lim_{n\to\infty} f_n$ *exists locally uniformly on $\mathbb{D}$.*

(ii) f *is the unique conformal map of $\mathbb{D}$ onto Ω_0 which satisfies $f(0) = 0$, $f'(0) > 0$.*

Hints: Since Ω_1 is bounded, the family $\{f_n\}$ is bounded. Therefore by 7.6 there are plenty of convergent subsequences. We shall show that the limit function of any one of them is conformal, takes the value 0 at the point 0, has positive derivative there and has range equal to Ω_0. Since the function answering to this description is unique by 9.8(i), we can invoke 7.8 to learn that the whole sequence $\{f_n\}$ converges to this function.

Thus suppose f is the local uniform limit of a subsequence of $\{f_n\}$. We may, for notational simplicity, suppose that subsequence is the whole sequence $\{f_n\}$. Choose $0 < r < 1$ so that $\bar{D}(0, r) \subset \Omega$. Such r exist by hypothesis. Then apply Schwarz' lemma, appropriately scaled, to f_n^{-1} in $D(0, r)$ to get $\left|(f_n^{-1})'(0)\right| \le 1/r$. From $f_n(0) = 0$ and the Chain Rule we have $(f_n^{-1})'(0) = 1/f_n'(0)$. Thus $f_n'(0) \ge r$ for all n. Since $f_n' \to f'$ (by 5.44(ii)), we have $f'(0) \ge r$. Hence f is not constant and so by 7.11 f is conformal.

Given $z_0 \in \mathbb{D}$, since f is one-to-one, $f - f(z_0)$ has a zero of order 1 in $\mathbb{D}$ at z_0. Since $f_n - f(z_0)$ converges to $f - f(z_0)$ locally uniformly in $\mathbb{D}$, it follows from 8.19 (or from 7.10) that for all sufficiently large n, $f_n - f(z_0)$ has a zero in $\mathbb{D}$, i.e., $f(z_0) \in f_n(\mathbb{D}) = \Omega_n$. This shows (because of the nesting) that $f(\mathbb{D}) \subset \bigcap \Omega_n$. Since $0 = f(0) \in f(\mathbb{D})$ and $f(\mathbb{D})$ is connected, we get $f(\mathbb{D}) \subset \Omega_0$.

Use the Basic Connectedness Lemma (1.27) to show that this inclusion is not proper. To this end, show that

(1) $$w_0 \in f(\mathbb{D}) \ \& \ D(w_0, 4\delta) \subset \Omega_0 \Rightarrow D(w_0, \delta) \subset f(\mathbb{D}).$$

Write $w_0 = f(z_0)$. For each n the function $f_n^{-1}(w_0 + \xi)$, for $\xi \in D(0, 4\delta)$, is holomorphic and maps into $\mathbb{D}$; for $\xi = 0$ it satisfies $\left|f_n^{-1}(w_0)\right| = \left|f_n^{-1} \circ f(z_0)\right| \le |z_0|$, by an application of 6.1 to $f_n^{-1} \circ f$. Therefore an application of 6.3(v) (and the Maximum-Modulus Principle) to the function $z \mapsto f_n^{-1}(w_0+z)$ shows that

$$\left|f_n^{-1}(w_0+\xi)\right| \le \frac{|z_0|+1/2}{1+|z_0|/2} =: r, \quad \text{for all } \xi \in D(0,2\delta).$$

$$(2) \qquad \left|f_n^{-1}(w)\right| \le r < 1 \qquad \forall w \in D(w_0,2\delta).$$

Set $R := \sqrt{r}$. For $z \in C(0,R)$, $f_n(z)$ cannot lie in $D(w_0,2\delta)$, since otherwise by (2), $|z| = \left|f_n^{-1}(f_n(z))\right| \le r < R$. Consequently,

$$|f_n(z) - w| \ge \delta \qquad \forall w \in D(w_0,\delta),\ z \in C(0,R)$$

and so

$$(3) \qquad \frac{1}{2\pi i}\int_{C(0,R)} \frac{f_n'}{f_n - w} \to \frac{1}{2\pi i}\int_{C(0,R)} \frac{f'}{f-w} \qquad \forall w \in D(w_0,\delta).$$

The left-hand expression in (3) is by 5.85 the number of zeros of $f_n - w$ in $D(0,R)$ and this is at least 1, since $f_n^{-1}(w) \in D(0,R)$ [by (2)] and $f_n\left(f_n^{-1}(w)\right) = w$. Consequently, the value of the expression on the right-hand side of (3) is at least 1. As this expression represents the number of zeros of $f - w$ in $D(0,R)$, we thus learn that $w \in f\left(D(0,R)\right) \subset f(\mathbb{D})$. This is the case for each $w \in D(w_0,\delta)$, thus confirming (1) and therewith the surjective character of f.

Theorem 9.23 (Carathéodory [1912a] and Courant [1914a]) *Let $\Gamma, \Gamma_n : \mathbb{T} \mapsto \mathbb{C}$ be homeomorphisms such that*

(i) $|\Gamma_n - \Gamma| \to 0$ *uniformly on* $\mathbb{T}$,

(ii) $0 \in \mathcal{J}(\Gamma) \subset \overline{\mathcal{J}(\Gamma)} \subset \mathcal{J}(\Gamma_{n+1}) \subset \mathcal{J}(\Gamma_n)$ *for all* $n \in \mathbb{N}$.

Let f_n be the unique [9.8(i)] *conformal map of $\mathbb{D}$ onto $\mathcal{J}(\Gamma_n)$ which satisfies $f_n(0) = 0$, $f_n'(0) > 0$. Its extension via* 9.14 *to a homeomorphism of $\overline{\mathbb{D}}$ onto $\overline{\mathcal{J}(\Gamma_n)}$ is also called f_n. Then the family $\{f_n\}$ is uniformly equicontinuous on $\mathbb{T}$.*

Proof: (T. Radó [1922/23d]) We argue by contradiction. If $\{f_n\}$ is not equicontinuous on $\mathbb{T}$, then, invoking compactness several times, passing to subsequences and re-labelling to assume that the final subsequence is the original one, we find

(1) distinct points $z_n,\ w_n \in \mathbb{T}$ such that

(2) $z_n,\ w_n \to z_0$,

(3) $|f_n(z_n) - f_n(w_n)| \ge 2\delta > 0 \qquad \forall n \in \mathbb{N}$,

(4) $f_n(z_n) \to a,\ f_n(w_n) \to b$.

Since $f_n(\mathbb{T}) = \Gamma_n(\mathbb{T})$, it follows readily from (4) and (i) that $a, b \in \Gamma(\mathbb{T})$. The points z_n, w_n divide $\mathbb{T}$ into two arcs, one of which, call it A_n, converges to z_0 in the obvious sense: every neighborhood of z_0 contains all but finitely many A_n. This one sees easily from (2) by using the exponential representation of $\mathbb{T}$. Now $f_n(A_n)$ is an arc which meets both the disjoint [by (3)] open sets $D\big(f_n(z_n), \delta\big)$ and $D\big(f_n(w_n), \delta\big)$ [in their centers]. Being connected, it cannot lie wholly in their union. Therefore there exists a point

$$(5) \qquad c_n \in f_n(A_n) \quad \text{such that} \quad |c_n - f_n(z_n)| \geq \delta \text{ and } |c_n - f_n(w_n)| \geq \delta.$$

As all c_n lie in compact $\overline{\mathfrak{I}(\Gamma_n)}$, passage to another subsequence lets us assume, in order to reach the ultimately sought contradiction, that for some $c \in \mathbb{C}$

$$(6) \qquad c_n \to c.$$

From (5) and (4) then we have

$$(7) \qquad |c - a| \geq \delta, \quad |c - b| \geq \delta.$$

As argued earlier, $c_n \in f_n(\mathbb{T}) = \Gamma_n(\mathbb{T})$ and (i) and (6) imply that $c \in \Gamma(\mathbb{T})$. From (3) and (4), $a \neq b$. Let t_a, t_b, t_c be the points of $\mathbb{T}$ which satisfy

$$(8) \qquad \Gamma(t_a) = a, \quad \Gamma(t_b) = b, \quad \Gamma(t_c) = c.$$

Let B denote that (compact) arc of $\mathbb{T}$ with endpoints t_a, t_b which does not contain t_c. Let A denote a (compact) non-degenerate arc of $\mathbb{T}$ symmetric about t_c but disjoint from B. (Use the exponential representation of $\mathbb{T}$ to effect all this.) Now Γ is one-to-one and continuous. Since A and B are compact and disjoint, so are $\Gamma(A)$ and $\Gamma(B)$; since t_c and $\overline{\mathbb{T} \setminus A}$ are disjoint, so are $\Gamma(t_c) = c$ and $\Gamma\big(\overline{\mathbb{T} \setminus A}\big) = \overline{\Gamma(\mathbb{T}) \setminus \Gamma(A)}$. Hence there is an η such that

$$(9) \qquad 0 < 3\eta < \operatorname{dist}\big(c, \Gamma(\mathbb{T}) \setminus \Gamma(A)\big)$$

$$(10) \qquad 0 < 2\eta < \operatorname{dist}\big(\Gamma(A), \Gamma(B)\big).$$

Using (i), (4) and (6) and discarding a few f_n, we can assume that

$$(11) \qquad |\Gamma_n - \Gamma| < \eta \quad \text{on } \mathbb{T} \text{ for all } n,$$

$$(12) \qquad |f_n(z_n) - a| < \eta, \ |f_n(w_n) - b| < \eta, \ |c_n - c| < \eta \quad \text{for all } n.$$

From these inequalities we want to deduce that

$$(13) \qquad D(c, 2\eta) \cap \Gamma_n(\mathbb{T}) \subset f_n(A_n) \qquad \forall n.$$

Suppose for some n this inclusion fails. Then

$$(14)\qquad |\Gamma_n(t) - c| < 2\eta \ \ \&\ \ \Gamma_n(t) \notin f_n(A_n) \quad \text{for some } t \in \mathbb{T}.$$

The point c_n belongs to $f_n(A_n) \subset f_n(\mathbb{T}) = \Gamma_n(\mathbb{T})$, so we can write

$$(15)\qquad c_n = \Gamma_n(t_n), \quad \text{some } t_n \in \mathbb{T}.$$

We have

$$(16)\ |\Gamma(t_n) - c| \le |\Gamma(t_n) - \Gamma_n(t_n)| + |\Gamma_n(t_n) - c| \overset{(11)}{<} \eta + |c_n - c| \overset{(12)}{<} 2\eta,$$

$$(17)\ |\Gamma(t) - c| \le |\Gamma(t) - \Gamma_n(t)| + |\Gamma_n(t) - c| \overset{(11)}{<} \eta + |\Gamma_n(t) - c| \overset{(14)}{<} 3\eta.$$

Because of (9) the inequalities (16) and (17) imply that $\Gamma(t_n), \Gamma(t) \in \Gamma(A)$. Since Γ is one-to-one, this means that $t_n, t \in A$. Let A' denote the subarc of A having these endpoints. Now $\Gamma_n(t_n) = c_n$ lies in the arc $f_n(A_n)$ but is not one of its endpoints $f_n(z_n)$, $f_n(w_n)$, by (5); and $\Gamma_n(t) \in \Gamma_n(\mathbb{T}) \setminus f_n(A_n)$ by (14). It follows from 4.47(v) that the arc $\Gamma_n(A')$, whose endpoints are $\Gamma_n(t_n)$ and $\Gamma_n(t)$, contains exactly one of $f_n(z_n)$, $f_n(w_n)$, say the former:

$$(18)\qquad f_n(z_n) = \Gamma_n(t_n') \quad \text{for some } t_n' \in A'.$$

But then we have

$$|\Gamma(t_n') - \Gamma(t_a)| \overset{(8)}{=} |\Gamma(t_n') - a| \overset{(18)}{\le} |f_n(z_n) - a| + |\Gamma_n(t_n') - \Gamma(t_n')| < 2\eta$$

by (12), (11). Since $t_a \in B$ and $t_n' \in A' \subset A$, this inequality violates (10). We have thus finally confirmed (13).

Now define

$$(19)\qquad \varepsilon_n := \sup\left\{\left|f_n^{-1}(w') - z_0\right| : w' \in f_n(A_n)\right\} = \sup\{|z - z_0| : z \in A_n\}.$$

Since $A_n \to z_0$, we have

$$(20)\qquad \varepsilon_n \to 0.$$

Our next goal is to use Lindelöf's estimate 5.14 to show that

$$(21)\qquad f_n^{-1}(w_0) - z_0 \to 0 \text{ as } n \to \infty \qquad \forall w_0 \in D(c, \eta) \cap \mathcal{I}(\Gamma).$$

Let $w_0 \in D(c, \eta) \cap \mathcal{I}(\Gamma)$ be given. Consider the disk $D(w_0, \eta)$. It contains $c \in \Gamma(\mathbb{T})$, the boundary of $\overline{\mathcal{I}(\Gamma)}$. Therefore $D(w_0, \eta)$ contains points outside

$\overline{\mathcal{J}(\Gamma)}$. If one such point has distance r $(< \eta)$ to w_0, then some non-degenerate arc of $C(w_0, r)$ lies outside $\overline{\mathcal{J}(\Gamma)}$. Let γ be such a (compact) arc with length $2\pi r/k$ for some positive integer k. Now γ is disjoint from $\overline{\mathcal{J}(\Gamma)}$ which by 9.21 equals $\bigcap_{n\in\mathbb{N}} \overline{\mathcal{J}(\Gamma_n)}$. Therefore by compactness γ is disjoint from $\overline{\mathcal{J}(\Gamma_n)}$ for all large n. Say,

$$(22) \qquad \gamma \cap \overline{\mathcal{J}(\Gamma_n)} = \varnothing \qquad \forall n \geq n_0.$$

Notice that $D(w_0, \eta) \subset D(c, 2\eta)$ and therefore by (13), $\Gamma_n \cap D(w_0, \eta) \subset f_n(A_n)$. It follows from this inclusion and from (19) that

$$(23) \quad \lim_{w \to w'} \left|f_n^{-1}(w) - z_0\right| = \left|f_n^{-1}(w') - z_0\right| \leq \varepsilon_n \qquad \forall w' \in \Gamma_n(\mathbb{T}) \cap D(w_0, \eta).$$

The function $f_n^{-1} - z_0$ is bounded by 2 and $\Gamma_n(\mathbb{T})$ is the boundary of $\mathcal{J}(\Gamma_n)$. Therefore (22) and (23) affirm that the hypotheses of Lindelöf's theorem 5.14 are met. We conclude that

$$\left|f_n^{-1}(w_0) - z_0\right| \leq \left(\varepsilon_n 2^{k-1}\right)^{1/k} \qquad \forall n \geq n_0.$$

If we let $n \to \infty$ and recall (20), the claim (21) stands confirmed.

Since each f_n^{-1} maps $\mathcal{J}(\Gamma)$ into $\mathbb{D}$, and $z_0 \in \overline{\mathbb{D}} \setminus \mathbb{D}$, it follows from (21) and 7.10 that $\{f_n^{-1}\}$ converges to z_0 throughout the connected set $\mathcal{J}(\Gamma)$. Since $f_n^{-1}(0) = 0$ for all n, this is a contradiction. Therefore (21), and therewith the assumption (of non-equicontinuity) from which (21) was deduced, is untenable. This proves the theorem.

Corollary 9.24 *Let the notation and hypotheses be as in the preceding theorem, and let f denote the unique continuous function on $\overline{\mathbb{D}}$ which maps $\mathbb{D}$ conformally onto $\mathcal{J}(\Gamma)$ and satisfies $f(0) = 0$, $f'(0) > 0$. Then $\{f_n\}$ converges to f uniformly on $\overline{\mathbb{D}}$.*

Proof: According to 9.21 and 4.45 the component Ω_0 of the interior of $\cap\mathcal{J}(\Gamma_n)$ which contains 0 is $\mathcal{J}(\Gamma)$. Therefore 9.22 insures that

$$(*) \qquad f_n \to f \text{ locally uniformly in } \mathbb{D}.$$

We first show that every subsequence of $\{f_n\}$ contains another which converges uniformly on $\overline{\mathbb{D}}$ to the function f. Then a short little argument like that in 7.8 shows that the sequence $\{f_n\}$ itself must converge uniformly on $\overline{\mathbb{D}}$. Here are the details.

For the first step, by a notational change it suffices to show that $\{f_n\}$ contains a subsequence which is uniformly convergent on $\overline{\mathbb{D}}$ to f. These functions all map

$\overline{\mathbb{D}}$ into the compact set $\overline{\mathcal{J}(\Gamma_1)}$. By the diagonal process we select a subsequence $\{f_{n_k}\}$ which converges at each point of some countable dense subset of $\mathbb{T}$. Since the family $\{f_n\}$ is uniformly equicontinuous on $\mathbb{T}$, Cauchyness of $\{f_{n_k}(z)\}$ for a dense set of $z \in \mathbb{T}$ and compactness lead via a finite-covering argument to the uniform Cauchyness of $\{f_{n_k}\}$ on $\mathbb{T}$. The Maximum-Modulus Principle propagates this into $\mathbb{D}$. Thus $\{f_{n_k}\}$ converges uniformly on $\overline{\mathbb{D}}$. The limit is a continuous function on $\overline{\mathbb{D}}$. On $\mathbb{D}$ this limit is f, by $(*)$. Consequently, by continuity, f is the limit everywhere in $\overline{\mathbb{D}}$.

Exercise 9.25 *Show that if* $\Gamma\colon \mathbb{T} \mapsto \mathbb{C}$ *is any homeomorphism with* $0 \in \mathcal{J}(\Gamma)$, *there exist* Γ_n *satisfying the hypotheses of* 9.23.

Hints: The result is trivial if Γ is the identity map. In the general case, extend Γ to a homeomorphism of $\mathbb{C}$ onto $\mathbb{C}$ [Schönflies 9.19] and thereby reduce the general case to this trivial one.

We are now in a position to prove a significant extension of the Polynomial Runge Theorem 8.9, still considerably short of the Mergelyan extension however.

Theorem 9.26 (J. L. WALSH *[1927]) Let* Ω, Ω_1, Ω_2 *be Jordan regions with* $0 \in \Omega_1 \cap \Omega$ *and* $\overline{\Omega}_1 \subset \Omega_2$. *Set* $A := \Omega_2 \setminus \overline{\Omega}_1$. *Then*

(i) *every continuous function* $F\colon \overline{\Omega} \mapsto \mathbb{C}$ *which is holomorphic in* Ω *is uniformly approximable on* $\overline{\Omega}$ *by polynomials in* z;

(ii) *every continuous function* $G\colon \bar{A} \mapsto \mathbb{C}$ *which is holomorphic in* A *is uniformly approximable on* $\bar{A}$ *by polynomials in* z *and* $1/z$.

Proof: (i) Let Γ be a homeomorphism of $\mathbb{T}$ onto $\partial\Omega$ [definition of Jordan region]. Then let Γ_n, f_n be the maps furnished by 9.25 and 9.23. According to 9.24 the functions f_n converge uniformly on $\overline{\mathbb{D}}$ to the unique conformal map f of $\mathbb{D}$ onto Ω which satisfies $f(0) = 0$, $f'(0) > 0$. Given $\varepsilon > 0$, choose $\delta > 0$ so that

(1) $$z_1, z_2 \in \overline{\mathbb{D}} \;\&\; |z_1 - z_2| < \delta \Rightarrow |F \circ f(z_1) - F \circ f(z_2)| < \varepsilon.$$

This is just the uniform continuity of $F \circ f$ on $\overline{\mathbb{D}}$. Similarly, choose $\eta > 0$ such that

(2) $$w_1, w_2 \in \overline{\Omega} \;\&\; |w_1 - w_2| < \eta \Rightarrow \left|f^{-1}(w_1) - f^{-1}(w_2)\right| < \delta.$$

Next choose n such that

(3) $$|f_n(z) - f(z)| < \eta \qquad \forall z \in \overline{\mathbb{D}}.$$

Apply f^{-1} in (3) to get via (2)

$$\left|f^{-1}(f_n(z)) - z\right| < \delta \qquad \forall z \in f_n^{-1}(\overline{\Omega}) \subset \overline{\mathbb{D}}.$$

That is, upon setting $w = f_n(z)$,

$$\left|f^{-1}(w) - f_n^{-1}(w)\right| < \delta \qquad \forall w \in \overline{\Omega}. \tag{4}$$

From (4) and (1) we have

$$\left|F \circ f \circ f^{-1}(w) - F \circ f \circ f_n^{-1}(w)\right| < \varepsilon \qquad \forall w \in \overline{\Omega},$$

i.e.,

$$\left|F(w) - F \circ f \circ f_n^{-1}(w)\right| < \varepsilon \qquad \forall w \in \overline{\Omega}. \tag{5}$$

Since $\overline{\Omega} = \overline{\mathcal{I}(\Gamma)}$ is a compact subset of the simply-connected region $\mathcal{I}(\Gamma_n)$, on which $F \circ f \circ f_n^{-1}$ is holomorphic, Runge's Polynomial Approximation Theorem 8.10 furnishes a polynomial P such that

$$\left|P - F \circ f \circ f_n^{-1}\right| < \varepsilon \quad \text{on } \overline{\Omega}. \tag{6}$$

Combining (5) and (6), we have finally $|F - P| < 2\varepsilon$ on $\overline{\Omega}$.

(ii) Theorem 8.24 supplies g holomorphic in $\mathbb{C} \setminus \overline{\Omega}_1$ and F holomorphic in Ω_2 such that

$$G = F + g \quad \text{in } A \text{ and} \tag{7}$$

$$g \text{ vanishes at } \infty. \tag{8}$$

Consider any $z_0 \in \partial(\mathbb{C} \setminus \Omega_1) = \partial\Omega_1$ and z_n in the open neighborhood Ω_2 of $\partial\Omega_1$ such that $z_n \to z_0$. Since G is by hypothesis continuous on $\bar{A} = \overline{\Omega}_2 \setminus \Omega_1$, we have $G(z_n) \to G(z_0)$, and since F is holomorphic in $\Omega_2 \supset \overline{\Omega}_1$, $F(z_n) \to F(z_0)$. Thus by (17), $g(z_n) \to G(z_0) - F(z_0)$. This shows that g extends to a continuous function on $\mathbb{C} \setminus \Omega_1$. Similarly, F extends to a continuous function on $\overline{\Omega}_2$. And then (7) becomes

$$G = F + g \quad \text{in } \bar{A}. \tag{9}$$

The function $z \mapsto 1/z$ maps $\partial\Omega_1$ onto a Jordan-curve $\widetilde{J}$ with $0 \in \mathcal{I}(\tilde{J}) =: \widetilde{\Omega}$ and maps $\mathbb{C} \setminus \Omega_1$ onto $\overline{\widetilde{\Omega}} \setminus \{0\}$ — see the hints to 9.19(ii). The function $H(z) := g(1/z)$ is holomorphic in $\widetilde{\Omega} \setminus \{0\}$ and due to (8), 0 is a removable singularity of H. Thus H is holomorphic in $\widetilde{\Omega}$ and continuous on $\overline{\widetilde{\Omega}}$. By part (i) there exist polynomials P_n, Q_n uniformly convergent to F, H in $\overline{\Omega}_2$, $\overline{\widetilde{\Omega}}$, respectively. Then $Q_n(1/z)$ converges to $g(z)$ uniformly in $\mathbb{C} \setminus \Omega_1$ and from (9) $P_n(z) + Q_n(1/z) \to G(z)$ uniformly for $z \in \overline{\Omega}_2 \cap [\mathbb{C} \setminus \Omega_1] = \bar{A}$.

Corollary 9.27 (Cauchy's Theorem on the boundary) *Let Ω be a Jordan region, Γ a piecewise smooth loop in $\overline{\Omega}$, $f\colon \overline{\Omega} \mapsto \mathbb{C}$ a continuous function which is holomorphic in Ω. Then $\int_\Gamma f = 0$.*

Proof: (J. L. Walsh [1933]) This follows from the preceding theorem just as 8.13 was deduced from the Polynomial Runge Theorem.

Remarks: (i) A casual reading belies the precision navigation through Scylla and Charybdis involved in 9.27. If that strait is closed, the ship founders: Consider the simply-connected region $\Omega := D(1,2) \setminus \overline{\mathbb{D}}$, whose boundary is not a Jordan-curve (having just one point, $z = -1$, that fails the 4.50 test), but is nevertheless parameterizable as a smooth loop, for example, $\Gamma(t) := -e^{2\pi i t}$ for $0 \le t \le 1$ and $\Gamma(t) := 1 + 2e^{\pi i t}$ for $1 \le t \le 2$. The function $f(z) := 1/z$ is continuous in $\overline{\Omega}$, holomorphic in Ω and $\int_\Gamma f = 2\pi i \operatorname{Ind}_\Gamma(0) = 4\pi i \neq 0$. Obviously this example also means that 9.26(i) may fail if Ω is a simply-connected region which is bounded by a loop that is *not* simple.

(ii) If in 9.27, Γ is a positively oriented and simple parametrization of $\partial\Omega$, then by the usual artifice the (iconic) Cauchy integral formula follows, unencumbered by any index factor, thanks to 4.43 and 4.44:

$$f(z) = \frac{1}{2\pi i}\int_\Gamma \frac{f(\xi)}{\xi - z}\,d\xi \qquad \forall z \in \Omega = \mathfrak{I}(\Gamma).$$

Exercise 9.28 (Painlevé [1887]) (i) *Let Γ be a piecewise smooth, simple loop, C a piecewise smooth cross-cut* (1.15) *in $\mathfrak{I}(\Gamma)$. Let f be continuous on $\overline{\mathfrak{I}(\Gamma)}$ and holomorphic in $\mathfrak{I}(\Gamma)\setminus C$. Show that in fact f is holomorphic in $\mathfrak{I}(\Gamma)$.*

Hints: Let Γ_1, Γ_2 be piecewise smooth arcs such that $\Gamma_1 \cap C = \Gamma_2 \cap C = \Gamma_1 \cap \Gamma_2$, $\Gamma_1 \cup \Gamma_2 = \Gamma$, and $\Gamma_j \cup C$ are Jordan-curves. See 4.47. If the arcs are consistently parameterized, meaning that initial point of C = terminal point of Γ_1, and terminal point of C = initial point of Γ_1 = terminal point of Γ_2, then by the preceding corollary

$$\int_{\Gamma_1} f + \int_C f = \int_C f - \int_{\Gamma_2} f = 0.$$

Subtract to get

$$\int_\Gamma f = \int_{\Gamma_1} f + \int_{\Gamma_2} f = 0.$$

For each $z \in \mathfrak{I}(\Gamma) \setminus C$ apply this conclusion to the function

$$F(\zeta) := \frac{f(\zeta) - f(z)}{\zeta - z}$$

in the role of f to get

$$(*) \qquad nf(z) = \frac{1}{2\pi i}\int_\Gamma \frac{f(\zeta)}{\zeta - z}\,d\zeta \qquad \forall z \in \mathfrak{I}(\Gamma)\setminus C,$$

where n is 1 or -1. Since C is a homeomorph of $[0,1]$, the Open-Map Theorem 4.61 ensures that C contains no open disk. [More elementarily, this is because removal of any three points destroys the connectedness of $[0,1]$ but not that of a disk.] In particular, every neighborhood of a point of $C \cap \mathfrak{I}(\Gamma)$ contains points of $\mathfrak{I}(\Gamma)\setminus C$. In other words, $\mathfrak{I}(\Gamma)\setminus C$ is dense in $\mathfrak{I}(\Gamma)$. Now both sides of $(*)$ are continuous functions of $z \in \mathfrak{I}(\Gamma)$, hence the equality there holds throughout $\mathfrak{I}(\Gamma)$. That formula proves that f is holomorphic in $\mathfrak{I}(\Gamma)$. (See 2.14.)

(ii) *Let U be any open subset of $\mathbb{C}$, γ a piecewise smooth arc which, except possibly for its endpoints, lies in U. Let $f\colon U \mapsto \mathbb{C}$ be continuous in U and holomorphic in $U \setminus \gamma$. Show that in fact f is holomorphic in U.*

Hints: We only have to check holomorphy in a neighborhood of each point of γ. So let the parameter interval for γ be $[0,1]$ and consider $t_0 \in (0,1)$. Select $r > 0$ so small that $\bar{D}(\gamma(t_0), r) \subset U \setminus \{\gamma(0), \gamma(1)\}$. Let t_1 be the largest $t \in [0, t_0]$ such that $\gamma(t) \in \mathbb{C} \setminus D(\gamma(t_0), r)$ and t_2 the smallest $t \in [t_0, 1]$ with this property. Evidently $t_1 < t_0 < t_2$ and

$$\gamma(t_1), \gamma(t_2) \in C(\gamma(t_0), r) \;\;\&\;\; \gamma(t_1, t_2) \subset D(\gamma(t_0), r).$$

Hence part (i) is applicable, with $\Gamma := C(\gamma(t_0), r)$ and $C := \gamma\,|_{[t_1,t_2]}$, and we infer that f is holomorphic in $D(\gamma(t_0), r)$. Since $t_0 \in (0,1)$ is arbitrary, the conclusion is that f is holomorphic throughout $U \setminus \{\gamma(0), \gamma(1)\}$. But then, being continuous in U, f is holomorphic throughout U by 5.41(i).

(iii) *Is the conclusion in* (ii) *valid if γ is not an arc but only a piecewise smooth curve?*

§ 7 Harmonic Functions and the General Dirichlet Problem

In this section we shall connect our definition (2.2) of harmonic function with the traditional one and then solve the Dirichlet problem (5.19) for (bounded) simply-connected regions. This is appropriate here because, as we shall see in 9.50, it is possible to use the solvability of the Dirichlet problem to provide yet another proof of the Riemann mapping theorem, since that solvability does not utilize Riemann's theorem. This approach to the Riemann mapping theorem is close to the historical one followed by Riemann, Schwarz, Neumann, Hilbert, Poincaré, Osgood and Lebesgue, i.e., via potential theory.

Exercise 9.29 *If U is an open subset of $\mathbb{C}$ and $h\colon U \mapsto \mathbb{R}$ is harmonic, then $D_{11}h$, $D_{22}h$ (exist and are continuous and) satisfy $D_{11}h + D_{22}h = \mathbf{0}$.*

Hints: The problem is a local one, so we can assume U is a disk and $h = \operatorname{Re} f$ for some $f \in H(U)$. Recall 5.3(ii) and differentiate the Cauchy-Riemann equation.

Remark: For $\mathbb{C}$-valued h we write Δh for $D_{11}h + D_{22}h$. This is called the *Laplacian of* h and $\Delta h = 0$ is called *Laplace's equation.* (See pp. 17, 18 of Osgood [1901].) Recall that the gradient of an $\mathbb{R}$-valued u is the function $\nabla u := D_1 u + iD_2 u = (D_1 u, D_2 u)$. If we think of ∇ as the "operator vector" (D_1, D_2), then the dot product $\nabla \bullet \nabla$ would be $D_1D_1 + D_2D_2 = D_{11} + D_{22} = \Delta$. Hence the suggestive notation ∇^2 is often used for Δ.

Theorem 9.30 *Let U be an open subset of $\mathbb{C}$, f a continuous real-valued function on U for which $D_{11}f$, $D_{22}f$ exist in U and satisfy $\Delta f \le 0$. Then f is superharmonic.*

Proof: Consider any disk $D := D(z_0, r), r > 0$, whose closure lies in U. Denote by h the solution of the Dirichlet problem in D with boundary data f. For $0 < \varepsilon < 1$ form

$$\phi_\varepsilon(z) := \varepsilon |z - z_0|^2 + h(z) - f(z), \quad z \in \bar{D}. \tag{1}$$

If ϕ_ε has a relative maximum at a point $z_1 = (x_1, y_1) \in D$, then the function $\phi_\varepsilon(x, y_1)$ has a relative maximum at x_1 so $D_{11}\phi_\varepsilon(x_1, y_1) \le 0$. Similarly $\phi_\varepsilon(x_1, y)$ has a relative maximum at y_1 so $D_{22}\phi_\varepsilon(x_1, y_1) \le 0$. Consequently,

$$4\varepsilon + \Delta h(z_1) - \Delta f(z_1) \overset{(1)}{=} D_{11}\phi_\varepsilon(z_1) + D_{22}\phi_\varepsilon(z_1) \le 0. \tag{2}$$

Since $\Delta h = 0$ in D by the preceding exercise, and $\Delta f \le 0$ by hypothesis, the contradiction $4\varepsilon \le 0$ follows from (2), rendering the existence of such $z_1 \in D$ untenable. Therefore the continuous function ϕ_ε takes its maximum over the compact set $\bar{D}$ at some $z_\varepsilon \in \partial D$. Since $h(z_\varepsilon) = f(z_\varepsilon)$, this means that

$$\varepsilon |z - z_0|^2 + h(z) - f(z) \le \varepsilon |z_\varepsilon - z_0|^2 = \varepsilon r^2 \quad \forall z \in \bar{D}.$$

Let $\varepsilon \downarrow 0$ to infer that $h - f \le 0$ throughout $\bar{D}$. From this inequality, 5.21(iii) and $f = h$ on ∂D follows

$$f(z_0) \ge h(z_0) = \frac{1}{2\pi} \int_0^{2\pi} h\left(z_0 + re^{i\theta}\right) d\theta = \frac{1}{2\pi} \int_0^{2\pi} f\left(z_0 + re^{i\theta}\right) d\theta.$$

Thus f satisfies the typical inequality in the definition of superharmonicity.

Corollary 9.31 (Converse of 9.29) *If f is a continuous real-valued function in the open subset U of $\mathbb{C}$ and $D_{11}f$, $D_{22}f$ exist and satisfy Laplace's equation in U, then f is harmonic.*

Proof: 5.23(iv) plus the preceding theorem.

Remarks: Together 9.29 and 9.31 say that the harmonic functions are the kernel of the operator Δ. On the one hand, this frees the concept of harmonicity from that of holomorphy (and serves as the *definition* of harmonicity in higher dimensions). On the other hand, it establishes a new connection between the two. Because by (C. R.) $f' = D_1 f = \frac{1}{2}(D_1 - iD_2)f$ for holomorphic f, the operator $\frac{1}{2}(D_1 - iD_2)$ is often denoted by ∂, and $\frac{1}{2}(D_1 + iD_2)$ by $\bar{\partial}$. 2.16 and 2.17 say that the holomorphic functions are just the kernel of the differential operator $\bar{\partial}$. Recalling 2.19(iii), one sees that Δ is essentially the composition of these two differential operators, in either order: $\Delta = 4\bar{\partial}\partial$. Thus harmonicity of h is equivalent to holomorphy of $g := \partial h$. (Does this mean that h is a primitive of g? See 10.8.) Note that on $\mathbb{R}$-valued functions, $2\bar{\partial} = D_1 + iD_2$ is the gradient operator ∇.

Corollary 9.32 (Converse of 9.30) *If f is a superharmonic function in an open subset U of $\mathbb{C}$ and $D_{11}f$, $D_{22}f$ exist and are continuous in U, then $\Delta f \le 0$.*

Proof: If $\Delta f(a) > 0$ for some $a \in U$, then by continuity there is a whole open disk $D \subset U$ in which $\Delta f > 0$. But then by 9.30, f is subharmonic in D. Being also superharmonic, it is harmonic there (by 5.23(iv)). Hence by 9.29 $\Delta f = 0$ in D, contradiction.

Exercise 9.33 (i) *Let f be holomorphic in the region Ω. Define $F(x, y) := |f(x + iy)|^2$ and show that*

$$D_{11}F(x, y) + D_{22}F(x, y) = 4|f'(x + iy)|^2$$

for all $x + iy \in \Omega$.

(ii) *Deduce from* (i) *that if the modulus of the holomorphic function h is harmonic, then h is constant.*

Hint: If h is not constant, go to a disk $D \subset \Omega$ in which h has no zeros and in D write $h = f^2$ for some $f \in H(D)$ (5.35).

(iii) *Show that if u, v are (real-valued) harmonic functions in a region Ω and $f := u + iv$ is holomorphic in some non-empty open subset of Ω, then f is holomorphic throughout Ω.*

Hints: First consider the special case where Ω is a disk. Cite 5.21(i) twice to come up with F, $G \in H(\Omega)$ such that $u = \operatorname{Re} F$, $v = \operatorname{Im} G$ in Ω. Let D

be a non-empty open disk in Ω in which f is holomorphic. Since $i(F-f)$ is real-valued in D, it is constant there (2.16(ii)) and by adding i times this constant to F we can retain $u = \operatorname{Re} F$ in Ω and have in addition $F = f$ in D. Do the same to G. Then $F = G = f$ in D, so by 5.62 $F = G$ in Ω. It follows that $f = u + iv = \operatorname{Re} F + i \operatorname{Im} G = \operatorname{Re} F + i \operatorname{Im} F = F$ in Ω. For the general region Ω denote by A the largest open subset of Ω in which f is holomorphic. If $a \in A$ and $D(a,r) \subset \Omega$, then we have f holomorphic in the neighborhood $A \cap D(a,r)$ of a, hence by the case already proven, f is holomorphic in $D(a,r)$. Thus $D(a,r) \subset A$ and so by 1.27, $A = \Omega$. A quite different proof can be based on 10.8(i), according to which the function $D_1\bar{f} - iD_2\bar{f}$ is holomorphic in Ω. Its complex conjugate $D_1 f + iD_2 f$ vanishes in D (2.16) and so by 5.62 it vanishes throughout Ω. Taking complex conjugates again says $D_1 f + iD_2 f = 0$ in Ω, which makes f holomorphic in Ω (2.17).

(iv) *Show that if u is a (real-valued) harmonic function in an open square Ω whose sides are parallel to the coordinate axes, and if the product $U(z) := u(z) \operatorname{Re} z$ is also harmonic in Ω, then $u(z) = a \operatorname{Im} z + b$ for some real constants a and b.*

Hints: Compute that

$$\Delta U(x+iy) = 2D_1 u(x+iy) + x\Delta u(x+iy) \qquad \forall x+iy \in \Omega.$$

Since both U and u satisfy Laplace's equation, we learn that $2D_1 u(x+iy) = 0$, so $u(x+iy)$ is a function only of y and $0 = \Delta u(x+iy) = D_{22} u(x+iy)$.

(v) *The product of two harmonic functions is seldom harmonic. Prove the following precise version of this assertion: if u and v are non-constant (real-valued) harmonic functions in a region Ω, then their product is harmonic in Ω if and only if there exists a non-zero real constant c such that $u + icv$ is holomorphic in Ω.*

Hints: One direction is trivial: if $f = u + icv$ is holomorphic in Ω and c is real and non-zero, then $uv = \operatorname{Re}(f^2/2ci)$ is harmonic in Ω. Now suppose that $U := uv$ is harmonic in Ω. If both $D_1 u$ and $D_2 u$ are identically 0, then by 2.17 u is holomorphic and $u' = D_1 u = 0$. The constancy of u in Ω would follow. Therefore not both $|D_1 u|$ and $|D_2 u|$ are identically 0 and there exists an open disk D in Ω where one of them is positive. In that disk there is a holomorphic function f such that $u = \operatorname{Re} f$. According to the Cauchy-Riemann equation $f' = D_1 u - iD_2 u$ and so f' is zero-free in D. In some smaller disk D_0, f is therefore one-to-one (5.79) and on the open set $V_0 := f(D_0)$ there is an inverse function g. Then

$$U \circ g = [v \circ g][u \circ g] = [v \circ g][(\operatorname{Re} f) \circ g] = [v \circ g][\operatorname{Re}(f \circ g)]$$

is harmonic in V_0. (Why?) Since $f \circ g(w) = w$ for all $w \in V_0$, this says that

$$U \circ g(w) = v \circ g(w) \operatorname{Re} w, \qquad w \in V_0,$$

is harmonic. By (iv) there are real constants a, b such that

$$v \circ g(w) = a \operatorname{Im} w + b, \qquad w \in V,$$

where V is an open square in V_0; that is,

$$v(z) = a \operatorname{Im} f(z) + b \qquad \forall z \in f^{-1}(V) \cap D_0.$$

Since $u = \operatorname{Re} f$ in D_0, the preceding equation gives

$$(*) \qquad au + iv = af + ib \quad \text{in } f^{-1}(V) \cap D_0.$$

The equation $(*)$ affirms that $au + iv$ is holomorphic in $f^{-1}(V) \cap D_0$, so by (iii) it is holomorphic throughout Ω. But then $a \neq 0$, else by 2.16(ii) v would be constant in Ω. Set $c := 1/a$.

There is another way to prove the "only if" claim in (v). The functions $U := D_1u - iD_2u$, $V := D_1v - iD_2v$ are holomorphic (cf. 10.8(i)), and computation reveals that $\Delta(uv) = U\overline{V} + \overline{U}V$, so harmonicity of uv means that

$$U\overline{V} + \overline{U}V = \mathbf{0}.$$

Since v is not constant, V is not $\mathbf{0}$ (proof?) and so $V^{-1}(0)$ has no cluster point in Ω. On the region (see 1.24) $\Omega \setminus V^{-1}(0)$ we have $U/V + \overline{U}/\overline{V} = 0$, that is, the holomorphic function U/V has real part 0 there, hence (2.16(ii)) is constant, say ic, with $c \in \mathbb{R}$:

$$U = icV \text{ in } \Omega \setminus V^{-1}(0), \text{ hence throughout } \Omega,$$

by continuity. Using the definitions of U and V, this equation translates into

$$-D_2(u - icv) = iD_1(u - icv),$$

which, after conjugation, says that $u + icv$ satisfies (C. R.).

(vi) *Show that if both $u \colon \Omega \mapsto \mathbb{R} \setminus \{0\}$ and $1/u$ are harmonic, then u is constant.*

Hints: Use (v) or compute that

$$\Delta(1/u) = \frac{2\big[(D_1u)^2 + (D_2u)^2\big] - u \cdot \Delta u}{u^3}.$$

We have all that is needed to solve the Dirichlet problem in a quite general setting. Start with

Definition 9.34 Let U be an open subset of $\mathbb{C}$, $b \in \partial U$. A *barrier* in U at b is a set $\{h_r : 0 < r < 1\}$ of non-negative superharmonic functions in U which each satisfy

(i) $\lim_{z\to b} h_r(z) = 0$,

(ii) $h_r\big(U \setminus D(b,r)\big) = 1$.

Exercise 9.35 *Show that if the Dirichlet problem is solvable for the bounded open set U, then there is a barrier in U at each $b \in \partial U$.*

Hints: Let h be the harmonic function in U such that $h(z) = |z - b|$ for all $z \in \partial U$. The function $|z - b|$ is subharmonic in U by 5.7, so the maximum principle 5.12 applied to the subharmonic function $|z - b| - h(z)$ shows that $h(z) \geq |z - b|$ for all $z \in U$. Thus $h > 0$ in U and $h(b) = 0$. Finally, for each $r > 0$ define $h_r := \min\{\mathbf{1}, h/r\}$.

The importance of the concept 9.34 (due to LEBESGUE [1912]) lies not in the trivial result 9.35 but in the fact that its converse is true. The proof odyssey of this begins with

Basic Construction: Let U be an open subset of $\mathbb{C}$, $f\colon U \mapsto \mathbb{R}$ superharmonic, $g\colon U \mapsto \mathbb{R}$ subharmonic and $g \leq f$ throughout U. Let $\{D_n\}_{n\in\mathbb{N}}$ be a sequence of open disks whose closures each lie in U. The sequence is to have one further property: each point z of U lies in an open disk D_z which lies inside D_n for infinitely many n. As an example of such a sequence, take all the open disks with rational (complex) centers and rational radii, whose closures lie in U. Now the Dirichlet problem is solvable in each D_n, so we can define a sequence of functions F_n on U inductively via Poisson modifications thus: F_1 equals f in $U \setminus D_1$ and coincides in D_1 with the solution of the Dirichlet problem having boundary function f. According to 5.25(i), F_1 is superharmonic and $f \geq F_1$. In $U \setminus D_1$ we have $F_1 = f \geq g$. In particular, the subharmonic function $g - F_1 \leq 0$ on ∂D_1. But then by the maximum principle 5.12, $F_1 \geq g$ throughout D_1. In summary,

$$f \geq F_1 \geq g \quad \text{in } U.$$

Repeat this construction with F_1 in the role of f, D_2 in the role of D_1, and continue. We get a sequence of superharmonic functions F_n on U that satisfy

(0) $F_n = F_{n-1}$ in $U \setminus D_n$,

(1) F_n is harmonic in D_n,

(2) $f \geq F_1 \geq F_2 \geq \cdots \geq g$ in U.

From (2) it follows that $\lim_{n\to\infty} F_n(z)$ exists (in $\mathbb{R}$) for every $z \in U$. Call this limit $F(z)$. Now let $z \in U$ be given and pick $1 \le n_1 < n_2 \cdots$ such that $D_z \subset \bigcap_{j\in\mathbb{N}} D_{n_j}$. From this and (1), each F_{n_j} is harmonic in D_z, so from (2) and 7.15 the convergence of F_{n_j} to F is locally uniform in D_z. We have proved that

(3) $\lim_{n\to\infty} F_n = F$ exists throughout U and defines a harmonic function there.

Moreover,

$$(4) \qquad f \ge F \ge g.$$

The construction of F is independent of which subharmonic minorant g of f is used. Therefore this all proves

Theorem 9.36 *If U open $\subset \mathbb{C}$, $f\colon U \mapsto \mathbb{R}$ is superharmonic and dominates some subharmonic function on U, then f has a* greatest harmonic minorant*; that is, a harmonic function F on U exists such that $f \ge F \ge g$ for all subharmonic g that are majorized on U by f.* (Cf. F. Riesz [1926].)

Remark: In contrast to 9.36, if the *sub*harmonic function f dominates the *super*harmonic function g, there need not be a harmonic function F satisfying $f \ge F \ge g$. See Edgar, Netuka and Veselý [1984].

Exercise 9.37 (Kjellberg [1950]) *Prove the following useful* Riesz-type decomposition: *If u, h_1, h_2 are non-negative harmonic functions on open $U \subset \mathbb{C}$ and $u \le h_1 + h_2$, then there exist non-negative harmonic functions u_1, u_2 on U such that $u_1 \le h_1$, $u_2 \le h_2$ and $u = u_1 + u_2$.*

Hints: The function $(u-h_2)^+ := \max\{0, u-h_2\}$ is subharmonic and satisfies $(u-h_2)^+ \le u, h_1$. The least harmonic majorant u_1 which 9.36 supplies for $(u-h_2)^+$ therefore satisfies

$$0 \le (u-h_2)^+ \le u_1 \le u, h_1, \text{ so}$$
$$0 \le u - u_1 \le u - (u-h_2)^+ \le u - (u-h_2) = h_2$$

and we may set $u_2 := u - u_1$.

Lemma 9.38 *Let U open $\subset \mathbb{C}$, $f\colon U \mapsto \mathbb{R}$ a bounded superharmonic function and F its greatest harmonic minorant* (9.36)*. Suppose that for some $z_0 \in \partial U$*

$$U \text{ has a barrier at } z_0 \text{ and } \lim_{z\to z_0} f(z) =: L \in \mathbb{R} \text{ exists.}$$

Then

$$\lim_{z\to z_0} F(z) = L.$$

Proof: By hypothesis $M := 2\sup|f(U)|$ is finite, so we may use $g := -M$ in the Basic Construction. Let $\{h_r\}$ be a barrier in U at z_0 and let $\varepsilon > 0$ be given. Choose $r > 0$ sufficiently small that

$$(1) \qquad L - \varepsilon < f(z) < L + \varepsilon \qquad \forall z \in U \cap D(z_0, r).$$

We shall first show that

$$(2) \qquad L - Mh_r(z) - \varepsilon \le f(z) \le L + Mh_r(z) + \varepsilon \qquad \forall z \in U.$$

For $z \in U \cap D(z_0, r)$ this is a trivial consequence of (1), h_r being non-negative. For $z \in U \setminus D(z_0, r)$ we have $|f(z) - L| \le Mh_r(z)$ by definition of M, L and property (ii) of a barrier, so again (2) follows. Let $\{F_n\}$ be the functions in the Basic Construction. We claim that

$$(2.1) \qquad L - Mh_r(z) - \varepsilon \le F_1(z) \le L + Mh_r(z) + \varepsilon \qquad \forall z \in U.$$

This holds in $U \setminus D_1$ by (2) and the fact that $F_1 = f$ in $U \setminus D_1$. In particular, it holds on ∂D_1. But the function on the left-hand side of (2.1) is subharmonic, that on the right-hand side is superharmonic. Therefore, F_1 being harmonic in D_1, both inequalities persist throughout D_1 (5.22(i)). It should be clear how to repeat this reasoning to establish inductively (via (0) in the Basic Construction)

$$(2.n) \qquad L - Mh_r(z) - \varepsilon \le F_n(z) \le L + Mh_r(z) + \varepsilon \qquad \forall z \in U.$$

Go to the limit here and get

$$(3) \qquad L - Mh_r(z) - \varepsilon \le F(z) \le L + Mh_r(z) + \varepsilon \qquad \forall z \in U.$$

Using the first property of a barrier, select $0 < r_\varepsilon < r$ sufficiently small that

$$(4) \qquad h_r(z) \le \frac{\varepsilon}{M+1} \qquad \forall z \in U \cap D(z_0, r_\varepsilon).$$

From (3) and (4) conclude that

$$L - 2\varepsilon \le F(z) \le L + 2\varepsilon \qquad \forall z \in U \cap D(z_0, r_\varepsilon),$$

which completes the proof.

Exercise 9.39 *Let U be an open subset of $\mathbb{C}$ which has a barrier at each point of ∂U. Show that if f is a bounded, continuous function on $\overline{U}$ which is superharmonic in U, then there exists a bounded, continuous function F on $\overline{U}$ which is harmonic in U and coincides on ∂U with f.*

Hints: Extend the F of the above construction to $\overline{U}$ by the decree $F := f$ on ∂U. Then $\lim_{z \to z_0, z \in U} F(z) = F(z_0)$ for each $z_0 \in \partial U$ and what is at issue is that $\lim_{z \to z_0, z \in \overline{U}} F(z) = F(z_0)$. This is easily proven from the continuity of f $(= F)$ on ∂U.

Lemma 9.40 *Let U be a bounded open subset of $\mathbb{C}$ which has a barrier at each point of ∂U, $C_{\mathbb{R}}(\partial U)$ the continuous, real-valued functions on ∂U and $\mathcal{H}$ the subset of all such functions which admit continuous extensions to $\overline{U}$ that are harmonic in U. Then $\mathcal{H}$ is a vector space which is closed in the uniform norm and contains all the functions $p_{n,m}(z) := (\operatorname{Re} z)^n(\operatorname{Im} z)^m$, n, m non-negative integers.*

Proof: It is obvious that $\mathcal{H}$ is a vector space. If $f_n \in \mathcal{H}$ and $f_n \to f$ uniformly on ∂U, let F_n be the corresponding extension of f_n. Then $F_n - F_m$ is a bounded harmonic function on U, so

$$\sup_{\overline{U}} |F_n - F_m| = \sup_{\partial U} |F_n - F_m| = \sup_{\partial U} |f_n - f_m| \to 0 \text{ as } m, n \to \infty.$$

Therefore $\{F_n\}$ is uniformly convergent on $\overline{U}$ to a continuous function F which coincides on ∂U with f. Thanks to 5.44(iii) the limit function F is also harmonic in U. Thus $f \in \mathcal{H}$. Finally, consider any non-negative integers n, m and let $P(z) := (\operatorname{Re} z)^n(\operatorname{Im} z)^m$, $P_1(z) := |z|^2 = |z^2|$, $z \in \mathbb{C}$. The function P_1 is subharmonic in $\mathbb{C}$, by 5.7. The continuous function ΔP has a maximum value M, say, on the compact set $\overline{U}$. Since $\Delta P_1 = 4$, it follows that $\Delta\big(P - \frac{1}{4}MP_1\big) \leq 0$ throughout $\overline{U}$. Therefore by 9.30 $P_2 := P - \frac{1}{4}MP_1$ is superharmonic in U. 9.39 ensures that both $P_2\,|_{\partial U}$ and $-P_1\,|_{\partial U}$ belong to $\mathcal{H}$. Since $\mathcal{H}$ is a vector space, it follows that $p_{n,m} = P_2\,|_{\partial U} + \frac{1}{4}MP_1\,|_{\partial U}$ also belongs to $\mathcal{H}$. This proves the lemma.

Remark: The proof really shows that *any* continuous real-valued function P for which $D_{11}P$, $D_{22}P$ exist and are bounded above in U can be expressed as the difference of two superharmonic functions in U.

Corollary 9.41 *The Dirichlet problem is solvable for any bounded open subset of $\mathbb{C}$ which has a barrier at each point of its boundary.*

Proof: In the language of 9.40 the problem is to see that $\mathcal{H} = C_{\mathbb{R}}(\partial U)$. But this follows from 9.40 and the two-dimensional Weierstrass approximation theorem 8.29(v)—after the boundary function has been extended to a rectangle via 1.44.

We next examine conditions which ensure that the hypotheses of 9.41 are fulfilled. I owe the elegant treatment of 9.43 below to Sadahiro Saeki. It is related to an idea of Georges Bouligand (1926), as expounded in the books of TSUJI [1959] and NARASIMHAN [1985]. A somewhat different treatment may be found in CONWAY [1978] (pp. 271–274).

Lemma 9.42 *Let D be a bounded open subset in $\mathbb{C}$, and C a closed connected set in $\mathbb{C}$ such that*

$$(*) \qquad C \text{ meets both } D \text{ and } \mathbb{C} \setminus \overline{D}.$$

Then each component of $C \cap \bar{D}$ *meets* ∂D.

Proof: Suppose contrariwise that $C \cap \bar{D}$ has a component K which lies wholly in D. It follows easily from 1.33 that then there is a compact subset L of $C \cap \bar{D}$ which is relatively open in $C \cap \bar{D}$ and satisfies $K \subset L \subset D$. That is,

(1) $$K \subset L \text{ compact} \subset C \cap D, \text{ and}$$

(2) $$C \cap \bar{D} \setminus L \text{ is compact.}$$

Then

(3) $$C = L \cup (C \setminus L),$$

$L \neq \varnothing$ and $C \setminus L \overset{(1)}{\supset} C \setminus D \overset{(*)}{\neq} \varnothing$. Moreover, $C \setminus L = [C \cap \bar{D} \setminus L] \cup [C \setminus D]$, so (2) shows that $C \setminus L$ is closed. We have therefore in (3) a disconnection of C —×—.

Theorem 9.43 *An open subset* U *of* $\mathbb{C}$ *has a barrier at each boundary point which does not constitute a component of* $\mathbb{C} \setminus U$. *In particular, an open simply-connected subset of* $\mathbb{C}$ *has a barrier at each of its boundary points.*

Proof: Let such an $a \in \partial U$ and $r > 0$ be given. We seek a non-negative superharmonic function $h = h_r$ on U which satisfies

(1) $$\lim_{z \to a} h(z) = 0, \quad \text{and}$$

(2) $$h = 1 \quad \text{on } U \setminus D(a, r).$$

By translation we may assume that $a = 0$. Let C be the component of $\mathbb{C} \setminus U$ which contains 0. Thus C is closed and by hypothesis $C \neq \{0\}$. Therefore, for $0 < \rho < r$ sufficiently small and $D := D(0, \rho)$, C meets $\mathbb{C} \setminus \bar{D}$. Let K denote the component of $C \cap \bar{D}$ which contains 0. This is a compact set and the preceding lemma insures that $K \cap C(0, \rho) \neq \varnothing$. After a rotation we may simply assume that $\rho \in K$. Let

(3) $$W := \mathbb{C} \setminus \big(K \cup [\rho, \infty)\big) = (\mathbb{C} \setminus K) \cap \big(\mathbb{C} \setminus (\rho, \infty)\big).$$

Since $K \subset \bar{D} \cap C \subset \bar{D} \cap (\mathbb{C} \setminus U)$, it follows that $\mathbb{C} \setminus K \supset U$, so

(4) $$W \cap \bar{D} \supset (U \cap \bar{D}) \cap \big(\mathbb{C} \setminus (\rho, \infty)\big) = U \cap \bar{D}.$$

Now $K \cup [\rho, \infty)$ is closed, connected, unbounded and contains 0. Hence in the open set $\mathbb{C} \setminus \big(K \cup [\rho, \infty)\big)$ the zero-free function z has a holomorphic logarithm $f(z)$, by 4.71 (or 4.28 + 4.67): $f \in H(W)$ and

(5) $$e^{f(z)} = z \qquad \forall z \in W.$$

Let L denote the Principal Branch of the Logarithm in $\mathbb{C}\setminus[0,\infty)$. Then for all $z \in \big(\mathbb{C}\setminus\bar{D}\big)\setminus[0,\infty) \subset W$, (5) says $e^{f(z)-L(z)} = 1$. Since the set $\big(\mathbb{C}\setminus\bar{D}\big)\setminus[0,\infty)$ is connected, there is a single integer n such that

$$f(z) - L(z) = 2\pi i n \qquad \forall z \in \big(\mathbb{C}\setminus\bar{D}\big)\setminus[0,\infty),$$

so that for these z we have $|\operatorname{Im} f(z)| \le 2\pi|n| + 2\pi$. Since W is open, every $w \in W \cap (\mathbb{C}\setminus D)$ is the limit of a sequence $\{z_k\} \subset W \cap [\mathbb{C}\setminus\bar{\mathbb{D}}\setminus[0,\infty)]$. The preceding inequality is valid for all z_k, so by continuity of f in W and L in $\mathbb{C}\setminus[0,\infty)$, these inequalities hold in the limit. In other words,

$$\text{(6)} \qquad |\operatorname{Im} f| \le M := (|n|+1)\,2\pi \quad \text{on } W\setminus D.$$

Now let

$$g := \mathbf{1} + \log\rho - f \in H(W).$$

(3) and (5) show that g is zero-free and that for $z \in W$

$$\operatorname{Re} g(z) = 1 + \log\rho - \log|z| = 1 + \log\left(\frac{\rho}{|z|}\right).$$

Since

$$\operatorname{Re}\left(\frac{1}{w}\right) = \operatorname{Re}\left(\frac{\overline{w}}{|w|^2}\right) = \frac{\operatorname{Re} w}{|w|^2}$$

for any non-zero complex number w, we therefore have

$$\operatorname{Re}\left(\frac{1}{g(z)}\right) = \frac{1+\log(\rho/|z|)}{\big[1+\log(\rho/|z|)\big]^2 + [\operatorname{Im} f(z)]^2} \qquad \forall z \in W.$$

Hence

$$\text{(7)} \qquad \lim_{z\to 0} \operatorname{Re}\left(\frac{1}{g(z)}\right) = 0,$$

$$\text{(8)} \qquad \operatorname{Re}\left(\frac{1}{g}\right) \ge 0 \quad \text{on } W\cap\bar{D} \overset{(4)}{\supset} U\cap\bar{D}, \text{ and}$$

$$\text{(9)} \qquad \operatorname{Re}\left(\frac{1}{g}\right) \overset{(6)}{\ge} \frac{1}{1+M^2} \quad \text{on } W\cap C(0,\rho) \overset{(4)}{\supset} U\cap C(0,\rho).$$

The function h defined in U by

$$h := \begin{cases} \min\left\{\operatorname{Re}\left(\dfrac{1+M^2}{g}\right), 1\right\} & \text{on } U\cap\bar{D} \\ 1 & \text{on } U\setminus D \end{cases}$$

is therefore well defined [by (9)], non-negative [by (8)], satisfies (1) [by (7)] and satisfies (2) [since $U\setminus D(a,r) \subset U\setminus D$]. Finally, according to 5.25(ii) [with $V := D$], h is superharmonic, and this finishes the proof.

Corollary 9.44 *The Dirichlet problem is solvable for any bounded, open subset of $\mathbb{C}$ whose complement contains no one-point components; in particular, for any bounded, open and simply-connected subset of $\mathbb{C}$. Moreover, the boundary function uniquely determines its harmonic extension.*

Proof: 9.41 and 9.43. Uniqueness is from 5.9.

Remark: The reasoning in remark (b) after 5.41 shows that solvability of the Dirichlet problem in U precludes *isolated* points in $\mathbb{C} \setminus U$. It is somewhat subtle that one-point components in $\mathbb{C}\setminus U$, however, are *not* precluded, so 9.44 involves sufficient but not necessary conditions: There is a bounded region U such that $\mathbb{C} \setminus U$ contains a one-point component, but a barrier exists at each point of ∂U, so (9.41) the Dirichlet problem for U is solvable. See CONWAY [1978], p. 274.

Exercise 9.45 *Formulate and prove versions of* 9.39 *and* 9.40 *allowing a finite proper subset E of ∂U at each point of which no barrier exists.*

Hints: The extension F in 9.39 will be continuous only in $\overline{U} \setminus E$. In 9.40 the new $\mathcal{H}$ should comprise those $f \in C_{\mathbb{R}}(\partial U)$ which have a *bounded*, continuous extension F to $\overline{U}\setminus E$ which is harmonic in U. The crucial equality $\sup_{\overline{U}\setminus E} |F| = \sup_{\partial U} |f|$ still holds, thanks to 7.17 and the boundedness of F.

With 9.45 in hand, more mileage can be gotten from the inversion device used in the proofs of 5.76, 7.17, 7.35 and 9.19, and a modest improvement of 9.44 can be eked out [but not needed in the sequel]:

Exercise 9.46 *Let U be an unbounded, open subset of $\mathbb{C}$ such that $\mathbb{C}\setminus U$ has no one-point components. Let $f\colon \partial U \mapsto \mathbb{R}$ be continuous and, in case ∂U is unbounded, satisfy the extra condition*

$$(*) \qquad \lim_{\substack{|z|\to\infty \\ z\in\partial U}} f(z) \quad \textit{exists in } \mathbb{R}.$$

(i) *Show that, if in addition U is not dense in $\mathbb{C}$, then f has a bounded, continuous extension to $\overline{U}$ which is harmonic in U.*

Hints: U not dense in $\mathbb{C}$ means that $\mathbb{C} \setminus U$ contains a non-empty open disk, say $\mathbb{D}$. Let $\psi(z) := 1/z$. The set $U_0 := \psi(U)$ is an open subset of $\mathbb{D} \setminus \{0\}$, $\mathbb{C}\setminus U_0 = \psi(\mathbb{C} \setminus \{0\} \setminus U) \cup \{0\}$ and $\partial U_0 = \psi(\partial U) \cup \{0\}$ (since U is unbounded). Now 0 belongs to just one component of $\mathbb{C} \setminus U$ and that component contains all of $\mathbb{D}$. Hence there are no one-point components in $\mathbb{C}\setminus\{0\}\setminus U$, consequently, none in $\psi(\mathbb{C} \setminus \{0\} \setminus U) = (\mathbb{C} \setminus U_0) \setminus \{0\}$ and so the only possible one-point component in $\mathbb{C} \setminus U_0$ is $\{0\}$. By 9.43 U_0 therefore has a barrier at every point of its boundary except possibly 0. We define f_0 on ∂U_0 by $f_0 := f \circ \psi^{-1}$

on $\psi(\partial U)$, $f_0(0)$ equal 0 if ∂U is bounded, and $f_0(0)$ equal the value of the limit in $(*)$ if ∂U is unbounded. Since 0 is an isolated point of ∂U_0 if ∂U is bounded, it follows that f_0 is continuous on ∂U_0. By the version of 9.40 stated in 9.45, coupled with 8.29(v), we are assured of the existence of a bounded, continuous function F_0 on $\overline{U}_0 \setminus \{0\}$ which is harmonic in U_0 and agrees with f_0 on $\partial U_0 \setminus \{0\}$. The sought-for extension of f has to be $F := F_0 \circ \psi$, although (as was true in the preceding) confirmation of details is tedious.

(ii) *Show that f has a bounded, continuous extension to $\overline{U}$ which is superharmonic in U.*

Hints: We may suppose that $\overline{\mathbb{D}} \subset U$. From $(*)$ it follows trivially that f is bounded. Let $M := \sup|f(\partial U)|$. Let $U_0 := U \setminus \mathbb{D}$ and notice that $\partial U_0 = \mathbb{T} \cup \partial U$ and no component of $\mathbb{C} \setminus U_0$ consists of a single point. Define f_0 on ∂U_0 by $f_0 := M$ on $\mathbb{T}$, $f_0 := f$ on ∂U and apply (i) to come up with a bounded, continuous function F_0 on $\overline{U}_0$ which is harmonic in U_0 and coincides with f_0 on ∂U_0. Extend F_0 to F on $\overline{U}$ by the decree $F := M$ on $\mathbb{D}$. Since $|F_0| = |f_0| \le M$ on ∂U_0 and F_0 is bounded in U_0, it follows from Remark (i) after 5.12 that in fact $|F_0| \le M$ throughout U_0. A direct appeal to the definition of superharmonicity in terms of circumferential mean-value inequalities (in the form 5.23(i)) then shows that F is superharmonic in U.

(iii) *Show that f has a bounded, continuous extension to $\overline{U}$ which is harmonic in U, and that there is only one such extension.*

Hints: Use (ii), 9.39 and 9.43. Remark (i) following 5.12 when applied to the difference of two f candidates implies that that difference is $\mathbf{0}$.

Remarks: CONWAY [1978] modifies the Basic Construction somewhat and utilizes the barrier in a slightly different way, so as to avoid appeal to the two-dimensional Weierstrass approximation theorem. Moreover, a careful reading of his proof establishes that in 9.46 the hypothesis $(*)$ can be replaced by the weaker requirement that f be bounded on ∂U and the conclusion (iii) still follows. But this is as far as the matter goes: GARDINER [1993] has given a topological characterization of the open sets for which the Dirichlet problem is solvable and he showed that some simply-connected regions do not qualify. For the region Ω_2 of his example it is not hard to exhibit explicitly a continuous boundary function that has no harmonic extension (see below). I thank David Armitage for bringing this to my attention. In unbounded regions solutions of the Dirichlet problem need not be unique; see 8.37.

Example 9.47 *There exists a simply-connected plane region Ω for which the Dirichlet problem is not solvable.*

Proof: (S. Saeki, adapted from GARDINER [1993]): For each $n \ge 3$ define the open sector

$$S_n := \left\{re^{i\theta} : 0 < r < n, \left|\theta - \tfrac{\pi}{n}\right| < \tfrac{\pi}{2n(n+1)}\right\}.$$

Since the restriction on θ here is

$$\frac{1+\frac{1}{2n}}{n+1} < \frac{\theta}{\pi} < \frac{1+\frac{1}{2(n+1)}}{n},$$

it is obvious that

$$\text{(1)} \qquad \bar{S}_m \cap \bar{S}_n = \{0\} \quad \text{whenever } m > n \geq 3.$$

Define

$$\Omega := D(0,2) \cup \bigcup_{n\geq 3} S_n.$$

This set is open and is starlike with respect to 0; hence it is connected. Hence too $\mathbb{C} \setminus \Omega$ is a union of half-lines, so $\mathbb{C} \setminus \Omega$ has no bounded component, making Ω simply-connected. Notice that

$$\text{(2)} \qquad \partial\Omega = \begin{cases} [2,\infty) \text{ together with arcs of } C(0,2) \text{ together} \\ \text{with the disjoint sets } (\partial S_n) \setminus D(0,2),\ n \geq 3. \end{cases}$$

Now it is clear that $z \mapsto (ze^{-i\pi/n})^{n(n+1)}$ maps S_n into $\mathbb{H}$ and the boundary rays of S_n into $i\mathbb{R}$. From this and (2) it is easy to see that a function $f : \partial\Omega \mapsto \mathbb{R}$ is well defined by

$$f(z) := \begin{cases} \operatorname{Re}\left[(ze^{-i\pi/n})^{n(n+1)}\right] & \text{for } z \in (\partial S_n) \setminus D(0,2),\ n \geq 3, \\ 0 & \text{for all other } z \in \partial\Omega \end{cases}$$

and is continuous. Suppose h is a continuous real-valued extension of f to $\overline{\Omega}$ which is harmonic in Ω. On the compact subset $\bar{D}(0,2) \subset \overline{\Omega}$, h is bounded below, say by $c \in (-\infty, 0)$. For each integer $n \geq 3$

$$\text{(3)} \qquad h(z) - \operatorname{Re}\left[(ze^{-i\pi/n})^{n(n+1)}\right] \geq c \qquad \forall z \in \partial S_n.$$

This is because for $z \in (\partial S_n) \setminus D(0,2) \subset \partial\Omega$ the left-hand side of (3) is $h(z) - f(z) = 0 > c$, while for $z \in (\partial S_n) \cap D(0,2)$ it is $h(z) - 0 \geq c$ by definition of c. By the Maximum Principle inequality (3) must hold as well at $z := 2e^{\pi i/n} \in S_n$, yielding

$$\text{(4)} \qquad h(2e^{i\pi/n}) - 2^{n(n+1)} \geq c.$$

The validity of (4) for all $n \geq 3$ makes h unbounded on the compact set $\bar{D}(0,2)$. This contradiction shows that no such h exists.

§ 8 The Dirichlet Problem and the Riemann Mapping Theorem

Definition 9.48 Let U be an open subset of $\mathbb{C}$, $u\colon U \mapsto \mathbb{R}$ a harmonic function. A function $v\colon U \mapsto \mathbb{R}$ such that $u + iv \in H(U)$ is called a *harmonic conjugate* of u. If for every such u there exists such a v, we shall say that U has *property* (H).

Thus U has property (H) if and only if each (real-valued) harmonic function on U is the real part of some holomorphic function on U.

Exercise 9.49 *Property* (H) *implies property* (HL).

Hints: Evidently either property is possessed by an open set if and only if it is possessed by each component thereof. So consider a region Ω with property (H) and any zero-free $f \in H(\Omega)$. Then $\log|f|$ is harmonic in Ω (see 5.7). Consequently by property (H) there is a function $g \in H(\Omega)$ such that $\log|f| = \operatorname{Re} g$ throughout Ω. That is, $|fe^{-g}| = 1$ throughout Ω. From the Maximum-Modulus Principle (or a glance back at 2.16(iii)) it follows that fe^{-g} is constant in Ω, say c. Since $c \neq 0$, it has the form e^a for some $a \in \mathbb{C}$. Then $g + a$ is a holomorphic logarithm for f in Ω.

Theorem 9.50 *Any proper subregion* Ω *of* $\mathbb{C}$ *which has property* (H) *is conformal to an open disk.*

Proof: According to 9.49, Ω has property (HL) and then, as we showed in the opening salvo of the proof of 9.7, Ω is conformal to a bounded region. Evidently that region inherits both property (H) and property (HL). Hence we may assume that Ω itself is bounded.

According to the implications (v) $\Rightarrow$ (vii) $\Rightarrow$ (i) of 4.71, which are elementary (cf. the remarks following 4.71), our region Ω is simply-connected, and so by 9.44 the Dirichlet problem is solvable for Ω. More in the present spirit, we can argue the solvability of the Dirichlet problem in Ω by showing the existence of barriers. Consider any point in $\partial\Omega$, which after a translation we may assume is 0, and any $r > 0$. Then the identity function on Ω has a holomorphic logarithm f. Now take $\rho := r$, m any real number larger than 1 and $\log(\sup|\Omega|) - \log\rho$, thus assuring that $g := m + \log\rho - f$ is zero-free, and for this g read the proof of 9.43 from equation (5) onward, with Ω for every occurrence of W and U. This produces a barrier at 0, i.e., at the typical point of $\partial\Omega$. So by 9.41 the Dirichlet problem is solvable in Ω.

Since for each $a \in \Omega$ the function $\log|z - a|$ is continuous on $\partial\Omega$, there is then a continuous function u_a on $\overline{\Omega}$ which is harmonic in Ω and satisfies

$$(1) \qquad u_a(z) = \log|z - a| \qquad \forall z \in \partial\Omega.$$

By property (H) of Ω we have

$$(2) \qquad u_a = \operatorname{Re} \phi_a \quad \text{for some } \phi_a \in H(\Omega).$$

Consider the holomorphic function

$$(3.a) \qquad g_a(z) := (z-a)e^{-\phi_a(z)}, \qquad z \in \Omega.$$

We have

$$(4) \qquad |g_a(z)| = |z-a|e^{-\operatorname{Re} \phi_a(z)} = |z-a|e^{-u_a(z)} \qquad \forall z \in \Omega.$$

From (4) and (1) we see that

$$(5.a) \qquad \lim_{z \to \xi} |g_a(z)| = 1 \qquad \forall \xi \in \partial\Omega.$$

It follows from this and 5.12 that

$$(6) \qquad |g_a(z)| \leq 1 \qquad \forall z \in \Omega.$$

But equation $(3.a)$ also shows that g_a is zero only at the point a and so, in particular, g_a is not constant. Therefore by the Maximum-Modulus Principle 5.10, strict inequality prevails everywhere in (6):

$$(6.a) \qquad |g_a(z)| < 1 \qquad \forall z \in \Omega.$$

We aim to show that g_a is one-to-one and $g_a(\Omega) = \mathbb{D}$. Consider any $b \in \Omega \backslash \{a\}$. Since $|g_a(b)| < 1$, we may form

$$\Phi(w) := \frac{w - g_a(b)}{1 - \bar{g}_a(b)w}, \qquad w \in \mathbb{D}.$$

This is a conformal map of $\mathbb{D}$ onto itself (see 2.6). Consequently

(7) Φ is a proper mapping, hence compactafugal (6.12)

Since by $(6.a)$, $g_a(\Omega) \subset \mathbb{D}$, we may form

$$(8) \qquad F := \Phi \circ g_a.$$

Now $\Phi(g_a(b)) = 0$, so $F(b) = 0$. However, equation $(3.b)$ shows that the holomorphic function g_b has a first-order zero at b and no other zeros. Consequently, F/g_b is holomorphic in Ω. Now for any $\xi \in \partial\Omega$

$$\begin{aligned} \lim_{z \to \xi} \left| \frac{F(z)}{g_b(z)} \right| &\overset{(5.b)}{=} \lim_{z \to \xi} |F(z)| \overset{(8)}{=} \lim_{z \to \xi} |\Phi(g_a(z))| \\ &= 1, \quad \text{by (7) and equation } (5.a). \end{aligned}$$

Therefore the Maximum-Modulus Principle in the form 5.12 implies that

$$(9) \qquad \left|\frac{F(z)}{g_b(z)}\right| \leq 1 \qquad \forall z \in \Omega.$$

Since $F(a) = -g_a(b)$, it follows from (9) that

$$(10) \qquad \frac{|g_a(b)|}{|g_b(a)|} = \left|\frac{F(a)}{g_b(a)}\right| \leq 1,$$

$$(11.a.b) \qquad |g_a(b)| \leq |g_b(a)|.$$

But the roles of a and b may be interchanged and the last inequality together with the corresponding inequality (11.b.a) gives us $|g_a(b)| = |g_b(a)|$. This, however, says that equality holds in (10) and so by the Maximum-Modulus Principle (9) becomes

$$(12) \qquad \left|\frac{F(z)}{g_b(z)}\right| = 1 \qquad \forall z \in \Omega.$$

Since g_b vanishes only at b (see equation (3.b)), it follows from (12) that F is never zero in $\Omega \setminus \{b\}$, i.e., recalling (8),

$$(13.b) \qquad g_a(z) \neq g_a(b) \qquad \forall z \in \Omega \setminus \{b\}.$$

Here b is any point of $\Omega \setminus \{a\}$. However, the inequality (13.a) is true too, as follows at once from (3.a). The inequalities (13.b) are thus valid for all $b \in \Omega$. They establish the univalence of g_a.

Now to prove the surjective character of g_a, suppose $w \in \mathbb{D} \setminus g_a(\Omega)$. We may then form the holomorphic function

$$G(z) := \frac{1 - \overline{w} g_a(z)}{g_a(z) - w}, \qquad z \in \Omega.$$

The function

$$f(\zeta) := (\zeta - w)/(1 - \overline{w}\zeta)$$

is a self-homeomorphism of $\overline{\mathbb{D}}$, so it is a proper map (6.12(i)). In the present context f "fleeing compacta" simply means that $|f(\zeta)| \to 1$ whenever $\zeta \to 1$. Since $G = 1/f \circ g_a$, it then follows from (5.a) the $\lim_{z \to \xi} |G(z)| = 1$ for every $\xi \in \partial\Omega$. Therefore, again by 5.12, $|-1/w| = |G(a)| \leq 1$, contrary to $w \in \mathbb{D}$.

Exercise 9.51 *Show how one can use* 6.15 *to complete the preceding proof from equation* (5.a) *more expeditiously.*

Hints: 6.15 says that the map g_a of Ω into $\mathbb{D}$ is exactly m-to-1 for some integer m, and that g_a is surjective. Since g_a assumes the value 0 only at a and with multiplicity 1 (see (3.a)), m must be 1.

Remarks 9.52: (*On Green's functions and a generalized Poisson formula*) The converse direction, deducing the solvability of the Dirichlet Problem from the Riemann Mapping Theorem, is evidently only possible when something is known about boundary behavior in the latter. In case the Riemann map extends to a homeomorphism f of the closure of the region Ω, i.e. (9.14), in the case of Jordan regions, the solvability of the Dirichlet Problem is a banality: one transfers the boundary data to $\mathbb{T}$ via f^{-1}, uses the Poisson formula, then transfers back to $\overline{\Omega}$ via f. Heavier hypotheses are needed to insure that the integral formula 5.17 itself can be transformed into an integral of the boundary function on $\partial\Omega$ times an appropriate positive kernel. Smoothness of f on $\partial\Omega$ is needed, and ideally should follow from the hypothesis that the Jordan-curve $\partial\Omega$ admits a sufficiently smooth parametrization Γ. (See, e.g., Proposition 4.7.22 of BERENSTEIN and GAY [1991] or BELL and KRANTZ [1987].) In that case the derivation of the integral formula parallels the proof of 5.17. In order to illustrate this without lengthy arguments to confirm needed boundary behavior, I will make the (unnecessarily strong) hypothesis that

$$(1) \qquad f \text{ extends to be conformal in a neighborhood of } \overline{\Omega}.$$

Each $a \in \Omega$ leads to another map of Ω onto $\mathbb{D}$,

$$(2) \qquad T_a := \frac{f - f(a)}{1 - \overline{f(a)}f}, \quad \text{also holomorphic in a neighborhood of } \overline{\Omega}.$$

Suppose $\partial\Omega$ is parameterized by a smooth, 2π-periodic function Γ which is one-to-one on $[0, 2\pi)$ and is positively oriented. $T_a \circ \Gamma$ being smooth, the discussion preceding 4.4 furnishes a smooth $\phi : [0, 2\pi] \mapsto \mathbb{C}$ such that

$$(3) \qquad T_a(\Gamma(t)) = e^{i\phi(t)} \qquad \forall t \in [0, 2\pi].$$

Since $T_a \circ \Gamma$ maps onto $\partial\mathbb{D}$, ϕ is $\mathbb{R}$-valued, and since $T_a \circ \Gamma$ is one-to-one on $[0, 2\pi)$, so is ϕ. According to 1.46, ϕ is therefore strictly monotone. The conformal map T_a has exactly one zero, namely a, in Ω. Consequently by 9.15

$$(4) \qquad 1 = \mathrm{Ind}_{T_a \circ \Gamma}(0) = [\phi(2\pi) - \phi(0)]/2\pi.$$

Thus the monotone function ϕ is an *increasing* function. Now if the continuous function $h : \overline{\Omega} \mapsto \mathbb{R}$ is harmonic in Ω, then $h \circ T_a^{-1}$ is continuous on $\overline{\mathbb{D}}$ and harmonic in $\mathbb{D}$, so Gauss' mean-value theorem gives

$$\begin{aligned} h(a) = h\big(T_a^{-1}(0)\big) &= \frac{1}{2\pi}\int_{\phi(0)}^{\phi(0)+2\pi} h\big(T_a^{-1}(e^{ix})\big)\,dx \\ &\overset{(4)}{=} \frac{1}{2\pi}\int_{\phi(0)}^{\phi(2\pi)} h(T_a^{-1}(e^{ix}))\,dx \\ &\overset{(3)}{=} \frac{1}{2\pi}\int_{\phi(0)}^{\phi(2\pi)} h(\Gamma(\phi^{-1}(x)))\,dx \\ &= \frac{1}{2\pi}\int_0^{2\pi} h(\Gamma(t))\phi'(t)\,dt. \end{aligned} \tag{5}$$

The chain rule on (3) gives $T_a'(\Gamma(t))\Gamma'(t) = ie^{i\phi(t)}\phi'(t)$, so

$$\phi'(t) = -i\frac{T_a'\big(\Gamma(t)\big)}{T_a\big(\Gamma(t)\big)}\Gamma'(t) \qquad \forall t \in [0, 2\pi]. \tag{6}$$

But ϕ increasing means $\phi' \geq 0$, so absolute values may be introduced on the right side of (6), giving, since T_a maps $\Gamma(\mathbb{R}) = \partial\Omega$ onto $\partial\mathbb{D}$,

$$\phi'(t) = |T_a'\big(\Gamma(t)\big)||\Gamma'(t)| \qquad \forall t \in [0, 2\pi],$$

whereupon (5) reads

$$h(a) = \frac{1}{2\pi}\int_0^{2\pi} h(\Gamma(t))|T_a'(\Gamma(t))||\Gamma'(t)|dt \qquad \forall a \in \Omega. \tag{7}$$

The form T_a'/T_a suggests the derivative of a logarithm. For each $z_0 \in \overline{\Omega}\setminus\{a\}$ there is a holomorphic logarithm L in a neighborhood of $T_a(z_0)$, so $T_a(z) = e^{L(T_a(z))}$ for all z near z_0. For such z, $\log|T_a(z)| = \operatorname{Re} L(T_a(z))$, so by the chain rule and (C. R.)

$$\begin{aligned} T_a' &= (L\circ T_a)'\cdot T_a = [D_1 \operatorname{Re}(L\circ T_a) - iD_2 \operatorname{Re}(L\circ T_a)]\cdot T_a \\ T_a'/T_a &= D_1 \log|T_a| - iD_2\log|T_a| = \overline{\nabla}(\log|T_a|). \end{aligned}$$

Defining

$$g_a := -\log|T_a| \tag{8}$$

and again recalling that $|T_a| = 1$ on $\partial\Omega$, we see that

$$|T_a'| = |\nabla g_a| \text{ on } \partial\Omega, \text{ for all } a \in \Omega, \tag{9}$$

enabling us to express (7) as

$$(10) \qquad h(a) = \frac{1}{2\pi}\int_0^{2\pi} h(\Gamma(t))|\nabla g_a(\Gamma(t))||\Gamma'(t)|dt \qquad \forall a \in \Omega,$$

often written as

$$(10)' \qquad h(a) = \frac{1}{2\pi}\int_\Gamma h(\xi)|\nabla g_a(\xi)||d\xi| \qquad \forall a \in \Omega.$$

Theorem 5.17 concerns the special case of all this where $\Omega := \mathbb{D}$ and $\Gamma(s) := e^{is}$. The map T_a was denoted there by T^{-1} and given explicitly as $w \mapsto \frac{w-a}{1-\bar{a}w}$, enabling the kernel (9) to be computed for the unit disk: $|T_a'(\xi)| = \frac{1-|a|^2}{|1-\bar{a}\xi|^2}$. Thus (9) is a generalized Poisson kernel and $(10)'$ a generalized Poisson formula.

The function g_a is called *Green's function for Ω with singularity at a.* It is harmonic and positive in $\Omega \setminus \{a\}$, vanishes on $\partial\Omega$, and approaches ∞ as $z \to a$ in the precise sense that $g_a(z) + \log|z-a|$ has a removable singularity (equivalently, a finite limit) at a. In fact, $-\log|T_a|$ is the unique function with these properties. For

$$G_a(z) := -\log\left|\frac{T_a(z) - T_a(a)}{z-a}\right| = -\log|T_a(z)| + \log|z-a|$$

is the unique solution of the Dirichlet problem in Ω with boundary data $\log|z-a|$. Thus $g_a(z) = G_a(z) - \log|z-a|$ can be alternatively accessed by solving a very special Dirichlet problem, in principle circumventing T_a and f. Moreover, $\partial\Omega$ being part of the level-set $g_a^{-1}(0)$, the vector $\nabla g_a(\xi)$ is normal to $\partial\Omega$ at each $\xi \in \partial\Omega$. In fact [as can be checked by re-examining (6)–(8)], the vector $\nabla g_a(\Gamma(t))$ is a positive multiple of the "inward" normal $i\Gamma'(t)$, so the directional derivative of g_a in that normal direction, denoted $D_{\mathbf{N}}g_a(\Gamma(t))$, equals $|\nabla g_a(\Gamma(t))|$; that is, *the Poisson kernel* (at a) *is the normal derivative of the Green's function* (with singularity at a). Therefore $(10)'$ is often written with $D_{\mathbf{N}}g_a(\xi)$ in place of $|\nabla g_a(\xi)|$.

If $l := \int_0^{2\pi}|\Gamma'|$ denotes the length of Γ, and if we re-parameterize Γ into a function Γ^* of arclength on $[0, l]$, then (cf. the hints to 10.8(iv)) (10) becomes

$$h(a) = \frac{1}{2\pi}\int_0^l h\big(\Gamma^*(s)\big)\big|\nabla g_a\big(\Gamma^*(s)\big)\big|\,ds \qquad \forall a \in \Omega,$$

s being the traditional arclength parameter.

It may be noted in passing that if Lebesgue integration is used, less smoothness has to be demanded of Γ for (10) to hold; see VERBLUNSKY [1951].

The formulas (10) are pregnant with physical significance. The reader will find it wonderfully explained, and formula (10) itself derived from thermodynamical considerations, in NEEDHAM [1997]. In order to utilize (10) to solve the Dirichlet problem, it is not necessary to be in possession of a Riemann map of Ω onto $\mathbb{D}$, but only of the Green's functions g_a, $a \in \Omega$. Their existence is assured by 9.44 and they can be found by various artifices—in the 19th century, by appeal to electrostatics!

Equation (10) can also be derived (and indeed under weaker regularity assumptions) from the classical Green's-Stokes' formula, which relates the integral of a certain differential form over a surface to that of another form over the boundary. But the machinery needed to carry this out rigorously (see **MR** 80i:58002 and 85j:26017) is formidable.

§ 9 Conformal Mapping of a Half-plane onto a Polygon

There is a remarkably explicit formula, due to Herman Amandus Schwarz and Elwin Bruno Christoffel for mapping $\mathbb{U}$ conformally onto the inside of a polygon. See LICHTENSTEIN [1919], pp. 274–275 for history and provenance. There are, as one would expect, many topological and boundary subtleties involved. A careful treatment in considerable generality will be found in PALKA [1991], pp. 450–465. I want to illustrate the beautiful geometric ideas involved by treating the special case of a rectangle in detail, roughly following SAKS and ZYGMUND [1971], pp. 233–237. The famous elliptic integrals come up naturally here. Mapping $\mathbb{U}$ onto an n-gon requires n pre-vertices on $\mathbb{R}$ and, instead of merely square-roots, involves other roots determined by the vertex angles of the polygon. The interested reader will find lots of explicit examples in HENRICI [1974], pp. 396–432, and a few cautionary "counterexamples" in JOHNSTON [1983] and PFALTZGRAFF [1986]. DRISCOLL and TREFETHEN [2002] is a thorough treatment of the many kinds of polygonal region (including doubly-connected ones) for which the method is successful, together with numerical analysis and computer techniques, many accurate diagrams, numerous applications, and a 180-item bibliography.

The fact that the principal branch of the logarithm in $\mathbb{C} \setminus [0, \infty)$ jumps by $2\pi i$ at the positive real axis is used to manufacture a product Φ of square-root functions that at four points on $\mathbb{R}$ abruptly changes direction by $\pi/2$. The integral f of Φ maps $\mathbb{R}$ to a rectangle R, and $\mathbb{U}$ onto the inside of R. That is the broad outline to be kept in mind as we wade into the details to follow. Most of the work consists in confirming that f maps $\mathbb{R}$ as claimed, and indeed bijectively. The proof is constructive and does not involve the Riemann mapping theorem at all.

Let $a \in (0,1)$ be given and set

$$a_1 := -1,\ a_2 := -a,\ a_3 := a,\ a_4 := 1. \tag{0}$$

Denote by L the holomorphic logarithm in $\mathbb{C} \setminus \{iy : y \le 0\}$ which satisfies

$$-\frac{\pi}{2} < \operatorname{Im} L < \frac{3\pi}{2},$$

and for each $j = 1, 2, 3, 4$ define

$$(z - a_j)^{-1/2} := \exp\bigl(-\tfrac{1}{2} L(z - a_j)\bigr), \quad z \in \mathbb{C} \setminus \{a_j + iy : y \le 0\}. \tag{s.j}$$

These functions are holomorphic, and all are defined at least in the region

$$\Omega := \mathbb{C} \setminus \{a_1, a_2, a_3, a_4\} \times (-\infty, 0]\,,$$

which is evidently simply-connected.

For $x \in \mathbb{R} \setminus \{a_j\}$

$$(x - a_j)^{-1/2} = \begin{cases} \dfrac{-i}{\sqrt{a_j - x}} & \text{if } x < a_j \\ \dfrac{1}{\sqrt{x - a_j}} & \text{if } x > a_j. \end{cases} \tag{1}$$

In particular

(2) $\left\{\begin{array}{l} \bigl|(x - a_j)^{-1/2}\bigr| = \dfrac{1}{\sqrt{|x - a_j|}} \\ \text{which is improperly Riemann integrable in a neighborhood of } a_j; \\ \text{since it has the bounded antiderivatives } -2\sqrt{a_j - x} \\ \text{on the interval } [-2, a_j) \text{ and } 2\sqrt{x - a_j} \text{ on the interval } (a_j, 2]. \end{array}\right.$

Define

$$\Phi(z) := \prod_{j=1}^{4} (z - a_j)^{-1/2}, \quad z \in \Omega. \tag{3}$$

Evidently

$$|z|^2 |\Phi(z)| \to 1 \text{ as } |z| \to \infty, \text{ so}$$

$$|\Phi(z)| < 2|z|^{-2} \text{ for large } |z|. \tag{4}$$

This means that $|\Phi|$ is integrable over the rest of $\mathbb{R}$ too:

$$\int_{\mathbb{R}} |\Phi(t)|\, dt < \infty. \tag{5}$$

Define

$$f(z) := \int_{[0,\operatorname{Re} z+i,z]} \Phi, \quad z \in \Omega. \tag{6}$$

Evidently each such polygonal path lies wholly in Ω and since Ω is simply-connected, 8.13 insures that the holomorphic function Φ in Ω satisfies the hypotheses of 2.11. Consequently f is holomorphic and

$$f' = \Phi \quad \text{in } \Omega. \tag{7}$$

Claim: For $k = 1, 2, 3, 4$

$$\lim_{\substack{z \to a_k \\ z \in \Omega}} f(z) =: v_k \quad \text{exists in } \mathbb{C}. \tag{8.k}$$

For notational ease consider only $k = 1$. The function $h_1(z) := \prod_{j=2}^{4} (z - a_j)^{-1/2}$ is holomorphic in a neighborhood $N_1 := D(a_1, 2\varepsilon)$ of a_1 with $0 < \varepsilon < (1-a)/2$. Let its Taylor series there be

$$h_1(w) = \sum_{n=0}^{\infty} c_n (w - a_1)^n. \tag{9.1}$$

The point $z_1 := a_1 + \varepsilon i \in N_1$ clearly satisfies $[z_1, z] \subset \Omega$ for every $z \in N_1 \cap \Omega$. By (7) then

$$f(z) = f(z_1) + \int_{[z_1,z]} \Phi \qquad \forall z \in N_1 \cap \Omega. \tag{10.1}$$

For any $z \in N_1 \cap \Omega$ the series (9.1) converges uniformly for $w \in [z_1, z]$ and multiplication by the bounded function $(w - a_1)^{-1/2}$ does not change that. Hence

$$\Phi(w) = (w - a_1)^{-1/2} h_1(w) = \sum_{n=0}^{\infty} c_n (w - a_1)^{-1/2} (w - a_1)^n$$

with convergence uniform in $w \in [z_1, z]$. Consequently, the integration in (10.1) can be performed termwise:

$$f(z) = f(z_1) + \sum_{n=0}^{\infty} c_n \int_{[z_1,z]} (w - a_1)^{-1/2} (w - a_1)^n \, dw.$$

Using the product differentiation rule and definition (s.1), we confirm that $\frac{(w-a_1)^{-1/2}(w-a_1)^{n+1}}{n+\frac{1}{2}}$ is an antiderivative for the nth integrand, so we get, after splitting into the difference of two series [allowable because by 3.1 and 3.3(iii) each series has the same radius of convergence as (9.1)],

$$f(z) = f(z_1)+(z-a_1)^{-\frac{1}{2}}\sum_{n=0}^{\infty}\frac{c_n}{n+\frac{1}{2}}(z-a_1)^{n+1}-(z_1-a_1)^{-\frac{1}{2}}\sum_{n=0}^{\infty}\frac{c_n}{n+\frac{1}{2}}(z_1-a_1)^{n+1}.$$

In this expression we let $z \in N_1 \cap \Omega$ converge to a_1, and the part on the right-hand side involving z converges to 0, thanks to the factorization

$$\left|(z-a_1)^{-1/2}\sum_{n=0}^{\infty}\frac{c_n}{n+\frac{1}{2}}(z-a_1)^{n+1}\right| = |z-a_1|^{-1/2}|z-a_1|\left|\sum_{n=0}^{\infty}\frac{c_n}{n+\frac{1}{2}}(z-a_1)^{n}\right|.$$

This confirms (8.1). Thus if we *define*

(11) $$f(a_k) := v_k, \quad k = 1, 2, 3, 4,$$

then f, so extended, is continuous in $\overline{\mathbb{U}}$.

Using (7) and the fundamental theorem of calculus leads to useful alternative integral representations of f. First, if $z = |z|e^{i\theta}$ with $\theta \in [0, \pi]$ and $|z| > 1$,

(12) $$f(z) - f(|z|) = \int_0^{\theta} \Phi(|z|e^{it})i|z|e^{it}\,dt$$

and, as well,

(13) $$f(s) - f(r) = \int_{[r,s]} \Phi \qquad \text{whenever real } s > r > 1.$$

From (12) and (4)

(14) $$\left|f(z) - f(|z|)\right| \le \int_0^{\theta} 2|z|^{-2}|z|\,dt \le 2\pi|z|^{-1} \qquad \text{for all large } z \in \overline{\mathbb{U}},$$

while from (13) and (4)

(15) $$|f(s) - f(r)| \le \int_r^s 2x^{-2}\,dx < \frac{2}{r} \qquad \text{whenever } s > r \text{ is sufficiently large.}$$

From (15) we learn that

(16) $$C := \lim_{r\uparrow\infty} f(r) \qquad \text{exists in } \mathbb{C}$$

and from this and (14) follows

$$\lim_{\substack{|z|\to\infty \\ z\in\overline{\mathbb{U}}}} f(z) = C. \tag{17}$$

According to (1), (3) and (7), for all $t \in \mathbb{R} \setminus \{a_1, a_2, a_3, a_4\}$

$$f'(t) = \prod_{j=1}^{4} (t - a_j)^{-1/2} = \frac{(-i)^N}{\prod_{j=1}^{4} \sqrt{|t - a_j|}}$$

$$= (-i)^N \, |f'(t)|, \qquad \text{with } N := \begin{cases} 4 & \text{if } t < a_1 \\ 3 & \text{if } a_1 < t < a_2 \\ 2 & \text{if } a_2 < t < a_3 \\ 1 & \text{if } a_3 < t < a_4 \\ 0 & \text{if } a_4 < t. \end{cases} \tag{18}$$

This shows that $|f'| > 0$ on $\mathbb{R} \setminus \{a_1, a_2, a_3, a_4\}$ and on each interval $(-\infty, a_1]$, $[a_1, a_2]$, $[a_2, a_3]$, $[a_3, a_4]$, $[a_4, +\infty)$, f takes values in a translate of $\mathbb{R}$ or of $i\mathbb{R}$, hence is one-to-one on each (by the mean-value theorem). Since $|f'| = |\Phi|$ is (improperly Riemann-) integrable on each of these intervals, by (2) and (5), we have

$$v_2 = f(a_2) - 0 = f(a_2) - f(0) = -\int_{a_2}^{0} f' \overset{(18)}{=} \int_{a_2}^{0} |f'| \in \mathbb{R}_+. \tag{19}$$

$$\begin{aligned} v_1 &= f(a_1) - f(a_2) + f(a_2) - f(0) = f(a_1) - f(a_2) + v_2 \\ &= -\int_{a_1}^{a_2} f' + v_2 \overset{(18)}{=} -i \int_{a_1}^{a_2} |f'| + v_2 \in \mathbb{C} \setminus \overline{\mathbb{U}}. \end{aligned} \tag{20}$$

$$v_3 = f(a_3) - f(0) = \int_0^{a_3} f' = -\int_0^{a_3} |f'| = -v_2.$$

The last equality is due to the fact that on $\mathbb{R}$, $|f'|$ is an even function, which (7), (2) and (0) make clear.

$$\begin{aligned} v_4 &= f(a_4) - f(a_3) + f(a_3) - f(0) = f(a_4) - f(a_3) + v_3 \\ &= \int_{a_3}^{a_4} f' + v_3 \overset{(18)}{=} -i \int_{a_3}^{a_4} |f'| + v_3 = -i \int_{a_1}^{a_2} |f'| - v_2 = v_1 - 2v_2 \end{aligned}$$

(again using evenness of $|f'|$ on $\mathbb{R}$). In summary,

$$v_3 = -v_2 \tag{21}$$

$$v_4 = v_1 - 2v_2 = -\bar{v}_1. \tag{22}$$

Recalling (20) that $v_1 \in \mathbb{C} \setminus \overline{\mathbb{U}}$ and $\operatorname{Re} v_1 = v_2$, we thus see that

(23) v_1, v_2, v_3, v_4 are the vertices of a rectangle in $\mathbb{C} \setminus \mathbb{U}$ with one side in $\mathbb{R}$.

For $a_2 < x < 0$

$$f(x) = f(x) - f(0) = -\int_x^0 f' \overset{(18)}{=} \int_x^0 |f'|$$

and so

(24) $$f \text{ maps } [a_2, 0] \text{ onto } [f(a_2), f(0)] = [v_2, 0] \subset \mathbb{R}.$$

For $a_1 < x < a_2$

$$f(x) = \int_{a_1}^x f' + f(a_1) \overset{(18)}{=} (-i)^3 \int_{a_1}^x |f'| + f(a_1) = i \int_{a_1}^x |f'| + v_1 \overset{(20)}{=} v_2 - i \int_x^{a_2} |f'|$$

and so

(25) $$f \text{ maps } [a_1, a_2] \text{ onto } [f(a_1), f(a_2)] = [v_1, v_2].$$

Similar analyses confirm that

(26) $$f \text{ maps } [0, a_3] \text{ onto } [f(0), f(a_3)] = [0, v_3] = [0, -v_2] \subset \mathbb{R}$$

and

(27) $$f \text{ maps } [a_3, a_4] \text{ onto } [f(a_3), f(a_4)] = [v_3, v_4] = [-v_2, -\bar{v}_1].$$

The change of variable $x := a/t$ shows that

$$\int_{a_4}^\infty |f'| = \int_1^\infty \frac{dt}{\sqrt{t^2 - 1}\sqrt{t^2 - a^2}} = \int_0^a \frac{dx}{\sqrt{a^2 - x^2}\sqrt{1 - x^2}} = \int_0^{-a_2} |f'| = \int_{a_2}^0 |f'|$$

since, as noted earlier, $|f'|$ is an even function on $\mathbb{R}$. Consequently,

(28) $$\int_{a_4}^\infty |f'| = v_2, \quad \text{by (19)}.$$

Now for $r > a_4$

$$f(r) = f(a_4) + \int_{a_4}^r |f'| = v_4 + \int_{a_4}^r |f'|$$

and letting $r \to \infty$,

$$C \overset{(16)}{=} \lim_{r\uparrow\infty} f(r) = v_4 + \int_{a_4}^{\infty} |f'| \overset{(28)}{=} v_4 + v_2,$$

$$C = \frac{v_4 + v_1}{2} = v_1 - v_2, \text{ by using (22).} \tag{29}$$

That is, C is the midpoint of the interval $[v_4, v_1]$. And we have

$$f \text{ maps } [a_4, +\infty) \text{ onto } [f(a_4), C) = \left[v_4, \frac{v_4 + v_1}{2}\right). \tag{30}$$

Similarly,

$$f \text{ maps } (-\infty, a_1] \text{ onto } (C, f(a_1)] = \left(\frac{v_4 + v_1}{2}, v_1\right]. \tag{31}$$

Together (25), (24) & (26), (27), (30) & (31) assert that

$$f(\mathbb{R}) = [v_1, v_2] \cup [v_2, v_3] \cup [v_3, v_4] \cup \left[v_4, \frac{v_4 + v_1}{2}\right) \bigcup \left(\frac{v_4 + v_1}{2}, v_1\right]. \tag{32}$$

We saw earlier that f is one-to-one on each of the $\mathbb{R}$-intervals determined by the a_j. In view of (23) the images of these intervals have only endpoints in common. It follows that

$$f \text{ is one-to-one on the whole real line.} \tag{33}$$

Recall that $\phi(w) := i\frac{1+w}{1-w}$ maps $\overline{\mathbb{D}} \setminus \{1\}$ homeomorphically onto $\overline{\mathbb{U}}$, $\mathbb{T} \setminus \{1\}$ homeomorphically onto $\mathbb{R}$, and $\mathbb{D}$ conformally onto $\mathbb{U}$. And obviously

$$\lim_{\substack{w\to 1 \\ w\in\overline{\mathbb{D}}\setminus\{1\}}} |\phi(w)| = \infty.$$

Therefore, in view of (17), the composite $f \circ \phi$ is holomorphic in $\mathbb{D}$ and continuous on $\overline{\mathbb{D}}$, with value C at $w = 1$. Moreover, $f \circ \phi$ maps $\mathbb{T}$ homeomorphically onto $\{C\} \cup f \circ \phi\,(\mathbb{T} \setminus \{1\}) = \{C\} \cup f(\mathbb{R}) = [v_1, v_2, v_3, v_4, v_1]$, by (33), (32) and (29):

$$f \circ \phi(\mathbb{T}) = [v_1, v_2, v_3, v_4, v_1] = [v_1, v_2, -v_2, -\bar{v}_1, v_1]. \tag{34}$$

At this point we can invoke the Darboux-Picard theorem (9.16) to assert that $f \circ \phi$ must map $\mathbb{D}$ conformally onto the inside of the Jordan-curve (34), that is, onto the interior of the rectangle

$$R := \operatorname{co}\{v_1, v_2, v_3, v_4\} = \operatorname{co}\{v_1, v_2, -v_2, -\bar{v}_1\}.$$

Then f itself maps $\mathbb{U}$ conformally onto the interior of R and $\overline{\mathbb{U}}$ homeomorphically onto $R \setminus \{C\} = R \setminus \{v_1 - v_2\}$—*mirabile dictu*!

Remarks 9.53: (i) It is important to observe that by proper choice of $a \in (0,1)$ the rectangle R can have any length-to-width ratio. To see this, first note that

$$|v_3 - v_2| \overset{(21)}{=} 2|v_2| \overset{(19)}{=} 2\int_{a_2}^{0} |f'| = 2\int_0^a \frac{dt}{\sqrt{1-t^2}\sqrt{a^2-t^2}}$$

$$(35) \qquad = 2\int_0^1 \frac{dx}{\sqrt{1-a^2x^2}\sqrt{1-x^2}} \quad (\text{via } x := t/a),$$

and

$$(36) \qquad |v_2 - v_1| \overset{(20)}{=} \int_{a_1}^{a_2} |f'| = \int_a^1 \frac{dt}{\sqrt{1-t^2}\sqrt{t^2-a^2}}.$$

The change of variable $x := \sqrt{\frac{1-t^2}{1-a^2}}$ and a heroic calculation transform (36) into

$$(37) \qquad |v_2 - v_1| = \int_0^1 \frac{dx}{\sqrt{1-(1-a^2)x^2}\sqrt{1-x^2}}.$$

From (35) and (37)

$$(38) \qquad \lim_{a\downarrow 0} |v_3 - v_2| = 2\lim_{a\uparrow 1} |v_2 - v_1| = 2\int_0^1 \frac{dx}{\sqrt{1-x^2}} = \pi,$$

$$\lim_{a\uparrow 1} |v_3 - v_2| = 2\lim_{a\downarrow 0} |v_2 - v_1| = 2\int_0^1 \frac{dx}{1-x^2}$$

$$(39) \qquad = 2\int_0^1 \frac{dx}{(1+x)(1-x)} \geq \int_0^1 \frac{dx}{1-x} = +\infty.$$

From (35) and (37) we also see that the length-to-width ratio $|v_3 - v_2|/|v_2 - v_1|$ is an increasing, continuous function of $a \in (0,1)$. In view of (38) and (39), this ratio therefore realizes every positive value.

(ii) (*Squaring the circle*) For a given ratio it is obviously a non-trivial task to find the appropriate a. Even confirming that the choice $a := \left(\sqrt{2}-1\right)^2$ produces a square is non-trivial. However, by adapting the arguments above, it is not hard to show that

$$z \mapsto \int_{[0,z]} w^{-1/2}(w+1)^{-1/2}(w-1)^{-1/2}\, dw$$

maps $\mathbb{U}$ conformally onto the interior of a square of side-length $\int_0^1 \frac{dx}{\sqrt{x(1-x^2)}}$, and

$$z \mapsto \int_{[0,z]} (1-w^4)^{-1/2}\,dw$$

maps $\mathbb{D}$ onto the interior of a square. Moreover, according to 9.14 this latter map extends to a homeomorphism of $\overline{\mathbb{D}}$ onto the image square.

(iii) Upon recalling definition (0) of the a_j, it is natural to inquire whether the definition of (3) of Φ can perhaps be simplified to

$$(3)_0 \qquad \Phi_0(z) := (z^2-1)^{-1/2}(z^2-a^2)^{-1/2}.$$

More precisely, are there some definitions of $(\ )^{-1/2}$ that make $(3)_0$ coincide with (3), at least for all $z \in \mathbb{U}$? Note that $z \in \mathbb{U} \Rightarrow z^2 \in \mathbb{C}_+ \Rightarrow z^2-1, z^2-a^2 \in \mathbb{C}_+$. Therefore the Principal Branch of the Logarithm in $\mathbb{C}_+$ (3.42(iii)) can be enlisted to create the holomorphic function Φ_0 in $\mathbb{U}$, which is obviously zero-free and satisfies $\Phi_0^2 = \Phi^2$. Moreover, $\Phi_0(i) = \Phi(i) = -1/\sqrt{2}\sqrt{1+a^2}$. From Exercise 1.7 follows $\Phi_0 = \Phi$ in $\mathbb{U}$, so Φ is a holomorphic extension of Φ_0 to Ω. The function Φ_0 suffices for the construction of the sought-for conformal map f (see (6)), although without (3) some of the geometric proof details about f may be less transparent or more troublesome to confirm. Is there a further $\big((z^2-1)(z^2-a^2)\big)^{-1/2}$ version?

(iv) The method applies to polygons which are not convex and to “polygons” with a vertex at infinity, even to situations where a side is traversed twice, in opposite directions. Thus, for example, it leads to conformal maps of $\mathbb{U}$ onto $\mathbb{C} \setminus [0,1] \times [0,\infty)$ or onto $\mathbb{H} \setminus [0,1]$. Such configurations are important in applications (fluid flow around an obstacle or in special channels).

(v) Since 9.14 we have known that the inside of any simple polygonal loop can be conformally mapped onto any other, and the mapping extends to a homeomorphism between their closures. Bak and Ding [2009] investigate what happens if it is required of such a mapping that it be *analytic* on the boundary (i.e., holomorphic in a neighborhood of the closure). They show that this is only possible if the two polygons are similar and the mapping in that case is affine (i.e., a polynomial of degree 1).

It is a remarkable fact, full of historical and mathematical significance, that a comparably explicit but formally very different recipe can be found for the inverse of the function f in (6), that is, for a conformal map of $\overset{\circ}{R}$ onto $\mathbb{U}$. Such a map is a natural avenue through which the famous *elliptic functions* can enter an exposition of complex analysis, so I propose to briefly outline here how that happens. However, a few ideas from Chapter XI are needed for any such undertaking, so finishing the present chapter might be better deferred until some basic concepts from XI have been studied. Briefly, for the function $\mathfrak{P}$ of 11.47(iv), with $p_1 := 2\alpha, p_2 := 2\beta i$ $(\alpha, \beta > 0)$ one can directly prove that it maps $(0, \alpha) \times (0, \beta)$ conformally onto $-\mathbb{U}$. This is done in section 22 of MARKUSCHEWITSCH [1967]. Here I will explore a different way. It is convenient to consider, not f^{-1}, but $(f \circ \phi)^{-1}$, that is, study the conformal mapping of $\mathring{R}$ onto $\mathbb{D}$, and to simplify the notation employed earlier with f. In the following it will be

$$(40) \qquad 0 < L < \infty, \quad R := [0, 2L] \times [0, 2].$$

The goal is to find explicitly, in terms of L only, a conformal map of $\mathring{R}$ onto $\mathbb{D}$. Interestingly, we need an existence theorem to confirm that our concrete candidate works! 9.14 supplies a homeomorphism $F : R \mapsto \overline{\mathbb{D}}$ which maps $\mathring{R}$ conformally onto $\mathbb{D}$. [Alternatively we could use for F the inverse of the function $f \circ \phi$ constructed by the Schwarz-Christoffel method.] We adjust F so that

$$(41) \qquad F(L + i) = 0.$$

The first phase of the program is to extend F (in three steps) to a *meromorphic function* in $\mathbb{C}$.

I: Setting $R_1 := [0, 2L] \times [-2, 2]$, we can use a variant of the (Schwarz) reflection principle embodied in 5.40(i) and 2.4 in which reflection in the range space is over $\mathbb{T}$ rather than $\mathbb{R}$, $z \mapsto 1/\bar{z}$ (more vividly, $re^{i\theta} \mapsto r^{-1}e^{i\theta}$), to extend F to a continuous function F_1 on $R_1 \setminus \{L - i\}$ which is holomorphic in $\mathring{R}_1 \setminus \{L - i\}$. The recipe for F_1 is

$$(42) \qquad F_1(w) := \frac{1}{\overline{F}(\bar{w})} \text{ for } w \text{ in the southern half of } R_1 \setminus \{L - i\}.$$

Due to (41), F_1 as just defined will experience a *pole* at $L - i$. That pole is simple because the zero of F at $L + i$ is simple (F being one-to-one):

$$(43) \qquad L - i \text{ is the unique pole of } F_1 \text{ in } \mathring{R}_1 \text{ and it is simple.}$$

The recipe (42) for F_1 and the fact that F maps ∂R onto $\mathbb{T}$ show that

$$(44)_1 \qquad F_1(x - 2i) = \frac{1}{\overline{F}(x + 2i)} = F(x + 2i) = F_1(x + 2i) \quad \forall x \in [0, 2L].$$

We next reflect over the western edge of R_1, thereby getting a function F_2 which is continuous in $R_2 \setminus \{L \pm i\}$ and holomorphic in its interior, where $R_2 := [-2L, 2L] \times [-2, 2]$. The recipe is

$$(45) \qquad F_2(w) := \frac{1}{\overline{F}_1(-\bar{w})} \text{ for } w \text{ in the western half of } R_2 \setminus \{-L \pm i\},$$

with $F_2(-L - i)$ being 0. Evidently

$$(46) \qquad -L + i =: p \text{ is a simple pole of } F_2, \text{ as is } L - i = -p.$$

Since $F_1(\partial R_1) \subset F(\partial R) = \mathbb{T}$ and F_2 extends F_1, we see that

$$\begin{aligned} F_2(-2L + iy) &\overset{(45)}{=} \frac{1}{\overline{F}_1(2L + iy)} \\ (47) \qquad &= F_1(2L + iy) = F_2(2L + iy) \quad \forall y \in [-2, 2]. \end{aligned}$$

Also, since F_2 extends F_1, $(44)_1$ now tell us that

$$(44)_2 \qquad F_2(x - 2i) = F_2(x + 2i) \qquad \forall x \in [0, 2L].$$

On the other hand, for $x \in [-2L, 0]$, $-x \in [0, 2L]$, so we also have

$$\begin{aligned} F_2(x - 2i) &\overset{(45)}{=} \frac{1}{\overline{F}_1(-x - 2i)} = F_1(-x - 2i) \overset{(42)}{=} \frac{1}{\overline{F}(-x + 2i)} \\ &= \frac{1}{\overline{F}_1(-x + 2i)}, \text{ since } F_1 \text{ extends } F, \\ &\overset{(45)}{=} F_2(x + 2i) \qquad \forall x \in [-2L, 0]. \end{aligned}$$

Combined with $(44)_2$ this gives

$$(48) \qquad F_2(x - 2i) = F_2(x + 2i) \qquad \forall x \in [-2L, 2L].$$

Since opposite edges of R_2 are oppositely oriented in the loop ∂R_2, it follows from (47) and (48) that $\int_{\partial R_2} F_2 = 0$. This equation and 11.29 tell us that the sum of the (two) residues of F_2 in $\overset{\circ}{R}_2$ is 0, that is, the poles being simple,

$$(49) \qquad r := \operatorname{Res}(F_2, p) = -\operatorname{Res}(F_2, -p) \neq 0.$$

[Strictly speaking, we need to apply 11.29 to $\tau \overset{\circ}{R}_2$ and let $\tau \uparrow 1$.] We make one final grand extension to the whole complex plane. In what follows, it is expedient to consider functions as taking values in $\mathbb{C}_\infty$ with value ∞ assigned at poles, and the conventions $1/0 := \infty$, $1/\infty := 0$. (See the Chapter I Notes.) Thereby F_2 becomes a continuous function on R_2. Define

(50) $$\Lambda := 4L\mathbb{Z} + 4i\mathbb{Z} \quad\text{and}\quad \Omega := \mathbb{C}\setminus(\pm p + \Lambda).$$

Thanks to (47) and (48), F_2 is $4L$-periodic and $4i$-periodic to the extent that this makes sense in R_2, that is, on its edges. This empowers us to extend F_2 to $\mathbb{C}$ by Λ-periodicity: Given $w \in \mathbb{C}$ there are unique $n, m \in \mathbb{Z}$ such that

$$-2L \le \operatorname{Re} w - 4nL < 2L \qquad\text{and}\qquad -2 \le \operatorname{Im} w - 4m < 2.$$

That is, there is a unique $\lambda := 4nL + 4mi \in \Lambda$ such that

(51) $$w - \lambda \in [-2L, 2L) \times [-2, 2), \text{ and we } \textit{define } F_3(w) := F_2(w-\lambda),$$

getting a function on $\mathbb{C}$ which is evidently Λ-periodic and coincides with F_2 on R_2. The *restriction* of F_3 to each Λ-translate of R_2 is thus continuous. Any $w \in \mathbb{C}$ lies in at most four Λ-translates of R_2, so the continuity of these restrictions at w implies the full continuity of F_3 at w. Evidently F_3 is also holomorphic in each Λ-translate of $\overset{\circ}{R}_2 \setminus \{\pm p\}$. The rest of Ω consists of the horizontal and vertical line segments determined by Λ. Therefore by 5.40(iii), F_3 is $\mathbb{C}$-differentiable at each point of these lines as well. And evidently too F_3 has simple poles at points of $\pm p + \Lambda$ and no others.

II: The next phase of the program is to find an *explicit* Λ-periodic meromorphic function with these poles and residues. Our candidate is the (elliptic) function rM, where for $w \in \Omega$

(52) $$M(w) := \frac{1}{w-p} - \frac{1}{w+p} + \sum_{\lambda\in\Lambda\setminus\{0\}} \left[\frac{1}{w-p-\lambda} - \frac{1}{w+p-\lambda} - \frac{2p}{\lambda^2}\right],$$

By this sum is meant the limit as $K \to \infty$ of the finite sums over λ with $0 < |\lambda| < K$. Granted this convergence, the function rM visibly has the same poles and *principal part* at each that F_3 does, so $rM - F_3$ is an entire function. If further M is Λ-periodic, then the Λ-periodic function $rM - F_3$ is bounded in $\mathbb{C}$, by its bound on the compactum R_2, hence by Liouville it is constant, say C. Since $0 \overset{(41)}{=} F(L+i) = F_3(L+i)$, we evaluate at $L+i$ to find C and therewith $F_3 = rM - rM(L+i)$. Since $F_3 = F$ maps R onto $\overline{\mathbb{D}}$, this last equation says $M - M(L+i)$ maps R onto $(1/r)\overline{\mathbb{D}}$. As boundary points map to boundary points, we have only to evaluate $|M - M(L+i)|$ at any boundary point of R, say at 0, to find the radius $1/|r|$ of this image disk. In summary, given $L > 0$, the function $M - M(L+i)$ defined by (52) explicitly, in terms of

L only (recall (46)), maps the rectangle $(0, 2L) \times (0, 2)$ conformally onto the disk of radius $|M(0) - M(L + i)|$ centered at 0.

In chapter VI, §3 NEHARI [1952] carries out the preceding program but realizes the function M as an infinite product, rather than an infinite series.

III: Let us confirm the needed properties of M. The role of the third term in the summand $[\cdots]$ in (52) is to guarantee convergence. To see how, first check that

$$\frac{\lambda^3[\cdots]}{2p} = \frac{2w + (p^2 - w^2)/\lambda}{1 - 2w/\lambda - (p^2 - w^2)/\lambda^2} \qquad \forall w \in \Omega,\ \lambda \in \Lambda \setminus \{0\}.$$

Clearly this quotient converges to $2w$ as $|\lambda| \to \infty$, and uniformly for w in any compact $K \subset \Omega$. Hence it is a bounded function of $w \in K$, $\lambda \in \Lambda \setminus \{0\}$. Consequently, the series (52) converges uniformly for $w \in K$; details in the 11.47(iv) hints. It should be clear that the meromorphic function thus defined has first-order poles of residue 1 at points of $p + \Lambda$ and residue -1 at points of $-p + \Lambda$. That M is Λ-periodic is trickier to prove, due to that convergence-producing third term in $[\cdots]$. Eliminate it by differentiating (termwise, via 5.44(ii)), to see that $M(w + 4i) - M(w)$ and $M(w + 4L) - M(w)$ both have derivative 0, and so are constant in the connected set Ω, say c_i and c_L, respectively. For example,

$$M'(w) = \sum_{\lambda \in \Lambda} [-(w - p - \lambda)^{-2} + (w + p - \lambda)^{-2}],$$

and $M'(w + 4i)$ is this same sum over λ in the index set $-4i + \Lambda$. As $-4i$ lies in the additive *group* Λ, this latter index set equals Λ also. (Bear in mind too that, convergence being absolute, order of terms is immaterial to the sum.) Expressing one sum as being over $\lambda \in \Lambda \setminus \{0\}$ and the other as being over $-\lambda \in \Lambda \setminus \{0\}$, shows that M is an even function. Take $w = -2i$ in the identity $M(w + 4i) - M(w) = c_i$ to conclude that $c_i = 0$. Then take $w = -2L$ in the identity $M(w + 4L) - M(w) = c_L$ to conclude that $c_L = 0$, thereby completing the proof that M is Λ-periodic, and has all the properties claimed!

Notes to Chapter IX

For some of the history of this famous theorem see WALSH [1973], A. SINCLAIR [1981], J. J. GRAY [1994], ULLRICH [1996], pp. 181–187 and 197–198 of REMMERT [1998], pp. 320–321 of HILLE [1962], §74 of DINGHAS [1961], pp. 49–53 of SIMONART [1931] and the encyclopedia articles of LICHTENSTEIN [1919] (chapter IV) and BIEBERBACH [1921]. In CARATHÉODORY [1912a] appears the method (used in treating the $\phi_{n,p}$ in the proof of 9.7, as well as in the proofs of 9.8(iii) and 9.22) of "variable regions", showing that when the regions converge geometrically, the Riemann maps from $\mathbb{D}$ onto them converge locally uniformly. For more on this see BIEBERBACH [1913a], [1914], COURANT [1914], GAIER [1956], and pp. 283–285 of POMMERENKE [1975] and the references cited there. The result 9.6, a special case of the above, affirms roughly that if Ω is close to a disk, then its Riemann map is close to the identity. CARATHÉODORY [1914] gives a version of 9.6 and uses it in a beautiful, self-contained proof of 9.7. (See pp. 61–72 of SIMONART [1931] for a French exposition.) Koebe's proof is in his paper [1915], having been announced and sketched in [1912]. He called this method the "Schmiegungsverfahren" (osculation method). For other constructive proofs see LEJA [1935], LEHTO [1949], POMMERENKE [1965], and JULIA [1927a]; compare also pp. 339 ff. of HILLE [1962] and the references there. OSTROWSKI [1929a] (also HENRICI [1983]) shows that not only does $R_n \to 1$ ((8) in the proof of 9.7), but in fact

$$1 - R_n < \frac{128|\log R_1|}{1 - R_1} \cdot \frac{1}{n}$$

for all $n \geq 512|\log R_1|$. See also VEECH [1967], p. 112. For another proof of 9.6 (based on area) see pp. 10–11 of BIEBERBACH [1931]; also VEECH [1967], pp. 99 ff., OSTROWSKI [1930], M. MÜLLER [1938], LAVRENTIEFF and KWESSELAWA [1940] and SAKASHITA [1956] for improvements on the majorant.

The proof in 9.9(i), (ii) appears in T. RADÓ [1922/23b]; see also OSTROWSKI [1929a]. The Riemann map is also the solution to many other extremum problems; for some of them see chapter VI of JULIA [1950]. For example (REMMERT [1998], pp. 178–179 or NARASIMHAN [1985], pp. 148–50), in the 9.9 proof we could choose to maximize $|f(p)|$ at some $p \neq 0$, instead of maximizing $|f'(0)|$, or among all conformal maps f of Ω into $\mathbb{C}$ with $f'(0) = 1$ we could choose one which minimizes the area of $f(\Omega)$ (BIEBERBACH [1914]—this will be treated in chapter 18). The Riemann map also maximizes the hyperbolic distance between $f(z_1)$ and $f(z_2)$ over all conformal maps f of Ω into $\mathbb{D}$, for each pair z_1, $z_2 \in \Omega$. As BEAR [1991] points out, this maximal spreading of the range explains intuitively why this map turns out to be surjective.

For other accounts of the fundamental theorem see BIEBERBACH [1914], LINDELÖF [1914], [1915], [1920], MONTEL [1917], FABER [1922], CARATHÉODORY [1928], [1948], [1952], chapter 10 of DE LA VALLÉE POUSSIN [1949], JULIA [1950], GARABEDIAN [1976] and HENRICI [1986]. For a constructive proof in the rigorous sense of that term, see BISHOP and BRIDGES [1985]. An unusual modern proof is that of William Thurston, as simplified by HE and SCHRAMM [1996].

The whole problem of conformal mapping, and Riemann's theorem in particular, can be treated from a Hilbert space [of areally square-integrable holomorphic functions] point of view. For a definitive account of this procedure see

BERGMAN [1970] together with LEHTO [1949], and for short and very readable introductions STONE [1960], §17.2 of HILLE [1962], or EPSTEIN [1965]. The latter is a completely *ab ovo* treatment requiring of the reader no knowledge of either Lebesgue integration or Hilbert space theory. From an orthonormal basis for the aforementioned Hilbert space the derivative of the mapping function can be constructed—*mirabile dictu*. A beautiful short proof of a different kind with a Hilbert-space flavor was given by GARABEDIAN [1991].

§48 of LICHTENSTEIN'S encyclopedia article [1919] is a good source for the history of the fundamental mapping theorem 9.14. There are several approaches: centered around CARATHÉODORY [1913a,b], [1952]; COURANT [1914] and KOEBE [1913a], [1915]; OSGOOD and TAYLOR [1913]; LINDELÖF [1914], [1915]. The first uses cross-cuts in Jordan regions and a famous theorem of Fatou that $\lim_{r\uparrow 1} f\left(re^{i\theta}\right)$ exists in $\mathbb{C}$ for almost all (Lebesgue measure) $\theta \in [0, 2\pi]$ whenever f is a bounded holomorphic function in $\mathbb{D}$. Actually only the case of univalent f is needed and for this CARATHÉODORY [1913b], §2 gives an elegant proof using area ideas that avoids Lebesgue integration. Area arguments also dominate in the second method above in which it is shown by *reductio ad absurdum* that the Riemann map must be uniformly continuous. (See also DE LA VALLÉE POUSSIN [1930], WOLFF [1930a], TSUJI [1930a], [1959], MCSHANE [1937], LELONG-FERRAND [1955], ARSOVE [1968a], NOVINGER [1975a], and HENRICI [1986]. When $\partial\Omega$ is a C^2-curve, the Riemann map is even Lipschitz; see RANGE [1976].) The third method is old-school potential-theory arguments. Lindelöf's is closer in spirit to Carathéodory's. (See pp. 372–373 of LICHTEINSTEIN, *loc. cit.*) In 1950 L. Ahlfors and A. Beurling used yet another method, involving a construct called the *conformal modulus* or *extremal length*. This is very clearly and elementarily set out in PALKA [1991], pp. 426–447. These ideas are made constructive (recipes provided for the various ε and δ) in MCNICHOLL [2016]. For other accounts of the mapping theorem see §§5–8 of STUDY [1913], MONTEL [1917] (also chapter IV of [1927]), FABER [1922], DE LA VALLÉE POUSSIN [1932], [1949], DOUGLAS [1931], FLOYD [1946], KLEINER [1955], [1962] and COLLINGWOOD and PIRANIAN [1964]. SCHWINGENDORF [1980] gives a modern functional analysis proof of 9.14. The treatment in the text is based on Carathéodory and on LINDELÖF's exposition [1920]. It uses, I believe, the fewest auxiliary results (e.g., no Lebesgue integration) and no geometric facts which cannot be precisely formulated and proven. Many of the older proofs abound with vague arguments which the demanding reader will doubtlessly deem deficient. But such a reader will probably also notice that elegance has taken a hit here.

Carathéodory himself was critical of the cavalier proof which Koebe offered for his fundamental lemma 9.10 and gave (in [1913b]) another (KOEBE rejoins on pp. 212–213 of [1915] and more on their competition will be found on p. 197 of REMMERT [1998]); that in the text is adapted from GOLUZIN [1969]. For other proofs see COURANT [1914b], MYLLER-LÉBÉDEFF [1938], TSUJI [1959], p. 302 and for generalizations see §19 of MONTEL [1917], MAITLAND [1939], FLETT [1954], §5 of BAGEMIHL and SEIDEL [1960], [1961], BAGEMIHL [1961], GAVRILOV [1965], CHOU [1967], RUNG [1968] and POMMERENKE [1975]. R. NEVANLINNA [1923], p. 283 claims that 9.10 can be derived from a version of his "two-constants" theorem (cf. 5.45). It should be noted that 9.10 has other uses; for example, it features in an alternative proof of Lindelöf's theorem 5.56 (cf. HERVÉ [1982]).

There are still other proofs that g is univalent (lemma 9.13). See, e.g., chapter 6 of HEINS [1962b], where the solubility of the Dirichlet problem is utilized. The elegant path followed by Montel and Lindelöf is that in the second proof in the text. Instead of 4.50 there, we can use the idea of locally sequentially accessible boundary point from ARSOVE [1968a]. He proves that every point of the Jordan-curve $\partial\Omega$ has this property, and the definition makes it clear that every such point is the endpoint of a curve in Ω.

The reader may note that the maps f and $g = f^{-1}$ are treated by different techniques in the proof of the Osgood-Taylor-Carathéodory theorem. It should be clear, however, that if lemma 9.10 is freed from the disk, then the technique of 9.11 could be used on g as well as f. This approach is offered in FLETT [1954]. It rests on an inequality of CARLEMAN [1926]. A symmetrical treatment of f and g is also offered by BEARDON [1979].

Carathéodory (see also STUDY [1913]) in his theory of "prime ends" also made definitive investigations into the boundary behaviour of non-Jordan regions. An account may be found in POMMERENKE [1975], [1992] or COLLINGWOOD and LOHWATER [1966]. DENJOY [1941], [1942] also investigated continuous, not necessarily univalent, extensions of the Riemann map to the boundary, in case the image set has finite area.

Another important problem is "conformality" of the Riemann map at the boundary (cf. 6.28) when various smoothness hypotheses are placed on the bounding curve. For this vast area of research I shall be content to cite chapter 10 of POMMERENKE [1975], his book [1992], LELONG-FERRAND [1955] and the two-volume work of GATTEGNO and OSTROWSKI [1949] and its bibliographies. Already in the 19th century the special case of regions bounded by C^∞-curves had been investigated (e.g., by P. Painlevé). A readable modern account of this theory will be found in BELL and KRANTZ [1987], and GREENE and KRANTZ [1997]. Finally there is the problem of constructibility of conformal maps, especially important in (extra-mathematical) applications. Again there is an enormous literature and I cite only five references: SEIDEL [1952], GAIER [1964], OPFER [1974], the special issue volume 14 (1986), no. 1–2 of the *Journal of Computational and Applied Mathematics*, and KYTHE [1998]. The bibliography of Gaier's book (consisting of 490 items, equipped with FM,

Zbl and MR coordinates) covers the field very well. It also will inform the reader on other approaches to the mapping problem not discussed here. See also **MR** 2005m:30008.

For more literature on boundary behavior see the references T. 62 on p. 130 of BERNARDI [1982], the book of COLLINGWOOD and LOHWATER [1966] and its bibliography. POMMERENKE [1980] is a nice survey and his book [1992] is an up-to-date monographic treatment.

The effect of 9.14 is to concentrate a lot of geometric difficulties in one place. Now to prove theorems involving boundary behavior of an arbitrary holomorphic function in an arbitrary Jordan region, instead of confronting all the unpleasantries of proof associated with an arbitrary Jordan region, one can prove the result for disks or some other geometrically simple region and transfer them to the general region by citing 9.14. E.g., the limiting behavior of a bounded function along two sides of an angular sector is easy to analyze but that along two abutting arcs of a Jordan-curve is not (see, e.g., pp. 226–228 of DINGHAS [1961]); so we reduce the latter to the former, as we did in 9.20. For an intriguing application of 9.14 to a problem in Fourier theory (theorem of Bohr and Pál), see p. 127 of ZALCMAN [1974].

9.14 also permits the introduction for a Jordan region Ω of a sequence of polynomials $\{P_n\}$, called *Faber polynomials*, which play for the region Ω a role analogous to that played by the monomials z^n in the unit disk $\mathbb{D}$: for certain Ω, each $f \in H(\Omega)$ has a unique expansion $\sum c_n P_n$ $(c_n \in \mathbb{C})$ which converges locally uniformly in Ω to f. The literature on Faber polynomials is considerable; the papers of SZEGÖ [1921], HEUSER [1934], J. H. CURTISS [1971], SUETIN [1964] and their bibliographies should get the interested reader started.

9.14 together with the Riemann mapping theorem provides the following important characterization of Jordan regions: a subset of $\mathbb{C}$ is a Jordan region if and only if it is the image of $\mathbb{D}$ under a homeomorphism of $\overline{\mathbb{D}}$. R. L. MOORE 1918] has given a wholly internal characterization: Ω is a Jordan region if and only if it is a bounded, simply-connected region which is uniformly connected *im kleinen*, i.e., for every $\varepsilon > 0$ there is a $\delta(\varepsilon) > 0$ such that any two points of Ω closer together than $\delta(\varepsilon)$ lie in a connected subset of Ω of diameter less than ε. (Cf. 4.50 and the Scholium following 9.14.) See also ARSOVE [1967], §9.3 of POMMERENKE [1975], and pp. 440–445 of PALKA [1991].

For more applications of the Argument Principle see exercises III. 181–194 of PÓLYA and SZEGÖ [1972]. The last of these is Rouché's theorem for Jordan regions (weaker than our 8.18). See also p. 193 of SAKS and ZYGMUND [1971]. For a generalization of it see MONTEL [1932b] and [1933a]; LIPKA [1929] and TERASAKA [1929]; KERÉKJÁRTÓ [1934/35]. A version of the argument principle for complex-valued harmonic functions is formulated and proved (along with some corollaries, like an open-mapping theorem) in DUREN, HENGARTNER, and LAUGESEN [1996].

If the non-constant holomorphic function f maps a to w, then $\mathrm{Ind}_{f\circ C(a,r)}(w)$ is non-zero and independent of $r > 0$ as long as r is sufficiently small, and in fact (5.86) it equals the (algebraic) multiplicity of a. This inspires a definition of multiplicity for continuous functions f which are not necessarily holomorphic: If a is an *isolated* zero of $f - w$, we can define the multiplicity of a to be $\mathrm{Ind}_{f\circ C(a,r)}(w)$ ($r > 0$, sufficiently small); this is an integer, but unlike the situation with holomorphic functions, it can be negative. A *topological* argument principle is easily formulated and proved. See T. RADÓ [1936] (cf. 4.59(iii) above) and NEEDHAM [1997].

For extensions of 9.16 to complex-valued harmonic functions see H. KNESER [1926], CHOQUET [1945], [1994] and BSHOUTY and HENGARTNER [1992], [1993]. Choquet proves the following: if Ω is a convex domain bounded by a Jordan-curve and $\phi\colon \mathbb{T} \mapsto \partial\Omega$ is an orientation-preserving homeomorphism, then

$$f(z) := (2\pi)^{-1} \int_0^{2\pi} (1 - |z|^2)|e^{it} - z|^{-2}\phi(e^{it})\, dt,$$

the unique harmonic extension of ϕ into $\mathbb{D}$, maps $\mathbb{D}$ one-to-one onto Ω; see esp. DUREN [2004]. For original credits on 9.16 see footnote 106 of OSGOOD [1901]. A significant topological extension (based on 4.61) was made by MEISTERS and OLECH [1963]; see also G. T. WHYBURN [1951], lemma 1 and STOÏLOW [1956]. For a different extension see BAGEMIHL [1967]. A very elementary, purely topological version of 9.16 may be found in DERRICK [1973]. T. SATÔ [1936] proves a variant of 9.16: if $f(\mathbb{T})$ is a Jordan-curve and f' is zero-free in $\mathbb{D}$, then f is one-to-one in $\mathbb{D}$; see also NOSHIRO [1937] and 10.12 below. ČERNÝ [1992], p. 219 extends 9.16 to meromorphic functions.

9.19 was proved in SCHÖNFLIES [1906], pp. 319–324 and in the longer work [1908], pp. 209–213, modulo some errors. See also BIEBERBACH [1913b], TIETZE [1913], J. R. KLINE [1920], ANTOINE [1921], R. L. MOORE [1926], GEHMAN [1926], NÖBELING [1950], CAIRNS [1951] (proving the Schönflies and Jordan Curve Theorems together), ČERNÝ [1977], CHOQUET [1945], and p. 198 of GUGGENHEIMER [1977]. A series of papers by W. Huebsch and M. Morse (circa 1960) treats the higher-dimensional case exhaustively. For more recent work see DOUADY and EARLE [1986], and **MR** 2007e:57011 for more history. A corresponding extension is possible for homeomorphisms of intervals in $\mathbb{C}$, i.e., for arcs. One first extends the arc to a simple loop (§7 of FEIGL [1928] or p. 164 of M. H. A. NEWMAN [1957]); see p. 214 of ČERNÝ [1992]. Extending the arc employs ideas similar to those in 4.50.

A variant of 9.22 with the weaker hypothesis that each f_{n+1} be subordinate to f_n is elegantly proved in POMMERENKE [1960].

For more material related to 9.26 see JULIA [1927], [1928], WALSH [1929], FARRELL [1932], LEJA [1936] and the monographs of WALSH. It is but a small step from 9.26 to Weierstrass' polynomial approximation theorem on an arbitrary Jordan-curve: If J is a Jordan-curve with $0 \in \mathfrak{I}(J)$, then every continuous function on J is uniformly approximable by polynomials in z and $1/z$. (See 13.8.) Of course the definitive result here is Mergelyan's theorem mentioned in the Chapter VIII notes. The reader may feel that we paid too high a price for theorem 9.26; especially if he is versed in abstract measure and integration theory, he could buy the full Mergelyan theorem with a comparable expenditure of energy. *Nolo contendere.* The functional analytic machinery for Mergelyan's theorem has even been adapted to give proofs of Walsh's theorem, e.g., by H. KÖNIG [1975] and SCHWINGENDORF [1980].

The *raison d'être* for Cauchy's theorem in the form 9.27 is mainly esthetical: one seldom needs this general a version of the theorem in applications (see, however, 18.46(iv)) and for the special configurations in which one does, more elementary *ad hoc* arguments usually suffice. However, if one does want the theorem and is not willing to go all the way to Walsh's result 9.26, then the proof of BECKENBACH [1943] or DELANGE [1956] is the best. The theorem proved there is a little less general than 9.27, however: the loop Γ must be the boundary of Ω. The most general Cauchy Theorem that I know of in the literature is NÖBELING [1949]: *If* Γ *is a rectifiable loop and* Ind_Γ *is bounded in* $\mathbb{C}\backslash\Gamma$, f *is holomorphic in* $\{z : \mathrm{Ind}_\Gamma(z) \neq 0\}$ *and continuous on the union of this set with* Γ, *then* $\int_\Gamma f = 0$; see also HARMSE [2008]. Many proofs of 9.27 have appeared in the literature; few are both artistically and logically satisfactory. Here is a partial roll call: PUCCIANO [1909], WATSON [1914], POLLARD [1923] (heavy), LEAU [1926], ČECH [1930], RIDDER [1930a], [1941], KAMKE [1932], HEILBRONN [1933], DENJOY [1933], [1955], ESTERMANN [1933], HU [1935], FEDERIGHI [1938], GONÇALVES [1940], [1941], [1942] (the *Zentralblatt* reviewer was never convinced), HESTENES [1941], W. T. REID [1941], MINAMI [1942], LOOMIS [1944], DEURING [1949], H. KNESER [1949], M. H. A. NEWMAN [1951], CRAVEN [1964], GRUNSKY [1966], ARSOVE [1968b], [1969]. For a more complete list (including some Chinese references) and some comments see FALLIN and GOULD [1965].

An inchoate version of 9.28 occurs in §2 of Riemann's dissertation. (But see §2.3 of ULLRICH [1994c], where work of Axel Harnack is described.) An example in §5 of ZALCMAN [1974] illustrates that some sort of smoothness hypothesis on the cross-cut in 9.28 is necessary. 9.28 is just the most prominent product of a big industry concerned with the study of "removable" sets (of singularities). See POMPEIU [1905a] and chapter V of ZORETTI [1911] for more on this theme. Cf. also BESICOVITCH [1931], VASILESCO [1930], the monograph of CARLESON [1967] and ZALCMAN [1968b].

The proof technique in 9.30 comes from ZAREMBA [1905]. In the 19th century, just as continuity of f' had to be hypothesized for holomorphic functions (see Chapter V notes), so too the continuity of the partial derivatives in 9.31 had to be hypothesized for harmonic functions. ZAREMBA [1905] showed that the latter hypothesis could be dropped. See WILKOSZ [1922], LOOMAN [1923–24], E. HOPF [1930] and ZAREMBA [1939]. However, WILKOSZ [1922] gives an example (due to Zaremba; see also ZAREMBA [1939]) of a discontinuous function h whose Laplacian exists and is identically zero: $h(x, y) := xy/(x^2 + y^2)^2$ if (x, y) is not $(0, 0)$, $h(0, 0) := 0$. If h is bounded, this cannot happen though (TOLSTOV [1951]).

On the other hand, relative to 9.31, the requirement that the Laplacian exist and equal zero throughout the region can be considerably weakened and the harmonicity of the function still inferred. See RIDDER [1940], [1946], BECKENBACH [1945] and ARSOVE [1955], as well as references cited in the Chapter II notes.

9.32 is due to F. RIESZ [1926]. An attractive relative of it involving no differentiability hypotheses is due to MONTEL [1928]: If f is continuous in open convex U and $x \mapsto f(x, y)$, $y \mapsto f(x, y)$ are all concave functions on the line segments in U on which they are defined, then f is superharmonic in U. In fact, the convexity hypotheses already insure that f is (jointly) continuous. See T. RADÓ [1937], §3.22.

The ultimate explication of the phenomenon in 9.33(vi) may be the results of E. REICH [1987]. The general (and elementary) fact that underlies it is this: If I is an open interval in $\mathbb{R}$, Ω a region in $\mathbb{C}$ and $f\colon I \mapsto \mathbb{R}$, then $f \circ u$ is subharmonic for every subharmonic $u\colon \Omega \mapsto I$ if and only if f is convex and non-decreasing on I, and $f \circ h$ is harmonic for every harmonic $h\colon \Omega \mapsto I$ if and only if f is affine. See SAKS [1930–32]. An alternative proof of 9.33(v) with a geometric flavor was given by AMDEBERHAN [1999], and the case of the complex-valued harmonic functions is treated by MORTINI [2002].

HEINS [1968], p. 308 shows in a very elegant and elementary way how 9.38 can be combined with the solvability of the Dirichlet problem in an annulus (§7 of chapter XI) to solve the Dirichlet problem in the complement of finitely many disjoint closed disks, without involving the approximation arguments in 9.40.

The solution of the Dirichlet problem has a remarkable extremality property which in many cases characterizes it. For a region Ω, let $C^{(2)}(\Omega)$ denote the class of continuous real-valued functions on $\overline{\Omega}$ which have continuous first and second partial derivatives in Ω. For such a function u the (possibly infinite) double integral $\int\int_\Omega [(D_1 u)^2 + (D_2 u)^2]$ is denoted $D(u)$ and called the *Dirichlet integral* of u. If $h \in C^{(2)}(\Omega)$ and h is harmonic in Ω, then $D(h) \le D(u)$ for all other u having the same boundary values as h, and (uniqueness) if $D(h)$ is finite, then the inequality is strict. A proof for Ω a rectangle appears as lemma 3 of COURANT [1914] and for Ω a disk in MARKUSCHEWITSCH [1965], II.172–173, by very different methods. Thanks to 9.14 either of these cases effectively covers all Jordan regions. Early in the game it was thought that a solution of the Dirichlet problem is always insured by this property (starkly put, a necessary condition was assumed to be sufficient): Given continuous $f : \partial\Omega \mapsto \mathbb{R}$, one had only to select from the $u \in C^{(2)}(\Omega)$ which agree on $\partial\Omega$ with f (do such u always exist?) one that minimizes $D(u)$. That a minimizing u is necessarily harmonic in Ω if $D(u) < \infty$ is pretty easy to prove; see pp. 279–280 of KELLOGG [1929]. Around 1870 examples of f and Ω were found for which all such $D(u)$ are infinite, and Weierstrass devised an analogous 1-dimensional extremum problem where no minimizing function exists. [The infimum over all continuously differentiable $u : [-1, 1] \mapsto \mathbb{R}$ satisfying $u(\pm 1) = \pm 1$ of the functional $J(u) := \int_{-1}^{1} x^2(u'(x))^2\,dx$ is 0, but $J(u) = 0$ for no such u.] This method, called the *Dirichlet Principle*, consequently fell out of favor! But it was later rehabilitated under stricter hypotheses by Hilbert and Courant. See MONNA [1975] or the Appendix in BOTTAZZINI [1986].

By now we have seen both integral and differential characterizations of harmonic functions. There are many more, including areal mean-value properties and mixtures of integral and differential properties. See, e.g., BLASCHKE [1916], [1918] and SAKS [1932].

Laplace's equation is very prevalent in classical physics and the Dirichlet problem is one of several kinds of boundary-value problem for partial-differential equations which are important there. Laplace's equation arises as follows: $h(x, y)$ measures a quantity (energy, fluid mass, heat) at the point (x, y) and the net flow into any region of the plane equals the net flow out, i.e., a steady state or an equilibrium state exists. This flow across a curve is often proportional to the normal derivative of h with respect to the curve. If one records this information in the form of an integral over a square of the normal derivative of h (in the case of a square parallel to the axes this is $\pm D_1 h$ or $\pm D_2 h$) being equal to 0, divides by the area of the square and lets the side-length decrease to 0, there emerges (as in the hints to 2.19(ix)) the equation $\Delta h = 0$. [An alternative derivation based on Morera's Theorem is sketched in 10.8.] Actually, as 10.8 shows, harmonicity is completely characterized by the vanishing of all such integrals or by the vanishing of the integral of the normal derivative over all (small) circles. See KOEBE [1906b], BÔCHER [1906], SEVERINI [1911], G. C. EVANS [1921], [1928], SAKS [1932], and RIDDER [1946].

Laplace's equation and the whole subject of potential theory were intensively investigated in the second half of the 19th century, but not until early in the

20th century were the methods of analytic function theory brought into the subject on a large scale. The pendulum has now swung even further and the subject is dominated by the methods of real analysis, probability and the theory of distributions. The Dirichlet problem for an arbitrary bounded, simply-connected region was finally disposed of by LEBESGUE [1907]. See also COURANT [1914b], LICHTENSTEIN [1916], pp. 734–739 of OSGOOD [1928], INOUE [1938], LEJA [1950]. The use of superharmonic functions as in the text (descendant from the "méthode de balayage" of Poincaré, 1890) was pioneered by PERRON [1923] and simplified by REMAK [1924], [1926], RADÓ and RIESZ [1925], and WHITNEY [1932]. Similar ideas were used by Wiener and Kellogg about this same time. The development in the text follows KELLOGG [1928], a similar account appearing in his book [1929]. The method of approximating the boundary data with polynomials (as in 9.40), due to Poincaré, also occurs in PICONE [1926a], WALSH [1930] and JULIA [1930a]. For fine expository accounts of the problem see CARATHÉODORY [1937a] and BRELOT [1933], [1938–40]. For a very detailed history of the Dirichlet problem up to about 1915, including outlines of the proofs, the encyclopedia article of LICHTENSTEIN is excellent; the survey article of KELLOGG [1926b] is also valuable. For later developments see ANGER [1961/62], with its bibliography of over 200 entries, and BRELOT [1952], [1972], also equipped with extensive bibliographies.

In AXLER, BOURDON and RAMEY [2001], pp. 226–228, a simpler definition of *barrier* for bounded U is adopted and our *Basic Construction* (following 9.35) is modified, to prove that for any continuous $f : \partial U \mapsto \mathbb{R}$, the function $F : \overline{U} \mapsto \mathbb{R}$ defined by $F(z) := \sup\{u(z) : u$ continuous on $\overline{U}$, subharmonic in U, and $u \leq f$ on $\partial U\}$ is harmonic in U and satisfies $\lim_{z \to b} F(z) = f(b)$ for every $b \in \partial U$ at which a barrier exists. The authors then reach 9.41 very quickly (without involving 9.40 and approximation arguments). Readers are urged to look at this extraordinarily elegant, concise and transparent treatment!

Ideas in 9.50 and 9.51 are from HEINS [1949]. To make the proof of 9.50 constructive, we need to be able, for a simply-connected region Ω and $a \in \Omega$, to construct the function u_a, which essentially means construct the Green's function for Ω with singularity at a. A very clever construction for smoothly bounded Jordan regions is offered in GARABEDIAN [1991]; it makes no appeal to the solvability of the Dirichlet problem in Ω. The last step, getting the function ϕ_a (that is, finding a harmonic conjugate of u_a), amounts to finding a primitive for the holomorphic function $D_1 u_a - i D_2 u_a$ (see 10.8(ii)), and that is just a matter of integrating (see the proof of 2.11). Green's function is utilized in a completely different way in M. LEVI [2007] to prove the mapping theorem.

Chapter X

Simple and Double Connectivity

§ 1 Simple Connectivity

We come now to an "enterprise of great pitch and moment," the climax of much of our previous work. To state the result in its most awesome form we introduce some convenient definitions to supplement those in 4.70 and 9.48.

Definition 10.1 Let U be an open subset of $\mathbb{C}$. Say U has

> *property* (*C*) if the conclusion of Cauchy's theorem holds in U, that is, $\int_\gamma f = 0$ for every $f \in H(U)$ and every piecewise smooth loop γ in U;
>
> *property* (*P*) if every holomorphic function in U has a primitive, that is, every function in $H(U)$ is the derivative of another;
>
> *property* (*AP*) if each holomorphic function in U is locally uniformly approximable on U by polynomials, that is, for each $f \in H(U)$ there exist polynomials p_n such that $p_n \to f$ locally uniformly on U.

Theorem 10.2 (Hauptsatz of the Cauchy Theory) *For any proper subregion Ω of $\mathbb{C}$ the following statements are equivalent:*

R. B. Burckel, *Classical Analysis in the Complex Plane*, Cornerstones,
https://doi.org/10.1007/978-1-0716-1965-0_11

(a) Ω *is conformal to an open disk*
(b) Ω *is homeomorphic to an open disk*
(c) Ω *is loophomotopically-connected*
(d) Ω *is homologically-connected*
(e) Ω *is simply-connected*
(f) Ω *contains the inside of every Jordan-curve in* Ω
(g) Ω *has property* (C)
(h) Ω *has property* (P)
(i) Ω *has property* (HL)
(j) Ω *has property* (HS)
(k) Ω *has property* (HS_1)
(ℓ) Ω *has property* (CL)
(m) Ω *has property* (CS)
(n) Ω *has property* (H)
(o) Ω *has property* (AP)

Proof: Here is one possible cyclic tour, touching all bases:

$$
\begin{array}{ccccccccccccccc}
(n) & \Longleftarrow & (a) & \overset{\text{trivial}}{\Longrightarrow} & (b) & \Longrightarrow & (c) & \overset{4.12}{\Longrightarrow} & (d) & \overset{4.71}{\underset{8.22}{\Longrightarrow}} & (e) & \overset{8.10}{\Longrightarrow} & (o) & \Longrightarrow & (g) \\
9.49\Downarrow & & 9.7\Uparrow & & & & & & & & & & & & \Downarrow 2.11 \\
(i) & \underset{\text{trivial}}{\Longrightarrow} & (k) & \underset{\text{trivial}}{\Longleftarrow} & (j) & \underset{4.67}{\Longleftarrow} & (m) & \underset{\text{trivial}}{\Longleftarrow} & (\ell) & \underset{4.71}{\Longleftarrow} & (f) & \underset{4.71}{\Longleftarrow} & (i) & \underset{5.35}{\Longleftarrow} & (h)
\end{array}
$$

Substantiation of the three unlabelled implications follows.

(a) $\Rightarrow$ (n) According to 5.21(i), every open disk has property (H) and that property is transmitted to Ω via the conformal map supplied by (a).

(b) $\Rightarrow$ (c) Open disks are loophomotopically-connected by 4.16 and homeomorphisms clearly preserve this property.

(o) $\Rightarrow$ (g) Examine the hint to 8.13.

Remarks 10.3: (i) Six other characterizations of simple-connectivity for open sets are provided by 1.38, 10.5, the Aumann-Carathéodory theorem (see §4 below and 12.49), 7.51 (see the discussion in 7.48 and 12.48), (iv) below (the Holomorphic Monodromy Theorem), and 4.38(v).

(ii) Part of the profundity of theorem 10.2 is that some of the properties which it establishes as equivalent (all but (d), (e) and (f)) are entirely internal to Ω, one is entirely external to Ω (namely (e), which concerns the connectedness of the complement of Ω) and one involves both the interior and the exterior of Ω (namely (d), which concerns how curves in Ω behave with respect to points outside Ω—see, however, the remark following 4.8). Moreover, properties (a), (n), (o) and (h)–(k) are "analytic," while the rest are purely "topological". Property (j) can be regarded as an "algebraic" characterization of simple-connectivity, because $f \in H(\Omega)$ zero-free just means that f is an invertible element of the ring $H(\Omega)$. In this connection it should also be noted that for any two regions Ω_1, Ω_2 the $\mathbb{C}$-algebras $H(\Omega_1)$ and $H(\Omega_2)$ are isomorphic if and only if Ω_1 and Ω_2 are conformally equivalent (see chapter XIII and 10.16 below).

(iii) It is desirable to establish the equivalence of all items in 10.2 save (f) via arguments that do not involve the Jordan Curve Theorem. To do this, just drop (f) out of the cycle of implications and use the modified form of 4.71 described in the remarks that follow it. An even better tactic, which eliminates the highly non-trivial 4.71 entirely, is to excise (f) $\Rightarrow$ (ℓ) $\Rightarrow$ (m) from the cycle and close it with the trivial implication (i) $\Rightarrow$ (j). Afterward confirm separately the very easy chain (b) $\Rightarrow$ (ℓ) $\Rightarrow$ (m) $\Rightarrow$ (j).

Note too that 10.9 below furnishes a direct proof that (h) $\Rightarrow$ (n), 9.50 a "direct" proof that (n) $\Rightarrow$ (a), and 1.42 together with 8.6 a direct proof that (o) $\Rightarrow$ (e). A direct proof of (b) $\Rightarrow$ (d) is easily constructed from 4.49(iv), 7.22 provides a direct proof of (d) $\Rightarrow$ (g), and 4.19(ii) one of (m) $\Rightarrow$ (ℓ). Also 7.23(i) and 2.11 together furnish a direct proof of (c) $\Rightarrow$ (h). 8.13 is a "direct" proof that (e) $\Rightarrow$ (g), 4.75(i) is a direct proof of (d) $\Rightarrow$ (ℓ), 5.44(vii) yields a direct proof of (o) $\Rightarrow$ (h), 1.38 contains (b) $\Rightarrow$ (e), and 2.12 proves (o) $\Rightarrow$ (h).

The Basic Integral Representation 8.3 can be obtained as an application of the homological Cauchy theorem 7.22, provided the needed family $\Gamma = (\Gamma_1, \ldots, \Gamma_m)$ of loops "surrounding" K can be produced without recourse to 8.3. Such a re-ordering of the proof of 10.2 is indeed possible. The combinatorial arguments in 8.1–8.3 disappear, but some non-trivial topological ones become necessary, and the Riemann mapping theorem itself is employed in a way similar to (but more complicated than) that in the third proof of 8.21; for details see pp. 111–114 of SCHMIEDER [1993].

(iv) There is another important analytic characterization of simple-connectivity, the (*Holomorphic*) *Monodromy Theorem*. Its enunciation requires a short definition: For a region Ω, curve $\Gamma\colon [0,1] \mapsto \Omega$ and open disk $D_0 \subset \Omega$ whose center is the initial point of Γ, say that a function $f_0 \in H(D_0)$ is *analytically continuable in* Ω *along* Γ if there exist open disks $D_1, \ldots, D_n \subset \Omega$ with $\Gamma(1)$ the center of D_n, $f_j \in H(D_j)$, and $0 = t_0 < t_1 < \cdots < t_n = 1$ such that $f_{j-1} = f_j$ in $D_{j-1} \cap D_j$ and $\Gamma[t_{j-1}, t_j] \subset D_{j-1}$ ($j = 1, \ldots, n$). The Monodromy Theorem asserts that a region Ω is simply-connected if and only if

(M) $\begin{cases} \text{for every open disk } D \subset \Omega, \text{ the set } H(\Omega)\big|D \text{ consists of all } f \in \\ H(D) \text{ which are analytically continuable in } \Omega \text{ along } \textit{every} \text{ curve} \\ \text{in } \Omega \text{ with initial point at the center of } D. \end{cases}$

Notice that if f_0 is the restriction to D_0 of some $f \in H(\Omega)$, then the only possible functions f_j are $f|D_j$. In particular, the continuation process depends only on the terminal point of Γ, and is thus "one-pathed," as suggested by the Greek constituents of the word "monodromic." In spite of the difficulty of verifying that a function is analytically continuable along *every* curve in a region (and the widespread carelessness in doing so—see §13 of ZALCMAN [1974] and §3 of STYER and MINDA [1974]), the property (M) of simply-connected regions can be quite useful for getting globally defined functions from locally defined ones. E.g., it is essential in the proofs described in 7.48 and in proving the appropriate global version of 5.95. If $f \in H(\Omega)$, then it follows from 3.6 that f has local primitives and that by the addition of appropriate constants these provide analytic continuations of one another along any curve in Ω. Therefore a region Ω which satisfies (M) has property (P). Thus, the Monodromy Theorem, born in Weierstrass' theory, enabled him to find (global) primitives while abjuring integrals, as his methodological convictions prompted (see pp. 239–241 of REMMERT [1991]); cf. also 2.12.

The usual proof that simple-connectivity implies (M) uses the homotopy definition of simple-connectivity. One first shows that the final function f_n in the continuation process is the same for all the curves in a fixed-endpoint homotopy. Then one proves (with a little work) that loophomotopic-connectivity of Ω implies that any two curves with common initial and common terminal points are part of such a homotopy. The continuation f_n therefore depends only on the terminal point of Γ and consequently well defines the desired holomorphic extension f of f_0 to Ω. Despite the ubiquity of this approach in textbooks, most presentations lack either precision (as this Remark 10.3 (iv) does) or perspicuity, but both will be found in the accounts in RUDIN [1966], [1987]. An advantage of this (notationally more intricate) approach is that the first of its two steps, the equality of analytic continuation along the curves in a fixed-endpoint homotopy in Ω, does not involve the simple-connectivity hypothesis. This useful fact is itself sometimes called the monodromy theorem. The proof is built on the uniform-continuity and paving ideas in the Remarks following 5.2.

The development of function theory in the present book, emphasizing the plane, has marginalized this Monodromy Theorem. But it becomes central when one does function theory on Riemann surfaces, where functions have to be synthesized out of parts brought over from $\mathbb{C}$ by the local coordinate maps, and where the myriad equivalent topological notions of simple-connectivity for plane sets in 10.2 must give way to a single intrinsic one, loophomotopic-connectivity. Specifically, this synthesizing often proceeds from the following easy consequence of the Monodromy Theorem: *The simply-connected region Ω in $\mathbb{C}$ (or Riemann surface Ω) is covered by subregions Ω_α and subsets $\mathcal{F}_\alpha \subset H(\Omega_\alpha)$ are given such that for any indices α, β and any $f_\alpha \in \mathcal{F}_\alpha$ there is an*

$f_\beta \in \mathcal{F}_\beta$ for which $f_\beta = f_\alpha$ in $\Omega_\beta \cap \Omega_\alpha$. Then there exists $f \in H(\Omega)$ satisfying $f|\Omega_\alpha \in \mathcal{F}_\alpha$ for every α.

Corollary 10.4 *For any open subset of $\mathbb{C}$ statements* (c) *through* (o) *in* 10.2 *are all equivalent.*

Proof: It is easy to see that the whole plane $\mathbb{C}$ satisfies (c) through (o)—the reader can search for the relevant previous theorem or exercise to cite in each case. Therefore we put aside the case $\Omega = \mathbb{C}$.

The equivalence of (d), (e), (f) and (i)–(m) is averred in 4.71, which does not have connectedness in its hypothesis. As to the other properties, we shall show that each is possessed by a non-void open set Ω if and only if it is possessed by each component of Ω. [Our corollary then follows from 10.2.] For (h) and (n) this is obvious. Since a curve in Ω and the range of any homotopy in Ω involving it are connected sets, they lie wholly in one component of Ω, so (g) and (c) are clear. For (o) we can pass back and forth from 8.10 and 10.2 thus: Ω has $(AP) \Rightarrow \Omega$ has (C) by the proof of 8.13 $\Rightarrow$ each component of Ω has $(C) \Rightarrow$ each component of Ω has (AP) by 10.2 $\Rightarrow$ each component of Ω simply-connected (10.2) $\Rightarrow \Omega$ simply-connected by 4.72 $\Rightarrow \Omega$ has (AP) by 8.10.

Exercise 10.5 (i) *Show that a region in $\mathbb{C}$ is simply-connected if and only if either* (a) *its boundary has no bounded component or* (b) *it is bounded and has a connected boundary.*

Hint: 10.2, 1.35 and 1.38(i). [Evidently, this complicated condition is equivalent to the boundary of the region *in* $\mathbb{C}_\infty$ being connected.]

(ii) *Show that an open subset U of $\mathbb{C}$ is simply-connected (if and) only if it contains every bounded (open) set V such that $\partial V \subset U$.*

Hints: 10.4(f) insures that this condition implies simple-connectivity. For the converse cite 1.38(ii).

(iii) *Use* (ii) *to prove that for any simply-connected open $U \subset \mathbb{C}$, $f \in H(U)$, $c > 0$, the* sublevel set $U_c := \{z \in U : |f(z)| < c\}$ *is simply-connected.*

Hints: Suppose the open set U_c has a hole C. Recall fact $(*)$ in the proof of 4.71. Its proof used the intricate 4.38. Here, instead of $(*)$, we need only

$$(**) \qquad \begin{cases} \text{For the hole } C \text{ in } U_c \text{ there is a polygonal loop} \\ \gamma \subset U_c \text{ such that } \operatorname{Ind}_\gamma \neq 0 \text{ in } C. \end{cases}$$

The same proof as before, but now calling on the simpler 4.18, works to confirm $(**)$. Having such a γ, let $V := \{z \in \mathbb{C} \setminus \gamma : \operatorname{Ind}_\gamma(z) \neq 0\}$. This is a bounded open subset of $\mathbb{C} \setminus \gamma$ and it lies in U, since U is simply connected.

Also, $\partial V \subset \gamma$. For if $z \in (\partial V) \setminus \gamma$, then some disk $D(z, r)$ lies in $\mathbb{C} \setminus \gamma$, while meeting V. Due to local constancy of Ind_γ in $\mathbb{C} \setminus \gamma$, this means $\mathrm{Ind}_\gamma \neq 0$ throughout $D(z, r)$; so $D(z, r) \subset V$ contrary to $z \in \partial V$. By the Boundary Maximum Principle 5.12

$$\sup |f(V)| \leq \max |f(\partial V)| \leq \max |f(\gamma)| < c,$$

the last inequality due to γ being a compact subset of U_c. These inequalities place V inside U_c, yet $C \subset V$ and C is a hole in U_c. This contradiction shows that no such C exists. [The requested proof via (ii) is, in essence, the same as this, but is shorter, at the expense of (implicitly) involving 4.38 and the Jordan Curve Theorem.]

Exercise 10.6 *Let n be a positive integer, U an open, simply-connected subset of $\mathbb{C}$, $f \in H(U)$. Show that f has a holomorphic n^{th} root in U if and only if the order of each of its zeros is divisible by n.*

Hints: One direction is trivial. For the other we may assume (4.72) U is connected and f is not identically 0. Then $A := f^{-1}(0)$ consists of isolated points and Weierstrass (7.35) provides an $h \in H(U)$ with a zero of order $n(a, f)/n$ at each $a \in A$ and no other zeros. Then f/h^n is zero-free and so by property (HL) it has the form e^F for some $F \in H(U)$. Then $he^{F/n}$ is a holomorphic n^{th} root of f. (A quite different proof will be found in 11.49(v).)

Exercise 10.7 *Use* 10.2 *to give a new proof of the simply-connected case of* 7.47*: If Ω is a simply-connected proper subregion of $\mathbb{C}$, $a \in \Omega$ and $\mathcal{G}_a$ is the group (under composition) of all conformal automorphisms of Ω which fix a, then $f \mapsto f'(a)$ is an isomorphism of $\mathcal{G}_a$ onto the multiplicative group $\mathbb{T}$.* (Cf. §4 below.)

Hints: By 10.2 there is a conformal map ψ of Ω onto $\mathbb{D}$. Following with the appropriate conformal automorphism of $\mathbb{D}$, we can suppose that $\psi(a) = 0$. Now 6.2(ii) shows that the conformal automorphisms of $\mathbb{D}$ which fix 0 all have the form $z \mapsto uz$ for some $u \in \mathbb{T}$. Consequently the typical function in $\mathcal{G}_a$ has the form $f_u(z) := \psi^{-1}(u\psi(z))$. By the chain rule $f_u'(a) = u$, so the map $f_u \mapsto u$ is well defined. It is evidently one-to-one and homomorphic (chain rule).

Exercise 10.8 (Cf. Remarks following 9.31) *Let U be an open subset of $\mathbb{C}$, $h \colon U \mapsto \mathbb{R}$ a continuous function for which all second-order partial derivatives exist and are continuous. Set $g := 2\partial h := D_1 h - iD_2 h$ and consider $f \in H(U)$ and piecewise smooth loops $\gamma \colon [0, 1] \mapsto U$.*

(i) *Show that h is harmonic in U if and only if g is holomorphic in U.*

Hints: Use 2.19(iii) to see that $D_1D_2h = D_2D_1h$. A calculation then shows that

$$D_1g + iD_2g = D_{11}h + D_{22}h.$$

From 2.16 & 2.17 together with 9.29 & 9.31 the claimed equivalence follows.

(ii) *Show that* $f' = g$ *if and only if* $h - \operatorname{Re} f$ *is locally constant.*

Hints: Either of these conditions makes h harmonic. Therefore, in any open disk $D \subset U$ there is a holomorphic function f_D with $\operatorname{Re} f_D = h$. From the (C. R.) equation it follows that $f_D' = g$ in D. If $f' = g$ in U, then $(f - f_D)' = 0$ in D, so $f - f_D$ is constant in D and $\operatorname{Re}(f_D - f) = h - \operatorname{Re} f$ is constant in D, i.e., $h - \operatorname{Re} f$ is locally constant in U. Conversely, if $h - \operatorname{Re} f = \operatorname{Re}(f_D - f)$ is constant in D, then by 2.16(ii), $f_D - f$ is also constant in D, so $0 = (f_D - f)' = f_D' - f' = g - f'$ in D. Since D is arbitrary, this means $g - f' = 0$ throughout U.

(iii) *Show that* $\operatorname{Re} \int_\gamma g = 0$.

Hints: For the case of smooth γ, consider $\phi := h \circ \gamma$. Since γ is a loop, $\phi(1) = \phi(0)$. And the chain rule in $\mathbb{R}^2$ (see §4 of chapter II) shows that

$$\phi' = (D_1h \circ \gamma)(\operatorname{Re} \gamma)' + (D_2h \circ \gamma)(\operatorname{Im} \gamma)' = \operatorname{Re}\big[(g \circ \gamma)\gamma'\big], \text{ by definition of } g.$$

Consequently, $\operatorname{Re} \int_\gamma g = \int_0^1 \operatorname{Re}\big[(g \circ \gamma)\gamma'\big] = \phi(1) - \phi(0) = 0$. For merely piecewise smooth γ, proceed as in the proof of 2.10(iv).

(iv) *Express* $\operatorname{Im} \int_\gamma g$ *in terms of* h *and interpret geometrically/physically.*

Hints: Suppose $0 = t_0 < t_1 < \cdots < t_n = 1$ mark the intervals in which γ is smooth. Using $g := D_1h - iD_2h$ we confirm that $-\operatorname{Im}[(g \circ \gamma)\gamma'] = (\nabla h \circ \gamma) \bullet i\gamma'$ in each interval $[t_{j-1}, t_j]$, where $\bullet$ means inner product in $\mathbb{R}^2$ and $\nabla h := (D_1h, D_2h)$ is the gradient of h. Consequently,

$$-\operatorname{Im} \int_\gamma g = \sum_{j=1}^n \int_{t_{j-1}}^{t_j} -\operatorname{Im}[(g \circ \gamma)\gamma'] = \sum_{j=1}^n \int_{t_{j-1}}^{t_j} (\nabla h \circ \gamma) \bullet i\gamma'.$$

If $\gamma'(t) \neq 0$, $i\gamma'(t)/|\gamma'(t)|$ is a unit vector orthogonal to the tangent vector $\gamma'(t)$, so its inner product with $\nabla h(\gamma(t))$ is the directional derivative of h in this *normal* direction to the curve γ at the point $\gamma(t)$. It is often denoted $D_{\mathbf{N}}h(\gamma(t))$, and then

$$\nabla h(\gamma(t)) \bullet i\gamma'(t) = D_{\mathbf{N}}h(\gamma(t))|\gamma'(t)|,$$

an equation which holds as well when $\gamma'(t) = 0$ however we choose to define $D_{\mathbf{N}}(\gamma(t))$ in such cases. In this language

$$-\operatorname{Im} \int_\gamma g = \sum_{j=1}^n \int_{t_{j-1}}^{t_j} D_{\mathbf{N}}h(\gamma(t))|\gamma'(t)|dt =: \int_\gamma D_{\mathbf{N}}h(\xi)|d\xi|.$$

Now suppose that γ is smooth and γ' is zero-free. The function $t \mapsto \int_0^t |\gamma'|$ represents arclength along γ. Its derivative is $|\gamma'| > 0$, so it is a strictly increasing map of $[0,1]$ onto $[0,L]$, L being the length of γ. Let γ^* denote the composition of γ with the inverse of this function. Then

$$\int_0^1 D_{\mathbf{N}}h(\gamma(t))|\gamma'(t)|\,dt = \int_0^L D_{\mathbf{N}}h(\gamma^*(s))\,ds$$

appears as an integral with respect to arclength. It represents the net amount per unit time entering or (depending on the orientation of γ) leaving the region bounded by γ of a (plane) fluid whose velocity at each point z is $D_1h(z) - iD_2h(z) = f'(z)$ (if $h = \operatorname{Re} f$). Thus, in the language of the remarks following 2.8, it is the *flux* of the Pólya vector field $-\overline{f'}$. The notation $D_{\mathbf{N}}h(\xi) = D_{\mathbf{N}}h(\gamma(t))$ is traditional but misleading, as the real number so designated depends on more than h and the point $\xi = \gamma(t)$: For two curves γ, Γ, the equality $D_{\mathbf{N}}h(\gamma(t_0)) = D_{\mathbf{N}}h(\Gamma(t_0))$ may fail even if $\gamma(t_0) = \Gamma(t_0)$! Better notation would be $\mathbf{N}_\gamma$, instead of simply $\mathbf{N}$, or $\gamma^\perp$, but that's proved to be unacceptably cumbersome.

(v) *Show that g is holomorphic if and only if*

(1) $$\operatorname{Im} \int_{\partial\triangle} g = 0$$

for every triangle $\triangle \subset U$. In the language of the preceding hints: h is harmonic if and only if

(2) $$\int_{\partial\triangle} D_{\mathbf{N}}h(\xi)\,|d\xi| = 0$$

for every triangle $\triangle \subset U$. The same conclusion follows if the hypothesis is modified to read that (1) *and* (2) *hold for $\int_{C(a,r)}$ whenever $\overline{D}(a,r) \subset U$.*

Hints: In view of (iii), (1) amounts to the formally stronger condition

(3) $$\int_{\partial\triangle} g = 0$$

for all triangles $\triangle \subset U$, and so the claim follows from Cauchy-Goursat 5.1 and Morera's converse 5.39. Under the alternative hypothesis, (3) becomes

$$\int_{C(a,r)} g = 0 \quad \text{whenever} \quad \overline{D}(a,r) \subset U,$$

and then the holomorphy of g follows from the polar-coordinate discussion preceding definition 3.17, which is applicable since here D_1g, D_2g exist and are continuous by hypothesis.

(vi) *Show that h is the real part of a holomorphic function in U if and only if*

$$\int_\gamma D_{\mathbf{N}}h(\xi)|d\xi| = 0 \tag{4}$$

for every piecewise smooth loop γ in U. Cf. 11.59(ii).

Hints: (Cf. MORTINI [2006]). Without loss of generality, assume U is connected. If (4) holds, then by (iii) and (iv), $\int_\gamma g = 0$ for all such γ, which enables 2.11 to produce $f \in H(U)$ with $f' = g$. Now cite (ii) and 1.6. Conversely, if $f \in H(U)$ and $h = \operatorname{Re} f$, then by (ii), $g = f'$, so $\int_\gamma g = 0$, which entails (see (iv) hints) that (4) holds.

Because of the interpretation made in the hints to (iv), (4) is a kind of conservation law (the "no-flux" or "sourceless-flow" condition). The reader may recall a companion conservation law, involving the work done going around loops in a force field, which is equivalent to the force being the gradient of some scalar function (a potential). See BRADEN [1987] and NEEDHAM [1997].

(vii) *Show that if $h = \log|F|$ for some zero-free $F \in H(U)$, then h is harmonic* (5.7) *and at least all the integrals* (4) *lie in $2\pi\mathbb{Z}$.* (See 10.28(ii) for the converse of this.)

Hints: For each open disk $D \subset U$ there exists $f_D \in H(D)$ such that

$$F = e^{f_D} \text{ in } D, \text{ and so}$$

$$h = \log|F| = \operatorname{Re} f_D \text{ in } D, \text{ and } f_D' = F'/F. \tag{5}$$

From (C. R.)

$$f_D' = D_1(\operatorname{Re} f_D) - iD_2(\operatorname{Re} f_D) \overset{(5)}{=} D_1h - iD_2h,$$

which with (5) again shows that

$$D_1h - iD_2h = F'/F \text{ in } D, \text{ hence throughout } U.$$

From this and 5.32(iii) it follows that all $\int_\gamma (D_1h - iD_2h)$ lie in $2\pi i\mathbb{Z}$.

(viii) (Removable curves) *Suppose* γ *is an arc in the open set* $V \subset \mathbb{R}^2$, $h\colon V \mapsto \mathbb{R}$ *is harmonic in* $V \setminus \gamma$ *and has continuous first-order partial derivatives at each point of* γ. *Show that* h *is harmonic throughout* V.

Hints: Consider $g := D_1h - iD_2h$. It is defined and continuous in V, and by (i) it is holomorphic in $U := V \setminus \gamma$. Hence by 9.28(ii) it is holomorphic throughout V. But then by (i) again h is harmonic throughout V. [The example $V := \mathbb{C}$, $\gamma(t) := t$, $h(z) := |\operatorname{Im} z|$ illustrates the need for the differentiability hypothesis on h at points of γ.]

Exercise 10.9 (i) *Use* 10.8 *to give a direct proof that property* (P) *implies property* (H).

Hints: Given region Ω with property (P) and any harmonic $h\colon \Omega \mapsto \mathbb{R}$, the holomorphic function $D_1h - iD_2h$ has a primitive f which satisfies $\operatorname{Re} f = h$ throughout Ω.

(ii) *Use* 10.8 *to prove a local maximum principle for harmonic functions; equivalently, show that a harmonic function which is constant in some non-empty open subset of a region is constant in the whole region.*

Hints: If the harmonic function h attains a local maximum in the region Ω, then it is constant in some disk (5.8). Therefore (5.62) the holomorphic function $D_1h - iD_2h$ is identically 0 in Ω, i.e., $D_1h = D_2h = 0$ throughout Ω. From 2.17 it follows that h is actually holomorphic in Ω and $h' = \mathbf{0}$. Cite 2.10(iv). An alternative one-line proof can be based on 9.33(iii).

(iii) *Part* (ii) *affirms that a* level-set *of a non-constant harmonic function has no interior. But much more can be said about such level-sets: locally each is a union of finitely many arcs. Formulate this precisely and prove it using* 5.83.

Hints: Suppose $h\colon D(0,\rho) \mapsto \mathbb{R}$ is harmonic and non-constant, and $h(0) = 0$. 5.21(i) furnishes $f \in H(D(0,\rho))$ with $\operatorname{Im} f = h$, and we can suppose $f(0) = 0$. This is an isolated zero of f because h, hence f, is not constant. Therefore, for some $n \in \mathbb{N}$, $r \in (0,\rho]$ and $F \in H(D(0,r))$ we have $f = F^n$ and F maps $D(0,r)$ conformally onto a neighborhood W of 0 (5.83). Pick $R > 0$ so that $\bar{D}(0,R) \subset W$, and consider $U := F^{-1}\big(\bar{D}(0,R)\big) \subset D(0,r)$ and the intervals

$$L_k := e^{\pi ik/n}\,[-R,R] \subset \bar{D}(0,R), \quad k \in \{1,\dots,n\}.$$

F^{-1} maps L_k onto an arc γ_k through $F^{-1}(0) = 0$ lying in the neighborhood U of 0, and any two of these arcs have only 0 in common. Finally, note that

$$\begin{aligned} h^{-1}(0) \cap U &= \{z \in U : \operatorname{Im} F^n(z) = 0\} = \{z \in U : F(z) \in L_k \text{ for some } k\} \\ &= \bigcup_{k=1}^{n} (U \cap \gamma_k) = \bigcup_{k=1}^{n} \gamma_k. \end{aligned}$$

(iv) (Extension of 5.38(iv)). *Let* $\{h_n\}$ *be harmonic and locally uniformly bounded in the region* Ω. *Suppose the sequence converges* (*in* $\mathbb{R}$) *at each point of a non-void open subset of* Ω. *Prove that convergence occurs throughout* Ω.

Hints: Use a version of 7.8, the compactness result 7.16(i), and the uniqueness result in (ii) above. Alternatively, reduce to 5.28(iii).

Exercise 10.10 (i) (HEINZ [1956]) $f\colon \overline{\mathbb{D}} \mapsto \mathbb{C}$ *is continuous and is holomorphic in* $U := \mathbb{D} \setminus f^{-1}(0)$; h *is the Poisson integral of* $f|\mathbb{T}$. *Show that* $h \in H(\mathbb{D})$ *and* $h = f$ *in* $\overline{U}$.

Hints: We may assume $|f| \leq 1$ and $U \neq \varnothing$. The $\mathbb{C}$-valued function h is continuous on $\overline{\mathbb{D}}$, harmonic in $\mathbb{D}$ and equal to f on $\mathbb{T}$ (see 5.20). For each $\varepsilon > 0$ the function $u = u_\varepsilon := \operatorname{Re}(h - f) + \varepsilon \log|f|$ is harmonic in U and satisfies

$$\varlimsup_{z \to \xi} u(z) \leq \begin{cases} 0 & \forall \xi \in \mathbb{T} \cap \partial U \\ -\infty & \forall \xi \in \mathbb{D} \cap \partial U \subset f^{-1}(0). \end{cases}$$

From 5.12 then $u \leq 0$ in U. Let $\varepsilon \downarrow 0$ to get $\operatorname{Re}(h - f) \leq 0$ in U. Treat $\operatorname{Re}(f - h)$, $\pm \operatorname{Im}(f - h)$ similarly. Thus $h = f$ in U, hence in $\overline{U}$ by continuity. Cite 9.33(iii) to get $h \in H(\mathbb{D})$.

(ii) *Under the hypotheses of* (i), $h = f$ *in* $\mathbb{D}$, *so* $f \in H(\mathbb{D})$.

Hints: Again assume $U \neq \varnothing$. The function $h \cdot f$ is not identically 0 and is holomorphic in U, as well as in the interior of $f^{-1}(0)$. Any $z_0 \in \mathbb{D}$ not in either of these sets lies in $\overline{U} \cap \mathbb{D} \setminus U \subset \mathbb{D} \setminus U = f^{-1}(0)$, so $h(z_0) = f(z_0) = 0$. Hence for $z \in \mathbb{D} \setminus \{z_0\}$, as $z \to z_0$

$$\frac{hf(z) - hf(z_0)}{z - z_0} = \frac{h(z) - h(z_0)}{z - z_0} \cdot f(z) \to h'(z_0) f(z_0) = 0,$$

showing that hf is ($\mathbb{C}$-)differentiable at z_0 too. In summary, $hf \in H(\mathbb{D})$. Its zero-set $f^{-1}(0)$ therefore has no limit point in $\mathbb{D}$, so its complement U is dense in $\mathbb{D}$.

(iii) (T. RADÓ [1924]) V *is an open subset of* $\mathbb{C}$, $f\colon V \mapsto \mathbb{C}$ *is continuous, and* f *is holomorphic in* $V \setminus f^{-1}(0)$. *Then* $f \in H(V)$.

Hint: Holomorphy being a local matter, apply (ii).

(iv) (T. RADÓ [1924]) Ω *is a region,* $b \in \partial\Omega$ *and* b *is not isolated in* $\mathbb{C} \setminus \Omega$. $F \in H(\Omega)$ *and for some neighborhood* U *of* b *satisfies*

(∗) $$\lim_{z \to \xi} F(z) = 0 \qquad \forall \xi \in U \cap \partial\Omega.$$

Use (iii) *to show that* $F = \mathbf{0}$. [This improves 5.76.]

Hints: We may assume $b = 0$ and $U \supset \overline{\mathbb{D}}$. Define f in $\mathbb{D}$ by

$$f := \begin{cases} F & \text{in } \mathbb{D} \cap \Omega \\ 0 & \text{in } \mathbb{D} \setminus \Omega. \end{cases}$$

Check that $(*)$ makes f continuous in $\mathbb{D}$. By (iii), $f \in H(\mathbb{D})$. By hypothesis on $b = 0$ there exist $z_n \in \mathbb{C} \setminus \Omega \setminus \{b\}$ which converge to b. As soon as $z_n \in \mathbb{D}$ it is a zero of f. By 5.62 then $f = \mathbf{0}$. In particular, $\{0\} = F(\mathbb{D} \cap \Omega)$. As $0 = b \in \partial\Omega$, the open set $\mathbb{D} \cap \Omega$ is not void. One more appeal to 5.62 gives $F = \mathbf{0}$.

Theorem 10.2 lets us significantly extend an earlier result affirming that maps that are boundary-preserving and locally univalent are in fact univalent.

Exercise 10.11 *Show that if Ω is a simply-connected region, V is any region in $\mathbb{C}$, and $f \colon \Omega \mapsto V$ is a continuous proper open map, then V is also simply-connected.*

Hints: According to 6.13 f is a closed mapping, hence $f(\Omega)$ is a clopen subset of V, hence $f(\Omega) = V$. According to 10.2(b) Ω is homeomorphic to $\mathbb{D}$. Let $\psi \colon \mathbb{D} \mapsto \Omega$ be a homeomorphism, apply 1.35 to $\phi := f \circ \psi$ and recall 1.38(i).

Theorem 10.12 *Let Ω_1, Ω_2 be regions in $\mathbb{C}$, $f \colon \Omega_1 \mapsto \Omega_2$ a proper local homeomorphism. If either Ω_1 or Ω_2 is simply-connected, then both are, and f is bijective.*

Proof: If Ω_1 is simply-connected, then so is Ω_2 by 10.11. If Ω_2 is simply-connected, then either $\Omega_2 = \mathbb{C} = D(0, \infty)$ or Ω_2 is conformal to $D(0, 1)$ by 10.2. Hence the desired conclusion is assured by theorem 6.17.

Remarks: The preceding result is quite sharp. Bijectivity of f can fail in the absence of simple-connectivity. The example $\Omega_1 = \Omega_2 := \mathbb{D} \setminus \{0\}$, $f(z) := z^2$ shows this. [Evidently f is a locally univalent, proper map.] On the other hand, even if both regions are simply-connected but f is only a locally univalent holomorphic surjection, the conclusion can also fail. Examples will be found in Greenstein [1958], Krzyż [1971] (problem 8.1.2) and Zalcman [1974] (p. 135); these use a deep result of E. Picard (12.18). A more complicated example which does not involve Picard will be offered in 18.15(iii). Ryff [1969] constructs a non-univalent but locally univalent holomorphic map of a rectangle onto a simply-connected region such that each point has at most three pre-images. In spite of all this, Ortel and Smith [1986] proved the following fact: If we hypothesize that both Ω_1 and Ω_2 are simply-connected, then any locally univalent mapping of Ω_1 onto Ω_2 which is continuous and for which there is some $N \in \mathbb{N}$ such that every point of Ω_2 has either 1 or N pre-images, is necessarily bijective.

§ 2 Double Connectivity

Any two disks are conformally equivalent and the conformal automorphisms of a disk are sufficiently abundant to act transitively. The situation is quite different with annuli, as we shall now see. Actually 10.14 below was explored (in the greater generality of proper maps) in 6.18, but the topic is so fascinating we give another treatment here. Notice that 10.13(iii) is an especially strong rigidity assertion. Finally, we prove in this section the appropriate form of the Riemann mapping theorem for doubly-connected regions: every one is conformal to some annulus. Two other proofs of this are offered in §5.

Theorem 10.13 *Let* $0 < r < R < \infty$, $A := \{z \in \mathbb{C} : r < |z| < R\}$, γ, Γ *piecewise smooth loops in* A. *Then*

(i) $\operatorname{Ind}_\gamma(0) \int_\Gamma f = \operatorname{Ind}_\Gamma(0) \int_\gamma f \qquad \forall f \in H(A)$.

(ii) *If* $\operatorname{Ind}_\Gamma(0) \neq 0$, $f \in H(A)$ *is zero-free and* $\operatorname{Ind}_{f\circ\Gamma}(0) = 0$, *then* f *has a holomorphic logarithm.*

(iii) *If* $f \in H(A)$, $f(A) \subset A$ *and* $\operatorname{Ind}_{f\circ\Gamma}(0) \neq 0$, *then for some constant* c *either* $f(z) \equiv cz$ *or* $f(z) \equiv c/z$. *Evidently* $|c| = 1$ *in the first case,* $|c| = rR$ *in the second. In either case* $f \in \operatorname{Aut}(A)$.

Proof: (i) $\operatorname{Ind}_\gamma$ and $\operatorname{Ind}_\Gamma$ each vanish identically in $\mathbb{C}\setminus D(0,R)$ and in $\bar{D}(0,r)$ have the constant values $\operatorname{Ind}_\gamma(0)$, $\operatorname{Ind}_\Gamma(0)$, respectively. Thus $\operatorname{Ind}_\gamma(0)\operatorname{Ind}_\Gamma - \operatorname{Ind}_\Gamma(0)\operatorname{Ind}_\gamma$ vanishes identically in $\bar{D}(0,r) \cup [\mathbb{C} \setminus D(0,R)] = \mathbb{C} \setminus A$. Therefore (i) follows from 8.14 (or from 7.22), more elementarily, from 11.2.

(ii) We have

$$(1) \qquad 2\pi i \operatorname{Ind}_{f\circ\Gamma}(0) = \int \frac{(f\circ\Gamma)'}{f\circ\Gamma} = \int \left(\frac{f'}{f}\circ\Gamma\right)\cdot\Gamma' = \int_\Gamma \frac{f'}{f}.$$

By hypothesis this is 0, so (i) with f'/f in the role of f gives

$$\operatorname{Ind}_\Gamma(0) \int_\gamma \frac{f'}{f} = 0.$$

Also, by hypothesis this latter index is non-zero, so we conclude that

$$\int_\gamma \frac{f'}{f} = 0.$$

As γ is any piecewise smooth loop in A, it follows from 2.11 that f'/f has a primitive F in A and then $f = e^{F-c}$ for some constant c, as shown in the proof of 5.35.

(iii) Let $\gamma := C(0, \sqrt{Rr})$. It follows from (i) that

$$\text{Ind}_\gamma(0) \int_\Gamma \frac{f'}{f} = \text{Ind}_\Gamma(0) \int_\gamma \frac{f'}{f}. \tag{2}$$

Put (1) (and its analog for γ) into (2) to get

$$\text{Ind}_\gamma(0)\text{Ind}_{f\circ\Gamma}(0) = \text{Ind}_\Gamma(0)\text{Ind}_{f\circ\gamma}(0)$$

or, since $\text{Ind}_\gamma(0) = 1$,

$$\text{Ind}_{f\circ\Gamma}(0) = \text{Ind}_\Gamma(0)\text{Ind}_{f\circ\gamma}(0). \tag{3}$$

Since the left-hand side of (3) is non-zero by hypothesis, it follows that $n := \text{Ind}_{f\circ\gamma}(0) \neq 0$ and therefore by 6.24(ii) (with $r_1 = r_2 = r$, $R_1 = R_2 = R$), f has the asserted form.

Corollary 10.14 *Let $0 < r_j < R_j < \infty$, $A_j := \{z \in \mathbb{C} : r_j < |z| < R_j\}$ ($j = 1, 2$) and f a conformal map of A_1 onto A_2. Then f has the form $f(z) \equiv cz$ or $f(z) \equiv c/z$ for some constant c. In particular, $R_2/r_2 = R_1/r_1$.*

Proof: Replacing f by $(r_1/r_2)f$, we can suppose that $r_1 = r_2$. Then, considering f^{-1} instead of f if necessary, we can suppose that $R_2 \leq R_1$, which insures that f maps A_1 into $A_2 \subset A_1$. Note that $\text{Ind}_{f\circ\Gamma}(0) \neq 0$, where $\Gamma := f^{-1} \circ C(0, \sqrt{r_2R_2})$, and consequently by 10.13(iii) f has the asserted form.

Theorem 10.15 *Let $0 < r < R < \infty$, $A := A(r, R)$. Let $\mathfrak{U}$ denote the invertible elements in the ring $H^\infty(A)$ and $\mathcal{E}$ the subset of $\mathfrak{U}$ consisting of functions which possess holomorphic n^{th} roots for every positive integer n. [It follows trivially from* 5.65(iii) *that $\mathcal{E}$ consists of the exponentials of functions in $H(A)$ having bounded real parts.] Let $\|f\| := \sup|f(A)|$ for $f \in H^\infty(A)$. Then we have*

$$\inf_{f\in\mathfrak{U}\setminus\mathcal{E}} \|f\|\|1/f\| = R/r. \tag{$*$}$$

Proof: According to 4.21 the function $f_1(z) := z$ has no holomorphic square root, so $f_1 \in \mathfrak{U} \setminus \mathcal{E}$ and

$$\inf_{f\in\mathfrak{U}\setminus\mathcal{E}} \|f\|\|1/f\| \leq \|f_1\|\|1/f_1\| = R/r. \tag{1}$$

Now consider any $f \in \mathfrak{U} \setminus \mathcal{E}$. We propose to show by *reductio ad absurdum* that

$$\|f\|\|1/f\| < R/r \tag{2}$$

is impossible. To this end, assume (2). Choosing $0 < R_0 < R$ such that $\|f\|\|1/f\| < R_0/r$, and replacing f by $(R_0/\|f\|)f$, lets us assume that

$$\|f\| < R \ \ \& \ \ \|1/f\| < 1/r.$$

Then f maps A into A. Setting $\gamma := C\big(0, \sqrt{Rr}\big)$, we note that $f \notin \mathcal{E}$ and 10.13(ii) imply that

$$\text{(3)} \qquad \operatorname{Ind}_{f\circ\gamma}(0) \neq 0.$$

It follows from (3) and 10.13(iii) that either $f(z) = cz$ or $f(z) = c/z$. In either case, therefore, a contradiction to (2) is reached.

Remarks 10.16: Let $Q := \{r_1 + r_2 i : r_1, r_2 \text{ rational}\}$ and for each $f \in H^\infty(A)$ let $S(f)$ denote the closure of $f(A)$. If f is non-constant, set

$$\text{(4)} \qquad \sigma(f) := \big\{\lambda \in Q : \text{ either } \ \ \lambda\mathbf{1} - f \notin \mathcal{U} \ \ \text{ or } \ \ \bar{\lambda}\mathbf{1} - f \notin \mathcal{U}\big\}.$$

For any $\lambda \in \mathbb{C}$ and any $f \in H^\infty(A)$, $\lambda\mathbf{1} - f \notin \mathcal{U}$ is clearly equivalent to $\lambda \in S(f)$. Moreover, if f is non-constant, $f(A)$ is open, so $Q \cap S(f)$ is dense in $S(f)$. Therefore, it follows from (4) that

$$\left.\begin{array}{r} \sup|\sigma(f)| = \sup|f(A)| \\ \inf|\sigma(f)| = \inf|f(A)| \end{array}\right\} \quad f \in H^\infty(A), \quad f \text{ non-constant.}$$

Since the functions in $\mathcal{U} \setminus \mathcal{E}$ are non-constant, $(*)$ then says

$$(**) \qquad \frac{R}{r} = \inf_{f \in \mathcal{U}\setminus\mathcal{E}} \left[\frac{\sup|\sigma(f)|}{\inf|\sigma(f)|}\right].$$

Next let us observe that $\mathcal{U}$, $\mathcal{E}$ and σ involve only the ring structure of $H^\infty(A)$. That is, if $A_j := \{z \in \mathbb{C} : r_j < |z - c_j| < R_j\}$, where $c_j \in \mathbb{C}$, $0 < r_j < R_j < \infty$ $(j = 1, 2)$ and Φ is a ring isomorphism of $H^\infty(A_1)$ onto $H^\infty(A_2)$, then with the obvious notation, $\Phi(\mathcal{U}_1) = \mathcal{U}_2$, $\Phi(\mathcal{E}_1) = \mathcal{E}_2$ and $\sigma(\Phi(f)) = \sigma(f)$. The first two of these equalities are obvious. For the third, note that $\Phi(\mathbf{1}) = \mathbf{1}$ and the additivity of Φ imply its rational homogeneity: $n\Phi(m/n\cdot\mathbf{1}) = \Phi(n\cdot m/n\cdot\mathbf{1}) = \Phi(m\cdot\mathbf{1}) = m\cdot\mathbf{1}$ for all integers m and n $(n \neq 0)$. Also $\big(\Phi(i\mathbf{1})\big)^2 = \Phi\big((i\mathbf{1})^2\big) = \Phi(-\mathbf{1}) = -\mathbf{1}$, so by 1.7 $\Phi(i\mathbf{1})$ is either the constant function $i\mathbf{1}$ or the constant function $-i\mathbf{1}$. Therefore either $\Phi(\lambda\mathbf{1}) = \lambda\mathbf{1}$ for all $\lambda \in Q$ or $\Phi(\lambda\mathbf{1}) = \overline{\lambda}\mathbf{1}$ for all $\lambda \in Q$. Together with $\Phi(\mathcal{U}_1) = \mathcal{U}_2$ and $\Phi(\mathcal{E}_1) = \mathcal{E}_2$, this implies via definition (4) that $\sigma(\Phi(f)) = \sigma(f)$ for all $f \in \mathcal{U}_1 \setminus \mathcal{E}_1$. Thus from $(**)$ $R_2/r_2 = R_1/r_1$. In particular, if ϕ is a conformal map of A_2 onto A_1, then $\Phi(f) := f\circ\phi$ clearly defines a ring isomorphism of $H^\infty(A_1)$ onto $H^\infty(A_2)$ and so we have a new proof (cf. 10.14 and 6.18(v)) that the existence of such a ϕ entails the equality

$R_2/r_2 = R_1/r_1$. However, we have done more than prove this equality; we have found the radius ratio R/r encoded in the ring structure of $H^\infty(A)$!

Next we take up a study of doubly-connected regions, culminating in the beautiful classification theorem 10.20. Our first result in this direction is a significant generalization of the conformal map $z \mapsto 1/z$ of $\mathbb{D} \setminus \{0\}$ onto $\mathbb{C} \setminus \bar{\mathbb{D}}$:

Theorem 10.17 *Let Ω be a region such that $\mathbb{C} \setminus \Omega$ is not a single point but is compact and connected. Then Ω is conformal to $\mathbb{D} \setminus \{0\}$ and there is a conformal bijection $F\colon \mathbb{D} \setminus \{0\} \mapsto \Omega$ of the form $F(z) = c/z + \sum_{n=0}^{\infty} c_n z^n$, $c \neq 0$.*

Proof: We may assume after a translation, that $0 \in K := \mathbb{C} \setminus \Omega$. Consider then $\phi\colon \mathbb{C}\setminus\{0\} \mapsto \mathbb{C}\setminus\{0\}$ given by $\phi(z) := 1/z$. $\phi(\Omega)$ is an open, connected set. If $0 < r < \infty$ is such that $K \subset D(0,r)$, then $D(0,1/r)\setminus\{0\} \subset \phi(\Omega)$. It follows that $\Omega_0 := \{0\} \cup \phi(\Omega)$ is open and connected. Moreover $\mathbb{C} \setminus \Omega_0 = \phi(K \setminus \{0\})$ and according to 1.36 each component of the latter is unbounded. This means that Ω_0 is simply-connected. By 10.2 there is therefore a conformal map f of $\mathbb{D}$ onto Ω_0. Preceding f with the appropriate conformal automorphism of $\mathbb{D}$, we can suppose $f(0) = 0$. Then $\phi \circ f = 1/f$ maps $\mathbb{D} \setminus \{0\}$ conformally onto Ω. Since f has a first-order zero at 0 and no other zeros in $\mathbb{D}$, we have $f(z) = zg(z)$, g holomorphic and zero-free in $\mathbb{D}$; so in $\mathbb{D}\setminus\{0\}$, $1/f(z) =: h(z)/z$, where h is holomorphic and zero-free in $\mathbb{D}$.

Lemma 10.18 *Let Ω be a region such that $\mathbb{C} \setminus \Omega$ has a unique bounded component C. Pick $z_0 \in C$ and let $E(w) := z_0 + e^w$ $(w \in \mathbb{C})$. Then $E^{-1}(\Omega)$ is connected.*

Proof: We obviously lose no generality by supposing $z_0 = 0$. Let $\Gamma\colon [0,1] \mapsto \Omega$ be a loop as provided by 8.21, so that

$$\text{(1)} \qquad \mathrm{Ind}_\Gamma(0) = 1.$$

In the present situation the third proof of 8.21 is the most appropriate, and simplifies considerably. In fact, in the only use to be made of this lemma, viz., in the proof of 10.20, Ω is $\mathbb{D} \setminus C$ for a compact set C with $0 \in C \subset \mathbb{D}$; in this case the circle $C(0,r)$ is obviously a suitable Γ whenever $1 > r$ is sufficiently near 1. According to 4.1 there exists a continuous ϕ: $[0,1] \mapsto \mathbb{C}$ such that $e^\phi = \Gamma$. We have then from (1) and definition 4.2

$$\text{(2)} \qquad 2\pi i = 2\pi i \mathrm{Ind}_\Gamma(0) = \phi(1) - \phi(0).$$

For each $z \in \Omega$ let $\Gamma_z\colon [0,1] \mapsto \Omega$ be a curve from $\Gamma(0)$ to z (see 1.28). Again 4.1 provides continuous $\phi_z\colon [0,1] \mapsto \mathbb{C}$ such that $e^{\phi_z} = \Gamma_z$. By adding an appropriate integer multiple of $2\pi i$ to ϕ_z, we may assume that it satisfies

$$\text{(3)} \qquad \phi_z(0) = \phi(0).$$

Now consider any $w \in E^{-1}(\Omega)$ and let $z := E(w) = e^w$. We have $e^w = \Gamma_z(1) = e^{\phi_z(1)}$, so

$$(4) \qquad w - \phi_z(1) = 2\pi i n, \quad \text{some integer } n.$$

In the following we consider $n \geq 0$; the changes to accommodate $n < 0$ are merely notational. Form the curve

$$\psi(t) := \begin{cases} \phi(t) & 0 \leq t \leq 1 \\ \phi(t-1) + 2\pi i & 1 \leq t \leq 2 \\ \quad\vdots & \\ \phi(t-n+1) + 2\pi i(n-1) & n-1 \leq t \leq n \\ \phi_z(t-n) + 2\pi i n & n \leq t \leq n+1. \end{cases}$$

Because of (2) and (3) ψ is well defined, hence continuous.

We have

$$e^{\psi(t)} = \left\{ \begin{array}{lll} e^{\phi(t)} & = \Gamma(t) & 0 \leq t \leq 1 \\ \quad\vdots & & \\ e^{\phi(t-n+1)} & = \Gamma(t-n+1) & n-1 \leq t \leq n \\ e^{\phi_z(t-n)} & = \Gamma_z(t-n) & n \leq t \leq n+1 \end{array} \right\} \in \Omega$$

and therefore $\psi(t) \in E^{-1}(\Omega)$ for all $t \in [0, n+1]$. Since $\psi(0) = \phi(0)$ and $\psi(n+1) = \phi_z(1) + 2\pi i n \overset{(4)}{=} w$, this shows that any point w of $E^{-1}(\Omega)$ can be joined by a curve in $E^{-1}(\Omega)$ to the point $\phi(0)$, proving (by 1.20 and 1.1) that $E^{-1}(\Omega)$ is connected.

Lemma 10.19 *Let C be a compact, connected subset of $\mathbb{D}$ with $0 \in C$. Let $E(w) := e^w$ $(w \in \mathbb{C})$. Then $E^{-1}(\mathbb{D} \setminus C)$ is simply-connected.*

Proof: We have $\mathbb{C} \setminus E^{-1}(\mathbb{D} \setminus C) = E^{-1}(\mathbb{C} \setminus (\mathbb{D} \setminus C)) = E^{-1}(\mathbb{C} \setminus \mathbb{D}) \cup E^{-1}(C)$ and

$$E^{-1}(\mathbb{C} \setminus \mathbb{D}) = \{w : |e^w| \geq 1\} = \{w : \operatorname{Re} w \geq 0\},$$

an unbounded connected set. Therefore it suffices to show that $E^{-1}(C)$ has no bounded component. Argue by contradiction: if $E^{-1}(C)$ has a bounded component, then there is a compact, relatively open subset K of $E^{-1}(C)$ which contains that component (1.34). If U is an open subset of $\mathbb{C}$ such that $K = U \cap E^{-1}(C)$, then $E(K) = E(U) \cap C$. Since E is an open map (5.77), the set $E(U)$ is open, so $E(U) \cap C$ is relatively open in C. That is, $E(K)$ is relatively open in C. But K is compact, so $E(K)$ is too. Thus $E(K)$ is a non-void, proper ($0 \in C$, $0 \notin E(K)$), relatively clopen subset of the connected set C, a contradiction.

Theorem 10.20 *For a doubly-connected region Ω one of the following three situations obtains:*

(i) Ω *equals* $\mathbb{C} \setminus \{c\}$ *for some* $c \in \mathbb{C}$,

(ii) Ω *is conformal to* $\mathbb{D} \setminus \{0\}$,

(iii) Ω *is conformal to* $A(r, 1)$, *some* $0 < r < 1$.

Proof: Let C be the unique bounded component of $\mathbb{C} \setminus \Omega$. The case where $C = \mathbb{C} \setminus \Omega$ is a single point, is (i) above. The case where $C = \mathbb{C} \setminus \Omega$ is not a single point, is dealt with in 10.17 and is (ii) above. So we suppose that

(1) $$\Omega \cup C \neq \mathbb{C}.$$

By 1.39 and 10.2 there is a conformal map of $\Omega \cup C$ onto $\mathbb{D}$. Following this with an appropriate conformal automorphism of $\mathbb{D}$, we see that we may simply assume in addition to (1) that

(2) $$0 \in C \text{ compact, connected} \subset \mathbb{D} \text{ and } \Omega = \mathbb{D} \setminus C.$$

Let $E \colon \mathbb{C} \mapsto \mathbb{C} \setminus \{0\}$ be the exponential map, $E(w) := e^w$. Cite the preceding two lemmas to assert that

(3) $$E^{-1}(\Omega) \text{ is open, connected and simply-connected.}$$

Evidently this is a proper subset of $\mathbb{C}$, so by (two applications of) 10.2 we can find a conformal map f of $E^{-1}(\Omega)$ onto $\mathbb{U}$. Now $T(w) := w + 2\pi i$ evidently maps $E^{-1}(\Omega)$ conformally onto itself and is fixed-point free. Then $\phi := f \circ T \circ f^{-1}$ is a fixed-point-free conformal map of $\mathbb{U}$ onto $\mathbb{U}$. According to 6.21(iii) there is a conformal map ψ of $\mathbb{U}$ onto $\mathbb{U}$ such that either for some $\beta \in \mathbb{R}$,

(4) $$\psi\big(\phi(\psi^{-1}(z))\big) = z + \beta \qquad \forall z \in \mathbb{U}$$

or for some $\alpha > 0$,

(5) $$\psi\big(\phi(\psi^{-1}(z))\big) = \alpha z \qquad \forall z \in \mathbb{U}.$$

Set then

(6) $$F := \psi \circ f,$$

a conformal map of $E^{-1}(\Omega)$ onto $\mathbb{U}$ such that either for some $\beta \in \mathbb{R}$,

(4)′ $$F\big(T(F^{-1}(z))\big) = z + \beta \qquad \forall z \in \mathbb{U}$$

or for some $\alpha > 0$

$$(5)' \qquad F\big(T\big(F^{-1}(z)\big)\big) = \alpha z \qquad \forall z \in \mathbb{U}.$$

(I) We shall treat the case $(4)'$ in detail. Notice that $\beta \neq 0$, since T is fixed-point free. Define

$$(7) \qquad A := \left\{e^{2\pi i w/\beta} : w \in \mathbb{U}\right\}.$$

Thus A is either $\mathbb{D} \setminus \{0\}$ (if $\beta > 0$) or the conformally equivalent (under $z \mapsto 1/z$) set $\mathbb{C}\backslash\overline{\mathbb{D}}$ (if $\beta < 0$). We shall show that Ω is conformal to A. Consider the function $g(v) := E(2\pi i F(v)/\beta)$ mapping $E^{-1}(\Omega)$ into A. Since F maps onto $\mathbb{U}$, g clearly maps onto A. Moreover, g is constant on each subset $E^{-1}(z)$, $z \in \Omega$. To see this, suppose v_1, $v_2 \in E^{-1}(z)$. Then $e^{v_1} = e^{v_2} \Rightarrow v_1 = v_2 + 2\pi i n$ for some integer n. Reversing the roles of v_1 and v_2 if necessary, we may suppose $n \geq 0$. Then $v_1 = v_2 + 2\pi i n = T^{[n]}(v_2)$, where $T^{[n]}$ is the nth iterate of T. It follows that

$$\begin{aligned} F(v_1) = F\Big(T^{[n]}(v_2)\Big) &= \Big(F \circ T^{[n]} \circ F^{-1}\Big)\big(F(v_2)\big) \\ &= \big(F \circ T \circ F^{-1}\big)^{[n]}\big(F(v_2)\big) \\ &= F(v_2) + n\beta \quad \text{by } (4)'. \end{aligned}$$

Therefore

$$g(v_1) = E\Big(\frac{2\pi i}{\beta}F(v_1)\Big) = E\Big(\frac{2\pi i}{\beta}F(v_2)\Big) = g(v_2).$$

With this confirmed, we can cite the Factorization Lemma 5.96 to find holomorphic $h \colon \Omega \mapsto A$ such that $g = h \circ E$. Then $A = g\big(E^{-1}(\Omega)\big) = h(\Omega)$ [since E covers Ω]. The proof that h is one-to-one is similar to the above analysis involving more unwinding of map compositions and iterations. Suppose $\omega_1, \omega_2 \in \Omega$ with $h(\omega_1) = h(\omega_2)$. Select any $v_1, v_2 \in E^{-1}(\Omega)$ such that $E(v_1) = \omega_1$, $E(v_2) = \omega_2$. Then

$$g(v_1) = h(E(v_1)) = h(\omega_1) = h(\omega_2) = h(E(v_2)) = g(v_2).$$

So the *definition* of g shows that for some $n \in \mathbb{Z}$ (and $n \geq 0$ can be assumed) $2\pi i F(v_1)/\beta = 2\pi i F(v_2)/\beta + 2\pi i n$. That is,

$$F(v_1) = F(v_2) + n\beta \overset{(4)}{=} (\psi \circ \phi \circ \psi^{-1})^{[n]}(F(v_2)).$$

So by (6)

$$\begin{aligned}
\psi \circ f(v_1) &= (\psi \circ \phi \circ \psi^{-1})^{[n]}((\psi \circ f)(v_2)) \\
f(v_1) &= (\psi^{-1} \circ (\psi \circ \phi \circ \psi^{-1})^{[n]} \circ \psi)(f(v_2)) = \phi^{[n]}(f(v_2)) \\
&= (f \circ T \circ f^{-1})^{[n]}(f(v_2)) \qquad \text{by definition of } \phi \text{, whence} \\
v_1 &= (f^{-1} \circ (f \circ T \circ f^{-1})^{[n]} \circ f)(v_2) \\
&= T^{[n]}(v_2) = v_2 + 2\pi i n, \qquad \text{by definition of } T.
\end{aligned}$$

Finally then

$$\omega_1 = E(v_1) = E(v_2) = \omega_2,$$

as was to be proved.

Proof: (II) The case $(5)'$ is not too different. Here is a sketch which leaves a few details to the reader. Let L be the Principal Branch of the Logarithm in $\mathbb{C} \setminus (-\infty, 0]$ (see 3.43). Then on $(0, \infty)$, L is real-valued and in fact is the natural logarithm of 3.17. Since T has no fixed-points, it follows from $(5)'$ that $\alpha \neq 1$. Therefore, if we define $S := \mathbb{R} \times (0, \pi)$ and $A := \{e^{2\pi i w/L(\alpha)} : w \in S\}$, then A is an annulus and $\mathbb{U}$ is mapped conformally onto S by L. Define $g(v) := E\big(2\pi i L(F(v))/L(\alpha)\big)$ for any $v \in E^{-1}(\Omega)$ and use the Factorization Lemma to find holomorphic $h \colon \Omega \mapsto A$ such that $g = h \circ E$. Prove, as before, that h is bijective.

Exercise 10.21 *Show that the three situations of the preceding theorem are mutually exclusive.*

Hint: If $f \colon \mathbb{D} \setminus \{0\} \mapsto A(r, 1)$ is conformal, use 5.41(i) and 5.35 to get $f = e^g$ for some $g \in H(\mathbb{D})$. Then $w = e^{g(f^{-1}(w))}$ for all $w \in A(r, 1)$, contradicting 3.42(v).

Exercise 10.22 *Here is an internal characterization of double-connectivity (for regions). Show that a region Ω is doubly-connected if and only if*

(i) *Ω is not simply-connected, and*

(ii) *For any two loops γ_0, γ_1 in Ω which are defined on the same parameter interval there exist integers n_0, n_1 not both 0 such that $n_0 {*} \gamma_0$ is loophomotopic in Ω to $n_1 {*} \gamma_1$.*

Hints: From 4.15(ii) and 10.20 we see that (i) and (ii) are necessary conditions. On the other hand, condition (i) insures that Ω has at least one hole and condition (ii) and 8.23 insure that it has at most one.

Exercise 10.23 *Let U, V be homeomorphic regions in $\mathbb{C}$. Show that one is doubly-connected if and only if the other is.*

Hints: 4.74 gives one proof. Here is another. If ϕ is a homeomorphism of U onto V and U is doubly-connected, then U has properties (i) and (ii) above and these pass over to V under ϕ, the latter because of the obvious fact that $\phi^{-1} \circ (n{*}\Gamma) = n{*}(\phi^{-1} \circ \Gamma)$ for any integer n and loop Γ in V.

Exercise 10.24 (Cf. 4.72) *Show that if U is an open, doubly-connected subset of $\mathbb{C}$ and V is a component of U, then V is either simply- or doubly-connected and either alternative is possible. What can you say about an open set if its components are each doubly-connected?*

Hints: If $\mathbb{C} \setminus V$ has at least two different bounded components C_1, C_2, use 8.21 to come up with loops γ_1, γ_2 in V such that

$$\mathrm{Ind}_{\gamma_1}(C_1)\mathrm{Ind}_{\gamma_2}(C_2) \neq 0 \ \ \& \ \ \mathrm{Ind}_{\gamma_1}(C_2) = \mathrm{Ind}_{\gamma_2}(C_1) = 0.$$

If neither $C_1 \subset U$ nor $C_2 \subset U$, say $c_1 \in C_1 \setminus U$ and $c_2 \in C_2 \setminus U$, we shall have

$$\mathrm{Ind}_{\gamma_1}(c_1)\mathrm{Ind}_{\gamma_2}(c_2) \neq 0, \tag{1}$$

and

$$\mathrm{Ind}_{\gamma_1}(c_2) = \mathrm{Ind}_{\gamma_2}(c_1) = 0. \tag{2}$$

Since Ind_γ is constant on components of $\mathbb{C} \setminus \gamma$, it follows from (1) and (2) that c_1, c_2 belong to different components of $\mathbb{C} \setminus U$ and from (1) and 4.3 that each of these components is bounded. This contradicts the double-connectivity of U. We conclude that one of C_1, C_2 lies in U, say $C_1 \subset U$. The connected set C_1 must be wholly in one component V_1 of U. Of course $V_1 \neq V$, so, distinct components being disjoint, $V_1 \cap V = \varnothing$. On the other hand, C_1 meets $\overline{V}$ (why?) and so V_1 meets $\overline{V}$. But then since V_1 is open, V_1 meets V, a contradiction.

Exercise 10.25 (i) *Given $0 < c < 1$ and $0 < r < 1 - c$, the region $\Omega := \mathbb{D} \setminus \bar{D}(c, r)$ is evidently doubly-connected. Find a conformal map of it onto $A(R, 1)$ for an appropriate $0 \leq R < 1$.*

Hint: Use the map discussed in the hints to 6.5(iii). What is the numerical value of R? (Example 10.3.3 of Beardon [1979] is a geometric construction and generalization of this result.)

(ii) *Show that the Joukowski transform $f(z) := \frac{1}{2}(z + 1/z)$ of 2.5 maps $\mathbb{T}$ onto $[-1, 1]$ and for each $0 < r < 1$ maps $r\mathbb{T}$ onto the ellipse with foci ± 1, major diameter $r^{-1} + r$ and minor diameter $r^{-1} - r$. Conclude that it maps $A(r, 1)$ conformally onto the inside of this ellipse minus $[-1, 1]$.*

(iii) *Let J_1, J_2 be Jordan-curves with $J_1 \subset \mathfrak{I}(J_2)$. Then any conformal map of $\Omega := \mathfrak{I}(J_2) \setminus \overline{\mathfrak{I}(J_1)}$ onto an annulus can be extended to a homeomorphism of $\overline{\Omega}$. Try to prove this by the methods of* §4, *chapter* IX.

(iv) *If $r = 1 - c$ is allowed in* (i), *then Ω is a crescent-shaped region which is not doubly-connected. Find a conformal map of Ω onto $\mathbb{D}$, and show that no such map extends continuously to $\Omega \cup \{1\}$.*

Hints: Consider instead $\Omega := D(1,1) \setminus \bar{D}(r,r)$, where $0 < r < 1$. Thus $\Omega = \{z \in \mathbb{C} : |z-1| < 1 < |z/r - 1|\}$. By squaring these inequalities and recalling that $1/z = \bar{z}/|z|^2$, show that Ω is also described by $1/2 < \operatorname{Re}(1/z) < 1/2r$. Hence, $\phi(z) := \frac{2\pi i r}{1-r}\left(\frac{1}{z} - \frac{1}{2}\right)$ maps Ω conformally onto the strip $\mathbb{R} \times (0,\pi)$, and following with the exponential E gives a conformal map onto $\mathbb{U}$. Set $f := \psi \circ E \circ \phi$, where $\psi(w) := \frac{w-i}{w+i}$ maps $\mathbb{U}$ conformally onto $\mathbb{D}$. Notice that as $z \in \Omega$ tends to 0 with $\operatorname{Im} z > 0$, $f(z) \to 1$, while if $\operatorname{Im} z < 0$, $f(z) \to -1$. The final form of the conformal map f of Ω onto $\mathbb{D}$ is

$$f(z) = i \tan\left(\frac{\pi r}{1-r}\left[\frac{1}{z} - \frac{1}{2}\right] - \frac{\pi}{4}\right).$$

If h is *any* conformal map of Ω onto $\mathbb{D}$, then $f \circ h^{-1} \in \operatorname{Aut}(\mathbb{D})$ and consequently extends continuously to $\overline{\mathbb{D}}$, by 6.2(ii). Therefore, since $f = (f \circ h^{-1}) \circ h$ does *not* extend continuously to 0, neither can h. Notice that in case $r = 1/2$ the map $E \circ \phi$ of Ω onto $\mathbb{U}$ is especially simple, just $-e^{2\pi i/z}$, and one can directly prove without great difficulty that f^{-1} extends to a continuous map of $\overline{\mathbb{D}}$ onto $\overline{\Omega}$. See 3.45(vi).

§ 3 Higher Connectivity

The question of canonical conformal types for multiply-connected regions and the appropriate analogs of the Riemann mapping theorem for them is briefly discussed in the chapter notes and will be taken up in chapter XVIII. Here we are going to present some (cf. 4.69) algebraic/analytic methods of counting holes (topology). These results bring us to the threshold of de Rham cohomology, but we do not cross it: That is material for a course on Riemann surfaces.

Theorem 10.26 *Let Ω be a region with only finitely many holes $C_1, C_2, \ldots, C_n$ and let $p_k \in C_k$ for $k = 1, 2, \ldots, n$. Then there exist simple polygonal loops $\gamma_1, \gamma_2, \ldots, \gamma_n$ in Ω with these properties:*

(i) $f(z) - \sum_{k=1}^{n} \frac{\gamma_k^*(f)}{z - p_k}$ *has a primitive in $H(\Omega)$, for every $f \in H(\Omega)$, where* $\gamma_k^*(f) := \frac{1}{2\pi i}\int_{\gamma_k} f$ $(k = 1, 2, \ldots, n)$.

(ii) $\int_\gamma f = \sum_{k=1}^n \mathrm{Ind}_\gamma(C_k) \int_{\gamma_k} f$ *for all* $f \in H(\Omega)$ *and all piecewise smooth loops* γ *in* Ω.

(iii) (cf. 4.66) *If* $F \in H(\Omega)$ *is zero-free and* $N_k := \mathrm{Ind}_{F\circ\gamma_k}(0)$ $(k = 1, 2, \ldots, n)$, *then there exists* $\phi \in H(\Omega)$ *such that*

$$F(z) = e^{\phi(z)} \prod_{k=1}^n (z - p_k)^{N_k} \qquad \forall z \in \Omega.$$

Proof: For each k, 8.21 supplies a simple polygonal loop γ_k in Ω such that

(1) $$\mathrm{Ind}_{\gamma_k}(C_k) = 1 \quad \text{and} \quad \mathrm{Ind}_{\gamma_k}(C_j) = 0 \qquad \forall j \neq k.$$

It follows that for any piecewise smooth loop γ in Ω

$$\mathrm{Ind}_\gamma - \sum_{k=1}^n \mathrm{Ind}_\gamma(C_k)\mathrm{Ind}_{\gamma_k} = 0$$

in each unbounded component of $\mathbb{C} \setminus \Omega$ by 4.3 and in each bounded component $C_1, C_2, \ldots, C_n$ by virtue of (1). Therefore 8.14 (or 7.22) applies and yields (ii). The chain in the cited theorems will be comprised of γ and an appropriate number of repetitions of each γ_k or $(-1) * \gamma_k$. Setting $F_k(z) := \frac{1}{z-p_k}$, $\mathrm{Ind}_\gamma(C_k) = \frac{1}{2\pi i}\int_\gamma F_k$ and so (ii) can be re-written as

$$\int_\gamma f = \sum_{k=1}^n \int_{\gamma_k} f \cdot \frac{1}{2\pi i}\int_\gamma F_k, \quad \text{that is,}$$

$$\int_\gamma \left[f - \sum_{k=1}^n \gamma_k^*(f) F_k \right] = 0.$$

As this holds for every piecewise smooth loop γ in Ω, (i) follows from 2.11.

To prove (iii), note that N_k is just $\gamma_k^*(F'/F)$ and so when (i) is applied to $f := F'/F$ we get a $\psi \in H(\Omega)$ such that

$$\frac{F'(z)}{F(z)} - \sum_{k=1}^n \frac{N_k}{z - p_k} = \psi'(z) \qquad \forall z \in \Omega,$$

that is, the zero-free holomorphic function $G(z) := F(z)\prod_{k=1}^n (z - p_k)^{-N_k}$ satisfies

$$\frac{G'(z)}{G(z)} = \psi'(z) \qquad \forall z \in \Omega.$$

In turn, this says that the function $Ge^{-\psi}$ has derivative 0 throughout Ω, so the desired holomorphic logarithm ϕ for G is ψ plus an appropriate constant.

Theorem 10.27 (Logarithmic Conjugation Theorem) *Let the notation and hypotheses be as in* 10.26*. Then for every real-valued harmonic function h on Ω there exist $a_k \in \mathbb{R}$ (uniquely determined by* 10.28(iii)*) and $g \in H(\Omega)$ such that*

$$h(z) = \operatorname{Re} g(z) + \sum_{k=1}^{n} a_k \log |z - p_k| \qquad \forall z \in \Omega. \tag{1}$$

Proof: Let $f := D_1 h - iD_2 h$. According to 10.8 this function is holomorphic in Ω. Define

$$a_k := \frac{1}{2\pi i} \int_{\gamma_k} f = \frac{1}{2\pi i} \int_{\gamma_k} (D_1 h - iD_2 h), \quad k = 1, 2, \ldots, n. \tag{2}$$

According to 10.8(iii) and (iv)

$$a_k = \frac{-1}{2\pi} \int_{\gamma_k} D_{\mathbf{N}} h(\xi) |d\xi| \in \mathbb{R}, \quad k = 1, 2, \ldots, n. \tag{3}$$

Let

$$G(z) := f(z) - \sum_{k=1}^{n} \frac{a_k}{z - p_k}, \qquad z \in \Omega. \tag{4}$$

Bearing in mind the definition of the a_k, 10.26(i) says that G has a primitive. That is, for some $F \in H(\Omega)$

$$F'(z) = f(z) - \sum_{k=1}^{n} \frac{a_k}{z - p_k}, \qquad z \in \Omega. \tag{5}$$

In any open disk $D \subset \Omega$ each of the zero-free holomorphic functions $z - p_k$ has a holomorphic logarithm L_k which, of course, satisfies

$$\operatorname{Re} L_k(z) = \log |z - p_k| \quad \text{and} \quad L_k'(z) = \frac{1}{z - p_k} \qquad \forall z \in D. \tag{6}$$

Hence in D equation (5) reads

$$\left[F + \sum_{k=1}^{n} a_k L_k \right]' = f \quad \text{in } D.$$

From 10.8(ii) therefore

$$\operatorname{Re} \left[F + \sum_{k=1}^{n} a_k L_k \right] - h \quad \text{is constant in } D,$$

that is, in view of (6) and (3)

$$\operatorname{Re} F(z) + \sum_{k=1}^{n} a_k \log|z - p_k| - h(z) \quad \text{is constant in } D.$$

Since such D cover Ω, this affirms that the function

$$\operatorname{Re} F(z) + \sum_{k=1}^{n} a_k \log|z - p_k| - h(z)$$

is locally constant in the connected set Ω. Hence it is constant, say a, in Ω. The function $g := F - a$ therefore satisfies (1).

Remarks 10.28: (i) The series $\sum_{k=1}^{\infty} 2^{-k} \log|z - 2^{-k}|$ converges locally uniformly in $\Omega := \mathbb{D} \setminus \{2^{-k} : k \in \mathbb{N}\}^-$ to a harmonic function $h(z)$ for which (as is not too hard to prove) no representation such as (1) in 10.27 exists. Hence the finite-connectivity hypothesis there is indispensable. However, it can be dispensed with if the representation (1) is only wanted on a compact subset K of Ω. To see this, we may assume that K is connected (why?) and then revisit the paving construction in 8.1 with $U := \Omega$. As noted there, the set $\mathcal{Q}$ of squares that meet K satisfies $K \subset \cup\mathcal{Q} \subset U$. Let V denote the interior of this union. Evidently V is the union of the $\overset{\circ}{Q}$ with $Q \in \mathcal{Q}$ together with any sides common to two such Q. Consequently, $K \subset V$ and the closed set $\mathbb{C} \setminus V$ consists of the closed squares $Q(n, m)$ that do not touch V (details?). Being bounded, V lies in $[-N\delta, N\delta] \times [-N\delta, N\delta] := S$ for some $N \in \mathbb{N}$. Hence the bounded components of $\mathbb{C} \setminus V$ all lie in S and each is a union of certain of these $Q(n, m)$. Thus $\mathbb{C} \setminus V$ can have only finitely many bounded components. Finally, note that since $Q \cap K \neq \varnothing$ for each $Q \in \mathcal{Q}$, the connectedness of K and 1.3 entail that $\overset{\circ}{Q} \cup K$ is connected, and so by 1.1 the set V is connected. In summary, 10.27 is applicable to V.

(ii) The considerations in (i) also lead to the converse of 10.8(vii): If $h : \Omega \mapsto \mathbb{R}$ is harmonic and $\int_\gamma D_{\mathbf{N}} h(\xi)|d\xi| \in 2\pi\mathbb{Z}$ for every piecewise smooth loop γ in Ω, then $h = \log|F|$ for some zero-free $F \in H(\Omega)$. To see this, suppose first that Ω has finite connectivity n. Then the integral hypothesis means that the numbers a_k in (3) of 10.27 are integers, and so $F(z) := e^{g(z)} \prod_{k=1}^{n} (z - p_k)^{a_k} \in H(\Omega)$, whence (1) in 10.27 says that $h = \log|F|$. In general, Ω can be exhausted by an increasing sequence of sets V_j like the V in the preceding construction. That is,

$$V_j \text{ is a finitely-connected region} \subset \overline{V_j} \text{ compact} \subset \Omega.$$

We then get zero-free $F_j \in H(V_j)$ such that $h = \log|F_j|$ in V_j. Using 2.16(iii) we can successively adjust the F_j so that each is an extension of its predecessor and thereby arrive at a single $F \in H(\Omega)$ such that $h = \log|F|$ throughout Ω.

(iii) There are interesting algebraic re-formulations of 10.26 and 10.27 which require a little notation. Let Ω be a region in $\mathbb{C}$, $C_1, C_2, \ldots, C_n$ (some of the) holes in Ω, $p_k \in C_k$ for $k = 1, 2, \ldots, n$. Let $H'(\Omega)$ denote the set $\{f' : f \in H(\Omega)\}$, let $\operatorname{Re} H(\Omega) := \{\operatorname{Re} f : f \in H(\Omega)\}$ and let $\operatorname{Har}(\Omega)$ denote the set of real-valued harmonic functions on Ω. Let $H_0(\Omega)$ denote the multiplicative group of zero-free functions in $H(\Omega)$ and $EH(\Omega)$ its subgroup $\{e^g : g \in H(\Omega)\}$. This latter set can be described purely algebraically within $H_0(\Omega)$ as the set of f such that the equation $f = g_n^n$ has a solution $g_n \in H_0(\Omega)$ for each $n \in \mathbb{Z}$ (see 4.19(i) and 4.67). The quotient $H(\Omega)/H'(\Omega)$ is a complex vector space, the quotient $\operatorname{Har}(\Omega)/\operatorname{Re} H(\Omega)$ is a real vector space and the quotient $H_0(\Omega)/EH(\Omega)$ is an abelian group. Let

$$F_k(z) := \frac{1}{z - p_k}, \qquad h_k(z) := \log|z - p_k|, \qquad z \in \Omega,\ k = 1, 2, \ldots, n$$

and notice that

(1) the cosets $F_k + H'(\Omega)$ are linearly independent in $H(\Omega)/H'(\Omega)$.

Indeed, if $\alpha_k \in \mathbb{C}$ and $\sum_{k=1}^n \alpha_k F_k = f' \in H'(\Omega)$, then for each j

$$0 = \int_{\gamma_j} f' = \sum_{k=1}^n \alpha_k \int_{\gamma_j} F_k = \sum_{k=1}^n 2\pi i \alpha_k \operatorname{Ind}_{\gamma_j}(C_k) = 2\pi i \alpha_j,$$

where γ_j are appropriate loops furnished by 8.21. Similarly,

(2) the cosets $h_k + \operatorname{Re} H(\Omega)$ are linearly independent in $\operatorname{Har}(\Omega)/\operatorname{Re} H(\Omega)$.

Indeed, suppose $a_k \in \mathbb{R}$ and $\sum_{k=1}^n a_k h_k = \operatorname{Re} f \in \operatorname{Re} H(\Omega)$. Proceed as in the proof of 10.27. Let D be any disk in Ω, L_k a holomorphic logarithm of $1/F_k$ in D. Then $\operatorname{Re} L_k(z) = -\log|F_k(z)| = h_k(z)$ for all $z \in D$, so the preceding equation says

$$\operatorname{Re}\left[\sum_{k=1}^n a_k L_k\right] = \sum_{k=1}^n a_k \operatorname{Re} L_k = \operatorname{Re} f \quad \text{in } D, \text{ whence}$$

$$\sum_{k=1}^n a_k L_k = f + ia \quad \text{in } D$$

for some (real) constant a, by 2.16(ii). Taking derivatives,

$$\sum_{k=1}^n a_k F_k = f' \quad \text{in } D.$$

By the uniqueness theorem this last equation must hold throughout Ω and therefore, as shown in (1), all a_k must be 0.

Finally we note that the group $H_0(\Omega)/EH(\Omega)$ is torsion-free (see the hint to 4.22(iv) and 4.67) and

(3) the cosets $F_k EH(\Omega)$ are independent in the group $H_0(\Omega)/EH(\Omega)$.

Indeed, if $N_k \in \mathbb{Z}$, $\phi \in H(\Omega)$ and $\prod_{k=1}^{n} F_k^{N_k} = e^\phi$, then upon differentiating and dividing by e^ϕ, we have $\sum_{k=1}^{n} -N_k F_k = \phi'$, so that from (1) it follows $N_1 = \cdots = N_n = 0$.

Now we have assembled most of the ingredients in the proof of

Corollary 10.29 *Let Ω be a region in $\mathbb{C}$. Then*

(i) *Ω has finite connectivity if and only if the complex vector space $H(\Omega)/H'(\Omega)$ is finite-dimensional, and in this case the dimension of this space equals the number of holes in Ω.*

(ii) *Ω has finite connectivity if and only if the real vector space $\mathrm{Har}(\Omega)/\mathrm{Re}\, H(\Omega)$ is finite-dimensional, and in this case the dimension of this space equals the number of holes in Ω.*

Notice that (e) $\Leftrightarrow$ (h) and (e) $\Leftrightarrow$ (n) from 10.2 are the special case of the preceding in which the number in question is 0.

Using 10.26(iii), (3) above and 4.32 the reader can easily prove the following attractive complement to the preceding results; essentially it is already contained in 4.69.

Exercise 10.30 *Let Ω be a region in $\mathbb{C}$. Then Ω has finite connectivity if and only if the torsion-free group $H_0(\Omega)/EH(\Omega)$ has finite rank, and in this case that rank equals the number of holes in Ω. Show in fact that the number of holes is $n \in \mathbb{N}$ if and only if $H_0(\Omega)/EH(\Omega)$ is isomorphic to $\mathbb{Z}^n$.*

Remarks: (i) Actually 10.30 can be inferred directly from 4.69(iii), by establishing the fact that the groups $H_0(\Omega)/EH(\Omega)$ and $\mathcal{G}(\Omega)/\mathcal{E}(\Omega)$ are isomorphic, the map being $fEH(\Omega) \mapsto f\mathcal{E}(\Omega)$. Surjectivity follows from 4.66* and injectivity from 4.67. Here $\mathcal{G}(\Omega)$ is the group of zero-free *continuous* functions on Ω and $\mathcal{E}(\Omega)$ is the subgroup of all exponentials of continuous functions on Ω.

(ii) In case Ω has infinitely many holes, then $H_0(\Omega)/EH(\Omega)$ is isomorphic to $\mathbb{Z}^{\mathbb{Z}}$, the (additive) group of all maps of $\mathbb{Z}$ into $\mathbb{Z}$ (cf. 4.33(ii)), but this is a deeper result (Rudin [1955b]; see also Kuratowski [1945], p. 347).

(iii) If ϕ is a conformal map of the region Ω_1 onto Ω_2, then it is very easy to see that ϕ induces natural isomorphisms between the associated vector spaces and groups discussed above. We thereby get a proof of the fact that the number of holes in a region is a conformal invariant. Although this is weaker than the homeomorphic invariance proved in 4.74, it is no less remarkable, because it too elicits information about $\mathbb{C}\setminus\Omega$ from information about Ω alone.

(iv) The rather intricate result 8.21 played a key role in proving 10.26–10.30, but the simpler version of it in the remark following its (third) proof could serve as well. We have only to re-formulate 10.26 with each γ_k not a single loop, but a tuple of loops (see definitions 2.8 and 4.2); proofs in this section suffer only notational modifications.

(v) There is yet another algebraic structure that counts the holes in Ω. The piecewise smooth loops in Ω can be formally organized into an abelian group, the free abelian group on these objects as generators, called the group of *cycles* on Ω. This group is *defined* to be the subset of $\mathbb{Z}$-valued functions in the vector space of all finitely non-zero, $\mathbb{C}$-valued functions on the set of piecewise smooth loops in Ω. The indicator functions of the individual loops clearly constitute a basis for this vector space. It is customary (if initially disquieting) to denote the indicator function $1_{\{\gamma\}}$ with the same symbol "γ" as the loop itself; so if $\gamma_1, \ldots, \gamma_k$ are the distinct loops for which a cycle is not 0, and its value at γ_j is $n_j \in \mathbb{Z}$, then that cycle would be written as $n_1\gamma_1 + \cdots + n_k\gamma_k$. [Thus, despite appearances, this is *not* a sum of scalar multiples of *curves* on a common domain.] The cycles $n_1\gamma_1 + \cdots + n_k\gamma_k$ for which $\sum n_j \mathrm{Ind}_{\gamma_j}$ vanishes identically in $\mathbb{C} \setminus \Omega$ constitute a subgroup, and the associated quotient group is called the *(first) homology group of* Ω *(with coefficients in* $\mathbb{Z}$*)*. It is torsion-free and its rank coincides with the number of holes in Ω. (In effect, 10.26 and 8.21 prove this.) See chapters 6 and 9 of FULTON [1995] or chapter 14 of REMMERT [1998]. This construction is arguably the historical cradle of algebraic topology. From the point of view of topology it might seem more natural and useful to make this construction using all (continuous) loops, not just the piecewise smooth ones. This point of view is the appropriate one for proving (via 4.49(iv)) that this quotient group is invariant of the homeomorphism class of Ω. However, the resulting quotient group is the same in either construction, since every loop is Ω-homologous to a piecewise smooth one, as follows easily from the 7.23(v) hint.

The first homology group can be realized as a group of functionals, and doing so establishes a connection with 10.29. Recall that the *dual* (or *conjugate*) *space* of a $\mathbb{C}$-vector space $\mathcal{V}$ is the $\mathbb{C}$-vector space of linear functionals (= $\mathbb{C}$-valued linear maps) on $\mathcal{V}$; its customary designation is $\mathcal{V}^*$. The cycle $\gamma := n_1\gamma_1 + \cdots + n_k\gamma_k$ well defines an element $\gamma^* \in H(\Omega)^*$ via $\gamma^*(f) := \sum n_j \frac{1}{2\pi i}\int_{\gamma_j} f$. Clearly, $\gamma \mapsto \gamma^*$ is a group homomorphism which, incidentally, is also compatible with the operation in 1.11(iv): $(n{*}\gamma)^* = n\gamma^*$ for every piecewise smooth loop γ, $n \in \mathbb{Z}$. According to 7.22 the kernel of this homomorphism consists of those cycles for which $\sum n_j \mathrm{Ind}_{\gamma_j}$ vanishes throughout $\mathbb{C} \setminus \Omega$. In this way the first homology group becomes isomorphic to an additive subgroup of $H(\Omega)^*$. But clearly every functional in this group lies in $H'(\Omega)^\perp$, the subspace of $H(\Omega)^*$ consisting of all functionals which annihilate $H'(\Omega)$. Suppose now that Ω has exactly $n \in \mathbb{N}$ holes. Then the $\mathbb{C}$-vector space $H(\Omega)/H'(\Omega)$ has dimension n (by 10.29), and so therefore (by elementary algebra) does its dual space, which is naturally isomorphic to $H'(\Omega)^\perp$. Let $\gamma_1, \ldots, \gamma_n$, $F_1, \ldots, F_n$ be as in the proof of 10.26, and consider any $\phi \in H'(\Omega)^\perp$. According to 10.26, for every $f \in H(\Omega)$ the function $f - \sum \gamma_k^*(f) F_k$ lies in $H'(\Omega)$, and consequently is annihilated by ϕ. It follows that $\phi = \sum \phi(F_k)\gamma_k^*$. Thus the $\mathbb{C}$-vector space spanned by the homology group is the whole n-dimensional space $H'(\Omega)^\perp$.

In a similar way, designating the first homology group by $\mathcal{H}(\Omega)$, we can consider its *dual group*, here defined as the set of all homomorphisms of $\mathcal{H}(\Omega)$ into $\mathbb{Z}$, itself a group in the obvious way, designated by $\mathcal{H}(\Omega)^*$. Consider $f \in H_0(\Omega)$. Evaluation at a fixed $F \in H(\Omega)$ is a linear (in particular, additive) functional on $H(\Omega)^*$, so the composite map $\gamma \mapsto \gamma^* \mapsto \gamma^*(f'/f)$ defines a homomorphism of $\mathcal{H}(\Omega)$ into $\mathbb{C}$, which we denote by f^*. For any piecewise smooth loop γ in Ω

$$f^*(\gamma) = \gamma^*(f'/f) = \frac{1}{2\pi i}\int_\gamma f'/f = \mathrm{Ind}_{f\circ\gamma}(0) \in \mathbb{Z}.$$

Thus f^* maps $\mathcal{H}(\Omega)$ into $\mathbb{Z}$ and hence is an element of $\mathcal{H}(\Omega)^*$. The last equation in conjunction with 4.14 shows that $(fg)^* = f^* + g^*$ for all $f, g \in H_0(\Omega)$; that is, $f \mapsto f^*$ is a homomorphism from the multiplicative group $H_0(\Omega)$ into $\mathcal{H}(\Omega)^*$. Finally, $0 = f^*(\gamma) = \mathrm{Ind}_{f\circ\gamma}(0)$ for all piecewise smooth loops γ in Ω is equivalent to $f \in EH(\Omega)$, by 4.18 and 3.41. Thus the homomorphism $f \mapsto f^*$ induces an isomorphism of the quotient group $H_0(\Omega)/EH(\Omega)$ *into* $\mathcal{H}(\Omega)^*$. Simple algebra, as in the preceding paragraph, shows that if Ω has just $n < \infty$ holes, then this isomorphism is surjective. In this case both $\mathcal{H}(\Omega)$ and $\mathcal{H}(\Omega)^*$ are clearly isomorphic to $\mathbb{Z}^n$, and the loops $\gamma_1, \ldots, \gamma_n$ are called a *homology basis*: their cosets in the group $\mathcal{H}(\Omega)$ are independent and generate the group.

§ 4 The Aumann-Carathéodory Theorem

We have assembled all of the considerable arsenal needed to prove the remarkable characterization of simple-connectivity discussed in 7.48. The arguments are very topological. Step by step we show that the orbits are Jordan-curves lying inside one another, like the concentric circles onto which they are mapped. The original proof of this theorem used comparably non-trivial facts about covering spaces and analytic continuation. The vector-valued integral and its functional equation used here are taken from H. CARTAN [1931], p. 70. I owe to Reinhold Remmert the suggestion that it might be possible to base a proof of the theorem of Aumann and Carathéodory on this integral. That proof turns out to be rather architectonic and instructive. Here is the

Set-up: *Ω is a bounded region in $\mathbb{C}$, $z_0 \in \Omega$ and the set $\mathcal{G}$ of conformal automorphisms of Ω which fix z_0 is infinite. We want to show that Ω is conformal to $\mathbb{D}$.* We can suppose

$$(0) \qquad 0 = z_0 \in \Omega \subset \mathbb{D}.$$

According to 7.47 there is a map $u \mapsto f_u$ of $\mathbb{T}$ onto $\mathcal{G}$ such that $f_u'(0) = u$. The set $\mathcal{G}$ is a group under composition and 7.47 shows that

(1) $\qquad u \mapsto f_u$ is a group isomorphism and a homeomorphism,

with the locally uniform convergence topology on $\mathcal{G}$. In particular,

(1)$'$ $\qquad (u, z) \mapsto f_u(z)$ is a continuous map of $\mathbb{T} \times \Omega$ into Ω.

Let us write f_θ for $f_{e^{i\theta}}$ and for each $z \in \Omega$ denote its *orbit* by

$$O_z := \{f_\theta(z) : \theta \in [0, 2\pi]\}.$$

Notice that thanks to (1) distinct orbits are disjoint. It is convenient to define loops Γ_z by

$$(2) \qquad \Gamma_z(\theta) := f_\theta(z), \quad 0 \le \theta \le 2\pi,\ z \in \Omega.$$

Now, following Cartan, we define

$$(3) \qquad F(z) := \frac{1}{2\pi}\int_0^{2\pi} e^{-i\theta} f_\theta(z)\, d\theta, \quad z \in \Omega.$$

Thus $F(0) = 0$. According to 7.21 this integral is a holomorphic function on Ω and

$$F'(z) = \frac{1}{2\pi}\int_0^{2\pi} e^{-i\theta} f_\theta'(z)\, d\theta \qquad \forall z \in \Omega.$$

In particular, since $f_\theta'(0) = e^{i\theta}$ for all θ, we have

$$(4) \qquad F'(0) = 1.$$

According to (0), $|f_\theta| < 1$ for all θ, so (3) shows that

$$(5) \qquad F(\Omega) \subset \mathbb{D}.$$

Our goal is to prove

Theorem 10.31 *F maps Ω conformally onto a disk.*

Proof: Note that for any $\theta \in \mathbb{R}$ and $z \in \Omega$

$$F\big(f_\theta(z)\big) = \frac{1}{2\pi}\int_0^{2\pi} e^{-i\psi} f_\psi\big(f_\theta(z)\big)\,d\psi \overset{(1)}{=} \frac{1}{2\pi}\int_0^{2\pi} e^{-i\psi} f_{\psi+\theta}(z)\,d\psi$$

$$= \frac{1}{2\pi}\int_0^{2\pi} e^{-i(t-\theta)} f_t(z)\,dt = e^{i\theta}F(z), \quad \text{that is,}$$

$$(6) \qquad F \circ f_\theta = e^{i\theta} \cdot F \qquad \forall \theta \in \mathbb{R}.$$

From this extraordinarily fecund relationship all the rest can be made to follow. First, we see from it that the range of F is closed under multiplication by all $e^{i\theta}$, that is, $F(\Omega) = \mathbb{T} \cdot F(\Omega)$. This means that $F(\Omega)$ is a union of circles centered at 0. Since F is continuous, this image-set is connected. It is also open by 5.77 and (4), and contains $0 = F(0)$. Hence for appropriate positive r [with $r \le 1$ by (5)]

$$(5)' \qquad F(\Omega) = D(0, r).$$

It remains to prove that F is univalent, but that is a bit arduous and calls for heavy machinery. It is convenient to introduce the notation

$$\gamma_z(\theta) := e^{i\theta}F(z), \quad 0 \le \theta \le 2\pi,\ z \in \Omega.$$

These are loops and by (2) and (6)

$$(6)' \qquad F \circ \Gamma_z = \gamma_z \qquad \forall z \in \Omega.$$

Step I: For each $z \in \Omega \setminus \{0\}$, O_z is a Jordan-curve and $0 \in \mathcal{J}(O_z)$.

To prove this, we first use (4) and 5.89 to see that F is one-to-one near 0. That is, there exists a neighborhood V_0 of 0 and an $r_0 \in (0, r)$ such that

$$F \text{ maps } V_0 \text{ conformally onto } D(0, r_0).$$

In particular, $F(z) \ne F(0) = 0$ for $z \in V_0 \setminus \{0\}$, so $(6)'$ shows that

$$(7) \qquad O_z \text{ is a Jordan-curve} \qquad \forall z \in V_0 \setminus \{0\}.$$

Let $G := (F \,|\, V_0)^{-1}$, with domain $D(0, r_0)$. Because of the joint continuity $(1)'$ there is a $\rho_0 > 0$ such that $D(0, \rho_0) \subset V_0$ and

$$f_\theta(z) \in V_0 \qquad \forall \theta \in [0, 2\pi],\ z \in D(0, \rho_0).$$

So for such θ and z

$$F\big(f_\theta(z)\big) = e^{i\theta} F(z) \in D(0, r_0).$$

Thus

$$\bar{D}\big(0, |F(z)|\big) \subset D(0, r_0) \quad \text{and} \quad G \circ \gamma_z = \Gamma_z.$$

We apply 5.85 to G and get

$$\begin{aligned} \operatorname{Ind}_{G \circ \gamma_z}(0) &= \text{number of zeros of } G \text{ inside } \gamma_z \\ &= \text{number of zeros of } G \text{ in } D\big(0, |F(z)|\big) \\ &= 1 \qquad \forall z \in D(0, \rho_0) \setminus \{0\}, \text{ since } G \text{ is injective.} \end{aligned}$$

Thus since $G \circ \gamma_z = \Gamma_z$,

$$\text{(8)} \qquad \operatorname{Ind}_{\Gamma_z}(0) = 1 \qquad \forall z \in D(0, \rho_0) \setminus \{0\}.$$

From the continuity of $(\theta, z) \mapsto f_\theta(z) = \Gamma_z(\theta)$ and 4.6 we know that $z \mapsto \operatorname{Ind}_{\Gamma_z}(0)$ is a continuous function on $\Omega \setminus \{0\}$. Since $\Omega \setminus \{0\}$ is connected, this integer-valued function must be constant, so (8) tells us that

$$\text{(8)}' \qquad \operatorname{Ind}_{\Gamma_z}(0) = 1 \qquad \forall z \in \Omega \setminus \{0\}.$$

In particular, no Γ_z with $z \neq 0$ reduces to a point. It follows that

$$\text{(9)} \qquad F(z) \neq 0 \qquad \forall z \in \Omega \setminus \{0\}.$$

Indeed, if $F(z) = 0$, then (6)′ and the definition of γ_z just preceding it show that F vanishes on $O_z = \Gamma_z[0, 2\pi]$. This is a connected compact subset of $\mathbb{C}$ and comprises more than a single point, by (8)′, so it must be infinite and then the uniqueness theorem 5.62 implies that $F = \mathbf{0}$, contrary to (4). With (9) it follows from (6) that

$$\theta \mapsto F\big(f_\theta(z)\big) \text{ is one-to-one on } [0, 2\pi) \qquad \forall z \in \Omega \setminus \{0\},$$

which means that $\theta \mapsto f_\theta(z) = \Gamma_z(\theta)$ is one-to-one on $[0, 2\pi)$, entailing that its image set

$$\text{(10)} \qquad O_z \text{ is a Jordan-curve} \qquad \forall z \in \Omega \setminus \{0\}$$

and that

$$\text{(11)} \qquad F \text{ is one-to-one on } O_z \qquad \forall z \in \Omega \setminus \{0\}.$$

Now (8)′ and (11) together with 4.43 tell us that

$$\text{(12)} \qquad 0 \in \mathfrak{I}(O_z) \qquad \forall z \in \Omega \setminus \{0\}.$$

This finishes the confirmation of step I.

Step II: $\mathcal{I}(O_z) \subset \Omega \qquad \forall z \in \Omega \setminus \{0\}$.

Fix $z \in \Omega \setminus \{0\}$ and suppose $z_1 \in \mathcal{I}(O_z) \setminus \Omega$. Let $\phi\colon [0,1] \mapsto \Omega$ be any curve in Ω joining $\phi(0) = 0$ to $\phi(1) = z$. Then $(\theta, t) \mapsto f_\theta\big(\phi(t)\big) = \Gamma_{\phi(t)}(\theta)$ is a loophomotopy in $\Omega \subset \mathbb{C} \setminus \{z_1\}$ of Γ_z with the constant loop Γ_0. Therefore by 4.12 $\mathrm{Ind}_{\Gamma_z}(z_1) = \mathrm{Ind}_{\Gamma_0}(z_1) = 0$. This contradicts 4.43, according to which the point $z_1 \in \mathcal{I}(O_z) = \mathcal{I}(\Gamma_z)$ must have index ± 1. Step II is thus secured.

Step III: For all z_1, $z_2 \in \Omega \setminus \{0\}$

$$\text{either} \quad \mathcal{I}(O_{z_1}) \subset \mathcal{I}(O_{z_2}) \quad \text{or} \quad \mathcal{I}(O_{z_2}) \subset \mathcal{I}(O_{z_1}).$$

Since either $O_{z_1} = O_{z_2}$ or $O_{z_1} \cap O_{z_2} = \varnothing$, we can assume that the latter is the case. Suppose $\mathcal{I}(O_{z_1})$ meets $\mathbb{C} \setminus \mathcal{I}(O_{z_2})$. Then since $\mathcal{I}(O_{z_1})$ is connected and also meets $\mathcal{I}(O_{z_2})$ [namely, in 0, by (12)], it must meet O_{z_2} [see 4.45], that is, $O_{z_2} \cap \mathcal{I}(O_{z_1}) \neq \varnothing$. But O_{z_2} is connected and does not meet O_{z_1}. Hence [by 4.45 again] $O_{z_2} \subset \mathcal{I}(O_{z_1})$. Since by 10.2(f) the simply-connected region $\mathcal{I}(O_{z_1})$ contains the inside of any Jordan-curve in it, the inclusion $\mathcal{I}(O_{z_2}) \subset \mathcal{I}(O_{z_1})$ now follows.

Step IV: F is one-to-one on Ω.

Consider z_1, $z_2 \in \Omega$ with $z_1 \neq z_2$. We need to show that $F(z_1) \neq F(z_2)$. If either $z_1 = 0$ or $z_2 = 0$, then $F(z_1) \neq F(z_2)$ follows from (9). If z_1, $z_2 \in \Omega \setminus \{0\}$, then by step III we can suppose $\mathcal{I}(O_{z_1}) \subset \mathcal{I}(O_{z_2})$. Since $\overline{\mathcal{I}(O_{z_2})} = O_{z_2} \cup \mathcal{I}(O_{z_2}) \subset \Omega$ by step II, (11) and the Darboux–Picard theorem (in the form 9.17(i)) insure that F is one-to-one on $\overline{\mathcal{I}(O_{z_2})}$, which contains both z_1 and z_2. So $F(z_1) \neq F(z_2)$, as claimed.

This finishes the proof of the whole theorem. 12.49 is an extension to unbounded regions. The ideas above are pushed somewhat further in BURCKEL and MINDA [1998]. For example, necessary and sufficient that a region Ω be *either* simply-connected *or* doubly-connected is the existence of a mapping of $\mathbb{T}$ into $\mathrm{Aut}(\Omega)$ which satisfies (1). For a more geometric proof of 10.31, see KRANTZ [2005].

§ 5 The Fundamental Mapping Theorem for Doubly-connected Regions Revisited

There are several other very different and very instructive ways to prove 10.20. As far as I can tell, the idea of the proof in §2 is due to KOEBE [1915]. He also indicated the desirability of basing a proof on Montel's compactness theorem and an argument like that of Fejér and Riesz in 9.9(i) and (ii). Recall that the essence of that argument is to consider a certain functional, $f \mapsto |f'(0)|$, and show that any (injective) function f that maximizes it (found via 7.13(iv)) must be surjective; because otherwise f could be followed by a Koebe map k so that $|(k \circ f)'(0)|$ is even larger, due to the dilating effect of Koebe maps (9.2(c) and (d)). CREMER [1930] outlined such a proof. As its details are non-trivial and illustrate many techniques of our subject, I want to present it here — for the really indefatigable reader.

The problem is that of conformally mapping a doubly-connected region Ω onto one of the three canonical types, $\mathbb{C} \setminus \{0\}$, $\mathbb{D} \setminus \{0\}$, $A(\rho, 1)$.

Initial Reductions. Let C denote the unique bounded component of $\mathbb{C} \setminus \Omega$, the (compact) hole in Ω. The case in which $C = \mathbb{C} \setminus \Omega$ is relatively easily dealt with — see 10.17. Furthermore, the case $C \neq \mathbb{C} \setminus \Omega$ is quickly reduced, in the opening phase of the 10.20 proof, to the situation where

$$(0) \qquad \{0\} \subsetneq C \subset \mathbb{D}, \quad \Omega = \mathbb{D} \setminus C.$$

In this case the inversion $\varphi(z) := 1/z$ maps Ω to a region Ω_1 lying in $\mathbb{C} \setminus \overline{\mathbb{D}}$ whose complement consists of $\varphi(\mathbb{C} \setminus \mathbb{D}) \cup \{0\} = \overline{\mathbb{D}}$ together with $\varphi(C \setminus \{0\}) \neq \varnothing$. According to 1.36 each component of $\varphi(C \setminus \{0\})$ is unbounded. Thus Ω_1 is a doubly-connected region and $\overline{\mathbb{D}}$ is its hole. Then the disjoint union $\Omega_1 \cup \overline{\mathbb{D}}$ is a simply-connected region (by 1.39) not equal to $\mathbb{C}$, so by 10.2 there is a conformal map ψ of $\Omega_1 \cup \overline{\mathbb{D}}$ onto $\mathbb{D}$. The image $\psi(\Omega_1)$ is a region whose complement consists of $\psi(\overline{\mathbb{D}}) \subset \mathbb{D}$ together with $\mathbb{C} \setminus \mathbb{D}$. Hence $\psi(\Omega_1)$ is doubly-connected, with hole $\psi(\overline{\mathbb{D}})$. Thus the hole contains the non-empty open set $\psi(\mathbb{D})$. The effect of following $\psi \circ \varphi$ by an appropriate automorphism of $\mathbb{D}$ (6.2) is therefore to replace the original doubly-connected region Ω with a conformal image, *still to be denoted by* Ω, for which (0) is sharpened to

$$(1) \quad \bar{D}(0, r) \subset C \text{ compact, connected} \subset \mathbb{D},\ \Omega = \mathbb{D} \setminus C \quad \text{for some } 0 < r < 1.$$

Clearly then if $\varepsilon > 0$ is sufficiently small, $\gamma := C(0, 1 - \varepsilon)$ lies in Ω. This loop is fixed for the sequel. Basic for all that follows is the fact

(∗) Every conformal image of Ω is doubly-connected.

It is a consequence of 4.74, but in the present context the proof indicated in Remark (iii) following 10.30 might be more appropriate.

The functional to be extremized. Consider conformal maps $f : \Omega \mapsto \mathbb{D} \setminus \bar{D}(0, r) = A(r, 1)$ such that

$$(2) \qquad \mathrm{Ind}_{f \circ \gamma}(0) \neq 0.$$

Let $\mathcal{F}$ denote this family and set

$$\rho := \sup\{\inf |f(\Omega)| : f \in \mathcal{F}\}. \tag{3}$$

Since the identity function is in $\mathcal{F}$, we have $\rho \geq r$. Pick $f_n \in \mathcal{F}$ with

$$\inf |f_n(\Omega)| \uparrow \rho. \tag{4}$$

Because of 7.13(iv) we may assume that $\{f_n\}$ converges locally uniformly to an $f \in H(\Omega)$. We have $|f_n| \geq \rho/2$ for all large n, by (4), and both $f_n \to f$ and $f_n' \to f'$ uniformly on γ. Consequently,

$$\mathrm{Ind}_{f_n \circ \gamma}(0) = \frac{1}{2\pi i} \int_\gamma \frac{f_n'}{f_n} \to \frac{1}{2\pi i} \int_\gamma \frac{f'}{f} = \mathrm{Ind}_{f \circ \gamma}(0).$$

As only non-zero integers appear on the left (by definition of $f_n \in \mathcal{F}$), it follows from this convergence that (2) holds. In particular, f is not constant, so f is one-to-one and, like all the f_n, maps into $A(r, 1)$. (See 7.11.) So $f \in \mathcal{F}$. We next confirm that

$$\inf |f(\Omega)| = \rho. \tag{5}$$

Since $f \in \mathcal{F}$, we have "$\leq$" in (5) by definition of ρ. Given $\delta > 0$, there exists $n_\delta \in \mathbb{N}$ such that $\inf |f_n(\Omega)| \geq \rho - \delta$ for all $n \geq n_\delta$, thanks to (4). That is, $|f_n(z)| \geq \rho - \delta$ for all $z \in \Omega$, $n \geq n_\delta$. It follows that $\inf |f(\Omega)| \geq \rho - \delta$, which proves "$\geq$" in (5) because $\delta > 0$ is arbitrary. The image $f(\Omega)$ is a region which by (5) lies in $A(\rho, 1)$. Recalling (∗), let K denote the unique hole in $f(\Omega)$. It is compact, and we have

$$\bar{D}(0, \rho) \subset K \subset K \cup f(\Omega) \subset \mathbb{D}. \tag{6}$$

Lemma 10.32 *The right-most inclusion in* (6) *is equality.*

Proof: Suppose not. According to 1.39, $K \cup f(\Omega)$ is a simply-connected region, which by (6) contains 0 and is a subset of $\mathbb{D}$. Our initial assumption means it's a *proper* subset. Thus $K \cup f(\Omega)$ is a simply-connected proper subregion of $\mathbb{D}$ which contains 0. After 10.2 it is therefore a Koebe region (9.2) and according to 9.4 possesses a Koebe map k. Let us show that

$$k \circ f \in \mathcal{F} \tag{7}$$

and that this leads to a contradiction of the maximality of ρ. The inequality $9.2(\mathrm{d})_0$ says

$$|k(w)| > |w| \qquad \forall\, w \in K \cup f(\Omega) \setminus \{0\}. \tag{8}$$

Since $0 \notin f(\Omega)$, we therefore have

$$(9) \qquad |k(f(z))| > |f(z)| \qquad \forall\, z \in \Omega.$$

From this and (5) follows $|k \circ f| > \rho \geq r$ on Ω, so

$$(10) \qquad k \circ f \text{ is a conformal map of } \Omega \text{ into } A(\rho, 1) \subset A(r, 1).$$

Since $f \circ \gamma$ is a simple loop in $K \cup f(\Omega)$ and k has a zero at 0, it follows from (2) and either 9.15 or 4.49(ii) that

$$(11) \qquad \operatorname{Ind}_{k \circ f \circ \gamma}(0) \neq 0.$$

Together (10) and (11) confirm (7). This means, by (3), that

$$\inf |k \circ f(\Omega)| \leq \rho.$$

Since (10) provides the reverse inequality, we get

$$(12) \qquad \inf |k \circ f(\Omega)| = \rho.$$

This equality means that there exist $z_n \in \Omega$ such that

$$(13) \qquad |k \circ f(z_n)| \to \rho.$$

We can assume that the bounded sequence $\{f(z_n)\}$ converges to some $w \in \overline{f(\Omega)}$. From (13), (5) and (9), $|w| = \rho > 0$. That is, $w \in C(0, \rho) \overset{(6)}{\subset} K \cup f(\Omega)$. On this set k is continuous, so $k \circ f(z_n) = k(f(z_n)) \to k(w)$. Hence from (13), $|k(w)| = \rho = |w|$, contradicting (8). This completes the proof of 10.32.

Theorem 10.33 *With notation and hypotheses as in the foregoing*

$$(14) \qquad f(\Omega) = \mathbb{D} \setminus \bar{D}(0, \rho) = A(\rho, 1).$$

Proof: As already noted, $f(\Omega) \subset A(\rho, 1)$. Consequently, the inversion $\varphi(w) := \rho/w$ maps $f(\Omega)$ to a doubly-connected region lying in $A(\rho, 1)$. It should be clear that therefore $\varphi \circ f \in \mathcal{F}$, for the mapping φ merely changes the sign of the index in (2). Since K is a compact subset of $\mathbb{D}$, the equality $K \cup f(\Omega) = \mathbb{D}$ of 10.32 implies the existence of $R > 1$ such that $A(1/R, 1) \subset f(\Omega)$. Hence

$$(15) \qquad A(\rho, \rho R) \subset \varphi(f(\Omega)) \subset A(\rho, 1)$$

which clearly entails that

$$(16) \qquad \rho = \inf |\varphi \circ f(\Omega)|.$$

But (15) also shows that the (unique) hole in the doubly-connected region $\varphi(f(\Omega))$ must in fact be $\bar{D}(0,\rho)$. Lemma 10.32 is about any function in $\mathcal{F}$ that realizes equality in (5), not just f. We have shown that $\varphi \circ f$ is such a function and that

(17) the hole in $\varphi \circ f(\Omega)$ is $\bar{D}(0,\rho)$.

Consequently, (6) for this function reads

$$\bar{D}(0,\rho) \subset \bar{D}(0,\rho) \cup \varphi \circ f(\Omega) \subset \mathbb{D}$$

and the conclusion of 10.32 for this function is

$$\bar{D}(0,\rho) \cup \varphi \circ f(\Omega) = \mathbb{D},$$

which together with (17) gives

$$\varphi \circ f(\Omega) = \mathbb{D} \setminus \bar{D}(0,\rho) = A(\rho, 1).$$

Applying $\varphi^{-1} = \varphi$ to this gives (14), and finishes our second proof of the fundamental mapping theorem 10.20.

There is yet a third way to get the mapping theorem 10.20. It too presents significant topological and analytic difficulties, surmounting which well illustrates many of the tools of the trade at work. This time solvability of the Dirichlet problem is central. There is an analogy with the proof of the Riemann mapping theorem in 9.50. However, the absence of harmonic conjugates here complicates matters, and the theory of boundary-preserving maps proves useful. The exposition to follow is inspired by §17.1 of HENRICI [1986], and leaves some details to the reader. It's helpful for keeping track of one's whereabouts in this complicated terrain, to put up a few global positioning satellites, i.e., to make some diagrams of regions and maps between them.

There are initial reductions like those in the preceding proof.

Third Proof: We start with (0) and apply the inversion φ, getting a doubly-connected region Ω_1, whose hole is $\bar{\mathbb{D}}$. Because C is a compact set, it follows from (0) that for appropriate $R_1 > 1$, $A(1/R_1, 1) \subset \Omega$, so

(18) $$A(1, R_1) \subset \Omega_1.$$

Finally, we again employ the conformal map ψ of the disjoint union $\Omega_1 \cup \bar{\mathbb{D}}$ onto $\mathbb{D}$, this time insisting that $\psi(0) = 0$. Thus

(19) $$\Omega_0 := \psi(\Omega_1) = \psi(\Omega_1 \cup \bar{\mathbb{D}}) \setminus \psi(\bar{\mathbb{D}}) = \mathbb{D} \setminus \psi(\bar{\mathbb{D}}).$$

Since $\psi(\overline{\mathbb{D}})$ is compact, this shows that there is an $0 < r_0 < 1$ such that

$$(20) \qquad A(r_0, 1) \subset \Omega_0.$$

For $\rho \geq 0$ define $C_\rho := C(0, \rho)$. According to the Argument Principle 5.86 and inclusion (18)

$$\mathrm{Ind}_{\psi \circ C_\rho}(0) = \text{number of zeros of } \psi \text{ in } D(0,\rho), \text{ namely, } 1 \qquad \forall\, 0 < \rho < R_1.$$

Since the conjugated function $\overline{C}_\rho$ is C_ρ parameterized in reverse, we get

$$(21) \qquad \mathrm{Ind}_{\psi \circ \overline{C}_\rho}(0) = -1 \qquad \forall\, 0 < \rho < R_1.$$

Given $z \in \Omega_0 = \mathbb{D} \setminus \psi(\overline{\mathbb{D}})$, if $1 < \rho < R_1$ is close enough to 1, we shall also have $z \in \mathbb{D} \setminus \psi(\overline{D}(0, \rho))$. Hence by the homotopy-invariance of index

$$(22) \qquad \mathrm{Ind}_{\psi \circ \overline{C}_\rho}(z) = \mathrm{Ind}_{\psi \circ \overline{C}_0}(z) = 0.$$

Similarly, if $r_0 < r < 1$ is close to 1, $z \in D(0, r)$, so

$$(23) \qquad \mathrm{Ind}_{C_r}(z) = 1.$$

Since 0 lies in the unique hole of Ω_0, it is clear from (21) and (23) that

(24) every pair $(\psi \circ \overline{C}_\rho, C_r)$ with $r_0 < r < 1 < \rho < R_1$ is Ω_0-nullhomologous.

In 10.34 below we will construct $0 < a < b < \infty$ and a mapping

$$F : \Omega_0 \mapsto A(a, b)$$

which is holomorphic and boundary-preserving (6.12), and has the form

$$(25) \qquad F(z) = z e^{\Phi(z)}, \qquad \Phi \in H(\Omega_0).$$

From 6.15 it follows that F is surjective and exactly m-to-1 for some $m \in \mathbb{N}$. We need to prove that $m = 1$ to complete this proof of the mapping theorem. It is clear from 4.14, (25) and (20) that

$$(26) \qquad \mathrm{Ind}_{F \circ C_r}(0) = \mathrm{Ind}_{C_r}(0) = 1 \qquad \text{for } r_0 < r < 1$$

and from 4.14, (25) and (21) that

$$(27) \qquad \mathrm{Ind}_{F \circ \psi \circ \overline{C}_\rho}(0) = \mathrm{Ind}_{\psi \circ \overline{C}_\rho}(0) = -1 \qquad \text{for } 1 < \rho < R_1.$$

Let us next show that for every compact $K \subset \Omega_0$

(**) $r_0 < r < 1 < \rho < R_1$ close enough to $1 \Rightarrow A(r,1), \psi(A(1,\rho)) \subset \mathbb{C} \setminus K$.

On one hand $K \subset \Omega_0 \subset \mathbb{D} \Rightarrow K \subset D(0,r)$ for all r sufficiently close to 1, whence $A(r,1) \subset \mathbb{D} \setminus K$. On the other hand, by (19), $\psi(\mathbb{T}) \subset \psi(\overline{\mathbb{D}}) \subset \mathbb{C} \setminus \Omega_0 \subset \mathbb{C} \setminus K$, so compact $\mathbb{T} \subset$ open $\psi^{-1}(\mathbb{C} \setminus K)$, whence $A(\rho^{-1}, \rho) \subset \psi^{-1}(\mathbb{C} \setminus K)$ for all ρ sufficiently close to 1, and in particular $\psi(A(1,\rho)) \subset \mathbb{C} \setminus K$. Now F being a boundary-preserving map implies (see 6.13) that for all $0 < \varepsilon < \frac{1}{2}(b-a)$, $F^{-1}(\bar{A}(a+\varepsilon, b-\varepsilon))$ is a compact subset of Ω_0. Consequently, (**) supplies $r_0 < r_\varepsilon < 1 < \rho_\varepsilon < R_1$ such that F maps $A(r_\varepsilon, 1)$ and $\psi(A(1,\rho_\varepsilon))$ into $F(\Omega_0) \setminus \bar{A}(a+\varepsilon, b-\varepsilon) = A(a, a+\varepsilon) \uplus_d A(b-\varepsilon, b)$. As all these F-images are connected sets, a dichotomy is manifest: either $F(A(r_\varepsilon, 1))$ lies in $A(a, a+\varepsilon)$ for every eligible ε, or $F(A(r_\varepsilon, 1))$ lies in $A(b-\varepsilon, b)$ for every eligible ε. Of course the sets $F(\psi(A(1,\rho_\varepsilon)))$ are distributed similarly. The (previously noted) surjectivity of F precludes two of these four possibilities. For if, say, $A(r_\varepsilon, 1) \cup \psi(A(1,\rho_\varepsilon))$ is mapped by F into $A(a, a+\varepsilon)$ for some eligible ε, then using the complement of this set in $\Omega_0 = \mathbb{D} \setminus \psi(\overline{\mathbb{D}})$, namely, the compact subset $\bar{D}(0, r_\varepsilon) \setminus \psi(D(0, \rho_\varepsilon))$, F must cover the rest of $A(a,b)$. But if z_j are in this set and $F(z_j) \to b$, compactness would entail $b \in F(\Omega_0) = A(a,b)$ $\rightarrow\!\leftarrow$. Thus either (28.1) or (28.2) prevails:

$$\lim_{r \uparrow 1} |F \circ C_r| = b \quad \text{and} \quad \lim_{\rho \downarrow 1} |F \circ \psi \circ \overline{C}_\rho| = a, \tag{28.1}$$

$$\lim_{r \uparrow 1} |F \circ C_r| = a \quad \text{and} \quad \lim_{\rho \downarrow 1} |F \circ \psi \circ \overline{C}_\rho| = b. \tag{28.2}$$

Consider a fixed $w \in A(a,b)$. As F is exactly m-to-1, w has m pre-images in Ω_0. Let them be, repeated according to multiplicity, $z_1, \ldots, z_m$. Use (22), (23) and either (28.1) or (28.2) to select $r < 1 < \rho$ sufficiently near 1 that

$$\mathrm{Ind}_{\psi \circ \overline{C}_\rho}(z_j) = 0 \text{ and } \mathrm{Ind}_{C_r}(z_j) = 1, \qquad j = 1, \ldots, m \tag{29}$$

and either (30.1) or (30.2) prevails:

$$|F \circ C_r| > |w| > |F \circ \psi \circ \overline{C}_\rho|, \tag{30.1}$$

$$|F \circ C_r| < |w| < |F \circ \psi \circ \overline{C}_\rho|. \tag{30.2}$$

In case (30.1), w clearly lies in the unbounded component of $\mathbb{C} \setminus F \circ \psi \circ \overline{C}_\rho$, so $\mathrm{Ind}_{F \circ \psi \circ \overline{C}_\rho}(w) = 0$; while the whole segment $[0, w]$ lies in $\mathbb{C} \setminus F \circ C_r$, so $\mathrm{Ind}_{F \circ C_r}(w) = \mathrm{Ind}_{F \circ C_r}(0) = 1$, by (26). Thus

$$\mathrm{Ind}_{F \circ \psi \circ \overline{C}_\rho}(w) + \mathrm{Ind}_{F \circ C_r}(w) = 1. \tag{31.1}$$

A similar argument (involving (27)) establishes that in case (30.2)

$$\text{Ind}_{F\circ\psi\circ\overline{C}_\rho}(w) + \text{Ind}_{F\circ C_r}(w) = -1. \tag{31.2}$$

There is a zero-free $F_0 \in H(\Omega_0)$ such that

$$F(z) - w = F_0(z)\prod_{j=1}^{m}(z - z_j), \text{ whence}$$

$$\frac{F'(z)}{F(z)-w} = \frac{F_0'(z)}{F_0(z)} + \sum_{j=1}^{m}\frac{1}{z-z_j}, \qquad z \in \Omega_0\backslash\{z_1, \ldots, z_m\}. \tag{32}$$

This is preparation for using a generalized argument principle. Set $\gamma_1 := C_r$, $\gamma_2 := \psi \circ \overline{C}_\rho$, $\gamma := (\gamma_1, \gamma_2)$, so that by (24)

$$\gamma \text{ is } \Omega_0\text{-nullhomologous.} \tag{33}$$

Apply 7.22 with F_0'/F_0 as the F there, to get from (32) and (33)

$$\text{Ind}_{(F-w)\circ\gamma}(0) = \sum_{j=1}^{m}\text{Ind}_\gamma(z_j).$$

(Cf. also 11.48.) In view of (29) each $\text{Ind}_\gamma(z_j)$ is 1, so this reads

$$m = \text{Ind}_{F\circ\gamma}(w) = \text{Ind}_{F\circ C_r}(w) + \text{Ind}_{F\circ\psi\circ\overline{C}_\rho}(w). \tag{34}$$

Since m is a positive integer, (34) decides for (31.1) over (31.2), so decides for (30.1) over (30.2) and (28.1) over (28.2). But the important thing is that we now see $m = 1$, making F injective and finishing, modulo 10.34, our third proof of the mapping theorem 10.20.

It remains to construct F, which we do by solving a Dirichlet problem and contriving a harmonic conjugate.

Lemma 10.34 *Given*

$$\{0\} \subsetneq C \textit{ compact, connected} \subset \mathbb{D} \textit{ for which } \Omega := \mathbb{D} \setminus C \textit{ is connected,}$$

there exist $0 < a < b < \infty$ *and a boundary-preserving holomorphic* $F : \Omega \mapsto A(a, b)$ *which has the form*

$$F(z) = ze^{\Phi(z)}, \qquad z \in \Omega,$$

for some $\Phi \in H(\Omega)$.

Proof: Since there are no one-point components in $\mathbb{C} \setminus \Omega$, 9.43 ensures that Ω has a barrier at each of its boundary points, whence (9.41) the Dirichlet problem is solvable for Ω. An easy consequence of C being a compact subset of the open set $\mathbb{D}$ is that

$$\partial\Omega = \partial(\mathbb{D} \setminus C) = \mathbb{T} \cup \partial C,$$

a disjoint union. Therefore there exists continuous $h : \overline{\Omega} \mapsto \mathbb{R}$ such that

(35) $$h = \begin{cases} 0 & \text{on } \mathbb{T} \\ 1 & \text{on } \partial C, \end{cases}$$

(36) $$h \text{ is harmonic in } \Omega.$$

The distance from the compact set C to the closed set $\mathbb{C} \setminus \mathbb{D}$ being positive, the loop $\Gamma := C(0, r)$ lies in Ω if $r < 1$ is sufficiently close to 1. Set

(37) $$g := D_1 h - i D_2 h \quad \text{in } \Omega.$$

According to 10.8(i) and (iii) $g \in H(\Omega)$ and

(38) $$c := \frac{1}{2\pi i} \int_\Gamma g \in \mathbb{R}.$$

The function $\ell(z) := \log|z|$ is harmonic in $\mathbb{C} \setminus \{0\}$ and, using the fact that $2\ell(z) = \log|z|^2$, one easily computes

(39) $$\frac{1}{z} = (D_1\ell - iD_2\ell)(z) \qquad \forall\, z \in \mathbb{C} \setminus \{0\}.$$

Form

(40) $$g_0(z) := \begin{cases} g(z) & \text{if } c = 0 \\ \frac{1}{c} g(z) - \frac{1}{z} & \text{if } c \neq 0, \end{cases} \qquad z \in \Omega.$$

That is, in view of (39) and (37)

(41) $$g_0 = \begin{cases} D_1 h - iD_2 h & \text{if } c = 0 \\ D_1\left(\frac{1}{c}h - \ell\right) - iD_2\left(\frac{1}{c}h - \ell\right) & \text{if } c \neq 0. \end{cases}$$

Now examine the (easy) proof of 10.13 to see that its part (i) remains valid with Ω in the role of the annulus A there: For any piecewise smooth loop γ in Ω

$$\mathrm{Ind}_\gamma(0) \int_\Gamma g = \mathrm{Ind}_\Gamma(0) \int_\gamma g = \int_\gamma g, \text{ so}$$

$$\frac{1}{2\pi i} \int_\gamma g = c \cdot \mathrm{Ind}_\gamma(0) \text{ by (38), and consequently from (40)}$$

$$\frac{1}{2\pi i} \int_\gamma g_0 = 0 \qquad \forall \text{ piecewise smooth loops } \gamma \text{ in } \Omega,$$

from which and 2.11 we infer that g_0 has a primitive $\Phi \in H(\Omega)$. In view of (41) and 10.8(ii) Φ may be adjusted by an additive real constant so that in addition

$$\text{(42)} \qquad \operatorname{Re}\Phi = \begin{cases} h & \text{if } c = 0 \\ \frac{1}{c}h - \ell & \text{if } c \neq 0. \end{cases}$$

For $z \in \Omega$ define

$$\text{(43)} \qquad F(z) := \begin{cases} e^{\Phi(z)} & \text{if } c = 0 \\ ze^{\Phi(z)} & \text{if } c \neq 0. \end{cases}$$

Then (42) implies

$$\text{(44)} \qquad |F(z)| = \begin{cases} e^{\operatorname{Re}\Phi(z)} = e^{h(z)} & \text{if } c = 0 \\ |z|e^{\operatorname{Re}\Phi(z)} = e^{\frac{1}{c}h(z)} & \text{if } c \neq 0. \end{cases}$$

By (35), (36) and the maximum principle

$$\text{(45)} \qquad 0 < h < 1 \quad \text{holds throughout } \Omega.$$

If we define $a := 1$, $b := e$ in case $c = 0$, and $a := \min\{1, e^{1/c}\}$, $b := \max\{1, e^{1/c}\}$ in case $c \neq 0$, then from (44) and (45) we see that the holomorphic function F maps Ω into $A(a, b)$. But then (44), (35) and continuity of h on $\overline{\Omega}$ further show that F is a boundary-preserving map. Hence by 6.15, F maps Ω *onto* $A(a, b)$ and is exactly m-to-1 for some $m \in \mathbb{N}$.

It remains only to prove that $c \neq 0$, which we do by *reductio ad absurdum*. If $c = 0$, then by (42)

$$\text{(46)} \qquad \operatorname{Re}\Phi = h,$$

so by (45), Φ maps Ω into the strip $S := (0, 1) \times \mathbb{R}$. As before, (46), (35) and the continuity of h on $\overline{\Omega}$ imply that Φ is a boundary-preserving map. Hence by 6.15, Φ maps Ω *onto* S. Given any $w = F(z)$ in $A(a, b)$, all the points $\Phi(z) + 2\pi i\mathbb{Z}$ lie in $S = \Phi(\Omega)$. Hence there exist $z_n \in \Omega$ with $\Phi(z_n) = \Phi(z) + 2\pi i n$, $n \in \mathbb{Z}$. These points are distinct and $w = F(z) = e^{\Phi(z)} = e^{\Phi(z_n)} = F(z_n)$ for all $n \in \mathbb{Z}$, contradicting the m-to-1 character of F.

This completes the proof of the lemma, and therewith our third proof of the fundamental mapping theorem.

The key ideas in both these proofs can be adapted to deal with regions of higher connectivity; see the chapter notes below.

Notes to Chapter X

As the reader can see, the *Hauptsatz* (my designation) 10.2 is just the assembly of many subsections proved elsewhere; this final *Beweisordnung* has however become fairly standard (see the texts of Rudin and Conway). For a somewhat different treatment involving the Jordan Curve Theorem see HERVÉ [1970].

For an open subset U of $\mathbb{C}$, any $f \in H(U)$ is the *pointwise* limit in U of a sequence of polynomials. (See, for example, MONTEL [1910], pp. 101–103, M. A. LAVRENTIEFF [1936], p. 13, REIJNIERSE [1938] or MARKUSCHEWITSCH [1967], p. 97. In view of 8.9 the proof comes down to a geometric construction: find simply-connected, compact subsets K_n of U such that $K_1 \subset K_2 \subset \cdots$ and $U = \bigcup_{n=1}^{\infty} K_n$. None of the cited references handles this construction very satisfactorily; but Lemma 1.1 in DAVIE [2007] does so in an elementary and reasonably transparent way.) This is to be contrasted to the equivalence (e) $\Leftrightarrow$ (o) in 10.2. 10.5(i) is from HAUSDORFF [1914], p. 345. (Cf. also MAZURKIEWICZ [1922a].) Other proofs of 10.2(f) and 10.5 may be found in BASYE [1935].

For more on the theme of level-lines 10.9(iii), see ELEKES [2002]. By quite different means the verbatim harmonic-function version of 10.10(iii) gets proved in GREEN [1960].

On 10.12 see CARATHÉODORY and RADEMACHER [1917], T. RADÓ [1922/23c], BANACH and MAZUR [1934] and NOSHIRO [1937]. For the purely topological version of 10.12 in the text, see CARTAN [1933], PLASTOCK [1974], HO [1975], KRASNOSEL'SKIĬ and OPOJTSEV [1978], ABBASOV [1982], and §2.7.5 of SHEIL-SMALL [2002]. For other nice extensions and applications of 10.11 and 10.12 see exercise 5.17 of FEYEL and PRADELLE [1973]. In GORDON [1972] the reader will find a proof of an n-dimensional analog of 10.12 due to Hadamard: If $f : \mathbb{R}^n \mapsto \mathbb{R}^n$ is continuously differentiable, has zero-free Jacobian, and is a proper map, then f is a bijection. (Using the Inverse-function Theorem in $\mathbb{R}^n$, it is easy to adapt the proof of 6.17 to obtain this.) In another direction, Radó (*loc. cit.*) shows that if the region Ω is not simply-connected but has only finitely many holes, none a single point, then any holomorphic, proper self-map of Ω is necessarily a conformal automorphism. See also MUELLER and RUDIN [1991].

10.13 is from HUBER [1951] (see also TSUJI [1960] and SCHIFFER [1946]). For far-reaching extensions of 10.13(iii) see LANDAU and OSSERMAN [1959–60] and MARDEN, RICHARDS and RODIN [1967]. The conclusion 10.14 that the radius ratio is a conformal invariant occurs already in SCHOTTKY [1877], p. 326. For a modern treatment of great generality (the length-area method and the *conformal modulus*) see pp. 220–224 of AHLFORS and SARIO [1960] or pp. 426–440 of PALKA [1991]. For more on the interesting phenomenon of ring equivalence of $H^\infty(\Omega_1)$ with $H^\infty(\Omega_2)$ implying the conformal or conjugate-conformal equivalence of Ω_1 with Ω_2 see BERS [1948], KAKUTANI [1955], RUDIN [1955a], NAGASAWA [1959], NAKAI [1963], [1975], ALLING [1968], SU [1972], ULLRICH [1983], the nice exposition in HEINS [1968], pp. 200–204 and in LUECKING and RUBEL [1984]. The beautiful results 10.15, 10.16 are from BECK [1964] and RICHARDS [1968].

There are a variety of proofs of 10.14; we saw another one in 6.18(v) and we shall see one due to TEICHMÜLLER [1938] in chapter XVIII. A fourth proof proceeds by normalizing the annuli appropriately and considering iterates of the given conformal map; another uses a reflection principle (see Chapter XIII) to extend the map to an entire function which is 0 at 0 and of constant modulus on some circle centered at 0. (Compare then 6.16(iii).) For these see exercises 24 and 25, p. 295 of RUDIN [1987] and RITT [1920–21], or GOLUZIN [1969], p. 208 or T. RADÓ [1924–26]. For a proof based on area (due to Carleman) see PÓLYA and SZEGÖ [1976], exercises IV.81–83 or MARKUSCHEWITSCH [1967], pp. 4–8. A very short proof using *extremal length* is given by FUCHS [1967], p. 92 and a sophisticated proof using covering spaces by RUBEL [1983]. CHERNOFF [1977] makes a nice application of 10.14 to show that there exists a foci-preserving conformal map between two ellipses if and only if they have the same eccentricity. NITSCHE [1962] proved that for each $0 \le \rho < 1$, $A(\rho, 1)$ can be mapped onto $A(r, 1)$ by a univalent function f with $\operatorname{Re} f$ and $\operatorname{Im} f$ each harmonic *if* $0 \le r \le 2(1+\rho^2)^{-1}\rho$. He conjectured that *only if* is also true and that was recently confirmed (**Zbl** 1214.31001).

The concept of a covering space (mentioned in the notes to Chapter IV) is again in the background in 10.18 and 10.19; these lemmas assert in effect that $E^{-1}(U)$ is a covering space of U. See, for example, JULIA [1955] or VEECH [1967] (esp. pp. 103–113) or AHLFORS and SARIO [1960]. On the inversion technique in 10.17 (and in 9.46) see KNASTER and KURATOWSKI [1924]. The proof of 10.20 is an elaboration of that in H. KNESER [1958], pp. 372–375 and should be compared with ideas used in 7.59. For an elementary treatment of 10.14 and 10.20 entirely in terms of covering surfaces see PESCHL [1967], pp. 226–236. For other treatments of the fundamental theorem on doubly-connected regions (10.20), the reader may look at LEVAVASSEUR [1902], GRAESER [1930], GIRONZA SOLANAS [1932], ZARANKIEWICZ [1934], KHAJALIA [1937], [1938], KOMATU [1945a], [1949], ROYDEN [1952], ALBRECHT [1954], LEJA [1954], HEINHOLD [1957], GAIER [1957], [1964] (chapter V), H. J. LANDAU [1961], FUCHS [1967] (via harmonic measure), and §17.2 of HENRICI [1986].

For regions with degree of connectivity n (i.e., $n-1$ bounded components in the complement) there are also canonical conformal types. Generally these involve the extended plane $\mathbb{C}_\infty$ (see the Chapter I notes), e.g., $\mathbb{C}_\infty$ with some parallel line segments [i.e., compact intervals in $\mathbb{C}$] removed (parallel slit regions); some intervals on lines through a common point removed (radial slit regions); or some arcs of concentric circles removed (circular slit regions). Here is a representative result, a version of which, along with more references to the literature, will be presented in Chapter XVIII. Theorem: (i) For any open, connected subset Ω of $\mathbb{C}_\infty$, there exists a conformal map of Ω onto a parallel slit domain. (ii) If $c \in \Omega \cap \mathbb{C}$, then among all conformal maps of Ω into $\mathbb{C}_\infty$ which near c have a representation of the form $1/(z-c) + \sum_{k=1}^{\infty} \alpha_k (z-c)^k$ there is one for which $\operatorname{Re} \alpha_1$ is maximal. (iii) The image of Ω under any such extremal function is a parallel slit region. (iv) If $\mathbb{C}_\infty \setminus \Omega$ has n components, then $\mathbb{C}_\infty \setminus f(\Omega)$ does too for any extremal f, and for a given $c \in \Omega \cap \mathbb{C}$ there is only one function of the form (ii) which maps Ω onto a parallel slit region. This result and its radial slit and circular slit avatars is due variously to Cecioni, Hilbert, Koebe, Courant, Grötzsch, Rengel, dePossel. The proof details are elegant and elementary. There is an excellent careful exposition in ČERNÝ [1982]. Distortion and area theorems from chapter XVIII are used to establish the local uniform boundedness of the functions in (ii) and then 7.6 is exploited to get an extremal function. This proof, reminiscent of Fejér and Riesz's proof of the Riemann Mapping Theorem, is from DEPOSSEL [1931]. (See also RENGEL [1934], the various papers of Koebe, and the long series of papers by H. Grötzsch in the *Proceedings of the Saxon Academy* at Leipzig between 1928 and 1935. GOLUZIN [1969], AHLFORS and SARIO [1960] contain the references on the latter.) In fact, at one point of the proof Riemann's theorem is invoked. Later DEPOSSEL [1939] found a constructive (and short) way around this (see also GARABEDIAN [1976]) and the theorem thus proved subsumes Riemann's completely. There is, not surprisingly, an intimate connection between the solvability of the Dirichlet problem and the existence of conformal maps (a special case of which is examined in §8 of Chapter IX). The mapping of a multiply-connected region onto one of the canonical ones above is treated via the Dirichlet problem in AHLFORS [1979], pp. 251–261. Several papers in the bibliography deal with mappings of multiply-connected regions but I have not tried to provide extensive coverage of it. I mention only the monographs of JULIA [1955], BERGMAN [1970], GOLUZIN [1969], CONWAY [1995] and the papers of LEHTO [1949], REICH and WARSCHAWSKI [1960], ROYDEN [1952], STONE [1962] and especially the survey GAIER [1978]. See also GAIER's book and its extensive bibliography as well as the annotated bibliography of SEIDEL [1952] and the notes of GRUNSKY [1978].

10.27 is a slight modification of the treatment by AXLER [1986]. It is proved somewhat differently (via the harmonic-function version of 8.24) in AXLER, BOURDON and RAMEY [2001]. For more on function theory in multiply-connected regions, see FISHER [1983] and GRUNSKY [1978].

There are several attractive relatives of the Aumann-Carathéodory characterizations of simple-connectivity (10.31). Let Ω be a region in $\mathbb{C}$. Then Aut (Ω) *acts transitively* (meaning that any point of Ω can be moved to any other by a map in Aut (Ω)) if and only if either Ω is simply-connected or Ω is $\mathbb{C}$ with some point removed. See KRANTZ [1983] (elementary) and BEARDON [1985]. Suppose Ω is bounded and its boundary consists of C^2-curves. Then Aut (Ω) is compact (with respect to local uniform convergence) if and only if Ω is *not* simply-connected. An elegant proof of this, together with a deduction of the transitivity theorem from it, will be found in KRANTZ [1990], pp. 137–145. Suppose Ω is a region in $\mathbb{C}$ whose boundary consists of C^1-curves. Instead of the point z_0 in 10.31 being fixed, we suppose there exist $\varphi_n \in$ Aut (Ω) such that the sequence $\{\varphi_n(z_0)\}$ converges to a boundary point of Ω. Under this hypothesis too it follows that Ω is conformal to $\mathbb{D}$; see Theorem 2.2 in KIM and KRANTZ [2005]. Finally, suppose only that the region Ω has finite connectivity. Then Aut (Ω) is infinite if and only if Ω is either simply- or doubly-connected; see HEINS [1946b], MUELLER and RUDIN [1991], and BEARDON and MINDA [2008a]. The region $\Omega := \mathbb{C} \setminus \mathbb{Z}$ has infinite connectivity and Aut (Ω) contains the translations $z \mapsto n + z$ $(n \in \mathbb{Z})$, hence is also infinite. However, for infinitely connected Ω the identity map is always an isolated point in Aut (Ω) (with respect to the local uniform convergence topology); this is a deep result due to F. Klein and H. Poincaré; see chapter 1.2 of ABATE [1989]. A different perspective is afforded by considering Aut (Ω) with the topology of uniform convergence on Ω; for this see GAIER [1984b] and VOLYNEC [1992].

In the course of this chapter we have seen many quite different characterizations of simple-connectivity. Not yet mentioned is an attractive and surprisingly elementary one due to J. H. SHAPIRO [1998]: For all $\lambda, z \in \mathbb{C}$ define $e_\lambda(z) := e^{\lambda z}$. The region Ω is simply-connected if and only if the linear span of $\{e_\lambda : \lambda \in \Lambda\}$ is dense in $H(\Omega)$ whenever the set Λ has a limit point in $\mathbb{C}$. An $(n+1)^{\text{st}}$ characterization of simple-connectivity is sequestered in the Remarks following 12.23 below.

Chapter XI
Isolated Singularities

§ 1 Laurent Series and Classification of Singularities

If f is holomorphic in $V \setminus \{z_0\}$ for some neighborhood V of z_0, we call z_0 an *isolated singularity* of f. The zeros of the denominator of a rational function are natural examples of such z_0, as is $z_0 := 0$ for $f(z) := F(1/z)$, with F entire. In this chapter we will study the behavior of f near z_0, rational functions serving as a prototype. The first step is the acquisition of a bilateral series expansion for f around such a point. Such an expansion, which is not surprising in view of 9.26(ii), flows naturally from the appropriate Cauchy integral formula (cf. the hints to 3.8), which we now derive. [But a proof from Fourier analysis is sketched in the chapter notes.] The appropriate series expansion will involve negative as well as positive exponents, so we need to clarify the nature of the convergence. For $c_n \in \mathbb{C}$ we adopt the definition

$$\sum_{n=-\infty}^{\infty} c_n z^n \qquad \text{means} \qquad \lim_{N\to\infty} \sum_{n=-N}^{N} c_n z^n$$

whenever this limit exists. The domain of convergence is always an annulus and inside the open annulus the asymmetric limit $\lim_{N,M\to\infty} \sum_{n=-M}^{N} c_n z^n$ exists and coincides with this symmetric one, as we soon show (11.8(i)).

Theorem 11.1 (A Cauchy Formula for the Annulus) *Let* $0 \le r < R \le \infty$, $c \in \mathbb{C}$, $A := \{z \in \mathbb{C} : r < |z-c| < R\}$, f *holomorphic in* A. *Then for any* $r < a < b < R$

$$f(z) = \frac{1}{2\pi i}\int_{C(c,b)} \frac{f(\xi)}{\xi - z}\, d\xi - \frac{1}{2\pi i}\int_{C(c,a)} \frac{f(\xi)}{\xi - z}\, d\xi, \qquad a < |z-c| < b.$$

R. B. Burckel, *Classical Analysis in the Complex Plane*, Cornerstones,
https://doi.org/10.1007/978-1-0716-1965-0_12

Proof: Of the many avenues open to us at this point (e.g., 10.13(i) or 7.22 or 8.14), we choose the most elementary. Let $\gamma := C(c, a)$, $\Gamma := C(c, b)$ and let z satisfying $a < |z - c| < b$ be given. Form

$$h(\xi) := \begin{cases} \dfrac{f(\xi) - f(z)}{\xi - z}, & \xi \in A \setminus \{z\} \\ f'(z), & \xi = z. \end{cases}$$

This function is continuous in A, holomorphic in $A \setminus \{z\}$. Apply 5.5(i) to $g(\xi) := (\xi - c)h(\xi)$ and split the integrands apart, keeping in mind that

$$\frac{1}{2\pi i}\int_\gamma \frac{d\xi}{\xi - z} = \mathrm{Ind}_\gamma(z) = 0, \quad \text{while} \quad \frac{1}{2\pi i}\int_\Gamma \frac{d\xi}{\xi - z} = \mathrm{Ind}_\Gamma(z) = 1.$$

Remark: By means of an artifice inspired by Carleman's formula (see the Notes to Chapter V) one of the integrals in 11.1 can be eliminated, and an integral formula for f found which only involves the values of f on one circle. Given $f \in H(A)$, apply the formula in 11.1 to the function $(z - c)^k f(z)$ and then divide by $(z - c)^k$ to get

$$f(z) = \frac{1}{2\pi i}\int_{C(c,b)} \left(\frac{\xi - c}{z - c}\right)^k \frac{f(\xi)}{\xi - z}\, d\xi - \frac{1}{2\pi i}\int_{C(c,a)} \left(\frac{\xi - c}{z - c}\right)^k \frac{f(\xi)}{\xi - z}\, d\xi$$

for all $a < |z - c| < b$ and all $k \in \mathbb{Z}$. Fix such a z and note that $\left|\frac{\xi - c}{z - c}\right|^k \to 0$ as $k \to +\infty$ uniformly for $\xi \in C(c, a)$. This limiting behavior "quenches" the second integral above, resulting in

$$f(z) = \lim_{k\to+\infty} \frac{1}{2\pi i}\int_{C(c,b)} \left(\frac{\xi - c}{z - c}\right)^k \frac{f(\xi)}{\xi - z}\, d\xi,$$

which, a having been arbitrary in the open interval (r, b), holds for all z satisfying $r < |z - c| < b$. On the other hand, for z satisfying $a < |z - c| < R$ the relevant formula for $f(z)$ involves a minus sign before the integral over the circle $C(c, a)$ and the limit as $k \to -\infty$.

The sufficiency of one circle is also encoded in the integral formulae for the Laurent coefficients (11.3), which let the preceding formula be reached via the Laurent series as well.

The decomposition created in the first part of the next theorem is just a special case of 8.24, but because it is so basic it is best not to cite that result with its rather involved proof. It is on the other hand revealing to adapt the idea behind the proof to the especially simple geometry of the present situation.

Theorem 11.2 (Laurent Decomposition and Expansion) *Let* $0 \leq r < R \leq \infty$, $c \in \mathbb{C}$,

$$A := \{z \in \mathbb{C} : r < |z - c| < R\},$$

f holomorphic in A. Then there exist functions F, G holomorphic in $D(0, R)$, $D(0, 1/r)$, respectively, such that

$$f(z) = F(z - c) + \frac{1}{z - c} G\left(\frac{1}{z - c}\right) \qquad \forall z \in A.$$

It follows that there exist $c_n \in \mathbb{C}$ such that

$$f(z) = \sum_{n=-\infty}^{\infty} c_n (z - c)^n = \lim_{N \to \infty} \sum_{n=-N}^{N} c_n (z - c)^n \qquad \forall z \in A$$

with convergence absolute and uniform on compact subsets of A.

Proof: Without loss of generality, $c = 0$. For $r < \rho < R$ define $\gamma_\rho := C(0, \rho)$ and

(1)
$$f_\rho(z) := \frac{1}{2\pi i} \int_{\gamma_\rho} \frac{f(\xi)}{\xi - z} \, d\xi, \qquad z \in \mathbb{C} \setminus \gamma_\rho.$$

If $\max\{|z|, r\} < \rho_1 < \rho_2 < R$, then apply 5.5(i) with $\xi \mapsto \xi f(\xi)/(\xi - z)$ as the function g, $b = R$ and a any number between $\max\{|z|, r\}$ and ρ_1 to conclude that $f_{\rho_1}(z) = f_{\rho_2}(z)$. We can therefore consistently define a holomorphic function F in $D(0, R)$ by

(2)
$$F(z) := f_\rho(z) \qquad \text{whenever } \max\{|z|, r\} < \rho < R.$$

We can apply these considerations to the function $g(w) := (1/w) f(1/w)$ in the annulus $\{w \in \mathbb{C} : 1/R < |w| < 1/r\}$. The analog of (2) says that the equation

(3)
$$G(w) := g_\rho(w) \qquad \text{whenever } \max\{|w|, 1/R\} < \rho < 1/r$$

defines a holomorphic function in $D(0, 1/r)$. Notice that if $r < \rho < \min\{|z|, R\}$,

$$\begin{aligned}
g_{1/\rho}\left(\frac{1}{z}\right) &= \frac{1}{2\pi i} \int_{\gamma_{1/\rho}} \frac{g(\xi)}{\xi - 1/z} \, d\xi = \int_0^1 \frac{g\left(\rho^{-1} e^{2\pi i t}\right)}{\rho^{-1} e^{2\pi i t} - 1/z} \cdot \frac{1}{\rho} e^{2\pi i t} \, dt \\
&= \int_0^1 \frac{f\left(\rho e^{-2\pi i t}\right)}{\rho^{-1} e^{2\pi i t} - 1/z} \, dt = z \int_0^1 \frac{f\left(\rho e^{-2\pi i t}\right) \rho e^{-2\pi i t}}{z - \rho e^{-2\pi i t}} \, dt \\
&= -z \int_0^1 \frac{f\left(\rho e^{2\pi i \tau}\right) \rho e^{2\pi i \tau}}{\rho e^{2\pi i \tau} - z} \, d\tau \quad (\text{setting } \tau := 1 - t) \\
&= -\frac{z}{2\pi i} \int_{\gamma_\rho} \frac{f(\xi)}{\xi - z} \, d\xi = -z f_\rho(z),
\end{aligned}$$

(4)
$$f_\rho(z) = -\frac{1}{z} g_{1/\rho}\left(\frac{1}{z}\right), \qquad r < \rho < \min\{|z|, R\}.$$

Finally, if $z \in A$ is given, pick $r < \rho_1 < |z| < \rho_2 < R$ and get

$$F(z) + \frac{1}{z}G\Big(\frac{1}{z}\Big) \overset{(2)}{=} f_{\rho_2}(z) + \frac{1}{z}G\Big(\frac{1}{z}\Big) \overset{(3)}{=} f_{\rho_2}(z) + \frac{1}{z}g_{1/\rho_1}\Big(\frac{1}{z}\Big)$$

$$\overset{(4)}{=} f_{\rho_2}(z) - f_{\rho_1}(z)$$

(5) $$= f(z), \quad \text{by 11.1.}$$

If the power series for F and G are, respectively,

$$F(z) = \sum_{n=0}^{\infty} c_n z^n \qquad \forall |z| < R, \qquad G(w) = \sum_{n=0}^{\infty} c_{-n-1} w^n \qquad \forall |w| < 1/r,$$

then (5) gives

$$f(z) = \sum_{n=-\infty}^{\infty} c_n z^n \qquad \forall z \in A.$$

Exercise 11.3 (i) *Show that if f is holomorphic in $A := \{z \in \mathbb{C} : r < |z-c| < R\}$ for some $0 \le r < R \le \infty$, then the functions F and G provided by* 11.2 *are unique. Find integral formulas for the coefficients c_n in* 11.2 *and conclude that these coefficients are unique* (CAUCHY [1841]) *and satisfy $|c_n| \le \rho^{-n} \sup\{|f(z)| : |z-c| = \rho\}$ for all $n \in \mathbb{Z}$, $r < \rho < R$.*

Hint: If $\widetilde{F}$, $\widetilde{G}$ also represent f in the same way, use the identity

$$(z-c)\big[F - \widetilde{F}\big](z-c) = \big[\widetilde{G} - G\big]\big((z-c)^{-1}\big), \quad z \in A,$$

to create an entire function that is 0 at c and has a finite limit at infinity, and then invoke Liouville—cf. the proof of 8.24. The inequalities also follow from 3.32(v).

(ii) *With hypotheses and notation as in* (i) *and $c = 0$, prove that*

$$\frac{1}{\pi}\int_0^{2\pi} \operatorname{Re} f(\rho e^{i\theta}) e^{-in\theta}\, d\theta = c_n \rho^n + \bar{c}_{-n}\rho^{-n} \qquad \forall n \in \mathbb{Z}, \quad r < \rho < R.$$

[In the hints to 11.54(ii) these equations are algebraically combined into integral formulae of this kind for the coefficients c_n themselves.]

(iii) (Cauchy from Weierstrass) *f is given in the annulus A above by a locally uniformly convergent bilateral series, as in* 11.2. *Derive from this the Cauchy integral formula in* 11.1.

Hint: Given $r < a < |z-c| < b < R$, insert into that series the integral formulae for the c_n from (i), using $\rho := a$ for $n < 0$ and $\rho := b$ for $n \geq 0$, and then interchange the summation with the integrations.

In view of 11.3(i) we can make

Definition 11.4 Let f be holomorphic in $A := \{z \in \mathbb{C} : r < |z-c| < R\}$ for some $c \in \mathbb{C}$, $0 \leq r < R \leq \infty$. Then

(i) The bilateral series $\sum_{n=-\infty}^{\infty} c_n(z-c)^n$ provided by 11.2 is called the *Laurent series* of f in A and the c_n are called the *Laurent coefficients* of f *in* A.

(ii) When $r = 0$ we call c_{-1} the *residue* of f at c, noted $\operatorname{Res}(f, c)$ or often $\operatorname{Res}(f(z), z = c)$, and the function $\sum_{n=1}^{\infty} c_{-n}(z-c)^{-n}$, that is, the function $(z-c)^{-1}G\big((z-c)^{-1}\big)$ is called the *principal part (=Hauptteil)* or the *singular part* of f at c. Note that by 11.2 this series converges for all $z \in \mathbb{C} \setminus \{c\}$. The power series $\sum_{n=0}^{\infty} c_n(z-c)^n$, that is, the function $F(z-c)$ is called the *regular part (=Nebenteil)* of f at c; this series converges throughout $D(c, R)$.

(iii) When $r = 0$ we call c a *removable singularity* of f if $c_{-n} = 0$ for all $n > 0$, a *polar singularity* or a *pole* of f of *order* or *multiplicity* m (m positive) if $c_{-m} \neq 0$ but $c_{-n} = 0$ for all $n > m$, and an *essential singularity* of f if $c_{-n} \neq 0$ for infinitely many $n > 0$. If $m > 1$, c_{-m} is called the *strength* of the pole; if $m = 1$, the pole is called *simple.*

Remarks: (i) According to 11.3

$$c_{-1} = \frac{1}{2\pi i} \int_{C(c,\rho)} f(z)\, dz = \frac{\rho}{2\pi} \int_0^{2\pi} f\big(c + \rho e^{i\theta}\big) e^{i\theta}\, d\theta$$

(for any $\rho \in (r, R)$). This is the sole surviving term when the Laurent series of f is integrated term-by-term. Hence the designation "residue". More generally, what remains after integrating a holomorphic function around a loop is the sum of the residues of f inside the loop, as will be (precised and) proved in 11.29 below. The rationale for the terminology in (iii) is contained in 11.5(i) and 11.7(ii). According to 11.7(ii) c is a polar singularity of f if and only if the graph of $|f|$ displays a pole at c, equivalently, if and only if f, thought of as a mapping into the Riemann sphere, maps c to the "north pole."

(ii) Termwise differentiation of Laurent series is valid; more precisely (5.44(ii)), if $\sum_{n\in\mathbb{Z}} c_n(z-c)^n$ is the Laurent series of f in A, then $\sum_{n\in\mathbb{Z}} nc_n(z-c)^{n-1}$ is that of f'. Thus the residue of every derivative is 0, a fact that is also clear from the integral formula in (i). Similarly, $\sum_{n\in\mathbb{Z}\setminus\{-1\}} (n+1)^{-1} c_n(z-c)^{n+1}$ is the Laurent series of a holomorphic function F in A such that $F'(z) =$

$f(z) - c_{-1}/(z-c)$. This fact, that every $f \in H(A)$ differs only by a scalar multiple of $1/(z-c)$ from a derivative, is part of 10.26, although there no recipe for the primitive was offered.

(iii) The relationship in 11.5(iv) below is one of several suggesting that greater unity and concision could be achieved by distinguishing (as we have not) between "order" and "multiplicity," the latter being the absolute value of the former and always non-negative. The integer m itself could be called "the order of f at c," denoted either $\text{ord}_f(c)$ or $\text{ord}_c(f)$, the point c being then a zero, a pole, or neither according as this integer is positive, negative, or zero. The right-hand side of 11.48 would then read simply $\sum\{\text{ord}_f(c)\text{Ind}_\gamma(c) : c \in \mathbb{U} \setminus \gamma\}$. The reader needs to be alert, as the literature is not uniform in this matter.

Exercise 11.5 *Let U be an open neighborhood of c, f holomorphic in $U \setminus \{c\}$. Then*

(i) *c is a removable singularity of f if and only if f has a holomorphic extension to U.*

(ii) *c is a pole of f if and only if for some non-constant polynomial P the function $f(z) - P(1/(z-c))$ has c as a removable singularity; c is a pole of order m if and only if $f(z) = F(z)/(z-c)^m$ for some $F \in H(U)$ having $F(c) \neq 0$, that is, if and only if $\lim_{z\to c}(z-c)^m f(z)$ exists in $\mathbb{C} \setminus \{0\}$. It follows that c is a pole of f of multiplicity m if and only if it is a removable singularity and a zero of $1/f$ of multiplicity m.*

(iii) *c is an essential singularity of f if and only if for some non-polynomial entire function g the function $f(z) - g(1/(z-c))$ has c as a removable singularity.*

Hint: Sufficiency comes from the uniqueness of Laurent series (11.3).

(iv) *If c is an inessential singularity of the non-constant function f and $m \in \mathbb{Z}$ is the smallest power of $z-c$ occurring in the Laurent expansion of f, then f has a pole (zero) of multiplicity $|m|$ if and only if $m < 0$ $(m \geq 0)$. Show that the integer m can be computed directly from f in three ways, as*

$$\lim_{r\to 0} \text{Ind}_{f\circ C(c,r)}(0) = \lim_{z\to c} \frac{(z-c)f'(z)}{f(z)} = \lim_{z\to c} \frac{\log|f(z)|}{\log|z-c|}.$$

Hints: According to 5.69 and (ii) above there is an $F \in H(U)$ which is zero-free near c, say in $D(c,r) \subset U$, and satisfies $f(z) = (z-c)^m F(z)$. Then for $z \in D(c,r) \setminus \{c\}$

$$\frac{(z-c)f'(z)}{f(z)} = m + (z-c)\frac{F'(z)}{F(z)}, \quad \frac{\log|f(z)|}{\log|z-c|} = m + \frac{\log|F(z)|}{\log|z-c|}.$$

Now recall 4.4 and the Argument Principle, 5.86.

(v) *Suppose f has a pole at c. After using* (iv) *to find its multiplicity m, we can use* (ii) *to learn that $(z-c)^m f(z)$ is (extendable to be) holomorphic in U. Show that we can then proceed to obtain* $\operatorname{Res}(f,c)$ *itself, as $\frac{1}{(m-1)!}$ times the $(m-1)^{\text{st}}$ derivative of this function at c.* [This theme culminates in 11.30(iv).]

Theorem 11.6 (Casorati-Weierstrass) *Let U be an open neighborhood of c, f holomorphic in $U \setminus \{c\}$. Then c is an essential singularity of f if and only if $f\big(D(c,R) \cap U \setminus \{c\}\big)$ is dense in $\mathbb{C}$ for every $R > 0$.* (Cf. 12.29!)

Proof: ($\Leftarrow$) If c is removable, then $\lim_{z\to c} |f(z)|$ exists and is finite, while if c is a pole, it is clear from 11.5(ii) that $\lim_{z\to c} |f(z)| = \infty$. In either case the density condition fails.

($\Rightarrow$) Suppose for some $R > 0$, $f\big(D(c,R) \cap U \setminus \{c\}\big)$ is not dense in $\mathbb{C}$. We may suppose $D := D(c,R) \subset U$. There is accordingly a $w \in \mathbb{C}$ and $d > 0$ such that $|f(z) - w| \geq d$ for all $z \in D \setminus \{c\}$. The function $h := 1/(f - w)$ is therefore bounded (by $1/d$) and holomorphic in $D \setminus \{c\}$, so by 5.41 may be extended to a holomorphic function, call it h also, in D. Thus

$$f = w + \frac{1}{h} \qquad \text{in } D \setminus \{c\}.$$

Therefore if $h(c) \neq 0$, $w + 1/h$ provides a holomorphic extension of f to D and by 11.5(i) c is a removable singularity. If $h(c) = 0$, then by 11.5(ii) c is a pole of $f - w$.

Remarks: (i) The denseness conclusion also follows when c is a limit point of poles, more precisely, under the following hypotheses: P is a sequence in U convergent to $c \in U \setminus P$ and f is holomorphic in $U \setminus \overline{P}$ with a pole at each $p \in P$. If for some $D := D(c,R) \subset U$ and some $w \in \mathbb{C}$, $f(D \setminus \overline{P})$ is disjoint from a neighborhood of w, then the function $h := 1/(f - w)$ is bounded and holomorphic in $D \setminus \overline{P}$. The points of $D \cap P$ are isolated from each other and by 11.7(ii) the definition $h(D \cap P) := 0$ extends h continuously to $D \setminus \{c\}$. By 5.41 this extension is holomorphic and further extends holomorphically to D. But then from 5.62 follows the contradiction $h(D) = 0$. Besicovitch [1931] is a significant extension of this.

(ii) The conclusion ($\Rightarrow$) of 11.6 can be strengthened (but still short of 12.29) to

The set $\{w \in \mathbb{C} : f^{-1}(w) \cap D(c,\varepsilon)$ is infinite$\}$ *is dense in $\mathbb{C}$, for every $\varepsilon > 0$.*

To prove it, first simplify notation by assuming $c = 0$, $\mathbb{D} \subset U$. For each $n \in \mathbb{N}$, set $A_n := D(0, 1/n) \setminus \{0\}$ and let a non-void open subset D_0 of $\mathbb{C}$ be given. $f(A_n)$ is open (5.77) and by 11.6 it is dense in $\mathbb{C}$. Hence $f(A_1) \cap D_0$ contains $\bar{D}_1$ for some open disk $D_1 \neq \varnothing$. But then $f(A_2) \cap D_1$ contains $\bar{D}_2$ for some open disk $D_2 \neq \varnothing$, etc. By compactness the subset $\bigcap \bar{D}_n$ of D_0 is not void. Any point w in it lies in each $f(A_n)$, that is, $w = f(z_n)$ for some $z_n \in A_n$. Whenever $n > 1/|z_k|$, the point z_n satisfies $|z_n| < 1/n < |z_k|$. Hence $\{z_n : n \in \mathbb{N}\}$ is an infinite subset of $f^{-1}(w)$, for the point $w \in D_0$.

(iii) Kulkarni [2007] pointed out an interesting consequence of (ii) [cf. 12.30(i)]:

If $F \in H(\mathbb{C})$ and $F^{-1}(w)$ is finite for each w in some non-void open subset of $\mathbb{C}$, then F is a polynomial.

Indeed, according to (ii) this property of F means that 0 is not an essential singularity of $f(z) := F(1/z)$. That is, the principal part of f has only finitely many terms, so the Taylor series of F has only finitely many terms.

Theorem 11.7 *Let U be an open neighborhood of c, f holomorphic in $U \setminus \{c\}$. Then*

(i) *c is a removable singularity of f if and only if f is bounded near c; if and only if* $\operatorname{Re} f$ *is bounded above (or below) near c. (Cf. 11.60(iii).)*

(ii) *c is a pole of f if and only if* $\lim_{z\to c} |f(z)| = \infty$.

Moreover, if c is a pole, then $f(U \setminus \{c\})$ is the complement of a compact set.

Proof: The implication ($\Rightarrow$) in both (i) and (ii) is clear from 11.5. On the other hand, if $\lim_{z\to c} |f(z)| = \infty$, then $f(D(c, R) \setminus \{c\})$ can be neither dense in $\mathbb{C}$ nor bounded, for small $R > 0$, so by 11.6 and 11.5(i), f has neither an essential nor a removable singularity at c. The other implications in (i) follow similarly from 11.6 and the last statement of the theorem, which we now prove (independently of them).

Suppose c is a pole. Then $\lim_{z\to c} |f(z)| = \infty$, so for $r > 0$ sufficiently small, $D(c, r) \subset U$ and f is zero-free in $V := D(c, r) \setminus \{c\}$. This means that $F := 1/f$ is holomorphic and zero-free in V and, by 5.41, has a removable singularity at c, removed by defining $F(c) := 0$. The extension (call it F too) is an open map; say, $D(0, R) \subset F(D(c, r))$, $0 < R < \infty$. This means that $f(V) = \{1/w : w \in F(V)\} \supset \{1/w : 0 < |w| < R\} = \mathbb{C} \setminus \bar{D}(0, 1/R)$. Hence $\mathbb{C} \setminus f(U \setminus \{c\})$ is a closed set (5.77) which lies in $\bar{D}(0, 1/R)$.

Remarks: (i) If f maps $U \setminus \{c\}$ onto a simply-connected region $\Omega \neq \mathbb{C}$, then c is a removable singularity. For the Riemann mapping theorem furnishes a conformal map φ of Ω onto $\mathbb{D}$ and 11.7(i) is applicable to $\varphi \circ f$. The extension F maps U onto an open subset of $\bar{\mathbb{D}}$ which contains $\mathbb{D}$, and hence must be $\mathbb{D}$, enabling us to form $\varphi^{-1} \circ F$, which is the holomorphic extension of f to U.

(ii) As indicated in the Chapter V notes, Weierstrass also proved the first "if" in 11.7(i). He did so by using the inequality in 11.3(i): If $|f|$ is bounded near c, and we let the radius ρ decrease to $r := 0$, this inequality forces $c_n = 0$ in the Laurent series expansion, for each $n < 0$, so the singular part of f is $\mathbf{0}$. Similarly, if $|\operatorname{Re} f|$ is bounded near c, and we let $\rho \downarrow 0$ in 11.3(ii), we learn that $c_n = 0$ for each $n < 0$.

Exercise 11.8 (i) *Suppose* $\lim_{N\to\infty} \sum_{n=-N}^{N} c_n z^n$ *exists in* $\mathbb{C}$ *for* $z = a \in \mathbb{C}$ *and for* $z = b \in \mathbb{C}$, *where* $0 < |a| < |b|$. *Prove that then* $\sum_{n=0}^{\infty} c_n z^n$ *converges absolutely for all* $|z| < b$ *and* $\sum_{n=1}^{\infty} c_{-n} z^{-n}$ *converges absolutely for all* $|z| > |a|$.

Hints: For $N \in \mathbb{N}$, $z \in \mathbb{C} \setminus \{0\}$ set $s_N(z) := \sum_{n=-N}^{N} c_n z^n$. Then

$$c_{-N} z^{-N} + c_N z^N = s_N(z) - s_{N-1}(z) \to 0$$

for $z \in \{a, b\}$. In $c_{-N} a^{-N} + c_N a^N \to 0$ multiply by $(ab^{-1})^N$, which also converges to 0, to get

$$c_{-N} b^{-N} + c_N (ab^{-1})^{2N} b^N \to 0.$$

Subtract this from $c_{-N} b^{-N} + c_N b^N \to 0$, getting $\left[1 - (ab^{-1})^{2N}\right] c_N b^N \to 0$. Since $(ab^{-1})^{2N} \to 0$, get $c_N b^N \to 0$. Thus $\{c_N b^N\}_{N\in\mathbb{N}}$ is bounded, so $\sum_{n=0}^{\infty} c_n z^n$ converges absolutely for every $|z| < |b|$ (as in the proof of 3.3).

(ii) *f is holomorphic in $A(c, r, R)$ and* $\lim_{N\to\infty} \sum_{n=-N}^{N} c_n (z - c)^n$ *exists uniformly for* $z \in C(c, \rho)$ *and equals* $f(z)$, *for some* $\rho \in (r, R)$. *Use the integral formula in 11.3(i) to show that the* c_n *must be the Laurent coefficients of* f. *Is this conclusion valid if the series only converges pointwise to* $f(z)$ *for all* $z \in C(c, \rho)$?

(iii) (Cf. 5.31) *Let* F, G *be holomorphic in* $A := \{z \in \mathbb{C} : r < |z| < R\}$ *with Laurent coefficients* f_n, g_n. *Show that for each* $n \in \mathbb{Z}$ *the series*

$$\sum_{k=-\infty}^{\infty} f_k g_{n-k} = \lim_{N\to\infty} \sum_{k=-N}^{N} f_k g_{n-k}$$

converges to the nth *Laurent coefficient of* FG.

Hints: Writing $F_N(z) := \sum_{k=-N}^{N} f_k z^k$, we have $F_N G \to FG$ locally uniformly in A, hence uniformly on $C(0, \rho)$ for each fixed $\rho \in (r, R)$. The nth

coefficient of FG is given (see 11.3) by

$$\begin{aligned}\frac{1}{2\pi i}\int_{C(0,\rho)}\frac{FG(z)}{z^{n+1}}\,dz &= \lim_{N\to\infty}\frac{1}{2\pi i}\int_{C(0,\rho)}\frac{F_N G(z)}{z^{n+1}}\,dz\\ &= \lim_{N\to\infty}\frac{1}{2\pi i}\int_{C(0,\rho)}\sum_{k=-N}^{N} f_k z^k\,\frac{G(z)}{z^{n+1}}\,dz\\ &= \lim_{N\to\infty}\sum_{k=-N}^{N} f_k\frac{1}{2\pi i}\int_{C(0,\rho)}\frac{G(z)}{z^{n-k+1}}\,dz\\ &= \lim_{N\to\infty}\sum_{k=-N}^{N} f_k g_{n-k}, \quad \text{by 11.3 again.}\end{aligned}$$

The next exercise is a useful work-out in the combinatorics of Laurent series, but it also has an ulterior motivation—these facts are needed in volume II.

Exercise 11.9 *Let U, V be open subsets of $\mathbb{C}$ with bounded complements. Let $g\colon U \mapsto V$ and $h\colon V \mapsto \mathbb{C}$ be holomorphic. Suppose that their Laurent expansions for large z are of the form*

$$g(z) = \sum_{j=-1}^{\infty} a_j z^{-j}, \quad a_{-1} \neq 0$$

$$h(z) = \sum_{j=-1}^{\infty} b_j z^{-j}.$$

(i) *Show that for the composite $h \circ g$ the Laurent expansion for large z is of the form*

$$h \circ g(z) = \sum_{j=-1}^{\infty} c_j z^{-j}, \qquad \textit{where}$$

$$c_{-1} = a_{-1}b_{-1},\; c_0 = b_0 + a_0 b_{-1},\; c_1 = \frac{b_1 + a_{-1}b_{-1}a_1}{a_{-1}}.$$

Hints: The hypotheses on g and h are that the functions

$$G(z) := g\left(\frac{1}{z}\right) - \frac{a_{-1}}{z}, \qquad H(z) := h\left(\frac{1}{z}\right) - \frac{b_{-1}}{z}$$

have removable singularities at 0, that $G(0) = a_0$ and $G'(0) = a_1$, while $H(0) = b_0$ and $H'(0) = b_1$. That is,

(1) $$\lim_{|z|\to\infty} [g(z) - a_{-1}z] \quad \text{exists and equals } a_0$$

(2) $$\lim_{|z|\to\infty} z[g(z) - a_{-1}z - a_0] = a_1$$

(3) $$\lim_{|z|\to\infty} [h(z) - b_{-1}z] \quad \text{exists and equals } b_0$$

(4) $$\lim_{|z|\to\infty} z[h(z) - b_{-1}z - b_0] = b_1.$$

Let $\phi(z) := z[g(z) - a_{-1}z - a_0]$ for large z. Then

$$g(z) = a_{-1}z + a_0 + \frac{\phi(z)}{z}$$

and

(2)′ $$\lim_{|z|\to\infty} \phi(z) = a_1.$$

Now for large z

$$h \circ g(z) - a_{-1}b_{-1}z = h\left(a_{-1}z + a_0 + \frac{\phi(z)}{z}\right) - b_{-1}\left(a_{-1}z + a_0 + \frac{\phi(z)}{z}\right) + \left(a_0 + \frac{\phi(z)}{z}\right)b_{-1}.$$

Since $|a_{-1}z + a_0 + \phi(z)/z| \to \infty$ as $|z| \to \infty$ [by (2)′ and the fact that $a_{-1} \neq 0$], it follows from (3) and (2)′ that

(5) $$\lim_{|z|\to\infty} [h \circ g(z) - a_{-1}b_{-1}z] \quad \text{exists and equals } b_0 + a_0b_{-1}.$$

Then for large z

$$\begin{aligned}
&z[h \circ g(z) - a_{-1}b_{-1}z - b_0 - a_0b_{-1}] \\
&= z\left[h\left(a_{-1}z + a_0 + \frac{\phi(z)}{z}\right) - b_{-1}\left(a_{-1}z + a_0 + \frac{\phi(z)}{z}\right) - b_0\right] + b_{-1}\phi(z) \\
&= \frac{1}{a_{-1}}\left(a_{-1}z + a_0 + \frac{\phi(z)}{z}\right) \\
&\quad \times \left[h\left(a_{-1}z + a_0 + \frac{\phi(z)}{z}\right) - b_{-1}\left(a_{-1}z + a_0 + \frac{\phi(z)}{z}\right) - b_0\right] \\
&\quad - \frac{1}{a_{-1}}\left(a_0 + \frac{\phi(z)}{z}\right) \\
&\quad \times \left[h\left(a_{-1}z + a_0 + \frac{\phi(z)}{z}\right) - b_{-1}\left(a_{-1}z + a_0 + \frac{\phi(z)}{z}\right) - b_0\right] + b_{-1}\phi(z).
\end{aligned}$$

Then let $|z| \to \infty$ and cite (4), (3) and (2)′ to see that

$$\lim_{|z|\to\infty} z[h \circ g(z) - a_{-1}b_{-1}z - b_0 - a_0 b_{-1}] = \frac{b_1}{a_{-1}} - \frac{a_0}{a_{-1}} \cdot 0 + b_{-1}a_1$$

(6)
$$= \frac{b_1 + a_{-1}b_{-1}a_1}{a_{-1}}.$$

From (5) and (6) we see that $h \circ g(1/z) - a_{-1}b_{-1}/z$ has a removable singularity at 0 and its Laurent series there looks like

$$h \circ g\left(\frac{1}{z}\right) - \frac{a_{-1}b_{-1}}{z} = \sum_{j=0}^{\infty} c_j z^j, \qquad \text{where}$$

$$c_0 = b_0 + a_0 b_{-1}, \quad c_1 = \frac{b_1 + a_{-1}b_{-1}a_1}{a_{-1}}.$$

Therefore for all large z

$$h \circ g(z) = \sum_{j=-1}^{\infty} c_j z^{-j}$$

where c_{-1}, c_0 and c_1 have the asserted values.

(ii) *Show that if g is a conformal map of U onto V and $\lim_{|z|\to\infty} |g(z)| = \infty$, then the Laurent expansion of g for large z is of the form*

$$g(z) = \sum_{j=-1}^{\infty} a_j z^{-j}, \qquad a_{-1} \neq 0$$

and that of g^{-1} has the form

$$g^{-1}(z) = \sum_{j=-1}^{\infty} b_j z^{-j}, \qquad \textit{with} \quad b_{-1} = \frac{1}{a_{-1}}, \quad b_0 = \frac{-a_0}{a_{-1}}, \quad b_1 = -a_1.$$

Hints: For small positive r, $\mathbb{C} \setminus \bar{D}(0, 1/r) \subset U$ and g is zero-free in this set. Then $f(w) := 1/g(1/w)$ is holomorphic in $D(0,r) \setminus \{0\}$ and $\lim_{w\to 0} f(w) = 0$. Thus f extends to be holomorphic in $D(0,r)$ with $f(0) = 0$. Of course, f is 0 nowhere else. Therefore the extended function is one-to-one. Consequently (5.78), $c := f'(0) \neq 0$, that is, 0 is a zero of multiplicity 1 and $w/f(w)$ is holomorphic in $D(0,r)$, with value $1/c$ at 0. Writing its Taylor series as $1/c + \sum_{n=0}^{\infty} a_n w^{n+1}$, we have

$$wg(1/w) = w/f(w) = 1/c + \sum_{n=0}^{\infty} a_n w^{n+1}, \quad w \in D(0,r) \setminus \{0\}$$

$$g(z) = \frac{1}{c} z + \sum_{n=0}^{\infty} a_n z^{-n}, \quad z \in \mathbb{C} \setminus D(0, 1/r),$$

which is of the form claimed, with $a_{-1} = 1/c$. The same arguments show that the Laurent series of $h := g^{-1}$ for large z has the form

$$h(z) = \sum_{j=-1}^{\infty} b_j z^{-j}, \quad b_{-1} \neq 0.$$

Therefore part (i) is applicable. Since the Laurent series of $(h \circ g)(z)$ is z, all the c_j in part (i) are 0 except $c_{-1} = 1$.

Exercise 11.10 *Let f be holomorphic in $\mathbb{R} \times (a, b)$ for some real a, b with $a < b$ and suppose that f is periodic of period 2π.*

(i) *Show that there exist $c_k \in \mathbb{C}$ such that*

$$f(z) = \sum_{k=-\infty}^{\infty} c_k e^{ikz} \qquad \forall z \in \mathbb{R} \times (a, b)$$

with convergence uniform in $\mathbb{R} \times K$ for any compact $K \subset (a, b)$.

Hints: The Factorization Lemma 5.96 is made to order for this set-up, but I shall outline an *ab ovo* treatment. Define F in $A := \{w \in \mathbb{C} : e^{-b} < |w| < e^{-a}\}$ by

$$F(re^{i\theta}) := f(\theta - i \log r), \qquad e^{-b} < r < e^{-a},\ \theta \in \mathbb{R}.$$

Then F is well defined due to the 2π-periodicity of f. Alternatively viewed, we have

(1) $$f(z) = F(e^{iz}) \qquad \forall z \in \mathbb{R} \times (a, b).$$

Given $w_0 \in A$, there is a holomorphic logarithm iL_0 defined in a neighborhood W_0 of w_0. Thus each $w \in W_0$ satisfies $w = e^{iL_0(w)}$ and so by (1), $F(w) = f(L_0(w))$ for all $w \in W_0 \cap A$. Therefore $F = f \circ L_0$ is holomorphic in $W_0 \cap A$, i.e., F is holomorphic in A. Write the Laurent expansion of F in this annulus

(2) $$F(w) = \sum_{k=-\infty}^{\infty} c_k w^k \qquad \forall w \in A.$$

From (1) and (2) the desired expansion follows.

(ii) *Find integral formulas for the c_k of part* (i).

Hint: Fix $c \in (a,b)$. Then the series converges uniformly on $\mathbb{R} \times \{c\}$, so for every integer n

$$\int_0^{2\pi} f(x+ic)e^{-in(x+ic)}\,dx = \sum_{k=-\infty}^{\infty} c_k \int_0^{2\pi} e^{i(k-n)(x+ic)}\,dx = 2\pi c_n.$$

(iii) *$F \in H(\mathbb{C})$ has two $\mathbb{R}$-linearly independent periods, p_1 and p_2. Show that F is constant.*

Hints: The entire function $f(z) := F(p_1 z/2\pi)$ has periods 2π and $p := 2\pi p_2/p_1$. Replacing $f(x+c)$ by $f(x+p+c)$ in the (ii) hint shows that

$$c_n = c_n e^{inp} \qquad \forall n \in \mathbb{Z}.$$

Since $p \notin \mathbb{R}$, $e^{inp} \neq 1$ if $n \neq 0$, so $c_n = 0$ for all $n \neq 0$. Liouville furnishes a more elementary proof: $F(\mathbb{C}) = F(P)$ for the compact set $\{\lambda_1 p_1 + \lambda_2 p_2 : 0 \leq \lambda_1, \lambda_2 \leq 1\} =: P$, due to periodicity and $\mathbb{R}$-independence.

Exercise 11.11 *Let f be an entire function such that $f(z)e^{-c|z|}$ is bounded for some $c > 0$. Suppose that f is periodic of period 2π. Show that there exist $c_k \in \mathbb{C}$ with $|k| \leq N \leq c$ such that $f(z) = \sum_{k=-N}^{N} c_k e^{ikz}$ for all $z \in \mathbb{C}$.*

Hints: According to 11.10

(1) $$f(z) = \sum_{k=-\infty}^{\infty} c_k e^{ikz} \qquad \forall z \in \mathbb{C},$$

where $c_k = \frac{1}{2\pi}\int_0^{2\pi} f(x)e^{-ikx}\,dx$. If M is a bound for $f(z)e^{-c|z|}$, then by 5.37(ii)

$$\left|f^{(n)}(z)\right| \leq (ce)^n M e^{c|z|}$$

for all $z \in \mathbb{C}$ and all positive integers n. Integrate by parts n-times to get

$$c_k = \frac{1}{2\pi} \cdot \frac{1}{(ik)^n} \int_0^{2\pi} f^{(n)}(x)e^{-ikx}\,dx.$$

Then

$$|c_k| \leq \frac{1}{|k|^n}(ce)^n M e^{2\pi c} = M e^{2\pi c}\left(\frac{ce}{|k|}\right)^n.$$

This holds for all positive integers n, so if $|k| > ce$, then $c_k = 0$. Thus if K is the greatest integer not greater than ce, we have from (1)

(2) $$f(z) = \sum_{k=-K}^{K} c_k e^{ikz} \qquad \forall z \in \mathbb{C}.$$

Because $f(z)e^{-c|z|}$ is bounded on $i\mathbb{R}$, it follows from (2) that in fact no c_k with $|k| > c$ can be non-zero.

§ 2 Rational Functions

We have already encountered rational functions in Chapter VIII. Here we are going to examine some interesting examples and derive a criterion for a power series to represent a rational function. Notice that the definition of pole for a rational function given in 8.4 coincides with that in 11.4, as a consequence of 11.7(ii), for example. Our first result here is purely algebraic and could be proved accordingly. For an *ab initio* algebraic proof see, e.g., HAMILTON [1972]. For history see §24 of NETTO and LEVAVASSEUR [1907]; an alternative proof based on the residue theorem is offered in 11.30(iii).

Theorem 11.12 (Partial-fraction decomposition) *If f is a rational function and its poles occur at $a_1, \ldots, a_n$ with orders $m_1, \ldots, m_n$, then there is a polynomial p and $c_{jk} \in \mathbb{C}$ such that*

$$f(z) = p(z) + \sum_{j=1}^{n} \sum_{k=1}^{m_j} \frac{c_{jk}}{(z-a_j)^k}, \qquad z \in \mathbb{C} \setminus \{a_1, \ldots, a_n\}.$$

Proof: The limit $\lim_{|z|\to\infty} |f(z)|$ exists in $[0, \infty]$. In fact, if P and Q are polynomials such that $f = P/Q$, then this limit is infinite if the degree of P exceeds that of Q, is zero if the degree of Q exceeds that of P and is the ratio of the moduli of the leading coefficients of P and Q when they have the same degree. Since $\lim_{|z|\to\infty} 1/|z-a_j| = 0$, it follows that if $f_j(z) := \sum_{k=1}^{m_j} c_{jk}/(z-a_j)^k$ is the principal part of f at a_j, then the function $p := f - \sum_{j=1}^{n} f_j$ can be extended to an entire function and the limit $\lim_{|z|\to\infty} |p(z)|$ exists in $[0, \infty]$ (and equals $\lim_{|z|\to\infty} |f(z)|$). If this limit is infinite, then p is a polynomial by 5.70, while if this limit is finite, it is obvious that p is bounded and so by Liouville (3.33) is constant.

Exercise 11.13 *If two rational functions agree on an infinite set, they are identical.* (Cf. 11.46.)

Hint: If P_1, P_2, Q_1, Q_2 are polynomials and $P_1/Q_1 = P_2/Q_2$ on a set S, then $P_1Q_2 - P_2Q_1$ is a polynomial which is zero on S. If S is infinite, this polynomial is identically zero.

Example 11.14 *If f is a rational function and $|f| = 1$ on an infinite subset of the unit circle, then f has the form*

$$f(z) = cz^m \prod_{k=1}^{n} \frac{z - a_k}{1 - \bar{a}_k z}$$

for some integer m and some $c \in \mathbb{T}$, $a_k \in \mathbb{C} \setminus \mathbb{T}$; that is, f is a quotient of two finite Blaschke products. (Cf. 11.47 and LANGER [2007].)

Proof: The proof is almost identical to that suggested for 6.16(i). By dividing away the appropriate (positive or negative) power of z, we may assume without loss of generality that f has neither a pole nor a zero at 0. We first prove by a "reflection argument" that in fact $|f| = 1$ everywhere on $\mathbb{T}$. To this end, let $\tilde{f}$ denote the rational function gotten from f by conjugating all its coefficients. Then $f(z)\tilde{f}(1/z)$ is a rational function $r(z)$ which for $z \in \mathbb{T}$ satisfies

$$|f(z)|^2 = f(z)\overline{f(z)} = f(z)\tilde{f}(\bar{z}) = f(z)\tilde{f}(1/z) = r(z).$$

Hence by hypothesis $r(z) = 1$ on an infinite set. Consequently by 11.13 r is the constant function $\mathbf{1}$ and the above chain of equalities means that $|f| = 1$ everywhere on $\mathbb{T}$.

Let $a_1, \ldots, a_N$ and $b_1, \ldots, b_M$ be the zeros and poles, respectively, of f in $\mathbb{D}$, each counted according to its multiplicity. Form

$$R(z) := \prod_{k=1}^{N} \frac{z - a_k}{1 - \bar{a}_k z} \prod_{l=1}^{M} \frac{1 - \bar{b}_l z}{z - b_l}.$$

If $a \in \mathbb{D}$ and $|z| = 1$, then

$$\left|\frac{z - a}{1 - \bar{a}z}\right| = \left|\frac{\bar{z}z - \bar{z}a}{1 - \bar{a}z}\right| = \left|\frac{\overline{1 - \bar{a}z}}{1 - \bar{a}z}\right| = 1,$$

so $|R(z)| = 1$. Moreover f/R and R/f are each (extendable to be) holomorphic in $\mathbb{D}$. Applying the Maximum-Modulus Principle to each, we learn that $|f/R| = 1$ in $\mathbb{D}$. Therefore for some $b \in \mathbb{T}$, $f/R = b$, $f = bR$ in $\mathbb{D}$. Thanks to 11.13 then $f = bR$ throughout $\mathbb{C}$. Finally, set $n = N + M$, $a_{N+l} = 1/\bar{b}_l$ $(l = 1, 2, \ldots, M)$ and $c = \left(\bar{b}_1/b_1\right) \cdots \left(\bar{b}_M/b_M\right) \cdot b$ in order to have

$$f(z) = bR(z) = c \prod_{k=1}^{n} \frac{z - a_k}{1 - \bar{a}_k z}.$$

Example 11.15 *If f is a rational function which is real and positive on the unit circle, then f has the form*

$$f(z) = c \prod_{k=1}^{n} \frac{(z - z_k)(1 - \bar{z}_k z)}{(z - p_k)(1 - \bar{p}_k z)}$$

for some p_k, $z_k \in \mathbb{D}$ and some positive c. (Cf. 11.47(ii).)

Proof: By considering the rational function $1/f$ instead of f if necessary, we may assume that 0 is a pole of f of order, say, l with $l \geq 0$ (and possibly $l = 0$). Form the rational function $F(z) := 1/\bar{f}(1/\bar{z})$.

We have $fF = \mathbf{1}$ on the boundary of $\mathbb{D}$. For if $|z| = 1$, then $1/\bar{z} = z$ and $f(1/\bar{z}) = f(z)$ is real, so $F(z) = 1/f(z)$. Conclude from 11.13 that $fF = \mathbf{1}$ throughout $\mathbb{C}$:

$$f(z) \equiv \bar{f}\left(\frac{1}{\bar{z}}\right).$$

It follows from this equality that if $z_1, \ldots, z_n$ are the zeros of f in $\mathbb{D} \setminus \{0\}$ and $p_1, \ldots, p_m$ are the poles of f in $\mathbb{D} \setminus \{0\}$, each repeated according to its multiplicity, then $1/\bar{z}_1, \ldots, 1/\bar{z}_n$ are the zeros of f in $\mathbb{C} \setminus \mathbb{D}$ and $1/\bar{p}_1, \ldots, 1/\bar{p}_m$ are the poles of f in $\mathbb{C} \setminus \mathbb{D}$. (Look at the concluding part of the proof of 7.35 to see that multiplicities are respected in this transition.) If therefore we form

$$R(z) := \frac{(z - z_1)\cdots(z - z_n)(1 - \bar{z}_1 z)\cdots(1 - \bar{z}_n z)}{(z - p_1)\cdots(z - p_m)(1 - \bar{p}_1 z)\cdots(1 - \bar{p}_m z)} \cdot \frac{1}{z^l}, \tag{1}$$

then f/R is a rational function with no zeros or poles, hence a constant:

$$f = cR. \tag{2}$$

Now for $z \neq 0$

$$R(z) = \frac{(z - z_1)\cdots(z - z_n)\left(\frac{1}{z} - \bar{z}_1\right)\cdots\left(\frac{1}{z} - \bar{z}_n\right)}{(z - p_1)\cdots(z - p_m)\left(\frac{1}{z} - \bar{p}_1\right)\cdots\left(\frac{1}{z} - \bar{p}_m\right)} \cdot z^{n-m-l},$$

so that if $|z| = 1$, that is, $1/z = \bar{z}$, we get

$$R(z) = \frac{|z - z_1|^2 \cdots |z - z_n|^2}{|z - p_1|^2 \cdots |z - p_m|^2} \cdot z^{n-m-l}. \tag{3}$$

Since $f(z)$ is positive for such z, it follows from (2) and (3) that

$$cz^{n-m-l} \quad \text{is positive if } |z| = 1. \tag{4}$$

It follows that $n - m - l = 0$ and $c > 0$. But set then

$$p_{m+1} = \cdots = p_{m+l} = p_n = 0$$

and the asserted result follows from (1) and (2).

Corollary 11.16 *If* $p(\theta) := \sum_{k=-n}^{n} a_k e^{ik\theta}$ $(a_k \in \mathbb{C})$ *and* $p(\theta) \geq 0$ *for all real* θ*, then there exists* $P(\theta) = \sum_{k=0}^{n} c_k e^{ik\theta}$ *such that* $p = |P|^2$.

Proof: Let $\epsilon > 0$ and consider

(1) $$R_\epsilon(z) := \epsilon + \sum_{k=-n}^{n} a_k z^k.$$

This is a rational function which satisfies

$$R_\epsilon\left(e^{i\theta}\right) = \epsilon + p(\theta) > 0 \quad \text{for all real } \theta.$$

Therefore R_ϵ has the form prescribed by the last example: For some integer $n(\epsilon)$, some $z_k(\epsilon)$, $p_k(\epsilon) \in \mathbb{D}$ and some $c(\epsilon) > 0$

(2) $$R_\epsilon(z) = c(\epsilon) \prod_{k=1}^{n(\epsilon)} \frac{\left(z - z_k(\epsilon)\right)\left(1 - \bar{z}_k(\epsilon)z\right)}{\left(z - p_k(\epsilon)\right)\left(1 - \bar{p}_k(\epsilon)z\right)}.$$

Since 0 is the only possible pole of R_ϵ, after appropriate cancellations have been made in (2) and notation changed accordingly, we shall have $p_k(\epsilon) = 0$ for all k:

(3) $$R_\epsilon(z) = c(\epsilon) \prod_{k=1}^{n(\epsilon)} \left(z - z_k(\epsilon)\right)\left(\frac{1}{z} - \bar{z}_k(\epsilon)\right).$$

A comparison of (3) with (1) shows that by ignoring or creating factors with $z_k(\epsilon) = 0$ we can simply suppose, as far as (3) is concerned, that the $n(\epsilon)$ there equals n. Then

$$\epsilon + p(\theta) = R_\epsilon\left(e^{i\theta}\right) = c(\epsilon) \prod_{k=1}^{n} \left|e^{i\theta} - z_k(\epsilon)\right|^2.$$

Letting

$$P_\epsilon(\theta) := \sqrt{c(\epsilon)} \prod_{k=1}^{n} \left[e^{i\theta} - z_k(\epsilon)\right] =: \sum_{k=0}^{n} c_k(\epsilon) e^{ik\theta},$$

we thus have

(4) $$\epsilon + p = |P_\epsilon|^2.$$

A simple computation shows that

$$\sum_{k=0}^{n} |c_k(\epsilon)|^2 = \frac{1}{2\pi} \int_0^{2\pi} |P_\epsilon|^2 = \epsilon + \frac{1}{2\pi} \int_0^{2\pi} p$$

and so $\{(c_0(\epsilon), \ldots, c_n(\epsilon)) : 0 < \epsilon < 1\}$ is a bounded set in $\mathbb{C}^{n+1}$ and compactness provides $\epsilon_j \downarrow 0$ such that $\lim_{j\to\infty} c_k(\epsilon_j)$ exists for each k. Calling this limit c_k and setting $P(\theta) := \sum_{k=0}^{n} c_k e^{ik\theta}$, it follows from (4) that $p = |P|^2$.

Exercise 11.17 *If f is a non-constant rational function such that $|f(z)| = 1$ whenever* $\operatorname{Re} z = 0$, *then f has the form*

$$f(z) = c \prod_{k=1}^{n} \frac{z + \bar{z}_k}{z - z_k}$$

for some $z_k \in \mathbb{C}$ and some unimodular c.

Hint: Let $f = P/Q$, P, Q being polynomials without non-constant common factors. Since $\operatorname{Re} z = 0$ implies $z = -\bar{z}$,

$$1 = f(z)\bar{f}(z) = f(z)\bar{f}(-\bar{z}),$$

whence

$$P(z)\overline{P}(-\bar{z}) = Q(z)\overline{Q}(-\bar{z}) \qquad \forall z \in i\mathbb{R}.$$

Since both sides of this equation are polynomials, it follows that the equality in fact prevails throughout $\mathbb{C}$. The zeros of the polynomial $P(z)\overline{P}(-\bar{z})$ are therefore just those of the polynomial $Q(z)\overline{Q}(-\bar{z})$. As $P(z)$, $Q(z)$ have no common zeros, z_k is a zero of $Q(z)$ if and only if $-\bar{z}_k$ is a zero of $P(z)$. Thus for some $c \in \mathbb{C}$, $N \in \mathbb{N}$, $P(z)/Q(z) = c\prod_{k=1}^{N}(z + \bar{z}_k) \,/\, \prod_{k=1}^{N}(z - z_k)$. (Why is $c \in \mathbb{T}$?)

Exercise 11.18 (i) *If f is one-to-one and holomorphic in an open set U, except for isolated singularities, then none of these is essential and at most one is a pole, which must be simple.*

Hints: Let the singularity-set be S. If $z_0 \in S$ is an essential singularity and $D(z_0, 2r) \setminus \{z_0\} \subset U \setminus S$, then the set $f\big(D(z_0, r) \setminus \{z_0\}\big)$ is dense in $\mathbb{C}$ (11.6) and so meets the non-void open (5.77) set $f\big(U \setminus \bar{D}(z_0, r) \setminus S\big)$. This violates the univalence of f in $U \setminus S$. If z_0, z_1 are distinct points of S which are poles, choose $0 < \epsilon < \frac{1}{2}|z_0 - z_1|$ so that each $D_j := D(z_j, \epsilon)$ lies in U and is disjoint from the rest of S. By 11.7 the f-images of the disjoint sets $D(z_j, \epsilon) \setminus \{z_j\}$ each contain the complement of a disk, which clearly violates the univalence of f. Using 11.5(ii), apply 5.88 to $1/f$ to prove that any pole must be simple.

(ii) *If f is holomorphic in the open set U and one-to-one in $U \setminus A$, where $A \subset U$ has no limit points in U, then f is one-to-one throughout U.*

Hints: Suppose $f(z_0) = f(z_1) = w$ for some $z_0 \in A$, $z_1 \in U \setminus \{z_0\}$. Choose disjoint open disks D_j in U centered at z_j so that $D_j \setminus \{z_j\}$ does not meet A. Then $f(D_j)$ are open neighborhoods of w. As in (i), this leads to a contradiction. [Notice that this is really a result about continuous open maps generally.]

(iii) *U, W are bounded open subsets of $\mathbb{C}$, neither has any isolated boundary points, $A \subset U$, $B \subset W$ have no limit points in these supersets, and f maps $U \setminus A$ conformally onto $W \setminus B$. Show that f extends to a conformal map of U onto W. Is the hypothesis about isolated boundary points necessary?*

Hints: Since f is bounded, there is no pole option in (i): every point of A is a removable singularity of f. So $F \in H(U)$ exists which extends f. Moreover, F is one-to-one in U, by (ii). Of course F maps U into $\overline{F(U \setminus A)} = \overline{f(U \setminus A)} = \overline{W \setminus B} \subset \overline{W}$. We need to show that in fact $F(U) \subset W$, that is, $F(a) \in W$ for each $a \in A$. There is an open disk D centered at a such that $D \setminus \{a\} \subset U \setminus A$, and (5.77) $F(D)$ is an open neighborhood of $F(a)$. The inclusion $F(D) \setminus \{F(a)\} \subset F\big(D \setminus \{a\}\big) = f\big(D \setminus \{a\}\big) \subset W$ shows that if $F(a) \in \overline{W} \setminus W = \partial W$, then $F(a)$ would be an isolated point in ∂W. There being no such points, we infer $F(a) \in W$. Similar conclusions apply to f^{-1}: there is a holomorphic map $G\colon W \mapsto U$ which extends f^{-1}. Since $F(G(w)) = f\big(f^{-1}(w)\big) = w$ for all $w \in W \setminus B$, continuity insures that $F(G(w)) = w$ for all $w \in W$, proving that $F(U) \supset W$.

As to the question, consider $U := D(0,2) \setminus \{1\}$, $W := D(0,2) \setminus \{0\}$, $A := \{0\}$, $B := \{1\}$ and f the identity map on $U \setminus A = W \setminus B$.

(iv) *Show that every $f \in \operatorname{Aut}(\mathbb{D} \setminus \{0\})$ is a rotation.*

Hint: f extends to $F \in H(\mathbb{D})$ (5.41). By the maximum principle F maps into $\mathbb{D}$ and by (ii) F is one-to-one, so $F(0)$ must be 0, and $F \in \operatorname{Aut}(\mathbb{D})$.

Exercise 11.19 *Show that a one-to-one entire function f is a first-degree polynomial.*

Hints: The coefficient of z^{-n} in the Laurent development of $F(z) := f(1/z)$ is $f^{(n)}(0)/n!$. If infinitely many of these were non-zero, F would have an essential singularity at 0 (11.4(iii)) and by 11.18(i) this would violate the one-to-oneness of F. Therefore f is in fact a polynomial, which one-to-oneness forces to be of degree 1.

Alternatively, notice that $f(\mathbb{C})$ is connected (1.4) and, indeed, simply-connected (4.75(ii)). Consequently if $f(\mathbb{C}) \neq \mathbb{C}$, 10.2 provides a conformal $g\colon f(\mathbb{C}) \mapsto \mathbb{D}$ and the composite $g \circ f$ violates Liouville. Thus f is surjective and $f^{-1}\colon \mathbb{C} \mapsto \mathbb{C}$ is continuous, so $f^{-1}(\overline{\mathbb{D}})$ is compact. Outside this compact set we have $|f| > 1$, and we finish by citing 5.70(i). (For yet another proof of the surjectivity of f, form the functions $f_r(z) := rf\big(z/(rf'(0))\big) - rf(0)$ with $z \in \mathbb{D}$, note that $\{f_r : r > 0\}$ is a subset of $\mathcal{S}$ and apply 7.28.) The result of this exercise is also a trivial corollary of the much deeper result 12.30, as well as a genuine corollary of the next exercise.

Exercise 11.20 (i) *Show that if f is holomorphic in $\mathbb{C}$ except for poles that have no limit point in $\mathbb{C}$, and $\lim_{|z|\to\infty} |f(z)|$ exists in $[0,\infty]$, then f is a rational function.*

Hints: Assume f is not identically $\mathbf{0}$. Let P and Z denote its pole-set and zero-set. Since f is holomorphic in $\mathbb{C} \setminus P$, which is connected (1.24), Z has no limit point in $\mathbb{C} \setminus P$ (5.62), hence none in $\mathbb{C}$ (why?). Now $1/f$ has a removable singularity at each point of P and a pole at each point of Z (11.5). Consequently, it satisfies the hypotheses that f does. Working with it instead of f, if need be, means that it suffices to treat the case in which the limit at ∞ is finite. In that case the function has only finitely many poles and the argument in 11.12 can be used.

(ii) *If f is one-to-one and holomorphic in $\mathbb{C}$ except for isolated singularities, then f is a Möbius Transformation: $f(z) = \frac{az+b}{cz+d}$ for some complex numbers a, b, c, d such that $ad - bc \neq 0$.*

Hints: By 11.18(i) f has no essential singularities and at most one pole. 11.18(i) can also be applied to $f(1/z)$, showing that this function has a removable singularity at $z = 0$, whence $\lim_{|z|\to\infty} f(z)$ exists in $\mathbb{C}$. Hence the hypotheses of (i) are met, and so f is rational: $f = P/Q$, where P and Q are polynomials without common zeros. Univalence means that P has at most one zero, hence the form $P(z) = a(z - z_0)^n$. Then f' zero-free (5.78) forces $n = 0$ or 1. Since $1/f$ satisfies the same hypotheses as does f, Q is also of degree 0 or 1. [Note that this proof is independent of 11.19, hence (ii) subsumes 11.19.]

(iii) *Show that all conformal mappings of $\mathbb{C} \setminus \{0\}$ into itself are of the form $z \mapsto c/z$ or $z \mapsto cz$.* (Cf. 10.14.)

(iv) *Each of the sets $A, B \subset \mathbb{C}$ has no cluster point in $\mathbb{C}$. Under what conditions is there a conformal map of $\mathbb{C} \setminus A$ onto $\mathbb{C} \setminus B$?*

Hints: If such a map exists, then from (ii) it has the form $f(z) = (az + b)/(cz + d)$. Hence we need to have $B \cup \{\infty\} = f\big(A \cup \{\infty\}\big)$ on the Riemann sphere. In other words, either B is an affine image of A $(c = 0)$, or the single pole p $(= -d/c)$ of f lies in A, the single pole q $(= a/c)$ of f^{-1} lies in B, and $B \setminus \{q\} = f\big(A \setminus \{p\}\big)$.

(v) *Show that if a_0, a_1 are any distinct complex numbers, then there are exactly six conformal maps of $\mathbb{C} \setminus \{0, 1\}$ onto $\mathbb{C} \setminus \{a_0, a_1\}$, namely*

$$z \mapsto a_j + (a_{1-j} - a_j)z, \quad z \mapsto a_j + (a_{1-j} - a_j)/z,$$
$$z \mapsto a_j + (a_{1-j} - a_j)/(1 - z) \quad (j = 0, 1).$$

On the other hand, there are exactly twenty conformal automorphisms of $\mathbb{C} \setminus \{-1, 0, 1\}$ *and exactly nineteen of* $\mathbb{C} \setminus \{-1, 0, 2\}$. *Hence there is no conformal map of one of these regions onto the other.*

(vi) *Use* (ii) *to find conditions on* $\theta_1, \theta_2, \theta_3 \in (0, 2\pi)$ *which insure that the region* $\mathbb{C} \setminus \{1, e^{i\theta_1}, e^{i\theta_2}, e^{i\theta_3}\}$ *is* rigid, *i.e., its only conformal automorphism is the identity map.*

Hints: 6.4 (iv) may be useful.

(vii) *Suppose* f *is non-constant and holomorphic in* $\mathbb{C}$ *except for a set of singularities* P *that has no limit points in* $\mathbb{C}$, *and suppose that* f *maps each circle in* $\mathbb{C} \setminus P$ *into a circle. Show that* f *is one-to-one in* $\mathbb{C} \setminus P$ *and conclude that it has the form specified in* (ii).

Hints: (HARUKI [1972]) Suppose a, $b \in \mathbb{C} \setminus P$ and

(1) $$f(a) = f(b).$$

Since f is not constant and $\mathbb{C} \setminus P$ is connected (1.24), there exists $c \in \mathbb{C} \setminus P \setminus \{a, b\}$ such that $f'(c) \neq 0$. Then there exists $r > 0$ such that $\bar{D}(c, r) \subset \mathbb{C} \setminus P \setminus \{a, b\}$ and f is one-to-one in $\bar{D}(c, r)$ (5.79). By performing appropriate linear transformations on z and $f(z)$, we can suppose that $c = 0$, $r = 1$ and that the circle into which f maps $\mathbb{T}$ is $\mathbb{T}$. In the open set $\Omega := \mathbb{C} \setminus P \setminus \{w : 1/\overline{w} \in P\} \setminus \{0\}$, which is connected (1.24 again), the function $F(z) := \bar{f}(1/\bar{z})$ is defined and holomorphic (2.4). Since $F(z)f(z) = \bar{f}(1/\bar{z})f(z) = \bar{f}(z)f(z) = 1$ for all $z \in \mathbb{T}$, the uniqueness theorem implies that

(2) $$F(z)f(z) = 1 \qquad \forall z \in \Omega.$$

Now a, $b \in \Omega$ (why?), so $F(a) \overset{(2)}{=} 1/f(a) \overset{(1)}{=} 1/f(b) \overset{(2)}{=} F(b)$. That is, in view of the definition of F, $f(1/\bar{a}) = f(1/\bar{b})$. Since $1/\bar{a}$, $1/\bar{b} \in \mathbb{D} = D(c, r)$ and f is one-to-one in that disk, we get $a = b$. This all shows that f is one-to-one in $\mathbb{C} \setminus P$.

The next five exercises develop the rationality criterion mentioned at the beginning of this section. Besides its intrinsic interest, it will be essential to certain Chapter XVI developments.

Exercise 11.21 *Let* $q_0, q_1, q_2, \ldots$ *be rational numbers and* $z_0, \ldots, z_n \in \mathbb{C}$ *not all* 0 *satisfy*

(1) $$z_0 q_k + z_1 q_{k+1} + \cdots + z_n q_{k+n} = 0 \qquad \forall k = 0, 1, 2, \ldots$$

Show that then there exist rational numbers $r_0, \ldots, r_n$ *not all* 0 *satisfying*

(2) $$r_0 q_k + r_1 q_{k+1} + \cdots + r_n q_{k+n} = 0 \qquad \forall k = 0, 1, 2, \ldots.$$

Hints: Let $Q_k := (q_k, q_{k+1}, \dots, q_{k+n}) \in \mathbb{C}^{n+1}$ $(k = 0, 1, 2, \dots)$. Then (1) says that the subspace spanned by these vectors has $\mathbb{C}$-dimension less than $n+1$, being orthogonal to the non-zero vector $(z_0, \dots, z_n)$. Let $Q_{k_1}, \dots, Q_{k_m}$ $(m \le n)$ be a maximal $\mathbb{C}$-linearly independent set. Then $Q_{k_1}, \dots, Q_{k_m}$ are also linearly independent over the rationals. As column rank (over the rationals) does not exceed row rank (over the rationals) in the $m \times (n+1)$ matrix

$$\begin{bmatrix} Q_{k_1} \\ \vdots \\ Q_{k_m} \end{bmatrix}$$

(for a painless and brief proof of which see LIEBECK [1966]), there exists a non-trivial linear relation among the $n+1$ columns of this matrix. That is, rational $r_0, \dots, r_n$ not all 0 exist such that

$$\begin{bmatrix} Q_{k_1} \\ \vdots \\ Q_{k_m} \end{bmatrix} \begin{bmatrix} r_0 \\ \vdots \\ r_n \end{bmatrix} = \begin{bmatrix} 0 \\ \vdots \\ 0 \end{bmatrix} \in \mathbb{C}^m. \tag{3}$$

Now, given Q_k, there exist $a_1, \dots, a_m \in \mathbb{C}$ such that

$$Q_k = a_1 Q_{k_1} + \cdots + a_m Q_{k_m}$$

(by maximality of $\{Q_{k_1}, \dots, Q_{k_m}\}$). Consequently

$$Q_k \begin{bmatrix} r_0 \\ \vdots \\ r_n \end{bmatrix} = a_1 Q_{k_1} \begin{bmatrix} r_0 \\ \vdots \\ r_n \end{bmatrix} + \cdots + a_m Q_{k_m} \begin{bmatrix} r_0 \\ \vdots \\ r_n \end{bmatrix} \overset{(3)}{=} 0 \in \mathbb{C}.$$

Exercise 11.22 *Let $f(z) = \sum_{n=0}^{\infty} a_n z^n$ and $g(z) = \sum_{n=0}^{\infty} b_n z^n$ each have positive radius of convergence. Suppose the a's are all integers and have no common integer factors save ± 1. Assume also that the b's are all integers. Show that any common integer factor of all the coefficients of $f(z)g(z)$ is also a common integer factor of all the b's.*

Hints: The nth coefficient of the product series is (5.31) $c_n := \sum_{j=0}^{n} a_j b_{n-j}$. Suppose $p > 1$ is a prime factor of all the c's which nevertheless is not a common factor of all the b's. There exists then a least $l \ge 0$ such that p is not a factor of a_l and a least $n \ge l$ such that p is not a factor of b_{n-l}. Then p is not a factor of the product $a_l b_{n-l}$. Yet by minimality of l and n, p is a factor of each term in the sum $\sum_{j=0}^{l-1} a_j b_{n-j} + \sum_{j=l+1}^{n} a_j b_{n-j}$, as well as being a factor of c_n. Since the difference of c_n and these sums is $a_l b_{n-l}$, we have a contradiction.

Exercise 11.23 (FATOU [1904]; cf. BOREL [1894]) *Let the power series $f(z) = \sum_{n=0}^{\infty} c_n z^n$ have positive radius of convergence and integer coefficients. Suppose that f is a rational function. Show that then there exist polynomials P, Q with integer coefficients such that $f = P/Q$, $Q(0) = 1$ and P, Q have no common non-constant polynomial factor.*

Hints: By hypothesis there exist $a_0, \ldots, a_m; b_0, \ldots, b_k \in \mathbb{C}$ such that

$$f(z) = \frac{a_0 + a_1 z + \cdots + a_m z^m}{b_0 + b_1 z + \cdots + b_k z^k} \qquad (z \text{ near } 0). \tag{1}$$

We may assume that $b_k \neq 0$ and that

$$(*) \qquad \left\{ \begin{array}{l} \text{No denominator in any representation of } f \text{ as a} \\ \text{quotient of polynomials has degree less than } k. \end{array} \right.$$

Multiplying away the denominator in (1) and equating coefficients, we have

$$a_j = \sum_{l=0}^{\min\{k,j\}} b_l c_{j-l}, \quad j = 0, 1, \ldots, m \qquad 0 = \sum_{l=0}^{\min\{k,j\}} b_l c_{j-l}, \quad \forall j > m.$$

If we simply define $c_j := 0$ for all $j < 0$, then these equations can be more conveniently written as

$$a_j = \sum_{l=0}^{k} b_l c_{j-l}, \qquad j = 0, 1, \ldots, m \tag{2}$$

$$0 = \sum_{l=0}^{k} b_l c_{j-l}, \qquad \forall j > m. \tag{3}$$

It follows from (3) and 11.21 that there exists rational numbers $q_0, \ldots, q_k$, not all 0, such that

$$0 = \sum_{l=0}^{k} q_l c_{j-l} \qquad \forall j > m. \tag{3$'$}$$

We may clear the denominators in (3)$'$ and so suppose simply that the q's are all integers. We may also divide away any common integer factors and so suppose that

(4) $\quad q_0, \ldots, q_k \quad$ are integers with no common integer factors save ± 1.

If we define integers $p_0, \ldots, p_m$ by

(2)′ $$p_j := \sum_{l=0}^{k} q_l c_{j-l} \qquad j = 0, 1, \ldots, m$$

and

(5) $$P(z) := \sum_{l=0}^{m} p_l z^l, \qquad Q(z) := \sum_{l=0}^{k} q_l z^l,$$

then using (2)′ and (3)′ in place of (2) and (3) we reason back to the equality

(6) $$f(z) = \frac{P(z)}{Q(z)}.$$

Notice that, because the degree of Q is k or less, it follows from (6) and $(*)$ that P and Q have no common non-constant polynomial factor.

Next, use 4.56 to select polynomials A, $B \in \mathfrak{R}$ (i.e., with rational coefficients) such that $AP + BQ = \mathbf{1}$. If d is the product of the denominators of the coefficients in A and B, then $U := dA$ and $V := dB$ are polynomials with integer coefficients such that $UP + VQ = d$, whence from (6)

(7) $$(Uf + V)Q = d.$$

Since f is a power series with integer coefficients and positive radius of convergence, so is $Uf + V$. Hence from (4), (7) and 11.22 it follows that d is a common integer factor of all the coefficients of $Uf + V$. If therefore we equate the constant terms in (7), we get $[U(0)f(0) + V(0)]q_0 = d$, where the square-bracketed term is an integer of which d is a factor. It follows that the square-bracketed term must in fact be $\pm d$ and so $q_0 = \pm 1$.

Exercise 11.24 *The power series $f(z) = \sum_{n=0}^{\infty} c_n z^n$ has positive radius of convergence and represents a rational function if and only if there exist non-negative integers p, N and complex numbers $a_0, \ldots, a_p$ not all 0 such that*

$(*)$ $$a_0 c_n + a_1 c_{n+1} + \cdots + a_p c_{n+p} = 0 \qquad \forall n \geq N.$$

Hint: $(\Leftarrow)$ We can suppose $p \geq 1$ and $a_p = 1$. Then $(*)$ reads

$$c_{n+p} = -\sum_{j=0}^{p-1} a_j c_{n+j} \qquad \forall n \geq N.$$

Setting $A := 1 + \sum_{j=0}^{p-1} |a_j|$ and $C := \max\{|c_N|, \ldots, |c_{N+p-1}|\}$, show by induction that

$$|c_{n+p}| \leq CA^{n+1} \leq CA^{n+p} \qquad \forall n \geq N, \text{whence}$$
$$|c_k|^{1/k} \leq C^{1/k} A \qquad \forall k \geq N + p.$$

The latter inequalities entail [3.1 and 3.3(iii)] that the series has a radius of convergence at least $1/A$. Set $P(z) := \sum_{j=0}^{p} a_j z^{p-j}$ and use $(*)$ to show that $f(z)P(z)$ is a polynomial of degree less than $N + p$.

Exercise 11.25 (Kronecker [1881]) *Let $c_0, c_1, \ldots$ be complex numbers and for each non-negative integer m form the* Hankel *matrix*

$$C_m := \begin{bmatrix} c_0 & c_1 & \cdots & c_m \\ c_1 & c_2 & \cdots & c_{m+1} \\ \vdots & \vdots & & \vdots \\ c_m & c_{m+1} & \cdots & c_{2m} \end{bmatrix}.$$

Show that the power series $\sum_{n=0}^{\infty} c_n z^n$ has positive radius of convergence and represents a rational function if and only if $\det C_m = 0$ for all sufficiently large m.

Hints: ($\Leftarrow$) Let p be the smallest non-negative integer with the property that

$$(1) \qquad \det C_m = 0 \qquad \forall m \geq p.$$

If $p = 0$, use induction on m to show that $c_m = 0$ for all $m \geq 0$ (hence $f = \mathbf{0}$, a rational function): we have $c_0 = \det C_0 = 0$. If $c_0 = c_1 = \cdots = c_{n-1} = 0$, then $0 \overset{(1)}{=} \det C_n = \pm c_n^{n+1}$, whence $c_n = 0$.

If $p > 0$, we have

$$(2) \qquad \det C_{p-1} \neq 0, \qquad \det C_p = 0.$$

It follows that the columns of C_{p-1} are linearly independent while the columns of C_p are linearly dependent. Therefore the last column of C_p is a linear combination of the others, that is, there exist $a_0, \ldots, a_{p-1} \in \mathbb{C}$, $a_p = 1$ such that

$$(3) \qquad \sum_{j=0}^{p} a_j c_{m+j} = 0, \qquad 0 \leq m \leq p.$$

Make the definition

$$(4) \qquad s_m := \sum_{j=0}^{p} a_j c_{m+j}, \qquad m = 0, 1, 2, \ldots .$$

We want to show that $s_m = 0$ for all $m \geq 0$. Go by induction on m and suppose, in view of (3), that for some $q > 0$

$$(5) \qquad s_m = 0 \qquad \forall\, 0 \leq m < p + q.$$

Now consider C_{p+q}. Perform on it the following column operations: For each $p \leq m \leq p+q$ add to the $(m+1)^{\text{st}}$ column (the column headed by c_m) each of the p immediately preceding columns multiplied left to right by $a_0, \ldots, a_{p-1}$. The resulting matrix has the same determinant as C_{p+q} and, thanks to (5), has the form

$$(*) \qquad \begin{bmatrix} c_0 & c_1 & \cdots & c_{p-1} & & & & & & \\ c_1 & c_2 & \cdots & c_p & & & \mathbf{0} & & & \\ \vdots & \vdots & & \vdots & & & & & & \\ c_{p-1} & c_p & \cdots & c_{2(p-1)} & & & & & & \\ & & & & 0 & 0 & \cdots & 0 & 0 & s_{p+q} \\ & & & & 0 & 0 & \cdots & 0 & s_{p+q} & \vdots \\ & & & & \vdots & \vdots & \iddots & & & \vdots \\ & & & & 0 & 0 & \iddots & & & \vdots \\ & & & & 0 & s_{p+q} & & & & \vdots \\ & & & & s_{p+q} & \cdots & \cdots & \cdots & \cdots & \vdots \end{bmatrix}.$$

Compute its determinant by expanding via minors in the column headed c_p. The result is $\pm s_{p+q}$ times the determinant of a $(p+q-1)$ square matrix which has essentially the same form as $(*)$, so the process can be repeated. After $q+1$ iterations we find that $0 \overset{(1)}{=} \det C_{p+q} = \pm s_{p+q}^{q+1} \det C_{p-1}$. Because of (2) this entails $s_{p+q} = 0$, thus completing the induction. We can now appeal to 11.24 (with $N = 0$) to conclude the proof.

($\Rightarrow$) Under these hypotheses we have the equations $(*)$ in 11.24. For $n = N, N+1, \ldots, N+m$ they constitute a non-trivial dependence relation among the columns of C_m headed by $c_N, \ldots, c_{N+p}$, whenever $m \geq N + p$.

§ 3 Isolated Singularities on the Circle of Convergence

Here we look at two beautiful classical results which enable one to decide when a point on the circle of convergence of a power series is the unique pole of a function and to find that pole in terms of the power series coefficients.

Example 11.26 *Let $z_0 \in \mathbb{C} \setminus \{0\}$, U a neighborhood of $\overline{D}(0, |z_0|)$ and f a function holomorphic in $U \setminus \{z_0\}$, with a pole at z_0. Then*

$$\lim_{n\to\infty} \frac{(n+1)f^{(n)}(0)}{f^{(n+1)}(0)} \quad \text{exists and equals } z_0.$$

Proof: Let $g(z) = \sum_{k=1}^{m} b_k/(z_0 - z)^k$ $(m \geq 1, b_m \neq 0)$ be the principal part of f at z_0. Thus $h := f - g$ is holomorphic in U and its power series $\sum_{n=0}^{\infty} a_n z^n$ converges at z_0 (in fact, in $D(0, |z_0| + \epsilon)$ for some $\epsilon > 0$). Therefore

$$(1) \qquad a_n z_0^n \to 0.$$

We have

$$g^{(n)}(0) = \sum_{k=1}^{m} b_k (k+n-1) \cdots (k+1) k z_0^{-k-n}$$

and so

$$\begin{aligned}
\frac{(n+1)f^{(n)}(0)}{f^{(n+1)}(0)} &= \frac{(n+1)h^{(n)}(0) + (n+1)g^{(n)}(0)}{h^{(n+1)}(0) + g^{(n+1)}(0)} \\
&= \frac{(n+1)n!a_n + (n+1)\sum\limits_{k=1}^{m} b_k (k+n-1)\cdots(k+1)k z_0^{-k-n}}{(n+1)!a_{n+1} + \sum\limits_{k=1}^{m} b_k (k+n)(k+n-1)\cdots(k+1)k z_0^{-k-n-1}}
\end{aligned}$$

$$(2) \qquad = \frac{\dfrac{(n+1)!}{(m+n)!} a_n z_0^n + \dfrac{n+1}{m+n} \sum\limits_{k=1}^{m} b_k z_0^{-k} \dfrac{(k+n-1)\cdots(k+1)k}{(m+n-1)!}}{\dfrac{1}{z_0}\dfrac{(n+1)!}{(m+n)!} a_{n+1} z_0^{n+1} + \sum\limits_{k=1}^{m} b_k z_0^{-k-1} \dfrac{(k+n)\cdots(k+1)k}{(m+n)!}}.$$

Now

$$0 \leq \frac{(n+1)!}{(m+n)!} \leq \frac{n+1}{n+m} \leq 1,$$

while

$$\frac{(k+n)\cdots(k+1)k}{(m+n)!} \leq \frac{1}{(m+n)(k-1)!} \to 0, \quad \text{as } n \to \infty,$$

if $k < m$. Therefore, recalling (1), we see that all the terms in both the numerator and the denominator of (2), save for $k = m$, converge to 0 as $n \to \infty$.

Exercise 11.27 *Let the power series $\sum_{n=0}^{\infty} c_n z^n$ have radius of convergence 1 and denote by f the holomorphic function it defines.*

(i) *Show that f has a holomorphic extension to $U \setminus \{1\}$ with a simple pole at 1, where U is a neighborhood of $\overline{\mathbb{D}}$, if and only if*

$$\overline{\lim_{n\to\infty}} \left|c_{n+1} - c_n\right|^{1/n} < 1. \tag{1}$$

Hints: Note that

$$(1-z)f(z) = (1-z)\sum_{n=0}^{\infty} c_n z^n = c_0 + \sum_{n=0}^{\infty}(c_{n+1} - c_n)z^{n+1} \quad \forall z \in \mathbb{D}. \tag{2}$$

Now suppose such an extension exists. Call it F. Since the pole which f has at 1 is supposed to be simple, the function $(1-z)F(z)$ is holomorphic in a neighborhood of $\overline{\mathbb{D}}$, say in $D(0,r)$, $r > 1$. The series on the right-hand side of (2) is the power series at 0 of the function $(1-z)F(z)$ and so by 5.30 it converges throughout $D(0,r)$. By 3.3 this means that $\overline{\lim}_{n\to\infty} \left|c_{n+1} - c_n\right|^{1/n} \leq 1/r < 1$.

Conversely suppose that (1) holds. Then by 3.3 the series on the right-hand side of (2) converges in a disk of radius greater than 1, say in $D(0,r)$. Let g denote the holomorphic function so defined. Then $g(z)/(1-z)$ is the desired holomorphic extension of f to $D(0,r) \setminus \{1\}$. The form of this function shows that 1 is either a first-order pole (if $g(1) \neq 0$) or a removable singularity (if $g(1) = 0$). The latter would mean that f has a holomorphic extension to $D(0,r)$, contrary to the radius of convergence of $\sum c_n z^n$ being 1.

(ii) *f has a holomorphic extension to $U \setminus \{1\}$ with a simple pole at 1, where U is a neighborhood of $\overline{\mathbb{D}}$, if and only if*

$$\overline{\lim_{n\to\infty}} \left|\frac{c_n}{c_{n+1}} - 1\right|^{1/n} < 1. \tag{3}$$

Hints: In view of (i) it suffices to establish the equivalence of (1) and (3). Suppose that (1) holds. Then using the notation of the hints to the second part of (i), we have

$$0 \neq g(1) = c_0 + \sum_{n=0}^{\infty}(c_{n+1} - c_n) = \lim_{N\to\infty} c_N.$$

Consequently

$$\lim_{n\to\infty} \frac{c_{n+1}}{c_n} = 1 \tag{4}$$

and so from 3.9 $\lim_{n\to\infty} |c_n|^{1/n} = 1$. It follows that

$$|c_{n+1}|^{1/n} = \left[|c_{n+1}|^{1/(n+1)}\right]^{1+1/n} \to 1, \quad \text{as } n \to \infty. \tag{5}$$

Since

$$|c_{n+1} - c_n|^{1/n} = |c_{n+1}|^{1/n} \left|1 - \frac{c_n}{c_{n+1}}\right|^{1/n}, \tag{6}$$

we deduce (3) from (1) and (5).

Conversely from (3) follows easily (4), thence (5); and then (3), (5) and (6) yield (1).

Exercise 11.28 *Show that if* $\lim_{n\to\infty} c_n = 0$, *then* $\sum_{n=0}^{\infty} c_n z^n$ *converges in* $\mathbb{D}$ *and no point of the unit circle is a pole for the function* f *so defined, i.e.,* f *does not admit extension to a holomorphic function in* $D(z_0, r) \setminus \{z_0\}$ *with a pole at* z_0 *for any* $z_0 \in \mathbb{T}$, $r > 0$.

Hint: It suffices to show that 1 is not a pole. The inequalities

$$|f(r)| \le \left|\sum_{n=0}^{N} c_n r^n\right| + \left(\sup_{n>N} |c_n|\right) \sum_{n=N+1}^{\infty} r^n = \left|\sum_{n=0}^{N} c_n r^n\right| + \sup_{n>N} |c_n| \frac{r^{N+1}}{1-r},$$

valid for all $0 \le r < 1$ and $N \in \mathbb{N}$, show that $\overline{\lim}_{r\uparrow 1}(1-r)|f(r)| \le \sup_{n>N} |c_n|$.

§ 4 The Residue Theorem and Some Applications

The main result here is the form which Cauchy's theorem takes when singularities are present, the integral formula $F(a) = \frac{1}{2\pi i}\int_\gamma \frac{F(z)}{z-a}\,dz$ being the special case in which $f(z)$ is the function $F(z)/(z-a)$ and F is holomorphic.

Theorem 11.29 (Residue Theorem) *Let* U *be an open subset of* $\mathbb{C}$, A *a non-empty subset of* U *with no limit points in* U, γ *a piecewise smooth loop in* $U \setminus A$ *which is* U*-nullhomologous,* f *holomorphic in* $U \setminus A$. *Then*

$$\frac{1}{2\pi i}\int_\gamma f = \sum_{a\in A} \operatorname{Res}(f, a)\operatorname{Ind}_\gamma(a)$$

and this sum contains only finitely many non-zero terms.

Proof: Let $N := \{z \in \mathbb{C} \setminus \gamma : \mathrm{Ind}_\gamma(z) \neq 0\}$. According to 4.3, $\gamma \cup N$ is compact, and by hypothesis it lies in U. Since A has no limit point in U, $A \cap N$ must be finite. If f_a denotes the principal part of f at a, then f_a is holomorphic in $\mathbb{C} \setminus \{a\}$ and $F := f - \sum_{a \in A \cap N} f_a$ has a removable singularity at each point of $A \cap N$ and so is (extendable to be) holomorphic in $U_0 := (U \setminus A) \cup (A \cap N) = U \setminus (A \setminus N)$. This set is open, since A has no limit points in U. By definition of N, Ind_γ vanishes in $A \setminus N$, hence throughout $(\mathbb{C} \setminus U) \cup (A \setminus N) = \mathbb{C} \setminus U_0$, due to γ being U-nullhomologous. That is, γ is U_0-nullhomologous. So by 7.22 (or by 8.14) $\int_\gamma F = 0$, which we write as

$$\int_\gamma f = \sum_{a \in A \cap N} \int_\gamma f_a. \tag{1}$$

By the definition of principal part, we know there is an entire function $g_a(z) = \sum_{k=1}^{\infty} c_k(a) z^k$ such that $f_a(z) = g_a(1/(z-a))$. The series for g_a converges uniformly on the compact set $\{1/(\xi - a) : \xi \in \gamma\}$, so
(2)

$$\int_\gamma f_a = \int_\gamma \sum_{k=1}^{\infty} c_k(a)(\xi - a)^{-k}\, d\xi = \sum_{k=1}^{\infty} c_k(a) \int_\gamma \frac{d\xi}{(\xi - a)^k} = c_1(a) \int_\gamma \frac{d\xi}{\xi - a},$$

using 2.15. Since $c_1(a)$ is $\mathrm{Res}\,(f, a)$, we have from (2) and 4.4

$$\int_\gamma f_a = 2\pi i \mathrm{Res}\,(f, a) \mathrm{Ind}_\gamma(a). \tag{3}$$

The claimed equality follows from (1) and (3), since of the $a \in A$ only those in $A \cap N$ contribute a non-zero summand. (Recall also the convention that a sum over the empty index set means 0.)

Exercise 11.30 (i) *Generalize Theorem* 11.29 *to the case of finitely many piecewise smooth loops* $\gamma_1, \ldots, \gamma_k$ *in* $U \setminus A$ *satisfying, for* $\gamma := (\gamma_1, \ldots, \gamma_k)$, $\mathrm{Ind}_\gamma = 0$ *throughout* $\mathbb{C} \setminus U$. *Use this and the technique in* 7.23(i) *to formulate a homotopy version of the residue theorem.*

(ii) *Deduce from* 11.29 *that if the open set* U *is simply-connected and* $A \subset U$ *has no limit point in* U, *then* $f \in H(U \setminus A)$ *has a primitive if and only if* $\mathrm{Res}(f, a) = 0$ *for all* $a \in A$.

Hint: By passing to its components we may assume that U is connected (4.72). Then $U \setminus A$ is also connected (1.24). According to (e) $\Rightarrow$ (d) in 10.2 all loops in U are U-nullhomologous. Consequently the equality in 11.29 is valid for all piecewise smooth loops γ in $U \setminus A$. If all residues are 0, it then follows from 2.11 that f has a primitive. Conversely, if f has a primitive, then every residue is 0 by the integral formula in Remark (i) following 11.4.

(iii) (Cauchy, 1826) *Suppose the rational function f has the form P/Q with a polynomial Q of degree greater than that of the polynomial P and let the poles of f be $a_1, \ldots, a_n$. Show that*

$$f(z) = \sum_{j=1}^{n} \operatorname{Res}\left(\frac{f(\xi)}{z-\xi}, \xi = a_j\right) \qquad \forall z \in \mathbb{C} \setminus \{a_1, \ldots, a_n\}$$

and use this to give a new derivation of the partial-fraction representation 11.12.

Hints: Fix such a z and consider the rational function

$$R(\xi) := \frac{f(\xi)}{z-\xi} = \frac{P(\xi)}{(z-\xi)Q(\xi)}.$$

Since $\deg(z-\xi)Q(\xi) \geq 2+\deg P(\xi)$, it is clear that $\lim_{r\to\infty} \int_{C(0,r)} R(\xi)\, d\xi = 0$. Hence by 11.29

$$0 = \sum_{\xi\in\mathbb{C}} \operatorname{Res}(R, \xi) = \operatorname{Res}(R, z) + \sum_{j=1}^{n} \operatorname{Res}(R, a_j). \tag{1}$$

Since f is holomorphic at z, the residue at z of $R(\xi) = -f(\xi)/(\xi - z)$ is $-f(z)$, so (1) reads

$$f(z) = \sum_{j=1}^{n} \operatorname{Res}(R, a_j). \tag{2}$$

Suppose P, Q have no common non-constant factor and that the multiplicity of a_j as a zero of Q is $m = m_j$. Then f has the form $f(\xi) = (\xi - a_j)^{-m} f_j(\xi)$, where f_j is holomorphic in a neighborhood of a_j, and $R(\xi) = f_j(\xi)(z-\xi)^{-1}(\xi - a_j)^{-m}$. This makes it clear that $\operatorname{Res}(R, a_j)$, the coefficient of $\frac{1}{\xi - a_j}$ in the Laurent expansion of R around a_j, is just the coefficient of $(\xi - a_j)^{m-1}$ in the Taylor expansion of the holomorphic function $f_j(\xi) \cdot (z-\xi)^{-1}$ around a_j. That is,

$$\operatorname{Res}(R, a_j) = \frac{1}{(m-1)!} \frac{d^{m-1}}{d\xi^{m-1}} \left[\frac{f_j(\xi)}{z-\xi}\right]\Bigg|_{\xi=a_j}.$$

Using Leibniz' product differentiation rule we see that for appropriate holomorphic functions $C_{jk}(\xi)$ this derivative has the form

$$\frac{d^{m-1}}{d\xi^{m-1}} \left[\frac{f_j(\xi)}{z-\xi}\right] = \sum_{k=0}^{m-1} \frac{C_{jk}(\xi)}{(z-\xi)^{k+1}},$$

so for appropriate numbers c_{jk}

$$\text{(3)} \qquad \operatorname{Res}(R, a_j) = \sum_{k=0}^{m_j-1} \frac{c_{jk}}{(z-a_j)^{k+1}}.$$

Equations (2) and (3) yield the partial-fraction representation of f.

(iv) *Extend the ideas in the preceding hints to show that if f is an arbitrary holomorphic function having a pole at a of multiplicity not exceeding $m \in \mathbb{N}$, then the principal part R of f at a is given, for each $z \neq a$, by*

$$R(z) = \lim_{\xi \to a} \frac{1}{(m-1)!} \frac{d^{m-1}}{d\xi^{m-1}} \left[\frac{(\xi - a)^m f(\xi)}{z - \xi} \right] \Bigg|_{\xi = a}.$$

What if the multiplicity of the pole exceeds m?

Notation: For the initial applications of 11.29 I have in mind, to Gauss sums, Fresnel integrals, the theta function and the Gaussian (normal) distribution, it is convenient to dispose of certain computations in advance; this is done in the following multi-part exercise. It is also convenient to set some notation for the immediate future. The positive integer n is fixed and

$$\begin{aligned} S_n(z) &:= \sum_{k=0}^{n-1} \exp\left[\frac{2\pi i}{n}(z+k)^2\right] && \forall z \in \mathbb{C}, \\ g(z) &:= e^{2\pi i z} - 1 && \forall z \in \mathbb{C}, \\ f &:= \frac{S_n}{g} \quad \text{on } \mathbb{C} \setminus \mathbb{Z}, \\ c &:= 1 + i. \end{aligned}$$

Exercise 11.31 (i) *For any $a, r \in \mathbb{R}$ with $r > 2|a|$ we have*

$$\left| \int_{-r}^{r} e^{2\pi i (a+tc)^2/n}\, dt - \int_{-r}^{r} e^{2\pi i (tc)^2/n}\, dt \right| \leq \sqrt{2}|a| e^{-2\pi r^2/n}.$$

Hints: Let $\Gamma_r := [a - rc, -rc, rc, a + rc, a - rc]$. The function $h(z) := e^{2\pi i z^2/n}$ is entire, so $\int_{\Gamma_r} h = 0$. That is,

$$\begin{aligned} &- \int_{[a+rc, a-rc]} h - \int_{[-rc, rc]} h = \int_{[rc, a+rc]} h + \int_{[a-rc, -rc]} h, \\ &- \int_0^1 h(a + rc - 2trc)\,(-2rc)\, dt - \int_0^1 h\,(-rc + 2trc)\,(2rc)\, dt \\ &= \int_0^1 h(rc + ta)(a)\, dt + \int_0^1 h(a - rc - ta)\,(-a)\, dt. \end{aligned}$$

Or, upon making the obvious changes of variable,

$$c\int_{-r}^{r} h(a+xc)\,dx - c\int_{-r}^{r} h(xc)\,dx = a\int_0^1 h(ta+rc)\,dt - a\int_0^1 h(ta-rc)\,dt.$$

Finally note that for $0 \le t \le 1$

$$\begin{aligned} |h(ta \pm rc)| &= \exp\left[\frac{-2\pi}{n}\operatorname{Im}(ta \pm rc)^2\right] \\ &= \exp\left[\frac{-4\pi}{n}(r \pm ta)r\right] \\ &< \exp\left[\frac{-4\pi}{n}\cdot\frac{r}{2}\cdot r\right] \qquad \text{if } r > 2|a|. \end{aligned}$$

(ii) $\lim_{r\to\infty}\int_{-1/2}^{1/2} f(t \pm rc)\,dt = 0.$

Hints: We have for all real t

$$\left|e^{2\pi i(t\pm rc)} - 1\right| = \left|e^{2\pi i(t\pm r)}e^{\mp 2\pi r} - 1\right|$$

and therefore

$$(1) \qquad \left|e^{2\pi i(t+rc)} - 1\right| \ge 1 - e^{-2\pi r} > \frac{1}{2} \qquad \forall r \ge \frac{1}{2\pi}$$

$$(1)' \qquad \left|e^{2\pi i(t-rc)} - 1\right| \ge e^{2\pi r} - 1 > 1 \qquad \forall r \ge \frac{1}{2\pi}.$$

For any integer $0 \le k \le n-1$ and all $-\frac{1}{2} \le t \le \frac{1}{2}$

$$\begin{aligned} \left|\exp\left[\frac{2\pi i}{n}(t \pm rc + k)^2\right]\right| &= \exp\left[-\frac{4\pi}{n}\operatorname{Re}(t \pm rc + k)\operatorname{Im}(t \pm rc + k)\right] \\ &= \exp\left[-\frac{4\pi}{n}(t \pm r + k)\,(\pm r)\right] \\ &= \exp\left[-\frac{4\pi}{n}r^2\left(1 \pm \frac{t+k}{r}\right)\right] \\ (2) \qquad &\le \exp\left[-\frac{2\pi}{n}r^2\right] \qquad \forall r \ge 2(k+1). \end{aligned}$$

From (1), (1)′ and (2)

$$(3) \qquad |f(t \pm rc)| \le 2ne^{-2\pi r^2/n} \qquad \forall r \ge 2n,\ t \in \left[-\tfrac{1}{2}, \tfrac{1}{2}\right].$$

The assertion (ii) follows.

(iii) $\quad f(z+1) - f(z) = e^{2\pi i z^2/n}\big[e^{2\pi i z} + 1\big] \qquad \forall z \in \mathbb{C} \setminus \mathbb{Z}.$

Hints:

$$\begin{aligned}
&f(z+1) - f(z) \\
&\quad = \frac{1}{e^{2\pi i z} - 1} \sum_{k=0}^{n-1} \left(\exp\left[\frac{2\pi i}{n}(z+1+k)^2\right] - \exp\left[\frac{2\pi i}{n}(z+k)^2\right]\right) \\
&\quad = \frac{1}{e^{2\pi i z} - 1} \left(\exp\left[\frac{2\pi i}{n}(z+n)^2\right] - \exp\left[\frac{2\pi i}{n}z^2\right]\right) \\
&\quad = e^{2\pi i z^2/n} \cdot \frac{\exp\left(\frac{2\pi i}{n}\left[(z+n)^2 - z^2\right]\right) - 1}{e^{2\pi i z} - 1} \\
&\quad = e^{2\pi i z^2/n} \cdot \frac{\exp\left(\frac{2\pi i}{n}\left[2nz + n^2\right]\right) - 1}{e^{2\pi i z} - 1} \\
&\quad = e^{2\pi i z^2/n} \cdot \frac{e^{4\pi i z} - 1}{e^{2\pi i z} - 1}.
\end{aligned}$$

(iv) *For any $r > n$*

$$\left| \int_{-r}^{r} \left[f(\tfrac{1}{2} + tc) - f\left(-\tfrac{1}{2} + tc\right)\right] dt - [1 + (-i)^n] \int_{-r}^{r} e^{2\pi i (tc)^2/n}\, dt \right| \le n e^{-2\pi r^2/n}.$$

Hints: Using (iii),

$$\begin{aligned}
&\int_{-r}^{r} \left[f(\tfrac{1}{2} + tc) - f\left(-\tfrac{1}{2} + tc\right)\right] dt \\
&= \int_{-r}^{r} \exp\left[\frac{2\pi i}{n}\left(-\tfrac{1}{2} + tc\right)^2\right] \left[\exp 2\pi i \left(-\tfrac{1}{2} + tc\right) + 1\right] dt \\
&= \int_{-r}^{r} \exp\left[\frac{2\pi i}{n}\left(-\tfrac{1}{2} + tc\right)^2\right] dt + \int_{-r}^{r} \exp\left[\frac{2\pi i}{n}\left(-\tfrac{1}{2} + tc\right)^2 + 2\pi i \left(-\tfrac{1}{2} + tc\right)\right] dt \\
&= \int_{-r}^{r} \exp\left[\frac{2\pi i}{n}\left(-\tfrac{1}{2} + tc\right)^2\right] dt + \int_{-r}^{r} \exp\left(\frac{2\pi i}{n}\left[\left(\frac{n}{2} - \frac{1}{2} + tc\right)^2 - \frac{n^2}{4}\right]\right) dt \\
&= \int_{-r}^{r} \exp\left[\frac{2\pi i}{n}\left(-\tfrac{1}{2} + tc\right)^2\right] dt + e^{-n\pi i/2} \int_{-r}^{r} \exp\left[\frac{2\pi i}{n}\left(\frac{n}{2} - \frac{1}{2} + tc\right)^2\right] dt.
\end{aligned}$$

Note that $e^{-n\pi i/2} = \left[e^{-\pi i/2}\right]^n = (-i)^n$. Now make two citations of part (i).

Example 11.32 (i) $I := \lim_{R\to\infty} \int_{-R}^{R} e^{-x^2}\,dx$ *exists and equals* $\sqrt{\pi}$.

(ii) $\lim_{R\to\infty} \int_{-R}^{R} \cos(x^2)\,dx$, $\lim_{R\to\infty} \int_{-R}^{R} \sin(x^2)\,dx$ *both exist and equal* $\sqrt{\pi/2}$.

(iii) $\sum_{k=0}^{n-1} e^{2\pi i k^2/n} = \sqrt{n}(1+(-i)^n)/(1-i)$ *for each positive integer* n.

Proof: We use the notation and results of the preceding exercise. Notice that the poles of f are at the zeros of its denominator g, namely at the integers. Set

$$\gamma_r := \left[-rc-\tfrac{1}{2}, -rc+\tfrac{1}{2}, rc+\tfrac{1}{2}, rc-\tfrac{1}{2}, -rc-\tfrac{1}{2}\right], \quad r>0.$$

Then a simple calculation reveals that $\gamma_r \cap \mathbb{R} = \left\{\pm\frac{1}{2}\right\}$. Therefore the unbounded connected sets $(-\infty,-1]$ and $[1,\infty)$ both lie in $\mathbb{C}\setminus\gamma_r$ and so by 4.3 Ind_{γ_r} is constantly 0 on each of these sets. In particular, $\mathrm{Ind}_{\gamma_r}(z)=0$ for all non-zero integers z and therefore

(1) $$\sum_{z\in\mathbb{Z}} \mathrm{Res}\,(f,z)\mathrm{Ind}_{\gamma_r}(z) = \mathrm{Res}\,(f,0)\mathrm{Ind}_{\gamma_r}(0).$$

We use the method of 11.5(v) [with $m=1$] to find this residue:

$$g(z) = e^{2\pi i z} - 1 = \sum_{k=1}^{\infty} \frac{(2\pi i z)^k}{k!} = 2\pi i z h(z),$$

where $h(z) := \sum_{k=1}^{\infty} (2\pi i)^{k-1} z^{k-1}/k!$. Since $h(0)=1$, S_n/h is holomorphic near 0, say $\frac{S_n(z)}{h(z)} =: S_n(0) + \sum_{k=0}^{\infty} c_k z^{k+1}$. Then

$$f(z) = \frac{S_n(z)}{g(z)} = \frac{1}{2\pi i z}\frac{S_n(z)}{h(z)} = \frac{S_n(0)}{2\pi i z} + \sum_{k=0}^{\infty} \frac{c_k}{2\pi i} z^k.$$

Therefore

(2) $$\mathrm{Res}\,(f,0) = \frac{S_n(0)}{2\pi i}.$$

By means of 4.17 we easily check that $\mathrm{Ind}_{\gamma_r}(0)=1$. Using this fact and (2) in equation (1),

(3) $$\sum_{z\in\mathbb{Z}} \mathrm{Res}\,(f,z)\mathrm{Ind}_{\gamma_r}(z) = \frac{S_n(0)}{2\pi i}.$$

According to the Residue Theorem (11.29) with $U := \mathbb{C}$, the left-hand side of (3) equals $\frac{1}{2\pi i}\int_{\gamma_r} f$. Using 2.9(ii) then, (3) becomes

$$\begin{aligned} S_n(0) &= \int_{[-rc-1/2,-rc+1/2]} f + \int_{[-rc+1/2,rc+1/2]} f \\ &\quad + \int_{[rc+1/2,rc-1/2]} f + \int_{[rc-1/2,-rc-1/2]} f \\ &= \int_0^1 f\left(-rc - \tfrac{1}{2} + t\right) dt + \int_0^1 f\left(-rc + \tfrac{1}{2} + 2trc\right)(2rc)\, dt \\ &\quad - \int_0^1 f(rc + \tfrac{1}{2} - t)\, dt + \int_0^1 f(rc - \tfrac{1}{2} - 2trc)\,(-2rc)\, dt. \end{aligned}$$

Whence, upon making the obvious changes of variable and combining the first with the third and the second with the fourth terms,

$$S_n(0) = \int_{-1/2}^{1/2} \left[f\left(-rc + t\right) - f(rc + t)\right] dt + c\int_{-r}^{r} \left[f\left(\tfrac{1}{2} + tc\right) - f\left(-\tfrac{1}{2} + tc\right)\right] dt.$$

Recalling 11.31(iv), we may then write, for some $\phi(r)$ with $\lim_{r\to\infty}\phi(r) = 0$,

$$S_n(0) = \int_{-1/2}^{1/2} \left[f(t - rc) - f(t + rc)\right] dt + c\left[1 + (-i)^n\right] \int_{-r}^{r} e^{2\pi i (tc)^2/n}\, dt + \phi(r).$$

Since $c^2 = 2i$, a change of variable here gives

$$\text{(4)} \qquad \begin{aligned} S_n(0) &= \int_{-1/2}^{1/2} \left[f(t - rc) - f(t + rc)\right] dt \\ &\quad + \frac{c}{2}\sqrt{\frac{n}{\pi}}\left[1 + (-i)^n\right] \int_{-2r\sqrt{\pi/n}}^{2r\sqrt{\pi/n}} e^{-x^2}\, dx + \phi(r). \end{aligned}$$

In (4) let $r \uparrow \infty$, taking account of 11.31(ii): We see that I exists (in $\mathbb{R}$) and satisfies

$$\text{(5)} \qquad S_n(0) = \frac{c}{2}\sqrt{\frac{n}{\pi}}\left[1 + (-i)^n\right] I \qquad \forall n \in \mathbb{N}.$$

Since $S_1(0) = 1$, we get for I the value

$$\text{(6)} \qquad I = \frac{2\sqrt{\pi}}{c(1 - i)} = \sqrt{\pi},$$

establishing (i); and then substitution of (6) into (5) gives (iii).

For the proof of (ii), apply Cauchy's Theorem for Convex Regions to the function $h(z) := e^{-z^2}$ in $\mathbb{C}$ and the piecewise smooth loops Γ_r defined by

$$\Gamma_r(t) := \begin{cases} re^{\pi it/4} & 0 \le t \le 1 \\ (2-t)re^{\pi i/4} & 1 \le t \le 2 \\ (t-2)r & 2 \le t \le 3. \end{cases}$$

Thus $\int_{\Gamma_r} h = 0$, which after the appropriate changes of variable says that

$$(7) \qquad e^{\pi i/4} \int_0^r e^{-ix^2}\,dx - \int_0^r e^{-x^2}\,dx = ir \int_0^{\pi/4} \exp\left[-r^2 e^{2ix}\right] e^{ix}\,dx.$$

For the right-hand side we have the bound

$$\begin{aligned} \left| ir \int_0^{\pi/4} \exp\left[-r^2 e^{2ix}\right] e^{ix}\,dx \right| &\le r \int_0^{\pi/4} \left|\exp\left[-r^2 e^{2ix}\right] e^{ix}\right| dx \\ &= r \int_0^{\pi/4} \exp\left[\operatorname{Re}\left(-r^2 e^{2ix}\right)\right] dx \\ &= r \int_0^{\pi/4} e^{-r^2 \cos 2x}\,dx \\ (8) \qquad &\le \frac{\pi}{4r}, \qquad \text{by 3.31.} \end{aligned}$$

Putting (8) into (7) gives

$$\begin{aligned} &\left| e^{\pi i/4} \int_0^r e^{-ix^2}\,dx - \int_0^r e^{-x^2}\,dx \right| \le \frac{\pi}{4r} \\ (9) \qquad &\left| \int_0^r e^{-ix^2}\,dx - e^{-\pi i/4} \int_0^r e^{-x^2}\,dx \right| \le \frac{\pi}{4r}. \end{aligned}$$

Let $r \uparrow \infty$ in (9), keeping in mind the result (i). It follows that

$$\lim_{r\to\infty} \int_0^r e^{-ix^2}\,dx \text{ exists and equals } e^{-\pi i/4} \cdot \frac{\sqrt{\pi}}{2} = \frac{1-i}{\sqrt{2}} \cdot \frac{\sqrt{\pi}}{2}$$

which is obviously equivalent to (ii).

Remarks: (i) For 11.32(ii) JAMET [1903] and OLDS [1968] use the triangle $[0, r, r(1+i), 0]$ instead of Γ_r; see also REMMERT [1991], p. 200.

(ii) If $\theta \in [0, \pi/4]$ and the loops Γ_r in 11.32 are modified by replacing $\pi/4$ with θ, then the proof there establishes that for c any positive multiple of $e^{i\theta}$

$$\lim_{R\to\infty} \int_{-R}^{R} e^{-c^2x^2}\,dx \quad \text{exists and equals} \quad \frac{\sqrt{\pi}}{c}.$$

All such integrals can also be evaluated by real-variable methods, as in 3.24.

(iii) In 1835 Peter Gustav Lejeune Dirichlet provided a Fourier series proof of 11.32(ii) and (iii). One needs to know that the limits in 11.32(ii) *exist*, a fact that follows easily from integration by parts: For $b > a > 0$

$$4\int_a^b e^{ix^2}dx = 2\int_{a^2}^{b^2} t^{-1/2}e^{it}dt = 2ia^{-1}e^{ia^2} - 2ib^{-1}e^{ib^2} - \int_{a^2}^{b^2} it^{-3/2}e^{it}dt,$$

whence

$$4\left|\int_a^b e^{ix^2}dx\right| \le 2a^{-1} + 2b^{-1} + \int_{a^2}^{b^2} t^{-3/2}dt = 4a^{-1}$$

The function $F_n(t) := S_n(t/2\pi), t \in [0, 2\pi]$, is continuously differentiable and, as is easily checked, $F_n(0) = F_n(2\pi)$. Therefore, according to Remark (iv) following 5.20, its Fourier series converges to it. In particular,

$$S_n(0) = F_n(0) = \sum_{j\in\mathbb{Z}} \widehat{F}_n(j). \tag{10}$$

Routine calculations yield

$$\widehat{F}_n(j) = n\int_0^1 e^{2\pi in(x^2 - jx)}dx, \qquad j \in \mathbb{Z}. \tag{11}$$

Now decompose $\int_{\mathbb{R}} e^{ix^2}dx$ in two ways:

$$\int_{-\infty}^{\infty} e^{2\pi inx^2}dx = \sum_{k\in\mathbb{Z}} \int_k^{k+1} e^{2\pi inx^2}dx = \sum_{k\in\mathbb{Z}} \int_0^1 e^{2\pi in(x^2+2kx)}dx \tag{12}$$

$$= \sum_{k\in\mathbb{Z}} \int_{k-1/2}^{k+1/2} e^{2\pi inx^2}dx = \sum_{k\in\mathbb{Z}} \int_0^1 e^{2\pi in(x+k-1/2)^2}dx$$

$$= \sum_{k\in\mathbb{Z}} i^n \int_0^1 e^{2\pi in(x^2+(2k-1)x)}dx, \tag{13}$$

the last equality thanks to $e^{2\pi i n(k-1/2)^2} = e^{\frac{\pi i n}{2}} = (e^{\frac{\pi i}{2}})^n$ for all $k \in \mathbb{Z}$. Adding (12) to i^{-n}(13) results in

$$(1+i^{-n})\int_{-\infty}^{\infty} e^{2\pi i n x^2}\,dx = \sum_{j\in\mathbb{Z}}\int_0^1 e^{2\pi i n(x^2+jx)}\,dx$$

$$\text{(14)}\qquad \overset{(11)}{=} \sum_{j\in\mathbb{Z}} \frac{1}{n}\widehat{F}_n(-j) \overset{(10)}{=} \frac{1}{n}S_n(0).$$

Taking $n := 1$ shows that $\int_{-\infty}^{\infty} e^{2\pi i x^2}\,dx = \frac{1}{2}(1+i)$, which is 11.32(ii); and with that fact a change of variable in the left-hand integral of (14) converts that equality into the Gauss sum 11.32(iii).

Exercise 11.33 *Let $a > 0$, F the entire function $F(z) := e^{-\pi a z^2}$ and f the holomorphic function $f(z) := F(z)/\big(e^{2\pi i z} - 1\big)$, $z \in \mathbb{C}\setminus\mathbb{Z}$.*

(i) *Show that for each real y*

$$\text{(1)}\qquad \lim_{r\to\infty}\int_{-r}^{r} e^{-\pi a(x+iy)^2}\,dx = \lim_{r\to\infty}\int_{-r}^{r} e^{-\pi a x^2}\,dx \in \mathbb{R}.$$

Call this number $I(a)$. Show that, in fact, uniformly in $y \in \mathbb{R}$ we have

$$\text{(2)}\qquad \lim_{r\to\infty} e^{-\pi a y^2}\left| I(a) - \int_{-r}^{r} e^{-\pi a(x+iy)^2}\,dx\right| = 0.$$

Hints: If R and t are real

$$\big|F(R+it)\big| = \big|e^{-\pi a(R^2+2iRt-t^2)}\big| = e^{-\pi a R^2}e^{\pi a t^2}, \quad \text{and so}$$

$$\text{(3)}\qquad \left|\int_{[R,R+iy]} F\right| \le \int_0^{|y|}\big|F(R\pm it)\big|\,dt \le e^{-\pi a R^2}\int_0^{|y|} e^{\pi a t^2}\,dt.$$

By Cauchy's theorem for rectangles

$$\int_{[-r+iy,r+iy]} F = \int_{[r,r+iy]} F - \int_{[-r,-r+iy]} F + \int_{[-r,r]} F.$$

By (3) the first two integrals on the right-hand side converge to 0 as $r \to \infty$. This gives (1), since from $e^{t^2} > t^2$ follows easily existence of $\lim \int_{-r}^{r} e^{-t^2}\,dt$.

With (1) thus secured, look at $R > r > 0$. Since for x, $y \in \mathbb{R}$

$$\big|e^{-\pi a(x+iy)^2}\big| = e^{\pi a y^2}e^{-\pi a x^2},$$

we have

$$e^{-\pi a y^2}\left|\int_{-R}^{R} e^{-\pi a(x+iy)^2}\,dx - \int_{-r}^{r} e^{-\pi a(x+iy)^2}\,dx\right|$$
$$\le \int_{-R}^{-r} e^{-\pi a x^2}\,dx + \int_{r}^{R} e^{-\pi a x^2}\,dx = \int_{-R}^{R} e^{-\pi a x^2}\,dx - \int_{-r}^{r} e^{-\pi a x^2}\,dx.$$

Let $R \to \infty$ and use (1) to get

$$e^{-\pi a y^2}\left|I(a) - \int_{-r}^{r} e^{-\pi a(x+iy)^2}\,dx\right| \le I(a) - \int_{-r}^{r} e^{-\pi a x^2}\,dx.$$

This proves (2) when we recall the definition of $I(a)$.

(ii) *Show that*

$$\lim_{r\to\infty}\int_{[-r+i,r+i]} f = -I(a)\sum_{n=0}^{\infty} e^{-\pi n^2/a}. \tag{4}$$

Hints: If $z \in i+\mathbb{R}$, then $\left|e^{2\pi i z}\right| = e^{-2\pi} < 1$ and so the series $1/\left(e^{2\pi i z}-1\right) = -\sum_{n=0}^{\infty}\left(e^{2\pi i z}\right)^n$ converges uniformly for such z. Therefore

$$f(z) = -\sum_{n=0}^{\infty} e^{-\pi a z^2}\left(e^{2\pi i z}\right)^n = -\sum_{n=0}^{\infty} e^{-\pi n^2/a} e^{-\pi a(z-in/a)^2}$$

with convergence uniform for z in any compact subset of $i+\mathbb{R}$. Consequently

$$\int_{[-r+i,r+i]} f = -\sum_{n=0}^{\infty} e^{-\pi n^2/a}\int_{[-r+i,r+i]} e^{-\pi a(z-in/a)^2}\,dz$$
$$= -e^{\pi a}\sum_{n=0}^{\infty}\left[e^{-\pi a(1-n/a)^2}\int_{-r}^{r} e^{-\pi a(x+i-in/a)^2}\,dx\right]e^{-2\pi n}. \tag{5}$$

The geometric series $\sum\left(e^{-2\pi}\right)^n$ is convergent and the square-bracketed term above converges to $I(a)e^{-\pi a(1-n/a)^2}$ as $r \to \infty$, uniformly in n, according to (2). It follows easily from this and (5) that

$$\lim_{r\to\infty}\int_{[-r+i,r+i]} f = -e^{\pi a}\sum_{n=0}^{\infty}\left[I(a)e^{-\pi a(1-n/a)^2}\right]e^{-2\pi n},$$

which is (4).

(iii) *Show that*

$$\lim_{r\to\infty}\int_{[-r-i,r-i]} f = I(a)\sum_{n=1}^{\infty} e^{-\pi n^2/a}.$$

Hints: If $z \in -i+\mathbb{R}$, then $\left|e^{-2\pi iz}\right| = e^{-2\pi} < 1$ and so the series

$$\frac{1}{e^{2\pi iz}-1} = \frac{-e^{-2\pi iz}}{e^{-2\pi iz}-1} = \sum_{n=0}^{\infty}\left(e^{-2\pi iz}\right)^{n+1} = \sum_{n=1}^{\infty} e^{-2\pi inz}$$

converges uniformly for such z. The rest is the same as in (ii).

(iv) *Prove that*

$$\sum_{n=-\infty}^{\infty} e^{-\pi an^2} = I(a)\sum_{n=-\infty}^{\infty} e^{-\pi n^2/a}.$$

Hints: Integrate f over the rectangle $[-r-i, r-i, r+i, -r+i, -r-i]$, r of the form $N+\frac{1}{2}$, N an integer. Apply the Residue Theorem, parts (ii) and (iii) and show as in (i) that

(6) $$\lim_{|N|\to\infty}\int_{[N+1/2-i,N+1/2+i]} f = 0.$$

To prove (6), note that for $z \in N+\frac{1}{2}+i\mathbb{R}$

$$1-e^{2\pi iz} = 1+e^{-2\pi\,\mathrm{Im}\,z} \geq 1$$

and so $|f(z)| \leq |F(z)|$ for such z. Therefore (6) follows from the second inequality in (3), with y there being ± 1. To calculate the needed residues, proceed as in the derivation of 11.32 (2).

(v) *Derive anew the identity* 11.32(i) *and show that for all* $z \in \mathbb{C}$ *with* $\mathrm{Re}\, z > 0$

(7) $$\theta(z) := \sum_{n=-\infty}^{\infty} e^{-\pi zn^2} = \frac{1}{\sqrt{z}}\sum_{n=-\infty}^{\infty} e^{-\pi n^2/z},$$

where $\sqrt{z}$ *denotes the (holomorphic) square-root in the right half-plane with positive real part.*

Hints: Take $a = 1$ in (iv) and get $I(1) = 1$. Then make a change of variable in the integral defining $I(a)$ to get $I(a) = \left(1/\sqrt{a}\right)I(1) = 1/\sqrt{a}$. Taking $a = 1/\pi$ gives 11.32(i). Next prove that both sides of (7) converge locally uniformly in the open right half-plane and so define holomorphic functions there. By (iv) and the fact $I(a) = 1/\sqrt{a}$, the two functions agree at all positive real numbers.

Remark: Of key importance in much of analysis and probability theory is the fact that the function $f(x) := e^{-\pi x^2}$, $x \in \mathbb{R}$, is its own *Fourier Transform.* That is, $f = \hat{f}$, where for any function g which is integrable over $\mathbb{R}$, $\hat{g}(y) := \int_{-\infty}^{\infty} e^{-2\pi ixy} g(x)\,dx$, $y \in \mathbb{R}$. To see that indeed $\hat{f}(y) = f(y)$ for all $y \in \mathbb{R}$, just take $a = 1$ in 11.33(i) and use 11.32(i). This real-variable result naturally also has real-variable proofs. One is outlined in 3.24(vi). Another consists in computing $\hat{f}$ by passing the integration in it through the exponential power series (how is that justified?), then using repeated integration by parts on the terms $\int_0^\infty x^n e^{-\pi x^2}\,dx$ in the resulting series.

The Fourier perspective leads to an alternative proof and extension of (v):

(vi) *Suppose that F is holomorphic in $S := \mathbb{R} \times (-\eta, \eta)$ for some $\eta > 0$ and that $z^2 F(z)$ is bounded in S. Show that $|F|$ is integrable over $\mathbb{R}$ and that the series*

$$f(z) := \sum_{n=-\infty}^{\infty} F(z+n) \tag{8}$$

converges locally uniformly in S. Then use 11.10 *and the above definition of the Fourier transform to show that*

$$\sum_{n=-\infty}^{\infty} F(z+n) = \sum_{n=-\infty}^{\infty} \hat{F}(n) e^{2\pi inz} \qquad \forall z \in S. \tag{9}$$

Note: Under less restrictive hypotheses on an integrable $F : \mathbb{R} \mapsto \mathbb{C}$, the instance $z = 0$ of (9) is known as the *Poisson summation formula.*

Hints: If $\sup\{|z^2 F(z)| : z \in S\} =: B < \infty$, then for $z \in S$, $n \in \mathbb{Z}$ and $m \in \mathbb{N}$ with $|z| \le m < |n|/2$

$$|F(z+n)| \le \frac{B}{|z+n|^2} \le \frac{B}{(|n|-m)^2} \le \frac{4B}{n^2},$$

proving convergence claim (8). The function f is thus holomorphic in S and 1-periodic. According to 11.10, appropriately re-scaled, we have

$$f(z) = \sum_{k=-\infty}^{\infty} c_k e^{2\pi ikz} \qquad \forall z \in S \tag{10}$$

with convergence uniform on $\mathbb{R} \times K$ for every compact $K \subset (-\eta, \eta)$, where for each $k \in \mathbb{Z}$

$$\begin{aligned} c_k &:= \int_0^1 f(x)e^{-2\pi ikx}dx \\ &= \sum_{n=-\infty}^{\infty} \int_0^1 F(x+n)e^{-2\pi ikx}dx, \\ &\qquad\qquad \text{because in (8) convergence is uniform over } x = z \in [0,1], \\ &= \sum_{n=-\infty}^{\infty} \int_n^{n+1} F(t)e^{-2\pi ikt}dt \\ &= \int_{-\infty}^{\infty} F(t)e^{-2\pi ikt}dt \qquad \text{(why?)} \\ &= \hat{F}(k), \quad \text{the Fourier transform of} \quad F|\mathbb{R}. \end{aligned}$$

When this is inserted into (10), (8) becomes (9).

(vii) *Show that for $a > 0$ the function $F(z) := e^{-\pi az^2}$ satisfies the hypotheses of* (vi), *for every $\eta > 0$, and infer the following extension of* (7):

(11) $$\sum_{n=-\infty}^{\infty} e^{-\pi a(z+n)^2} = \frac{1}{\sqrt{a}} \sum_{n=-\infty}^{\infty} e^{2\pi inz}e^{-\pi n^2/a} \qquad \forall z \in \mathbb{C}.$$

Hint: As noted in the preceding Remark, $e^{-\pi x^2}$ is its own Fourier transform. A simple change of variable then shows that the Fourier transform of $F|\mathbb{R}$ is

$$\hat{F}(y) = \frac{1}{\sqrt{a}}e^{-\pi y^2/a} \qquad \forall\ y \in \mathbb{R}.$$

Remarks: (i) In 11.35(ii) it is shown that the function $\operatorname{sech}(\pi z) := 2/(e^{\pi z} + e^{-\pi z})$ is also its own Fourier transform. Consequently, there is a relation analogous to (11) for this function; its formulation is left to the reader, along with the easy proof that sech satisfies the requisite boundedness condition in $\mathbb{R} \times (-\eta, \eta)$ (for each $0 < \eta < \infty$).

(ii) If p, r are positive real numbers, then $\int_{-r}^{r} e^{-p(t+z)^2}\,dt$ is an entire function of z (by 7.21). It is easy to see that this integral has a limit in $\mathbb{C}$ as $r \to \infty$ and that convergence is uniform on any compact set of z. Hence (by 5.44) $\int_{\mathbb{R}} e^{-p(t+z)^2}\,dt$ defines an entire function. It is constant on $\mathbb{R}$ because of the translation-invariance of integration over $\mathbb{R}$. Therefore it is constant in $\mathbb{C}$ (5.62). This observation (made by Cauchy in 1823) proves 11.33(1). It also proves that the left-hand side of 11.31(i) converges to 0 as $r \to \infty$, which is the only use made of that inequality in the 11.33 developments.

(iii) The drastically disparate rates of convergence of the two sides of (7) make for some subtle pitfalls ("high-precision fraud"—see **MR** 93f:40001). For example, taking $z := \pi 10^{10}$ in (7) gives

$$\frac{1}{\sqrt{\pi}10^5}\sum_{n\in\mathbb{Z}} e^{-n^2/10^{10}} = \sum_{n\in\mathbb{Z}} e^{-\pi^2 n^2 10^{10}} = 1 + 2\sum_{n=1}^{\infty} e^{-\pi^2 n^2 10^{10}}$$

and easy estimates (without resort to calculators) show that this number lies in the open interval $(1, 1+10^{-4\cdot 10^{10}})$. Consequently, computer evidence would strongly suggest the spurious identity

$$\frac{1}{10^5}\sum_{n\in\mathbb{Z}} e^{-n^2/10^{10}} = \sqrt{\pi}.$$

Similarly, the mathematician's esthetic impulse, seduced by calculating agreement to 16 decimal places, may suggest the specious equality

$$\int_{\mathbb{R}} e^{-x^2/4}\,dx = \sum_{\mathbb{Z}} e^{-n^2/4}.$$

Exercise 11.34 In this multi-part exercise the reader is taken through the steps which LINDELÖF [1902] indicated would culminate in the remarkable "prolongement analytique"

$$(0) \qquad f(z) = \lim_{\alpha\downarrow 0}\sum_{n=0}^{\infty}\frac{1}{n^{\alpha n}}\frac{f^{(n)}(0)}{n!}z^n$$

for all $f \in H(\Omega)$, all $z \in \Omega$ and all regions Ω which are starlike with respect to 0, which naturally includes all convex regions containing 0. Convergence is uniform on compact subsets of Ω.

Note that for $\alpha > 0$ the series converges for all $z \in \mathbb{C}$ (apply the root test), but for $\alpha = 0$ perhaps only converges for z in the largest $D(0, r)$ that lies in Ω. Most of the work is concentrated on proving the special case $\Omega := \mathbb{C} \setminus [1, \infty)$, $f(z) := 1/(1 - z)$, where $f^{(n)}(0)/n! = 1$ for all n. After that the proof of 8.25 can simply be repeated. The heavy lifting consists in proving that the limit *exists*, and that is accomplished by using the residue theorem to represent the series as an (improper) integral. (In the literature one finds several different contours used.) The account below is based on that in HARDY [1949].

Notation: L is the Principal Branch of the Logarithm in $\mathbb{C}\setminus(-\infty,0]$, l that in $\mathbb{C}\setminus[0,\infty)$. (See 3.43.)

$$(1) \qquad 0<\eta<\pi/2; \quad 0\le\alpha\le\eta/\pi; \quad T,R>1; \quad N\in\mathbb{N}.$$

We write

$$(2) \qquad z:=re^{i\theta}, \quad \frac{1}{2}<r<R, \quad 2\eta<\theta<2\pi-2\eta.$$

$$(3) \qquad u:=\tau+it, \quad \tau\ge 1/2, \quad t\in\mathbb{R}, \text{ and } u\notin\mathbb{N}.$$

Denote the set of all such z by $\Delta(\eta,R)$, and let

$$(4) \qquad R_{N,T}:=\left[\frac{1}{2},N+\frac{1}{2}\right]\times[-T,T],$$

$$(5) \qquad F(\alpha,z,u):=\frac{z^u e^{-\alpha uL(u)}e^{-\pi iu}}{\sin\pi u},$$

where, of course, z^u means $e^{u(\log r+i\theta)}$.

F is designed so that for fixed z and α, as a function of u it is holomorphic and has residue $z^n e^{-\alpha n\log n}=z^n n^{-\alpha n}$ at each $n\in\mathbb{N}$. The factor $e^{-\pi iu}$ does not affect this but is inserted to cause the convergence (as N, $T\to\infty$) of the integral of F over $\partial R_{N,T}$, which, via the residue theorem, will sum the series $\sum z^n n^{-\alpha n}$. Various estimates have to be uniform in the parameters and limits taken in a particular order, which is all a bit intricate. But let us begin.

In the sequel any unqualified equation is being asserted of *all* its variables, subject only to (1)–(3).

(i) *Prove that*

$$4|\sin w|\ge e^{|\operatorname{Im} w|} \quad \textit{whenever } w\in\mathbb{C} \textit{ satisfies } |\operatorname{Im} w|\ge\log\sqrt{2}.$$

Hint: $2|\sin w| = |e^{iw}-e^{-iw}| \ge \max\{|e^{\pm iw}|\}-\min\{|e^{\pm iw}|\} = e^{|\operatorname{Im} w|}-e^{-|\operatorname{Im} w|}$.

(ii) *Confirm that*

$$|F(\alpha,z,u)|=\frac{r^\tau\exp\big(t(\pi-\theta+\alpha\tan^{-1}(\tau^{-1}t))\big)}{|\sin\pi u|(\tau^2+t^2)^{\alpha\tau/2}}.$$

(iii) *Using* (i), (ii) *and the fact* $\theta \in (2\eta, 2\pi - 2\eta)$ *deduce that*

$$|F(\alpha, z, u)| \le \frac{4r^\tau e^{-\eta|t|}}{(\tau^2 + t^2)^{\alpha\tau/2}} \quad \textit{whenever } |t| = |\operatorname{Im} u| \ge \tfrac{\log 2}{2\pi}.$$

(iv) *Show that for* $\tau = N + \frac{1}{2}$

$$|F(\alpha, z, u)| \le 2r^{N+\frac{1}{2}} \left(N + \frac{1}{2}\right)^{-\alpha(N+\frac{1}{2})} e^{\pi|t|}.$$

Hint: Use (ii) and 3.30(ii), noting that $\pi - \theta + \alpha \tan^{-1}(\tau^{-1}t) \in (-\pi, \pi)$ for all eligible θ, α, τ, t.

(v) *Show that for* $T^* > T > 1$

$$\left| \int_{[\tau \pm iT, \tau \pm iT^*]} F(\alpha, z, u)\, du \right| < \frac{4r^\tau}{\eta} e^{-\eta T}, \tag{6}$$

and for $r \ne 1$

$$\left| \int_{[\frac{1}{2} \pm iT, N+\frac{1}{2} \pm iT]} F(\alpha, z, u)\, du \right| < 4r^{1/2} \left| \frac{r^N - 1}{\log r} \right| e^{-\eta T}, \tag{7}$$

while for $r = 1$, $4Ne^{-\eta T}$ *is a majorant.*

Hint: Minorize the denominator in (iii) by 1 before integrating.

(vi) *From* (6) *deduce that*

$$\lim_{T\to\infty} \int_{[\tau - iT, \tau + iT]} F(\alpha, z, u)\, du \quad \textit{exists in } \mathbb{C} \tag{8}$$

and from (7) *deduce that*

$$\lim_{T\to\infty} \int_{[\frac{1}{2} \pm iT, N+\frac{1}{2} \pm iT]} F(\alpha, z, u)\, du = 0. \tag{9}$$

Hint: (for (8)) Inequality (6) is a "Cauchy condition" on the "tails."

(vii) *From* (8) *[for both* $\tau := \frac{1}{2}$ *and* $\tau := N + \frac{1}{2}$*] and* (9) *infer that the three limits below exist and satisfy the asserted equality:*

$$\lim_{T\to\infty} \int_{\partial R_{N,T}} F(\alpha, z, u)\, du = \lim_{T\to\infty} \int_{[N+\frac{1}{2} - iT, N+\frac{1}{2} + iT]} F(\alpha, z, u)\, du - \lim_{T\to\infty} \int_{[\frac{1}{2} - iT, \frac{1}{2} + iT]} F(\alpha, z, u)\, du.$$

(viii) *Prove that if* $\alpha > 0$, *then for any* $T > 1$

$$(10)\qquad \left| \int_{[N+\frac{1}{2}-iT,N+\frac{1}{2}+iT]} F(\alpha, z, u)\, du \right| < \left[\frac{4}{\pi} e^{\pi} + \frac{8}{\eta}\right] r^{N+\frac{1}{2}} \left(N + \frac{1}{2}\right)^{-\alpha(N+\frac{1}{2})}.$$

Hint: (iv) and (iii) say that the integral is majorized by

$$2\int_0^1 2r^{N+\frac{1}{2}}\left(N+\frac{1}{2}\right)^{-\alpha(N+\frac{1}{2})} e^{\pi t}\, dt + 2\int_1^T 4r^{N+\frac{1}{2}}\left(N+\frac{1}{2}\right)^{-\alpha(N+\frac{1}{2})} e^{-\eta t}\, dt.$$

(ix) *Infer from* (viii) *and* (8) *that if* $\alpha > 0$, *then*

$$\lim_{N\to\infty} \lim_{T\to\infty} \left| \int_{[N+\frac{1}{2}-iT,N+\frac{1}{2}+iT]} F(\alpha, z, u)\, du \right| \quad \textit{exists and equals } 0.$$

Hint: On the right-hand side of (10) the factor $\left[\frac{r}{(N+\frac{1}{2})^{\alpha}}\right]^{N+\frac{1}{2}}$ is less than $\left(\frac{1}{2}\right)^{N+\frac{1}{2}}$ if N is sufficiently large.

(x) *From* (vii) *and* (ix) *infer that for* $\alpha > 0$

$$\lim_{N\to\infty} \lim_{T\to\infty} \int_{\partial R_{N,T}} F(\alpha, z, u)\, du = \lim_{T\to\infty} \int_{[\frac{1}{2}-iT,\frac{1}{2}+iT]} -F(\alpha, z, u)\, du.$$

(xi) *From the residue theorem deduce that*

$$\frac{1}{2i} \int_{\partial R_{N,T}} F(\alpha, z, u)\, du = \sum_{n=1}^{N} n^{-\alpha n} z^n \qquad \forall T > 1, N \in \mathbb{N}.$$

Hints: With α, z fixed, $u \mapsto F(\alpha, z, u)$ is holomorphic in $\mathbb{H} \setminus \mathbb{N}$ and has a simple pole of residue $\pi^{-1} n^{-\alpha n} z^n$ at each $n \in \mathbb{N}$ [see the hints to 11.37(ii)]. None of these poles lies on $\partial R_{N,T}$, and those inside this positively oriented Jordan-curve are $n = 1, 2, \ldots, N$.

(xii) *By inserting* (xi) *into the left side of* (x), *infer that for* $(\alpha, z) \in (0, \eta/\pi) \times \Delta(\eta, R)$

$$\sum_{n=1}^{\infty} n^{-\alpha n} z^n = \lim_{T\to\infty} \frac{1}{2i} \int_{[\frac{1}{2}-iT,\frac{1}{2}+iT]} -F(\alpha, z, u)\, du.$$

(xiii) *Prove that for each* $T \geq 1$

$$(\alpha, z) \mapsto \int_{[\frac{1}{2}-iT, \frac{1}{2}+iT]} F(\alpha, z, u)\, du$$

is a uniformly continuous function on $[0, \eta/\pi] \times \Delta(\eta, R)$ *which is holomorphic in* z *for each fixed* α.

Hints: According to (5)

$$F(\alpha, z, u) = \frac{e^{-\alpha u L(u)} e^{-\pi i u}}{\sin \pi u} \cdot e^{u l(z)}.$$

Hence F is a continuous function of $(\alpha, z, u) \in [0, \eta/\pi] \times \bar{\Delta}(\eta, R) \times [\frac{1}{2} - iT, \frac{1}{2} + iT]$. It is moreover a holomorphic function of $z \in \Delta(\eta, R)$ for each fixed α, u. Hence its integral with respect to u is a holomorphic function of z (by 7.21).

(xiv) *According to* (8)

$$G(\alpha, z) := \lim_{T \to \infty} \frac{1}{2i} \int_{[\frac{1}{2}-iT, \frac{1}{2}+iT]} -F(\alpha, z, u)\, du$$

exists in $\mathbb{C}$ *for every* $(\alpha, z) \in [0, \eta/\pi] \times \Delta(\eta, R)$. *Show that the convergence is uniform in these parameters.*

Hints: Inequality (6) insures that for all such α, z and all $T^* > T > 1$

$$\left| \int_{[\frac{1}{2}-iT^*, \frac{1}{2}+iT^*]} F(\alpha, z, u)\, du - \int_{[\frac{1}{2}-iT, \frac{1}{2}+iT]} F(\alpha, z, u)\, du \right| < 2 \cdot \frac{4r^{1/2}}{\eta} e^{-\eta T} \leq \frac{8R^{1/2}}{\eta} e^{-\eta T}.$$

(xv) *From* (xiv) *and* (xiii) *infer that for each* $\alpha \in [0, \eta/\pi]$, $G(\alpha, z)$ *is a holomorphic function of* $z \in \Delta(\eta, R)$.

Hint: For each fixed $T > 1$ and fixed α the function in (xiii) is holomorphic in $z \in \Delta(\eta, R)$, and these functions converge uniformly in z to $-2iG(\alpha, z)$ as $T \to \infty$, according to (xiv).

(xvi) *Prove that*

$$(11) \qquad \lim_{\alpha \downarrow 0} \sum_{n=1}^{\infty} n^{-\alpha n} w^n = \frac{w}{1-w} \quad \textit{locally uniformly for } w \in \mathbb{D}.$$

Hint: This follows from the inequalities

$$\left|\frac{w}{1-w} - \sum_{n=1}^{\infty} n^{-\alpha n} w^n\right| = \left|\sum_{n=1}^{\infty}(1-n^{-\alpha n})w^n\right| \le \sum_{n=1}^{N}(1-n^{-\alpha n}) + \sum_{n=N+1}^{\infty} |w|^n$$
$$= \sum_{n=1}^{N}(1-n^{-\alpha n}) + |w|^{N+1}\bigl(1-|w|\bigr)^{-1},$$

valid for all $w \in \mathbb{D}$, $N \in \mathbb{N}$, $\alpha > 0$.

(xvii) *Show that*

$$\lim_{\alpha\downarrow 0} \sum_{n=1}^{\infty} n^{-\alpha n} z^n = \frac{z}{1-z} \quad \textit{uniformly for } z \in \Delta(\eta, R).$$

Hints: From (xiv) and (xii) we have

$$G(\alpha, z) = \sum_{n=1}^{\infty} n^{-\alpha n} z^n \qquad \forall z \in \Delta(\eta, R),\ \alpha \in (0, \eta/\pi]\,. \tag{12}$$

Because of (xiii) and the fact convergence in (xiv) is uniform, the function $G(\alpha, z)$ is uniformly continuous on $[0, \eta/\pi] \times \Delta(\eta, R)$. Therefore

$$\lim_{\alpha\downarrow 0} G(\alpha, z) = G(0, z) \qquad \text{uniformly in} \quad z \in \Delta(\eta, R).$$

But for $z \in \mathbb{D} \cap \Delta(\eta, R)$

$$G(0, z) = \lim_{\alpha\downarrow 0} G(\alpha, z) \overset{(12)}{=} \lim_{\alpha\downarrow 0} \sum_{n=1}^{\infty} n^{-\alpha n} z^n \overset{(11)}{=} \frac{z}{1-z}. \tag{13}$$

By the uniqueness theorem for holomorphic functions (5.62) the equality in (13) persists for all $z \in \Delta(\eta, R)$. That is, claim (xvii) holds.

(xviii) *Prove the assertion* (0).

Hints: Evidently any compact subset C of $\mathbb{C} \setminus [1, \infty)$ lies in $\rho\mathbb{D} \cup \Delta(\eta, R)$ for some choice of $0 < \eta < \pi/2$ and some $R > 1 > \rho$, since these open sets cover $\mathbb{C} \setminus [1, \infty)$. Setting $0^0 := 1$, it then follows from (xvi) and (xvii) that

$$\lim_{\alpha\downarrow 0} \sum_{n=0}^{\infty} n^{-\alpha n} w^n = \frac{1}{1-w} \quad \text{uniformly for } w \in C.$$

Now adapt the proof of 8.25: Replace the polynomial $P_n(w)$ there by the series $\sum_{n=0}^{\infty} n^{-\alpha n} w^n$, whose convergence is clearly uniform in any compact set of w, for fixed $\alpha > 0$, thus letting $\int_\gamma$ commute with $\sum$, as in 8.25.

Exercise 11.35 (i) (Euler, 1735) *Prove that* $\sum_{n=0}^{\infty} (-1)^n / (2n+1)^3 = \pi^3/32$.

Hints: There is a quick Fourier series derivation: Take $m = 1, t = 1/4$ in equation (14) of 3.38. More in the spirit of this chapter, let

$$f(z) := \frac{1}{(2z+1)^3}, \quad g(z) := \frac{\pi}{\sin(\pi z)}, \quad \gamma_N := C\left(0, N + \tfrac{1}{2}\right), N \in \mathbb{N}.$$

Note that $\mathrm{Res}\,(fg, n) = f(n)\mathrm{Res}\,(g, n)$ for $n \in \mathbb{Z}$, cull $\mathrm{Res}\,(g, n)$ from the 11.37(ii) hints, then use the Residue Theorem to write

$$\frac{1}{2\pi i} \int_{\gamma_N} fg = \sum_{n=-N}^{N} (-1)^n f(n) + \mathrm{Res}\,(fg, -\tfrac{1}{2}).$$

This residue equals $g''(-1/2)/16$ (why?). Finally make use of the fact (3.30(ii)) that $\sin(\pi z)$ is bounded away from 0 in $\mathbb{C} \setminus \bigcup_{n \in \mathbb{Z}} D(n, 1/4)$ to show that $\int_{\gamma_N} fg$ converges to 0 as $N \to \infty$.

(ii) *Use the Residue Theorem to prove that the hyperbolic secant function* $\mathrm{sech}(\pi z) := 2/(e^{\pi z} + e^{-\pi z})$ *is its own* Fourier Transform, *that is,*

$$(*) \qquad \int_{-\infty}^{\infty} e^{-2\pi i x y}\, \mathrm{sech}(\pi x)\, dx = \mathrm{sech}(\pi y) \qquad \forall y \in \mathbb{R}.$$

Hints: First note that the zeros of the entire function $f(z) := e^{\pi z} + e^{-\pi z}$ are all the numbers $p_n := (n + 1/2)i$, $n \in \mathbb{Z}$. Since $f'(p_n) = (-1)^n 2\pi i \neq 0$, each zero is simple, and so, upon factoring the Taylor expansion at p_n, we see that $f(z) = (z - p_n) f_n(z)$, where f_n is a power series in $(z - p_n)$ with constant term $f_n'(p_n) = (-1)^n 2\pi i$. Consequently, for $z \in \mathbb{C} \setminus (\mathbb{Z} + 1/2)i$

$$\frac{1}{f(z)} = \frac{1}{z - p_n} \cdot \frac{1}{f_n(z)}.$$

This tells us that

$$(1) \qquad \mathrm{Res}\,(1/f, p_n) = 1/f_n(p_n) = (-1)^n / 2\pi i, \qquad n \in \mathbb{Z}.$$

Fix *positive* $y \in \mathbb{R}$ and introduce the holomorphic function

$$(2) \qquad F(z) := \frac{e^{2\pi i y z}}{e^{\pi z} + e^{-\pi z}}, \qquad z \in \mathbb{C} \setminus (\mathbb{Z} + 1/2)i$$

and the simple loops

$$(3)\qquad \Gamma_N := [N, N+iN, -N+iN, -N, N] \subset \mathbb{C}\setminus(\mathbb{Z}+1/2)i, \qquad n\in\mathbb{N}.$$

Check that $|w \pm w^{-1}| \geq |w|/2$ whenever $|w| \geq \sqrt{2}$, so

$$(*)\qquad |e^{\pi z} \pm e^{-\pi z}| \geq (1/2)e^{\pi|\operatorname{Re} z|} \quad \text{whenever} \quad |\operatorname{Re} z| \geq (1/2\pi)\log 2.$$

Consequently

$$(4)\qquad |F(z)| \leq 2e^{-2\pi y \operatorname{Im} z}e^{-\pi|\operatorname{Re} z|} \quad \text{if} \quad |\operatorname{Re} z| \geq (1/2\pi)\log 2.$$

The poles of F are the p_n, and since the factor $e^{2\pi i y z}$ is a zero-free entire function, we get from (1)

$$(5)\quad \operatorname{Res}(F, p_n) = \operatorname{Res}(1/f, p_n)e^{2\pi i y p_n} = ((-1)^n/2\pi i)e^{-(2n+1)\pi y}, \quad n\in\mathbb{Z}.$$

Note that the poles in $\mathcal{I}(\Gamma_N)$ are p_n with $0 \leq n \leq N-1$. Consequently, the Residue Theorem delivers

$$\frac{1}{2\pi i}\int_{\Gamma_N} F = \sum_{n=0}^{N-1}\operatorname{Res}(F, p_n) = \sum_{n=0}^{N-1}\frac{(-1)^n}{2\pi i}e^{-(2n+1)\pi y},$$

$$\int_{\Gamma_N} F = e^{-\pi y}\sum_{n=0}^{N-1}(-e^{-2\pi y})^n = e^{-\pi y}\frac{1-(-e^{-2\pi y})^N}{1-(-e^{-2\pi y})} = \frac{1-(-e^{-2\pi y})^N}{e^{\pi y}+e^{-\pi y}}.$$

It follows that

$$(6)_+\qquad \lim_{N\to\infty}\int_{\Gamma_N} F \quad \text{exists and equals} \quad \frac{1}{e^{\pi y}+e^{-\pi y}} \qquad \forall y > 0.$$

Inequality $(*)$ is valid for all $z \in [N, N+iN] \cup [-N+iN, -N]$ since $|\operatorname{Re} z| = N > (1/2\pi)\log 2$ for such z. It follows that $|F(z)| \leq 2e^{-\pi N}e^{-2\pi y \operatorname{Im} z} \leq 2e^{-\pi N}$ (y and $\operatorname{Im} z$ being non-negative).

For $z \in [-N+iN, N+iN]$ we have $z = t+iN$, $t\in[-N, N]$, so

$$|e^{\pi z}+e^{-\pi z}| = |e^{\pi t + i\pi N} + e^{-\pi t - i\pi N}| = e^{\pi t}+e^{-\pi t} \geq 1,$$

whence

$$|F(z)| = \frac{|e^{2\pi i y z}|}{|e^{\pi z}+e^{-\pi z}|} \leq e^{-2\pi y N}.$$

The last two inequalities for $|F(z)|$ give the crude bounds

$$\left|\int_{[N, N+iN, -N+iN, -N]} |F|\right| \leq 4Ne^{-\pi N} + 2Ne^{-2\pi y N} \qquad \forall N\in\mathbb{N}.$$

Since $y > 0$, these integrals converge to 0 as $N \to \infty$. Therefore

$$\lim_{N\to\infty} \int_{\Gamma_N} F \text{ (proved to exist at } (6)_+) \;=\; \lim_{N\to\infty} \int_{[-N,N]} F =: \int_{-\infty}^{\infty} F$$

and $(6)_+$ itself now reads

$$\int_{-\infty}^{\infty} F(x)\,dx = \frac{1}{e^{\pi y} + e^{-\pi y}} \qquad \forall y > 0.$$

Upon phasing out F via (2), this says

$$\int_{-\infty}^{\infty} e^{-2\pi i y z} \operatorname{sech}(\pi x)\,dx = \operatorname{sech}(\pi y) \qquad \forall y > 0.$$

Changing the variable of integration here to $-x$ gives the same equality for $y < 0$. (What about $y = 0$?)

Remarks: Readers are encouraged to supply *critical* details in the following short(er) alternative derivation of $(*)$: Note that the integral in $(*)$ equals $2\operatorname{Re}\int_0^\infty e^{-2\pi i x y} \operatorname{sech}(\pi x)\,dx$, expand $\operatorname{sech}(\pi x) = 2e^{-\pi x}/\big(1 - \big[-e^{-2\pi x}\big]\big)$ into a geometric series with ratio $-e^{-2\pi x}$ and integrate termwise. Call the integrated series $s(y)$ and change the summation index $k \in \mathbb{Z}_+$ involved in $\overline{s(y)}$ to $n := -k-1 \in -\mathbb{N}$. The resulting summand is $\overline{(-1)^{-n-1}/\pi(-iy + 1/2 - n - 1)} = (-1)^{-n}/\pi(iy - 1/2 + n + 1) = (-1)^n/\pi(iy + 1/2 + n)$. Thus $\pi s(y) + \pi\overline{s(y)}$ becomes the bilateral series in 11.37(vii) at $z := iy + 1/2$. Ramanujan [1915] proved a generalization of $(*)$ by yet a third method. Note that equality $(*)$ is valid as well for all complex y with $|\operatorname{Im} y| < 1/2$, because each side is a holomorphic function of such y.

Exercise 11.36 *Let f be entire and satisfy for some finite positive M*

$$(*) \qquad |f(z)| \le M e^{|\operatorname{Im} z|} \qquad \forall z \in \mathbb{C}.$$

Define a holomorphic function F by

$$F(z) := \frac{f(z)}{z^2 \cos z}$$

for all z not equal to 0 or an odd multiple of $\pi/2$. For each positive integer n define $v_n := n\pi(1+i)$ and

$$\Gamma_n := [v_n, -\bar{v}_n, -v_n, \bar{v}_n, v_n].$$

(i) *Show that*

$$\frac{1}{2\pi i}\int_{\Gamma_n} F = f'(0) - \sum_{k=-n}^{n-1}(-1)^k \frac{f\left(\left(k+\frac{1}{2}\right)\pi\right)}{\left(k+\frac{1}{2}\right)^2\pi^2}.$$

Hints: Use the Residue Theorem, remember that $\cos(z+(k+1/2)\pi)$ equals $(-1)^{k+1}\sin z$, and compute residues as in the 11.35(i) hints.

(ii) *Show that* $|\cos z| \geq \frac{1}{4}e^{|\operatorname{Im} z|}$ *for all* $z \in \Gamma_n$.

Hint: If $\operatorname{Re} z = \pm n\pi$, then $2|\cos z| = e^{-\operatorname{Im} z} + e^{\operatorname{Im} z}$. If $\operatorname{Im} z = \pm n\pi$, then $2|\cos z| \geq e^{|n\pi|} - e^{-|n\pi|}$.

(iii) *Show that*

$$|F(z)| \leq \frac{4M}{|z|^2} \qquad \forall z \in \Gamma_n.$$

(iv) *From* (i) *and* (iii) *deduce that*

$$f'(0) = \sum_{k=-\infty}^{\infty}(-1)^k \frac{f\left(\left(k+\frac{1}{2}\right)\pi\right)}{\left(k+\frac{1}{2}\right)^2\pi^2}.$$

(v) *Apply* (iv) *to the function* $f := \sin$ *to deduce that*

$$\sum_{k=-\infty}^{\infty}\frac{1}{(2k+1)^2} = \frac{\pi^2}{4}. \qquad \text{(Cf. 3.37 (vi).)}$$

(vi) *From* (iv) *and* (v) *deduce that in general for any* f *which satisfies* $(*)$ *we have*

$$|f'(0)| \leq \sup|f(\mathbb{R})|.$$

(vii) (S. N. Bernstein) *Let* g *be an entire function and for some finite positive* c, M *satisfy*

$$|g(z)| \leq Me^{c|\operatorname{Im} z|} \qquad \forall z \in \mathbb{C}.$$

Show that

$$g'(x) = \frac{4c}{\pi^2}\sum_{k=-\infty}^{\infty}\frac{(-1)^k}{(2k+1)^2}g\left(x+\frac{2k+1}{2c}\pi\right) \qquad \forall x \in \mathbb{R},$$

and

$$|g'(x)| \leq c\sup|g(\mathbb{R})| \qquad \forall x \in \mathbb{R}.$$

Hint: Given $x \in \mathbb{R}$, the function $f(z) := g(x + z/c)$ satisfies the hypothesis $(*)$ above. Cite the conclusions (iv) and (vi).

(viii) *The boundedness hypothesis*

$$(**) \qquad |f(z)| \le Me^{|z|} \qquad \forall z \in \mathbb{C} \quad \& \quad |f(x)| \le M \qquad \forall x \in \mathbb{R}$$

is formally weaker than $(*)$. *With the help of* 5.16′ *show that it is in fact equivalent to* $(*)$. *Formulate a corresponding "strengthening" of* (vii).

Hints: Suppose $(**)$ holds, and consider the sector $S := \mathbb{H} \cap \mathbb{U}$ and in $\bar{S}$ the function

$$f^*(z) := f(z)e^{iz}.$$

For $x \in \mathbb{R}$

$$|f^*(x)| = |f(x)| \le M,$$

and for $y > 0$

$$|f^*(iy)| = |f(iy)|e^{-|iy|} \le M \quad \text{by } (**),$$

whereas for all $z \in \bar{S}$

$$|f^*(z)| = |f(z)|e^{-\operatorname{Im} z} \le |f(z)| \overset{(**)}{\le} Me^{|z|}, \qquad z \in \bar{S}.$$

The discussion immediately preceding 5.16′ (with $\theta_1 := 0$, $\theta_2 := \pi/2$, $\gamma := 1$) pointed out that as a consequence of 5.16′ the conclusion of 5.16 remains valid with its boundedness hypothesis weakened to exponential boundedness. Thus the conclusion of 5.16(i) (the m there being the M here) is valid for f^* in $\bar{S}$, which clearly entails that f^* is bounded in $\bar{S}$. The same argument works in $\mathbb{U} \setminus \bar{S}$. Thus f^* is bounded in $\bar{\mathbb{U}}$, say by M^*. By definition of f^* then

$$(***) \qquad |f(z)| = |f^*(z)e^{-iz}| = |f^*(z)|e^{\operatorname{Im} z} \le M^* e^{|\operatorname{Im} z|} \qquad \forall z \in \mathbb{U}$$

Finally, this whole argument applied to the entire function $\bar{f}(\bar{z})$, elevates the validity of inequality $(***)$ to the whole of $\mathbb{C}$.

Exercise 11.37 *Let (recalling* 3.30(iii)*)*

$$f(z) := \frac{\pi^2}{\sin^2 \pi z}, \quad g(z) := \sum_{n=-\infty}^{\infty} \frac{1}{(z-n)^2} \qquad \text{for } z \in \mathbb{C} \setminus \mathbb{Z}.$$

(i) *Show that the series converges locally uniformly in* $\mathbb{C} \setminus \mathbb{Z}$.

(ii) *Show that the principal part of f at n is $(z-n)^{-2}$ for each $n \in \mathbb{Z}$.*

Hint: Due to periodicity it suffices to consider $n = 0$. By 3.27

$$\sin \pi z = \pi z \left[1 + \sum_{n=1}^{\infty} \frac{(-1)^n \pi^{2n} z^{2n}}{(2n+1)!} \right] =: \pi z [1 - z^2 \psi(z)].$$

For all sufficiently small z (i.e., small enough that $|z^2\psi(z)| < 1$) we then have

$$\frac{\pi z}{\sin \pi z} = \frac{1}{1 - z^2\psi(z)} = \sum_{n=0}^{\infty} \left[z^2 \psi(z) \right]^n = 1 + \sum_{n=1}^{\infty} z^{2n} \psi^n(z),$$

$$\frac{\pi}{\sin \pi z} = \frac{1}{z} + z \sum_{n=1}^{\infty} z^{2(n-1)} \psi^n(z) =: \frac{1}{z} + z\phi(z),$$

$$\frac{\pi^2}{\sin^2 \pi z} = \frac{1}{z^2} + 2\phi(z) + z^2 \phi^2(z).$$

(iii) *Conclude from* (i) *and* (ii) *that the function* $h := f - g$, *which is holomorphic in* $\mathbb{C} \setminus \mathbb{Z}$, *actually extends to be entire.*

Hint: For each n there are functions f_n, g_n holomorphic near n such that $f(z) = (z-n)^{-2} + f_n(z)$, $g(z) = (z-n)^{-2} + g_n(z)$. Thus $h = f_n - g_n$ near n.

(iv) *Let* h *be extended as in* (iii). *Show that*

$$h\left(\frac{z}{2}\right) + h\left(\frac{z+1}{2}\right) = 4h(z) \qquad \forall z \in \mathbb{C}.$$

Hint: Calculate. At one point the "double angle" formula $\sin(\frac{\pi z}{2})\cos(\frac{\pi z}{2}) = \frac{1}{2}\sin(\pi z)$ will be needed.

(v) *Deduce from* (iv) *that* $h = \mathbf{0}$, *that is,*

$$\frac{\pi^2}{\sin^2 \pi z} = \sum_{n=-\infty}^{\infty} \frac{1}{(z-n)^2} \qquad \forall z \in \mathbb{C} \setminus \mathbb{Z}.$$

Hint: The extended h is continuous on $\overline{\mathbb{D}}$, so $M := \sup |h(\mathbb{D})| < \infty$. By (iv) and convexity of $\mathbb{D}$, we have

$$|h(z)| \le \frac{1}{4}\left| h\left(\frac{z}{2}\right)\right| + \frac{1}{4}\left| h\left(\frac{z+1}{2}\right)\right| \le \frac{1}{2}M \qquad \forall z \in \mathbb{D}.$$

Thus $M \le \frac{1}{2}M$, so $M = 0$. By 5.62, $h = 0$ throughout $\mathbb{C}$. Most of us, to whom the muse did not vouchsafe to reveal this "1/4-trick," would probably turn to Liouville, 5.34(iii): First of all, $|\sin \xi| > \frac{1}{4}e^{|\operatorname{Im}\xi|} > \frac{1}{4}|\operatorname{Im}\xi|$ whenever

$|\operatorname{Im}\xi| \geq 1$ (see $(*)$ in the 11.35 hints). Therefore if $x := \operatorname{Re} z$, $y := \operatorname{Im} z$, $N \in \mathbb{N}$ and $|y| \geq N$, $\pi^2|\sin^2 \pi z|^{-1} < \pi^2(\frac{1}{4}|\pi y|)^{-2} \leq 16N^{-1}$. Next, if $x \in [0,1]$, $\sum_{|n|\geq 0} |z-n|^{-2} \leq \sum_{|n|\leq N+1}(0+y^2)^{-1} + \sum_{|n|>N+1}((n-x)^2+0)^{-1} < (2N+3)N^{-2} + \sum_{|n|>N+1}(|n|-1)^{-2} = (2N+3)N^{-2} + 2\sum_{n>N} n^{-2} < 5N^{-1} + 2\sum_{n>N}((n-1)^{-1} - n^{-1}) = 5N^{-1} + 2N^{-1}$. Consequently, for such z, $|h(z)| < 23N^{-1}$. Thus h is bounded in $[0,1] \times \mathbb{R}$. As the entire function h is evidently 1-periodic, the latter means that h is bounded in $\mathbb{C}$, hence is constant. The preceding inequalities show that constant can only be 0.

Readers who fought through the preceding arguments deserve to experience a brilliant "two-line" proof of R. M. YOUNG [1979]: For any $z \in \mathbb{R} \setminus \mathbb{Z}$, $n \in \mathbb{Z}$ the nth Fourier coefficient of the function $h_z(t) := e^{izt}$ is trivially computed to be $ie^{\pi i(z-n)} \cdot \sin \pi(z-n)/\pi(z-n)$, whereas $\int_{-\pi}^{\pi} |h_z|^2 = 2\pi$. Consequently, the version $(14)_0$ of *Parseval's identity* in Remark (iii) following 5.20 asserts that

$$1 = \sum_{n\in\mathbb{Z}} |\sin \pi(z-n)/\pi(z-n)|^2 = \sin^2 \pi z \sum_{n\in\mathbb{Z}} 1/\pi^2(z-n)^2 \qquad \forall z \in \mathbb{R} \setminus \mathbb{Z}.$$

Since this is a holomorphic function of z in the region $\mathbb{C} \setminus \mathbb{Z}$, the identity persists for all such z (5.62). See also (xv) below.

(vi) *Deduce from* (v) *that* [cf. (29) in 3.38]

$$\pi \cot \pi z = \frac{1}{z} + \sum_{k=1}^{\infty} \frac{2z}{z^2-k^2} = \lim_{N\to\infty} \sum_{n=-N}^{N} \frac{1}{z-n} \qquad \forall z \in \mathbb{C} \setminus \mathbb{Z}.$$

Hints: For $z \in \mathbb{C} \setminus \mathbb{Z}$ define

(1) $$F(z) := \pi \cot \pi z - \frac{1}{z},$$

(2) $$F_N(z) := \sum_{\substack{n=-N \\ n\neq 0}}^{N} \frac{1}{z-n}, \qquad N \in \mathbb{N}.$$

Combine the terms $n = k$ and $n = -k$ in the sum for F_N and get

(3) $$F_N(z) = \sum_{k=1}^{N} \frac{2z}{z^2-k^2}.$$

We have

$$F'(z) = \frac{1}{z^2} - \frac{\pi^2}{\sin^2 \pi z} \quad \text{and by (2)} \quad F_N'(z) = \sum_{\substack{n=-N \\ n\neq 0}}^{N} \frac{-1}{(z-n)^2}.$$

Therefore by (v)

$$F'_N \to F' \quad \text{locally uniformly in } \mathbb{C} \setminus \mathbb{Z}. \tag{4}$$

It is clear from (3) that $\{F_N\}$ converges locally uniformly in $\mathbb{C} \setminus \mathbb{Z}$, say to G. Then by 5.44(ii), $F'_N \to G'$ and we have from (4) that $F' = G'$ in $\mathbb{C} \setminus \mathbb{Z}$. As this set is connected (1.24), we infer from 2.10(iv) that $F - G$ is constant there. By (1) and (3), F and each F_N is an odd function. The constant $F - G = F - \lim F_N$ is therefore an odd function, hence equal to $\mathbf{0}$.

(vii) *Deduce from* (vi) *that*

$$\frac{\pi}{\sin \pi z} = \frac{1}{z} + \sum_{k=1}^{\infty} \frac{(-1)^k\, 2z}{z^2 - k^2} = \lim_{N\to\infty} \sum_{n=-N}^{N} \frac{(-1)^n}{z-n} \qquad \forall z \in \mathbb{C} \setminus \mathbb{Z}.$$

Hints: The right-hand side evidently converges; call it $H(z)$. For $N = 2K+1$, separate even and odd indices into groups and get

$$\frac{1}{z} + \sum_{k=1}^{N} \frac{(-1)^k\, 2z}{z^2 - k^2} = \sum_{n=-N}^{N} \frac{(-1)^n}{z-n} = \sum_{n=-K}^{K} \frac{1}{z-2n} - \sum_{n=-K-1}^{K} \frac{1}{z-1-2n}.$$

From (vi) then $H(z) = \frac{1}{2}\pi \cot(\pi z/2) - \frac{1}{2}\pi \cot(\pi(z-1)/2) = \pi/\sin(\pi z)$.

(viii) *For non-zero real a, express $\sum_{k=1}^{\infty} \frac{1}{k^2+a^2}$ and $\sum_{k=1}^{\infty} \frac{(-1)^k}{k^2+a^2}$ in terms of $e^{\pi a}$ by taking $z = ia$ in* (vi) *and* (vii). *Can $\sum_{n=1}^{\infty} 1/n^2 = \pi^2/6$ be deduced from* (v)? [Note that (vii) with $z := 1/2$ gives 3.22.]

Hints: Take $z = 1/2$ in (v) to get $\pi^2/8 = \sum (2n-1)^{-2}$. Use $\sum k^{-2} = \sum (2n)^{-2} + \sum (2n-1)^{-2}$ to get $(3/4)\sum k^{-2} = \sum (2n-1)^{-2}$.

(ix) *From* (vi) *and* 5.32 *deduce* Euler's product *representation of the sine function:*

$$\sin \pi z = \lim_{N\to\infty} \pi z \prod_{n=1}^{N} \left(1 - \frac{z^2}{n^2}\right) \qquad \forall z \in \mathbb{C},$$

with convergence uniform on compact sets of z.

Hints: Let $\Omega := \mathbb{C} \setminus (\mathbb{Z} \setminus \{0\})$. The function $s(z) := \frac{\sin \pi z}{\pi z}$ is entire and is zero-free in Ω, and a little calculation gives

$$\frac{s'(\xi)}{s(\xi)} = \pi \cot \pi\xi - \frac{1}{\xi}$$

$$= \lim_{N\to\infty} \sum_{\substack{n=-N \\ n\neq 0}}^{N} \frac{1}{\xi - n} \quad \text{locally uniformly in } \Omega, \tag{5}$$

by (vi). For all $z \in \mathbb{C}$ set $P_z := [0, i, z]$ and then define

$$\gamma_z := \begin{cases} P_z & \text{if } \operatorname{Im} z \geq 0 \\ \overline{P}_{\bar{z}} & \text{if } \operatorname{Im} z < 0. \end{cases}$$

Clearly

(6) $$\gamma_z \subset \Omega \quad \text{and} \quad l(\gamma_z) \leq |z| + 2 \qquad \forall z \in \Omega.$$

Notice that

$$(t, z) \mapsto P_z(t) \quad \text{and} \quad (t, z) \mapsto \overline{P}_{\bar{z}}(t)$$

are continuous functions. Hence for compact $K \subset \Omega$

$$\{P_z(t) : t \in [0,2], z \in K, \operatorname{Im} z \geq 0\}, \quad \{\overline{P}_{\bar{z}}(t) : t \in [0,2], z \in K, \operatorname{Im} z \leq 0\}$$

are compact subsets of Ω. Clearly their union contains

$$K^* := \{\gamma_z(t) : t \in [0,2], z \in K\}.$$

Therefore convergence in (5) is uniform on K^*, which entails not only that

(7) $$\int_{\gamma_z} \frac{s'(\xi)}{s(\xi)}\, d\xi = \lim_{N\to\infty} \sum_{\substack{n=-N \\ n\neq 0}}^{N} \int_{\gamma_z} \frac{1}{\xi - n}\, d\xi$$

for each $z \in K$, but, thanks to 2.9(i) and the length condition in (6), this convergence is actually uniform in $z \in K$. It follows (cf. 7.1(ii)) that

(8) $$\exp\left(\int_{\gamma_z} \frac{s'(\xi)}{s(\xi)}\, d\xi\right) = \lim_{N\to\infty} \prod_{\substack{n=-N \\ n\neq 0}}^{N} \exp\left(\int_{\gamma_z} \frac{d\xi}{\xi - n}\right) \qquad \forall z \in \Omega.$$

Now according to 5.32(i)

$$\exp\left(\int_{\gamma_z} \frac{s'(\xi)}{s(\xi)}\, d\xi\right) = \frac{s(z)}{s(0)} = s(z) \quad \text{and} \quad \exp\left(\int_{\gamma_z} \frac{d\xi}{\xi - n}\right) = \frac{z-n}{0-n} = 1 - \frac{z}{n}.$$

Therefore (8) reads

(9) $$s(z) = \lim_{N\to\infty} \prod_{\substack{n=-N \\ n\neq 0}}^{N} \left(1 - \frac{z}{n}\right) = \lim_{N\to\infty} \prod_{n=1}^{N} \left(1 - \frac{z^2}{n^2}\right)$$

(upon combining the $(-n)^{\text{th}}$ and n^{th} factors), in which convergence is uniform on every compact subset of $z \in \Omega$. In particular, for each $k \in \mathbb{N}$ convergence in (9) is uniform on the circle $C\big(0, k+\frac{1}{2}\big)$, hence by the maximum-modulus principle uniform on $\bar{D}\big(0, k+\frac{1}{2}\big)$.

(ix)* (PRINGSHEIM [1915]) *If the symmetric bilateral products in* (9) *are replaced by non-symmetric ones, the limit may change or cease to exist. Show that if* $p, q \in \mathbb{N}$, *then*

$$(9)^* \qquad \lim_{M\to\infty} \pi z \prod_{\substack{n=-Mq\\ n\neq 0}}^{Mp} \Big(1-\frac{z}{n}\Big) = e^{z\log(q/p)}\sin(\pi z) \qquad \forall z \in \mathbb{C}.$$

Hints: Consider $q > p$ and write

$$\pi z \prod_{\substack{n=-Mq\\ n\neq 0}}^{Mp} \Big(1-\frac{z}{n}\Big) = \Bigg[\pi z \prod_{\substack{n=-Mq\\ n\neq 0}}^{Mq} \Big(1-\frac{z}{n}\Big)\Bigg]\Bigg[\prod_{n=Mp+1}^{Mq} \Big(1-\frac{z}{n}\Big)e^{z/n}\Bigg]^{-1} \exp\Bigg(z\sum_{n=Mp+1}^{Mq}\frac{1}{n}\Bigg),$$

$M \in \mathbb{N}$, $z \in \mathbb{C} \setminus \mathbb{Z}$. As $M \to \infty$ the first bracketed product is governed by (9) and the last sum by 3.46(ii). For the central factor, use (2) in the proof of 3.19 to write

$$1-\frac{z}{n} = \exp\Bigg(-\sum_{k=1}^{\infty}\frac{1}{k}\Big(\frac{z}{n}\Big)^k\Bigg),$$

$$\Big[\Big(1-\frac{z}{n}\Big)e^{z/n}\Big]^{-1} = \exp\Bigg(\sum_{k=2}^{\infty}\frac{1}{k}\Big(\frac{z}{n}\Big)^k\Bigg), \qquad n \in \mathbb{N}, z \in D(0,n),$$

and note that for $M > 2|z|/p$

$$\Bigg|\sum_{n=Mp+1}^{Mq}\sum_{k=2}^{\infty}\frac{1}{k}\Big(\frac{z}{n}\Big)^k\Bigg| \le \sum_{n=Mp}^{\infty}\sum_{k=2}^{\infty}\Big|\frac{z}{n}\Big|^k = \sum_{n=Mp}^{\infty}\Big|\frac{z}{n}\Big|^2\frac{1}{1-|z|/n} \le \sum_{n=Mp}^{\infty}\frac{2|z|^2}{n^2}$$

which clearly converges to 0 as $M \to \infty$. [Inserting the (mutually cancelling) factors $e^{z/n}$, then engaging 3.19, as above, is another way to confirm the *existence* of the limit in (9). Cf. the use of "convergence factors" in the proof of 7.35.]

(x) *Deduce from* (ix) *and the double-angle formula,* $2\sin w\cos w = \sin 2w$, Euler's product *representation of the cosine function:*

$$\cos \pi z = \lim_{N\to\infty}\prod_{n=1}^{N}\Bigg(1-\frac{z^2}{(n-\frac{1}{2})^2}\Bigg) \qquad \forall z \in \mathbb{C}.$$

Hint: For all $z \in \mathbb{C} \setminus \mathbb{Z}, N \in \mathbb{N}$

$$\prod_{n=1}^{N}\left(1-\frac{(2z)^2}{(2n-1)^2}\right)=\prod_{m=1}^{2N}\left(1-\frac{(2z)^2}{m^2}\right) \Big/ \prod_{n=1}^{N}\left(1-\frac{(2z)^2}{(2n)^2}\right).$$

The quotient on the right side converges locally uniformly to $\frac{\sin(2\pi z)}{2\pi z} / \frac{\sin(\pi z)}{\pi z} = \cos(\pi z)$. Thus the polynomials on the left side converge to $\cos(\pi z)$ uniformly on $C(0, k+\frac{1}{2})$, hence by the maximum modulus principle, uniformly on $\bar{D}(0, k+\frac{1}{2})$, for each $k \in \mathbb{N}$.

(xi) *From two applications of* (ix) *deduce the product representation*

$$1-\sin \pi z=\prod_{n \in \mathbb{Z}}\left(1+\frac{2z}{4n-1}\right)^2 \qquad \forall z \in \mathbb{C}.$$

Hint: $1-\sin \pi z = 2\sin^2\left(\frac{\pi}{4}-\frac{\pi z}{2}\right) = \left[\frac{\sin\left(\frac{\pi}{4}-\frac{\pi z}{2}\right)}{\sin\frac{\pi}{4}}\right]^2.$

(xii) *Use* (ix) *and* 5.44(ii) *to get yet another proof (Euler's third) of the identity* $\pi^2/6 = \sum k^{-2}$.

Hints: From 3.27 we can read off the coefficient of z^2 in the Taylor expansion of the entire function $s(z) := \frac{\sin(\pi z)}{\pi z}$, that is, read off $\frac{s''(0)}{2}$. But we can also use (ix) and 5.44(ii) to compute $s''(0)$ as a limit, because by induction on N there are polynomials P_N with zero constant term such that $\prod_{n=1}^{N}(1-z^2/n^2) = 1-\left(\sum_{n=1}^{N}\frac{1}{n^2}\right)z^2 + z^2 P_N(z)$.

(xiii) *From* (ix) *deduce Wallis' product representation of* π*:*

(w) $$\frac{\pi}{2}=\lim_{N\to\infty}\prod_{n=1}^{N}\left(\frac{2n}{2n-1}\cdot\frac{2n}{2n+1}\right)=\lim_{N\to\infty}\prod_{n=1}^{N}\frac{4n^2}{4n^2-1}.$$

By writing the Nth *product of quotients as a quotient of products and introducing the missing even factors in the denominator, deduce from this that*

(a) $$\sqrt{\pi}=\lim_{N\to\infty}\frac{1}{\sqrt{N}}\frac{2^{2N}(N!)^2}{(2N)!}.$$

Then from (w) *deduce the remarkable identities*

(b) $$\prod_{n=1}^{\infty}\left(1-\frac{1}{(2n)^2}\right)=\frac{2}{\pi},$$ (c) $$\prod_{n=2}^{\infty}\left(1-\frac{1}{(2n-1)^2}\right)=\frac{\pi}{4},$$

(d)
$$\prod_{n=2}^{\infty}\left(1-\frac{1}{n^2}\right)=\frac{1}{2}=\prod_{n=2}^{\infty}\left(1-\frac{(-1)^n}{n}\right).$$

Hints: Take reciprocals in (w) to get (b). For (c) shift indices to get

$$\prod_{n=1}^{N}(2n-1)(2n+1)=\prod_{n=1}^{N}(2n-1)\prod_{n=1}^{N}(2n+1)=\prod_{n=1}^{N}(2n-1)\prod_{m=2}^{N+1}(2m-1)$$
$$=(2N+1)\prod_{n=2}^{N}(2n-1)^2, \text{ and}$$
$$\prod_{n=1}^{N}(2n)(2n)=\prod_{n=1}^{N}2n\prod_{m=2}^{N+1}2(m-1)=2\prod_{n=2}^{N}(2n)2(n-1)\cdot 2N, \text{ so}$$
$$\prod_{n=1}^{N}\frac{(2n)(2n)}{(2n-1)(2n+1)}=\frac{4N}{2N+1}\prod_{n=2}^{N}\frac{(2n)2(n-1)}{(2n-1)^2}=\frac{4N}{2N+1}\prod_{n=2}^{N}\left(1-\frac{1}{(2n-1)^2}\right).$$

(c) thus follows from (w). [A derivation of (c) from (x) is also possible: Divide away the factor $n=1$ in (x) and do some rearranging to get, for all $z\in\mathbb{C}\setminus\{1/2\}$

$$\frac{\cos(\pi z)-\cos(\pi/2)}{\pi z-\pi/2}\cdot\frac{-\pi}{2(1+2z)}=\lim_{N\to\infty}\prod_{n=2}^{N}\left(1-\frac{4z^2}{(2n-1)^2}\right).$$

These polynomials converge uniformly on $\mathbb{T}$, hence uniformly on $\overline{\mathbb{D}}$, and so define a function which is continuous at $1/2$. The left side too has a limit as $z\to 1/2$, namely, $\cos'(\pi/2)\cdot(-\pi/4)$.] The first equality in (d) is the product of (b) and (c). For the second equality, note that

$$\prod_{n=2}^{2N}\left(1-\frac{(-1)^n}{n}\right)=\prod_{m=1}^{N}\left(1-\frac{1}{2m}\right)\prod_{m=1}^{N}\left(1+\frac{1}{2m+1}\right)=$$
$$\prod_{m=1}^{N}\frac{2m-1}{2m}\Bigg/\prod_{m=1}^{N}\frac{2(m+1)-1}{2(m+1)}=\frac{1}{2}\Bigg/\frac{2(N+1)-1}{2(N+1)}.$$

(xiv) *Use* (xiii)(a) *to show that the constant* S *in Stirling's formula of* 3.47 *is* $\sqrt{2\pi}$.

Hint: In (xiii) replace $N!$ by $S\sqrt{N}\left(\frac{N}{e}\right)^N e^{\theta_N/12N}$ and replace $(2N)!$ by the corresponding expression from 3.47, where all the θ's lie in $[0,1]$.

(xv) [Alternative Fourier derivation of (v), (vi), (vii)] *Fix* $z \in \mathbb{C} \setminus \mathbb{Z}$ *and set* $h_z(t) := \cos(zt)$ *for* $t \in [-\pi, \pi]$ *extended by* 2π*-periodicity to* $\mathbb{R}$. *Show that*

$$\text{(10)} \qquad \frac{1}{2\pi}\int_{-\pi}^{\pi} h_z(t)e^{-int}dt = \frac{(-1)^n \sin \pi z}{\pi}\frac{z}{z^2 - n^2} \qquad \forall n \in \mathbb{Z}$$

and deduce from the latter part of Remark (ii) *following* 5.20 *that the Fourier series of* h_z *converges to it, that is,*

$$\text{(11)} \quad \cos(zt) = \frac{\sin \pi z}{\pi}\left[\frac{1}{z} + 2z\sum_{n=1}^{\infty}\frac{(-1)^n}{z^2 - n^2}\cos(nt)\right] \qquad \forall\, t \in [-\pi, \pi].$$

Take $t = 0$ *to get* (vii) *and* $t = \pi$ *to get* (vi). *Similarly, find the Fourier series for* $t\sin(zt)$ *and evaluate it at* $t = \pi$ *to get* (v)*; alternatively, get* (v) *from differentiating* (vi) *via* 5.44(ii).

Hints: Since h_z is an even function,

$$\widehat{h}_z(n) := \frac{1}{2\pi}\int_{-\pi}^{\pi} h_z(t)e^{-int}dt = \frac{1}{\pi}\int_0^{\pi} h_z(t)\cos(nt)dt,$$

which yields (10) upon writing $2h_z(t)\cos(nt)$ as $\cos(z - n)t + \cos(z + n)t$. For z in any fixed compact subset of $\mathbb{C} \setminus \mathbb{Z}$, (10) shows that $\sum |\widehat{h}_z(n)| < \infty$, so the (symmetric, bilateral) Fourier series of h_z converges to it (uniformly in such z). When the terms with index $-n$ and n are combined, the series looks like (11).

(xvi) *From* (ix) *deduce the following remarkable relation between the sine function and the generalized factorial* Γ *of* 3.48*:*

$$\Gamma(z)\Gamma(1 - z) = \pi/\sin(\pi z) \qquad \forall z \in \mathbb{C} \setminus \mathbb{Z}.$$

Hint: For the auxiliary function f in 3.48 re-write (9) in (ix) using

$$\prod_{0 \neq n = -N}^{N}\left(1 - \frac{z}{n}\right) = \prod_{n=1}^{N}\left(\frac{n^2 - z^2}{n^2}\right) = \frac{1}{N!}\prod_{n=1}^{N}(n - z)\frac{1}{N!}\prod_{n=1}^{N}(n + z)$$

to account for all the 1-degree factors in the product $f(z)f(1 - z)$.

Some of the ideas in 11.37 and 11.36 can be used to derive a general (partial-fraction) representation which subsumes (vii) [and (vi)?] above; namely

Exercise 11.38 *If f is an entire function and $f(z)e^{-c|\operatorname{Im} z|}$ is bounded for some $c \in [0,\pi)$, then*

$$f(z) = \frac{\sin \pi z}{\pi} \lim_{N\to\infty} \sum_{n=-N}^{N} \frac{(-1)^n f(n)}{z-n} \qquad \forall z \in \mathbb{C}\setminus\mathbb{Z}.$$

Hints: Fix $z \in \mathbb{C}\setminus\mathbb{Z}$ and look at the function $F(\zeta) := \frac{f(\zeta)}{(\zeta-z)\sin\pi\zeta}$ in $\mathbb{C}\setminus\mathbb{Z}\setminus\{z\}$. When the integer N exceeds $|z|$, the sum of the residues of F inside the square with northeast vertex $v_N := (N+1/2)(1+i)$ is

$$\frac{f(z)}{\sin \pi z} + \sum_{n=-N}^{N} \frac{f(n)}{n-z}\cdot\frac{(-1)^n}{\pi}.$$

So (11.29) we want to show that for $\Gamma_N := [v_N, -\bar{v}_N, -v_N, \bar{v}_N, v_N]$ we have

$$\lim_{N\to\infty} \int_{\Gamma_N} F = 0.$$

We can suppose that $|f(\zeta)|e^{-c|\operatorname{Im}\zeta|} \le 1$ for all ζ. Then for $N \ge 2|z|$ and $\zeta \in \Gamma_N$ with $|\operatorname{Im}\pi\zeta| \ge 1$ we have

$$|F(\zeta)| \le \frac{1}{N/2}\cdot\frac{e^{c|\operatorname{Im}\zeta|}}{\frac{1}{4}e^{|\operatorname{Im}\pi\zeta|}} = \frac{8}{N}e^{(c-\pi)|\operatorname{Im}\zeta|}. \tag{1}$$

This is because of the elementary estimate

$$2|\sin\xi| = \left|e^{i\xi} - e^{-i\xi}\right| \ge e^{|\operatorname{Im}\xi|} - e^{-|\operatorname{Im}\xi|} > \frac{1}{2}e^{|\operatorname{Im}\xi|}$$

which is valid whenever $|\operatorname{Im}\xi| \ge 1$ (see $(*)$ in the 11.35 hints). On the other hand, for $N \ge 2|z|$ and $\zeta \in \Gamma_N$ we have the estimate

$$|F(\zeta)| \le \frac{1}{N/2}\cdot\frac{e^{c|\operatorname{Im}\zeta|}}{1/2} = \frac{4}{N}e^{c|\operatorname{Im}\zeta|} \tag{2}$$

thanks to 3.30(ii). Inequality (1) insures that for $\delta \in (1/\pi, N)$ and $N \ge 2|z|$ the integral of F over the part of Γ_N where $|\operatorname{Im}\zeta| \ge \delta$ is bounded by the quantity $4(2N+1)\cdot\frac{8}{N}e^{(c-\pi)\delta} \le 96e^{(c-\pi)\delta}$. (Here we use the fact that $c < \pi$.) And if $N \ge 2|z|$, inequality (2) insures that the integral of F over the remaining part of Γ_N is bounded by $4\delta\cdot\frac{4}{N}e^{c\delta}$. It follows that

$$\left|\int_{\Gamma_N} F\right| \le 96e^{(c-\pi)\delta} + \frac{16\delta e^{c\delta}}{N}.$$

Letting $N \to \infty$ we see that

$$\varlimsup_{N\to\infty} \left| \int_{\Gamma_N} F \right| \le 96 e^{(c-\pi)\delta}$$

and this holds for every $\delta > \pi^{-1}$. Let $\delta \to +\infty$.

Remarks: Useful formulas (for example, 11.36(iv)) also emerge when the argument in the hints above is applied at a point $z \in \mathbb{Z}$. If we recall that $\sin \pi z = (-1)^n \sin \pi(z-n)$ and introduce the entire function $s(z) := \frac{\sin \pi z}{\pi z}$, (often designated $\operatorname{sinc}(\pi z)$), then the formula in 11.38 reads

$$f(z) = \lim_{N\to\infty} \sum_{n=-N}^{N} f(n) s(z-n),$$

valid for $z \in \mathbb{Z}$ as well, since $\operatorname{sinc}(0) = 1$. This series is sometimes called the *cardinal series* of f. See the monograph of J. M. WHITTAKER [1935]. Its resemblance to convolution is no accident and in fact this representation, under other names like the *Whittaker-Shannon-Kotel′nikov Sampling Theorem*, is very important in the Fourier theory of band-limited signals and is derivable from the Poisson summation formula 11.33(vi). For history and a beautiful survey see HIGGINS [1985]. Also valuable are BUTZER [1983] and BUTZER, *et al.* [1984], [1988]. Generalizations of a different kind (still in the realm of entire functions) due to VALIRON [1925c] will be examined in chapter XIV. When the boundedness hypothesis in 11.38 is relaxed to allow $c = \pi$, the appropriate conclusion is

$$f(z) = \frac{\sin \pi z}{\pi} \left[f'(0) + \frac{f(0)}{z} + z \lim_{N\to\infty} \sum_{|n|=1}^{N} \frac{(-1)^n f(n)}{n(z-n)} \right].$$

This can be derived from the cotangent expansion 11.37(vi) and, in fact, is equivalent to it; see HIGGINS [2003] and HIGGINS, SCHMEISSER AND VOSS [2000].

§ 5 Specifying Principal Parts—Mittag-Leffler's Theorem

Here we prove an analog for principal parts of Weierstrass' theorem 7.35 on prescribing the zeros of a holomorphic function and give some of the more prominent applications of this famous theorem. The relation to Weierstrass is more than mere analogy, as 11.40 illustrates. Moreover, a grand synthesis theorem allowing both poles and zeros to be specified can easily be fabricated from the two results (11.41).

Theorem 11.39 (MITTAG-LEFFLER [1884a]) *Suppose that U is an open subset of $\mathbb{C}$, A a subset of U with no limit point in U and for each $a \in A$, f_a is an entire function which is zero at 0. Then there exists a holomorphic function f in $U \setminus A$ whose principal part at each $a \in A$ is the function $g_a(z) := f_a\big(1/(z-a)\big)$.*

Proof: If the series $\sum g_a$ converges in $U \setminus A$, it pretty clearly defines an f with the desired properties. By the same token, any series $\sum(g_a - h_a)$, with all $h_a \in H(U)$, that converges in $U \setminus A$ would serve as well. So we will look for such convergence-inducing summands. Let $K_1, K_2, \ldots$ be compact subsets of U as per 1.31. Thus

(1) $$K_n \subset \overset{\circ}{K}_{n+1}, \qquad U = \bigcup_{n \in \mathbb{N}} K_n.$$

(2) Each bounded component of $\mathbb{C} \setminus K_n$ contains a component of $\mathbb{C} \setminus U$.

In case $U = \mathbb{C}$, let us explicitly take each $K_n := \bar{D}(0, n)$. Set $K_0 := \varnothing$ and define

(3) $$A_n := A \cap \big(K_n \setminus K_{n-1}\big), \quad n \in \mathbb{N}.$$

Since A has no limit point in U, it meets each compact subset of U in only a finite set. Thus in particular each A_n is a finite set. Define

(4) $$F_n(z) := \sum_{a \in A_n} f_a\left(\frac{1}{z-a}\right), \qquad z \in \mathbb{C} \setminus A_n, n \in \mathbb{N},$$

a sum over the empty set being 0. Thus

(5) $$F_n \text{ is holomorphic in } \mathbb{C} \setminus A_n \supset K_{n-1}.$$

Each function $f_a\left(\frac{1}{z-a}\right)$ in (4) may be uniformly approximated on the compact set $K_{n-1} \subset \mathbb{C} \setminus \{a\}$ by polynomials in $\frac{1}{z-a}$, just by using partial sums of the power series of f_a. In turn, each positive power of $\frac{1}{z-a}$ can be uniformly approximated on K_{n-1} by polynomials in z in case a lies in the unbounded component of $\mathbb{C} \setminus K_{n-1}$ (by 8.6), and otherwise by polynomials in $\frac{1}{z-w}$ for some $w \in \mathbb{C} \setminus U$, thanks to (2) and 8.5. In summary

(6) $$\text{there exists (rational) } R_n \in H(U)$$

such that

(7) $$|F_n(z) - R_n(z)| < 2^{-n} \qquad \forall z \in K_{n-1}.$$

We shall show that the following series converges throughout $U \setminus A$ and defines a function f with the desired properties:

$$(8) \qquad f(z) := \sum_{n=1}^{\infty} [F_n(z) - R_n(z)].$$

Fix $N \in \mathbb{N}$. Then by (5) and (6), $F_n - R_n$ is holomorphic in a neighborhood of K_N for every $n \geq N+1$ and on K_N we have $|F_n - R_n| < 2^{-n}$ for such n by (7). Therefore the series $\sum_{n=N+1}^{\infty} [F_n - R_n]$ converges uniformly on K_N, to a function g_N which is holomorphic in $\overset{\circ}{K}_N$. Consequently, (8) converges for all $z \in K_N \setminus A$, since the N initial terms there have singularities in K_N only at points of A. As $N \in \mathbb{N}$ is arbitrary, (8) therefore converges for all $z \in \bigcup (K_N \setminus A) = U \setminus A$, and for each N

$$f(z) = \sum_{n=1}^{N} [F_n(z) - R_n(z)] + g_N(z), \quad z \in \overset{\circ}{K}_N \setminus A$$

$$(9) \quad f(z) - F_1(z) - \cdots - F_N(z) = g_N(z) - R_1(z) - \cdots - R_N(z), \; z \in \overset{\circ}{K}_N \setminus A.$$

Since $g_N - R_1 - \cdots - R_N$ is holomorphic in $\overset{\circ}{K}_N$, it follows from (9) that $f - F_1 - \cdots - F_N$ is also. Therefore, recalling (5), f is holomorphic in $\overset{\circ}{K}_N \setminus (A_1 \cup \cdots A_N) = \overset{\circ}{K}_N \setminus A$ and f has precisely the principal part $f_a\big(1/(z-a)\big)$ at each $a \in \overset{\circ}{K}_N$. This is because, recalling (4), every $f_{a'}$ with $a' \neq a$ which occurs in (9) is holomorphic near a. Since $U = \bigcup_{N \in \mathbb{N}} \overset{\circ}{K}_N$, the proof is complete.

Exercise 11.40 *Use Mittag-Leffler's theorem* 11.39 *to deduce the theorem of Weierstrass* 7.35 *on prescribing the zeros of a holomorphic function.*

Hints: Let U, A, $m(a)$ be as in 7.35. 11.39 provides $h \in H(U \setminus A)$ with principal part $m(a)/(z-a)$ at each $a \in A$. This function may be chosen so that in addition it satisfies

$$(1) \qquad \frac{1}{2\pi i} \int_{\gamma} h \in \mathbb{Z}$$

for every piecewise smooth loop $\gamma \subset U \setminus A$. To see this examine the genesis of h in the proof of 11.39: $h = \sum_{n=1}^{\infty} (F_n - R_n)$ [local uniform convergence in $U \setminus A$, whence uniform convergence on γ], where the functions $F_n(z)$ are finite sums of the desired principal parts $\frac{m(a)}{z-a}$ and R_n are rational approximations to F_n having all poles in $\mathbb{C} \setminus U$. For each function $\frac{1}{z-a}$ being approximated here, the approximants are either polynomials in z or (according to the Scholium to 8.5) can be chosen to be polynomials in $\frac{1}{z-w}$, $w \in \mathbb{C} \setminus U$, in which the term $\frac{1}{z-w}$ has coefficient 1. Hence $\frac{1}{2\pi i}\int_\gamma R_n$ is the sum of $m(a)\mathrm{Ind}_\gamma(w)$ over the relevant w and a, and so is an integer; while of course $\frac{1}{2\pi i}\int_\gamma F_n$ is the sum of $m(a)\mathrm{Ind}_\gamma(a)$ over the relevant a, hence also an integer. From (1) and 5.32(iii) we get a zero-free $F \in H(U \setminus A)$ such that

$$h = F'/F \qquad \text{in } U \setminus A. \tag{2}$$

Now to investigate F near an $a \in A$, choose $R > 0$ so small that $D(a, R) \subset U \setminus \big[A \setminus \{a\}\big]$. By definition of h there exists a function h_a holomorphic in $D(a, R)$ such that

$$h(z) = \frac{m(a)}{z-a} + h_a(z) \qquad \forall z \in D(a, R) \setminus \{a\}.$$

If we let H_a denote a primitive for h_a and $P_a(z) := (z-a)^{m(a)}$ $(z \in \mathbb{C})$, this last equation reads

$$h = \frac{P_a'}{P_a} + H_a' \qquad \text{in } D(a, R) \setminus \{a\}.$$

In conjunction with (2) this implies that

$$\big(e^{-H_a} F/P_a\big)' = 0 \qquad \text{in } D(a, R) \setminus \{a\}.$$

That is, for some (non-zero) constant c_a

$$\begin{aligned} F(z) &= P_a(z) c_a e^{H_a(z)} \\ &= (z-a)^{m(a)} c_a e^{H_a(z)} \qquad \forall z \in D(a, R) \setminus \{a\}. \end{aligned} \tag{3}$$

Since H_a is holomorphic in $D(a, R)$, equation (3) shows that F extends to be holomorphic in $D(a, R)$ with a zero at a of order $m(a)$.

Remarks: The case of 11.39 and 11.40 in which $U = \mathbb{C}$ can be treated more simply. In fact, WEIERSTRASS [1876] proved 11.40 in that case by a method that, true to his philosophical convictions, involved no integration (and logarithms only superficially) and his inauguration of convergence-inducing terms there was Mittag-Leffler's inspiration for 11.39 when $U = \mathbb{C}$. Poincaré considered that idea to be Weierstrass' principal contribution to function theory, so it is worthwhile looking at this proof.

We may assume $0 \notin A$ and that A is infinite. It is, of course countable and can be enumerated as $\{a_n\}_{n\in\mathbb{N}}$ with $|a_n| \uparrow \infty$. Write m_k for the positive integer $m(a_k)$ giving the desired multiplicity of a_k. We have first to manufacture the function h used in the 11.40 proof. This is done by the technique used in the 11.39 proof, but here is simpler and more explicit. Notice that for $a \neq 0$ and $|z| < |a|$

$$\frac{1}{z-a} = -\frac{1}{a} \cdot \frac{1}{1-z/a} = \sum_{k=0}^{\infty} -a^{-k-1} z^k$$

and therefore for $|z| \leq |a|/2$ and all $K \in \mathbb{N}$

$$\left| \frac{1}{z-a} + \sum_{k=0}^{K} a^{-k-1} z^k \right| \leq |a|^{-1} \sum_{k=K+1}^{\infty} |z/a|^k = \frac{|z/a|^{K+1}}{|a|-|z|} \leq |a|^{-1} 2^{-K}.$$

In particular, polynomials p_n exist satisfying

$$\left| \frac{m_n}{z-a_n} + p_n(z) \right| \leq 2^{-n} \qquad \forall |z| \leq |a_n|/2,\ n \in \mathbb{N}.$$

This entails that the series $\sum_{n=N}^{\infty} [m_n (z-a_n)^{-1} + p_n(z)]$ converges uniformly in $D(0, |a_N|/2)$, for each $N \in \mathbb{N}$. Hence the series

$$h(z) := \sum_{n=1}^{\infty} \left[\frac{m_n}{z-a_n} + p_n(z) \right]$$

converges locally uniformly in $\mathbb{C} \setminus A$. Clearly h is a holomorphic function and for each n the

(1) $$\text{principal part of } h \text{ at } a_n \text{ is } \frac{m_n}{z-a_n}.$$

Let H be a primitive for h in $D(0, |a_1|) \subset \mathbb{C} \setminus A$. H comes directly from the Taylor series of h at $0 \in \mathbb{C} \setminus A$. Define

$$F := e^H, \text{ which is holomorphic in } D(0, |a_1|).$$

Notice that

(2) $$F'/F = H' = h \quad \text{in } D(0, |a_1|).$$

The proof is finished by establishing the

Claim:

> F admits holomorphic extension to $\mathbb{C}$ with zero-set A, the multiplicity of a_n being m_n, for each n.

Consider any real $R > |a_1|$ and let $a_1, a_2 \ldots, a_r$ be all of A that lies in $D(0,R)$. Introduce the polynomial

(3)
$$P(z) := \prod_{n=1}^{r} (1 - z/a_n)^{m_n}.$$

From the product differentiation rule

$$P'(z)/P(z) = \sum_{n=1}^{r} \frac{m_n}{z - a_n} \qquad \forall z \in \mathbb{C} \setminus \{a_1, a_2, \ldots, a_r\}.$$

In view of (1) therefore

(4)
$$h_R := h - P'/P \text{ is holomorphic in } D(0,R).$$

Let H_R be a primitive of h_R in $D(0,R)$ with $H_R(0) = H(0)$, and define

(5)
$$F_R := Pe^{H_R} \qquad \text{in } D(0,R).$$

Then in $D(0,R) \setminus \{a_1, a_2, \ldots, a_r\}$

$$F_R'/F_R = P'/P + H_R' = P'/P + h_R \overset{(4)}{=} h.$$

Together with (2) this says that $F_R'/F_R = F'/F$ in $D(0, |a_1|)$. That is, in this disk

$$\big(F_R/F\big)' = \big(FF_R' - F_R F'\big)/F^2 = 0$$

and so in this disk $F_R/F = F_R(0)/F(0) = P(0)e^{H_R(0)}/e^{H(0)} = 1$, since $H_R(0) = H(0)$ and $P(0) = 1$, by (3). In summary,

(6)
$$F = F_R \text{ in } D(0, |a_1|),$$

that is, F_R is a holomorphic extension to $D(0,R)$ of F. As $R \uparrow \infty$, the F_R define a single holomorphic function in $\mathbb{C}$, thanks to (6) and 5.62. Or, as Weierstrass put it, the Taylor series of F at 0 has radius of convergence at least R, for every real R. The form (5) and (3) of F_R shows that the zeros of F_R are exactly $a_1, a_2, \ldots, a_r$, with multiplicities $m_1, m_2, \ldots, m_r$. The claim is thus confirmed.

Theorem 11.41 (Mittag-Leffler's "Anschmiegungssatz") *Let $U \subset \mathbb{C}$ be open, $A \subset U$ have no limit points in U. For each $a \in A$ let a neighborhood $U_a \subset U$ of a, a function $g_a \in H\big(U_a \setminus \{a\}\big)$ and a positive integer n_a be given. Then there exists $F \in H(U \setminus A)$ such that for each $a \in A$, the Laurent coefficients of F at a coincide with those of g_a for all indices less than n_a.*

Proof: 7.35 (or 11.40) provides an $f \in H(U)$ which at each $a \in A$ has a zero of order n_a and no others. Therefore $1/f$ is holomorphic in $U \setminus A$. Let then $h_a \in H\big(\mathbb{C} \setminus \{a\}\big)$ be the principal part at a of the function g_a/f. Use 11.39 to come up with a function $h \in H(U \setminus A)$ which for each $a \in A$ has principal part h_a at a. Then set $F := fh$. By definition, about each $a \in A$ there is an open disk D_a inside U_a (and disjoint from $A \setminus \{a\}$) and $f_a,\ F_a,\ H_a \in H(D_a)$ such that

$$f(z) = (z-a)^{n_a} f_a(z), \quad \frac{g_a(z)}{f(z)} = h_a(z) + F_a(z), \quad h(z) = h_a(z) + H_a(z)$$

for all $z \in D_a \setminus \{a\}$. For these z then a little calculation shows that

$$F(z) = g_a(z) + (z-a)^{n_a} \big[f_a H_a - f_a F_a\big](z).$$

Since the square-bracketed expression is holomorphic in the neighborhood D_a of a, it is clear that the Laurent series of F at a has the form claimed.

Remark: The relations $\operatorname{sinc}\pi(z-n)+\operatorname{sinc}\pi(z+n) = 2(-1)^n z\sin(\pi z)/\pi(z^2-n^2)$ lead to an *explicit* solution to certain special cases of 11.41. Namely, for any bounded sequence $\{a_n : n \in \mathbb{Z}^+\}$ the limit

$$\lim_{N\to\infty} \sum_{n=-N}^{N} a_{|n|} \operatorname{sinc}(\pi(z-n)) = \frac{\sin \pi z}{\pi} \left[\frac{a_0}{z} + \sum_{n=1}^{\infty} a_n \frac{2(-1)^n z}{z^2-n^2}\right]$$

exists locally uniformly in $\mathbb{C}$ and defines an entire function $f(z)$ that satisfies $f(n) = a_n$ for all $n \in \mathbb{Z}^+$: For any $K \in \mathbb{N}$ the limit on the right exists uniformly for $z \in C(0, K+1/2)$, since such z are never in $\mathbb{Z}$ and they satisfy $|z^2 - n^2|^{-1} \le 2n^{-2}$ whenever $n \ge \sqrt{2}(K+1/2)$. Thus the series of entire functions on the left converges uniformly on each $C(0, K+1/2)$.

Corollary 11.42 *If U is an open subset of $\mathbb{C}$ and $f_1, \ldots, f_n$ are holomorphic in U and have no common zero, then there exist $g_1, \ldots, g_n$ holomorphic in U such that* (Bézout's equation) $f_1 g_1 + \cdots + f_n g_n = \mathbf{1}$.

Proof: Passing to components, we may assume without loss of generality that U is connected. We go by induction on n and so suppose that no f_j is identically $\mathbf{0}$. The case $n=1$ being trivial, we look at $n=2$. If a is a zero of f_2 of order $n(a)$, then $f_1(a) \neq 0$, and so $1/f_1$ is holomorphic near a. Now $f_2^{-1}(0)$ has no limit point in U, since U is connected and f_2 is not identically $\mathbf{0}$. Therefore 11.41 provides a function g_1 holomorphic in U such that $g_1 - 1/f_1$ has a zero at a of order not less than $n(a)$ for each zero a of f_2. If

$$g_1(z) - \frac{1}{f_1(z)} = (z-a)^{n(a)} h_a(z), \quad z \text{ near } a,$$

then

$$1 - f_1(z)g_1(z) = -(z-a)^{n(a)}h_a(z)f_1(z), \quad z \text{ near } a,$$

so $(\mathbf{1} - f_1g_1)/f_2$ has a removable singularity at a. Thus $(\mathbf{1} - f_1g_1)/f_2$ extends to a holomorphic function g_2 in U. We have evidently $\mathbf{1} - f_1g_1 = f_2g_2$.

Now suppose the result true for $n = N$ and consider $n = N+1$. Let $A := f_1^{-1}(0) \cap \cdots \cap f_N^{-1}(0)$ and for each $a \in A$ let $m(a)$ be the minimum of the orders of a as a zero of $f_1, \dots, f_N$. Weierstrass provides an $h \in H(U)$ with a zero of order $m(a)$ at each $a \in A$ and no others. Evidently then $F_j := f_j/h$ is holomorphic in U for each $1 \le j \le N$ and $F_1, \dots, F_N$ have no common zero. By the induction hypothesis there exist $G_1, \dots, G_N$ in $H(U)$ such that

$$F_1G_1 + \cdots + F_NG_N = \mathbf{1},$$

$$(*) \qquad f_1G_1 + \cdots + f_NG_N = h.$$

We have $h^{-1}(0) \cap f_{N+1}^{-1}(0) = A \cap f_{N+1}^{-1}(0) = f_1^{-1}(0) \cap \cdots \cap f_N^{-1}(0) \cap f_{N+1}^{-1}(0) = \varnothing$ and so by the case $n = 2$ considered above there exists g, $g_{N+1} \in H(U)$ such that

$$hg + f_{N+1}g_{N+1} = \mathbf{1},$$

that is, recalling $(*)$ and setting $g_j := gG_j$ $(j = 1, 2, \dots, N)$,

$$(f_1g_1 + \cdots + f_Ng_N) + f_{N+1}g_{N+1} = \mathbf{1}.$$

Exercise 11.43 *Deduce from* 11.42 *that every finitely-generated ideal in* $H(U)$ *is principal, that is, of the form* $F \cdot H(U)$ *for some* $F \in H(U)$.

Hints: The problem reduces at once to connected U. Given $F_1, \dots, F_N \in H(U)$, none identically 0, the set $A = F_1^{-1}(0) \cap \cdots \cap F_N^{-1}(0)$ has no limit point in U (5.62) and 7.35 provides an $F \in H(U)$ such that at each $a \in A$, F has a zero of order $\min\{n(a, F_j) : 1 \le j \le N\}$ and no other zeros in U. Then the functions $f_j := F_j/F \in H(U)$ and 11.42 applies, providing $g_j \in H(U)$ such that $\sum_{j=1}^N f_jg_j = \mathbf{1}$. Then $\sum_{j=1}^N F_jg_j = F$, so F belongs to the ideal generated by $F_1, \dots, F_N$.

Exercise 11.44 *Let* U *be an open subset of* $\mathbb{C}$, $\varnothing \neq \mathcal{F} \subset H(U)$ *a family of functions with no common zero in* U.

(i) (Extension of 11.42) *Prove that there exist distinct* $f_0, f_1, f_2, \dots \in \mathcal{F}$ *and* $g_0, g_1, g_2, \dots \in H(U)$ *such that* $\sum_n f_ng_n = \mathbf{1}$ *holds locally uniformly throughout* U.

Hints: Consider the case of connected U. Now $\mathcal{F}$ contains an $f_0 \neq \mathbf{0}$ and its zero-set A must be isolated (5.62), hence countable. The case A finite reduces at once to 11.42, so we assume A is infinite. Let $\{a_j\}_{j \in \mathbb{N}}$ be an enumeration of it and let n_j be the multiplicity of a_j. Then $f_j^*(z) := (z - a_j)^{-n_j} f_0(z)$ is holomorphic in U and its zero-set is $A \setminus \{a_j\}$. By hypothesis $\mathcal{F}$ contains an F_j with $F_j(a_j) \neq 0$. Let $\{K_n\}$ be an increasing compact exhaustion of U (1.31) and choose $0 < c_j < \infty$ so that

$$(1) \qquad c_j \sup \left| f_j^*(K_j) \right| \leq 2^{-j} \left[1 + \sup |F_j(K_j)| \right]^{-1}.$$

Let $\{f_1, f_2, \ldots\}$ be a (possibly finite, but one-to-one) enumeration of $\{F_1, F_2, \ldots\}$. Then (1) shows that for each $n > 0$ the series

$$h_n := \sum_{\{j : F_j = f_n\}} c_j f_j^*$$

converges locally uniformly in U. Moreover, (1) also shows that throughout U

$$\infty > \sum_{j=1}^{\infty} \left| c_j f_j^* F_j \right| = \sum_{n>0} \sum_{\{j : F_j = f_n\}} \left| c_j f_j^* F_j \right| \geq \sum_{n>0} |f_n h_n|.$$

It follows that

$$(2) \qquad h := \sum_{n>0} f_n h_n \quad \text{exists and equals} \quad \sum_{j=1}^{\infty} c_j f_j^* F_j \quad \text{in } U,$$

with both series converging locally uniformly. Notice that if $k \neq j$, then $f_j^*(a_k) = 0$ and so by the second formula in (2), $h(a_k) = c_k f_k^*(a_k) F_k(a_k) \neq 0$, for every $k \in \mathbb{N}$. Since $\{a_1, a_2, \ldots\}$ are all the zeros of f_0, this means that f_0 and h have no common zeros in U. Therefore by 11.42 there exist g_0, $g \in H(U)$ such that $f_0 g_0 + hg = \mathbf{1}$. We can therefore take $g_n := h_n g$, for each $n > 0$.

(ii) *Show that any ideal in $H(U)$ which is closed with respect to uniform-on-compacta convergence is a principal ideal, i.e., has the form $h \cdot H(U)$ for some $h \in H(U)$.*

Hints: Let I be such an ideal, $I \neq \{\mathbf{0}\}$. Again assume without loss of generality that U is connected. Then the set A of common zeros of the functions in I is an isolated set in U. For each $a \in A$ let $n(a)$ be the smallest order of a as a zero of any function in I and cite 7.35 to produce an $h \in H(U)$ which has a zero of order $n(a)$ at each $a \in A$ and no other zeros. Then $f/h \in H(U)$ for each $f \in I$ and the set of these quotients is an ideal J which has no common zeros. It is also easy to see that J is closed with respect to uniform-on-compacta convergence. Therefore an application of (i) to $\mathcal{F} := J$ shows that $\mathbf{1} \in J$, so $J = H(U)$.

(iii) *Prove the converse of* (ii)*: every principal ideal in $H(U)$ is closed with respect to uniform-on-compacta convergence.*

Hint: $h \cdot H(U)$ is the set of all $f \in H(U)$ which vanish at each zero of h to at least the order of that zero in h.

(iv) *Suppose U is connected. Prove that every closed maximal ideal in $H(U)$ has the form $\{f \in H(U) : f(a) = 0\}$ for some $a \in U$ but that there are maximal ideals not of this form.*

Hints: The claim about maximal ideals which are closed is immediate from (ii) [cf. the hints to (iii)]. Let A be an infinite subset of U with no limit point in U. Clearly the set $I := \{f \in H(U) : A \setminus f^{-1}(0) \text{ is finite}\}$ is an ideal. Use the Axiom of Choice to get a maximal ideal $M \supset I$. For each $a \in U$ there is (7.35 or 11.40) an $f_a \in I$ whose zero-set is exactly $A \setminus \{a\}$. Therefore it is clear that M is not of the form described.

§ 6 Meromorphic Functions

If U is an open subset of $\mathbb{C}$, A a subset of U with no limit points in U and f is holomorphic in $U \setminus A$ with each point of A a pole or a removable singularity, we say f is *meromorphic* in U. The uniqueness theorem 5.62 leads routinely to an equivalent definition, which does not mention an exceptional set: A non-zero function f is meromorphic in the region Ω if and only if to each $p \in \Omega$ are associated $m_p \in \mathbb{Z}$, neighborhood $U_p \subset \Omega$ and zero-free $h_p \in H(U_p)$ such that $f(z) = (z-p)^{m_p} h_p(z)$ for all $z \in U_p \setminus \{p\}$. The rationale for this terminology (compare its Greek origins) is

Theorem 11.45 *A function f is meromorphic in an open set U if and only for some holomorphic functions g, h in U, where h does not vanish identically in any component of U, and some set A of isolated points in U, we have $f = g/h$ in $U \setminus A$. In this case, g and h can be chosen without common zeros.*

Proof: ($\Leftarrow$) The set $A := h^{-1}(0)$ has no limit points in U, since h is not identically zero in any component of U. Then $f = g/h$ is holomorphic in $U \setminus A$ and every point $a \in A$ is clearly either a removable singularity or a pole of f. Indeed if $h(z) = (z-a)^n H(z)$, where H is holomorphic in U and $H(a) \neq 0$, then near a

$$f(z) = \frac{1}{(z-a)^n} \frac{g(z)}{H(z)} = \sum_{k=0}^{\infty} c_k (z-a)^{k-n},$$

where $\sum_{k=0}^{\infty} c_k (z-a)^k$ is the power series for g/H near a.

(⇒) If f is meromorphic in U and A is its set of (isolated) singularities, we may suppose each $a \in A$ is in fact a pole and we let $n(a)$ denote its order. Weierstrass' theorem (7.35 or 11.40) provides an $h \in H(U)$ with a zero of order $n(a)$ at each $a \in A$ and no others. Consider fh in $U \setminus A$. Evidently each point of A is a removable singularity for this function, so it extends to be holomorphic in U. By definition of $n(a)$, the extension g of fh is non-zero at each point of A; as the zeros of h are all in A, this means g and h have no common zero. Of course, $f = g/h$ in $U \setminus A$.

Corollary 11.46 *If f is meromorphic in the region Ω and $f^{-1}(0)$ has a limit point in Ω, then f is identically $\mathbf{0}$. The set of all functions which are meromorphic in Ω thus constitutes a field, the field of quotients of the integral domain $H(\Omega)$.*

Proof: This is immediate from 11.45 and 5.62. The first statement can also be deduced without the deeper results on which 11.45 is based by simply recalling 1.24 and applying 5.62 to Ω minus the pole-set of f, because in removing the pole-set the hypothesized limit point of $f^{-1}(0)$ in Ω is not lost, thanks to 11.7(ii).

Remarks: (i) The second statement in 11.46 is perhaps too succinct and needs amplification. First, we agree that each function which is meromorphic in Ω has all its removable singularities removed. For two such functions f_1, f_2 we understand $f_1 + f_2$ and $f_1 \cdot f_2$ to be defined algebraically (pointwise) off the two pole-sets and then extended over any removable singularities. When f_2 is not the $\mathbf{0}$ function, we understand f_1/f_2 to be defined algebraically off the union of $f_2^{-1}(0)$ and the two pole-sets, then extended over any removable singularities. Note that the latter will include any poles of f_2 which are not also poles of f_1.

(ii) The issue of convergence of meromorphic functions is somewhat subtle and will not be explored. There are at least two possible definitions of a sequence $\{f_n\}$ of meromorphic functions in a region Ω converging to a $\mathbb{C}$-valued limit function f on Ω minus a limit-point-free set (of potential poles). They are:

(1) Every $z \in \Omega$ has a neighborhood U_z such that almost every f_n is pole-free in $U_z \setminus \{z\}$ and $|f_n - f| \to 0$ uniformly in $U_z \setminus \{z\}$. [f is then holomorphic in U_z if almost all f_n have no pole in U_z, and otherwise f has a pole at z.] This is perhaps more transparent in terms of series, where it reads (without specifying the limit): a series $\sum g_n$ of meromorphic functions g_n in region Ω converges in Ω if each $z \in \Omega$ has a neighborhood U_z and an $N \in \mathbb{N}$ such that g_n is pole-free in U_z for all $n \geq N$ and the series $\sum_{n=N}^{\infty} g_n$ converges uniformly in U_z to a (holomorphic) limit function G_N. The sum of the original series

has then to be consistently defined from the various functions $\sum_{n=0}^{N-1} g_n + G_N$. For a brief clear discussion, see §1, Chapter 11 of REMMERT [1991].

(2) Consider all functions as taking values on the Riemann sphere $\mathbb{C}_\infty$ where the chordal metric χ (see the Chapter VII Notes) is used and convergence of functions is to be locally uniform in this metric: Each $z \in \Omega$ has a neighborhood U_z such that $\sup_{w \in U_z} \chi(f_n(w), f(w)) \to 0$. In (1) only meromorphic f show up as limits, but in (2) the limit function f is either meromorphic or identically ∞. (See, for example, Theorem 15.2.1 in HILLE [1962].) Moreover, (2) does not harmonize with the algebraic structure of the field of meromorphic functions. For example, one can have $f_n \to f$ but $f_n - f$ fail to converge to $\mathbf{0}$.

For other topologies on spaces of meromorphic functions see GROSSE-ERDMANN [1995] and its appendix and bibliography.

Exercise 11.47 (i) *Does the conclusion of* 11.14 *hold for all f which are meromorphic in $\mathbb{D}$ and satisfy* $\lim_{|z|\to 1} |f(z)| = 1$*?*

Hint: f has only finitely many poles in $\mathbb{D}$, say $p_1, \ldots, p_n$, repeated according to multiplicity. Let $B(z) := \prod_{k=1}^{n} \frac{z-p_k}{1-\bar{p}_k z}$ and apply 6.16(i) to $f \cdot B$.

(ii) *Does the conclusion of* 11.15 *hold for all f which are meromorphic in $\mathbb{C}$ and real and positive on $\mathbb{T}$?*

Hint: The c in the proof of 11.15 is now a zero-free entire function. Consider $F(z) := z^{n-m-l} \cdot c(z)$ if $n - m - l \geq 0$, otherwise $F(z) := z^{m-n+l}/c(z)$. Then F is real and positive on $\mathbb{T}$. From 5.17 F is real in $\mathbb{D}$, hence (2.16(ii)) constant there, hence constant in $\mathbb{C}$. The constant is positive, so $0 < F(0)$ implies $n - m - l = 0$.

(iii) *Does the conclusion of* 11.15 *hold for all f which are meromorphic in $\mathbb{D}$ and have a finite positive limit at each point of $\mathbb{T}$?*

(iv) *Does the conclusion of* 11.10(iii) *hold if the hypothesis $F \in H(\mathbb{C})$ is relaxed to F meromorphic in $\mathbb{C}$?*

Hints: Consider $\mathbb{R}$-linearly independent $p_1, p_2 \in \mathbb{C}$ and form $\Lambda := \mathbb{Z}p_1 + \mathbb{Z}p_2$. Independence means that $(x, y) \in \mathbb{R}^2 \mapsto |xp_1 + yp_2|$ is a continuous function which is 0 only at $(0, 0)$. On the compact set $\{(x, y) \in \mathbb{R}^2 : |x| + |y| = 1\}$ it has a positive minimum B. Hence

$$|xp_1 + yp_2| \geq (|x| + |y|)B \qquad \forall (x, y) \in \mathbb{R}^2. \tag{1}$$

In particular,

$$\text{every bounded subset of } \Lambda \text{ is finite,} \tag{2}$$

so Λ has no limit points in $\mathbb{C}$. For $N \in \mathbb{N}$ and $z, \omega \in \mathbb{C}$, $|\omega| > 2N \geq 2|z| \Rightarrow$

$$(3) \qquad \left| \frac{1}{(z-\omega)^2} - \frac{1}{\omega^2} \right| \leq \left| \frac{N(2|\omega| + N)}{|\omega|^2 (|\omega| - |z|)^2} \right| \leq \frac{N(2|\omega| + \frac{1}{2}|\omega|)}{|\omega|^2 (|\omega| - \frac{1}{2}|\omega|)^2} = \frac{10N}{|\omega|^3}.$$

Choose integer $p > |p_1| + |p_2|$. Since $|xp_1 + yp_2| \leq (|x| + |y|)(|p_1| + |p_2|) \leq (|x| + |y|)p$, $\omega := n_1 p_1 + n_2 p_2$ satisfies $|n_1| + |n_2| \geq p^{-1}|\omega|$. If $N, K \in \mathbb{N}$ and $K \geq 2Np^{-1}$, then (3) will hold whenever $|z| \leq N$ and $|\omega| > pK$. The sum in (3) over such ω will involve only $n_1, n_2 \in \mathbb{Z}$ with $|n_1| + |n_2| \geq p^{-1}|\omega| > K$ and hence will be majorized by

$$10N \sum_{k \geq K} \sum_{|n_1|+|n_2|=k} |n_1 p_1 + n_2 p_2|^{-3} \leq 10N \sum_{k \geq K} \sum_{|n_1|+|n_2|=k} B^{-3}(|n_1| + |n_2|)^{-3}$$

$$\leq 10N \sum_{k \geq K} 4kB^{-3}k^{-3} = 40NB^{-3} \sum_{k=K}^{\infty} k^{-2} \to 0 \text{ as } K \to \infty.$$

Thus the finite [by (2)] sums

$$(4) \qquad \sum \{(z-\omega)^{-2} - \omega^{-2} : \omega \in \Lambda, \quad 0 < |\omega| \leq pK\}$$

converge as $K \to \infty$, absolutely and uniformly for z in any bounded subset of $\mathbb{C} \setminus (\Lambda \setminus \{0\})$. Their limit defines a holomorphic function $\mathfrak{P}_0(z)$ in $\mathbb{C} \setminus (\Lambda \setminus \{0\})$.

The function $\mathfrak{P}(z) := z^{-2} + \mathfrak{P}_0(z)$, $z \in \mathbb{C} \setminus \Lambda$, is thus holomorphic, with a second-order pole at 0. Replacing ω by $-\omega$ in the sums (4) shows that $\mathfrak{P}(z) = \mathfrak{P}(-z)$. $\mathfrak{P}'(z)$ can be computed termwise (by 5.44(ii)) and is therefore the sum of the terms $-2(z-\omega)^{-3}$ over $\omega \in \Lambda = -p_j + \Lambda$. Hence $\mathfrak{P}'(z) = \mathfrak{P}'(z + p_j)$ for $z \in \mathbb{C} \setminus \Lambda$, $j = 1, 2$. Thus $\mathfrak{P}(z + p_j) - \mathfrak{P}(z)$ is constant in the connected set $\mathbb{C} \setminus \Lambda$. Evaluate at $z := -p_j/2$ to see that this constant is 0. Thus both p_1 and p_2, hence all points of Λ, are periods. Conversely, since translation by any period p must permute the pole-set Λ, $p = 0 + p \in \Lambda$.

Just as the Residue Theorem is the final generalization of Cauchy's Theorem, there are meromorphic function generalizations of the various zero-counting integral theorems (argument principles). Here is one; the reader can easily adapt its proof to produce the appropriate generalization of 5.84.

Theorem 11.48 *Let U be an open subset of $\mathbb{C}$, F a function meromorphic in U and not identically zero in any component of U. Let P be the pole-set and Z the zero-set of F in U, $n(z, F)$ and $n(p, F)$ the multiplicities of z and p as zeros and poles of F, respectively. Let $\gamma_1, \ldots, \gamma_k$ be piecewise smooth* *loops in* $U \setminus (P \cup Z)$ *such that* $\gamma := (\gamma_1, \ldots, \gamma_k)$ *is* U*-nullhomologous. Write* $F \circ \gamma := (F \circ \gamma_1, \ldots, F \circ \gamma_k)$. *Then*

$$\mathrm{Ind}_{F\circ\gamma}(0) = \frac{1}{2\pi i}\int_\gamma \frac{F'}{F} = \sum_{z\in Z} n(z,F)\mathrm{Ind}_\gamma(z) - \sum_{p\in P} n(p,F)\mathrm{Ind}_\gamma(p),$$

both sums having only a finite number of non-zero terms. (See also 11.49(vi).)

Proof: We note that $P \cap Z = \varnothing$ and the set $A := P \cup Z$ has no limit point in U. Let $f := F'/F$ in $U \setminus A$. This function is holomorphic and we shall show that each point of A is a pole such that

$$(*) \qquad \mathrm{Res}\left(\frac{F'}{F}, a\right) = \begin{cases} n(z,F) & \text{if } a = z \in Z \\ -n(p,F) & \text{if } a = p \in P. \end{cases}$$

The assertion of the theorem then follows from the Residue Theorem (in the form of 11.30(i)). If $a \in A$, then there is a non-zero integer n such that $(z-a)^n F(z) =: H_a(z)$ has a removable singularity and no zero at a: $|n| = n$ is the multiplicity of a as a pole if $a \in P$, $|n| = -n$ is the multiplicity of a as a zero if $a \in Z$. Then

$$n(z-a)^{n-1}F(z) + (z-a)^n F'(z) = H_a'(z),$$

$$\frac{F'(z)}{F(z)} = \frac{-n}{z-a} + \frac{H_a'(z)}{H_a(z)},$$

which confirms the assertion $(*)$, since H_a'/H_a is holomorphic near a.

Exercise 11.49 (i) (Argument Principle *redivivus*). *Let* U *be an open subset of* $\mathbb{C}$, K *a non-void compact subset of* U. *Show that there exist polygonal loops* $\gamma_1, \ldots, \gamma_k$ *in* $U \setminus K$ *such that, with* $\gamma := (\gamma_1, \ldots, \gamma_k)$ *and* $(F \circ \gamma_1, \ldots, F \circ \gamma_k) =: F \circ \gamma$

$$(*) \qquad \mathrm{Ind}_{F\circ\gamma}(0) = \frac{1}{2\pi i}\int_\gamma \frac{F'}{F} = Z(F) - P(F)$$

for all functions F *which are meromorphic in* U, *not identically* 0 *in any component of* U, *and have no zeros or poles outside* K. *Here* $Z(F)$ [*resp.*, $P(F)$] *denotes the number of zeros* [*resp., poles*] *of* F *in* U *counted according to multiplicity, and is finite.*

Hints: Let $\gamma_1, \ldots, \gamma_k$ be the closed polygons supplied by 8.3 such that $\gamma := (\gamma_1, \ldots, \gamma_k)$ satisfies

$$\mathrm{Ind}_\gamma(w) = 0 \qquad \forall w \in \mathbb{C} \setminus U, \text{ that is,}$$

(0) γ is U-nullhomologous,

and

$$(1) \qquad \mathrm{Ind}_\gamma(w) = 1 \qquad \forall w \in K \supset P \cup Z.$$

From (0), (1) and 11.48 claim $(*)$ follows.

(ii) (Rouché's theorem for meromorphic functions). Φ,Ψ *are meromorphic in the open set* U *and there is a compact* $K \subset U$ *which contains all their poles and outside of which the strict inequality*

$$(**) \qquad |\Phi + \Psi| < |\Phi| + |\Psi|$$

prevails. Letting $Z(F)$, $P(F)$ *be as in* (i), *prove that*

$$Z(\Phi) - P(\Phi) = Z(\Psi) - P(\Psi).$$

Hints: Let $\gamma_1, \dots, \gamma_k$ be as supplied by (i). Since $(**)$ holds in $U \setminus K$, for no $z \in U \setminus K$ can $\Phi(z)$ be a non-negative real multiple of $\Psi(z)$. That is, Φ/Ψ maps $U \setminus K$ into $\mathbb{C} \setminus [0, \infty)$. In the same way that (6) in the proof of 8.18 was established, we can use this fact to show that

$$\frac{1}{2\pi i} \int_\gamma \frac{\Phi'}{\Phi} = \frac{1}{2\pi i} \int_\gamma \frac{\Psi'}{\Psi},$$

and the desired conclusion then follows from (i).

(iii) *Deduce from* 11.48 *the following result on conformal invariance of the index (which is really just a special case of* 4.49(ii)): U *is an open subset of* $\mathbb{C}$, $f \in H(U)$ *is injective,* γ *is a piecewise smooth loop in* U *which is* U*-nullhomologous, and* $a \in U \setminus \gamma$. *Then*

$$\mathrm{Ind}_{f\circ\gamma}(f(a)) = \mathrm{Ind}_\gamma(a).$$

Hint: Take $F := f - f(a)$ in 11.48, noting that $\mathrm{Ind}_{F\circ\gamma}(0) = \mathrm{Ind}_{f\circ\gamma}(f(a))$.

(iv) *Deduce from* 11.48 *the effect of inversion on indexes, namely, the following fact: If* γ *is a piecewise smooth loop in* $\mathbb{C} \setminus \{0\}$, $0 \neq a \in \mathbb{C} \setminus \gamma$, $a^* := 1/a$ *and* $\gamma^* := 1/\gamma$, *then*

$$\mathrm{Ind}_{\gamma^*}(a^*) = \mathrm{Ind}_\gamma(a) - \mathrm{Ind}_\gamma(0).$$

Hint: Take $U := \mathbb{C}$, $P := \{0\}$, $Z := \{a\}$, $F(z) := 1/z - 1/a$ in 11.48 to get

$$\mathrm{Ind}_{\gamma^*}(a^*) = \mathrm{Ind}_{F\circ\gamma}(0) = 1 \cdot \mathrm{Ind}_\gamma(a) - 1 \cdot \mathrm{Ind}_\gamma(0).$$

(A simple proof based directly on the definition of the index is also possible.)

(v) *Show how* 11.48 *can be combined with* 5.65(ii) *to give a new proof of* 10.6.

Hints: Consider n, f, U as in 10.6, and any piecewise smooth loop γ in $U \setminus f^{-1}(0)$. Since U is simply-connected, Ind_γ vanishes in $\mathbb{C} \setminus U$. Therefore by 11.48

$$\mathrm{Ind}_{f\circ\gamma}(0) = \sum_{z \in U \setminus \gamma} n(z, f)\mathrm{Ind}_\gamma(z),$$

which is a multiple of n, since by hypothesis each $n(z, f)$ is. Now cite 5.65(ii).

(vi) *The piecewise-smoothness hypothesis in* 11.48 *is really only there to support the integration. Prove the version of* 11.48 *in which both are suppressed. Extend* (iii) *and* (iv) *accordingly.*

Hints: As shown in the proof of 11.29, there is a compact subset K of U which contains γ and outside which $\mathrm{Ind}_\gamma = 0$. There exists an open set V with compact closure such that $K \subset V \subset \overline{V} \subset U$. Both $Z \cap \overline{V}$ and $P \cap \overline{V}$ are finite. Use 4.6(i) to approximate each γ_j by a polygonal loop $\Gamma_j \subset V$ well enough that the indexes of γ_j and Γ_j agree at the finitely many points in $(Z \cup P) \cap \overline{V}$. If this approximation can be done so that in addition $\mathrm{Ind}_{\Gamma_j} = 0$ throughout $\mathbb{C} \setminus V$, then the present claim follows from the original 11.48. Since each component of $\mathbb{C} \setminus V$ meets ∂V (1.30(v)), it suffices to achieve the equality in the *compact* set ∂V. And 4.6(ii) is adequate to that task. [An alternative proof can be based on the 7.23(v) hint.]

(vii) (Cauchy at the North Pole). *U is an open subset of $\mathbb{C}$ whose complement is bounded, say by $r \in \mathbb{N}$, γ is a piecewise smooth loop in U whose index is constant, say n, in $\mathbb{C} \setminus U$, and $f \in H(U)$. Show that*

$$(*) \qquad \frac{1}{2\pi i}\int_\gamma \frac{f(\zeta)}{\zeta - z}\,d\zeta = [\mathrm{Ind}_\gamma(z) - n]f(z) + nF(z) \qquad \forall z \in \mathbb{C} \setminus \gamma,$$

where F is the regular part of f in the annulus $\mathbb{C} \setminus \overline{D}(0, r) \subset U$. How does $(*)$ simplify if $f(z)$ is bounded for large z?

Hints: According to 11.4(ii) F is an entire function and there exists $G \in H(D(0, 1/r))$ such that

$$(1) \qquad f(z) = F(z) + \frac{1}{z}G\left(\frac{1}{z}\right) \qquad \forall z \in \mathbb{C} \setminus \overline{D}(0, r).$$

Given $z \in \mathbb{C} \setminus \gamma$, for every $R > |z| + r$ define $\gamma_R(t) := Re^{-int}$. Then

$$\mathrm{Ind}_{\gamma_R}(z) = \mathrm{Ind}_{\gamma_R}(w) = -n \qquad \forall w \in \mathbb{C} \setminus U.$$

From the hypothesis on γ therefore, $\mathrm{Ind}_\gamma(w)+\mathrm{Ind}_{\gamma_R}(w)=0$ for all $w\in\mathbb{C}\setminus U$; that is, the pair (γ,γ_R) is U-nullhomologous. Form the function $H(\zeta):=\frac{f(\zeta)}{\zeta-z}$ and compute from (1) that

$$\frac{1}{2\pi i}\int_{\gamma_R} H=\mathrm{Ind}_{\gamma_R}(z)F(z)-\frac{n}{2\pi}\int_0^{2\pi}\frac{G(R^{-1}e^{int})}{Re^{-int}-z}\,dt.$$

Since G is continuous at 0, the last integral converges to 0 as $R\to\infty$. Hence, $\mathrm{Ind}_{\gamma_R}(z)$ being $-n$ for all $R>|z|$,

$$\lim_{R\to\infty}\frac{1}{2\pi i}\int_{\gamma_R} H=-nF(z). \tag{2}$$

If $z\in U\setminus\gamma$, H is meromorphic in U with only pole z and with residue $f(z)$ there. According to 11.30(i) therefore

$$\frac{1}{2\pi i}\int_\gamma H+\frac{1}{2\pi i}\int_{\gamma_R} H=\mathrm{Res}\,(H,z)[\mathrm{Ind}_\gamma(z)+\mathrm{Ind}_{\gamma_R}(z)]=f(z)[\mathrm{Ind}_\gamma(z)-n]. \tag{3}$$

From this and (2) equation $(*)$ follows for such z. On the other hand, if $z\in\mathbb{C}\setminus U$, H is holomorphic in U, the singularity set A is empty, and so in applying 11.30(i) we find the left-hand side of (3) equals 0. As before, appeal to (2) now yields $(*)$ for such z.

Finally, suppose $f(z)$ is bounded for large z. Then $F(z)$ is too, by (1) and the fact that G is continuous at 0. Hence (Liouville) F is constant. Evidently, (1) then implies that $\lim_{|z|\to\infty}f(z)$ exists and equals this constant.

(viii) *$f:\mathbb{C}\mapsto\mathbb{C}$ is bounded and continuous, $\varnothing\neq K$ compact $\subset\mathbb{C}$, and f is holomorphic in $U:=\mathbb{C}\setminus K$. Show that $f(\mathbb{C})=f(K)$.*

Hints: Due to the boundedness of f, application of the removable singularity criterion (5.41) to $z\mapsto f(1/z)$ ensures that

$$f(\infty):=\lim_{|z|\to\infty}f(z)\text{ exists in }\mathbb{C}. \tag{1}$$

We will suppose that $f(u)\notin f(K)$ for some $u\in U$ and deduce a contradiction. If Ω is the component of U which contains u, then $\overline{\Omega}\cap K\neq\varnothing$ (1.30(v)), so $f(u)\notin f(K)$ entails f is not constant in Ω. By the open-mapping theorem 5.77, $f(\Omega)$ is then a neighborhood of $f(u)\in\mathbb{C}\setminus f(K)$. Consequently, $f(\Omega)$ contains a point $w\in\mathbb{C}\setminus f(K)$ different from $f(\infty)$. Replacing f with $f-w$ lets us assume that

$$0\in f(U)\setminus f(K)\quad\text{and}\quad f(\infty)\neq 0. \tag{2}$$

The definition (1) and $f(\infty) \neq 0$ mean that for some (finite) $R > 0$

$$f(\mathbb{C} \setminus D(0,R)) \subset D(f(\infty), |f(\infty)|) \subset \mathbb{C} \setminus \{0\}. \tag{3}$$

Moreover, by (2) $f^{-1}(\mathbb{C} \setminus \{0\})$ is an open neighborhood V of K. Thus all the zeros of f lie in the compact subset $\bar{D}(0,R) \setminus V$ of U. Only finitely many of the (disjoint, open) components of U meet this compact subset of U. If any of these, say C, contains infinitely many zeros, then these zeros have an accumulation point $c \in \bar{C}$. Either $c \in C$, whence $f(C) = \{0\}$, by 5.62, and consequently $f(\partial C) = \{0\}$; or $c \in \partial C$ and again $0 \in f(\partial C)$. Since $\partial C \subset K$ [by 1.30(v)], we get in either case $0 \in f(K)$, contrary to (2). Thus f has only finitely many zeros, and at least one by (2). Let them be $z_1, \ldots, z_k$ repeated according to multiplicity. Then $g(z) := f(z) \prod_{j=1}^{k} (z - z_j)^{-1}$ well defines a continuous, zero-free function on $\mathbb{C}$. For any loop $\gamma : [0,1] \mapsto \mathbb{C}$ the mapping $(\tau, t) \in [0,1] \times [0,1] \mapsto g(\tau\gamma(t))$ is a loophomotopy in $\mathbb{C} \setminus \{0\}$ with the constant $g(0)$. Hence by 4.12, $\mathrm{Ind}_{g\circ\gamma}(0) = 0$. Taking $\gamma(t) = Re^{2\pi i t}$, and recalling the additivity of index (4.14), we get from $f \circ \gamma = (g \circ \gamma) \prod_{j=1}^{k} (\gamma - z_j)$

$$\mathrm{Ind}_{f\circ\gamma}(0) = \sum_{j=1}^{k} \mathrm{Ind}_{\gamma - z_j}(0). \tag{4}$$

According to (3) each $z_j \in D(0,R)$, so $\mathrm{Ind}_{\gamma - z_j}(0) = \mathrm{Ind}_{\gamma}(z_j) = 1$. But (3) also says that 0 lies in the unbounded component of $\mathbb{C} \setminus f \circ \gamma$, making $\mathrm{Ind}_{f\circ\gamma}(0) = 0$. Together these last two facts contradict (4).

Remark: An interesting variant of the above proof of (viii) is offered in YOUNSI [2019], where a continuous logarithm of the continuous function $g : \mathbb{C} \mapsto \mathbb{C} \setminus \{0\}$ is recruited to take over the index arguments here.

Exercise 11.50 *Let* $0 < r < R < \infty$, $A := \{z \in \mathbb{C} : r < |z| < R\}$, $a \in A$. *Let* f *be holomorphic in* $A \setminus \{a\}$ *with a first-order pole at* a, *be continuous on* $\bar{A} \setminus \{a\}$ *and satisfy* $|f| \leq 1$ *on* ∂A.

(i) *Show that* f *assumes every value in* $\mathbb{C} \setminus \bar{\mathbb{D}}$ *exactly once in* A.

Hints: Let $\epsilon > 0$. Since f is continuous on $\bar{A} \setminus \{a\}$ and $1 + \epsilon > 1 \geq |f(\partial A)|$, we have

$$1 + \epsilon > \big|f\big(C(0, r_\epsilon)\big)\big| \quad \text{and} \quad 1 + \epsilon > \big|f\big(C(0, R_\epsilon)\big)\big|$$

whenever $r_\epsilon - r > 0$ and $R - R_\epsilon > 0$ are sufficiently small. Choose and fix such r_ϵ, R_ϵ so that $a \in A_\epsilon := \{z \in \mathbb{C} : r_\epsilon < |z| < R_\epsilon\}$. If $|w| \geq 1 + \epsilon$, an application of 11.48 to $F := f - w$ shows that

$$N(w) := \frac{1}{2\pi i} \int_{C(0,R_\epsilon)} \frac{f'}{f - w} - \frac{1}{2\pi i} \int_{C(0,r_\epsilon)} \frac{f'}{f - w}$$

is the number of zeros of $f - w$ in A_ϵ minus the number of its poles. But N is evidently a continuous function of $w \in \mathbb{C} \setminus f\big(C(0, R_\epsilon) \cup C(0, r_\epsilon)\big) \supset \mathbb{C} \setminus D(0, 1+\epsilon)$ and is arbitrarily small when $|w|$ is large. Therefore $N(w) = 0$ for all w in the connected set $\mathbb{C} \setminus D(0, 1+\epsilon)$. Since by hypothesis $f - w$ has exactly one pole in A_ϵ and that of order 1, it follows that $f - w$ has exactly one zero in A_ϵ and that of order 1. This is the case for every $w \in \mathbb{C} \setminus D(0, 1+\epsilon)$. And $\epsilon > 0$ is arbitrary.

(ii) *Suppose, in addition to the hypothesis preceding* (i), *that the pole a is on the negative real axis. Show that $|f(x)| < 1$ for all $r < x < R$.*

Hints: Let $F(z) := \frac{1}{2} f(z) + \frac{1}{2}\bar{f}(\bar{z})$. Then F is meromorphic in A, its only possible pole being at a, F is real on the real axis and $|F(\partial A)| \le 1$. First, suppose that the singularity at a is removable; then $|F(z)| \le 1$ for all $z \in A$, with equality for some such z only if F is constant, by the Maximum-Modulus Principle. Since $F = \operatorname{Re} f$ on (r, R), it follows that $|\operatorname{Re} f(x)| \le 1$ for all $x \in (r, R)$ with equality for some such x only if F is a unimodular constant, hence only if $\operatorname{Re} f$ is constantly 1 or -1 on (r, R). Suppose this happens, say $\operatorname{Re} f = 1$ on (r, R). Since $|f| \le 1$ on ∂A, we get $|f| = 1$ at r and at R, whence $f(r)$, $f(R)$ are real and equal 1. Thus $f[r, R]$ is a loop lying in $\{1\} \times \mathbb{R}$. This loop must be constantly 1, since otherwise some value in $\{1\} \times \mathbb{R} \setminus \{1\} \subset \mathbb{C} \setminus \overline{\mathbb{D}}$ would be assumed more than once by f on (r, R), contrary to (i). From f constant on (r, R) follows the contradiction f constant throughout $A \setminus \{a\}$.

On the other hand, suppose that F has a pole at a. Write $f(z) = c/(z - a) + g(z)$, for some $c \in \mathbb{C}$ and some g holomorphic in A. Then $F(z) = (\operatorname{Re} c)/(z - a) + \frac{1}{2} g(z) + \frac{1}{2}\bar{g}(\bar{z})$ and $\operatorname{Re} c \neq 0$. Now $\operatorname{Re} g(x)$ is bounded for x near a. Therefore, taking $x \in (-R, -r)$ near a, we see that $F(x)$ assumes arbitrarily large positive and arbitrarily large negative values. Since $|F(-r)| \le 1$ and $|F(-R)| \le 1$, it follows from connectedness considerations that $F(-R, a) \cup F(a, -r) \supset (-\infty, -1) \cup (1, \infty)$. But then by (i), none of the values in $(-\infty, -1) \cup (1, \infty)$ is assumed by F outside $(-R, -r)$. Since F is not constant, by the Open-Map Theorem $F\big(A \setminus (-R, -r)\big)$ is an open set. It is disjoint from $(-\infty, -1) \cup (1, \infty)$ and therefore disjoint from $(-\infty, -1] \cup [1, \infty)$ too. In particular, $F(x) = \operatorname{Re} f(x) \in (-1, 1)$ for all $x \in (r, R)$.

To get the inequality for $|f(x)|$, given $x_0 \in (r, R)$, multiply f by $e^{i\theta_0}$ for appropriate real θ_0 so that $|f(x_0)| = \operatorname{Re}\big(e^{i\theta_0} f\big)(x_0)$ and apply the foregoing considerations to $e^{i\theta_0} f$ in the role of f to conclude that $|f(x_0)| = \big|\operatorname{Re}\big(e^{i\theta_0} f\big)(x_0)\big| < 1$.

(iii) (Polar analog of (9.16)) *Suppose $f : \overline{\mathbb{D}} \setminus \{0\} \mapsto \mathbb{C}$ is continuous, holomorphic in $\mathbb{D} \setminus \{0\}$ with a simple pole at 0, is one-to-one on $\partial\mathbb{D}$, and the simple loop $f \circ \partial\mathbb{D}$ is negatively oriented* (4.44). *Prove that f maps $\mathbb{D} \setminus \{0\}$ conformally onto the outside of the Jordan-curve $f(\partial\mathbb{D})$.*

Hints: For $0 < r \le 1$ set $\gamma_r := C(0,r)$. Then $f \circ \gamma_1$ is a simple loop and its range $f(\partial\mathbb{D})$ is a Jordan-curve, which (as usual) we also denote $f \circ \gamma_1$. Recall [4.44 and 4.45] that

$$(1) \qquad \mathcal{O}(f \circ \gamma_1) := \{w \in \mathbb{C} \setminus f \circ \gamma_1 : \mathrm{Ind}_{f\circ\gamma_1}(w) = 0\}.$$

For $r < 1$ and $w \in \mathbb{C} \setminus f \circ \gamma_r$, 11.48 asserts that

$$\begin{aligned} \mathrm{Ind}_{f\circ\gamma_r}(w) &= \#\text{ zeros of } (f-w) \text{ in } D(0,r) - \#\text{ poles of } (f-w) \text{ in } D(0,r) \\ (2) \qquad &= \#\text{ zeros of } (f-w) \text{ in } D(0,r) - 1, \end{aligned}$$

since by hypothesis the number of poles of f in $D(0,r)$ is 1. If $w \in \mathcal{O}(f \circ \gamma_1)$, then $w \in \mathbb{C} \setminus f \circ \gamma_1$ and by continuity, for all $r < 1$ but close to 1, $w \in \mathbb{C} \setminus f \circ \gamma_r$ and $\mathrm{Ind}_{f\circ\gamma_r}(w) = \mathrm{Ind}_{f\circ\gamma_1}(w) = 0$ [see 4.6(i)]. Hence (2) says that $f - w$ has exactly one zero in $D(0,r)$—for all r near 1:

$$(3) \qquad w \in \mathcal{O}(f \circ \gamma_1) \Rightarrow f - w \text{ has exactly one zero in } \mathbb{D}.$$

The hypothesis that $f \circ \gamma_1$ is negatively oriented means [4.44] that for every $w \in \mathcal{I}(f \circ \gamma_1)$, $\mathrm{Ind}_{f\circ\gamma_1}(w) = -1$. Continuity considerations as above show that for all r near 1, $\mathrm{Ind}_{f\circ\gamma_r}(w) = -1$. It follows then from (2) that $f - w$ has no zeros in $D(0,r)$. That is,

$$(4) \qquad w \in \mathcal{I}(f \circ \gamma_1) \Rightarrow f - w \text{ has no zeros in } \mathbb{D}.$$

Summarizing,

$$(5) \qquad \mathcal{O}(f \circ \gamma_1) \subset f(\mathbb{D} \setminus \{0\}) \subset \mathbb{C} \setminus \mathcal{I}(f \circ \gamma_1).$$

Due to 4.42 the open set $f(\mathbb{D} \setminus \{0\})$ cannot meet the Jordan-curve $f \circ \gamma_1$ without meeting its inside. Therefore (5) further entails that

$$\mathcal{O}(f \circ \gamma_1) \subset f(\mathbb{D} \setminus \{0\}) \subset \mathbb{C} \setminus \mathcal{I}(f \circ \gamma_1) \setminus f \circ \gamma_1 = \mathcal{O}(f \circ \gamma_1), \quad \text{that is,}$$
$$\mathcal{O}(f \circ \gamma_1) = f(\mathbb{D} \setminus \{0\}),$$

which, with (3), proves the conformality claim.

There is an interesting class of meromorphic functions in $\mathbb{C}$ which arise in connection with the quantum theory of collisions (Wigner [1951]) and which admit a very elegant description that we now explore, following Wigner [1952]. Actually, the problem had been raised and solved before Wigner without any such "practical" motivation.

Consider in what follows functions f which are meromorphic in $\mathbb{C}$ with pole-set P and satisfy

(a) $$f(\mathbb{R} \setminus P) \subset \mathbb{R}$$

(b) $$f(\mathbb{U} \setminus P) \subset \mathbb{U}.$$

Following Wigner we shall call these *R-functions*; this stands for "reaction matrix" but will serve to remind us of the condition (a). Our goal is a very explicit partial-fraction representation of such functions—see 11.52(ii) and compare with 11.38—which generalizes 8.37. A short and elegant alternative development, which uses techniques like those in 8.37, will be found in BOAS and BOAS [1988] and a version for functions meromorphic in $\mathbb{D}$ in A. W. GOODMAN [1956].

Exercise 11.51 *Let $a \in \mathbb{C}$, $r > 0$ and F be holomorphic in $D(a,r) \setminus \{a\}$ with a pole at a. Further, let f be an R-function with pole-set P.*

(i) *Show that F cannot map the punctured disk $D(a,r) \setminus \{a\}$ entirely into $\overline{\mathbb{U}}$.*

Hints: This is immediate from 11.7(i), but the following construction (cf. the hint to 2.21(ii)) is useful in the second part of this problem as well. If m (≥ 1) denotes the order of the pole, then there is a holomorphic function g in $D(a,r)$ such that $g(a) \neq 0$ and $F(z) = (z-a)^{-m} g(z)$. Write $g(a) := r_0 e^{i\theta_0}$ ($r_0 > 0$, $0 \leq \theta_0 < 2\pi$) and look at z of the form $a + \rho \exp\big(i(\theta_0 + \pi/2) \,/\, m\big)$. Then $F(z) = -i\rho^{-m} e^{-i\theta_0} g(z)$. As $\rho \downarrow 0$, we have $-ie^{-i\theta_0} g(z) \to -ir_0 \in \mathbb{C} \setminus \overline{\mathbb{U}}$. So for small $\rho > 0$, $-ie^{-i\theta_0} g(z) \in \mathbb{C} \setminus \overline{\mathbb{U}}$.

(ii) *Show that if $a \in \mathbb{R}$ and either the pole there is not simple or the residue of F there is not real and negative, then F cannot map $D(a,r) \cap \mathbb{U}$ entirely into $\overline{\mathbb{U}}$.*

Hints: With the notation of the preceding Hints, choose a neighborhood N of a and in it a holomorphic function G such that $g = e^G$ there and $G(a) = \log r_0 + i\theta_0$. Consider two scenarios:

Case I: $m \in \mathbb{N} \setminus \{1\}$ & $0 \leq \theta_0 < 2\pi$, or $m = 1$ & $0 \leq \theta_0 < \pi$.

One checks easily that $0 \leq \theta_0/m < \pi$, so $(\theta_0/m, (\theta_0 + \pi)/m) \cap (0,\pi) \neq \varnothing$. Select and fix θ in this intersection. Thus $\theta \in (0,\pi)$ and $\theta_0 - m\theta \in (-\pi, 0)$, so ($a$ being real)

(∗) $$z := a + \rho e^{i\theta} \in \mathbb{U} \quad \forall \rho > 0 \text{ and } e^{i(\theta_0 - m\theta)} \in \mathbb{C} \setminus \overline{\mathbb{U}}, \text{ an open set.}$$

Since G is continuous and $\operatorname{Im} G(a) = \theta_0$, if z is close enough to a (that is, ρ is small), then it follows from (∗) that

(∗∗) $$e^{i(\operatorname{Im} G(z) - m\theta)} \in \mathbb{C} \setminus \overline{\mathbb{U}}.$$

Therefore

$$F(z) = (z-a)^{-m} g(z) = (z-a)^{-m} |g(z)| e^{i \operatorname{Im} G(z)}$$

$$(***) \qquad = \rho^{-m} |g(z)| e^{i(\operatorname{Im} G(z) - m\theta)} \in \mathbb{C} \setminus \overline{\mathbb{U}}, \text{ by } (**).$$

Case II: $m = 1$ & $0 < \theta_0 < 2\pi$.

Then $2\pi < (\theta_0 + \pi)/m$, so $(\theta_0/m, (\theta_0+\pi)/m) \cap (2\pi, 3\pi) \neq \varnothing$. Select any θ in this intersection. The rest of the argument is the same as before, and culminates in $(***)$.

Notice that I and II cover all the cases addressed in (ii).

(iii) *Show that at every* $x \in \mathbb{R} \setminus P$, $f'(x)$ *is real and non-negative.*

Hints: Write $v := \operatorname{Im} f$. From (a) and (b)

$$\frac{v(x+iy) - v(x)}{y} = \frac{v(x+iy)}{y} \geq 0 \qquad \forall y > 0.$$

It follows that $D_2 v(x) \geq 0$, that is, (C. R.) $\operatorname{Re} f'(x) \geq 0$. But by computing $f'(x)$ as $(f \,|\, \mathbb{R})'(x)$ we also see from (a) that $f'(x)$ is real.

(iv) *Prove that each* $p \in P$ *is simple and real, that* $\operatorname{Res}(f, p)$ *is real and negative, and that* $f(z) - \dfrac{\operatorname{Res}(f,p)}{z-p}$ *is also an R-function.*

Hints: The meromorphic functions $f(z)$ and $\bar{f}(\bar{z})$ agree on $\mathbb{R} \setminus P$ (by(a)) and so by 11.46 they are identical. It follows then from (b) and (i) that $P \subset \mathbb{R}$ and then from (b) and (ii) that each pole is simple and each residue real and negative. Let $p_1 \in P$ be given and set $c_1 := \operatorname{Res}(f, p_1)$, $F_1(z) := f(z) - \frac{c_1}{z-p_1}$. By (a) and the fact that c_1 is real we have

$$(\text{a})_1 \qquad \operatorname{Im} F_1 = 0 \quad \text{on } \mathbb{R} \setminus P.$$

Let $z_0 \in \mathbb{U}$ and $\epsilon > 0$ be given. Choose $R > \epsilon^{-1}|c_1|$ and $R > |z_0|$. For $z \in C(p_1, R) \cap \mathbb{U}$ we have

$$\left| \frac{c_1}{z - p_1} \right| = \frac{|c_1|}{R} < \epsilon,$$

and so (b) yields

$$(1) \qquad \operatorname{Im} F_1(z) > -\epsilon \qquad \forall z \in C(p_1, R) \cap \mathbb{U}.$$

Consider a typical pole p_2 of F_1, i.e., a point $p_2 \in P\backslash\{p_1\} \subset \mathbb{R}$ and the residue c_2 of f there. We shall find $r > 0$ sufficiently small that

$$\text{(2)} \qquad \operatorname{Im} F_1 > 0 \quad \text{on } C(p_2, r) \cap \mathbb{U}.$$

First, choose $r < |z_0|$ small enough that $\bar{D}(p_2, r)$ contains no other poles. Then the function

$$\begin{aligned} F_2(z) &:= F_1(z) - \frac{c_2}{z - p_2} \\ &= f(z) - \frac{c_1}{z - p_1} - \frac{c_2}{z - p_2} \end{aligned}$$

is holomorphic in a neighborhood of $\bar{D}(p_2, r)$ and because of (a) and the fact c_1, c_2 are real, F_2 is real-valued on $(p_2 - r, p_2 + r)$. Hence for $z \in D(p_2, r) \cap \mathbb{U}$ the mean-value theorem gives

$$\begin{aligned} \operatorname{Im} F_2(z) &= \operatorname{Im} F_2(z) - 0 \\ &= \operatorname{Im} F_2(z) - \operatorname{Im} F_2(\operatorname{Re} z) \\ &= D_2(\operatorname{Im} F_2)(\xi) \cdot \operatorname{Im} z \\ &= \operatorname{Re} F_2'(\xi) \cdot \operatorname{Im} z \end{aligned}$$

for some $\xi \in [\operatorname{Re} z, z]$. Let M denote the (finite) minimum of $\operatorname{Re} F_2'$ in $\bar{D}(p_2, r)$. Then the last equation implies that

$$\operatorname{Im} F_2(z) \geq M \operatorname{Im} z = M \operatorname{Im}(z - p_2) \qquad \forall z \in \bar{D}(p_2, r) \cap \mathbb{U}.$$

Hence

$$\begin{aligned} \operatorname{Im} F_1(z) &= \operatorname{Im} F_2(z) + c_2 \operatorname{Im}\left(\frac{1}{z - p_2}\right) \\ &= \operatorname{Im} F_2(z) - \frac{c_2}{|z - p_2|^2} \operatorname{Im}(z - p_2) \\ &\geq M \operatorname{Im}(z - p_2) - \frac{c_2}{|z - p_2|^2} \operatorname{Im}(z - p_2) \\ \text{(3)} \qquad &\geq \left(M - \frac{c_2}{r^2}\right) \operatorname{Im}(z - p_2) \qquad \forall z \in \bar{D}(p_2, r) \cap \mathbb{U}. \end{aligned}$$

In the last inequality use was made of the fact $c_2 < 0$; which fact also means that if $r > 0$ is small enough, $M - \dfrac{c_2}{r^2} > 0$ and (3) will yield (2).

If we treat each of the finitely many points of $\big(P \setminus \{p_1\}\big) \cap [-R, R]$ this way, we find a region Ω in $\mathbb{U}$ containing z_0, bounded by $C(p_1, R) \cap \mathbb{U}$ and by a crenulation along the real axis, which as a result of $(a)_1$, (1) and the analogs of (2), satisfies

$$\operatorname{Im} F_1(z) > -\epsilon \qquad \forall z \in \partial\Omega.$$

But then by the maximum principle for harmonic functions, the point $z_0 \in \Omega$ satisfies $\operatorname{Im} F_1(z_0) > -\epsilon$. As $\epsilon > 0$ and $z_0 \in \mathbb{U}$ are arbitrary, we have proved

$$(b)_1 \qquad \operatorname{Im} F_1 \geq 0 \quad \text{in } \mathbb{U}.$$

Exercise 11.52 (i) *Suppose f is an R-function with pole-set P. This set is finite or countable and it is notationally convenient to suppose the latter (by adjoining "poles" at which f has singular part $\mathbf{0}$, if necessary). Enumerate $P \setminus \{0\}$ as $\{p_j\}_{j\in\mathbb{N}}$ and let $c_j := \operatorname{Res}(f, p_j)$.*

Show that the series

$$(1) \qquad \sum_{j=1}^{\infty}\left(\frac{c_j}{z-p_j}+\frac{c_j}{p_j}\right)$$

converges locally uniformly in $\mathbb{C} \setminus P$.

Hints: According to 11.51(iv), for each n

$$f_n(z) := f(z) - \sum_{j=1}^{n}\frac{c_j}{z-p_j}$$

defines another R-function. Therefore 11.51(iii) gives, for each $x \in \mathbb{R} \setminus P$

$$0 \leq f_n'(x) = f'(x) + \sum_{j=1}^{n}\frac{c_j}{(x-p_j)^2}, \quad \text{that is,}$$

$$(2) \qquad 0 \leq \sum_{j=1}^{n}\frac{-c_j}{(x-p_j)^2} \leq f'(x) \qquad \forall x \in \mathbb{R} \setminus P.$$

Since the poles do not accumulate, $|p_j| \to \infty$ as $j \to \infty$, so $|x - p_j| \leq 2|p_j|$ if j is large enough; whence

$$(3) \qquad \left|\frac{c_j}{x-p_j}+\frac{c_j}{p_j}\right| = \left|\frac{c_j x}{p_j(x-p_j)}\right| \leq \frac{2|c_j x|}{|x-p_j|^2} = \frac{-2c_j|x|}{|x-p_j|^2}.$$

From (2) and (3) we infer that the series (1) converges absolutely for every $x \in \mathbb{R} \setminus P$. Now

$$(4) \qquad \frac{c_j}{z-p_j}+\frac{c_j}{p_j} = \frac{c_j}{p_j^2}\cdot z\left(\frac{z}{z-p_j}-1\right) \qquad \forall z \in \mathbb{C} \setminus P.$$

So if we choose a non-zero $x \in \mathbb{R} \setminus P$, the absolute convergence of (1) at $z = x$ and the fact that $\left|\frac{x}{x-p_j} - 1\right| \geq 1/2$ for all large j show that

(5) $$0 \leq \sum_{j=1}^{\infty} \frac{-c_j}{p_j^2} < \infty.$$

Returning to (4), we infer via (5) that the series (1) converges (absolutely) uniformly on each compact subset of $\mathbb{C} \setminus P$.

(ii) (Theorem of ČEBOTAREV and MEĬMAN [1949]). *Deduce from* (i) *that if f is an R-function, then there exist $p_j \in \mathbb{R} \setminus \{0\}$ and real numbers b, c, c_j with $c \geq 0$ and all $c_j \leq 0$, such that* (5) *holds and*

(6) $$f(z) = cz + b + \frac{c_0}{z} + \sum_{j=1}^{\infty} \left(\frac{c_j}{z - p_j} + \frac{c_j}{p_j} \right),$$

with convergence uniform on compact subsets of $\mathbb{C} \setminus \{0, p_1, p_2, \ldots\}$.

Hints: Let p_j be the non-zero poles of f, c_j the residue of f at p_j, c_0 the residue at 0. According to 11.51(iv) the p_j are real and simple and the c_j are real and non-positive. As in (i), we may suppose the set $\{p_j\}$ infinite (countable) and $|p_j| \to \infty$, by allowing some $c_j = 0$. Denote this set by P. The functions $F_n(z) := \frac{c_0}{z} + \sum_{j=1}^{n} \left(\frac{c_j}{z-p_j} + \frac{c_j}{p_j} \right)$, $n \in \mathbb{N}$, converge locally uniformly in $\mathbb{C} \setminus P$ to a function F which is holomorphic there, by (i). Evidently the principal part of F at p_j is $c_j/(z - p_j)$, which is also the principal part of f at p_j, since p_j is either a removable singularity of f and $c_j = 0$ or p_j is a first-order pole of f with residue c_j. Thus the function $f - F$ is holomorphic in $\mathbb{C} \setminus P$ and each $p \in P$ is a removable singularity, that is, $f - F$ is entire. According to 11.51(iv) each $f - F_n$ is an R-function, hence maps $\mathbb{U}$ into $\overline{\mathbb{U}}$, $\mathbb{R} \setminus P$ into $\mathbb{R}$. The limit function $f - F$ clearly inherits these properties. And therefore by continuity and the Open Map Theorem, it maps $\mathbb{R}$ into $\mathbb{R}$ and $\mathbb{U}$ into $\mathbb{U}$. It is then an easy consequence of 8.37 that this function has the form $cz + b$ for some constants $c \geq 0$, $b \in \mathbb{R}$.

(iii) *Prove the converse of* (ii): *Every f of the form* (6), *with $|p_j| \to \infty$ and the series locally uniformly convergent in $\mathbb{C} \setminus \{0, p_1, p_2, \ldots\}$, is an R-function.*

(iv) (Extension of 8.37) *Let $P = \{p_j\}_{j \in \mathbb{N}}$ be a (finite or countable) subset of $\mathbb{R}$ which has no limit points in $\mathbb{R}$. Suppose u is harmonic and positive in $\mathbb{U}$ and satisfies*

(1)′ $$\lim_{z \to x} u(z) = 0 \qquad \forall x \in \mathbb{R} \setminus P,$$

(2)′ *for each $p \in P$ there is an $n_p \in \mathbb{N}$ such that*
$$\lim_{z \to p} (z - p)^{n_p} u(z) = 0.$$

[Regarding hypothesis (2)′, see the remarks below.]

Show that there exist $c_j \le 0$ *and* $c \ge 0$ *such that*

$$u(z) = c \operatorname{Im} z + \sum c_j \operatorname{Im}\left(\frac{1}{z - p_j}\right) \qquad \forall z \in \mathbb{U}.$$

Hints: Extend u to h defined in $\mathbb{C} \setminus P$ by

$$h(z) := \begin{cases} u(z) & z \in \mathbb{U} \\ 0 & z \in \mathbb{R} \setminus P \\ -u(\bar{z}) & z \in \mathbb{C} \setminus \overline{\mathbb{U}}. \end{cases}$$

There is an $F \in H(\mathbb{U})$ with $\operatorname{Im} F = u$. Consider two adjacent points $p < q$ in P. Set $r := \frac{1}{2}(q - p)$, $a := \frac{1}{2}(q + p)$. By 8.36, h is harmonic in $D(a, r)$. Let g be holomorphic in $D(a, r)$ with $\operatorname{Im} g = h$. Then $\operatorname{Im}(F - g) = 0$ in $D(a, r) \cap \mathbb{U}$, so $F - g$ equals a real constant there (2.16). By adding this constant to g we can suppose that

$$g = F \quad \text{in } D(a, r) \cap \mathbb{U}. \tag{7}$$

Now $z \mapsto \bar{g}(\bar{z})$ is holomorphic in $D(a, r)$ and satisfies $\bar{g}(\bar{z}) = g(\bar{z}) = g(z)$ for all $z \in D(a, r) \cap \mathbb{R}$, since $\operatorname{Im} g(\bar{z}) = h(\bar{z}) = 0$ for such z. Therefore by the uniqueness theorem $\bar{g}(\bar{z}) = g(z)$ for all $z \in D(a, r)$. In view of (7) then

$$g(z) = \begin{cases} F(z) & \text{in } D(a, r) \cap \mathbb{U}; \\ \overline{F}(\bar{z}) & \text{in } D(a, r) \setminus \overline{\mathbb{U}}. \end{cases}$$

This shows that the extension of F to $\mathbb{C} \setminus \mathbb{R}$ by reflection has a further holomorphic extension to the region $(\mathbb{C} \setminus \mathbb{R}) \cup (p, q)$. As $p \in P$ is arbitrary, we infer that F has a holomorphic extension, call it f, to the whole of $\mathbb{C} \setminus P$, and $f(\mathbb{R} \setminus P) \subset \mathbb{R}$.

Finally, we need to show that any $p \in P$ which is not removable for f is a pole. Hypothesis (2)′ says that

$$\sup_{z \in \mathbb{U} \cap D(p,r)} |z - p|^{n_p} \operatorname{Im} f(z) \to 0 \quad \text{as} \quad r \downarrow 0.$$

Since $\operatorname{Im} f = 0$ in $\mathbb{R} \setminus P$ and f is defined in $\mathbb{C} \setminus \overline{\mathbb{U}}$ by reflection, we get

$$\sup_{p \neq z \in D(p,r)} |z - p|^{n_p} |\operatorname{Im} f(z)| \to 0 \quad \text{as} \quad r \downarrow 0.$$

That is,

$$\lim_{z\to p} |z-p|^{n_p} \operatorname{Im} f(z) = 0.$$

Now scrutinize the hints to 11.60(i). The idea there can be adapted to infer from this last equality that f has a pole at p of order less than n_p. In summary, f is an R-function. From its representation (6) follows the claimed form of $u = \operatorname{Im} f$.

(v) *Show that for any distinct* $z_1, \ldots, z_n \in \mathbb{R}$ *and any* $w_1, \ldots, w_n \in \mathbb{R}$ *there exists a rational* R*-function* f *such that* $f(z_k) = w_k$ *for each* $k = 1, \ldots, n$.

Hints: (Younis [1980]) For each $k = 1, \ldots, n$ define

$$g_k(z) := \frac{-w_k^2}{z}, \quad r_k(z) := \sum_{\substack{j=1\\ j\neq k}}^{n} \frac{-1}{z-z_j} - w_k - \sum_{\substack{j=1\\ j\neq k}}^{n} \frac{1}{z_j - z_k}.$$

According to (ii) these are all R-functions. Since $-\overline{w} \in \mathbb{U}$ whenever $w \in \mathbb{U}$, property (b) of an R-function f implies that $-1/f(z) = -\overline{f(z)}/|f(z)|^2 \in \mathbb{U}$ for all $z \in \mathbb{U}$, showing that $-1/f$ is also an R-function. Thus each $f_k := g_k \circ r_k$ is an R-function. Note that if $l \neq k$, then r_k has a pole at z_l, so $f_k = g_k \circ r_k$ has a zero at z_l, while $f_k(z_k) = w_k$. Finally, set $f := \sum_{k=1}^n f_k$.

Remarks: If P is a non-empty subset of $\mathbb{C}$ with no limit points and $f \in H(\mathbb{C}\setminus P)$ has properties (a) and (b) of an R-function, is f an R-function? That is, are the points of P *inessential* singularities? An affirmative answer was indicated to the author by Harold Boas. This yields a modest but esthetically attractive improvement of 11.52(iv), as it renders hypothesis (2)′ there superfluous. The proof uses a deeper result from the next chapter, but otherwise belongs here. Without loss of generality, suppose no point of P is a removable singularity. The hints to 11.51(i) show that no pole can lie in $\mathbb{U}$, whereas property (b) and 11.6 show that no essential singularity can lie in $\mathbb{U}$. Thus $P \subset \mathbb{C} \setminus \mathbb{U}$. Let $P^* := \{\bar{p} : p \in P\}$. From 1.24 and 5.62 we learn that $f(z) = \overline{f(\bar{z})}$ for all $z \in \mathbb{C} \setminus (P \cup P^*)$, since these two holomorphic functions coincide in $\mathbb{R} \setminus P$. It follows that P^* does not meet $\mathbb{U}$ either, so $P \subset \mathbb{R}$. Since f is not constant we now have $f(\mathbb{U}) \subset \mathbb{U}$ and $f(\mathbb{C} \setminus \overline{\mathbb{U}}) \subset \mathbb{C} \setminus \overline{\mathbb{U}}$, so $f^{-1}(\mathbb{R}) \subset \mathbb{R} \setminus P$. Consider $p \in P$ and $r > 0$ sufficiently small that $D(p,r) \cap P = \{p\}$. In 11.51(iii) [which uses only (a) and (b)] we saw that f' is real and non-negative in $\mathbb{R} \setminus P$. Hence f is non-decreasing on every interval in $\mathbb{R} \setminus P$. If p is an essential singularity, it follows from 12.29 that in $D(p,s) \setminus \{p\}$, for every $0 < s < r$, f assumes, with at most one exception, every value in $\mathbb{C}$. Since f is only real-valued in $\mathbb{R}$, real values can only be taken in $D(p,s) \cap \mathbb{R}$. In particular, some real number is assumed twice by f in one of the intervals $(p-r,p)$, $(p,p+r)$. Since f is monotone on that interval, this fact makes f constant on a whole interval in $\mathbb{R} \setminus P$. Hence f is constant throughout the connected set $\mathbb{C} \setminus P$ —⋊—.

§ 7 Poisson's Formula in an Annulus and Isolated Singularities of Harmonic Functions

There are several interesting techniques for solving the Dirichlet problem in an annulus. I want to begin by outlining two of them, and then presenting a detailed development of a different kind. The first method goes back to Carl Neumann and H. A. Schwarz in the 19th century.

Consider $0 < r < 1$, $A := \{z \in \mathbb{C} : r < |z| < 1\}$, and continuous $f : \partial A \mapsto \mathbb{R}$. In 5.20 we constructed continuous $h : \overline{\mathbb{D}} \mapsto \mathbb{R}$ which is harmonic in $\mathbb{D}$ and equal to f on $\mathbb{T}$. In equation (10) of the subsequent remarks h appeared as a bilateral sum of functions of the form

$$\rho e^{i\theta} \mapsto \hat{f}(n)\rho^{|n|}e^{in\theta} \qquad (n \in \mathbb{Z}).$$

Consequently, when the (complex-valued) harmonic functions

$$z \mapsto \begin{cases} \frac{\hat{f}(n)r^{|n|}}{r^n - r^{-n}}(z^n - (\bar{z})^{-n}) & , \qquad 0 \neq n \in \mathbb{Z} \\ \frac{\hat{f}(0)}{\log r}\log|z| & , \qquad n = 0 \end{cases}$$

are summed over $n \in \mathbb{Z}$ (the series converges absolutely for $r \leq |z| \leq 1$ because $\max\{|\rho^n - \rho^{-n}| : r \leq \rho \leq 1\}$ is $|r^n - r^{-n}|$ for each $n \in \mathbb{Z}$), the resulting (harmonic) function agrees with h on $r\mathbb{T}$ and is 0 on $\mathbb{T}$. When this function is subtracted from h, one gets a continuous function h_{01} on $\bar{A}$ which equals f on $\mathbb{T}$ and 0 on $r\mathbb{T}$. Carry out this construction with $f^*(z) := f(r/\bar{z})$ in the role of f, getting a harmonic function h_{01}^*. The harmonic function $h_{10}(z) := h_{01}^*(r/\bar{z})$ then equals f on $r\mathbb{T}$ and 0 on $\mathbb{T}$. So, finally, $h_{01} + h_{10}$ solves the Dirichlet problem in A for boundary data f. When the series involved are commuted with the integration defining the Fourier coefficients, the solution appears in the form of a kernel integrated against the boundary function f. Details of all this are routine (pp. 227–230 of Henrici [1986], e.g.). However, to identify this kernel by direct calculation with the real part of the one we will generate in 11.53(ii) is a bit challenging, as is the task of expressing it in terms of Jacobi elliptic functions.

An alternative solution of the Dirichlet problem in an annulus A is furnished by the conformal mapping technique of 8.43 and 9.52. In fact, with just a twist, the exponential and logarithm functions used in 8.43 to map $S := \left(\frac{-\pi}{2}, \frac{\pi}{2}\right) \times \mathbb{R}$ onto $\mathbb{H}$ can be employed to map S onto A and thereby reduce the Dirichlet problem on A to 8.43. Here is an outline:

Again let E denote the exponential function and L the Principal Branch of the Logarithm in $\mathbb{C}_-$. E maps the strip S holomorphically onto $A := A(e^{-\pi/2}, e^{\pi/2})$ and maps ∂S onto ∂A. [By changing scale in S the most general annulus $A(r^{-1}, r)$ can be realized, so treating this special case suffices.] Given continuous $f : \partial A \mapsto \mathbb{R}$, the function $g := f \circ E$ is continuous and $2\pi i$-periodic (hence bounded) on ∂S. 8.43 supplies a continuous function $G : \overline{S} \mapsto \mathbb{R}$ which is harmonic in S and agrees with g on ∂S. Moreover, 8.43(vi) insures that G is $2\pi i$-periodic. Because of this periodicity and the fact that $E(\overline{S}) = \overline{A}$, a function F is well defined in $\overline{A}$ by declaring $F(E(z)) := G(z)$ for each $z \in \overline{S}$. Clearly F agrees on $\partial A = E(\partial S)$ with f. Using local holomorphic logarithms in $\overline{A}$ shows that F is continuous on $\overline{A}$ and harmonic in A. This endgame is tantamount to citing the factorization lemma 5.96. A Poisson formula in A can now be derived from the one in 8.43(v) and a logarithmic mapping of $\overline{A}$ back to $\overline{S}$, but the calculations required appear formidable.

Exercise 11.53 (Poisson Formula for the Annulus). *Let* $A = A(r, 1)$, $0 < r < 1$.

(i) *Show that for all non-zero complex numbers* z, w *which satisfy*

(1) $$r^2 < \left|\frac{z}{w}\right| < r^{-2}$$

the series

(2) $$H(w,z) = 2\sum_{k=1}^{\infty} \frac{r^{2k}}{1-r^{2k}} \left[\left(\frac{w}{z}\right)^k - \left(\frac{z}{w}\right)^k\right]$$

converges absolutely, and that convergence is uniform on $\overline{A} \times \overline{A}$ *and defines a continuous function there.*

Hint: $\displaystyle\sum_{k=1}^{\infty} \frac{r^{2k}}{1-r^{2k}} \left|\frac{z}{w}\right|^k \le \sum_{k=1}^{\infty} \frac{1}{1-r^2} \left|r^2\frac{z}{w}\right|^k$, and $|r^2\frac{z}{w}| < 1$. If $(w,z) \in \overline{A} \times \overline{A}$, then $r^2\left|\frac{z}{w}\right| \le r$ and $r^2\left|\frac{w}{z}\right| \le r$.

(3) *If* $|z| = |w| \neq 0$, *then* $\operatorname{Re} H(w,z) = 0$.

Hint: $\displaystyle\left(\frac{w}{z}\right)^k - \left(\frac{z}{w}\right)^k = \frac{(w\bar{z})^k}{|z|^{2k}} - \frac{(z\bar{w})^k}{|w|^{2k}} = \frac{1}{|z|^{2k}} 2i \operatorname{Im}(w\bar{z})^k$.

(4) *If* $|z| = 1$ *and* $|w| = r$, *then* $\operatorname{Re}\left[H(w,z) - \dfrac{w+z}{w-z} - 1\right] = 0$.

Hint: $\displaystyle \frac{w+z}{w-z}+1=-\frac{2w/z}{1-w/z}=-2\sum_{k=1}^{\infty}\left(\frac{w}{z}\right)^k.$

Consequently

$$\begin{aligned} H(w,z)-\frac{w+z}{w-z}-1 &= 2\sum_{k=1}^{\infty}\frac{r^{2k}}{1-r^{2k}}\left[\left(\frac{w}{z}\right)^k-\left(\frac{z}{w}\right)^k+\frac{1-r^{2k}}{r^{2k}}\left(\frac{w}{z}\right)^k\right] \\ &= 2\sum_{k=1}^{\infty}\frac{r^k}{1-r^{2k}}\left[\left(\frac{w}{rz}\right)^k-\left(\frac{rz}{w}\right)^k\right]. \end{aligned}$$

Now apply to w and rz the equality in the hint to (3).

(5) *If* $|z|=r$ *and* $|w|=1$, *then* $\displaystyle \operatorname{Re}\left[H(w,z)-\frac{w+z}{w-z}+1\right]=0.$

(ii) *Let* $A:=\{z\in\mathbb{C}: r<|z|<1\}$, f *a continuous real-valued function on* ∂A *which satisfies*

(6) $$\int_{-\pi}^{\pi} f\big(re^{i\theta}\big)\,d\theta=\int_{-\pi}^{\pi} f\big(e^{i\theta}\big)\,d\theta=:2\pi c.$$

For $z\in A$ *define*

(7) $$\begin{aligned} F(z):=&\frac{1}{2\pi}\int_{-\pi}^{\pi}\left[\frac{e^{i\theta}+z}{e^{i\theta}-z}-H\big(e^{i\theta},z\big)\right]f\big(e^{i\theta}\big)\,d\theta \\ &-\frac{1}{2\pi}\int_{-\pi}^{\pi}\left[\frac{re^{i\theta}+z}{re^{i\theta}-z}-H\big(re^{i\theta},z\big)\right]f\big(re^{i\theta}\big)\,d\theta-c. \end{aligned}$$

Show that F *is holomorphic in* A.

Hint: H is continuous on the compact set $\partial A\times\bar{A}$. Cite 5.43(iii).

(iii) $\lim_{z\to z_0, z\in A}\operatorname{Re}F(z)=f(z_0)$ *for each* $z_0\in\partial A$.

Hints: From (3), (4) and the continuity of H on $\partial A\times\bar{A}$

$$\begin{aligned} \lim_{\substack{z\to e^{i\phi}\\ z\in A}}\operatorname{Re}F(z) &= \lim_{\substack{z\to e^{i\phi}\\ z\in A}}\frac{1}{2\pi}\int_{-\pi}^{\pi}\operatorname{Re}\left[\frac{e^{i\theta}+z}{e^{i\theta}-z}\right]f\big(e^{i\theta}\big)\,d\theta+\frac{1}{2\pi}\int_{-\pi}^{\pi}f\big(re^{i\theta}\big)\,d\theta-c \\ &= \lim_{\substack{z\to e^{i\phi}\\ z\in A}}\frac{1}{2\pi}\int_{-\pi}^{\pi}\operatorname{Re}\left[\frac{e^{i\theta}+z}{e^{i\theta}-z}\right]f\big(e^{i\theta}\big)\,d\theta \quad \text{by (6)} \\ &= f\big(e^{i\phi}\big) \quad \text{by 5.20.} \end{aligned}$$

Whereas (3) and (5) yield

$$
\begin{aligned}
&\lim_{\substack{z\to re^{i\phi}\\ z\in A}} \operatorname{Re} F(z)\\
&\quad = \frac{1}{2\pi}\int_{-\pi}^{\pi} f\big(e^{i\theta}\big)\,d\theta + \lim_{\substack{z\to re^{i\phi}\\ z\in A}} \frac{1}{2\pi}\int_{-\pi}^{\pi} \operatorname{Re}\left[-\frac{re^{i\theta}+z}{re^{i\theta}-z}\right] f\big(re^{i\theta}\big)\,d\theta - c\\
&\quad = \lim_{\substack{z\to re^{i\phi}\\ z\in A}} \frac{1}{2\pi}\int_{-\pi}^{\pi} \operatorname{Re}\left[\frac{e^{-i\theta}+r/z}{e^{-i\theta}-r/z}\right] f\big(re^{i\theta}\big)\,d\theta \quad \text{by (6)}\\
&\quad = \lim_{\substack{w\to e^{-i\phi}\\ w\in A}} \frac{1}{2\pi}\int_{-\pi}^{\pi} \operatorname{Re}\left[\frac{e^{it}+w}{e^{it}-w}\right] f\big(re^{-it}\big)\,dt \quad (w := r/z, t := -\theta)\\
&\quad = f\big(re^{i\phi}\big) \quad \text{by 5.20 applied to the function } h(e^{it}) := f\big(re^{-it}\big).
\end{aligned}
$$

9.41 together with 9.43 insure that the Dirichlet problem is solvable in every annulus $A(r,R)$, $0 < r < R < \infty$, and 10.27 even settles the question of harmonic conjugates. But the Poisson formula just developed provides an independent, more constructive route to these results:

Exercise 11.54 (Dirichlet Problem for the Annulus). *Let* $0 < r < R < \infty$ *and* f *be a continuous real-valued function on* $C(0,r)\cup C(0,R)$. *Define*

$$
a_0 := \frac{1}{2\pi}\int_0^{2\pi} f\big(re^{i\theta}\big)\,d\theta, \quad A_0 := \frac{1}{2\pi}\int_0^{2\pi} f\big(Re^{i\theta}\big)\,d\theta, \quad b := \frac{A_0 - a_0}{\log R - \log r}.
$$

(i) *Show that there is a holomorphic function* F *in*

$$
A(r,R) := \{z\in\mathbb{C} : r < |z| < R\}
$$

such that $\operatorname{Re} F$ *is* (*extendable to be*) *continuous on* $\bar{A}(r,R)$ *and satisfies*

$$
f(z) = b\log|z| + \operatorname{Re} F(z) \qquad \forall z\in C(0,r)\cup C(0,R).
$$

Hints: Define f_0 on $\partial A(r,R)$ by

$$
f_0(z) := f(z) - b\log|z|.
$$

Then the definition of b is just such as to insure that

$$
\frac{1}{2\pi}\int_{-\pi}^{\pi} f_0\big(re^{i\theta}\big)\,d\theta = \frac{1}{2\pi}\int_{-\pi}^{\pi} f_0\big(Re^{i\theta}\big)\,d\theta.
$$

Apply 11.53, after a change of scale, to come up with a function F which is holomorphic in $A(r,R)$ and such that $\operatorname{Re} F$ is continuous on $\bar{A}(r,R)$ with

$$\operatorname{Re} F = f_0 \quad \text{on } \partial A(r,R).$$

(ii) *Express the Laurent coefficients of F directly in terms of f. (F is only determined by the data up to an imaginary constant, so for the 0th Laurent coefficient only the real part is expressible in terms of f.)*

Hints: Let c_n be the nth Laurent coefficient of F:

$$F(z) = \sum_{n=-\infty}^{\infty} c_n z^n, \qquad z \in A(r,R).$$

Then for $r < \rho < R$

$$\begin{aligned}\operatorname{Re} F\big(\rho e^{i\theta}\big) &= \frac{1}{2}\sum_{n=-\infty}^{\infty} c_n \rho^n e^{in\theta} + \frac{1}{2}\sum_{n=-\infty}^{\infty} \bar{c}_n \rho^n e^{-in\theta} \\ &= \sum_{n=-\infty}^{\infty} \tfrac{1}{2}\big(c_n\rho^n + \bar{c}_{-n}\rho^{-n}\big)e^{in\theta},\end{aligned}$$

whence

$$\frac{1}{2\pi}\int_0^{2\pi} \operatorname{Re} F\big(\rho e^{i\theta}\big)e^{-in\theta}\,d\theta = \tfrac{1}{2}\big(c_n\rho^n + \bar{c}_{-n}\rho^{-n}\big), \quad n \in \mathbb{Z},\ r < \rho < R.$$

By continuity of $\operatorname{Re} F$ in $\bar{A}(r,R)$, these equations continue to hold for $\rho = r$ and $\rho = R$ and lead, for $n \in \mathbb{Z}$ and $\rho \in \{r, R\}$, to

$$(*) \qquad \frac{1}{2\pi}\int_0^{2\pi} \big[f\big(\rho e^{i\theta}\big) - b\log\rho\big]e^{-in\theta}\,d\theta = \tfrac{1}{2}\big(c_n\rho^n + \bar{c}_{-n}\rho^{-n}\big).$$

Since $\int_0^{2\pi} e^{-in\theta}\,d\theta = 0$ if $n \neq 0$, this gives

$$\left.\begin{aligned} A_n &:= \frac{1}{2\pi}\int_0^{2\pi} f\big(Re^{i\theta}\big)e^{-in\theta}\,d\theta = \tfrac{1}{2}\big(c_n R^n + \bar{c}_{-n}R^{-n}\big) \\ a_n &:= \frac{1}{2\pi}\int_0^{2\pi} f\big(re^{i\theta}\big)e^{-in\theta}\,d\theta = \tfrac{1}{2}\big(c_n r^n + \bar{c}_{-n}r^{-n}\big)\end{aligned}\right\} \quad n = \pm1, \pm2, \ldots.$$

Therefore

$$c_n = \frac{2A_nR^n - 2a_nr^n}{R^{2n} - r^{2n}}, \quad n = \pm1, \pm2, \ldots.$$

Take $n = 0$ and $\rho = r$ in $(*)$ to get in addition

$$\operatorname{Re} c_0 = \frac{1}{2\pi}\int_0^{2\pi} f\left(re^{i\theta}\right) d\theta - b\log r = \frac{a_0 \log R - A_0 \log r}{\log R - \log r}.$$

Notice that A_n, a_n are the Fourier coefficients of the two boundary functions.

(iii) *Use* (ii) *to explain the origin of the Poisson formula in* 11.53.

Hint: Take $R = 1$ and assume (6), as before, as well as c_0 real. Then justify all the steps necessary to derive the Poisson formula (7) for F from its Laurent series by substituting the integral formulas for A_n, a_n into the expression for c_n and interchanging the sum in the Laurent series with the resulting integrals.

Exercise 11.55 (Cf. 10.27.) *Let* $0 \le r < R \le \infty$, $h\colon A(r,R) \mapsto \mathbb{R}$ *be harmonic. Show that there exists a unique constant* b *and a holomorphic function* F *in* $A(r,R)$, *unique to within a purely imaginary additive constant, such that*

$$h(z) = \operatorname{Re} F(z) + b\log|z| \qquad \forall z \in A(r,R).$$

Hints: Choose $r_n \downarrow r$ and $r_1 < R_n \uparrow R$ and set $A_n := A(r_n, R_n)$. 11.54 supplies $F_n \in H(A_n)$, $b_n \in \mathbb{R}$ such that $\operatorname{Re} F_n(z) + b_n \log|z| = h(z)$ for all $z \in C(0,r_n) \cup C(0,R_n)$, hence for all $z \in A_n$ by the maximum principle. In $A_n \subset A_{n+1}$ we have $\operatorname{Re}(F_{n+1} - F_n)(z) = (b_n - b_{n+1})\log|z|$. Since $\log|z|$ has no harmonic conjugate in A_n (5.21(ii)), it follows that $b_n - b_{n+1} = 0$. Thus $b_1 = b_2 = \cdots = b$, say, and we can add purely imaginary constants successively to the F_n in order to have $F_{n+1} = F_n$ in A_n for all n. (Cf. hints to 5.21.)

Alternatively, the proof of 10.27 [with $n := 1$, $p_1 = 0$, $C_1 := \{0\}$] can be applied, made more elementary by using the Laurent series of $f := D_1 h - iD_2 h$ to eliminate appeal to 10.26. Such a proof is a bit more constructive: It shows that b [$= a_1$ in 10.27] is the residue at 0 of the holomorphic function f and F is a primitive of the function $f(z) - b/z$, which can be found by termwise integration, since the Laurent series of $f(z) - b/z$ has no $1/z$ term. Note that if we differentiate with respect to ρ in the equation $h(\rho e^{i\theta}) = \operatorname{Re} F(\rho e^{i\theta}) + b\log\rho$ and then integrate with respect to θ [using the polar form of (C. R.) with F], we can express b directly in terms of h as

$$b = \frac{\rho}{2\pi}\int_0^{2\pi} D_\rho h(\rho e^{i\theta})\, d\theta \quad \text{for any } \rho \in (r,R).$$

A more basic formula for b in terms of h is encoded in 11.54. Namely, for any unequal $\rho', \rho \in (r,R)$

$$b = \frac{\int_0^{2\pi}\left[h(\rho' e^{i\theta}) - h(\rho e^{i\theta})\right] d\theta}{2\pi[\log\rho' - \log\rho]}.$$

Evidently this formula converges to the previous one as $\rho' \to \rho$.

There is yet a third instructive approach to this problem: Consider the Principal Branch L of the Logarithm in $\mathbb{C}_+ := \mathbb{C} \setminus [0, \infty)$, $A := A(r, R)$. The regions $A_+ := A \cap \mathbb{C}_+$ and $A_- := A \cap \mathbb{C}_-$ are each simply-connected, and so there exist $f_+ \in H(A_+)$, $f_- \in H(A_-)$ satisfying

$$h = \begin{cases} \operatorname{Re} f_+ & \text{in } A_+ \\ \operatorname{Re} f_- & \text{in } A_-. \end{cases} \tag{1}$$

The function $f_+ - f_-$ is holomorphic in $A_+ \cap A_- = A \setminus \mathbb{R}$ and has real part 0. Hence in each component of $A_+ \cap A_-$ it is constant. Call these purely imaginary constants ic_+ and ic_-. The function

$$f(z) := f_+(z) + \frac{c_+ - c_-}{2\pi} L(z), \qquad z \in A_+ \tag{2}$$

is holomorphic and since by (1)

$$\operatorname{Re} f(z) = h(z) + \frac{c_+ - c_-}{2\pi} \log |z| \qquad \forall z \in A_+, \tag{3}$$

$\operatorname{Re} f$ extends continuously to A. By definition of c_+ and c_- we have, for all $z \in A \setminus \mathbb{R}$

$$\begin{aligned} \operatorname{Im} f_+(z) &= \operatorname{Im} f_-(z) + \operatorname{Im}(f_+ - f_-)(z) \\ &= \begin{cases} \operatorname{Im} f_-(z) + c_+ & \text{if } \operatorname{Im} z > 0 \\ \operatorname{Im} f_-(z) + c_- & \text{if } \operatorname{Im} z < 0. \end{cases} \end{aligned}$$

Since f_- is continuous at each $x \in (r, R)$, it follows that

$$\lim_{\substack{z \to x \\ \operatorname{Im} z > 0}} \operatorname{Im} f_+(z) = \operatorname{Im} f_-(x) + c_+ \tag{4}$$

$$\lim_{\substack{z \to x \\ \operatorname{Im} z < 0}} \operatorname{Im} f_+(z) = \operatorname{Im} f_-(x) + c_-. \tag{5}$$

Because of the limiting behavior of $\operatorname{Im} L$ at each real $x > 0$, it follows from (2) that

$$\begin{aligned} \lim_{\substack{z \to x \\ \operatorname{Im} z < 0}} \operatorname{Im} f(z) &\overset{(5)}{=} \operatorname{Im} f_-(x) + c_- + \frac{c_+ - c_-}{2\pi} \cdot 2\pi \\ &= \operatorname{Im} f_-(x) + c_+ \overset{(4)}{=} \lim_{\substack{z \to x \\ \operatorname{Im} z > 0}} \operatorname{Im} f(z). \end{aligned}$$

Therefore $\operatorname{Im} f$ too extends continuously to A. In summary, f has a continuous extension F to A. By 5.40(i), F is also holomorphic in A. Thanks to (3) and continuity, this F does all that is wanted.

Exercise 11.56 (i) *Show that if h is harmonic in $A(a,b)$ and*

$$H(r) := \frac{1}{2\pi}\int_0^{2\pi} h\big(re^{i\theta}\big)\, d\theta \quad \text{for } a < r < b,$$

then there are two real constants α, β such that $H(r) \equiv \alpha + \beta \log r$.

Hint: Integrate the equality in 11.55 and recall 5.5(i).

(ii) *Show that if h is harmonic in $A(a,b)$ and $h(z)$ depends only on $|z|$, then h has the form $h(z) = \alpha + \beta \log |z|$.*

Hint: In this case $H(r) = h(re^{i\theta})$ for all $\theta \in \mathbb{R}$.

(iii) (F. Riesz [1922/23]) *Show that if u is subharmonic in $A(a,b)$ and $U(r) := \frac{1}{2\pi}\int_0^{2\pi} u\big(re^{i\theta}\big)\, d\theta$ for $a < r < b$, then $U(r)$ is a convex function of $\log r$.*

Hints: Given $a < r_1 < r_2 < b$, 11.54 provides a function h which is harmonic in $A(r_1, r_2)$, continuous in $\overline{A}(r_1, r_2)$ and coincides on the boundary of $A(r_1, r_2)$ with u. By the harmonic-majorant property of u (5.22) we have $u \leq h$ in $A(r_1, r_2)$, hence $U \leq H$ in (r_1, r_2). Therefore for each $0 \leq \lambda \leq 1$

$$\begin{aligned} U\Big(e^{\lambda \log r_1 + (1-\lambda)\log r_2}\Big) &\leq H\Big(e^{\lambda \log r_1 + (1-\lambda)\log r_2}\Big) \overset{\text{(i)}}{=} \lambda H(r_1) + (1-\lambda) H(r_2) \\ &= \lambda U(r_1) + (1-\lambda) U(r_2). \end{aligned}$$

The last equality is because $u = h$ on $C(0, r_j)$ entails $U(r_j) = H(r_j)$, $j = 1, 2$.

Remarks: If $a = 0$ in 11.56(iii), it follows from the most elementary properties of convex functions (see, e.g., Stromberg [1981], p. 202 and H. König [1992]) that $U(r)$ and $U(r)/\log r$ are monotonic for small r and consequently have limits in $[-\infty, \infty]$ as $r \downarrow 0$. In fact, one can even assert that $\lim_{r\downarrow 0} u\big(re^{i\theta}\big)/\log r$ exists and equals $\lim_{r\downarrow 0} U(r)/\log r$ for most θ. (See Arsove and Huber [1973].)

Exercise 11.57 *Let $b > 0$ and $h\colon A(0,b) \mapsto \mathbb{R}$ be harmonic.*

(i) *Suppose h satisfies the (weaker than boundedness near 0) condition*

$$(*) \qquad \varliminf_{r\downarrow 0} \frac{\int_0^{2\pi} \big|h\big(re^{i\theta}\big)\big|\, d\theta}{-\log r} = 0.$$

Show that h has a harmonic extension to $D(0,b)$. (Cf. 5.41(ii).)

Hints: According to 11.55 there is a holomorphic function F in $A(0,b)$ and a real constant B such that

(1) $$h(z) = \operatorname{Re} F(z) + B\log|z| \qquad \forall z \in A(0,b).$$

Moreover, 11.56(i) and 5.5(i) show that

(2) $$B = \frac{H(R) - H(r)}{\log R - \log r}$$

for any pair $0 < r < R < b$, where

(3) $$H(r) := \frac{1}{2\pi}\int_0^{2\pi} h\big(re^{i\theta}\big)\, d\theta, \quad 0 < r < b.$$

If c_n is the nth Laurent coefficient of F, the uniform convergence of the Laurent series to F on $C(0,r)$ permits an interchange of summation and integration, which leads to

(4) $$\frac{1}{2\pi}\int_0^{2\pi} 2\big[\operatorname{Re} F\big(re^{i\theta}\big)\big]e^{-in\theta}\, d\theta = c_n r^n + \bar{c}_{-n}r^{-n}, \quad n \in \mathbb{Z}.$$

Use $(*)$ to select $b > R > r_1 > r_2 > \cdots$ convergent to 0 such that

(5) $$\lim_{k\to\infty} \frac{\int_0^{2\pi} \big|h\big(r_k e^{i\theta}\big)\big|\, d\theta}{-\log r_k} = 0.$$

From (3) and (5) we evidently have

$$\lim_{k\to\infty} \frac{H(r_k)}{\log R - \log r_k} = 0,$$

from which it follows that $B = 0$. Thus $\operatorname{Re} F = h$, so (4) yields

(6) $$\big|c_n r^n + \bar{c}_{-n}r^{-n}\big| \le \frac{1}{\pi}\int_0^{2\pi} \big|h\big(re^{i\theta}\big)\big|\, d\theta.$$

Now $r^N/(-\log r)$ converges to 0 as $r \downarrow 0$ if $N > 0$, but converges to ∞ if $N < 0$ (3.18). Therefore if we consider $n > 0$, divide by $-\log r$, take $r = r_k$ in (6) and let $k \to \infty$, we are forced to conclude via (5) that $c_{-n} = 0$.

(ii) *Suppose that h is bounded below [above] near 0. Show that there is a non-positive [non-negative] constant B and a holomorphic function F in $D(0,b)$ such that*

$$h(z) = \operatorname{Re} F(z) + B\log|z| \qquad \forall z \in A(0,b).$$

Hints: Without loss of generality, $h \geq 0$. In the above reasoning (4) yields $(n \neq 0)$

$$\begin{aligned}\pi\left|c_n r^n + \bar{c}_{-n} r^{-n}\right| &= \left|\int_0^{2\pi} \operatorname{Re}\left[F\left(re^{i\theta}\right)\right] e^{-in\theta}\, d\theta\right| \\ &= \left|\int_0^{2\pi} \left[h\left(re^{i\theta}\right) - B\log r\right] e^{-in\theta}\, d\theta\right| \\ &= \left|\int_0^{2\pi} h\left(re^{i\theta}\right) e^{-in\theta}\, d\theta\right|, \quad \text{since } n \neq 0, \\ &\leq \int_0^{2\pi} \left|h\left(re^{i\theta}\right) e^{-in\theta}\right| d\theta = \int_0^{2\pi} h\left(re^{i\theta}\right) d\theta \quad \text{since } h \geq 0, \\ &= \int_0^{2\pi} \left[\operatorname{Re} F\left(re^{i\theta}\right) + B\log r\right] d\theta = 2\pi \operatorname{Re} c_0 + 2\pi B \log r.\end{aligned}$$

It follows as before that $c_{-n} = 0$ for all positive n.

(iii) *Draw a similar conclusion as in* (ii) *under the* (*weaker than non-negativity*) *hypothesis that*

$$\overline{\lim_{r\to 0}} \frac{\min h\big(C(0,r)\big)}{-\log r} > -\infty.$$

Hints: Choose a real c less than this limit superior and then $\min\{b,1\} > r_1 > r_2 > \cdots \to 0$ such that

$$\frac{\min h\big(C(0,r_k)\big)}{-\log r_k} > c, \quad k \in \mathbb{N}.$$

Then $h(z) + c\log|z| > 0$ on $\bigcup_{k=1}^{\infty} C(0,r_k)$. Apply the Maximum Principle (5.11) to the harmonic function $h^*(z) := -h(z) - c\log|z|$ in each annulus $A(r_{k+1}, r_k)$ to conclude that it is negative throughout $A(0,r_1)$. Thus (ii) may be applied to $-h^*$.

(iv) (Generalized Liouville theorem for harmonic functions). *$S \subset \mathbb{C}$ has no limit points in $\mathbb{C}$ and h is harmonic and non-negative in $\mathbb{C} \setminus S$, with every point of S a non-removable (isolated) singularity. Use* 7.17 *to prove that $S = \varnothing$ and h is constant.*

Hints: For each $s \in S$ part (ii) supplies numbers $b > 0 > B$ and holomorphic F in $D(s,b)$ such that

$$h(w) = \operatorname{Re} F(w) + B\log|w-s| \qquad \forall\, w \in D(s,b) \setminus \{s\} \subset \mathbb{C} \setminus S.$$

It follows that

$$\text{(0)} \qquad \lim_{w \to s} h(w) = +\infty \qquad \forall\, s \in S.$$

Fix $w \in \mathbb{C} \setminus S$ and consider the harmonic function $u := h(w) - h$ in $U := \mathbb{C} \setminus S$, an open set whose boundary is S. Thus (0) implies that

$$\text{(1)} \qquad \lim_{z \to \xi} u(z) = -\infty \qquad \forall\, \xi \in \partial U,$$

and since $h \geq 0$ we also have

$$\text{(2)}' \qquad \overline{\lim_{\substack{|z| \to \infty \\ z \in U}}} \frac{u(z)}{\log|z|} = \overline{\lim_{\substack{|z| \to \infty \\ z \in U}}} \frac{-h(z)}{\log|z|} \leq 0.$$

Now cite the Remark following 7.17 to infer that $u \leq 0$. That is,

$$h(w) \leq h \quad \text{ in } \quad \mathbb{C} \setminus S.$$

In summary,

$$h(w) \leq h(z) \qquad \forall\, w, z \in \mathbb{C} \setminus S,$$

which means that h is constant in $\mathbb{C} \setminus S$, and from (0) follows that $S = \varnothing$.

(v) *From* (ii) *deduce the following analog of the Casorati-Weierstrass theorem for harmonic functions: If $h(z)$ does not converge in $[-\infty, \infty]$ as $z \to 0$, then $h\big(D(0,r) \setminus \{0\}\big) = \mathbb{R}$ for each $0 < r < b$.*

Hint: For each $0 < r < b$, the continuous-image set $h\big(D(0,r) \setminus \{0\}\big)$ is an interval. If it were either bounded above or bounded below, then $\lim_{z \to 0} h(z)$ would exist in $[-\infty, \infty]$ by (ii).

Remarks 11.58: Another hypothesis under which the conclusion of 11.57(ii) holds is that $\iint_{A(0,r)} |h(x+iy)|^2\, dxdy < \infty$ for some $0 < r < b$. See, e.g., pp. 15–20 of HERVÉ [1970]. On the other hand, finiteness of the integral of $(D_1 h)^2 + (D_2 h)^2$ over $A(0,r)$ for some $0 < r < b$ is equivalent to 0 being a removable singularity. See HILBERT [1904], §10. For a stronger form of (iv) involving subharmonic h, see p. 70 of RANSFORD [1995].

Exercise 11.59 *Suppose $h : A(r,R) \mapsto \mathbb{R}$ is harmonic, $0 \leq r \leq R \leq \infty$.*

(i) *Show that for some zero-free holomorphic function f in $A(r,R)$ and some real number c*

$$h = c \log|f|.$$

Hint: Use 11.55. If the b there is 0, take $c := 1$ and $f := e^F$; if $b \neq 0$, take $c := b$ and $f(z) := z e^{F(z)/b}$.

(ii) *Show that for $U := A(r,R)$ the conclusion in* 10.8(vi) *that $h = \operatorname{Re} f$ for some $f \in H(U)$ follows even when the "no-flux" hypothesis*

$$\int_\gamma D_{\mathbf{N}} h(\xi)|d\xi| = 0 \tag{4}$$

there is only required to hold for at least one $\gamma := C(0,\rho)$, $r < \rho < R$.

Hints: As noted in the 10.8 development, the function $g := 2\partial h = D_1 h - iD_2 h$ is holomorphic and the integral in (4) equals $i\int_\gamma g$. Hence (4) is saying that g has residue 0 at 0. Consequently, as noted in Remark (ii) following 11.4, g has a primitive F. But then according to 10.8(ii), h differs from $\operatorname{Re} F$ only by a constant.

Exercise 11.60 (i) *Establish the following improvement of* 5.41(i)*: If f is holomorphic in $U \setminus \{a\}$ for some neighborhood U of a, and if it satisfies*

$$\varliminf_{z\to a} |z-a| \operatorname{Re} f(z) \geq 0$$

then the singularity at a is removable.

Hints: Without loss of generality, $U = \mathbb{D}$, $a = 0$. 11.2 furnishes $F \in H(\mathbb{D})$, $G \in H(\mathbb{C})$ such that

$$f(z) = F(z) + \frac{1}{z}G\left(\frac{1}{z}\right) \qquad \forall z \in \mathbb{D}\setminus\{0\}.$$

Since F is continuous at 0, our hypothesis implies that

$$\varliminf_{z\to 0} |z| \operatorname{Re}\left(\frac{1}{z}G\left(\frac{1}{z}\right)\right) \geq 0, \quad \text{that is,}$$

$$\varliminf_{|w|\to\infty} \frac{1}{|w|}\operatorname{Re}\big(wG(w)\big) \geq 0. \tag{$*$}$$

From 6.37 it follows that the entire function $w \mapsto wG(w)$ is a polynomial of degree at most 1. Since $wG(w)$ has no constant term, it therefore has the form cw, $c \in \mathbb{C}$. If $c \neq 0$, we could consider $w := -r/c$ and have

$$\frac{1}{|w|}\operatorname{Re}(wG(w)) = \frac{1}{|w|}\operatorname{Re}(cw) = -|c|.$$

As $r \to +\infty$ this is incompatible with $(*)$.

(ii) *Suppose $h\colon \mathbb{D}\setminus\{0\} \mapsto \mathbb{R}$ is harmonic and satisfies $\varliminf_{z\to 0}|z|h(z) \geq 0$. Use* (i) *and* 11.55 *to show that h has the form*

$$h(z) = \operatorname{Re} F(z) + B\log|z|$$

for some $B \in \mathbb{R}$, $F \in H(\mathbb{D})$. This improves 11.57(ii).

Hint: We need to know that $\lim_{z\to 0} |z| \log |z| = 0$ and that was proved in 3.18(4).

(iii) *Establish the following improvement of* 11.7(i)*: If* f *is holomorphic in* $U \setminus \{a\}$ *for some neighborhood* U *of* a *and if for some* $c \in \mathbb{R}$ *the inequality* $\operatorname{Re} f(z) \le c \log |z-a|$ *holds for all* $z \in U \setminus \{a\}$ *sufficiently near* a*, then* a *is a removable singularity.*

Hints: Without loss of generality, $a = 0$, $U = \mathbb{D}$. Apply part (ii) to $h(z) := c\log|z| - \operatorname{Re} f(z)$ to learn that $\operatorname{Re} f(z) = \operatorname{Re} F(z) + b\log|z|$ for some $b \in \mathbb{R}$, $F \in H(\mathbb{D})$. It follows from this and the preceding hint that $\lim_{z\to 0} |z| \operatorname{Re} f(z) = 0$, so (iii) follows from (i).

(iv) *Use* (i) *to prove the following* Liouville-type *result: If* $u\colon \mathbb{C} \mapsto \mathbb{R}$ *is harmonic and* $\underline{\lim}_{|w|\to\infty} |w|^{-1} u(w) \ge 0$*, then* u *is constant.*

Hints: $u = \operatorname{Re} F$ for some $F \in H(\mathbb{C})$. The function $f(z) := F(1/z)$ is holomorphic in $\mathbb{C} \setminus \{0\}$ and satisfies

$$\varliminf_{z\to 0} |z| \operatorname{Re} f(z) = \varliminf_{|w|\to\infty} |w|^{-1} u(w) \ge 0.$$

Hence by (i), f extends to be entire. In particular, f is bounded in $\mathbb{D} \setminus \{0\}$, so F is bounded in $\mathbb{C} \setminus \overline{\mathbb{D}}$, hence bounded in $\mathbb{C}$, consequently constant, by Liouville.

Remark: A skillful use of Harnack's inequalities leads to an improvement of 11.60(iv):

$$\varlimsup_{r\to\infty} \frac{1}{r} \min_{\theta} u(re^{i\theta}) \ge 0 \Rightarrow u \text{ constant.}$$

See p. 36 of Hayman and Kennedy [1976] or p. 21 of Ransford [1995].

We close with a significant extension of 11.60(iv), achieved with a most elegant and elementary proof, from Nakai and Tada [1997].

Theorem 11.61 $u : \mathbb{C} \mapsto \mathbb{R}$ *is harmonic and for some* $n \in \mathbb{N}$ *satisfies*

$$(*) \qquad \varliminf_{z\to\infty} \frac{u(z)}{|z|^n} \ge 0.$$

Then u *is the real part of a polynomial of degree less than* n.

Proof: There is a power series $\sum_{k=0}^{\infty} c_k z^k =: f(z)$, convergent throughout $\mathbb{C}$, such that $u = \operatorname{Re} f$. We may suppose $u(0) = 0$ and $f(0) = 0$. Thus

$$u(z) = \sum_{k=1}^{\infty} \operatorname{Re}(c_k z^k) = \sum_{k=1}^{\infty} (a_k \operatorname{Re} z^k + b_k \operatorname{Im} z^k) \qquad \forall z \in \mathbb{C},$$

where $a_k := \operatorname{Re} c_k$, $b_k := -\operatorname{Im} c_k$. Thus

(1) it suffices to prove that $a_k = b_k = 0$ for all $k \geq n$.

Given $\varepsilon > 0$, (∗) says that $\exists r_\varepsilon \in \mathbb{R}$ such that $u(z)|z|^{-n} \geq -\varepsilon$ whenever $|z| \geq r_\varepsilon$. Thus

(2) $$0 \leq \varepsilon r^n + u(re^{i\theta}) = \varepsilon r^n + \sum_{k=1}^{\infty}(a_k \cos k\theta + b_k \sin k\theta)r^k \qquad \forall r \geq r_\varepsilon, \theta \in \mathbb{R}.$$

Fix $j \in \mathbb{N}$, multiply (2) by $1 \pm \cos(j\theta)$, integrate over $\theta \in [0, 2\pi]$, and use the orthogonality of the family $\{\cos(k\theta), \sin(k\theta) : k \geq 0\}$ to get

$$0 \leq 2\pi\varepsilon r^n \pm \int_0^{2\pi} a_j(\cos j\theta)^2 r^j \, d\theta = 2\pi\varepsilon r^n \pm \pi a_j r^j, \text{ whence}$$

(3) $$0 \leq 2\varepsilon r^{n-j} \pm a_j \qquad \forall r \geq r_\varepsilon, j \in \mathbb{N}.$$

For $j > n$ let $r \uparrow \infty$ and get $0 \leq 0 \pm a_j$, that is,

(4) $$a_j = 0 \qquad \forall j > n.$$

If $j = n$, (3) says that $-2\varepsilon \leq \pm a_n$, with $\varepsilon > 0$ arbitrary, so we also learn that

(5) $$a_n = 0.$$

Similarly we prove that

(6) $$b_j = 0 \qquad \forall j \geq n.$$

According to (1), the conjunction of (4), (5) and (6) completes the proof.

Corollary 11.62 (cf 6.37) *If $u : \mathbb{C} \mapsto \mathbb{R}$ is harmonic and for some $0 < C < \infty$, $n \in \mathbb{N}$ satisfies $u(z) \leq C|z|^{n-1}$ for all large z, then u is the real part of a polynomial of degree less than n.*

Proof: We have $-u(z)|z|^{-n} \geq -C|z|^{-1}$ for all large z, hence $-u$ satisfies hypothesis (∗) of 11.61.

Notes to Chapter XI

The bilateral series expansion in 11.2 was known to WEIERSTRASS in 1841 but not published then. (See his *Werke* I, pp. 51–66 and ULLRICH [1994b].) It was announced in a short note by P. A. LAURENT [1843], in which he remarked only that it followed easily from Cauchy's earlier work. (A proof of his appeared posthumously in 1863.) On pp. 938–940 of the same journal Cauchy and Liouville comment on this communiqué and then CAUCHY [1843] gives a detailed proof. (See pp. 34–35 of MITTAG-LEFFLER [1923].) The treatment of 11.1 and 11.2 in the text is Cauchy's. For other treatments of 11.2 see pp. 1028–1030 of BURKHARDT [1914], MITTAG-LEFFLER [1884b], SCHEEFFER [1884], and O. BIERMANN [1908]. The proof by Weierstrass' student Scheeffer avoids integrals (a desideratum that had eluded the master) by utilizing the Joukowski transform $z \mapsto \frac{1}{2}(z + 1/z)$ (see 2.5) and the factorization lemma 5.96 in a clever way and is very clearly and detailedly presented in ULLRICH [1990] and on pp. 352–354 of REMMERT [1991a]. The treatment in PRINGSHEIM [1896], [1925] is also interesting; as remarked in the Chapter V notes, he replaces integrals with limits of special sums, a device which Cauchy himself had on occasion used. See also BEESACK [1972] and SHISHA [1989]. Another interesting approach is STACHÓ [1994]. There is an interesting alternative proof of 11.2 using Fourier analysis (which can also be freed of integration!). Here is a sketch: Suppose $0 \le a < 1 < b$ and f is holomorphic in $A(a,b)$ so f' exists and is continuous. For each $r \in (a,b)$ the function $f_r(e^{i\theta}) := f(re^{i\theta})$ is continuously differentiable and so, according to Remark (iv) following 5.20, its Fourier series converges absolutely and uniformly to it. For fixed $n \in \mathbb{Z}$ look at the function $F(z) := z^{-n}f(z)$. We have $r^{-n}\hat{f}_r(n) = (2\pi)^{-1}\int_0^{2\pi} F(re^{i\theta})d\theta$, and according to 5.5 this is independent of r. Thus $\hat{f}_r(n) = r^n\hat{f}_1(n)$ for all r, n; which means that the family of Fourier series of the f_r, $a < r < b$, is actually a bilateral series representation of f in $A(a,b)$. In part II of LELAND [1971] an ingenious alternative definition of the Fourier coefficients is employed which is based on Weierstrass's approximation theorem in $C(\mathbb{T})$ and avoids integration. The above proof is then executed without appeal to 5.5 and without using the Fourier convergence theorem, but rather proving the needed version of it too. Moreover, the removable singularity theorem for bounded functions in $A(0,b)$ is a corollary of the proof.

Theorem 11.6 (CASORATI [1868], WEIERSTRASS [1876], §8) was independently discovered by the Polish mathematician J. W. Sochozki in his St. Petersburg masters thesis. The history of this theorem and the priority issue is presented in great detail by NEUENSCHWANDER [1978] (see also chapter 7 of MITRINOVIĆ and KEČKIĆ [1993]). It seems likely that a special case was known to Cauchy: If f is a non-polynomial entire function, then for every $c \in \mathbb{C}$, $r \in \mathbb{R}_+$, f takes values arbitrarily close to c in $\mathbb{C} \setminus D(0,r)$. [Just apply 5.70 to $F := f - c$.] For a different treatment of the theorem itself see HÖLDER [1882]. More on 11.9 in HENRICI [1986], 507–513. On 11.10, 11.11 see §11, chapter III of BEHNKE and SOMMER [1965]. For an even shorter proof of 11.11, proceeding independently of 5.37, see the development in BOAS [1964a] (which is only sixteen lines long!).

On 11.15 see KAKEYA [1923]. 11.16, which is due to Fejér and F. Riesz (FEJÉR [1915], pp. 55–59; SZÁSZ [1919], pp. 166–167), is an important ingredient in the classical proof of the Spectral Theorem for unitary operators in a Hilbert space. See p. 282 of RIESZ and SZ.-NAGY [1955]. There is a nice generalization of 11.16 to certain classes of entire functions of exponential type. See, e.g., ACHIEZER [1956], p. 154, pp. 437–443 of B. LEVIN [1980] and BRICKMAN [1964]. The ultimate generalization may be KREIN and NUDEL'MAN [1985].

The compactness argument in 11.16 is rather heavy-handed and the reader may simply prefer to give an *ab initio* proof modelled on 11.15. (See, e.g., PÓLYA and SZEGÖ [1976], exercise VI.40.) Alternatively, he can extend 11.15 to cover the case where zeros (and poles) are permitted on the unit circle. [The proof of this is essentially the same as that of 11.15 but the pairing off of poles and zeros is slightly more complicated.] There is also an obvious version of 11.16 involving polynomials which are non-negative on $\mathbb{R}$ (see BRICKMAN and STEINBERG [1962]), and both this and 11.16 itself have operator analogs. E.g., if P_j are bounded linear operators on Hilbert space $\mathcal{H}$ and $P(x) := \sum_{j=0}^{2n} x^j P_j$ is a positive operator for each real x, then there exist bounded linear operators Q_j on $\mathcal{H}$ such that $P(x) = Q^*(x)Q(x)$ for all real x, where $Q(x) := \sum_{j=0}^{n} x^j Q_j$. See ROSENBLUM and ROVNYAK [1971], pp. 312–313, and MIAMEE and SALEHI [1976].

For more on 11.14 and 11.47(i) see GRONWALL [1912–13a] and on this general theme, PÓLYA and SZEGÖ [1976], exercise VI.44–VI.49. For material related to 11.19 see STYER and MINDA [1974], and R. K. WILLIAMS [1976] for a variety of other proofs. E.g., 11.19 is a corollary of 12.30. Another of Williams' proofs occurs in Chapter XVIII. An important feature of the second proof of 11.19 was the deduction of surjectivity from injectivity. For more on that interesting theme see BANACH and MAZUR [1934] and the references given therein, as well as REICHBACH [1958], GORDON [1972], PLASTOCK [1974] and HO [1975]. 11.20(i) occurs in CAUCHY [1844] (perhaps earlier too) and MÉRAY [1855]. The analog for $\mathbb{D}\backslash\{a_0, a_1\}$ of 11.20(v) is detailed on p. 314 of REMMERT [1991](a) and an amazing extension in the Remarks following 12.50 below. For a much stronger version of 11.20(vii) see CARATHÉODORY [1937b] and HUA [2008]: If f is an injective self-map of the Riemann sphere $\mathbb{C}_\infty$ (no continuity hypothesis) and f maps every circle or line onto a circle or a line, then either f or $\bar{f}$ is a Möbius transformation; the same conclusion if $f : \mathbb{D} \mapsto \mathbb{C}$ is an injection which maps each circle in $\mathbb{D}$ onto a circle. Cf. 6.5(iv).

The suggested proof of 11.23 is from PÓLYA [1916] who attributes it to Hurwitz. A new proof and generalization are given by VAN DER WAERDEN [1986].

J. KÖNIG [1884] was my source for 11.26 (a faulty proof of which occurs in LECORNU [1887]). The proof in the text can be adapted to cover the case of finitely many other poles on $C(0,|z_0|)$, each having order less than that of z_0. HADAMARD [1892b], p. 109 gave the following example to show that the converse of 11.26 fails:

$$\frac{1}{1-z} + \log(1+z) = 1 + \sum_{n=1}^{\infty}\left[1 + \frac{(-1)^{n+1}}{n}\right] z^n.$$

Here $c_n/c_{n+1} \to 1$ yet there are two singular points (1 and -1) on the circle of convergence. Moreover, other examples show that even if there is only one singularity z_0 on the circle of convergence, but it is not a pole, then the ratio c_n/c_{n+1} need not converge to z_0. See, e.g., pp. 1049–1100 of PRINGSHEIM [1932]. 11.27(i) is due to PRINGSHEIM [1929], pp. 115–124 and 11.27(ii) to HADAMARD [1892b]. For the case where the (unique) pole is not simple see E. W. MILLER [1935]. Notice that by a change of variable 11.27(ii) reads as follows: a necessary and sufficient condition for f to have a holomorphic extension to $U \setminus \{z_0\}$ with a simple pole at z_0, for some neighborhood U of $\overline{\mathbb{D}}$ and some $z_0 \in \partial\mathbb{D}$, is that

$$\overline{\lim_{n\to\infty}} \left|\frac{c_n}{c_{n+1}} - z_0\right|^{1/n} < 1.$$

HADAMARD [1892b] also provides the following related criterion which does not explicitly involve the unique pole z_0:

$$\overline{\lim_{n\to\infty}} |c_{n-1}c_{n+1} - c_n^2| < 1.$$

For a perspicuous proof see OSTROWSKI [1926b]. There is a related (deeper) result due to FABRY [1896a] (On this attribution see, however, pp. 97 and 298 of PRINGSHEIM [1929].): If c_n/c_{n+1} converges to some $z_0 \in \mathbb{C} \setminus \{0\}$, then the series has radius of convergence $|z_0|$ (by 3.3(iv)) and f does not have a holomorphic extension to any neighborhood of z_0. In §131 of PRINGSHEIM [1932] the reader will find a proof of the following stronger form of this theorem: If r (non-zero and finite) is the radius of convergence of $\sum c_n z^n$ and $(c_n/c_{n+1})|c_{n+1}/c_n|$ has the (unimodular) limit u_0, then $z_0 = ru_0$ is a singular point in the above sense. For other proofs of (various stronger forms of) this Fabry result, see pp. 307–315 of BIEBERBACH [1931], pp. 64 ff. of BIEBERBACH [1955], pp. 76–86 of E. LANDAU [1929a], pp. 377 ff. of DIENES [1931], S. K. SINGH [1964/65] and ARAKÉLIAN and MARTIROSIAN [1988]. See also AGMON [1948], BROGGI [1935], V. BERNSTEIN [1935], and DUFRESNOY and

Pisot [1951] (p. 122). (Singh's proof seems incomplete to me—and to the *Zentralblatt* reviewer as well.) For the more elementary version of Fabry's theorem when the series represents a rational function see Broggi [1931]. 11.28 is from Pólya and Szegö [1972], exercise III.246 and is also contained, along with generalizations, in Tschakaloff [1948]. For a good survey of results and history on questions of singularities on the circle of convergence see §45 of Bieberbach's encyclopedic article.

I have followed Rudin [1987] in the proof of 11.29 (for its history see §35 of Burkhardt [1914]); for a recent Russian history see Kirsanov [1985]. A brief survey of Cauchy's papers on the theorem and some of the secondary literature will be found in Mitrinović and Kečkić [1984], [1993]. This book together with Lindelöf [1905] is also an exhaustive source of applications of the residue theorem, especially to the evaluation of sums and integrals. A typical and fruitful example of the latter is the computation of the integral $\int_0^{2\pi} R(\cos t, \sin t)\,dt$, $R(x,y)$ a rational function of $x, y \in \mathbb{R}$, by re-writing it as the contour integral $\frac{1}{i}\int_{C(0,1)} R(\frac{1}{2}(z+\frac{1}{z}), \frac{1}{2i}(z-\frac{1}{z}))\frac{1}{z}\,dz$. The reader intrigued by contour integrals *per se* can be sated by looking into Garnir and Gobert [1965].

The treatment of 11.31, 11.32 is due to Mordell [1933], pp. 351–352. For a somewhat different one see pp. 285–286 of Dieudonné [1971]. That the parallelogram γ_r employed in the proof of 11.32 is no *deus ex machina* but is "natural," even "inevitable" is argued by Desbrow [1998]. The integral in 11.32(i) seems to have been first computed by Euler; see van Veen [1943] and van Yzeren [1979]. The most popular derivation of 11.32(i) (due to Poisson) involves iterated integrals, their conversion to a double integral and the polar coordinate change of variable in this integral. A more elementary variant of this uses Fubini's theorem only over a compact rectangle (i.e., uses 2.19(ii)); see Gomes Teixeira [1889] and Sandham [1946]. Other interesting proofs are those in Krafft [1955], [1956], Spivak [1967], pp. 328–329 (see also **MR** 97j:26001), Jameson [1994] = Nicholas [1957] (perhaps the most elementary), Weinstock [1990], R. M. Young [1991] (reproduced in 3.24), [2005], Niizeki and Sasaki [1997], Kortram [1993] (counting lattice points), and Lord [2003b]. Levrie and Daems [2009] is a recent elementary derivation based on a refinement of 3.21 and proves Wallis' product formula 11.37(xiii) en route. Other references are provided by Mitrinović and Kečkić [1984], pp. 158–168. Because the (Gaussian) distribution function $F(x) := \frac{1}{\sqrt{2\pi}}\int_{-\infty}^{x} e^{-t^2/2}\,dt$ plays a central role in probability, the integral 11.32(i) is called the probability integral (also the [Gauss] error integral). The integrals 11.32(ii) are known in the literature as *Fresnel Integrals* and are important in optics; they were, however, first evaluated by ? — of course, Euler! As with the probability integral above, they are susceptible to treatment by a wide range of different techniques. Some of these are CAYLEY [1873], GLAISHER [1875], FABRY [1896b], GODEFROY [1898], BLOCH [1919], [1922], FLANDERS [1982], LEONARD [1988], WEINSTOCK [1990] (elementary), VÝBORNÝ [1991], R. M. YOUNG [1991], and CHEN [2009]. Recall also 3.24(ix).

The sum in 11.32(iii) first occurred in an 1811 paper of C. F. Gauss and is called a Gauss sum; the related expression $\sum_{k=0}^{n-1} e^{2\pi i h k^2/n}$ is called a generalized Gauss sum. For the contributions of Gauss, Cauchy and Dirichlet to their evaluation see the notes in KRONECKER [1889a]; the evaluation of the Gauss sums via residues first occurs here; see also p. 243 of LANDSBERG [1893]. For a discussion of Kronecker's proof as well as proofs due to SCHUR [1921] and MERTENS [1896] and the one in the text due to Dirichlet, see pp. 197–218 of E. LANDAU [1958]. See also LERCH [1903a] (who corrects errors in Kronecker and Bachmann), SALVADORI [1904], §38 of LINDELÖF [1905] and 5.1 of MITRINOVIĆ and KEČKIĆ [1993]. For an algebraic treatment, close to Gauss' original, which also evaluates the generalized Gauss sums (as does SCHUR *op. cit.*), see chapter 11 of RADEMACHER [1964], §7 of P. BACHMANN [1894] and BRESSOUD [1981]. (Another algebraic treatment occurs in SHANKS [1958].) These sums are related to the important law of quadratic reciprocity and there is a reciprocity relationship for the generalized Gauss sums too. This is also elucidated in Rademacher's and Bachmann's books, in KRONECKER [1890], in LANDSBERG [1893], in KRAZER [1903], pp. 184–186 and in BELLMAN [1961], pp. 38–39. In turn there are important relationships between the Gauss sums and the formula 11.33(v). For these see KRAZER [1903], LANDSBERG [1893] and pp. 222–226 of DYM and MCKEAN [1972]; for more on the connection of this formula with the important special functions θ, Γ, ζ see also HAMBURGER [1922]. The proof of 11.33(v) in the text is from Landsberg. A different proof is offered in PÓLYA [1927] (somewhat demystified in BELLMAN[1961]); the Fourier proof in (vi) is apparently due to Cauchy. See DYM and MCKEAN [1972], pp. 65–66 for the relation of the identity in (vii) to the heat equation of physics.

The history of Gauss sums and their many generalizations is surveyed and a very complete 191-entry bibliography is provided in the paper of BERNDT and EVANS [1981]; see especially its §2 and §10. The Gauss sums involve characters of finite groups and their connection with Fourier analysis on finite groups is set out in the very readable article of AUSLANDER and TOLIMIERI [1985]. (See also **MR** 2011e:43001.)

In 1950 Paul Lévy observed that the probability distribution sech(πx) plays a role in Brownian motion, which makes the computation of its characteristic function (=Fourier transform) important. That was done, using the series method sketched in 11.35(ii), by BASS and LÉVY [1950], although the result itself occurs already in LINDELÖF [1905], p. 72.

11.36 is from S. N. BERNSTEIN [1923]; the equivalence in 3.36 provides the nexus. The proof in the text is attributed by DONOGHUE [1969] to Lars Gårding; another proof occurs in Chapter XIV. For (finite) exponential polynomials see O'HARA [1973]. See also BOAS [1969]. For a converse see p. 139 of ACHIEZER [1956].

I think the treatment of 11.37 is pretty economical. These facts are more or less equivalent to one another and can be derived in many different ways, some involving only real-variable methods (cf. (xv)). The device in parts (iv) and (v) is a modification of one used by LERCH [1903b] and is due to G. Herglotz; it appeared in the 1931 first edition of the booklet by his student ARTIN [1964]. For other applications of it see ZACHARIAS [1993], CIPRA and RUPP [1995], and ELSTRODT [1998]. REMMERT [1991a] is a good source of history of (v) and (vi) (key ideas appear in SCHRÖTER [1868], elaborated by KNOPP [1951], pp. 205–207, and in SCHOTTKY [1892]). In addition the reader will find there a beautiful and unusual treatment of these matters via Eisenstein series. Compare also WIGNER [1952], where the partial-fraction expansion of the tangent analogous to 11.37(vi) is taken as the definition of the tangent function and the whole theory of the exponential and trigonometric functions is derived from it. Another unusual real-variable proof of 11.37(v) is offered by HOFBAUER [2002], and there are popular proofs via the residue theorem (e.g., p. 136 of HERVÉ [1982], p. 122 of CONWAY [1978]). Euler had yet another way of deriving the cotangent partial-fraction representation 11.37(vi), by evaluating $\int_0^1 (x^{p-1} + x^{q-p-1})/(1 + x^q)\, dx$ in two different ways! See STÄCKEL [1907/1908] or pp. 521–523 of V. S. VARADARAJAN [2007].

The product representation of the sine function in 11.37(ix) is more brilliant work of Euler. He thought of series as infinite polynomials, so by 3.27 $(\pi z)^{-1} \sin(\pi z)$ is a "polynomial" with constant term 1 and zero-set $\mathbb{Z} \setminus \{0\}$ (each zero simple), for which a product representation like 4.52(ii) is to be expected! This product is treated by other methods in E. H. MOORE [1895], PADÉ [1898], PRINGSHEIM [1925] (pp. 485 ff.), NEVILLE [1951], MOHR [1959], VENKATACHALIENGAR [1962], MACKEY [1967] (p. 148 — based on LERCH [1903b]), FELLER [1967] (elementary, real-variables arguments), EBERLEIN [1977] (rigorous version of Euler's original proof — a different such in STROMBERG [1981]), WALTER [1982] (very elementary, discusses several different approaches), BUTZER, *et al.* [1984], HARUKI [1985a], KORTRAM [1996], BAK and NEWMAN [1997], pp 221–223, and PRIBITKIN [2001]. The method in the hints to 11.37(xii) was further exploited by Euler to deduce 11.37(v), (vii) and 3.22 from the product formula 11.37(ix); see §4 of AYOUB [1974]. There is an analog of Stirling's formula 11.37(xiv) with the generalized factorial $\Gamma(z)$ in the role of $n!$ and $z \in \mathbb{C}_-$. The exponent $\theta_n/12n$ in the final factor is replaced by a more complicated function of z, which still converges to 0 like

$1/|z|$ whenever $|z| \to \infty$ and $\operatorname{Re} z > -\lambda|z|$ for some $\lambda < 1$ (that is, convergence is inside some closed angular sector in $\mathbb{C}_-$). See REMMERT [1998], p. 58. For history of the whole of 11.37 see §23–25 of PRINGSHEIM, FABER and MOLK [1911].

The technique in 11.39 is from RUNGE [1885], pp. 243–244; in fact it appears that Runge proved 11.39 independently of Mittag-Leffler. MITTAG-LEFFLER's work originated in papers written in Swedish and in the French Academy notes of [1882]. The paper [1884a] represents the grand synthesis and generalization; 11.41 appears there (as theorems B, p. 43; D, p. 53 and A′, p. 72). On page 20 and at other points throughout this paper he records this evolution and the contribution of others (some of whom had read his Swedish papers), notably SCHERING [1881], WEIERSTRASS [1880b], DINI [1881], HERMITE [1881], CAZZANIGA [1882], CASORATI [1882], GUICHARD [1884] and RUNGE [1885]. See also STÄCKEL [1890], FABER [1905], JOURDAIN [1905a], ARONSZAJN [1935], PÓLYA [1941] and GERMAY [1946], [1948]. For more detailed history, see TURNER [2013]. GELFOND [1967] proves the following: If Ω is open and simply-connected, $A \subset \Omega$, then there exists a non-polynomial $f \in H(\Omega)$ which together with all its derivatives takes only integer values in A if and only if A has no limit point in Ω. See also MAHLER [1971], FRANKLIN [1926] and ISRAEL [1989].

Theorem 11.41 (also known suggestively as the "osculation" theorem) can even be sharpened a bit more: we can arrange that F has no zeros outside A. The derivation 11.40 of Weierstrass' theorem on zeros from 11.39 was given by Mittag-Leffler himself (after a suggestion of HERMITE [1881]). Yet another derivation of 11.40, using a version of Runge's theorem in which the approximating functions are zero-free, is given by NARASIMHAN [1985], p. 127; on p. 123 he also offers a wholly different proof of 11.39.

The use of 8.5 in the proof of 11.39 is only a convenience and can be avoided. The classical proof (nicely set out in HERVÉ [1982], PALKA [1990] and REMMERT [1991b]) constructs the rational functions R_n from judiciously selected Laurent expansions of the g_a and is very beautiful. One consequence of the proof of 11.39 is the following attractive "partial-fraction representation" of meromorphic functions: *If f is meromorphic in $\mathbb{C}$ with pole-set P, then there exist rational functions r_n such that $\sum_{n=1}^{\infty} r_n$ converges locally uniformly in $\mathbb{C} \setminus P$ to f and for each $n \in \mathbb{N}$, r_n has at most one pole, at some point in P.* Similarly, from the proof of 7.35 it follows easily that every $F \in H(U)$ has the form $F = \lim_{n\to\infty} \prod_{k=1}^{n} f_k$, where each $f_k \in H(U)$ has at most one zero and the convergence is locally uniform in U. These two representations are very useful: e.g., they permit the ideal theory of 11.42–11.44 to be done a bit more perspicuously. For this see the treatment in REMMERT [1991b].

On the other hand, there is an abstract version of Mittag-Leffler's theorem in the setting of projective limits of complete metric spaces. Among its consequences is 11.39 (using Runge to confirm its hypotheses), as well as the important P. J. Cohen factorization theorem in Banach algebras having bounded approximate identities. A nice exposition is ESTERLE [1984]. A version for operator-valued holomorphic functions will be found in GOHBERG and RODMAN [1983].

11.42 is due to WEDDERBURN [1915], who re-discovered an idea of HURWITZ [1897] and it was in turn re-discovered by HELMER [1940], to whom is due 11.43. 11.44 is due to SCHILLING [1946] (with different proofs in NARASIMHAN [1985] and REMMERT [1991b]). See also KAKUTANI [1955], COHEN [1961], chapters 6 and 10 of HOFFMAN [1962], AGUILÓ FUSTER [1965], ALLING [1968] (esp., pp. 25–28 for history), RUBEL [1970], the nice treatment in §2 of H. S. SHAPIRO [1970], VON RENTELN [1976], [1981], FORELLI [1982], §4 chapter 6 and its notes in NARASIMHAN [1985], and chapter 13 of LUECKING and RUBEL [1984]. The analogs of 11.42 for some other rings of holomorphic functions besides the full ring $H(U)$ are discussed in some of these references. See especially the work of Raymond Mortini and co-authors.

Problem 10, p. 317 of RUDIN [1987] is relevant to 11.43.

WEIERSTRASS [1876] proved 11.45, for $U = \mathbb{C}$, via his zero-specifying theorem 7.35 and, according to ULLRICH [1989], p. 167, establishing 11.45 was in fact Weierstrass' impetus for proving 7.35. Ullrich also remarks (p. 166) on an earlier proof of 11.45 by Enrico Betti. The general case for arbitrary U is due to MITTAG-LEFFLER [1884a].

The function in 11.47(iv) is known as the Eisenstein-Weierstrass pe-function (being universally designated by the Gothic letter $\mathfrak{P}$ or by $\mathfrak{P}_\Lambda$ or $\mathfrak{P}(z; p_1, p_2)$ if the lattice Λ needs to be specified). A non-constant meromorphic function with $\mathbb{R}$-linearly independent periods p_1 and p_2 is called an *elliptic function* and, remarkably, every such function has the form $g \circ \mathfrak{P} + \mathfrak{P}' \cdot h \circ \mathfrak{P}$ for some rational functions g and h. For two ingenious constructions of elliptic functions quite different from that (of Weierstrass) in 11.47(iv), see ALPAY [2011], Exer. 3.6.13 (based on infinite products) and Exer. 7.6.3 (using a series like 11.10); see also REMMERT [1998], pp. 83–85. The study of elliptic functions is a deep chapter in complex analysis and was the driving force for many of its pioneers in the 19th century, as chapter 4 in BOTTAZZINI and GRAY [2013] delineates. See Chapter VIII of SAKS and ZYGMUND [1971] for the tip of the iceberg.

Theorem 4, p. 102 of NARASIMHAN [2001] and Theorem 10 of MORTINI and RUPP [2014] are versions of 11.48 based on area integrals instead of line integrals and indices of loops.

Not only can smoothness of γ be eliminated from 11.48 (see 11.49(vi)), but even the meromorphy of F! See the argument principle for *continuous* F on

p. 44 of SHEIL-SMALL [2002].

11.50 is from ROBINSON [1943], with extensions by ABE [1958].

There is a beautiful multiplicative analog of 11.52(ii) from which 11.52(ii) itself can be deduced: For each R-function f there exist $c > 0$ and real sequences $\{a_n\}$, $\{b_n\}$, which may be finite or indexed by $\mathbb{Z}^+$ or $\mathbb{Z}$, but which satisfy $a_{-1} < 0 < b_1$, $b_k < a_k < b_{k+1}$ for all k and

$$f(z) = c\frac{z - a_0}{z - b_0} \prod \left(1 - \frac{z}{a_n}\right)\left(1 - \frac{z}{b_n}\right)^{-1}$$

for all $z \in \mathbb{C} \setminus \{b_n\}$. Proofs of this may be found in WIGNER [1952] and B. LEVIN [1980], p. 308 (where it is said to be an unpublished result of M. G. Krein, sometime before 1943). See also **MR** 2012h:30190. A trivial modification of the formula in 11.37(vi) yields $\pi\cot(\pi z) = 1/z + \sum[1/(z - n) + 1/n]$, so that according to 11.52(iii), $-\cot(\pi z)$ is an R-function. WIGNER [1952] showed that this equation can be taken as the definition of the cotangent and the whole theory of the trigonometric and exponential functions developed from properties of R-functions. A similar approach to the circular functions in the 19th century was devised by Gotthold Eisenstein; see §4*, chapter 11 of REMMERT [1991a], and WEIL [1976].

On the Dirichlet problem for an annulus see DINI [1870–71], [1913], H. A. SCHWARZ (pp. 207–209 of his *Gesammelte Mathematische Abhandlungen*), VILLAT [1912], [1921], GRONWALL [1912/13b] (the sources of 11.53), KOEBE [1913b], PICONE [1926a], JULIA [1930a], DEMTCHENKO [1931], CISOTTI [1931], KOMATU [1945b], CONSIGLIO [1950], GOLUZIN [1951], ZAYED [1986] and §14 of LICHTENSTEIN [1919]. When the boundary data in 11.54 are trigonometric polynomials, the problem reduces to an easy algebraic one. Whenever trigonometric polynomials converge to a continuous boundary function f, the harmonic functions so constructed converge, thanks to the maximum principle, to the solution of the Dirichlet problem with boundary data f. Theorems of Weierstrass and Fejér insure that every continuous function f is approximable by trigonometric polynomials, so this proof also solves the Dirichlet problem in a disk, independent of Poisson-formula considerations (although, as indicated in Remark (ii) following 5.20, one of the many proofs of Weierstrass' theorem is via the Poisson formula). See pp. 300–302 of HEINS [1968]. See also pp. 155–158 of GAIER [1964] and the references there. The whole theory in §7 can be derived *ab initio* by Fourier methods, as hinted in 11.54(iii). In fact, by differentiating under the integral sign (in the definition of a_n in 11.54(ii)) and using the polar form of the Laplace's Equation, differential equations for these coefficients (as functions of r) are found which can be solved by inspection, leading to the appropriate form of a_n ($n \neq 0$) and a_0. For the details see pp. 56–58 of HEINS [1962b]. This method can actually be used to "realize" the foundations of complex analysis, as was done by HARNACK [1883] and W. H. YOUNG [1911]. For if one shows by real-variable methods (e.g., convolution approximate identities) that the Fourier series of every continuously differentiable 2π-periodic function converges absolutely to it, then the argument just outlined derives the existence of Taylor and Laurent series expansions of a function f that has continuous partial derivatives which satisfy (C. R.) directly from the Fourier series of the periodic functions $f_r(\theta) := f(re^{i\theta})$. Cf. BEESACK [1972].

For the literature on removable singularities of harmonic functions (11.57(i)) see the Chapter V notes.

For a different treatment and extension of 11.57(ii) to $\mathbb{R}^n$ see AXLER, BOURDON, and RAMEY [1992]. The result 11.59(ii) appears in BÔCHER [1902/03] and 11.60(ii) was reached by a more circuitous route in ANANDAM and DAMLAKHI [1993/94], while in [1997/98] they give a proof of 11.60(iv) directly from the Poisson integral formula. See also 9.10 and 9.11 in AXLER, BOURDON and RAMEY [2001]. KURAN [1966b] contains a generalization and alternative proof of 11.61.

We have by now seen Poisson formulas for disks, annuli, strips and half-planes. There are also simple and useful ones for ellipses and half-disks. In DINGHAS [1954] (or p. 69 of [1961]) the formula for half-disks is gotten by a simple application of the residue theorem; to get it from the conformal equivalence of $\mathbb{D}^+$ and $\mathbb{D}$ in 6.16(vi) is rather daunting. A Poisson formula for ellipses is given by ROYSTER [1964] and HENRICI [1986], p. 231. See also KOEBE [1913b].

Chapter XII
Omitted Values and Normal Families

The center-piece of this chapter is the following remarkable result due to Friedrich Schottky: if f is holomorphic in $\mathbb{D}$ and assumes neither of the values 0 or 1, then f is bounded in $D(0,r)$ by a bound depending on r and $|f(0)|$ only, for each $r < 1$. [In fact, we prove a generalization in which f does not assume 0 and $f^{(n)}$ (for some $n \geq 0$) does not assume 1.] The principal application is the almost immediate fact that a family of holomorphic functions on a common domain, none of which has 0 or 1 in its range, if bounded at a point, is uniformly bounded in a neighborhood of that point. The compactness theorems of Chapter VII then come into play with astounding consequences. Of course "0" and "1" here are convenient normalizations: any two distinct complex numbers would serve as well.

To scale these peaks we start with some modest preparation of gear.

§ 1 Logarithmic Means and Jensen's Inequality

Lemma 12.1 *The function* $\log^+$ *defined on* $[0,\infty)$ *by*

$$\log^+(x) := \log(\max\{1, x\})$$

has the following properties:

(i) $\log^+$ *is continuous, non-decreasing and non-negative*

(ii) $\log^+(x_1 \cdots x_n) \leq \log^+(x_1) + \cdots + \log^+(x_n) \qquad \forall x_1, \ldots, x_n \in [0,\infty)$

(iii) $\log^+(x_1 + \cdots + x_n) \leq \log^+(x_1) + \cdots + \log^+(x_n) + \log(n)$

(iv) $\log^+(x) \leq \log(1+x) \leq \sqrt{x}$ *for all* $x \geq 0$

(v) $|\log x| = \log^+(x) + \log^+(1/x)$ *and*
$\log x = \log^+(x) - \log^+(1/x)$ *for all* $x > 0$.

R. B. Burckel, *Classical Analysis in the Complex Plane*, Cornerstones,
https://doi.org/10.1007/978-1-0716-1965-0_13

Proof: The function $\log^+$, being the composite of the continuous non-decreasing function $x \mapsto \max\{1, x\}$ of $[0, \infty)$ onto $[1, \infty)$ and the continuous non-decreasing function $\log\colon [1, \infty) \mapsto [0, \infty)$, is itself continuous, non-decreasing and non-negative.

Now (ii) is trivial if the left-hand side is 0. Otherwise, some $x_j > 1$ and we may suppose the notation chosen so that

$$0 \leq x_1 \leq x_2 \leq \cdots \leq x_{j-1} \leq 1 < x_j \leq \cdots \leq x_n.$$

Then

$$x_1 \cdots x_n \leq x_j \cdots x_n.$$

Therefore

$$\begin{aligned}\log^+(x_1 \cdots x_n) \leq \log^+(x_j \cdots x_n) &= \log(x_j \cdots x_n) = \log(x_j) + \cdots + \log(x_n) \\ &\leq \log^+(x_1) + \cdots + \log^+(x_j) + \cdots + \log^+(x_n).\end{aligned}$$

Similarly, (iii) is trivial if each $x_j < 1/n$. Otherwise, let the points be monotone in their indices as before and have $x_n \geq 1/n$ and

$$\begin{aligned}\log^+(x_1 + \cdots + x_n) \leq \log^+(nx_n) &= \log(nx_n) = \log(n) + \log(x_n) \\ &\leq \log(n) + \log^+(x_1) + \cdots + \log^+(x_n).\end{aligned}$$

Since $\log^+$ is non-decreasing, we have

$$\log^+(x) \leq \log^+(1 + x) = \log(1 + x).$$

Also the function $f(x) := \sqrt{x} - \log(1 + x)$ has derivative

$$f'(x) = \frac{\left(1 - \sqrt{x}\right)^2}{2\sqrt{x}(1 + x)} \geq 0$$

on $(0, \infty)$ and so f is non-decreasing. Since $f(0) = 0$, we have $f(x) \geq 0$ for all $x \geq 0$, completing the proof of (iv).

For (v) consider the two cases $x \geq 1$ or $x \leq 1$.

Definition 12.2 If $r \geq 0$ and f is a continuous complex-valued function on $C(0, r)$ define the *means*

$$M(r, f) := \sup_{\theta \in \mathbb{R}} \left|f\left(re^{i\theta}\right)\right|$$

$$L(r, f) := \frac{1}{2\pi} \int_0^{2\pi} \log^+ \left|f\left(re^{i\theta}\right)\right| d\theta.$$

If, moreover, f is zero-free, define

$$V(r, f) := \frac{1}{2\pi} \int_0^{2\pi} \left|\log\left|f\left(re^{i\theta}\right)\right|\right| d\theta.$$

Then by 12.1(v) we see that

$$V(r, f) = L(r, f) + L\left(r, \frac{1}{f}\right).$$

Theorem 12.3 *If f is holomorphic in $D(0, r)$, $0 < R < r$ and f has no zeros in $\overline{D}(0, R)$, then*

$$\log|f(z)| = \frac{1}{2\pi} \int_0^{2\pi} \log\left|f\left(Re^{i\theta}\right)\right| \operatorname{Re}\left[\frac{Re^{i\theta} + z}{Re^{i\theta} - z}\right] d\theta \qquad \forall |z| < R.$$

Proof: Let F be a holomorphic logarithm for f in some neighborhood of $\overline{D}(0, R)$, apply the Cauchy-Schwarz formula (5.18 appropriately scaled to the disk $\overline{D}(0, R)$) to F and take real parts, remembering that $\operatorname{Re} F = \log|f|$.

Theorem 12.4 *If F is holomorphic in $\mathbb{D}$ and $0 < R < 1$, then*

(i) $$\log^+|F(z)| \le \frac{1}{2\pi} \int_0^{2\pi} \log^+\left|F\left(Re^{i\theta}\right)\right| \operatorname{Re}\left[\frac{Re^{i\theta} + z}{Re^{i\theta} - z}\right] d\theta \quad \forall z \in D(0, R).$$

If in addition F is zero-free on $C(0, R)$, then

(ii) $$\log|F(z)| \le \frac{R + |z|}{R - |z|} L(R, F) - \frac{R - |z|}{R + |z|} L\left(R, \frac{1}{F}\right) \quad \forall z \in D(0, R) \setminus F^{-1}(0)$$

and if F is not identically 0, then

(iii) $$\log M(r, F) \le \frac{R + r}{R - r} L(R, F) \quad \forall 0 < r < R.$$

Proof: First consider the case where F has no zeros on $C(0, R)$. Then F is not identically zero in $\overline{D}(0, R)$, hence has only finitely many zeros there (5.62). Let these zeros counted according to multiplicity be $z_1, \ldots, z_n$. We have $|z_j| < R$ by assumption, so each of the functions

$$\phi_j(z) := \frac{R(z - z_j)}{R^2 - \bar{z}_j z}$$

is holomorphic in a neighborhood of $\bar{D}(0,R)$. Thus $f := F/\phi_1 \cdots \phi_n$ is holomorphic in a neighborhood of $\bar{D}(0,R)$ and has no zeros in $\bar{D}(0,R)$. We apply the preceding theorem to this function, noting that $|\phi_j(z)| = 1$ when $|z| = R$:

$$(1) \quad \log\left|\frac{F(z)}{\phi_1(z)\cdots\phi_n(z)}\right| = \frac{1}{2\pi}\int_0^{2\pi} \log\left|F(Re^{i\theta})\right| \operatorname{Re}\left[\frac{Re^{i\theta}+z}{Re^{i\theta}-z}\right] d\theta \quad \forall |z| < R.$$

Since $|\phi_j(z)| < 1$ for all $z \in D(0,R)$, we have $\log|\phi_j(z)| < 0$ whenever $z \in D(0,R) \setminus F^{-1}(0)$, so for such z the left-hand side of (1) dominates $\log|F(z)|$ and we get the so-called *Jensen's Inequality*:

$$(\mathrm{J}) \qquad \log|F(z)| \le \frac{1}{2\pi}\int_0^{2\pi} \log\left|F(Re^{i\theta})\right| \operatorname{Re}\left[\frac{Re^{i\theta}+z}{Re^{i\theta}-z}\right] d\theta.$$

Recall that $\log(x) = \log^+(x) - \log^+(1/x)$ for all $x > 0$, to see from (J) that

$$(2) \qquad \log|F(z)| \le \frac{1}{2\pi}\int_0^{2\pi} \log^+\left|F(Re^{i\theta})\right| \operatorname{Re}\left[\frac{Re^{i\theta}+z}{Re^{i\theta}-z}\right] d\theta \\ - \frac{1}{2\pi}\int_0^{2\pi} \log^+\left|\frac{1}{F(Re^{i\theta})}\right| \operatorname{Re}\left[\frac{Re^{i\theta}+z}{Re^{i\theta}-z}\right] d\theta.$$

Now (equation (2) in the proof of 5.20)

$$\operatorname{Re}\left[\frac{Re^{i\theta}+z}{Re^{i\theta}-z}\right] = \frac{R^2 - |z|^2}{\left|Re^{i\theta}-z\right|^2}$$

and therefore

$$(3) \qquad \frac{R+|z|}{R-|z|} \ge \operatorname{Re}\left[\frac{Re^{i\theta}+z}{Re^{i\theta}-z}\right] \ge \frac{R-|z|}{R+|z|} > 0 \quad \text{for } |z| < R.$$

Since $\log \le \log^+$, it follows from (3) that

$$(4) \qquad \frac{1}{2\pi}\int_0^{2\pi} \log\left|F(Re^{i\theta})\right| \operatorname{Re}\left[\frac{Re^{i\theta}+z}{Re^{i\theta}-z}\right] d\theta \\ \le \frac{1}{2\pi}\int_0^{2\pi} \log^+\left|F(Re^{i\theta})\right| \operatorname{Re}\left[\frac{Re^{i\theta}+z}{Re^{i\theta}-z}\right] d\theta.$$

But also $\log^+ \ge 0$, so the right-hand side of (4) is non-negative and it follows from (J) and (4) that

$$\log^+|F(z)| = \max\{0, \log|F(z)|\} \le \frac{1}{2\pi}\int_0^{2\pi} \log^+\left|F(Re^{i\theta})\right| \operatorname{Re}\left[\frac{Re^{i\theta}+z}{Re^{i\theta}-z}\right] d\theta$$

for all $z \in D(0,R) \setminus F^{-1}(0)$, giving (i). While (2) and the inequalities (3) lead to (ii) and (iii).

Now consider the general case and a fixed $z \in \mathbb{D}$. An easy uniform continuity argument shows that the right-hand side of (i), call it $\Phi(R)$, is a continuous function of $R \in (|z|, 1)$. If F is identically 0, then (i) is trivial. Otherwise the zeros of F do not accumulate in $\mathbb{D}$, hence there are only finitely many in each $\bar{D}(0, 1 - 1/n)$, hence only countably many in $\mathbb{D}$. Thus for a dense set of $R \in [0,1)$, F has no zeros on $C(0,R)$. Given $R \in (|z|, 1)$, pick R_n from this dense set with $|z| < R_n < R$ and $R_n \to R$. Then $\log^+ |F(z)| \leq \Phi(R_n)$ by the result of the first paragraph and $\Phi(R_n) \to \Phi(R)$ by continuity of Φ. Thus $\log^+ |F(z)| \leq \Phi(R)$, which is (i).

Similarly, $L(R,F)$ is a continuous function of $R \in [0,1)$ and (ii), (iii) are already established for the dense set of R for which F has no zeros on $C(0,R)$. Therefore they follow for all $R \in [0,1)$ by continuity.

Remark: Notice that (i) also follows from 5.22(i) and the subharmonicity of $\log^+ |F|$ (5.26(ii)).

Theorem 12.5 *If f is holomorphic in $\mathbb{D}$, then $L(r,f)$ is a non-decreasing function of $r \in [0,1)$.*

Proof: 5.26(ii) shows that $\log^+ |f|$ is a subharmonic function and so the conclusion follows from 5.26(i). Here is an alternative, more prosaic proof. Let $0 < r < R < 1$. We may apply 12.4(i) to write

$$\log^+ |f(re^{it})| \leq \frac{1}{2\pi} \int_0^{2\pi} \log^+ |f(Re^{i\theta})| \operatorname{Re} \left[\frac{Re^{i\theta} + re^{it}}{Re^{i\theta} - re^{it}} \right] d\theta, \quad t \in \mathbb{R}.$$

Integrate this:

$$\begin{aligned} L(r,f) &= \frac{1}{2\pi} \int_0^{2\pi} \log^+ |f(re^{it})| \, dt \\ &\leq \frac{1}{2\pi} \int_0^{2\pi} \frac{1}{2\pi} \int_0^{2\pi} \log^+ |f(Re^{i\theta})| \operatorname{Re} \left[\frac{Re^{i\theta} + re^{it}}{Re^{i\theta} - re^{it}} \right] d\theta \, dt. \end{aligned}$$

Since the integrand on the right-hand side is a continuous function of $(\theta, t) \in [0, 2\pi] \times [0, 2\pi]$, it is possible to invert the orders of integration in this iterated integral (see 2.19(ii) or 8.29(vi)) to get

$$L(r,f) \leq \frac{1}{2\pi} \int_0^{2\pi} \log^+ |f(Re^{i\theta})| \left(\frac{1}{2\pi} \int_0^{2\pi} \operatorname{Re} \left[\frac{Re^{i\theta} + re^{it}}{Re^{i\theta} - re^{it}} \right] dt \right) d\theta.$$

But (formula (4) in the proof of 5.20) this inner integral is 1 and so we get

$$L(r, f) \leq \frac{1}{2\pi} \int_0^{2\pi} \log^+ \left| f\left(Re^{i\theta}\right) \right| d\theta = L(R, f).$$

Theorem 12.6 *Let F be a holomorphic function in $\mathbb{D}$ which is not identically 0 but satisfies*

$$(*) \qquad \sup_{r<1} \int_0^{2\pi} \log^+ \left| F\left(re^{i\theta}\right) \right| d\theta =: M < \infty.$$

Then F has at most countably many zeros. Let them be, with due regard to multiplicity, $z_1, z_2, \ldots$. Then

$$\sum_{j=1}^{\infty} (1 - |z_j|) < \infty.$$

Proof: Since for each $R < 1$, F has only finitely many zeros in $\overline{D}(0, R)$, we may suppose the notation chosen so that $|z_1| \leq |z_2| \leq \cdots$. Moreover, by dividing off an appropriate power of z, we may suppose without loss of generality that 0 is not a zero of F. Given n, choose R so that $|z_n| < R < 1$ and $R \neq |z_k|$ for each k. Let $z_1, \ldots, z_N$ $(N \geq n)$ be the zeros of F in $\overline{D}(0, R)$ and for $1 \leq j \leq N$ define

$$\phi_j(z) := \frac{R(z - z_j)}{R^2 - \bar{z}_j z}, \qquad f := \frac{F}{\phi_1 \cdots \phi_N}.$$

Then f is holomorphic and zero-free in a neighborhood of $\overline{D}(0, R)$. Apply 12.3 to f with $z = 0$, remembering that $|\phi_j| = 1$ on $C(0, R)$, to see that

$$\text{(JJ)} \qquad \log |F(0)| + \sum_{j=1}^{N} \log \frac{R}{|z_j|} = \frac{1}{2\pi} \int_0^{2\pi} \log \left| F\left(Re^{i\theta}\right) \right| d\theta.$$

Consequently, from $(*)$

$$\sum_{j=1}^{n} \log \frac{R}{|z_j|} \leq \sum_{j=1}^{N} \log \frac{R}{|z_j|} \leq \frac{M}{2\pi} - \log |F(0)|.$$

Let $R \uparrow 1$ through admissible values to have

$$\sum_{j=1}^{n} \log \frac{1}{|z_j|} \leq \frac{M}{2\pi} - \log |F(0)|.$$

Since for any $x > 1$ we have

$$\log x = \int_1^x \frac{1}{t}\,dt > \int_1^x \frac{1}{x}\,dt = \frac{x-1}{x} = 1 - \frac{1}{x},$$

it follows that

$$\sum_{j=1}^{n}(1 - |z_j|) \le \sum_{j=1}^{n} \log \frac{1}{|z_j|} \le \frac{M}{2\pi} - \log|F(0)|,$$

holding for *all* positive integers n.

Remark: The hypothesis $(*)$ is certainly fulfilled if F is bounded, so this result generalizes 6.7(iii). The equality (JJ) is known as the *Jacobi-Jensen* formula. See E. LANDAU [1914] and KOWALEWSKI [1918] for the priorities and exact references. It is valid even if R is the modulus of a zero of F, due to the fact that the integral always exists as an improper Riemann integral (see, e.g., pp. 151-152 of SANSONE and GERRETSEN [1960]).

Exercise 12.7 (i) (PRIVALOV [1924a]) *Let F be holomorphic in $\mathbb{D}$ and satisfy*

$$(*) \qquad \sup_{\rho<1} \int_0^{2\pi} \log^+ \left|F\left(\rho e^{i\theta}\right)\right| d\theta =: K < \infty.$$

Show that F satisfies

$$|F(z)| \le e^{K/(\pi(1-|z|))} \qquad \forall z \in \mathbb{D}.$$

Hints: We may assume that F is not identically 0, the conclusion being trivial otherwise. For each $R < 1$, 12.4 gives

$$\log|F(z)| \le \frac{R+|z|}{R-|z|} L(R,F) \le \frac{2}{R-|z|} L(R,F) \qquad \forall z \in D(0,R) \setminus F^{-1}(0).$$

Using hypothesis $(*)$, we get

$$\log|F(z)| \le \frac{2}{R-|z|} \cdot \frac{K}{2\pi},$$
$$|F(z)| \le e^{K/(\pi(R-|z|))} \qquad \forall z \in D(0,R) \setminus F^{-1}(0).$$

The restriction $z \notin F^{-1}(0)$ is now unnecessary. Given $z \in \mathbb{D}$, pick $R \in (|z|, 1)$ and let $R \uparrow 1$.

(ii) *Use* (i) *to show that the boundedness hypothesis in* 7.4 *can be weakened to* $\sup_n \sup_{r<1} \int_0^{2\pi} \log^+ |f_n(re^{i\theta})|\, d\theta =: M < \infty$ *and the conclusion of* 7.4 *still follows.*

Hints: Under the present hypothesis (i) insures that $\{f_n\}$ is locally uniformly bounded in $\mathbb{D}$, hence by 7.6 possesses locally uniformly convergent subsequences. Consider any pair g_1, g_2 of limits of such subsequences. Fixing $0 < r < 1$, we infer from the present hypothesis that

$$(1) \qquad \int_0^{2\pi} \log^+ |g_k(re^{i\theta})|\, d\theta \leq M < \infty \qquad \forall\, 0 < r < 1, k = 1, 2.$$

If $\{z_j\}$ satisfy hypothesis 7.4(i), then $g_1(z_j) = g_2(z_j)$ for all j:

$$(2) \qquad (g_1 - g_2)(z_j) = 0 \qquad \forall\, j \in \mathbb{N}.$$

Now (1) shows that $g_1 - g_2$ fulfills the hypothesis of 12.6. Consequently, (2) and the divergence hypothesis 7.4(i) mean that $g_1 - g_2$ must be $\mathbf{0}$. Therefore 7.8 can be invoked to get the conclusion in 7.4. (See PRIVALOV [1924d] for a related extension of 7.4.)

§ 2 Miranda's Theorem

This section is somewhat long and a bit arduous. It culminates in a generalization of Schottky's theorem mentioned in the introductory remarks. Here the function f omits 0 but one of its derivatives $f^{(n)}$ omits 1. The case $n = 0$ is Schottky's theorem and that is adequate for all the applications we make here, save one (12.19). Moreover, a self-contained proof of Schottky's theorem is offered in the last section of this chapter. Therefore the reader can skip this section and do §7 instead. He can then cite Schottky instead of Miranda in the subsequent sections and get the case $n = 0$ of all subsequent theorems. As noted, this will secure for him all the applications except 12.19.

Lemma 12.8 *If f is holomorphic and zero-free in the open set U and $F := f'/f$, then for every $n \geq 1$ the quotient $f^{(n)}/f$ is a sum of $n!$ or fewer terms, each of which is a product of n or fewer factors of the form $F^{(j)}$ with $0 \leq j < n$.*

Proof: By induction on n. Trivial for $n = 1$. If true for $n = N$, consider $n = N + 1$. We have that $f^{(N)}$ is a sum of $N!$ or fewer terms of the form $f \cdot F^{(j_1)} \cdots F^{(j_k)}$ $(0 \leq j_1, \ldots, j_k < N,\ k \leq N)$. Since

$$\frac{\left[f F^{(j_1)} \cdots F^{(j_k)}\right]'}{f} = F^{(0)} F^{(j_1)} \cdots F^{(j_k)} + \sum_{l=1}^{k} F^{(j_1)} \cdots F^{(j_l+1)} \cdots F^{(j_k)},$$

a sum of $N+1$ or fewer terms, each of which is a product of $N+1$ or fewer factors of the form $F^{(j)}$ with $0 \le j < N+1$, the assertion for $n = N+1$ follows.

Theorem 12.9 *For each integer $n \ge 1$ there exists a finite positive constant A_n such that for any f which is holomorphic and zero-free in $\mathbb{D}$*

$$L\left(r, \frac{f^{(n)}}{f}\right) \le A_n + n(n+2)\log\frac{1}{R-r} + n\log^+ V(R,f), \quad 0<r<R<1.$$

Proof: Fix such an n and R, and let such an f be given. Let h be a holomorphic logarithm of f in $\mathbb{D}$ and apply the Cauchy-Schwarz formula (5.18) to h in $D(0,R)$:

$$h(z) = -\overline{h(0)} + \frac{1}{\pi}\int_0^{2\pi} \frac{\operatorname{Re} h\big(Re^{i\theta}\big)\, Re^{i\theta}}{Re^{i\theta}-z}\, d\theta.$$

Apply 2.14 to this integral to conclude that

$$h'(z) = \frac{1}{2\pi}\int_0^{2\pi} \operatorname{Re} h\big(Re^{i\theta}\big)\frac{2Re^{i\theta}}{\big(Re^{i\theta}-z\big)^2}\, d\theta, \quad |z|<R.$$

Since $f = e^h$, $h' = f'/f$ and $\operatorname{Re} h = \log|f|$, so

$$(1) \qquad \frac{f'(z)}{f(z)} = \frac{1}{2\pi}\int_0^{2\pi} \log\big|f\big(Re^{i\theta}\big)\big|\frac{2Re^{i\theta}}{\big(Re^{i\theta}-z\big)^2}\, d\theta, \quad |z|<R.$$

Let $F := f'/f$. Repeated differentiations under the integral in (1) (i.e., appeals to 2.14) give

$$(2) \qquad F^{(j)}(z) = \frac{(j+1)!}{2\pi}\int_0^{2\pi} \log\big|f\big(Re^{i\theta}\big)\big|\frac{2Re^{i\theta}}{\big(Re^{i\theta}-z\big)^{j+2}}\, d\theta, \qquad |z|<R,\ j=0,1,\ldots.$$

It follows that

$$\left|F^{(j)}(z)\right| \le \frac{(j+1)!}{2\pi}\int_0^{2\pi} \big|\log\big|f\big(Re^{i\theta}\big)\big|\big|\frac{2R}{(R-|z|)^{j+2}}\, d\theta, \quad |z|<R,\ j=0,1,\ldots$$

$$\begin{aligned} \left|F^{(j)}\big(re^{it}\big)\right| &\le \frac{2R(j+1)!}{(R-r)^{j+2}}\frac{1}{2\pi}\int_0^{2\pi} \big|\log\big|f\big(Re^{i\theta}\big)\big|\big|\, d\theta \\ (3) \qquad &= \frac{2R(j+1)!}{(R-r)^{j+2}} V(R,f), \quad t\in\mathbb{R},\ r<R,\ j=0,1,\ldots. \end{aligned}$$

Thus, since $0 < R - r < R < 1$,

$$\text{(4)} \qquad \left|F^{(j)}\left(re^{it}\right)\right| \le \max\{1, 2(n+1)!(R-r)^{-n-2}V(R,f)\}, \qquad t \in \mathbb{R},\ r < R,\ j = 0, 1, \ldots, n.$$

Use the fact that $\log^+$ is a non-decreasing function and

$$\log^+(n!x) \le \log(n!) + \log^+(x) \quad (12.1(\text{ii}))$$

together with the preceding lemma and (4), to see that

$$\log^+ \left|\frac{f^{(n)}\left(re^{it}\right)}{f\left(re^{it}\right)}\right| \le \log(n!) + \log^+\left[2(n+1)!(R-r)^{-n-2}V(R,f)\right]^n, \quad t \in \mathbb{R},\ r < R.$$

Then use 12.1(ii) again to get

$$\begin{aligned} \log^+ \left|\frac{f^{(n)}\left(re^{it}\right)}{f\left(re^{it}\right)}\right| & \\ &\le \log(n!) + n\left[\log(2(n+1)!) + (n+2)\log\frac{1}{R-r} + \log^+ V(R,f)\right] \\ \text{(5)} \qquad &= \log(n!(2(n+1)!)^n) + n(n+2)\log\frac{1}{R-r} + n\log^+ V(R,f), \\ & \qquad\qquad t \in \mathbb{R},\ r < R. \end{aligned}$$

Let $A_n := \log(n!(2(n+1)!)^n)$ and integrate in (5) to get

$$L\left(r, \frac{f^{(n)}}{f}\right) \le A_n + n(n+2)\log\frac{1}{R-r} + n\log^+ V(R,f), \quad r < R,$$

as desired.

Lemma 12.10 (Bureau's Bootstrap Lemma) *Let $0 \le a < b \le 1$ and f be a non-negative, non-decreasing function on (a,b). Suppose that for some finite positive constants A and B, f satisfies*

$$\text{(1)} \qquad f(r) \le A\log^+ f(R) + A\log\frac{1}{R-r} + B \qquad \forall a < r < R < b.$$

If $B \ge 2A(A+2)$, then f also satisfies

$$\text{(2)} \qquad f(r) \le 2A\log\frac{1}{b-r} + 2B \qquad \forall a < r < b.$$

Hence, regardless of the size of B, f satisfies

$$f(r) \le 2A \log \frac{1}{b-r} + 2B + 4A(A+2) \qquad \forall a < r < b. \tag{3}$$

Proof: Suppose that

$$B \ge 2A(A+2) \tag{4}$$

and that nevertheless (2) fails:

$$f(r_0) > 2A \log \frac{1}{b-r_0} + 2B \quad \text{for some } a < r_0 < b. \tag{5}$$

Then (5) also holds with b replaced by $b' < b$, if b' is sufficiently close to b. Moreover, f is bounded on (a, b'), by $f(b')$. Therefore without loss of generality we can suppose, in trying to deduce a contradiction from (1), (4) and (5), that f is bounded on (a, b). Set

$$r_1 := \frac{r_0 + b}{2} = r_0 + \frac{b - r_0}{2}$$

and apply (1) with $r = r_0$, $R = r_1$ to get

$$\begin{aligned} f(r_0) &\le A \log^+ f(r_1) + A \log \frac{2}{b-r_0} + B \\ \log^+ f(r_1) &\ge \frac{1}{A} f(r_0) + \log \frac{b-r_0}{2} - \frac{B}{A}, \end{aligned}$$

whence, recalling 12.1(iv),

$$\begin{aligned} \sqrt{f(r_1)} &\ge \frac{1}{A} f(r_0) + \log \frac{b-r_0}{2} - \frac{B}{A} \\ &> 2\log \frac{1}{b-r_0} + \frac{2B}{A} + \log \frac{b-r_0}{2} - \frac{B}{A} \quad \text{by (5)} \\ &= \log \frac{2}{b-r_0} + \frac{B}{A} - \log 4 \qquad (6) \\ &> \frac{B}{A} - \log 4 \quad (\text{since } b - r_0 < b - a \le 1). \qquad (6)' \end{aligned}$$

Now from (4)

$$\begin{aligned} &B \ge 2A(A+2) = 2A^2 + 4A \\ &B - 4A \ge 2A^2 \\ &\frac{B}{A} - \log 4 > \frac{B}{A} - 4 \ge 2A \qquad (7) \end{aligned}$$

and also from (4)

$$
\begin{aligned}
&B \geq 2A(A+2) \geq 2A(A+\log 4) \\
&B^2 - 2AB(\log 4 + A) + A^2 \log^2 4 \geq 0 \\
&(B - A\log 4)^2 \geq 2A^2 B \\
\text{(8)} \qquad &\left(\frac{B}{A} - \log 4\right)^2 \geq 2B.
\end{aligned}
$$

Now, multiply inequality (6) by (6)′, keeping (7) and (8) in mind:

$$
\begin{aligned}
f(r_1) &> \left(\frac{B}{A} - \log 4\right)\left(\log \frac{2}{b - r_0} + \frac{B}{A} - \log 4\right) \\
&\geq 2A \log \frac{2}{b - r_0} + 2B \\
\text{(9)} \qquad &= 2A \log \frac{1}{b - r_1} + 2B.
\end{aligned}
$$

From the (5) to (9) implication it follows that if we define

$$
r_n := \frac{r_{n-1} + b}{2}, \qquad n = 1, 2, \ldots,
$$

then

$$
f(r_n) > 2A \log \frac{1}{b - r_n} + 2B.
$$

Since $b - r_n = (b - r_0)/2^n$, this gives

$$
f(r_n) > 2A \log \frac{2^n}{b - r_0} + 2B
$$

and this makes f unbounded on (a, b), a contradiction.

Thus (2) is established. For (3), just replace B by $B + 2A(A+2)$ in (1) and quote (2) for this new constant.

Lemma 12.11 *For each non-negative integer n there exists a finite positive constant B_n such that every f which is holomorphic and zero-free in $\mathbb{D}$ and satisfies*

$$
\text{(i)} \qquad \left|f^{(n)}(z)\right| \geq 1 \qquad \forall z \in \mathbb{D}
$$

also satisfies

$$
\text{(ii)} \qquad \log M(r, f) \leq \frac{1}{1-r}\left[8 \log^+ |f(0)| + B_n \log \frac{2}{1-r}\right] \qquad \forall 0 \leq r < 1.
$$

Proof: We have from 12.2 and 12.1(v)

$$V(r,f) = L(r,f) + L\left(r, \frac{1}{f}\right) \tag{1}$$

and from 12.3 and 12.1(v)

$$\log|f(0)| = \frac{1}{2\pi}\int_0^{2\pi} \log\left|f\left(re^{i\theta}\right)\right| d\theta = L(r,f) - L\left(r, \frac{1}{f}\right).$$

Therefore

$$V(r,f) = 2L\left(r, \frac{1}{f}\right) + \log|f(0)|. \tag{2}$$

By hypothesis

$$\left|\frac{1}{f}\right| \leq \left|\frac{f^{(n)}}{f}\right|$$

and therefore from 12.9

$$L\left(r, \frac{1}{f}\right) \leq L\left(r, \frac{f^{(n)}}{f}\right) \leq A_n + n(n+2)\log\frac{1}{R-r} + n\log^+ V(R,f) \tag{3}$$

holds for $0 < r < R < 1$. Use (3) in (2) to get

$$V(r,f) \leq 2A_n + \log^+|f(0)| + 2n(n+2)\log\frac{1}{R-r} + 2n\log^+ V(R,f), \tag{4}$$

holding for $0 < r < R < 1$. Now by (1) and 12.5, $V(r,f)$ is a non-decreasing function of r, so we can apply Bureau's lemma 12.10 to conclude that

$$\begin{aligned} V(r,f) &\leq 4A_n + 2\log^+|f(0)| + \\ &\qquad 4(2n(n+2))(2n(n+2)+2) + 4n(n+2)\log\frac{1}{1-r} \\ &= 2\log^+|f(0)| + a_n + b_n\log\frac{1}{1-r}, \quad 0 < r < 1 \end{aligned} \tag{5}$$

(with the obvious definitions of a_n and b_n). Next apply 12.4(iii) with $R = (1+r)/2$ to see that

$$\log M(r,f) \leq \frac{4}{1-r}L\left(\frac{1+r}{2}, f\right) \overset{(1)}{\leq} \frac{4}{1-r}V\left(\frac{1+r}{2}, f\right), \quad 0 < r < 1. \tag{6}$$

Apply (5) and (6) together to get

$$(7)\quad \log M(r,f) \le \frac{1}{1-r}\left[8\log^+|f(0)| + 4a_n + 4b_n \log\frac{2}{1-r}\right], \qquad 0<r<1.$$

Choose for B_n any number which satisfies

$$B_n \ge 8b_n \quad \text{and} \quad B_n \ge \frac{8a_n}{\log 2}.$$

Then

$$\begin{aligned} B_n \log\frac{2}{1-r} &\ge \frac{1}{2}B_n\log 2 + \frac{1}{2}B_n \log\frac{2}{1-r} \\ &\ge 4a_n + 4b_n\log\frac{2}{1-r}, \qquad 0\le r<1 \end{aligned}$$

and (7) gives

$$(8)\qquad \log M(r,f) \le \frac{1}{1-r}\left[8\log^+|f(0)| + B_n\log\frac{2}{1-r}\right], \qquad 0<r<1.$$

(Trivially the inequality (8) holds as well at $r=0$.)

Lemma 12.12 *For each non-negative integer n there exists a finite positive constant C_n such that every f which is holomorphic and zero-free in $\mathbb{D}$ and satisfies*

$$\text{(i)}\qquad \left|f^{(n)}(z)\right| \le 1 \qquad \forall z\in\mathbb{D}$$

also satisfies

$$\text{(ii)}\qquad \log M(r,f) \le \frac{1}{1-r}\left[16\log^+|f(0)| + C_n\log\frac{2}{1-r}\right] \qquad \forall 0\le r<1.$$

Proof: We go by induction on n. If $n=0$, (i) says that any $C_0\ge 0$ will work in (ii). Suppose the result true for $0\le n<N$ and consider $n=N$. Let such a function and an $r\in[0,1)$ be given and set $r' := (1+r)/2$. Consider the two cases:

Case I: There exists $z'\in D(0,r')$ such that

$$(1)\qquad \left|f^{(N-1)}(z')\right| \le 1.$$

Then for all $z \in \mathbb{D}$

$$\begin{aligned}\left|f^{(N-1)}(z)\right| &\le \left|f^{(N-1)}(z')\right| + \left|f^{(N-1)}(z) - f^{(N-1)}(z')\right| \\ &\le 1 + \left|\int_{[z',z]} f^{(N)}\right| \le 1 + |z-z'| \quad \text{by (i)} \\ &< 3.\end{aligned}$$

Apply the induction hypothesis to the function $\frac{1}{3}f$:

$$\log M\left(r, \tfrac{1}{3}f\right) \le \frac{1}{1-r}\left[16\log^+\left|\tfrac{1}{3}f(0)\right| + C_{N-1}\log\frac{2}{1-r}\right],$$

whence, since $\log^+\left|\frac{1}{3}f(0)\right| \le \log^+|f(0)|$,

$$\begin{aligned}\log M(r,f) &= \log M\left(r, \tfrac{1}{3}f\right) + \log 3 \\ &\le \frac{1}{1-r}\left[16\log^+|f(0)| + C_{N-1}\log\frac{2}{1-r}\right] + \log 3 \\ &\le \frac{1}{1-r}\left[16\log^+|f(0)| + \left(C_{N-1} + \frac{\log 3}{\log 2}\right)\log\frac{2}{1-r}\right]. \qquad (2)\end{aligned}$$

Case II:

$$\left|f^{(N-1)}(z)\right| > 1 \qquad \forall z \in D(0,r'). \qquad (3)$$

In this case the function $F(z) := f(r'z)/(r')^{N-1}$ ($z \in \mathbb{D}$) satisfies the hypotheses of the preceding lemma with $n = N-1$ and so we have

$$\log M(R,F) \le \frac{1}{1-R}\left[8\log^+|F(0)| + B_{N-1}\log\frac{2}{1-R}\right], \quad 0 \le R < 1. \qquad (4)$$

Also for $0 \le R < 1$

$$\sup_{|z|=R}|F(z)| = \frac{1}{(r')^{N-1}}\sup_{|z|=r'R}|f(z)|.$$

In particular then, with $R := r/r'$,

$$\log M\left(r, \frac{1}{(r')^{N-1}}f\right) = \log M\left(\frac{r}{r'}, F\right)$$

and (4) gives

$$\begin{aligned}
&\log M(r,f)\\
&\quad\le \log M\left(r,\frac{1}{(r')^{N-1}}f\right)\\
&\quad\le \frac{1}{1-r/r'}\left[8\log^+\left|\frac{f(0)}{(r')^{N-1}}\right| + B_{N-1}\log\frac{2}{1-r/r'}\right]\\
&\quad\le \frac{1+r}{1-r}\left[8\log^+|f(0)| + (N-1)8\log\frac{2}{1+r} + B_{N-1}\log\frac{2(1+r)}{1-r}\right],\\
&\qquad\text{by definition of } r' \text{ and by 12.1(ii)},\\
&\quad\le \frac{1}{1-r}\left[16\log^+|f(0)| + 16(N-1)\log\frac{2}{1+r} + 2B_{N-1}\log\frac{4}{1-r}\right]\\
&\quad\le \frac{1}{1-r}\Bigg[16\log^+|f(0)| + (16(N-1)+2B_{N-1})\log 2 +\\
&\qquad\qquad 2B_{N-1}\log\frac{2}{1-r}\Bigg]\\
(5)\quad&\quad\le \frac{1}{1-r}\left[16\log^+|f(0)| + (16(N-1)+4B_{N-1})\log\frac{2}{1-r}\right].
\end{aligned}$$

From (2) and (5) we are lead to define

$$C_N := \max\left\{C_{N-1} + \frac{\log 3}{\log 2},\ 16(N-1)+4B_{N-1}\right\}$$

and thereby have (ii) valid for $n = N$.

Lemma 12.13 *For each non-negative integer n there exists a finite positive constant K_n such that every f which is holomorphic and zero-free in $\mathbb{D}$ and satisfies*

(i) $$f^{(n)} - \mathbf{1} \quad \textit{is zero-free and} \quad \left|\frac{f^{(n+1)}}{f^{(n)}-\mathbf{1}}\right| \le 1 \quad \textit{in } \mathbb{D}$$

also satisfies

(ii) $$\log M(r,f) \le \frac{1}{1-r}\left[32\log^+|f(0)| + K_n\log\frac{2}{1-r}\right] \qquad \forall 0 \le r < 1.$$

Proof: Let such a function and an $r \in [0,1)$ be given and set $r' := (1+r)/2$. Consider the two cases as before:

Case I: There exists $z' \in D(0, r')$ such that

$$\left|f^{(n)}(z')\right| \le 1. \tag{1}$$

Let F be a holomorphic logarithm for $f^{(n)} - \mathbf{1}$ in $\mathbb{D}$:

$$f^{(n)} - \mathbf{1} = e^F, \tag{2}$$

$$\operatorname{Re} F(z') = \log\left|f^{(n)}(z') - 1\right| \overset{(1)}{\le} \log 2. \tag{3}$$

As in the preceding proof

$$\begin{aligned} \operatorname{Re} F(z) &\le \operatorname{Re} F(z') + \left|\operatorname{Re} \int_{[z,z']} F'\right| \le \log 2 + \left|\operatorname{Re} \int_{[z,z']} \frac{f^{(n+1)}}{f^{(n)} - \mathbf{1}}\right| \\ &\le \log 2 + |z' - z| \quad \text{by (i)} \\ &< \log 2 + 2 \qquad \forall z \in \mathbb{D}. \end{aligned} \tag{4}$$

It follows that

$$\left|f^{(n)}\right| \le \mathbf{1} + \left|f^{(n)} - \mathbf{1}\right| \overset{(2)}{=} \mathbf{1} + e^{\operatorname{Re} F} \overset{(4)}{\le} \mathbf{1} + 2e^2. \tag{5}$$

Let $c := 1 + 2e^2$ and form

$$\widetilde{F}(z) := \frac{f(r'z)}{c(r')^n}, \qquad z \in \mathbb{D}. \tag{6}$$

Because of (5) we may apply the preceding lemma to $\widetilde{F}$:

$$\log M\big(R, \widetilde{F}\big) \le \frac{1}{1-R}\left[16 \log^+\left|\widetilde{F}(0)\right| + C_n \log \frac{2}{1-R}\right] \qquad \forall 0 \le R < 1. \tag{7}$$

We have for $0 \le R < 1$

$$\sup_{|z|=R} \left|\widetilde{F}(z)\right| = \frac{1}{c(r')^n} \sup_{|z|=r'R} |f(z)|.$$

In particular then, with $R = r/r'$,

$$\log M\left(r, \frac{1}{c(r')^n} f\right) = \log M\left(\frac{r}{r'}, \widetilde{F}\right)$$

and (7) gives

$$\begin{aligned}\log M(r,f) &\le \log c + \log M\left(r, \frac{1}{c(r')^n} f\right) = \log c + \log M\left(\frac{r}{r'}, \widetilde{F}\right)\\ &\le \log c + \frac{1}{1-r/r'}\left[16\log^+\left|\frac{f(0)}{c(r')^n}\right| + C_n \log\frac{2}{1-r/r'}\right]\\ &\le \log c + \frac{1}{1-r/r'}\left[16\log^+\left|\frac{f(0)}{(r')^n}\right| + C_n \log\frac{2}{1-r/r'}\right], \qquad \text{since } c>1.\end{aligned}$$

Appraisals like those in the proof of inequality (5) of the preceding lemma then lead us to

$$\begin{aligned}\log M(r,f) &\le \log c + \frac{1}{1-r}\left[32\log^+|f(0)| + (32n+4C_n)\log\frac{2}{1-r}\right]\\ \text{(8)}\qquad &\le \frac{1}{1-r}\left[32\log^+|f(0)| + \left(32n+4C_n+\frac{\log c}{\log 2}\right)\log\frac{2}{1-r}\right].\end{aligned}$$

Case II:

$$\text{(9)}\qquad \left|f^{(n)}(z)\right| > 1 \qquad \forall z \in D(0,r').$$

This is exactly case II of the preceding proof and so, as was deduced there,

$$\text{(10)}\qquad \log M(r,f) \le \frac{1}{1-r}\left[16\log^+|f(0)| + (16n+4B_n)\log\frac{2}{1-r}\right].$$

From (8) and (10) it appears that the choice

$$K_n := \max\left\{16n+4B_n,\ 32n+4C_n+\frac{\log c}{\log 2}\right\}$$

will validate (ii).

Theorem 12.14 (MIRANDA [1935]) *For each non-negative integer n there exists a finite positive constant M_n such that every f which is holomorphic in $\mathbb{D}$ and for which f and $f^{(n)} - \mathbf{1}$ are zero-free satisfies*

$$\log M(r,f) \le \frac{1}{1-r}\left[64\log^+|f(0)| + \frac{M_n}{(1-r)^4}\right] \qquad \forall 0 \le r < 1.$$

Proof: Let such a function f and an $r \in [0,1)$ be given and set $r' := (1+r)/2$. Three cases arise, of which the first two are trivial reductions to previous lemmas:

Case I:

$$(1) \qquad \left|f^{(n)}(z)\right| \le 2 \qquad \forall z \in D(0,r').$$

Look at the argument in Case I of the 12.13 proof, from its inequality (5) onward. The function F used there to get (5) is not needed, for now we have (1) above to replace (and refine) inequality (5); that is, in defining the function $\tilde{F}$ in (6), one may employ $c := 2$ instead of the $c := 1 + 2e^2$ used there, and then go on to complete the Case I proof, getting now, in place of (8) there, the inequality

$$(2) \qquad \log M(r,f) \le \frac{1}{1-r}\left[32\log^+|f(0)| + (32n + 4C_n + 1)\log\frac{2}{1-r}\right].$$

Case II:

$$(3) \qquad \left|\frac{f^{(n+1)}(z)}{f^{(n)}(z)-1}\right| \le 1 \qquad \forall z \in D(0,r').$$

Then for the function $F(z) := f(r'z)/(r')^n$ on $\mathbb{D}$ we have $F^{(n)}(z) = f^{(n)}(r'z)$, whence $F^{(n)} - \mathbf{1}$, F are zero-free in $\mathbb{D}$ and

$$\left|\frac{F^{(n+1)}(z)}{F^{(n)}(z)-1}\right| = r'\left|\frac{f^{(n+1)}(r'z)}{f^{(n)}(r'z)-1}\right| \le r' < 1 \qquad \forall z \in \mathbb{D},$$

and so 12.13 is applicable to F. After the canonical maneuvers there results

$$\begin{aligned} \log M(r,f) &\le \frac{1}{1-r}\left[64\log^+|f(0)| + 64n\log\frac{2}{1+r} + 2K_n\log\frac{4}{1-r}\right] \\ (4) \qquad &< \frac{1}{1-r}\left[64\log^+|f(0)| + (64n + 4K_n)\log\frac{2}{1-r}\right]. \end{aligned}$$

Case III:

$$(5) \qquad \left|f^{(n)}(z')\right| > 2 \quad \text{for some } z' \in D(0,r')$$

and

$$(6) \qquad \left|\frac{f^{(n+1)}(z'')}{f^{(n)}(z'')-1}\right| > 1 \quad \text{for some } z'' \in D(0,r').$$

From (6) we have in particular that $f^{(n+1)}$ is not identically 0 in $\mathbb{D}$ and so, as noted before (penultimate paragraph of the proof of 12.4), $f^{(n+1)}$ has a zero on $C(0,\rho)$ for at most countably many $\rho \in [0,1)$. In what follows consider

$$(7) \qquad \rho > \frac{1+r'}{2} \quad \text{such that } 0 \notin f^{(n+1)}(C(0,\rho)).$$

Then on $C(0,\rho)$ we may write

$$\frac{1}{f} = \frac{f^{(n)}}{f} - \frac{f^{(n)} - \mathbf{1}}{f^{(n+1)}} \cdot \frac{f^{(n+1)}}{f}$$

and apply 12.1(ii) and (iii) to see that

$$L\left(\rho, \frac{1}{f}\right) \le \log 2 + L\left(\rho, \frac{f^{(n)}}{f}\right) + L\left(\rho, \frac{f^{(n+1)}}{f}\right) + L\left(\rho, \frac{f^{(n)} - \mathbf{1}}{f^{(n+1)}}\right).$$

As noted in the opening lines of the proof of 12.11,

$$V(\rho, f) = 2L\left(\rho, \frac{1}{f}\right) + \log|f(0)|,$$

and so

(8)

$$V(\rho, f) \le \log|f(0)| + \log 4 + 2L\left(\rho, \frac{f^{(n)}}{f}\right) + 2L\left(\rho, \frac{f^{(n+1)}}{f}\right) + 2L\left(\rho, \frac{f^{(n)} - \mathbf{1}}{f^{(n+1)}}\right).$$

Apply 12.4(ii) to $F := f^{(n+1)}/\big(f^{(n)} - \mathbf{1}\big)$, $z = z''$ and $R = \rho$:

$$0 \overset{(6)}{<} \log\left|\frac{f^{(n+1)}(z'')}{f^{(n)}(z'') - 1}\right| \le \frac{\rho + |z''|}{\rho - |z''|} L\left(\rho, \frac{f^{(n+1)}}{f^{(n)} - \mathbf{1}}\right) - \frac{\rho - |z''|}{\rho + |z''|} L\left(\rho, \frac{f^{(n)} - \mathbf{1}}{f^{(n+1)}}\right),$$

$$\begin{aligned} L\left(\rho, \frac{f^{(n)} - \mathbf{1}}{f^{(n+1)}}\right) &\le \left(\frac{\rho + |z''|}{\rho - |z''|}\right)^2 L\left(\rho, \frac{f^{(n+1)}}{f^{(n)} - \mathbf{1}}\right) \\ &\le \frac{4}{(\rho - r')^2} L\left(\rho, \frac{f^{(n+1)}}{f^{(n)} - \mathbf{1}}\right) \\ &\le \frac{64}{(1-r)^2} L\left(\rho, \frac{f^{(n+1)}}{f^{(n)} - \mathbf{1}}\right), \quad \text{by (7) and definition of } r'. \qquad (9) \end{aligned}$$

Apply 12.9 with the n there equal 1 and the f there equal $f^{(n)} - \mathbf{1}$:

$$(10) \quad L\left(\rho, \frac{f^{(n+1)}}{f^{(n)} - \mathbf{1}}\right) \le A_1 + 3\log\frac{1}{R' - \rho} + \log^+ V\big(R', f^{(n)} - \mathbf{1}\big), \ \rho < R' < 1.$$

Similarly appraise the terms $L\big(\rho, f^{(n)}/f\big)$ and $L\big(\rho, f^{(n+1)}/f\big)$ in (8) via 12.9 and use (9) and (10) to get

$$\begin{aligned} V(\rho, f) \le{} & \log|f(0)| + (\log 4 + 2A_n + 2A_{n+1}) \\ & + 2n(n+2)\log\frac{1}{R-\rho} + 2(n+1)(n+3)\log\frac{1}{R-\rho} \\ & + 2n\log^+ V(R,f) + 2(n+1)\log^+ V(R,f) \\ & + \frac{128}{(1-r)^2}\left[A_1 + 3\log\frac{1}{R'-\rho} + \log^+ V\big(R', f^{(n)} - \mathbf{1}\big)\right], \end{aligned}$$

holding for all $\rho < R' < R < 1$. Take $R' := \frac{1}{2}(\rho + R)$ and get

$$\begin{aligned} V(\rho, f) \leq \log|f(0)| &+ (\log 4 + 2A_n + 2A_{n+1}) \\ &+ 2(n(n+2) + (n+1)(n+3)) \log \frac{1}{R-\rho} \\ &+ 2(2n+1) \log^+ V(R, f) + \frac{128A_1}{(1-r)^2} + \frac{384}{(1-r)^2} \log \frac{2}{R-\rho} \\ &+ \frac{128}{(1-r)^2} \log^+ V\big(R', f^{(n)} - \mathbf{1}\big), \end{aligned}$$

holding for all $\rho < R < 1$. Set

(11) $H_n := \log 4 + 2A_n + 2A_{n+1} + 2n(n+2) + 2(n+1)(n+3) + 128A_1 + 384$

and have then *a fortiori*

$$\begin{aligned} V(\rho, f) \leq \log|f(0)| + H_n &+ \frac{H_n}{(1-r)^2} \log \frac{2}{R-\rho} \\ &+ \frac{H_n}{(1-r)^2} \Big[\log^+ V(R, f) + \log^+ V\big(R', f^{(n)} - \mathbf{1}\big)\Big], \end{aligned} \tag{12}$$

holding for all $\rho < R < 1$, $R' = \frac{1}{2}(\rho + R)$. As in the derivation of (9), use the fact $\big|f^{(n)}(z') - 1\big| > 1$ (from (5)) and 12.4(ii) to deduce that

$$\begin{aligned} L\left(R', \frac{1}{f^{(n)} - \mathbf{1}}\right) &\leq \left(\frac{R' + |z'|}{R' - |z'|}\right)^2 L\big(R', f^{(n)} - \mathbf{1}\big) \\ &\leq \frac{4}{(R'-\rho)^2} L\big(R', f^{(n)} - \mathbf{1}\big) = \frac{16}{(R-\rho)^2} L\big(R', f^{(n)} - \mathbf{1}\big). \end{aligned}$$

Therefore

$$\begin{aligned} V\big(R', f^{(n)} - \mathbf{1}\big) &= L\big(R', f^{(n)} - \mathbf{1}\big) + L\left(R', \frac{1}{f^{(n)} - \mathbf{1}}\right) \\ &\leq L\big(R', f^{(n)} - \mathbf{1}\big) + \frac{16}{(R-\rho)^2} L\big(R', f^{(n)} - \mathbf{1}\big) \\ &\leq \frac{17}{(R-\rho)^2} L\big(R', f^{(n)} - \mathbf{1}\big) \\ &\leq \frac{17}{(R-\rho)^2} \Big[L\big(R', f^{(n)}\big) + \log 2\Big], \quad \text{by 12.1(iii)} \\ &\leq \frac{17}{(R-\rho)^2} \left[L\left(R', \frac{f^{(n)}}{f}\right) + L(R', f) + \log 2\right], \quad \text{by12.1(ii)} \\ &\leq \frac{17}{(R-\rho)^2} \left[L\left(R', \frac{f^{(n)}}{f}\right) + V(R, f) + \log 2\right], \quad \text{since } R' < R. \end{aligned} \tag{13}$$

Now apply $\log^+$ to (13), again making use of 12.1:

$$\begin{aligned}
&\log^+ V\big(R', f^{(n)} - \mathbf{1}\big) \\
&\quad\le \log \frac{17}{(R-\rho)^2} + \log^+ L\Big(R', \frac{f^{(n)}}{f}\Big) + \log^+ V(R,f) + \log^+ \log 2 + \log 3 \\
&\quad\le \log \frac{17}{(R-\rho)^2} + L\Big(R', \frac{f^{(n)}}{f}\Big) + \log^+ V(R,f) + \log 3,
\end{aligned}$$

using the obvious facts $\log^+ x \le x$ and $\log 2 < 1$.

An application of 12.9 then gives

$$\begin{aligned}
\log^+ V\big(R', f^{(n)} - \mathbf{1}\big) &\le \log \frac{17}{(R-\rho)^2} + A_n + n(n+2)\log \frac{1}{R-R'} \\
&\quad + n\log^+ V(R,f) + \log^+ V(R,f) + \log 3 \\
&= A_n + \log \frac{51}{4} + 2\log \frac{2}{R-\rho} + n(n+2)\log \frac{2}{R-\rho} \qquad (14)\\
&\quad + (n+1)\log^+ V(R,f).
\end{aligned}$$

We set

$$H'_n := A_n + \log \frac{51}{4} + (n+1)(n+2) \qquad (15)$$

and have *a fortiori* from (14)

$$\log^+ V\big(R', f^{(n)} - \mathbf{1}\big) \le H'_n + H'_n \log \frac{2}{R-\rho} + H'_n \log^+ V(R,f). \qquad (16)$$

If we insert (16) into (12),

$$\begin{aligned}
V(\rho, f) &\le \log|f(0)| + H_n + \frac{H_n H'_n}{(1-r)^2} + \frac{H_n(H'_n+1)}{(1-r)^2}\log \frac{2}{R-\rho} \\
&\quad + \frac{H_n(H'_n+1)}{(1-r)^2}\log^+ V(R,f) \\
&\le \log^+|f(0)| + H_n + \frac{H_n(2H'_n+1)}{(1-r)^2} + \frac{H_n(H'_n+1)}{(1-r)^2}\log \frac{1}{R-\rho} \\
&\quad + \frac{H_n(H'_n+1)}{(1-r)^2}\log^+ V(R,f) \\
&\le \log^+|f(0)| + \frac{2H_n(H'_n+1)}{(1-r)^2} + \frac{H_n(H'_n+1)}{(1-r)^2}\log \frac{1}{R-\rho} \qquad (17)\\
&\quad + \frac{H_n(H'_n+1)}{(1-r)^2}\log^+ V(R,f).
\end{aligned}$$

This inequality is valid for a dense set of $\rho \in ((1+r')/2, 1)$ and for each such ρ it is valid for all $\rho < R < 1$. But an easy uniform continuity argument on the integrand defining $V(\rho, f)$ shows that it is a continuous function of ρ and so it follows that (17) is valid for all $\rho \in ((1+r')/2, 1)$ and all $\rho < R < 1$. Since (12.5) $V(\rho, f)$ is a non-decreasing function of ρ, we are in a position to apply Bureau's lemma 12.10. We get

$$\begin{aligned}(18)\qquad V(\rho, f) \leq 2\log^+|f(0)| + \frac{4H_n(H_n'+1)}{(1-r)^2} &+ \\ \frac{4H_n(H_n'+1)}{(1-r)^2}\left[\frac{H_n(H_n'+1)}{(1-r)^2} + 2\right] &+ \\ \frac{2H_n(H_n'+1)}{(1-r)^2}\log\frac{1}{1-\rho}, &\quad \frac{1+r'}{2} < \rho < 1.\end{aligned}$$

Let

$$(19)\qquad L_n := 4H_n(H_n'+1)$$

and have *a fortiori* from (18)

$$\begin{aligned}(20)\qquad V(\rho, f) \leq 2\log^+|f(0)| &+ \frac{L_n}{(1-r)^2} + \frac{L_n}{(1-r)^2}\left[\frac{L_n}{(1-r)^2} + 2\right] \\ &+ \frac{L_n}{(1-r)^2}\log\frac{1}{1-\rho}, \qquad \frac{1+r'}{2} < \rho < 1.\end{aligned}$$

An eligible ρ is

$$(21)\qquad \rho := \frac{3+r'}{4} = \frac{7+r}{8}.$$

We have $\rho - r = \frac{7}{8}(1-r) > (1-r)/2$, so we may apply 12.4(iii) to see that

$$\begin{aligned}\log M(r, f) &\leq \frac{\rho+r}{\rho-r}L(\rho, f) \leq \frac{\rho+r}{\rho-r}V(\rho, f) \\ (22)\qquad &\leq \frac{2}{\rho-r}V(\rho, f) \leq \frac{4}{1-r}V(\rho, f).\end{aligned}$$

Use (21) in (20) and insert the result into (22):

$$\begin{aligned}\log M(r, f) &\leq \frac{8}{1-r}\log^+|f(0)| + \frac{4L_n}{(1-r)^3} + \frac{4L_n}{(1-r)^3}\left[\frac{L_n}{(1-r)^2} + 2\right] \\ &\quad + \frac{4L_n}{(1-r)^3}\log\frac{8}{1-r} \\ &\leq \frac{8}{1-r}\log^+|f(0)| + \frac{12L_n}{(1-r)^3} + \frac{4L_n^2}{(1-r)^5} + \frac{32L_n}{(1-r)^4} \\ &\quad \text{(since } \log x \leq x \text{ for all } x > 0\text{)} \\ (23)\qquad &\leq \frac{8}{1-r}\log^+|f(0)| + \frac{48L_n^2}{(1-r)^5} \quad \text{(since } L_n \geq 1\text{)}.\end{aligned}$$

From (2), (4) and (23) we see that we may take for M_n the number

$$M_n := \max\{64n + 8C_n + 2,\ 128n + 8K_n,\ 48L_n^2\}.$$

Exercise 12.15 (Schottky) *When $n = 0$, the dependence on r in Miranda's theorem can be sharpened:*

$$\log M(r, f) \le \frac{1}{1-r}\left[S_0 \log^+ |f(0)| + S_1 \log \frac{2}{1-r}\right], \qquad 0 \le r < 1$$

for some finite positive constants S_0, S_1 which are independent of f. (Cf. 12.55.)

Hints: Refine the argument for *Case III* in the proof of Miranda's theorem, notably by citing 12.4(ii) (with $F := f'/(f - \mathbf{1})$) to assert that

$$(9)' \qquad L\left(r, \frac{f-\mathbf{1}}{f'}\right) \le L\left(r, \frac{f'}{f-\mathbf{1}}\right) + \log\left|\frac{f(0)-1}{f'(0)}\right| \quad \text{if } f'(0) \ne 0,$$

and using (9)$'$ instead of (9) in the rest of the proof. Similarly, get a simple estimate in (13) free of the factor $1/(R-\rho)^2$. In general, if f is not constant, there exist $z_n \to 0$ such that $f'(z_n) \ne 0$ and we apply the inequality already secured to the functions $f_n(z) := f(z_n + (1 - |z_n|)z)$. Eliminating the term $|f'(0)|$ is troublesome; help with that can be found on pp. 13–15 of VALIRON [1937].

§ 3 Immediate Applications of Miranda

Corollary 12.16 *If F is a zero-free entire function and for some non-negative integer n the range of $F^{(n)}$ does not contain all non-zero complex numbers, then F is a (non-zero) constant.*

Proof: If $\mathbb{C}\backslash\{0\} \not\subset F^{(n)}(\mathbb{C})$, then by scaling we may suppose that $1 \notin F^{(n)}(\mathbb{C})$. Consider $R \ge \frac{1}{2}$. Set $f(z) := F(2Rz)/(2R)^n$. Then F is holomorphic and zero-free in $\mathbb{D}$ and $f^{(n)}(z) = F^{(n)}(2Rz)$, so $f^{(n)} - \mathbf{1}$ is also zero-free. Take $r = \frac{1}{2}$ in Miranda's theorem to conclude that

$$\log M(\tfrac{1}{2}, f) \le 2\left[64 \log^+ |f(0)| + 16M_n\right] = 128 \log^+ \left|\frac{F(0)}{(2R)^n}\right| + 32M_n$$

$$(1) \qquad \le 128 \log^+ |F(0)| + 32M_n,$$

since $\log^+$ is non-decreasing and $2R \geq 1$. Let B denote the number on the right-hand side of (1). Then (1) says that $\sup_{|z|\leq 1/2} |f(z)| \leq e^B$, that is,

$$\sup_{|z|\leq R} \frac{|F(z)|}{(2R)^n} \leq e^B$$

$$\sup_{|z|\leq R} |F(z)| \leq \left(2^n e^B\right) R^n, \tag{2}$$

holding for all $R \geq 1/2$. It follows from this and 3.34(ii) that F is a polynomial of degree not greater than n. Since F has no zeros, it must in fact be constant.

Exercise 12.17 (Cf. 7.28) *Let $\mathcal{S}_0$ denote the set of all holomorphic functions f in $\mathbb{D}$ with $f(0) = 0$, $f'(0) = 1$ and $0 \notin f(\mathbb{D} \setminus \{0\})$. Show that there exists a real number $\alpha > 0$ such that $D(0, \alpha) \subset f(\mathbb{D})$ for every $f \in \mathcal{S}_0$.*

Hints: Suppose contrariwise that there exist $\alpha_n \in \mathbb{C} \setminus \{0\}$ with $\alpha_n \to 0$ and $f_n \in \mathcal{S}_0$ such that $\alpha_n \notin f_n(\mathbb{D})$. Then $\mathbf{1} - f_n/\alpha_n$ is zero-free and so there exists for it a holomorphic logarithm L_n in $\mathbb{D}$, which may be normalized by $L_n(0) = 0$. The numbers $\pm 2\pi i$ are then absent from the range of each L_n and so the function $(L_n + 2\pi i)/4\pi i$ omits 0 and 1. Consequently, 12.14 entails, as in the preceding proof, that $\{L_n\}$ is uniformly bounded in $D\left(0, \frac{1}{2}\right)$. From this and 5.37(i) follows the boundedness of the set $\left\{(\mathbf{1} - f_n/\alpha_n)'(0)\right\} = \{-1/\alpha_n\}$.

Corollary 12.18 (Picard's Little Theorem) *At most one complex number is absent from the range of any non-constant entire function.* (Cf. 7.29(v) and 12.30(i).) *At most two complex numbers are absent from the range of any non-constant function which is meromorphic in $\mathbb{C}$.*

Proof: If f is entire and a, $b \notin f(\mathbb{C})$, where $a \neq b$, then the entire function $F := f - a$ is zero-free and $0 \neq b - a \notin F(\mathbb{C})$, so by 12.16 F is constant. (A different proof is outlined in 7.29(v).)

Next consider a function G which is meromorphic in $\mathbb{C}$ and suppose the distinct complex numbers a, b, c are absent from its range. Let P denote the (isolated) set of poles of G. If $P = \varnothing$, then G is constant, by the preceding paragraph. So assume $P \neq \varnothing$. Look at the function

$$F := \frac{1}{G - a}, \quad \text{which is holomorphic in } \mathbb{C} \setminus P, \tag{$*$}$$

since $G - a$ is holomorphic and zero-free in $\mathbb{C} \setminus P$. We have $\lim_{z\to p} |G(z)| = \infty$ for each $p \in P$, by 11.7(ii). Therefore $\lim_{z\to p} F(z) = 0$ for all $p \in P$ and F extends (5.41) to be holomorphic in all of $\mathbb{C}$, with the value 0 at each point of P. Since b, $c \notin G(\mathbb{C} \setminus P)$, we see that the (distinct) complex numbers $1/(b-a)$, $1/(c - a) \notin F(\mathbb{C})$. By the result of the first paragraph, F is constant. Since

$F(p) = 0$ for all $p \in P \neq \varnothing$, this constant is 0, which contradicts $(*)$, according to which F is zero-free in $\mathbb{C} \setminus P$. Thus no such G exists.

Remarks: The non-constant functions e^z and $1/(1 - e^z)$ are examples where one or two values, respectively, are omitted. PICARD [1880], p. 153 noted the following amusing derivation of the fundamental theorem of algebra from 12.18: If the non-constant polynomial P has no zeros, write $P = e^f$ with f entire and not constant. Then f assumes infinitely many of the values $\{2\pi i n : n \in \mathbb{N}\}$, so $P - \mathbf{1}$ assumes 0 infinitely often, which is not possible for a non-zero polynomial. Alternatively (R. P. Boas), $P^{-1}[0, 1]$ is closed and bounded (=compact), so $P(P^{-1}[0, 1])$ is a compact subset of $[0, 1]$, hence cannot be $(0, 1]$. In particular, if the range of P misses 0, it also misses some *other* number in $[0, 1]$. Another amusing application: every periodic entire function has a fixed-point. For if f is entire and $p \neq 0$ is a period, then the range of $f(z) - z$ contains either 0 or p.

Corollary 12.19 (Pólya-Saxer) *If f is an entire function and $f \cdot f' \cdot f''$ is zero-free, then $f(z) = e^{az+b}$ for some constants a and b. If F is an entire function and none of F, F', F'' is surjective, then $F(z) = c + e^{az+b}$ for some constants a, b, c.*

Proof: f never vanishes, hence is of the form $f = e^g$ for some entire g and $g' = f' \cdot e^{-g}$ never vanishes, hence has the form $g' = e^h$ for some entire h. Then $f' = g'e^g = e^{g+h}$, so $g' + h' = f'' \cdot e^{-(g+h)}$ never vanishes, that is, $e^h + h'$ never vanishes. Therefore $e^h\big(\mathbf{1} + h'e^{-h}\big) = e^h\big(\mathbf{1} - (e^{-h})'\big)$ never vanishes. It follows that $(e^{-h})'$ does not take the value 1. 12.16 then implies that e^{-h} must be a constant. Since $g' = e^h$, it follows that g is linear and since $f = e^g$, the desired conclusion about the form of f is obtained.

If F is constant, it has the stated form, so suppose F is not constant. Then F' is not constant either: For if $F' \equiv \alpha$, then $\alpha \neq 0$ and $F(z) \equiv \alpha z + \beta$ would be surjective. The hypothesis supplies $c \in \mathbb{C} \setminus F(\mathbb{C})$. Then $f := F - c$ is non-constant and zero-free, so according to 12.16 the (only) value omitted by $F' = f'$ must be 0. Since f' is not constant, another appeal to 12.16 insures that the (only) value omitted by $F'' = f''$ must also be 0. In summary, $f \cdot f' \cdot f''$ is zero-free, so f, hence also F, has the stated form.

Corollary 12.20 *Let f, g be entire functions, each of which is fixed-point-free and g is not a translation. Then $f \circ g$ has infinitely many fixed-points.*

Proof: (Saeki) The entire functions $f(z) - z$ and $g(z) - z$ are zero-free and so have the form $e^{\phi(z)}$, $e^{\psi(z)}$ for some ϕ, $\psi \in H(\mathbb{C})$. Let $n \in \mathbb{N}$ be given and set $F := e^{\phi/n}$, $G := e^{\psi/n}$. In terms of these functions

$$(f \circ g)(z) = z + G^n(z) + F^n(g(z))$$

and so z_0 is a fixed-point of $f \circ g$ if and only if

$$(1) \qquad G^n(z_0) = -F^n(g(z_0)).$$

Suppose that there is a constant C such that

$$(2) \qquad G = C\,F \circ g.$$

Then g is a one-to-one function. For from $g(z_1) = g(z_2)$ and (2) follows $G(z_1) = G(z_2)$ and then $z_1 = g(z_1) - G^n(z_1) = g(z_2) - G^n(z_2) = z_2$. According to 11.19 then g has the form $g(z) \equiv az + b$ for certain constants a, b. Since g is fixed-point free, this means $a = 1$, contrary to the hypothesis that g is not a translation. We conclude that no constant fulfilling (2) exists. That is, the zero-free entire function $G/(F \circ g)$ is not constant. Therefore Picard's Little Theorem insures that for each $k \in \{1, 2, \ldots, n\}$ there exists $z_k \in \mathbb{C}$ such that

$$(3) \qquad G(z_k) = e^{\pi i(2k+1)/n} F(g(z_k)).$$

That is, each z_k is a z_0 satisfying (1) and so each z_k is a fixed-point of $f \circ g$. Clearly because of (3), the z_k are all distinct.

Lemma 12.21 (BOREL [1896–97]) *If C_1, C_2, $C_3 \in \mathbb{C} \setminus \{0\}$, f_1, f_2, f_3 are entire and $C_1 e^{f_1} + C_2 e^{f_2} + C_3 e^{f_3} = \mathbf{0}$, then $f_3 - f_1$ and $f_3 - f_2$ are each constant.*

Proof: Writing $C_j := e^{c_j}$ and replacing f_j with $f_j + c_j$, we can suppose that $C_1 = C_2 = C_3 = 1$. Divide by e^{f_3} and set $g_1 := f_1 - f_3$, $g_2 := f_2 - f_3$. Then $e^{g_1} + e^{g_2} + \mathbf{1} = \mathbf{0}$. Thus

$$(*) \qquad e^{g_1} = -\mathbf{1} - e^{g_2}.$$

Since e^{g_1} is zero-free, $(*)$ shows that g_2 assumes none of the values $(2n+1)\pi i$, $n \in \mathbb{Z}$. Consequently by Picard's Little Theorem g_2 is constant. By symmetry, g_1 is also constant.

Theorem 12.22 *If $m \geq 3$, $n \geq 2$ are integers and f, g are entire functions satisfying $f^n + g^m = \mathbf{1}$, then f and g are each constant.*

Proof: (MONTEL [1927], p. 136.)

$$(1) \qquad f^n - \mathbf{1} = -g^m = \left(g e^{\frac{\pi i}{m}}\right)^m,$$

from which it follows that the order of every zero of $f^n - \mathbf{1}$ is divisible by m. Now consider a zero α of $f - \mathbf{1}$, with order k, say. Write

$$f(z) - 1 = (z - \alpha)^k h(z), \quad \text{where } h(\alpha) \neq 0.$$

Then

$$f^n(z) = 1 + n(z-\alpha)^k h(z) + \sum_{j=2}^{n} \binom{n}{j} (z-\alpha)^{kj} h^j(z)$$

$$f^n(z) - 1 = (z-\alpha)^k \left[nh(z) + (z-\alpha)^k \sum_{j=2}^{n} \binom{n}{j} (z-\alpha)^{k(j-2)} h^j(z) \right].$$

Since the value at $z = \alpha$ of the bracketed term is $nh(\alpha) \neq 0$, we see that the order of α as a zero of $f^n - \mathbf{1}$ is k, which is therefore divisible by m, as noted above. From 10.6 we therefore know that there is an entire function ϕ such that

$$f - \mathbf{1} = \phi^m. \tag{2}$$

Now let $\varepsilon \neq 1$ be an nth root of 1. Then $\varepsilon^{-1} f$ satisfies the same equation that f does, hence $\varepsilon^{-1} f - \mathbf{1} = \varepsilon^{-1}(f - \varepsilon)$ is the mth power of some entire function. Thus there is an entire function ψ such that

$$f - \varepsilon = \psi^m. \tag{3}$$

From (2) and (3) we see that $\phi(z) \neq \psi(z)$ for all z, so

$$\phi - \psi = e^{h_0} \tag{4}$$

for some entire h_0. Letting $\eta := e^{\frac{2\pi i}{m}}$, the functions $\eta\psi$ and $\eta^2\psi$ in the role of ψ also satisfy (3) and so we infer that

$$\phi - \eta\psi = e^{h_1} \tag{5}$$

$$\phi - \eta^2\psi = e^{h_2} \tag{6}$$

for a pair of entire functions h_1, h_2. Then

$$e^{h_2} = \phi - \eta^2\psi \overset{(5)}{=} \phi + \eta\left[e^{h_1} - \phi\right] = (1-\eta)\phi + \eta e^{h_1} \overset{(4),(5)}{=} \left(e^{h_1} - \eta e^{h_0}\right) + \eta e^{h_1}$$

and so

$$e^{h_2} - (1+\eta)e^{h_1} + \eta e^{h_0} = \mathbf{0}. \tag{7}$$

Notice that $1 + \eta \neq 0$, since $m \neq 2$. Hence the preceding lemma is applicable to (7), and we learn that $h_2 - h_0 =: c_2$, $h_1 - h_0 =: c_1$ are constants. Returning to (4) and (5), this says that

$$\frac{\phi - \eta\psi}{\phi - \psi} = e^{c_1}, \quad \text{whence}$$

$$(1 - e^{c_1})\phi = (\eta - e^{c_1})\psi. \tag{8}$$

First suppose $e^{c_1} \neq 1$. Then, with $C := \big(\eta - e^{c_1}\big)/\big(1 - e^{c_1}\big)$, we have from (8) that $\phi = C\psi$. Combined with (2) and (3) this gives

$$\mathbf{1} - \varepsilon = \psi^m - \phi^m = (1 - C^m)\psi^m.$$

Since $\varepsilon \neq 1$, this means ψ^m is constant. On the other hand, if $e^{c_1} = 1$, then $\psi = \mathbf{0}$ by (8). In either case therefore, f is constant by (3) and then by (1), g^m is also constant. Therefore the set $g(\mathbb{C})$ is finite, yet by 1.4 it is connected; hence it's a single point.

Corollary 12.23 *Let f and g be entire functions and for some positive integer n satisfy the identity $f^n + g^n = \mathbf{1}$.*

(i) *If $n = 2$, then there is an entire function h such that $f = \cos \circ h$ and $g = \sin \circ h$.*

(ii) *If $n > 2$, then f and g are each constant.*

Proof: (i) When $n = 2$, the identity is

$$(f + ig)(f - ig) = f^2 + g^2 = \mathbf{1}. \tag{1}$$

Thus the entire function $f + ig$ has no zeros and consequently is of the form e^{ih} for some entire h. It follows then from (1) that $f - ig = 1/(f + ig) = e^{-ih}$. Therefore

$$f = \frac{1}{2}(f + ig) + \frac{1}{2}(f - ig) = \frac{e^{ih} + e^{-ih}}{2} = \cos \circ h,$$

$$g = \frac{1}{2i}(f + ig) - \frac{1}{2i}(f - ig) = \frac{e^{ih} - e^{-ih}}{2i} = \sin \circ h.$$

(ii) is immediate from the preceding theorem, but it admits a more transparent proof:

Suppose $n \geq 3$, f and g entire, and $f^n + g^n = \mathbf{1}$. Since $u_k := e^{(2k+1)\pi i/n}$ $(k = 1, 2, \dots, n)$ are all the nth roots of -1, the polynomial $X^n + 1$ coincides with $\prod_{k=1}^n (X - u_k)$. Therefore $\mathbf{1} = f^n + g^n = \prod_{k=1}^n (f - u_k g)$. If g is not the function $\mathbf{0}$, then f/g is a meromorphic function in $\mathbb{C}$ and this factorization shows that no u_k lies in its range. It follows from 12.18 that f/g is a constant c, different from all the u_k, i.e., $c^n \neq -1$. Then $\mathbf{1} = f^n + g^n = (c^n + 1)g^n$ means that g^n is constant, etc.

Remarks: The claim that for integers $n > 2$ the equation $f^n + g^n = h^n$ has no non-constant entire solutions f, g, h might be called Fermat's Last Theorem for entire functions, save for the fact that it is not quite a theorem. This assertion is equivalent (with $F := f/h$, $G := g/h$) to the statement that there exist no non-constant *meromorphic* solutions of the equation $F^n + G^n = \mathbf{1}$. This is in fact the case for $n > 3$ but false for $n = 3$. See GROSS and OSGOOD [1981] and REMMERT [1998], Chapter 10, §2. In the same vein, for every positive integer n there exist non-constant f, g, $h \in H(\mathbb{D})$ such that $f^n + g^n = h^n$ (see p. 24 of GROSS and OSGOOD). For reasonably elementary accounts of the special case $n = 4$ see VAN DER POL [1955], BELLMAN [1961], C. Y. LIN [1979], and F. GROSS [1966a]. In the latter paper there is also an explicit example of three non-constant polynomials f, g, h such that $f^3 + g^3 + h^3 = \mathbf{1}$. (But see **MR** 2004f:30019 and 2005e:30041.)

If Ω is any simply-connected region and f, $g \in H(\Omega)$ satisfy $f^2 + g^2 = \mathbf{1}$, then $f = \cos \circ h$ and $g = \sin \circ h$ for some $h \in H(\Omega)$; the proof of (i) given above establishes this. Jürgen Elstrodt has observed that this property is yet another characterization of simple-connectivity: If Ω is a region and every pair f, $g \in H(\Omega)$ satisfying $f^2 + g^2 = \mathbf{1}$ has the form specified in (i), then Ω is simply-connected. For if $\phi \in H(\Omega)$ is zero-free, then $f := \frac{1}{2}(\phi + 1/\phi)$ and $g := \frac{1}{2i}(\phi - 1/\phi)$ satisfy $f^2 + g^2 = \mathbf{1}$. It follows that $\phi = f + ig = \cos \circ h + i \sin \circ h = \exp \circ (ih)$ for the relevant $h \in H(\Omega)$. In other words, Ω has property (HL).

§ 4 Normal Families and Julia's Extension of Picard's Great Theorem

Definition 12.24 A set $\mathfrak{F}$ of functions holomorphic in an open subset V of $\mathbb{C}$ is called *normal* (in V) if every sequence in $\mathfrak{F}$ contains a subsequence which either converges locally uniformly in V or diverges to ∞ locally uniformly in V.

Remark: There is an alternative definition that is easily proved equivalent to this one and is sometimes more convenient: $\mathfrak{F}$ is normal if every sequence in $\mathfrak{F}$ either diverges to ∞ locally uniformly or contains a subsequence which converges locally uniformly.

Lemma 12.25 (MONTEL [1916], p. 227) *Let Ω be a region, $\mathfrak{F}$ a family of functions holomorphic in Ω with the property that about each point $z \in \Omega$ there is an open disk D_z in which $\mathfrak{F}$ is normal. Then $\mathfrak{F}$ is normal in Ω.*

Proof: Let $\{f_n\}$ be a sequence in $\mathfrak{F}$. Since Ω is a countable union of compacta (e.g., the sets $\{z \in \mathbb{C} : |z| \leq n \text{ and } d(z, \mathbb{C} \setminus \Omega) \geq 1/n\}$), it is easy to see that countably many of the disks D_z cover Ω. Since $\mathfrak{F}$ is normal in each of these, a diagonal argument produces a single subsequence $\{f_{n_j}\}$ of $\{f_n\}$ with the property that every point z of Ω lies in some open disk on which $\{f_{n_j}\}$ either converges uniformly or diverges uniformly to ∞. If the two classes of points

z are denoted by C and D respectively, then by their very definitions each is open, $C \cap D = \varnothing$ and $C \cup D = \Omega$. It follows from the connectedness of Ω that either $C = \Omega$ or $D = \Omega$. A simple covering-compactness argument shows that in the first case $\{f_{n_j}\}$ converges uniformly on each compact subset of Ω and in the second case $\{f_{n_j}\}$ diverges uniformly to ∞ on each compact subset of Ω.

Theorem 12.26 (Montel [1911]) *If U is an open subset of $\mathbb{C}$, $\mathcal{F} \subset H(U)$ is a normal family and for some $a \in U$ the set $\mathcal{F}(a) := \{f(a) : f \in \mathcal{F}\}$ is bounded, then $\mathcal{F}$ is locally uniformly bounded on U.*

Proof: Suppose contrariwise that $\mathcal{F}$ is not uniformly bounded on some compact $K \subset U$. So there exist $f_n \in \mathcal{F}$ with

$$(*) \qquad \sup |f_n(K)| > n.$$

A subsequence $\{f_{n_j}\}$ of $\{f_n\}$ either converges locally uniformly on U or diverges locally uniformly to ∞ on U. Since $\mathcal{F}(a)$ is bounded, the sequence $\{f_{n_j}(a)\}$ is bounded. Hence the second part of the alternative is untenable, the first part prevails and $(*)$ is contradicted.

Theorem 12.27 *Let Ω be a region, N a non-negative integer, $a, b \in \mathbb{C}$, $b \neq 0$ if $N \neq 0$, and $a \neq b$ if $N = 0$. Let $\mathcal{F}$ be a family of holomorphic functions in Ω such that for each $f \in \mathcal{F}$, $a \notin f(\Omega)$ and $b \notin f^{(N)}(\Omega)$. Then $\mathcal{F}$ is a normal family.*

Proof: For each $f \in \mathcal{F}$ define

$$\tilde{f} := \begin{cases} \dfrac{f-a}{b-a}, & \text{if } N = 0; \\ \dfrac{f-a}{b}, & \text{if } N \neq 0, \end{cases}$$

and let $\widetilde{\mathcal{F}} := \{\tilde{f} : f \in \mathcal{F}\}$. Manifestly $\mathcal{F}$ is normal if and only if $\widetilde{\mathcal{F}}$ is. Moreover, for each $\tilde{f} \in \widetilde{\mathcal{F}}$ we have $0 \notin \tilde{f}(\Omega)$ and $1 \notin \tilde{f}^{(N)}(\Omega)$. We are therefore reduced to considering the case $a = 0$, $b = 1$.

We first establish that, for each $R > 0$ there is a constant $M_N(R)$ such that

$$\begin{aligned} &D(w, 2R) \subset \Omega \Rightarrow \\ (1) \qquad &\log|f(z)| \leq 128 \log^+ |f(w)| + M_N(R) \quad \forall z \in D(w,R),\ f \in \mathcal{F}. \end{aligned}$$

Indeed, given that $D(w, 2R) \subset \Omega$, it is possible to define for each $f \in \mathcal{F}$ a function F on $\mathbb{D}$ by $F(z) := f(2Rz + w)/(2R)^N$. This function meets the hypotheses of Miranda's theorem and so, taking $r = \frac{1}{2}$ in that theorem,

$$\sup_{|z| \leq 1/2} \log|F(z)| \leq 2\big[64 \log^+ |F(0)| + 16 M_N\big],$$

that is (using 12.1(ii), (iii)),

$$\begin{aligned} \log \frac{|f(z)|}{(2R)^N} &\leq 2\left[64 \log^+ \left|\frac{f(w)}{(2R)^N}\right| + 16M_N\right] \\ \log|f(z)| &\leq 128 \log^+|f(w)| + 32M_N + 129\left|\log(2R)^N\right| \qquad \forall z \in D(w,R). \end{aligned}$$

We set $M_N(R) := 32M_N + 129N|\log(2R)|$ and have (1). It follows at once from (1) that

$$\text{(2)}\ \ D(w,2R) \subset \Omega \Rightarrow \sup_{z\in D(w,R)} \log|f(z)| \leq 128\log^+|f(w)| + M_N(R) \quad \forall f \in \mathcal{F}$$

and also that

$$\text{(3)}\ \ D(z,3R) \subset \Omega \Rightarrow \log|f(z)| \leq 128 \inf_{w\in D(z,R)} \log^+|f(w)| + M_N(R) \quad \forall f \in \mathcal{F},$$

since $D(z,3R) \subset \Omega$ implies $D(w,2R) \subset \Omega$ for every $w \in D(z,R)$.

Given any point $u \in \Omega$, choose $R = R(u) > 0$ so that $D(u,3R) \subset \Omega$. We shall show that $\mathcal{F}$ is normal in $D(u,R)$. Then we can cite 12.25 to assert that $\mathcal{F}$ is normal in Ω, as desired. To prove that $\mathcal{F}$ is normal in $D(u,R)$, let a sequence $\{f_n\}$ in $\mathcal{F}$ be given. We want to show that either there is a subsequence of $\{f_n\}$ locally uniformly convergent on $D(u,R)$ or there is a subsequence of $\{f_n\}$ divergent to ∞ uniformly on $D(u,R)$. In case $\{f_n(u)\}$ is bounded, take $w = u$ in (2) and learn that $\{f_n\}$ is uniformly bounded in $D(u,R)$. Then cite 7.6 to come up with a subsequence of $\{f_n\}$ locally uniformly convergent on $D(u,R)$. In case $\{f_n(u)\}$ is unbounded, there exist $1 \leq n_1 < n_2 < \cdots$ such that $\lim_{k\to\infty}\left|f_{n_k}(u)\right| = \infty$. Then (3) with $z = u$ shows that $\lim_{k\to\infty}\left|f_{n_k}(w)\right| = \infty$ uniformly for $w \in D(u,R)$.

Lemma 12.28 *Let $U := \{z \in \mathbb{C} : 0 < |z| < 1\}$ and let f be holomorphic in U and have an essential singularity at 0. Define $f_n(z) := f(2^{-n}z)$ $(z \in U)$. Then $\{f_n\}_{n=1}^{\infty}$ is not a normal family in U.*

Proof: (Montel [1927], pp. 79–80) Consider the compact annulus

$$A := \left\{z \in \mathbb{C} : 1/4 \leq |z| \leq 1/2\right\} \subset U.$$

(1) No subsequence of $\{f_n\}$ is uniformly bounded in A.

Indeed, if $1 \leq n_1 < n_2 < \cdots$ and $\{f_{n_k}\}_{k\in\mathbb{N}}$ is bounded by $M < \infty$ in A, then for every $z \in \mathbb{T}$

$$f\left(2^{-n_k-1}z\right) = f_{n_k}\left(\tfrac{1}{2}z\right) \in f_{n_k}(A) \subset \overline{D}(0,M).$$

That is, $|f| \le M$ on $2^{-n_k-1}\mathbb{T}$ for all k. By the Maximum-Modulus Principle then $|f| \le M$ on $\bigcup_{k\in\mathbb{N}} \{z : 2^{-n_{k+1}-1} \le |z| \le 2^{-n_k-1}\} = \bar{D}(0, 2^{-n_1-1}) \setminus \{0\}$, which by 5.41 makes 0 a removable singularity. This contradiction proves (1).

By 11.7(ii) or by 11.6 there exist $z_k \in \frac{1}{4}U$ with $|z_k| \downarrow 0$ such that

$$\{f(z_k) : k \in \mathbb{N}\} \quad \text{is bounded.} \tag{2}$$

There exist integers $n_k \uparrow \infty$ such that $2^{-n_k-2} \le |z_k| < 2^{-n_k-1}$. Thus $2^{n_k} z_k \in A$, so

$$f(z_k) = f_{n_k}(2^{n_k} z_k) \in f_{n_k}(A) \qquad \forall k \in \mathbb{N}. \tag{3}$$

From (2) and (3) it is clear that $\{f_{n_k}\}_{k\in\mathbb{N}}$ contains no subsequence which is uniformly divergent to ∞ on A, and by (1) no subsequence which is uniformly convergent on A. Thus $\{f_n\}$ is not a normal family in U.

Theorem 12.29 (Gaston Julia) *If the holomorphic function f has an essential singularity at a, then there exists a real number θ_0 such that the set*

$$\bigcap_{\varepsilon>0} f\big(\{a + re^{i\theta} : |\theta - \theta_0| < \varepsilon,\ 0 < r < \varepsilon\}\big)$$

contains all complex numbers with at most one exception.

Remark: Any such $e^{i\theta_0}$ is called a *Julia direction* and the associated ray $\{a + re^{i\theta_0} : r \in \mathbb{R}^+\}$ is called a *Julia line* (for the function f). The same terminology is used if f is defined in a sector with vertex a and the condition "$0 < r < \varepsilon$" is changed to "$r > 1/\varepsilon$".

Proof: We may suppose $a = 0$ and that f is holomorphic in $U := \mathbb{D}\setminus\{0\}$. Let $f_n(z) := f(2^{-n}z)$, $n \in \mathbb{N}$, $z \in U$. Then by 12.28 and 12.25 there is a $z_0 \in U$ such that $\{f_n\}$ is not normal in any neighborhood of z_0. According to 12.27 this means that for each $m \in \mathbb{N}$ and each $\rho > 0$ the family $\{f_n\}_{n\ge m}$ does not omit two complex numbers on $D(z_0, \rho)$, that is,

$$(*) \qquad \begin{cases} a,\ b \in \mathbb{C}\ \&\ a \ne b \Rightarrow \exists\ n \ge m \text{ such that either } a \\ \text{or } b \text{ lies in } f_n(D(z_0,\rho)) = f\big(D(2^{-n}z_0, 2^{-n}\rho)\big). \end{cases}$$

Write $z_0/|z_0| = e^{i\theta_0}$ and for each $0 < \varepsilon < 1$

$$S_\varepsilon := \{re^{i\theta} : |\theta - \theta_0| < \varepsilon,\ 0 < r < \varepsilon\}.$$

This is an open set which contains $\varepsilon z_0/2|z_0| = \varepsilon e^{i\theta_0}/2$, so for some $\rho > 0$

$$D\left(\frac{\varepsilon z_0}{2|z_0|}, \frac{\varepsilon}{2|z_0|}\rho\right) \subset S_\varepsilon.$$

Since S_ε is obviously closed under multiplication by positive numbers less than 1, this entails that if $m \in \mathbb{N}$ is large enough

$$D\big(2^{-n}z_0, 2^{-n}\rho\big) \subset S_\varepsilon \quad \text{for all } n \geq m.$$

Given $a, b \in \mathbb{C}$ with $a \neq b$, $(*)$ therefore shows that

$$\text{either } a \text{ or } b \text{ lies in } f(S_\varepsilon).$$

Thus at most one complex number is absent from $f(S_\varepsilon)$, and this is true of every $\varepsilon > 0$. Since the S_ε decrease with ε, at most one complex number is absent from $\bigcap_{\varepsilon>0} f(S_\varepsilon)$.

Exercise 12.30 (i) *Deduce from* 12.29 *the following stronger form of Picard's Little Theorem: a non-polynomial entire function assumes infinitely often every complex value with at most one exception*—PICARD [1879c].

Hints: If $a \neq b$ and a, b are each assumed only finitely many times by the entire function F, then neither is assumed outside $D(0, 1/r)$ for sufficiently small positive r. Consequently the function $f(z) := F(1/z)$, $0 < |z| < r$, does not have an essential singularity at 0, i.e., its Laurent series contains only finitely many negative powers of z. Evidently then, F is a polynomial.

(ii) *P is a sequence in $\mathbb{C}$ convergent to $c \in D \setminus P$, D an open disk; f is holomorphic in $D \setminus \overline{P}$ with a pole at each $p \in P \cap D$. Then at most two complex numbers are absent from $f\big(D \setminus \overline{P}\big)$.*

Hints: Suppose the distinct complex numbers $w, a, b \notin f\big(D \setminus \overline{P}\big)$ and adapt the strategy in Remark (i) following 11.6: Since h is not the zero function but c is a limit of zeros of h, c is neither a removable singularity nor a pole of h. But since h omits the distinct complex numbers $(a - w)^{-1}$ and $(b - w)^{-1}$, c is not essential either, by 12.29 $\longrightarrow\!\!\!\!\!\times$. In the example $D := \mathbb{D}$, $c := 0$, $P := \{(2\pi i n)^{-1} : n \in \mathbb{Z} \setminus \{0\}\}$, $f(z) := 1/(1 - e^{1/z})$ the two numbers $0, 1$ are both absent from $f(D \setminus \overline{P})$.

Exercise 12.31 *Let Ω be a region, $\{f_n\}$ be a sequence of* univalent *holomorphic functions in Ω. The object of* (i)–(iv) *below is to show that there exists a point $a \in \Omega$ and a subsequence $\{h_k\}$ of $\{f_n\}$ which is normal in $\Omega \setminus \{a\}$;* (v) *and* (vi) *are applications of this fact.*

(i) *There is a point $a \in \Omega$ and a subsequence $\{f_{n_j}\}$ of $\{f_n\}$ such that in each compact $K \subset \Omega \setminus \{a\}$, for only finitely many j does the function f_{n_j} have a zero.*

Hints: If the set of subsequential limits of the sequence $\{f_n^{-1}(0)\}$ meets Ω, let a be any point in the intersection. Otherwise, let a be any point whatsoever in Ω. Let $\{K_j\}$ be a compact exhaustion of $\Omega \setminus \{a\}$ as furnished by 1.31. For each j let N_j be a neighborhood of a in $D(a, 1/j) \cap \Omega$ which is disjoint from K_j. In the first case there exists a subsequence $\{f_{n_j}\}$ such that $f_{n_j}^{-1}(0) \in N_j$ for each j. In the second case no set $\{f_n^{-1}(0)\}_{n=k}^{\infty}$ $(k = 1, 2, 3, \dots)$ lies wholly in any K_j and we can choose the n_j successively such that $1 \le n_1 < n_2 < \cdots$ and $f_{n_j}^{-1}(0) \notin K_j$. Thus in either case $K_j \cap \{f_{n_l}^{-1}(0)\}_{l=j}^{\infty} = \varnothing$, for each j. Since any compact subset K of $\Omega \setminus \{a\}$ lies in some K_j, the claim (i) is established.

(ii) *There is a point $b \in \Omega \setminus \{a\}$ and a subsequence $\{h_k\}$ of $\{f_{n_j}\}$ such that in any compact $K \subset \Omega \setminus \{a, b\}$, for only finitely many k does the function h_k take either of the values* 0 *or* 1.

Hints: Apply the argument of part (i) to the sequence $\{\mathbf{1} - f_{n_j}\}$ of univalent functions in the region $\Omega \setminus \{a\}$.

(iii) *The sequence $\{h_k\}$ of* (ii) *is normal in $\Omega \setminus \{a, b\}$.*

Hints: By (ii) and 12.27 $\{h_k\}$ is normal in every open disk D such that $\bar{D} \subset \Omega \setminus \{a, b\}$. Since the latter is connected (1.24), it follows from 12.25 that $\{h_k\}$ is normal therein.

(iv) *The sequence $\{h_k\}$ of* (iii) *is normal in $\Omega \setminus \{a\}$.*

Hints: Since $\{h_k\}$ is normal in $\Omega \setminus \{a, b\}$, any subsequence of $\{h_k\}$ contains a further subsequence $\{g_l\}$ such that either (1): $\{g_l\}$ converges locally uniformly in $\Omega \setminus \{a, b\}$ or (2): $\{|g_l|\}$ diverges to ∞ locally uniformly in $\Omega \setminus \{a, b\}$. If (1) prevails, then $\{g_l\}$ converges uniformly on the boundaries of two disjoint closed disks D_a, D_b centered at a and b and lying in Ω (since such circles are compact subsets of $\Omega \setminus \{a, b\}$). Then by the Maximum-Modulus Principle $\{g_l\}$ converges uniformly in D_a and D_b. Thus (1) implies (1)′: $\{g_l\}$ converges locally uniformly throughout Ω. If (2) prevails, select $r > 0$ so that $\bar{D}(b, r) \subset \Omega \setminus \{a\}$. Then $|g_l| \to \infty$ uniformly on $C(b, r)$. Bearing in mind the definition of the f_{n_j} from (i) and the fact that $\{g_l\}$ is a subsequence of $\{f_{n_j}\}$, we know that for only finitely many l does g_l have a zero in $\bar{D}(b, r)$, so $\{1/g_l\}$ is a sequence of holomorphic functions in $D(b, r)$. From $|1/g_l| \to 0$ uniformly on $C(b, r)$ and the Maximum-Modulus Principle it follows that $|1/g_l| \to 0$ uniformly in $D(b, r)$, so $|g_l| \to \infty$ uniformly in $D(b, r)$. Thus (2) implies (2)′: $\{|g_l|\}$ diverges to ∞ locally uniformly throughout $\Omega \setminus \{a\}$.

(v) (*Extension of* 7.32(i)) *Show that if*

(∗) *$\{f_n(c), f_n(d) : n \in \mathbb{N}\}$ is bounded for two distinct points c, $d \in \Omega$,*

then $\{f_n\}$ is locally uniformly bounded in Ω.

Hints: Consider any subsequence of $\{f_n\}$. Apply (iv) to this subsequence to come up with a point $a \in \Omega$ and a further subsequence $\{g_l\}$ of the given subsequence, such that either (1): $\{g_l\}$ converges locally uniformly in $\Omega \setminus \{a\}$ or (2): $\{|g_l|\}$ diverges to ∞ in $\Omega \setminus \{a\}$. Since one of the points c, d must be in $\Omega \setminus \{a\}$, alternative (2) cannot prevail because of $(*)$. Now argue as in (iv) that $\{g_l\}$ converges uniformly in some disk centered at a, hence locally uniformly throughout Ω. We have shown that $\{f_n\}$ is normal in Ω. Now conclude by citing 12.26.

(vi) *Prove the following extension of* 7.11 (*due to* MONTEL [1925], p. 253): *If* $\lim_{n\to\infty} f_n(c)$, $\lim_{n\to\infty} f_n(d)$ *exist in* $\mathbb{C}$ *and coincide, equal* L, *say, for two distinct points* c, $d \in \Omega$, *then* $\{f_n\}$ *converges to* L *locally uniformly in* Ω.

Hints: Use (v) above, 7.11 and 7.8.

(vii) *Prove that if* $0 \notin f_n(\Omega)$ *for all* n, *then* $\{f_n\}$ *is a normal family.*

Hints: Given a subsequence of $\{f_n\}$, part (i) supplies $a \in \Omega$ and a further subsequence $\{f_{n_j}\}$ such that in each compact $K \subset \Omega \setminus \{a\}$, for only finitely many j does $\mathbf{1} - f_{n_j}$ have a zero. The ideas in the hints to (iv) then show that $\{f_{n_j}\}$ is normal in Ω.

Exercise 12.32 *Let* Ω *be a region,* $\mathcal{F}$ *a family of* univalent *holomorphic functions in* Ω.

(i) *Show that if* $\{f'(a) : f \in \mathcal{F}\}$ *is bounded for some* $a \in \Omega$, *then* $\mathcal{F}$ *is a normal family.*

Hints: Set $M(z) := \sup\{|f'(z)| : f \in \mathcal{F}\}$. From Koebe's Distortion Theorem 7.31(iv) and the hypothesis $M(a) < \infty$ it follows that $M(z) < \infty$ for each $z \in \Omega$. Now by 12.25 it suffices to show that each point $z_0 \in \Omega$ has a neighborhood in which $\mathcal{F}$ is normal. Pick $r > 0$ such that $D(z_0, r) \subset \Omega$ and infer via 7.27 that for some finite constant B

$$\left|\frac{f(z) - f(z_0)}{rf'(z_0)}\right| \le B \qquad \forall z \in D(z_0, r/2), \quad f \in \mathcal{F}$$

whence

$$(1) \qquad |f(z)| \le |f(z_0)| + rB|f'(z_0)| \le |f(z_0)| + rBM(z_0) \qquad \forall z \in D(z_0, r/2),\ f \in \mathcal{F}$$

and

$$(2) \qquad |f(z)| \ge |f(z_0)| - rB|f'(z_0)| \ge |f(z_0)| - rBM(z_0) \qquad \forall z \in D(z_0, r/2),\ f \in \mathcal{F}.$$

Let now a sequence $\{f_n\} \subset \mathcal{F}$ be given. If $\{f_n(z_0)\}$ is bounded, then by (1) $\{f_n\}$ is uniformly bounded in $D(z_0, r/2)$ and so by Montel (7.6) has a subsequence which is locally uniformly convergent in $D(z_0, r/2)$. If on the other hand $\{f_n(z_0)\}$ is not bounded, then there exist $1 \le n_1 < n_2 < \cdots$ such that $\lim_{j\to\infty} |f_{n_j}(z_0)| = \infty$. In this case (2) shows that $\lim_{j\to\infty} |f_{n_j}| = \infty$ uniformly in $D(z_0, r/2)$. Thus $\mathcal{F}$ is normal in $D(z_0, r/2)$.

(ii) *Show that $\{f' : f \in \mathcal{F}\}$ is a normal family.*

Hints: Fix $a \in \Omega$. We may suppose $f(a) = 0$ for all $f \in \mathcal{F}$. According to 5.78 all f' are zero-free. Consider any sequence $\{f_n\}$. 7.31(i) and 7.6 supply a subsequence $g_{n_j} := f'_{n_j}/f'_{n_j}(a)$ which converges locally uniformly to a $g \in H(\Omega)$. Since $g_{n_j}(a) = 1$ and $|g_{n_j}| > 0$ for all j, it follows from 7.11 that g is zero-free. Now $\{f'_{n_j}(a)\}$ has a subsequence which either converges to $c \in \mathbb{C}$ or diverges in modulus to ∞. The corresponding subsequence of $\{f'_{n_j}\}$ either converges to cg locally uniformly or diverges in modulus to ∞ locally uniformly.

§ 5 Sectorial Limit Theorems

Montel [1912] used normal families in a simple but ingenious way to investigate boundary behavior of holomorphic functions in angular domains. (Later he used these ideas to treat the boundary problem for conformal maps which we analysed in Chapter IX.) We have already explored this theme, on the boundary itself in 5.16 and internally in 5.56. Most of the theorems of this section look at internal approach. The bounded version of 12.33 is, of course, contained in 5.56. It is interesting to note that, in spite of the diversity of proof techniques involved, the bounded 12.33 plus 5.16 is equivalent to 12.34.

Theorem 12.33 *Let $\alpha < \gamma < \beta$, $0 < R < \infty$, $S := \{re^{i\theta} : r > R,\ \alpha < \theta < \beta\}$ and let f be holomorphic in S and for some complex c satisfy $\lim_{r\to\infty} f(re^{i\gamma}) = c$. Suppose that there are two distinct complex numbers absent from the range of f. (For example, f might be bounded.) Then $\lim_{r\to\infty} f(re^{i\theta}) = c$ for every $\theta \in (\alpha, \beta)$ and uniformly in any compact subinterval.*

Proof: After a change of scale we can assume $R < 1$. Note that S is an open, connected subset of $\mathbb{C}$. Define for each positive integer n functions f_n in S by

$$(1) \qquad f_n(z) := f(nz), \quad z \in S.$$

Notice that

$$(2) \qquad \lim_{n\to\infty} f_n(re^{i\gamma}) = c \qquad \forall r > R.$$

In particular, the set $\{f_n(e^{i\gamma})\}$ is bounded. Since also the two distinct numbers absent from the range of f are both absent from the range of each f_n, it follows from 12.27 and 12.26 that $\{f_n\}$ is uniformly bounded on each compact subset of S. But then from (2) and the Vitali-Porter theorem 7.5 we know that f_n converges to a holomorphic function f_0 uniformly on each compact subset of S. On the ray $(R,\infty)e^{i\gamma}$ we have $f_0 = c$, so by the fundamental uniqueness theorem, $f_0 \equiv c$. If I is a compact subinterval of (α,β), let $K(I) := \{re^{i\theta} : 1 \le r \le 2,\, \theta \in I\}$, a compact subset of S. Given $\varepsilon > 0$, there exists then $N = N(I,\varepsilon)$ such that

$$(3) \qquad |f_n(z) - c| < \varepsilon \qquad \forall n \ge N, \quad \forall z \in K(I).$$

In particular,

$$(4) \qquad \left|f\big(re^{i\theta}\big) - c\right| < \varepsilon \qquad \forall r \ge N, \quad \forall \theta \in I.$$

Indeed, given such θ and r, let n be the integer such that $n \le r < n+1$. Then the number $z := (r/n)e^{i\theta}$ belongs to $K(I)$ and $n \ge N$. Since $f\big(re^{i\theta}\big) = f_n(z)$, the inequality in (4) follows from (3) and the proof is finished.

Exercise 12.34 (Lindelöf [1915]) *Let* $0 < \alpha \le \pi$,

$$S := \{re^{i\theta} : r > 0,\, -\alpha < \theta < \alpha\}, \quad \widetilde{S} := \{re^{i\theta} : r > 0,\, -\alpha \le \theta < \alpha\}$$

and suppose that $f \colon \widetilde{S} \mapsto \mathbb{C}$ *is bounded, continuous, holomorphic in* S *and satisfies*

$$(1) \qquad \lim_{r\to\infty} f\big(re^{-i\alpha}\big) = 0.$$

Show that

$$(2) \quad \lim_{r\to\infty} f\big(re^{i\theta}\big) = 0 \quad \text{uniformly for } \theta \text{ in any compact subset of } [-\alpha,\alpha).$$

Hints: Introduce

$$S_0 := \{re^{i\theta} : r > 0,\, -\alpha/2 < \theta < \alpha/2\}$$

and form

$$F(z) := f\big(ze^{-i\alpha/2}\big)\bar{f}\big(\bar{z}e^{-i\alpha/2}\big), \quad z \in \bar{S}_0.$$

This function is continuous and bounded in $\bar{S}_0$, holomorphic in S_0. As $F\big(re^{-i\alpha/2}\big)$ equals $f\big(re^{-i\alpha}\big)$ times a bounded factor and $F\big(re^{i\alpha/2}\big)$ equals $\bar{f}\big(re^{-i\alpha}\big)$ times a bounded factor, we see from (1) that

$$\lim_{r\to\infty} F\big(re^{\pm i\alpha/2}\big) = 0.$$

Therefore by 5.16, $\lim_{r\to\infty} F\big(re^{i\theta}\big) = 0$ for each $\theta \in [-\alpha/2, \alpha/2]$. In particular,

$$0 = \lim_{r\to\infty} F(r) = \lim_{r\to\infty} \big|f\big(re^{-i\alpha/2}\big)\big|^2$$

and so

(3) $$\lim_{r\to\infty} f\big(re^{-i\alpha/2}\big) = 0.$$

From (1), (3) and 5.16 it follows that

(4) $$\lim_{r\to\infty} f\big(re^{i\theta}\big) = 0 \quad \text{uniformly for } \theta \in [-\alpha, \alpha/2],$$

while from (3) and 12.33 it follows that

(5) $$\lim_{r\to\infty} f\big(re^{i\theta}\big) = 0 \quad \text{uniformly for } \theta \text{ in any compact subset of } (-\alpha, \alpha).$$

From (4) and (5) the claim (2) follows.

Exercise 12.35 *Give an* ab initio *proof of* 12.34 *using the harmonic-majorant technique.*

Hints: It is convenient to put the problem in the following form: Let $f\colon \mathbb{U} \cup (0, \infty) \mapsto \mathbb{C}$ be bounded and continuous there and holomorphic in $\mathbb{U}$. Suppose that $f(x) \to 0$ as real $x \to +\infty$. We are to show that

(∗) $$\lim_{\substack{|z|\to\infty \\ z\in S_\varepsilon}} f(z) = 0$$

for each set $S_\varepsilon := \big\{re^{i\theta} : r > 0,\ 0 \le \theta < (1-\varepsilon)\pi\big\}$, $\varepsilon > 0$. We may assume that $|f| \le 1$. Let $\varepsilon > 0$ and $0 < \delta < 1$ be given. Choose $x_\delta > 0$ so large that

(1) $$|f(x)| < \delta \qquad \forall x \ge x_\delta.$$

Let L denote the holomorphic logarithm in $\mathbb{C} \setminus \{iy : y \le 0\}$ which satisfies

(2) $$L\big(re^{i\theta}\big) = \log r + i\theta \quad \text{for } r > 0,\ -\frac{\pi}{2} < \theta < \frac{3\pi}{2}.$$

(Cf. 3.43.) Look at the function

(3) $$h(z) := 1 - \frac{1}{\pi} \operatorname{Im} L(z - x_\delta), \quad z \in \overline{\mathbb{U}} \setminus \{x_\delta\}.$$

It is harmonic and bounded in $\mathbb{U}$, equals 0 on $(-\infty, x_\delta)$ and equals 1 on (x_δ, ∞). Consequently the function

$$\log\left(\max\{|f|, \delta\}\right) - (\log\delta)h = \log\delta + \log^+\left|\frac{f}{\delta}\right| - (\log\delta)h$$

is bounded and subharmonic in $\mathbb{U}$ (by 5.26(ii)), and has non-positive limit superior at each point of $\partial\mathbb{U} \setminus \{x_\delta\} = (-\infty, x_\delta) \cup (x_\delta, \infty)$. Thus, thanks to 7.17 this function is non-positive throughout $\mathbb{U}$:

$$\log\left(\max\{|f|, \delta\}\right) \le (\log\delta)h \quad \text{on } \mathbb{U}. \tag{4}$$

Since $\operatorname{Im} L\big(re^{i\theta} - x_\delta\big) = \operatorname{Im} L\big(e^{i\theta} - x_\delta/r\big)$ [for all $r > x_\delta$, $0 \le \theta \le \pi$] and since there exists [by uniform continuity of L on the compact set $\overline{\mathbb{U}} \cap \mathbb{T}$] an $x'_\delta > x_\delta$ such that

$$\operatorname{Im} L\big(e^{i\theta} - x_\delta/r\big) \le \operatorname{Im} L\big(e^{i\theta}\big) + \frac{\varepsilon\pi}{2} = \theta + \frac{\varepsilon\pi}{2} \qquad \forall r \ge x'_\delta, \quad 0 \le \theta \le \pi,$$

it follows that for such r and θ

$$\operatorname{Im} L(z - x_\delta) \le (1-\varepsilon)\pi + \frac{\varepsilon\pi}{2} = \left(1 - \frac{\varepsilon}{2}\right)\pi,$$

and so the definitions of h and S_ε give

$$h(z) \ge \frac{\varepsilon}{2} \qquad \forall z \in S_\varepsilon \text{ with } |z| \ge x'_\delta. \tag{5}$$

From (4) and (5), remembering that $\log\delta < 0$, we get

$$\log\left(\max\{|f|, \delta\}\right) \le (\log\delta)\frac{\varepsilon}{2} \qquad \forall z \in S_\varepsilon \text{ with } |z| \ge x'_\delta.$$

Exponentiating gives us finally

$$|f(z)| \le \delta^{\varepsilon/2} \qquad \forall z \in S_\varepsilon \text{with } |z| \ge x'_\delta.$$

Since $\lim_{\delta\downarrow 0} \delta^{\varepsilon/2} = 0$ for each fixed $\varepsilon > 0$, the claim $(*)$ is established.

Exercise 12.36 (i) *Deduce both* 12.33 *and* 5.16 *from* 12.34.

(ii) *Try to deduce* 5.56 *from* 9.20 *and* 12.33 *by a reflection argument like that used in the proof of Milloux's theorem* 5.50. (*See* p. 26 *of* Cartwright [1935].)

Exercise 12.37 (Seidel and Littauer [1932]–cf. 5.16(ii)) *Let* $\alpha_1 < \alpha_2 < \alpha_1 + 2\pi$, $S := \{re^{i\theta} : r > 0,\ \alpha_1 < \theta < \alpha_2\}$ *and* f *holomorphic in some open* *sector which contains* $\bar{S} \setminus \{0\}$ *be given. Suppose that* $\lim_{r\to\infty} f\left(re^{i\alpha_j}\right) = L_j \in \mathbb{C}$ *exist. Show that then*

either

(a) $\quad L_1 = L_2$ *and* $\lim_{r\to\infty} f\left(re^{i\theta}\right) = L_1$ *uniformly for* $\theta \in [\alpha_1, \alpha_2]$

or

(b) $\quad f$ *has at least one Julia direction in* $[\alpha_1, \alpha_2]$.

Hints: Suppose (b) fails. This means that each $\theta \in [\alpha_1, \alpha_2]$ lies in some open interval I_θ such that in some sector $S_\theta := \left\{re^{it} : r > r_\theta,\, t \in I_\theta\right\}$ f omits two distinct complex numbers. A finite subset $\mathcal{C}$ of $\{I_\theta : \alpha_1 \le \theta \le \alpha_2\}$ suffices to cover $[\alpha_1, \alpha_2]$. Suppose $I_{\theta_1} \in \mathcal{C}$ contains α_1. If $\alpha_2 \in I_{\theta_1}$, set $\theta_2 := \theta_1$. If $\alpha_2 \notin I_{\theta_1}$, then the right endpoint of I_{θ_1} lies in some $I_{\theta_2} \in \mathcal{C}$. Continue in this fashion to produce $I_{\theta_1}, I_{\theta_2}, \ldots, I_{\theta_n} \in \mathcal{C}$ with $\alpha_2 \in I_{\theta_n}$ and $I_{\theta_{j-1}} \cap I_{\theta_j} \ne \varnothing$ for $j = 2, 3, \ldots, n$. Apply 12.33 *successively* in S_{θ_1}, S_{θ_2}, ... to reach the conclusion (a).

If we restrict to "non-tangential" approach, then the method of 12.33 can be further exploited to yield

Exercise 12.38 *Let* $\alpha < \beta$, $0 < R < \infty$, $S := \left\{re^{i\theta} : r > R,\, \alpha < \theta < \beta\right\}$, f *a holomorphic function in* S *for which two distinct complex numbers are absent from* $f(S)$. *Suppose that there exist* $\alpha < \alpha' < \beta' < \beta$ *such that*

$$(*) \qquad \sup_{\rho > r} \inf\left\{\left|f\left(\rho e^{i\theta}\right)\right| : \alpha' < \theta < \beta'\right\} \to 0 \text{ as } r \to \infty.$$

Show that $\lim_{r\to\infty} f\left(re^{i\theta}\right) = 0$ *uniformly for* θ *in any compact subset of* (α, β).

Hints: (Montel [1917], p. 20) We can suppose $R < 1$. Define $f_n(z) := f(2^n z)$, $z \in S$, $n = 1, 2, \ldots$ and $K := \left\{re^{i\theta} : 1 \le r \le 2,\, \alpha' \le \theta \le \beta'\right\}$. For each $r \in [1, 2]$ select $\theta_n(r) \in [\alpha', \beta']$ such that

$$(**) \qquad \begin{aligned} \left|f_n\left(re^{i\theta_n(r)}\right)\right| &= \left|f\left(2^n re^{i\theta_n(r)}\right)\right| = \inf\left\{\left|f(2^n re^{i\theta})\right| : \alpha' < \theta < \beta'\right\} \\ &\to 0 \quad \text{as } n \to \infty, \text{ by } (*). \end{aligned}$$

This shows that no subsequence of $\{|f_n|\}$ can diverge to ∞ *uniformly* in the compact set K. However, by 12.27 $\{f_n\}$ is a normal family in S. Now from 7.2 and $(**)$ it follows easily that if g is the limit of a convergent subsequence $\{f_{n_j}\}$ of $\{f_n\}$, then $g(z) = 0$ for each cluster point z of $\left\{re^{i\theta_n(r)} : n = n_1, n_2, \ldots\right\}$. We have this for each $r \in [1, 2]$, so such z constitute an infinite subset of K and by 5.62, g is the 0 function. By 7.8 $\{f_n\}$ therefore converges to 0 locally uniformly in S. The proof is concluded from this point just as was 12.33.

Exercise 12.39 *Use the method of* 12.33 *to prove this result of* Lindelöf [1909]: *if f is holomorphic in an open sector S, never* 0 *or* 1, *and is bounded on some half-line in S, then f is bounded in every closed subsector of S.*

Next we are going to investigate the analog of 12.33 when $\left|f\left(re^{i\theta}\right)\right|$ converges as $r \to \infty$, rather than $f\left(re^{i\theta}\right)$ itself. (An easy result along these lines is 7.34.) Here it is more convenient to work with strips and half-strips rather than sectors, so we begin with

Exercise 12.40 *If $a, b, c \in \mathbb{R}$ with $a < b$, then there is an exponential which maps the semi-strip $(a,b) \times (c,\infty)$ conformally onto a truncated sector $S \setminus \bar{D}(0,R)$. Using it, formulate a semi-strip version of* 12.33.

Exercise 12.41 *Let $\alpha, \beta, a, b, c \in \mathbb{R}$ with $\alpha < a < b < \beta$ and form $S_c := (\alpha,\beta) \times (c,\infty)$. Let $f\colon S_c \mapsto \mathbb{C}$ be holomorphic and bounded, by K, say.*

(i) *Suppose that*

(1) $$\lim_{y\to+\infty} |f(a+iy)| = \lim_{y\to+\infty} |f(b+iy)| = 1,$$

(2) $$b - a < \frac{1}{2}(\beta - \alpha),$$

(3) $$a - \alpha \le b - a \quad \textit{and} \quad \beta - b \le b - a.$$

Show that for any $\delta > 0$ there exists $c(\delta) > c$ such that

(4) $$|f(x+iy)| \ge \frac{1}{2K} \quad \textit{whenever } \alpha + \delta \le x \le \beta - \delta,\ y \ge c(\delta).$$

Hints: For $z \in S'_c := (\alpha, 2a - \alpha) \times (c, \infty)$ define $g(z) := \bar{f}(\overline{2a - z})$. Note that

$$\overline{2a - z} = 2a - \bar{z} \in S'_c \subset S_c,$$

so g is well defined and holomorphic (cf. 2.4). Then $F := fg$ is holomorphic and bounded in S'_c and satisfies

$$F(a+iy) = |f(a+iy)|^2 \to 1 \quad \text{as } y \to +\infty, \text{ by (1)}.$$

Since $\alpha < a < 2a - \alpha$, we may cite 12.40 to conclude that $\lim_{y\to+\infty} F(x+iy) = 1$ uniformly for x in any compact subset of $(\alpha, 2a-\alpha)$. In particular, for each $0 < \delta < a - \alpha$ there exists $c_1(\delta) > c$ such that

(5) $$|F(x+iy)| > \frac{1}{2} \qquad \forall (x,y) \in (\alpha+\delta, 2a-\alpha-\delta) \times (c_1(\delta), \infty).$$

Because $|f| \leq K$, (5) and the definition of F show that

$$(6) \qquad |f(x+iy)| > \frac{1}{2K} \qquad \forall (x,y) \in (\alpha+\delta, 2a-\alpha-\delta) \times (c_1(\delta), \infty).$$

Argue similarly to produce $c_2(\delta) > c$, for each $0 < \delta < \beta - b$, such that

$$(7) \qquad |f(x+iy)| > \frac{1}{2K} \qquad \forall (x,y) \in (2b-\beta+\delta, \beta-\delta) \times (c_2(\delta), \infty).$$

Now by (2), if δ is sufficiently small, the strips (6) and (7) overlap and so for $c(\delta) := \max\{c_1(\delta), c_2(\delta)\}$ the inequalities (6) and (7) yield (4).

(ii) *Under hypotheses* (1), (2), (3) *show that*

$$(8) \qquad \lim_{y \to +\infty} |f(x+iy)| = 1 \quad \textit{uniformly for } x \textit{ in any compact subset of } (\alpha, \beta).$$

Hints: Let $\varepsilon > 0$ be given. There is then by (1) a $y(\varepsilon) > c$ such that

$$(9) \qquad |f(a+iy)| \leq 1+\varepsilon, \quad |f(b+iy)| \leq 1+\varepsilon \qquad \forall y \geq y(\varepsilon).$$

The function

$$h_\varepsilon(z) := \frac{z-(a+iy(\varepsilon))}{z-(a+iy(\varepsilon))+K(b-a)}, \qquad z \in [a,b] \times [y(\varepsilon), \infty)$$

is bounded by 1 and on $[a,b] \times \{y(\varepsilon)\}$ it is bounded by $1/K$. Consequently, the holomorphic function fh_ε is bounded in this strip by K and because of (9) bounded by $1+\varepsilon$ on its boundary. From 5.13 it follows then that fh_ε is bounded by $1+\varepsilon$ throughout this set. Since $\lim_{y\to+\infty} h_\varepsilon(x+iy) = 1$ uniformly for $x \in [a,b]$, we may select $y_1(\varepsilon) > y(\varepsilon)$ sufficiently large that $|h_\varepsilon(x+iy)| > (1+\varepsilon)/(1+2\varepsilon)$ for all $y \geq y_1(\varepsilon)$ and all $x \in [a,b]$. We shall then have

$$|f(x+iy)| \leq 1+2\varepsilon \qquad \forall (x,y) \in [a,b] \times [y_1(\varepsilon), \infty).$$

Because of (4), the same argument may be applied to $1/f$. Conclude that

$$(10) \qquad \lim_{y\to+\infty} |f(x+iy)| = 1 \quad \text{uniformly for } x \in [a,b].$$

Now consider $a - \alpha > \delta > 0$. In the hints to (i) we saw that $\lim_{y\to+\infty} F(x+iy) = 1$ uniformly for $x \in [\alpha+\delta, a]$. Thus, by definition of F,

$$(11) \qquad \lim_{y\to+\infty} |f(x+iy)||f(2a-x+iy)| = 1 \quad \text{uniformly for } x \in [\alpha+\delta, a].$$

Since $a \le 2a - x \le 2a - \alpha \overset{(3)}{\le} b$ for $x \in [\alpha + \delta, a]$, it follows that for these x the behavior of the second factor on the left-hand side of (11) is governed by (10). It follows therefore from (11) that $\lim_{y\to+\infty} |f(x+iy)| = 1$ uniformly for $x \in [\alpha+\delta, a]$, and similarly for $x \in [b, \beta-\delta]$. As this holds for all sufficiently small $\delta > 0$, (8) is established.

(iii) *Assuming only hypotheses* (1) *and* (2), *prove that for some* $\alpha \le \alpha_0 < a < b < \beta_0 \le \beta$

$(8)_0$ $\lim_{y\to+\infty} |f(x+iy)| = 1$ *uniformly for* x *in any compact subset of* (α_0, β_0).

Hints: Check that

$$\alpha_0 := a - \lambda(b-a), \qquad \beta_0 := b + \gamma(b-a)$$

satisfy the above chain of inequalities as well as

$(2)_0$ $b - a < \frac{1}{2}(\beta_0 - \alpha_0)$, and

$(3)_0$ $a - \alpha_0 \le b - a, \qquad \beta_0 - b \le b - a,$

whenever

$$0 < \lambda, \gamma \le 1 \quad \text{and} \quad \lambda \le \frac{a-\alpha}{b-a}, \quad \gamma \le \frac{\beta - b}{b-a} \quad \text{and} \quad \lambda + \gamma > 1.$$

Therefore the choices

$$\lambda := \min\left\{1, \frac{a-\alpha}{b-a}\right\}, \qquad \gamma := \min\left\{1, \frac{\beta-b}{b-a}\right\},$$

enable α_0, β_0 to satisfy *all* the relevant inequalities needed in the proof of (ii). So that proof delivers the conclusion $(8)_0$.

(iv) *Prove that if* f *is holomorphic and bounded in the strip* $S := (\alpha, \beta) \times \mathbb{R}$ *and satisfies* (1) *and* (2), *then*

(12) $\lim_{y\to+\infty} |f(x+iy)| = 1$ *uniformly for* x *in any compact subset of* (α, β).

Hints: (HARDY [1926]) the function $f^*(z) := \bar{f}(\bar{z})$ is holomorphic in S (see 2.4) and bounded. Consider any real sequence $y_n \to +\infty$ and the holomorphic functions $g_n(z) := f(z+iy_n)f^*(z-iy_n)$, $z \in S$. For $x \in (\alpha, \beta), y \in \mathbb{R}$

(13) $|f(x+iy)|^2 = f(x+iy)\bar{f}(x+iy) = f(x+iy)f^*(x-iy)$ $[= g_n(x)$ if $y = y_n]$.

For $x \in [a,b]$ we have $\lim_{n\to\infty} |f(x+iy)|^2 = 1$, by (iii). Hence (13) says that the bounded holomorphic functions g_n in S satisfy

$$(14) \qquad \lim_{n\to\infty} g_n(x) = 1 \quad \forall x \in [a,b], \text{ a set with a limit point in } S.$$

It follows then from 7.5 that (14) holds as well uniformly for x in any compact subset of S. In particular, via (13),

$$\lim_{n\to\infty} |f(x+iy_n)| = 1 \text{ uniformly for } x \text{ in any compact subset of } (\alpha,\beta).$$

From this holding for *every* real sequence $y_n \to +\infty$, claim (12) follows easily.

Exercise 12.42 *Let α, β, a, $b \in \mathbb{R}$ with $\alpha < a < b < \beta$, $F\colon (\alpha,\beta)\times\mathbb{R} \mapsto \mathbb{C}$ a bounded holomorphic function. Suppose that*

$$(1) \qquad \lim_{y\to+\infty} |F(a+iy)| =: A > 0, \qquad \lim_{y\to+\infty} |F(b+iy)| =: B > 0,$$

$$(2) \qquad b-a < \frac{1}{2}(\beta-\alpha).$$

Show that $\lim_{y\to+\infty} |F(x+iy)|$ exists for each $x \in (\alpha,\beta)$, uniformly in compact subsets of such x. Find the value of this limit.

Hint: The function $f(z) := A^{-(b-z)/(b-a)} B^{-(z-a)/(b-a)} F(z)$ satisfies the hypotheses of 12.41(iv).

Exercise 12.43 *Let α, $\beta \in \mathbb{R}$ with $\alpha < \beta$, $F\colon (\alpha,\beta)\times\mathbb{R} \mapsto \mathbb{C}$ be a bounded holomorphic function. Suppose that there are $x_1 < x_2 < x_3 \in (\alpha,\beta)$ such that $\lim_{y\to+\infty} |F(x_k+iy)|$ exists and is finite for $k = 1,2,3$. Prove that then $\lim_{y\to+\infty} |F(x+iy)|$ exists uniformly for x in any compact subset of (α,β) and has the form be^{ax} for appropriate real constants a and b.*

Hint: If one limit is 0, use a strip version of 12.33. If each limit is positive, then 12.42 is applicable, because at least one of $x_2 - x_1$ and $x_3 - x_2$ must be less than $\beta - \alpha$.

Exercise 12.44 *Suppose $\alpha < a < b < \beta$ and $h\colon S := (\alpha,\beta)\times\mathbb{R} \mapsto \mathbb{R}$ is a bounded harmonic function satisfying*

$$\lim_{y\to+\infty} h(a+iy) =: A \in \mathbb{R}, \qquad \lim_{y\to\infty} h(b+iy) =: B \in \mathbb{R}.$$

Show that

$$\lim_{y\to+\infty} h(x+iy) = B\frac{x-a}{b-a} + A\frac{b-x}{b-a}$$

uniformly for x in any compact subset of (α,β).

Hints: Replace $h(z)$ by $h(z) - B(\operatorname{Re} z - a)/(b-a) - A(b - \operatorname{Re} z)/(b-a)$ to achieve $A = B = 0$. Select (10.2) $F \in H(S)$ with $h = \operatorname{Re} F$ and form $f := e^F$. Since *both* f and $1/f$ are then bounded, the first part of the proof of 12.41(ii) [down to (10)] shows, without appeal to 12.41(i) or its hypotheses (2), (3), that

$$\lim_{y \to +\infty} |f(x+iy)| = 1 \text{ for all } x \in [a,b].$$

The rest of the argument here is verbatim the same as that for 12.41(iv).

Exercise 12.45 *Let f be entire and suppose that for each real θ we have*

$$(*) \qquad \begin{cases} \text{either} & \lim_{r\to\infty} \left|f\left(re^{i\theta}\right) - c(\theta)\right| = 0 \quad \text{for some } c(\theta) \in \mathbb{C} \\ \text{or} & \lim_{r\to\infty} \left|f\left(re^{i\theta}\right)\right| = \infty. \end{cases}$$

(Non-constant examples of such functions will be produced in Chapter XV and according to BERNAL-GONZÁLEZ [1996] the set of them is dense in $H(\mathbb{C})$.)

(i) *Show that in any non-degenerate closed interval I there is a non-degenerate sub-interval J and a real number R such that*

$$\begin{array}{llll} \textit{either} & (1.J) & \left|f\left(re^{i\theta}\right)\right| \geq 1 & \forall\theta \in J \quad \textit{and} \quad \forall r \geq R \\ \textit{or} & (2.J) & \left|f\left(re^{i\theta}\right)\right| \leq 2 & \forall\theta \in J \quad \textit{and} \quad \forall r \geq R. \end{array}$$

Hints: Suppose that no such J exists. Take $J_0 := I$ and construct non-degenerate closed intervals $J_n \subset I$ and $r_n > n$ inductively as follows: given J_{2n-1}, the failure of $(1.J_{2n-1})$ implies $\theta_{2n} \in J_{2n-1}$ and $r_{2n} > 2n$ exist such that $\left|f(r_{2n}e^{i\theta_{2n}})\right| < 1$. By continuity there is then a whole (non-degenerate) subinterval J_{2n} of J_{2n-1} such that

$$(3.n) \qquad \left|f\left(r_{2n}e^{i\theta}\right)\right| < 1 \qquad \forall\theta \in J_{2n}.$$

Similarly, given J_{2n}, the failure of $(2.J_{2n})$ implies $\theta_{2n+1} \in J_{2n}$ and $r_{2n+1} > 2n+1$ exist such that $\left|f(r_{2n+1}e^{i\theta_{2n+1}})\right| > 2$. By continuity there is then a whole (non-degenerate) subinterval J_{2n+1} of J_{2n} such that

$$(4.n) \qquad \left|f\left(r_{2n+1}e^{i\theta}\right)\right| > 2 \qquad \forall\theta \in J_{2n+1}.$$

For any $\theta \in \bigcap_{n=1}^{\infty} J_n$ (and there is at least one, by compactness) $(3.n)$ and $(4.n)$ each hold for all n. It follows that $\lim_{n\to\infty} \left|f(r_n e^{i\theta})\right|$ does not exist in $[0,\infty]$. Since $r_n \to \infty$, this contradicts $(*)$.

(ii) *Call a non-degenerate interval I a* convergence sector *(for f) if*

$$(**) \qquad \begin{cases} \textit{either} & \lim_{r\to\infty} \sup_{\substack{\rho > r \\ \theta \in I}} \left|f\left(\rho e^{i\theta}\right) - c\right| = 0 \quad \textit{for some } c \in \mathbb{C} \\ \textit{or} & \lim_{r\to\infty} \inf_{\substack{\rho > r \\ \theta \in I}} \left|f\left(\rho e^{i\theta}\right)\right| = \infty. \end{cases}$$

Show that every non-degenerate interval contains a convergence sector.

Hint: Because of (i) and (∗), 12.33 is applicable to one of f or $1/f$.

Exercise 12.46 *Let $a < b$, $c < d$ be real numbers, $R := (a,b) \times (c,d)$ and f continuous on $\overline{R}$, holomorphic in R. Prove that if*

$$\min_{c \le y \le d} [\operatorname{Re} f(b+iy) - \operatorname{Re} f(a+iy)] \ge b-a,$$

then

$$\max_{a \le x \le b} [\operatorname{Im} f(x+id) - \operatorname{Im} f(x+ic)] \ge d-c.$$

Hints: According to 5.2, $\int_{\partial R} f = 0$. Writing out this integral and equating its imaginary part to 0 yields

$$\int_c^d [\operatorname{Re} f(b+iy) - \operatorname{Re} f(a+iy)]\, dy = \int_a^b [\operatorname{Im} f(x+id) - \operatorname{Im} f(x+ic)]\, dx,$$

whence

$$(d-c) \min_{c \le y \le d} [\operatorname{Re} f(b+iy) - \operatorname{Re} f(a+iy)] \le (b-a) \max_{a \le x \le b} [\operatorname{Im} f(x+id) - \operatorname{Im} f(x+ic)].$$

Exercise 12.47 *Let f be holomorphic in $S := (a,b) \times (0,\infty)$, continuous on $\overline{S}$ and suppose that $\overline{\lim}_{y\to\infty} \operatorname{Re} f(a+iy) < \underline{\lim}_{y\to\infty} \operatorname{Re} f(b+iy)$. Then f is unbounded in S.*

Hints: (PÓLYA [1933]) By translating and scaling we may suppose that

$$\overline{\lim_{y\to\infty}} \operatorname{Re} f(a+iy) < a < b < \underline{\lim_{y\to\infty}} \operatorname{Re} f(b+iy).$$

There exists then finite $c > 0$ such that

$$(*) \qquad \operatorname{Re} f(a+iy) < a \quad \text{and} \quad \operatorname{Re} f(b+iy) > b \qquad \forall y \ge c.$$

Set $m := \max_{a \le x \le b} |f(x+ic)|$ and let real $M > 0$ be given. Consider $d := c + m + M$. By (∗) we have

$$\min_{c \le y \le d} [\operatorname{Re} f(b+iy) - \operatorname{Re} f(a+iy)] \ge b-a$$

and therefore from the preceding exercise

$$\max_{a \le x \le b} [\operatorname{Im} f(x+id) - \operatorname{Im} f(x+ic)] \ge d - c.$$

It follows then from the definition of m that

$$\max_{a \le x \le b} |f(x+id)| \ge d - c - m = M.$$

§ 6 Applications to Iteration Theory

Exercise 12.48 *Prove that in* 7.43–7.51 [*except* 7.44(ii)] *the hypothesis that the region* Ω *be bounded can be weakened to the simple assumption that* $\mathbb{C} \setminus \Omega$ *contain at least two points.*

Hints: Examine the proofs of these results to see that boundedness of Ω is only used to ensure that the iterates of the self-map f constitute a normal family. Since our assumption is that there are at least two distinct points $a,\ b \in \mathbb{C} \setminus \Omega$, we have $a,\ b \notin f^{[n]}(\Omega)$ for all n, so the desired normality is a consequence of 12.27. Moreover, whenever f has a fixed-point, $\{f^{[n]}\}$ is locally uniformly bounded, by 12.26. One subtle point occurs in the proof of 7.43. We use normality to choose m_{j_k} so that $\{f^{[m_{j_k}]}\}$ either diverges to ∞ locally uniformly on Ω or else converges locally uniformly on Ω to a holomorphic limit ϕ. We have to show that it is the second alternative which occurs. Since g is assumed non-constant, $g(\Omega) \subset \Omega$, by 7.12. Now we may fix a $z \in \Omega$ and then $\{g(z)\} \cup \{f^{[n_j]}(z) : j = 1, 2, \ldots\}$ is a compact subset K of Ω. The identity $f^{[m_{j_k}]}(f^{[n_{j_k}]}(z)) = f^{[n_{j_{k+1}}]}(z)$ then shows that $\{f^{[m_{j_k}]}\}$ does not diverge *uniformly* to ∞ on K. Hence the first alternative for $\{f^{[m_{j_k}]}\}$ does not occur and the proof proceeds as in Chapter VII. To get the needed local boundedness of the sequence of iterates in 7.46(i), (iii) and 7.47(ii), cite 12.26.

Remarks: 7.44(ii) is just a little subtler when the set $\{f^{[n]}\}$ is not locally bounded. Then we need to allow the constant function ∞ as a subsequential limit, and closed and connected need to be referred to the Riemann sphere. To argue wholly in $\mathbb{C}$, apply to the set C in the 7.44(ii) hints the argument in the third paragraph of the 1.35 proof. That shows that if C has any compact component at all, then C itself is connected.

Exercise 12.49 (Extension of 10.31) *If* Ω *is a region in* $\mathbb{C}$, $a \in \Omega$ *and the group of conformal automorphisms of* Ω *that fix* a *is infinite, then* Ω *is simply-connected.*

Hints: The form of conformal automorphisms of $\mathbb{C} \setminus \{0\}$ given in 11.20(iii) shows that the present hypothesis is incompatible with $\mathbb{C} \setminus \Omega$ being a single point, and if $\Omega = \mathbb{C}$, there is nothing to prove. Hence we can suppose that $\mathbb{C} \setminus \Omega$ contains at least two points. The proof of 10.31 can then be repeated. In its preliminaries we use the strengthened version of 7.47 covered by 12.48 above, and (5) in the proof must be dropped. The proof then shows that F is a conformal map of Ω and that $F(\Omega)$ is a union of circles centered at 0. As $F(\Omega)$ is open and connected, it follows that either $F(\Omega)$ is an open disk or $F(\Omega) = \mathbb{C}$.

We've now accounted for part (a) of the following grand summary.

Theorem 12.50 (HEINS [1941b]) *Let Ω be a region in $\mathbb{C}$ whose complement in $\mathbb{C}$ contains at least two points and let f be a holomorphic self-map of Ω.*

(a) *If f has a fixed-point a in Ω, then*

either (i) $f \in \mathrm{Aut}(\Omega)$ *and has finite order* [*i.e., $f^{[n]}$ is the identity map on Ω for some $n \in \mathbb{N}$*];

or (ii) $f \in \mathrm{Aut}(\Omega)$ *is not of finite order, Ω is conformal to $\mathbb{D}$, the subgroup of* $\mathrm{Aut}(\Omega)$ *fixing a is compact, isomorphic to $\mathbb{T}$, and $\{f^{[n]}\}$ is dense in it;*

or (iii) $|f'(a)| < 1$ *and $\{f^{[n]}\}$ converges locally uniformly in Ω to a.*

(b) *If f fixes no point of Ω, then*

either (i) $f \in \mathrm{Aut}(\Omega)$ *has finite order;*

or (ii) $f \in \mathrm{Aut}(\Omega)$ *is not of finite order and the non-constant subsequential limits of $\{f^{[n]}\}$ constitute a commutative subgroup of* $\mathrm{Aut}(\Omega)$ *containing f;*

or (iii) *every convergent subsequence of $\{f^{[n]}\}$ has a constant limit, which lies in $\partial\Omega$. The totality C of such limits is a closed subset of $\partial\Omega$. C is bounded and connected if $\{f^{[n]}\}$ is pointwise bounded in Ω and otherwise C has no bounded component; that is, $C \cup \{\infty\}$ is a connected subset of the Riemann sphere.*

Proof: Only part (b) remains to be dealt with. Also, the supplemental claims in its (iii) follow from reviewing the hints to 7.44(ii) in the light of the remarks preceding 12.49.

So let us suppose that f fixes no point of Ω and that (iii) of (b) fails, and prove that then (ii) or (i) must hold. Failure of (iii) means that the limit g of some (convergent in $\mathbb{C}$) subsequence of $\{f^{[n]}\}$ is either non-constant or is a constant that lies in Ω. In the latter case the constant g would be a fixed-point of f in Ω, of which there are none under hypothesis (b). Hence g is not constant, and the contradiction $f \in \mathrm{Aut}(\Omega)$ arises this time from the 12.48 version of 7.43.

Remarks: The apparent missing symmetry between (a)(ii) and (b)(ii) above is supplied by the following fact:

In case (b)(ii), Ω *is conformal to an annulus and* $\{f^{[n]} : n \in \mathbb{N}\}$ *is dense in the connected component of* $\mathrm{Aut}(\Omega)$ *which contains the identity map, a compact subgroup of* $\mathrm{Aut}(\Omega)$ *isomorphic to* $\mathbb{T}$. (Cf. 6.18(v).)

The techniques needed to prove this are beyond the present book. Interested readers may consult p. 82 of ABATE [1989], §2, chapter IV of CARLESON and GAMELIN [1993], or p. 58 of MILNOR [1999]. However, in an awesome *tour de force* of "elementary" methods (accessible to anyone reading this book) BARGMANN [1998] established the annulus conclusion by moving a small part of its conclusion into its hypotheses: $f \in \mathrm{Aut}(\Omega)$ is fixed-point free, not of finite order, and *every* subsequential limit of $\{f^{[n]}\}$ is non-constant.

Notice that the theorem fails if $\mathbb{C} \setminus \Omega$ consists of fewer than two points: For $\Omega = \mathbb{C}$ and for $\Omega = \mathbb{C} \setminus \{0\}$ the holomorphic self-map $f(z) := z^2$ fixes the point 1, but all three alternatives in (a) fail.

The situation where $\mathbb{C} \setminus \Omega$ consists of *exactly* two points is simple (but surprising) and is *completely* described by 11.20(v) together with the following:

Amazing Fact: *If* Ω *is a region in* $\mathbb{C}$ *whose complement consists of exactly two points, then every non-constant holomorphic self-map of* Ω *is a conformal automorphism, that is, one of the six functions in* 11.20(v).

Here is the beautiful proof of BEHNKE and ZUMBUSCH [1936]. Let Ω be $\mathbb{C} \setminus \{-1, 1\}$ and ϕ be a non-constant holomorphic self-map. Consider $f(z) := \phi(1/z)$. Since $f(\Omega \setminus \{0\}) \subset \Omega$ misses two complex numbers, none of $0, \pm 1$ can be an essential singularity of f, by 12.29. That is, each is either a pole or a removable singularity. Also $\lim_{|z|\to\infty} f(z)$ exists $[= \phi(0)]$ in $\mathbb{C}$. By 11.20(i) f is therefore a rational function, hence so is ϕ, say $\phi = P/Q$, where the polynomials P and Q have no common zero. There are two cases:

$$\text{(I)} \qquad \lim_{|z|\to\infty} |\phi(z)| = \infty.$$

In this case $\deg P > \deg Q$ and $\phi = P/Q$ therefore assumes every complex value. (Why?) As ± 1 are not assumed in $\Omega = \mathbb{C} \setminus \{-1, 1\}$, it must be that ± 1 are removable singularities and that ϕ, now a polynomial since it has no poles, satisfies

$$\phi\{-1, 1\} = \{-1, 1\}.$$

Let the degree of ϕ be n; so $n \geq 1$. Replacing ϕ by $-\phi$ if necessary, we can suppose that

$$\phi(-1) = -1, \quad \phi(1) = 1.$$

Since the polynomial $\phi - 1$ has no zero except 1, it has the form

$$\phi(z) - 1 = a(z-1)^n \qquad \forall z \in \mathbb{C} \tag{1}$$

for some constant $a \neq 0$. Similarly for some $b \neq 0$

$$\phi(z) + 1 = b(z+1)^n \qquad \forall z \in \mathbb{C}. \tag{2}$$

Differentiation yields

$$na(z-1)^{n-1} = nb(z+1)^{n-1} \qquad \forall z \in \mathbb{C},$$

whence $n - 1 = 0$ and $a = b$. But then by subtracting (1) from (2) we infer $a = 1$. Thus ϕ is the identity map.

$$\lim_{|z|\to\infty} \phi(z) =: c \in \mathbb{C}. \tag{II}$$

In this case $\deg P \leq \deg Q$ and $\phi = P/Q$ assumes every complex value except possibly c. (Why?) As before, one of ± 1 must therefore be a removable singularity at which one of the values ± 1 is assumed. Replacing $\phi(z)$ by $\phi(-z)$ or $-\phi(z)$ or $-\phi(-z)$, as need be, we can suppose that -1 is removable and $\phi(-1) = -1$. Note that now 1 *must* be a pole. For otherwise the rational function ϕ is poleless, hence a (non-constant) polynomial, which (II) precludes. As ϕ is now a holomorphic self-map of $\mathbb{C} \setminus \{1\}$, we may iterate it. $\phi \circ \phi$ is a self-map of $\mathbb{C} \setminus \{1\}$ which sends Ω into Ω. Since 1 is a pole of ϕ, it follows from (II) that

$$\lim_{w\to 1} \phi \circ \phi(w) = \lim_{|z|\to\infty} \phi(z) = c. \tag{3}$$

Hence $\phi \circ \phi$ extends to an entire function. If $c \neq 1$, then from (II) we would get

$$\lim_{|z|\to\infty} \phi \circ \phi(z) = \phi(c) \in \mathbb{C}.$$

This would make $\phi \circ \phi$ constant (by Liouville), and in turn that would make ϕ constant. (Why?) Consequently it must be that $c = 1$. But then since ϕ has a pole at 1, we get

$$\lim_{|z|\to\infty} |\phi \circ \phi(z)| = \lim_{w\to 1} |\phi(w)| = \infty.$$

This shows that, as a self-mapping of Ω, $\phi \circ \phi$ comes under Case (I). The conclusion, bearing in mind the normalization made in Case (I), is that $\pm\phi\circ\phi$ is the identity function on Ω. But that clearly makes ϕ itself a bijection of

Ω, i.e., a conformal automorphism of Ω—the form of which is specified by 11.20(v).

Working on the Riemann sphere $\mathbb{C}_\infty$, Ted Gamelin, BEARDON and MINDA [2008a,b], and (independently) GROSSE-ERDMANN and MORTINI [2009] extended and simplified the ideas above to prove that if S is a finite subset of $\mathbb{C}_\infty$ and ϕ is a non-constant holomorphic self-mapping of $\mathbb{C}_\infty \setminus S$, then ϕ is a Möbius transformation which permutes $S \cup \{\infty\}$. BEARDON [2010] is an exhaustive account. Interestingly, for $S := \{-1,\ 1,\ i,\ (1+i)\ /\ \sqrt{2}\}$ the only non-constant holomorphic self-map of $\mathbb{C} \setminus S$ is the identity.

Exercise 12.51 (i) (Extension of 7.45.) *Let Ω be a region whose complement contains more than one point, $f\colon \Omega \mapsto \Omega$ holomorphic. Show that if f fixes two distinct points $a, b \in \Omega$, then f is a conformal automorphism of finite period, that is, $f^{[n]}$ is the identity map for some $n \in \mathbb{N}$.*

Hints: One of (i)–(iii) in part (a) of 12.50 prevails. Clearly (iii) cannot, because $f^{[n]}(b) = b$ for all n and $b \neq a$. But (ii) cannot prevail either, in view of 6.4(ii).

(ii) *If $\Omega \neq \mathbb{C}$ and $f \in \operatorname{Aut}(\Omega)$ fixes some $a \in \Omega$ and $f'(a) > 0$, then f is the identity map.*

Hints: If $\mathbb{C} \setminus \Omega$ contains more than one point, this fact is contained in 7.46 as extended in 12.48. Otherwise, $\mathbb{C} \setminus \Omega$ consists of exactly one point, which we may suppose to be 0, and the conclusion is a trivial consequence of 11.20(iii). [For a somewhat different proof see SAKS and ZYGMUND [1971], p. 232.]

Remark: The example $\Omega := \mathbb{C} \setminus \{0\}$, $f(z) := ze^z$, which has infinitely many fixed-points but is not univalent (because $f'(-1) = 0$), shows that the hypothesis on Ω in (i) cannot be weakened. The example $\Omega := \mathbb{C} \setminus \{-i, i\}$, $f(z) := 1/z$, $a := -1$, $b := 1$ shows that the holomorphy hypothesis in (i) cannot be weakened to meromorphy. But one could hope to relate n to the number of holes in Ω when Ω has finite connectivity.

§ 7 Ostrowski's Proof of Schottky's Theorem and an Application

Exercise 12.52 *Let f be holomorphic in $D(0,r)$, $f(0) = 1$, and $0 < |f| \leq M < \infty$. Then for all $0 < \rho < r$*

$$|f(z)| \geq M^{-2\rho/(r-\rho)} \qquad \forall z \in D(0,\rho).$$

Hint: Let F be a holomorphic logarithm for M/f and apply to $\operatorname{Re} F$ the upper Harnack inequality (5.28)(i).

Lemma 12.53 *Let f be holomorphic in a neighborhood of $\overline{\mathbb{D}}$ with $f \cdot (f - \mathbf{1})$ zero-free and for all $0 < r < 1$ define*

$$M(r) := \max_{|z| \le r} |f(z)|, \qquad m(r) := \max_{|z|\le r} \frac{1}{|f(z)|}$$

$$\widetilde{M}(r) := \max\{e, M(r)\}, \qquad \widetilde{m}(r) := \max\{e, m(r)\}$$

$$c := \max\{1, \big|\log|f(0)|\big|\}.$$

Then for any $0 < \rho < r < 1$

$$(*) \qquad \widetilde{m}(\rho) \le e^{10c/(r-\rho)} \big[\log \widetilde{M}(r)\big]^{2/(r-\rho)}.$$

Proof: Fix such an r and ρ. Notice that $\widetilde{M}(r) \ge e$ and $c \ge 1$, so $10c/(r-\rho) \ge 10/(r-\rho) > 10$. Consequently the right-hand side of $(*)$ exceeds e and the proposed inequality is true if $\widetilde{m}(\rho) \le e$, that is, if $m(\rho) \le e$. So we need only address ourselves to the case where

$$(1) \qquad \widetilde{m}(\rho) = m(\rho) > e.$$

Select z_ρ with $|z_\rho| = \rho$ such that

$$(2) \qquad \widetilde{m}(\rho) = \frac{1}{|f(z_\rho)|}.$$

Let n be the positive integer specified by

$$(3) \qquad n \le \log \widetilde{M}(r) < n + 1.$$

Now f is holomorphic and $\mathbf{1} - f$ is zero-free in $V := D(0, R)$ for some $R > 1$. Therefore by 5.35 there is a holomorphic function F such that

$$(4) \qquad F^n = \mathbf{1} - f \quad \text{in } V.$$

We set $f^* := n(\mathbf{1} - F)$. Let s denote the holomorphic n^{th} root in $D(1,1)$ which is positive on $(0, 2)$, that is, $s = e^{L/n}$ where L is the Principal Branch of the Logarithm in $\mathbb{C} \setminus (-\infty, 0]$ (3.43). We shall show that F can be chosen so that

$$(5) \qquad F = s \circ (\mathbf{1} - f) \quad \text{near } z_\rho.$$

Indeed we have

$$|f(z_\rho)| = \frac{1}{\widetilde{m}(\rho)} < \frac{1}{e} < \frac{1}{2}, \tag{6}$$

so for some connected neighborhood U of z_ρ in V we have

$$|f(z)| < 1 \qquad \forall z \in U.$$

For these z then $F(z)$ and $s(1-f(z))$ are each (non-zero) n^{th} roots of $1-f(z)$ and so their quotient is an nth root of 1. As $s \circ (\mathbf{1} - f)/F$ is continuous on the connected set U, it is therefore constantly equal to some one n^{th} root of 1, say $e^{2\pi ij/n}$. We may replace F by $e^{2\pi ij/n}F$ without altering (4) and upon so doing we shall have (5) valid throughout U.

Computing $s^{(k)}$ as $\big[s\,|\,(0,2)\big]^{(k)}$ and remembering that on $(0,2)$, s is just the unique positive n^{th} root function, we see

$$\begin{aligned} s^{(k)}(1) &= \frac{1}{n}\left(\frac{1}{n}-1\right)\cdots\left(\frac{1}{n}-k+1\right) \\ &= \frac{1}{n}(-1)^{k-1}\left(1-\frac{1}{n}\right)\left(2-\frac{1}{n}\right)\cdots\left(k-1-\frac{1}{n}\right), \end{aligned}$$

$$s(w)-1 = \sum_{k=1}^{\infty} \frac{1}{n}\left(\frac{1}{n}-1\right)\cdots\left(\frac{1}{n}-k+1\right)\frac{1}{k!}(w-1)^k \qquad \forall w \in D(1,1).$$

It follows that $\left|s^{(k)}(1)\right| \le (k-1)!/n$ and so

$$\begin{aligned} |1-s(w)| &\le \sum_{k=1}^{\infty} \frac{1}{n}|w-1|^k = \frac{1}{n}\frac{|w-1|}{1-|w-1|} \\ &< \frac{2}{n}|w-1| \qquad \forall w \in D(1, \tfrac{1}{2}). \end{aligned} \tag{7}$$

Thus for z near z_ρ we have from (5), (6) and (7)

$$|f^*(z)| < 2|f(z)|.$$

In particular

$$|f^*(z_\rho)| < 2|f(z_\rho)| = \frac{2}{\widetilde{m}(\rho)} \quad \text{by (2)}$$

and so

$$\widetilde{m}(\rho) < \frac{2}{|f^*(z_\rho)|} \le 2 \max_{|z|\le\rho} \frac{1}{|f^*(z)|} =: 2m^*(\rho),$$

$$\frac{\widetilde{m}(\rho)}{2} < m^*(\rho). \tag{8}$$

Moreover, from the definitions of F and f^* and with

$$M^*(r) := \max\{|f^*(z)| : |z| \le r\},$$

we have

$$\begin{aligned}
M^*(r) &\le n\Big[1 + \big(1 + \widetilde{M}(r)\big)^{1/n}\Big] \\
&< \big[\log \widetilde{M}(r)\big]\Big[1 + \big(1 + e^{n+1}\big)^{1/n}\Big] \quad \text{by definition (3) of } n \\
&= \Big[1 + e\big(e^{-n} + e\big)^{1/n}\Big] \log \widetilde{M}(r) \\
&\le \big[1 + e\big(e^{-n} + e\big)\big] \log \widetilde{M}(r) \\
&\le \big(e^2 + 2\big) \log \widetilde{M}(r) \\
&< 10 \log \widetilde{M}(r). \qquad (9)
\end{aligned}$$

In the transition to the final inequality we used the fact that $e^2 < 8$. [See the Remark following 3.15.] Next note that $|F(0)| \le (1 + |f(0)|)^{1/n}$ by (4), so by (4) and the definition of f^*

$$\begin{aligned}
\frac{1}{|f^*(0)|} &= \frac{1}{n|f(0)|}\left|\frac{1 - F^n(0)}{1 - F(0)}\right| \\
&\le \frac{1}{n|f(0)|}\Big(1 + |F(0)| + \cdots + |F(0)|^{n-1}\Big) < \frac{n(1 + |f(0)|)}{n|f(0)|} \\
&= 1 + \frac{1}{|f(0)|} \le 1 + e^c \quad \text{since } c \ge -\log|f(0)| \text{ by definition,} \\
&< e^{2c}. \qquad (10)
\end{aligned}$$

Now 12.52 applied to $f^*/f^*(0)$ [note that f^* is zero-free] says that

$$|f^*(0)|m^*(\rho) = \sup_{|z|\le\rho} \frac{1}{\left|\dfrac{f^*(z)}{f^*(0)}\right|} \le \left[\sup_{|z|\le r}\left|\frac{f^*(z)}{f^*(0)}\right|\right]^{2\rho/(r-\rho)} = \left[\frac{M^*(r)}{|f^*(0)|}\right]^{2\rho/(r-\rho)},$$

$$m^*(\rho) \le \left[\frac{1}{|f^*(0)|}\right]^{(r+\rho)/(r-\rho)} M^*(r)^{2\rho/(r-\rho)}.$$

Using this and (8), (9), (10)

$$\frac{\widetilde{m}(\rho)}{2} \le \big[e^{2c}\big]^{2/(r-\rho)} \big[10 \log \widetilde{M}(r)\big]^{2/(r-\rho)},$$

$$\widetilde{m}(\rho) \le 2(10)^{2/(r-\rho)} e^{4c/(r-\rho)} \big[\log \widetilde{M}(r)\big]^{2/(r-\rho)}. \qquad (11)$$

But $0 < r - \rho < 1$, $c \geq 1$ and $e^6 > \sum_{k=0}^{6} 6^k/k! > 200$, so (11) yields

$$\widetilde{m}(\rho) \leq e^{10c/(r-\rho)} \left[\log \widetilde{M}(r)\right]^{2/(r-\rho)},$$

as claimed.

Lemma 12.54 *Let the notation and hypotheses be those of the preceding lemma. Then*

$$(*) \qquad M(1-\theta) \leq e^{168c(1/\theta)\log(e/\theta)} \qquad \forall 0 < \theta < 1.$$

Proof: First consider only $\theta \in (0, 1/e]$ and suppose, with a view towards ultimately reaching a contradiction, that for some $\theta \in (0, 1/e]$

$$(1) \qquad M(1-\theta) > e^{56c(1/\theta)\log(1/\theta)}.$$

Now apply the preceding lemma to $1/f$ in the role of f. The conclusion is then

$$(2) \qquad \widetilde{M}(\rho) \leq e^{10c/(r-\rho)} [\log \widetilde{m}(r)]^{2/(r-\rho)}.$$

Here let us take $\rho = 1 - \theta$ and $r = 1 - \theta/2$, $(r-\rho)/2 = \theta/4$. We get

$$\begin{aligned}
\log \widetilde{m}(r) &\geq e^{-5c} \widetilde{M}(\rho)^{(r-\rho)/2} = e^{-5c} \widetilde{M}(1-\theta)^{\theta/4} \quad \text{by (2)} \\
&> e^{-5c} e^{-14c \log\theta} \quad \text{by (1)} \\
&= \left(\frac{1}{\theta}\right)^{14c} \left(\frac{1}{e}\right)^{5c} = \left(\frac{1}{\theta}\right)^{2c} \left(\frac{1}{\theta e}\right)^{5c} \left(\frac{1}{\theta}\right)^{7c} \\
&\geq \left(\frac{1}{\theta}\right)^{2} e^{7c} \quad (\text{since } c \geq 1 \text{ and } \theta \leq 1/e) \\
&> \frac{1}{\theta^2}\left(e^{7c} - 1\right) = \frac{7c}{\theta^2} \cdot \frac{e^{7c}-1}{7c} > \frac{7c}{\theta^2} \cdot \frac{e^7 - 1}{7} \\
&\quad (\text{since } (e^x - 1)/x \text{ is an increasing function of } x \in [1, \infty)) \\
&= \left(\frac{2}{\theta}\right)^2 c \frac{e^7 - 1}{4} > \left(\frac{2}{\theta}\right)^2 c \cdot 100 \\
&\quad (\text{from } e^6 > 200, \text{ above, follows } e^7 > 401) \\
&> 100c \frac{2}{\theta} \log \frac{2}{\theta} \quad (\text{since } x > \log x).
\end{aligned}$$

Thus

$$\widetilde{m}(1-\theta/2) > e^{100c(2/\theta)\log(2/\theta)}.$$

The right-hand side exceeds e, so the definition of $\widetilde{m}$ shows that $\widetilde{m}(1-\theta/2) = m(1-\theta/2)$ and so our inequality says that

$$m(1-\theta/2) > e^{100c(2/\theta)\log(2/\theta)} > e^{56c(2/\theta)\log(2/\theta)}. \tag{3}$$

Now just as (3) was deduced from (1) via 12.53 applied to $1/f$, we can deduce from (3) via 12.53 applied to f that

$$M(1-\theta/4) > e^{56c(4/\theta)\log(4/\theta)}. \tag{4}$$

The whole implication "(1) ⇒ (4)" can now simply be iterated n times (for any positive integer n) and there results the inequality

$$M(1-\theta/4^n) > e^{56c(4^n/\theta)\log(4^n/\theta)}, \quad n \in \mathbb{N}. \tag{5}$$

Now $(4^n/\theta)\log(4^n/\theta) \to \infty$ as $n \to \infty$ and so for large values of n, (5) contradicts the boundedness of f in $\mathbb{D}$. (Remember, f is holomorphic in a neighborhood of $\overline{\mathbb{D}}$.) Thus (1) is untenable and we have proved that

$$M(1-\theta) \le e^{56c(1/\theta)\log(1/\theta)}, \quad 0 < \theta \le 1/e. \tag{6}$$

Now $(1/\theta)\log(e/\theta)$ is clearly a decreasing function of $\theta \in (0,1]$, so its smallest value on this interval occurs at $\theta = 1$ and is 1. It follows that for $\theta \in [1/e, 1)$

$$M(1-\theta) \le M(1-1/e) \overset{(6)}{\le} e^{56ce} < e^{168c} \le e^{168c(1/\theta)\log(e/\theta)}. \tag{7}$$

Since the right-hand side of (6) is not greater than the right-hand side of $(*)$, the conjunction of (6) and (7) yields $(*)$.

Theorem 12.55 (Schottky) *Let f be holomorphic in $\mathbb{D}$ and omit 0 and 1 from its range. Set $d := \max\{1, \log|f(0)|\}$. Then*

$$\sup_{|z|\le 1-\theta} |f(z)| \le e^{169d(1/\theta)\log(e/\theta)} \qquad \forall 0 < \theta < 1. \tag{$*$}$$

Proof: For each $r \in (0,1)$ the function $f_r(z) := f(rz)$ is holomorphic in the neighborhood $D(0,1/r)$ of $\overline{\mathbb{D}}$ and so the preceding lemma applies to it and affirms that

$$|f(rz)| \le e^{168c(1/\theta)\log(e/\theta)} \qquad \forall z \in D(0, 1-\theta).$$

This holding for all $r < 1$, we let $r \uparrow 1$ and conclude that

$$|f(z)| \le e^{168c(1/\theta)\log(e/\theta)} \qquad \forall z \in D(0, 1-\theta). \tag{$**$}$$

Here $c := \max\{1, |\log|f(0)||\}$ and of course $d \le c$. If $d = c$, we have in (∗∗) (a bit more than) the desired inequality (∗).

If $d < c$, then $d = 1$ and $c > 1$. Thus $\log|f(0)| \ne |\log|f(0)||$, so $\log|f(0)| = -|\log|f(0)|| = -c < -1$, $|f(0)| < 1/e < 1/2$. Then certainly $e > |1 - f(0)| > 1/e$. Consequently the appropriate c-constant for the function $\mathbf{1} - f$ is 1, and (∗∗) for this function reads

$$|1 - f(z)| \le e^{168(1/\theta)\log(e/\theta)} \qquad \forall z \in D(0, 1-\theta),$$

whence

$$|f(z)| \le 1 + e^{168(1/\theta)\log(e/\theta)} < e^{169(1/\theta)\log(e/\theta)} \qquad \forall z \in D(0, 1-\theta)$$

(since $(1/\theta)\log(e/\theta) > 1$, as noted in the preceding proof, and $1 + e^t < e^{t+1}$ for $t \ge 0$). This confirms (∗) in the complementary case $d < c$ and finishes the proof.

Remarks: This is perhaps the simplest of the elementary derivations of Schottky's theorem which nevertheless give quantitative estimates for the majorant, free of unspecified constants. It is from OSTROWSKI [1925a]. He also presents the same proof in the appendix of his monograph [1931] but with minor modifications which lead to the better inequality

$$\sup_{|z| \le 1-\theta} |f(z)| \le e^{20d(1/\theta)\log(e/\theta) - 16}.$$

On the other hand, by sacrificing the explicit dependence on r, a bound which is sensitive to small values of $f(0)$ can easily be found. This is done in the next exercise; 12.57 is an illustration of the usefulness of such a bound. Let $H_{0,1}(\mathbb{D})$ denote the set of functions in $H(\mathbb{D})$ which map into $\mathbb{C} \setminus \{0, 1\}$, and for $r \in [0, 1)$, $a > 0$ define

(1) $$S(a, r) := \sup\{|f(z)| : f \in H_{0,1}(\mathbb{D}),\ |f(0)| \le a,\ |z| \le r\}.$$

It is clear from Schottky's theorem that this is always finite, and in fact from 12.55

$$S(a, r) \le \exp\left(\frac{169}{1-r}\log\left(\frac{e}{1-r}\right) \cdot \max\{1, \log a\}\right).$$

It is also clear that for each fixed r, $S(a, r)$ is a non-decreasing function of a.

Exercise 12.56 *Show that* $\lim_{a \downarrow 0} S(a, r) = 0$ *for each* $r \in [0, 1)$.

Hints: (OSTROWSKI [1933], p. 58) Suppose that for some $r \in [0, 1)$ this is false. That is, there exists $\varepsilon > 0$ such that

(2) $$S(a, r) > \varepsilon \qquad \forall a > 0.$$

These inequalities and definition (1) mean that there exist $f_n \in H_{0,1}(\mathbb{D})$ with $|f_n(0)| \leq 1/n$ such that

$$(3) \qquad \sup\{|f_n(z)| : |z| \leq r\} > \varepsilon \qquad \forall n \in \mathbb{N}.$$

Set $R := (1+r)/2 \in (r,1)$. The sequence $\{f_n\}$ is bounded in $D(0,R)$ by $S(1,R)$, these functions are zero-free, and $f_n(0) \to 0$. It follows from 7.10 then that $|f_n| \to 0$ uniformly in $\overline{D}(0,r)$, contradicting (3).

Example 12.57 (Mañé, Sad and Sullivan [1983]) *Let Ω be a bounded region in $\mathbb{C}$. Suppose that*

(1) *f_λ maps Ω conformally into $\mathbb{C}$* $\quad \forall \lambda \in \mathbb{D}$;

(2) *$\lambda \mapsto f_\lambda(z)$ is holomorphic in $\mathbb{D}$* $\quad \forall z \in \Omega$;

(3) $f_0(z) = z \quad \forall z \in \Omega$.

Then each f_λ extends to a homeomorphism of $\overline{\Omega}$ and (2) holds for $z \in \partial\Omega$ as well. Moreover, $(\lambda, z) \mapsto f_\lambda(z)$ is continuous on $\mathbb{D} \times \overline{\Omega}$.

Proof: For any three distinct points z, w, $v \in \Omega$, the function

$$h_{z,w,v}(\lambda) := \frac{f_\lambda(z) - f_\lambda(w)}{f_\lambda(v) - f_\lambda(w)}, \qquad \lambda \in \mathbb{D},$$

lies in $H_{0,1}(\mathbb{D})$, thanks to (1) and (2), and by (3)

$$h_{z,w,v}(0) = \frac{z-w}{v-w}.$$

Consider $r \in (0,1)$, $\lambda \in \overline{D}(0,r)$. By definition of S above

$$|h_{z,w,v}(\lambda)| \leq S\left(\left|\frac{z-w}{v-w}\right|, r\right), \qquad \text{that is,}$$

$$(4) \qquad |f_\lambda(z) - f_\lambda(w)| \leq |f_\lambda(v) - f_\lambda(w)| S\left(\left|\frac{z-w}{v-w}\right|, r\right).$$

Let d denote the diameter of Ω. Then for each $w \in \Omega$ there exists $w^* \in \Omega$ with $|w - w^*| > d/3$. With $v = w^*$ (4) yields

$$(5) \qquad |f_\lambda(z) - f_\lambda(w)| \leq |f_\lambda(w^*) - f_\lambda(w)| S\big(3|z-w|d^{-1}, r\big),$$

$$(6) \qquad \leq |f_\lambda(w^*) - f_\lambda(w)| S(3, r),$$

holding for all $z \in \Omega \setminus \{w, w^*\}$. Fix $w_0 \in \Omega$ and note that by (2) the number

$$A_r := \sup\{|f_\lambda(w_0^*)| + |f_\lambda(w_0)| : |\lambda| \leq r\}$$

is finite. Taking $w = w_0$ in (6) yields

$$|f_\lambda(z)| \le A_r + A_r \cdot S(3, r) =: M_r,$$

holding for all $\lambda \in \bar{D}(0, r)$ and for all $z \in \Omega$ (even for $z = w, w^*$ because of continuity of f_λ). Now (5) yields

$$(7) \qquad |f_\lambda(z) - f_\lambda(w)| \le 2M_r \cdot S(3|z - w|d^{-1}, r) \quad \forall \lambda \in \bar{D}(0, r),$$

holding for all $w \in \Omega$, $z \in \Omega \setminus \{w, w^*\}$. Given $\varepsilon > 0$, 12.56 provides $\delta > 0$ such that $S(3\delta d^{-1}, r) \le \varepsilon/(1+2M_r)$. Then (7) says that for $w \in \Omega$, $z \in \Omega \setminus \{w, w^*\}$

$$(8) \qquad |z - w| < \delta \Rightarrow |f_\lambda(z) - f_\lambda(w)| \le \varepsilon \quad \forall \lambda \in \bar{D}(0, r).$$

Continuity again lets the restriction $z \notin \{w, w^*\}$ be dropped, whereupon (8) says that each f_λ is *uniformly* continuous on Ω, and consequently extends to a continuous function (call it also f_λ) on $\overline{\Omega}$. Evidently (8) remains valid for $z, w \in \overline{\Omega}$, from which it follows routinely that $(\lambda, z) \mapsto f_\lambda(z)$ is continuous at each point $(\lambda_0, z_0) \in \bar{D}(0, r) \times \overline{\Omega}$, $r < 1$; that is, at each point of $\mathbb{D} \times \overline{\Omega}$.

If we fix $z \in \partial\Omega$ and let $w_n \in \Omega$ converge to z, we learn from (8) that $f_\lambda(w_n)$ converges to $f_\lambda(z)$ uniformly for λ in the open set $D(0, r)$. Since $r < 1$ is arbitrary and each map $\lambda \mapsto f_\lambda(w_n)$ is holomorphic in $\mathbb{D}$ by (2), it follows that $\lambda \mapsto f_\lambda(z)$ is also holomorphic in $\mathbb{D}$.

Since f_λ is conformal on Ω, $f_\lambda(\partial\Omega)$ is disjoint from $f_\lambda(\Omega)$. To see this, suppose $z_0 \in \partial\Omega$ and $f_\lambda(z_0) = f_\lambda(w_0)$ for some $w_0 \in \Omega$. Take disjoint open neighborhoods $W_0 \subset \Omega$ of w_0 and V_0 of z_0 such that the continuous function f_λ on $\overline{\Omega}$ maps the neighborhood $\overline{\Omega} \cap V_0$ of z_0 into the open (5.77) neighborhood $f_\lambda(W_0)$ of $f_\lambda(w_0) = f_\lambda(z_0)$. For any point $v \in \Omega \cap V_0$ (such exist, since $v_0 \in \partial\Omega$) we have $v \in \Omega \setminus W_0$, yet $f_\lambda(v) \in f_\lambda(W_0)$, in clear violation of the univalence of f_λ in Ω. In particular, if we fix $w_0 \in \Omega$, the number

$$d_\lambda := \text{dist}\,(f_\lambda(w_0), f_\lambda(\partial\Omega)) \text{is positive.}$$

Consider any v, $w \in \Omega \setminus \{w_0\}$ with $|v - w| > \varepsilon > 0$ and set $z = w_0$ in (4). We get

$$\begin{aligned} |f_\lambda(w_0) - f_\lambda(w)| &\le |f_\lambda(v) - f_\lambda(w)| S\left(\left| \frac{w_0 - w}{v - w} \right|, r \right) \\ &\le |f_\lambda(v) - f_\lambda(w)| S(d\varepsilon^{-1}, r). \end{aligned}$$

By continuity this inequality holds too for all v, $w \in \overline{\Omega}$ satisfying $|v - w| > \varepsilon$. Thus

$$0 < d_\lambda \le |f_\lambda(v) - f_\lambda(w)| S(d\varepsilon^{-1}, r) \qquad \forall v, w \in \overline{\Omega} \text{ with } |v - w| > \varepsilon.$$

Here $\varepsilon > 0$ is arbitrary. Consequently, if $|f_\lambda(v) - f_\lambda(w)| < d_\lambda / S(d\varepsilon^{-1}, r)$ for a pair of points $v, w \in \overline{\Omega}$, then these points cannot satisfy $|v - w| > \varepsilon$; that is, they must satisfy $|v - w| \le \varepsilon$. This proves that f_λ is one-to-one on $\overline{\Omega}$ and f_λ^{-1} is (uniformly) continuous on $f_\lambda(\overline{\Omega})$.

Final Remarks: (i) The continuity of $(\lambda, z) \mapsto f_\lambda(z)$ in 12.57 is a special instance of a very general phenomenon (cf. 7.21(ii)). If U, W are open subsets of $\mathbb{C}$, $F\colon U \times W \mapsto \mathbb{C}$ and $z \mapsto F(z, w)$ is holomorphic in U for each $w \in W$, while $w \mapsto F(z, w)$ is holomorphic in W for each $z \in U$, then F is continuous on $U \times W$. In fact, in any bidisk $\bar{D}(z_0, r) \times \bar{D}(w_0, \rho)$ lying in $U \times W$, $F(z, w)$ coincides with a uniformly convergent double series $\lim_{K,N\to\infty} \sum_{k=0}^{K} \sum_{n=0}^{N} c_{k,n}(z - z_0)^n (w - w_0)^k$. For this important theorem of Friedrich Hartogs see §§356–361 of CARATHÉODORY [1960].

(ii) E. LANDAU [1906] proved Schottky's theorem in the form

$$|f(z)| \le \big[A|f(0)| + A\big]^{\frac{1+r}{1-r}} \qquad \forall |z| \le r, \tag{9}$$

where $A \ge 1$ is a universal constant. VALIRON [1927b] pointed out how this yields a quantitative version of 12.56: Fix $z \in D(0, r)$ and apply (9) to the function $w \mapsto 1/f\big(\frac{w+z}{1+\bar{w}z}\big)$, getting

$$|f(z)|^{\frac{1+r}{1-r}} \le \left| f\left(\frac{w+z}{1+\bar{w}z}\right)\right| [A|f(z)| + A]^{\frac{1+r}{1-r}} \quad \forall |w| \le r.$$

In particular, for $w := -z$

$$|f(z)| \le |f(0)|^{\frac{1-r}{1+r}} A[1 + |f(z)|], \text{ whence from (9) again}$$

$$|f(z)| \le |f(0)|^{\frac{1-r}{1+r}} A\left[1 + \big(A|f(0)| + A\big)^{\frac{1+r}{1-r}}\right] \qquad \forall |z| \le r. \tag{10}$$

Therefore there is a universal constant B such that

$$|f(z)| \le B^{\frac{1}{1-r}} |f(0)|^{\frac{1-r}{1+r}} \quad \forall |z| \le r, \text{ whenever } |f(0)| \le 1.$$

For example (since $A \ge 1$), $B := (2A)^2$ works. In the notation of 12.56 this means that

$$S(a, r) \le B^{\frac{1}{1-r}} a^{\frac{1-r}{1+r}} \qquad \forall 0 < a,\ r < 1.$$

Incidentally, as Valiron points out, this inequality easily implies the result 7.10 which features in our proof of 12.56.

Notes to Chapter XII

The trek to the *Hauptsatz* 12.14 (MIRANDA [1935]) is laborious but, I think, worth the trouble. For one thing, these techniques have a quite broad applicability; they represent the elementary phase of a very fruitful and deep chapter of function theory inaugurated in the 20th century by R. NEVANLINNA. See his book [1953], [1970] and HAYMAN [1964]. My account follows VALIRON [1937] (a related, less self-contained, treatment being that by LIVERPOOL and NNADI [1983]). I have, however, tried to watch over the constants more carefully and to keep before the reader at all the times their independence from the particular functions being analyzed. The inequality we obtain is not the best. It can be improved, as regards its dependence on $1/(1-r)$, for example, by using techniques from MILLOUX [1940] (see also BUREAU [1996]). HIONG [1958a] proves that for each non-negative integer n there are positive constants H_n, K_n such that

$$(*)\qquad \log M(r,f) \le \frac{1}{1-r}\left[H_n|\log|f(0)|| + K_n \log\frac{2}{1-r}\right], \qquad 0 \le r < 1$$

for all f as in 12.14. In its dependence on $1/(1-r)$ this is a marked improvement over the inequality in the text. Contrary to that author's claim (p. 987, *op. cit.*), I found that when the proof details and supporting material had all been supplied, the proof of $(*)$ was longer than that in the text. However, a more serious disadvantage of $(*)$ is the fact that the bound on f which it provides gets coarser as $|f(0)| < 1$ gets smaller, and because of this I was unable, despite the author's claim (p. 994, *op. cit.*), to deduce from $(*)$ the important normality criterion 12.27. [For the history and priorities on the latter see p. 987, *op. cit.*] So I settled for the version of Miranda's Theorem in the text. The reader may wish to investigate this matter on his own. He will find in HIONG [1968] an English translation of (a Chinese translation of) HIONG [1958a] and interesting related matter in HIONG and HO [1961], SHIEH [1962] and YANG and CHANG [1965]. In fact, the Chinese school has made many important contributions to this theory and for a survey of them, together with all the relevant literature, the reader is referred to CHUANG and YANG [1985]. Two attractive such results are: (1) the verbatim generalization by Yong Xing Ku (= Y.-X. Gu) of 12.27 to meromorphic functions and (2) a normality criterion involving just *one* omitted value: if for some non-negative integer $n \neq 1$ all the products $f \cdot f^{(n)}$ omit 1, the family is normal. See also the monographs of SCHIFF [1993] and CHUANG [1993].

The special case of this theorem in 12.15 and 12.55 is due to SCHOTTKY [1904], [1906], [1917] (and BOUTROUX [1905], [1906] independently, though his derivation contains a non-trivial lacuna), and is already a very deep and striking result. His original theorem asserted a bound on $\log M(r,f)$ dependent only on $f(0)$ and $(1-r)^{-4}$. E. LANDAU [1906] (and LÉVY [1912]) showed that the dependence on r was of order $(1-r)^{-1}$. For Schottky's theorem there exist very polished direct proofs, the products of many hands: E. LANDAU [1926], BLOCH [1924], [1925], VALIRON [1926], [1930], H. KÖNIG [1957]. See, for example, the account in SAKS and ZYGMUND [1971], pp. 341–350, CONWAY [1978] or E. LANDAU [1940] or [1929a]. (See especially the notes in the 3rd ed. of the latter for history.) For a somewhat different proof see TITCHMARSH [1939], CARTWRIGHT [1956] and HOLLAND [1973]. For an exhaustive account of Schottky's theorem exploring many of its applications see the treatise of OSTROWSKI [1931], chapters 1 and 4. For extensions of a different kind from Miranda's see BIEBERBACH [1922], E. LANDAU [1922b], FEKETE [1922], SAXER [1934] and KRAJKIEWICZ [1977].

In our version neither the constants nor the order of dependence on $1/(1-r)$ are best possible. In HAYMAN [1947] (and [1989], pp. 685 ff.) occurs one of the sharpest forms of the inequality: $\log M(r, f) \leq \left[\pi + \log^+ |f(0)|\right] \frac{1+r}{1-r}$. JENKINS [1955] gives a short proof of this and some improvements which depend on the size of $|f(0)|$. Earlier AHLFORS [1938] had proved a corresponding inequality with the (larger) constant $4 + \log 10$ instead of π. See also ROBINSON [1939]. HEMPEL [1980] succeeded in reducing the right-hand side of Hayman's inequality by the additive term $-\log 16$. In Chapter XVIII we shall prove a sharp Schottky-type inequality for *univalent* functions. OSTROWSKI [1933] finds asymptotic expressions for $M(r, f)$ as $r \uparrow 1$.

The amazing lemma 12.10 is from BUREAU [1932]; the ideas in it go back to BOREL [1896], [1896–97] and BLUMENTHAL [1910]. A version of it features also in Ostrowski's proof. (See the proof of 12.54.)

12.16 is due to BUREAU [1932] and 12.18 to PICARD [1879a]. The beautiful corollary 12.19 is asserted in PÓLYA [1921] and [1922] and proved in SAXER [1923], VAROPOULOS [1928] and CSILLAG [1928], [1935]. This result has a remarkable generalization due to TUMURA [1937] (whose proof contained gaps) and CLUNIE [1962]: if f is entire and for some $k \geq 2$, $f \cdot f^{(k)}$ is zero-free, then f has the form $f(z) = e^{az+b}$ for some constants a and b. See pp. 67 ff. of HAYMAN [1964]. For a short, but not elementary, proof when $k = 2$, (first proved by Hayman) see C.-C. YANG [1970]. See also S. S. MACINTYRE [1949] and EDREI [1955]. Finally, LANGLEY [1993] and G. FRANK [1976] have proved a conjecture of Hayman that if meromorphic f are admitted, the only new members of the club are $f(z) = (az + b)^{-n}$, n a positive integer.

12.17 is *Satz* VIII in HURWITZ [1904], with $\alpha \geq 1/58$. E. LANDAU [1906], p. 285 gave a proof of this also. 12.17 was rediscovered by BOCHNER [1926] and FEKETE [1927a]. CARATHÉODORY [1907] and Bochner each confirmed Hurwitz' conjecture that $1/16$ is the best value for α. (See also NEHARI [1952], p. 323 or SANSONE and GERRETSEN [1969], p. 502.) The proofs use a modular function; see below. For a related result see E. LANDAU [1922a], BOHR [1923], MONTEL [1929] and p. 453 of HAYMAN [1989].

A brilliant elementary proof of 12.18 (using essentially only 5.28) was discovered by J. L. LEWIS [1994]. A nice exposition of it appears in chapter 1 of RANSFORD [1995]. GIRELA [2004] was able to extend these ideas to a proof of Picard's Great Theorem (see below); GRAHL and ROTH [2009] did so as well, and in addition captured 12.27.

For 12.20 see ROSENBLOOM [1948], [1952]. BERGWEILER [1990] proves that the conclusion of 12.20 is true of any two non-polynomial entire functions f and g. If $f \in H(\mathbb{C})$ is not a polynomial, I. N. BAKER [1960] proved: For all but at most one $n > 1$, f has infinitely many fixed-points of exact order n (meaning points fixed by $f^{[n]}$ but not by $f^{[n-1]}$); BERGWEILER [1991] showed that in fact there are no exceptional $n > 1$. By combining 12.20 with some of the ideas in the proof of 7.46 one can deduce the following interesting result: If f, g are non-constant entire functions and $f \circ g = f$, then g has the form $g(z) = az + b$, where a is a root of unity. For details see FLATTO, GROSS and GOLDSTEIN [1966]. This theme of fixed-points and "factorization" via composition is exhaustively explored in CHUANG and YANG [1990].

12.23 is from chapter IV of MONTEL [1916]. (See the last paragraph of the review of this paper, and also p. 80 of MONTEL [1927].) Later elementary proofs were given by VALIRON [1929a], GANAPATHY IYER [1938–39], JATEGAONKAR [1965] and F. GROSS [1966]. For $n > 2$ the equation $f^n + g^n = h^n$ was found to have no polynomial (hence no rational) solutions, other than f and g constant multiples of h, by LIOUVILLE [1879] and KORKINE [1880]; see also GREENLEAF [1969], RIBENBOIM [1979] and p. 39 of LANG [1990]. The modern proofs are algebraic and elementary.

The important special case of 12.27 when $N = 0$ is due to MONTEL [1912] (but is adumbrated in *Satz* VI of CARATHÉODORY and LANDAU [1911]; see also the footnotes there). For other proofs in that case see DE LA VALLÉE POUSSIN [1915–16] and p. 22 of RANSFORD [1995]. This theorem is the principal ingredient in our development of the Julia theorem (12.29); in turn its principal ingredient is Schottky's theorem. The definition of normality for a family $\mathcal{F}$ of functions which are meromorphic in a region Ω is simply that every sequence in $\mathcal{F}$ contain a subsequence which converges locally uniformly in the chordal metric (see the notes to chapters VII and the Remarks before 11.47). In modern terms, this means that $\mathcal{F}$ has compact closure in the appropriately topologized space of $\mathbb{C}_\infty$-valued functions, and in turn that is equivalent to $\mathcal{F}$ being equicontinuous (cf. 7.13(i)) on each compact subset of Ω with respect to the chordal metric. This was first pointed out by OSTROWSKI [1925b], p. 231. If $\mathcal{F}$ is a family of meromorphic functions, the appropriate analog of the sufficient condition for normality in 12.27 ($N = 0$ case) is that there be *three* distinct values in $\mathbb{C}_\infty$ not in the range of any of the functions. Its reduction to 12.27 is via the idea in the proof of 12.18. Also in terms of the chordal metric a very useful normality criterion was given in 1931 by Felix Marty. See HAYMAN [1964], pp. 158–160 for references and a statement and proof of the result and ROYDEN [1985], SCHIFF [1993] for an extension. The important paper of CARATHÉODORY [1929b] should also be mentioned in this context.

MANDELBROJT [1929] provided a boundedness criterion for a family $\mathcal{F}$ of zero-free holomorphic functions on a region Ω to be normal: For each compact $K \subset \Omega$, $f \in \mathcal{F}$ define $M(f, K) := \max\{|f(z)|/|f(w)| : z, w \in K\}$ and $L(f, K) := \max\{\log|f(z)|/\log|f(w)| : z, w \in K\}$ if $f(K) \cap \mathbb{T} = \varnothing$, otherwise $L(f, K) := +\infty$. Then $\mathcal{F}$ is normal if and only if for each K the set $\{\min\{M(f, K), L(f, K)\} : f \in \mathcal{F}\}$ is bounded. Another proof of this was given by DOVBUSH [2014].

For an elementary proof of a slightly weakened version of 12.30 see ZALCMAN [1978]. Also of interest in this connection is GETHNER [1992].

Before 12.29, PICARD [1879b] had proved that there exists $b \in \mathbb{C}$ such that $\mathbb{C} \setminus \{b\} \subset f(D(a, R))$ for all (sufficiently small) $R > 0$, i.e., in every neighborhood of a, f assumes all complex values with at most one exception. This is the so-called Big (or Great) Picard Theorem [in contrast to the little (or lesser) one in 12.18]. See FRANKLIN [1925a] for a modest extension and CARATHÉODORY [1912b] (also OSTROWSKI [1929b], BERNAYS [1913] (p. 220) and E. LANDAU [1929d]) for a quantitative extension: there exists a constant $r \in (0, 1)$ such that each f which is holomorphic in $A(0, 1)$ and omits 0 and 1 from its range satisfies either $f(A(0, r)) \subset D(0, 2)$ or $f(A(0, r)) \subset \mathbb{C} \setminus D(0, \frac{1}{2})$. 12.30(ii) may be regarded as the extension of Picard's Great Theorem to meromorphic functions. For Julia's extension of Picard's theorem see p. 102 of JULIA [1924]. Interesting extensions of Julia's theorem may be found in OSTROWSKI [1931]. An extension to harmonic functions appears in MONTEL [1932c]. The "cercles de remplissage" $D_{k,m}$ constructed in the proof of Julia's theorem depended on the function f. Much research has been done to find large sets $E \subset \mathbb{C}$ (so-called "Picard sets") such that in $\mathbb{C} \setminus E$ *every* non-constant entire function takes all complex values with at most one exception. I shall mention only the attractive paper of ANDERSON and CLUNIE [1980] where sets E of the form $\bigcup_{n=1}^{\infty} D(a_n, \rho_n)$, with $|a_n|$ and ρ_n^{-1} rapidly diverging to ∞ are constructed.

There is an entirely different, rather geometric, development of Schottky's theorem and many of its corollaries like the Picard theorems, based on the Monodromy Theorem (10.3(iv)) and a modular function. The latter is a holomorphic map of the upper half-plane onto $\mathbb{C} \setminus \{0, 1\}$ which is invariant under a certain group of conformal self-maps of the upper half-plane. For a careful treatment of it see pp. 328–331 of RUDIN [1987] or chapter 23 of ESTERMANN [1962]. This approach was the historically first one, used for example by Picard himself, and it is via a modular function that the sharpest inequalities have been obtained. See E. LANDAU [1906] and the works referenced there, CARATHÉODORY [1960], DINGHAS [1961], §§71, 72; HILLE [1962], §14.5; VEECH [1967], pp. 126–136 and LEHNER [1969]. The proof given by PERRON [1929] is a more elementary version of the modular function proofs; he manufactures on the spot a substitute for the modular function. For a more detailed exposition of Perron's work see §140 of PRINGSHEIM [1932]. Cf. also DE LA VALLÉE POUSSIN [1915–16]. There is a differential geometry treatment of the theorems of Schottky, Montel (12.27 with $N = 0$) and Picard which uses something like the modular function, but can be made entirely elementary. For a beautiful account of it the reader is urged to look at MINDA and SCHOBER [1983] or KRANTZ [1990], also chapter 4 of NARASIMHAN [1985]. The case $N = 0$ of 12.27 is proved using a version of 12.10 and a few ideas from Nevanlinna theory (like those in §2) by SCHWICK [1992].

After Picard proved his results with the modular function, a search began for an elementary proof (meaning one which did not use the modular function or any analytic results of comparable depth). In [1896] BOREL achieved this for the little theorem (see also the appendix to his book [1921]) and in [1904], building on Borel's technique, SCHOTTKY did it for the great theorem. For a nice treatment of their work (including contributions of Carathéodory and Landau) see the congress lecture of LINDELÖF [1910]. The next phase occurred with the arrival of Bloch's theorem 7.29(iii) (see Chapter VII notes), already presaged in the work of Landau, Hurwitz and Koebe. From this follows a relatively easy proof of Schottky's theorem, thence the Picard theorems. This is the approach followed in almost all textbooks which want to avoid the modular function. The account in E. LANDAU [1929a] (together with ESTERMANN [1971]) is especially recommended. 7.29(v) is a sample of this elegant reasoning; the function F constructed there plays a further role in the proof of Schottky's theorem (see p. 298 of CONWAY [1978]). However, the majorant that such proofs produce is less transparent in its dependence on r and is orders of magnitude greater than those found in the proofs of 12.15 and 12.55. An elementary proof of a different kind (using Harnack-type inequalities) was given by GIRELA [2004]; this is an extraordinarily elegant paper.

If one only wants the Great Picard Theorem and not Julia's, then normal families can be avoided and a constructive proof based directly on Schottky; see BRIDGES [1981] and the book of BISHOP and BRIDGES [1985]. At the other non-constructive extreme there is a proof of Picard's theorem using Brownian motion and probability theory; see DAVIS [1979].

For a detailed history of the Schottky and Picard theorems up to 1920, see pp. 409–417 of BIEBERBACH [1921]; for related history see VALIRON [1932].

In 12.31 (taken from pp. 66–70 of MONTEL [1927]) the reader has a brief introduction to Montel's theory of quasi-normal families. See his book just cited, VALIRON [1929b] and SCHIFF [1993] for more on this topic. Notice that the functions $f_n(z) := nz$, $z \in \Omega := \mathbb{D}$ show that the hypotheses of 12.31(v) cannot be weakened, and that the exceptional point in 12.31(iv) may be genuine.

12.33 is due to MONTEL [1912]. Nice generalizations in which we only hypothesize that $f(z_n) \to c$ for a certain sequence of points $z_n \in S$ may be found in EGGLESTON [1951], A. EVANS [1953] and BOWEN and MACINTYRE [1954] (cf. 7.34). For a good summary statement of the classical results, containing $5.16 \cup 5.56 \cup 12.39 \cup 12.34$, plus references to important work of W. Gross and F. Iverson on the subject, see p. 420 of the encyclopedia article of BIEBERBACH [1921]. For a unified treatment of all these results via harmonic-majorants see pp. 301–308 of TSUJI [1959]. For other proofs of 12.34 see §21 of MONTEL [1917], exercise III. 277 of PÓLYA and SZEGÖ [1972] and pp. 12–13 of F. and R. NEVANLINNA [1922]. For 12.39 see p. 518 of MONTEL [1912] and OU [1957].

12.41–12.43 are from HARDY, INGHAM and PÓLYA [1928]. They actually state 12.41(iv) for semi-strips $(\alpha, \beta) \times (c, \infty)$ instead of strips S. Unable to fully confirm their proof, I settled for the strip version. These authors show that equality can be tolerated in hypothesis (2) of 12.41 and 12.42 and provide a counterexample to the conclusion when $b-a > \frac{1}{2}(\beta-\alpha)$. But they show how to eliminate all hypotheses on a, b (beyond $a \neq b$, of course) when the function has no zeros. See CARTWRIGHT [1962] and HAYMAN [1962] for extensions. 12.44 is from HARDY [1926]; for a generalization see TSUJI [1930b], SHIMIZU [1931a], BONSALL [1949] and BEARDON [1971] (subharmonic functions); see also pp. 392 ff. of OSTROWSKI [1937]. A related result due to PRIVALOV [1936] says that if u is subharmonic and bounded above in $(\alpha, \beta) \times \mathbb{R}$, then $\sup_y u(x+iy)$ is a convex function of x.

I took 12.45 from ROTH [1938–39], though it is adumbrated in BOHR [1930] and [1927]. Compare also LAURITZEN [1950]. There is a more extensive treatment in GAIER [1987], pp. 164–168. Bohr proves an analog of 12.45 for entire functions f which are bounded on every ray: given any non-degenerate closed interval $J \subset [0, 2\pi]$, there is a non-degenerate closed interval $I \subset J$ such that $\{f(re^{i\theta}) : \theta \in I, r \geq 0\}$ is bounded. He also shows that if $\{I_n\}$ are disjoint open intervals in $[0, 2\pi]$ whose union is dense in $[0, 2\pi]$, then there exists a non-constant entire f which is bounded on every ray and uniformly bounded in each set $\{re^{i\theta} : \theta \in I_n, r \geq 0\}$. Without the uniformity feature 12.45(ii) is essentially contained in BOHR [1927]. Non-constant entire functions bounded on every ray through 0 are not too hard to construct: see D. J. NEWMAN [1976] or ULLRICH [1994a]. However, if $f : [0, 2\pi] \mapsto [0, \infty]$ is an integrable function, then any entire function that is bounded on the half-line $\{re^{i\theta} : r \geq 0\}$ by $\exp(\exp(f(\theta)))$ for every $\theta \in [0, 2\pi]$ is necessarily constant (CARLEMAN

[1926a]). As to non-constant entire functions which have finite limits at infinity along every ray, these too exist. The easiest examples I know of are the function $f(z)f(iz)$, where f is the function Newman constructs, *op. cit.*, and the F in Example 2.1 of BEARDON and MINDA [2003], which actually vanish at ∞ along every line through the origin; further refinements by ARMITAGE [2007].

Both 12.46 and 12.47 are taken from PÓLYA [1933]. See also AHLFORS [1933], MILLOUX [1935], KERÉKJÁRTÓ [1934/35], A. J. MACINTYRE [1939] and JENKINS [1953]. An even briefer proof of a related result can be found in LANDAU and OSSERMAN [1959]. They use the Cauchy-Riemann equations and a double integral instead of Cauchy's theorem.

12.48 may fail if only one point is omitted. Specifically, for $\Omega := \mathbb{C} \setminus \{-2\}$ and $f(z) := 2(e^z - 1)$, we have $f(0) = 0$ and $f(\Omega) = \Omega$. Since, however, $f'(0) = 2$, the conclusion of 7.46(i) does not hold there.

12.6 was improved by OSTROWSKI [1926a] as follows: Let $\mathcal{L} = \{f \in H(\mathbb{D}) : \sup_{r<1} \int_0^{2\pi} \log|f(re^{i\theta})|\, d\theta < \infty\}$, the integral being in Lebesgue's sense. A necessary and sufficient condition on points $z_n \in \mathbb{D}$ that there exist $f \in \mathcal{L}$ whose zero set is $\{z_1, z_2, \ldots\}$ is that $\sum_{n=1}^{\infty}(1 - |z_n|) < \infty$. For a related result see DENJOY [1929] (cf. also the Chapter VII notes). Proofs of 12.6 and 12.7(ii) may also be found on pp. 26–28 of F. and R. NEVANLINNA [1922]. They show that the hypothesis $(*)$ of 12.6 is equivalent to F being a quotient of two bounded holomorphic functions in the disk (this follows easily from 7.16(iii) and the Jacobi-Jensen formula), so the generality of 12.6 over 6.7(iii) is illusory. Ostrowski's result above (resp., 6.7(iii)) together with 7.36 enable one to factor any function of class $\mathcal{L}$ (resp., $H^\infty(\mathbb{D})$) into the product of a (bounded) zero-free holomorphic function in $\mathbb{D}$, i.e., the exponential of a holomorphic function in $\mathbb{D}$, and a Blaschke product. This representation is extremely useful. See the elementary discussion on pp. 40–45 of MONTEL [1927]. For a comprehensive treatment see chapter VII of R. NEVANLINNA [1970].

If hypothesis $(*)$ in 12.6 is weakened to $\sup_{r<1} \int_0^r \int_0^{2\pi} \log^+ |F(\rho e^{i\theta})|\, d\theta\, d\rho < \infty$, then the weaker conclusion $\sum_{j=1}^{\infty}(1 - |z_j|)^2 < \infty$ can be inferred; see OPIAL [1957], KABAILA [1971] and HEILPER [1979].

The theorems of this chapter led André Bloch to formulate the following "heuristic principle": *A family of holomorphic* (*meromorphic*) *functions which have a property P in common in a domain D is* [apt to be] *a normal family in D if P cannot be possessed by non-constant holomorphic* (*meromorphic*) *functions in the whole plane* $\mathbb{C}$. [An example of such a P is "uniform boundedness."] A rigorous formulation and proof of this was given by ZALCMAN [1975], with extensions and counterexamples by MINDA [1985], RUBEL [1986] and PANG [1989]. It gives a new approach to the main theorems here, showing, for example, how Montel's normality theorem and Picard's Big Theorem are consequences of Picard's Little Theorem. MINDA [1992] shows how in turn to get Bloch's covering theorem (7.29(iii)) and thence Picard's Little Theorem from Zalcman's rigorous formulation and 7.19. ZALCMAN [1993], [1998], [2016] give an especially attractive overview of this program, and it is used by GRAHL [2003] to prove the important normality criterion 12.27 independently of 12.14. See also GRAHL and ROTH [2009].

SCHIFF [1993] is an excellent (classical in flavor) treatment of many of the ideas in this chapter and chapter VII, including sections on normality for harmonic functions and normality for meromorphic functions. See also CHUANG [1993] and BERGWEILER [2006].

Epilog

And each man hears as the twilight nears, to the beat of
his dying heart,
The Devil drum on the darkened pane: "You did it,
but was it Art?"

Rudyard Kipling (1865–1936)

Bibliography

In spite of its bulk, this bibliography is rather selective. E.g., hardly any papers on Riemann surfaces, elliptic functions, Dirichlet series, the order and type theory of entire functions or Nevanlinna theory appear because these topics are not discussed in the text. In general too the references and notes do not attempt to lead the reader into the differentiable manifold, global analysis or several complex variable extensions of the material presented in the text.

Most of the entries carry a reference to one or two of the review journals: **MR** = *Mathematical Reviews* (published by the American Mathematical Society, Providence), **Zbl** = *Zentralblatt für Mathematik und ihre Grenzgebiete* (published by Springer-Verlag, Berlin), **FM** = *Jahrbuch über die Fortschritte der Mathematik* (published by Verlag Georg Reimer, Berlin, prior to 1913 and thereafter by Walter de Gruyter and Co., Berlin). The first covers 1940 to the present, the second covers 1931 to the present, and the third covers 1868 to 1942. Every paper which was reviewed is equipped with its review coordinates. In addition, many of the books and monographs here have been accorded longer, more critical reviews in the *Jahresbericht der Deutschen Mathematiker Vereinigung* or in **BAMS** = *Bulletin of the American Mathematical Society*. For the reader's further convenience, books reviewed in the latter are also supplied with the volume and page numbers where that review can be found. The reader should always consult the several reviews of each paper (some being more detailed than others) because the reviews point out contact with related results in the literature and frequently discuss questions of priority or indicate errors. I hope, too, that reference to the review will compensate for any bibliographic inadequacies or inaccuracies in the individual books and papers listed here. "See also **MR** … **Zbl** …" after an entry signals a closely related work that could not be accommodated in this bibliography. Since its founding in 1953 *Referativnyĭ Žurnal Matematika* has also provided comprehensive reviews (in Russian) of the world's mathematical literature and the reader might wish to check there works published after 1953. For more details about these review journals, all of which are now accessible electronically, see Lohwater; [1966–67] and **MR** 95f:01041, **MR** 95k:01033 and **Zbl** 1288.01021.

With books I try to list (only) the latest edition and rely on the review to mention ISBN numbers, all previous editions, reprints and translations and the coordinates of their reviews; almost always the review supplies this information. The only disadvantage of this is a chronological distortion when, for example, one cites "the great classic Saks and Zygmund [1971]." Some journals (like the *Bulletin of the American Mathematical Society* and the *Comptes*

R. B. Burckel, *Classical Analysis in the Complex Plane*, Cornerstones,
https://doi.org/10.1007/978-1-0716-1965-0

Rendus de l'Académie des Sciences Paris) carry research announcements with no proof details. Often (but not always!) when an author announces results like this and follows with a later detailed paper elsewhere, my bibliography only lists the latter paper.

If a recent book or paper on a certain result supersedes several older ones and mentions them in its bibliography, I sometimes list only the former here and refer my readers to it (and, implicitly, to its bibliography).

For some of the authors cited here collected (or selected) works have been published and it was my original intention to list these and give references to them as well as to the original works, but limitations of space have precluded that. Also I had hoped to include whenever available an obituary notice, since these usually provide a short biography, an overview and assessment of the person's work, a complete list of publications and sometimes a photograph. This project also proved unfeasible and must therefore be commended to the initiative of the interested reader. A big help locating obituaries, biographies and collected works is MAY [1973]. The 4-volume *Biographical Dictionary of Mathematicians* (Charles Scribner's Sons (1991), New York; **MR**97h:01001) contains scholarly biographies of some of the principal actors in this big drama. See also **MR** 99f:01002. There are also several large photo collections the curious reader might like to consult: For the late 19th century, BÖLLING [1994] and the 30-year index published by the journal *Acta Mathematica* in 1913; for the 20th century, PÓLYA [1987] and HALMOS [1987]. The University of St. Andrews in Scotland has a rich on-line source for all aspects of the history of mathematics at `http://www-history.mcs.st-and.ac.uk`.

To exploit the full potential of **MR** goldmine the reader should consult *Notices of the American Mathematical Society* 66, No. 8 (2019), 1320–1324.

ABATE, M.

Iteration theory of holomorphic maps on taut manifolds, Research and Lecture Notes in Mathematics, Mediterranean Press (1989), Commende di Rende. **Zbl** 747.32002. **MR** 92i:32032.

ABBASOV, A. G.

"Sufficient conditions for homeomorphism of mappings," *Izv. Akad. Nauk Azerbaĭdzhan SSR Ser. Fiz.-Tekhn. Mat. Nauk* 3 (1982), No. 4, 13–17 (Russian). **Zbl** 527.54010. **MR** 85m:26016.

ABE, H.

"On some analytic functions in an annulus," *Kōdai Math. Sem. Reports* 10 (1958), 38–45. **Zbl** 82, p. 61. **MR** 20 #3294.

ABIAN, A.

"A proof and extension of Brouwer's fixed point theorem for the closed 2-cell," *Boll. della Un. Mat. Ital.* (3) 16 (1961), 281–284. **Zbl** 147, p. 419. **MR** 25 #1546.

"The identity theorem for analytic functions," *Jour. Math. Anal. and Applic.* 45 (1974), 682–683. **Zbl** 275.30001. **MR** 49 #9168.

(a) "The coefficients of the Laurent expansion of analytic functions," *Archivum Mathematicum* , Brno 13 (1977), 65–68. **Zbl** 373.30026. **MR** 57 #6437.

(b) "Hurwitz' theorem implies Rouché's theorem," *Jour. Math. Anal. and Applic.* 61 (1977), 113–115. **Zbl** 409.30005. **MR** 57 #12823.

"Unique determination of any harmonic function from its values given on the points of two convergent sequences," *Acta Math. Vietnamica* 9 (1984), 33–40. **Zbl** 595.31001. **MR** 86i:31001.

ABIAN, A. and BROWN, A. B.

"A fixed point theorem for the closed 2-cell," *Portugaliae Math.* 21 (1962), 93–98. **Zbl** 107, p. 166. **MR** 26 #3032.

ACHIEZER, N. I.

Theory of Approximation (translated from Russian by C. J. Hyman), Frederick Ungar Publishing Co. (1956), New York. **Zbl** 152, p. 253. **MR** 36 #5567. **BAMS** 63, p. 163. Reprinted by Dover Publications, Inc. (1992), New York. **MR** 94b:01041.

ACKER, F.

"The missing link," *Math. Intelligencer* 18, no. 3 (1996), 4–9 and 19, no. 3 (1997), 7. **Zbl** 881.26002. **MR** 97i:01014.

ADEL'SON-VEL'SKIĬ, G. M. and KRONROD, A. S.

"On a direct proof of the analyticity of a monogenic function," *Dokl. Akad. Nauk SSSR (N.S.)* 50 (1945), 7–9 (Russian). **MR** 14, p. 546.

AGMON, S.

"Sur deux théorèmes de Fabry," *C. R. Acad. Sci. Paris* 226 (1948), 1673–1674. **Zbl** 30, p. 153. **MR** 9, p. 576.

AGOSTINI, A.

"Il teorema fondamentale dell' algebra," *Periodico di Mat.* (4) 4 (1924), 307–327. **FM** 50, p. 8.

AGOSTON, M. K.

Algebraic Topology, A First Course, Pure & Applied Mathematics, Vol. 32. Marcel Dekker, Inc. (1976), New York. **Zbl** 337.55001. **MR** 56 #3825.

AGRANOVSKIĬ, M. L.

"The Fourier transform on $SL_2(\mathbb{R})$ and theorems of Morera-type," *Soviet Math. Dokl.* 19 (1978), 1522–1526. **Zbl** 416.30001. **MR** 80j:22014.

AGUILÓ FUSTER, R.

"Untersuchung der Ideale des Ringes der ganzen Funktionen," *Collect. Math.* 17 (1965), 105–134. Correction, *ibid.*, 297 (Spanish). **Zbl** 163, p. 35. **MR** 34 #2608 and 35 #4204.

AHLFORS, L. V.

"Untersuchungen zur Theorie der konformen Abbildung und der ganzen Funktionen," *Acta Soc. Sci. Fennicae* (2) A1, No. 9 (1930). **FM** 56, p. 984.

(a) "Sur une généralisation du théorème de Picard," *C. R. Acad. Sci. Paris* 194 (1932), 245–247. **FM** 58, p. 343. **Zbl** 3, p. 407.

(b) "Ein Satz über die charakteristische Funktion und den Maximalmodul einer meromorphen Funktion," *Soc. Sci. Fennicae Comment. phys.-math.* (=*Finska Vetenskaps-Societeten*) 6, No. 9 (1932). **FM** 58, p. 333. **Zbl** 5, p. 300.

"Sur les domaines dans lesquels une fonction méromorphe prend des valeurs appartenant à une région donnée," *Acta Soc. Sci. Fennicae* (2) A2, No. 2 (1933). **FM** 59, p. 1033. **Zbl** 8, p. 262.

"On Phragmén-Lindelöf's principle," *Trans. Amer. Math. Soc.* 41 (1937), 1–8. **FM** 63, p. 286. **Zbl** 6, p. 32.

"An extension of Schwarz's Lemma," *Trans. Amer. Math. Soc.* 43 (1938), 359–364. **FM** 64, p. 315. **Zbl** 18, p. 410.

"Classical and contemporary analysis," *SIAM Review* 3 (1961), 1–9. **Zbl** 117, p. 36. **MR** 23 #A291.

"Remarks on Carleman's formula for functions in a half-plane," *SIAM Jour. on Num. Anal.* 3 (1966), 183–187. **Zbl** 145, p. 150. **MR** 34 #2874.

Conformal Invariants: Topics in Geometric Function Theory, McGraw-Hill Book Co. (1973), New York. **Zbl** 272.30012. **MR** 50 #10211.

Complex Analysis, McGraw-Hill Book Co. (3rd ed., 1979), New York. **Zbl** 395.30001. **MR** 14, p. 857 and 80c:30001. **BAMS** 59, p. 464.

AHLFORS, L. V. and HEINS, M.

"Questions of regularity connected with the Phragmén-Lindelöf Principle," *Annals of Math.* (2) 50 (1949), 341–346. **Zbl** 36, p. 47. **MR** 10, p. 522.

AHLFORS, L. V. and SARIO, L.

Riemann Surfaces, Princeton Mathematical Series, No. 26, Princeton University Press (1960). **Zbl** 196, p. 338. **MR** 22 #5729.

AIGNER, A.

Tangenten an den Frohsinn, Berichte der Mathematisch-statistischen Sektion im Forschungszentrum Graz (1978). **Zbl** 376.00006.

AIGNER, M. and ZIEGLER, G. M.

Proofs from THE BOOK, Springer-Verlag (4th ed., 2010), Berlin. **Zbl** 1185.00001. **MR** 2010m:00001.

AIZENBERG, L.

Carleman's Formulas in Complex Analysis, Theory and Applications, Kluwer Academic Publishers (1993), Dordrecht. **Zbl** 783.32002. **MR** 95d:32004.

ÅKERBERG, B.

"Proof of Poisson's formula," *Proc. Camb. Phil. Soc.* 57 (1961), 186. **Zbl** 91, p. 97. **MR** 22 #8234.

ALBRECHT, R.

"Iterationsverfahren zur konformen Abbildung eines Ringgebietes auf einen konzentrischen Kreisring," *Sitzungsber. Bayer. Akad. Wissen. München* (1954), 169–178. **Zbl** 64, p. 76. **MR** 17, p. 26.

ALEKSANDROV, I. A.

"Conditions for convexity of the image region under mapping by functions regular and univalent in the unit circle," *Izv. Vysš. Učebn. Zaved. Matematika* (1958), No. 6 (7), 3–6. **Zbl** 118, p. 73. **MR** 23 #A3834.

ALEXANDER, D. S.

A History of Complex Dynamics, from Schröder to Fatou and Julia, Aspects of Mathematics vol. E 24, Vieweg Publ. Co. (1994), Wiesbaden. **Zbl** 788.30001. **MR** 95d:01014. (See also **Zbl** 1179.01017, 1244.37002 and **MR** 2011k:01008, 2012k:37103.)

ALEXANDER, J. W., II,

"Functions which map the interior of the unit circle upon simple regions," *Annals of Math.* (2) 17 (1915–16), 12–22. **FM** 45, p. 672.

"A proof and extension of the Jordan-Brouwer separation theorem," *Trans. Amer. Math. Soc.* 23 (1922), 333–349. **FM** 49, p. 403.

ALLING, N.

"The valuation theory of meromorphic function fields," pp. 8–29 of *Entire Functions and Related Parts of Analysis,* Proceedings of Symposia in Pure

Mathematics, Vol. 11, American Mathematical Society (1968), Providence. **Zbl** 182, p. 63. **MR** 38 #4700.

ALMIRA, J. M., JIMÉNEZ, M. and DEL TORO, N.
"Another topological proof of the Fundamental Theorem of Algebra," *Elemente der Math.* 57 (2002), 32–37. **Zbl** 1076.12001. **MR** 1894589.

ALPAY, D.
A Complex Analysis Problem Book, Birkhäuser-Springer (2011), Basel. **Zbl** 1226.00001. **MR** 2012g:30001.

ALVAREZ, C. and DHOMBRES, J.
Une histoire de l'imaginaire mathematique. Vers le théorème fondamental de l'algèbre et sa démonstration par Laplace en 1795, Hermann (2011), Paris. **Zbl** 1243.01005.

Une histoire de l'invention mathèmatique. Les démonstrations du théorème de l'algèbre dans le cadre de l'analyse réele et de l'analyse complexe de Gauss à Liouville, Vol. II. Hermann (2012), Paris. **Zbl** 1243.01004.

AMDEBERHAN, T.
"Harmonic products of harmonic functions," *Amer. Math. Monthly* 106 (1999), 782.

AMES, L. D.
"An arithmetic treatment of some problems in *analysis situs*," *Amer. Jour. of Math.* 27 (1905), 343–380. **FM** 36, p. 529.

ANANDAM, V. and DAMLAKHI, M.
"Bôcher's theorem in $\mathbb{R}^2$ and Carathéodory's inequality," *Real Anal. Exchange* 19 (1993/94), 537–539. **Zbl** 803.30003. **MR** 95f:31001 and 95a:31004.

"Harmonic singularity at infinity in $\mathbb{R}^n$," *Real Anal. Exchange* 23 (1997/98), 471–476. **Zbl** 938.31003. **MR** 99f:31008.

ANDERSON, J.
"Iterated exponentials," *Amer. Math. Monthly* 111 (2004), 668–679. **Zbl** 1187.40001. **MR** 2005j:40001.

ANDERSON, J. M. and CLUNIE, J.
"Entire functions of finite order and lines of Julia," *Math. Zeit.* 112 (1969), 59–73 **Zbl** 216, p. 101. **MR** 40 #4456.

"Picard sets of entire and meromorphic functions," *Ann. Acad. Scient. Fennicae, Ser. A. I. Math.* 5 (1980), No. 1, 24–43. **Zbl** 456.30026. **MR** 81m:30026.

ANDÔ, T. and FAN, K.
"Pick-Julia theorems for operators," *Math. Zeit.* 168 (1979), 23–34. **Zbl** 408.47007. **MR** 80h:47018.

ANDREWS, J. J. and LACHER, R. C.
"Stacked exponents," *Aequationes Math.* 16 (1977), 137–147. **Zbl** 418.40001. **MR** 57 #6331.

ANGER, G.

"Die Entwicklung der Potentialtheorie im Hinblick auf ihre grundlegenden Existenzsätze," *Jahresber. Deutsch. Math. Verein.* 64 (1961/62), 1te Abteilung, 51–78 and 100–134. **Zbl** 106, p. 77. **MR** 26 #5170 (a), (b).

ANGHELUTZA, TH.

"Sur une classe de développements en série d'une fonction holomorphe," *Mathematica* (Cluj) 2 (1929), 10–15. **FM** 55, p. 182.

ANKENY, N. C.

"One more proof of the fundamental theorem of algebra," *Amer. Math. Monthly* 54 (1947), 464. **Zbl** 31, p. 351. **MR** 9, p. 90.

ANONYME, UN

"Sur l'intégrale $\int_0^\infty e^{-x^2}dx$," *Bull. Sci. Math.* (2) 13 (1889), 84. **FM** 21, p. 277.

ANTOINE, L.

"Sur l'homéomorphie de deux figures et de leurs voisinages," *Jour. de Math. Pures et Appliq.* (8) 4 (1921), 221–325. **FM** 48, p. 650.

"Sur les voisinages de deux figures homéomorphes," *Fund. Math.* 5 (1924), 264–287. **FM** 50, p. 369.

APOSTOL, T. M.

"Term-wise differentiation of power series," *Amer. Math. Monthly* 59 (1952), 323–326. (See also **MR** 40 #270.)

"Another elementary proof of Euler's formula for $\zeta(2n)$," *Amer. Math. Monthly* 80 (1973), 425–431. **Zbl** 267.10050. **MR** 47 #3330.

"An elementary view of Euler's summation formula," *Amer. Math. Monthly* 106 (1999), 409–418. **Zbl** 1076.41509.

APOSTOL, T. M. and YOUNG, R. L.

"Famous formula," *Math. Mag.* 53 (1980), 181–183.

APPELL, P.

"Développements en série d'une fonction holomorphe dans une aire limitée par des arcs de cercles," *Math. Annalen* 21 (1883), 118–124. **FM** 15, p. 324.

ARAKÉLIAN, N. U.

"The asymptotic approximation to entire functions in infinite regions," *Mat. Sbornik (N. S.)* 53 (95) (1961), 515–538 (=*American Mathematical Society Translations* (2) 43 (1964), 169–193.) **Zbl** 143, p. 295. **MR** 24 #A830. (See also **Zbl** 592.30042.)

"Uniform approximation on closed sets by entire functions," *Izv. Akad. Nauk SSSR. Ser. Mat.* 28 (1964), 1187–1206 (Russian). **Zbl** 143, p. 296. **MR** 30 #258.

"Certain questions of approximation theory and the theory of entire functions," *Math. Notes* 9 (1971), 267–271. **MR** 44 #440. (See also **Zbl** 234.30029 and **MR** 54 #10609.)

"Uniform and tangential approximation on the real axis by entire functions with an estimate of their growth," *Math. USSR-Sb.* 41 (1982), No. 1, 1–32. **Zbl** 478.30029. **MR** 83e:30041.

Arakélian, N. U. and Martirosian, V. A.

"The location of the singularities of power series on the boundary of the circle of convergence, II," *Soviet Jour. Contemp. Math. Analysis* 23 (1988), No. 2, 17–34. **Zbl** 662.30001. **MR** 90c:30001.

Arantes Sad, L., Vieira Teixeira, M. and Baldino, R. R.

"Cauchy and the problem of point-wise convergence," *Arch. Internat. Histoire Sci.* 51 (2001), 277–308. **Zbl** 1064.01011. **MR** 2004a:01016.

Arens, R

"On sandwich slicing," *Colloq. Math. Soc. János Bolyai* 23 (1980), 57–60. **Zbl** 431.55001. **MR** 82e:55002.

Armitage, D. H.

"Entire functions that tend to zero on every line," *Amer. Math. Monthly* 114 (2007), 251–256. **Zbl** 1132.30334. **MR** 2007j;30018.

Armitage, D. H. and Gauthier, P. M.

"Recent developments in harmonic approximation, with applications," *Results in Math.* 29 (1996), 1–15. **Zbl** 859.31001. **MR** 97f:31007.

Classical Potential Theory, Springer-Verlag (2001), London. **Zbl** 972.31101. **MR** 201m:31001.

Armitage, D. H. and Goldstein, M.

"Better than uniform approximation on closed sets by harmonic functions with singularities," *Proc. Lon. Math. Soc.* (3) 60 (1990), 319–343. **Zbl** 702.31003. **MR** 91a:31005. (See also **Zbl** 770.31003.)

"Quadrature and harmonic approximation of subharmonic functions in strips," *Jour. Lon. Math. Soc.* (2) 46 (1992), 171–179. **Zbl** 757.31002. **MR** 93j:31003. (See also **Zbl** 770.31004.)

"Non-uniqueness for the Radon transform," *Proc. Amer. Math. Soc.* 117 (1993), 175–178. **Zbl** 765.44001. **MR** 93c:44004.

Armitage, D. H. and Kuran, Ü.

"The convexity of a domain and the superharmonicity of the signed distance function," *Proc. Amer. Math. Soc.* 93 (1985), 598–600. **Zbl** 568.31002. **MR** 86k:31005.

Armitage, D. H. and Nelson, C. S.

"A harmonic quadrature formula characterizing open strips," *Math. Proc. Camb. Phil. Soc.* 113 (1993), 147–151. **Zbl** 783.31002. **MR** 93h:31004.

Armstrong, M. A.

Basic Topology, Undergraduate Texts in Mathematics, Springer-Verlag (1983), New York. **Zbl** 514.55001. **MR** 84f:55001.

ARNOLD, D. N. and ROGNESS, J.

"Möbius transformations revealed" *Notices Amer. Math. Soc.* 55 (2008), 1226–1231. **Zbl** 1200.30036. **MR** 2009j:30100.

ARONSZAJN, N.

"Sur les décompositions des fonctions analytiques uniformes et sur leurs applications," *Acta Math.* 65 (1935), 1–156. **FM** 61, p. 303. **Zbl** 12, p. 262. (See also **MR** 2002m:31018, 2005d:30081.)

ARSOVE, M. G.

(a) "On the definition of an analytic function," *Amer. Math. Monthly* 62 (1955), 22–25. **Zbl** 64, p. 66. **MR** 16, p. 683.

(b) "The Looman-Menchoff theorem and some subharmonic function analogues," *Proc. Amer. Math. Soc.* 6 (1955), 94–105. **Zbl** 65, p. 336. **MR** 16, p. 1108.

"Intrinsic characterization of regions bounded by closed curves," *Duke Math. Jour.* 34 (1967), 425–429. **Zbl** 165, p. 97. **MR** 36 #367.

(a) "The Osgood-Taylor-Carathéodory theorem," *Proc. Amer. Math. Soc.* 19 (1968), 38–44. **Zbl** 165, p. 97. **MR** 36 #3966.

(b) "Some boundary properties of the Riemann mapping function," *Proc. Amer. Math. Soc.* 19 (1968), 560–568. **Zbl** 162, p. 103. **MR** 37 #1572.

"A correction to 'Some boundary properties of the Riemann mapping function'," *Proc. Amer. Math. Soc.* 22 (1969), 711–712. **Zbl** 186, p. 400. **MR** 39 #4368.

ARSOVE, M. G. and HUBER, A.

"Local behavior of subharmonic functions," *Indiana Univ. Math. Jour.* 22 (1973), 1191–1199. **Zbl** 264.31001. **MR** 48 #11536.

ARTIN, E.

The Gamma Function (translated from German by M. Butler), Holt, Rinehart and Winston, Inc. (1964), New York. **FM** 57, p. 419. **Zbl** 144, p. 68. **MR** 29 #2437. (See also **MR** 2007k:01017.)

The Collected Papers of Emil Artin (S. Lang and J. T. Tate, editors), Addison-Wesley Publishing Co. (1965), Reading, Massachusetts. **Zbl** 146, p. 1. **MR** 31 #1159. (See also **MR** 2007k:01017.)

ARTIN, E. and SCHREIER, O.

"Algebraische Konstruktion reeller Körper," *Abhand. Math. Sem. Univ. Hamburg* 5 (1926), 85–99. **FM** 52, p. 120. (See also **MR** 83j:01083 and 2007k:01017.)

ARZELÀ, C.

"Note on a series of analytic functions," *Annals of Math.* (2) 5 (1903–04), 51–63. **FM** 35, p. 396.

ASCOLI, G.

(a) "Sulle singolarità isolate delle funzioni armoniche," *Boll. della Un. Mat. Ital.* (2) 7 (1928), 230–237. **FM** 54, p. 510.

(b) "Sulla unicità della soluzione nel problema di Dirichlet," *Rend. Accad. d. Lincei Roma* (6) 8 (1928), 348–351. **FM** 54, p. 509.

ASH, J. M. and BROWN, R.

"Uniqueness and non-uniqueness for harmonic functions with zero non-tangential limits," pp. 30–40 of *Harmonic Analysis* (*Sendai*, 1990), (S. Igari, editor), Springer-Verlag (1991), Tokyo. **Zbl** 778.31002. **MR** 94m:31002.

AUMANN, G. and CARATHÉODORY, C.

"Ein Satz über die konforme Abbildung mehrfach zusammenhängender ebener Gebiete," *Math. Annalen* 109 (1934), 756–763. **FM** 60, p. 285. **Zbl** 9, p. 26.

AUSLANDER, L. and TOLIMIERI, R.

"Ring structure and the finite Fourier transform," *Math. Intelligencer* 7, No. 3 (1985), 49–52. **Zbl** 567.12013. **MR** 87b:11123. (See also **MR** 2011e:43001.)

AVKHADIEV, F. G. and AKSENT'EV, L. A.

"The main results on sufficient conditions for an analytic function to be schlicht," *Russian Math. Surveys* 30 (1975), No. 4, 1–64. **Zbl** 331.30016. **MR** 53 #5848. Continued in *Soviet Math.* 30, No. 10 (1986), 1–20. **Zbl** 624.30031. **MR** 88j: 30037. (See also **Zbl** 711.30017.)

AVKHADIEV, F. G. and WIRTHS, K.-J.

Schwarz-Pick Type Inequalities, Birkhäuser Verlag AG (2009), Basel-Boston-Berlin. **Zbl** 1168.30001. **MR** 2010j:30049.

AXLER, S.

"Harmonic functions from a complex analysis viewpoint," *Amer. Math. Monthly* 93 (1986), 246–258. **Zbl** 604.31001. **MR** 88h:31002.

AXLER, S., BOURDON, P. and RAMEY, W.

"Bôcher's theorem," *Amer. Math. Monthly* 99 (1992), 51–55. **Zbl** 758.31002. **MR** 92m:31003.

Harmonic Function Theory, Graduate Texts in Mathematics, Vol. 137, Springer-Verlag (2nd ed. 2001), New York. **Zbl** 765.31001. **MR** 2001j:31001.

AXLER, S. and RAMEY, W.

"Harmonic polynomials and Dirichlet-type problems," *Proc. Amer. Math. Soc.* 123 (1995), 3765–3773. **Zbl** 848.31002. **MR** 96b:31003.

AYOUB, R.

"Euler and the Zeta function," *Amer. Math. Monthly* 81 (1974), 1067–1086. **Zbl** 293.10001. **MR** 50 #12566. (See also **MR** 51 #7778.)

"What is a Napierian logarithm?", *Amer. Math. Monthly* 100 (1993), 351–364. **Zbl** 779.01004. **MR** 94f:01014.

B

BACHMANN, F.

"Aufbau des Zahlensystems," *Encyklopädie der Mathematischen Wissenschaften* Bd. I, 1st Part, Heft 2, A (1939). **FM** 65, p. 33. **Zbl** 21, p. 205. **MR** 1, p. 46.

BACHMANN, P.

Zahlentheorie: Zweiter Teil: Die Analytische Zahlentheorie, B. G. Teubner (1894), Leipzig. Reprinted by Johnson Reprint Corp. (1968), New York and London. **FM** 25, p. 249.

BAGEMIHL, F.

"Concerning non-continuable, transcendentally transcendental power series," *Proc. Nat. Acad. Sci. U. S. A.* 37 (1951), 211–213. **Zbl** 44, p. 298. **MR** 13, p. 221.

"Some identity and uniqueness theorems for normal meromorphic functions," *Ann. Acad. Scient. Fennicae, Ser. A. I. Math.* No. 299 (1961). **Zbl** 100, p. 71. **MR** 23 #A302.

"A boundary condition for a holomorphic function in a Jordan region to be schlicht," *Proc. Nat. Acad. Sci. U. S. A.* 58 (1967), 1102–1103. **Zbl** 155, p. 400. **MR** 36 #1633.

BAGEMIHL, F. and PIRANIAN, G.

"Absolutely convergent power series," *Ann. Univ. Sci. Budapest. Eötvös Sect. Math.* 3–4 (1960/61), 27–34. **Zbl** 108, p. 71. **MR** 24 #A3273.

BAGEMIHL, F. and SEIDEL, W. P.

"Sequential and continuous limits of meromorphic functions," *Ann. Acad. Scient. Fennicae, Ser. A. I. Math.* No. 280 (1960). **Zbl** 95, p. 58. **MR** 22 #12226.

"Koebe arcs and Fatou points of normal functions," *Comm. Math. Helv.* 36 (1961), 9–18. **Zbl** 125, p. 317. **MR** 25 #5183.

BAILLIE, R. BORWEIN, D. and BORWEIN, J. M.

"Surprising sinc sums and integrals," *Amer. Math. Monthly* 115 (2008), 888–901. **Zbl** 1187.33001. **MR** 2010d:33001. (See also **Zbl** 1297.26002, **MR** 3182260.)

BAK, J. and DING, P.

"Shape distortion by analytic functions," *Amer. Math. Monthly* 116 (2009), 143–150. **Zbl** 1178.30003. **MR** 2010c:30011.

BAK, J. and NEWMAN, D. J.

Complex Analysis, Undergraduate Texts in Mathematics, Springer-Verlag (1982), New York. (3rd edition, 2010), **Zbl** 482.30001 and 906.30001. **MR** 84b:30001 and 2011d:30001.

BAKER, A.

Transcendental Number Theory, Cambridge University Press (1975), London. **Zbl** 297.10013. **MR** 54 #10163. **BAMS** 84, p. 1370.

BAKER, I. N.

"Zusammensetzungen ganzer Funktionen," *Math. Zeit.* 69 (1958), 121–163. **Zbl** 178, p. 75. **MR** 20 #4000.

"The existence of fixpoints of entire functions," *Math. Zeit.* 73 (1960), 280–284. **Zbl** 129, p. 291. **MR** 22 #4838b.

"The iteration of polynomials and transcendental entire functions," *Jour. Austral. Math. Soc.* (Ser. A) 30 (1981), 483–495. **Zbl** 474.30023. **MR** 82g: 30043.

"Iteration of entire functions: An introductory survey," pp. 1–17 of *Lectures on Complex Analysis* (Chi-Tai Chuang, editor), World Scientific Publishing Co. (1988), Singapore. **Zbl** 746.30003.

BAKER, I. N. and POMMERENKE, CH.

"On the iteration of analytic functions in a half-plane, II," *Jour. Lon. Math. Soc.* (2) 20 (1979), 255–258. **Zbl** 419.30025. **MR** 83j:30024.

BAKER, I. N. and RIPPON, P. J.

"Convergence of infinite exponentials," *Ann. Acad. Scient. Fennicae, Ser. A. I. Math.* 8 (1983), 179–186. **Zbl** 513.30018. **MR** 85g:30039.

"Iteration of exponential functions," *Ann. Acad. Scient. Fennicae, Ser. A. I. Math.* 9 (1984), 49–77. **Zbl** 558.30029. **MR** 86d:58065.

"A note on complex iteration," *Amer. Math. Monthly* 92 (1985), 501–504. **Zbl** 605.30027. **MR** 86m:30025.

"Towers of exponents and other composite maps," *Complex Variables* 12 (1989), 181–200. **Zbl** 678.30017. **MR** 91b:30068. (See also **MR** 97k:30030 and 2003g: 30042.)

"On compositions of analytic self-mappings of a convex domain," *Arch. der Math.* 55 (1990), 380–386. **Zbl** 715.30022. **MR** 91k:30062.

BAKER, J. A.

"Plane curves, polar coordinates and winding numbers," *Math. Mag.* 64 (1991), 75–91. **Zbl** 742.30001. (See also **Zbl** 1330.55001.)

BAKER, J. A. and SLUIS, W.

"Arg," *Math. Mag.* 70 (1997), 297.

BALK, M. B.

"A theorem on entire functions," *Moskov. Oblast. Pedagog. Inst. Uč. Zap.* 57 (1957), 51–53 (Russian). **Zbl** 90, p. 291. **MR** 20 #5283.

BALTUS, C.

"D'Alembert's proof of the fundamental theorem of algebra," *Historia Math.* 31 (2004), 414–428. **Zbl** 1060.01006. **MR** 2005g:01019.

BANACH, S.

"Über die Baire'sche Kategorie gewisser Funktionenmengen," *Studia Math.* 3 (1931), 174–179. **FM** 57, p. 305. **Zbl** 3, p. 297.

BANACH, S. and MAZUR, S.

"Über mehrdeutige stetige Abbildungen," *Studia Math.* 5 (1934), 174–178. **FM** 60, p. 1227. **Zbl** 13, p. 82.

BANK, S. B. and ORLAND, G. H.

"A note on Rouché's theorem," *Fund. Math.* 63 (1968), 137–141. **Zbl** 165, p. 401. **MR** 39 #1656.

BARGMANN, D.

"Conjugations on rotation domains as limit functions of the geometric means of the iterates," *Ann. Acad. Scient. Fennicae, Ser. A. I. Math.* 23 (1998), 507–524. **Zbl** 910.30024. **MR** 99g:30029.

BARTH, K. F. and SCHNEIDER, W. J.

"On a problem of Erdös concerning the zeros of the derivatives of an entire function," *Proc. Amer. Math. Soc.* 32 (1972), 229–232. **Zbl** 236.30032. **MR** 45 #2168.

"A Julia line which is not a Milloux line," *Complex Variables* 13 (1989), 57–65. **Zbl** 764.30023. **MR** 91b:30081.

BAŠMAKOVA, I. G.

"On the proof of the fundamental theorem of algebra," *Istoriko-mat. Issledovanija* 10 (1957), 257–304 (Russian). **Zbl** 101, p. 246.

"Le théorème fondamental de l'algèbre et la construction des corps algébriques," *Arch. Internat. Histoire Sci.* 13 (1960), 211–222. **Zbl** 126, p. 258.

BASHMAKOVA, I. G. and SMIRNOVA, G. S.

The Beginnings and Evolution of Algebra, (translated from Russian by Abe Shenitzer) Dolciani Mathematical Expositions, No. 23, The Mathematical Association of America (2000), Washington, D. C. **Zbl** 942.01001. **MR** 2000h: 01002.

BASS, J. and LÉVY, P.

"Propriétés des lois dont les fonctions caractéristiques sont $1/ch\, z$, $z/sh\, z$, $1/ch^2\, z$," *C. R. Acad. Sci. Paris* 230 (1950), 815–817. **Zbl** 36, p. 86. **MR** 11, p. 443.

BASYE, R. E.

"Simply connected sets," *Trans. Amer. Math. Soc.* 38 (1935), 341–356. **FM** 61, p. 636. **Zbl** 12, p. 250.

BEAR, H. S.

"A new look at the three circles theorem," *Amer. Math. Monthly* 81 (1974), 487–490. **Zbl** 288.30019. **MR** 49 #9186.

"Part metric and hyperbolic metric," *Amer. Math. Monthly* 98 (1991), 109–123 and 99 (1992), 669. **Zbl** 742.30039. **MR** 92k:51029.

"Distance-decreasing functions on the hyperbolic plane," *Mich. Math. Jour.* 39 (1992), 271–279. **Zbl** 758.30035. **MR** 93g:30034.

An Introduction to Mathematical Analysis, Academic Press (1997), San Diego. Reprinted by Blackburn Press (2004), Caldwell, N. J. **Zbl** 927.26001. **MR** 98h:00001.

Beardon, A. F.

"Montel's theorem for subharmonic functions and solutions of partial differential equations," *Proc. Camb. Phil. Soc.* 69 (1971), 123–150. **Zbl** 207, p. 110. **MR** 42 #4756.

Complex Analysis: The Argument Principle in Analysis and Topology, John Wiley and Sons, Ltd. (1979), Chichester. **Zbl** 399.30001. **MR** 80e:30001.

"Conformal automorphisms of plane domains," *Jour. Lon. Math. Soc.* (2) 32 (1985), 245–253. **Zbl** 605.30008. **MR** 86k:30052.

"Iteration of contractions and analytic maps," *Jour. Lon. Math. Soc.* (2) 41 (1990), 141–150. **Zbl** 703.30022. **MR** 91i:51025.

Iteration of Rational Functions: Complex Analytic Dynamical Systems, Graduate Texts in Mathematics, Vol. 132, Springer-Verlag (1991), New York. **Zbl** 742.30002. **MR** 92j:30026.

"Sums of powers of integers," *Amer. Math. Monthly* 103 (1996), 201–213 and 104 (1997), 686. **Zbl** 851.11012. **MR** 97f:11020.

"Analytic contraction of the unit disc," *Bull. Hong Kong Math. Soc.* 1 (1997), 209–218. **Zbl** 913.30014. **MR** 99b:30030.

"Meromorphic self-maps of regions," *Computational Methods and Function Theory* 10 (2010), 441–453. **Zbl** 1230.30015. **MR** 2012d:37102.

Beardon, A. F. and Carne, T. K.

"A strengthening of the Schwarz-Pick inequality," *Amer. Math. Monthly* 99 (1992), 216–217. **Zbl** 753.30015. **MR** 94f:30055. (See also **MR** 97m:30062.)

Beardon, A. F., Carne, T. K., Minda, D. and Ng, T. W.

"Random iteration of analytic maps," *Ergod. Th. & Dynam. Sys.* 24 (2004), 659–675. **Zbl** 1055.37053. **MR** 2005b:37086.

Beardon, A. F. and Minda, D.

"On the pointwise limit of complex analytic functions," *Amer. Math. Monthly* 110 (2003), 289–297.

"A multi-point Schwarz-Pick Lemma," *Jour. d'Analyse Math.* 92 (2004), 81–104. **Zbl** 1055.30038. **MR** 2005f:30044.

(a) "Conformal automorphisms of finitely connected regions," pp. 37–73 of *Transcendental Dynamics and Complex Analysis,* P. J. Rippon and G. M. Stallard, editors, London Mathematical Society Lecture Note Series 348, Cambridge University Press (2008), Cambridge. **Zbl** 1156.30007. **MR** 2010c:30061.

(b) "Dieudonné points of holomorphic self-maps of regions," *Comput. Methods Funct. Theory* 8 (2008), 409–432. **Zbl** 1153.30022. **MR** 2009h:30078.

BECK, A.

"On rings on rings," *Proc. Amer. Math. Soc.* 15 (1964), 350–353. **Zbl** 135, p. 289. **MR** 29 #1347.

BECKENBACH, E. F.

"The stronger form of Cauchy's integral theorem," *Bull. Amer. Math. Soc.* 49 (1943), 615–618. **MR** 5, p. 35.

"Concerning the definition of harmonic functions," *Bull. Amer. Math. Soc.* 51 (1945), 240–245. **MR** 6, p. 227.

BECKENBACH, E. F. and GRAHAM, E. W.

"On subordination in complex function theory," pp. 247–254 of *Construction and Applications of Conformal Maps,* National Bureau of Standards Applied Math. Series No. 18, E. F. Beckenbach, editor, (1952), Washington, D. C. **Zbl** 49, p. 175. **MR** 14 p. 632.

BEESACK, P. R.

"The Laurent expansion without Cauchy's theorem," *Canad. Math. Bull.* 15 (1972), 473–480. **Zbl** 266.30040. **MR** 46 #9301.

BEHAN, D. F.

"Commuting analytic functions without fixed points," *Proc. Amer. Math. Soc.* 37 (1973), 114–120. **Zbl** 251.30009. **MR** 46 #7492.

BEHNKE, H.

"Karl Weierstrass und seine Schule," *Arbeitsgemeinschaft Nordrhein-Westfalen. Wissen. Abhand.* 33 (1966), 13–40. **Zbl** 138, p. 252. **MR** 34 #2412.

BEHNKE, H. and SOMMER, F.

Theorie der analytischen Funktionen einer komplexen Veränderlichen, Die Grundlehren der mathematischen Wissenschaften in Einzeldarstellungen, Bd. 77. Springer-Verlag (3rd ed., 1965), Berlin. **Zbl** 273.30001 and 324.30001. **MR** 26 #5137.

BEHNKE, H. and ZUMBUSCH, H.

"Konforme Abbildung von Bereichen auf in ihnen liegende Bereiche," *Semester-Ber. Math. Sem. Münster* 8 (1936), 100–121. **FM** 62, p. 380.

BEHREND, F. A.

"A contribution to the theory of magnitudes and the foundations of analysis," *Math. Zeit.* 63 (1956), 345–362. **Zbl** 70, p. 47. **MR** 17, p. 702.

BELHOSTE, B.

Augustin-Louis Cauchy: A Biography (translated from French by Frank Ragland), Springer-Verlag (1991), New York. **Zbl** 726.01015. **MR** 92a:01062 and 92m, p. xxvi.

BELL, S. R.

The Cauchy transform, potential theory, and conformal mapping, CRC Press (1992), Boca Raton. (2nd ed., 2015). **MR** 94k:30013.

BELL, S. R. and KRANTZ, S. G.

"Smoothness to the boundary of conformal maps," *Rocky Mountain Jour. Math.* 17 (1987), 23–40. **Zbl** 626.30005. **MR** 88e:30023.

BELLMAN, R.

A Brief Introduction to Theta Functions, Holt, Rinehart and Winston, Inc. (1961), New York. **Zbl** 98, p. 283. **MR** 23 #A2556.

BENDIXSON, J.

"Sur une extension à infinité de la formule d'interpolation de Gauss," *Acta Math.* 9 (1886/87), 1–34. **FM** 18, p. 208.

BENKO, D.

"The Basel problem as a telescoping series," *The College Math. Jour.* 43 (2012), 244–250. **MR** 2916494. (See also **Zbl** 1338.40006, **MR** 3453538.)

BERENŠTEĬN, M. N.

"Certain sufficient conditions for the analyticity of functions," *Izv. Vysš. Učebn. Zaved. Matematika* No. 11 (114) (1971), 11–18 (Russian). **Zbl** 234.30002. **MR** 45 #5322.

BERENSTEIN, C. A. and GAY, R.

"A local version of the two-circles theorem," *Israel Jour. of Math.* 55 (1986), 267–288. **Zbl** 624.31002. **MR** 88f:42044. (See also **Zbl** 668.30037 and 888.30002.)

Complex Variables, An Introduction, Graduate Texts in Mathematics, Vol. 125, Springer-Verlag (1991), New York. **Zbl** 741.30001. **MR** 92f:30001.

BERENSTEIN, C. A. and TAYLOR, B. A.

"The 'three squares' theorem for continuous functions," *Arch. Rational Mech. and Anal.* 63 (1977), 253–259. **Zbl** 353.46027. **MR** 56 #3335.

BERGGREN, L., BORWEIN, J. and BORWEIN, P.

Pi: a source book, Springer-Verlag (3rd ed., 2004), New York. **Zbl** 1054.11001. **MR** 2005h:11001.

BERGMAN, S.

The Kernel Function and Conformal Mapping, Mathematical Surveys, Vol. V, American Mathematical Society (2nd ed. 1970), Providence. **Zbl** 473.30006. **MR** 12, p. 402 and 58 #22502. **BAMS** 58, p. 76.

BERGWEILER, W.

"Proof of a conjecture of Gross concerning fix-points," *Math. Zeit.* 204 (1990), 381–390. **Zbl** 703.30024. **MR** 92j:30028. (See also **MR** 2007g:30043.)

"Periodic points of entire functions: proof of a conjecture of Baker," *Complex Variables* 17 (1991), 57–72. **Zbl** 748.30020. **MR** 92m:30048.

"Bloch's principle," *Comput. Methods Funct. Theory* 6 (2006), 77–108. **Zbl** 1101.30034. **MR** 2007f:30045.

BERNAL-GONZÁLEZ, L.

"A lot of 'counter-examples' to Liouville's theorem," *Jour. Math. Anal. and Applic.* 210 (1996), 1002–1009. **Zbl** 855.30030. **MR** 97d:30036.

BERNARD, A. GARNETT, J. B. and MARSHALL, D. E.

"Algebras generated by inner functions," *Jour. of Fct. Anal.* 25 (1977), 275–285. **Zbl** 348.46044. **MR** 56 #1072.

BERNARDI, S. D.

Bibliography of Schlicht Functions, Mariner Publishing Co. (1982), Tampa. **Zbl** 515.30009. **MR** 84k:00014.

BERNAYS, P.

"Zur elementaren Theorie der Landauschen Funktion $\phi(\alpha)$," *Vierteljahrsschr. Naturforsch. Gesell. Zürich* 58 (1913), 203–238. **FM** 44, pp. 247 and 508.

BERNDT, B. C.

"On Gaussian sums and other exponential sums with periodic coefficients," *Duke Math. Jour.* 40 (1973), 145–156. **Zbl** 255.10042. **MR** 47 #1757.

"Elementary evaluation of $\zeta(2n)$," *Math. Mag.* 48 (1975), 148–154. **Zbl** 303.10038. **MR** 51 #3078.

BERNDT, B. C. and EVANS, R. J.

"The determination of Gauss sums," *Bull. Amer. Math. Soc.* (2) 5 (1981), 107–129. **Zbl** 471.10028. **MR** 82h:10051. (See also **MR** 99d:11092.)

BERNSTEIN, F.

"Über einen Schönflies'schen Satz der Theorie der stetigen Funktionen zweier reeller Veränderlichen," *Göttingen Nachr.* (1900), 98–102. **FM** 31, p. 478.

BERNSTEIN, S. N.

"Sur le principe de Dirichlet et le développement des fonctions harmoniques en séries de polynômes," *C. R. Acad. Sci. Paris* 148 (1909), 1306–1308. **FM** 40, p. 450.

"Sur une propriété des fonctions entières," *C. R. Acad. Sci. Paris* 176 (1923), 1603–1605. **FM** 49, p. 215.

BERNSTEIN, V.

"Alcune osservazioni sopra un teorema di Fabry," *Rend. Accad. d. Lincei Roma* (6) 21 (1935), 780–785. **FM** 61, p. 312. **Zbl** 12, p. 261.

BERS, L.

"On rings of analytic functions," *Bull. Amer. Math. Soc.* 54 (1948), 311–315. **Zbl** 32, p. 203. **MR** 9, p. 575.

BESICOVITCH, A. S.

"Quelques théorèmes générales sur l'interversion des intégrations et l'intégration des séries," *Perm, Jour. Soc. Phys. et Math.* 1 (1918–19), 99–139. **FM** 50, p. 632.

"On sufficient conditions for a function to be analytic, and on behaviour of analytic functions in the neighbourhood of non-isolated singular points," *Proc. Lon. Math. Soc.* (2) 32 (1931), 1–9. **FM** 56, p. 272.

BEURLING, A.

Études sur un problème de majoration, (thesis), Almqvist and Wiksell (1933), Upsala. **FM** 59, p. 1042. **Zbl** 8, p. 318. (See also **Zbl** 732.01042 and **MR** 92k:01046.)

BHATIA, R. and MUKHERJEA, K. K.

"The space of unordered tuples of complex numbers," *Linear Alg. and its Appl.* 52/53 (1983), 765–768. **Zbl** 525.32009.

BIEBERBACH, L.

(a) "Über einen Satz des Herrn Carathéodory," *Göttingen Nachr.* (1913), 552–560. **FM** 44, p. 760.

(b) "Über den Jordanschen Kurvensatz, die Schönfliesschen Sätze von Erreichbarkeit and Unbewalltheit und den Satz von der Invarianz des ebenen Gebietes," *Jahresber. Deutsch. Math. Verein.* 22 (1913), 144–153. **FM** 44, p. 559.

"Zur Theorie und Praxis der konformen Abbildung," *Rend. Circ. Mat. Palermo* 38 (1914), 98–112. **FM** 45, p. 670. (See also **Zbl** 922.30006.)

"Über einige Extremalprobleme im Gebiete der konformen Abbildung," *Math. Annalen* 77 (1916), 153–172. **FM** 46, p. 549.

"Über eine Vertiefung des Picardschen Satzes bei ganzen Funktionen endlicher Ordnung," *Math. Zeit.* 3 (1919), 175–190. **FM** 47, p. 299.

"Neuere Untersuchungen über Funktionen von komplexen Variablen," *Encyklopädie der Mathematischen Wissenschaften* Bd. II, 3rd Part, 1st Half, Heft 4, 379–532. B. G. Teubner (1921), Leipzig. **FM** 48, p. 313.

"Über die Verteilung der Null- und Einsstellen analytischer Funktionen," *Math. Annalen* 85 (1922), 141–148. **FM** 48, p. 324. (See also **MR** 83i:01003.)

"Eine hinreichende Bedingung für schlichte Abbildungen des Einheitskreises," *Jour. für Reine und Angew. Math.* 157 (1927), 189–192. **FM** 53, p. 322.

Lehrbuch der Funktionentheorie I (4th ed., 1934), II (2nd ed., 1931), B. G. Teubner, Leipzig. Reprinted by Chelsea Publishing Co. (1945), New York and by Johnson Reprint Corp. (1968), New York. **FM** 56, p. 258 and 57, p. 340. **Zbl** 1, p. 211. **MR** 6, p. 261 and 40 #2823 (a), (b). **BAMS** 28, p. 467; 34, p. 244; 38, pp. 19, 790.

"Über einen Satz Pólyascher Art," *Arch. der Math.* 4 (1953), 23–27. **Zbl** 51, p. 57. **MR** 14, p. 1074.

Analytische Fortsetzung, Ergebnisse der Mathematik und ihrer Grenzgebiete, Neue Folge, Heft 3. Springer-Verlag (1955), Berlin. **Zbl** 174, p. 368. **MR** 35 #5585. **BAMS** 62, p. 184.

Einführung in die konforme Abbildung, Sammlung Göschen Band 768/768a, Walter de Gruyter and Co. (6th ed., 1967), Berlin. **FM** 63, p. 294. **Zbl** 158, p. 323. **MR** 38 #4661. **BAMS** 34, p. 244. English translation of 4th ed. by F. Steinhardt, Chelsea Publishing Co. (1953), New York. **Zbl** 50, p. 84. **MR** 14, p. 462.

Biehler, Ch.

"Sur la limite de $(1+\frac{x}{m})^m$ quand m augmente indéfiniment," *Nouvelles Annales de Math.* (3) 6 (1887), 60–67. **FM** 19, p. 235.

Biermann, K.-R.

"Karl Theodor Wilhelm Weierstrass," pp. 219–224, Vol. XIV of *Dictionary of Scientific Biography* (C. C. Gillispie, editor), Charles Scribner's Sons (1971), New York. **MR** 97h:01001d.

"Karl Weierstrass. Ausgewählte Aspekte seiner Biographie," *Jour. für Reine und Angew. Math.* 223 (1966), 191–220. **Zbl** 196, p. 288. **MR** 33 #1210.

Biermann, O.

"Eine formale Ableitung der Laurentschen Reihe einer Funktion aus einer ihrer rationalgebrochenen Näherungsfunktionen," *Jour. für Reine und Angew. Math.* 135 (1908), 142–145. **FM** 39, p. 466.

Birkhoff, G.

A Source Book in Classical Analysis, Harvard University Press (1973), Cambridge, Massachusetts. Paperback reprint by Princeton University Press (1986). **Zbl** 275.01009. **MR** 57 #9395.

Bishop, E.

Foundations of Constructive Analysis, McGraw-Hill Book Co. (1967), New York. **Zbl** 183, p. 15. **MR** 36 #4930. **BAMS** 76, pp. 301–323.

Bishop, E. and Bridges, D.

Constructive Analysis, Grundlehren der mathematischen Wissenschaften 279, Springer-Verlag (1985), New York. **Zbl** 656.03042. **MR** 87d:03172.

Blaschke, W.

"Eine Erweiterung des Satzes von Vitali über Folgen analytishcer Funktionen," *Ber. Verhandl. Kön. Sächs. Gesell. Wissen. Leipzig* 67 (1915), 194–200. **FM** 45, p. 638.

"Ein Mittelwertsatz und eine kennzeichnende Eigenschaft des logarithmischen Potentials," *Ber. Verhandl. Kön. Sächs. Gesell. Wissen. Leipzig* 68 (1916), 3–7. **FM** 46, p. 742.

"Mittelwertsätze der Potentialtheorie," *Jahresber. Deutsch. Math. Verein.* 27 (1918), 157–160. **FM** 46, p. 742.

Blåsjö, V.

"How to find the logarithm of any number using nothing but a piece of string," *The College Math. Jour.* 47 (2016), 95–100 and 48 (2017), 391–392. **Zbl** 06920804. **MR** 3519055.

Błaszczyk, P., Katz, M. G. and Sherry, D.

"Ten misconceptions from the history of analysis and their debunking," *Foundations of Science* 18 (2013), 43–74. **Zbl** 06201725. **MR** 3031794. (See also **MR** 3530988.)

Bloch, A.

"Sur les intégrales de Fresnel," *Bull. Sci. Math.* (2) 43 (1919), 179–180 and 46 (1922), 34–35. **FM** 47, p. 226 and 48, p. 259.

(a) "Démonstration directe des théorèmes de M. Picard," *C. R. Acad. Sci. Paris* 178 (1924), 1593–1595. **FM** 50, p. 217.

(b) "Les théorèmes de M. Valiron sur les fonctions entières, et la théorie de l'uniformisation," *C. R. Acad. Sci. Paris* 178 (1924), 2051–2053. **FM** 50, p. 217.

"Les théorèmes de M. Valiron sur les fonctions entières, et la théorie de l'uniformisation," *Annales de la Faculté des Sciences de Toulouse* (3) 17 (1925), 1–22. **FM** 52, p. 324.

Les fonctions holomorphes et méromorphes dans le cercle-unité, Mémorial des Sciences Mathématiques, Fasc. XX, Gauthier-Villars (1926), Paris. **FM** 52, p. 324.

Blumenthal, O.

Principes de la théorie des fonctions entières d'ordre infini, Gauthier-Villars (1910), Paris. **FM** 41, p. 462. **BAMS** 21, p. 36.

Blutel, E.

"Sur une méthode d'approximation," *Bull. Soc. Math. France* 39 (1911), 155–159. **FM** 42, p. 113.

Boas, H. P.

"Julius and Julia: mastering the art of the Schwarz lemma," *Amer. Math. Monthly* 117 (2010), 770–785. **Zbl** 1205.30021. **MR** 2012d:30063.

"Mocposite functions," *Amer. Math. Monthly* 123 (2016), 427–438. **Zbl** 1391.30006. **MR** 3508963.

"Cauchy's residue sore thumb," *Amer. Math. Monthly* 125 (2018), 16–28. **Zbl** 1395.30040. **MR** 3756325.

Boas, H. P., and Boas, R. P.

"Short proofs of three theorems on harmonic functions," *Proc. Amer. Math. Soc.* 102 (1988), 906–908. **Zbl** 664.31004. **MR** 89d:31008.

BOAS, R. P., JR.

Entire Functions, Pure and Applied Mathematics, Vol. V, Academic Press (1954), New York. **Zbl** 58, p. 302. **MR** 16, p. 914. **BAMS** 62, p. 57.

(a) "Periodic entire functions," *Amer. Math. Monthly* 71 (1964), 782. **Zbl** 128, p. 72.

(b) "Yet another proof of the fundamental theorem of algebra," *Amer. Math. Monthly* 71 (1964), 180. **Zbl** 117, p. 38.

"Inequalities for the derivatives of polynomials," *Math. Mag.* 42 (1969), 165–174. **Zbl** 179, p. 378. **MR** 41 #4067.

BOAS, R. P., JR. and BUCK, R. C.

Polynomial Expansions of Analytic Functions, Ergebnisse der Mathematik und ihrer Grenzgebiete, Neue Folge, Heft 19. Springer-Verlag (2nd printing 1964), Berlin. **Zbl** 116, p. 281. **MR** 29 #218. **BAMS** 65, p. 150.

BOAS, R. P., JR. and POLLARD, H.

"Continuous analogues of series," *Amer. Math. Monthly* 80 (1973), 18–25. **Zbl** 252.42019. **MR** 47 #3903.

BOBOC, N.

Complex Functions (Roumanian), Editura Didactică si Pedagogică (1969), Bucureşti. **Zbl** 212, p. 97 and 220, p. 433.

BÔCHER, M.

"Singular points of functions which satisfy partial differential equations of elliptic type," *Bull. Amer. Math. Soc.* 9 (1902/03), 455–456. **FM** 34, p. 386.

(a) "On harmonic functions in two dimensions," *Proc. Amer. Acad. Arts and Sci.* 41 (1906), 577–583. **FM** 37, p. 481.

(b) "Another proof of the theorem concerning artificial singularities," *Annals of Math.* (2) 7 (1906), 163–164. **FM** 37, p. 416.

"On semi-analytic functions of two variables," *Annals of Math.* (2) 12 (1910–11), 18–26. **FM** 41, p. 485.

BOCHNER, S.

"Remarks on the theorems of Picard-Landau and Picard-Schottky," *Jour. Lon. Math. Soc.* 1 (1926), 100–103. **FM** 52, p. 325.

(a) "Functions of one complex variable as viewed from the theory of functions in several variables," pp. 315–333 of *Lectures on Functions of a Complex Variable,* W. Kaplan, editor; University of Michigan Press (1955), Ann Arbor. **Zbl** 68, p. 284. **MR** 17, p. 250.

(b) "Green-Goursat theorem," *Math. Zeit.* 63 (1955), 230–242. **Zbl** 65, p. 410. **MR** 17, p. 250.

BOGGIO, T.

"Sulle funzioni di variabile complessa in un'area circolare," *Atti della R. Accad. d. Scienze di Torino* 47 (1911–12), 22–37. **FM** 43, p. 487.

Bohl, P.

"Über die Bewegung eines mechanischen Systems in der Nähe einer Gleichgewichtslage," *Jour. für Reine und Angew. Math.* 127 (1904), 179–276. **FM** 35, p. 720.

Bohr, H.

"Über streckentreue und konforme Abbildung," *Math. Zeit.* 1 (1918), 403–420. **FM** 46, p. 558.

"Über einen Satz von Edmund Landau," *Scripta Univ. atque Biblio. Hierosolymitanarum* 1, No. 2 (1923). **FM** 49, p. 711.

"On the limit values of analytic functions," *Jour. Lon. Math. Soc.* 2 (1927), 180–181. **FM** 53, p. 284.

"Über ganze transzendente Funktionen, die auf jeder durch den Nullpunkt gehenden Geraden beschränkt sind," pp. 39–46 of *Opuscula Mathematica Andreae Wiman Dedicata,* Lundequist (1930), Upsala.

Bölling, R.

Das Fotoalbum für Weierstrass/A Photo Album for Weierstrass, Friedr. Vieweg & Sohn (1994), Braunschweig. **Zbl** 829.01028. **MR** 95h:01028.

Bonk, M.

"On Bloch's constant," *Proc. Amer. Math. Soc.* 110 (1990), 889–894. **Zbl** 713.30033. **MR** 91c:30011. (See also **MR** 97j:30002.)

Bonnet, O.

"Démonstration de la continuité des racines d'une équation algébrique," *Bull. Sci. Math.* 2 (1871), 215–221. **FM** 3, p. 19.

Bonsall, F. F.

"Note on a theorem of Hardy and Rogosinski," *Quart. Jour. of Math.* (*Oxford Series*) 20 (1949), 254–256. **Zbl** 34, p. 363. **MR** 11, p. 358.

Borel, É.

"Sur une application d'un théorème de M. Hadamard," *Bull. Sci. Math.* (2) 18 (1894), 22–25. **FM** 25, p. 705.

"Démonstration élémentaire d'un théorème de M. Picard sur les fonctions entières," *C. R. Acad. Sci. Paris* 122 (1896), 1045–1048. **FM** 27, p. 321.

"Sur les zéros des fonctions entières," *Acta Math.* 20 (1896–97), 357–396. **FM** 28, p. 360.

"Addition au mémoire sur les séries divergentes," *Ann. Sci. École Norm. Sup.* (3) 16 (1899), 132–136. **FM** 30, p. 230.

Leçons sur les fonctions entières, Gauthier-Villars (2nd ed. 1921), Paris. **FM** 48, p. 316.

Leçons sur les fonctions de variables réelles et les développements en séries de polynômes, Gauthier-Villars (2nd ed. 1928), Paris. **FM** 54, p. 268. **BAMS** 12, p. 399.

BOREL, É. and ROSENTHAL, A.

"Neuere Untersuchungen über Funktionen reeller Veränderlichen," *Encyklopädie der Mathematischen Wissenschaften* Bd. II, 3rd Part, 2nd Half, Heft 7, 851–1187. B. G. Teubner (1924), Leipzig. **FM** 50, p. 176. **BAMS** 31, p. 453.

BORSUK, K.

"Über Schnitte der n-dimensionalen Euklidischen Räume," *Math. Annalen* 106 (1932), 239–248. **FM** 58, p. 628. **Zbl** 4, p. 73.

(a) "Drei Sätze über die n-dimensionale euklidische Sphäre," *Fund. Math.* 20 (1933), 177–190. **FM** 59, p. 560. **Zbl** 6, p. 424.

(b) "Über stetige Abbildungen der euklidischen Räume," *Fund. Math.* 21 (1933), 236–243. **FM** 59, p. 1254. **Zbl** 8, p. 133.

"Sur les prolongements des transformations continues," *Fund. Math.* 28 (1937), 99–110. **FM** 63, p. 1168. **Zbl** 15, p. 321.

"Set theoretical approach to the disconnection theory of the Euclidean space," *Fund. Math.* 37 (1950), 217–241. **Zbl** 40, p. 102. **MR** 13, p. 150.

BOTTAZZINI, U.

The "Higher Calculus": A History of Real and Complex Analysis from Euler to Weierstrass, (translated from Italian by W. Van Egmond). Springer-Verlag (1986), New York. **Zbl** 597.01011. **MR** 88a:01016. **BAMS** (2) 17 (1987), 186.

"Remarques sur les fondements de la théorie de fonctions d'une variable complexe au milieu du XIX siècle," *Cahiers Hist. Philos. Sci.* (Nouv. Sér.) 20 (1987), 152–182. **Zbl** 652.01016.

"Three traditions in complex analysis: Cauchy, Riemann and Weierstrass," pp. 419–431 of *Companion Encyclopedia of the History and Philosophy of the Mathematical Sciences,* vol. I (edited by I. Grattan-Guinness), Routledge (1994), London and New York. **Zbl** 833.01001. **MR** 98g:01003.

BOTTAZZINI, U. and GRAY, J. J.

"Complex function theory from Zurich (1897) to Zurich (1932)," *Rend. Circ. Mat. Palermo* (2) Supplement no. 44 (1996), 85–111. **Zbl** 895.01005. **MR** 98d:01028. **Zbl** 1276.30006. **MR** 3099398.

Hidden Harmony–Geometric Fantasies: The Rise of Complex Function Theory, Springer (2013), New York. **Zbl** 1276.30006. **MR** 3099398.

BOULIGAND, G.

"Démonstration élémentaire d'un théorème de détermination à un facteur constant près d'une fonction harmonique," *Mathematica* (Cluj) 6 (1932), 80–85. **FM** 58, p. 503. **Zbl** 4, p. 395.

BOURBAKI, N.

Elements of Mathematics, General Topology. Part 2, Hermann et Cie (1966), Paris and Addison-Wesley Publishing Co. (1966), Reading. Reprinted by Springer-Verlag (1989), New York. **Zbl** 145, p. 193; 197, p. 304; 301.54002 and 337.54001. **MR** 34 #5044(b), 40, p. 1704, 41 #984, 58 #18315 and 90a: 54001b.

BOUTROUX, P.

"Propriétés d'une fonction holomorphe dans un cercle où elle ne prend pas les valeurs zéro et un," *C. R. Acad. Sci. Paris* 141 (1905), 305–307 and *Bull. Soc. Math. France* 34 (1906), 30–39. **FM** 36, p. 466 and 37, p. 421.

BOWEN, N. A.

(a) "On the limit of the modulus of a bounded regular function," *Proc. Edin. Math. Soc.* (2) 14 (1964–65), 21–24. **Zbl** 145, p. 84. **MR** 29 #1309.

(b) "On mean value limits for bounded regular functions," *Proc. Edin. Math. Soc.* (2) 14 (1964–65), 103–108. **Zbl** 146, p. 99. **MR** 31 #5981.

BOWEN, N. A. and MACINTYRE, A. J.

"Interpolatory methods for theorems of Vitali and Montel type," *Proc. Royal Soc. Edin.* A 64 (1954), 71–79. **Zbl** 56, p. 72. **MR** 16, p. 232.

BRADEN, B.

"Pólya's geometric picture of complex contour integrals," *Math. Mag.* 60 (1987), 321–327. **Zbl** 636.30038. **MR** 89m:30003.

BRAWN, F. T.

"Positive harmonic majorization of subharmonic functions in strips," *Proc. Lon. Math. Soc.* (3) 27 (1973), 261–289. **Zbl** 265.31008. **MR** 48 #8819.

BRELOT, M.

"Über die Singularitäten der Potentialfunktionen und der Integrale der Differentialgleichungen vom elliptischen Typus," *Sitzungsber. Berlin Math. Gesell.* 31 (1932), 46–54. **FM** 58, p. 492. **Zbl** 5, p. 399.

"Einige neuere Untersuchungen über das Dirichletsche Problem," *Jahresber. Deutsch. Math. Verein.* 42 (1933), 111–129. French translation: **FM** 59, p. 481. **Zbl** 6, p. 349.

Étude des fonctions sousharmoniques au voisinage d'un point, Actualités Scientifiques et Industrielles, No. 139, Hermann et Cie. (1934), Paris. **FM** 60, p. 1134. **Zbl** 9, p. 19. **BAMS** 43, p. 458.

"Familles de Perron et problème de Dirichlet," *Acta Litt. ac Scient. Univ. Hung.* (Szeged) 9 (1938–40), 133–153. **FM** 65, p. 418. **Zbl** 23, p. 233. **MR** 1, p. 121.

(a) "Quelques applications aux fonctions holomorphes de la théorie moderne du potentiel et du problème de Dirichlet," *Bull. Soc. Roy. Sci. Liége* 8 (1939), 385–391. **FM** 65, p. 419. **Zbl** 21, p. 319. **MR** 1, p. 114.

(b) "Sur l'allure à la frontière des fonctions harmoniques, sousharmoniques ou holomorphes," *Bull. Soc. Roy. Sci. Liége* 8 (1939), 468–477. **FM** 65, p. 419. **Zbl** 23, p. 234. **MR** 1, p. 122.

"La théorie moderne du potentiel," *Ann. Inst. Fourier* (Grenoble) 4 (1952), 113–140. **Zbl** 55, p. 89. **MR** 15, p. 527. (See also **Zbl** 201, p. 139 and **MR** 42 #6260.)

Éléments de la théorie classique du potentiel, Centre de Documentation Universitaire (4th ed. 1969), Paris. **Zbl** 84, p. 309 and 116, p. 75. **MR** 31 #2412 and #1387.

"Les étapes et les aspects multiples de la théorie du potentiel," *L'Enseignement Mathématique* (2) 18 (1972), 1–36. **Zbl** 235.31002. **MR** 51 #8436 and 53, p. 2382.

Brenner, J. L. and Lyndon, R. C.

"Proof of the fundamental theorem of algebra," *Amer. Math. Monthly* 88 (1981), 253–256. **Zbl** 472.30010. **MR** 82e:30010.

Bressoud, D. M.

"On the value of Gaussian sums," *Jour. of Number Theory* 13 (1981), 88–94. **Zbl** 443.10026. **MR** 82b:10046.

Brickman, L.

"Non-negative entire and meromorphic functions," *Jour. Lon. Math. Soc.* 39 (1964), 15–18. **Zbl** 123, p. 268. **MR** 28 #4112.

Brickman, L. and Steinberg, L.

"On non-negative polynomials," *Amer. Math. Monthly* 69 (1962), 218–221.

Bridges, D. S.

"On Montel's proof of the Great Picard Theorem," *Amer. Math. Monthly* 88 (1981), 45–47 and 89 (1982), 239. **Zbl** 476.30021 and 492.30022. **MR** 82e:30032 and 83e:30033.

Brill, A. and Noether, M.

"Die Entwicklung der Theorie der algebraischen Functionen in älterer und neuerer Zeit," *Jahresber. Deutsch. Math. Verein.* 3 (1892–93), 107–566. **FM** 25, p. 70.

Brillinger, D. R.

"The analyticity of the roots of a polynomial as functions of the coefficients," *Math. Mag.* 39 (1966), 145–147. **Zbl** 164, p. 375. **MR** 34 #4455.

Brisse, Ch.

"Démonstration du théorème de d'Alembert," *Jour. École Polyt.* 56 (1886), 163–169. **FM** 18, p. 60.

Brodovich, M. T.

"Monogeneity conditions for discontinuous mappings," *Teor. Funkciĭ Funkcional. Anal. i Priložen Vyp.* 12 (1970), 94–103 (Russian). **Zbl** 218, p. 169. **MR** 45 #5323.

"A certain sufficient condition for the conformality of an arbitary one-to-one mapping," *Dopovidī Akad. Nauk Ukraïn. RSR, Ser. A* (1974), 489–492, 572 (Ukrainian). **Zbl** 283.30008. **MR** 51 #8388. (See also **MR** 55 #3269 and 56 #5855. **Zbl** 446.30003 and 447.30007.)

"Holomorphy of an arbitrary unbounded mapping of a plane domain into the plane that preserves angles along a system of rays," *Siberian Math. Jour.* 32 (1991), no. 1, 21–28. Errata, *ibid.* 43 (2002), 194. **Zbl** 739.30001. **MR** 92j: 30008 and 2003a:30007.

"Holomorphy of an arbitrary approximately holomorphic mapping from a plane domain into the plane," *Siberian Math. Jour.* 34 (1993), 412–418. **Zbl** 813.30006. **MR** 95e:30002.

"On a holomorphy criterion for an arbitrary mapping of a planar domain into the plane," *Siberian Math. Jour.* 38 no. 5 (1997), 868–870. Errata, *ibid.* 43 (2002), 194. **Zbl** 880.30004. **MR** 99c:30001 and 2003a:30007. (See also **MR** 2005f:30002.)

BROGGI, U.

"Un teorema sulle serie di potenze che rappresentano funzioni razionali," *Rend. del R. Istituto Lombardo di Scienze e Lettere* (2) 64 (1931), 238-243. **FM** 57, p. 416. **Zbl** 4, p. 10.

"Su di un teorema concernente le serie di potenze," *Rend. del R. Istituto Lombardo di Scienze e Lettere* (2) 68 (1935), 859–862. **FM** 61, p. 1134. **Zbl** 14, p. 23.

BROMWICH, T. J. I'A.,

An Introduction to the Theory of Infinite Series, Cambridge University Press (2nd ed. revised, 1926), London. Reprinted by Chelsea Publishing Co. (1991), New York. **FM** 52, p. 208. **Zbl** 133, p. 8. **BAMS** 34, p. 244.

BROUWER, L. E. J.

"Zur Analysis Situs," *Math. Annalen* 68 (1910), 422–434. **FM** 41, p. 543.

"Beweis der Invarianz der Dimensionszahl," *Math. Annalen* 70 (1911), 161–165. **FM** 42, p. 416.

(a) "Beweis der Invarianz des n-dimensionalen Gebiets," *Math. Annalen* 71 (1912), 305–313 and 72 (1912), 55–56. Correction, *ibid.* 82 (1921), 286. **FM** 42, p. 418; 43, p. 479.

(b) "Über Abbildung von Mannigfaltigkeiten," *Math. Annalen* 71 (1912), 97-115. Correction, *ibid.* 71 (1912), 598 and 82 (1921), 286. **FM** 42, p. 417.

(c) "Beweis der Invarianz der gechlossenen Kurve," *Math. Annalen* 72 (1912), 422–425. **FM** 43, p. 569.

Collected Works, vol II: Geometry, Analysis, Topology and Mechanics, H. Freudenthal, editor, North-Holland Publishing Co. (1976), Amsterdam. **Zbl** 328.01020. **MR** 58 #21337.

BROWDER, A.

"Topology in the complex plane," *Amer. Math. Monthly* 107 (2000), 393–401 and 113 (2006), 935–937. **Zbl** 1076.30500, 1151.30330. **MR** 2001g:30001, 2007f: 30044.

BROWDER, F. E.

"Covering spaces, fibre spaces, and local homeomorphisms," *Duke Math. Jour.* 21 (1954), 329–336. **Zbl** 56, p. 166. **MR** 15, p. 978.

"On the proof of Mergelyan's approximation theorem," *Amer. Math. Monthly* 67 (1960), 442–444. **Zbl** 106, p. 45. **MR** 22 #6894.

BROWN, A. B.

"Extensions of the Brouwer fixed point theorem," *Amer. Math. Monthly* 69 (1962), 643. **Zbl** 134, p. 189.

BROWN, J. E.

"Images of disks under convex and starlike functions," *Math. Zeit.* 202 (1989), 457–462. **Zbl** 679.30011. **MR** 91a:30010.

BROWN, L., GAUTHIER, P. M. and SEIDEL, W.

"Possibility of complex asymptotic approximation on closed sets," *Math. Annalen* 218 (1975), 1–8. **Zbl** 304.30026. **MR** 54 #10610.

BROWN, L., SCHREIBER, B. and TAYLOR, A.

"Spectral synthesis and the Pompeiu problem," *Ann. Inst. Fourier* (Grenoble) 23 (1973), Fasc. 3, 125–154. **Zbl** 265.46044. **MR** 50 #4979.

BROWN, R. F.

"Elementary consequences of the noncontractibility of the circle," *Amer. Math. Monthly* 81 (1974), 247–252. **Zbl** 284.55009. **MR** 48 #9629.

BRUCKNER, A. M., LOHWATER, A. J. and RYAN, F.

"Some non-negativity theorems for harmonic functions," *Ann. Acad. Scient. Fennicae, Ser. A. I. Math.* No. 452 (1969). **Zbl** 187, p. 358. **MR** 42 #529. (See also **MR** 46 #3795.)

BRÜCKNER, H.

"Ein elementarer Beweis der algebraischen Abgeschlossenheit des Körpers $\mathbb{C}$ der komplexen Zahlen," *Mitt. Math. Gesell. Hamburg* 12 (1991), 487–489. **Zbl** 753.12001. **MR** 92j:12001.

BRUNÉ, C.

"Sulla rappresentazione conforme dei campi piani semplici pluriconnessi," *Atti del Reale Istituto Veneto di Scienze, Lettere ed Arti* 91 (1932), 359–380. **FM** 58, p. 1092.

BSHOUTY, D. and HENGARTNER, W.

"Boundary correspondence of univalent harmonic mappings from the unit disc onto a Jordan domain," pp. 51–60 of *Approximation by Solutions of Partial*

Differential Equations, NATO Advanced Science Institutes, Ser. C, Math. Phys. Sci. 365, Kluwer Academic Publishers (1992), Dordrecht. **Zbl** 777.30009. **MR** 93f:31004.

"Univalent solutions of the Dirichlet problem for ring domains," *Complex Variables* 21 (1993), 159–169. **Zbl** 794.30014. **MR** 95h:30018.

BUHL, A.

Séries analytiques. Sommabilité, Mémorial des Sciences Mathématiques, Fasc. 7, Gauthier-Villars (1925), Paris. **FM** 51, p. 242. **BAMS** 33, p. 120.

BURCKEL, R. B., MARSHALL, D. E., MINDA, D., POGGI-CORRADINI, P. and RANSFORD, T. J.

"Area, capacity and diameter versions of Schwarz's Lemma," *Conformal Geometry and Dynamics* 12 (2008), 133–152. **Zbl** 1233.30016. **MR** 2010j:30050.

BURCKEL, R. B. and MINDA, D.

"A characterization of simple and double connectivity" *Expositiones Math.* 16 (1998), 133–144. **Zbl** 909.30006. **MR** 99d:30011.

BURDICK, D. and LESLEY, F. D.

"Some uniqueness theorems for analytic functions," *Amer. Math. Monthly* 82 (1975), 152–155. **Zbl** 306.30004. **MR** 50 #10234.

BUREAU, F.

"Mémoire sur les fonctions uniformes à point singulier essentiel isolé," *Mém. Soc. Roy. Sci. Liége* (3) 17 No. 3 (1932). **FM** 58, p. 1085. **Zbl** 6, p. 408.

"Analytic functions and their derivatives," *Acad. Roy. Belg. Cl. Sci. Mém. Collect* 8° (3) 12 (1997). **Zbl** 873.30001. **MR** 98f:30031.

BURGERS, W. A. M.

Radiale randlimieten van holomorfe functies, Centrale Drukkerij (1929), Nijmegen. **FM** 55, p. 770.

BURGHELEA, A.

"About the numbering of the roots of a family of generic polynomials," *Bul. Inst. Politehn. Bucureşti Ser. Mec.* 39, No. 3 (1977), 11–15 (Roumanian). **Zbl** 434.30006. **MR** 57 #12469.

BURKHARDT, H.

"Trigonometrische Reihen und Integrale," *Encyklopädie der Mathematischen Wissenschaften* Bd. II, 1st Part, 2nd Half, Heft 7, 819–1354. B. G. Teubner (1914), Leipzig. **FM** 45, p. 374. (See also **FM** 36, p. 56 and 39, p. 62.)

BURRILL, C. W.

Foundations of Real Numbers, McGraw-Hill Book Co. (1967), New York. **Zbl** 175 p. 339. **MR** 36 #1585.

BUTLER, R.

"On the evaluation of $\int_0^\infty (\sin^m t)/t^m\, dt$ by the trapezoidal rule," *Amer. Math. Monthly* 67 (1960), 566–569. **Zbl** 100, p. 56. **MR** 22 #4841.

BUTZER, P. L.

"The Shannon sampling theorem and some of its applications; an overview," *Jour. Math. Res. and Expo.* 3 (1983), 185–212. **Zbl** 523.94003. **MR** 85h:94007.

BUTZER, P. L., RIES, S. and STENS, R. L.

"Shannon's sampling theorem, Cauchy's integral formula, and related results," pp. 363–377 of *Anniversary volume on approximation theory and functional analysis,* International Series in Numerical Mathematics, Vol. 65. Birkhäuser Verlag (1984), Basel. **Zbl** 557.30034. **MR** 87g:30047.

BUTZER, P. L., SPLETTSTÖSSER, W. and STENS, R. L.

"The sampling theorem and linear prediction in signal analysis," *Jahresber. Deutsch. Math. Verein.* 90 (1988), 1–70. **Zbl** 633.94002. **MR** 89b:94006.

C

CAIRNS, S. S.

"An elementary proof of the Jordan-Schoenflies theorem," *Proc. Amer. Math. Soc.* 2 (1951), 860–867. **Zbl** 44, p. 198. **MR** 13, p. 764.

CAJORI, F.

"History of the exponential and logarithm concepts," *Amer. Math. Monthly* 20 (1913), 5–14; 35–47; 75–84; 107–117; 148–151; 173–182; 205–210; 23 (1916), 71–72. **FM** 44, p. 60; 45, p. 95; 46, p. 51.

CANTOR, D. G. and PHELPS, R.

"An elementary interpolation theorem," *Proc. Amer. Math. Soc.* 16 (1965), 523–525. **Zbl** 146, p. 96. **MR** 31 #358.

CARATHÉODORY, C.

"Sur quelques applications du théorème de Landau-Picard," *C. R. Acad. Sci. Paris* 144 (1907), 1203–1206. **FM** 38, p. 432.

"Über den Variabilitätsbereich der Fourier'schen Konstanten von positiven harmonischen Funktionen," *Rend. Circ. Mat. Palermo* 32 (1911), 193–217. **FM** 42, p. 429.

(a) "Untersuchungen über die konformen Abbildungen von festen und veränderlichen Gebieten," *Math. Annalen* 72 (1912), 107–144. **FM** 43, p. 524.

(b) "Sur le théorème de M. Picard," *C. R. Acad. Sci. Paris* 154 (1912), 1690–1693. **FM** 43, p. 506.

(a) "Über die gegenseitige Beziehung der Ränder bei der konformen Abbildung des Inneren einer Jordanschen Kurve auf einen Kreis," *Math. Annalen* 73 (1913), 305–320. **FM** 44, p. 757.

(b) "Zur Ränderzuordnung bei konformer Abbildung," *Göttingen Nachr.* (1913), 509–518. **FM** 44, p. 758.

"Elementarer Beweis für den Fundamentalsatz der konformen Abbildungen," pp. 19–41 of *Mathematische Abhandlungen H. A. Schwarz Gewidmet,* (O. Bolza, *et al.*, editors) Verlag Julius Springer (1914), Berlin. **FM** 45, p. 667. Reprinted by Chelsea Publishing Co. (1974), New York.

"Über die Studysche Rundungsschranke," *Math. Annalen* 79 (1919), 402. **FM** 47, p. 327.

"Bemerkungen zu den Existenztheoremen der konformen Abbildung," *Bull. Calcutta Math. Soc.* 20 (1928), 125–134. **FM** 56, p. 295.

(a) "Über die Winkelderivierten von beschränkten analytischen Funktionen," *Sitzungsber. Preuss. Akad. Wissen. Berlin, Phys. -math. Kl.* (1929), 39–54. **FM** 55, p. 209.

(b) "Stetige Konvergenz und normale Familien von Funktionen," *Math. Annalen* 101 (1929), 515–533. **FM** 55, p. 198.

"Über die Abbildungen, die durch Systeme von analytischen Funktionen von mehreren Veränderlichen erzeugt werden," *Math. Zeit.* 34 (1932), 758–792. **FM** 58, p. 349. **Zbl** 3, p. 407.

"Eine Verschärfung des Schwarzschen Lemmas," *Praktika de l'Acad. d'Athens* 11 (1936), 276–286. **FM** 62, p. 1211. **Zbl** 15, p. 306.

(a) "On Dirichlet's problem," *Amer. Jour. of Math.* 59 (1937), 709–731. **FM** 63, p. 454. **Zbl** 17, p. 260.

(b) "The most general transformations of plane regions which transform circles into circles," *Bull. Amer. Math. Soc.* 43 (1937), 573–579. **FM** 63, p. 294. **Zbl** 17, p. 229.

"A proof of the first principal theorem on conformal representation," pp. 75–83 of *Studies and Essays Presented to R. Courant on his 60th Birthday,* Interscience Publishers, Inc. (1948), New York. **Zbl** 38, p. 52. **MR** 9, p. 232.

Conformal Representation, (translated from German by B. M. Wilson and M. Kennedy) Cambridge Tracts in Mathematics and Physics, No. 28, Cambridge University Press (2nd ed. 1952), London; reprinted by Dover Publications, Inc. (1998), New York. **FM** 58, p. 354. **Zbl** 47, p. 79. **MR** 13, p. 734. **BAMS** 60, p. 281.

Theory of Functions of a Complex Variable I, (2nd ed. 1958), II (2nd ed. 1960), (translated from German by F. Steinhardt) Chelsea Publishing Co. New York. **Zbl** 96, p. 49. **MR** 15, p. 612; 16, p. 346 and 24 #A209. **BAMS** 57, p. 190.

Carathéodory, C. and Fejér, L.

"Remarques sur le théorème de M. Jensen," *C. R. Acad. Sci. Paris* 145 (1907), 163–165. **FM** 38, p. 427.

Carathéodory, C. and Landau, E.

"Beiträge zur Konvergenz von Funktionenfolgen," *Sitzungsber. Kön. Preuss. Akad. Wissen. Berlin* (1911), 587–613. **FM** 42, p. 275.

CARATHÉODORY, C. and RADEMACHER, H.

"Über die Eineindeutigkeit im Kleinen und im Grossen stetiger Abbildungen von Gebieten," *Arch. der Math. und Physik* (3) 26 (1917), 1–9. **FM** 46, p. 834.

CARGO, G. T.

"Blaschke products and singular functions with prescribed boundary values," *Jour. Math. Anal. and Applic.* 71 (1979), 287–296. **Zbl** 428.30021. **MR** 80i: 30053. (See also **MR** 2003g:30066.)

CARLEMAN, T.

"Über die Approximation analytischer Funktionen durch lineare Aggregate von vorgegebenen Potenzen," *Arkiv för Mat., Astro. och Fysik* 17, No. 9 (1922–23). **FM** 49, p. 708.

(a) "Extension d'un théorème de Liouville," *Acta Math.* 48 (1926), 363–366. **FM** 52, p. 316.

(b) *Les Fonctions Quasi Analytiques,* Gauthier-Villars (1926), Paris. **FM** 52, p. 255.

"Sur un théorème de Weierstrass," *Arkiv för Mat., Astro. och Fysik* 20 B, No. 4 (1927). **FM** 53, p. 237. (See also **MR** 82j:31001.)

CARLESON, L.

"Mergelyan's theorem on uniform approximation," *Math. Scand.* 15 (1965), 167–175. **Zbl** 163, p. 86. **MR** 33 #6368.

Selected Problems on Exceptional Sets, Van Nostrand Mathematical Studies #13, D. Van Nostrand Co. (1967), Princeton. Reprinted by Wadsworth International Group (1982), Belmont California. **Zbl** 224.31001. **MR** 50 #4932.

CARLESON, L. and GAMELIN, T.

Complex Dynamics, Springer-Verlag (1993), New York. **Zbl** 782.30022. **MR** 94h:30033.

CARLSON, F.

"Sur le module maximum d'une fonction analytique uniforme. I," *Arkiv för Mat., Astro. och Fysik* 26A, No. 9 (1938). **FM** 64, p. 1071. **Zbl** 19, p. 221.

"Quelques inégalités concernant les fonctions analytiques," *Arkiv för Mat., Astro. och Fysik* 29B, No. 11 (1943). **MR** 6, p. 205.

CARTAN, H.

"Les fonctions de deux variables complexes et le problème de la représentation analytique," *Jour. de Math. Pures et Appliq.* (9) 10 (1931), 1–114. **FM** 57, p. 387. **Zbl** 1, p. 285.

"Sur les fonctions de plusieurs variables complexes. L'itération des transformations intérieures d'un domaine borné," *Math. Zeit.* 35 (1932), 760–773. **FM** 58, p. 349. **Zbl** 4, p. 406.

"Sur les transformations localement topologiques," *Acta Litt. ac Scient. Univ. Hung.* (Szeged) 6 (1933), 85–104. **FM** 59, p. 563. **Zbl** 7, p. 370.

CARTWRIGHT, M. L.

"Some generalizations of Montel's theorem," *Proc. Camb. Phil. Soc.* 31 (1935), 26–30. **FM** 61, p. 307. **Zbl** 11, p. 29.

"On certain integral functions of order one," *Quart. Jour. of Math.* (*Oxford Series*) 7 (1936), 46–55. **FM** 62, p. 354. **Zbl** 13, p. 358.

Integral Functions, Cambridge Tracts in Mathematics, No. 44, Cambridge University Press (1956), London. **Zbl** 75, p. 59. **MR** 17, p. 1067. **BAMS** 67, p. 454.

"A generalization of Montel's theorem," *Jour. Lon. Math. Soc.* 37 (1962), 179–184. **Zbl** 137, p. 50. **MR** 25 #4106.

CASORATI, F.

Teorica delle funzioni di variabili complesse, Vol I, Tipografia dei fratelli Fusi (1868), Pavia. **FM** 1, p. 128.

"Un teorema fondamentale nella teorica delle discontinuità delle funzioni," *Rend. del R. Istituto Lombardo di Scienze e Lettere* (2) 1 (1868), 123–125. **FM** 1, p. 128. (See also **MR** 13, p. 612 and **Zbl** 49, p. 291.)

"Aggiunte a recenti lavori dei signori Weierstrass e Mittag-Leffler sulle funzioni di una variabile complessa," *Annali Mat. Pura Appl.* (2) 10 (1882), 261–278. **FM** 14, p. 327. (See also **FM** 25, p. 707.)

CASTELLANOS, D.

"The ubiquitous π," *Math. Mag.* 61 (1988), 67–98 and 148–163, and 62 (1989), 148. **Zbl** 654.10001,2. **MR** 89c:01025, 11184.

CATER, F. S.

"Differentiable, nowhere analytic functions," *Amer. Math. Monthly* 91 (1984), 618–624. **Zbl** 598.26035. **MR** 86b:26034.

"An elementary proof that analytic functions are open mappings," *Real Anal. Exchange* 27 (2001/2002), 389–392. **Zbl** 1075.30500. **MR** 2002k:30011.

CAUCHY, A. -L.

Cours d'analyse de l'École Royale Polytechnique. Première partie: analyse algébrique (1821), Paris (=Vol. 3, 2nd ser. of *Oeuvres complètes*). Reprinted by Wissenschaftliche Buchgesellschaft (1968), Darmstadt. German translation by C. Itzigsohn, Verlag Julius Springer (1885), Berlin. **FM** 17, p. 200. (See also **FM** 28, p. 12.) Reprinted by Éditions Jacques Gabay (1989), Seaux. **MR** 94a:01019. Reprinted (with 170-page English introduction by U. Bottazzini) by Cooperativa Libraria Universitaria Editrice Bologna (1992), Bologna. **Zbl** 806.01030. **MR** 95a:01026. English translation by R. E. Bradley and C. E. Sandifer, Springer-Verlag (2009), New York. **Zbl** 1189.26001. **MR** 2010h: 01011.

"Extrait d'une lettre à M. Coriolis," *C. R. Acad. Sci. Paris* 4 (1837), 216–218.

"Méthode simple et nouvelle pour la détermination complète des sommes alternées formées avec les racines primitives des équations binômes," *C. R. Acad. Sci. Paris* 10 (1840), 560–572 and *Jour. de Math. Pures et Appliq.* 5 (1840), 154–168.

"Note sur le développement des fonctions en séries," *C. R. Acad. Sci. Paris* 13 (1841), 909–914.

"Note sur le développement des fonctions en séries convergentes ordonnées suivant les puissances entières des variables," *C. R. Acad. Sci. Paris* 17 (1843), 940–942.

"Mémoire sur quelques propositions fondamentales du calcul des résidus, et sur la théorie des intégrales singulières," *C. R. Acad. Sci. Paris* 19 (1844), 1337–1344.

"Mémoire sur les fonctions complémentaires," *C. R. Acad. Sci. Paris* 19 (1844), 1377–1384.

"Sur les intégrales qui s'étendent à tous les points d'une courbe fermée," *C. R. Acad. Sci. Paris* 23 (1846), 251–255 and 689–704.

"Sur les intégrales dans lesquelles la fonction sous le signe $\int$ change brusquement de valeur," *C. R. Acad. Sci. Paris* 23 (1846), 537 and 557–563.

"Mémoire sur une nouvelle théorie des imaginaires et sur les racines symboliques des équations et des équivalences," *C. R. Acad. Sci. Paris* 24 (1847), 1120–1130.

"Sur les fonctions de variables imaginaires," *C. R. Acad. Sci. Paris* 32 (1851), 160–162.

"Note sur les séries convergentes dont les divers termes sont des fonctions continues d'une variable réelle ou imaginaire entre des limites données," *C. R. Acad. Sci. Paris* 36 (1853), 454–459.

"Abhandlung über bestimmte Integrale zwischen imaginären Grenzen," translation (from French) and commentary by P. Stäckel in *Ostwald's Klassiker der exakten Wissenschaften,* No. 112. Verlag von W. Engelmann (1900), Leipzig. **FM** 31, p. 314. (See also **FM** 7, p. 155, **MR** 57 #12151 and **Zbl** 274.01026 for reprintings of original (1825) French and Iacobacci [1965] for an English translation.)

Œuvres complètes d'Augustin Cauchy. Publiée sous la direction scientifique de l'Académie des Sciences. I^e^ sér. Tome 1–12. *II*^e^ sér. Tome 1–15. Gauthier-Villars (1882–1974), Paris. See **Zbl** 297.01012 and **MR** 58#21338.

Cauer, W.

"The Poisson integral for functions with positive real part," *Bull. Amer. Math. Soc.* 38 (1932), 713–717. **FM** 58, p. 1087. **Zbl** 5, p. 361.

Cayley, A.

"Note on the integrals $\int_0^\infty \cos x^2\,dx$ and $\int_0^\infty \sin x^2\,dx$," *Quart. Jour. of Math.* 12 (1873), 118–126. **FM** 4, p. 139.

Cazzaniga, P.

"Espressione di funzioni intere che in posti dati arbitrariamente prendono valori prestabiliti," *Annali Mat. Pura Appl.* (2) 10 (1880–82), 279–290. **FM** 14, p. 324.

Čebotarev [=Tschebotareff], N. G. and Meĭman, N. N.

"The Routh-Hurwitz Problem for polynomials and entire functions," *Trudy Mat. Inst. Steklov.* 26 (1949) (Russian). **Zbl** 41, pp. 198, 662. **MR** 11, p. 509.

Čech, E.

"Une démonstration du théorème de Cauchy et de la formule de Gauss," *Rend. Accad. d. Lincei Roma* (6) 11 (1930), 884–887. **FM** 56, p. 958.

"Encore sur le théorème de Cauchy," *Rend. Accad. d. Lincei Roma* (6) 12 (1930), 286–289. **FM** 56, p. 958.

Point Sets (translated from 1936 Czech original by A. Pultr), Academic Press (1969), New York. **Zbl** 191, p. 532. **MR** 41 #1000.

Černý, I.

Foundations of Analysis in the Complex Domain (Czech), Academia-Nakladatelství Československé akademie věd (1967), Prague. **Zbl** 168, p. 34. **MR** 36 #5309.

"A simple proof of Cauchy theorem," *Časopis pro Pěstování Matematiky a Fysiki* 101 (1976), 366–369. **Zbl** 353.30001. **MR** 56 #5900.

"Extension of a homeomorphism of a topological circumference," *Časopis pro Pěstování Matematiky a Fysiki* 102 (1977), 105–127. **Zbl** 366.57001. **MR** 56 #6628.

Analysis in the Complex Plane (Czech), Academia-Nakladatelství Československé akademie věd (1983), Prague. **MR** 86d:30001. English translation published by Ellis Horwood (1992), Chichester. **Zbl** 774.30001. **MR** 93g:30001.

Challener, D. and Rubel, L.

"A converse of Rouché's theorem," *Amer. Math. Monthly* 89 (1982), 302–305. **Zbl** 489.30030. **MR** 83e:30035.

Chaundy, T. H.

"On a trigonometric series," *Jour. Lon. Math. Soc.* 12 (1937), 296–298. **FM** 63, p. 293. **Zbl** 17, p. 302.

Chen, Hongwei

"The Fresnel integrals revisited," *The College Math. Jour.* 40 (2009), 259–262. **MR** 2010h:26002.

Chernoff, P.

"The boundary values of a holomorphic function," *Amer. Math. Monthly* 74 (1967), 843.

"Conformal maps of ellipses onto ellipses," *Amer. Math. Monthly* 84 (1977), 396.

"Pointwise convergence of Fourier series," *Amer. Math. Monthly* 87 (1980), 399–400. **Zbl** 439.42005. **MR** 81b:42025. (See also **MR** 87g:42012 and 85j:42008.)

CHINN, W. G. and STEENROD, N. E.

Five concepts of topology. The geometry of mappings of segments, curves, circles, and disks, New Mathematical Library, vol. 18. Random House (1966), New York. **Zbl** 201, p. 553. **MR** 34 #6705.

CHESSIN, A. S.

"On the singularities of single-valued and generally analytic functions," *Annals of Math.* 11 (1895–96), 52–56. **FM** 27, p. 301.

CHOQUET, G.

"Sur un type de transformation analytique généralisant la représentation conforme et définie au moyen de fonctions harmoniques," *Bull. Sci. Math.* (2) 69 (1945), 156–165. **MR** 8, p. 93.

"Sur les homéomorphies harmoniques d'un disque D sur D," *Complex Variables* 24 (1994), 47–48. **Zbl** 793.30012. **MR** 95k:31002.

CHOU, HSIN-TI

"On the uniqueness theorem for analytic functions and its applications," *Chinese Math.* 9 (1967), 433–438. **Zbl** 172, p. 373. **MR** 21 #5717.

CHUANG, CHI-TAI

Normal families of meromorphic functions, World Scientific Publishing Co. (1993), Singapore. **Zbl** 878.30026. **MR** 95d:30065. **BAMS** (2) 32, p. 257.

CHUANG, CHI-TAI and YANG, LO

"Distributions of the values of meromorphic functions," pp. 21–63 of *Analytic Functions of One Complex Variable* (Chung Chun Yang and Chi-Tai Chuang, editors), Contemporary Mathematics 48 (1985). **Zbl** 578.30019. **MR** 87e: 30039.

Fix-Points and Factorization of Meromorphic Functions, World Scientific Publishing Co. (1990), Teaneck, New Jersey. **Zbl** 743.30027. **MR** 91b:30072.

CIMA, J. A. and STRESSIN, M.

"On the recovery of analytic functions," *Canad. Jour. Math.* 48 (1996), 288–301. **Zbl** 851.30018. **MR** 97g:30035.

CIMMINO, G.

"Formole di maggiorazione nel problema di Dirichlet per le funzioni armoniche," *Rend. Seminario Mat. Padova* 3 (1932), 46–66. **FM** 58, p. 508. **Zbl** 5, p. 16.

CIORANESCO, N.

"Le problème de Dirichlet harmonique et les médiations successives," *Bull. Sci. Math.* (2) 54 (1930), 191–200. **FM** 56, p. 1068.

CIPRA, B. and RUPP, R.

"A functional equation," *Math. Mag.* 68 (1995), 227.

CISOTTI, U.

"Determinazione della funzione di variabile complessa regolare in una corona circolare note: La parte reale sulla circonferenza esterna e la parte immaginaria sull'interna," *Rend. Accad. d. Lincei Roma* (6) 13 (1931), 395–399. **FM** 57, p. 348. **Zbl** 2, p. 33.

CLARK, P. L.

"The instructor's guide to real induction," *Math. Mag.* 92 (2019), 136–150 + 400. **MR** 3929271.

CLARKE, R. J.

"Some integrals of the form $\int_0^\infty \frac{\sin^m x \cos^n x}{x^p}\,dx$," *Math. Gazette* 82 (1998), 290–293.

CLUNIE, J.

"On integral and meromorphic functions," *Jour. Lon. Math. Soc.* 37 (1962), 17–27. **Zbl** 104, p. 295. **MR** 26 #1456.

"Three-circles theorems," *Jour. d'Analyse Math.* 46 (1986), 65–79. **Zbl** 603.30043. **MR** 87m:30005.

CLUNIE, J., HALLENBECK, D. J. and MACGREGOR, T. H.

"A peaking and interpolation problem for univalent functions," *Jour. Math. Anal. and Applic.* 111 (1985), 559–570. **Zbl** 583.30006. **MR** 87b:30057. (See also **MR** 2000a:30018.)

COHEN, L. W. and EHRLICH, G.

The Structure of the Real Number System, D. Van Nostrand Co. (1963), Princeton. Reprinted by Robert E. Krieger Publishing Co., Inc. (1977), Melbourne, Florida. **Zbl** 118, p. 285.

COHEN, P. J.

"On Green's theorem," *Proc. Amer. Math. Soc.* 10 (1959), 109–112. **Zbl** 90, p. 35. **MR** 21 #3004.

"A note on constructive methods in Banach algebras," *Proc. Amer. Math. Soc.* 12 (1961), 159–163. **Zbl** 97, p. 106. **MR** 23 #A1827.

COLLINGWOOD, E. F.

"On three circles theorems (I)," *Jour. Lon. Math. Soc.* 7 (1932), 162–166. **FM** 58, p. 331. **Zbl** 5, p. 17.

COLLINGWOOD, E. F. and LOHWATER, A. J.

The Theory of Cluster Sets, Cambridge Tracts in Mathematics and Mathematical Physics, No. 56, Cambridge University Press (1966), London. **Zbl** 212, p. 423. **MR** 50 #13528. (See also the Russian supplement: **MR** 53 #3311 and **Zbl** 283.30032.)

COLLINGWOOD, E. F. and PIRANIAN, G.

"The mapping theorems of Carathéodory and Lindelöf," *Jour. de Math. Pures et Appliq.* (9) 43 (1964), 187–199. **Zbl** 133, p. 327. **MR** 29 #6022.

COLWELL, P.

Blaschke Products. Bounded Analytic Functions, The University of Michigan Press (1985), Ann Arbor. **MR** 86f:30033.

CONNELL, E. H.

"A classical theorem in complex variable," *Amer. Math. Monthly* 72 (1965), 729–732. **Zbl** 152, p. 60. **MR** 32 #1360.

CONNELL, E. H. and PORCELLI, P.

"An algorithm of J. Schur and the Taylor series," *Proc. Amer. Math. Soc.* 13 (1962), 232–235. **Zbl** 108, p. 71. **MR** 24 #A3300.

CONSIGLIO, A.

"Sulla risoluzione in termini finiti dei problemi di Villat e Dirichlet relativi alla corona circolare," *Atti Accad. Gioenia Sci. natur. Catania* (6) 6, Mem. 7 (1950). **Zbl** 39, p. 324. **MR** 12, p. 88.

CONWAY, J. B.

Functions of One Complex Variable, Graduate Texts in Mathematics, Vol. 11, Springer-Verlag (2nd ed. 1978), New York. **Zbl** 277.30001. **MR** 56 #5843 and 80c:30003.

Functions of One Complex Variable II, Graduate Texts in Mathematics, Vol. 159, Springer-Verlag (1995), New York. **Zbl** 887.30003. **MR** 96i:30001.

COROMINAS, F. and SUNYER I BALAGUER, F.

"Conditions for an infinitely differentiable function to be a polynomial," *Revista Mat. Hisp.-Amer.* (4) 14 (1954), 26–43 (Spanish). **Zbl** 55, p. 288. **MR** 15, p. 942. (See also **MR** 2006g:26028.)

CORPUT, J. G., VAN DER,

"Le théorème fondamental de l'algèbre," *Actual. Math. Centrum, Amsterdam, Scriptum* 2, l. Teil (1950). **Zbl** 39, p. 11. **MR** 12, p. 475.

COURANT, R.

(a) "Über eine Eigenschaft der Abbildungsfunktionen bei konformer Abbildung," *Göttingen Nachr.* (1914), 101–109 and (1922), 69–70. **FM** 45, p. 668 and 48, p. 1235.

(b) "Über die Existenztheoreme der Potential-und Funktionentheorie," *Jour. für Reine und Angew. Math.* 144 (1914), 190–211. **FM** 45, p. 668.

COWEN, C.

"Iteration and the solution of functional equations for functions analytic in the unit disc," *Trans. Amer. Math. Soc.* 265 (1981), 69–95. **Zbl** 476.30017. **MR** 82i:30036.

"Commuting analytic functions," *Trans. Amer. Math. Soc.* 283 (1984), 685–695. **Zbl** 542.30030. **MR** 85i:30054.

COWEN, C. and POMMERENKE, CH.

"Inequalities for the angular derivative of an analytic function in the unit disk," *Jour. Lon. Math. Soc.* (2) 26 (1982), 271–289. **Zbl** 488.30001. **MR** 84a:30006. (See also **MR** 92h:30002.)

COWLING, V. F.

"On analytic functions having a positive real part in the unit circle," *Amer. Math. Monthly* 63 (1956), 329–330. **Zbl** 70, p. 72. **MR** 17, p. 1070.

CRAIG, C., JR. and MACINTYRE, A. J.

"Inequalities for functions regular and bounded in a circle," *Pac. Jour. of Math.* 20 (1967), 449–454. **Zbl** 182, p. 408. **MR** 34 #7806.

CRAVEN, B. D.

"A note on Green's theorem," *Jour. Austral. Math. Soc.* 4 (1964), 289–292. **Zbl** 135, p. 114. **MR** 32 #5807.

CREMER, H.

"Über die Iteration rationaler Funktionen," *Jahresber. Deutsch. Math. Verein.* 33 (1925), 185–210. **FM** 51, p. 262.

"Ein Existenzbeweis der Kreisringabbildung zweifach zusammenhängender schlichter Bereiche," *Ber. Verhandl. Sächs. Akad. Wissen., math.-phys. Kl.* 82 (1930), 190–192. **FM** 56, p. 296.

Carmina mathematica und andere poetische Jugendsünden (7th ed.), Verlag J. A. Meyer (1982), Aachen. **Zbl** 625.00031.

CREUTZ, M. and STERNHEIMER, R. M.

"On the convergence of iterated exponentiation - I," *Fibonacci Quart.* 18 (1980), 341–347. **Zbl** 493.26004. **MR** 82b:26006.

CROWE, J. and SAMPERI, D.

"On open maps," *Amer. Math. Monthly* 96 (1989), 242–243. **Zbl** 663.54007. **MR** 90f:54035.

CSÁSZÁR, Á.

"Sur les auxiliaires topologiques des éléments de la théorie des fonctions analytiques," *Mat. Lapok* 13 (1962), 73–94 (Hungarian). **Zbl** 122, p. 75. **MR** 26 #3918.

CSILLAG, P.

"Untersuchungen über die Borelschen Verallgemeinerungen des Picardschen Satzes," *Math. Annalen* 100 (1928), 367–383. **FM** 54, p. 347.

"Über ganze Funktionen, welche drei nicht verschwindende Ableitungen besitzen," *Math. Annalen* 110 (1935), 745–752. **FM** 61, p. 341. **Zbl** 10, p. 265.

CUCKER, F. and CORBALAN, A. G.

"An alternate proof of the continuity of the roots of a polynomial," *Amer. Math. Monthly* 96 (1989), 342–345. **Zbl** 708.30010. **MR** 90h:12003.

CUNNINGHAM, F., JR.

"Taking limits under the integral sign," *Math. Mag.* 40 (1967), 179–186 and *Amer. Math. Monthly* 75 (1968), 827. **Zbl** 156, p. 62.

ĆURGUS, B. and MASCIONI, V.

"Roots and polynomials as homeomorphic spaces," *Expositiones Math.* 24 (2006), 81–95. **Zbl** 1101.30008. **MR** 2006m:30012.

CURTISS, D. R.

"Note on the sufficient conditions for an analytic function," *Bull. Amer. Math. Soc.* 8 (1901/02), 329–331. **FM** 33, p. 395.

"A proof of the theorem concerning artificial singularities," *Annals of Math.* (2) 7 (1905-06), 161–162. **FM** 37, p. 416.

"On certain theorems of mean value for analytic functions of a complex variable," *Annals of Math.* (2) 8 (1906-07), 118–126. **FM** 38, p. 337.

CURTISS, J. H.

"Faber polynomials and Faber series," *Amer. Math. Monthly* 78 (1971), 577–596. Correction, *ibid.* 79 (1972), 363. **Zbl** 215, p. 415; 237.30001. **MR** 45 #2183, #8893.

D

DAIA, L.

"On a theorem of Shizuo Kakutani," *Real Anal. Exchange* 12 (1986/87), 361–365. **Zbl** 642.26010.

DALZELL, D. P.

"On 22/7," *Jour. Lon. Math. Soc.* 19 (1944), 133–134. **Zbl** 60, p. 153. **MR** 7, p. 152.

DANERS, D.

"A short elementary proof of $\sum 1/k^2 = \pi^2/6$," *Math. Mag.* 85 (2012), 361–364. **Zbl** 1274.97037. **MR** 3007217.

DANIKAS, N.

"On an identity theorem in the Nevanlinna class $\mathcal{N}$" *Jour. Approx. Theory* 77 (1994), 184–190, **Zbl** 804.30031. **MR** 95d:30067. (See also **Zbl** 922.30005.)

DANTSCHER, V. VON

"Bemerkung zur logarithmischen Reihe," *Monatshefte für Math. und Physik* 6 (1895), 390–392. **FM** 26, p. 287.

DARBOUX, J. G.

"Mémoire sur les fonctions discontinues," *Ann. Sci. École Norm. Sup.* (2) 4 (1875), 57–112. **FM** 7, p. 243 and 11, p. 274.

"Mémoire sur l'approximation des fonctions de très-grands nombres, et sur une classe étendue de développements en série," *Jour. de Math. Pures et Appliq.* (3) 4 (1878), 5–56 and 377–416. **FM** 8, p. 304 and 10, p. 279.

"Un peu de géométrie à propos de l'intégrale de Poisson," *Bull. Sci. Math.* (2) 34 (1910), 287–300. **FM** 41, p. 319.

DARST, R.

"Most infinitely differentiable functions are nowhere analytic," *Canad. Math. Bull.* 16 (1973), 597–598. **Zbl** 285.26015. **MR** 49 #10838. (See also **MR** 52 #6685.)

DAUBEN, J. W.

"The invariance of dimension: Problems in the early development of set theory and topology," *Historia Math.* 2 (1975), 273–288. **Zbl** 313.01007. **MR** 57 #15888. (See also **Zbl** 833.01002.)

DAVID, J. -M.

"Notice sur les oeuvres de Cauchy," *Mém. Acad. Sci. de Toulouse* (8) 4 (premier semestre) (1882), 178–201. **FM** 14, p. 18.

"Sur une formule de Cauchy," *Mém. Acad. Sci. de Toulouse* (8) 7 (1885), 193–215. **FM** 17, p. 271.

DAVIDSON, K. R.

"Pointwise limits of analytic functions," *Amer. Math. Monthly* 90 (1983), 391–394 and 95 (1988), 535–536. **Zbl** 518.30003 and 657.30003. **MR** 84h:30049 and 89g:30062.

DAVIE, A. M.

"Pointwise limits of analytic functions," *Jour. Lon. Math. Soc.* (2) 75 (2007), 133–145 and 79 (2009), 272. **Zbl** 1127.30015, 1157.30320. **MR** 2009b:30073, 2010b:30057.

DAVIS, B.

"Brownian motion and analytic functions," *Ann. of Probability* 7 (1979), 913–932. **Zbl** 421.60072. **MR** 80j:30001.

DAVIS, B. and LEWIS, J. L.

"Paths for subharmonic functions," *Proc. Lon. Math. Soc.* (3) 48 (1984), 401–427. **Zbl** 541.31001. **MR** 85m:31002.

DEAUX, R. AND DELCOURTE, M.,

"Calcul des intégrales $(m,n) = \int_0^\infty \frac{\sin^m x}{x^n}\,dx$, m et n entiers positifs, $m \geq n$," *Mathesis* 66 (1957), 16–22. **Zbl** 85, p. 43. **MR** 19, p. 29.

DE JONG, T.

"Lagrange multipliers and the fundamental theorem of algebra," *Amer. Math. Monthly* 116 (2009), 828–830. **Zbl** 1229.12003. **MR** 2572089.

DE OLIVEIRA, O. R. B

"The fundamental theorem of algebra: An elementary and direct proof," *Math. Intelligencer* 33, no. 2 (2011), 1–2. **Zbl** 1225.00027. **MR** 2813254.

"The fundamental theorem of algebra: From the four basic operations," *Amer. Math. Monthly* 119 (2012), 753–758. **Zbl** 1295.12002. **MR** 2990933.

DE POSSEL, R.

"Zum Parallelschlitztheorem unendlich-vielfach zusammenhängender Gebiete," *Göttingen Nachr.* (1931), 199–202. **FM** 57, p. 400. **Zbl** 3, p. 314. (See also **Zbl** 77, p. 58.)

"Sur la représentation conforme d'un domaine à connexion infinie sur un domaine à fentes parallèles," *Jour. de Math. Pures et Appliq.* (9) 18 (1939), 285–290. **FM** 65, p. 1242. **Zbl** 23, p. 55. **MR** 1, p. 111. (See also **Zbl** 77, p. 58.)

DE SOUZA, P. N. and SILVA, J.-N.

Berkeley Problems in Mathematics, Springer-Verlag (3rd ed., 2004), New York. **Zbl** 1041.00003. **MR** 2004i:00004.

DE VILLIERS, J. M. and ROBINSON, P. N.

"The interval of convergence and limiting functions of a hyperpower sequence," *Amer. Math. Monthly* 93 (1986), 13–23. **Zbl** 649.40001. **MR** 87d:26005. (See also **MR** 2005k:30053.)

DE VITO, L.

"Sulla connessione dei campi piani," *Rend. Mat. e Appl.* (5) 16 (1957), 297–314. **Zbl** 105, p. 167. **MR** 19, p. 1186.

DELANGE, H.

"Sur la forme forte du théorème de Cauchy relatif à l'intégrale d'une fonction de variable complexe," *Bull. Sci. Math.* (2) 80 (1956), 156–160. **Zbl** 72, p. 290. **MR** 19, p. 23.

DELSARTE, J.

"Note sur une propriété nouvelle des fonctions harmoniques," *C. R. Acad. Sci. Paris* 246 (1958), 1358–1360. **Zbl** 84, p. 94. **MR** 20 #2548.

Lectures on Topics in Mean Periodic Functions and the Two Radius Theorem, Notes by K. B. Vedak, Tata Institute of Fundamental Research (1961), Bombay.

DEMIN, E.

"Remarque sur une formule de Laurent," *Bull. Soc. Roy. Sci. Liége* 15 (1946), 84–86. **Zbl** 61, p. 148. **MR** 8, p. 507.

DEMTCHENKO, B.

"Sur la formule de M. H. Villat résolvant le problème de Dirichlet dans un anneau circulaire," *Jour. de Math. Pures et Appliq.* (9) 10 (1931), 201–211. **FM** 57, p. 1422.

DENJOY, A.

"Sur les fonctions définies par des séries de fractions rationnelles," *C. R. Acad. Sci. Paris* 174 (1922), 95–98. **FM** 48, p. 320. (See also **Zbl** 77, p. 58.)

"Sur l'itération des fonctions analytiques," *C. R. Acad. Sci. Paris* 182 (1926), 255–257. **FM** 52, p. 309. (See also **Zbl** 77, p. 58.)

"Sur une classe de fonctions analytiques," *C. R. Acad. Sci. Paris* 188 (1929), 140–142 and 1084–1086. **FM** 55, pp. 180, 181. (See also **Zbl** 77, p. 58.)

"Sur les polygones d'approximation d'une courbe rectifiable," *C. R. Acad. Sci. Paris* 196 (1933), 29–32. **FM** 59, p. 313. **Zbl** 6, p. 62. (See also **Zbl** 77, p. 58.)

"Sur l'intégration des differentielles totales," *Bull. Soc. Sci. Cluj* 7 (1934), 580–587 and *Mathematica* (Cluj) 10 (1935), 117–124. **FM** 61, p. 241. **Zbl** 9, p. 256 and 10, p. 259. (See also **MR** 17, p. 698.)

"Représentation conforme des aires limitées par des continus cycliques," *C. R. Acad. Sci. Paris* 213 (1941), 975–977. **FM** 67, p. 285. **Zbl** 26, p. 218. **MR** 5, p. 115. (See also **Zbl** 77, p. 58.)

"Les continus cycliques et la représentation conforme," *Bull. Soc. Math. France* 70 (1942), 97–124. **Zbl** 28, p. 403. **MR** 6, p. 207. (See also **Zbl** 17, p. 698.)

"Le théorème de Cauchy-Goursat," *C. R. Acad. Sci. Paris* 240 (1955), 386–389 and 473–476. **Zbl** 64, p. 67. **MR** 16, p. 683. (See also **Zbl** 77, p. 58.)

DERKSEN, H.

"The fundamental theorem of algebra and linear algebra," *Amer. Math. Monthly* 110 (2003), 620–623. **Zbl** 1082.12002. **MR** 2004g:12002. (See also **MR** 2543717.)

DERRICK, W. R.

"A condition under which a mapping is a homeomorphism," *Amer. Math. Monthly* 80 (1973), 554–555. **Zbl** 285.57001. **MR** 48 #2977.

DESBROW, D.

"On the irrationality of π^2," *Amer. Math. Monthly* 97 (1990), 903–906. **Zbl** 735.11029. **MR** 91j:11055.

"On zero derivatives," *Amer. Math. Monthly* 103 (1996), 410–411. **Zbl** 853.30001.

"On evaluating $\int_{-\infty}^{\infty} e^{ax(x-2b)}\,dx$ by contour integration round a parallelogram," *Amer. Math. Monthly* 105 (1998), 726–731. **Zbl** 923.26002. **MR** 99h:26002.

DEURING, M.

"Eine Bemerkung zum Cauchyschen Integralsatz," *Arch. der Math.* 1 (1949), 321–322. **Zbl** 31, p. 352. **MR** 11, p. 91.

DIAMOND, H. G., POMERANCE, C. and RUBEL, L.

"Sets on which an entire function is determined by its range," *Math. Zeit.* 176 (1981), 383–398. **Zbl** 473.30022. **MR** 82e:30031. (See also **MR** 85b:30038.)

DIENES, P.

The Taylor Series: An Introduction to the Theory of Functions of a Complex Variable, Clarendon Press (1931), Oxford. **FM** 57, p. 339. **Zbl** 3, p. 155.

BAMS 38, p. 792. Reprinted by Dover Publications, Inc. (1957), New York. **Zbl** 78, p. 59. **MR** 19, p. 735.

DIEUDONNÉ, J.

"Recherches sur quelques problèmes relatifs aux polynômes et aux fonctions bornées d'une variable complexe," *Ann. Sci. École Norm. Sup.* (3) 48 (1931), 247–358. **FM** 57, p. 355. **Zbl** 3, p. 119.

Foundations of Modern Analysis I, Pure and Applied Mathematics, Vol. 10-I, Academic Press (1969), New York. **Zbl** 176, p. 5, 264.26001 and 504.46001. **MR** 50 #1781 and #1782. **BAMS** 67, p. 246 and (2) 3 (1980), pp. 719–724.

Infinitesimal Calculus, Houghton Mifflin Co. (1971), Boston; Kershaw Publishing Co., Ltd. (1973), London. **Zbl** 497.26004. **MR** 50 #1780.

DINEEN, S.

The Schwarz lemma, Oxford Mathematical Monographs, Clarendon Press (1989), Oxford. **Zbl** 708.46046. **MR** 91f:46064.

DINGHAS, A.

"Über das Phragmén-Lindelöfsche Prinzip und den Julia-Carathéodory'schen Satz," *Sitzungsber. Preuss. Akad. Wissen. Berlin, Phys.-math. Kl.* (1938), 32–48. **FM** 64, p. 301. **Zbl** 18, p. 262.

"A simple proof of a formula in the theory of functions," *Math. Student* 22 (1954), 101–102. **Zbl** 57, p. 307. **MR** 16, p. 231.

Vorlesungen über Funktionentheorie, Die Grundlehren der mathematischen Wissenschaften in Einzeldarstellungen, Bd. 110. Springer-Verlag (1961), Berlin. **Zbl** 102, p. 293. **MR** 31 #3577. **BAMS** 68, p. 311.

DINI, U.

"Sopra le funzioni di una variabile complessa," *Annali Mat. Pura Appl.* (2) 4 (1870–71), 159–174. **FM** 3, p. 179.

"Alcuni teoremi sulle funzioni di una variabile complessa," pp. 258–276 of *In memoriam Dominici Chelini. Collectanea mathematica,* (L. Cremona and E. Beltrami, editors), Hoepli (1881), Milan. **FM** 13, p. 310.

"Il problema di Dirichlet in un' area anulare, e nello spazio compreso fra due sfere concentriche," *Rend. Circ. Mat. Palermo* 36 (1913), 1–28.

DISNEY, S. A. R., GRAY, J. D. and MORRIS, S. A.

"Is a function that satisfies the Cauchy-Riemann equations necessarily analytic?", *Austral. Math. Soc. Gaz.* 2 (1975), 67–81. **MR** 57 #594. (See also **MR** 57 #9940.)

DIXON, J. D.

"A brief proof of Cauchy's Integral Theorem," *Proc. Amer. Math. Soc.* 29 (1971), 625–626. **Zbl** 205, p. 378. **MR** 43 #3432. (See also **Zbl** 748.30030 and **MR** 92d:30001.)

Doetsch, G.

"Über die obere Grenze des absoluten Betrages einer analytischen Funktion auf Geraden," *Math. Zeit.* 8 (1920), 237–240. **FM** 47, p. 274.

Dolzhenko, E. P.

"D. E. Men'shov's work on the theory of analytic functions and the current state of monogeneity theory," *Russian Math. Surveys* 47 no. 5 (1992), 71–102. **Zbl** 794.30001. **MR** 94d:30002. (See also **Zbl** 784.01011 and **MR** 94d:30001.)

Donoghue, W. F., Jr.

Distributions and Fourier Transforms, Pure and Applied Mathematics, Vol. 32, Academic Press (1969), New York. **Zbl** 188, p. 181.

Monotone Matrix Functions and Analytic Continuation, Die Grundlehren der mathematischen Wissenschaften in Einzeldarstellungen, Bd. 207. Springer-Verlag (1974), New York. **Zbl** 278.30004. **MR** 58 #6279. **BAMS** 81, p. 847.

Doran, R. S.

"An application of idempotents in the classifiction of complex algebras," *Elemente der Math.* 35 (1980), 16–17. **Zbl** 427.46030. **MR** 81b:46065.

Dörge, K.

"Über den Fundamentalsatz der Algebra," *Sitzungsber. Preuss. Akad. Wissen. Berlin, Phys. -math. Kl.* (1928), 87–89. **FM** 54, p. 117.

Dörrie, H.

Unendliche Reihen, Verlag von R. Oldenbourg (1951), München. **Zbl** 45, p. 333. **MR** 13, p. 737.

Douady, A. and Earle, C. J.

"Conformally natural extension of homeomorphisms of the circle," *Acta Math.* 157 (1986), 23–48. **Zbl** 615.30005. **MR** 87j:30041. (See also **Zbl** 712.30040 and **MR** 91i:30015.)

Douglas, J.

"Solution of the problem of Plateau," *Trans. Amer. Math. Soc.* 33 (1931), 263–321. **FM** 57, pp. 661 and 1542. **Zbl** 1, p. 141.

Dovbush, P. V.

"On a normality criterion of Mandelbrojt," *Complex Variables & Elliptic Equations* 59 (2014), 1388–1394. **Zbl** 1377.32002. **MR** 3219684.

Drexler, J. and Kryster, M.

"Was ist ein ebener Winkel?", *Mitteilungen der DMV* 23 (2015), 204.

Driscoll, T. A. and Trefethen, L. N.

Schwarz-Christoffel Mapping, Cambridge Monographs on Applied and Computational Mathematics 8, Cambridge University Press (2002). **Zbl** 1003.30005. **MR** 2003e:30012.

DUFFIN, R. J.
"A note on Poisson's integral," *Quarterly of Applied Math.* 15 (1957), 109–111. **Zbl** 79, p. 122. **MR** 19, p. 261.

DUFRESNOY, J. and PISOT, C.
"Prolongement analytique de la série de Taylor," *Ann. Sci. École Norm. Sup.* (3) 68 (1951), 105–124. **Zbl** 43, p. 294. **MR** 13, p. 221.

DUGUNDJI, J. and GRANAS, A.
Fixed Point Theory, Vol. I, Monografie Matematyczne 61, PWN-Polish Scientific Publishers (1982), Warsaw. **Zbl** 483.47038. **MR** 83j:54038. **BAMS** (2) 41 (2004), 267. 2nd edition, (2003), **Zbl** 1025.47002. **MR** 2004d:58012.

DULST, D., VAN
"A functional analytic proof of Rouché's Theorem," *Amer. Math. Monthly* 78 (1971), 770–771. **Zbl** 224.30004. **MR** 45 #2504.

DUNHAM, W.
Euler The Master of Us All, The Dolciani Mathematical Expositions No. 22, (1999), The Mathematical Association of America, Washington, D. C. **Zbl** 951.01012. **MR** 2000b:01007.

DUNMORE, C.
"Meta-level revolutions in mathematics," pp. 209–225 of *Revolutions in Mathematics* (Donald Gillies, editor), Oxford University Press (1992). **Zbl** 967.00506. **MR** 94c:00018.

DUREN, P. L.
"Subordination," pp. 22–29 of *Proceedings of a Conference on Complex Analysis, Lexington Kentucky, 1976.* Lecture Notes in Mathematics 599, Springer-Verlag (1977), New York. **Zbl** 372.30013. **MR** 56 #12252.
Univalent Functions, Grundlehren der mathematischen Wissenschaften 259, Springer-Verlag (1983), New York and Berlin. **Zbl** 514.30001. **MR** 85j:30034. **BAMS** (2) 12 (1985), 158.
Harmonic Mappings in the Plane, Cambridge Tracts in Mathematics, No. 156, Cambridge University Press (2004), Cambridge & New York. **Zbl** 1055.31001. **MR** 2005d:31001.

DUREN, P. L., HENGARTER, W. and LAUGESEN, R. S.
"The argument principle for harmonic functions," *Amer. Math. Monthly* 103 (1996), 411–415. **Zbl** 863.31001. **MR** 97f:30002.

DVORETSKY, A. and ERDÖS, P.
"On power series diverging everywhere on the circle of convergence," *Mich. Math. Jour.* 3 (1955), 31–35. **Zbl** 73, p. 62. **MR** 17, p. 138.

DYM, H. and MCKEAN, H. P.
Fourier Series and Integrals, Probability and Mathematical Statistics, Vol. 14. Academic Press (1972), New York. **Zbl** 242.42001. **MR** 56 #945. **BAMS** 79, p. 641.

DYSON, F. J.
"Continusous functions on spheres," *Annals of Math.* (2) 54 (1951), 534–536. **Zbl** 45, p. 29. **MR** 13, p. 450.

DZEWAS, J.
"Konstruktion der Winkelfunktionen mittels der Theorie der topologischen Gruppen," *Math.-Phys. Semesterber.* (2) 15 (1968), 163–171. **Zbl** 164, p. 80. **MR** 38 #3473.

E

EATON, J. R.
"The fundamental theorem of algebra," *Amer. Math. Monthly* 67 (1960), 578–579.

EBBINGHAUS, H.-D., *et al.*
Numbers, Graduate Texts in Mathematics, Vol. 123 (translated from German by H. L. S. Orde), Springer-Verlag (1990), New York. **Zbl** 705.00001. **MR** 91h:00005.

EBERLEIN, W. F.
"The elementary transcendental functions," *Amer. Math. Monthly* 61 (1954), 386–392. **Zbl** 56, p. 281. **MR** 15, p. 791. (See also **MR** 40 #270.)
"The circular function(s)," *Math. Mag.* 39 (1966), 197–201. **Zbl** 147, p. 60.
"The Gauss-Green and Cauchy integral theorems," *Amer. Math. Monthly* 82 (1975), 625–629. **Zbl** 311.30036. **MR** 51 #3465. (See also *Math. Intelligencer* 9 no. 1 (1987), 58–59.)
"On Euler's infinite product for the sine," *Jour. Math. Anal. and Applic.* 58 (1977), 147–151. **Zbl** 349.33001. **MR** 55 #3345.

EDELSTEIN, M.
"An extension of Banach's contraction principle," *Proc. Amer. Math. Soc.* 12 (1961), 7–10. **Zbl** 96, p. 171. **MR** 22A#11375.

EDGAR, G. A., NETUKA, I., and VESELY, J.
"Solution of Advanced Problem 6393," *Amer. Math. Monthly* 91 (1984), 61–62. **MR** 1540343.

EDREI, A.
"On the zeros of successive derivatives," *Proc. Amer. Math. Soc.* 6 (1955), 386–391. **Zbl** 64, p. 70. **MR** 16, p. 914.

EDWARDS, A. W. F.
"Sums of powers of integers: a little of the history," *Math. Gazette* 66 (1982), 22–28. **Zbl** 493.10004. **MR** 83e:10012. (See also **MR** 87h:11099 and **Zbl** 605.40004 and 806.11015.)

EDWARDS, C. H., JR.

The Historical Development of the Calculus, Springer-Verlag (1979), New York. **Zbl** 452.01001. **MR** 81c:01003.

EGAN, M. F.

"The integrals $\int_0^{\frac{\pi}{2}} \frac{\sin^2 mx}{\sin^2 x}\,dx$, $\int_0^{\infty} \frac{\sin^2 x}{x^2}\,dx$ and $\int_0^{\infty} \frac{\sin x}{x}\,dx$," *Math. Gazette* 8 (1915–16), 299–300.

EGGLESTON, H. G.

"A Tauberian lemma," *Proc. Lon. Math. Soc.* (3) 1 (1951), 28–45. **Zbl** 45, p. 335. **MR** 13, p. 22.

EILENBERG, S.

"Sur les transformations d'espaces métriques en circonférence," *Fund. Math.* 24 (1935), 160–176. **FM** 61, p. 621. **Zbl** 10, p. 277.

"Transformations continues en circonférence et la topologie du plan," *Fund. Math.* 26 (1936), 61 112. **FM** 62, p. 680. **Zbl** 13, p. 420.

"An invariance theorem for subsets of S^n," *Bull. Amer. Math. Soc.* 47 (1941), 73–75. **FM** 67, p. 738. **Zbl** 25, p. 234. **MR** 2, p. 179.

Lectures on Topology, mimeographed lecture notes, New York University (1948).

EISENACK, G. and FENSKE, C.

Fixpunkttheorie, Bibliographisches Institut (1978), Mannheim. **Zbl** 369.47001. **MR** 80g:58001.

EISENBERG, M. and GUY, R.

"A proof of the hairy ball theorem," *Amer. Math. Monthly* 86 (1979), 572–574. **Zbl** 433.57011. **MR** 80i:57018.

EISERMANN, M.

"The fundamental theorem of algebra made effective: An elementary real-algebraic proof via Sturm chains," *Amer. Math. Monthly* 119 (2012), 715–752. **Zbl** 1267.26016. **MR** 2990932.

ELEKES, M.

"Level sets of differentiable functions of two variables with non-vanishing gradient," *Jour. Math. Anal. and Applic.* 270 (2002), 369–382. **Zbl** 1006.26006. **MR** 2003d:26015.

ELJOSEPH, N.

"On the iteration of linear fractional transformations," *Amer. Math. Monthly* 75 (1968), 362–366. **Zbl** 159, p. 367.

ELKINS, J. M.

"A Borel-Carathéodory inequality and approximation of entire functions by polynomials with restricted zeros," *Jour. Approx. Theory* 4 (1971), 274–278. **Zbl** 234.30030. **MR** 44 #4191.

ELSTRODT, J.

"Iterierte Potenzen," *Math. Semesterber.* 41 (1994), 167–178. **Zbl** 817.40001. **MR** 96e:11032.

"Partialbruchentwicklung des Kotangens, Herglotz-Trick und die Weierstrass'sche stetige, nirgends differenzierbare Funktion," *Math. Semesterber.* 45 (1998), 207–220. **Zbl** 949.39010. **MR** 2000c:39021.

Ely, G. S.

"Bibliography of Bernoulli's Numbers," *Amer. Jour. of Math.* 5 (1882), 228–235. **FM** 15, p. 21. (See also **Zbl** 741.11001 and **MR** 89i:11002.)

Epstein, B.

Orthogonal Families of Analytic Functions, The Macmillan Company (1965), New York. **Zbl** 145, p. 89. **MR** 31 #4916.

Erdös, P. and Radó, R.

"Advanced problems and solutions: Solution 4229," *Amer. Math. Monthly* 55 (1948), 171. **MR** 1526929.

Essén, M.

The $\cos \pi\lambda$ *Theorem,* Lecture Notes in Mathematics 467, Springer-Verlag (1975), New York. **Zbl** 335.31001. **MR** 57 #6464. (See also **MR** 2004d:30050.)

Esterle, J.

"Mittag-Leffler methods in the theory of Banach algebras and a new approach to Michael's problem," *Contemporary Math.* 32 (1984), 107–129. **Zbl** 569.46031. **MR** 86a:46056.

Estermann, T.

"Über die totale Variation einer stetigen Funktion und den Cauchyschen Integralsatz," *Math. Zeit.* 37 (1933), 556–560 and 38 (1934), 641. **FM** 59, p. 313 and 60, p. 243. **Zbl** 7, p. 213 and 9, p. 24.

"On the fundamental theorem of algebra," *Jour. Lon. Math. Soc.* 31 (1956), 238–240. **Zbl** 71, p. 249. **MR** 18, p. 4.

Complex Numbers and Functions, Athlone Press (1962), London. **Zbl** 115, p. 60.

"Notes on Landau's proof of Picard's 'Great' Theorem," pp. 101–106 of *Studies in Pure Mathematics,* L. Mirsky, editor, Academic Press (1971), New York. **Zbl** 215, p. 420. **MR** 44 #1809.

Ettlinger, H. J.

"Cauchy's paper of 1814 on definite integrals," *Annals of Math.* (2) 23 (1921–22), 255–270. **FM** 49, p. 209. (See also **FM** 16, p. 25.)

"On the inversion of the order of integration of a two fold integral," *Annals of Math.* (2) 28 (1926–27), 65–68. **FM** 52, p. 245.

Euler, L.

Zur Theorie Komplexer Funktionen, German translation by W. Purkert, *et al.*, remarks by A. P. Yuškevič in *Ostwald's Klassiker der exakten Wissenschaften,* Bd. 261, Akademische Verlagsgesellschaft Geest & Portig K.-G. (1983), Leipzig. **Zbl** 597.01019 and 883.01050. **MR** 86k:01049.

EUSTICE, D. J.
"Holomorphic idempotents and common fixed points on the 2-disk," *Mich. Math. Jour.* 19 (1972), 347–352. **Zbl** 254.32008. **MR** 47 #2040.

EVANS, A.
"The application of complex variable methods to Tauberian theorems," *Jour. Lon. Math. Soc.* 28 (1953), 94–102. **Zbl** 50, p. 67. **MR** 14, p. 551.

EVANS, G. C.
"Fundamental points of potential theory," pp. 252–329 of *The Rice Institute Pamphlet* 7, No. 4 (1920), Houston. **FM** 48, p. 1268.
"Problems of potential theory," *Proc. Nat. Acad. Sci. U. S. A.* 7 (1921), 89–98. **FM** 48, p. 568.
"Sur l'intégrale de Poisson," *C. R. Acad. Sci. Paris* 177 (1923), 241–242. **FM** 49, p. 339.
"Note on a theorem of Bôcher," *Amer. Jour. of Math.* 50 (1928), 123–126. **FM** 54, p. 508.

EVANS, G. C. and BRAY, H. E.
"Sur l'intégrale de Poisson generalisée," *C. R. Acad. Sci. Paris* 176 (1923), 1042–1044 and 1368–1371. **FM** 49, p. 339.

EVARD, J.-CL. and JAFARI, F
"A complex Rolle's theorem," *Amer. Math. Monthly* 99 (1992), 858–861. **Zbl** 772.30003. (See also **MR** 96h:30009.)

EVARD, J.-CL., JAFARI, F. and MOORHOUSE, G. E.
"A sufficient condition for an entire function to be a polynomial of degree one," *Nieuw Archief voor Wiskunde* (4) 12 (1994), 9–12. **Zbl** 821.30027. **MR** 95d: 30047.

EVGRAFOV, M. A., *et al.*
Recueil de problèmes sur la théorie des fonctions analytiques, Éditions Mir (1974), Moscow. **Zbl** 286.30001. **MR** 51 #3406.

EYMARD, P. and LAFON, J.-P.
Autor du nombre π, Hermann et Cie (1999), Paris. **Zbl** 932.00001. **MR** 2001a: 11001. English translation, by Stephen S. Wilson, American Mathematical Society (2004), Providence. **Zbl** 1038.11001. **MR** 2004j:11152.

F

FABBRI, E.
Il Teorema dell' Integrale di Cauchy, Contributo Alla Storia Critica dell' Analisi, Zamorani e Albertazzi (1900), Bologna. **FM** 31, p. 43.

FABER, G.

"Über polynomische Entwicklungen," *Math. Annalen* 57 (1903), 389–408 and 64 (1907), 116–135. **FM** 34, p. 430 and 38, p. 439.

"Über arithmetische Eigenschaften analytischer Funktionen," *Math. Annalen* 58 (1904), 545–557. **FM** 35, p. 407.

"Über analytische Funktionen mit vorgeschriebenen Singularitäten," *Math. Annalen* 60 (1905), 379–397. **FM** 36, p. 457.

"Neuer Beweis eines Koebe-Bieberbachschen Satzes über konforme Abbildung," *Sitzungsber. Kön. Bayer. Akad. Wissen. München* (1916), 39–42. **FM** 46, p. 550.

"Über den Hauptsatz aus der Theorie der konformen Abbildung," *Sitzungsber. Bayer. Akad. Wissen. München* (1922), 91–100. **FM** 48, p. 1231.

FABRY, E.

(a) "Sur les points singuliers d'une fonction donnée par son développement en série et l'impossibilité du prolongement analytique dans des cas très généraux," *Ann. Sci. École Norm. Sup.* (3) 13 (1896), 367–399. **FM** 27, p. 303.

(b) "Sur les intégrales de Fresnel," *Nouvelles Annales de Math.* (3) 15 (1896), 504–505. **FM** 27, p. 235.

FALK, M.

"Démonstration du théorème de Cauchy sur l'intégrale d'une fonction complexe," *Nova Acta Regiae Societatis Sci. Upsaliensis* (3) 12 (1883) and *Bull. Sci. Math.* (2) 7 (1883), 137–139. **FM** 15, p. 317 and 16, p. 365.

FALLIN, H. I., JR. and GOULD, H. W.

"Chronological bibliography of the Cauchy integral theorem," *Mathematica Monongaliae* No. 10 (1965, 2nd ed., 1972), Department of Mathematics, West Virginia University, Morgantown, West Virginia.

FAN, K.

"Julia's lemma for operators," *Math. Annalen* 239 (1979), 241–245. **Zbl** 389.47003. **MR** 80f:47009.

"Iteration of analytic functions of operators," *Math. Zeit.* 179 (1982), 293–298. **Zbl** 475.47013. **MR** 83d:47024.

"Iteration of analytic functions of operators. II," *Linear and Multilinear Alg.* 12 (1983), 295–304. **Zbl** 564.47006. **MR** 84c:47018.

"The angular derivative of an operator-valued analytic function," *Pac. Jour. of Math.* 121 (1986), 67–72. **Zbl** 588.47018. **MR** 87h:47034.

FANG, QIQIN

"A theorem of Milloux's type," *Acta Sci. Nat. Univ. Peking* 28, no. 3 (1992), 262–266 (Chinese). **Zbl** 765.30010.

FARKAS, J.

"Sur les fonctions itératives," *Jour. de Math. Pures et Appliq.* (3) 10 (1884), 101–108. **FM** 16, p. 375.

FARRELL, O. J.

"On approximation to a mapping function by polynomials," *Amer. Jour. of Math.* 54 (1932), 571–578. **FM** 58, p. 356. **Zbl** 4, p. 403.

FATOU, P.

"Sur les séries entières à coéfficients entiers," *C. R. Acad. Sci. Paris* 138 (1904), 342–344. **FM** 35, p. 257.

"Séries trigonométriques et séries de Taylor," *Acta Math.* 30 (1906), 335–400. **FM** 37, p. 283.

"Sur les substitutions rationnelles," *C. R. Acad. Sci. Paris* 164 (1917), 806–808. **FM** 46, p. 519.

"Sur les équations fonctionnelles," *Bull. Soc. Math. France* 47 (1919), 161–271; 48 (1920), 33–94; 208–314; 50 (1922), 37–41 and 51 (1923), 16–22. **FM** 47, p. 921; 48, p. 1223 and 49, p. 144.

"Sur les fonctions holomorphes et bornées à l'intérieur d'un cercle," *Bull. Soc. Math. France* 51 (1923), 191–202. **FM** 49, p. 221.

"Sur l'itération des fonctions transcendantes entières," *Acta Math.* 47 (1926), 337–370. **FM** 52, p. 309.

FEDERIGHI, U.

"Sul teorema fondamentale di Cauchy per le funzioni analitiche," *Rend. del R. Istituto Lombardo, Cl. Sci. mat. natur.* (3) 2 (1938), 186–192. **FM** 64, p. 1048. **Zbl** 18, p. 260.

FEIGL, G.

"Über einige Eigenschaften der einfachen stetigen Kurven, Teil I," *Math. Zeit.* 27 (1928), 161–186. **FM** 53, p. 567.

FEJÉR, L.

"Untersuchungen über Fouriersche Reihen," *Math. Annalen* 58 (1904), 51–69. **FM** 34, p. 287. (See also **MR** 23 #A789.)

"Über die Wurzel vom kleinsten absoluten Betrage einer algebraischen Gleichung," *Math. Annalen* 65 (1908), 413–423 and *Math. és Fiz. Lapok* 17 (1908), 308–324. **FM** 39, p. 123.

"Über gewisse Potenzreihen an der Konvergenzgrenze," *Sitzungsber. Kön. Bayer. Akad. Wissen. München* 40 (1910), 1–17. **FM** 41, p. 284.

"La convergence sur son cercle de convergence d'une série de puissances effectuant une représentation conforme du cercle sur le plan simple," *C. R. Acad. Sci. Paris* 156 (1913), 46–49. **FM** 44, p. 290.

"Über die Konvergenz der Potenzreihe an der Konvergenzgrenze in Fällen der konformen Abbildung auf die schlichte Ebene," pp. 42–53 of *Mathematische Abhandlungen H. A. Schwarz Gewidmet,* (O. Bolza, *et al.*, editors) Verlag Julius Springer (1914), Berlin. **FM** 45, p. 670. Reprinted by Chelsea Publishing Co. (1974), New York.

"Über trigonometrische Polynome," *Jour. für Reine und Angew. Math.* 146 (1915), 53–82. **FM** 45, p. 406.

"Über Potenzreihen, deren Summe im abgeschlossenen Konvergenzkreise überall stetig ist," *Sitzungsber. Kön. Bayer. Akad. Wissen. München* (1917), 33–50. **FM** 46, p. 480.

"Interpolation und konforme Abbildung," *Göttingen Nachr.* (1918), 319–331. **FM** 46, p. 517.

FEKETE, M.

"Eine Bemerkung zu der Arbeit des Herrn Bieberbach 'Über die Verteilung der Null-und Einsstellen analytischer Funktionen'," *Math. Annalen* 88 (1922), 166–168. **FM** 48, p. 325.

"Zum Koebeschen Verzerrungssatz," *Göttingen Nachr.* (1925), 142–150. **FM** 51, p. 272.

(a) "Über die Wurzelverteilung analytischer Funktionen, deren Wert an zwei Stellen gegeben ist," *Jahresber. Deutsch. Math. Verein.* 36 (1927), 216–222. **FM** 53, p. 279.

(b) "Zur Theorie der konformen Abbildung," *Acta Litt. ac Scient. Univ. Hung.* (Szeged) 3 (1927), 25–31. **FM** 53, p. 322.

FELLER, W.

"A direct proof of Stirling's Formula," *Amer. Math. Monthly* 74 (1967), 1223–1225. Correction: *ibid.* 75 (1968), 518. **Zbl** 168, p. 298. **MR** 36 #5585 and 37 #5586.

FENCHEL, W.

"Elementare Beweise und Anwendungen einiger Fixpunktsätze," *Mat. Tidsskr. B* (1932), 66–87. **FM** 58, p. 1204. **Zbl** 5, p. 415.

FENTON, P. C.

(a) "Functions having the restricted mean value property," *Jour. Lon. Math. Soc.* (2) 14 (1976), 451–458. **Zbl** 344.31002. **MR** 55 #10703.

(b) "On conformal maps with starlike images," *Math. Scand.* 39 (1976), 102–112. **Zbl** 371.30010. **MR** 54 #13050.

"On sufficient conditions for harmonicity," *Trans. Amer. Math. Soc.* 253 (1979), 139–147. **Zbl** 404.31001. **MR** 80m:31001.

FERRAR, W. L.

"Note on a paper of L. Neder," *Math. Zeit.* 31 (1930), 519–520. **FM** 56, p. 261.

FERRARO, G.

The Rise and Development of the Theory of Series up to the Early 1820's, Springer (2008), New York. **Zbl** 1141.01007. **MR** 2008k:01003.

FEYEL, D. and PRADELLE, A. DE LA,

Exercices sur les fonctions analytiques. Maîtrise de mathématiques. Série "Mathématiques," Librairie Armand Colin (1973), Paris. **Zbl** 487.30001. **MR** 82m:30001.

FICHTENHOLZ, G.

"Un théorème sur l'intégration sous le signe intégrale," *Rend. Circ. Mat. Palermo* 36 (1913), 111–114. **FM** 44, p. 346.

"Sur l'intégrale de Poisson et quelques questions qui s'y rattachent," *Fund. Math.* 13 (1929), 1–33. **FM** 55, p. 889.

FINE, B. and ROSENBERGER, G.

The Fundamental Theorem of Algebra, Undergraduate Texts in Mathematics, Springer-Verlag (1997), New York. **Zbl** 902.12001. **MR** 98h:30001.

FINKELSTEIN, M. and SCHEINBERG, S.

"Kernels for solving problems of Dirichlet type in a half-plane," *Adv. in Math.* 18 (1975), 108–113. **Zbl** 309.31001. **MR** 52 #3559.

FISCHER, H.

"Die Geschichte des Integrals $\int_0^\infty \frac{\sin x}{x}\,dx$. Eine Geschichte der Analysis in der Nussschale," *Math. Semesterber.* 54 (2007), 13–30. **Zbl** 1118.01014. **MR** 2007m:01018.

FISCHER, W. and LIEB, I.

Ausgewählte Kapitel aus der Funktionentheorie, Viewig Studium: Aufbaukurs Mathematik 48, Friedr. Vieweg & Sohn (1988), Braunschweig. **Zbl** 657.30001. **MR** 90d:30001.

FISHER, J. C. and SHILLETO, J.

"Three aspects of Fubini's theorem," *Math. Mag.* 59 (1986), 40–42. **Zbl** 638.26008.

FISHER, S. D.

"The convex hull of the finite Blaschke products," *Bull. Amer. Math. Soc.* 74 (1968), 1128–1129. **Zbl** 164, p. 83. **MR** 38 #2316.

"Approximation by unimodular functions," *Canad. Jour. Math.* 23 (1971), 257–269 and 1093. **Zbl** 211, p. 387 and 225.30040. **MR** 43 #888 and 44 #6969.

FISHER, S. D. and FRANKS, J.

"The fixed points of an analytic self-mapping," *Proc. Amer. Math. Soc.* 99 (1987), 76–78. **Zbl** 611.30009. **MR** 87m:30069.

FLAMENT, D.

Histoire Des Nombres Complexes, entre algèbre et géométrie, CNRS Editions (2003), Paris. **Zbl** 1089.01001. **MR** 2009d:01001. (See also **MR** 99b:01003.)

FLANDERS, H.

"On the Fresnel integrals," *Amer. Math. Monthly* 89 (1982), 264–266. **Zbl** 599.26012. **MR** 83g:26001.

FLATTO, L., GROSS, F. and GOLDSTEIN, R.

"Integral functions of an integral function," *Amer. Math. Monthly* 73 (1966), 904–905.

FLATTO, L. and SHISHA, O.

"A proof of Cauchy's Integral Theorem," *Jour. Approx. Theory* 7 (1973), 386–390. **Zbl** 258.30033. **MR** 49 #7456. (See also **Zbl** 317.30038 and **MR** 49 #9166.)

FLETT, T. M.

"A note on conformal mapping," *Jour. Lon. Math. Soc.* 29 (1954), 118–121. **Zbl** 55, p. 72. **MR** 15, p. 303.

FLOYD, E. E.

"On the extension of homeomorphisms on the interior of a two cell," *Bull. Amer. Math. Soc.* 52 (1946), 654–658. **Zbl** 60, p. 401. **MR** 8, p. 50.

FORD, L. R.

"On properties of regions which persist in the subregions bounded by level curves of the Green's function," *Duke Math. Jour.* 1 (1935), 103–104. **FM** 62, p. 1216. **Zbl** 11, p. 261.

FORELLI, F.

"A note on ideals in the disc algebra," *Proc. Amer. Math. Soc.* 84 (1982), 389–392. **Zbl** 501.30041. **MR** 83c:46041.

FORT, M. K., JR.

"Some properties on continuous functions," *Amer. Math. Monthly* 59 (1952), 372–375. **Zbl** 49, p. 338. **MR** 13, p. 925.

"The complements of bounded, open, connected subsets of euclidean space," *Bull. Acad. Polon. Sci.* 9 (1961), 457–460. **Zbl** 103, p. 394. **MR** 24 #A2953.

"Continuous square roots of mappings," pp. 169–180 of *Lectures on Calculus*, K. O. May, editor, Holden-Day, Inc. (1967), San Francisco. **Zbl** 204, p. 77.

FOUËT, ÉD. A.

Leçons sur la théorie des fonctions analytiques. Seconde partie. Théorèmes d'existence. – Étude des fonctions analytiques au point de vue de Cauchy, de Weierstrass, de Riemann. Gauthier-Villars (1904), Paris. **FM** 35, p. 375. (See also **FM** 38, p. 415; 40, p. 435 and **BAMS** 18, p. 30.)

FRAENKEL, A.

"Der Zusammenhang zwischen dem ersten und dem dritten Gauss'schen Beweis des Fundamentalsatzes der Algebra," *Jahresber. Deutsch. Math. Verein.* 31 (1922), 234–238.

FRANK, G.

"Eine Vermutung von Hayman über Nullstellen meromorpher Funktionen," *Math. Zeit.* 149 (1976), 29–36. **Zbl** 312.30032. **MR** 54 #10601.

FRANKLIN, P.

"A qualitative definition of the potential functions," *Bull. Amer. Math. Soc.* 30 (1924), 41–50. **FM** 50, p. 332.

(a) "Functions with an essential singularity," *Bull. Amer. Math. Soc.* 31 (1925), 157–162. **FM** 51, p. 261.

(b) "Analytic transformations of everywhere dense point sets," *Trans. Amer. Math. Soc.* 27 (1925), 91–100. **FM** 51, p. 166.

"Functions of a complex variable with assigned derivatives at an infinite number of points, and an analogue of Mittag-Leffler's theorem," *Acta Math.* 47 (1926), 371–385. **FM** 52, p. 303.

"Analytic functions with assigned values," *Bull. Amer. Math. Soc.* 33 (1927), 461–466. **FM** 53, p. 279.

"A qualitative definition of the sub- and superharmonic functions," *Jour. of Math. and Physics of Massachusetts Institute of Technology* 7 (1928), 86–92. **FM** 54, p. 518.

FREUDENTHAL, H.

"The cradle of modern topology, according to Brouwer's inedita," *Historia Math.* 2 (1975), 495–502. **Zbl** 334.01027. (See also **Zbl** 275.01011 and **MR** 58 #30949.)

(a) "Louis-Augustin Cauchy," pp. 131–148, Vol. III of *Dictionary of Scientific Biography* (C. C. Gillispie, editor), Charles Scribner's Sons (1971), New York. **MR** 97h:01001a.

(b) "Georg Friedrich Bernhard Riemann," pp. 447–456, Vol. XI of *Dictionary of Scientific Biography* (C. C. Gillispie, editor), Charles Scribner's Sons (1971), New York. **MR** 97h:01001d.

FRIED, E. and VARGA, T.

"Über die Frage der Einführung der komplexen Zahlen," *Publ. Math. Debrecen* 4 (1956), 484–487. **Zbl** 71, p. 49. **MR** 17, p. 1169.

FROBENIUS, G.

"Über die Entwickelung analytischen Functionen in Reihen, die nach gegebenen Functionen fortschreiten," *Jour. für Reine und Angew. Math.* 73 (1871), 1–31. **FM** 3, p. 188.

FROSTMAN, O.

"Potentiel d'équilibre et capacité des ensembles avec quelques applications à la théorie des fonctions," *Medd. Lunds Univ. Mat. Sem.* 3 (1935). **FM** 61, p. 262. **Zbl** 13, p. 63.

FUCHS, W. H. J.

"A uniqueness theorem for mean values of analytic functions," *Proc. Lon. Math. Soc.* (2) 48 (1943), 35–47. **Zbl** 28, p. 402. **MR** 5, p. 36.

(a) "On the zeros of power series with Hadamard gaps," *Nagoya Math. Jour.* 29 (1967), 167–174. **Zbl** 172, p. 369. **MR** 35 #1780.

(b) *Topics in the Theory of Functions of One Complex Variable,* Manuscript prepared with the collaboration of Alan Schumitsky. Van Nostrand Mathematical Studies #12, D. Van Nostrand Co. (1967), Princeton. **Zbl** 155, p. 115. **MR** 36 #3954.

Théorie de l'approximation des fonctions d'une variable complexe, Séminaire de Mathématiques Supérieures, No. 26, The University of Montreal Press (1968). **Zbl** 199, p. 397. **MR** 41 #5630.

FUCHS, L. and SZELE, T.

"Introduction of complex numbers as vectors in the plane," *Amer. Math. Monthly* 59 (1952), 628–631. **Zbl** 47, p. 53.

FUCHSSTEINER, B. and LUSKY, W.

Convex Cones, North-Holland Mathematics Studies Vol. 56. North-Holland Publishing Co. (1981), Amsterdam. **Zbl** 478.46002. **MR** 83m:46018.

FULTON, W.

Algebraic Topology, A First Course, Graduate Texts in Mathematics, Vol. 153, Springer-Verlag (1995), New York. **Zbl** 852.55001. **MR** 97b:55001.

G

GAIER, D.

"Schlichte Potenzreihen, die auf $|z| = 1$ gleichmässig, aber nicht absolut konvergieren," *Math. Zeit.* 57 (1953), 349–350. **Zbl** 50, p. 78. **MR** 14, p. 737.

"Schlichte Potenzreihen an der Konvergenzgrenze," *Math. Zeit.* 58 (1953), 456–458. **Zbl** 52, p. 80. **MR** 15, p. 113.

"Über die konforme Abbildung veränderlichen Gebiete," *Math. Zeit.* 64 (1956), 385–424. **Zbl** 73, p. 68. **MR** 17, p. 1191.

"Über ein Iterationsverfahren von Komatu zur konformen Abbildung von Ringgebieten," *Jour. Math. and Mech.* 6 (1957), 865–883. **Zbl** 79, p. 298. **MR** 20 #105.

Konstruktive Methoden der konformen Abbildung, Springer Tracts in Natural Philosophy, Vol. 3, Springer-Verlag (1964), Berlin. **Zbl** 132, p. 367. **MR** 33 #7507.

"Konforme Abbildung mehrfach zusammenhängender Gebiete," *Jahresber. Deutsch. Math. Verein.* 81 (1978), 25–44. **Zbl** 425.30001. **MR** 58 #22499.

(a) "Approximation in Komplexen," *Jahresber. Deutsch. Math. Verein.* 86 (1984), 151–159. **Zbl** 544.30034. **MR** 86c:30074.

(b) "Über Räume konformer Selbstabbildungen ebener Gebiete," *Math. Zeit.* 187 (1984), 227–257. **Zbl** 543.30009. **MR** 85h:30009.

Lectures on Complex Approximation (translated from German by R. McLaughlin), Birkhäuser Boston, Inc. (1987), Cambridge, Massachusetts. **Zbl** 612.30003. **MR** 88i:30059.

"Über die Entwicklung der Funktionentheorie in Deutschland von 1890 bis 1950," pp. 361–420 of *Ein Jahrhundert Mathematik 1890–1990, Festschrift Zum Jubiläum der DMV,* (Gerd Fischer, *et al.*, editors) Friedr. Vieweg & Sohn (1990), Wiesbaden. **Zbl** 807.30000. **MR** 92b:01041.

"Calculus on arcs in the complex plane," *Analysis* 18 (1998), 291–302. **Zbl** 919.30002. **MR** 99m:30002.

GAIER, D. and POMMERENKE, CH.

"On the boundary behavior of conformal maps," *Mich. Math. Jour.* 14 (1967), 79–82. **Zbl** 182, p. 102. **MR** 34 #4470.

GAMELIN, T. W.

Uniform Algebras, Prentice-Hall, Inc. (1969), Englewood Cliffs; 2nd ed., Chelsea Publishing Co. (1984), New York. **Zbl** 213, p. 404. **MR** 53 #14137. **BAMS** 76, p. 1226.

GAMELIN, T. W. and GREENE, R. E.

Introduction to Topology, Saunders College Publishing (1983), Philadelphia. Reprinted by Dover Publications, Inc. (1999), New York. **Zbl** 625.54002 and 935.54001. **MR** 84h:54001 and 2000c:54001.

GANAPATHY IYER, V.

"On certain functional equations," *Jour. Indian Math. Soc.* (2) 3 (1938–39), 312–315. **FM** 65, p. 1326. **Zbl** 22, p. 214.

GARABEDIAN, P. R.

"Univalent functions and the Riemann mapping theorem," *Proc. Amer. Math. Soc.* 61 (1976), 242–244. **Zbl** 346.30008. **MR** 54 #13052.

"A simple proof of a simple version of the Riemann mapping theorem by simple functional analysis," *Amer. Math. Monthly* 98 (1991), 824–826. **Zbl** 741.30007. **MR** 92j:30008.

GARDINER, S. J.

"The Dirichlet and Neumann problems for harmonic functions in half-space," *Jour. Lon. Math. Soc.* (2) 24 (1981), 502–512. **Zbl** 466.31013. **MR** 82m:31006.

"The Dirichlet problem with non-compact boundary," *Math. Zeit.* 213 (1993), 163–170. **Zbl** 789.31006. **MR** 94e:31001.

"Harmonic approximation," pp. 51–65 of *Potential Theory — ICPT 94 (Kouty, 1994)* (J. Král, *et al.*, editors). W. de Gruyter (1996), Berlin. **Zbl** 858.31006. **MR** 97h:31003.

GARNETT, J. B.

"On a theorem of Mergelyan," *Pac. Jour. of Math.* 26 (1968), 461–468. **Zbl** 165, p. 169. **MR** 38 #1532.

Analytic Capacity and Measure, Lecture Notes in Mathematics 297, Springer-Verlag (1972), New York. **Zbl** 253.30014. **MR** 56 #12257.

GARNETT, J. B. and MARSHALL, D. E.

Harmonic measure, New Mathematical Monographs, 2. Cambridge University Press (2005), Cambridge. **Zbl** 1077.31001. **MR** 2006g:31002.

GARNIR, H. G. and GOBERT, J.

Fonctions d'une variable complexe, Dunod (1965), Paris. **Zbl** 146, p. 298.

GASAPINA, U.

"Il teorema fondamentale dell' algebra," *Periodico di Mat.* (4) 35 (1957), 149–163. **Zbl** 79, p. 242. **MR** 19, p. 1034.

GATTEGNO, C. and OSTROWSKI, A.

Représentation conforme à la frontière: domaines généraux, Mémorial des Sciences Mathématiques, Fasc. 109, Gauthier-Villars (1949), Paris. **Zbl** 37, p. 180. **MR** 11, p. 425.

Représentation conforme à la frontière: domaines particuliers, Mémorial des Sciences Mathématiques, Fasc. 110, Gauthier-Villars (1949), Paris. **Zbl** 40, p. 331. **MR** 11, p. 426.

GAUGHAN, E. D.

"An isometry of the plane," *Amer. Math. Monthly* 69 (1962), 546.

GAUSS, C. F.

"Die vier Beweise für die Zerlegung ganzer algebraischer Functionen in reelle Factoren ersten oder zweiten Grades (1799–1849)," translation and commentary by E. Netto in *Ostwald's Klassiker der exakten Wissenschaften,* Nr. 14. Verlag von W. Engelmann (1890), Leipzig. **FM** 22, p. 105. (3rd ed. 1913) **FM** 44, p. 128.

GAUTHIER, P.

"*Cercles de remplissage* and asymptotic behaviour," *Canad. Jour. Math.* 21 (1969), 447–455. **Zbl** 181, p. 355. **MR** 38 #6074.

"Uniform approximation," pp. 235–271 of *Complex Potential Theory* NATO Advanced Science Institutes, Ser. C, Math. Phys. Sci. 439, Kluwer Academic Publishers (1992), Dordrecht. **Zbl** 810.30001. **MR** 96d:30048.

GAUTHIER, P. and HENGARTNER, W.

"Complex approximation and simultaneous interpolation on closed sets," *Canad. Jour. Math.* 29 (1977), 701–706. **Zbl** 352.30028. **MR** 58 #28520.

GAVRILOV, V. I.

"On holomorphic functions bounded on point sequences," *Sibirsk. Mat. Žurnal* 6 (1965), 1227–1233 (Russian). **Zbl** 163, p. 31. **MR** 33 #285.

GEHMAN, H. M.

"On extending a continuous (1–1) correspondence of two plane continuous curves to a correspondence of their planes," *Trans. Amer. Math. Soc.* 28 (1926), 252–265. **FM** 52, p. 604.

GEHRING, F. W. and LOHWATER, A. J.

"On the Lindelöf theorem," *Math. Nachr.* 19 (1958), 165–170. **Zbl** 89, p. 53. **MR** 21 #4246.

GELBAUM, B. R.

Problems in Real and Complex Analysis, Springer-Verlag (1992), New York. **Zbl** 749.00003. **MR** 94d:00003.

GELFOND, A. O.

"Sur quelques questions concernant les fonctions entières à valeurs entières," *C. R. Acad. Sci. Paris* 264 (1967), A932-A934. **Zbl** 183, p. 72. **MR** 37 #414.

Residues and their applications (translated from Russian by V. M. Volosov), Mir Publishers (1971), Moscow. **Zbl** 229.30003. **MR** 50 #2449.

GERBER, L.

"Complex iterated radicals," *Proc. Amer. Math. Soc.* 41 (1973), 205–210. **Zbl** 271.40002. **MR** 47 #7267.

GERMAY, R. H. J.

(a) "Sur une application des théorèmes de Weierstrass et de Mittag-Leffler de la théorie générale des fonctions," *Annales de la Soc. Sci. de Bruxelles* (1) 60 (1946), 190–195. **MR** 8, p. 324.

(b) "Extension du théorème d' E. Picard sur la décomposition en facteurs primaires des fonctions uniformes ayant une ligne de points singuliers essentiels," *Bull. Soc. Roy. Sci. Liége* 15 (1946), 9–13. **MR** 8, p. 507.

(a) "Sur une proposition de la théorie générale des fonctions analytiques," *Bull. Soc. Roy. Sci. Liége* 17 (1948), 62–65. **Zbl** 34, p. 47. **MR** 11, p. 21.

(b) "Sur une application d'un théorème de E. Picard relatif aux produits indéfinis de facteurs primaires," *Bull. Soc. Roy. Sci. Liége* 17 (1948), 138–143. **Zbl** 34, p. 48. **MR** 11, p. 21.

(c) "Extension d'un théorème d' E. Picard sur les produits indéfinis de facteurs primaires," *Bull. Soc. Roy. Sci. Liége* 17 (1948), 180–185. **Zbl** 34, p. 48. **MR** 11, p. 21.

GERVAIS, R. and RUBEL, L. A.

"Yet another characterization of the sine function," *Internat. Jour. of Math. & Math. Sci.* 4 (1981), 371–381. **Zbl** 475.30008. **MR** 82h:30022.

GETHNER, R. M.

"Two relatives of Picard's theorem on entire functions," *Amer. Math. Monthly* 99 (1992), 13–19. **Zbl** 760.30001. **MR** 92m:30059.

GEVIRTZ, J.

"On the mean value theorem for analytic functions," *Mich. Math. Jour.* 33 (1986), 365–375. **Zbl** 608.30024. **MR** 88a:30091.

GILAIN, CH.

"Sur l'histoire du théorème fondamental de l'algèbre: théorie des équations et calcul intégral," *Arch. Hist. Exact Sci.* 42 (1991), 91–136. **Zbl** 766.01018. **MR** 92j:01034.

"Le théorème fondamental de l'algèbre et la théorie géométrique des nombres complexes au XIXe siècle," pp. 50–73 of *Le nombre, une hydre à n visages* (D. Flament, editor), Maison des sciences de l'homme (1997), Paris. **Zbl** 884.00024.

GILLESPIE, D. C.

"Repeated integrals," *Annals of Math.* (2) 20 (1918–19), 224–228. **FM** 46, p. 396.

GIRELA, D.

"A proof of the great Picard theorem," *Bull. Belg.Math. Soc. Simon Stevin* 11 (2004), no. 2, 271–280. **Zbl** 1069.30047. **MR** 2005g:30028.

GIRONZA SOLANAS, J. A.

"Representación conforme de recintos doblemente conexos sobre la corona circular," *Universidad* (Zaragoza) (1932), 11–30 and 41–69. **FM** 58, p. 1093.

GIUDICE, F.

"Esistenza, calcolo, e differenze de radici d'equazioni numeriche," *Rend. Circ. Mat. Palermo* 16 (1902), 180–184. **FM** 33, p. 109.

"Sulla dimonstrazione di Clifford del teorema fondamentale d'algebra," *Giornale di Mat. di Battaglini* 45 [(2), 14] (1907), 7–15. **FM** 38, p. 117.

GLAISHER, J. W. L.

"On the integrals $\int_0^\infty \sin x^2\,dx$ and $\int_0^\infty \cos x^2\,dx$," *Quart. Jour. of Math.* 13 (1875), 343–349. **FM** 7, p. 161.

"On some methods of calculating logarithms without the use of series," *Quart. Jour. of Pure and Applied Math.* 47 (1916), 249–301. **FM** 45, p. 100.

GLAISTER, P.

"Integrals, infinite series and closed forms," *Internat. Jour. of Math. Ed. Sci. Tech* 32, no. 5 (2001), 765–774. **Zbl** 1007.40004. **MR** 2002h:40001.

GLICKSBERG, I.

"A remark on Rouché's theorem," *Amer. Math. Monthly* 83 (1976), 186–187. **Zbl** 318.30002. **MR** 53 #779 and 54, p. 2365.

GLOBEVNIK, J.

"Integrals over circles passing through the origin and a characterization of analytic functions," *Jour. d'Analyse Math.* 52 (1989), 199–209. **Zbl** 676.30003. **MR** 90d:30003.

"Zero integrals on circles and characterizations of harmonic and analytic functions," *Trans. Amer. Math. Soc.* 317 (1990), 313–330. **Zbl** 696.31001 and 701.30041. **MR** 90e:44001.

GLOBEVNIK, J. and QUINTO, E. T.

"Morera theorems via microlocal analysis," *Jour. Geom.Anal.* 6 (1996), 19–30. **Zbl** 866.30031. **MR** 98c:30004.

GLOBEVNIK, J. and RUDIN, W.

"A characterization of harmonic functions," *Indag. Math.* 50 (1988), 419–426. **Zbl** 705.31001. **MR** 90a:31004. (See also **Zbl** 778.31001.)

GLUCHOFF, A.

"A simple interpretation of the complex contour integral," *Amer. Math. Monthly* 98 (1991), 641–644. **Zbl** 741.30033.

GODEFROY, M.

"Sur les intégrales de Fresnel," *Nouvelles Annales de Math.* (3) 17 (1898), 205–206. **FM** 29, p. 245.

"Sur la dérivation des séries uniformément convergentes," *L'Enseignement Mathématique* 6 (1904), 294–296. **FM** 35, p. 254.

GOEBEL, K. and REICH, S.

Uniform Convexity, Hyperbolic Geometry, and Nonexpansive Mappings, Pure & Applied Mathematics, Vol. 83. Marcel Dekker, Inc. (1984), New York. **Zbl** 537.46001. **MR** 86d:58012.

GOHBERG, I. and RODMAN, L.

"Analytic operator valued functions with prescribed local data," *Acta Sci. Math.* (Szeged) 45 (1983), 189–199. **Zbl** 545.47011. **MR** 85g:47020.

GOLDBERG, J. L.

"Functions with positive real part in a half-plane," *Duke Math. Jour.* 29 (1962), 333–339. **Zbl** 101, p. 297. **MR** 29 #1340.

"Bounds on the derivatives of positive functions," *SIAM Review* 8 (1966), 343–345. **Zbl** 147, p. 69. **MR** 35 #381.

GOLDSTINE, H. H.

A History of Numerical Analysis from the 16th *through the* 19th *Century,* Studies in the History of Mathematics & Physical Sciences 2, Springer-Verlag (1977), New York. **Zbl** 402.01005. **MR** 58 #4774.

GOLUZIN, G. M.

"On the parametric representation of functions univalent in a ring," *Mat. Sbornik (N. S.)* 29 (71) (1951), 469–476 (Russian). **Zbl** 44, p. 307. **MR** 13, p. 930.

Geometric Theory of Functions of a Complex Variable, Translations of Mathematical Monographs, Vol. 26, American Mathematical Society (1969), reprinted, 1982, Providence. **Zbl** 148, p. 306. **MR** 40 #308.

GOLUZIN, G. M. and KRYLOW, W. J.

"Verallgemeinerung einer Formel von Carleman und ihre Anwendung zur analytischen Fortsetzung," *Recueil Math. Moscou* (=*Mat. Sbornik*) 40 (1933), 144–149 (Russian). **FM** 59, p. 1021. **Zbl** 7, p. 416.

GOMES TEIXEIRA, F.

"Sur l'intégrale $\int_0^\infty e^{-x^2} dx$," *Sitzungsber. Kön. Böhm. Gesell. der Wissen. Math. -naturwissen. Cl.* 1 (1889), 118–120. **FM** 21, p. 277.

GOMEZ-CALDERON, J. and WELLS, D. M.

"Why polynomials have roots," *The College Math. Jour.* 27 (1996), 90–94.

GONÇALVES, J. V.

"Sur l'intégrale prise sur un contour variable," *Portugaliae Math.* 1 (1940), 343–345. **FM** 66, p. 328. **Zbl** 24, p. 25. **MR** 2, p. 78.

"Contours de Jordan et intégrale de Cauchy," *Portugaliae Math.* 2 (1941), 166–172 and 3 (1942), 253–257. **FM** 67, p. 256. **Zbl** 25, p. 151 and 27, p. 396. **MR** 3, p. 76 and 4, p. 241.

GONZÁLEZ-VELASCO, E. A.

"The homotopic proof of Cauchy's integral theorem," *Int. Jour. Math. Educ. Sci. Technol.* 11 (1980), 189–191. **Zbl** 452.30004. **MR** 81g:30001.

GOODMAN, A. W.

"Functions typically-real and meromorphic in the unit circle," *Trans. Amer. Math. Soc.* 81 (1956), 92–105. **Zbl** 72, p. 293. **MR** 17, p. 724.

Univalent Functions, vols I, II, Mariner Publishing Co. (1983), Tampa. **Zbl** 1041.30500, 30501.

GORDON, W. B.

"On the diffeomorphisms of euclidean space," *Amer. Math. Monthly* 79 (1972), 755–759 and 80 (1973), 674–675. **Zbl** 263.57015–6. **MR** 46 #4548, 47 #7758.

GOULD, H. W.

"Explicit formulas for Bernoulli numbers," *Amer. Math. Monthly* 79 (1972), 44–51. **Zbl** 227.10010. **MR** 46 #5229. (See also *ibid.* 109 (2002), 205 and 111 (2004), 432–434.)

GOURSAT, É.

"Sur la définition générale des fonctions analytiques, d'après Cauchy," *Trans. Amer. Math. Soc.* 1 (1900), 14–16. **FM** 31, p. 398. (See also **FM** 48, p. 318.)

"Sur un théorème de M. Jensen," *Bull. Sci. Math.* (2) 26 (1902), 298–302. **FM** 33, p. 401.

(a) "Sur la théorie des fonctions implicites," *Bull. Soc. Math. France* 31 (1903), 184–192. **FM** 34, p. 426.

(b) "Sur quelques développements de $\frac{1}{1-x}$ en séries de polynômes," *Bull. Sci. Math.* (2) 27 (1903), 226–232. **FM** 34, p. 305.

GRABINER, S.

"A short proof of Runge's theorem," *Amer. Math. Monthly* 83 (1976), 807–808. **MR** 55 #670.

GRAESER, E.

"Über konforme Abbildung des allgemeinen zweifach zusammenhängenden schlichten Bereichs auf die Fläche eines Kreisrings," *Dissertation* (1930), Leipzig. **FM** 56, p. 987.

GRAHL, J.

"A short proof of Miranda's theorem and some extensions using Zalcman's lemma," *Journal of Analysis* 11 (2003), 105–113. **Zbl** 1132.30337. **MR** 2006m: 30052.

GRAHL, J. and ROTH, O.

"An extension of Lewis's lemma, renormalization of harmonic and analytic functions, normal families," *Math. Nachr.* 282 (2009), 540–551. **Zbl** 1163.30019. **MR** 2010f:30073.

GRANAS, A.

"The theory of compact vector fields and some of its applications to topology of functional spaces (I)," *Rozprawy Matematyczne* (=*Dissertationes Mathematicae*) 30 (1962). **Zbl** 111, p. 110. **MR** 26 #6743.

GRATTAN-GUINNESS, I.,

The Development of the Foundations of Mathematical Analysis from Euler to Riemann, MIT Press (1970), Cambridge, Massachusetts. **Zbl** 215, p. 44. **MR** 58 #15948. **BAMS** 78, p. 385.

"The Cauchy-Stokes-Seidel story on uniform convergence again: Was there a fourth man?", *Bull. Soc. Math. Belgique* 38 (1986), 225–235. **Zbl** 625.01011. **MR** 88d:01019. (See also **MR** 2008g:01013.)

GRAY, J. D. and MORRIS, S. A.

"When is a function that satisfies the Cauchy-Riemann equations analytic?," *Amer. Math. Monthly* 85 (1978), 246–256. **Zbl** 416.30002. **MR** 57 #9940. (See also **MR** 57 #594.)

GRAY, J. J.

"The nineteenth-century revolution in mathematical ontology," pp. 226–248 of *Revolutions in Mathematics* (Donald Gillies, editor), Oxford University Press (1992). **Zbl** 967.00508. **MR** 94c:00018.

"On the history of the Riemann mapping theorem," *Rend. Circ. Mat. Palermo* (2) Supplement no. 34 (1994), 47–94. **Zbl** 810.01005. **MR** 95k:30017.

"Goursat, Pringsheim, Walsh, and the Cauchy integral theorem," *Math. Intelligencer* 22, no. 4 (2000), 60–66 and 77. **Zbl** 1052.30503. **MR** 1796768.

The Real and the Complex: A History of Analysis in the 19th *Century,* Springer (2015). **Zbl** 1330.01001. **MR** 3380969.

Gray, W. L.

"A note on functions with a natural boundary," *Amer. Math. Monthly* 72 (1965), 40–41. **Zbl** 147, p. 62. **MR** 30 #3982.

Green, J. W.

"Mean values of harmonic functions on homothetic curves," *Pac. Jour. of Math.* 6 (1956), 279–282. **Zbl** 71, p. 320. **MR** 18, p. 295.

Greene, R. E. and Krantz, S. G.

Function Theory of One Complex Variable, John Wiley and Sons, Inc. (1997), New York. 3rd ed., Graduate Studies in Mathematics, vol. 40 (2006), American Mathematical Society, Providence. **Zbl** 1114.30001. **MR** 2006m:30001.

Greenleaf, N.

"On Fermat's equation in $\mathcal{C}(t)$," *Amer. Math. Monthly* 76 (1969), 808–809. **MR** 40 #4204.

Greenlee, W. M.

"On Green's theorem and Cauchy's theorem," *Real Anal. Exchange* 30 (2004/05), 703–717. **Zbl** 1098.53012. **MR** 2006f:26022.

Greenstein, D. S.

"Locally schlicht mapping," *Amer. Math. Monthly* 65 (1958), 214.

Greiter, G.

"A simple proof for a theorem of Kronecker," *Amer. Math. Monthly* 85 (1978), 756–757. **MR** 80a:12003.

Grinshpan, A. Z.

"The Bieberbach conjecture and Milin's functionals," *Amer. Math. Monthly* 106 (1999), 203–214. **MR** 2000b:30027.

Gronwall, T. H.

(a) "On analytic functions of constant modulus on a given contour," *Annals of Math.* (2) 14 (1912–13), 72–80. **FM** 43, p. 491.

(b) "Some special boundary problems in the theory of harmonic functions," *Bull. Amer. Math. Soc.* 19 (1912/13), 227–233. **FM** 44, p. 433.

"On the maximum modulus of an analytic function," *Annals of Math.* (2) 16 (1914–15), 77–81. **FM** 45, p. 641.

"Sur la déformation dans la représentation conforme," *C. R. Acad. Sci. Paris* 162 (1916), 249–252 and 316–318. **FM** 46, pp. 555, 556.

GROSS, F.

(a) "On the functional equation $f^n + g^n = h^n$," *Amer. Math. Monthly* 73 (1966), 1093–1096. **Zbl** 154, p. 401. **MR** 34 #4494. (See also **MR** 2004f: 30019.)

(b) "On the equation $f^n + g^n = 1$," *Bull. Amer. Math. Soc.* 72 (1966), 86–88; erratum (*ibid.*) 74 (1968), 767. **Zbl** 131, p. 136. **MR** 32 #2595, #5882 and 37 #1613.

"On the equation $f^n + g^n = 1$. II," *Bull. Amer. Math. Soc.* 74 (1968), 647–648. **Zbl** 159, p. 202. **MR** 37 #2990.

Factorization of Meromorphic Functions, Mathematics Research Center Naval Research Laboratory (1972), Washington, D. C. **Zbl** 266.30006. **MR** 53 #11030.

GROSS, F. and OSGOOD, C. F.

"On the functional equation $f^n + g^n = h^n$ and a new approach to a certain class of more general functional equations," *Indian Jour. of Math.* 23 (1981), 17–39. **Zbl** 513.30017. **MR** 84k:30027.

GROSSE-ERDMANN, K.-G.

"The locally convex topology on the space of meromorphic functions," *Jour. Austral. Math. Soc.* (Ser. A) 59 (1995), 287–303. **Zbl** 853.30030. **MR** 96j: 46020.

GROSSE-ERDMANN, K.-G. and MORTINI, R.

"Universal functions for composition operators with non-automorphic symbol," *Jour. d'Analyse Math.* 107 (2009), 355–376. **Zbl** 05646622. **MR** 2010c:30073.

GROSSMAN, N.

"A C^∞ Lagrange inversion theorem," *Amer. Math. Monthly* 112 (2005), 512–514. **Zbl** 1128.26007. **MR** 2006a:26017.

GRUNSKY, H.

"Der Hauptsatz der komplexen Analysis," *Jahresber. Deutsch. Math. Verein.* 68 (1966), 1te Abteilung, 1–6. **Zbl** 131, p. 304. **MR** 33 #2800.

Lectures on Theory of Functions in Multiply Connected Domains, Studia Mathematica, Skript 4. Vandenhoeck & Ruprecht (1978), Göttingen. **Zbl** 371.30015. **MR** 57 #3365.

GRÜNWALD, G. and TURÁN, P.

"Über den Blochschen Satz," *Acta Litt. ac Scient. Univ. Hung.* (Szeged) 8 (1936–1937), 236–240. **FM** 63, p. 300. **Zbl** 17, p. 23.

GUGGENHEIMER, H.

"The Jordan Curve Theorem and an unpublished manuscript by Max Dehn," *Arch. Hist. Exact Sci.* 17 (1977), 193–200. **Zbl** 357.01023. **MR** 58 #27069.

GUICHARD, C.

"Sur les fonctions entières," *Ann. Sci. École Norm. Sup.* (3) 1 (1884), 427–432. **FM** 16, p. 186.

GUICHARDET, A.

Analyse harmonique commutative, Dunod (1968), Paris. **Zbl** 159, p. 186. **MR** 39 #1901.

GUINARD, A. P.

"The umbral method: a survey of mnemonic and manipulative uses," *Amer. Math. Monthly* 86 (1979), 187–195. **Zbl** 406.10013. **MR** 80e:05001.

GUTZMER, A.

"Sur certaines moyennes arithmétiques des fonctions d'une variable complexe," *Teixeira Jour.* 8 (1887), 147–156 and *Nouvelles Annales de Math.* (3) 8 (1889), 101–111. **FM** 19, p. 370 and 21, p. 233.

"Ein Sätz über Potenzreihen," *Math. Annalen* 32 (1888), 596–600. **FM** 20, p. 248.

H

HAAR, A.

"Über analytische Funktionen mit singulärer Linie," *Göttingen Nachr.* (1914), 115–123. **FM** 45, p. 659.

HADAMARD, J.

"Sur le rayon de convergence des séries ordonnées suivant les puissances d'une variable," *C. R. Acad. Sci. Paris* 106 (1888), 259–262. **FM** 20, p. 247.

(a) "Sur les fonctions entières de la forme $e^{G(x)}$," *C. R. Acad. Sci. Paris* 114 (1892), 1053–1055. **FM** 24, p. 400.

(b) "Essai sur l'étude des fonctions données par leur développement de Taylor," *Jour. de Math. Pures et Appliq.* (4) 8 (1892), 101–186. **FM** 24, p. 359 and 33, p. 448. (See also **FM** 62, p. 1039 and **Zbl** 13, p. 1.)

"Étude sur les propriétés des fonctions entières et en particulier d'une fonction considérée par Riemann," *Jour. de Math. Pures et Appliq.* (4) 9 (1893), 171–215. **FM** 25, p. 698 and 33, p. 448. (See also **FM** 62, p. 1039 and **Zbl** 13, p. 1.)

"Sur les fonctions entières," *Bull. Soc. Math. France* 24 (1896), 186–187. **FM** 27, p. 321. (See also **FM** 62, p. 1039 and **Zbl** 13, p. 1.)

Notice sur les travaux scientifiques de M. Jacques Hadamard, II: 1901–1912, Libraire Scientifique A. Hermann et Fils (1912), Paris. **FM** 43, p. 61.

HADWIGER, H.

"Elementare Begründung ausgewählter stetigkeitsgeometrischer Sätze für Kreis und Kugelfläche," *Elemente der Math.* 14 (1959), 49–60. **Zbl** 90, p. 393. **MR** 21 #5188.

"Ein Satz über stetige Funktionen auf der Kugelfläche," *Arch. der Math.* 11 (1960), 65–68. **Zbl** 117, p. 170. **MR** 22A #4050.

"Elementare Kombinatorik und Topologie," *Elemente der Math.* 15 (1960), 49–60. **Zbl** 96, p. 173. **MR** 22A #7132.

HALES, T. C.

(a) "The Jordan curve theorem, formally and informally," *Amer. Math. Monthly* 114 (2007), 882–894. **Zbl** 1137.03305. **MR** 2008m:03027.

(b) "Jordan's proof of the Jordan curve theorem," pp. 45–60 of *From Insight to Proof: Festschrift in Honour of Andrzej Trybulec* (ed. H. Swieczkowska), *Studies in Logic, Grammar and Rhetoric* 10 (23) 2007.

HALL, T.

"Sur la mesure harmonique de certains ensembles," *Arkiv för Mat., Astro. och Fysik* 25A, No. 28 (1937). **FM** 63, p. 287. **Zbl** 16, p. 216.

HALMOS, P. R.

Naive Set Theory, Van Nostrand Reinhold Co. (1960), New York. Reprinted by Springer-Verlag (1975), New York. **Zbl** 321.04001. **MR** 55 #102 and 56 #11794.

I have a photographic memory, American Mathematical Society (1987), Providence. **Zbl** 658.01017. **MR** 89f:01067.

HAMBURGER, H.

"Über einige Beziehungen, die mit der Funktionalgleichung der Riemannschen ζ-Funktion äquivalent sind," *Math. Annalen* 85 (1922), 129–140. **FM** 48, p. 1214. (See also **MR** 83i:01003.)

HAMEL, G.

"Über die Umkehrung einer Potenzreihe," *Deutsche Math.* 5 (1940), 338–339. **FM** 66, p. 329. **Zbl** 24, p. 214. **MR** 7, p. 193.

HAMILTON, H. J.

"The partial fraction decomposition of a rational function," *Math. Mag.* 45 (1972), 117–119. **Zbl** 241.12103.

HANCHE-OLSEN, H.

"On Goursat's proof of Cauchy's integral theorem," *Amer. Math. Monthly* 115 (2008), 648–652 and 116 (2009), 381–382. **Zbl** 1167.30018. **MR** 2009f:30093.

HANČL, J.

"A simple proof of the irrationality of π^4," *Amer. Math. Monthly* 93 (1986), 374–375. **Zbl** 597.10029. **MR** 87g:11084.

HANSEN, D. J.

"A note on the operation of multiplication in the complex plane," *Amer. Math. Monthly* 71 (1964), 185–186. **Zbl** 122, p. 247.

HANSEN, W.

"Restricted mean value property and harmonic functions," pp. 67–90 of *Potential Theory — ICPT 94 (Kouty, 1994)* (J. Král, *et al.*, editors). W. de Gruyter (1996), Berlin. **Zbl** 857.31001. **MR** 97g:31005.

"Liouville's theorem and the restricted mean value property in the plane," *Jour. de Math. Pures et Appliq.* (9) 77 (1998), 943–947. **Zbl** 919.31002. **MR** 96b:31006.

"A strong version of Liouville's theorem," *Amer. Math. Monthly* 115 (2008), 583–595. **Zbl** 1161.31001. **MR** 2010b:31002.

HANSEN, W. and NADIRASHVILI, N.

(a) "Mean values and harmonic functions," *Math. Annalen* 297 (1993), 157–170. **Zbl** 794.31002. **MR** 94m:31004b.

(b) "A converse to the mean value theorem for harmonic functions," *Acta Math.* 171 (1993), 139–163. **Zbl** 808.31004. **MR** 94m:31004a.

"Littlewood's one-circle problem," *Jour. Lon. Math. Soc.* (2) 50 (1994), 349–360. **Zbl** 804.31001. **MR** 95j:31002. (See also **MR** 2010c:31003.)

HANTOUT, Y.

"Une preuve géométrique du théorème fondamental de l'algèbre," *Publ. I. R. M. A. de Lille* 26, No. V (1991).

HARDY, G. H.

"Some theorems connected with Abel's theorem on the continuity of power series," *Proc. Lon. Math. Soc.* (2) 4 (1907), 247–265. **FM** 37, p. 429.

"The integral $\int_0^\infty \frac{\sin x}{x}\,dx$," *Math. Gazette* 5 (1909–11), 98–103 and 8 (1915–16), 301–303. **FM** 40, p. 341.

"Notes on some points in the integral calculus XXXVIII. On the definition of an analytic function by means of a definite integral," *Messenger of Math.* 43 (1914), 29–33. **FM** 44, p. 348.

"The mean value of the modulus of an analytic function," *Proc. Lon. Math. Soc.* (2) 14 (1915), 269–277. **FM** 45, p. 1331.

"A theorem concerning harmonic functions," *Jour. Lon. Math. Soc.* 1 (1926), 130–131. **FM** 52, p. 497.

Divergent Series, Clarendon Press (1949), Oxford and Chelsea (1991), New York. **Zbl** 897.01044. **MR** 11, p. 25.

HARDY, G. H., INGHAM, A. E. and PÓLYA, G.

"Theorems concerning mean values of analytic functions," *Proc. Royal Soc. Lon.* (A) 113 (1927), 542–569. **FM** 53, p. 304.

"Notes on moduli and mean values," *Proc. Lon. Math. Soc.* (2) 27 (1928), 401–409. **FM** 54, p. 331.

HARDY, G. H. and ROGOSINSKI, W.

"Theorems concerning functions subharmonic in a strip," *Proc. Royal Soc. Lon.* (A) 185 (1946), 1–14. **Zbl** 61, p. 232. **MR** 7, p. 448.

HARMSE, J. E.

"Extensions of the Cauchy-Goursat theorem," *Jour. Math. Anal. and Applic.* 339 (2008), 429–437. **Zbl** 1177.30004. **MR** 2009b:30078.

HARNACK, A.

"Anwendung der Fourier'schen Reihe auf die Theorie der Functionen einer complexen Veränderlichen," *Math. Annalen* 21 (1882), 305–326. **FM** 15, p. 185.

"Existenzbeweise zur Theorie des Potentiales in der Ebene und im Raume," *Ber. Verhandl. Kön. Sächs. Gesell. Wissen. Leipzig* (1886), 144–169, reproduced in *Math. Annalen* 35 (1890), 19–40. **FM** 18, p. 919 and 21, p. 994.

HARRIS, G. and MARTIN, C.

"The roots of a polynomial vary continuously as a function of the coefficients," *Proc. Amer. Math. Soc.* 100 (1987), 390–392. **Zbl** 619.30008. **MR** 88h:30006. Addendum *ibid.* 102 (1988), 993–994. **Zbl** 646.30011. **MR** 89c:26026.

HARTE, R. and KEOGH, G.

"Touché Rouché," *Bull. Korean Math. Soc.* 40 (2003), 215–221. **MR** 2004d:30038.

HARTOGS, F.

"Über die Grenzfunktionen beschränkter Folgen von analytischen Funktionen," *Math. Annalen* 98 (1928), 164–178. **FM** 53, p. 281.

HARTOGS, F. and ROSENTHAL, A.

"Über Folgen analytischer Funktionen," *Math. Annalen* 100 (1928), 212–263. **FM** 54, p. 358.

"Über Folgen analytischer Funktionen," *Math. Annalen* 104 (1931), 606–610. **FM** 57, p. 371. **Zbl** 1, p. 213.

HARUKI, H.

"On a functional equation for the exponential function of a complex variable," *Glasgow Math. Jour.* 12 (1971), 31–34. **Zbl** 224.30016. **MR** 45 #5359.

"A proof of the principle of circle-transformation by use of a theorem on univalent functions," *L'Enseignement Mathématique* (2) 18 (1972), 145–146. **Zbl** 241.30017. **MR** 48 #4312. (See also **MR** 43 #492 and **Zbl** 223.30003.)

"On the equivalence of Hille's and Robinson's functional equations," *Annales Polonici Mathematici* 28 (1973), 261–264. **Zbl** 268.30002. **MR** 49 #7430.

(a) "A proof of Euler's identity by a functional equation," *Aequationes Math.* 28 (1985), 138–143. **Zbl** 561.30019. **MR** 86c:30045.

(b) "A generalization of Hille's and Robinson's functional equations," *Portugaliae Math.* 40 (1985), 335–343. **Zbl** 582.39010. **MR** 87d:30030.

HASELEN, A. VAN

"Sur la représentation conforme," *Kon. Akad. Wetensch. Amsterdam, Proc.* 35 (1932), 867–869. **FM** 58, p. 360. **Zbl** 5, p. 169.

HATHAWAY, D.

"Using continuity induction," *The College Math. Jour.* 42 (2011), 229–231. See also **MR** 2009b:03011.

HAUSDORFF, F.

Grundzüge der Mengenlehre, Veit and Company (1914), Leipzig. **FM** 45, p. 123. **BAMS** 27, p. 116. Reprinted by Chelsea Publishing Co. (1955), New York. **MR** 11, p. 88. This is vol. II of his *Gesammelte Werke,* Springer (2006), Berlin.

"Über halbstetige Funktionen und deren Verallgemeinerung," *Math. Zeit.* 5 (1919), 292–309. **FM** 47, p. 240.

Set Theory (translated from German by J. Aumann, *et al.*), Chelsea Publishing Co. (1957), New York. **FM** 61, p. 60. **Zbl** 12, p. 203. **MR** 25 #4999. **BAMS** 33, p. 778.

HAVIL, J.

Gamma: Exploring Euler's Constant, Princeton University Press (2003), Princeton, N.J. **Zbl** 1023.11001. **MR** 2004k:11145.

HAYASHI, T.

"Relation between the zeros of a rational integral function and its derivate," *Annals of Math.* (2) 15 (1914–15), 112–113. **FM** 45, p. 167.

"Evaluation of some integrals," *Nieuw Archief voor Wiskunde* (2) 13 (1919/1921), 324–333. **FM** 47, p. 225.

"On the integral $\int_0^\infty \frac{\sin^n x}{x^m}\,dx$," *Nieuw Archief voor Wiskunde* (2) 14 (1922/1924), 13–18. **FM** 48, p. 257.

"On the integral $\int_0^\infty x^{-m} \sin^n x\,dx$, with an appendix to an application to a theory of approximation of a function," *Tôhoku Math. Jour.* 22 (1923), 165–170. **FM** 48, p. 258.

HAYMAN, W. K.

"Some remarks on Schottky's theorem," *Proc. Camb. Phil. Soc.* 43 (1947), 442–454. **Zbl** 29, p. 123. **MR** 9, p. 84.

"On the limits of moduli of analytic functions," *Annales Polonici Mathematici* 12 (1962), 143–150. **Zbl** 107, p. 283. **MR** 26 #324 and p. 1544.

Meromorphic Functions, Clarendon Press (1964), reprinted with appendix, 1975, Oxford. **Zbl** 149, p. 30. **MR** 35 #5617.

"On a non-regular parametric integral," *Proc. Royal Soc. Edin.* A83 (1979), 185–188. **Zbl** 417.30001. **MR** 81d:30074. (For part II see **MR** 2001d:30069.)

Subharmonic Functions, Vol. 2, London Mathematical Society Monographs No. 20, Academic Press (1989), New York and London. **Zbl** 699.31001. **MR** 91f:31001. **BAMS** (2) 25 (1991), 458–469.

"Function theory 1900–1950," pp. 369–384 of *Development of Mathematics 1900–1950* (Jean-Paul Pier, editor), Birkhäuser Verlag (1994), Basel and Boston. **Zbl** 807.01016. **MR** 95j:01022.

HAYMAN, W. K. and KENNEDY, P. B.

Subharmonic Functions, Vol. 1, London Mathematical Society Monographs No. 9, Academic Press (1976), New York and London. **Zbl** 521.31001. **MR** 82a:31001. **BAMS** (2) 1, p. 376.

HAYMAN, W. K. and KORENBLUM, B.

"A critical growth rate for functions regular in a disk," *Mich. Math. Jour.* 27 (1980), 21–30. **Zbl** 428.30020. **MR** 81b:30059.

HE, Z.-X. and SCHRAMM, O.

"On the convergence of circle packings to the Riemann map," *Invent. Math.* 125 (1996), 285–305. **Zbl** 868.30010. **MR** 97i:30009. (See also **Zbl** 960.44461 and **MR** 97i:30008.)

HEFFTER, L.

"Zur Theorie der reellen Kurvenintegrale," *Göttingen Nachr.* (1902), 115–140. **FM** 33, p. 310.

"Zum Beweis des Cauchy-Goursatschen Integralsatzes," *Göttingen Nachr.* (1903), 312–316. **FM** 34, p. 338.

Begründung der Funktionentheorie auf alten und neuen Wegen, Springer-Verlag (2nd ed. 1960), Berlin. **Zbl** 89, p. 47. **MR** 22 #9587. **BAMS** 62, p. 271.

HEILBRONN, H.

"Zu dem Integralsatz von Cauchy," *Math. Zeit.* 37 (1933), 37–38. **FM** 59, p. 312. **Zbl** 6, p. 260.

"On the representation of homotopic classes by regular functions," *Bull. Acad. Polon. Sci.* 6 (1958), 181–184. **Zbl** 80, p. 160. **MR** 20 #5481.

HEILPER, A.

"The zeros of functions in Nevanlinna's area class," *Israel Jour. of Math.* 34 (1979), 1–11. **Zbl** 498.30040. **MR** 82e:30046.

HEINHOLD, J.

"Zur konformen Abbildung schlichter Gebiete," *Math. Zeit.* 67 (1957), 133–138. **Zbl** 79, p. 298. **MR** 19, p. 401.

HEINS, M. H.

(a) "A generalization of the Aumann-Carathéodory 'Starrheitssatz'," *Duke Math. Jour.* 8 (1941), 312–316. **FM** 67, p. 283. **Zbl** 25, p. 173. **MR** 3, p. 81.

(b) "On the iteration of functions which are analytic and single-valued in a given multiply-connected region," *Amer. Jour. of Math.* 63 (1941), 461–480. **FM** 67, p. 283. **Zbl** 25, p. 46. **MR** 2, p. 275.

(c) "A note on a theorem of Radó concerning the $(1, m)$ conformal maps of a multiply-connected region into itself," *Bull. Amer. Math. Soc.* 47 (1941), 128–130. **FM** 67, p. 283. **Zbl** 25, p. 173. **MR** 2, p. 186.

"The problem of Milloux for functions analytic throughout the interior of the unit circle," *Amer. Jour. of Math.* 67 (1945), 212–234. **Zbl** 60, p. 217. **MR** 6, p. 262.

(a) "On the Phragmén-Lindelöf Principle," *Trans. Amer. Math. Soc.* 60 (1946), 238–244. **Zbl** 60, p. 221. **MR** 8, p. 371.

(b) "On the number of 1–1 directly conformal maps which a multiply-connected plane region of finite connectivity p (> 2) admits onto itself," *Bull. Amer. Math. Soc.* 52 (1946), 454–457. **MR** 8, p. 21.

"Entire functions with bounded minimum modulus; subharmonic function analogues," *Annals of Math.* (2) 49 (1948), 200–213. **Zbl** 29, p. 298. **MR** 9, p. 341.

"The conformal mapping of simply connected Riemann surfaces," *Annals of Math.* (2) 50 (1949), 686–690. **Zbl** 37, p. 55. **MR** 11, p. 93.

(a) "On a class of conformal metrics," *Nagoya Math. Jour.* 21 (1962), 1–60. **Zbl** 113, p. 56. **MR** 26 #1451.

(b) *Selected topics in the Classical Theory of Functions of a Complex Variable,* Holt, Rinehart and Winston, Inc. (1962), New York. **MR** 29 #217.

Complex Function Theory, Pure and Applied Mathematics, Vol. 28. Academic Press (1968), New York. **Zbl** 155, p. 115. **MR** 39 #413. **BAMS** 76, p. 968.

"On a theorem of Study concerning conformal maps with convex images," pp. 171–176 of *Mathematical Essays Dedicated to A. J. Macintyre,* Ohio University Press (1970), Athens, Ohio. **Zbl** 211, p. 100. **MR** 42 #7878.

"Some remarks concerning nonnegative harmonic functions," *Jour. Approx. Theory* 5 (1972), 118–122. **Zbl** 225.31001. **MR** 49 #9225.

"On a theorem of Study concerning conformal maps with convex images. II," *Math. Scand.* 32 (1973), 245–257. **Zbl** 271.30009. **MR** 49 #551.

"Some characterizations of finite Blaschke products of positive degree," *Jour. d'Analyse Math.* 46 (1986), 162–166. **Zbl** 607.30031. **MR** 87k:30063.

"A note concerning the lemma of Julia-Wolff-Carathéodory," *Ann. Acad. Scient. Fennicae, Ser. A. I. Math.* 14 (1989), 133–136. **Zbl** 642.30025. **MR** 90d:30096.

HEINZ, E.

"Ein elementarer Beweis des Satzes von Radó-Behnke-Stein-Cartan über analytische Funktionen," *Math. Annalen* 131 (1956), 258–259. **Zbl** 71, p. 294. **MR** 18, p. 149.

"An elementary analytic theory of the degree of mapping in n-dimensional space," *Jour. Math. and Mech.* 8 (1959), 231–247. **Zbl** 85, p. 171. **MR** 21 #1370.

HEJHAL, D. A.

"On the loci $|f(z)| = R$, $f(z)$ entire," *Glasgow Math. Jour.* 10 (1969), 94–102. **Zbl** 189, p. 362. **MR** 41 #470.

HELMER, O.

"Divisibility properties of integral functions," *Duke Math. Jour.* 6 (1940), 345–356. **FM** 66, p. 105. **Zbl** 23, p. 239. **MR** 1, p. 307.

HELSON, H. and SARASON, D.

"Past and future," *Math. Scand.* 21 (1967), 5–16. **Zbl** 241.60029. **MR** 38 #5282. (See also **MR** 40 #7814.)

HEMPEL, J. A.

"Precise bounds in the theorems of Schottky and Picard," *Jour. Lon. Math. Soc.* (2) 21 (1980), 279–286. **Zbl** 431.30017 and 501.30019. **MR** 81i:30053 and 82h:30017. (See also **MR** 2009d:30046.)

HENLE, J.

"Classical mathematics . . . ," *Amer. Math. Monthly* 103 (1996), 18–29. **Zbl** 962.01507. **MR** 1369148.

HENRICI, P.

Applied and Computational Complex Analysis, Vol. I (1974), Vol. II (1977), Vol. III (1986), Wiley-Interscience, New York. **Zbl** 313.30001, 363.30001, 578.30001. **MR** 51 #8378, 56 #12235, 87h:30002.

"A general theory of osculation algorithms for conformal mapping," *Linear Alg. and its Appl.* 52/53 (1983), 361–382. **Zbl** 515.30005. **MR** 85b:30011.

HERGLOTZ, G.

"Über Potenzreihen mit positivem, reellen Teil im Einheitskreis," *Ber. Verhandl. Kön. Sächs. Gesell. Wissen. Leipzig* 63 (1911), 501–511. **FM** 42, p. 438.

HERMITE, C.

"Sur la formule d'interpolation de *Lagrange.* Extrait d'une lettre à M. Borchardt," *Jour. für Reine und Angew. Math.* 84 (1877), 70–79. **FM** 9, p. 312.

"Sur quelques points de la théorie des fonctions. Extrait d'une lettre à M. Mittag-Leffler," *Jour. für Reine und Angew. Math.* 91 (1881), 54–78. **FM** 13, p. 307 and 310.

"Sur les fonctions holomorphes," *Jour. de Math. Pures et Appliq.* (4) 1 (1885), 9–10. **FM** 17, p. 391.

HERVÉ, M.

"Sur l'itération dans un domaine multiplement connexe," *C. R. Acad. Sci. Paris* 230 (1950), 707–708. **Zbl** 35, p. 52. **MR** 11, p. 589.

"Quelques propriétés des transformations intérieures d'un domaine borné," *Ann. Sci. École Norm. Sup.* (3) 68 (1951), 125–168. **Zbl** 44, p. 303. **MR** 13, p. 734.

Compléments sur les fonctions d'une variable complexe (lecture notes), Faculté des Sciences de Paris. Mathématiques approfondies 1965/66. Secrétariat mathématique (1970), Paris. **Zbl** 195, p. 81. **MR** 42 #1980.

Les fonctions analytiques, Presses Universitaires de France (1982), Paris. **Zbl** 564.32001. **MR** 84h:30001.

Herz, A.

Repetitorium Funktionentheorie, Friedr. Vieweg & Sohn (1996), Braunschweig/ Wiesbaden. **Zbl** 1074.30001.

Herzig, A.

"Die Winkelderivierte und das Poisson-Stieltjes-Integral," *Math. Zeit.* 46 (1940), 129–156. **FM** 66, p. 362. **Zbl** 22, p. 238. **MR** 1, p. 213.

Herzog, F.

"A note on power series which diverge everywhere on the unit circle," *Mich. Math. Jour.* 2 (1953–54), 175–177. **Zbl** 58, p. 60. **MR** 16, p. 578.

Hessenberg, G.

Transzendenz von e und π. Ein Beitrag zur höheren Mathematik vom elementaren Standpunkt aus. B. G. Teubner (1912), Leipzig and Berlin. **FM** 43, p. 530. **BAMS** 20, p. 421. Reprinted by Johnson Reprint Corp. (1965), New York and London. **MR** 32 #1169.

Hestenes, M. R.

"An analogue of Green's theorem in the calculus of variations," *Duke Math. Jour.* 8 (1941), 300–311. **FM** 67, p. 366. **Zbl** 25, p. 181. **MR** 3, p. 53.

Heuser, P.

"Über Entwicklungen analytischer Funktionen nach gebietsabhängigen Polynomen," *Math. Annalen* 110 (1934), 1–11. **FM** 60, p. 248. **Zbl** 9, p. 118.

Higgins, J. R.

"Five short stories about the cardinal series," *Bull. Amer. Math. Soc.* (2) 12 (1985), 45–89. **Zbl** 562.42002. **MR** 86k:42045.

"Two basic formulae of Euler and their equivalence to Tschakolov's sampling theorem," *Sampl. Theory Signal Image Process.* 2 (2003), 259–270. **MR** 2005c: 01013.

Higgins, J. R., Schmeisser, G. and Voss, J. J.

"The sampling theorem and several equivalent results in analysis," *Jour. Comp. Analysis and Appl.* 2 (2000), 333–371. **Zbl** 1030.41010. **MR** 2001i:41012.

Hilbert, D.

"Über die Entwickelung einer beliebigen analytischen Function einer Variabeln in eine unendliche nach ganzen rationalen Functionen fortschreitende Reihe," *Göttingen Nachr.* (1897), 63–70. **FM** 28, p. 359.

Über das Dirichletsche Prinzip, Weidmannsche Buchhandlung (1901), Berlin and *Math. Annalen* 59 (1904), 161–186. **FM** 32, p. 423 and 35, p. 436.

"Zur Theorie der konformen Abbildung," *Göttingen Nachr.* (1909), 314–323. **FM** 40, p. 732.

Hille, E.

"Über die Variation der Bogenlänge bei konformer Abbildung von Kreisbereichen," *Arkiv för Mat., Astro. och Fysik* 11, No. 27 (1916–17). **FM** 46, p. 537.

"A Pythagorean functional equation," *Annals of Math.* (2) 24 (1922–23), 175–180. **FM** 49, p. 327. (See also **MR** 2001g:39057.)

"Essai d'une bibliographie de la représentation analytique d'une fonction monogène," *Acta Math.* 52 (1929), Appendix, 1–80. **FM** 55, p. 172.

Analytic Function Theory, I, II, Ginn and Co. (1959, 1962), Boston; 2nd ed. Chelsea Publishing Co. (1973), New York. **Zbl** 88, p. 52, 273.30002, 102, p. 294. **MR** 21 #6415, 34 #1490. **BAMS** 66, p. 253.

HIONG, KING-LAI

"Sur l'impossibilité de quelques relations indentiques entre des fonctions entières," *C. R. Acad. Sci. Paris* 243 (1956), 222–225. **Zbl** 70, p. 297. **MR** 18, p. 27.

Sur les fonctions méromorphes et les fonctions algébroïdes. Extensions d'un théorème de M. R. Nevanlinna, Mémorial des Sciences Mathématiques, Fasc. 139, Gauthier-Villars (1957), Paris. **Zbl** 79, p. 296. **MR** 19, p. 950.

"Sur les fonctions holomorphes sans zéros dont les dérivées admettent une valeur exceptionnelle," *C. R. Acad. Sci. Paris* 244 (1957), 2125–2127. **Zbl** 80, p. 57. **MR** 19, p. 130.

"Sur la limitation d'une fonction holomorphe sans zéro et admettant une valeur exceptionnelle *B.*," *Bull. Sci. Math.* (2) 81 (1957), 71–77. **Zbl** 78, p. 65. **MR** 19, p. 1170.

(a) "Sur le cycle de Montel-Miranda dans la théorie des familles normales," *Scientia Sinica* 7 (1958), 987–1000. **Zbl** 86, p. 62. **MR** 21 #4243 a, b.

(b) "Sur un problème de M. Montel concernant la théorie des familles normales des fonctions," *Sci. Record* (N. S.) 2 (1958), 189–192. **Zbl** 82, p. 291. **MR** 21 #3562 a.

"On the Montel-Miranda cycle in the theory of normal families," *Chinese Math.* 9 (1967), 390–399 (1968). **Zbl** 173, p. 320.

HIONG, KING-LAI and HO, YU-TSAIM,

"Sur les valeurs multiples des fonctions méromorphes et de leurs dérivées," *Scientia Sinica* 10 (1961), 267–285 and *Chinese Math.* 3 (1962), 156–168. **Zbl** 104, p. 296 and 163, p. 92. **MR** 27 #5910 and #2629.

HIRSCH, M. W. and SMALE, S.

"On algorithms for solving $f(x) = 0$," *Comm. in Pure and Applied Math.* 32 (1979), 281–312. **Zbl** 408.65032. **MR** 80b:65061. (See also **MR** 83i:65044.)

HO, C.-W.

"A note on proper maps," *Proc. Amer. Math. Soc.* 51 (1975), 237–241. **Zbl** 303.54002. **MR** 51 #6698.

HOFBAUER, J.

"A simple proof of $1 + \frac{1}{2^2} + \frac{1}{3^2} + \cdots = \frac{\pi^2}{6}$ and related identities," *Amer. Math. Monthly* 109 (2002), 196–200. **Zbl** 1022.40002. (See also 1043 #40500.)

HOFFMAN, K. M.

Banach Spaces of Analytic Functions, Prentice-Hall, Inc. (1962), Englewood Cliffs. Reprinted by Dover Publications, Inc. (1989), New York. **Zbl** 734.46033. **MR** 92d:46066. **BAMS** 69, p. 687.

HOISCHEN, L.

"A note on the approximation of continuous functions by integral functions," *Jour. Lon. Math. Soc.* 42 (1967), 351–354. **Zbl** 146, p. 300. **MR** 35 #2038.

"Eine Verschärfung eines Approximationssatzes von Carleman," *Jour. Approx. Theory* 9 (1973), 272–277. **Zbl** 271.30035. **MR** 51 #3459.

"Approximation und Interpolation durch ganze Funktionen," *Jour. Approx. Theory* 15 (1975), 116–123. **Zbl** 318.30034. **MR** 52 #14309.

"Asymptotische Approximationseigenschaften der Ableitungen analytischer Funktionen," *Jour. Approx. Theory* 35 (1982), 86–93. **Zbl** 496.30034. **MR** 83f: 30036.

HÖLDER, O.

"Beweis des Satzes, dass eine eindeutige analytische Function in unendlicher Nähe einer wesentlich singulären Stelle jedem Werth beliebig nahe kommt," *Math. Annalen* 20 (1882), 138–143. **FM** 14, p. 337.

HOLLAND, A. S. B.

Introduction to the Theory of Entire Functions, Pure and Applied Mathematics, Vol. 56. Academic Press (1973), New York. **Zbl** 278.30001. **MR** 56 #5882.

HOLMANN, H.

"Merkwürdige Eigenschaften der Kugeloberfläche," *Math.-Phys. Semesterber.* (2) 10 (1964), 99–108. **Zbl** 117, p. 406.

HOLZER, L.

"Zwei Beispiele zum Verhalten von Potenzreihen auf dem Konvergenzkreis," *Deutsche Math.* 4 (1939), 190–193. **FM** 65, p. 307. **Zbl** 21, p. 37.

HOPF, E.

"Bemerkungen zur Aufgabe 49," *Jahresber. Deutsch. Math. Verein.* 39 (1930), 2te Abteilung, 4–6. **FM** 56, p. 1071.

HOPPE, R.

"Summirung einer Reihe," *Arch. der Math. und Physik* 61 (1877), 224. **FM** 9, p. 180.

HOROWITZ, L.

"A proof of the 'fundamental theorem of algebra' by means of Galois theory and 2-Sylow groups," *Nieuw Archief voor Wiskunde* (3) 14 (1966), 95–96. **Zbl** 143, p. 58. **MR** 33 #5617.

HOWARD, E. J.

"Analyticity of almost everywhere differentiable functions," *Proc. Amer. Math. Soc.* 110 (1990), 745–753. **Zbl** 705.30001. **MR** 91b:26015. (See also **MR** 34 #346.)

HOWSON, A. G.

A handbook of terms used in algebra and analysis, Cambridge University Press (1972). **Zbl** 233.00027. **MR** 50 #1783. (See also **Zbl** 522.00017.)

HU, K.

"On Cauchy's integral theorem," *Sci. Reports of Nat. Tsing Hua Univ. Peking* A3 (1935), 179–183. **FM** 61, p. 1132. **Zbl** 12, p. 261.

HUA, XIA

"A converse to a theorem on linear fractional transformations," *Math. Mag.* 81 (2008), 367–369. **Zbl** 1175.30008.

HUBER, H.

"Ein Mittelwertsatz für Funktionen einer komplexen Veränderlichen," *Comm. Math. Helv.* 21 (1948), 58–66. **Zbl** 29, p. 393. **MR** 9, p. 506.

"Über analytische Abbildungen von Ringgebieten in Ringgebiete," *Compositio Math.* 9 (1951), 161–168. **Zbl** 43, p. 302. **MR** 13, p. 337.

HUMMEL, J. A., SCHEINBERG, S. and ZALCMAN, L.

"A coeffieient problem for bounded nonvanishing functions," *Jour. d'Analyse Math.* 31 (1977), 169–190. **Zbl** 346.30013. **MR** 58 #11358.

HUREWICZ, W.

"Ueber ein topologisches Theorem," *Math. Annalen* 101 (1929), 210–218. **FM** 55, p. 313.

HURWITZ, A.

"Ueber die Nullstellen der Bessel'schen Function," *Math. Annalen* 33 (1889), 246–266. **FM** 20, p. 502.

"Ueber beständig convergierende Potenzreihen mit rationalen Zahlencoefficienten und vorgeschriebenen Nullstellen," *Acta Math.* 14 (1891), 211–215. **FM** 23, p. 250.

"Sur l'intégrale finie d'une fonction entière," *Acta Math.* 20 (1897), 285–312. **FM** 28, p. 251.

"Über die Entwickelung der allgemeinen Theorie der analytischen Funktionen in neuer Zeit," pp. 91–112 of *Verhandlungen des Ersten Internationalen Mathematiker–Kongresses in Zürich 1897.* B. G. Teubner (1898), Leipzig. **FM** 29, p. 338.

"Über die Anwendung der elliptischen Modulfunktionen auf einen Satz der allgemeinen Funktionentheorie," *Vierteljahrsschr. Naturforsch. Gesell. Zürich* 49 (1904), 242–253. **FM** 35, p. 401.

"Über die Einführung der elementaren transzendenten Funktionen in der algebraischen Analysis," *Math. Annalen* 70 (1911), 33–47. **FM** 41, p. 495.

HUYLEBROUCK, D.

"Similarities in irrationality proofs for π, $\ln 2$, $\zeta(2)$, $\zeta(3)$," *Amer. Math. Monthly* 108 (2001), 222–231. **Zbl** 986.11045. **MR** 2002b:11095.

I

IACOBACCI, R. F.

Augustin-Louis Cauchy and the development of mathematical analysis, thesis (1965), New York University.

IGLISCH, R.

"Über den Fundamentalsatz der Algebra," *Deutsche Math.* 5 (1940), 339–340. **FM** 66, p. 58. **Zbl** 23, p. 400. **MR** 8, p. 127.

"Zur Stetigkeit der Wurzeln einer algebraischen Gleichung," *Deutsche Math.* 7 (1942), 520–521. **Zbl** 60, p. 52. **MR** 8, p. 266.

INOUE, M.

"Sur un procédé pour construire la solution du problème de Dirichlet," *Proc. Imper. Acad. Tokyo* 14 (1938), 368–372. **FM** 64, p. 1161. **Zbl** 20, p. 130.

ISEKI, K.

"On the fundamental theorem of algebra," *Jour. Math. Soc. of Japan* 6 (1954), 129–130. **Zbl** 56, p. 253. **MR** 16, p. 328.

ISRAEL, R.

"Making nothing from something," *Amer. Math. Monthly* 95 (1988), 266–267.

"Polynomial or not," *Math. Intelligencer* 11 (1989), No. 2, 32.

IVERSEN, F.

Recherches sur les fonctions inverses des fonctions méromorphes, thesis (1914), Helsingfors. **FM** 45, p. 656.

IZUMI, S.

"Analytic function with assigned values at the given point set," *Tôhoku Math. Jour.* 30 (1929), 182–184. **FM** 55, p. 178.

J

JACK, I. S.

"Functions starlike and convex of order α," *Jour. Lon. Math. Soc.* (2) 3 (1971), 469–474. **Zbl** 224.30026. **MR** 43 #7611.

JACOBSTHAL, E.

"Bemerkungen zum Vitalischen Satze," *Jahresber. Deutsch. Math. Verein.* 36 (1927), 361–363. **FM** 53, p. 279.

JAHNKE, H. N., *et al.*

Geschichte der Analysis, Spektrum Akademischen Verlag (1999), Heidelberg & Berlin. **Zbl** 931.01002. English translation by the authors, London Mathematical Society and American Mathematical Society (2003). **Zbl** 1088.01001. **MR** 2004h:01005.

JAMESON, T. P.

"The probability integral by volume of revolution," *Math. Gazette* 78 (1994), 339–340.

JAMET, V.

"Sur le théorème de d'Alembert," *Mathesis* (2) 4 (1894), 5–14. **FM** 25, p. 142.

"Sur les intégrales de Fresnel," *Nouvelles Annales de Math.* (3) 15 (1896), 372–376 and (4) 3 (1903), 357–359. **FM** 27, p. 235 and 34, p. 335.

JANIEC, E.

"Some sufficient conditions for univalence of holomorphic functions," *Demonstratio Math.* 22 (1989), 717–727. **Zbl** 702.30015. **MR** 91b:30054. (See also **MR** 94d:30017.)

JANISZEWSKI, Z.

"Über die durch Kontinua bewirkte Zerschneidung der Ebene," *Prace Mat.-Fiz.* 26 (1913), 11–63. (Polish, French résumé). **FM** 45, p. 132. (See also **MR** 25 #3803.)

JATEGAONKAR, A. V.

"Elementary proof of a theorem of P. Montel on entire functions," *Jour. Lon. Math. Soc.* 40 (1965), 166–170. **Zbl** 134, p. 54. **MR** 30 #248.

JENKINS, J.

"Some results related to extremal length," *Annals of Mathematics Studies,* No. 30 (1953), 87–94. **Zbl** 52, p. 81. **MR** 15, p. 115.

"On explicit bounds for Schottky's theorem," *Canad. Jour. Math.* 7 (1955), 76–82. **Zbl** 64, p. 73. **MR** 16, p. 579.

JENSEN, J. L. W. V.

"Sur un nouvel et important théorème de la théorie des fonctions," *Acta Math.* 22 (1898–99), 359–364. **FM** 30, p. 364.

"Sur les fonctions convexes et les inégalites entre les valeurs moyennes," *Acta Math.* 30 (1906), 175–193. **FM** 37, p. 422.

"Investigation of a class of fundamental inequalities in the theory of analytic functions," Authorized translation from the Danish by T. H. Gronwall, *Annals of Math.* (2) 21 (1919–20), 1–29. **FM** 47, p. 271.

Jentzsch, R.
"Untersuchungen zur Theorie der Folgen analytischer Funktionen," *Acta Math.* 41 (1918), 219–251. **FM** 46, p. 516.

Jessen, B.
"Elementare Beweise einiger Sätze über stetige Abbildungen einer Kugel auf sich," *Mat. Tidsskr. B* (1933), 55–58. **FM** 59, p. 1255. **Zbl** 8, p. 181.

Jiang, Runrong
"A generalization of the Grace theorem," *Jour. Math. Wuhan Univ.* 11 (1991), 61–63 (Chinese). **Zbl** 748.30003.

Johansson, S.
"Über einige neuere Fragen aus der Theorie der konformen Abbildung," *Acta Soc. Sci. Fennicae* 40, Nr. 1 (1910). **FM** 41, p. 748.

Johnson, D. M.
"The problem of the invariance of dimension in the growth of modern topology, Part I," *Arch. Hist. Exact Sci.* 20 (1979), 97–188. **Zbl** 424.54020. **MR** 80k:01034. Part II, *ibid.* 25 (1981), 85–267. **Zbl** 532.55001. **MR** 83e:01050.

Johnsonbaugh, R.
"Evaluation of trigonometric sums," *Amer. Math. Monthly* 87 (1980), 497. (See also 113 (2006), 597.)

Johnston, E. H.
"The mean value theorem for analytic functions," *Proc. Edin. Math. Soc.* (2) 26 (1983), 289–295. **Zbl** 559.30006. **MR** 84k:30007.

Jones, G. A.
"$\frac{6}{\pi^2}$," *Math. Mag.* 66 (1993), 290–298. **Zbl** 808.11001. **MR** 94m:11002.

Jones, P. W.
"Ratios of interpolating Blaschke products," *Pac. Jour. of Math.* 95 (1981), 311–321. **Zbl** 479.30021. **MR** 82m:30032.

Jones, T. W.
"Discovering and proving that π is irrational," *Amer. Math. Monthly* 117 (2010), 553–557. **Zbl** 1235.11069. **MR** 2011d:11170.

Jones, W. B. and Ruscheweyh, S.
"Blaschke product interpolation and its application to the design of digital filters," *Constr. Approx.* 3 (1987), 405–409. **Zbl** 634.41007. **MR** 88m:30082. (See also **MR** 2003g:30066 and 2009d:30088.)

Jørgensen, V.
"Über die Randableitung einer beschränkten analytischen Funktion nebst einer Anwendung auf den Picardschen Satz," *Math. Annalen* 116 (1939), 658–663. **FM** 65, p. 336 and 1682. **Zbl** 21, p. 418.
"Elementary proof of a theorem of the Picard type," *Mat. Tidsskr. B* (1946), 129–134 (Danish). **Zbl** 60, p. 225. **MR** 7, p. 427.

JOURDAIN, P. E. B.

(a) "The theory of functions with Cauchy and Gauss," *Biblio. Math.* (3) 6 (1905), 190–207. **FM** 36, p. 59.

(b) "On the general theory of functions," *Jour. für Reine und Angew. Math.* 128 (1905), 169–210. **FM** 36, p. 443.

JULIA, G.

"Sur les lignes singulières de certaines fonctions analytiques," *Bull. Soc. Math. France* 41 (1913), 351–366. **FM** 44, p. 470.

"Mémoire sur l'itération des fonctions rationnelles," *Jour. de Math. Pures et Appliq.* (8) 1 (1918), 47–245 and *Bull. Sci. Math.* (2) 43 (1919), 106–109. **FM** 46, p. 520 and 47, pp. 47, 311.

(a) "Une propriété générale des fonctions entières liée au théorème de M. Picard," *C. R. Acad. Sci. Paris* 168 (1919), 502–504 and 598–600. **FM** 47, p. 312.

(b) "Sur quelques problèmes relatifs à l'itération des fractions rationnelles," *C. R. Acad. Sci. Paris* 168 (1919), 147–149. **FM** 47, p. 311.

"Sur quelques propriétés nouvelles des fonctions entières ou méromorphes," *Ann. Sci. École Norm. Sup.* (3) 36 (1919), 93–125; 37 (1920), 165–218; 38 (1921), 165–181. **FM** 47, p. 312 and 48, p. 363.

"Extension nouvelle d'un lemme de Schwarz," *Acta Math.* 42 (1920), 349–355. **FM** 47, p. 272.

Leçons sur les fonctions uniformes à point singulier essentiel isolé, Gauthier-Villars (1924), Paris. **FM** 50, p. 254. **BAMS** 31, p. 359.

(a) "Développement en série de polynômes ou de fonction rationnelle de la fonction que fournit la représentation conforme d'une aire simplement connexe sur un cercle," *Ann. Sci. École Norm. Sup.* (3) 44 (1927), 289–316. **FM** 53, p. 323.

(b) "Sur la représentation conforme des aires simplement connexes," *C. R. Acad. Sci. Paris* 184 (1927), 1107–1109. **FM** 53, p. 323.

(c) "Sur une classe de polynômes," *C. R. Acad. Sci. Paris* 184 (1927), 1227–1228. **FM** 53, p. 323.

(d) "Sur les moyennes des modules de fonctions analytiques," *Bull. Sci. Math.* (2) 51 (1927), 198–214. **FM** 53, p. 306.

"Sur une suite double de polynômes liée à la représentation conforme des aires planes simplement connexes," *Jour. de Math. Pures et Appliq.* (9) 7 (1928), 381–407. **FM** 54, p. 377.

(a) "Remarques sur les problèmes de 'meilleure approximation' et sur le problème de Dirichlet," pp. 95–101 of Vol. 3 of *Atti del Congresso Internazionale dei Matematici Bologna, Settembre 1928,* N. Zanichelli (1930), Bologna. Reprinted by Kraus Reprint, Ltd. (1967), Nedeln/Liechtenstein. **FM** 56, p. 423.

(b) *Principes géométriques d'analyse,* Part I (1930), Part II (1932), Cahiers scientifiques, fasc. 6 and 11. Gauthier-Villars , Paris. **FM** 56, p. 294 and 58, p. 355. **Zbl** 4, p. 10. **BAMS** 36, p. 789 and 39, p. 15.

"Sur quelques majorantes utiles dans la théorie des fonctions analytiques ou harmoniques," *Ann. Sci. École Norm. Sup.* (3) 48 (1931), 15–64. **FM** 57, p. 349. **Zbl** 1, p. 279.

"Essai sur le développement de la théorie des fonctions de variables complexes," pp. 102–125 of Vol. I of *Verhandlungen des Internationalen Mathematiker-Kongresses Zürich 1932.* Orell Füssli, Verlag (1932), Zürich and Leipzig. Reprinted by Kraus Reprint, Ltd. (1967), Nedeln/Liechtenstein. **FM** 58, p. 298. **Zbl** 6, p. 260. **MR** 53 #7713. Reprinted with a Foreward by Gauthier-Villars et Cie. (1933), Paris. **FM** 59, p. 311. **Zbl** 6, p. 211. **BAMS** 40, p. 521.

"La représentation conforme des aires multiplement connexes," *L'Enseignement Mathématique* 33 (1935), 137–168. **FM** 61, p. 363. **Zbl** 11, p. 169.

Leçons sur la représentation conforme des aires simplement connexes, Cahiers scientifiques, fasc. 8. Gauthier-Villars (2nd ed. 1950), Paris. **FM** 57, p. 397. **Zbl** 41, p. 50.

Leçons sur la représentation conforme des aires multiplement connexes, Cahiers scientifiques, fasc. 14. Gauthier-Villars (2nd ed. 1955), Paris. **FM** 60, p. 285. **Zbl** 11, p. 312.

Exercices d'analyse. II. Fonctions analytiques. Développements en série. Résidues. Transformations analytiques. Représentation conforme, Gauthier-Villars (2nd ed. 1958), reprinted 1969, Paris. **FM** 59, p. 1018. **Zbl** 169, p. 90. **MR** 39 #414. **BAMS** 40, p. 23.

Jürgens, E.

Allgemeine Sätze über Systeme von zwei eindeutigen und stetigen reellen Funktionen von zwei reellen Veränderlichen, B. G. Teubner (1879), Leipzig. **FM** 11, p. 269.

K

Kabaila, V. P.

"An analogue of the Poisson-Jensen formula with a double integral," *Litovsk. Mat. Sb.* 11 (1971), 241–253 (Russian). **Zbl** 224.30055. **MR** 45 #3718.

Kac, M.

"Une remarque sur les équations fonctionnelles," *Comm. Math. Helv.* 9 (1936–37), 170–171. **FM** 63, p. 408. **Zbl** 16, p. 249.

KAKEYA, S.

"On the mean modulus of polynomial," *Proc. Phys.-Math. Soc. Japan* (3) 5 (1923), 77–89.

KAKUTANI, S.

"A proof that there exists a circumscribing cube around any bounded closed convex set in $\mathbb{R}^3$," *Annals of Math.* (2) 43 (1942), 739–741. **Zbl** 61, p. 383. **MR** 4, p. 111. (See also **Zbl** 609.01049 and **MR** 92g:01109.)

"Rings of analytic functions," pp. 71–83 of *Lectures on Functions of a Complex Variable,* W. Kaplan, editor; University of Michigan Press (1955), Ann Arbor. **Zbl** 67, p. 54. **MR** 16, p. 1125. (See also **Zbl** 609.01049 and **MR** 92g:01109.)

KALB, K. G.

"Zur Entwicklung der Kompaktheitsschlussweisen," *Revista Colombiana de Mat.* 21 (1987), 107–134. **Zbl** 657.01009. **MR** 90c:01026.

KALMAN, D.

"Six ways to sum a series," *The College Math. Jour.* 24 (1993), 402–421 and 26 (1995), 110. **Zbl** 1266.40006. **MR** 3287877.

KAMIYA, H.

"An elementary proof of a theorem of L. Moore's on continuous curves," *Japan. Jour. of Math.* 7 (1931), 301–304. **FM** 57, p. 742. **Zbl** 1, p. 407.

KAMKE, E.

"Verallgemeinerung des Jordanschen Kurvensatzes und stetige Winkelfunktionen," *Sitzungsber. Preuss. Akad. Wissen. Berlin, Phys. -math. Kl.* (1928), 283–299. **FM** 54, p. 625. (See also **MR** 2003h:54041.)

"Zu dem Integralsatz von Cauchy," *Math. Zeit.* 35 (1932), 539–543. **FM** 58, p. 300. **Zbl** 4, p. 247.

KAMOWITZ, H.

"The spectra of endomorphisms of algebras of analytic functions," *Pac. Jour. of Math.* 66 (1976), 433–442. **Zbl** 358.46038. **MR** 58 #2426.

KAPLAN, W.

"Approximation by entire functions," *Mich. Math. Jour.* 3 (1955–56), 43–52. **Zbl** 70, p. 62. **MR** 17, p. 31.

KAUNZNER, W.

"Logarithms," pp. 210–228 of *Companion Encyclopedia of the History and Philosophy of the Mathematical Sciences,* vol. I (edited by I. Grattan-Guinness), Routledge (1994), London and New York. **Zbl** 833.01001. **MR** 98g:01003.

KAWAKAMI, Y.

"On Montel's theorem," *Nagoya Math. Jour.* 10 (1956), 125–127. **Zbl** 71, p. 71. **MR** 18, p. 292.

KAZARINOFF, N. D.

Analytic Inequalities, Holt, Rinehart and Winston, Inc. (1964), New York. **MR** 37 #3451.

KELDYCH, M. and LAVRENTIEFF, M.

"Sur un problème de M. Carleman," *C. R. Acad. Sci. URSS* (2) 23 (1939), 746–748. **FM** 65, p. 1227. **Zbl** 21, p. 335. **MR** 2, p. 82.

KELLOGG, O. D.

(a) "On some theorems of Bôcher concerning isolated singular points of harmonic functions," *Bull. Amer. Math. Soc.* 32 (1926), 664–668. **FM** 52, p. 496.

(b) "Recent progress with the Dirichlet problem," *Bull. Amer. Math. Soc.* 32 (1926), 601–625 and 34 (1928), 154. **FM** 52, p. 491 and 54, p. 507.

"Les moyennes arithmétiques dans la théorie du potentiel," *L'Enseignement Mathématique* 27 (1928), 14–26. **FM** 54, p. 502.

Foundations of Potential Theory, Verlag Julius Springer (1929), Berlin. Reprinted by Dover Publications, Inc. (1953), New York and by Springer-Verlag (1967), New York. **FM** 55, p. 282. **Zbl** 53, p. 73 and 152, p. 313. **MR** 36 #5369. **BAMS** 37, p. 141.

"Converses of Gauss' theorem on the arithmetic mean," *Trans. Amer. Math. Soc.* 36 (1934), 227–242. **FM** 60, p. 430. **Zbl** 9, p. 112.

KEMENY, J.

"The exponential function," *Amer. Math. Monthly* 64 (1957), 158–160. **Zbl** 77, p. 2. **MR** 18, p. 887.

KERÉKJÁRTÓ, B. V.

Vorlesungen über Topologie I, Flächentopologie. Verlag Julius Springer (1923), Berlin. **FM** 49, p. 396. **BAMS** 31, p. 176.

"Sur l'indice des transformations analytiques," *Acta Litt. ac Scient. Univ. Hung.* (Szeged) 7 (1934/35), 163–172. **FM** 61, p. 626 and 1353. **Zbl** 11, p. 407.

KHAJALIA, G.

"Sur la représentation conforme des aires doublement connexes sur l'anneau circulaire," *Trav. Inst. Math. Tbilissi* [also known as *Trudy Tbiliss. Mat. Inst.*] 1 (1937), 89–107 (Russian). **FM** 63, p. 982. **Zbl** 16, p. 407.

"Sur la représentation conforme des domaines doublement connexes," *Trav. Inst. Math. Tbilissi* [also known as *Trudy Tbiliss. Mat. Inst.*] 4 (1938), 123–133 (Russian). **FM** 64, p. 1074. **Zbl** 19, p. 126.

KHINTCHINE, A.

"Sur les suites de fonctions analytiques bornées dans leur ensemble," *Fund. Math.* 4 (1923), 72–75. **FM** 49, p. 212.

"On sequences of analytic functions," *Recueil Math. Moscou* (=*Mat. Sbornik*) 31 (1922–24), 147–151 (Russian). **FM** 48, p. 374.

KIM, K.-T. and KRANTZ, S. G.

"The automorphism groups of domains," *Amer. Math. Monthly* 112 (2005), 585–601. **Zbl** 1138.32302. **MR** 2006c:32024.

KIRSANOV, V. S.

"V. Ya Bunyakovskij's thesis and A. Cauchy's theory of residues," *Istoriko-mat. Issledovanija* 28 (1985), 261–266 (Russian). **Zbl** 584.01007. **MR** 87e:01016c.

KJELLBERG, B.

"On the growth of minimal positive harmonic functions in a plane region," *Ark. för Mat.* 1 (1950), 347–351. **Zbl** 40, p. 55. **MR** 12, p. 410.

KLEINER, W.

"Démonstration du théorème de Osgood-Carathéodory par la méthode des points extrémaux," *Annales Polonici Mathematici* 2 (1955), 67–72. **Zbl** 65, p. 310. **MR** 17, p. 473.

"Démonstration du théorème de Carathéodory par la méthode des points extrémaux," *Annales Polonici Mathematici* 11 (1962), 217–224. **Zbl** 111, p. 78. **MR** 25 #202.

KLEPPNER, A.

"Another proof of Ascoli's theorem," *Amer. Math. Monthly* 74 (1967), 315–316. **Zbl** 145, p. 162. **MR** 34 #8371.

KLINE, J. R.

"A new proof of a theorem due to Schönflies," *Proc. Nat. Acad. Sci. U. S. A.* 6 (1920), 529–531. **FM** 47, p. 519.

"What is the Jordan curve theorem?", *Amer. Math. Monthly* 49 (1942), 281–286. **Zbl** 61, p. 402. **MR** 3, p. 318.

KLINE, M.

Mathematical Thought from Ancient to Modern Times, Oxford University Press (2nd printing 1974). New York. **Zbl** 277.01001 and 864.01001. **MR** 57 #12010 and 91i: 01003a, b, c. **BAMS** 80, p. 805.

KLOOSTERMAN, H. D.

"Über die Pole auf dem Rande des Konvergenzgebiets gewisser Potenzreihen," *Math. Zeit.* 26 (1927), 733–743. **FM** 53, p. 288.

KNASTER, B. and KURATOWSKI, K.

"Sur les continus non-bornés. (Application de la méthode d'inversion)," *Fund. Math.* 5 (1924), 23–58. **FM** 50, p. 139.

KNASTER, B., KURATOWSKI, K. and MAZURKIEWICZ, S.

"Ein Beweis des Fixpunktsatzes für n-dimensionale Simplexe," *Fund. Math.* 14 (1929), 132–137. **FM** 55, p. 972. (See also **MR** 55 #5357 and **Zbl** 352.01010.)

KNESER, A.

"Arithmetische Begründung einiger algebraischer Fundamentalsätze," *Jour. für Reine und Angew. Math.* 102 (1887), 20–55. **FM** 19, p. 66.

"Die elementare Theorie der analytischen Funktionen und die komplexe Integration," *Sitzungsber. Bayer. Akad. Wissen. München* (1920), 65–81. **FM** 47, p. 267.

KNESER, H.

"Lösung der Aufgabe 41," *Jahresber. Deutsch. Math. Verein.* 35 (1926), 2te Abteilung, 123–124. **FM** 52, p. 498. (See also **Zbl** 692.57016.)

"Laplace, Gauss und der Fundamentalsatz der Algebra," *Deutsche Math.* 4 (1939), 318–322. **FM** 65, p. 49. **Zbl** 21, p. 196.

"Der Fundamentalsatz der Algebra und der Intuitionismus," *Math. Zeit.* 46 (1940), 287–302. **FM** 66, p. 58. **Zbl** 22, p. 301. **MR** 1, p. 322.

"Zur Stetigkeit der Wurzeln einer algebraischen Gleichung," *Math. Zeit.* 48 (1942), 101–104. **FM** 68, p. 34. **Zbl** 26, p. 385. **MR** 4, p. 273.

"Über den Beweis des Cauchyschen Integralsätzes bei streckbarer Randkurve," *Arch. der Math.* 1 (1949), 318–321. **Zbl** 31, p. 352. **MR** 11, p. 91.

Funktionentheorie, Studia Mathematica Bd. 13, Vandenhoeck & Ruprecht (1958; 2nd ed. 1966), Göttingen. **Zbl** 135, p. 283. **MR** 26 #6361. **BAMS** 65, p. 337.

KNESER, H. and ULLRICH, E.

"Funktionentheorie," chapter 12 of Part I of Pure Mathematics section of *FIAT Review of German Science 1939–1946,* (W. Süss, editor). Dieterich'sche Verlagsbuchhandlung (1948), Wiesbaden. **Zbl** 30, p. 152. **MR** 11, pp. 91 and 148.

KNESER, M.

"Ergänzung zu einer Arbeit von Hellmuth Kneser über den Fundamentalsatz der Algebra," *Math. Zeit.* 177 (1981), 285–287. **Zbl** 452.12007. **MR** 82i:03072. (See also **Zbl** 1054.03041.)

"Ein etwas anderer Zugang zur Exponentialfunktion," *Math. Semesterber.* 51 (2004), 225–229. **Zbl** 1057.26003. **MR** 2109654.

KNOEBEL, R. A.

"Exponentials reiterated," *Amer. Math. Monthly* 88 (1981), 235–252. **Zbl** 493.26007. **MR** 82e:26004. (See also **MR** 2005k:30053.)

KNOPP, K.

"Einheitliche Erzeugung und Darstellung der Kurven von Peano, Osgood und v. Koch," *Arch. der Math. und Physik* (3) 26 (1917), 103–115. **FM** 46, p. 400.

Theory and Application of Infinite Series, Blackie and Son, Ltd. (2nd ed. 1951), Glasgow (translated from German by R. C. H. Young). Reprinted by Dover Publications, Inc. (1989), New York. **FM** 54, p. 222. **Zbl** 1, p. 392, 42 p. 292 and 124, p. 283. **MR** 17, p. 1074 and 32 #1473. **BAMS** 36, p. 614 and 37, p. 810.

Problem Book in the Theory of Functions, Vol. I (translated from German by L. Bers, 1948), Vol. II (translated from German by F. Bagemihl, 1953), Dover Publications, Inc. New York. **FM** 57, p. 341 and 68, p. 159. **Zbl** 3, p. 62; 32, p. 65 and 27, p. 305. **MR** 10, p. 288 and 21 #4225. **BAMS** 31, p. 183; 35, p. 415; 38, p. 471.

KOCH, H., VON

"Quelques théorèmes sur les fonctions entières," *Öfversigt af Kongl. Svenska Vetenskaps-Akademiens Förhandlingar* 58 (1901), 405–413. **FM** 32, p. 411.

"Sur une classe remarquable de fonctions entières et transcendantes," *Arkiv för Mat., Astro. och Fysik* 1 (1903–04), 205–208. **FM** 34, p. 456 and 35, p. 417.

KOCHOL, M.

"An elementary proof of the fundamental theorem of algebra," *Internat. Jour. of Math. Ed. Sci. Tech* 30 (1999), 614–615. **Zbl** 1061.30034.

KOEBE, P.

(a) "Über konforme Abbildung mehrfach zusammenhängender ebener Bereiche, insbesondere solcher Bereiche, deren Begrenzung von Kreisen gebildet wird," *Jahresber. Deutsch. Math. Verein.* 15 (1906), 142–153. **FM** 37, p. 684.

(b) "Herleitung der partiellen Differentialgleichung der Potentialfunktion aus deren Integraleigenschaft," *Sitzungsber. Berlin Math. Gesell.* 5 (1906), 39–42. **FM** 37, p. 384.

"Über die Uniformisierung beliebiger analytischer Kurven," *Göttingen Nachr.* (1907), 191–210. Zweite Mitteilung, *ibid.* (1907), 633–669. Dritte Mitteilung, *ibid.* (1908), 337–358. Vierte Mitteilung, *ibid.* (1909), 324–361. **FM** 38, pp. 454 and 455; 40, pp. 467 and 468.

(a) "Über die Uniformisierung der algebraischen Kurven durch automorphe Funktionen mit imaginärer Substitutionsgruppe," *Göttingen Nachr.* (1909), 68–76. **FM** 40, p. 468.

(b) "Über die Uniformisierung der algebraischen Kurven. I," *Math. Annalen* 67 (1909), 145–224. **FM** 40, p. 470. II, *ibid.* 69 (1910), 1–81. **FM** 41, p. 480.

"Über die Hilbertsche Uniformisierungsmethode," *Göttingen Nachr.* (1910), 59–74. **FM** 41, p. 479.

"Über eine neue Methode der konformen Abbildung und Uniformisierung (Voranzeige)," *Göttingen Nachr.* (1912), 844–848. **FM** 43, p. 520.

(a) "Ränderzuordnung bei konformer Abbildung," *Göttingen Nachr.* (1913), 286–288. **FM** 44, p. 760.

(b) "Lösung der Randwertaufgabe der Potentialtheorie für Kreisring, Ellipse und Rechteck mittels des Poissonschen Integrals," *Ber. Verhandl. Kön. Sächs. Gesell. Wissen. Leipzig* 65 (1913), 210–213. **FM** 44, p. 759.

"Abhandlungen zur Theorie der konformen Abbildung. I. Die Kreisabbildung des allgemeinsten einfach und zweifach zusammenhängenden schlichten Bereichs und die Ränderzuordnung bei konformer Abbildung," *Jour. für Reine und Angew. Math.* 145 (1915), 177–223. **FM** 45, p. 667.

"Kontinuitätsbeweis des Fundamentalsatzes der Algebra," *Göttingen Nachr.* (1918), 45–53. **FM** 46, p. 558.

"Über das Schwarzsche Lemma und einige damit zusammenhängende Ungleichheitsbeziehungen der Potentialtheorie und Funktionentheorie," *Math. Zeit.* 6 (1920), 52–84. **FM** 47, p. 271.

"Hydrodynamische Potentialströmungen in mehrfach zusammenhängenden ebenen Bereichen im Zusammenhang mit der konformen Abbildung solcher Bereiche," *Ber. Verhandl. Sächs. Akad. Wissen., math.-phys. Kl.* 87 (1935), 287–318. **FM** 61, p. 1528. **Zbl** 14, p. 61.

KOENIGS, G.

"Recherches sur les substitutions uniformes," *Bull. Sci. Math.* (2) 7 (1883), 340–357. **FM** 15, p. 114.

KOETSIER, T.

"Etappen in der Entwicklungsgeschichte des sogenannten Vertauschbarkeitssatzes von Schwarz," pp. 59–78 of *Rechnen mit dem Unendlichen* (D. D. Spalt, editor), Birkhäuser Verlag (1990), Basel. **Zbl** 729.01003. **MR** 95a:01003.

KOMATU, Y.

(a) "Ein alternierendes Approximationsverfahren für konforme Abbildung von einem Ringgebiete auf einen Kreisring," *Proc. Japan Acad.* 21 (1945), 146–155. **Zbl** 60, p. 236. **MR** 11, p. 341.

(b) "Sur la représentation de Villat pour les fonctions analytiques définies dans un anneau circulaire concentrique," *Proc. Imper. Acad. Tokyo* 21 (1945), 94–96. **Zbl** 60, p. 236. **MR** 7, p. 286.

"Existence theorem of conformal mapping of double-connected domains," *Kōdai Math. Sem. Reports,* No. 5–6 (1949), 3–4. **Zbl** 45, p. 41. **MR** 11, p. 590.

"On angular derivative," *Kōdai Math. Sem. Reports* 13 (1961), 167–179. **Zbl** 119, p. 62. **MR** 25 #2210.

"Topics from classical theory of conformal mapping," pp. 165–183 of *Complex Analysis,* Banach Center Publications, Vol. 11. PWN-Polish Scientific Publishers (1983), Warsaw. **Zbl** 549.30005. **MR** 85j:30005.

KOMMERELL, K.

"Der Fundamentalsatz der Algebra," *Arch. der Math. und Physik* (3) 13 (1908), 99–102. **FM** 39, p. 119.

KÖNIG, H.

"Über die Landausche Verschärfung des Schottkyschen Satzes," *Arch. der Math.* 8 (1957), 112–114. **Zbl** 77, p. 285. **MR** 19, p. 1045.

"Ein funktionalanalytischer Beweis des Approximationssatzes von Walsh," *Jour. für Reine und Angew. Math.* 274/275 (1975), 158–163. **Zbl** 327.30037. **MR** 51 #10643.

"On a property of convex functions," p. 471 of *General Inequalities 6* (W. Walter, editor), International Series in Numerical Mathematics, Vol. 103. Birkhäuser Verlag (1992), Basel. **Zbl** 765.26003. **MR** 93k:26001.

KÖNIG, J.

"Die Factorenzerlegung ganzer Functionen und damit zusammenhängende Eliminationsprobleme," *Math. Annalen* 15 (1879), 161–173. **FM** 11, p. 62.

"Über eine Eigenschaft der Potenzreihen," *Math. Annalen* 23 (1884), 447–449. **FM** 16, p. 202.

KÔNO, N.

"Démonstration probabiliste du théorème de d'Alembert," pp. 207–208 of *Séminaire de Probabilités XIX 1983/84 Proceedings,* Lecture Notes in Mathematics 1123, Springer-Verlag (1985), New York. **Zbl** 558.60012. **MR** 88h:30005. (See also **MR** 2006i:30006.)

KOOSIS, P.

Introduction to H_p Spaces, London Mathematical Society Lecture Note Series 40, Cambridge University Press (1980). **Zbl** 435.30001. **MR** 81c:30062.

KOPFERMANN, K.

"Ein kurzer Beweis des Cauchyschen Integralsatzes," *Math. Semesterber.* 28 (1981), 1–10. **Zbl** 501.30002. **MR** 82d:30036.

KOREVAAR, J.

"Polynomial and rational approximation in the complex domain," pp. 251–292 of *Aspects of Contemporary Complex Analysis* (D. A. Brannan and J. G. Clunie, editors), Academic Press (1980), London and New York. **Zbl** 496.30033. **MR** 82j:30053.

"On Newman's quick way to the prime number theorem," *Math. Intelligencer* 4 (1982), 108–115. **Zbl** 496.10027. **MR** 84b:10063.

KOREVAAR, J. and MEYERS, J. L. H.

"Logarithmic convexity for supremum norms of harmonic functions," *Bull. Lon. Math. Soc.* 26 (1994), 353–362. **Zbl** 819.31001. **MR** 95m:31008.

KORKINE, A.

"Sur l'impossibilité de la relation algébrique $X^n + Y^n + Z^n = 0$," *C. R. Acad. Sci. Paris* 90 (1880), 303–304. **FM** 12, p. 134. (See also **FM** 14, p. 134.)

KÖRNER, T. W.

"The behavior of power series on their circle of convergence," pp. 54–94 of *Banach Spaces, Harmonic Analysis and Probability Theory (Storrs, Conn. 1980/1981),* Lecture Notes in Mathematics 995, Springer-Verlag (1983), New York. **Zbl** 518.30005. **MR** 84j:30005.

"On the fundamental theorem of algebra," *Amer. Math. Monthly* 113 (2006), 347–348. **Zbl** 1158.26306.

KORTRAM, R. A.

"Another computation of $\int_0^\infty e^{-u^2}\,du$," *Elemente der Math.* 48 (1993), 170–172. **Zbl** 875.26007.

"Simple proofs for $\sum \frac{1}{k^2} = \frac{\pi^2}{6}$ and $\sin x = x \prod(1 - \frac{x^2}{k^2\pi^2})$," *Math. Mag.* 69 (1996), 122–125. **Zbl** 860.40001.

KOSMODEM'YANSKII, A. A., JR.

"A converse of the mean value theorem for harmonic functions," *Russian Math. Surveys* 36, No. 5 (1981), 159–160. **Zbl** 548.31002. **MR** 84d:31001.

KOVANKO, A.

"Sur les suites de fonctions à une variable complexe," *C. R. Acad. Sci. Paris* 179 (1924), 873–875. **FM** 50, p. 245.

KOWALEWSKI, G.

"Ein funktionentheoretischer Satz Jacobis," *Jahresber. Deutsch. Math. Verein.* 27 (1918), 53–55 and 160. **FM** 46, p. 537.

KOWALSKY, H.-J.

"Zum Fundamentalsatz der Algebra," *Abhand. Braunsch. Wissen. Gesell.* 35 (1983), 111–120. **Zbl** 538.12010. **MR** 86d:12004.

KRAFFT, M.

"Der Satz von der Gebietstreue," *Jour. für Reine und Angew. Math.* 167 (1932), 388–389. **FM** 58, p. 356. **Zbl** 3, p. 210.

"Elementare Ermittlung des Wertes des Integrals $\int_0^\infty e^{-x^2} dx$," *Jahresber. Deutsch. Math. Verein.* 57 (1955), 31–33. **Zbl** 57, p. 292. **MR** 15, p. 609.

"Elementare Bestimmung des Wertes des Wahrscheinlichkeitsintegrals," *Math.-Phys. Semesterber.* 5 (1956), 120–122. **Zbl** 74, p. 290. **MR** 18, p. 368.

KRAJKIEWICZ, P.

"A new proof of a theorem of Saxer," *Annales Polonici Mathematici* 33 (1977), 293–309. **Zbl** 348.30011. **MR** 55 #8344.

KRANTZ, S. G.

"Characterization of smooth domains in $\mathbb{C}$ by their biholomorphic self-maps," *Amer. Math. Monthly* 90 (1983), 555–557. **Zbl** 524.30007. **MR** 84m:30014.

Complex Analysis: The Geometric Viewpoint, Carus Mathematical Monographs, No. 23, Mathematical Association of America (1990), Washington, D. C. **Zbl** 743.30002. **MR** 2004m:30001.

"Two results on uniqueness of conformal mappings," *Complex Variables* 50 (2005), 427–432. **Zbl** 1083.30010. **MR** 2005m:30009.

KRASNOSEL'SKIĬ, M. A. and OPOJTSEV, V. I.

"Theorem on global homeomorphism," *Teor. Funkts. Funkts. Anal. Prilozh.* 30 (1978), 83–85 (Russian). **Zbl** 449.54009. **MR** 80g:54023.

KRAZER, A.

Lehrbuch der Thetafunktionen, B. G. Teubner (1903), Leipzig. **FM** 34, p. 492. **BAMS** 11, p. 375. Reprinted by Chelsea Publishing Co. (1970), New York. **Zbl** 212, p. 429.

KREIN, M. G. and NUDEL'MAN, A. A.

"On the representation of entire functions positive on the real axis, or on a semi-axis, or outside a finite interval," *American Mathematical Society Translations. Series 2,* Vol. 127 (1985), 17–32. American Mathematical Society, Providence. **Zbl** 583.30026. **MR** 82g:30050. (See also **Zbl** 698.26005 and **MR** 92b:41012.)

KRONECKER, L.

"Zwei Sätze über Gleichungen mit ganzzahligen Coefficienten," *Jour. für Reine und Angew. Math.* 53 (1857), 173–175.

"Zur Theorie der Elimination einer Variabeln aus zwei algebraischen Gleichungen," *Monatsber. Kön. Preuss. Akad. Wiss. Berlin* (1881), 535–600. **FM** 13, p. 114.

(a) "Summirung der Gauss'schen Reihen $\sum_{h=0}^{h=n-1} e^{2h^2\pi i/n}$," *Jour. für Reine und Angew. Math.* 105 (1889), 267–268. **FM** 21, p. 251.

(b) "Bemerkung über die Darstellung von Reihen durch Integrale," *Jour. für Reine und Angew. Math.* 105 (1889), 157–159 and 345–354. **FM** 21, p. 281.

"Über die Dirichlet'sche Methode der Werthbestimmung der Gauss'schen Reihen," *Festschrift der Math. Gesell. Hamburg* (1890), 32–36. **FM** 22, p. 205.

Vorlesungen über die Theorie der einfachen und der vielfachen Integrale (E. Netto, editor), B. G. Teubner (1894), Leipzig. **FM** 25, p. 34.

KRYLOV, V. I.

"On functions regular in a half-plane," *Recueil Math. Moscou* (=*Mat. Sbornik*) N. S. 6 (1939), 95–137; *American Mathematical Society Translations. Series 2,* Vol. 32 (1963), 37–81. **FM** 65, p. 337. **Zbl** 22, p. 363. **MR** 1, p. 308.

KRZY.Z, J. G.

Problems in Complex Variable Theory, Modern Analytic and Computational Methods in Science and Mathematics, Number 36. American Elsevier Publishing Co. (1971), New York. **Zbl** 152, p. 59 and 239.30001. **MR** 56 #5844.

KULKARNI, S. H.

"A relatively elementary proof of a characterization of polynomials," *Indian Jour. Pure & Appl. Math.* 38 (2007), 229–230. **Zbl** 1137.30303. **MR** 2008f:30012.

KUNUGUI, K.

"Sur l'allure d'une fonction analytique uniforme au voisinage d'un point frontière de son domaine de définition," *Japan. Jour. of Math.* 18 (1941–43), 1–39. **Zbl** 60, p. 229. **MR** 8, p. 24.

Kuran, Ü.

(a) "Classes of subharmonic functions in $\mathbf{R}^n \times (0, \infty)$," *Proc. Lon. Math. Soc.* (3) 16 (1966), 473–492. **Zbl** 142, p. 87. **MR** 34 #2917.

(b) "Generalizations of a theorem on harmonic functions," *Jour. Lon. Math. Soc.* 41 (1966), 145–152. **Zbl** 138, p. 364. **MR** 33 #298.

"Harmonic majorization in half-balls and half-spaces," *Proc. Lon. Math. Soc.* (3) 21 (1970), 614–636. **Zbl** 207, p. 416. **MR** 47 #3697.

"On the mean-value property of harmonic functions," *Bull. Lon. Math. Soc.* 4 (1972), 311–312. **Zbl** 257.31007. **MR** 47 #8887. (See also **MR** 52 #8464.)

"Some extension theorems for harmonic, superharmonic and holomorphic functions," *Jour. Lon. Math. Soc.* (2) 22 (1980), 269–284. **Zbl** 446.31004. **MR** 82a:31006.

Kuratowski, K.

"Sur les continus de Jordan et le théorème de M. Brouwer," *Fund. Math.* 8 (1926), 137–150. **FM** 52, p. 599.

"Théorèmes sur l'homotopie des fonctions continues de variable complexe et leurs rapports à la théorie des fonctions analytiques," *Fund. Math.* 33 (1945), 316–367. **Zbl** 60, p. 415. **MR** 8, p. 50.

"Remark on an invariance theorem," *Fund. Math.* 37 (1950), 251–252. **Zbl** 39, p. 394. **MR** 13, p. 150.

"Sur quelques invariants topologiques dans l'espace euclidien," *Jour. de Math. Pures et Appliq.* (9) 36 (1957), 191–200. **Zbl** 78, p. 365. **MR** 19, p. 970.

"Sur l'extension de la notion de fonction rationnelle à l'espace euclidien n-dimensionnel," *Bull. Acad. Polon. Sci.* 6 (1958), 281–287. **Zbl** 89, p. 389. **MR** 20 #3532.

"Un critère de coupure de l'espace euclidien par un sous-ensemble arbitraire," *Math. Zeit.* 72 (1959), 88–94. **Zbl** 88, p. 383. **MR** 21 #6574.

Topology, Vol. II (translated from French by A. Kirkor) Academic Press (1968), New York. **Zbl** 102, p. 376. **MR** 41 #4467. **BAMS** 58, p. 265.

Introduction to Set Theory and Topology (translated from Polish by L. F. Boron), Pergamon Press (2nd edition, 1972), New York. **Zbl** 247.54001 and 267.54002. **MR** 58 #30948a, b.

Kythe, P. K.

Computational Conformal Mapping, Birkhäuser Verlag (1998), Boston & Basel. **Zbl** 918.30001. **MR** 99k:65027.

L

LABELLE, G.

"On the theorems of Gauss-Lucas and Grace," *Ann. Univ. Mariae Curie-Skłodowska, Sect. A* 20 (1966), 5–21 (1971). **Zbl** 266.30002. **MR** 46 #7487.

LACZKOVICH, M.

"On Lambert's proof of the irrationality of π," *Amer. Math. Monthly* 104 (1997), 439–443. **Zbl** 876.11032. **MR** 98a:11090.

LAGRANGE, J.

"Calcul des intégrales $I_m = \int_0^\infty \frac{\sin x^m}{x^a} dx$, $J_m = \int_0^\infty \frac{\cos x^m}{x^a} dx$, m entier positif, a quelconque," *Mathesis* 66 (1957), 363–369. **Zbl** 85, p. 43. **MR** 20 #3240.

LAI, W.-T.

"Über den Satz von Schottky," *Sci. Record* (N. S.) 3 (1959), 381–384. **Zbl** 102, p. 296. **MR** 23 #A1802.

LAMMEL, E.

"Eine Bemerkung zum Satz von Vitali über Konvergenz von Funktionenfolgen," *Math. Nachr.* 18 (1958), 309–312. **Zbl** 81, p. 297. **MR** 20 #2423.

LANDAU, E.

"Über eine Verallgemeinerung des Picardschen Satzes," *Sitzungsber. Kön. Preuss. Akad. Wissen. Berlin* (1904), 1118–1133. **FM** 35, p. 401.

"On a familiar theorem of the theory of functions," *Bull. Amer. Math. Soc.* 12 (1905/06), 155–156. **FM** 37, p. 416.

"Über den Picard'schen Satz," *Vierteljahrsschr. Naturforsch. Gesell. Zürich* 51 (1906), 252–318. **FM** 37, p. 418.

(a) "Über die Approximation einer stetigen Funktion durch eine ganze rationale Funktion," *Rend. Circ. Mat. Palermo* 25 (1908), 337–346. **FM** 39, p. 472.

(b) "Beiträge zur analytischen Zahlentheorie," *Rend. Circ. Mat. Palermo* 26 (1908), 169–302. **FM** 39, p. 267.

"Über eine Aufgabe aus der Funktionentheorie," *Tôhoku Math. Jour.* 5 (1914), 97–116. **FM** 45, p. 640.

"Neuer Beweis eines Hardyschen Satzes," *Arch. der Math. und Physik* (3) 25 (1917), 173–178. **FM** 46, p. 536.

"Über die Blaschkesche Verallgemeinerung des Vitalischen Satzes," *Ber. Verhandl. Sächs. Gesell. Wissen. Leipzig* 70 (1918), 156–159. **FM** 46, p. 517 and 47, p. 311.

(a) "Zum Koebeschen Verzerrungssatz," *Rend. Circ. Mat. Palermo* 46 (1922), 347–348. **FM** 48, p. 406.

(b) "Bemerkung zu der Arbeit des Herrn Bieberbach 'Über die Verteilung der Null - und Einsstellen analytischer Funktionen'," *Math. Annalen* 86 (1922), 158–160. **FM** 48, p. 324.

"Der Picard-Schottkysche Satz und die Blochsche Konstante," *Sitzungsber. Preuss. Akad. Wissen. Berlin, Phys. -math. Kl.* (1926), 467–474. **FM** 52, p. 324.

(a) *Darstellung und Begründung einiger neuerer Ergebnisse der Funktionentheorie,* Verlag Julius Springer (2nd ed. 1929), Berlin. Reprinted by Chelsea Publishing Co. (1946 and 1960), New York. Third edition with two appendices by D. Gaier, Springer-Verlag (1986), New York. **FM** 55, p. 171. **Zbl** 601.30001. **MR** 88d:01046. **BAMS** 37, p. 656.

(b) "Über die Blochsche Konstante und zwei verwandte Weltkonstanten," *Math. Zeit.* 30 (1929), 608–634. **FM** 55, p. 770.

(c) "Über schlichte Funktionen," *Math. Zeit.* 30 (1929), 635–638. **FM** 55, p. 187.

(d) "Über die Carathéodorysche Verschärfung des grossen Picardschen Satzes," *Math. Zeit.* 30 (1929), 208–210 and 796. **FM** 55, p. 193.

(e) "Bestimmung der genauen Konstanten in einem Koebeschen Hilfssatz," *Jour. für Reine und Angew. Math.* 161 (1929), 135–136. **FM** 55, p. 211.

"Über den Millouxschen Satz," *Göttingen Nachr.* (1930), 1–9. **FM** 56, p. 270.

"Ausgewählte Kapitel der Funktionentheorie," *Trav. Inst. Math. Tbilissi* [also known as *Trudy Tbiliss. Mat. Inst.*] 8 (1940), 23–68. **FM** 66, p. 350. **Zbl** 24, p. 420. **MR** 3, p. 78. (See also **Zbl** 669.01024.)

Differential and Integral Calculus (translated from German by M. Hausner and M. Davis), Chelsea Publishing Co. (1950), New York. **FM** 60, p. 167. **Zbl** 8, p. 303 and 42, p. 55. **MR** 2002j:01021. **BAMS** 41, p. 317.

Foundations of Analysis (translated from German by F. Steinhardt), Chelsea Publishing Co. (1951), New York. **FM** 56, p. 191. **Zbl** 42, p. 278 and 198, p. 385. **MR** 22 #661. (See also **Zbl** 319.00001 and **MR** 57 #6314; **Zbl** 349.04002 and **MR** 54 #9987; **Zbl** 402.68064 and **MR** 58#32124.)

Elementary Number Theory (translated from German by J. E. Goodman), Chelsea Publishing Co. (1958), New York. **FM** 53, p. 123. **Zbl** 79, p. 62. **MR** 19, p. 1159.

LANDAU, E. and VALIRON, G.

"A deduction from Schwarz's Lemma," *Jour. Lon. Math. Soc.* 4 (1929), 162–163. **FM** 55, p. 769.

LANDAU, E. and TOEPLITZ, O.

"Über die grösste Schwankung einer analytischen Funktion in einem Kreise," *Arch. der Math. und Physik* (3) 11 (1907), 302–307. **FM** 38, p. 433.

LANDAU, H. J.

"On canonical conformal maps of multiply connected domains," *Trans. Amer. Math. Soc.* 99 (1961), 1–20. **Zbl** 102, p. 68. **MR** 22 #12212.

LANDAU, H. J. and OSSERMAN, R.

"Some distortion theorems for multivalent mappings," *Proc. Amer. Math. Soc.* 10 (1959), 87–91. **Zbl** 99, p. 59. **MR** 21 #3553.

"On analytic mappings of Riemann surfaces," *Jour. d'Analyse Math.* 7 (1959–60), 249–279. **Zbl** 101, p. 55. **MR** 23 #A311.

LANDSBERG, G.

"Zur Theorie der Gauss'schen Summen und der linearen Transformation der Thetafunctionen," *Jour. für Reine und Angew. Math.* 111 (1893), 234–253. **FM** 25, p. 283.

LANG, S.

"Old and new conjectured Diophantine inequalities," *Bull. Amer. Math. Soc.* (2) 23 (1990), 37–75. **Zbl** 714.11034. **MR** 90k:11032. (See also **Zbl** 794.11015, **MR** 94g:11044.)

LANGE, R. AND WALSH, R. A.

"A heuristic for the Poisson integral for the half-plane and some caveats," *Amer. Math. Monthly* 92 (1985), 356–358.

LÄNGER, H.

"An elementary proof of the convergence of iterated exponentials," *Elemente der Math.* 51 (1996), 75–77. **Zbl** 865.26005.

LANGER, J.

"Rational diffeomorphisms of the circle," *Applied Math. Letters* 20 (2007), 876–879. **Zbl** 1142.30333. **MR** 2008b:30057.

LANGLEY, J. K.

"Proof of a conjecture of Hayman concerning f' and f''," *Jour. Lon. Math. Soc.* (2) 48 (1993), 500–514. **Zbl** 789.30020. **MR** 94k:30075. (See also **MR** 2001c:30035.)

LAUGWITZ, D.

Bernhard Riemann 1826–1866: Turning Points in the Conception of Mathematics, (translated from German by A. Shenitzer), Birkhäuser Verlag (1999), Boston-Basel-Berlin. **Zbl** 931.01027. **MR** 2000b:01012. **BAMS** (2) 37, p. 477.

LAUGWITZ, D. and NEUENSCHWANDER, E.

"Riemann and the Cauchy-Hadamard formula for the convergence of power series," *Historia Math.* 21 (1994), 64–70 and *Math. Semesterber.* 40 (1993), 115–120. **Zbl** 797.01005 and 798.01005. **MR** 94b:01013 and 96g:01025.

LAURENT, H.

"Sur les séries de polynômes," *Jour. de Math. Pures et Appliq.* (5) 8 (1902), 309–328. **FM** 33, p. 410.

LAURENT, P. A.

"Extension du théorème de M. Cauchy relatif à la convergence du développement d'une fonction suivant les puissances ascendantes de la variable," *C. R. Acad. Sci. Paris* 17 (1843), 348–349.

LAURITZEN, S

"On entire transcendental functions which approach a definite limit along every ray from the origin," *Mat. Tidsskr. B* (1950), 42–48 (Danish). **Zbl** 40, p. 35. **MR** 12, p. 326.

LAVRENTIEFF, M. A.

Sur les fonctions d'une variable complexe représentables par des séries de polynômes, Actualités Scientifiques et Industrielles, No. 441, Hermann et Cie. (1936), Paris. **FM** 62, p. 1205. **Zbl** 17, p. 206. **BAMS** 44, p. 22.

LAVRENTIEFF, M. A. and KWESSELAWA, D.

"Über einen Ostrowskischen Satz," *Mitt. Georg. Abt. Akad. Wiss. USSR* 1 (1940), 171–174 (Russian). **FM** 66, p. 367. **Zbl** 23, p. 139. **MR** 2, p. 83.

LAVRENTIEFF, M. M.

"Numerical estimates in interior theorems of uniqueness," *Dokl. Akad. Nauk SSSR (N.S.)* 110 (1956), 731–734 (Russian). **Zbl** 72, p. 70. **MR** 18, p. 728. (See also **Zbl** 81, p. 69 and **MR** 20 #1759.)

LAX, P.

"The Cauchy integral theorem," *Amer. Math. Monthly* 114 (2007), 725–727 and 115 (2008), 276. **Zbl** 1134.30027.

LAZER, A. C.

"From the Cauchy-Riemann equations to the fundamental theorem of algebra," *Math. Mag.* 79 (2006), 210–213. **Zbl** 1230.30005.

LEAU, L.

"Nouvelle démonstration du théorème de d'Alembert," *L'Enseignement Mathématique* 11 (1909), 110–118. **FM** 40, p. 119.

"Cas extrême d'une intégrale de Cauchy et limite de certaines intégrales," *Jour. de Math. Pures et Appliq.* (9) 5 (1926), 211–218. **FM** 52, p. 289.

Les suites de fonctions en général (domaine complexe), Mémorial des Sciences Mathématiques, Fasc. 59, Gauthier-Villars (1932), Paris. **FM** 58, p. 301. **Zbl** 6, p. 352.

LEBESGUE, H.

"Sur l'approximation des fonctions," *Bull. Sci. Math.* (2) 22 (1898), 278–287. **FM** 29, p. 352.

"Sur le problème de Dirichlet," *Rend. Circ. Mat. Palermo* 24 (1907), 371–402 and 412. **FM** 38, p. 392.

"Sur la représentation approchée des fonctions," *Rend. Circ. Mat. Palermo* 26 (1908), 325–328. **FM** 29, p. 473.

"Sur le problème de Dirichlet," *C. R. Acad. Sci. Paris* 154 (1912), 335–337. **FM** 43, p. 368.

"Sur le théorème de la moyenne et le problème de Dirichlet," *C. R. des Séances Soc. Math. France* 48 (1920), 26.

"Sur les singularités des fonctions harmoniques," *C. R. Acad. Sci. Paris* 176 (1923), 1097–1099 and 1270–1271. **FM** 49, p. 338.

LEBORGNE, D.

Calcul différentiel complexe, Presses Universitaires de France (1991), Paris. **Zbl** 731.30001. **MR** 1091546.

LECORNU, L.

"Sur les séries entières," *C. R. Acad. Sci. Paris* 104 (1887), 349–352. **FM** 19, p. 228.

LEHMER, D. H.

"A new approach to Bernoulli polynomials," *Amer. Math. Monthly* 95 (1988), 905–911 and 96 (1989), 510. **Zbl** 663.10009. **MR** 90c:11014.

LEHNER, J.

"The Picard theorems," *Amer. Math. Monthly* 76 (1969), 1005–1012. **Zbl** 186, p. 394. **MR** 40 #2866.

LEHTO, O.

"Anwendung orthogonaler Systeme auf gewisse funktionentheoretische Extremal- und Abbildungsprobleme," *Ann. Acad. Scient. Fennicae, Ser. A. I.* Nr. 59 (1949). **Zbl** 35, p. 56. **MR** 11, p. 170.

LEIBMAN, G.

"A nonstandard proof of the Fundamental Theorem of Algebra," *Amer. Math. Monthly* 112 (2005), 705–712. **Zbl** 1159.12303. **MR** 2167770.

LEINFELDER, H.

"Eine Ergänzung zu meiner Note 'Zum Fundamentalsatz der Algebra'," *Didaktik der Math.* 11 (1983), 329–331.

LEJA, F.

"Construction de la fonction analytique effectuant la représentation conforme d'un domaine plan quelconque sur le cercle," *Math. Annalen* 111 (1935), 501–504. **FM** 61, p. 356. **Zbl** 12, p. 214.

"Sur une suite de polynômes et la représentation conforme d'un domaine plan quelconque sur le cercle," *Ann. Soc. Polon. Math.* 14 (1936), 116–134. **FM** 62, p. 1216. **Zbl** 18, p. 260.

"Une méthode élémentaire de résolution du problème de Dirichlet dans le plan," *Ann. Soc. Polon. Math.* 23 (1950), 230–245. **Zbl** 39, p. 325. **MR** 12, p. 703.

"Polynômes extrémaux et la représentation conforme des domaines doublement connexes," *Annales Polonici Mathematici* 1 (1954), 13–28. **Zbl** 56, p. 74. **MR** 16, p. 348.

"Construction of the function mapping conformally an arbitrary simply connected domain upon a circle," *Zastosowania Mat.* 2 (1955), 117–122 (Polish). **Zbl** 64, p. 325. **MR** 16, p. 917.

LELAND, K. O.

"Topological analysis of analytic functions," *Fund. Math.* 56 (1964–65), 157–182. **Zbl** 145, p. 80. **MR** 30 #2157.

"A polynomial approach to topological analysis, II, III," *Jour. Approx. Theory* 4 (1971), 6–12; 433–441. **Zbl** 212, p. 98; 235.30047. **MR** 55 #8374, 8375.

LELONG-FERRAND, J.

Représentation conforme et transformations à l'intégrale de Dirichlet bornée, Cahiers scientifiques, fasc. XXII. Gauthier-Villars (1955), Paris. **Zbl** 64, p. 322. **MR** 16, p. 1096.

LENARD, A.

"A note on Liouville's theorem," *Amer. Math. Monthly* 93 (1986), 200–201. **Zbl** 604.30032. **MR** 87f:30072.

"A nonzero complex sequence with vanishing power-sums," *Proc. Amer. Math. Soc.* 108 (1990), 951–953. **Zbl** 685.40002. **MR** 90i:40007. (See also **MR** 92m: 40003.)

LEONARD, I. E.

"More on Fresnel integrals," *Amer. Math. Monthly* 95 (1988), 431–433. **Zbl** 645.26003. **MR** 89m:26002.

LERCH, M.

"Über die Gauss'schen Summen," *Časopis pro Pěstování Matematiky a Fysiki* 28 (1898), 1–24. (Böhmisch). **FM** 29, p. 168.

(a) "Zur Theorie der Gauss'schen Summen," *Math. Annalen* 57 (1903), 554–567. **FM** 34, p. 228.

(b) "Démonstration élémentaire de la formule $\frac{\pi^2}{\sin^2 \pi x} = \sum_{\nu=-\infty}^{\infty} \frac{1}{(x+\nu)^2}$," *L'Enseignement Mathématique* 5 (1903), 450–453. **FM** 34, p. 311.

LEVAVASSEUR, R.

"Sur la représentation conforme de deux aires planes à connexion multiple, d'après M. Schottky," *Annales de la Faculté des Sciences de Toulouse* (2) 4 (1902), 45–100. **FM** 33, p. 707.

LEVEQUE, W. J.

Reviews in Number Theory, 6 volumes, American Mathematical Society (1974), Providence. **Zbl** 287.10001. **MR** 50 #2040–2045.

LEVI, E. E.

"Sopra una proprietà caratteristica delle funzioni armoniche," *Rend. Accad. d. Lincei Roma* (5) 18, part 1 (1909), 10–15. **FM** 40, p. 452.

LEVI, M.

"Riemann mapping theorem by steepest descent," *Amer. Math. Monthly* 114 (2007), 246–251. **Zbl** 1132.30300. **MR** 2007j:30001.

LEVI-CIVITA, T.

"Sulla continuazione analitica," *Atti e Mem. della R. Accad. Scienze, Lettere ed Arti Padova* 28 (1912), 3–5. **FM** 43, p. 490.

LEVIN, B. YA.

Distribution of Zeros of Entire Functions, Translations of Mathematical Monographs, Vol. 5, American Mathematical Society (rev. ed. 1980), Providence. **Zbl** 152, p. 67. **MR** 81k:30011.

Lectures on Entire Functions, Translations of Mathematical Monographs, Vol. 150, American Mathematical Society (1996), Providence. **Zbl** 856.30001. **MR** 97j:30001.

LEVIN, V. I.

"Ein Beitrag zu dem Milloux-Landauschen Satz," *Jahresber. Deutsch. Math. Verein.* 44 (1934), 262–265. **FM** 60, p. 255. **Zbl** 10, p. 308.

LEVRIE, P. and DAEMS, W.

"Evaluating the probability integral using Wallis's product formula for π," *Amer. Math. Monthly* 116 (2009), 538–541. **Zbl** 1229.26022. **MR** 2010h:60066.

LÉVY, P.

"Sur une généralisation des théorèmes de MM. Picard, Landau et Schottky," *C. R. Acad. Sci. Paris* 153 (1911), 658–660. **FM** 42, p. 427.

"Remarques sur le théorème de M. Picard," *Bull. Soc. Math. France* 40 (1912), 25–39. **FM** 43, p. 506.

LEWIN, J.

"A truly elementary approach to the bounded convergence theorem," *Amer. Math. Monthly* 93 (1986), 395–397. See **Zbl** 0195.34302 and 0156.06201. (See also **MR** 1535791 and **MR** 97e:26006.)

"Some applications of the bounded convergence theorem for an introductory course in analysis," *Amer. Math. Monthly* 94 (1987), 988–993. **MR** 1541210.

LEWIS, C. J.

"The problem of Milloux for functions analytic in an open annulus," *Duke Math. Jour.* 25 (1958), 591–600. **Zbl** 116, p. 60. **MR** 20 #1783.

LEWIS, J. L.

"Picard's theorem and Rickman's theorem by way of Harnack's inequality," *Proc. Amer. Math. Soc.* 122 (1994), 199–206. **Zbl** 807.30010. **MR** 94k:30052. (See also **MR** 2005g:30028 and 2010f:30073.)

LICHTENSTEIN, L.

"Über die Integration eines bestimmten Integrals in Bezug auf einen Parameter," *Göttingen Nachr.* (1910), 468–475. **FM** 41, p. 331.

"Über die zweimalige Integration von Funktionen zweier reellen Veränderlichen," *Sitzungsber. Berlin Math. Gesell.* 10 (1911), 55–69. **FM** 42, p. 318.

"Über die erste Randwertaufgabe der Potentialtheorie," *Sitzungsber. Berlin Math. Gesell.* 15 (1916), 92–96. **FM** 46, p. 738.

"Neuere Entwicklung der Potentialtheorie. Konforme Abbildung," *Encyklopädie der Mathematischen Wissenschaften* Bd. II, 3rd Part, 1st Half, Heft 3, 181–377. B. G. Teubner (1919), Leipzig. **FM** 47, p. 450.

LIEBECK, H.

"The convergence of sequences with linear fractional recurrence relation," *Amer. Math. Monthly* 68 (1961), 353–355. **Zbl** 115, p. 277. **MR** 23 #A1181.

"A proof of the equality of column and row rank of a matrix," *Amer. Math. Monthly* 73 (1966), 1114. **Zbl** 143, p. 50. (See also **MR** 57 #2802 and **Zbl** 357.00001.)

LIN, C. Y.

"An elementary proof of the identity $\theta_{00}^4 = \theta_{10}^4 + \theta_{01}^4$," *Chinese Jour. of Math.* 7 (1979), 113–116. **Zbl** 429.14015. **MR** 84b:10036.

LINDELÖF, E.

"Une application de la théorie des résidus au prolongement analytique des séries de Taylor," *C. R. Acad. Sci. Paris* 135 (1902), 1315–1318 and *Jour. de Math. Pures et Appliq.* (5) 9 (1903), 213–221. **FM** 33, p. 409 and 34, p. 441.

"Sur la détermination de la croissance des fonctions entières définies par un développement de Taylor," *Bull. Sci. Math.* (2) 27 (1903), 213–226. **FM** 34, p. 440.

Le calcul des résidus et ses applications à la théorie des fonctions, Gauthier-Villars (1905), Paris. **FM** 36, p. 468. **BAMS** 12, p. 134. Reprinted by Chelsea Publishing Co. (1947), New York and Éditions Jacques Gabay (1989), Sceaux. **MR** 93f:30001.

"Sur un théorème de M. Hadamard dans la théorie des fonctions entières," *Rend. Circ. Mat. Palermo* 25 (1908), 228–234. **FM** 39, p. 466.

"Mémoire sur certaines inégalités dans la théorie des fonctions monogènes et sur quelques propriétés nouvelles de ces fonctions dans le voisinage d'un point singulier essentiel," *Acta Soc. Sci. Fennicae* 35, Nr. 7 (1909). **FM** 40, p. 439.

"Sur le théorème de M. Picard dans la théorie des fonctions monogènes," pp. 112–136 of *Compte rendu du congrès des mathématiciens tenu à Stockholm 22–25 septembre 1909.* B. G. Teubner (1910), Berlin and Leipzig. **FM** 41, p. 465.

"Démonstration nouvelle d'un théorème fondamental sur les suites de fonctions monogènes," *Bull. Soc. Math. France* 41 (1913), 171–178. **FM** 44, p. 489.

"Sur la représentation conforme," *C. R. Acad. Sci. Paris* 158 (1914), 245–247. **FM** 45, p. 665.

"Sur un principe général de l'analyse et ses applications à la théorie de la représentation conforme," *Acta Soc. Sci. Fennicae* 46, Nr. 4 (1915). **FM** 45, p. 665.

"Sur la représentation conforme d'une aire simplement connexe sur l'aire d'un cercle," pp. 59–90 of *Quatrième congrès des mathématiciens scandinaves à Stockholm 1916,* Almqvist & Wiksell (1920), Upsala. **FM** 47, p. 322.

LINDWART, E. and PÓLYA, G.

"Über einen Zusammenhang zwischen der Konvergenz von Polynomfolgen und der Verteilung ihrer Wurzeln," *Rend. Circ. Mat. Palermo* 37 (1914), 297–304. **FM** 45, p. 650.

LIOUVILLE, R. J.

"Sur l'impossibilité de la relation algébrique $X^n + Y^n - Z^n = 0$," *C. R. Acad. Sci. Paris* 89 (1879), 1108–1110. **FM** 11, p. 138.

LIPKA, S.

"Eine Verallgemeinerung des *Rouché*schen Satzes," *Jour. für Reine und Angew. Math.* 160 (1929), 143–150. **FM** 55, p. 175.

LIPMAN, J.

"A lemma on continuous functions," *Canad. Math. Bull.* 3 (1960), 186–187. **Zbl** 92, p. 154. **MR** 23 #A4115.

LITTLEWOOD, J. E.

"On inequalities in the theory of functions," *Proc. Lon. Math. Soc.* (2) 23 (1925), V-IX and 481–519; (2) 25 (1926), 337. **FM** 51, p. 247.

"Mathematical notes (6): On the definition of subharmonic function," *Jour. Lon. Math. Soc.* 2 (1927), 189–192. **FM** 53, p. 467.

"Mathematical notes (7): On functions subharmonic in a circle," *Jour. Lon. Math. Soc.* 2 (1927), 192–196. **FM** 53, p. 468.

"Mathematical notes (8): On functions subharmonic in a circle (II)," *Proc. Lon. Math. Soc.* (2) 28 (1928), 383–394. **FM** 54, p. 516.

"Mathematical notes (14): 'Every polynomial has a root'," *Jour. Lon. Math. Soc.* 16 (1941), 95–98. **FM** 67, p. 976. **Zbl** 28, p. 198. **MR** 3, p. 110.

Some Problems in Real and Complex Analysis, D. C. Heath and Co. (1968), Lexington, Mass. **Zbl** 185, p. 115. **MR** 39 #5777.

LIVERPOOL, L. S. O. and NNADI, J.

"The theorems of Schottky and Miranda –a novel approach," *Jour. Nigerian Math. Soc.* 2 (1983), 17–21. **Zbl** 558.30033. **MR** 86c:30065.

LLOYD, N. G.

Degree Theory, Cambridge Tracts in Mathematics, No. 73, Cambridge University Press (1978), Cambridge. **Zbl** 367.47001. **MR** 58 #12558. **BAMS** (2) 1, p. 632. (See also **MR** 85b:47004.)

LOEB, P. A.

"A note on Dixon's proof of Cauchy's Integral Theorem," *Amer. Math. Monthly* 98 (1991), 242–244 and 100 (1993), 680–681. **Zbl** 748.30030 and 790.30024. **MR** 92d:30001.

LOHWATER, A. J.

(a) "A uniqueness theorem for a class of harmonic functions," *Proc. Amer. Math. Soc.* 3 (1952), 278–279. **Zbl** 46, p. 326. **MR** 13, p. 743.

(b) "The boundary values of a class of meromorphic functions," *Duke Math. Jour.* 19 (1952), 243–252. **Zbl** 46, p. 300. **MR** 14, p. 34.

"Mathematical literature," *Library Trends* 15 (1966–67), 852–867.

LOOMAN, H.

"Über die Cauchy-Riemannschen Differentialgleichungen," *Göttingen Nachr.* (1923), 97–108. **FM** 49, p. 709.

"Bemerkung zur Definition der harmonischen Funktionen," *Christiaan Huygens* 3 (1923–24), 41–42. **FM** 50, p. 646.

"Über eine Erweiterung des Cauchy-Goursatschen Integralsatzes," *Nieuw Archief voor Wiskunde* (2) 14 (1922/1924), 234–239. **FM** 50, p. 250.

LOOMIS, L. H.

"The converse of Fatou's theorem for positive harmonic functions," *Trans. Amer. Math. Soc.* 53 (1943), 239–250. **Zbl** 61, p. 233. **MR** 4, p. 199.

"An elementary proof of the strong form of the Cauchy theorem," *Bull. Amer. Math. Soc.* 50 (1944), 831–833. **MR** 6, p. 122.

LOOMIS, L. H. and WIDDER, D. V.

"The Poisson integral representation of functions which are positive and harmonic in a half-plane," *Duke Math. Jour.* 9 (1942), 643–645. **Zbl** 61, p. 233. **MR** 4, p. 101.

LORD, N.

(a) "Generalising an integration 'joke'," *Math. Gazette* 87 (2003), 123–125.

(b) "An elementary single-variable proof of $\int_{-\infty}^{\infty} e^{-x^2/2}\,dx = \sqrt{2\pi}$," *Math. Gazette* 87 (2003), 308–311.

"Beyond the Basel problem: Euler's derivation of the general formula for $\zeta(2n)$," *Math. Gazette* 98 (2014), 459–474. **Zbl** 06766747.

"The most elementary proof that $\sum_{k=1}^{\infty} \frac{1}{k^2} = \frac{\pi^2}{6}$?", *Math. Gazette* 100 (2016), 429–434. **Zbl** 06990099. **MR** 3563583.

LORENTZEN, L. [= JACOBSEN, L.]

"Compositions of contractions," *Jour. Comput. & Appl. Math.* 32 (1990), 169–178. **Zbl** 713.30005. **MR** 92f:30035.

"Continued compositions of linear fractional transformations," *Jour. Math. Anal. and Applic.* 199 (1996), 130–137. **Zbl** 852.30017. **MR** 97d:30031.

LORIA, G.

"Il teorema fondamentale della teoria delle equazioni algebriche," *Rivista di Mat.* 1 (1891), 185–248 and 3 (1893), 105–108. **FM** 23, p. 34 and 25, p. 61.

"Esame di alcune ricerche concernenti l'esistenza di radici nelle equazioni algebriche," *Biblio. Math.* (2) 5 (1891), 99–112 and 7 (1893), 47–50. **FM** 23, p. 35 and 25, p. 61.

"L'Enigma dei numeri immaginari attraverso i secoli," *Scientia* 21 (1917), 101–121; French translation by S. Jankelevitch, *op. cit.*, supplement, 47–65. **FM** 46, p. 40.

LÖWNER, K. [=LOEWNER, C.]

"Untersuchungen über die Verzerrung bei konformen Abbildungen des Einheitskreises $|z| < 1$, die durch Funktionen mit nicht verschwindender Ableitung geliefert werden," *Ber. Verhandl. Kön. Sächs. Gesell. Wissen. Leipzig* 69 (1917), 89–106. **FM** 46, p. 556.

LÖWNER, K. and RADÓ, T.

"Bemerkung zu einem Blaschkeschen Konvergenzsatze," *Jahresber. Deutsch. Math. Verein.* 32 (1923), 198–200. **FM** 49, p. 231.

LUCAS, S. K.

"Approximations to π derived from integrals with nonnegative integrands," *Amer. Math. Monthly* 116 (2009), 166–172. **Zbl** 1179.26008.

LUECKING, D. H. and RUBEL, L. A.

Complex Analysis: A Functional Analysis Approach, Springer-Verlag (1984), New York and Berlin. **Zbl** 546.30002. **MR** 86d:30001. **BAMS** (2) 12, p. 283.

LÜNEBURG, H.

Von Zahlen und Größen: Dritthalbtausend Jahre Theorie und Praxis, Birkhäuser (2008), Basel and Berlin (2 volumes). **Zbl** 1158.00002, 3.

LÜTZEN, J. (ed.)

"Around Caspar Wessel and the geometric representation of complex numbers," *Proceedings of the Wessel Symposium at the Royal Danish Academy of Sciences and Letters Copenhagen, August* 11–15, 1998. **Zbl** 970.00008. **MR** 2003k:01012, 01014. (See also **Zbl** 957.01034, **MR** 2003j:01010.)

M

MACGREGOR, T. H.

"An inequality concerning analytic functions with a positive real part," *Canad. Jour. Math.* 21 (1969), 1172–1177. **Zbl** 185, p. 147. **MR** 40 #2878.

"Geometric problems in complex analysis," *Amer. Math. Monthly* 79 (1972), 447–468. **Zbl** 242.30016. **MR** 42 #7034.

MacGregor, T. H. and Tepper, D. E.

"Finite boundary interpolation by univalent functions," *Jour. Approx. Theory* 52 (1988), 315–321. **Zbl** 662.41002. **MR** 89e:30062.

Macintyre, A. J.

(a) "On Bloch's theorem," *Math. Zeit.* 44 (1939), 536–540. **FM** 94, p. 1075. **Zbl** 19, p. 419.

(b) "On a theorem concerning functions regular in an annulus," *Recueil Math. Moscou* (=*Mat. Sbornik*) (2) 5 (1939), 307–308 (Russian). **FM** 65, p. 337. **Zbl** 22, p. 238. **MR** 1, p. 308.

"Euler's limit for e^x and the exponential series," *Edin. Math. Notes* No. 37 (1949), 26–28. **Zbl** 36, p. 35. **MR** 10, p. 446.

"Convergence of $i^{i^{i\cdots}}$," *Proc. Amer. Math. Soc.* 17 (1966), 67. **Zbl** 133, p. 324. **MR** 32 #5855.

Macintyre, A. J. and Rogosinski, W. W.

"Some elementary inequalities in function theory," *Edin. Math. Notes* No. 35 (1945), 1–3. **Zbl** 60, p. 217. **MR** 7, p. 150.

Macintyre, S. S.

"On the zeros of successive derivatives of integral functions," *Trans. Amer. Math. Soc.* 67 (1949), 241–251. **Zbl** 35, p. 49. **MR** 11, p. 340.

Mackey, G.

Lectures on the Theory of Functions of a Complex Variable, Van Nostrand Mathematical Studies #9, D. Van Nostrand Co. (1967), Princeton. Reprinted with corrections by Robert E. Krieger Publishing Co., Inc. (1977), Huntington, New York. **Zbl** 381.30002. **MR** 36 #2781.

Macki, J. W.

"On Julia's corollary to Picard's great theorem," *Amer. Math. Monthly* 75 (1968), 655–656. **Zbl** 157, p. 398. **MR** 38 #324.

Mačys, J.

"On the fundamental theorem of algebra," *Lithuanian Math. Jour.* 42 (2002), 364–372. **Zbl** 1020.12005. **MR** 1968740.

Mahler, K.

"An elementary existence theorem for entire functions," *Bull. Austral. Math. Soc.* 5 (1971), 415–419. **Zbl** 219.30014. **MR** 45 #515.

Lectures on Transcendental Numbers, Lecture Notes in Mathematics 546, Springer-Verlag (1976), New York. **Zbl** 332.10019. **MR** 58 #10772. **BAMS** 84, p. 1370.

Maitland, B. J.

"A note on functions regular and bounded in the unit circle and small at a set of points near the circumference of the circle," *Proc. Camb. Phil. Soc.* 35 (1939), 382–388. **FM** 65, p. 333. **Zbl** 21, p. 240. **MR** 1, p. 112.

MANDELBROJT, S.

"Sur les suites de fonctions holomorphes. Les suites correspondantes des fonctions dérivées. Fonctions entières," *Jour. de Math. Pures et Appliq.* (9) 8 (1929), 173–195. **FM** 55, p. 762.

MAÑÉ, R., SAD, P. and SULLIVAN, D.

"On the dynamics of rational maps," *Ann. Sci. École Norm. Sup.* (4) 16 (1983), 193–217. **Zbl** 524.58025. **MR** 85j:58089. (See also **MR** 88i:30044.)

MANNING, K. R.

"The emergence of the Weierstrassian approach to complex analysis," *Arch. Hist. Exact Sci.* 14 (1975), 297–383. **Zbl** 338.01009. **MR** 56 #2746.

MANDELKERN, M.

"On the uniform continuity of Tietze extensions," *Arch. der Math.* 55 (1990), 387–388. **Zbl** 712.54009. **MR** 92c:54034.

"A short proof of the Tietze-Urysohn extension theorem," *Arch. der Math.* 60 (1993), 364–366. **Zbl** 778.54004. **MR** 94b:54050.

MANSION, P.

"Toute équation algébrique a une racine," *Annales de la Soc. Sci. de Bruxelles* 4 (B) (1880), 99–124. Corrections, 17(A) (1892–93), 65–66. **FM** 12, p. 63 and 25, p. 143.

MARDEN, A., RICHARDS, I. and RODIN, B.

"Analytic self-mappings of Riemann surfaces," *Jour. d'Analyse Math.* 18 (1967), 197–225. **Zbl** 152, p. 274. **MR** 35 #3057.

MARDEN, M.

Geometry of Polynomials, Mathematical Surveys No. 3, American Mathematical Society (2nd ed. 1966), Providence. **Zbl** 162, p. 371. **MR** 37 #1562. **BAMS** 56, p. 78.

"On the zeros of the derivative of an entire function," *Amer. Math. Monthly* 75 (1968), 829–839. **Zbl** 165, p. 401. **MR** 38 #3436.

"Much ado about nothing," *Amer. Math. Monthly* 83 (1976), 788–798. **Zbl** 352.30001. **MR** 54 #7762.

"The search for a Rolle's Theorem in the complex plane," *Amer. Math. Monthly* 92 (1985), 643–650, **Zbl** 627.30002. **MR** 87c:30008.

MARKUSCHEWITSCH (also MARKUSHEVICH), A. I.,

Skizzen zur Geschichte der Analytischen Funktionen (translated from Russian by W. Ficker), Hochschulbücher für Mathematik, Bd. 16, VEB Deutscher Verlag der Wissenschaften (1955), Berlin. **Zbl** 64, p. 310. **MR** 17, p. 445.

Theory of Functions of a Complex Variable (translated from Russian and edited by R. A. Silverman), I (1965), II (1965), III (1967), Prentice-Hall, Inc., Englewood Cliffs. Second edition (3 volumes bound as one), Chelsea Publishing

Co. (1977), New York. **Zbl** 357.30002. **MR** 30 #2125; 31 #5965; 35 #6799; 56 #3258. (See also **MR** 35 #5581 and 38 #304.)

Selected chapters in the theory of analytic functions (Russian), Izdat. "Nauka" (1976), Moscow. **Zbl** 459.30001. **MR** 55 #8324.

"Theory of analytic functions," (Russian) pp. 115–269 of *Mathematics of the XIXth Century. Geometry, Theory of Analytic Functions* (A. N. Kolmogorov and A. P. Yushkevich, editors), Izdat. "Nauka" (1981), Moscow. **Zbl** 506.01013. English translation by Roger L. Cooke, Birkhäuser Verlag (1996), Basel and Boston. **Zbl** 849.01048. **MR** 98m:01017.

Marshall, D. E.

"An elementary proof of the Pick-Nevanlinna interpolation theorem," *Mich. Math. Jour.* 21 (1974), 219–223. **Zbl** 281.30032. **MR** 51 #903.

"Blaschke products generate H^∞," *Bull. Amer. Math. Soc.* 82 (1976), 494–496. **Zbl** 327.30029. **MR** 53 #5877.

Mathew, P. and Patyi, I.

"The fundamental theorem of calculus implies that of algebra," *Math. Gazette* 95 (2011), 88–89.

Matsumoto, K.

"An elementary summation of the Leibniz series," *Math. Gazette* 79 (1995), 567–568.

Mattner, L.

"Complex differentiation under the integral," *Nieuw Archief voor Wiskunde* (5) 2 (2001), 32–35. **MR** 2002a:30064.

May, K. O.

Bibliography and Research Manual of the History of Mathematics, University of Toronto Press (1973), Toronto and Buffalo. **Zbl** 254.01001. **MR** 58 #26679.

Mazur, B.

Imagining Numbers (Particularly the Square Root of Minus Fifteen), Ferrar, Straus and Giroux (2003), New York. **Zbl** 1049.00502. **MR** 2003m:11001.

Mazurkiewicz, S.

"Sur les lignes de Jordan," *Fund. Math.* 1 (1920), 166–209. **FM** 47, p. 521.

(a) "Extension du théorème de Phragmén-Brouwer aux ensembles non bornés," *Fund. Math.* 3 (1922), 20–25. **FM** 48, p. 213.

(b) "Sur les séries de puissances," *Fund. Math.* 3 (1922), 52–58. **FM** 48, p. 337.

"Sur les fonctions non-dérivables," *Studia Math.* 3 (1931), 92–93. **FM** 57, p. 305. **Zbl** 3, p. 297.

"Sur le théorème de Rouché," *Sprawozdania Z Posiedzeń Towarzystwa Naukowego Warszawskiego* (=*C. R. Soc. Sci. Lett. Varsovie*) 28 (1936), 78–79. **FM** 62, p. 1199. **Zbl** 13, p. 311.

"Ultraconvergence et espace fonctionnel," *Fund. Math.* 28 (1937), 289–293. **FM** 63, p. 361. **Zbl** 16, p. 125.

McCall, J. D.
"Mean growth, H_p spaces and subharmonic functions in the upper half-plane," *SIAM Jour. of Math. Anal.* 14 (1983), 982–986. **Zbl** 538.30032. **MR** 86g:42039.

McNicholl, T. H.
"Carathéodory's theorem and moduli of local connectivity," *Complex Variables & Elliptic Equations* 61 (2016), 76–85. **Zbl** 1337.30011. **MR** 3428853.

McShane, E. J.
"On the Osgood-Carthéodory theorem," *Amer. Math. Monthly* 44 (1937), 288–291. **FM** 63, p. 299. **Zbl** 16, p. 266.

Meier, K. E.
"Über die Randwerte meromorpher Funktionen und hinreichende Bedingungen für Regularität von Funktionen einer komplexen Variablen," *Comm. Math. Helv.* 24 (1950), 238–259. **Zbl** 41, pp. 52 and 661. **MR** 12, p. 490.
"Zum Satz von Looman-Menchoff," *Comm. Math. Helv.* 25 (1951), 181–195. **Zbl** 44, p. 78. **MR** 14, p. 150.
"Eine hinreichende Bedingung für die Regularität einer komplexen Funktion," *Comm. Math. Helv.* 34 (1960), 67–70. **Zbl** 90, p. 47. **MR** 22 #5714.
"Hinreichende Bedingungen für die Regularität einer komplexen Funktion," *Comm. Math. Helv.* 45 (1970), 256–264. **Zbl** 194, p. 375. **MR** 43 #6406.

Meisters, G. and Olech, C.
"Locally one-to-one mappings and a classical theorem on schlicht functions," *Duke Math. Jour.* 30 (1963), 63–80. **Zbl** 112, p. 377. **MR** 26 #1471.

Menchoff, D. E.
"Sur les fonctions monogènes," *Bull. Soc. Math. France* 59 (1931), 141–182. **FM** 57, p. 344. **Zbl** 4, p. 117.
"Sur la généralisation des conditions de Cauchy-Riemann," *Fund. Math.* 25 (1935), 59–97. **FM** 61, p. 305. **Zbl** 12, p. 82.
Les conditions de monogénéité, Actualités Scientifiques et Industrielles, No. 329, Hermann et Cie. (1936), Paris. **FM** 62, p. 318. **Zbl** 14, p. 167. **BAMS** 44, p. 18.
"Sur une généralisation d'un théorème de M. H. Bohr," *Recueil Math. Moscou* (=*Mat. Sbornik*) (2) 2 (1937), 339–356. **FM** 63, p. 965. **Zbl** 17, p. 215. (See also **MR** 2007k:30002.)

Menger, K.
"Gulliver's return to the land without one, two, three," *Amer. Math. Monthly* 67 (1960), 641–648. **MR** 23A #760 and 22A #9427.

Méray, Ch.
"Mémoire sur les fonctions doublement périodiques, monogènes et monodromes," *C. R. Acad. Sci. Paris* 40 (1855), 787–789.

"Observations sur la légitimité de l'interpolation," *Ann. Sci. École Norm. Sup.* (3) 1 (1884), 165–176. **FM** 16, p. 196.

"Démonstration analytique de l'existence et des propriétés essentielles des racines des équations binômes," *Ann. Sci. École Norm. Sup.* (3) 2 (1885), 337–356. **FM** 17, p. 60.

"Valeur de l'intégrale définie $\int_0^\infty e^{-x^2}dx$ déduite de la formule de Wallis," *Bull. Sci. Math.* (2) 12 (1888), 174–176. **FM** 20, p. 300.

"Théorie analytique du logarithme népérien et de la fonction exponentielle," *Annales de la Faculté des Sciences de Toulouse* 4 (1890), 1–35. Gauthier-Villars et Fils (1891), Paris. **FM** 22, p. 259; 23, p. 444.

(a) "Méthode directe fondée sur l'emploi des séries pour prouver l'existence des racines des équations entières à une inconnue par la simple exécution de leur calcul numérique," *Bull. Sci. Math.* (2) 15 (1891), 236–252. **FM** 23, p. 95.

(b) "Démonstration élementaire d'un lemme fondamental de Cauchy," *Bull. Sci. Math.* (2) 15 (1891), 86–96 and (2) 19 (1895), 154–159. **FM** 23, p. 413 and 26 p. 428.

Mercer, P. R.

"Another look at Julia's lemma," *Complex Variables* 43 (2000), 129–138. **Zbl** 1021.30021. **MR** 2001i:30022.

Mergelyan, S. N.

"Uniform approximation to functions of a complex variable," *Uspehi Mat. Nauk* (N. S.) 7, No. 2 (48) (1952), 31–122. American Mathematical Society Translation No. 101 (1954), and pp. 294–391 of *Translations. Series* 1, Vol. 3: *Series and Approximation,* (1962). American Mathematical Society, Providence. **Zbl** 49, p. 327. **MR** 15, p. 612 and 38 #1979.

Mertens, F.

"Der Fundamentalsatz der Algebra," *Monatshefte für Math. und Physik* 3 (1892), 293–308 and *Sitzungsber. Kaiserl. Akad. Wiss. Wien* 101 (1892), Zweite Abteilung, 415–424. **FM** 24, p. 87.

"Über die Gaussischen Summen," *Sitzungsber. Kön. Preuss. Akad. Wissen. Berlin* (1896), 217–219. **FM** 27, p. 144.

"Über die Zerfällung einer ganzen Funktion einer Veränderlichen in zwei Faktoren," *Sitzungsber. Kaiserl. Akad. Wiss. Wien* 120 (1911), Zweite Abteilung, 1485–1502. **FM** 42, p. 114.

Miamee, A. and Salehi, H.

"Positive operators on a Banach space and the Fejér-Riesz theorem," *Proc. Amer. Math. Soc.* 56 (1976), 189–192. **Zbl** 324.30044. **MR** 53 #3788.

Mignosi, G.

"Sul teorema fondamentale dell'algebra nell'algebra classica e nell'algebra moderna," *Esercitazioni Mat. Catania* (2) 13 (1941), 28–47. **FM** 67, p. 61. **Zbl** 24, p. 145.

MILLER, E. W.

"On singularities of an analytic function," *Bull. Amer. Math. Soc.* 41 (1935), 561–565. **FM** 61, p. 310. **Zbl** 12, p. 170.

MILLER, S. S. and MOCANU, P.

"Second order differential inequalities in the complex plane," *Jour. Math. Anal. and Applic.* 65 (1978), 289–305. **Zbl** 379.34003. **MR** 80f:30009. (See also **MR** 80m:34006 and **Zbl** 423.30016.)

MILLOUX, H.

"Le théorème de M. Picard, suites de fonctions holomorphes, fonctions méromorphes et fonctions entières," *Jour. de Math. Pures et Appliq.* (9) 3 (1924), 345–401. **FM** 50, p. 211.

"Les cercles de remplissage des fonctions méromorphes ou entières et le théorème de Picard-Borel," *Acta Math.* 52 (1928), 189–255. **FM** 55, p. 193.

"Sur certaines fonctions holomorphes et bornées dans un cercle," *Mathematica* (Cluj) 4 (1930), 182–185. **FM** 56, p. 971.

"Sur une inégalité de la théorie des fonctions et ses applications," *C. R. Acad. Sci. Paris* 194 (1932), 587–589. **FM** 58, p. 332. **Zbl** 3, p. 406.

"Sur les valeurs asymptotiques des fonctions entières d'ordre infini," *Compositio Math.* 1 (1935), 305–313. **FM** 60, p. 1019. **Zbl** 10, p. 364.

Les fonctions méromorphes et leurs dérivées. Extensions d'un théorème de M. R. Nevanlinna. Applications. Actualités Scientifiques et Industrielles, No. 888, Hermann et Cie. (1940), Paris. **FM** 66, p. 1249. **Zbl** 26, p. 316. **MR** 7, p. 427.

(a) "Une application de la théorie des familles normales," *Bull. Sci. Math.* (2) 72 (1948), 12–16. **Zbl** 33, p. 116. **MR** 10, p. 289.

(b) "Le problème de la distribution des valeurs d'une fonction uniforme," *Mathematica, Timişoara* 23 (1948), 76–84. **Zbl** 31, p. 22. **MR** 10, p. 28.

Principes. Méthodes générales, Fasc. I. Avec la collaboration de Charles Pisot. (Traite de théorie des fonctions. Tome I.), Gauthier-Villars (1953), Paris. **Zbl** 53, p. 43. **MR** 15, pp. 300 and 1139.

MILNER, E. C.

"Hadamard's three circles theorem," *Amer. Math. Monthly* 75 (1968), 525–527. **Zbl** 162, p. 95.

MILNOR, J.

Dynamics in One Complex Variable, Friedr. Vieweg & Sohn (1999), Braunschweig & Wiesbaden. **Zbl** 946.30013. **MR** 2002i:37057. **BAMS** (2) 38, p. 495. (Third edition, 2006, Princeton University Press)

MINAMI, U.

"Sur un théorème de Phragmén-Lindelöf," *Tôhoku Math. Jour.* 44 (1937), 85–93. **FM** 63, p. 978. **Zbl** 18, p. 141.

"An extension of Phragmén-Lindelöf's theorem," *Proc. Imper. Acad. Tokyo* 13 (1937), 241–243. **FM** 63, p. 978. **Zbl** 19, p. 419.

"On the Cauchy's integral theorem," *Proc. Imper. Acad. Tokyo* 18 (1942), 440–445. **Zbl** 60, p. 200. **MR** 7, p. 284.

MINDA, C. D.

"Analytic functions on nonopen sets," *Math. Mag.* 46 (1973), 223–224. **MR** 48 #2349.

(a) "Fixed points of analytic self-mappings of Riemann surfaces," *Manuscripta Math.* 27 (1979), 391–399. **Zbl** 405.30035. **MR** 81a:30024.

(b) "The Aumann-Carathéodory rigidity constant for doubly connected regions," *Kōdai Math. Jour.* 2 (1979), 420–426. **Zbl** 426.30019. **MR** 81d:30043.

(c) "Conformal automorphisms of circular regions," *Amer. Math. Monthly* 86 (1979), 684–686. **Zbl** 422.30004. **MR** 80i:30015.

"A heuristic principle for a nonessential isolated singularity," *Proc. Amer. Math. Soc.* 93 (1985), 443–447. **Zbl** 552.30015. **MR** 86g:30040.

"The Dirichlet problem for a disk," *Amer. Math. Monthly* 97 (1990), 220–223. **Zbl** 738.31002. **MR** 91e:31005.

"Another approach to Picard's theorem and a unifying principle in function theory," pp. 186–200 of *Current Topics in Analytic Function Theory* (H. M. Srivastava and Shigeyoshi Owa, editors), World Scientific Publishing Co. (1992), Singapore. **Zbl** 982.30301. **MR** 94k:30086.

MINDA, C. D. and SCHOBER, G.

"Another elementary approach to the theorems of Landau, Montel, Picard and Schottky," *Complex Variables* 2 (1983), 157–164. **Zbl** 572.30021. **MR** 86h:30059.

MINSKER, S.

"A familiar combinatorial identity proved by complex analysis," *Amer. Math. Monthly* 80 (1973), 1051. **Zbl** 277.05005.

MIODUSZEWSKI, J.

"On two-to-one continuous functions," *Rozprawy Matematyczne* (=*Dissertationes Mathematicae*) 24 (1961). **Zbl** 104, p. 173. **MR** 26 #3021.

MIRANDA, C.

"Sur un nouveau critère de normalité pour les familles de fonctions holomorphes," *Bull. Soc. Math. France* 63 (1935), 185–196. **FM** 61, p. 1155. **Zbl** 13, p. 272.

MITCHELL, T

"Invariant means and analytic actions," *Pac. Jour. of Math.* 85 (1979), 145–153. **Zbl** 447.43003. **MR** 81m:43001.

MITRINOVIĆ, D. S. and KEČKIĆ, J. D.

The Cauchy Method of Residues: Theory and Applications, D. Reidel Publishing Co. (1984), Dordrecht and Boston. **Zbl** 546.30004. **MR** 85e:30002. Vol. 2,

Kluwer Academic Publishing Co. (1993), Dordrecht and Boston. **Zbl** 786.30001. **MR** 94k:30001.

MITTAG-LEFFLER, G.

"Sur la théorie des fonctions uniformes d'une variable. Extrait d'une lettre adressée à M. Hermite," *C. R. Acad. Sci. Paris* 94 (1882), 414–416, 511–514, 713–715, 781–783, 938–941, 1040–1042, 1105–1108, 1163–1166; 95 (1882), 335–336. **FM** 14, p. 325.

(a) "Sur la représentation analytique des fonctions monogènes uniformes d'une variable indépendante," *Acta Math.* 4 (1884), 1–79. **FM** 16, p. 351.

(b) "Démonstration nouvelle du théorème de Laurent," *Acta Math.* 4 (1884), 80–88 and *Mém. Soc. Roy. Sci. Liége* (2) 11 (1885). **FM** 16, pp. 218, 350 and 17, 383.

"Sur la représentation analytique des fonctions d'une variable réelle," *Rend. Circ. Mat. Palermo* 14 (1900), 217–224. **FM** 31, p. 409.

"Die ersten 40 Jahre des Lebens von Weierstrass," *Acta Math.* 39 (1923), 1–57. **FM** 49, p. 8.

MOHR, E.

"Beweis des sogenannten Fundamentalsatzes der Algebra im reellen Gebiete," *Jour. für Reine und Angew. Math.* 184 (1942), 175–177. **FM** 68, p. 34. **Zbl** 27, p. 7. **MR** 5, p. 169. Corrections, *ibid.* 189 (1952), 250–252. **Zbl** 46, p. 245. **MR** 12, p. 938.

"Der sogenannte Fundamentalsatz der Algebra als Satz der reellen Analysis," *Math. Nachr.* 6 (1951/52), 65–69. **Zbl** 44, p. 8. **MR** 12, p. 938. Corrections, *ibid.* 6 (1951–52), 385–386. **Zbl** 46, p. 14. **MR** 14, p. 465.

(a) "Ein elementarer Beweis für den Fundamentalsatz der Algebra in Reellen," *Annali Mat. Pura Appl.* (4) 34 (1953), 407–410. **Zbl** 51, p. 10. **MR** 15, p. 418.

(b) "Elementarer Beweis für die Partialbruchzerlegung des Cotangens," *Zeit. für Math. and Mech.* 33 (1953), 247–248. **Zbl** 50, p. 288. **MR** 14, p. 1080.

"Elementarer Beweis für die Produktentwicklung des Sinus und die Partialbruchzerlegung des Cotangens," *Zeit. für Math. and Mech.* 39 (1959), 78–80. **Zbl** 90, p. 46. **MR** 21 #1404.

MONNA, A. F.

Dirichlet's principle: a mathematical comedy of errors and its influence on the development of analysis, Oosthoek, Scheltema and Holkema (1975), Utrecht. **Zbl** 312.31001.

MONTEL, P.

"Sur les suites infinies de fonctions," *Ann. Sci. École Norm. Sup.* (3) 24 (1907), 233–334. **FM** 38, p. 440.

Leçons sur les séries de polynômes à une variable complexe, Gauthier-Villars (1910), Paris. **FM** 41, p. 277.

"Sur l'indétermination d'une fonction uniforme dans les voisinages de ses points essentiels," *C. R. Acad. Sci. Paris* 153 (1911), 1455–1456. **FM** 42, p. 426.

"Sur les familles de fonctions analytiques, qui admettent des valeurs exceptionnelles dans un domaine," *Ann. Sci. École Norm. Sup.* (3) 29 (1912), 487–535. **FM** 43, p. 509.

"Sur le théorème de d'Alembert et la continuité des fonctions algébriques," *Nouvelles Annales de Math.* (4) 13 (1913), 481–492. **FM** 44, p. 116.

"Sur les familles normales de fonctions analytiques," *Ann. Sci. École Norm. Sup.* (3) 33 (1916), 223–302. **FM** 46, p. 519.

"Sur la représentation conforme," *Jour. de Math. Pures et Appliq.* (7) 3 (1917), 1–54. **FM** 46, p. 548.

"Sur les modules des zéros des polynômes," *Ann. Sci. École Norm. Sup.* (3) 40 (1923), 1–34. **FM** 49, p. 46 and 48, p. 87.

"Sur les suites de fonctions analytiques qui ont pour limite une constante," *Bull. Soc. Math. France* 53 (1925), 246–257. **FM** 52, p. 306.

Leçons sur les familles normales de fonctions analytiques et leurs applications, Gauthier-Villars (1927), Paris. **FM** 53, p. 303. Reprinted by Chelsea Publishing Co. (1974), New York.

"Sur les fonctions convexes et les fonctions sousharmoniques," *Jour. de Math. Pures et Appliq.* (9) 7 (1928), 29–60. **FM** 54, p. 517. (See also **Zbl** 556.31001.)

"Sur les domaines formés par les points représentant les valeurs d'une fonction analytique," *Ann. Sci. École Norm. Sup.* (3) 46 (1929), 1–23. **FM** 55, p. 198.

(a) "Sur les séries de fractions rationnelles," *Publications Math. Univ. Belgrade* 1 (1932), 157–169. **FM** 58, p. 302. **Zbl** 6, p. 211.

(b) "Sur un théorème de Rouché," *C. R. Acad. Sci. Paris* 195 (1932), 1214–1216. **FM** 58, p. 300. **Zbl** 6, p. 62.

(c) "Sur les fonctions harmoniques qui admettent des valeurs exceptionnelles," *C. R. Acad. Sci. Paris* 194 (1932), 40–41. **FM** 58, p. 503. **Zbl** 3, p. 263.

(d) "Sur une formule de Darboux et les polynômes d'interpolation," *Ann. Scuola Norm. Sup. Pisa* (2) 1 (1932), 371–384. **FM** 58, p. 299. **Zbl** 5, p. 290. (See also **MR** 5, p. 92 and **Zbl** 60, p. 200.)

(a) "Sur un théorème de M. Pompeiu," *Bull. Math. Soc. Roum. Sci.* 35 (1933), 179–181. **FM** 59, p. 1023. **Zbl** 8, p. 317.

(b) *Leçons sur les fonctions univalentes ou multivalentes,* Gauthier-Villars (1933), Paris. **FM** 59, p. 346. **Zbl** 6, p. 351. **BAMS** 42, p. 463.

(c) "Sur les fonctions méromorphes limites de fractions rationnelles à termes entrelacés," *Ann. Sci. École Norm. Sup.* (3) 50 (1933), 171–196. **FM** 59, p. 331. **Zbl** 8, p. 20.

"Le rôle des familles normales," *L'Enseignement Mathématique* 33 (1934), 5–21. **FM** 60, p. 1007. **Zbl** 10, p. 264.

"Sur les relations de Cauchy," *Bull. Math. Soc. Roum. Sci.* 38 (1936), 97–99. **FM** 62, p. 1222. **Zbl** 16, p. 264.

"Sur les fonctions localement univalentes ou multivalentes," *Ann. Sci. École Norm. Sup.* (3) 54 (1937), 39–54. **FM** 63, p. 290. **Zbl** 17, p. 24.

"Harmonische und subharmonische Funktionen," *Publ. Inst. Mat. Rosario* 2 (1940), 1–23 (Spanish). **FM** 66, p. 448. **Zbl** 24, p. 116. **MR** 2, p. 76.

"Sur le rôle des familles de fonctions dans l'analyse moderne," *Bull. Soc. Roy. Sci. Liége* 15 (1946), 262–278. **Zbl** 61, p. 149. **MR** 8, p. 507. (Italian translation: **MR** 12, p. 490.)

Leçons sur les récurrences et leurs applications, Gauthier-Villars (1957), Paris. **Zbl** 77, p. 66. **MR** 19, p. 427.

Moore, E. H.

"Concerning the definition by a system of functional properties of the function $f(z) = \frac{\sin \pi z}{\pi z}$," *Annals of Math.* 9 (1895), 43–49. **FM** 26, p. 470.

"A simple proof of the fundamental Cauchy-Goursat theorem," *Trans. Amer. Math. Soc.* 1 (1900), 499–506. **FM** 31, p. 398.

Moore, R. L.

"A theorem concerning continuous curves," *Bull. Amer. Math. Soc.* 23 (1916–17), 233–236. **FM** 46, pp. 308, 829.

"A characterization of Jordan regions by properties having no reference to their boundaries," *Proc. Nat. Acad. Sci. U. S. A.* 4 (1918), 364–370. **FM** 46, p. 1451.

"Conditions under which one of two given closed linear point sets may be thrown into the other one by a continuous transformation of a plane into itself," *Amer. Jour. of Math.* 48 (1926), 67–72. **FM** 52, p. 604.

Morant, J.

(a) "Sur certains développements de fonctions holomorphes dans une aire limitée par des arcs de cercle," *Bull. Soc. Roy. Sci. Liége* 7 (1938), 159–164. **FM** 64, p. 276. **Zbl** 18, p. 260.

(b) "Quelques applications d'une formule de Laurent à certains développements de fonctions holomorphes," *Bull. Soc. Roy. Sci. Liége* 7 (1938), 470–476. **FM** 64, p. 276. **Zbl** 20, p. 137.

Mordell, L. J.

"On a simple summation of the series $\sum_{s=0}^{n-1} e^{2s^2\pi i/n}$," *Messenger of Math.* 48 (1918), 54–56. **FM** 46, p. 538.

"The value of the definite integral $\displaystyle\int_{-\infty}^{\infty} \frac{e^{at^2+bt}dt}{e^{ct}+d}$", *Quart. Jour. of Pure and Applied Math.* 48 (1920), 329–342. **FM** 59, p. 197.

"The definite integral $\int_{-\infty}^{\infty} \frac{e^{ax^2+bx}dx}{e^{cx}+d}$ and the analytic theory of numbers," *Acta Math.* 61 (1933), 323–360. **FM** 59, p. 197. **Zbl** 8, p. 55.

MORENO, S. G.

"A short and elementary proof of the Basel problem," *The College Math. Jour.* 47 (2016), 134–135. **MR** 3519066.

MORERA, G.

(a) "Un teorema fondamentale nella teorica delle funzioni di una variabile complessa," *Rend. del R. Istituto Lombardo di Scienze e Lettere* (2) 19 (1886), 304–307. **FM** 18, p. 338.

(b) "Sulla rappresentazione delle funzioni di una variabile complessa per mezzo di espressioni analitiche infinite," *Atti della R. Accad. d. Scienze di Torino* 21 (1886), 892–899. **FM** 18, p. 339.

"Sulla definizione di funzione di una variabile complessa," *Atti della R. Accad. d. Scienze di Torino* 37 (1901–02), 99–102. **FM** 33, p. 396.

MORGENSTERN, D.

"Unendlich oft differenzierbare nicht-analytische Funktionen," *Math. Nachr.* 12 (1954), 74. **Zbl** 58, p. 289. **MR** 16, p. 342.

MORTINI, R.

"When is the product of two planar harmonic mappings harmonic?", *Lecturas Matemáticas* 23 (2002), 5–10. **Zbl** 1085.30003. **MR** 2008g:31006.

"A classroom note: Harmonic conjugates in arbitrary planar domains," *Matematički Bilten* 30 (LVI), (2006), 37–42. **Zbl** 1208.31001. **MR** 2356767.

"The Riemann-Cantor uniqueness theorem for unilateral trigonometric series via a special version of the Lusin-Privalov theorem," *Ann. Univ. Mariae Curie-Skłodowska, Sect. A* (2017), 71, A73–76. **Zbl** 1373.30052. **MR** 3668788.

MORTINI, R. and RUPP, R.

"Mappings into the Euclidean sphere," *Amer. Math. Monthly* 119 (2012), 485–494. **Zbl** 1260.54088. **MR** 2928661.

"The symmetric versions of Rouché's theorem via $\bar{\partial}$-calculus," *Jour. of Complex Analysis* (2014), Article ID 260953. **Zbl** 1310.30004. **MR** 3166802.

MOTZKIN, TH. and OSTROWSKI, A.

"Über den Fundamentalsatz der Algebra," *Sitzungsber. Preuss. Akad. Wissen. Berlin* (1933), 255–258. **FM** 59, pp. 120 and 121. **Zbl** 6, pp. 243 and 338.

MUELLER, C. and RUDIN, W.

"Proper holomorphic self-maps of plane regions," *Complex Variables* 17 (1991), 113–121. **Zbl** 745.30024. **MR** 93e:30013. (See also **Zbl** 805.30030.)

MÜLLER, C.

"Über die Umkehrung des Cauchyschen Integralsatzes," *Arch. der Math.* 6 (1954), 47–51. **Zbl** 59, p. 65. **MR** 16, p. 346.

MÜLLER, J. and WENGENROTH, J.

"Separating singularities of holomorphic functions," *Canad. Math. Bull.* 41 (1998), 473–477. **Zbl** 917.30026. **MR** 99g:30050.

MÜLLER, M.

"Zur konformen Abbildung angenähert kreisförmiger Gebiete," *Math. Zeit.* 43 (1938), 628–636. **FM** 64, p. 1073. **Zbl** 17, p. 408.

MULLIKIN, A. M.

"Certain theorems related to plane connected point sets," *Trans. Amer. Math. Soc.* 24 (1922), 144–162. **FM** 49, p. 143. (See also **Zbl** 1195.01048.)

MURAKAMI, M.

"A nonstandard proof of the continuity of zeros of polynomials," *Jour. Fac. Sci. Univ. Tokyo* Sect. IA, Math. 40 (1993), 53–54. **Zbl** 781.12004. **MR** 94c: 12001.

MYLLER-LÉBÉDEFF, V.

"Sur une application du lemme de M. Carleman," *Ann. Sci. Univ. Jassy, I: Math.* 24 (1938), 1–4. **FM** 64, p. 313. **Zbl** 18, p. 141.

N

NABETANI, K.

"On Study's theorem in the theory of conformal representation," *Tôhoku Math. Jour.* 41 (1935–36), 406–410. **FM** 62, p. 374. **Zbl** 14, p. 24.

NAGASAWA, M.

"Isomorphisms between commutative Banach algebras with an application to rings of analytic functions," *Kōdai Math. Sem. Reports* 11 (1959), 182–188. **Zbl** 166, p. 400. **MR** 22 #12379.

NAGEL, E.

"'Impossible numbers': A chapter in the history of modern logic," being chapter 8 of his anthology *Teleology Revisited and Other Essays in the Philosophy and History of Science,* Columbia Universtiy Press (1979), New York.

NAKAI, M.

"On rings of analytic functions on Riemann surfaces," *Proc. Japan Acad.* 39 (1963), 79–84. **Zbl** 113, p. 58. **MR** 27 #295.

"Divisors on meromorphic function fields," *Proc. Japan Acad.* 51 (1975), 507–509. **Zbl** 368.46030. **MR** 52 #5956.

NAKAI, M. and TADA, T.

"A form of classical Liouville theorem," *Proc. Japan Acad. Ser.* A 73 no. 9 (1997), 166–167. **Zbl** 903.31003. **MR** 99a:31004.

NARASIMHAN, R.

Complex Analysis in One Variable, Birkhäuser Verlag (1985), Basel and Boston. **Zbl** 561.30001. **MR** 87h:30001. 2nd ed. with Y. NIEVERGELT (2001). **Zbl** 1009.30001. **MR** 2002e:30001.

NAUX, C.

Histoire des Logarithmes de Neper à Euler. Tome I: *La découverte des logarithmes et le calcul des premières tables.* Tome II: *La promotion des logarithmes au rang de valeur analytique,* Librairie Scientifique et Technique A. Blanchard (1966, 1971), Paris. **Zbl** 151, p. 3 and 232.01001. **MR** 36 #15 and 43 #7285.

NEDER, L.

"Konvergenzdefekte der Potenzreihen stetiger Funktionen auf dem Rande des Konvergenzkreises," *Math. Zeit.* 6 (1920), 262–269. **FM** 47, p. 276.

NEEDHAM, T.

"The geometry of harmonic functions," *Math. Mag.* 67 (1994), 92–108 and 68 (1995), 268. **Zbl** 805.31001.

Visual Complex Analysis, Clarendon Press (1997), Oxford. **Zbl** 1095.30500. **MR** 98g:30001 and 2002d:30001.

NEHARI, Z.

Conformal Mapping, McGraw-Hill Book Co. (1952), New York. **Zbl** 48, p. 315. **MR** 13, p. 640. **BAMS** 58, p. 515. Reprinted by Dover Publications, Inc. (1975), New York. **MR** 51 #13206.

NELSON, E.

"A proof of Liouville's theorem," *Proc. Amer. Math. Soc.* 12 (1961), 995. **Zbl** 124, p. 312. **MR** 41 #3791.

NERSESJAN, A. A.

"Carleman sets," *Izv. Akad. Nauk Armjan. SSR Ser. Mat.* 6, No. 6 (1971), 465–471 (Russian); American Mathematical Society Translations. Series 2, 122 (1984), 85–97. **Zbl** 552.30029. **MR** 46 #367. (See also **MR** 51 #3460, 3461 and **Zbl** 669.30027, 292.30037.)

NETTO, E.

"Beweis der Wurzelexistenz algebraischer Gleichungen," *Jour. für Reine und Angew. Math.* 88 (1880), 16–21. **FM** 11, p. 61.

NETTO, E. and LEVAVASSEUR, R.

"Les fonctions rationnelles," *Encyclopédie des sciences mathématiques pures et appliquées,* t. I, v. 2, fasc. 1, 1–232. Gauthier-Villars (1907), Paris. **FM** 30, p. 95. **MR** 95f:00029a.

NETUKA, I and VESELÝ, J.

"Mean value property and harmonic function," pp. 359–398 of *Classical and Modern Potential Theory and Applications,* NATO Advanced Science Institutes,

Ser. C, Math. Phys. Sci. 430, Kluwer Academic Publishers (1994), Dordrecht. **Zbl** 863.31012. **MR** 96c:31001.

NEUENSCHWANDER, E.

"The Casorati-Weierstrass theorem (studies in the history of complex function theory I)," *Historia Math.* 5 (1978), 139–166. **Zbl** 374.01010. **MR** 58 #10109.

"Augustin Cauchy: ein Wendepunkt in der Geschichte der Analysis," *Nieuw Archief voor Wiskunde* (3) 27 (1979), 50–69 and pp. 275–294 of *Proceedings of the Bicentennial Congress Wiskundig Genootschap,* Part II, Mathematical Centre Tracts #101, Amsterdam (1979). **Zbl** 398.01008 and 419.01014. **MR** 80f:01013, 01014.

"Studies in the history of complex function theory II: Interactions among the French school, Riemann, and Weierstrass," *Bull. Amer. Math. Soc.* (2) 5 (1981), 87–105. **Zbl** 483.01008. **MR** 82j:01050. (See also **Zbl** 476.01002 and **MR** 83a:01030.)

Riemanns Einführung in die Funktionentheorie. Eine quellenkritische Edition seiner Vorlesungen mit einer Bibliographie zur Wirkungsgeschichte der Riemannschen Funktionentheorie, Abhandlungen der Akademie der Wisenschaften in Göttingen, Mathematisch-physikalische Klasse, Dritte Folge, Nr. 44. Vandenhoeck & Ruprecht (1996), Göttingen. **Zbl** 844.01020. **MR** 97d:01041. (See also **Zbl** 922.01018.)

NEUMANN, C.

"Revision einiger allgemeinen Sätze aus der Theorie des logarithmischen Potentials," *Math. Annalen* 3 (1871), 325–349. **FM** 3, p. 491.

NEVANLINNA, F. and NEVANLINNA, R.

"Über die Eigenschaften analytischer Funktionen in der Umgebung einer singulären Stelle oder Linie," *Acta Soc. Sci. Fennicae* 50, No. 5 (1922). **FM** 48, p. 358.

NEVANLINNA, F. and NIEMINEN, T.

"Das Poisson-Stieltjes'sche Integral und seine Anwendung in der Spektraltheorie des Hilbert'schen Raumes," *Ann. Acad. Scient. Fennicae, Ser. A. I.* Nr. 207 (1955). **Zbl** 67, p. 90. **MR** 17, p. 648.

NEVANLINNA, R.

"Über beschränkte Funktionen, die in gegebenen Punkten vorgeschriebene Werte annehmen," *Ann. Acad. Scient. Fennicae, Ser. A. I.* 13, No. 1 (1919). **FM** 46, p. 1466 and 47, p. 271.

(a) "Asymptotische Entwicklungen beschränkter Funktionen und das Stieltjessche Momentenproblem," *Ann. Acad. Scient. Fennicae, Ser. A. I.* 18, Nr. 5 (1922). **FM** 48, p. 1226.

(b) "Kriterien für die Randwerte beschränkter Funktionen," *Math. Zeit.* 13 (1922), 1–9. **FM** 48, p. 322.

"Über die Anwendung des Poisson'schen Integrals zur Untersuchung der Singläritaten analytischer Funktionen," pp. 273–289 of *Wissenschaftliche Vorträge Gehalten auf dem Fünften Kongress der Skandinavischen Mathematiker in Helsingfors vom 4. bis 7. Juli 1922.* Akademische Buchhandlung (1923), Helsingfors. **FM** 48, p. 358 and 49, p. 213.

(a) "Untersuchungen über den Picard'schen Satz," *Acta Soc. Sci. Fennicae* 50, Nr. 6 (1924). **FM** 50, p. 219.

(b) "Beweis des Picard-Landauschen Satzes," *Göttingen Nachr.* (1924), 151–154. **FM** 50, p. 219.

"Über eine Erweiterung des Poissonschen Integrals," *Ann. Acad. Scient. Fennicae, Ser. A. I.* 24, No. 4 (1925). **FM** 51, p. 360.

(a) "Remarques sur le lemme de Schwarz," *C. R. Acad. Sci. Paris* 188 (1929), 1027–1029. **FM** 55, p. 768.

(b) *Le théorème de Picard-Borel et la théorie des fonctions méromorphes,* Gauthier-Villars (1929), Paris. Reprinted by Chelsea Publishing Co. (1974), New York. **FM** 55, p. 773. **Zbl** 357.30019. **MR** 54 #5468.

(c) "Über beschränkte analytische Funktionen," *Ann. Acad. Scient. Fennicae, Ser. A. I.* 32 (Lindelöf-Festschrift), Nr. 7 (1929). **FM** 55, p. 768.

(d) "Sur un problème d'interpolation," *C. R. Acad. Sci. Paris* 188 (1929), 1224–1226. **FM** 55, p. 768.

"Über eine Minimumaufgabe in der Theorie der konformen Abbildung," *Göttingen Nachr.* (1933), 103–115. **FM** 59, p. 348. **Zbl** 6, p. 408.

"Das harmonische Mass von Punktmengen und seine Anwendung in der Funktionentheorie," pp. 116–133 of *Comptes rendus du huitième Congrès des mathématiciens scandinaves tenu à Stockholm 14–18 août 1934.* Håkon Ohlssons Boktryckeri (1935), Lund. **FM** 61, p. 306. **Zbl** 12, p. 78.

"Beweis des Satzes über die Vertauschbarkeit der Differentiationen," *Math. Zeit.* 56 (1952), 121–122. **Zbl** 47, p. 293. **MR** 14, p. 257.

Eindeutige Analytische Funktionen, Die Grundlehren der mathematischen Wissenschaften in Einzeldarstellungen, Bd. 46. Springer-Verlag (2nd ed. 1953), Berlin. **FM** 62, p. 315. **Zbl** 278.30002. **MR** 15, p. 208 and 49 #9165. **BAMS** 45, p. 52.

"Entwicklung der Theorie der eindeutigen analytischen Funktionen einer komplexen Veränderlichen seit Weierstrass," *Arbeitsgemeinschaft Nordrhein-Westfalen. Wiss. Abhand.* 33 (1966), 97–122. **Zbl** 146, p. 297. **MR** 33 #4284.

Analytic Functions (translated from German by P. Emig), Die Grundlehren der mathematischen Wissenschaften in Einzeldarstellungen, Bd. 162. Springer-Verlag (1970), Berlin. **Zbl** 199, p. 125. **MR** 43 #5003.

"On the minimal modulus of an entire function," pp. 418–420 of *Complex analysis and its applications* (Russian), N. N. Bogoljubov, M. A. Lavent'ev and A. V. Bicadze, editors, Izdat. "Nauka" (1978), Moscow. **Zbl** 435.30023. **MR** 80a:30028.

NEVILLE, E. H.

"A trigonometrical identity," *Proc. Camb. Phil. Soc.* 47 (1951), 629–632. **Zbl** 43, p. 285. **MR** 13, p. 19.

NEWMAN, D. J.

"Entire functions," *Amer. Math. Monthly* 67 (1960), 597.

"An entire function bounded in every direction," *Amer. Math. Monthly* 83 (1976), 192–193. **Zbl** 321.30033. **MR** 52 #8433.

NEWMAN, D. J. and BARLAZ, J.

"Arbitrarily large entire functions," *Amer. Math. Monthly* 60 (1953), 427.

NEWMAN, M. H. A.

Elements of the Topology of Plane Sets of Points, Cambridge University Press (2nd ed. 1951), London, reprinted by Dover (1992), New York. **FM** 65, p. 873. **Zbl** 123, p. 343. **MR** 24 #A2374. **BAMS** 58, p. 101.

NICHOLAS, C. P.

"Another look at the probability integral," *Amer. Math. Monthly* 64 (1957), 739–741.

NICOLESCO, M.

"Sur un critère d'harmonicité de Volterra et Vitali," *C. R. Acad. Sci. Roum.* 7 (1945), 16–19. **MR** 9, p. 142.

NIELAND, L. W.

"Over $\sum_{n=1}^{m} \frac{1}{n^{2h}}$ en $\sum_{n=1}^{\infty} \frac{1}{n^{2h}} = \zeta(2h)$," *Nieuw Archief voor Wiskunde* (2) 16 (1929), 1–13. **FM** 55, p. 130.

NIELSEN, N.

Traité élémentaire des nombres de Bernoulli, Gauthier-Villars (1923), Paris. **FM** 49, p. 99 and 50, p. 170.

NIEVERGELT, Y.

"Yet a stronger generalization of Rouché's theorem," *Int. Jour. Math. Educ. Sci. Technol.* 26 (1995), 156–158. **Zbl** 880.30002.

NIIZEKI, S. and SASAKI, M.

"Remarks on the integral of $\exp(-x^2)$ over $\mathbb{R}$," *Mem. Fac. Sci. Kochi Univ. (Math.)* 18 (1997), 105–110. **Zbl** 876.26002. **MR** 97j:26001.

NITSCHE, J. C. C.

"On the module of double connected regions under harmonic mappings," *Amer. Math. Monthly* 69 (1962), 781–782. **Zbl** 109, p. 305. (See also **Zbl** 1214.31001.)

NIVEN, I.

Irrational Numbers, Carus Mathematical Monographs, No. 11, John Wiley and Sons (1956), New York. **Zbl** 70, p. 271. **MR** 18, p. 195. **BAMS** 64, p. 68.

"Formal power series," *Amer. Math. Monthly* 76 (1969), 871–889. **Zbl** 184, p. 296. **MR** 40 #5606.

NOAILLON, P.

"Point singulier isolé non critique d'une fonction harmonique," *C. R. Acad. Sci. Paris* 184 (1927), 360–362. **FM** 53, p. 464.

NÖBELING, G.

"Eine allgemeine Fassung des Hauptsatzes der Funktionentheorie von Cauchy," *Math. Annalen* 121 (1949), 54–66. **Zbl** 33, p. 360. **MR** 10, p. 691.

"Ein gemeinsamer Beweis für den Jordanschen Kurvensatz und zwei damit zusammenhängende Sätze," *Jour. für Reine und Angew. Math.* 188 (1950), 22–39. **Zbl** 38, p. 362. **MR** 14, p. 192.

NOSHIRO, K.

"Some theorems on a cluster-set of an analytic function," *Proc. Imper. Acad. Tokyo* 13 (1937), 27–29. **FM** 63, p. 283. **Zbl** 17, p. 407.

Cluster Sets, Ergebnisse der Mathematik und ihrer Grenzgebiete, Neue Folge, Heft 28. Springer-Verlag (1960), Berlin. **Zbl** 90, p. 288. **MR** 24 #A3295.

NOVINGER, W. P.

(a) "An elementary approach to the problem of extending conformal maps to the boundary," *Amer. Math. Monthly* 82 (1975), 279–282. **Zbl** 313.30007. **MR** 51 #8390.

(b) "Some theorems from geometric function theory: applications," *Amer. Math. Monthly* 82 (1975), 507–510. **Zbl** 313.30008. **MR** 51 #858.

O

O'HARA, P. J.

"Another proof of Bernstein's theorem," *Amer. Math. Monthly* 80 (1973), 673–674. **Zbl** 271.30037. **MR** 48 #507.

OLDS, C. D.

"The Fresnel integrals," *Amer. Math. Monthly* 75 (1968), 285–286. **Zbl** 161, p. 248.

OMAN, G.

"An independent axiom system for the real numbers," *The College Math. Jour.* 40 (2009), 78–86. **MR** 2010d:03018.

ONICESCU, O.

"Sur le théorème fondamental de l'algèbre," *Mathematica, Timişoara* 22 (1946), 208–214. **Zbl** 60, p. 46. **MR** 8, p. 127.

OPFER, G.

"Konforme Abbildungen und ihre numerische Behandlung," *Überblicke Math.* Band 7 (1974), 33–113. **Zbl** 305.30011. **MR** 51 #871.

OPIAL, Z.

"Sur une famille de fonctions analytiques," *Annales Polonici Mathematici* 3 (1957), 312–318. **Zbl** 80, p. 57. **MR** 18, p. 885.

ORTEL, M and SMITH, W.

"A covering theorem for continuous locally univalent maps of the plane," *Bull. Lon. Math. Soc.* 18 (1986), 359–363. **Zbl** 584.30042. **MR** 88b:30013.

OSGOOD, W. F.

"Some points in the elements of the theory of functions," *Bull. Amer. Math. Soc.* 2 (1895/96), 296–302. **FM** 27, p. 301.

"Über einen Satz des Herrn Schönflies aus der Theorie der Functionen zweier reeller Veränderlichen," *Göttingen Nachr.* (1900), 94–97. **FM** 31, p. 478.

"Allgemeine Theorie der analytischen Funktionen a) einer und b) mehrerer komplexen Grössen," *Encyklopädie der Mathematischen Wissenschaften* Bd. II, 2nd Part, Heft 1, 1–114. B. G. Teubner (1901), Leipzig. **FM** 33, p. 389. (See also **MR** 93m:00053b.)

"Note on the functions defined by infinite series whose terms are analytic functions of a complex variable; with corresponding theorems for definite integrals," *Annals of Math.* (2) 3 (1901–02), 25–34. **FM** 32, p. 399. (See also ARZELÀ [1903–04].)

"On the Neumann's existence proof," *Annals of Math.* (2) 25 (1923–24), 238–240. **FM** 50, p. 648.

Lehrbuch der Funktionentheorie I, B. G. Teubner (5th ed. 1928), Leipzig and Berlin. **FM** 54, p. 326. **BAMS** 37, p. 330. Reprinted by Chelsea Publishing Co. (1965), New York. **Zbl** 184, p. 297. **MR** 33 #4233(a).

OSGOOD, W. F. AND TAYLOR, E. H.

"Conformal transformations on the boundaries of their regions of definition," *Trans. Amer. Math. Soc.* 14 (1913), 277–298. **FM** 44, p. 758.

OSSA, E.

"A simple proof of the Tietze-Urysohn extension theorem," *Arch. der Math.* 71 (1998), 331–332. **Zbl** 909.54018. **MR** 99h:54023.

OSSERMAN, R.

"A sharp Schwarz inequality on the boundary," *Proc. Amer. Math. Soc.* 128 (2000), 3513–3517. **Zbl** 963.30014. **MR** 2001b:30039.

Ostrowski, A.

"Über vollständige Gebiete gleichmässiger Konvergenz von Folgen analytischer Funktionen," *Abhand. Math. Sem. Univ. Hamburg* 1 (1922), 327–350. **FM** 48, p. 372.

"Über die Bedeutung der Jensenschen Formel für einige Fragen der komplexen Funktionentheorie," *Acta Litt. ac Scient. Univ. Hung.* (Szeged) 1 (1922/23), 80–87. **FM** 49, p. 713.

"Über allgemeine Konvergenzsätze der komplexen Funktionentheorie," *Jahresber. Deutsch. Math. Verein.* 32 (1923), 185–194. **FM** 49, p. 231.

(a) "Über den Schottkyschen Satz und die Borelschen Ungleichungen," *Sitzungsber. Preuss. Akad. Wissen. Berlin, Phys. -math. Kl.* (1925), 471–484. **FM** 51, p. 262.

(b) "Über Folgen analytischer Funktionen und einige Verschärfungen des Picardschen Satzes," *Math. Zeit.* 24 (1925), 215–258 and 38 (1934), 642. **FM** 51, p. 260 and 60, p. 243.

(a) "Mathematische Miszellen III: Über Nullstellen gewisser im Einheitskreis regulärer Funktionen und einige Sätze zur Konvergenz unendlicher Reihen," *Jahresber. Deutsch. Math. Verein.* 34 (1926), 161–171. **FM** 51, p. 246.

(b) "Mathematische Miszellen VI: Über einen Satz von Herrn Hadamard," *Jahresber. Deutsch. Math. Verein.* 35 (1926), 179–182. **FM** 52, p. 292.

(a) "Mathematische Miszellen, XV: Zur konformen Abbildung einfach zusammenhängender Gebiete," *Jahresber. Deutsch. Math. Verein.* 38 (1929), 168–182. **FM** 55, p. 788.

(b) "Über Schwankungen analytischer Funktionen, die gegebene Werte nicht annehmen," *Sitzungsber. Preuss. Akad. Wissen. Berlin, Phys. -math. Kl.* (1929), 276–289. **FM** 55, p. 194.

"Über konforme Abbildungen annähernd kreisförmiger Gebiete," *Jahresber. Deutsch. Math. Verein.* 39 (1930), 78–81. **FM** 56, p. 297.

Studien über den Schottkyschen Satz, B. Wepf et Cie. (1931), Basel. **FM** 57, p. 350. **Zbl** 3, p. 211.

"Asymptotische Abschätzung des absoluten Betrages einer Funktion, die die Werte 0 und 1 nicht annimmt," *Comm. Math. Helv.* 5 (1933), 55–87. **FM** 59, p. 327. **Zbl** 5, p. 251.

"Zur Randverzerrung bei konformer Abbildung," *Prace Mat.-Fiz.* 44 (1937), 371–471. **FM** 62, p. 381. **Zbl** 20, p. 238.

"Sur la continuité relative des racines d'équations algébriques," *C. R. Acad. Sci. Paris* 209 (1939), 777–779. **FM** 65, p. 1119. **Zbl** 26, p. 53. **MR** 1, p. 97.

"Recherches sur la méthode de Graffe et les zéros des polynômes et des séries de Laurent," *Acta Math.* 72 (1940), 99–257 and 75 (1942), 183–186. **FM** 66, p. 578. **Zbl** 23, p. 334. **MR** 1, p. 323; 2, p. 342; 7, p. 58.

(a) "Sur l'inverse d'une transformation continue et biunivoque," *C. R. Acad. Sci. Paris* 223 (1946), 229–230. **Zbl** 60, p. 406. **MR** 8, p. 49.

(b) "Nouvelle démonstration du théorème de Schoenflies pour les espaces à n dimensions," *C. R. Acad. Sci. Paris* 223 (1946), 530–531. **Zbl** 60, p. 406. **MR** 8, p. 164.

OTHMAN, S. I.

"Harmonic functions with asymptotic growth," *Jour. of Analysis* 4 (1996), 17–21. **Zbl** 873.31001. **MR** 97g:31002.

OU, ŠO-MO

"Some properties of analytic function omitting two values," *Rev. Roum. Math. Pures et Appliq.* 2 (1957), 145–150. **Zbl** 88, p. 286. **MR** 20 #3284.

P

PACE, L.

"Probabilistically proving that $\zeta(2) = \pi^2/6$," *Amer. Math. Monthly* 118 (2011), 641–643. **Zbl** 1227.00039. **MR** 2012g:11216.

PADÉ, H.

"Note sur la formule $\sin x = x\prod_{n=1}^{\infty}(1 - \frac{x^2}{n^2\pi^2})$," *Nouvelles Annales de Math.* (3) 17 (1898), 312–314. **FM** 29, p. 376.

PAINLEVÉ, P.

"Sur le développement en série de polynômes d'une fonction holomorphe dans une aire quelconque," *C. R. Acad. Sci. Paris* 102 (1886), 672–675. **FM** 18, p. 346.

"Sur les lignes singulières des fonctions analytiques," *Annales de la Faculté des Sciences de Toulouse* (1) 2 (1887), 1–130. **FM** 20, p. 404.

"Sur la représentation des fonctions analytiques uniformes," *C. R. Acad. Sci. Paris* 126 (1898), 200–202. **FM** 29, p. 361.

"Sur le développement des fonctions uniformes ou holomorphes dans un domaine quelconque," *C. R. Acad. Sci. Paris* 126 (1898), 318–321. **FM** 29, p. 361.

PÁL, J.

"Zur Topologie der Ebene," *Acta Litt. ac Scient. Univ. Hung.* (Szeged) 1 (1922/23), 226–239. **FM** 49, p. 409.

PALKA, B. P.

An Introduction to Complex Function Theory, Springer-Verlag (1991), New York. **Zbl** 726.30001. **MR** 92b:30001.

PANG, XUE CHENG

"Bloch's principle and normal criterion," *Sci. China Ser. A* 32 (1989), 782–791. **Zbl** 687.30023. **MR** 91i:30031. (See also **MR** 90i:30051.)

PARAMEŚWARAN, S.
"Whish's showroom revisited," *Math. Gazette* 76 (1992), 28–36. **Zbl** 869.01004.

PARKER, M. J.
"Mean values and distance functions in potential theory," Ph. D. thesis, University of Liverpool (1987).

PARTINGTON, J. R.
Interpolation, Identification and Sampling, London Mathematical Society Monographs new series 17, Oxford University Press (1997). **Zbl** 892.93004. **MR** 99m:41001.

PARVATHAM, R.
"Generalized Jack-Miller-Mocanu lemma," *Math. Student* 57 (1989), 145–152. **Zbl** 723.30013. **MR** 91e:30046.

PASCH, M.
"Über die Einführung des Imaginären," *Arch. der Math. und Physik* (3) 7 (1904), 102–108. **FM** 35, p. 196.

PATIL, D. J.
"Representation of H^p-functions," *Bull. Amer. Math. Soc.* 78 (1972), 617–620. **Zbl** 255.30010. **MR** 45 #7069. (See also **Zbl** 317.60027 and **MR** 54 #8863.)

PATYI, I.
"On some elementary functions," *Math. Gazette* 99 (2015), 263–275. **Zbl** 06783723. **MR** 3414148.

PEANO, G.
"Sulle funzioni interpolari," *Atti della R. Accad. d. Scienze di Torino* 18 (1884), 573–580. **FM** 16, p. 197.

PEIFFER, J.
"Joseph Liouville (1809–1882): ses contributions à la théorie des fonctions d'une variable complexe," *Rev. Hist. Sci.* 36 (1983), 209–248. **Zbl** 531.01003. **MR** 87j:01031.

PELLING, M. J.
"On the singularity set of complex functions satisfying the Cauchy-Riemann equations," *Mich. Math. Jour.* 26 (1979), 225–229. **Zbl** 452.30005. **MR** 80i:30041.

PERETZ, R.
"Applications of subordination theory to the class of bounded nonvanishing functions," *Complex Variables* 17 (1992), 213–222. **Zbl** 738.30025. **MR** 93c:30050.

PERKINS, F. W.
"On the oscillation of harmonic functions," *Annals of Math.* (2) 27 (1925–26), 159–170. **FM** 51, pp. 359, 365.

"An intrinsic treatment of Poisson's integral," *Amer. Jour. of Math.* 50 (1928), 389–414. **FM** 54, p. 326.

PERRON, O.

"Eine neue Behandlung der ersten Randwertaufgabe für $\Delta u = 0$," *Math. Zeit.* 18 (1923), 42–54. **FM** 49, p. 340.

"Über die Picard-Landauschen Sätze," *Göttingen Nachr.* (1929), 65–72. **FM** 55, p. 192.

"Ein Beweis des Fundamentalsatzes der Algebra im Reellen," *Annali Mat. Pura Appl.* (4) 28 (1949), 183–187. **Zbl** 39, p. 11. **MR** 12, p. 4.

PERRY, A. D. and YOUNGS, J. W. T.

"Remarks on analyticity and integration," *Amer. Math. Monthly* 54 (1947), 313–318. **Zbl** 29, p. 31. **MR** 9, p. 138.

PESCHL, E.

Funktionentheorie. I, B. I. Hochschultaschenbücher. Band 131/131a, Bibliographisches Institut (1967), [2nd edition (1983)], Mannheim. **Zbl** 533.30001. **MR** 86b:30003.

PETERSEN, J.

"Quelques remarques sur les fonctions entières," *Acta Math.* 23 (1900), 85–90. **FM** 30, p. 378.

PETROVA, S. S.

"Über den ersten Beweis des Fundamentalsatzes der Algebra," *Istor. Metodolog. estestv. Nauk* 11 (1971), 123–127 (Russian). **Zbl** 261.01013. (See also **Zbl** 263.01022 and **MR** 57 #15879.)

"Aus der Geschichte der analytischen Beweise des Fundamentalsatzes der Algebra," *Istor. Metodolog. estestv. Nauk* 14 (1973), 167–172 (Russian). **Zbl** 285.01013. **MR** 58 #21227.

"Sur l'histoire des démonstrations analytiques du théorème fondamental de l'algèbre," *Historia Math.* 1 (1974), 255–261. **Zbl** 281.01008. **MR** 57 #5609.

PFALTZGRAFF, J. A.

"Univalence constraints on the Schwarz-Christoffel parameters," *SIAM Jour. of Math. Anal.* 17 (1986), 231–235. **Zbl** 588.30008. **MR** 87b:30007.

PFLUGER, A.

"Über numerische Schranken im Schottkyschen Satz," *Comm. Math. Helv.* 7 (1934–35), 159–170. **FM** 61, p. 355. **Zbl** 11, p. 119.

"Operatorwertige analytische Funktionen und das Lemma von Julia," *Festband 70. Geburtstag R. Nevanlinna,* pp. 82–88. Springer-Verlag (1966), Berlin. **Zbl** 168, p. 118. **MR** 34 #6133.

Lectures on Conformal Mapping, Department of Mathematics, Indiana University (1969), Bloomington.

PHRAGMÉN, E.

"Über die Begrenzung von Continua," *Acta Math.* 7 (1885/86), 43–48. **FM** 17, p. 505.

"Remarques sur la théorie de la représentation conforme," *Acta Math.* 14 (1890–91), 225–232. **FM** 23, p. 890.

"Ein elementarer Beweis des Fundamentalsatzes der Algebra," *Öfversigt af Kongl. Svenska Vetenskaps-Akademiens Förhandlingar* 48 (1891), 113–129. **FM** 23, p. 86.

"Sur la représentation analytique des fonctions réelles, données sur un ensemble quelconque de points," *Rend. Circ. Mat. Palermo* 14 (1900), 256–261. **FM** 31, p. 410.

PHRAGMÉN, E. and LINDELÖF, E.

"Sur une extension d'un principe classique de l'analyse et sur quelques propriétés des fonctions monogènes dans le voisinage d'un point singulier," *Acta Math.* 31 (1908), 381–406. **FM** 39, p. 465.

PICARD, E.

(a) "Sur une propriété des fonctions entières," *C. R. Acad. Sci. Paris* 88 (1879), 1024–1027. **FM** 11, p. 267.

(b) "Sur les fonctions analytiques uniformes dans le voisinage d'un point singulier essentiel," *C. R. Acad. Sci. Paris* 89 (1879), 745–747. **FM** 11, p. 267.

(c) "Sur les fonctions entières," *C. R. Acad. Sci. Paris* 89 (1879), 662–665. **FM** 11, p. 268.

"Mémoire sur les fonctions entières," *Ann. Sci. École Norm. Sup.* (2) 9 (1880), 145–166. **FM** 12, p. 327.

"Sur la décomposition en facteurs primaires des fonctions uniformes ayant une ligne de points singuliers essentiels," *C. R. Acad. Sci. Paris* 92 (1881), 690–692. **FM** 13, p. 316.

Traité d'analyse Vol. II, Gauthier-Villars (2nd ed. 1905), Paris. **FM** 36, p. 355.

"Deux théorèmes élémentaires sur les singularités des fonctions harmoniques," *C. R. Acad. Sci. Paris* 176 (1923), 933–935. **FM** 49, p. 338.

"Quelques théorèmes élémentaires sur les fonctions harmoniques," *Bull. Soc. Math. France* 52 (1924), 162–166. **FM** 50, p. 333.

PICK, G.

(a) "Über eine Eigenschaft der konformen Abbildung kreisförmiger Bereiche," *Math. Annalen* 77 (1916), 1–6. **FM** 45, p. 671.

(b) "Über die Beschränkungen analytischer Funktionen, welche durch vorgegebene Funktionswerte bewirkt werden," *Math. Annalen* 77 (1916), 7–23. **FM** 45, p. 642.

"Über beschränkte Funktionen mit vorgeschriebenen Wertzuordnungen," *Ann. Acad. Scient. Fennicae, Ser. A. I.* 15, Nr. 3 (1920).

PICONE, M.

(a) "Sul problema di Dirichlet per la corona circolare," *Boll. della Un. Mat. Ital.* 5 (1926), 114–118. **FM** 52, p. 493.

(b) "Sulle singolarità delle funzioni armoniche," *Rend. Accad. d. Lincei Roma* (6) 3 (1926), 655–660. **FM** 52, p. 496.

"Sulle singolarità isolate delle funzioni armoniche," *Rend. Accad. d. Lincei Roma* (6) 9 (1929), 1067–1073. **FM** 55, p. 889.

"Sul calcolo delle funzioni olomorfe di una variabile complessa," pp. 118–126 of *Studies in Mathematics and Mechanics presented to Richard von Mises*, Academic Press (1954), New York. **Zbl** 58, p. 301. **MR** 16, p. 459.

PIEPER, H.

Die Komplexen Zahlen: Theorie-Praxis-Geschichte, Deutscher Verlag der Wissenschaften (3rd ed. 1991), Berlin. **Zbl** 741.00001. **MR** 92j:12003. (See also **Zbl** 941.01037, 629.01001 and **MR** 91i:01152, 01153.)

PIER, J.-P.

"Historique de la notion de compacité," *Historia Math.* 7 (1980), 425–443. **Zbl** 451.01005. **MR** 82f:01068.

PINGYA, XU

"The sufficient condition for the differentiability of functions of several variables," *Math. Mag.* 71 (1998), 297–299.

PINKUS, A.

"Weierstrass and approximation theory," *Jour. Approx. Theory* 107 (2000), 1–66. **Zbl** 968.41001. **MR** 2001k:01039. (See also **MR** 2006m:41002.)

PIRANIAN, G. and THRON, W.

"Convergence properties of sequences of linear fractional transformations," *Mich. Math. Jour.* 4 (1957), 129–135. **Zbl** 80, p. 282. **MR** 20 #102. (See also **MR** 2005g:30051.)

PLANCHEREL, M.

"Les problèmes de Cantor et de du Bois-Reymond dans la théorie des séries de polynômes de Legendre," *Ann. Sci. École Norm. Sup.* (3) 31 (1914), 223–262. **FM** 45, p. 704.

PLASTOCK, R.

"Homeomorphisms between Banach spaces," *Trans. Amer. Math. Soc.* 200 (1974), 169–183. **Zbl** 291.54009. **MR** 50 #8593.

PLEMELJ, J.

"Die Grenzkreis-Uniformisierung analytischer Gebilde," *Monatshefte für Math. und Physik* 23 (1912), 297–304. **FM** 43, p. 519.

"Ein Abschätzungssatz der Potentialtheorie," *Publications Math. Univ. Belgrade* 2 (1933), 150–153. **FM** 59, p. 1133. **Zbl** 8, p. 357.

PLESSNER, A.

"Zur Theorie der konjugierten trigonometrischen Reihen," *Mitt. Math. Sem. Univ. Giessen* 10 (1923). **FM** 49, p. 204.

POINCARÉ, H.

"Sur les fonctions à espaces lacunaires," *Amer. Jour. of Math.* 14 (1892), 201–221. **FM** 24, p. 388.

"Sur l'uniformisation des fonctions analytiques," *Acta Math.* 31 (1907), 1–63. **FM** 38, p. 452.

POL, VAN DER B.

"Démonstration élémentaire de la relation $\Theta_3^4 = \Theta_0^4 + \Theta_2^4$ entre les differéntes fonctions de Jacobi," *L'Enseignement Mathématique* (2) 1 (1955), 258–261 and 2 (1956), 60. **Zbl** 71, p. 80. **MR** 17, p. 1082.

POLLARD, S.

"On the conditions for Cauchy's theorem," *Proc. Lon. Math. Soc.* (2) 21 (1923), 456–482. **FM** 49, p. 209.

PÓLYA, G.

"Ueber ganzwertige ganze Funktionen," *Rend. Circ. Mat. Palermo* 40 (1915), 1–16. **FM** 45, p. 655.

"Über Potenzreihen mit ganzzahligen Koeffizienten," *Math. Annalen* 77 (1916), 497–513. **FM** 46, p. 481.

"Bestimmung einer ganzen Funktion endlichen Geschlechts durch viererlei Stellen," *Mat. Tidsskr. B* (1921), 16–21. **FM** 48, p. 354.

(a) "Arithmetische Eigenschaften und analytischer Charakter," *Jahresber. Deutsch. Math. Verein.* 31 (1922), 107–115. **FM** 48, p. 327.

(b) "Über die Nullstellen sukzessiver Derivierten," *Math. Zeit.* 12 (1922), 36–60. **FM** 48, p. 370.

"Elementarer Beweis einer Thetaformel," *Sitzungsber. Preuss. Akad. Wissen. Berlin, Phys.-math. Kl.* (1927), 158–161. **FM** 53, p. 344.

"Untersuchungen über Lücken und Singularitäten von Potenzreihen," *Math. Zeit.* 29 (1929), 549–640; French résumés in *Boll. della Un. Mat. Ital.* 8 (1929), 211–214 and *Atti del Congresso Internazionale dei Matematici Bologna, Settembre 1928,* Vol. III, 243–247. Part II, *Annals of Math.* (2) 34 (1933), 731–777. **FM** 55, p. 186; 56, p. 271; 59, p. 319. **Zbl** 8, p. 62.

"Stelle der grössten Beanspruchung an der Oberfläche?", *Zeit. für Math. and Mech.* 10 (1930), 353–360. **FM** 56, p. 1228.

"Über analytische Deformationen eines Rechtecks," *Annals of Math.* (2) 34 (1933), 617–620. **FM** 59, p. 348. **Zbl** 7, p. 169.

"Sur les séries entières lacunaires non-prolongeables," *C. R. Acad. Sci. Paris* 208 (1939), 709–711. **FM** 65, p. 310. **Zbl** 20, p. 234.

"Sur l'existence de fonctions entières satisfaisant à certaines conditions linéaires," *Trans. Amer. Math. Soc.* 50 (1941), 129–139. **FM** 67, p. 269. **Zbl** 25, p. 166. **MR** 2, p. 356.

The Pólya Picture Album: Encounters with a Mathematician (G. L. Alexanderson, editor), Birkhäuser Verlag (1987), Boston and Basel. **Zbl** 614.01028. **MR** 88i: 01125.

PÓLYA, G. and SZEGÖ, G.

Aufgaben und Lehrsätze aus der Analysis, Bd. I, II. Die Grundlehren der mathematischen Wissenschaften in Einzeldarstellungen, Bd. 19, 20. Springer-Verlag (3rd ed. 1964), Berlin. **FM** 51, p. 173. **Zbl** 201, p. 381. **MR** 42 #6160 and 49 #8781. **BAMS** 34, p. 233.

Problems and Theorems in Analysis, Vol. I (translated from German by D. Aeppli), Die Grundlehren der mathematischen Wissenschaften in Einzeldarstellungen, Bd. 193. Springer-Verlag (1972), Berlin and New York. Vol. II (translated from German by C. E. Billigheimer), *ibid.* Bd. 216 (1976). **Zbl** 338.00001; 311.00002 and 359.00003. **MR** 81e:00002 and 57 #5529. **BAMS** 84, p. 53.

POMEY, É.

"Nouvelles démonstrations du théorème de Dalembert," *Nouvelles Annales de Math.* (4) 5 (1905), 388–394. **FM** 36, p. 121.

POMMERENKE, CH.

"On sequences of subordinate functions," *Mich. Math. Jour.* 7 (1960), 181–185. **Zbl** 91, p. 252. **MR** 27 #1598.

"Images of convex domains under conformal mappings," *Mich. Math. Jour.* 9 (1962), 257–269. **Zbl** 114, p. 279. **MR** 26 #6378.

"Lacunary power series and univalent functions," *Mich. Math. Jour.* 11 (1964), 219–223. **Zbl** 135, p. 124. **MR** 29 #4883.

"Polynome und konforme Abbildung," *Monatshefte für Math.* 69 (1965), 58–61. **Zbl** 151, p. 93. **MR** 30 #4919.

Univalent Functions, Studia Mathematica Bd. 25, Vandenhoeck & Ruprecht (1975), Göttingen. **Zbl** 298.30014. **MR** 58 #22526.

"On the iteration of analytic functions in a half-plane, I," *Jour. Lon. Math. Soc.* (2) 19 (1979), 439–447. **Zbl** 405.30018. **MR** 83j:30023.

"Boundary behavior of conformal mappings," pp. 313–331 of *Aspects of Contemporary Complex Analysis* (D. A. Brannan and J. G. Clunie, editors), Academic Press (1980), London and New York. **Zbl** 491.30009. **MR** 82i:30008.

"On asymptotic iteration of analytic functions in the disk," *Analysis* 1 (1981), 45–61. **Zbl** 477.30019. **MR** 82g:30048.

"On the boundary continuity of conformal maps," *Pac. Jour. of Math.* 120 (1985), 423–430. **Zbl** 573.30007. **MR** 87c:30011.

Boundary Behaviour of Conformal Maps, Grundlehren der mathematischen Wissenschaften 299, Springer-Verlag (1992), New York. **Zbl** 762.30001. **MR** 95b:30008.

POMPEIU, D.

(a) "Sur la continuité des fonctions de variables complexes," *Annales de la Faculté des Sciences de Toulouse* (2) 7 (1905), 264–315. **FM** 36, p. 454.

(b) "Sur l'extension du théorème des accroissements finis aux fonctions analytiques d'une variable complexe," *Rend. Circ. Mat. Palermo* 19 (1905), 309–313. **FM** 36, p. 454.

"Einige Sätze über monogene Funktionen," *Sitzungsber. Kaiserl. Akad. Wiss. Wien* 120 (1911), Zweite Abteilung, 1249–1252. **FM** 42, p. 422. (See also **FM** 57, p. 346 and **Zbl** 3, p. 300.)

"Sur une propriété des fonctions holomorphes," *Bull. Math. Section Sci. Acad. Roumaine* 30 (1927), 67–68. **FM** 53, p. 318.

"Sur une propriété des fonctions holomorphes," *Ann. Scuola Norm. Sup. Pisa* (2) 2 (1933), 227–230. **FM** 59, p. 1023. **Zbl** 6, p. 316.

PONOMAREV, S. P.

"On a condition of analyticity," *Siberian Math. Jour.* 11 (1970), 360–363. **Zbl** 203, p. 72. **MR** 41 #8636.

PORTER, M. B.

(a) "On functions defined by an infinite series of analytic functions of a complex variable," *Annals of Math.* (2) 6 (1904–1905), 45–48. **FM** 35, p. 397.

(b) "Concerning series of analytic functions," *Annals of Math.* (2) 6 (1904–1905), 190–192. **FM** 36, p. 459.

POUKKA, K. A.

"Über die grösste Schwankung einer analytischen Funktion auf einer Kreisperipherie," *Arch. der Math. und Physik* (3) 12 (1907), 251–254. **FM** 33, p. 433. (See also **FM** 62, p. 368 and **Zbl** 13, p. 211.)

PRACHAR, K.

"Über einige einfache Folgen und Reihen im Schulunterricht," *Elemente der Math.* 30 (1975), 36–39. **MR** 51 #1188. (See also *Amer. Math. Monthly* 108 (2001), 357.)

PRAWITZ, H.

"Über Mittelwerte analytischer Funktionen," *Arkiv för Mat., Astro. och Fysik* 20A, No. 6 (1927–1928). **FM** 53, p. 307.

PRIBITKIN, W. DE AZEVEDO

"A simpler proof of $\sin \pi z = \pi z \prod(1 - z^2/k^2)$," *Amer. Math. Monthly* 108 (2001), 767–768. **Zbl** 1129.30300.

PRIESTLEY, W. M.

"Complex sequences whose 'moments' all vanish," *Proc. Amer. Math. Soc.* 116 (1992), 437–444. **Zbl** 757.40002. **MR** 92m:40003.

PRINGSHEIM, A.

"Zur Theorie der Taylor'schen Reihe und der analytischen Functionen mit beschränktem Existenzbereich," *Sitzungsber. Kön. Bayer. Akad. Wissen. München* 22 (1892), 211–245 and *Math. Annalen* 42 (1893), 153–184. **FM** 24, p. 356 and 25, p. 386.

"Über Functionen, welche in gewissen Punkten endliche Differentialquotienten jeder endlichen Ordnung, aber keine Taylor'sche Reihenentwickelung besitzen," *Math. Annalen* 44 (1894), 41–56. **FM** 25, p. 389.

"Über die Entwickelung eindeutiger analytischer Functionen in Potenzreihen," *Sitzungsber. Kön. Bayer. Akad. Wissen. München* 25 (1895), 75–92. **FM** 26, p. 270.

(a) "Zur Theorie der synektischen Functionen," *Sitzungsber. Kön. Bayer. Akad. Wissen. München* 26 (1896), 167–182. **FM** 27, p. 300.

(b) "Über Vereinfachungen in der elementaren Theorie der analytischen Functionen," *Math. Annalen* 47 (1896), 121–154. **FM** 27, p. 300.

"Zur Geschichte des Taylor'schen Lehrsatzes," *Biblio. Math.* (3) 1 (1900), 433–479. **FM** 31, p. 42.

"Über den Goursat'schen Beweis des Cauchy'schen Integralsatzes," *Trans. Amer. Math. Soc.* 2 (1901), 413–421. **FM** 32, p. 308.

"Zur Theorie der ganzen transcendenten Functionen," *Sitzungsber. Kön. Bayer. Akad. Wissen. München* 32 (1902), 163–192 and 295–304. **FM** 33, p. 416.

"Der Cauchy-Goursat'sche Integralsatz und seine Übertragung auf reelle Kurven-Integrale," *Sitzungsber. Kön. Bayer. Akad. Wissen. München* 33 (1903), 673–682. **FM** 34, p. 339.

"Elementare Theorie der ganzen transcendenten Funktionen von endlicher Ordnung," *Math. Annalen* 58 (1904), 257–342. **FM** 35, p. 405.

"Über die Weierstrass'sche Produktdarstellung ganzer transzendenter Funktionen und über bedingt konvergente unendliche Produkte," *Sitzungsber. Kön. Bayer. Akad. Wissen. München* (1915), 387–400. **FM** 45, p. 652.

"Elementare Funktionentheorie und komplexe Integration," *Sitzungsber. Bayer. Akad. Wissen. München* (1920), 145–182 and (1921), 255–258. **FM** 47, p. 267 and 48, p. 317.

"Kritisch-historische Bemerkungen zur Funktionentheorie," *Sitzungsber. Bayer. Akad. Wissen. München* (1929), 95–112, 113–124, 281–306. **FM** 55, p. 760.

Vorlesungen über Zahlen- und Funktionenlehre Vol. II: *Funktionenlehre* (Part 1: *Grundlagen der Theorie der analytischen Funktionen einer komplexen*

Veränderlichen. Part 2: *Eindeutige analytische Funktionen.*) B. G. Teubner (1925), (1932), Leipzig and Berlin. **FM** 51, p. 237 and 58, p. 296. **Zbl** 5, p. 199. **BAMS** 39, p. 487. Reprinted by Johnson Reprint Corp. (1968), New York. **MR** 40 #5407b.

PRINGSHEIM, A., FABER, G. and MOLK, J.

"Analyse algébrique," *Encyclopédie des sciences mathématiques pures et appliquées,* t. II, v. 2, fasc. 1, 1–93. Gauthier-Villars (1911), Paris. **FM** 40, p. 300 and 42, p. 458. **MR** 93m:00053a.

PRINGSHEIM, A. and FRÉCHET, M.

"Algorithmes illimités de nombres complexes," *Encyclopédie des sciences mathématiques pures et appliquées,* t. I, v. 1, fasc. 3, 469–488. Gauthier-Villars (1908), Paris. **FM** 39, p. 50. **MR** 93m:00053a.

PRINGSHEIM, A. and MOLK, J.

"Algorithmes illimités," *Encyclopédie des sciences mathématiques pures et appliquées,* t. I, v. 1, fasc. 2, 209–328. Gauthier-Villars (1907), Paris. **FM** 38, p. 53. **MR** 93m:00052a.

"Principes fondamentaux de la théorie des fonctions," *Encyclopédie des sciences mathématiques pures et appliquées,* t. II, v. 1, fasc. 1, 1–112. Gauthier-Villars (1909), Paris. **FM** 40, p. 44.

PRIVALOV, I. I.

(a) "Eine Erweiterung des Satzes von Vitali über Folgen analytischer Funktionen," *Math. Annalen* 93 (1924), 149–152. **FM** 50, p. 247.

(b) "Sur les fonctions, qui donnent la représentation conforme bi-univoque," *Recueil Math. Moscou* (=*Mat. Sbornik*) 31 (1924), 350–365 (Russian). **FM** 50, p. 251.

(c) "Über einen Mittelwertsatz in der Theorie der analytischen Funktionen," *Recueil Math. Moscou* (=*Mat. Sbornik*) 32 (1924), 50–53. **FM** 50, p. 251.

(d) "Über die Folgen von analytischen Funktionen," *Recueil Math. Moscou* (=*Mat. Sbornik*) 32 (1924), 46–49 (Russian). **FM** 50, p. 247.

(e) "Sur les suites de fonctions analytiques," *C. R. Acad. Sci. Paris* 178 (1924), 178–180. **FM** 50, p. 246.

"Sur certaines questions de la théorie des fonctions subharmoniques et des fonctions analytiques," *Recueil Math. Moscou* (=*Mat. Sbornik*) 41 (1935), 527–550 (Russian). **FM** 60, p. 1135. **Zbl** 11, p. 314.

"Sur la théorie générale des fonctions harmoniques et subharmoniques," *Recueil Math. Moscou* (=*Mat. Sbornik*) (2) 1 (1936), 103–122 (Russian). **FM** 62, p. 556. **Zbl** 14, p. 21.

Randeigenschaften Analytischer Funktionen (translated from Russian by J. Auth), VEB Deutscher Verlag der Wissenschaften (1956), Berlin. **Zbl** 73, p. 65. **MR** 18, p. 727.

PUCCIANO, G.

"Sulle condizioni di validità del teorema di Cauchy," *Giornale di Mat. di Battaglini* 47 [(2), 16] (1909), 55–64. **FM** 40, p. 440.

PUCCIO, L.

"Su di una dimostrazione puramente algebrica ed elementare del teorema fondamentale dell' algebra," *Bolletino di Mat.* (2) 12 (1933), 110–118. **FM** 59, p. 121.

PULFORD, J. H. and WALDEN, B. L.

"Angle distortion and conformality," *Expositiones Math.* 21 (2003), 33–45. **Zbl** 1032.30005. **MR** 2003j:30018.

R

RADEMACHER, H.

(a) "Über streckentreue und winkeltreue Abbildung," *Math. Zeit.* 4 (1919), 131–138. **FM** 47, p. 268. (See also **Zbl** 311.01022.)

(b) "Bemerkungen zu den Cauchy-Riemannschen Differentialgleichungen und zum Moreraschen Satz," *Math. Zeit.* 4 (1919), 177–185. **FM** 47, p. 268. (See also **Zbl** 311.01022.)

Lectures on Elementary Number Theory, Blaisdell Publishing Co. (1964), New York. Reprinted by Robert E. Krieger Publishing Co., Inc. (1977), Huntington, New York. **Zbl** 119, p. 278. **MR** 58 #10677.

RADO, R.

"Über stetige Fortsetzung reeller Funktionen," *Sitzungsber. Bayer. Akad. Wissen. München* (1931), 81–84. **FM** 57, p. 300. **Zbl** 3, p. 154.

RADÓ, T.

(a) "Bemerkung zu einem Unitätssatze der konformen Abbildung," *Acta Litt. ac Scient. Univ. Hung.* (Szeged) 1 (1922/23), 101–103. **FM** 49, p. 247.

(b) "Über die Fundamentalabbildungen schlichter Gebiete," *Acta Litt. ac Scient. Univ. Hung.* (Szeged) 1 (1922/23), 240–251. **FM** 49, p. 712.

(c) "Zur Theorie der mehrdeutigen konformen Abbildungen," *Acta Litt. ac Scient. Univ. Hung.* (Szeged) 1 (1922/23), 55–64. **FM** 48, p. 1235.

(d) "Sur la représentation conforme de domaines variables," *Acta Litt. ac Scient. Univ. Hung.* (Szeged) 1 (1922/23), 180–186. **FM** 49, p. 247.

"Über die konformen Abbildungen schlichter Gebiete," *Acta Litt. ac Scient. Univ. Hung.* (Szeged) 2 (1924–26), 47–60. **FM** 50, p. 255 and 640.

"Über eine nicht fortsetzbare Riemannsche Mannigfaltigkeit," *Math. Zeit.* 20 (1924), 1–6. **FM** 50, p. 255.

"Bemerkung über die konformen Abbildungen konvexer Gebiete," *Math. Annalen* 102 (1930), 428–429. **FM** 55, p. 208 and 57, p. 1425.

"On continuous transformations in the plane," *Fund. Math.* 27 (1936), 201–211. **FM** 62, p. 809. **Zbl** 15, p. 418.

Subharmonic Functions, Ergebnisse der Mathematik und ihrer Grenzgebiete, Bd. 5. Verlag Julius Springer (1937), Berlin. Reprinted by Chelsea Publishing Co. (1949), New York and Springer-Verlag (1971), New York. **FM** 63, p. 458. **Zbl** 211, p. 138. **MR** 49 #9718. **BAMS** 43, p. 758.

Length and Area, American Mathematical Society Colloquium Publications, Vol. 30 (1948), Providence. **Zbl** 33, p. 170. **MR** 9, p. 505. **BAMS** 54, p. 861.

RADÓ, T. and RIESZ, F.

"Über die erste Randwertaufgabe für $\Delta u = 0$," *Math. Zeit.* 22 (1925), 41–44. **FM** 51, p. 361.

RÅDSTRÖM, H.

"On the iteration of analytic functions," *Math. Scand.* 1 (1953), 85–92. **Zbl** 50, p. 300. **MR** 15, p. 115.

RAJAGOPAL, C. T.

"On inequalities for analytic functions," *Amer. Math. Monthly* 60 (1953), 693–695. **Zbl** 53, p. 46. **MR** 15, p. 412.

"On an absolute constant for a class of power series," *Math. Scand.* 5 (1957), 267–270. **Zbl** 80, p. 281. **MR** 20 #5276. (See also **MR** 2008c:30003.)

RAMANKUTTY, P.

"Extensions of Liouville theorems," *Jour. Math. Anal. and Applic.* 90 (1982), 58–63. **Zbl** 505.31001. **MR** 84b:31001.

RAMANUJAN, S.

"Some definite integrals," *Messenger of Math.* 44 (1915), 10–18. **FM** 45, p. 1288.

RANGE, R. M.

"On a Lipschitz estimate for conformal maps in the plane," *Proc. Amer. Math. Soc.* 58 (1976), 375–376. **Zbl** 337.30004. **MR** 53 #11033.

RANSFORD, T. J.

Potential Theory in the Complex Plane, London Mathematical Society Student Texts 28, Cambridge University Press (1995), Cambridge and New York. **Zbl** 828.31001. **MR** 96e:31001.

RAO, N. V.

"A generalization of the Looman-Menchoff theorem," *Israel Jour. of Math.* 70 (1990), 93–103. **Zbl** 706.30002. **MR** 91k:30004. (See also **Zbl** 863.30001 and **MR** 96g:30002.)

RAYNOR, G. E.

"Isolated singular points of harmonic functions," *Bull. Amer. Math. Soc.* 32 (1926), 537–544. **FM** 52, p. 495.

"Note on the expansion of harmonic functions in the neighborhood of isolated singular points," *Annals of Math.* (2) 31 (1930), 35–42. **FM** 56, p. 1066.

READE, M. O.

"Some remarks on subharmonic functions," *Duke Math. Jour.* 10 (1943), 531–536. **MR** 5, p. 7.

REDHEFFER, R.

"The fundamental theorem of algebra," *Amer. Math. Monthly* 64 (1957), 582–585. **Zbl** 77, p. 25. **MR** 19, p. 537.

"What! Another note just on the fundamental theorem of algebra?", *Amer. Math. Monthly* 71 (1964), 180–185. **Zbl** 117, p. 38.

"The homotopy theorems of function theory," *Amer. Math. Monthly* 76 (1969), 778–787. **Zbl** 194, p. 375. **MR** 41 #7126.

REICH, E.

"Elementary proof of a theorem on conformal rigidity," *Proc. Amer. Math. Soc.* 17 (1966), 644–645. **Zbl** 153, p. 100. **MR** 33 #4247.

"The composition of harmonic mappings," *Ann. Acad. Scient. Fennicae, Ser. A. I. Math.* 12 (1987), 47–53. **Zbl** 636.31002. **MR** 88a:31001.

REICH, E. and WARSCHAWSKI, S. E.

"On canonical conformal maps of regions of arbitrary connectivity," *Pac. Jour. of Math.* 10 (1960), 965–985. **Zbl** 91, p. 255. **MR** 22 #8120.

"Canonical conformal maps onto a circular slit annulus," *Scripta Math.* 25 (1960), 137–146. **Zbl** 99, p. 60. **MR** 22 #12228.

REICHBACH, M.

"Generalizations of the fundamental theorem of algebra," *Bulletin Research Council of Israel,* vol. 7F (1958), 155–164. **Zbl** 98, p. 309. **MR** 23 #A2195.

REID, W. T.

"Green's lemma and related results," *Amer. Jour. of Math.* 63 (1941), 563–574. **FM** 67, p. 185. **Zbl** 25, p. 151. **MR** 3, p. 75.

REIFF, R.

Geschichte der unendlichen Reihen, H. Laupp Verlag (1889), Tübingen; reprinted by Martin Sändig (1969), Wiesbaden. **FM** 21, p. 31.

REIJNIERSE, J. M.

"Sur la limite d'une suite de polynômes," *Nieuw Archief voor Wiskunde* (2) 19 (1938), 241–248. **FM** 64, p. 1056. **Zbl** 18, p. 207.

REMAK, R.

"Über winkeltreue und streckentreue Abbildung an einem Punkte und in der Ebene," *Rend. Circ. Mat. Palermo* 38 (1914), 193–246. **FM** 45, p. 674.

"Über potentialkonvexe Funktionen," *Math. Zeit.* 20 (1924), 126–130. **FM** 50, p. 333.

"Über die erste Randwertaufgabe der Potentialtheorie," *Jour. für Reine und Angew. Math.* 156 (1926), 227–230. **FM** 53, p. 460.

Remmert, R.

(a) *Theory of Complex Functions* (translated from German by R. B. Burckel), Graduate Texts in Mathematics, Vol. 122. Springer-Verlag (1991), New York. **Zbl** 780.30001. **MR** 91m:30001.

(b) *Funktionentheorie II,* Grundwissen Mathematik 6, Springer-Verlag (1991), 3rd ed. (2007), Berlin. **Zbl** 748.30002. **MR** 93f:30002.

Classical Topics in Complex Function Theory (translated from German by Leslie Kay), Graduate Texts in Mathematics, Vol. 172. Springer-Verlag (1998), New York. **Zbl** 895.30001. **MR** 98g:30002. **BAMS** (2) 35, p. 334.

"Vom Fundamentalsatz der Algebra zum Satz von Gelfand-Mazur," *Math. Semesterber.* 40 (1993), 63–71. **Zbl** 789.30001. **MR** 96h:01024.

Rengel, E.

"Existenzbeweise für schlichte Abbildungen mehrfach zusammenhängender Bereiche auf gewisse Normalbereiche," *Jahresber. Deutsch. Math. Verein.* 44 (1934), 51–55. **FM** 60, p. 286. **Zbl** 9, p. 173.

Renggli, H.

"On holomorphic mappings of annuli," pp. 345–348 of *Romanian-Finnish Seminar on Complex Analysis, Proceedings Bucharest 1976.* Lecture Notes in Mathematics 743, Springer-Verlag (1979), New York. **Zbl** 431.30012. **MR** 81c: 30019.

Rennie, B. C.

"Repeated Riemann integrals," *Proc. Camb. Phil. Soc.* 76 (1974), 187–189. **Zbl** 283.26005. **MR** 50 #4857.

Renteln, M. von

"Divisibility structure and finitely generated ideals in the disc algebra," *Monatshefte für Math.* 82 (1976), 51–56. **Zbl** 339.30035. **MR** 54 #13576.

"A simple constructive proof of an analogue of the corona theorem," *Proc. Amer. Math. Soc.* 83 (1981), 299–303. **Zbl** 479.30037. **MR** 82h:46073.

Ribenboim, P.

13 Lectures on Fermat's Last Theorem, Springer-Verlag (1979), New York and Berlin. **Zbl** 456.10006. **MR** 81f:10023. **BAMS** (2) 4, p. 218.

Richards, I.

"More on rings on rings," *Bull. Amer. Math. Soc.* 74 (1968), 677. **Zbl** 169, p. 409. **MR** 37 #1980.

"Axioms for analytic functions," *Adv. in Math.* 5 (1970), 311–338. **Zbl** 205, p. 136. **MR** 55 #10701. (See also **Zbl** 298.46052 and **MR** 51 #8413.)

RICHMOND, D. E.

"Complex numbers and vector algebra," *Amer. Math. Monthly* 58 (1951), 622–628.

RIDDER, J.

(a) "Über den Cauchyschen Integralsatz für reelle und komplexe Funktionen," *Math. Annalen* 102 (1930), 132–156. **FM** 55, p. 761.

(b) "Ein Satz über iterierte Integrale (R)," *Nieuw Archief voor Wiskunde* (2) 16 (1930), 43–45. **FM** 56, p. 914.

(c) "Ein Satz über iterierte Integrale und seine Anwendung zur Untersuchung der Analytizität von komplexen Funktionen," *Math. Zeit.* 31 (1930), 141–148. **FM** 55, p. 139.

"Harmonische, subharmonische und analytische Funktionen," *Ann. Scuola Norm. Sup. Pisa* (2) 9 (1940), 277–287. **FM** 66, p. 448. **Zbl** 24, p. 206. **MR** 3, p. 125.

"Über den Greenschen Satz in der Ebene," *Nieuw Archief voor Wiskunde* (2) 21 (1941), 28–32. **FM** 67, p. 185. **Zbl** 26, p. 207. **MR** 7, p. 376.

"Über harmonische Funktionen," *Nieuw Archief voor Wiskunde* (2) 22 (1946), 162–170. **Zbl** 61, p. 233. **MR** 8, p. 461.

"Ein einfacher Eindeutigkeitssatz für analytische Funktionen," *Kon. Nederl. Akad. Wetensch., Proc. Ser. A* 70 (=*Indag. Math.* 29) (1967), 373–374. **Zbl** 185, p. 147. **MR** 36 #3992.

RIEMANN, B.

Gesammelte Mathematische Werke und Wissenschaftlicher Nachlass, edited by H. Weber with supplement (1902) by M. Noether and W. Wirtinger, B. G. Teubner (2nd ed., 1892), Berlin and Leipzig. **FM** 24, p. 21 and 33, p. 25. Reprinted, with English introduction by H. Lewy, by Dover Publications, Inc. (1953), New York. Reprinted with more supplements by Springer-Verlag (1990), New York and Berlin. **Zbl** 53, p. 194 and 703.01020. **MR** 14, p. 610 and 91j:01070. French translation by L. Laugel with a preface by C. Hermite and an address by F. Klein, Gauthier-Villars et Fils (1898), Paris. **FM** 29, p. 9. Reprinted by Librairie Scientifique et Technique Albert Blanchard (1968), Paris. **Zbl** 188, p. 301. **MR** 36 #4952. English translation by Roger Baker, Charles Christenson, and Henry Orde, Kendrick Press, Heber City (2004). **Zbl** 1101.01013. **MR** 2005m:01028. (See also **Zbl** 957.01031.)

Riemanns Einführung in die Funktionentheorie, Eine quellenkritische Edition seiner Vorlesungen mit einer Bibliographie zur Wirkungsgeschichte der Riemannschen Funktionentheorie (E. Neuenschwander, editor) Vandenhoeck & Ruprecht (1996), Göttingen. **Zbl** 844.01020. **MR** 97d:01041.

RIESZ, F.

"Sur certains systèmes singuliers d' équations intégrales," *Ann. Sci. École Norm. Sup.* (3) 28 (1911), 33–62. **FM** 42, p. 374.

(a) "Sur les suites de fonctions analytiques," *Acta Litt. ac Scient. Univ. Hung.* (Szeged) 1 (1922/23), 88–97. **FM** 49, p. 713.

(b) "Sur les valeurs moyennes du module des fonctions harmoniques et des fonctions analytiques," *Acta Litt. ac Scient. Univ. Hung.* (Szeged) 1 (1922/23), 27–32. **FM** 48, p. 1268. (See also **MR** 2002m:31001.)

"Über subharmonische Funktionen und ihre Rolle in der Funktionentheorie und der Potentialtheorie," *Acta Litt. ac Scient. Univ. Hung.* (Szeged) 2 (1924–26), 87–100. **FM** 51, p. 363.

"Sur une inégalité de M. Littlewood dans la théorie des fonctions," *Proc. Lon. Math. Soc.* (2) 23 (1925), XXXVI–XXXIX. **FM** 51, p. 247.

"Sur les fonctions subharmoniques et leur rapport à la théorie du potentiel, I," *Acta Math.* 48 (1926), 329–343. **FM** 52, p. 497.

"Sur le théorème de Jordan," *Acta Litt. ac Scient. Univ. Hung.* (Szeged) 9 (1939), 154–162. **FM** 65, p. 867. **Zbl** 22, p. 84.

RIESZ, F. and SZ.-NAGY, B.

Functional Analysis (translated from French by Leo F. Boron), Frederick Ungar Publishing Co. (1955), New York. Reprinted by Dover Publications, Inc. (1990), New York. **Zbl** 732.47001. **MR** 91g:00002. **BAMS** 62, p. 423.

RIESZ, M.

"Über einen Satz des Herrn Serge Bernstein," *Acta Math.* 40 (1916), 337–347. **FM** 46, p. 472.

"Sur le principe de Phragmén-Lindelöf," *Proc. Camb. Phil. Soc.* 20 (1920–1921), 205–209 and 21 (1922), 6. **FM** 47, p. 924.

"Sur certaines inégalités dans la théorie des fonctions avec quelques remarques sur les géométries non-euclidiennes," *Kungl. Fysiografiska Sällskapets i Lund Förhandlingar* Bd. 1, No. 4 (1931), 18–38. **FM** 56, p. 983 and 57, p. 247. **Zbl** 3, p. 259. (See also **MR** 90a:01109.)

RIPPON, P. J.

"Infinite exponentials," *Math. Gazette* 67 (1983), 189–196 and *Irish Math. Soc. Newsletter* No. 5 (1982), 8–19 and No. 6 (1982), 46. **MR** 84b:30028.

RITT, J. F.

"On the conformal maping of a region into a part of itself," *Annals of Math.* (2) 22 (1920–21), 157–160. **FM** 48, p. 363.

ROBERTS, J. H.

"Two-to-one transformations," *Duke Math. Jour.* 6 (1940), 256–262. **FM** 66, p. 982. **Zbl** 27, p. 96. **MR** 1, p. 319.

ROBERTSON, M. I. S.

"On the theory of univalent functions," *Annals of Math.* (2) 37 (1936), 374–408. **FM** 62, p. 373. **Zbl** 14, p. 165.

"Bounds for the coefficient a_1 in certain series expansions," *Amer. Math. Monthly* 60 (1953), 131–132.

"Coefficient theorems for analytic functions," *Rev. Roum. Math. Pures et Appliq.* 23 (1978), 479–486. **Zbl** 374.30012. **MR** 57 #12838.

ROBINSON, R. M.

"On numerical bounds in Schottky's theorem," *Bull. Amer. Math. Soc.* 45 (1939), 907–910. **FM** 65, p. 332. **Zbl** 23, p. 51. **MR** 1, p. 112.

"Analytic functions in circular rings," *Duke Math. Jour.* 10 (1943), 341–354. **Zbl** 60, p. 218. **MR** 4, p. 241.

"Hadamard's three circles theorem," *Bull. Amer. Math. Soc.* 50 (1944), 795–802. **Zbl** 60, p. 218. **MR** 6, p. 122.

"Univalent majorants," *Trans. Amer. Math. Soc.* 61 (1947), 1–35. **Zbl** 32, p. 156. **MR** 8, p. 370.

"A curious trigonometric identity," *Amer. Math. Monthly* 64 (1957), 83–85. **Zbl** 81, p. 340. **MR** 18, p. 568. (See also **MR** 2001g:39057.)

ROBISON, G. R.

"A new approach to circular functions, π and $\lim \left(\frac{\sin x}{x}\right)$," *Math. Mag.* 41 (1968), 66–70. **Zbl** 165, p. 400.

ROGALSKI, M.

"De Leibniz à Euler: Cartier, Dumont, Krivine, Titchmarsh et les autres ... La relation $1 + \frac{1}{3^2} + \frac{1}{5^2} + \frac{1}{7^2} + \cdots = 2(1 - \frac{1}{3} + \frac{1}{5} - \frac{1}{7} + \cdots)^2$," *Gazette des Mathématiciens,* No. 68 (1996), 47–61. **MR** 97g:01010.

ROGERS, J. T., JR. and ZAME, W.

"Extension of analytic functions and the topology in spaces of analytic functions," *Indiana Univ. Math. Jour.* 31 (1982), 809–818. **Zbl** 559.30001. **MR** 83k:30050.

ROGOSINSKI, W.

"Zum Schwarzchen Lemma," *Jahresber. Deutsch. Math. Verein.* 44 (1934), 258–261 + 45 (1935), 88 and 243. **FM** 60, p. 254; 61, p. 348. **Zbl** 10, p. 307; 12, p. 359.

"On subordinate functions," *Proc. Camb. Phil. Soc.* 35 (1939), 1–26. **FM** 65, p. 332. **Zbl** 20, p. 140.

"On the coefficients of subordinate functions," *Proc. Lon. Math. Soc.* (2) 48 (1943), 48–82. **Zbl** 28, p. 355. **MR** 5, p. 36.

"On the order of the derivatives of a function analytic in an angle," *Jour. Lon. Math. Soc.* 20 (1945), 100–109. **MR** 8, p. 324.

ROHDE, H.-W.

"Complex iterated radicals," *Amer. Math. Monthly* 81 (1974), 14–21. **Zbl** 282.30008. **MR** 49 #3098.

ROSAY, J.-P. and RUDIN, W.

"Arakelian's approximation theorem," *Amer. Math. Monthly* 96 (1989), 432–434. **Zbl** 678.30025. **MR** 90h:30091.

ROSENBLATT, A.
"On the modulus of functions analytic in the unit circle," *Actas Acad. Ci. Lima* 8 (1945), 27–44 (Spanish). **MR** 8, p. 19.

ROSENBLOOM, P. C.
"An elementary constructive proof of the fundamental theorem of algebra," *Amer. Math. Monthly* 52 (1945), 562–570. **Zbl** 60, p. 47. **MR** 7, p. 295.
"L'itération des fonctions entières," *C. R. Acad. Sci. Paris* 227 (1948), 382–383. **Zbl** 30, p. 251. **MR** 10, p. 187.
"The fix-points of entire functions," *Medd. Lunds Univ. Mat. Sem.* (1952), tome supplémentaire, 186–192. **Zbl** 47, p. 316. **MR** 14, p. 546.

ROSENBLUM, M. and ROVNYAK, J.
"The factorization problem for non-negative operator valued functions," *Bull. Amer. Math. Soc.* 77 (1971), 287–318. **Zbl** 214, p. 121. **MR** 42 #8315.

ROSENHOLTZ, I.
"A topological mean value theorem for the plane," *Amer. Math. Monthly* 98 (1991), 149–154. **Zbl** 741.26003. **MR** 91m:26014.

ROSS, K. A.
Elementary Analysis: The Theory of Calculus, Undergraduate Texts in Mathematics, Springer (2nd ed. 2013), New York. **Zbl** 1282.26002. **MR** 3076698.

ROSSI, J.
"A sharp result concerning cercles de remplissage," *Ann. Acad. Scient. Fennicae, Ser. A. I. Math.* 20 (1995), 179–185. **Zbl** 817.30016. **MR** 95k:30062.

ROTH, A.
"Approximationseigenschaften und Strahlengrenzwerte meromorpher und ganzer Funktionen," *Comm. Math. Helv.* 11 (1938–39), 77–125. **FM** 64, p. 1067. **Zbl** 20, p. 235.
"Meromorphe Approximationen," *Comm. Math. Helv.* 48 (1973), 151–176. **Zbl** 275.30035. **MR** 57 #16615.
"Uniform and tangential approximation by meromorphic functions on closed sets," *Canad. Jour. Math.* 28 (1976), 104–110. **Zbl** 322.30035. **MR** 57 #9978. (See also **MR** 80d:30037.)

ROTHE, E. H.
Introduction to Various Aspects of Degree Theory in Banach Spaces, Mathematical Surveys and Monographs, Vol. 23, American Mathematical Society (1986), Providence. **Zbl** 597.47040. **MR** 87m:47145. **BAMS** (2) 17, p. 340.

ROUCHÉ, E.
"Mémoire sur la série de Lagrange," *Jour. École Polyt.* 22 (1862), 193–224.

ROUX, D.
"Una dimostrazione del teorema fondamentale dell'algebra," *Boll. della Un. Mat. Ital.* (3) 14 (1959), 563–567. **Zbl** 93, p. 18. **MR** 22 #5721.

ROY, R.

"The discovery of the series formula for π by Leibniz, Gregory and Nilakantha," *Math. Mag.* 63 (1990), 291–306. **Zbl** 727.01007. **MR** 92a:01029.

Sources in the Development of Mathematics: Infinite Series and Products from the Fifteenth to the Twenty-first Century, Cambridge University Press (2011). **Zbl** 1228.01001. **MR** 2012f:01002.

ROYDEN, H. L.

"A modification of the Neumann-Poincaré method for multiply connected regions," *Pac. Jour. of Math.* 2 (1952), 385–394. **Zbl** 47, p. 79. **MR** 14, p. 182.

"A generalization of Morera's theorem," *Annales Polonici Mathematici* 12 (1962), 199–202. **Zbl** 105, p. 281. **MR** 25 #5163.

"A criterion for the normality of a family of meromorphic functions," *Ann. Acad. Scient. Fennicae, Ser. A. I. Math.* 10 (1985), 499–500. **Zbl** 588.30034. **MR** 86i:30040. (See also **Zbl** 824.30022 and **MR** 85g:30047.)

ROYSTER, W. C.

"A Poisson integral formula for the ellipse and some applications," *Proc. Amer. Math. Soc.* 15 (1964), 661–670. **Zbl** 156, p. 297. **MR** 34 #2895.

RUBEL, L. A.

"A complex-variables proof of Hölder's inequality," *Proc. Amer. Math. Soc.* 15 (1964), 999. **Zbl** 133, p. 7. **MR** 30 #4882.

"Some applications of the Gauss-Lucas Theorem," *L'Enseignement Mathématique* (2) 12 (1966), 33–39. Errata, *ibid.* (2) 16 (1970), 113. **Zbl** 144, p. 331. **MR** 34 #4457.

"Nonmaximal prime ideals in the ring of holomorphic functions," pp. 341–342 of *Mathematical Essays Dedicated to A. J. Macintyre,* Ohio University Press (1970), Athens, Ohio. **Zbl** 209, p. 152. **MR** 42 #7919.

"How to use Runge's theorem," *L'Enseignement Mathématique* (2) 22 (1976), 185–190. Errata, (2) 23 (1977), 149. **Zbl** 347.30033 and 357.30027. **MR** 55 #3223 and 57 #9942.

"Conformal inequivalence of annuli and the first-order theory of subgroups of $PSL(2,\mathbf{R})$," *Proc. Amer. Math. Soc.* 88 (1983), 679–683. **Zbl** 512.03008. **MR** 85a:30018.

"Four counterexamples to Bloch's principle," *Proc. Amer. Math. Soc.* 98 (1986), 257–260. **Zbl** 602.30040. **MR** 87i:30064.

"Prescribed zero-sets for an entire function and its derivatives," *Amer. Math. Monthly* 97 (1990), 254–255.

RUBEL, L. A. and VENKATESWARAN, S.

"Simultaneous approximation and interpolation by entire functions," *Arch. der Math.* 27 (1976), 526–529. **Zbl** 348.30029. **MR** 55 #5869.

RUBINSTEIN, Z.

"On analytic functions satisfying the mean value theorem and a conjecture of W. G. Dotson," *Math. Mag.* 42 (1969), 256–259. **Zbl** 184, p. 297. **MR** 41 #435.

RUDIN, W.

"A theorem on subharmonic functions," *Proc. Amer. Math. Soc.* 2 (1951), 209–212. **Zbl** 43, p. 104. **MR** 12, p. 825.

"Analyticity and the maximum modulus principle," *Duke Math. Jour.* 20 (1953), 449–457. **Zbl** 52, p. 76. **MR** 15, p. 21.

(a) "Some theorems on bounded analytic functions," *Trans. Amer. Math. Soc.* 78 (1955), 333–342 and *Bull. Amer. Math. Soc.* 61 (1955), 543. **Zbl** 64, p. 312. **MR** 16, p. 685.

(b) "Multiplicative groups of analytic functions," *Proc. Amer. Math. Soc.* 6 (1955), 83–87. **Zbl** 64, p. 312. **MR** 16, p. 578.

Principles of Mathematical Analysis, McGraw-Hill Book Co. (3rd ed. 1976), New York. **Zbl** 449.26002. **MR** 83e:26004. **BAMS** 59, p. 572.

Real and Complex Analysis, McGraw-Hill Book Co. (1966, 3rd ed., 1987), New York. **Zbl** 333.28001, 461.0006 and 613.26001. **MR** 58 #15765, 87f:00009, 88k:00002 and 2000i:00003. **BAMS** 74, p. 79.

Functional Analysis, McGraw-Hill Book Co. (1973, 2nd ed., 1991), New York. **Zbl** 253.46001, 741.43007. **MR** 56 #16309, 92k:46001. **BAMS** 81, p. 842.

"Proper holomorphic maps and finite reflection groups," *Indiana Univ. Math. Jour.* 31 (1982), 701–720. **Zbl** 506.32009. **MR** 84d:32038.

"Power series with zero-sum on countable sets," *Complex Variables* 18 (1992), 283–284. **Zbl** 766.30002. **MR** 93h:30002.

"Preservation of level-sets by automorphisms of $\mathbb{C}^n$," *Indag. Math.* (2) 4 (1993), 489–497. **Zbl** 801.32002. **MR** 94i:32030.

RUNG, D. C.

"Behavior of holomorphic functions in the unit disk on arcs of positive hyperbolic diameter," *Jour. Math. Kyoto Univ.* 8 (1968), 417–464. Corrigendum, *ibid.* 9 (1969), 337. **Zbl** 176, p. 29 and 184, p. 305. **MR** 39 #4403 and 40 #2875.

RUNGE, C.

"Zur Theorie der eindeutigen analytischen Functionen," *Acta Math.* 6 (1885), 229–244. **FM** 17, p. 379.

"Über die Darstellung willkürlicher Functionen," *Acta Math.* 7 (1885/86), 387–392. **FM** 18, p. 344.

RUSCHEWEYH, S.

"Neighborhoods of univalent functions," *Proc. Amer. Math. Soc.* 81 (1981), 521–527. **Zbl** 458.30008. **MR** 82c:30016.

"Two remarks on bounded analytic functions," *Serdica* 11 (1985), 200–202. **Zbl** 581.30009. **MR** 87h:30006. (See also **MR** 2003g:30038 and 2009c:30063.)

RYFF, J. V.

"Remarks on the monodromy theorem," *Univ. Beograd. Publ. Elektrotehn. Fak. Ser. Mat. Fiz.* (1969), 274–301; (1970), 23–24. **Zbl** 226.30004. **MR** 42 #1981.

RZADKOWSKI, G.

"Remarks on the formulae of Stirling and Wallis," *Math. Gazette* 81 (1997), 429–431 and 84 (2000), 306–307. (See also **MR** 98m:26008.)

S

SAAKIAN, R. SH.

"Some applications of theorems on approximation by an analytic function," *Soviet Jour. Contemp. Math. Analysis* 24, no. 3 (1989), 51–60. **Zbl** 691.30031. **MR** 90k:30079.

SAALSCHÜTZ, L.

Vorlesungen über die Bernoullischen Zahlen, ihren Zusammenhang mit den Secanten-Coefficienten und ihre wichtigeren Anwendungen, Verlag Julius Springer (1893), Berlin. **FM** 24, p. 236.

SAFF, E. B.

"Polynomial and rational approximation in the complex domain," *Proceedings of Symposia in Applied Math.* 36 (1986), 21–49. **Zbl** 593.41010. **MR** 88h:30061.

SAGAN, H.

Space-filling Curves, Springer-Verlag (1994), New York and Berlin. **Zbl** 806.01019. **MR** 95h:00001.

SAKASHITA, H.

"On the conformal mapping of nearly circular domains," *Bull. Kyoto Gakugei Univ., Ser B.,* No. 8 (1956), 10–14. **MR** 20 #5278.

SAKS, S.

"Sur une inégalité de la théorie des fonctions," *Acta Litt. ac Scient. Univ. Hung.* (Szeged) 4 (1928/29), 51–55. **FM** 54, p. 334.

"On subharmonic functions," *Acta Litt. ac Scient. Univ. Hung.* (Szeged) 5 (1930–32), 187–193. **FM** 58, p. 518. **Zbl** 4, p. 116.

"Note on defining properties of harmonic functions," *Bull. Amer. Math. Soc.* 38 (1932), 380–382. **FM** 58, p. 1141. **Zbl** 5, p. 17.

Theory of the Integral (translated from Polish by L. C. Young), G. E. Stechert (2nd ed. 1937), New York. **FM** 63, p. 183. **Zbl** 17, p. 300. **MR** 29 #4850. **BAMS** 44, p. 615. Reprinted by Dover Publications, Inc. (1964), New York.

SAKS, S. and ZYGMUND, A.

Analytic Functions (translated from Polish by E. J. Scott), American Elsevier Publishing Co. (3rd ed. 1971), New York. **FM** 64, p. 1048. **Zbl** 199, p. 126. **MR** 50 #2456. **BAMS** 60, p. 495.

SALVADORI, M.

Esposizione della teoria delle somme di Gauss, Tipografia fratelli Nistri (1904), Pisa. **FM** 35, pp. 210 and 977.

SALZMANN, H. and ZELLER, K.

"Singularitäten unendlich oft differenzierbarer Funktionen," *Math. Zeit.* 62 (1955), 354–367. **Zbl** 64, p. 299. **MR** 17, p. 134.

SAMUELSSON, Å.

"A local mean value theorem for analytic functions," *Amer. Math. Monthly* 80 (1973), 45–46. **Zbl** 261.30002. **MR** 47 #8822.

SANDERSON, D. E.

"Advanced plane topology from an elementary standpoint," *Math. Mag.* 53 (1980), 81–89 and 313. **Zbl** 454.54003. **MR** 81j:55002. (See also **Zbl** 376.57003, **MR** 56 #13207.)

SANDHAM, H. F.

"A well-known integral," *Amer. Math. Monthly* 53 (1964), 587.

SANDIFER, C. E.

"Euler's solution of the Basel problem — the longer story," pp. 105–117 of *Euler at 300, an Appreciation* (R. E. Bradley, *et al.*, editors) Mathematical Association of America (2007), Washington, D. C. **Zbl** 1177.01021. **MR** 2008i: 01006.

SANSONE, G. and GERRETSEN, J.

Lectures on the Theory of Functions of a Complex Variable I (1960), II (1969), P. Noordhoff Ltd., Groningen. **Zbl** 93, p. 268; 188, p. 381. **MR** 22 #4819; 41 #3714. **BAMS** 69, p. 39.

SANTOS, J. C.

"Another proof of the fundamental theorem of algebra," *Amer. Math. Monthly* 112 (2005), 76–78. **Zbl** 1138.12301. **MR** 2110116.

"The fundamental theorem of algebra deduced from elementary calculus," *Math. Gazette* 91 (2007), 302–303.

SARASON, D.

"Angular derivatives via Hilbert space," *Complex Variables* 10 (1988), 1–10. **Zbl** 635.30024. **MR** 89f:30045.

Complex Function Theory, American Mathematical Society (2nd ed. 2007), Providence. **Zbl** 940.30001, 1138.30001. **MR** 2008i:30001.

SATÔ, T.

"Sur une condition nécessaire et suffisante pour l'univalence d'une fonction holomorphe," *Proc. Imper. Acad. Tokyo* 12 (1936), 332–334. **FM** 62, p. 1215. **Zbl** 17, p. 216.

"On the proof of Carleman's extension theorem of Liouville theorem," *Memoirs of College of Science Kyoto Imperial Univ. A* 21 (1938), 47–49. **FM** 64, p. 1066. **Zbl** 19, p. 313.

SAXER, W.

"Über die Picardschen Ausnahmewerte sukzessiver Derivierten," *Math. Zeit.* 17 (1923), 206–227. **FM** 49, p. 219.

"Sur les valeurs exceptionnelles des dérivées successives des fonctions méromorphes," *C. R. Acad. Sci. Paris* 182 (1926), 831–833. **FM** 52, p. 323.

"Über eine Verallgemeinerung des Satzes von Schottky," *Compositio Math.* 1 (1934), 207–216. **FM** 60, p. 255. **Zbl** 9, p. 216.

SCARF, H.

"Group invariant integration and the fundamental theorem of algebra," *Proc. Nat. Acad. Sci. U. S. A.* 38 (1952), 439–440. **Zbl** 46, p. 244. **MR** 14, p. 126.

SCHARM, R.

"On the importance of harmonic majorization for a systematization of the theory of PHRAGMÉN-LINDELÖF," *Math. Nachr.* 133 (1987), 343–357. **Zbl** 638.30023. **MR** 88j:30051.

SCHATTE, P.

"Über eine Erweiterung des Cauchyschen Integralsatzes auf nichtrektifizierbare Kurven," *Wissen. Zeit. der Univ. Rostock* 16 (1967), 783–786. **Zbl** 237.30040. **MR** 53 #3270.

SCHEEFFER, L.

"Beweis des Laurent'schen Satzes," *Acta Math.* 4 (1884), 375–380. **FM** 16, p. 350.

SCHEINBERG, S.

"Uniform approximation by entire functions," *Jour. d'Analyse Math.* 29 (1976), 16–18. **Zbl** 343.41022. **MR** 58 #22664.

"Approximation and non-approximation on Riemann surfaces," pp. 110–118 of *Complex Approximation* (B. Aupetit, editor), Progress in Math. 4, Birkhäuser Verlag (1980), Boston. **Zbl** 447.30021. **MR** 81i:30066.

SCHEMPP, W.

"Approximation in the hardy space $H^1(D)$ with respect to the norm topology," pp. 291–300 of *Quantitative Approximation, Proceedings International Symposium, Bonn, 1979* (R. A. DeVore and K. Scherer, editors). Academic Press (1980), London. **Zbl** 489.41020. **MR** 81j:30053.

SCHERING, E.

"Das Anschliessen einer Function an algebraische Functionen in unendlich vielen Stellen," *Göttingen Abhand.* 27 (1881). **FM** 12, p. 315.

SCHIFF, J. L.

Normal Families, Springer-Verlag (1993), New York and Berlin. **Zbl** 770.30002. **MR** 94f:30046. **BAMS** (2) 32, p. 257.

SCHIFF, J. L. and WALKER, W. J.

"A sampling theorem for analytic functions," *Proc. Amer. Math. Soc.* 99 (1987), 737–740. **Zbl** 621.30001. **MR** 88b:30003.

SCHIFFER, M. M.

"On the modulus of doubly-connected domains," *Quart. Jour. of Math.* (*Oxford Series*) 17 (1946), 197–213. **Zbl** 60, p. 237. **MR** 8, p. 325.

SCHILLING, O. F. G.

"Ideal theory on open Riemann surfaces," *Bull. Amer. Math. Soc.* 52 (1946), 945–963. **MR** 8, p. 454.

SCHIMMACK, R.

"Axiomatische Untersuchungen über die Vektoraddition," *Nova Acta Deutsch. Akad. der Naturforschen zu Halle* 90 (1908), 1–104. **FM** 40, pp. 148, 620, 1027.

SCHIRMER, H.

"A Brouwer-type coincidence theorem," *Canad. Math. Bull.* 9 (1966), 443–446. **Zbl** 142, p. 213. **MR** 34 #2002. (See also **Zbl** 276.54043, **MR** 49 #6215.)

SCHLÖMILCH, O.

"Ueber das bestimmte Integral $\int_0^\infty \frac{\sin^p x}{x^q} dx$," *Zeit. für Math. und Physik* 5 (1860), 286–292. See **MR** 2005g:26012.

SCHMIDT, H.

"Zur Faktorenzerlegung reeller Polynome einer Veränderlichen," *Jahresber. Deutsch. Math. Verein.* 46 (1936), 2te Abteilung, 64–67. **FM** 62, p. 68. **Zbl** 19, p. 101.

SCHMIEDER, G.

Grundkurs Funktionentheorie, B. G. Teubner (1993), Stuttgart. **Zbl** 783.30002. **MR** 96e:30083.

"Punktweise Konvergenz und Holomorphie," *Math. Semesterber.* 41 (1994), 77–80. **Zbl** 804.30003. **MR** 96e:30083.

"Der Satz von Rolle im Komplexen," *Math. Semesterber.* 44 (1997), 109–120. **Zbl** 911.30005. **MR** 99i:30009.

SCHNEIDER, W. J.

"Some extensions of the Nevanlinna 2-constant theorem and the Hadamard 3-circle theorem," *Jour. Math. Anal. and Applic.* 17 (1967), 280–291. **Zbl** 152, p. 70. **MR** 35 #4416.

"Approximation and harmonic measure," pp. 333–349 of *Aspects of Contemporary Complex Analysis* (D. A. Brannan and J. G. Clunie, editors), Academic Press (1980), London and New York. **Zbl** 505.30025. **MR** 82g:30058.

SCHÖNFLIES, A.

"Über einen Satz der Analysis Situs," *Göttingen Nachr.* (1899), 282–290. **FM** 30, p. 434.

"Über einen grundlegenden Satz der Analysis Situs," *Göttingen Nachr.* (1902), 185–192. **FM** 33, p. 502.

(a) "Über die geometrischen Invarianten der Analysis Situs," *Göttingen Nachr.* (1904), 514–525. **FM** 35, p. 505.

(b) "Beiträge zur Theorie der Punktmengen, I," *Math. Annalen* 58 (1904), 195–234. II, *Math. Annalen* 59 (1904), 129–160. III, *Math. Annalen* 62 (1906), 286–328. **FM** 34, p. 74; 35, p. 88; 37, p. 73. Correction *ibid.* 65 (1908), 431–432. **FM** 39, p. 96.

"Die Entwickelung der Lehre von den Punktmannigfaltigkeiten, II," *Jahresber. Deutsch. Math. Verein.* Der Ergänzungsbände II. Bd. (1908), 1–331. **FM** 39, p. 95.

"Bemerkung zu dem vorstehenden Aufsatz des Herrn L. E. J. Brouwer," *Math. Annalen* 68 (1910), 435–444. **FM** 41, p. 543.

SCHÖNHAGE, A.

"Zur Ränderzuordnung beim Riemannschen Abbildungssatz," *Math. Annalen* 179 (1969), 97–100. **Zbl** 167, p. 65. **MR** 39 #1662.

SCHOTTKY, F.

"Über die conforme Abbildung mehrfach zusammenhängender ebener Flächen," *Jour. für Reine und Angew. Math.* 83 (1877), 300–351. **FM** 9, p. 584.

"Über das Additionstheorem der Cotangente und der Function $\zeta(u) = \sigma'(u)/\sigma(u)$," *Jour. für Reine und Angew. Math.* 110 (1892), 324–337. **FM** 24, p. 427.

"Über den Picard'schen Satz und die Borel'schen Ungleichungen," *Sitzungsber. Kön. Preuss. Akad. Wissen. Berlin* (1904), 1244–1262 and (1906), 32–36. **FM** 35, p. 401 and 37, p. 417.

"Über zwei Beweise des allgemeinen Picard'schen Satzes," *Sitzungsber. Kön. Preuss. Akad. Wissen. Berlin* (1907), 823–840. **FM** 38, p. 432.

"Über das Cauchysche Integral," *Jour. für Reine und Angew. Math.* 146 (1916), 234–244. **FM** 46, p. 470. (See also **FM** 45, p. 638.)

"Problematische Punkte und die elementaren Sätze, die zum Beweis des Picard'schen Theorems dienen," *Jour. für Reine und Angew. Math.* 147 (1917), 161–173. **FM** 46, p. 515.

SCHOU, E.

"Sur la théorie des fonctions entières," *C. R. Acad. Sci. Paris* 125 (1897), 763–764. **FM** 28, p. 361.

Schröder, E.

"Über unendlich viele Algorithmen zur Auflösung der Gleichungen," *Math. Annalen* 2 (1870), 317–363. **FM** 2, p. 42.

Schröter, H.

"Ableitung der Partialbruch-und Produkt-Entwicklungen für die trigonometrische Functionen," *Zeit. für Math. und Physik* 13 (1868), 254–259. **FM** 1, p. 88.

Schubring, G.

Conflicts between Generalization, Rigor, and Intuition: Number Concepts Underlying the Development of Analysis in 17–19th *Century France and Germany,* Springer (2005), New York. **Zbl** 1086.01001. **MR** 2006d:26001. (See also **MR** 3530988.)

Schur, I.

"Über Potenzreihen, die im Innern des Einheitskreises beschränkt sind. I, II," *Jour. für Reine und Angew. Math.* 147 (1917), 205–232; 148 (1918), 122–145. **FM** 46, p. 475. English translation in *Operator Theory: Advances & Applications* 18 (1986), 31–88. **Zbl** 618.30001, 2. **MR** 89c:00052. (See also **Zbl** 741.01029.)

"Über die Gaussschen Summen," *Göttingen Nachr.* (1921), 147–153. **FM** 48, p. 130.

Schuske, G. and Thron, W.

"Infinite radicals in the complex plane," *Proc. Amer. Math. Soc.* 12 (1961), 527–532. **Zbl** 103, p. 44. **MR** 27 #1570.

"On periodic infinite radicals," *Ann. Acad. Scient. Fennicae, Ser. A. I. Math.* , Nr. 307 (1962). **Zbl** 103, p. 44. **MR** 25 #1385.

Schwarz, H. A.

Gesammelte Mathematische Abhandlungen, (2 vols.) Springer-Verlag (1890), Berlin. Reprinted by Chelsea Publishing Co. (1972), New York. **FM** 22, p. 31 and 23, p. 438. **MR** 52 #13287. See *Sitzungsber. Berlin Math. Gesell.* 21 (1922), 47–52 (**FM** 48, p. 17) for works not included here.

Schwatt, I. J.

"Expressions for the Euler numbers obtained by expanding $\sec x$ by means of Maclaurins Theorem," *Math. Zeit.* 31 (1930), 151–158. **FM** 55, p. 135.

Schwick, W.

"A new joint proof of the theorems of Picard and Montel," *Resultate der Math.* 21 (1992), 403–407. **Zbl** 759.30014. **MR** 93e:30069.

Schwingendorf, K. E.

"Uniform polynomial approximation," *Jour. Math. Anal. and Applic.* 75 (1980), 81–101. **Zbl** 452.30022. **MR** 82d:30034.

SEARCÓID, M. Ó.

Elements of Abstract Analysis, Springer-Verlag (2002), London. **Zbl** 999.00002. **MR** 2002k:00001.

SEELEY, R. T.

"Fubini implies Leibniz implies $F_{yx} = F_{xy}$," *Amer. Math. Monthly* 68 (1961), 56–57.

SEGAL, S. L.

Nine Introductions in Complex Analysis, North-Holland Mathematics Studies, Vol. 53. North-Holland Publishing Co. (1981 revised ed. 2008), Amsterdam. **Zbl** 482.30002. **MR** 83b:30001 and 2008m:30002.

SEID, H. A.

"A corollary of the Gelfand-Mazur theorem," *Amer. Math. Monthly* 77 (1970), 282–283. **Zbl** 189, p. 445. **MR** 41 #4253.

SEIDEL, W.

"Über die Ränderzuordnung bei konformen Abbildungen," *Math. Annalen* 104 (1931), 182–243. **FM** 57, p. 398. **Zbl** 1, p. 19.

"Bibliography of numerical methods in conformal mapping," pp. 269–280 of *Construction and Applications of Conformal Maps*, National Bureau of Standards Applied Math. Series No. 18, E. F. Beckenbach, editor, (1952), Washington, D. C. **Zbl** 49, p. 335. **MR** 14, p. 589.

SEIDEL, W. P. and LITTAUER, S. B.

"Lines of Julia of integral functions," *Proc. Nat. Acad. Sci. U. S. A.* 18 (1932), 90–93. **FM** 58, p. 341. **Zbl** 3, p. 405.

SELBERG, A.

"Løste opgaver," *Norsk Mat. Tidsskrift* 15 (1933), Nr. 1, 24.

SEN, ANINDYA

"Fundamental theorem of algebra — yet another proof," *Amer. Math. Monthly* 107 (2000), 842–843. **Zbl** 1028.12001. **MR** 1543753.

SERRIN, J.

"A note on harmonic functions defined in a halfplane," *Duke Math. Jour.* 23 (1956), 523–526. **Zbl** 73, p. 82. **MR** 18, p. 120.

SEVERINI, G.

"Alcune ricerche sulla teoria delle funzioni analitiche," *Rend. del R. Istituto Lombardo di Scienze e Lettere* (2) 34 (1901), 891–904. **FM** 32, p. 395.

"Sulle serie di funzioni analitiche," *Atti del Reale Istituto Veneto di Scienze, Lettere ed Arti* [(8), 7] 64 (1905), 1609–1613. **FM** 36, p. 476.

"Sopra una proprietà caratteristica delle funzioni armoniche," *Boll. Accad. Gioenia* (2) fasc. 16 (1911), 2–5. **FM** 42, p. 495.

SHALIT, O. M.

"A sneaky proof of the maximum modulus principle," *Amer. Math. Monthly* 120 (2013), 359–362. **Zbl** 1272.30080. **MR** 3035130.

SHANKS, D.

"Two theorems of Gauss," *Pac. Jour. of Math.* 8 (1958), 609–612. **Zbl** 84, p. 60. **MR** 20 #5994.

SHAPIRO, H. S.

"Some function-theoretic problems motivated by the study of Banach algebras," pp. 95–113 of *Proceedings of the NRL Conference on Classical Function Theory* (1970), Washington, D. C., F. Gross, editor. **Zbl** 286.30030. **MR** 48 #2775.

SHAPIRO, J. H.

Composition Operators and Classical Function Theory. Springer-Verlag (1993), New York & Berlin. **Zbl** 791.30033. **MR** 94k:47049. **BAMS** (2) 32 (1995), p. 150.

"Simple connectivity and linear chaos," *Rend. Circ. Mat. Palermo* (2) Supplement no. 56 (1998), 27–48. **Zbl** 996.47009. **MR** 2001b:47051.

SHASHKEVICH, M.

"Remark on Fejér's inequality which is used in the Weierstrass factorization theorem," *L'Enseignement Mathématique* (2) 8 (1962), 279–280. **Zbl** 111, p. 273. **MR** 26 #6375. (See also **FM** 34, p. 477.)

SHEFFER, I. M.

"A proof of the fundamental theorem of algebra," *Bull. Amer. Math. Soc.* 35 (1929), 227–230. **FM** 55, p. 69.

SHEIL-SMALL, T.

Complex Polynomials, Cambridge Studies in Advanced Mathematics 73, Cambridge University Press (2002), Cambridge. **Zbl** 1012.30001. **MR** 2004b:30001.

SHELL, D. L.

"On the convergence of infinite exponentials," *Proc. Amer. Math. Soc.* 13 (1962), 678–681. **Zbl** 109, p. 299. **MR** 25 #5307.

SHELUPSKY, D.

"Derivation of the Gibbs phenomenon," *Amer. Math. Monthly* 87 (1980), 210–212. **Zbl** 439.42003.

SHIEH, HUI-CHUN

"On functions holomorphic in the unit circle with two exceptional values B," *Chinese Math.* 1 (1962), 226–234. **Zbl** 152, p. 68. **MR** 28 #223.

SHIELDS, A. L.

"On fixed points of commuting analytic functions," *Proc. Amer. Math. Soc.* 15 (1964), 703–706. **Zbl** 129, p. 291. **MR** 29 #2790.

SHIH, M.-H.

"An analog of Bolzano's theorem for functions of a complex variable," *Amer. Math. Monthly* 89 (1982), 210–211. **Zbl** 494.30010. **MR** 83b:30007. (See also **Zbl** 713.30002 and **MR** 91e:30010.)

SHIH, M.-H. and YEH, C.-C.

"Successive approximations in complex analysis," *Jour. Math. Anal. and Applic.* 81 (1981), 182–188. **Zbl** 456.30034. **MR** 82f:30040.

SHIMIZU, T.

(a) "On the domain of indetermination of a regular function," *Japan. Jour. of Math.* 7 (1931), 275–300. **FM** 57, p. 358. **Zbl** 1, p. 397.

(b) "On equi-modular functions and their applications," *Proc. Phys.-Math. Soc. Japan* (3) 13 (1931), 79–92. **FM** 57, p. 369. **Zbl** 1, p. 344.

SHISHA, O.

"On sequences of power series with restricted coefficients," *Amer. Math. Monthly* 72 (1965), 533–537. **Zbl** 158, p. 66. **MR** 36 #5316.

"Proof of power series and Laurent expansions of complex differentiable functions without use of Cauchy's integral formula or Cauchy's integral theorem," *Jour. Approx. Theory* 57 (1989), 117–135 and 58 (1989), 246. **Zbl** 676.30004. **MR** 90m:30002.

SHOIKHET, D.

Semigroups in Geometrical Function Theory, Kluwer Academic Publishers (2001), Dordrecht and Boston. **Zbl** 980.30001. **MR** 2002g:30012. (See also **MR** 2012c:30042.)

SHOLANDER, M.

"On defining the sine and cosine," *Math. Mag.* 43 (1970), 72–75. **Zbl** 191, p. 349.

SHORT, L.

"Iteration and bilinear transformations," *Int. Jour. Math. Educ. Sci. Technol.* 24 (1993), 391–411. **Zbl** 865.30037. **MR** 94c:39003.

SIDON, S.

"Über einen Satz von Herrn Bohr," *Math. Zeit.* 26 (1927), 731–732. **FM** 53, p. 281.

SIEGBERG, H. W.

"Some historical remarks concerning degree theory," *Amer. Math. Monthly* 88 (1981), 125–139. **Zbl** 463.55002. **MR** 82e:55001.

SIERPIŃSKI, W.

"Sur une série potentielle qui, étant convergente en tout point de son cercle de convergence, représente sur ce cercle une fonction discontinue," *Rend. Circ. Mat. Palermo* 41 (1916), 187–190. **FM** 46, p. 1466.

SIMMONS, J. G.

"Analytic functions, ideal fluid flow and Bernoulli's equation," *SIAM Review* 38 (1996), 666–667. **Zbl** 866.30001. **MR** 97i:76001.

SIMONART, F.

"Sur les transformations ponctuelles et leurs applications géométriques. Première partie: les transformations ponctuelles," *Annales de la Soc. Sci. de Bruxelles* 49A (1929), 121–149 and 50A (1930), 35–51. "Deuxième partie: la représentation conforme," *ibid.* 50A (1930), 81–96 and 51A (1931), 49–72. **FM** 55, p. 994; 56, pp. 1145, 982; 57, p. 1425. **Zbl** 2, p. 195.

"Limitations du module d'une fonction holomorphe et de sa dérivée sur un cercle quelconque," *Annales de la Soc. Sci. de Bruxelles* 50A (1930), 128–131. **FM** 56, p. 974.

SINCLAIR, A.

"A general solution for a class of approximation problems," *Pac. Jour. of Math.* 8 (1958), 857–866. **Zbl** 93, p. 65. **MR** 21 #2746.

"On development of the Riemann mapping theorem and related results," *Math. Scientist* 6 (1981), 27–34. **Zbl** 452.30002. **MR** 82k:30001.

"Strong Carleman and strong uniform approximation," *Pac. Jour. of Math.* 117 (1985), 417–428. **Zbl** 555.30026. **MR** 86g:30050.

SINDALOVSKIĬ, G. KH.

"Cauchy-Riemann conditions in a class of functions with summable modulus, and certain boundary properties of analytic functions," *Math. USSR-Sb.* 56 (1987), No. 2, 359–377. **Zbl** 607.30004. **MR** 87m:30052.

SINE, R.

"Behavior of iterates in the Poincaré metric," *Houston Jour. of Math.* 15 (1989), 273–289. **Zbl** 712.47049. **MR** 91a:32034.

SINGH, A. N.

"On the evaluation of a class of definite integrals," *Bull. Calcutta Math. Soc.* 15 (1925), 139–158. **FM** 51, p. 196.

SINGH, D.

"The spectrum in a Banach algebra," *Amer. Math. Monthly* 113 (2006), 756–758. **Zbl** 1146.46022. **MR** 2007c:46048.

SINGH, S. K.

"A note on a theorem of Fabry," *Jour. Karnatak Univ. Sci.* 9–10 (1964/65), 163–164. **Zbl** 192, p. 169. **MR** 32 #2563.

SLAUGHT, H. E.

"The evolution of numbers–an historical drama in two acts," *Amer. Math. Monthly* 35 (1928), 146–151. and *The Mathematics Teacher* 21 (1928), 305–315. **FM** 54, p. 5.

SMIRNOV, V. I. and LEBEDEV, N. A.

Functions of a Complex Variable, Constructive Theory (translated from Russian by Scripta Technica, Ltd.), MIT Press (1968), Cambridge, Massachusetts. **Zbl** 164, p. 375. **MR** 37 #5369.

SMITH, J. D.

"Harmonic analysis of scalar and vector fields in $\mathbb{R}^n$," *Proc. Camb. Phil. Soc.* 72 (1972), 403–416. **Zbl** 247.42021. **MR** 46 #7789.

"Determination of polynomials and entire functions," *Amer. Math. Monthly* 82 (1975), 822–825. **Zbl** 317.30002. **MR** 52 #5966.

SMITHIES, F.

Cauchy and the Creation of Complex Function Theory, Cambridge University Press (1997), Cambridge. **Zbl** 883.01020. **MR** 99b:01013.

SMORYŃSKI, C.

Adventures in Formalism, Texts in Mathematics 2. London: College Publications (2012). **Zbl** 1254.00013. **MR** 2919787.

SOKAL, A.

"A ridiculously simple and explicit implicit function theorem," *Sémin. Lothar. Comb.* 61A, B61Ad (2009). **Zbl** 1182.30006. **MR** 2010i:30003.

SONDOW, J.

"An antisymmetric formula for Euler's constant," *Math. Mag.* 31 (1998), 219–220. **MR** 1573325.

SØRENSEN, H. K.

"Exceptions and counterexamples: understanding Abel's comment on Cauchy's theorem," *Historia Math.* 32 (2005), 453–480. **Zbl** 1080.01005. **MR** 2006k:01011. (See also **MR** 2011a:01006.)

SPALT, D. D.

"Cauchys Kontinuum. Eine historiografische Annährung via Cauchys Summensatz," *Arch. Hist. Exact Sci.* 56 (2002), 285–338. **Zbl** 998.01013. **MR** 2003e:01024.

SPECHT, W.

"Algebraische Gleichungen mit reellen oder komplexen Koeffizienten," *Encyklopädie der Mathematischen Wissenschaften* Bd. I, 1, Heft 3, Teil II, 1–76. B. G. Teubner (2nd ed., 1958), Stuttgart. **Zbl** 82, p. 245. **MR** 21 #5008. **BAMS** 66, p. 63.

"Eine Bemerkung zum Satze von Gauss-Lucas," *Jahresber. Deutsch. Math. Verein.* 62 (1959), 85–92. **Zbl** 87, p. 20. **MR** 22 #4821.

SPERNER, E.

"Neuer Beweis für die Invarianz der Dimensionszahl und des Gebietes," *Abhand. Math. Sem. Univ. Hamburg* 6 (1928), 265–272. **FM** 54, p. 614.

SPIEGEL, M. R.

"On the evaluation of $\int_0^\infty \frac{\sin^n x}{x^n} dx$," *Delta* (Waukesha) (1971), 1–9. **Zbl** 244.30001.

SPINDLER, K.

"A short proof of the formula of Faà di Bruno," *Elemente der Math.* 60 (2005), 33–35. **Zbl** 1074.05014. **MR** 2188344. (See also **MR** 2003d:01019 and 2010i: 26003.)

SPIVAK, M.

Calculus, W. A. Benjamin, Inc. (1967), New York and Amsterdam; 2nd ed. Publish or Perish, Inc. (1980), Berkeley. **Zbl** 458.26001.

SPRINGER, G.

"On Morera's theorem," *Amer. Math. Monthly* 64 (1957), 323–331. **Zbl** 77, p. 75. **MR** 19, p. 22.

SPRINGER, T. A.

"Der Satz von Cauchy," *Simon Stevin* 32 (1958), 68–79 (Dutch). **Zbl** 87, p. 285. **MR** 20 #4629.

STACHÓ, L. L.

"Direct approach to Laurent expansions," *Nieuw Archief voor Wiskunde* (4) 12 (1994), 31–33. **Zbl** 821.30003.

STÄCKEL, P.

"Zur Theorie der eindeutigen Functionen," *Jour. für Reine und Angew. Math.* 106 (1890), 189–192. **FM** 22, p. 396.

"Integration durch imaginäres Gebiet. Ein Beitrag zur Geschichte der Funktionentheorie," *Biblio. Math.* (3) 1 (1900), 109–128. **FM** 31, p. 43.

"Beiträge zur Geschichte der Funktionentheorie im achtzehnten Jahrhundert," *Biblio. Math.* (3) 2 (1901), 111–121. **FM** 32, p. 48.

"Eine vergessene Abhandlung Leonhard Eulers über die Summe der reziproken Quadrate der natürlichen Zahlen," *Biblio. Math.* (3) 8 (1907/08), 37–60. **FM** 38, p. 61.

STANKIEWICZ, J. and STANKIEWICZ, Z.

"On classes of functions regular in a halfplane," *Bull. Polish Math. Soc.* 39 (1991), 49–56. **Zbl** 757.30015. **MR** 94b:30019. (See also **MR** 94g:30014, 96g: 30036, 2003m:30029.)

STARK, E. L.

"A new method of evaluating the sums of $\sum_{k=1}^\infty (-1)^{k+1} k^{-2p}$, $p = 1, 2, 3, \ldots$ and related series," *Elemente der Math.* 27 (1972), 32–34. **Zbl** 225.40002. **MR** 45 #3996. (See also **MR** 57 #16483.)

"The series $\sum_{k=1}^\infty k^{-s}$, $s = 2, 3, 4, \ldots$, once more," *Math. Mag.* 47 (1974), 197–202. **Zbl** 291.40004. **MR** 50 #5261.

STEEN, L. and SEEBACH, J.

Counterexamples in Topology, Holt, Rinehart and Winston, Inc. (1970), New York. 2nd ed. Springer-Verlag (1978), New York, reprinted by Dover Publ. Inc. (1995), Mineola NY. **Zbl** 386.54001. **MR** 80a:54001 and 96k:54001.

STEINITZ, E.

"Bedingt konvergente Reihen und konvexe Systeme," *Jour. für Reine und Angew. Math.* 143 (1913), 128–175. **FM** 44, p. 287.

"Die allgemeine Theorie der Körper und der Fundamentalsatz der Algebra," pp. 420–426 of *Mathematische Abhandlungen H. A. Schwarz Gewidmet,* (O. Bolza, *et al.*, editors) Verlag Julius Springer (1914), Berlin. Reprinted by Chelsea Publishing Co. (1974), New York.

STEINLEIN, H.

"Borsuk's antipodal theorem and its generalizations and applications: a survey," pp. 166–235 of *Topological Methods in Nonlinear Analysis,* Séminaire de Mathématiques Supérieures 95, Montréal (1985). **Zbl** 573.55003. **MR** 86k:55002.

"Spheres and symmetry: Borsuk's antipodal theorem," *Topological Methods in Nonlinear Analysis* 1 (1993), 15–33. **Zbl** 795.55004. **MR** 94b:55008.

STEINMETZ, N.

Rational Iteration: Complex Analytic Dynamical Systems, deGruyter Studies in Mathematics 16, Walter deGruyter (1993), Berlin. **Zbl** 773.58010. **MR** 94h:30035.

ŠTĚPÁNEK, F.

"On certain class of power series," *Functiones et Approximatio. Comm. Math.* 12 (1982), 13–15. **Zbl** 504.30001. **MR** 86m:30003.

STIELTJES, T. J.

"Recherches sur les fractions continues," *Annales de la Faculté des Sciences de Toulouse* 8 (1894), 1–122. **FM** 25, p. 326 and 35, p. 978. **MR** 97i:01054. (For an English translation see **MR** 95g:01033.)

STOÏLOW, S.

"Sur les transformations continues des espaces topologiques," *Bull. Math. Soc. Roum. Sci.* 35 (1933), 229–235. **FM** 59, p. 1254. **Zbl** 8, p. 326.

"Sur une extension topologique du principe du maximum du module et ses applications à la théorie des fonctions," *Bull. Section Sci. Acad. Roumaine* 23 (1940), 28–30. **FM** 66, p. 978. **Zbl** 24, p. 127. **MR** 2, p. 79.

Leçons sur les principes topologiques de la théorie des fonctions analytiques, Gauthier-Villars (2nd ed. 1956), Paris. **FM** 64, p. 309. **Zbl** 121, p. 61. **MR** 32 #5899. **BAMS** 44, p. 758.

STONE, M. H.

"On a theorem of Pólya," *Jour. Indian Math. Soc.* (N. S.) 12 (1948), 1–7. **Zbl** 31, p. 57. **MR** 10, p. 308.

"Hilbert space methods in conformal mapping," pp. 409–425 of *Proceedings of the International Symposium on Linear Spaces,* Jerusalem Academic Press, Jerusalem and Pergamon Press, New York (1960), **Zbl** 118, p. 299. **MR** 25 #178.

"Topological aspects of conformal mapping theory," pp. 343–346 of *General Topology and its Relations to Modern Analysis and Algebra,* Academic Press, New York and Publishing House of the Czechoslovakian Academy of Science, Prague (1962). **Zbl** 109, p. 303. **MR** 31 #323.

STOUT, E. L.

The Theory of Uniform Algebras, Bogden and Quigley (1971), Tarrytown-on-Hudson. **Zbl** 286.46049. **MR** 54 #11066.

STO.ZEK, W.

"Sur l'allure de fonctions harmoniques dans le voisinage d'un point exceptionnel," *C. R. Acad. Sci. Paris* 180 (1925), 727–728. **FM** 51, p. 362.

"Sur l'allure de fonctions harmoniques dans le voisinage d'un point exceptionnel," *Ann. Soc. Polon. Math.* 4 (1926), 52–58. **FM** 52, p. 497.

STRASZEWICZ, S.

"Über eine Verallgemeinerung des Jordan'schen Kurvensatzes," *Fund. Math.* 4 (1923), 129–135. **FM** 49, p. 404.

STROMBERG, K.

An Introduction to Classical Real Analysis, Wadsworth International Group and Prindle, Weber and Schmidt (1981), Belmont, California, reprinted by AMS Chelsea (2015), Providence, Rhode Island. **Zbl** 454.26001 and 1331.00003. **MR** 82c:26002 and 3444105.

STROMBERG, K. and TSENG, S.

"Simple plane arcs of positive area," *Expositiones Math.* 12 (1994), 31–52. **Zbl** 797.28003. **MR** 95b:28004.

STUDY, E.

Vorlesungen über ausgewählte Gegenstände der Geometrie, Heft 2, "Konforme Abbildung einfach-zusammenhängender Bereiche," B. G. Teubner (1913), Leipzig and Berlin. **FM** 44, p. 755. **BAMS** 20, p. 443.

STUDY, E. and CARTAN, E.

"Nombres complexes," *Encyclopédie des sciences mathématiques pures et appliquées,* t. I, v. 1, fasc. 3, 329–468. Gauthier-Villars (1908), Paris. **FM** 39, p. 50. **MR** 93m:00053a.

STYER, D. AND MINDA, C. D.

"The use of the Monodromy Theorem and entire functions with nonvanishing derivative," *Amer. Math. Monthly* 81 (1974), 639–642. **Zbl** 288.30003. **MR** 49 #5348.

SU, LI PI

"Rings of analytic functions on any subset of the complex plane," *Pac. Jour. of Math.* 42 (1972), 535–538. **Zbl** 245.30038. **MR** 47 #2076. (See also **Zbl** 854.30035.)

SUETIN, P. K.

"The basic properties of Faber polynomials," *Uspehi Mat. Nauk* (N. S.) 19 (1964), No. 4 (118), 125–154. English translation in *Russian Math. Surveys* 19 (1964), No. 4, 121–149. (Published by the London Mathematical Society). **Zbl** 138, p. 294. **MR** 29 #6029. (See also **MR** 56 #15896 and **Zbl** 313.42018.)

Series of Faber Polynomials (translated from Russian by E. V. Pankratiev), Gordon & Breach (1998), Reading. **Zbl** 936.30027. **MR** 99m:30001.

SUFFRIDGE, T. J.

"Common fixed points of commuting holomorphic maps of the hyperball," *Mich. Math. Jour.* 21 (1974), 309–314. **Zbl** 333.47026. **MR** 51 #3903.

SUITA, K.

"On fixed points of conformal self-mappings," *Hokkaido Math. Jour.* 10 (Special Issue) (1981), 667–671. **Zbl** 502.30040. **MR** 83f:30006.

ŠURA-BURA, M. R.

"Zur Theorie der bikompakten Räume," *Recueil Math. Moscou* (=*Mat. Sbornik*) N. S. 9 (51) (1941), 385–388 (Russian). **FM** 67, p. 756. **Zbl** 25, p. 95. **MR** 3, p. 137.

SUZUKI, J.

"Lagrange's proof of the fundamental theorem of algebra," *Amer. Math. Monthly* 113 (2006), 705–714. **Zbl** 1132.01302. **MR** 2007g:26014.

ŚWIĄTKOWSKI, T.

"On the holomorphism of the integral with respect to a complex parameter," *Colloq. Math.* 16 (1967), 61–65. **Zbl** 191, p. 369. **MR** 35 #1765.

SZÁSZ, O.

"Über Potenzreihen und Bilinearformen," *Math. Zeit.* 4 (1919), 163–176. **FM** 47, p. 273.

"Über Funktionen, die den Einheitskreis schlicht abbilden," *Jahresber. Deutsch. Math. Verein.* 42 (1933), *73–75*. **FM** 58, p. 1095. **Zbl** 5, p. 108.

SZEGÖ, G.

"Über orthogonale Polynome, die zu einer gegebenen Kurve der komplexen Ebene gehören," *Math. Zeit.* 9 (1921), 218–270. **FM** 48, p. 374.

"Conformal mapping of the interior of an ellipse onto a circle," *Amer. Math. Monthly* 57 (1950), 474–478. **Zbl** 41, p. 413. **MR** 12, p. 401.

T

TAHA, T. S.

"A note on a class of rational functions," *Math. Japonica* 36 (1991), 1041–1042. **Zbl** 745.30008. **MR** 92k:30009.

TAKAHASHI, S.

"Bemerkung zu einer Arbeit von Herrn Radó," *Japan. Jour. of Math.* 7 (1930), 161–162. **FM** 56, p. 986.

TAKENAKA, S.

"On the approximation of a function of two variables by polynomials," *Tôhoku Math. Jour.* 16 (1919), 16–25. **FM** 47, p. 251.

TAMMI, O.

"Note on Gutzmer's coefficient theorem," *Rev. Fac. Sci. Univ. Istanbul A* 22 (1957), 9–12. **Zbl** 83, p. 66. **MR** 20 #5293.

TANAKA, C.

"On E. Lindelöf's theorem on the meromorphic function of bounded characteristic in the unit circle," *Mem. School Sci. Engineering Waseda Univ.* No. 28 (1964), 57–61. **Zbl** 284.30022. **MR** 42 #6242.

"On the asymptotic values for regular functions with bounded characteristic," *Kōdai Math. Sem. Reports* 27 (1976), 94–115. **Zbl** 329.30026. **MR** 53 #3313.

TAO, ZHIGUANG

"Analytic operator functions," *Jour. Math. Anal. and Applic.* 103 (1984), 293–320. **Zbl** 608.47023. **MR** 86a:47011.

"Iteration of analytic normal function of matrices," *Chinese Annals Math. Ser. B* 6 (1985), 97–102. **Zbl** 557.47012. **MR** 87h:47039.

"On the operator equation $f(A) = A$," *Acta Math. Sinica* (N. S.) 1 (1985), 327–334. Part II, *Chinese Annals Math. Ser. B* 12 (1991), 175–180. **Zbl** 729.47014. **MR** 92h:47023.

TAR, M. M.

"On a criterion for the analyticity of a function of a complex variable," *Ukrain. Math. Jour.* 21 (1969), 354–357 (1970). **Zbl** 201, p. 87. **MR** 40 #310.

"Über gewisse hinreichende Bedingungen der Analytizität von Funktionen einer komplexen Veränderlichen," *Dopovīdī Akad. Nauk Ukraïn. RSR, Ser. A* (1971), 212–214, 284 (Ukrainian). **Zbl** 234.30003. **MR** 45 #510.

"Analyticity of functions of a complex variable," *Ukrain. Math. Jour.* 25 (1973), 59–65. **Zbl** 261.30001. **MR** 47 #3642.

"Conditions for monogenicity of complex functions," (Russian) pp. 153–157 of *Complex analysis and manifolds* (Ju A. Mitropol'skii and Ju Ju Trohimčuk, editors), Izdanie Inst. Mat., Akad. Nauk Ukrain. SSR (1978), Kiev. **Zbl** 408.30001. **MR** 81i:30001.

TAUVEL, P.

Exercices d'analyse complexe, Masson (1994), Paris. **Zbl** 798.30001.

TAYLOR, A. E.

(a) "A note on the Poisson kernel," *Amer. Math. Monthly* 57 (1950), 478–479. **Zbl** 37, p. 347. **MR** 12, p. 411.

(b) "New proofs of some theorems of Hardy by Banach space methods," *Math. Mag.* 23 (1950), 115–124. **Zbl** 35, p. 172. **MR** 11, p. 507.

TAYLOR, E. H.

"An extension of a theorem of Painlevé," *Bull. Amer. Math. Soc.* 19 (1913), 403–406. **FM** 44, p. 470.

TEICHMÜLLER, O.

"Untersuchungen über konforme und quasikonforme Abbildung," *Deutsche Math.* 3 (1938), 621–678. **FM** 64, p. 313. **Zbl** 20, p. 238.

TEISSIER DU CROS, F.

"Sur la convergence d'une série entière dont le terme général a sa partie réelle bornée en deux points de la circonférence-unité," *C. R. Acad. Sci. Paris* 219 (1944), 44–45. **Zbl** 60, p. 214. **MR** 7, p. 200.

TELYAKOVSKIĬ, D. S.

"On the holomorphy of functions specifying angle-preserving transformations," *Math. Notes* 56 (1994), 1200–1204. **Zbl** 863.30004. **MR** 97c:30002.

TERASAKA, H.

"Über eine Verschärfung des Rouché-Lipkaschen Satzes," *Proc. Phys.-Math. Soc. Japan* (3) 11 (1929), 90–94. **FM** 55, p. 175.

THOMAS, J. M.

"The resolvents of a polynomial," *Amer. Math. Monthly* 47 (1940), 686–694. **FM** 66, p. 1199. **Zbl** 61, p. 19. **MR** 2, p. 242.

THOMPSON, K. W. and SCHONBEK, T.

"A problem of the Pompeiu type," *Amer. Math. Monthly* 87 (1980), 32–36. **Zbl** 449.30001. **MR** 81d:26008.

TIDEMAN, M.

"Elementary proof of a uniqueness theorem for positive harmonic functions," *Nordisk Mat. Tidskrift* 2 (1954), 95–96. **Zbl** 55, p. 332. **MR** 16, p. 129.

TIETZE, H.

"Über einfach zusammenhängende Flächen und ihre Deformationen in sich," *Sitzungsber. Kaiserl. Akad. Wiss. Wien* 122 (1913), Zweite Abteilung, 1653–1658. **FM** 44, p. 562.

"Über Funktionen, die auf einer abgeschlossenen Menge stetig sind," *Jour. für Reine und Angew. Math.* 145 (1914), 9–14. **FM** 45, p. 628.

"Über stetige Kurven, Jordansche Kurvenbögen und geschlossene Jordansche Kurven," *Math. Zeit.* 5 (1919), 284–288. **FM** 47, p. 520.

TIMTSCHENKO, I. J.

"Grundlagen der Theorie der analytischen Functionen. Bd I. Geschichte der Entwickelung der Begriffe und Methoden, die der Theorie der analytischen Functionen zu Grunde liegen," (Russian), *Denkschriften der math. Abteilung der neurussischen Gesellschaft der Naturforscher* 12 (1892), 1–236; 16 (1896), 1–216, 257–472; 19 (1899), 1–183, 473–655. **FM** 30, p. 48, 383.

TITCHMARSH, E.

The Theory of Functions, Clarendon Press (2nd ed. 1939; reprinted, 1975), Oxford. **FM** 65, p. 302. **Zbl** 336.30001. **MR** 33 #5850. **BAMS** 39, p. 650.

TOLSTOV, G. P.

"Über beschränkte Funktionen, die der Laplaceschen Differentialgleichung genügen," *Mat. Sbornik (N. S.)* 29 (71) (1951), 559–564 (Russian). **Zbl** 44, p. 100. **MR** 13, p. 943. (See also **Zbl** 865.3102 and **MR** 98f:31004.)

TOMIĆ, M.

"Sur un théorème de H. Bohr," *Math. Scand.* 11 (1962), 103–106. **Zbl** 109, p. 302. **MR** 31 #316.

TONELLI, L.

"Sopra una proprietà caratteristica delle funzioni armoniche," *Rend. Accad. d. Lincei Roma* (5) 18, part 1 (1909), 577–582. **FM** 40, p. 452.

(a) "Sulla rappresentazione analitica delle funzioni di più variabili reali," *Rend. Circ. Mat. Palermo* 29 (1910), 1–36. **FM** 41, p. 490.

(b) "Su gli zeri del limite di una successione di funzioni analitiche," *Rend. Accad. d. Lincei Roma* (5) 19, part 1 (1910), 5–10. **FM** 41, p. 459.

TOTIK, V.

"Recovery of H^p-functions," *Proc. Amer. Math. Soc.* 90 (1984), 531–537. **Zbl** 544.30030. **MR** 85j:30073.

"A tale of two integrals," *Amer. Math. Monthly* 106 (1999), 227–240 (see also *ibid.* 92 (1985), 433). **Zbl** 1021.26001. **MR** 2000d:26002.

TRAININ, J.

"Integrating expressions of the form $\frac{\sin^n x}{x^m}$ and others," *Math. Gazette* 94 (2010), 216–223 and 95 (2011), 212.

TROKHIMCHUK, Y. Y.

"Applications continues et fonctions analytiques," pp. 7–29 of *Fonctions d'une variable complexe. Problèmes contemporains,* A. I. Markouchevitch, editor (translated from Russian by L. Nicolas), Gauthier-Villars (1962), Paris. **Zbl** 116, pp. 288 and 281. **MR** 22 #11099.

Continuous Mappings and Conditions of Monogeneity, (translated from Russian by R. Mandl), Israel Program for Scientific Translations, Ltd. (1964), Jerusalem. **Zbl** 133, p. 38. **MR** 33 #2801.

TROKHIMCHUK, Y. Y. and SAFONOV, V. M.

"On one criterion for the constancy of a complex function," *Ukrain. Math. Jour.* 51 (1999), 1237–1245. **Zbl** 998.30036. **MR** 2001j:30001.

TSARPALIAS, A.

"A version of Rouché's theorem for continuous functions," *Amer. Math. Monthly* 96 (1989), 911–913. **Zbl** 704.30007. **MR** 91b:30023.

TSCHAKALOFF, L.

"Sur les singularités polaires des séries entières," *C. R. Acad. Bulgare Sci. Mat. Nat.* 1 (1948), 9–12. **Zbl** 36, p. 331. **MR** 10, p. 691. (See also **MR** 15, p. 514 and **Zbl** 53, p. 375.)

TSCHEBOTAREFF [=ČEBOTAREV], N. G.

"Über die Realität von Nullstellen ganzer transzendenter Funktionen," *Math. Annalen* 99 (1928), 660–686. **FM** 54, p. 351.

TSUJI, M.

"On Blaschke's theorem," *Japan. Jour. of Math.* 3 (1926), 65–68. **FM** 52, p. 312.

(a) "On the theorems of Carathéodory and Lindelöf in the theory of conformal representation," *Japan. Jour. of Math.* 7 (1930), 91–99. **FM** 56, p. 983.

(b) "Theorems concerning Poisson integrals," *Japan. Jour. of Math.* 7 (1930), 227–253. **FM** 56, p. 1065.

"On a positive harmonic function in a half-plane," *Japan. Jour. of Math.* 15 (1939), 277–285. **FM** 65, p. 1239. **Zbl** 33, p. 54.

"A remark on Schottky's Theorem," *Jour. Math. Soc. of Japan* 1 (1949), 266–269. **Zbl** 38, p. 53. **MR** 11, p. 718.

"On a positive harmonic function in a half-plane," *Jour. Math. Soc. of Japan* 7 (1955), 76–78. **Zbl** 66, p. 87. **MR** 16, p. 819.

"On a non-negative subharmonic function in a half-plane," *Kōdai Math. Sem. Reports* 8 (1956), 134–141. **Zbl** 72, p. 315. **MR** 18, p. 885.

Potential Theory in Modern Function Theory, Maruzen Co., Ltd. (1959), Tokyo. Reprinted by Chelsea Publishing Co. (1975), New York. **Zbl** 322.30001. **MR** 54 #2990.

"Huber's theorem on analytical mappings of a ring domain in a ring domain," *Comm. Math. Univ. Sancti Pauli* 8 (1960), 41–43. **Zbl** 91, p. 255. **MR** 22 #5727.

TUCCIARONE, J.

"The development of the theory of summable divergent series from 1880 to 1925," *Arch. Hist. Exact Sci.* 10 (1973), 1–40. **Zbl** 275.01010. **MR** 56#5168.

TUCKER, A. W.

"Some topological properties of disk and sphere," pp. 285–309 of *Proceedings of the First Canadian Mathematical Congress,* Montreal, 1945. University of Toronto Press (1946), Toronto. **Zbl** 61, p. 403. **MR** 8, p. 525.

TUMURA, Y.

"On extensions of Borel's theorem and Saxer-Csillag's theorem," *Proc. Phys.-Math. Soc. Japan* (3) 19 (1937), 29–35. **FM** 63, p. 275. **Zbl** 16, p. 217.

TURNER, L. E.

"The Mittag-Leffler theorem: the origin, evolution, and reception of a mathematical result, 1876–1884," *Historia Math.* 40 (2013), no. 1, 36–83. **Zbl** 1261.01008. **MR** 3007859.

U

ULLRICH, P.

"Automorphismen analytischer Banach-Algebren," *Arch. der Math.* 40 (1983), 266–272. **Zbl** 498.46040. **MR** 84f:30060.

"Weierstrass' Vorlesung zur 'Einleitung in die Theorie der analytischen Funktionen'," *Arch. Hist. Exact Sci.* 40 (1989), 143–172. **Zbl** 761.01007. **MR** 91b:01056.

"Wie man beim Weierstraßschen Aufbau der Funktionentheorie das Cauchysche Integral vermeidet," *Jahresber. Deutsch. Math. Verein.* 92 (1990), 89–110. **Zbl** 707.30002. **MR** 92b:30002.

(a) "Punktweise und lokal gleichmässige Konvergenz von Folgen holomorpher Funktionen. I," *Math. Semesterber.* 41 (1994), 69–75. Part II, *ibid.* 81–87. **Zbl** 815.30002, 30003. **MR** 96e:30089.

(b) "The proof of the Laurent expansion by Weierstrass," pp. 139–153 of *The History of Modern Mathematics,* vol. III (edited by E. Knobloch and D. Rowe) Academic Press (1994), San Diego and London. **Zbl** 806.01012. **MR** 95c:01023.

(c) "The Riemann removable singularity theorem from 1841 onwards," pp. 155–178 of *The History of Modern Mathematics,* vol. III (edited by E. Knobloch and D. Rowe) Academic Press (1994), San Diego and London. **Zbl** 808.01004. **MR** 95c:01024.

"The Riemann mapping problem," *Rend. Circ. Mat. Palermo* (2) Supplement no. 44 (1996), 9–42. **Zbl** 899.01013. **MR** 98b:01024.

UNKELBACH, H.

"Über die Randverzerrung bei konformer Abbildung," *Math. Zeit.* 43 (1938), 739–742. **FM** 64, p. 313. **Zbl** 18, p. 224.

"Über die Randverzerrung bei schlichter konformer Abbildung," *Math. Zeit.* 46 (1940), 329–336. **FM** 66, p. 364. **Zbl** 23, p. 50. **MR** 2, p. 83.

URYSOHN, P.

"Sur une fonction analytique partout continue," *Fund. Math.* 4 (1923), 144–150. **FM** 49, p. 213.

"Über die Mächtigkeit der zusammenhängenden Mengen," *Math. Annalen* 94 (1925), 262–295. **FM** 51, p. 452.

V

VAGGIONE, D.

"On the fundamental theorem of algebra," *Colloq. Math.* 73 (1997), 193–194 and 77 (1998), 321. **Zbl** 876.12001. **MR** 97m:30003.

VAHLEN, K. TH.

"Der Fundamentalsatz der Algebra und die Auflösung der Gleichungen durch Quadratwurzeln," *Acta Math.* 21 (1897), 287–299. **FM** 28, p. 91.

VALIRON, G.

(a) "Remarque sur un théorème de M. Julia," *Bull. Sci. Math.* (2) 49 (1925), 68–73 and 270–275. **FM** 51, p. 261.

(b) *Fonctions entières et fonctions méromorphes d'une variable,* Mémorial des Sciences Mathématiques, Fasc. 2, Gauthier-Villars (1925), Paris. **FM** 51, p. 249.

(c) "Sur la formule d'interpolation de Lagrange," *Bull. Sci. Math.* (2) 49 (1925), 181–192 and 203–224. **FM** 51, p. 250.

"Sur les théorèmes de MM. Bloch, Landau, Montel et Schottky," *C. R. Acad. Sci. Paris* 183 (1926), 728–730. **FM** 52, p. 324.

(a) "Sur un théorème de MM. Koebe et Landau," *Bull. Sci. Math.* (2) 51 (1927), 34–42. **FM** 53, p. 303.

(b) "Compléments au théorème de Picard-Julia," *Bull. Sci. Math.* (2) 51 (1927), 167–183. **FM** 53, p. 301.

(a) "Le théorème de M. Picard et le complément de M. Julia," *Jour. de Math. Pures et Appliq.* (9) 7 (1928), 113–126. **FM** 54, p. 347.

(b) "Remarque sur le principe de la symétrie de Schwarz," *Bull. Sci. Math.* (2) 52 (1928), 289–292. **FM** 54, p. 340.

(a) "Sur le théorème de M. Picard," *L'Enseignement Mathématique* 28 (1929), 55–59. **FM** 55, p. 193.

(b) *Familles normales et quasi-normales de fonctions méromorphes,* Mémorial des Sciences Mathématiques, Fasc. 38, Gauthier-Villars (1929), Paris. **FM** 55, p. 762.

(c) "Sur un théorème de M. Julia étendant le lemme de Schwarz," *Bull. Sci. Math.* (2) 53 (1929), 70–76. **FM** 55, p. 769.

"Sur le théorème de Bloch," *Rend. Circ. Mat. Palermo* 54 (1930), 76–82. **FM** 56, p. 269.

"Sur l'itération des fonctions holomorphes dans un demi-plan," *Bull. Sci. Math.* (2) 55 (1931), 105–128. **FM** 57, p. 381. **Zbl** 1, p. 281. (See also **MR** 83j: 30023.)

"Le théorème de Borel-Julia dans la théorie des fonctions méromorphes," pp. 270–279 of Vol. I of *Verhandlungen des Internationalen Mathematiker–Kongresses Zürich 1932.* Orell Füssli, Verlag (1932), Zürich and Leipzig. Reprinted by Kraus Reprint, Ltd. (1967), Nedeln/Liechtenstein. **FM** 58, p. 335. **Zbl** 6, p. 409. **MR** 53 #7713.

Sur les valeurs exceptionnelles des fonctions méromorphes et de leurs dérivées, Actualités Scientifiques et Industrielles, No. 570, Hermann et Cie. (1937), Paris. **FM** 63, p. 977. **Zbl** 19, p. 419. **BAMS** 47, p. 7.

"Sur l'approximation des nombres réels et un théorème de M. Teissier du Cros," *C. R. Acad. Sci. Paris* 219 (1944), 45–47. **Zbl** 60, p. 214. **MR** 7, p. 200.

"Directions de Julia et directions de Picard des fonctions entières," *Revista Unión Math. Argentina* 12 (1946), 49–54. **Zbl** 61, p. 150. **MR** 9, p. 84.

"Remarques sur la représentation approchée par des polynomes des fonctions continues de plusiers variables," *Bull. Sci. Math.* (2) 72 (1948), 9–12. **Zbl** 33, p. 56. **MR** 10, p. 295.

Fonctions analytiques, Presses Universitaires de France (1954), Paris. **Zbl** 55, p. 67. **MR** 15, p. 861. **BAMS** 61, p. 85.

Théorie des fonctions, Cours d'analyse mathématique. I. Masson et Cie. (2nd ed. 1955), Paris. **FM** 68, p. 99. **Zbl** 28, p. 208. **MR** 7, p. 283.

VALLÉE POUSSIN, C. DE LA,

"Sur les applications de la notion de convergence uniforme dans la théorie des fonctions d'une variable complexe," *Annales de la Soc. Sci. de Bruxelles* 17 (B) (1892–93), 323–330. **FM** 25, p. 463.

"Sur la définition de la différentielle totale et sur les intégrales curvilignes qui ne dépendent que de leurs limites," *Annales de la Soc. Sci. de Bruxelles* 38 (1914), 67–72. **FM** 45, p. 452.

"Démonstration simplifiée du théorème fondamental de M. Montel sur les familles normales de fonctions," *Annals of Math.* (2) 17 (1915–16), 5–11. **FM** 45, p. 637.

"Application de l'intégral de Lebesgue au problème de la représentation d'une aire simplement connexe sur un cercle," *Annales de la Soc. Sci. de Bruxelles* 50 (A) (1930), 23–34. **FM** 56, p. 983.

"Utilisation de la méthode du balayage dans la théorie de la représentation conforme," *Bull. Acad. Bruxelles* (5) 18 (1932), 385–400. **FM** 58, p. 1091. **Zbl** 4, p. 355.

Le potentiel logarithmique. Balayage et représentation conforme, Louvain Librairie Universitaire, Gauthier-Villars (1949), Paris. **Zbl** 37, p. 346.

VAMANAMURTHY, M. K.

"A binomial inequality and application to power series," *Math. Chronicle* 18 (1989), 79–83. **Zbl** 704.30002. **MR** 91e:05011.

VANDEN EYNDE, R.

"Historical evolution of the concept of homotopic paths," *Arch. Hist. Exact Sci.* 45 (1992), 127–188. **Zbl** 766.01015. **MR** 93m:01048. (See also **Zbl** 944.55001. **MR** 2000h:55005.)

VARADARAJAN, V. S.

"Euler and his work on infinite series," *Bull. Amer. Math. Soc.* (2) 44 (2007), 515–539. **Zbl** 1135.01010. **MR** 2008g:01012.

VAROPOULOS, TH.

"Sur les valeurs exceptionnelles des fonctions dérivées et le théorème de M. Saxer," *Acta Math.* 51 (1928), 23–29. **FM** 53, p. 296.

VASILESCO, F.

"Sur les singularités des fonctions harmoniques," *Jour. de Math. Pures et Appliq.* (9) 9 (1930), 81–111. **FM** 56, p. 1067.

VAUGHAN, H. E.

"On two theorems of plane topology," *Amer. Math. Monthly* 60 (1953), 462–468. **Zbl** 51, p. 146. **MR** 15, p. 146.

VEECH, W.

A Second Course in Complex Analysis, W. A. Benjamin, Inc. (1967), New York. **Zbl** 145, p. 299. **MR** 36 #3955.

VEEN, S. C., VAN

"Historische Besonderheiten über $\int_0^\infty e^{-x^2}dx$," *Mathematica, Zutphen. B.* (=*Simon Stevin*) 12 (1943), 1–4 (Dutch). **Zbl** 28, p. 194. **MR** 7, p. 354.

VENEZIANI, C.

"Extrait d'une lettre de M. Hermite," *Amer. Jour. of Math.* 7 (1885), 168–169. **FM** 17, p. 374.

VENKATACHALIENGAR, K.

"Elementary proofs of the infinite product for sin z and allied formulae," *Amer. Math. Monthly* 69 (1962), 541–545. **Zbl** 100, p. 61.

VERBLUNSKY, S.

"On positive harmonic functions in a half-plane," *Proc. Camb. Phil. Soc.* 31 (1935), 482–507. **FM** 61, p. 523. **Zbl** 13, p. 158.

"A theorem on positive harmonic functions," *Proc. Camb. Phil. Soc.* 45 (1949), 207–212. **Zbl** 33, p. 63. **MR** 10, p. 533.

"On a fundamental formula of potential theory," *Jour. Lon. Math. Soc.* 26 (1951), 25–30. **Zbl** 42, p. 106. **MR** 12, p. 703.

VERHEUL, E. R.

"Elementary proofs concerning results about functions on the n-sphere," *Topology and its Applic.* 40 (1991), 101–116. **Zbl** 732.55001. **MR** 92i:55004.

VERLEY, J.-L.

"Les fonctions analytiques," pp. 129–163 vol. I of *Abrégé d'histoire des mathematiques 1700–1900* (J. Dieudonné, editor) Hermann (1978), Paris. **Zbl** 383.01009. **MR** 80k:01002. German translation (1985). **Zbl** 558.01003. **MR** 86e:01003.

VIGNATI, M.

"A geometric property of functions harmonic in a disk," *Elemente der Math.* 47 (1992), 33–38. **Zbl** 758.31001. **MR** 92k:31002. (See also **FM** 60, p. 1133.)

VIJAYARAGHAVAN, T.

"A power series that converges and diverges at everywhere dense sets of points on its circle of convergence," *Jour. Indian Math. Soc.* (N. S.) 11 (1947), 69–72. **Zbl** 30, p. 153. **MR** 10, p. 25.

VILLAT, H.

"Sur le problème de Dirichlet relatif au cercle," *Bull. Soc. Math. France* 39 (1911), 443–456.

"Le problème de Dirichlet dans une aire annulaire," *Rend. Circ. Mat. Palermo* 33 (1912), 134–174. **FM** 43, p. 490.

"Sur la résolution de certaines équations intégrales, et sur quelques problèmes qui s'y rattachent," *Acta Math.* 40 (1916), 101–178. **FM** 45, p. 526.

"Sur la représentation conforme des aires doublement connexes," *Ann. Sci. École Norm. Sup.* (3) 38 (1921), 183–227. **FM** 48, p. 406.

VISSER, C.

"Sur les limites des fonctions bornées à la frontiere de leur domaine d'existence," *Nieuw Archief voor Wiskunde* (2) 17 (1931–1932), 147–150. **FM** 58, p. 329. **Zbl** 3, p. 402.

"Über beschränkte analytische Funktionen und die Randverhältnisse bei konformen Abbildungen," *Math. Annalen* 107 (1933), 28–39. **FM** 58, p. 361. **Zbl** 4, p. 405.

VITALI, G.

"Sopra le serie di funzioni analitiche," *Rend. del R. Istituto Lombardo di Scienze e Lettere* (2) 36 (1903), 772–774. **FM** 34, p. 419.

"Sopra le serie di funzioni analitiche," *Annali Mat. Pura Appl.* (3) 10 (1904), 65–82. **FM** 35, p. 397.

"Sopra una proprietà caratteristica delle funzioni armoniche," *Rend. Accad. d. Lincei Roma* (5) 21, part 2 (1912), 315–320. **FM** 43, p. 492.

"Sul teorema fondamentale dell' algebra," *Periodico di Mat.* (4) 8 (1928), 102–105. **FM** 54, p. 117.

VIVANTI, G.

"Sul *valor medio* di Pringsheim e sulla sua applicazione alla teoria delle funzioni analitiche," *Math. Annalen* 58 (1904), 457–468. **FM** 35, p. 393.

VIZZINI, G.

"Su di un importante teorema sulle funzioni continue," *Bolletino di Mat.* (2) 11 (1932), 121–126. **FM** 58, p. 245.

VOLCHKOV, V. V.

"On a problem of Zalcman and its generalizations," *Math. Notes* 53 (1993), 134–138. **Zbl** 798.30025. **MR** 94i:43008.

"Morera type theorems on the unit disk," *Analysis Mathematica* 20 (1994), 49–63. **Zbl** 792.30027. **MR** 95h:30052. (See also **MR** 2005e:30071.)

VOLKOVYSKII, L., LUNTS, G. and ARAMANOVICH, I.

A Collection of Problems on Complex Analysis (translated from Russian by J. Berry), Addison-Wesley Publishing Co. (1965), Reading; also Mir Publishers (1977), Moscow and Dover Publications, Inc. (1991), New York. **Zbl** 357.30001 and 272.30002. **MR** 49 #9164, 56 #3259 and 93e:30003.

VOLPI, R.

"Osservazioni per una teoria puramente analitica ed elementare delle funzioni circolari ed iperboliche e loro relazioni coll' esponenziale," *Giornale di Mat. di Battaglini* 41 (1903), 33–46. **FM** 34, p. 474.

VOLTERRA, V.

"Alcune osservazioni sopra proprietà atte ad individuare una funzione," *Rend. Accad. d. Lincei Roma* (5) 18, part 1 (1909), 263–266. **FM** 40, p. 453.

VOLYNEC, I. A.

"Groups of conformal automorphisms of plane domains in the uniform metric," *Complex Variables* 19 (1992), 195–203. **Zbl** 769.30005. **MR** 95b:30009.

VÝBORNÝ, R.

"On the use of a differentiable homotopy in the proof of the Cauchy theorem," *Amer. Math. Monthly* 86 (1979), 380–382. **Zbl** 423.30036. **MR** 80f:30033.

"Differentation of power series," *Amer. Math. Monthly* 94 (1987), 369–370.

"Elementary evaluation of Fresnel's integrals," *Math. Bohemica* 116 (1991), 401–404. **Zbl** 739.26001. **MR** 92m:26002.

W

WAERDEN, B L. VAN DER

"Verallgemeinerung eines Satzes von Fatou," *Österreich. Akad. Wiss. Math.-Natur. Kl. Sitzungsber. II* 195 (1986), no. 1–3, 191–194. **Zbl** 619.13001. **MR** 88i:11072.

WALECKI, F.

"Démonstration du théorème de d'Alembert," *Nouvelles Annales de Math.* (3) 2 (1883), 241–248. **FM** 15, p. 45.

WALKER, P. L.

"Complex iterated radicals," *Math. Gazette* 67 (1983), 269–273. **Zbl** 528.30014.

WALL, C. T. C.

A Geometric Introduction to Topology, Addison-Wesley Publishing Co. (1972), Reading. Reprinted by Dover Publications, Inc. (1994), New York. **Zbl** 261.55001. **MR** 57 #17617.

WALLEN, L. J.

"The pth root of an analytic function," *Math. Mag.* 65 (1992), 260–261. **Zbl** 781.30001.

WALSH, J. L.

(a) "Über die Entwicklung einer analytischen Funktion nach Polynomen," *Math. Annalen* 96 (1927), 430–436. **FM** 52, p. 299.

(b) "Über die Entwicklung einer Funktion einer komplexen Veränderlichen nach Polynomen," *Math. Annalen* 96 (1927), 437–450. **FM** 52, p. 300.

"On the expansion of analytic functions in series of polynomials and in series of other analytic functions," *Trans. Amer. Math. Soc.* 30 (1928), 307–332. **FM** 54, p. 360.

"On approximation by rational functions to an arbitrary function of a complex variable," *Trans. Amer. Math. Soc.* 31 (1929), 477–502. **FM** 55, p. 762.

"On the overconvergence of sequences of polynomials of best approximation," *Trans. Amer. Math. Soc.* 32 (1930), 794–816. **FM** 56, p. 263.

"The Cauchy-Goursat theorem for rectifiable Jordan curves," *Proc. Nat. Acad. Sci. U. S. A.* 19 (1933), 540–541. **FM** 59, p. 312.

Approximation by polynomials in the complex domain, Mémorial des Sciences Mathématiques, Fasc. 73, Gauthier-Villars (1935), Paris. **FM** 61, p. 319. **Zbl** 11, p. 298.

Interpolation and Approximation by Rational Functions in the Complex Domain, American Mathematical Society Colloquium Publications, Vol. 20 (5th ed., 1969), Providence. **FM** 61, p. 315. **Zbl** 146, p. 299. **MR** 36 #1672(b). **BAMS** 42, p. 604.

"History of the Riemann Mapping Theorem," *Amer. Math. Monthly* 80 (1973), 270–276. **Zbl** 273.30003. **MR** 48 #2348.

WALTER, W.

"Old and new approaches to Euler's trigonometric expansions," *Amer. Math. Monthly* 89 (1982), 225–230. **MR** 83e:01032.

WALTHER, A.

"Maximum und Minimum einer harmonischen Funktion auf Kreisen," *Math. Zeit.* 11 (1921), 157–160. **FM** 48, p. 557.

WATSON, G. N.

Complex Integration and Cauchy's Theorem, Cambridge Tracts in Mathematics and Physics, No. 15, Cambridge University Press (1914), London. Reprinted by Hafner Publishing Co. (1960), New York. **FM** 45, p. 664. **MR** 22 #12210.

WAVRE, R.

"Sur la réduction des domaines par une substitution à m variables complexes et l'existence d'un seul point invariant," *L'Enseignement Mathématique* 25 (1926), 218–234. **FM** 53, p. 554.

WAVRE, R. AND BRUTTIN, A.

"Sur une transformation continue et l'existence d'un point invariant," *C. R. Acad. Sci. Paris* 183 (1926), 843–845. **FM** 52, p. 574.

WEDDERBURN, J. H. M.

"On matrices whose coefficients are functions of a single variable," *Trans. Amer. Math. Soc.* 16 (1915), 328–332. **FM** 45, p. 264.

WEIERSTRASS, K. T. W.

"Zur Theorie der eindeutigen analytischen Functionen," *Abhandl. Kön. Preuss. Akad. Wiss. Berlin* (1876), 11–60. **FM** 10, p. 282. French translation by É. Picard in *Ann. Sci. École Norm. Sup.* (2) 8 (1879), 111–150. **FM** 11, p. 264.

(a) "Zur Functionenlehre," *Monatsber. Kön. Preuss. Akad. Wiss. Berlin* (1880), 719–743. **FM** 12, p. 310. French translation by J. Tannery in *Bull. Sci. Math.* (2) 5 (1881), 157–183. **FM** 13, p. 306.

(b) "Über einen functionstheoretischen Satz des Herrn G. Mittag-Leffler," *Monatsber. Kön. Preuss. Akad. Wiss. Berlin* (1880), 707–717. **FM** 12, p. 311. French translation in *Bull. Sci. Math.* (2) 5 (1881), 113–124. **FM** 13, p. 307.

"Über die analytische Darstellbarkeit sogenannter willkürlicher Functionen reeller Argumente," *Sitzungsber. Kön. Preuss. Akad. Wissen. Berlin* (1885), 633–640, 789–806. **FM** 17, p. 384. French translation by L. Laugel in *Jour. de Math. Pures et Appliq.* (4) 2 (1886), 105–138. **FM** 18, p. 344.

"Neuer Beweis des Satzes, dass jede ganze rationale Function einer Veränderlichen dargestellt werden kann als ein Product aus linearen Functionen derselben Veränderlichen," *Sitzungsber. Kön. Preuss. Akad. Wissen. Berlin*

(1891), 1085–1101. **FM** 23, p. 83. French translation by C. Bourlet in *Ann. Sci. École Norm. Sup.* (3) 12 (1895), 317–335. **FM** 26, p. 114.

Mathematische Werke I (1894); II (1895); III (1903, Mayer and Müller, Berlin). **FM** 25, p. 49; 26, p. 41; 34, p. 23. Reprinted by Georg Olms Verlagsbuchhandlung, Hildesheim and Johnson Reprint Corp. , New York (1967). **MR** 36 #1279–1281.

"Zur Funktionentheorie," *Acta Math.* 45 (1925), 1–10. **FM** 50, p. 210. (This extract from an 1884 seminar is not included in the *Werke.*)

Einleitung in die Theorie der analytischen Funktionen, Vorlesung Berlin 1878. Lecture notes by A. Hurwitz. Dokumente zur Geschichte der Math. 4. Friedr. Vieweg & Sohn (1988), Braunschweig. **Zbl** 652.01039. **MR** 90a: 01110.

Ausgewählte Kapitel aus der Funktionenlehre. Vorlesung, gehalten in Berlin 1886, (R. Siegmund-Schultze, editor) Teubner-Archiv zur Mathematik 9 (1988), Vienna. **Zbl** 679.01010. **MR** 91i:01146.

WEIL, A.

Elliptic Functions according to Eisenstein and Kronecker, Ergebnisse der Mathematik und ihrer Grenzgebiete, Bd. 88. Springer-Verlag (1976), New York. **Zbl** 318.33004. **MR** 58 #27769.

WEINSTOCK, R.

"Elementary evaluations of $\int_0^\infty e^{-x^2}\,dx$, $\int_0^\infty \cos(x^2)\,dx$ and $\int_0^\infty \sin(x^2)\,dx$," *Amer. Math. Monthly* 97 (1990), 39–42. **Zbl** 704.26007. **MR** 91a:26004.

WEN, GUO CHUN

Conformal mapping and boundary value problems (translated from Chinese by Kuniko Weltin), Translations of Mathematical Monographs, Vol. 106, American Mathematical Society (1992), Providence. **Zbl** 778.30011. **MR** 93h:30059.

WERMUTH, E. M. E.

"Derivatives eventually zero," *Math. Intelligencer* 13, no. 4 (1991), 52.

WHITNEY, H.

"Note on Perron's solution of the Dirichlet problem," *Proc. Nat. Acad. Sci. U. S. A.* 18 (1932), 68–70. **FM** 58, p. 509. **Zbl** 3, p. 349.

WHITTAKER, J. M.

Interpolatory Function Theory, Cambridge Tracts in Mathematics and Physics, No. 33, Cambridge University Press (1935), London. Reprinted by Stechert-Hafner Service Agency, Inc. (1972), New York. **FM** 61, p. 331. **Zbl** 12, p. 155. **MR** 32 #2798. **BAMS** 42, p. 305. (See also **Zbl** 38, p. 228, **MR** 11, p. 344.)

(a) "An inequality involving derivatives," *Bull. Lon. Math. Soc.* 2 (1970), 297–300. **Zbl** 206, p. 87. **MR** 43 #488.

(b) "The basis of Vitali's theorem," pp. 353–357 of *Mathematical Essays Dedicated to A. J. Macintyre,* (H. Shankar, editor) Ohio University Press (1970), Athens, Ohio. **Zbl** 209, p. 100. **MR** 43 #2200.

WHYBURN, G. T.

"On the construction of simple arcs," *Amer. Jour. of Math.* 54 (1932), 518–524. **FM** 32, p. 640. **Zbl** 5, p. 27.

"An open mapping approach to Hurwitz's theorem," *Trans. Amer. Math. Soc.* 71 (1951), 113–119. **Zbl** 43, p. 178. **MR** 13, p. 149.

(a) "Developments in topological analysis," *Fund. Math.* 50 (1962), 305–318. **Zbl** 100, p. 67. **MR** 29 #3651.

(b) "The Cauchy inequality in topological analysis," *Proc. Nat. Acad. Sci. U.S.A.* 48 (1962), 1335–1336. **Zbl** 108, p. 71. **MR** 32 #4281.

Topological Analysis, Princeton Mathematical Series, No. 23, Princeton University Press (2nd ed. 1964). **Zbl** 186, p. 559. **MR** 29 #2758. **BAMS** 66, p. 255.

"What is a curve?," pp. 23–38 of *Studies in Modern Topology,* MAA Studies in Mathematics, Vol. 5, (P. J. Hilton, editor). Mathematical Association of America and Prentice-Hall, Inc. (1968), Englewood Cliffs. **Zbl** 187, p. 448. **MR** 37 #895.

WHYBURN, W. M.

"A connectedness theorem in abstract sets," *Bull. Amer. Math. Soc.* 41 (1935), 365–366. **FM** 61, p. 630. **Zbl** 12, p. 57.

WIDDER, D. V.

"Functions harmonic in a strip," *Proc. Amer. Math. Soc.* 12 (1961), 67–72. **Zbl** 96, p. 77. **MR** 24 #A2674.

WIGERT, S.

"Sur un théorème concernant les fonctions entières," *Arkiv för Mat., Astro. och Fysik* 11, Nr. 21 (1916). **FM** 46, p. 505.

"Sur le théorème fondamental de l'algèbre," *Arkiv för Mat., Astro. och Fysik* 25 B, No. 17 (1936). **FM** 63, p. 856. **Zbl** 15, p. 97.

WIGNER, E. P.

"On a class of analytic functions from the quantum theory of collisions," *Annals of Math.* (2) 53 (1951), 36–67. **Zbl** 42, p. 452. **MR** 12, p. 490.

"Simplified derivation of the properties of elementary transcendentals," *Amer. Math. Monthly* 59 (1952), 669–683. **Zbl** 48, p. 53. **MR** 14, p. 460.

WILDBERGER, N. J.

"Real fish, real numbers, real jobs," *Math. Intelligencer* 21, no. 2 (1999), 4–7.

WILDER, R. L.

"Evolution of the topological concept of 'connected'," *Amer. Math. Monthly* 85 (1978), 720–726 and 87 (1980), 31–32. **Zbl** 394.54009. **MR** 80f:01019(a) and 81f:01023.

WILKOSZ, W.

"Sur un point fondamental de la théorie du potentiel," *C. R. Acad. Sci. Paris* 174 (1922), 435–437. **FM** 48, p. 556.

Les propriétés topologiques du plan euclidien, Mémorial des Sciences Mathématiques, Fasc. 45, Gauthier-Villars (1931), Paris. **FM** 56, p. 1142. **Zbl** 1, p. 171.

"Sur le théorème intégral de Cauchy," *Ann. Soc. Polon. Math.* 11 (1933), 19–27 and 56–57. **FM** 59, p. 1020. **Zbl** 7, p. 105.

WILLIAMS, K. S.

(a) "On $\sum_{n=1}^{\infty} \frac{1}{n^{2k}}$," *Math. Mag.* 44 (1971), 273–276. **Zbl** 224.40008. **MR** 45 #3997.

(b) "Note on $\int_0^\infty (\sin x/x)dx$," *Math. Mag.* 44 (1971), 9–11.

WILLIAMS, R. K.

"A note on conformality," *Amer. Math. Monthly* 80 (1973), 299–300. **Zbl** 273.30004. **MR** 48 #506.

"On the linearity of one-to-one entire functions," *Delta* (Waukesha) 6 (1976), 72–77. **Zbl** 347.30011. **MR** 55 #3256.

"Some conditions for one-to-one-ness," *Pi Mu Epsilon Jour.* 7 (1980), 115–116. **Zbl** 439.30004.

WINTERNITZ, A.

"Beweis für die Invarianz des ebenen Gebiets," *Math. Zeit.* 26 (1927), 165–169. **FM** 53, p. 564.

WŁODARCZYK, K.

"Julia's lemma and Wolff's theorem for J^*-algebras," *Proc. Amer. Math. Soc.* 99 (1987), 472–476. **Zbl** 621.46041. **MR** 88a:46049.

WOLF, F.

"Ein Eindeutigkeitssatz für analytische Funktionen," *Math. Annalen* 117 (1940/1941), 383. **FM** 66, p. 363. **Zbl** 23, p. 49. **MR** 2, p. 81.

"On majorants of subharmonic and analytic functions," *Bull. Amer. Math. Soc.* 48 (1942), 925–932. **Zbl** 61, p. 231. **MR** 4, p. 144.

"Extension of analytic functions," *Duke Math. Jour.* 14 (1947), 877–887. **Zbl** 29, p. 123. **MR** 9, p. 420.

WOLFENSTEIN, S.

"Proof of the fundamental theorem of algebra," *Amer. Math. Monthly* 74 (1967), 853–854. **Zbl** 153, p. 96. (See also **FM** 44, p. 116.)

WOLFF, J.

"Sur les séries $\sum \frac{A_k}{z - a_k}$," *C. R. Acad. Sci. Paris* 173 (1921), 1327–1328 and 1057–1059. **FM** 48, p. 320. (See also **Zbl** 385.30004, 715.30003 and **MR** 58 #11339.)

"Über die Loomansche Erweiterung eines Satzes von Pompeiu," *Nieuw Archief voor Wiskunde* (2) 14 (1922/1924), 337–339 and pp. 457–459 of Vol. 1 of *Proceedings of the International Mathematical Congress Toronto, 1924,* University of Toronto Press. Reprinted by Kraus Reprint, Ltd. (1967), Nedeln/Liechtenstein. **FM** 50, p. 250 and 54, p. 328.

(a) "Sur l'itération des fonctions holomorphes dans une région, et dont les valeurs appartiennent à cette région," *C. R. Acad. Sci. Paris* 182 (1926), 42–43. **FM** 52, p. 309.

(b) "Sur l'itération des fonctions bornées," *C. R. Acad. Sci. Paris* 182 (1926), 200–201. **FM** 52, p. 309.

(c) "Sur une généralisation d'un théorème de Schwarz," *C. R. Acad. Sci. Paris* 182 (1926), 918–920 and 183 (1926), 500–502. **FM** 52, p. 309.

"Über die Iteration derjenigen in einem Gebiete regulären Funktionen, deren Werte dem Gebiete angehören," *Math. Zeit.* 26 (1927), 125–127. **FM** 53, p. 304.

"Sur les limites radiales d'une fonction holomorphe dans un cercle," *Bull. Soc. Math. France* 56 (1928), 167–173. **FM** 54, p. 346.

"Sur l'itération des fonctions holomorphes dans un demi-plan," *Bull. Soc. Math. France* 57 (1929), 195–203. **FM** 55, p. 769. (See also **MR** 83j:30023.)

(a) "Sur la représentation conforme," *Kon. Akad. Wetensch. Amsterdam, Proc.* 33 (1930), 96–97. **FM** 56, p. 983.

(b) "Quelques propriétés des fonctions holomorphes dans un demi-plan dont toutes les valeurs sont dans ce demi-plan," *Kon. Akad. Wetensch. Amsterdam, Proc.* 33 (1930), 1185–1188. **FM** 56, p. 983.

WOLFF, J. and DE KOK, F.

"Les fonctions holomorphes à partie réelle positive et l'intégral de Stieltjes," *Bull. Soc. Math. France* 60 (1932), 221–227. **FM** 59, p. 324. **Zbl** 6, p. 171.

WOLFF, J. and DENJOY, A.

"Sur la division d'une sphère en trois ensembles," *L'Enseignement Mathématique* 32 (1933), 66–68. **FM** 59, p. 1256. **Zbl** 8, p. 181.

"Über die bibliographischen Hilfsmittel der Mathematik," *Jahresber. Deutsch. Math. Verein.* 12 (1903), 408–426. **FM** 34, p. 47.

WOLTERMANN, M.
"Convergence of an iterated exponential," *Math. Mag.* 70 (1997), 303–304.

WULF, TH.
"Die Grenzwerte $\lim_{n=+\infty}(1+\frac{b}{n})^n$ und $\lim_{n=+\infty}(\sqrt[n]{a}-1)n$ als Grundlage für die Theorie der reellen Potenz und des reellen Logarithmus," *Monatshefte für Math. und Physik* 8 (1897), 43–53. **FM** 28, p. 245.

WYLER, O.
"The Cauchy integral formula," *Amer. Math. Monthly* 72 (1965), 50–53. **Zbl** 135, p. 123.

X

XU, C.
"A GM-AM ratio," *Math. Mag.* 83 (2010), 49–50. **Zbl** 1227.97036.

Y

YAMABE, H.
"On an extension of Helly's theorem," *Osaka Math. Jour.* 2 (1950), 15–17. **Zbl** 36, p. 362. **MR** 12, p. 616.

YAMASHITA, S.
"Conformality and semiconformality of a function holomorphic in the disk," *Trans. Amer. Math. Soc.* 245 (1978), 119–138. **Zbl** 398.30005. **MR** 80h:30018.

YANAGIHARA, N.
"On the difference of positive harmonic functions," *Jour. College Arts Sci. Chiba Univ.* 4 (1965), 235–239. **MR** 36 #3993.

YANG, CHUNG-CHUN
"On the zeros of an entire function and its second derivative," *Rend. Accad. d. Lincei Roma* (8) 49 (1970), 27–29. **Zbl** 218, p. 181. **MR** 45 #3705.

YANG, LE
"Normality for families of meromorphic functions," *Scientia Sinica* Ser. A 29 (1986), 1263–1274. **Zbl** 629.30032. **MR** 88f:30049.

YANG, LO
Value Distribution Theory, Springer-Verlag Berlin and Science Press Beijing (1993). **Zbl** 790.30018. **MR** 95h:30039.

YANG, LO and CHANG, KUAN-HEO
"Recherches sur la normalité des familles de fonctions analytiques à des valeurs multiples. I. Un nouveau critère et quelques applications," *Scientia Sinica* 14 (1965), 1258–1271. **Zbl** 166, p. 326. **MR** 33 #7550. "II. Généralisations," *ibid.* 15 (1966), 433–453. **Zbl** 173, p. 320. **MR** 34 #2843.

YOUNIS, R.

"Interpolation by a finite Blaschke product," *Proc. Amer. Math. Soc.* 78 (1980), 451–452. **Zbl** 439.30020. **MR** 80k:30035.

YOUNG, R. M.

"An elementary proof of a trigonometric identity," *Amer. Math. Monthly* 86 (1979), 296–297. **Zbl** 417.42002. **MR** 81a:42023.

"When is $\mathbb{R}^n$ a field," *Math. Gazette* 72 (1988), 128–129. **MR** 89h:12004.

"On the evaluation of certain improper integrals," *Math. Gazette* 75 (1991), 88–89.

"On evaluating the probability integral: a simple one-variable proof," *Math. Gazette* 89 (2005), 252–254.

YOUNG, W. H.

"On parametric integration," *Monatshefte für Math. und Physik* 21 (1910), 125–149. **FM** 41, p. 325.

"On the fundamental theorem in the theory of functions of a complex variable," *Proc. Lon. Math. Soc.* (2) 10 (1911), 1–6. **FM** 42, p. 421.

YOUNSI, M.

"Peano curves in complex analysis," *Amer. Math. Monthly* 126 (2019), 635–640. **Zbl** 1422.30059. **MR** 3979739.

YÛJÔBÔ, Z.

"A theorem concerning subharmonic functions defined in a strip domain," *Comm. Math. Univ. Sancti Pauli* 1 (1952), 1–4. **Zbl** 48, p. 341. **MR** 15, p. 526.

YUŠKEVIČ, A. P.

"On the origin of Cauchy's definition of the integral," *Akad. Nauk SSSR. Trudy Inst. Istorii Estestvoznaniya* 1 (1947), 373–411 (Russian). **MR** 11, p. 572.

YZEREN, J. VAN

"A rehabilitation of $(1+\frac{z}{n})^n$," *Amer. Math. Monthly* 77 (1970), 995–998. **Zbl** 204, p. 77. **MR** 1536096. (See also **Zbl** 746.30006.)

"Moivre's and Fresnel's integrals by simple integration," *Amer. Math. Monthly* 86 (1979), 691–693. **Zbl** 446.26003. **MR** 1539141.

Z

ZACHARIAS, K.

"The Herglotz trick applies," *Amer. Math. Monthly* 100 (1993), 194–195.

ZAGIER, D.

"Newman's short proof of the prime number theorem," *Amer. Math. Monthly* 104 (1997), 705–708. **Zbl** 887.11039. **MR** 98j:11069.

Zalcman, L.

(a) *Analytic Capacity and Rational Approximation,* Lecture Notes in Mathematics 50, Springer-Verlag (1968), New York. **Zbl** 171, p. 37. **MR** 37 #3018.

(b) "Null sets for a class of analytic functions," *Amer. Math. Monthly* 75 (1968), 462–470. **Zbl** 181, p. 351. **MR** 37 #6448.

"Analyticity and the Pompeiu problem," *Arch. Rational Mech. and Anal.* 47 (1972), 237–254. **Zbl** 251.30047. **MR** 50 #582.

"Real proofs of complex theorems (and vice-versa)," *Amer. Math. Monthly* 81 (1974), 115–137 and 83 (1976), 84–85. **Zbl** 279.30001. **MR** 48 #6370.

"A heuristic principle in complex function theory," *Amer. Math. Monthly* 82 (1975), 813–817. **Zbl** 315.30036. **MR** 52 #757. (See also **MR** 90i:30051 and 91i:30031.)

"Picard's theorem without tears," *Amer. Math. Monthly* 85 (1978), 265–268. **Zbl** 392.30020. **MR** 58 #1152.

"Offbeat integral geometry," *Amer. Math. Monthly* 87 (1980), 161–175. **Zbl** 433.53048. **MR** 83f:53052.

"Modern perspectives on classical function theory," *Rocky Mountain Jour. Math.* 12 (1982), 75–92. **Zbl** 558.30001 and 657.30002. **MR** 86d:30003 and 86j:30002.

"Normal families revisited," pp. 149–164 of *Complex Analysis and Related Topics* (J. J. O. O. Wiegerinck, editor), University of Amsterdam (1993), Amsterdam.

"Normal families: new perspectives," *Bull. Amer. Math. Soc.* (2) 35 (1998), 215–230. **Zbl** 1037.30021. **MR** 99g:30048.

"A tale of three theorems," *Amer. Math. Monthly* 123 (2016), 643–656 and 124 (2017), 636. **MR** 3539851.

Zarankiewicz, K.

"Sur la représentation conforme d'un domaine doublement connexe sur un anneau circulaire," *C. R. Acad. Sci. Paris* 198 (1934), 1347–1349. **FM** 60, p. 286. **Zbl** 9, p. 26.

"Über ein numerisches Verfahren zur konformen Abbildung zweifach zusammenhängender Gebiete," *Zeit. für Math. and Mech.* 14 (1934), 97–104. **FM** 60, p. 1207. **Zbl** 9, p. 26.

Zaremba, S.

"Contributions à la théorie d'une équation fonctionnelle de la physique," *Rend. Circ. Mat. Palermo* 19 (1905), 140–150. **FM** 36, p. 822.

"Sur l'unicité de la solution du problème de Dirichlet," *Bulletin Internat. de l'Acad. des Sciences de Cracovie* 1 (1909), 561–564. **FM** 40, p. 452.

"Sur le principe de Dirichlet," *Acta Math.* 34 (1911), 293–316. **FM** 42, p. 393.

"Sur une propriété générale des fonctions harmoniques," pp. 171–176 of *Conférences de la Réunion Internationale de Mathématiciens, Paris, 1937.* Gauthier-Villars (1939), Paris. **FM** 64, p. 469.

ZASSENHAUS, H.

"On the fundamental theorem of algebra," *Amer. Math. Monthly* 74 (1967), 485–497 and 75 (1968), 827. **Zbl** 145, p. 279. **MR** 36 #2605. (See also **MR** 57 #2802 and **Zbl** 357.00001.)

ZAYED, A. I.

"Recoverability of some classes of analytic functions from their boundary values," *Proc. Amer. Math. Soc.* 87 (1983), 493–498. **Zbl** 506.30030. **MR** 84c:30062 and 84e:30050.

"On the Poisson and conjugate Poisson kernels of an annulus," *Tamkang Jour. Math.* 14, No. 3 (1986), 69–81 and *Glasnik Mat.* (3) 22 (42) (1987), 353–363. **Zbl** 616.42004 and 661.31003. **MR** 88c:31001 and 89j:42008.

ZHENG, L.

"An elementary proof for two basic alternating series," *Amer. Math. Monthly* 109 (2002), 187–188. **Zbl** 1022.40003.

ZHOU, L.

"Irrationality proofs à la Hermite," *Math. Gazette* 95 (2011), 407–413.

ZHOU, L. and MARKOV, L.

"Recurrent proofs of the irrationality of certain trigonometric values," *Amer. Math. Monthly* 117 (2010), 360–362. **Zbl** 1222.11093. **MR** 2647819.

ZIN, G.

"Esistenza e rappresentazione di funzioni analitiche, le quali, su una curva di Jordan, si reducono a una funzione assegnata," *Annali Mat. Pura Appl.* (4) 34 (1953), 365–405. **Zbl** 51, p. 309. **MR** 14, p. 1073.

ZINTERHOF, P.

"Konstruktion von schlichten Funktionen mit unendlich vielen Fixpunkten," *Rend. Ist. Math. Univ. Trieste* 3 (1971), 125–134 (1972). **Zbl** 225.30018. **MR** 46 #9331. (See also **MR** 52 #8405 and **Zbl** 353.30015.)

ZORETTI, L.

Leçons sur le prolongement analytique, Gauthier-Villars (1911), Paris. **FM** 42, p. 414. **BAMS** 20, p. 321.

ZORN, M.

"Approximating sums," *Amer. Math. Monthly* 54 (1947), 148–151. **Zbl** 29, p. 31. **MR** 8, p. 450.

Name Index

R. B. Burckel, *Classical Analysis in the Complex Plane*, Cornerstones,
https://doi.org/10.1007/978-1-0716-1965-0

Subject Index

R. B. Burckel, *Classical Analysis in the Complex Plane*, Cornerstones,
https://doi.org/10.1007/978-1-0716-1965-0

Symbol Index

R. B. Burckel, *Classical Analysis in the Complex Plane*, Cornerstones,
https://doi.org/10.1007/978-1-0716-1965-0

Series summed, Integrals evaluated

$$\frac{\pi}{4} = \int_0^1 \frac{dx}{1+x^2} = \sum_{n=0}^{\infty} \frac{(-1)^n}{2n+1} \qquad 127$$

$$\frac{\pi}{6} = \frac{1}{\sqrt{3}} \sum_{n=0}^{\infty} \frac{(-1)^n}{2n+1} \cdot \frac{1}{3^n} \qquad 129$$

$$\frac{\pi}{2} = 1 + \sum_{k=1}^{\infty} \frac{1 \cdot 3 \cdots (2k-1)}{2 \cdot 4 \cdots (2k)} \cdot \frac{1}{2k+1} \qquad 369$$

$$\int_0^1 x^{-x}\, dx = \sum_{n=1}^{\infty} n^{-n} \qquad 130$$

$$\int_0^{\infty} e^{-t^2}\, dt = \frac{\sqrt{\pi}}{2}, \quad \int_0^{\infty} e^{it^2}\, dt = \sqrt{\frac{\pi}{2}} \left(\frac{1+i}{2} \right) \qquad 130, 132, 748$$

$$\sum_{k=1}^{\infty} (-1)^{k+1}/k = \sum_{k=1}^{\infty} 1/k2^k = \log 2 \qquad 124$$

$$\sum_{k=1}^{\infty} \frac{1}{k^2} = \frac{\pi^2}{6}, \quad \sum_{k=1}^{\infty} \frac{1}{k^4} = \frac{\pi^4}{90}, \quad \sum_{k=1}^{\infty} \frac{1}{k^6} = \frac{\pi^6}{945}, \quad \sum_{k=1}^{\infty} \frac{1}{k^8} = \frac{\pi^8}{9450} \qquad 157, 773$$

$$\sum_{k=1}^{\infty} \frac{1}{k^2+1}, \qquad \sum_{k=1}^{\infty} \frac{(-1)^k}{k^2+1} \qquad 770$$

$$\sum_{k=-\infty}^{\infty} \frac{1}{(2k+1)^2} = \frac{\pi^2}{4} \qquad 155, 766$$

$$\sum_{k=0}^{\infty} \frac{(-1)^k}{(2k+1)^3} = \frac{\pi^3}{32} \qquad 763$$

$$\sum_{n=-\infty}^{\infty} \frac{1}{(2n+1)^2} = \left(\sum_{n=-\infty}^{\infty} \frac{(-1)^n}{2n+1} \right)^2 \qquad 156$$

R. B. Burckel, *Classical Analysis in the Complex Plane*, Cornerstones,
https://doi.org/10.1007/978-1-0716-1965-0

$$\frac{\pi}{2} = \int_{-\infty}^{\infty} \frac{\sin^4 x}{x^2}\,dx, \quad \frac{2\pi}{3} = \int_{-\infty}^{\infty} \left(\frac{\sin x}{x}\right)^4 dx \qquad 266,\ 267$$

$$\int_{-\infty}^{\infty} \left(\frac{\sin x}{x}\right)^m dx, \quad m \text{ integer} \geq 2 \qquad 268$$

$$\sum_{k=-\infty}^{\infty} \frac{\sin k}{k} = \int_{-\infty}^{\infty} \frac{\sin x}{x}\,dx = \int_{-\infty}^{\infty} \left(\frac{\sin x}{x}\right)^2 dx = \sum_{k=-\infty}^{\infty} \left(\frac{\sin k}{k}\right)^2 = \pi \qquad 268$$

$$\sum_{k=1}^{\infty} \left(\frac{\sin k\pi x}{k\pi x}\right)^2 = \frac{1-x}{2x} = \sum_{k=1}^{\infty} \frac{\sin k\pi x}{k\pi x}, \quad x \in (0,1] \qquad 160$$

$$\sum_{k=1}^{\infty} \frac{\sin(2k-1)x}{2k-1} = \frac{\pi}{4}, \quad x \in (0,\pi) \qquad 154$$

$$\sum_{n=1}^{\infty} \frac{\sin(nx)}{n} = \frac{\pi - x}{2}, \quad x \in (0, 2\pi) \qquad 154$$

$$\sum_{n=1}^{\infty} \frac{\cos(nx)}{n} = -\log(2\sin(x/2)), \quad x \in (0, 2\pi) \qquad 154$$

$$\sum_{n=1}^{\infty} (-1)^{n+1} \frac{1-\cos(nx)}{n^2} = \frac{x^2}{4}, \quad x \in [-\pi, \pi] \qquad 155$$

$$\sum_{n=1}^{\infty} \frac{\cos(nx)}{n^2} = \frac{3x^2 - 6\pi x + 2\pi^2}{12}, \quad x \in [0, 2\pi] \qquad 155$$

$$\frac{\pi}{\sin \pi z} = \frac{1}{z} + \sum_{k=1}^{\infty} \frac{(-1)^k 2z}{z^2 - k^2}, \quad z \in \mathbb{C} \setminus \mathbb{Z} \qquad 770$$

$$\frac{\pi^2}{\sin^2 \pi z} = \sum_{k=-\infty}^{\infty} \frac{1}{(z-k)^2}, \quad z \in \mathbb{C} \setminus \mathbb{Z} \qquad 768$$

$$\pi \cot \pi z = \frac{1}{z} + \sum_{k=1}^{\infty} \frac{2z}{z^2 - k^2}, \quad z \in \mathbb{C} \setminus \mathbb{Z} \qquad 769$$

$$\tan z = \sum_{n=1}^{\infty} \frac{(-1)^{n-1}\, 2^{2n}(2^{2n}-1)B_{2n}}{(2n)!} z^{2n-1}, \quad |z| < \frac{\pi}{2} \qquad 159$$

$$\arctan(x) = \frac{x}{\sqrt{1+x^2}} \sum_{k=0}^{\infty} \frac{(2k)!}{2^{2k}(k!)^2} \frac{1}{2k+1} \left(\frac{x^2}{1+x^2}\right)^k, \quad x \in \mathbb{R} \qquad 370$$

$$\arctan(x) = \frac{x}{1+x^2} \sum_{k=0}^{\infty} \frac{2^{2k}(k!)^2}{(2k)!} \frac{1}{2k+1} \left(\frac{x^2}{1+x^2}\right)^k, \quad x \in \mathbb{R} \qquad 129$$

$$\arctan(x) = \sum_{n=0}^{\infty} \frac{(-1)^n x^{2n+1}}{2n+1}, \quad x \in [-1,1] \qquad 128$$

$$\arcsin(z) = \sum_{k=0}^{\infty} \frac{(2k)!}{2^{2k}(k!)^2} \frac{1}{2k+1} z^{2k+1}, \quad z \in \mathbb{D} \qquad 367$$

$$\frac{(\pi-1)^2}{6} = \sum_{k=1}^{\infty} \frac{\sin^2 k}{k^4} \qquad 160$$

$$\sum_{n=-\infty}^{\infty} e^{-\pi n^2 z} = \frac{1}{\sqrt{z}} \sum_{n=-\infty}^{\infty} e^{-\pi n^2/z}, \quad z \in \mathbb{H} \qquad 754$$

$$\int_{-\infty}^{\infty} e^{-2\pi i x y} e^{-\pi x^2}\, dx = e^{-\pi y^2}, \quad y \in \mathbb{R} \qquad 131, 755$$

$$\int_{-\infty}^{\infty} e^{-2\pi i x y} \operatorname{sech}(\pi x)\, dx = \operatorname{sech}(\pi y), \quad y \in \mathbb{R} \qquad 763$$

$$1 = \sum_{n=1}^{\infty} \frac{n^{n-1}}{n!\, e^n} \qquad 378$$

$$\frac{\pi}{2} = \prod_{n=1}^{\infty} \frac{4n^2}{4n^2-1} \qquad 773$$

$$\frac{2}{\pi} = \prod_{n=1}^{\infty} \left(1 - \frac{1}{(2n)^2}\right), \quad \frac{\pi}{4} = \prod_{n=2}^{\infty} \left(1 - \frac{1}{(2n-1)^2}\right) \qquad 773$$

$$\prod_{n=2}^{\infty} \left(1 - \frac{1}{n^2}\right) = \frac{1}{2} = \prod_{n=2}^{\infty} \left(1 - \frac{(-1)^n}{n}\right) \qquad 773$$

MIX
Papier aus verantwortungsvollen Quellen
Paper from responsible sources
FSC® C105338

If you have any concerns about our products,
you can contact us on
ProductSafety@springernature.com

In case Publisher is established outside the EU,
the EU authorized representative is:
Springer Nature Customer Service Center GmbH
Europaplatz 3, 69115 Heidelberg, Germany

Printed by Libri Plureos GmbH
in Hamburg, Germany